Handbuch der experimentellen Pharmakologie

Vol. 45/I Heffter-Heubner New Series

Handbook of Experimental Pharmacology

Drug Addiction I

Morphine, Sedative/Hypnotic and Alcohol Dependence

Contributors

H.T. Conrad · H.F. Fraser · C.W. Gorodetzky · D.R. Jasinski
P. Kielholz · D. Ladewig · W.R. Martin · N.K. Mello
J.H. Mendelson · J.W. Sloan · C.M. Smith

Editor

William R. Martin

Springer-Verlag Berlin Heidelberg New York 1977

William R. Martin, M.D., U.S. Department of Health, Education, and Welfare, Public Health Service, National Institute on Drug Abuse, Addiction Research Center, Lexington, KY 40511, USA

With 24 Figures

ISBN-13: 978-3-642-66614-8 e-ISBN-13: 978-3-642-66612-4
DOI: 10.1007/978-3-642-66612-4

Library of Congress Cataloging in Publication Data Main entry under title Drug addiction (Handbook of experimental pharmacology New series, v 45, pts 1—2) Contributors, v 2 E Anggård and others Includes bibliographies and indexes CONTENTS 1 Morphine, sedative-hypnotic, and alcohol dependence — 2 Amphetamine, psychotogen, and marihuana dependence 1 Narcotic habit 2 Drug abuse 3 Mental illness—Physiological aspects I Conrad, Harold, T II Martin, William R, 1921- III Anggård, E IV Series Handbuch der experimentellen Pharmakologie New series, v 45, pts 1—2 QP905 H3 vol 45, pts. 1—2 [RC566] 615' 1'08s
ISBN 0-387-08170-4 (v 1) [615' 782] 77-24381

Softcover reprint of the hardcover 1st edition 1977

Preface

This volume addresses the general problem of drug addiction from several points of view, which are in some ways quite unique and different from other areas of pharmacology. Drug addiction is closely associated with criminal behavior. One of the great and noble edifices of civilization is the philosophic and ethical view that man is perfectible, and some believe that this can be achieved by providing the appropriate circumstance or environment in which man can mature and be educated. Some have postulated that drug abuse is a consequence of an inadequate or pathologic set of socializing experiences or is a consequence of basic conflicts between the values and accepted patterns of behavior of a subculture and that of a larger culture. The degree to which man is malleable and perfectible by social forces is not known nor do we know the true desirability of socializing individuals to the extent that their behavior does not deviate from social norms. Some deviancy is essential for innovation and creativity, and at times there may be difficulties in determining whether an innovator or creator is exhibiting sociopathic behavior or not. This aspect of drug addiction is inherently a matter of social values and ethics.

These problems have not been treated in depth in this volume. Rather the major thrust of this volume is to explore the possibility that many drug abusers have an organic dysfunction of the brain. In recent decades, the possibility that mental illness is a consequence of improper functioning of the brain has gained increasing acceptance. The discovery of malfunctioning brain mechanisms that are responsible for behavioral pathology seems to be a goal achievable in the foreseeable future. With the advances in the understanding of mental illness, pathologic brain function and chemotherapy, it is unfortunate that the most prevalent and most costly of all mental diseases, psychopathy, has received the least attention of the neurobiologists.

It is hoped that this volume will stimulate the experimental therapists to become more involved in the general problems of psychopathy and the particular problem of drug abuse. The greater portion of this volume will be devoted to the description of the effects of drugs of abuse on the function of the brain and body. Although the state of the art at the present time does not permit definitive medical treatment of either drug abusers or criminals, another major portion of this volume will be devoted to methods and to description of endeavors that the pharmacologists can make and have made in both treating and limiting drug abuse.

There are two conspicuous omissions in this volume, namely, chapters on tobacco and cocaine. Tobacco was omitted because it was felt that it could not be treated comprehensively within the limitation of this volume. Cocaine was omitted because at the time this volume was formulated it was felt that knowledge about its modes and sites of action in producing euphoria and psychotic behavior was not well understood or even vigorously investigated and that most of our current speculations about its central nervous system actions would be treated in the sections on amphetamines and LSD-like hallucinogens.

The support that the National Institute of Mental Health, the National Institute on Drug Abuse, and the President's Special Action Office for Drug Abuse Prevention have given to the research of myself and other scientists at the Addiction Research Center, as well as to many other scientists who have had an interest in problems of drug abuse and whose works are cited in this volume, is gratefully acknowledged. The views that are expressed in this volume do not, however, necessarily represent the positions or views of these agencies or the Department of Health, Education and Welfare.

Many people have contributed to this volume both directly and indirectly. Mrs. BARBARA W. SMITH and Mrs. LULA H. MOORE, my dedicated secretaries, have provided invaluable assistance in preparing and editing the chapters and bibliographies. There will probably not be another quite so appropriate circumstance where I will be able to express my gratitude to my teachers and friends, Drs. K. R. UNNA, A. WIKLER, H. F. FRASER, and H. ISBELL, for shaping my interests in pharmacology, drug abuse, and diseases of the mind.

This volume has evolved more slowly than I thought it would when I undertook the task of editing it. I wish to express my gratitude to Prof. ALFRED FARAH for selecting me as its editor and to him and the publisher for patiently allowing this volume to evolve. It is only with the forebearance and encouragement of my wife Catherine and my children Catherine, David, and Douglas that this task could have been completed.

I would like also to apologize to our colleagues whose works have been omitted on behalf of myself and the other contributors. In some areas the literature is so vast that it could not have been comprehensively reviewed in this volume. Further, the literature in the area of drug addiction is expanding rapidly and many of the omissions of the more current works may reflect the difference in times that the various chapter were completed.

Lexington, Kentucky, July 1977 WILLIAM R. MARTIN

Contents

SECTION I

Problems of Drug Dependence

CHAPTER 1

General Problems of Drug Abuse and Drug Dependence. W. R. MARTIN. With 1 Figure

SECTION II

Morphine Dependence

CHAPTER 1

Neuropharmacology and Neurochemistry of Subjective Effects, Analgesia, Tolerance, and Dependence Produced by Narcotic Analgesics. W. R. MARTIN and J. W. SLOAN. With 9 Figures

CHAPTER 3

Assessment of the Abuse Potentiality of Morphinelike Drugs (Methods Used in Man).
D. R. Jasinski. With 1 Figure

CHAPTER 6

Detection of Drugs of Abuse in Biological Fluids. C. W. GORODETZKY

SECTION III

Sedative/Hypnotics and Alcohol Dependence

CHAPTER 1

The Pharmacology of Sedative/Hypnotics, Alcohol, and Anesthetics: Sites and Mechanisms of Action. C. M. SMITH

CHAPTER 2

The Assessment of the Abuse Potentiality of Sedative/Hypnotics (Depressants). (Methods Used in Animals and Man). H. F. Fraser and D. R. Jasinski. With 3 Figures

CHAPTER 3

Clinical Aspects of Alcohol Dependence. N. K. MELLO and J. H. MENDELSON. With 8 Figures

CHAPTER 4

Abuse of Non-Narcotic Analgesics. P. Kielholz and D. Ladewig. With 1 Figure

Contents
Part II: Amphetamine, Psychotogen and Marihuana Dependence

List of Contributors

H. T. Conrad, M. D., Consultant in Psychiatry, H. E. W. Regional Office IX, San Francisco, CA 94102, USA

H. F. Fraser, M. D., Consultant, National Institute on Drug Abuse, Lexington, KY 40502, USA

C. W. Gorodetzky, M. D., NIDA Addiction Research Center, Lexington, KY 40511, USA

D. R. Jasinski, M. D., NIDA Addiction Research Center, Lexington, KY 40511, USA

P. Kielholz, Prof. Dr. med., Psychiatrische Universitätsklinik 4025 Basel, Switzerland

D. Ladewig, PD Dr. med., Psychiatrische Universitätsklinik, 4025 Basel, Switzerland

W. R. Martin, M. D., NIDA Addiction Research Center, Lexington, KY 40511, USA

N. K. Mello, Ph. D., Associate Professor, Alcohol and Drug Abuse Research Center, McLean Hospital, Belmont, MA 02178, USA

J. H. Mendelson, Prof. Dr., Director Psychiatry Service, Boston City Hospital, Boston MA 02118, USA

J. W. Sloan, NIDA Addiction Research Center, Lexington, KY 40511, USA

C. M. Smith, M. D., Professor of Pharmacology and Therapeutics, State University of New York at Buffalo AND Director, Research Institute on Alcoholism, New York State Department of Mental Hygiene, Buffalo, NY 14203, USA

SECTION I

Problems of Drug Dependence

CHAPTER 1

General Problems of Drug Abuse and Drug Dependence

W. R. MARTIN

A. Introduction

In the past decade, there has been emerging interest in and increasing concern about drug abuse that is related in part to a modest increase in drug use and increased demands for legislative changes reducing the penalties for illegal drug use and legalizing the use of marihuana. There are many aspects to the ongoing public debates about drug use: (1) The encroachment on individual freedom, a consequence of legal efforts to minimize drug use, is deplored by many. Counterbalancing this position is the concern of others about the adverse effects of drug use on social functioning as well as individual and public health. (2) The emergence of a complex urban and technological society, particularly in the Western world, has created unique stresses and increased leisure and affluence to which mankind has not adjusted either biologically or socially. It is felt by some that there is a legitimate need for mind-affecting drugs which facilitate this adjustment. With the increase in complexity of our society and technology there has been a decrease in society's capacity to tolerate irrational, incompetent, and irresponsible actions including those induced by drug use. (3) The cost of drug abuse is becoming increasingly apparent even to the most affluent societies, and its competition with other individual and societal needs for resources is forcing a reassessment of our thinking about this problem. (4) The question of whether drug abusers are "normal" individuals who have because of their individual experiences evolved antisocial patterns of behavior, or alternatively, have disordered thinking processes which are responsible for their antisocial behavior occupies the concern of therapists. Correlatively, the solutions to these types of disorders have been directed toward changing behavior patterns through punishment and reward, or alternatively, viewed as a medical disease which can be treated with definitive therapeutic intervention.

It will be the purpose of this chapter to identify the issues that bear upon these debates. Despite the fact that there are strong protagonists on both sides of most of these issues, it is important to emphasize that nothing that approaches a consensus has been reached.

The use of drugs and chemicals to alter feelings, moods, and sensations was practiced even in antiquity by both primitive and advanced cultures for a variety of purposes, some of which were in keeping with the acceptable patterns of behavior, others not. Some agents have been freely available, others not, and some available licitly only through practitioners of magical arts and medicine. Until quite recently, there have been for the most part only limited, regional, and sporadic attempts to deal with problems of drug abuse. In the last century a number of changes have occurred that have led to increasing control of drugs at both national and international levels. With this control has come the inevitable encroachment on the freedom of individuals to manufacture, cultivate, process, dispense, prescribe, sell, and use a variety of agents that alter the functioning of the nervous system.

It is true that where money is to be made in satisfying useful and useless human needs, laws will be broken and treaties ignored. At best, national and international legal systems may to a degree limit the size and scope of drug abuse but this limitation has come at a cost that is hard to estimate but is very large.

There are several substantive reasons for believing that a complex industrial society can neither afford nor tolerate large drug abuse problems. It is not clear how drug abuse can be held to tolerable dimensions without intolerable and repressive legislation and controls. This dilemma is currently giving rise to anxieties and desperate, irrational attempts to cope with the problem.

In this chapter an attempt will be made to define the problem of drug abuse, to review the evolution of ideas concerning drug abuse, to analyze the issues that are involved, and to suggest courses of action that may allow our society to regulate drug abuse in a humane way with a minimum of loss of personal liberties.

B. Cost of Abuse of Major Drugs (Economic, Social, and Health)

I. Narcotic Analgesics

The narcotic analgesics relieve pain and diminish anxiety. In some subjects under certain conditions, they have a mood-elevating effect, consisting of feelings of efficiency, social acceptability, and other feelings of well-being. In other patients, the narcotic analgesics produce nausea, sedation, and feelings of apathy (see Sect. II, Chap. 1). When these agents are taken chronically, tolerance develops to many effects; however, contrary to popular opinion, increasing the dose does not allow the patient to recapture the euphorigenic properties of the drugs; rather, patients chronically intoxicated with narcotics become withdrawn or sedated, apathetic, narcissistic, and hypochondriacal. The consequence of chronic administration of narcotic analgesics on subjective effects is opposite to that effect sought by drug abusers. With the development of tolerance, subjects become physically dependent on narcotic analgesics, and when withdrawn, a discomforting abstinence syndrome develops. This is followed by physiologic abnormalities which persist for over 6 months after withdrawal, and which are associated with continuing feelings of apathy, inefficiency, and social withdrawal (MARTIN et al., 1973a).

It is difficult to estimate the cost of narcotic dependence with any precision. It is generally assumed that there were probably in the vicinity of 500000 narcotic addicts in the United States in the early 1970s, and one can assume that about half of them are at any one time using narcotics. The average cost of the drugs that active addicts use is probably at least $5000 per year resulting in an estimated cost of $1.25 billion annually. It can be further assumed that at any one time half of the addicts are not gainfully employed, and this represents approximately a minimum loss of productivity of about $5000 per year for those unemployed for a total of another $1.25 billion annually. Another very major cost is the decrease in life expectancy. A variety of estimates indicate that the mortality rate among heroin users is approximately 1 to

2% per annum and that the life of the addict, once he has become addicted, may be reduced by 50% (O'DONNELL, 1969). Because of the young age of addicts, this decrease in life expectancy comes during their potentially most productive years, and a rough calculation estimates the cost of this source of loss of productivity as something on the order of $2–2.5 billion per year. Further, the cost to the public of providing treatment to addicts currently probably is of the order of $0.2–0.3 billion per year (Strategy Council on Drug Abuse, 1973). Thus, the total cost of narcotic addiction in the United States at the present time probably exceeds $5 billion per year. In addition, the disruptive effect of the criminality associated with drug abuse which interferes with normal social intercourse would add to this cost.

II. Sedative/Hypnotics and Minor Tranquilizers (Depressants)

These drugs produce sedation, somnolence, and, in large doses, coma, overt sensory motor incoordination (drunkenness); a variety of mood changes which, depending on both the dose and situation, include feelings of well-being, self-confidence, talkativeness, grandiosity, garrulousness, apathy, sedation, and stupor (HILL et al., 1963a, b). Repeated administration of these agents leads to some degree of tolerance and physical dependence that is similar to that produced by alcohol. Withdrawal signs associated with barbiturate dependence include grand mal convulsions, tremulousness, fever, and delirium. The mortality of untreated alcohol dependence, depending on its severity, may be as high as 15–20%, and it may be assumed that untreated sedative/hypnotic dependence may have a comparable mortality. No systematic investigation has been made of the mood changes associated with chronic administration of these drugs; however, ISBELL et al. (1950) reported that subjects who were given increasing doses of barbiturates chronically after an initial period of elation became disheveled, confused, irritable, quarrelsome, and belligerent.

The incidence of the abuse of barbiturates as well as the individual harm that they may cause can only be roughly approximated. Based on data obtained on 6500 State of New York employes, CHAMBERS and HECKMAN (1972) found that approximately 4, 3, and 7% used respectively barbiturates, nonbarbiturate sedative/hypnotics, and minor tranquilizers, while 3, 1, and 3% used them regularly. A survey conducted by the Special Action Office for Drug Abuse Prevention (ROBINS, 1973, 1974) indicated that approximately 25% of enlisted men gave a history of barbiturate use alone or with other drugs while in Vietnam. This compares with an incidence of narcotic use of 34–38%. Approximately 50% of the narcotic abusers felt that they were addicted. Urine tests conducted by the U. S. Army of servicemen in Europe indicated that about 1% were positive for drugs and one-third of these were positive for barbiturates. Of all servicemen hospitalized for problems related to drug abuse, approximately 12% were for sedative-hypnotic abuse (TENNANT, 1972). PARRY et al. (1973) estimated that over 5% of the population are using depressants regularly. Thus, in all probability the incidence of sedative/hypnotic abuse in the United States is probably greater than abuse of narcotics, but is less than 1% of the population.

The societal and medical costs of sedative/hypnotic abuse are also unknown. Death is known to occur during abstinence in patients dependent on barbiturates (FRASER et al., 1953) and other depressants (see ESSIG, 1964). Although the incidence of death in patients dependent on depressants is unknown, it has been presumed that patients dependent on sedative/hypnotics may have a mortality approximating or somewhat less than that of alcohol dependence which may be as high as 15–20% when withdrawn without adequate treatment (VICTOR and ADAMS, 1953). Indeed, studies of barbiturate dependence in animals support this assumption (ESSIG and CARTER, 1962).

In a study of drug use in individuals arrested in several major cities (Chicago, New Orleans, New York, Los Angeles, San Antonio, and St. Louis), approximately 25% of those arrested had used barbiturates and 16% were using them at the time of their arrest (ECKERMAN et al., 1971). The overall trend was for criminality against persons to be somewhat greater than for other types of drug users but less than for non–drug-using criminals. In this study the incidence of barbiturate use was somewhat less than the use of narcotics. TINKLENBERG et al. (1974) found that secobarbital and alcohol, but not other barbiturate use, was associated with assaultive behavior.

The cost of producing sedative/hypnotics is less than the cost of equieffective doses of alcohol. Finally, the primary reinforcing property of sedative/hypnotics is probably as great or greater than than that of alcohol, as is their ability to produce physical dependence.

Thus, if it is assumed that the number of barbiturate abusers in the United States is between 200000 and 2000000 and that the cost of barbiturate abuse on a per capita basis is the same as alcohol abuse (ca. $2000 per year), the overall cost would be $1–4 billion a year.

BALTER and LEVINE (1969) estimated that approximately 30% of all prescriptions for psychoactive drugs were for barbituratelike drugs and almost 35% were minor tranquilizers. The total sales of ethical psychoactive pharmaceutical agents were approximately $1 billion in 1971. It has further been estimated that approximately 50% of all barbiturates are diverted into illicit traffic. The total cost of this illicit traffic cannot be accurately estimated, but it is unlikely that the cost of abused sedative/hypnotics and minor tranquilizers would exceed $0.5 billion per annum in the United States.

III. Alcohol

The effects of alcohol are very similar to the effects of sedative/hypnotics and minor tranquilizers and, like these drugs, repeated usage of alcohol leads to tolerance and physical dependence. The use of alcohol in the United States is by and large uncontrolled, and the size of the alcohol abuse problem is large compared to other abuse problems. The first report of the Department of Health, Education and Welfare to the U.S. Congress on alcohol and health (1972b) estimates that there are more than 90 million drinkers in the United States and approximately 9 million alcohol abusers and alcoholics. It is estimated that the cost of alcoholism due to loss of work and health problems, to welfare services provided to alcoholics, to property damage, and to medical expenses is approximately $15 billion per year. This estimate does not include the cost of the alcohol that is consumed by alcoholics, which would be approximately $5 billion per year. In 1971 total retail sales of alcoholic beverages were estimated to be $24 billion, and sales have increased $7 billion from 1966 to 1971 (National Commission on Marihuana and Drug Abuse, 1973).

Chronic use of alcohol reduces longevity by approximately 10 years. In addition, over 20000 highway fatalities annually are related to alcoholic use. Chronic use of alcohol produces serious pathology of the gastrointestinal, cardiovascular, and nervous systems.

IV. Amphetaminelike Agents (Stimulants)

Stimulant drugs cause or maintain wakefulness, increase and facilitate motor activity and performance, and stimulate the sympathetic nervous system. This group of agents includes not only amphetaminelike drugs (methamphetamine, methylphenidate, and phenmetrazine) but cocaine. They increase blood pressure, pulse rate, and pupillary diameter. In moderate doses (5–30 mg of *d*-amphetamine) they produce feelings of well-being and self-confidence; in larger doses, they produce nervousness, restlessness, tremulousness, anxiety, and delusions including paranoia, hallucinations, and disorientation. Stimulants also reduce appetite and food consumption. Repeated use leads to tolerance to some of these effects (see Part II, Sect. I, Chaps. 1 and 2), and exaggeration of others. Withdrawal of these agents results in depression and hypersomnolence which may be appropriately termed an abstinence syndrome. The euphoriant effects of amphetamines are similar to those produced by narcotic analgesics in narcotic addict subjects (HILL et al., 1963b; HAERTZEN, 1966; MARTIN et al., 1971) and are greater than those produced by narcotic analgesics in nonaddict populations (LASAGNA et al., 1955). Chronic administration of amphetamine produces a toxic psychosis, disorientation, and paranoid delusion (CONNELL, 1958; GRIFFITH et al., 1970, 1972; ANGRIST and GERSHON, 1972) as well as other less well-characterized changes including loss of motivation and good work habits, fatigue, violent behavior, compulsive stereotypy, unconsciousness, aphasia, paralysis, anorexia, and insomnia (KRAMER et al., 1967; GOLDBERG, 1968a, b, 1972; KRAMER, 1972).

JOHNSTON (1973) surveyed drug use in high schools and found that about 10% of high school students had used amphetamines and approximately 2% were using them regularly. The incidence of regular use was several times that of heroin, barbiturates, and hallucinogens. The incidence of amphetamine use in the armed services (TENNANT, 1972; ROBINS, 1973) is approximately the same as the incidence of barbiturate abuse and less than that of narcotic use. The second report of the National Commission on Marihuana and Drug Abuse (1973) indicates that there has been a progressive increase in the use of stimulants by high school and college students since 1967 with an incidence of about 20% in 1972. The incidence of current amphetamine use by criminals is approximately 10% and the percentage of serious crimes committed by amphetamine abusers is greater than that committed by barbiturate abusers, about the same as that committed by other types of drug abusers, but about half that committed by non–drug-abusing criminals (ECKERMANN et al., 1971). The incidence of amphetamine abuse based on a study of their use and abuse in Sweden is highest in the adolescent and young adult and in the criminal populations (GOLDBERG, 1968a, b), whereas the use and abuse of depressants become greater with increasing age.

The cost of amphetamines abused can only be roughly estimated. In 1969 over 45000 kg of amphetamine and methamphetamine were synthesized or imported into the United States, which would represent over 4.5 billion 10-mg doses. With the establishment of quotas, only 6000 kg were synthesized or imported in 1972. These figures represent only licit production and importation of the two most commonly used and abused stimulants and anorexigenics, and do not include many of the other agents which share their properties. A conservative estimate of the cost of stimulants abused in 1969 would be $0.25 billion a year. Further, there are no data that would allow an estimate of the social cost of amphetamine abuse, but clinical impressions indicate that it is less pernicious than alcohol abuse and that its incidence is greater than hypnotic abuse. Assuming then that there are currently 1 million stimulant abusers in the United States and that the social cost is $1000 an abuser-year, the total annual cost would be $1 billion.

V. Hallucinogens

Although there are several distinct classes of hallucinogens other than LSD-like hallucinogens which have had sporadic abuse (atropinelike agents, phencyclidine, ketamine, cannabis, and the stimulants, which have been previously discussed), LSD-like hallucinogens were used in antiquity in religious ceremonies and rites (mescaline and plants containing psilocin) and have become a distinct current drug abuse problem of some magnitude. These drugs produce feelings of well-being and euphoria in modest dose levels, and anxiety states, panic reactions, sensory distortions, delusions, hallucinations, flashbacks, and disorientation in higher dose levels. These changes in subjective states are accompanied by physiologic changes which include increased blood pressure and pulse rate, dilated pupils, and increased somatic reflex activity. When these drugs are administered chronically, tolerance develops rapidly to both subjective and physiologic changes. When use of the agents is discontinued, no abstinence syndrome has been observed.

The prevalence of hallucinogen abuse is probably of the same order of magnitude or slightly lower than amphetamine abuse in high school students (JOHNSTON, 1973; National Commission on Marihuana and Drug Abuse, 1973) and about one-half that of amphetamine abuse among college students (National Commission on Marihuana and Drug Abuse, 1973). These drugs differ from all other drugs of abuse in that they have no established role in medical practice and are not available legally. The incidence of serious crimes committed by users of hallucinogens (psychedelics) is less than that for barbiturate and amphetamine users and non–drug-using criminals.

Although the incidence of hallucinogen abuse is somewhat less than that of amphetamine abuse, the pattern of drug ingestion is different and the dosage consumed is considerably less. Existing clinical impressions suggest that there are undesirable consequences of hallucinogen abuse; however, their social and economic cost cannot be estimated.

VI. Cannabis, Marihuana, Hashish

Marihuana, hashish, and other preparations of *Cannabis sativa* have been used and possibly abused for their central effects over 2500 years, for the Indian Vedas recognized their inebriating qualities. Until the 1960s marihuana abuse had attracted scant and intermittent public attention, and its use or abuse was limited in Western

World civilization. In the 1960s marihuana use became well established in North America, particularly among young adults and adolescents. This has given rise to several serious attempts to formulate a rational public policy concerning the use and regulation of marihuana (National Commission on Marihuana and Drug Abuse, 1972; Commission of Inquiry into the Non-Medical Use of Drugs, 1972). Further, the United States Government has committed many millions of dollars to the investigation of the properties of marihuana and the tetrahydrocannabinols, which are responsible for the major central actions of marihuana, and their effects on behavior and social function with the hope that a rational public policy about marihuana could evolve. These endeavors have been summarized in the annual reports of the Department of Health, Education, and Welfare (1971, 1972a). To date, these intensive efforts have not resolved the social, public health, or medical issues.

The bulk of evidence suggests that marihuana's effects are caused by Δ^9-tetrahydrocannabinol (THC) although other cannabinols may have activity and interact with Δ^9-THC. The mood changes produced by cannabis and Δ^9-THC include feelings of well-being, silliness, drunkenness, sensory distortions, panic reactions, delusions, and hallucinations. The question as to whether tolerance develops to the action of cannabis cannot be broadly answered for it appears that tolerance develops to some but not other effects. Whether tolerance develops to the reinforcing effects of marihuana is not known. Existing information suggests that if marihuana or related compounds produce physical dependence the associated abstinence is difficult to detect. No hard estimates of the incidence of marihuana abuse have been made. A survey conducted by the National Commission on Marihuana and Drug Abuse (1972) indicated that 15% of adults and 14% of youths had used marihuana; however, only 2% of the adults and 4% of the youths who ever used marihuana were engaged in heavy use. These figures would suggest that several million individuals are abusing marihuana in the United States and indicate that only alcohol and tobacco abuse are more common.

There is little agreement about the public health and social implications of chronic marihuana use. The National Commission on Marihuana and Drug Abuse (1972) reported that there was "little proven danger of physical or psychological harm from ... intermittent use of natural preparations of cannabis ..." On the other hand, the Commission of Inquiry into the Non-Medical Use of Drugs (1973) was more cautious and felt that there were four major areas of concern about chronic cannabis use in which more information was necessary: the effects of cannabis on (1) adolescent maturation, (2) psychomotor function as it relates to driving and operating other machinery, (3) mental health, and (4) a predisposition to other drug use. In addition, our experience with tobacco with regard to pulmonary pathology should give rise to caution about the pulmonary complications of marihuana use. The relative insensitivity of epidemiologic methods in identifying chronic toxicity argues against the overinterpretation of negative data.

The cost of marihuana use and abuse cannot be estimated. There is good reason to believe that the extent of marihuana use could reach the same proportions as tobacco use if it were legalized. Thus, the National Commission on Marihuana and Drug Abuse (1972) survey indicated that experimentation and use of marihuana was approximately one-third that of tobacco among youths 12–17 years old. The Drug Enforcement Administration has estimated that 6 million pounds of marihuana and nearly 500000 pounds of hashish are consumed annually. The cost of cannabis to the user probably exceeds $4 billion annually. (This estimate has been arrived at by assuming the cost of marihuana is $250 a pound and hashish $5000 a pound.) In the event that marihuana were to be legalized, the cost of producing and merchandising it would approach that of tobacco and could range between $5 and 10 billion a year.

VII. Summary and Conclusions

Table 1 summarizes the estimates of costs of drug use and abuse and its incidence in the United States. These figures are not precise, for there is no way that all of the costs of drug use and abuse can be accurately estimated; however, these estimates do not represent extreme calculations. It is apparent that drug use and abuse probably cost in excess of $40 billion annually in the United States. This is a huge commitment of resources that inevitably competes with other social priorities.

Further, strong legal deterrents are used in the United States to minimize the abuse of marihuana, amphetamine, LSD-like hallucinogens, and sedative-hypnotics. In the absence of these deterrents, the magnitude of drug abuse problems would probably be greater and its cost greater.

The first and second reports of the National Commission on Marihuana and Drug Abuse (1972, 1973) have identified three aspects of the social impact of drug abuse: (1) public safety, (2) public health and welfare, and (3) normative social order or social values. Each of these dimensions of the drug abuse problem can in turn be viewed from several perspectives. Thus, assaults on public safety, health, and welfare have both an economic and personal vector. The economic consequence of industrial and automobile accidents and the loss of health and production in a rough way have been previously assessed. The personal unhappiness and the decrease in quality of life caused by broken families, poverty, stunted social development of children, crime, related personal grief, and anxiety associated with drug abuse cannot be quantitated but do enter into the politics of advocacy which influence public policy

Table 1. Estimates of the incidence and total cost of abuse and nontherapeutic use of psychoactive drugs

	Probable no. of abusers	Frequency[a] of drug abuse % of total	Cost in billions of dollars
Alcohol	9000000		20–25
Tobacco			10
Narcotic analgesics	500000	63.9	5
Sedative/hypnotics	200000–2000000	6.5	0.5–4
Amphetamines and cocaine	200000	5.6	1
Hallucinogens	100000	2.6	0.5
Cannabis	2000000	12.3	4
Total	12–13.8 million[b]		41.0–49.5

[a] These figures are from the April 24, 1974 report of the Client Oriented Data Acquisition Process (National Institute on Drug Abuse) based on information obtained from federally funded drug abuse treatment programs.
[b] This figure excludes users of tobacco.

and attitude. The question of the influence of drug abuse on social values and public morality cannot be discussed objectively. One of the major problems of Western World civilization is its inability to define major long-term social goals and to achieve a consensus concerning them. Certain aspects of viewing drug abuse must be deferred until social trial-and-error experiences give us a more intuitive feel for the dimensions of a heavily populated agricultural and industrial world. It seems inevitable that if we are to enjoy the advantages of life, longevity, and freedom from want, these ends will come at a cost of the loss of some personal freedom. Consensus on the issues of whether drug abuse will significantly affect the achievement of social goals and whether the decision to abuse drugs can remain a personal one has not been reached. The fact that present laws and mores are inconsistent and that there is an intensive public dialogue about drug abuse attest to this lack of consensus.

Although much of the current dialogue concerning the abuse of drugs has been concerned with individual freedoms and rights, little solid scientific effort has been devoted to learning if social drug use serves a useful function, an equally important problem.

C. Legal and Regulatory Approaches to Minimizing Drug Abuse

I. Forces Influencing Federal Legislation

Laws concerned with drug abuse have had several intents including minimizing availability of drugs of abuse, discouragement of drug abuse by punishing drug abusers, institutional rehabilitation of drug abusers, and removal of addicts from normal social intercourse by imprisonment. It is unfortunate that only the crudest of measures are available for the assessment of the efficacy, the monetary cost, and the social cost of these legal measures. Nevertheless, until effective alternatives are found to these traditional ways of dealing with problems of drug abuse, they will continue to be used.

Discussion of legal measures to minimize drug abuse will be restricted primarily to those endeavors of the United States Government, with only tangential references to international endeavors for the following reasons. Although China's concern about the regulation of opium traffic dates to the latter part of the sixteenth century (KOLB, 1958), preceding that of all other nations, problems of drug abuse in Western World civilization seem to have reached socially significant proportions during the twentieth century, earlier in the United States than in other Western World countries. The issue of individual freedom as opposed to societal need has been most vehemently articulated by protagonists in the United States to the end of evolving humanistic solutions to drug use and abuse problems. Finally, states and the federal government of the United States have provided continuing leadership in legislative, social, and medical endeavors to attempt to successfully cope with drug abuse problems for over 50 years.

The origins of the United States federal effort to limit drug abuse through legal measures are complex and are a consequence of a variety of influences and advocates who had several concerns. One major influence was the prohibitionists who felt that alcohol and drug use was morally wrong. This movement, which achieved its maximum political influence during the first two decades of the twentieth century, was characterized by an authoritarian intolerance of nonmedical drug use, a realistic assessment of the harm caused by drug and alcohol abuse, and an unrealistic assessment of the popular support for rigid prohibition. In the end a grand social experiment was conducted to clearly demonstrate that prohibition, although partially successful in decreasing the harm caused by alcohol abuse, was unacceptable and unworkable legally and politically.

Political purposes related to international trade and relations may also have exerted a significant influence, through a circuitous set of circumstances, on the federal laws relating to narcotic abuse. ANSLINGER and TOMPKINS (1953) and STEVENSON et al. (1956) have argued that China's concern about the opium problem during the nineteenth century was to a large degree due to the loss of silver reserves to the East India Company and hence to the British government and shareholders. Further, the defeat of China by the British in the Opium War increased feelings of nationalism and the rejection of foreign influences in China. MUSTO (1973) suggests that one of the motives of the State Department of the United States in taking a position for the strong international regulation of opium traffic was to decrease tensions between the United States and China, thus providing a more favorable circumstance for increasing United States trade with China. A second motive was to curtail the increasing spread of opium use in the Philippines following their cession to the United States following the Spanish-American War (ANSLINGER and TOMPKINS, 1953; MUSTO, 1973). A partial solution to this problem was to operate a government opium monopoly in the Philippines and to restrict its use to the Chinese, thus reinstituting the policy of the Spanish government. Legislation providing a statutory basis for this arrangement was introduced and then withdrawn because of the feeling of American prohibitionists that such an arrangement would be immoral. Governor Taft formed the Opium Committee which investigated the opium problem in the Philippines and made recommendations concerning its management. The testimony before this committee regarding the harmful consequences of opium use was ambiguous. Many witnesses felt that moderate opium use caused little or no harm (STEVENSON et al., 1956). In 1902 "Congress made it a punishable offense to give, sell or otherwise supply opium other than for medicinal purposes to an aboriginal native..." of the Philippines (U.S. Senate, 1956). With these ends in view, the State Department requested that Congress provide funds for the establishment of a commission (American Opium Commission) to arrange an international meeting of nations with Far East interests. To make the United States position more tenable in international negotiations, domestic legislation controlling narcotics was needed.

During the last two decades of the nineteenth century and the first two decades of the twentieth, there was a concurrent increase in professionalism and organization in medicine and pharmacy as well as an increased concern in public health about the safety of food and drugs. During this period the health services and sciences began the formulation of principles of sound practices in prescribing and dispensing narcotic analgesics, sedative/hypnotics, and cocaine (stimulants).

Dr. HARVEY W. WILEY was concerned about narcotics in freely available over-the-counter preparations whose constituents were unknown and unlabeled. The American Pharmaceutical Association also advocated restricted dispensing of narcotics and other drugs. Limitation of cocaine use as such and in beverages (Coca Cola) was advocated because it was feared that excessive use by Southern Negroes might lead to attacks on white society (cf. MUSTO, 1973).

Table 2. Major drug abuse laws

1902	Embargo on Certain Pacific Islands Opium Trade	32 Stat 33
1909	Opium Exclusion Act	35 Stat 614
1914	Narcotic Drug Import and Export Act	38 Stat 275
1914	Harrison Narcotic Act	38 Stat 785
1924	Prohibition of Heroin Use	43 Stat 657
1930	Establishment of the Bureau of Narcotics	46 Stat 585
1937	Marihuana Tax Act	50 Stat 551
1937	Penalties for Repeated Offenses	50 Stat 627
1944	Isonipecaine Tax Act	58 Stat 721
1946	Synthetic Drug Act	60 Stat 38
1951	Boggs Act	65 Stat 767
1960	Narcotic Manufacturing Act	74 Stat 55
1965	Drug Abuse Control Amendment	79 Stat 226
1968	Establishment of Bureau of Narcotic and Dangerous Drugs	82 Stat 1367
1970	Comprehensive Drug Abuse Prevention and Control Act	84 Stat 1236

In the formulation of the Harrison Narcotic Law (see below) and the subsequent execution of an international agreement which governed and regulated the production and dispensation of drugs of abuse, a number of compromises needed to be struck to reconcile a variety of interests. Manufacturers of drugs and pharmaceutical preparations opposed provisions that would put them in a disadvantaged competitive position, adversely affect profits, or make commercial transactions difficult. Similarly, physicians and pharmacists who were involved in formulating and dispensing medication objected to recordkeeping.

Out of such a concatenation of views, interests, and influences, the laws of the United States have emerged. Many of these same factors are still operating.

II. Summary of Laws of the United States Regulating Drugs of Abuse (Table 2)

The earliest federal legislation was a tax passed in 1890 which placed a duty of \$1 a pound on raw opium and \$6 a pound on smoking opium. This modest tax was neither a strong deterrent to opium use nor an especially important source of revenue. In 1890, a tax of \$10 a pound was levied on opium manufactured in the United States, and \$12 a pound on smoking opium.

As previously indicated, a variety of laws have been passed that were a consequence of international treaties and relations. According to TERRY and PELLENS (1928) there were in all four treaties with China dating from 1844 regulating opium trade in China by American citizens. In 1880 a federal law was passed which prohibited the import of opium by Chinese and which made it a misdemeanor for a U.S. citizen to traffic in opium in China.

As previously discussed, efforts to control rapidly growing abuse of opium in the Philippines not only gave rise to the passage of a law in 1902 which placed an embargo on the selling of opium to Philippine natives but was in part responsible for the meeting of the International Opium Commission (Shanghai Conference) in 1909. This conference adopted nine resolutions aimed at suppressing opium smoking. To permit the American delegates to take a strong stand at this conference, Congress passed the Opium Exclusion Act (1909) which prohibited the importation of opium into this country except for medicinal purposes. The Hague Convention (1912) resulted from a meeting of representatives from the powers which participated in the Shanghai Conference. At this convention, the principles of control and regulation of raw opium, morphine, heroin, and cocaine manufacture and commerce were established. The Narcotic Drugs Import and Export Act became law in 1914 and prohibited the import, export, and transshipment of opium except for medicinal purposes. Exportation of smoking opium was prohibited by this act and it stipulated that the export of opium, cocaine, and their derivatives be in accord with the laws of the importing country. A prohibitive domestic tax ($300 a pound) was placed upon the manufacture of all smoking opium (1914). The Narcotic Drug Import and Export Act of 1922 strengthened the Import and Export Act of 1914 and created the Federal Narcotic Control Board which established import quotas based on medical and other legitimate needs for crude opium and coca leaves.

Until the passage of the Harrison Licensing Act (1914), none of the federal legislation concerning the use of opium and cocaine had as its primary goal the regulation of domestic use of narcotics. It is not known with any certainty that there was a mandate for control of these drugs. There can be no doubt that the greater portion of opium imported and manufactured was not prescribed by physicians. Nor were the indications for the medical use of narcotics as rigorous as they are now. Nevertheless, it was uncertain, despite the magnitude of nonmedical and nonindicated use of morphine and opium, what the dimension of the problem was. Terry and Pellens (1928) have summarized both data and their interpretation concerning the magnitude of the opium abuse problems before and in the decade following passage of the Harrison Narcotic Act (Table 3). The range of estimates of the incidence is from 0.1% (Kolb and Dumez, 1924) to 1% of the population. The most prevalent estimate of the incidence of identifiable opium (narcotic) abuse was about 0.2%. Terry and Pellens (1928) present arguments that as many as one-half or two-thirds of addicts may not have been identified. Although this argument has a high degree of validity, it is complicated by the fact that addicts may be inactive as much as half the time. At the time of the passage of the Harrison Licensing Act, the major concern of health-oriented professionals was the harm that narcotic abuse caused to the individual but there were no vigorous attempts to examine its effects on social function. The fact that opium and cocaine were used and abused primarily by minorities such as Chinese and Negroes and by sporting class whites (Isbell, 1958) may have facilitated the passage of the Harrison Act in that users and abusers of these drugs had no strong political voice. The Harrison Act (1914) required licensing of and imposed a tax on all individuals who produced, imported, processed, manufactured, distributed, sold, prescribed, or dispensed opium and cocaine or their derivatives.

In 1924 the Narcotic Drugs Import and Export Act of 1922 was amended to prohibit the import of opium for the manufacture of heroin. This law had the support of the AMA (Musto, 1973). The purpose of this law was to prevent the spread of heroin abuse because it was felt that it was more dangerous than other narcotic drugs, had an especially pernicious effect on moral responsibility, and had no purpose that could not be served by other available drugs. From the vantage of nearly 50 years of experience, several conclusions can be made about this law and its passage: (1) It has not achieved its end. (2) It was based on false premises. Subsequent studies have shown that heroin does not differ significantly from morphine

Table 3. Summary of surveys and estimates

Locality	Year	Author	Source	Number of addicts recorded or estimated	Estimated number of addicts in continental U.S. for given year	Percentage of population affected
Michigan (96 cities)	1877	MARSHALL	Letters to physicians, one in each locality	1313	251936[a]	0.58%
Iowa (territory covered by 50 reporting druggists)	1884	HULL	Letters to druggists	235	182215[b]	0.32%
Jacksonville, Fla	1913	TERRY	Clinic and duplicate prescriptions of physicians	541	782118[c]	0.80%
Tennessee	1915	BROWN	Registration by state law	5000	269000[c]	0.27%
New York City	1919	HUBBARD	Clinic	7464	140554	0.13%
New York state	1920	HERRICK	Registration by state law	3900	396978	0.37%
Los Angeles, Cal	1920	BUCHER	Clinic	564	102005	0.10%
Shreveport, La	1920	BUTLER	Clinic	211	264276	0.25%

[a] Based on Marshall's estimate of 7763 addicts in state of Michigan.
[b] Based on Hull's estimate of 5732 addicts in state of Iowa.
[c] Not corrected for race composition.
[Reproduced from TERRY and PELLENS (1928) with the permission of the publisher (reprint edition).]

and other narcotic analgesics (MARTIN and FRASER, 1961) in its analgesic, euphorigenic, or dependence-producing properties. (3) Despite the fact that adequate comparisons between morphine and heroin were not done at the time of its introduction into clinical medicine, it was promoted by Bayer as an agent superior to morphine and codeine. It was even advocated by some physicians as a treatment for morphinism (Bulletin on Narcotics, 1953). In the intervening years, many drugs have been synthesized which have an abuse potentiality comparable to that of heroin, but because they offered no therapeutic advantage and because their abuse potentiality was recognized early, there was little commercial advantage to be gained by their promotion and thus they were not introduced into clinical medicine and have not been abused. This issue will be discussed subsequently in this chapter and volume.

Legislation was passed in 1930 establishing the Bureau of Narcotics in the Department of the Treasury which centralized the enforcement of an increasing number of tax acts aimed at suppressing narcotic and cocaine addiction. In 1930 the Commissioner of Narcotics was authorized to pay informers for information about the violation of narcotic laws.

In 1937 the Marihuana Tax Act was passed with the intent of controlling marihuana traffic in the United States. The history of this law has been recounted by

MUSTO (1973). The major force leading to this law appears to have originated among law enforcement officials and politicians of the West and Southwest who contended that the use of marihuana by Mexicans gave rise to violent behavior. Testimony on this law was not consistent. Some experts felt that marihuana use was especially pernicious, causing insanity; while others felt that the harmfulness of its effects was greatly exaggerated. It must be remembered that objective assessment of the effects of marihuana had not been made at this time. It was not until 10 years later that WILLIAMS et al. (1946) studied the effects of marihuana under controlled circumstances. Thirty years elapsed before the chemical Δ^9-tetrahydrocannabinol was isolated and shown to be responsible for most of marihuana's subjective and mood-changing effects in man (ISBELL et al., 1967). Further, at the time there were no effective vocal protagonists and propagandists for marihuana, and no segment of our economy was heavily dependent on cannabis production. Again, exaggeration of the health and social hazards of drug abuse and the absence of strong protagonists for marihuana use allowed a punitive law to come into effect that proved to be ineffective in accomplishing its goals. In 1937 a law was passed which increased penalties for repeated offenses of narcotic laws. The wisdom of the marihuana tax has recently become a subject of vigorous debate (KAPLAN, 1970) and it is unlikely that the issues concerning its use and control will be resolved in the immediate future.

During World War II the abuse of heroin and other narcotic drugs decreased markedly in the United States; however, following the war there was a great increase in the incidence of heroin addiction, particularly in young adults who were members of disadvantaged minority groups of large cities. Again, control of the problem was sought by increasing the severity of the penalties for violation of the tax laws. In 1951 the Boggs Law provided for mandatory penalties that increased in severity with repeated offenses. In 1956 the Narcotic Control Act was passed which further increased the severity of mandatory penalties. The penalties for sale were increased for a first offense to a mandatory 5–20 years; for a second offense to a mandatory 10–40 years; and for sale of heroin to a minor by an adult to 10 years to life or death at the discretion of the jury. Suspension or probation was allowed only for the first offense.

The rapidly developing science of pharmacology and experimental therapeutics has had an enormous impact on drug abuse in general and narcotic abuse in particular. The systematic investigation of chemicals on body function has led to the discovery of many drugs with psychopharmacologic activity. Using animal screening procedures, a very large number of compounds having pain-relieving properties have been discovered which have chemical structures quite different from the opium derivatives morphine and codeine. The Harrison Licensing Act controlled opium, cocaine, and their derivatives. In 1939 a new spasmolytic, meperidine (isonipecaine, Demerol), was synthesized which was shown to produce analgesia and had a chemical structure quite different from morphine or other alkaloids contained in opium. Studies by HIMMELSBACH (1943) and ISBELL (1955) clearly demonstrated that it had euphoric and physical dependence-producing properties similar to morphine. Because meperidine was not an opiate it was necessary to enact the Isonipecaine Tax Act of 1944 to control it. During World War II, pharmacologists and chemists in Germany developed yet another synthetic analgesic, methadone, which was shown to have an addiction liability similar to morphine (ISBELL et al., 1947, 1948). Methadone also differed chemically from opium alkaloids. In 1946 legal provisions were made which allowed substances with an addiction-forming or addiction-sustaining liability similar to that of morphine to be subject to narcotic law controls after public hearing. Under the authority of this law, a number of synthetic and semisynthetic agents were brought under control including methadone, alpha-acetyl methadol, ketobemidone (a meperidine congener) and drugs of the levorphanol series (dromoran). The Narcotics Manufacturing Act of 1960 further facilitated the control of morphine and cocainelike agents. Under this act, the Secretary of the Treasury or his delegate (Bureau of Narcotics) "... (after considering the technical advice of the Secretary of Health, Education, and Welfare or his delegate, on the subject) ... could, after due notice and public hearing declare substances with an ... addiction-forming or addiction-sustaining liability similar to morphine or cocaine..." or precursors opiates, and make them subject to narcotic laws control.

During the 1950s and 1960s there was increased use of and illicit traffic in barbiturates, amphetamines, and hallucinogenic drugs. The abuse of depressants and sedatives was by no means a new phenomenon. Anecdotal accounts of the abuse of chloral hydrate, paraldehyde, bromides, barbital, and phenobarbital were discussed by LEWIN (1964) and others. The perniciousness of the abuse of barbiturates was explicitly demonstrated by the studies of ISBELL and his collaborators (ISBELL et al., 1950; FRASER et al., 1953; FRASER et al., 1954), who showed experimentally that barbiturates when used chronically in large quantities induced tolerance and physical dependence. Similarly, there had been sporadic episodes of abuse of amphetamines in the United States, Europe, particularly Scandinavian countries, and Japan. Further, clinical and experimental evidence (cf., CONNELL, 1958; GRIFFITH et al., 1970, 1972) had demonstrated that chronic use of amphetamines produced a type of insanity. Perhaps the only new abuse problem that emerged during the fifties and sixties was the use of LSD-like hallucinogens. Even this type of drug abuse was not new. Peyote (mescaline), which resembles LSD in its action, was used by the Indians of the southwestern part of the United States. Although peyote use was an indication for admission to the Public Health Service Hospitals at Lexington and Fort Worth, there were no admissions of peyote users to these hospitals until the 1960s. The Federal Food, Drug and Cosmetic Act was amended by the Drug Abuse Control Amendment of 1965 and gave the Food and Drug Administration the responsibility of controlling the manufacturing, processing, dispensing, and prescribing of depressant, stimulant, and hallucinogenic drugs. The Secretary of the Department of Health, Education, and Welfare had the responsibility of controlling depressants, stimulants, and hallucinogens and was empowered to appoint expert committees to advise him. Penalties in the form of prison sentences and fines for violation of this law were quite moderate compared to the penalties of laws regulating narcotics, cocaine, and marihuana.

The final report of the President's Advisory Commission on Narcotic and Drug Abuse (1963) recommended that the responsibilities of both the Secretary of the Treasury and the Secretary of Health, Education, and Welfare with regard to investigation of illicit traffic in narcotics, marihuana, and dangerous drugs be transferred to the Department of Justice. In 1968 the Bureau of Narcotics and Dangerous Drugs was established in the Department of Justice, and functions concerned with drug traffic that had been previously under the Bureau of Narcotics and the FDA were transferred to this bureau. The formation of the Bureau of Narcotic and Dangerous

Drugs was part of a governmental effort to increase activities against organized crime in the United States. The Bureau of Narcotics and Dangerous Drugs felt that its most important mission was to attack organized interstate and international drug traffic. These administrative changes represented a part of a larger movement away from attempting to decrease drug abuse by making illicit use a criminal offense and by prosecuting the abuser. This change in attitude and tactic was a consequence of political pressures from the governments of large cities during the late 1950s and early 1960s where criminality and drug abuse were becoming increasingly disruptive and costly. The prosecution of the drug user was no longer able to contain the problem and other solutions were sought.

During the 1960s drug abuse problems and crime continued to increase. In 1968 there were increased efforts to redirect the federal drug enforcement effort again at the addict. In 1972 the office of Drug Abuse Law Enforcement was established in the Department of Justice to prosecute middle and street-level (users) pushers. In 1973 the Bureau of Narcotics and Dangerous Drugs was abolished by Executive Order and its functions as well as those of the office for Drug Abuse Law Enforcement and the office of National Narcotic Intelligence were placed in the Drug Enforcement Administration. The latter office had been formed to coordinate drug enforcement efforts of federal, state, and local law enforcement agencies and to provide a central coordinating point for intelligence concerning illicit drug traffic (see Strategy Council on Drug Abuse, 1973).

The continuing and growing concern about drug abuse during the 1960s led to the enactment of the Comprehensive Drug Abuse Prevention and Control Act of 1970. The scope of this act was broad, bringing under its control in a single act narcotic drugs, cocaine, marihuana, stimulants, depressants, and hallucinogens, and used interstate commerce instead of taxation as the major basis for its authority. Penalties for sale were reduced for narcotic drugs and marihuana. Although the language defining the various categories of controlled drugs was taken from previous legislation, the Attorney General was given a more rational and flexible set of criteria for considering drugs for scheduling. In regard to scheduling, the attorney general was required to seek the advice of the Secretary of Health, Education and Welfare. Five schedules were provided under the law: Schedule I contained drugs of high abuse potentiality and no or limited medical usefulness. Schedules II, III, IV, and V contained drugs of progressively lower abuse potentiality and increasing medical utility.

D. Medical Treatment and Research Efforts

Medical concern about problems related to drug abuse has undergone progressive evolution. TERRY and PELLENS (1928) and MUSTO (1973) have reviewed early efforts to treat narcotic and opium users. These endeavors were aimed for the most part at detoxifying users. Although gradual dose reduction, using any one of a number of potent narcotic analgesics including morphine and methadone, has been shown to be effective and relatively nondiscomforting, there are still recurring proposals and concerns about the development of nonnarcotic detoxifying modalities. Increasing sophistication in evaluating detoxification procedures as a consequence of the report of the Mayor's Committee on Drug Addiction (1930) and techniques devised by

HIMMELSBACH (1937) and HIMMELSBACH and ANDREWS (1943) allowed an objective assessment for the effectiveness of detoxification procedures, and to date no procedure other than dose reduction using narcotic analgesics has been proved to be more effective and safe.

Attempts to define and systematically characterize the nature of narcotic addiction and to define its medical aspects have had several origins in the United States: The Bureau of Social Hygiene, Inc., New York, established a Committee on Drug Addiction in 1921, which initiated and supported studies of the dimensions of the narcotic problems (TERRY and PELLENS, 1928), a clinical study of narcotic addiction at the Philadelphia General Hospital, and a study of the effects of narcotics in dogs at Iowa State University. (Papers emanating from these latter two studies are described and referred to in Sect. II, Chaps. 1 and 2.) To provide greater continuity of administrative leadership, this committee was taken over by the National Research Council in 1929, and signal accomplishments of this committee are described below.

A second major medical influence and driving force was the involvement of the federal government in problems of narcotic addiction which began with the establishment of two treatment facilities for narcotic addicts. These hospitals not only provided leadership in the treatment of narcotic addiction at a time when neither the medical nor scientific establishment had any great interest in this problem but led to the establishment of the first intramural research laboratory, the Addiction Research Center, in the Division of Mental Hygiene of the Treasury Department, the forerunner of the National Institute of Mental Health.

I. National Research Council

The history of the National Research Council's Committee on Drug Addiction, which subsequently was renamed the Committee on Problems of Drug Dependence, has recently been reviewed by EDDY (1973). The major thrust of this committee's research plan was to develop substitutes for all of the legitimate therapeutic uses of narcotics which would be devoid of addicting properties. To this end, a notable group of investigators collaborated in an effort whose main objective was the development of a nonaddicting, nontoxic analgesic such that there would be no legitimate need for narcotic analgesics. The beginning efforts in synthesizing new compounds were under the direction of Dr. LYNDON SMALL, a chemist of the University of Virginia and later the National Institutes of Health. Following World War II, several members of the pharmaceutical industry mounted major analgesic synthesizing programs. Animal studies of the analgesic and other actions of these compounds were assessed in Dr. EDDY's laboratory and in the pharmaceutical industry. Perhaps one of the most signal accomplishments of this committee was the support and encouragement of investigators who developed methods and facilities for measuring pain and analgesia in man. Major investigators involved in this effort include Drs. BEECHER, DE KORNFELD, EDDY, FORREST, HOUDE, KANTOR, KEATS, LASAGNA, and LEE (see Sect. II, Chaps. 2 and 3). Efforts to assess the abstinence-suppressing and for some drugs dependence-producing abilities were conducted in the monkey at the University of Michigan facility under the direction of Dr. SEEVERS (see Sect. II, Chaps. 2). These activities were supported by funds of the National Research Council derived for the most part from contributions from the pharmaceutical industry.

Those drugs that showed the potential for therapeutic use were assessed in man for their abuse potential at the Addiction Research Center.

The endeavor to find a nonaddicting analgesic has, to this time, met with modest success. Further, it has become clear that making the use of abused narcotics illegal has not and will not provide an adequate and complete solution to drug abuse problems. However, this is not to say that this endeavor has not had positive effects on narcotic abuse problems. The demonstration and objective characterization of the nature of narcotic addiction have been in part responsible for the development of a conservative physician attitude toward unnecessary and unjustified prescribing of narcotics and to a lesser degree other psychoactive drugs. Further, the systematic study of new analgesics for their abuse potentiality, prior to their being introduced into clinical medicine, has had several beneficial effects: (1) It has prevented the advertisement and promotion of newly developed narcotics as nonaddicting and thus has prevented their introduction into the drug subculture by the medical profession. (2) It has permitted the application of appropriate legal controls to new analgesics. As a consequence, no analgesic has been introduced into clinical medicine without appropriate controls since the late 1940s that has led to a significant abuse problem. The collaborative efforts of assessing the efficacy and abuse potentiality which have been conducted under the auspices of the National Research Council's committee have been a most successful endeavor in preventive medicine and have provided a proven model by which other psychoactive drugs with abuse potential can be assessed.

II. The Federal Government (Table 4)

1. Treatment

Federal laws that authorize the treatment of addicts and the conduct of research on problems of narcotic dependence have also had a complex and dynamic evolution. In 1929 the establishment of two narcotic farms for the treatment of both prisoner and volunteer addicts was authorized. The purpose of the establishment of these farms was to relieve the congestion in federal prisons, to minimize the effect that addicts might have on other prisoners, and to provide a constructive environment for the treatment and rehabilitation of addicts. Two large facilities were constructed at Lexington, Kentucky (1935) and Fort Worth, Texas (1938), which were subsequently to become U.S. Public Health Service Hospitals.

Table 4. Laws authorizing treatment of and research on narcotic addiction

1929	Establishment of Narcotic Farm (USPHS hospitals)	45 Stat 1085
1930	Establishment of Bureau of Narcotics	46 Stat 585
1944	Public Health Service Act	58 Stat 682
1966	The Narcotic Addict Rehabilitation Act	80 Stat 1438
1968	Alcoholic and Narcotic Addict Rehabilitation Amendments of 1968	82 Stat 1006
1970	The Community Mental Health Centers Amendment of 1970	84 Stat 54
1970	Comprehensive Alcohol Abuse and Alcoholism Prevention, Treatment and Rehabilitation Act of 1970	84 Stat 1848
1972	Drug Abuse Office and Treatment Act of 1972	86 Stat 65

These represented the first endeavors of the federal government in the treatment of narcotic dependence. These hospitals provided a variety of desirable services such as vocational therapy, continuing education, medical, surgical, and dental care, planned recreational opportunities, and psychiatric treatment to voluntary and prisoner patients who habitually used habit-forming drugs including opium, coca leaves and the alkaloids derived therefrom, Indian hemp, meperidine, and peyote which endangered "public morals, health, safety, or welfare" (Public Health Service Acts, 1929 and 1944). All of these services were offered in a nearly drug-free prison environment at pastoral sites. Although both hospitals incorporated what at the time were enlightened innovations in the treatment of addiction, they did not fulfill their expectations. A number of investigations of the usefulness and effectiveness of the Lexington and Fort Worth hospitals have been undertaken (PESCOR, 1943a, b; BRILL et al., 1963; DUVALL et al., 1963; MADDUX et al., 1971; O'DONNELL, 1969; VAILLANT, 1966, 1973), and no clear conclusions were reached concerning the value of the hospitals or the value of incarceration in the treatment of narcotic addiction in comparison to other treatment facilities and programs.

Although voluntary patients were eligible for admission to these hospitals, the professional staffs did not favor their admission for two reasons: (1) Most voluntary patients spent only a short time at the facilities, frequently leaving against medical advice after they had been detoxified. (2) The staffs felt that the institutional efforts devoted to the care and treatment of voluntary patients not only served no useful therapeutic end but diluted staff efforts. Between 1946 and 1956, the U.S. Public Health Service Hospital at Lexington used a Kentucky statute which made the habitual use of narcotic drugs a misdemeanor as a basis for the involuntary commitment of voluntary patients who sought admission to the hospital. Addicts were convicted of a misdemeanor and the sentence probated on the condition that they accept treatment at the Lexington hospital (KAY, 1973). Many of the staff members of the hospitals felt, on the basis of these experiences, that an involuntary commitment procedure would provide a better circumstance for treatment than a purely voluntary procedure.

During the 1960s civil commitment gained increasing acceptance as a modality of treating and managing narcotic addicts. California enacted a state civil commitment procedure in 1961 and New York in 1963. The President's Advisory Commission on Narcotics and Drug Abuse (1963) recommended that the federal government also enact a civil commitment statute. The Public Health Service also felt that civil commitment might be a way of civilly compelling patients to remain in therapy and supported federal legislation to this end. The Narcotic Addict Rehabilitation Act of 1966 was enacted and allowed the civil commitment of addicts, (1) in place of prosecution (Title I), (2) in place of sentencing (Title II), and (3) after the petitioning of the court by the addict or a related individual (Title III) (see Sect. II, Chap. 4). In addition, this law authorized the appropriation of funds for training of personnel for treating addicts as well as for developing and assessing treatment programs in communities through grants.

Even as the Narcotic Addict Rehabilitation Act was being implemented at the USPHS Hospitals, evidence was already accumulating in California and New York about the limitations of the civil commitment procedure. Judges had difficulty in identifying efficacious treatment programs to which addicts could be committed.

Many patients who were eligible for commitment would not accept it and when committed would be disruptive and would elope from the program. Finally, only one-third of the patients committed remained in good standing during their commitment (KRAMER et al., 1968). These same difficulties were encountered at the USPHS Hospital in Lexington. Further, the Public Health Service and the National Institute of Mental Health (NIMH) felt that addicts should be treated in their communities. In 1968, Congress passed legislation which placed the responsibility of narcotic treatment within community mental health centers which were administered locally but received their support from the NIMH. This change in emphasis in federal involvement in the treatment of narcotic addicts probably had several underlying purposes: (1) The USPHS and NIMH felt that treatment of the addict could be most effectively carried out in the community and that large centralized institutions were less than optimal treatment facilities. (2) The NIMH and the medical and legal communities at large felt that addiction was a disease and should not be regarded or dealt with as a crime (Joint Committee of the American Bar Association and the American Medical Association on Narcotic Drugs, 1961). (3) The NIMH wished to strengthen community mental health centers through the use of federal funds for their operation. Although support for NIMH's position for the use of federal funds for the operation of community mental health centers and the provision of social services by either the medical profession or Congress was not unanimous, nevertheless, the community mental health treatment facilities for alcoholics and narcotic addicts were continued from 1970 to 1973.

In 1970 the National Institute of Alcohol Abuse and Alcoholism was established to develop and conduct comprehensive health, education, training, research, and planning programs for the prevention and treatment of alcoholism and for the rehabilitation of alcohol abusers and alcoholics. This institute was placed in the National Institute of Mental Health. Further, the scope of the treatment responsibilities of the community mental health centers was broadened to include the treatment of all forms of drug abuse and dependence under the provisions of the Comprehensive Drug Abuse Prevention and Control Act of 1970.

The decade of the 1970s started with drug abuse and crime continuing to be issues of high political visibility and concern in America which was heightened by the outbreaks of heroin addiction in the armed forces, particularly among soldiers stationed in Vietnam. The problems of large cities continued to mount and were compounded by drug addiction. The USPHS and the NIMH, which had provided leadership in the treatment of narcotic addicts since the 1930s, had come under increasing criticism for their failure to mount extensive and effective regional treatment facilities. There were several reasons for this failure: (1) The NIMH did not view drug abuse as a major mental health problem to which a major commitment of resources should be made. (2) There was a legitimate question as to whether effective treatment modalities were available. The disappointments associated with the operation of two USPHS hospitals for the treatment of addicts probably was in part responsible for this pessimistic attitude. (3) There were few trained professionals who had both a knowledge of and an interest in drug abuse to mount and administer new treatment programs. (4) The philosophy of the mental health community concerning the nature of behavioral problems greatly restricted therapeutic and control measures that were acceptable to them.

Regardless of the validity of these reasons, the failure to develop regional treatment programs and effective civil commitment programs by NIMH led to political efforts to decrease its responsibility in the federal drug abuse treatment program. These factors together with the explosive political impact of the Vietnam heroin epidemic led a presidential task force to recommend the formation of a Special Action Office for Drug Abuse Prevention which was established in 1971. In 1972, the Drug Abuse Office and Treatment Act was enacted which had several major provisions: (1) A statutory basis was provided for the temporary establishment of the Special Action Office giving it authority to make recommendations about programs and budgets of agencies that had drug abuse programs. (2) Funds were appropriated for continuation of treatment under the auspices of community mental health centers. In additon, funds were also appropriated which enabled the Secretary of Health, Education, and Welfare to make grants to communities for the operation of drug abuse programs. (3) It established the National Institute on Drug Abuse which would become active in 1974. (4) Funds were appropriated for the development of nonaddictive analgesics, narcotic antagonists, and detoxification agents. (5) It established National Advisory Councils. In 1973 the Alcohol, Drug Abuse and Mental Health Administration was established in DHEW which was comprised of the National Institute of Alcohol Abuse and Alcoholism, the National Institute on Drug Abuse, and the National Institute of Mental Health.

By 1973 the annual federal commitment to drug abuse was approximately $0.75 billion. Nearly 90% of those funds were spent for treatment and rehabilitation, drug traffic prevention (law enforcement), and training and education. Most of these funds were spent for treatment and rehabilitation (49%) and law enforcement (33%). At the present time, there can be no hard evaluation of the effectiveness of this large commitment. Much of the training and education (6%) funds were committed to drug information efforts which have probably not been effective in decreasing drug experimentation, and some data suggest that it may have increased it. There has been little effort to determine the efficacy of training and educational efforts.

In 1972 approximately 60000 patients were being treated by federally funded programs with over two-thirds of them on methadone maintenance. The cost of treatment including administration costs has not been precisely calculated, but during 1972 it probably exceeded $2500 for a patient year of treatment. Several lines of evidence suggest that the rate of growth of the heroin abuse problem declined during the latter part of 1972. Thus, in Washington, D.C., DuPont and Greene (1973) and Greene and DuPont (1974) found that the number of narcotic overdose deaths, opiate offenses, percent positive urines, new admissions to treatment, and criminal offenses showed a decrease in 1972 and 1973. DuPont and Greene (1973) feel that the availability of treatment and decreased availability of heroin were largely responsible for this decrease. At the time of the writing of this chapter, there is evidence that heroin availability is again increasing.

The magnitude of the drug abuse problem has fluctuated over the years, and it has been difficult to identify with any certainty the factors that are responsible for this variation; however, it is apparent that definitive social and medical remedies have not been found and that the degree of success that has been obtained, which may have contained the problem, has been far short of a socially desirable goal.

2. Research

The involvement of the federal government in research on drug abuse problems probably started with the work of Dr. LAWRENCE KOLB, a psychiatrist in the Hygienic Laboratory of the USPHS, the forerunner of the NIMH. Kolb's major areas of research were epidemiologic studies which attempted to assess the size of the narcotic abuse problem, studies of the personality characteristics of the narcotic addict, and studies of morphine physical dependence in the monkey. Dr. C. K. HIMMELSBACH, another USPHS officer, initiated a clinical research program at the U.S. Penitentiary Annex at Leavenworth, Kansas in 1933. When the USPHS Hospital at Lexington was opened with Dr. KOLB as its head, Dr. Himmelsbach's laboratory was transferred to it and this laboratory subsequently became the Addiction Research Center (ARC). In 1948 with the formation of the NIMH, the research unit was transferred to the NIMH and became the NIMH Addiction Research Center. The ARC was the major intramural research facility dealing with problems of drug dependence of the federal government, although both USPHS hospitals and the NIMH intramural research program also made important contributions to the field. In 1967 the Center for Studies of Narcotics and Drug Abuse was formed in the NIMH to administer a rapidly growing grants and contracts extramural program whose sole concern was problems related to drug abuse. The NIMH Division of Narcotic Addiction and Drug Abuse was formed in 1968 and assumed administrative responsibilities for all of the NIMH intra- and extramural treatment and research activities. All of these activities were transferred from the NIMH to the NIDA in 1974.

Accompanying these administrative changes, there was rapid growth of research in drug abuse which had reached a figure of $51 million in 1973. Approximately two-thirds of these funds were administered by the NIMH for biomedical and behavioral research. Simultaneously there was a rapid growth of drug abuse research in the Department of Defense (12% of funds) and the Department of Justice (7%) as well as other governmental offices and departments. Figure 1 shows the types of research that were supported by these funds in 1972.

III. Medical and Psychiatric Concepts of Addiction

Although there has been increasing acceptance of the idea that drug abuse may be a medical or psychiatric disease, the characterization of the disease processes responsible for drug abuse and the evolution of definitive therapeutic modalities has not made adequate progress. The commitment of the medical and research communities to the disease states associated with the addictions has been quite modest. This is paradoxical when it is considered that psychopathy (personality or character disorders), of which drug addiction is one sociopathic manifestation, is the most prevalent of all mental disease and is the most costly. In the United States the incidence of clinically significant psychopathy probably exceeds 5% of the population and as has been previously estimated, the cost of drug abuse alone in the United States probably exceeds $40 billion a year. Despite the high prevalence and cost of psychopathy and drug addiction, there has been little speculation concerning its underlying pathology.

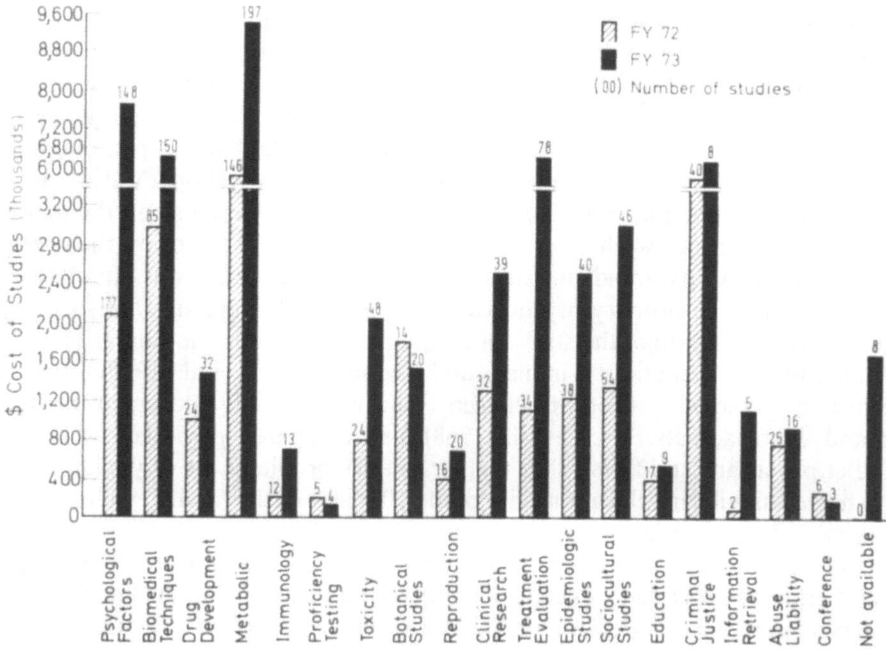

Fig. 1. Allocation of all federal drug abuse research funds to various types of experimentation. The total expenditures during 1972 were $30864256 and during 1973 were $51424976. These data are those of the Special Action Office for Drug Abuse Prevention's analysis of the commitment of federal research funds. [Reproduced from Strategy Council on Drug Abuse (1973) with permission of the Special Action Office for Drug Abuse Prevention]

1. Psychopathology

Many physicians have recognized the possibility that both hereditary and psychiatric factors could be related to drug abuse (cf. FELIX, 1939). KOLB (1925a) began classifying addicts and attempted to understand the processes that were responsible for their addiction. He recognized that a certain portion of addicts were psychoneurotic or psychotic and that other addicts, who were neither psychoneurotic or psychotic, had either a psychopathic diathesis, were constitutional psychopaths, had an inebriate personality, or were normal people who were "accidentally" addicted. KOLB (1925b) felt that narcotics were pleasurable only to certain individuals and that they produced deterioration only in the psychopath. FELIX (1939) recognized many of the problems associated with the term and concept of psychopathy and that narcotic addicts were for the most part either psychoneurotic, well-crystallized psychopaths or patients with a psychopathic diathesis. In the framework of KOLB's and FELIX's formulations of the problem, the psychoneurotic used narcotics to relieve anxiety with associated discomforting thoughts and feelings (negative euphoria), while the psychopath, who was morally defective and hedonistic, used narcotics to obtain a pleasurable state (positive euphoria) (KOLB, 1925b).

HILL et al. (1962) using the Minnesota Multiphasic Personality Inventory (MMPI) compared 200 addicts with a group of 199 hospitalized alcoholics and a group of 200 prisoners institutionalized in a state penitentiary. Using a factor analytic technique, he identified three primary addict personality types: (1) The undifferentiated psychopath, (2) the primary psychopath, and (3) the neurotic psychopath. All three types had a marked elevation on the psychopathic deviate scale of the MMPI and the psychopathic deviate scale was the only scale that was pathologically elevated in the undifferentiated psychopath. The primary psychopath also had an elevation on the hypomania and the paranoid scales and a decrease on the depression scale. The neurotic psychopath was characterized by elevations on the depression, psychoasthenia, paranoid, and masculinity-feminity scales. It was HILL's (1962) feeling that it was the primary psychopath who exhibited aggressive antisocial behavior. A striking feature of the study of HILL et al. (1962) was the similarity of the personality profiles of addicts, criminals, and alcoholics. HILL et al. (1960) found no differences in personality profiles between teenage and adult addicts or between black and white addicts. HILL et al. (1968) also compared physician addicts to nonaddict physicians and found that the personality profile of the physician addict group showed significant elevation on 8 of the 10 scales of the MMPI, including the psychopathic deviate scale.

The epidemic of heroin use by American soldiers in Vietnam during 1970 and 1971 also provided a circumstance to identify important variables in the genesis of drug dependence. ROBINS (1974) found that a prior history of marihuana, amphetamine, narcotic, or alcohol use and a history of arrests, truancy, or unemployment were the best predictors of drug use by American soldiers in Vietnam. These observations also lend support to the concept that psychopathy and consequent sociopathic behavior are a common characteristic of drug abusers.

The observations of HILL and HAERTZEN that narcotic addicts share many personality characteristics with other types of psychopaths have gone largely unattended. Further suggesting that there is a defect common to drug abusers and criminals is the observation of GUZE et al. (1974) that the most common psychiatric diagnoses in patients convicted of a felony were sociopathy, alcoholism, and drug dependence. There is general agreement about most of the clinical characteristics of the psychopath. Table 5 presents a summary of these characteristics (HILL et al., 1960, 1962; HILL 1962; CLECKLEY, 1964; GRAY and HUTCHISON, 1964; ROBINS, 1966; American Psychiatric Association, 1968; HARE, 1970; MARTIN et al., 1973a). ROBINS (1966) prefers the term "sociopathic personality." Although sociopathic or antisocial behavior is the reason that psychopaths come before the courts, are institutionalized, and are referred to treatment facilities, it speaks to the consequences, not the causes, of the disease. Sociopathic behavior does provide an objective measure of the disease process. In addition to sociopathic behavior, the psychopath has four other salient characteristics: (1) impulsivity and immaturity, (2) egocentricity or narcissism, (3) hypophoric feeling states [1], and (4) increased need states. Each of these characteristics can be: (a) conceptualized in terms of the psychic or physiologic deficits or disease states, or (b) operationally defined in terms of related or consequent behavior. It has been postulated that an increase in need or want can account for the characteristics

[1] Hypophoria is characterized by feelings of ineptness, inefficiency, unpopularity, and low esteem as well as a poor self-image. It differs from dysphoria in that uncontrollable unpleasant thoughts do not occur and from depression in that there are no associated feelings of unworthiness, loss of sleep or appetite or an inability to experience joy.

Table 5. Characteristics of psychopaths

Conceptual characteristics	Operational characteristics
A. Impulsiveness and immaturity	
1. Failure to learn by experience	1. Failure to follow a life plan
2. Poor judgment	2. Impulsive behavior
3. Failure to profit from punishment	3. Recidivism
B. Egocentricity, narcissism	
1. Incapacity to love	1. Poor marital adjustment
2. No meaningful interpersonal relationships	2. Social isolation
3. Impersonal and trivial sex life	3. Callous behavior
4. Self-centeredness	
5. Superficiality	
6. Lack of empathy	
C. Hypophoria	
1. Poverty in affective reactions	
2. Poor self-image, insecurity	
3. Thinks other people do not like them	
D. Increased need state	
1. Frustration threshold is low	1. Precocious and irresponsible sexual relations
2. Selfishness	2. Excessive interest in sex
	3. Thievery
	4. Embezzlement
	5. Indebtedness
E. Sociopathic Behavior	
1. Unreliable	1. Unemployed
2. Untruthfulness, lying	2. On welfare
3. Insincerity, lack of loyalty	3. Arrests
4. Impersonal and trivial sex life	4. Incarcerations
5. Lack of moral sense	5. Divorced or separated
6. Unable to experience guilt	6. Broken home
7. Remorse or shame	7. Truancy
8. Aggressiveness	8. School discipline problem
	9. School dropout
	10. Poor military adjustment
	11. Deviant sexual behavior
	12. Vagrancy
	13. Belligerency
	14. Use of aliases
	15. Alcohol abuse
	16. Drug abuse
	17. Wild adolescence
	18. Pathologic lying

of the addict psychopath including impulsivity, egocentricity, and hypophoria (MARTIN et al., 1973a; MARTIN et al., 1974). The enhanced need or want states may have several dimensions: (1) Because of increased need states, the psychopath is less likely to delay gratification, and this behavior is interpreted as impulsivity and immaturity. The fact that many need states are greatest in adolescence and early adulthood and that the highest incidence of both drug abuse and criminality occurs in this age range also is in keeping with this formulation. Finally, the possible relationship between egocentricity and need state is obvious. (2) The addict seems to have increased responsivity to strongly stressful and nociceptive stimuli (HIMMELSBACH, 1941; MARTIN et al., 1974). As a consequence their need to avoid unpleasant stimuli is increased and the addict may perceive the world as being more hostile and unpleasant than nonaddicts and this perception may be responsible for the addict's hypophoric feeling states. On less certain grounds are speculations concerning the strength of other types of need states. However, the fact that many psychopaths commit impulsive sex acts would be in keeping with this formulation. Further, crimes against property such as theft may be a consequence of an increased need state.

The role of needs and wants has occupied a central place in thinking about the addictions. RADO (1933) felt that addicting drugs reduced psychic discomfort associated with ungratified wants, and in suppressing wants and needs (libidinal) in turn decreased the need for objects of love. WIKLER and RASOR (1953) in summarizing analytic formulations of addiction concluded that the pathology was a deficit on the part of the addict in learning that all wants cannot be fulfilled. They also recognized that the narcissism was associated with an impairment of self-esteem. WIKLER (1953) has also emphasized the need-reducing properties of narcotics. The theory proposed concerning need states in psychopathy differs from previous formulations in that it suggests that the basic pathology is an increase in need state rather than a defect in the way with which need states are dealt.

The concept of need and want requires some discussion, for there are several conceptual frameworks from which it may be viewed. Some needs and wants are related to the maintenance of a congenial and the avoidance of a hostile environment; however, there are many toxins and hostile environmental circumstances that the central nervous system cannot recognize and hence cannot initiate the appropriate adaptive behavior. Further, there are needs and wants that have no unique physiologic basis. Although certain homeostatic regulatory mechanisms such as those governing tissue oxygen and CO_2 concentrations have clear and obvious adaptive utility, the role of other homeostats and need states is less clear and precise. It is not clear, as an example, that a temperature of 98–99° F is optimal or near optimal for function, species survival, or longevity in man. The sex drive in man is neither efficient, precise, nor many times appropriate in meeting the procreative needs of the species. It is possible that at some past time in both the biological and social evolution of man certain needs and wants were more adaptive than they now are. Thus, when man was ravaged by disease and more threatened by a hostile environment, and his life expectancy was less than it is now, impulsive sexual activity may have been highly adaptive. Not only did it allow the species to be replenished but procreation was initiated earlier in life so as to permit the offspring to grow to maturity during the life of the parent. In today's overpopulated, highly industrialized world, this type of impulsive behavior has apparently become less adaptive and in some instances may be harmful. Thus, it may be that at one time in the history of man not only was the immaturity and impulsivity caused by high need states not a social detriment, as it is now in Western World civilization, but a desirable characteristic. The characterization of psychopathy as a disease state may be in part a social decision based on social specific goals and needs. However, this does not mean that it is not a medical disease nor that need states do not have a biologic basis that can be therapeutically altered.

The study of addictions may provide insight into the problem of psychopathy. DOLE (DOLE et al., 1966; DOLE and NYSWANDER, 1968; DOLE, 1970) has speculated about the possibility of a neurologic or biochemical defect responsible for the pathologic hunger for narcotics of the addict and has suggested that narcotics may meet a metabolic need. Further, it has been shown that the chronic administration of narcotics induces long-persisting physiologic and behavioral changes (MARTIN et al., 1963; MARTIN et al., 1968; MARTIN and JASINSKI, 1969; MARTIN et al., 1973a; MARTIN et al., 1974) and that these changes are associated with exacerbation of characteristics of the psychopathic personality. All drugs of abuse can relieve hypophoric feelings of the addict (MARTIN, 1973) producing what has been called a feeling of well-being or euphoria. As many investigators have observed and reported, drugs of abuse do not commonly produce hyperphoria, exhilaration, or ecstasy. With chronic use of drugs of abuse, tolerance develops to their euphorigenic actions. Chronic administration of narcotics not only results in tolerance to euphoric effects but to the exacerbation of hypophoria (MARTIN et al., 1973a). The neurochemical correlates of hypophoria and euphoria have not been clearly identified (see Sect. II, Chap. 1 and Part II, Sect. I, Chaps. 1–3); however, this is an exciting and developing area of research.

2. Sociologic Theories

Although psychopathy is in all probability the major underlying defect in the genesis and continuation of addiction, other processes may be of great importance. There are a number of social circumstances that contribute to the initial addiction processes (CHEIN, 1956; CLAUSEN, 1957; CHEIN et al., 1964). There is now general agreement that most addicts obtain their first heroin from friends, peers, or relatives. Probably less than half of those who experiment with narcotics become addicted and use them on a regular basis. In the United States most addicts are found in the ghettos of the large cities. The social aspirations of the ghetto differ from those of the middle class and appear to place less value on the control of emotions, sexuality, and obtaining property. Thus, local social controls aimed at maintaining middle-class values are less effective. In addition to drug addiction, other types of deviant behavior are prevalent in the run-down neighborhoods. CHEIN (1956) found that environmental deprivation did not play a major role in addiction. BALL's (1965) analysis of admissions to the USPHS hospitals at Lexington and Fort Worth reemphasized the fact that narcotic addiction has been a significant problem among middle-class, middle-aged whites as well as among poor, young metropolitan minorities. The idea that peer and regional cultured attitudes towards narcotics and their availability are general influences that determine the probability that members of a subculture will experiment with and become addicted to drugs has face validity. These types of influences can be manipulated to some degree by social pressures but not by medical therapy. Further, drug use may itself be one of the activities characteristic of groups, cults, or subcultures and an activity necessary for social acceptance. Further, these subcultures have their own mythologies about the psychic, and in some cases physiologic actions of drugs, which in the subculture have the reputation of being desirable actions. Through some vague process which may be called suggestion or social reinforcement these socially induced imagined desirable properties cause the initiation and continuation of drug use (WIKLER, 1973a).

3. Tolerance, Dependence, and Conditioning

The phenomena of tolerance and dependence also play an important role in the addiction process (see Sect. II, Chap. 1). One of the primary reinforcing properties of drugs of abuse is their ability to elevate feeling state (euphoria). The nature of this change has been conceptualized in a variety of ways including the relief of anxiety, the activation of a pleasure center, the relief of hypophoria, and the diminution of needs and wants that give rise to unpleasant internal conflicts and hypophoria. These effects in themselves would be desirable if they were not associated with drug toxicity and tolerance. With the chronic administration of narcotics, not only does tolerance develop to the euphorigenic effects, anxieties and hypophoria are increased (WIKLER, 1952; MARTIN et al., 1973a). Tolerance is also known to develop to the euphorigenic actions of alcohol, depressants, stimulants, and hallucinogens. Whether chronic use of these agents is associated with an exacerbation of hypophoric feeling states has not been systematically studied and is not known. Clinical impressions of patients chronically intoxicated with alcohol, depressants, and stimulants would suggest that these patients are hypophoric. The psychopathology of the addict is not only not improved, it is worsened as a consequence of chronic drug use and abuse.

In addition, the narcotic analgesics, alcohol, and sedative-hypnotics induce physical dependence. The physically dependent patient has not only increased mental discomfort when stabilized but a dramatic exacerbation of negative feeling states when withdrawn. The drug of abuse is capable of relieving the discomfort associated with abstinence. In subjects self-administering drugs, this action has been termed *indirect primary pharmacologic reinforcement* by WIKLER (1973a).

WIKLER first recognized the role of conditioning in drug addiction and relapse and has, in a series of papers, presented several formulations of his thinking (WIKLER, 1948, 1955, 1961, 1965, 1973a, b). Tables 6 and 7 present his theory in a modified form. According to WIKLER, several conditioning processes occur. The relief of needs, hypophoria, and the abstinence syndrome (unconditioned and conditioned) constitute the primary reinforcing properties of the narcotic analgesics. The response to these needs is a learned behavior, drug "hustling," and through the association of "hustling" and the pharmacologic reinforcing properties of the drug, "hustling" behavior in itself becomes secondarily reinforcing according to an operant conditioning paradigm. Further, narcotics and probably other drugs stimulate both extero- (bitter taste, itching) and entero- (nausea) sensory receptors and through the association of these stimuli with the primary reinforcing properties, they too become secondary reinforcers or conditioned stimuli according to the classical conditioning paradigm. Thus, heroin-induced discomforting nausea in the novice becomes a "pleasant sick" in the addict. Presumably the stimulation of other enteroceptors, although not perceived, could also have secondary reinforcing properties. Yet another group of associations or type of conditioning occurs. A number of commonly recurring environmental stimuli that are associated with "hustling" and abstinence become conditioned stimuli which evoke conditioned "hustling" and abstinence both in the actively dependent patient and the postaddicted patient. In an extensive series of experiments, WIKLER and his collaborators have demonstrated that abstinence can be conditioned (WIKLER and PESCOR, 1967), that postaddict rats show a persistent relapse tendency (WIKLER and PESCOR, 1970), and that secondary rein-

Table 6. Wikler's conditioning theory of addiction and relapse (WIKLER, 1975)

Antecedent needs	**Primary pharmacologic reinforcement**						
	Direct			Indirect			
0 →	US →	UR →	Reinforcing event (or reinforcer) - - →	US →	UR →	New need →	Reinforcing event (or reinforcer)
Social pressure by peer group ("belonging") Hypophoria Anxiety	Receptor actions of opioid	Reflex re-sponses to receptor actions of opioid (signs of opioid effects)	Acceptance by peer group Reduction of hypophoria and anxiety ("euphoria")	Withholding of opioid	Abstinence changes in CNS (signs of opioid abstinence)	Abstinence distress	Hustling ↓ 0 ↓ Suppression of abstinence

New needs (after opioid withdrawal)	**Secondary pharmacologic reinforcement**				
	→ CS →	CR →	New need →	Reinforcing event (or reinforcer) →	0 Relapse → Suppression of abstinence
Protracted abstinence with dysphoria	Exteroceptive: Street associates Neighborhood characteristics Strung-out addict or dealer Dope talk Interoceptive: Receptor actions of drugs resembling those of opioids	Conditioned abstinence changes in CNS (fragmentary signs of opioid abstinence)	Conditioned abstinence distress ("craving")	Conditioned hustling	

0 = self-administration of an opioid (e.g., heroin);
US = unconditioned stimulus; UR = unconditioned response; CS = conditioned stimulus; CR = conditioned response. This scheme represents an attempt to combine the concepts and terminologies of classical and operant conditioning, assuming both can be explained by a single neural model (see WIKLER, 1973b).

forcement will persist for months (WIKLER et al., 1971). These concepts are presented in Table 6 (WIKLER, 1973b).

The potential role and importance of conditioned abstinence in relapse to narcotic use following detoxification is obvious. WIKLER (1965) recognized the necessity of extinguishing conditioned abstinence and drug-seeking behavior in the treatment of the detoxified addict and suggested that unsuccessful drug-seeking behavior be stimulated by a prolongation of the abstinence syndrome in an institutional setting. Subsequently, the use of narcotic antagonists such as cyclazocine and later naltrexone for the extinction of conditioned drug-seeking behavior and abstinence was proposed (MARTIN et al., 1966, 1973b). In an extensive series of experiments, it was shown that cyclazocine had a long duration of action, was effective orally, and that although tolerance developed to its agonistic actions, no tolerance developed to its antagonistic actions when it was administered chronically (MARTIN et al., 1966). It was further demonstrated that chronic administration of cyclazocine prevents the euphoric effects of morphine from becoming manifest and inhibits the development of physical dependence when morphine is administered chronically. Table 7 presents a reformulation of WIKLER'S concepts as they relate to the use of antagonists in the

Table 7. A reformulation of Wikler's conditioning theory of addiction and relapse

Need	Unconditioned stimuli		Unconditioned response	Conditioned stimuli	Conditioned response
	Reinforcers	Reinforcing event			
1. Hypophoria, anxiety, etc. 2. Neuronal alterations associated with abstinence and associated feeling states	Primary 1. Narcotic	1. Euphoria 2. Relief of abstinence	There may be no unlearned reflex-conditioned response to the conditioned stimulus which facilitates either the reduction of hypophoria or abstinence symptoms. Hustling for narcotics is a learned response that facilitates the reduction of these need states	1. Street associates 2. Neighborhood characteristics 3. Seeing a strungout addict or a dealer 4. Dope talk	1. Conditioned hustling 2. Conditioned abstinence
3. Conditioned abstinence	Secondary or aquired 1. Hustling 2. Entero and extero stimuli induced by drug of abuse	3. Hustling for narcotics			

Placing of addiction into a classic conditioning paradigm presents some difficulties. This table presents one formulation.

treatment of narcotic addiction. As can be seen, antagonists will prevent the reinforcing events of the narcotic from occurring and thus the conditioned stimulus will not be reinforced and will be extinguished. This formulation differs from WIKLER's in two respects. (1) The symptoms of the narcotic abstinence syndrome, which are discomforting, are regarded as an induced need state as are the presumed symptoms that are associated with conditioned abstinence. (2) In the classic and operant conditioning paradigm, the unconditioned response is an unlearned behavior that facilitates need reduction. Because of the artificial and unique nature of the needs associated with physical dependence, all responses that facilitate the reduction of the abstinence syndrome are probably learned.

4. Heredity

The dominant theories about drug abuse are that much of the addict's sociopathic behavior is learned and that this learned behavior can be treated with behavior-modifying techniques. An equally viable hypothesis is that there is an underlying organic disorder in brain function that is responsible for antisocial behavior of the psychopath and that this disorder can be rectified by appropriate chemotherapy. Determining the role of familial and genetic factors in drug abuse and psychopathy may assist in establishing the nature of the basic pathology in these disease processes. The role of heredity in criminality has not been established (McCORD and McCORD, 1964). Studies of the relationships between the XYY karyotype body physique and criminality have been recently reviewed by KESSLAR and MOOS (1970). Although data suggest that there may be relationship between the presence of a 47 XYY karyotype, tall body size, and aggressive criminality, it is not clear whether these three variables are causally related.

There is a relatively large body of data on the familial nature of alcoholism (cf. GOODWIN, 1971). One-fourth to one-half of fathers and brothers of alcoholics are alcoholics. The incidence of alcoholism in mothers and sisters of alcoholics is much lower and is similar to the incidence in the general population (ca. 5%). The concordance of alcoholism and heavy drinking is higher in monozygotic than in dizygotic twins. Studies of adopted and nonadopted male twins born of alcoholic parents suggest that environmental factors are of little importance in determining the incidence of alcoholism in offspring (GOODWIN et al., 1974). Further, a study of the incidence of alcoholism in adopted siblings indicates a familial influence. Through inbreeding in rats, it has been possible to obtain strains which have a high preference for alcohol as well as other strains which have a low preference (RODGERS and McCLEARN, 1962).

The role of familial influence in narcotic addiction has not been vigorously studied. POHLISCH (1933) as cited by GOODWIN (1971) found that the incidence of opium addiction in the siblings and parents of narcotic addicts was higher than the incidence of alcoholism. BROWN (1940) made anthropometric measurements of 400 white addict prisoners and found they were similar in height and weight to other control populations. O'DONNELL (1969) found that nearly 30% of 266 Kentucky addicts studied had a father, mother, or sibling who was either a narcotic addict, alcoholic, or criminal. NICHOLS and HSIAO (1967) found that one strain of selectively bred rats had a high preference for a morphine solution, while another showed a low preference. FRIEDLER and COCHIN (1972) and FRIEDLER (1974a, b) have shown in rats and mice that offspring born of parents who were postaddicts were tolerant to morphine and were smaller than appropriate controls. In studies conducted in mice, the decrease in body size was seen in the first and second generation of inbred offspring of postaddict males.

Although these observations do not prove that there is a genetic factor in drug abuse, they are consistent with this hypothesis. Further, these observations as well as others (see Sect. II, Chap. 1) suggest that through some unknown mechanism long-persisting physiologic and behavioral changes can be induced by chronic use of narcotic analgesics which may be transmittable to offspring.

E. Conclusions

The cost of licit and illicit drug use and abuse is large. In the United States it is estimated to cost between $40 and $50 billion a year, approximately 5% of the gross national product. It is not clear that drug abuse serves any useful purpose, and there is ample evidence that it causes harm. If the licit use of more drugs of abuse were permitted, probably the social, economic, and health costs of drug abuse would increase. Despite the cost and harm of drug abuse, it is highly probable that the thrust of social pressures in Western World civilization will be toward a permissive attitude toward the use of at least certain drugs. It is possible, however, that even affluent societies may at sometime in the future find that the competition for resources between the use and abuse of psychoactive drugs and other social needs will become so acute that strongly repressive measures may be employed to restrict their use. It is also possible that other individual freedoms may also be curtailed in this process.

Although it is argued by some that drug abuse is caused by social deprivation and assimilation into deviant subcultures, an alternate hypothesis is that drug abusers share with other social deviants a disorder in their thinking processes characterized by impulsivity, immaturity, egocentricity, hypophoric and increased need states. It is proposed that this disorder may have a hereditary basis or be a consequence of or exacerbated by drug abuse and may be biologically transmitted. These fundamental defects in turn give rise to other types of pathologic behavior such as conditioned abstinence and conditioned drug-seeking behavior which may produce further psychic debility and sociopathic behavior.

Although psychopathy is probably the most prevalent and most costly of all mental illnesses, it has received only modest attention from the psychiatric and medical community. This neglect has probably been a consequence of several factors: (1) Psychopaths are difficult and manipulative patients to treat and to deal with. (2) The medical profession has had little in the way of practical and useful remedies to offer the psychopath. For these reasons the neglect of these patients by the therapist is understandable. The failure of the psychiatric research community to vigorously attack this mental illness is less understandable but may be in part a consequence of the fact that physicians and biomedical scientists have had little contact with the psychopath. The vogues of current psychiatric therapy and theory have not been concerned with psychopathy. Criminality has been considered the domain of law and law enforcement.

With the development of biologic theories about psychopathy, rational therapeutic approaches to these disease processes will be conceived and in time definitive treatment modalities will evolve. To provide an optimal circumstance for the development of biologic theories and definitive chemotherapeutic modalities for the understanding and treatment of addicts it is essential that the experimental biologist and therapist work in a situation where they have frequent and intimate interactions with psychopaths. It is only in such a circumstance that knowledge about and insight into the disease process can develop unobscured by commonly held erroneous biases and stereotypes.

The search for potentially useful chemotherapeutic modalities for the treatment of psychopathy must have a new impetus. There are several reasons why, at the present time, conventional ways of developing new drugs will not be adequate in this search. (1) Psychopathy is not recognized as a medical disease. (2) There are no accepted hypotheses concerning the disease process that will provide a model for the search for and identification of new drugs. (3) Because many psychopaths are indigent, it is felt that a research investment will not result in a profitable drug product. (4) Because of the controversial nature of psychopathy and because of the ambivalence of the way the disease is viewed, there is the feeling that a venture into this area of research may lead to adverse publicity.

Although the United States government is making a large commitment to drug abuse, only a small portion of the funds are committed to basic research whose purpose is to understand the disease and to treat it. In order to develop psychotherapeutic agents for the treatment of psychopathy, a concerted effort must be made to synthesize and test new agents for their efficacy. There is every reason to be optimistic about the development of drugs for the treatment of psychopathy. The neuropharmacology and neurochemistry of moods and feeling states as well as need and drive states are making rapid progress.

Most Usual Abbreviations

AMA	American Medical Association
ARC	Addiction Research Center
DHEW	Department of Health, Education, and Welfare
FDA	Food and Drug Administration
MMPI	Minnesota Multiphasic Personality Inventory
NIDA	National Institute on Drug Abuse
NIMH	National Institute of Mental Health
USPHS	United States Public Health Service

References

American Psychiatric Association: Diagnostic and statistical manual of mental disorders, 2nd Ed. Washington, D.C.: Amer. Psychiat. Ass. 1968

Angrist, B., Gershon, S.: Possible dose-response relationships in amphetamine psychosis. In: Zarafonetis, C.J.D. (Ed.): Drug Abuse; Proceedings of the International Conference. Philadelphia: Lea & Febiger 1972

Anslinger, H.J., Tompkins, W.F.: The traffic in narcotics. New York: Funk & Wagnalls Co. 1953

Ball, J. C.: Two patterns of narcotic drug addiction in the United States. J. Criminal Law Criminol. Police Sci. **56**, 203—211 (1965)
Balter, M. B., Levine, J.: The nature and extent of psychotropic drug usage in the United States. Psychopharmacol. Bull. **5**, 3—13 (1969)
Brill, L., Rosenbloom, M. N., Farber, L., Gellert, A., Hill, R., Katz, E., Mangin, L.: Rehabilitation in drug addiction. A report on a five-year community experiment of the New York Demonstration Center. PHS Publication No. 1013. Washington, D. C.: U.S. Government Printing Office 1963
Brown, R. R.: The relation of body build to drug addiction. Publ. Hlth Rep. (Wash.) **55**, 1954—1963 (1940)
Bull. Narcot., **5**, No. 3 (1953)
Chambers, C. D., Heckman, R. D.: Employee drug abuse. York, Pa.: The Maple Press Co. 1972
Chein, I.: Narcotics use among juveniles. Social Work **1**, 50—60 (1956)
Chein, I., Gerard, D. L., Lee, R. S., Rosenfeld, E.: The road to H: Narcotics, delinquency and social policy. New York-London: Basic Books Inc 1964
Clausen, J. A.: Social and psychological factors in narcotics addiction. Law Contemporary Probl. **22**, 34—51 (1957)
Cleckley, H.: The mask of sanity, 4th Ed. Saint Louis, Missouri: C. V. Mosby Co. 1964
Commission of Inquiry into the Non-Medical Use of Drugs: Cannabis: A Report of the Commission of Inquiry into the Non-Medical Use of Drugs. Ottawa: Information Canada 1972
Commission of Inquiry into the Non-Medical Use of Drugs: Final Report of the Commission of Inquiry into the Non-Medical Use of Drugs. Ottawa: Information Canada 1973
Connell, P. H.: Amphetamine psychosis. Maudsley Monograph No. 5. London: Institute of Psychiatry 1958
Department of Health, Education, and Welfare: Marihuana and health. A Report to the Congress. Washington, D. C.: U.S. Government Printing Office 1971
Department of Health, Education, and Welfare: Marihuana and health. Second Annual Report to Congress, DHEW Publ. (HSM) 72—9113. Washington, D. C.: U.S. Government Printing Office 1972a
Department of Health, Education, and Welfare: First Special Report to the U.S. Congress on Alcohol and Health. DHEW Publ. 72—9099. Washington, D. C.: U. S. Government Printing Office 1972b
Dole, V. P.: Biochemistry of addiction. Ann. Rev. Biochem. **39**, 821—840 (1970)
Dole, V. P., Nyswander, M. E.: Methadone maintenance and its implication for theories of narcotic addiction. In: Wikler, A. (Ed.): The addictive states. Res. Publ. Ass. nerv. ment. Dis., Vol. 46. Baltimore: Williams & Wilkins 1968
Dole, V. P., Nyswander, M. E., Kreek, M. J.: Narcotic blockade. Arch. intern. Med. **118**, 304—309 (1966)
DuPont, R. L., Greene, M. H.: The dynamics of a heroin addiction epidemic. Science **181**, 716—722 (1973)
Duvall, H. J., Locke, B. Z., Brill, L.: Follow-up study of narcotic drug addicts five years after hospitalization. Pub. Hlth Rep. (Wash.) **78**, 185—193 (1963)
Eckerman, W. C., Bates, J. D., Rachel, J. V., Poole, W. K.: Drug usage and arrest charges; a study of drug usage and arrest charges among arrestees in six metropolitan areas of the United States. Bureau of Narcotics and Dangerous Drugs, U.S. Department of Justice, Final Report, BNDD Contract No. J-70-35 (SCID-TR-4). Washington, D. C. 1971
Eddy, N. B.: The National Research Council involvement in the opiate problem, 1928—1971. Washington, D. C.: National Academy of Sciences 1973
Essig, C. F.: Addiction to nonbarbiturate sedative and tranquilizing drugs. Clin. Pharmacol. Ther. **5**, 334—343 (1964)
Essig, C. F., Carter, W. W.: Failure of diphenylhydantoin in preventing barbiturate withdrawal convulsions in the dog. Neurology (Minneap.) **12**, 481—484 (1962)
Felix, R. H.: Some comments on the psychopathology of drug addiction. Ment. Hyg. (N.Y.) **23**, 567—582 (1939)
Fraser, H. F., Isbell, H., Eisenman, A. J., Wikler, A., Pescor, F. T.: Chronic barbiturate intoxication. Further studies. Arch. intern. Med. **94**, 34—41 (1954)

Fraser, H. F., Shaver, M. R., Maxwell, E. S., Isbell, H.: Death due to withdrawal of barbiturates. Ann. intern. Med. **38**, 1319—1325 (1953)
Friedler, G.: Morphine administration to male mice: Effects on subsequent progeny. Fed. Proc. **33**, 515 (1974a)
Friedler, G.: Long-term effects of opiates. In: Dancis, J., Hwang, J. C. (Eds.): Perinatal pharmacology: Problems and priorities. New York: Raven Press 1974b
Friedler, G., Cochin, J.: Growth retardation in offspring of female rats treated with morphine prior to conception. Science **175**, 654—656 (1972)
Goldberg, L.: Drug abuse in Sweden. Bull. Narcot. **20** (1), 1—31 (1968a)
Goldberg, L.: Drug abuse in Sweden (II). Bull. Narcot. **20** (2), 9—36 (1968b)
Goldberg, L.: Epidemiology of drug abuse in Sweden. In: Zarafonetis, C. J. D. (Ed.): Drug abuse; proceedings of the International Conference. Philadelphia: Lea & Febiger 1972
Goodwin, D. W.: Is alcoholism hereditary? A review and critique. Arch. gen. Psychiat. **25**, 545—549 (1971)
Goodwin, D. W., Schulsinger, F., Moller, N., Hermansen, L., Winokur, G., Guze, S. B.: Drinking problems in adopted and nonadopted sons of alcoholics. Arch. gen. Psychiat. **31**, 164—169 (1974)
Gray, K. G., Hutchison, H. C.: The psychopathic personality: A survey of Canadian psychiatrists' opinions. Canad. psychiat. Ass. J. **9**, 452—461 (1964)
Greene, M. H., DuPont, R. L.: Heroin addiction trends. Amer. J. Psychiat. **131**, 545—550 (1974)
Griffith, J. D., Cavanaugh, J., Held, J., Oates, J. A.: Dextroamphetamine: Evaluation of psychotomimetic properties in man. Arch. gen. Psychiat. **26**, 97—100 (1972)
Griffith, J. D., Cavanaugh, J., Oates, J. A.: Experimental psychosis induced by the administration of *d*-amphetamine. In: Costa, E. D., Garattini, S. (Eds.): International Symposium on Amphetamine and Related Compounds. New York: Raven Press 1970
Guze, S. B., Woodruff, R. A., Clayton, P. J.: Psychiatric disorders and criminality. J. Amer. med. Ass. **227**, 641—642 (1974)
Haertzen, C. A.: Development of scales based on patterns of drug effects, using the Addiction Research Center Inventory (ARCI). Psychol. Rep. **18**, 163—194 (1966)
Hare, R. D.: Psychopathy: Theory and research. New York: John Wiley & Sons, Inc. 1970
Hill, H. E.: The social deviant and initial addiction to narcotics and alcohol. Quart. J. Stud. Alcohol **23**, 562—582 (1962)
Hill, H. E., Haertzen, C. A., Davis, H.: An MMPI factor analytic study of alcoholics, narcotic addicts and criminals. Quart. J. Stud. Alcohol **23**, 411—431 (1962)
Hill, H. E., Haertzen, C. A., Glaser, R.: Personality characteristics of narcotic addicts as indicated by the MMPI. J. gen. Psychol. **62**, 127—139 (1960)
Hill, H. E., Haertzen, C. A., Wolbach, A. B., Jr., Miner, E. J.: The Addiction Research Center Inventory: Standardization of scales which evaluate subjective effects of morphine, amphetamine, pentobarbital, alcohol, LSD-25, pyrahexyl, and chlorpromazine. Psychopharmacologia (Berl.) **4**, 167—183 (1963a)
Hill, H. E., Haertzen, C. A., Wolbach, A. B., Jr., Miner, E. J.: The Addiction Research Center Inventory: Appendix. I. Items comprising empirical scales for seven drugs. II. Items which do not differentiate placebo from any drug condition. Psychopharmacologia (Berl.) **4**, 184—205 (1963b)
Hill, H. E., Haertzen, C. A., Yamahiro, R. S.: The addict physician: A Minnesota Multiphasic Personality Inventory study of the interaction of personality characteristics and availability of narcotics. In: Wikler, A. (Ed.): The addictive states. Res. Publ. Ass. nerv. ment. Dis., Vol. 46. Baltimore: Williams & Wilkins 1968
Himmelsbach, C. K.: Clinical studies of drug addiction. II. "Rossium" treatment of drug addiction. Publ. Hlth Rep. (Wash.) Suppl. **125**, (1937)
Himmelsbach, C. K.: Studies on the relation of drug addiction to the autonomic nervous system: Results of cold pressor tests. J. Pharmacol. exp. Ther. **73**, 91—98 (1941)
Himmelsbach, C. K.: Further studies on the addiction liability of Demerol (1-methyl-4-phenyl-piperidine-4-carboxylic acid ethyl ester hydrochloride). J. Pharmacol. exp. Ther. **79**, 5—9 (1943)
Himmelsbach, C. K., Andrews, H. L.: Studies on modification of the morphine abstinence syndrome by drugs. J. Pharmacol. exp. Ther. **77**, 17—23 (1943)

Isbell, H.: Withdrawal symptoms in "primary" meperidine addicts. Fed. Proc. **14**, 354 (1955)
Isbell, H.: Clinical research on addiction in the United States. In: Livingston, R.B. (Ed.): Narcotic drug addiction problems; proceedings of the Symposium on the History of Narcotic Drug Addiction Problems, Bethesda, Maryland, March 27—28, 1958. PHS Publication No. 1050. Washington, D.C.: U.S. Government Printing Office 1958
Isbell, H., Altschul, S., Kornetsky, C.H., Eisenman, A.J., Flanary, H.G., Fraser, H.F.: Chronic barbiturate intoxication. An experimental study. Arch. Neurol. Psychiat. (Chic.) **64**, 1—28 (1950)
Isbell, H., Gorodetzky, C.W., Jasinski, D.R., Claussen, U., von Spulak, F., Korte, F.: Effects of (−) Δ^9-trans-tetrahydrocannabinol in man. Psychopharmacologia (Berl.) **11**, 184—188 (1967)
Isbell, H., Wikler, A., Eddy, N.B., Wilson, J.L., Moran, C.F.: Tolerance and addiction liability of 6-dimethylamino-4-4-diphenyl-heptanone-3 (methadon). J. Amer. med. Ass. **135**, 888—894 (1947)
Isbell, H., Wikler, A., Eisenman, A.J., Frank, K., Daingerfield, M.: Liability of addiction to 6-dimethylamino-4-4-diphenyl-3-heptanone (methadon, Amidone or 10820) in man. Arch. intern. Med. **82**, 362—392 (1948)
Joint Committee of the American Bar Association and the American Medical Association on Narcotic Drugs: Drug addiction: Crime or disease? Interim and final reports of the Joint Committee of the American Bar Association and the American Medical Association on Narcotic Drugs. Bloomington, Indiana: Indiana University Press 1961
Johnston, L.: Drugs and American youth. Ann Arbor, Mich.: Institute for Social Research, University of Michigan 1973
Kaplan, J.: Marijuana—the new prohibition. New York-Cleveland: World Publishing Co. 1970
Kay, D.C.: Civil commitment in the Federal medical program for opiate addicts. In: Brill, L., Harms, E. (Eds.): Yearbook of Drug Abuse. New York: Behavioral Publications 1973
Kessler, S., Moos, R.H.: The XYY karyotype and criminality: A review. J. psychiat. Res. **7**, 153—170 (1970)
Kolb, L.: Types and characteristics of drug addicts. Ment. Hyg. (N.Y.) **9**, 300—313 (1925a)
Kolb, L.: Pleasure and deterioration from narcotic addiction. Ment. Hyg. (N.Y.) **9**, 699—724 (1925b)
Kolb, L.: Factors that have influenced the management and treatment of drug addicts. In: Livingston, R.B. (Ed.): Narcotic drug addiction problems; proceedings of the Symposium on the History of Narcotic Drug Addiction Problems. PHS Publication No. 1050. Washington, D.C.: U.S. Government Printing Office 1958
Kolb, L., DuMez, A.G.: The prevalence and trend of drug addiction in the United States and factors influencing it. Publ. Hlth Rep. (Wash.) **39**, 1179—1204 (1924)
Kramer, J.C.: Some observations on and a review of the effects of high-dose use of amphetamines. In: Zarafonetis, C.J.D.: Drug abuse; proceedings of the International Conference. Philadelphia: Lea & Febiger 1972
Kramer, J.C., Bass, R.A., Berecochea, J.E.: Civil commitment for addicts: The California program. Amer. J. Psychiat. **125**, 128—136 (1968)
Kramer, J.C., Fischman, V.S., Littlefield, D.C.: Amphetamine abuse. Pattern and effects of high doses taken intravenously. J. Amer. med. Ass. **201**, 305—309 (1967)
Lasagna, L., von Felsinger, J.M., Beecher, H.K.: Drug-induced mood changes in man. 1. Observations on healthy subjects, chronically ill patients, and "postaddicts." J. Amer. med. Ass. **157**, 1006—1020 (1955)
Lewin, L.: Phantastica: Narcotic and stimulating drugs; their use and abuse. New York: E. P. Dutton & Co. 1964
Maddux, J.F., Berliner, A., Bates, W.F.: Engaging opioid addicts in a continuum of services. A community-based study in the San Antonio area. Behavioral Science Monographs, Number 71—1. Fort Worth, Texas: Texas Christian University Press 1971
Martin, W.R.: Drug abuse—The need for a rational pharmacologic approach. In: Brill, L., Harms, E. (Eds.): Yearbook of Drug Abuse. New York: Behavioral Publications 1973
Martin, W.R., Eades, C.G., Thompson, W.O., Thompson, J.A., Flanary, H.G.: Morphine physical dependence in the dog. J. Pharmacol. exp. Ther. **189**, 759—771 (1974)
Martin, W.R., Fraser, H.F.: A comparative study of physiological and subjective effects of heroin and morphine administered intravenously in postaddicts. J. Pharmacol. exp. Ther. **133**, 388—399 (1961)

Martin, W. R., Gorodetzky, C. W., McClane, T. K.: An experimental study in the treatment of narcotic addicts with cyclazocine. Clin. Pharmacol. Ther. **7**, 455—465 (1966)

Martin, W. R., Jasinski, D. R.: Physiological parameters of morphine dependence in man—tolerance, early abstinence, protracted abstinence. J. psychiat. Res. **7**, 9—17 (1969)

Martin, W. R., Jasinski, D. R., Haertzen, C. A., Kay, D. C., Jones, B. E., Mansky, P. A., Carpenter, R. W.: Methadone—a reevaluation. Arch. gen. Psychiat. **28**, 286—295 (1973a)

Martin, W. R., Jasinski, D. R., Mansky, P. A.: Naltrexone, an antagonist for the treatment of heroin dependence. Arch. gen. Psychiat. **28**, 784—791 (1973b)

Martin, W. R., Jasinski, D. R., Sapira, J. D., Flanary, H. G., Kelly, O. A., Thompson, A. K., Logan, C. R.: The respiratory effects of morphine during a cycle of dependence. J. Pharmacol. exp. Ther. **162**, 182—189 (1968)

Martin, W. R., Sloan, J. W., Sapira, J. D., Jasinski, D. R.: Physiologic, subjective, and behavioral effects of amphetamine, methamphetamine, ephedrine, phenmetrazine and methylphenidate in man. Clin. Pharmacol. Ther. **12**, 245—258 (1971)

Martin, W. R., Wikler, A., Eades, C. G., Pescor, F. T.: Tolerance to and physical dependence on morphine in rats. Psychopharmacologia (Berl.) **4**, 247—260 (1963)

Mayor's Committee on Drug Addiction: Report of the Mayor's Committee on Drug Addiction to the Hon. Richard C. Patterson, Jr., Commissioner of Correction, New York City. Amer. J. Psychiat. **10**, 433—538 (1930)

McCord, W., McCord, J.: The psychopath: An essay on the criminal mind. New York: Van Nostrand Reinhold Co. 1964

Musto, D. F.: The American disease; origins of narcotic control. New Haven and London: Yale University Press 1973

National Commission on Marihuana and Drug Abuse: Marihuana: A signal of misunderstanding; First Report of the National Commission on Marihuana and Drug Abuse. Washington, D. C.: U.S. Government Printing Office 1972

National Commission on Marihuana and Drug Abuse: Drug use in America: Problem in perspective; Second Report of the National Commission on Marihuana and Drug Abuse. Washington, D. C.: U.S. Government Printing Office 1973

Nichols, J. R., Hsiao, S.: Addiction liability of albino rats: Breeding for quantitative differences in morphine drinking. Science **157**, 561—563 (1967)

O'Donnell, J. A.: Narcotic addicts in Kentucky. PHS Publication No. 1881. Washington, D. C.: U.S. Government Printing Office 1969

Parry, H. J., Balter, M. B., Mellinger, G. D., Cisin, I. H., Manheimer, D. I.: National patterns of psychotherapeutic drug use. Arch. gen. Psychiat. **28**, 769—783 (1973)

Pescor, M. J.: A statistical analysis of the clinical records of hospitalized drug addicts. Publ. Hlth Rep. (Wash.) Suppl. **143** (1943a)

Pescor, M. J.: Follow-up study of treated drug addicts. Publ. Hlth Rep. (Wash.) Suppl. **170** (1943b)

Pohlisch, K.: Soziale und persönliche Bedingungen des chronischen Alkoholismus. In: Sammlung psychiatrischer und neurologischer Einzeldarstellungen. Leipzig: G. Thieme Verlag 1933

President's Advisory Commission on Narcotic and Drug Abuse: Final Report. Washington, D. C.: U.S. Government Printing Office 1963

Radó, S.: The psychoanalysis of pharmacothymia (drug addiction). Psychoanal. Quart. **2**, 1—23 (1933)

Robins, L. N.: Deviant children grown up; a sociological and psychiatric study of sociopathic personality. Baltimore: Williams & Wilkins 1966

Robins, L. N.: A follow-up of Vietnam drug users. Special Action Office Monograph, Series A, No. 1. Washington, D. C.: Special Action Office for Drug Abuse Prevention 1973

Robins, L. N.: The Vietnam drug user returns. Final Report, September 1973. Special Action Office Monograph, Series A, No. 2. Washington, D. C.: Special Action Office for Drug Abuse Prevention 1974

Rodgers, D. A., McClearn, G. E.: Alcohol preference of mice. In: Bliss, E. L. (Ed.): Roots of behavior. New York: Harper & Row 1962

Stevenson, G. H., Lingley, L. R. A., Trasov, G. E., Stansfield, H.: Drug addiction in British Columbia; a research survey. Vols. I and II. Vancouver: University of British Columbia 1956

Strategy Council on Drug Abuse: Federal strategy for drug abuse and drug traffic prevention. Washington, D.C.: Strategy Council on Drug Abuse 1973

Tennant, F.S., Jr.: Drug abuse in the U.S. Army, Europe. J. Amer. med. Ass. **221**, 1146—1149 (1972)

Terry, C.E., Pellens, M.: The opium problem. New York: Bureau of Social Hygiene 1928. (Reprint edition: Publ. No. 115, Patterson Smith Reprint Series in Criminology, Law Enforcement and Social Problems. Montclair, N.J.: Patterson Smith 1970)

Tinklenberg, J.R., Murphy, P.L., Murphy, P., Darley, C.F., Roth, W.T., Kopell, B.S.: Drug involvement in criminal assaults by adolescents. Arch. gen. Psychiat. **30**, 658—689 (1974)

United States Senate: Laws controlling illicit narcotics traffic; summary of Federal legislation, statutes, executive orders, regulations, and agencies for control of the illicit narcotics traffic in the United States, including international, state, and certain municipal regulations (through 1st session, 84th Congress). 84th Congress, 2nd Session, Document No. 120, 1956

Vaillant, G.E.: A twelve-year follow-up of New York narcotic addicts: I. The relation of treatment to outcome. Amer. J. Psychiat. **122**, 727—737 (1966)

Vaillant, G.E.: A 20-year follow-up of New York narcotic addicts. Arch. gen. Psychiat. **29**, 237—241 (1973)

Victor, M., Adams, R.D.: The effect of alcohol on the nervous system. In: Merritt, H.H., Hare, C.C. (Eds.): Metabolic and toxic diseases of the nervous system. Res. Publ. Ass. Res. nerv. ment. Dis., Vol. 32. Baltimore: Williams & Wilkins 1953

Wikler, A.: Recent progress in research on the neurophysiologic basis of morphine addiction. Amer. J. Psychiat. **105**, 329—338 (1948)

Wikler, A.: A psychodynamic study of a patient during experimental self-regulated re-addiction to morphine. Psychiat. Quart. **26**, 279—293 (1952)

Wikler, A.: Opiate addiction. Psychological and neurophysiological aspects in relation to clinical problems. Springfield, Ill.: Charles C. Thomas 1953

Wikler, A.: Rationale of the diagnosis and treatment of addictions. Conn. med. J. **19**, 560—569 (1955)

Wikler, A.: On the nature of addiction and habituation. Brit. J. Addict. **57**, 73—79 (1961)

Wikler, A.: Conditioning factors in opiate addiction and relapse. In: Wilner, D.M., Kassebaum, G.G. (Eds.): Narcotics. New York: McGraw-Hill 1965

Wikler, A.: Sources of reinforcement for drug using behavior—a theoretical formulation. In: Cochin, J. (Ed.): Drug abuse and contraception. Pharmacology and the future of man; proceedings of the 5th Int. Cong. on Pharmacology, Vol. 1. Basel: Karger 1973a

Wikler, A.: Conditioning of successive adaptive responses to the initial effects of drugs. Conditional Reflex **8**, 193—210 (1973b)

Wikler, A.: Opioid antagonists and deconditioning in addiction treatment. In: Bostrum, H. (Ed.): Drug dependence—treatment and treatment evaluation. Skandia International Symposium, Oct. 15—17, 1974. Stockholm: Almqvist & Wiksell International 1975

Wikler, A., Pescor, F.T.: Classical conditioning of a morphine abstinence phenomenon, reinforcement of opioid-drinking behavior and "relapse" in morphine-addicted rats. Psychopharmacologia (Berl.) **10**, 255—284 (1967)

Wikler, A., Pescor, F.T.: Persistence of "relapse-tendencies" of rats previously made physically dependent on morphine. Psychopharmacologia (Berl.) **16**, 375—384 (1970)

Wikler, A., Pescor, F.T., Miller, D., Norrell, H.: Persistent potency of a secondary (conditioned) reinforcer following withdrawal of morphine from physically dependent rats. Psychopharmacologia (Berl.) **20**, 103—117 (1971)

Wikler, A., Rasor, R.W.: Psychiatric aspects of drug addiction. Amer. J. Med. **14**, 566—570 (1953)

Williams, E.G., Himmelsbach, C.K., Wikler, A., Ruble, D.C., Lloyd, B.J., Jr.: Studies on marihuana and pyrahexyl compound. Publ. Hlth Rep. (Wash.) **61**, 1059—1083 (1946)

SECTION II

Morphine Dependence

CHAPTER 1

Neuropharmacology and Neurochemistry of Subjective Effects, Analgesia, Tolerance, and Dependence Produced by Narcotic Analgesics

W. R. MARTIN and J. W. SLOAN

A. Introduction

The purposes of this chapter are: (1) to discuss the subjective effects and behavioral correlates, analgesia, and the tolerance and dependence-producing properties of the narcotic analgesics and to relate these changes to their abuse, (2) to review the neurophysiologic and neurochemical literature which bear upon the site and mode of action of narcotic analgesics, and (3) to speculate on the nature of the heroin abuse problem as it relates to the actions of the narcotic analgesics, as well as on areas of research which may lead to the identification of salient psychopathologic processes in the addict.

The importance of pharmacologic factors has occupied a central position in theories of narcotic addiction and dependence and has been concerned with two probably basic properties of narcotic analgesics: (1) their habituating or primary reinforcing actions, and (2) their ability to induce tolerance and dependence. Neither of these properties has been defined conceptually with any degree of sharpness, and thinking about them has been basically phenomenological. The analgesic actions of narcotic analgesics as well as their ability to relieve anxieties, suffering, apprehensions, and insecurities have been thought to be related in some way to their habituating properties. Thus, the relief of the unpleasant states which may be of clinical significance in certain psychopathologic and neurotic states has been called negative euphoria (KOLB, 1925). In other circumstances where there is no overt evidence of pain or anxiety, the narcotic drugs produce pleasurable effects. The degrees of pleasantness of sensation (CABANAC, 1971) and of liking of drugs (FRASER et al., 1961) have been measured in man. It is presumed that such descriptive terms as pleasantness and liking are in some way related to what KOLB (1925) referred to as "positive euphoria." It has been further presumed that the same attributes can be assessed in animals by measuring drug-seeking behavior.

Physical dependence has been conceptualized as an induced need state (WIKLER, 1961), which in turn adds another dimension to the reinforcing properties of narcotic analgesics. As CABANAC (1971) has shown, the pleasantness of many sensations that are associated with homeostatic and need mechanisms increases as need becomes greater and decreases with satiation. Similarly, the reinforcing properties of narcotics are greatest when dependent subjects are acutely abstinent. The assumption of a relationship between physical dependence and a need state provides a basis for conditioning to occur and WIKLER (1961, 1965) has postulated that conditioned abstinence and conditioned drug-seeking behavior may be of importance in relapse.

HILL (1962) and HILL et al. (1962, 1968) found that institutionalized and physician addicts, alcoholics, and criminals had elevated scores on the psychopathic deviate (Pd) scale of the MMPI. More recently, OLSON (1964), GILBERT and LOMBARDI (1967), ELLINWOOD (1967), COX and SMART (1972), and HELLER and MORDKOFF (1972) have also found that not only narcotic addicts but amphetamine and other drug abusers also have elevated scores on the MMPI Pd scale. The Pd scale contains items with many conceptual implications; however, over half the items are concerned with an inadequate or poor self-image, and disapproval on the part of others. Individuals who have elevated scores on the MMPI Pd scale have feelings of unhappiness, inefficiency, ineptness, unpopularity, and of being unfortunate, and they recognize that others disapprove of them. Similarly, the psychopathic scale developed by HAERTZEN and PANTON (1967) from the Addiction Research Center Inventory (ARCI) has many items relating to disapproval by others and poor self-image. HILL et al. (1963a) and HAERTZEN (1966) developed a series of scales that measure the subjective effects of narcotic analgesics (M, morphine), amphetamines (A, amphetamine), LSD (L), pyrahexyl (Py), pentobarbital (P), and alcohol (Al). These drugs, which comprise prototypes of all major drugs of abuse except nicotine, produce a type of euphoria which shares the common characteristics of producing a better self-image, feelings of improved performance and competence, as well as feelings of contentment, happiness, and harmony and decreased feelings of discouragement and worry (MARTIN, 1973). Table 1 summarizes some of the responses to items of the MBG (Morphine-Benzedrene Group Scale) (HAERTZEN, 1966) obtained for these drugs of abuse in inactive narcotic addicts.

Table 1. Questions from the MBG Scale of the Addiction Research Center Inventory and the level of significance of responses in subjects receiving morphine (M), amphetamine (A), LSD (L), pyrahexyl (Py), pentobarbital (P), and alcohol (Al)

MBG scale Items	M	A	L	Py	P	Al
3	S	S	M	S		
91	S	S	S	S	M	
98	S	S	S		M	S
102	S	S	M	S	M	
155		M			M	
299	S	S	M	S	M	
325	S	S	M			
345	S	S	S		S	S
396	S	S			M	
398	S	S	S		S	S

S = significant; M = marginally significant.

3. I feel as if I would be more popular with people today.
91. I feel as if something pleasant had just happened to me.
98. I am in the mood to talk about the feeling I have.
102. I feel less discouraged than usual.
155. Other people are likely to regard me as stimulating.
299. Right now I feel as if all my needs are satisfied.
325. Things around me seem more pleasing than usual.
345. I feel so good that I know other people can tell it.
396. I would be happy all the time if I felt as I feel now.
398. This is one day I'll put my troubles on the shelf.

Important positive euphoria-producing properties of these agents are related to feelings of pleasantness, happiness, enhanced self-image, and of a lack of trouble. It is highly probable that these subjective effects may have a higher positive valence to addicts in particular and psychopaths in general than they do to normal subjects and may be causally related to their abusing drugs. During chronic use of morphine and methadone, not only does tolerance develop to these euphoric effects but negative feeling states associated with attitudes of poor self-image, unpopularity, and apathy are enhanced (HAERTZEN and HOOKS, 1969; MARTIN et al., 1973b). These negative feeling states are further enhanced during abrupt withdrawal and persist for many months following withdrawal as part of the protracted abstinence syndrome (HAERTZEN and HOOKS, 1969; MARTIN et al., 1973b).

These feeling states may best be designated as *hypophoric* to distinguish them from both dysphoric and depressive states. Drug-induced dysphoria shares in common with hypophoria feelings of inadequacy and inefficiency, but also exhibits the characteristics of anxiety, apprehension, and the inability to suppress unwanted and disturbing thoughts. Similarly, hypophoria may be associated with depression but differs in that feelings of despair and worthlessness usually are not present, and disturbances of sleep and appetite are uncommon. Thus, one of the major differences between the psychopath and the depressed patient may be with regard to need and drive states. Whereas many need states are decreased in the depressed patient, they are not overtly decreased in the psychopath, and may be exaggerated (see Sect. I, Chap. 1). It has been suggested that these persisting negative feeling states may be the fundamental pathologic disorder in addicts and other psychopaths and may be responsible for other characteristics seen in psychopaths including narcissism and the lack of concern for others, immaturity and the inability to delay gratification (MARTIN et al., 1973b).

The analysis of the basis of hypophoria may provide insight into the pathophysiology of psychopathy. The understanding of the mode of action of drugs that produce euphoria and relieve hypophoria may not only assist in this analysis but may allow the understanding of the pathogenesis of hypophoria and psychopathy.

B. Subjective States and Their Correlates

I. Introduction

The subjective effects produced by narcotics have been regarded as one of the primary reasons for their abuse, and it is highly probable that the nature of their subjective effects is causally related to the pathology of narcotic addiction. For this reason, the elucidation of not only the mechanisms of action of narcotic analgesics in producing subjective changes but the delineation of the problem is central to the problem of narcotic addiction. To understand the neurophysiologic and neurochemical basis of subjective effects produced by narcotic analgesics, a variety of techniques have been employed, including the development of techniques to objectively measure and characterize subjective changes produced by narcotic analgesics in man. There have been attempts to relate neurophysiologic phenomena and behavioral signs both in man and animals to subjective effects of narcotic analgesics in man.

Morphine is alleged to have both excitant and depressant properties. A variety of criteria, which depend at least in part on the theoretical framework employed by the classifier, have been used to decide whether a given effect is excitant or depressant. Among the criteria that have been employed are naturalistic characterizations; thus, sleep, depressed respiration, and analgesia are depressant effects, and increased motor activity and convulsions are excitatory effects. Another criterion has been the character of neuronal activity; thus, miosis and vagal slowing have been regarded as excitant because they reflect increased activity of the third and tenth nerves respectively. Yet another criterion has been related to the development of tolerance with the assumption that tolerance develops to depressant but not to excitatory actions of narcotic analgesics (TATUM et al., 1929). Finally, opioid antagonists have been postulated to antagonize the depressant but not the stimulant effects of narcotic analgesics (SEEVERS and DENEAU, 1963). Despite the fact that it is apparent that these criteria are not mutually consistent, the terms excitant and depressant will continue to be used to categorize the various actions of narcotic analgesics, for they recognize the apparent paradoxical effects of these agents and in all probability their multiple modes of actions.

II. Subjective Effects in Man

1. Single Doses

The dominant subjective effects produced by morphine are euphoria or feelings of well-being and sedation. There have been several systematic investigations of these subjective changes. LASAGNA et al. (1955) found that morphine and heroin produced euphoria and some feelings of improved mentation in postaddicts but dysphoria, sedation, and feelings of impaired mentation in nonaddict subjects.

HILL and HAERTZEN using the Addiction Research Center Inventory (ARCI) (HILL et al., 1963a; 1963b) have studied the subjective effects produced by a number of psychoactive drugs, including morphine, on prisoner patients with a history of narcotic dependence. The ARCI contains 550 items which include items relating to a variety of subjective states (HAERTZEN et al., 1963). The Morphine Specific Scale (M) of HILL et al. (1963b) consisted of items which were responded to significantly more frequently when patients had received morphine than when they had received a placebo. The types of items varied; some were quite specific and related to the emetic and histamine-releasing actions of morphine; while another large group of items was concerned with enhanced self-image (e.g., popularity), feeling of efficiency, and an absence of concern (e.g., euphoria or well-being). Among those items that were marginally significant were ones related to sedation, sleepiness, and drowsiness. Subsequently, HAERTZEN (1966), using a factor analytic technique, derived the MBG scale, which contained items to which patients responded positively while under the influence of morphine and amphetamine but not other drugs. This scale contains for the most part items related to feelings of well-being, self-esteem, satisfaction, and competence. Although morphine may on some occasions produce sedativelike subjective changes manifested by increases in the Pentobarbital-Chlorpromazine-Alcohol Group Scale (PCAG), this is unusual (see Sect. II, Chap. 3).

SMITH and BEECHER (1959) studied the effects of morphine in college and graduate students. The predominant and most common effect of morphine was to produce feelings of drowsiness, tiredness, and sleepiness, as well as specific effects such as nausea and itching (SMITH and BEECHER, 1959). Other symptoms such as friendliness, geniality, peacefulness, satisfaction, and detachment were reported, but less frequently. In contrast to the findings of HILL and HAERTZEN in addicts, students report feelings of indifference and lack of efficiency and industry. In a comparison of the effects of heroin and morphine in student subjects, SMITH and BEECHER (1962) found that both drugs produced drowsiness, tiredness, and sleepiness. These subjective changes were associated with a decrement in performance (SMITH et al., 1962).

The studies of HILL and HAERTZEN and of SMITH and BEECHER are not as disparate as they may seem at first reading. Both groups of investigators have found evidence of both euphoric and sedative effects of narcotics. The euphoric effects predominate in nondependent addicts, while the sedative effects predominate in nonaddicts. The reasons for these differences are not understood. Although different test instruments were used by the investigators, it is unlikely that this is the important factor in accounting for the differences in results of these two groups of investigators. Postaddicts may have a residual tolerance to the sedative effects of morphine; however, this variable has not been systematically studied. As suggested in Section I, Chapter 1, the addict's personality and psychopathologic process may make the addict more susceptible to the euphorigenic actions of narcotic analgesics and hence make them more reinforcing, thus creating a greater need for them. Finally, the set under which the narcotic is given may have an effect upon the types of subjective effects reported; however, this variable also has not been systematically investigated.

2. Effects of Chronic Administration and Withdrawal

When addict patients become tolerant to narcotics, the subjective effects that they produce change from feelings of well-being and euphoria to an apathetic lethargic sedation (HAERTZEN and HOOKS, 1969; MARTIN et al., 1973b). Specifically, scores on the PCAG scale (HAERTZEN, 1966) are significantly elevated and subjects have feelings of tiredness, are hypochondrial and withdrawn. These changes are also associated with elevations on the hypochondriasis and schizophrenic scales of the MMPI (Minnesota Multiphasic Personality Inventory) (HAERTZEN and HOOKS, 1969; MARTIN et al., 1973b).

There have been only limited efforts to analyze or study the central nervous system mechanism of the subjective responses produced by narcotic analgesics. The narcotic antagonists antagonize the euphorigenic subjective effects of narcotic analgesics (MARTIN, 1967) in nontolerant nondependent subjects. The overwhelming body of evidence suggests that these agents are competitive antagonists of the narcotic analgesics, and provide little evidence as to the neuronal events that are responsible for the feeling states. Lithium, which decreases pathologic mania, does not antagonize either morphine's subjective or pupillary effects (JASINSKI et al., 1973), suggesting that mania and morphine-induced euphoria are different types of phenomena. Morphine and amphetamine produce similar subjective effects in man (HILL et al., 1963a; HAERTZEN, 1966). Like amphetamine, morphine decreases sleep (KAY et al., 1969). These observations suggest that catecholamines are involved in the production of euphoria; however, this must be a complex relationship, for methadone, which is equieuphoric to morphine, does not produce arousal, but rather sleep

(MARTIN et al., 1973b; NUTT and JASINSKI, 1973). The facts that β-phenethylamine is present in the central nervous system (MOSNAIM and INWANG, 1973; SAAVEDRA, 1974), produces amphetaminelike stereotypic behavior, is reinforcing in the dog (RISNER, 1975), and that there may be a specific β-phenethylamine receptor in the central nervous system (MARTIN and EADES, 1974), suggest that it may be a neurotransmitter that is intimately involved in feeling states. Tryptamine may also be a neurotransmitter that modulates feeling states (see Part II, Sect. I, Chap. 3).

III. Mouse

Three behavioral measures of the actions of narcotic analgesics have been studied in the mouse: (1) stereotyped motor activity, (2) Straub effect, and (3) analgesia.

KRUEGER et al. (1941) eloquently described the effects of morphine on motor activity:

> Shortly after the administration of a moderate dose of morphine to mice, 0.1 mg per g subcutaneously, there is a brief decrease in spontaneous movement; then the tail is erected over the back in an S-shape, commonly spoken of as the Straub reaction, and the animal begins to show excitement, characterized by a rapid and regular circular turning movement. The turning is either clockwise or counter-clockwise. Reflex excitability may be increased. During the early part of the excitement, the back is arched giving the animal a crouched appearance even though it is in motion. With small doses the motor excitement and Straub reaction may be the only effects shown. With larger doses, as the turning continues, the hind quarters sag, the hind limbs are more or less extended and become less active so that the animal drags itself around with its forefeet, appearing to pivot on one extended hind leg. The turning gradually becomes slower and irregular, the flattening increases and the animal comes to rest on its belly, practically unresponsive to external stimulation.

A number of morphinelike analgesics increase motor activity in the mouse, including morphine, meperidine, levorphanol, ketobemidone, and *l*-methadone (MILOSEVIC, 1955; BROWNLEE and WILLIAMS, 1963; GOLDSTEIN and SHEEHAN, 1969; SHARKAWI and GOLDSTEIN, 1969; HOLLINGER, 1969; SHELDON and SMITH, 1971; RETHY et al., 1971). The stereotyped motor activity induced can be antagonized by nalorphine, cyclazocine, levallorphan, diprenorphine, naltrexone, and naloxone (GOLDSTEIN and SHEEHAN, 1969; RETHY et al., 1971; PARKER, 1974). Thebaine does not increase motor activity (SLOAN et al., 1962; RETHY et al., 1971) nor does naloxone or *d*-methadone (RETHY et al., 1971). The data concerning dextrorphan are conflicting in that RETHY et al. (1971) found that dextrorphan produced a nonsignificant increase in motor activity, whereas GOLDSTEIN and SHEEHAN (1969) did not observe an increase. Interestingly, the increase in motor activity induced by dextrorphan was not antagonized by naloxone.

CHENEY and GOLDSTEIN (1971a) found that pCPA enhanced levorphanol-induced circling behavior, whereas neither HOLLINGER (1969) nor ESTLER (1973) observed enhancement. MAO inhibitors enhance meperidine and levorphanol-induced hyperactivity (BROWNLEE and WILLIAMS, 1963; HOLLINGER, 1969), whereas α-MT inhibits levorphanol-induced hyperactivity (HOLLINGER, 1969). MENON et al. (1967) found that α-MT, which itself does not alter activity, antagonized the hyperactivity produced by morphine as well as that produced by cocaine, *d*-amphetamine, and mescaline, but not LSD. ESTLER (1973) found that phenoxybenzamine but not the β-blockers propranolol and INPEA (N-isopropyl-p-nitrophenyl-ethanolamine) antagonized morphine-enhanced motor activity. [–]-5α, 9α-Diethyl-2′-hydroxy-2-methylene-cyclopropyl-methyl-6,7-benzmorphan, a potent analgesic, which produces physical dependence and increases the incorporation of tyrosine into catecholamines in the mouse brain, does not increase motor activity (VILLARREAL et al., 1971). Epinephrine (i.p.) has been reported to decrease motor activity and produce the Straub tail reaction and to antagonize morphine-induced hyperactivity (MILOSEVIC, 1955). Physostigmine inhibits levorphanol circling (SHARKAWI and GOLDSTEIN, 1969).

A number of investigators have shown that direct tolerance develops to the increased motor activity with chronic administration of morphine (HUIDOBRO et al., 1968; CHENEY and GOLDSTEIN, 1971b; MARSHALL and GRAHAME-SMITH, 1971; RETHY et al., 1971), dihydromorphinone (SHUSTER et al., 1963), and levorphanol (GOLDSTEIN and SHEEHAN, 1969), and that mice tolerant to these drugs are cross-tolerant to morphine, codeine, meperidine, methadone, and ketobemidone but not to amphetamine (SHUSTER et al., 1963; GOLDSTEIN and SHEEHAN, 1969; RETHY et al., 1971). pCPA does not alter tolerance to levorphanol (CHENEY and GOLDSTEIN, 1971a).

IV. Rat

The behavioral effects of morphine in the rat are quite complex. Small doses of morphine (1 mg/kg) increase grooming, motor activity, eating, and drinking, while larger doses of morphine and methadone decrease activity and produce depression and catalepsy (SLOAN et al., 1962; EIDELBERG and SCHWARTZ, 1970; FOG, 1970; YARBROUGH et al., 1971; DATTA et al., 1971; KUSCHINSKY and HORNYKIEWICZ, 1972; SASAME et al., 1972; AYHAN and RANDRUP, 1973a). Morphine has also been shown to produce sterotype jumping in the rat (WEISSMAN, 1971) and gnawing (VEDERNIKOV, 1970). Increased locomotor activity has also been observed after pentazocine, methadone, and levorphanol but not after naloxone and meperidine. Naloxone does not antagonize this effect of pentazocine (HOLTZMANN and JEWETT, 1972; DAVIS and BRISTER, 1973). The initial depression in motor activity seen with intermediate to large doses of morphine is followed within an hour or two by increased activity (JOEL and ETTINGER, 1926; SLOAN et al., 1962) and the increase in activity is associated with an increase in body temperature. Some of these effects are generally referred to as excitant and include walking, standing, scratching, preening, exploring, hostility on handling, increased body temperature and metabolic rate, and convulsions, while others are referred to as depressant such as loss of righting, apparent sleep, analgesia, and depressed respiration.

Morphine and methadone-induced catalepsy and rigidity are antagonized by naloxone and by L-DOPA and apomorphine, dopaminergic agonists (KUSCHINSKY and HORNYKIEWICZ, 1972; SASAME et al., 1972; WAND et al., 1973), whereas methadone-induced catalepsy is enhanced by α-MT (SASAME et al., 1972). AHTEE and KÄÄRIÄINEN (1973b) attempted to relate rat caudate HVA levels to catalepsy and found that etorphine, piminodine, methadone, and morphine produced marked catalepsy and increased HVA levels. Amantadine, but not trihexyphenidyl or diethazine, partially antagonized methadone-induced rigidity and the associated increase in HVA (AHTEE et al., 1972). Thebaine and cyclazocine did not produce catalepsy and did not elevate HVA levels. Codeine, meperidine, propoxyphene, and pentazocine produced a modest level of catalepsy and a modest to a marked increase in HVA. COSTALL and

NAYLOR (1974) have shown that lesions of the central amygdaloid nucleus antagonized the cataleptic action of morphine.

The stimulant effects of morphine and other analgesics are antagonized by the catecholamine depleters, α-MT and FLA-63 (AYHAN and RANDRUP, 1973a; DAVIS and BRISTER, 1973; KHALSA and DAVIS, 1973; BERNEY and BUXBAUM, 1973), apomorphine, L-DOPA, and amphetamine (AYHAN and RANDRUP, 1973b), but not by 6-OHDA. Morphine (25 mg/kg) antagonizes spontaneous as well as amphetamine- and apomorphine-induced stereotype behavior (PURI et al., 1973). Further, PURI and LAL (1973a, 1974a) and PURI et al. (1973) found that morphine, like haloperidol, increased DA turnover, produced catalepsy in the rat, and antagonized the decrement of DA turnover produced by apomorphine. In this regard, DRESSLER et al. (1974) have shown that morphine antagonized the dog renal vasodilator and the cat vasodepressor effect of DA. Morphine, however, does not antagonize apomorphine-induced stereotypic behavior in the rat (MCKENZIE and SADOF, 1974). Morphine has been reported to increase turnover of DA and NE during both the cataleptic and hyperactive states (BERNEY and BUXBAUM, 1973). In rats pretreated with pCPA, morphine does not induce catalepsy but produces hyperactivity (EIDELBERG and SCHWARTZ, 1970; BERNEY and BUXBAUM, 1973). pCPA and to a lesser degree α-MT abolish EEG slow-wave activity and stupor produced by morphine (COLASANTI and KHAZAN, 1973a). Although these findings suggest that serotonin may be in part responsible for the cataleptic effects of morphine, the fact that 5-HTP and iproniazid decrease morphine-induced catalepsy (EIDELBERG and SCHWARTZ, 1970) is not easily reconciled with this hypothesis. With chronic administration of α-MT, tolerance developed to its ability to antagonize morphine, but not amphetamine-induced hyperactivity, and the activity of morphine, but not amphetamine, was enhanced when α-MT was discontinued (KHALSA and DAVIS, 1973).

VEDERNIKOV (1970) found that morphine (2–5 mg/kg) increased the incidence and prolonged the duration of amphetamine and apomorphine stereotypy, findings which have been confirmed by MCKENZIE and SADOF (1974). The interpretation of these data is complicated by the fact that the effect of morphine on stereotypic behavior was not assessed. Tolerance develops to this effect with chronic administration of morphine.

Pentazocine, a weak narcotic antagonist which may have agonistic activity of both the morphine and nalorphine type, stimulates motor activity which is antagonized by α-MT (HOLTZMANN and JEWETT, 1972). Levallorphan, a narcotic antagonist which has predominantly nalorphine-like agonistic activity, also stimulates motor activity which is antagonized by naloxone. Larger doses of levallorphan decrease brain NE and DA levels (STEINERT et al., 1973).

When morphine is administered chronically, it increases a variety of types of motor activity including preening, walking, running, standing, sniffing, licking, and paw biting (JOEL and ETTINGER, 1926; FICHTENBERG, 1951; KAYMAKCALAN and WOODS, 1956; GUNNE, 1963; MARTIN et al., 1963; EIDELBERG and SCHWARTZ, 1970; KUMAR et al., 1971; AYHAN and RANDRUP, 1972; BUXBAUM et al., 1973). Phenoxybenzamine, spiramide, aceperone, reserpine, FLA-63, and α-MT antagonize certain stereotypic behavior induced by morphine in the morphine-tolerant and -dependent rat (AYHAN and RANDRUP, 1972). In general, agents which deplete the brain of NE (reserpine), inhibit the synthesis of NE or DA (FLA-63, diethylthiocarbamate, α-MT) or antagonize the effects of NE or DA (phenoxybenzamine, aceperone, and pimozide), decrease stereotypy and increase motor activity (EIDELBERG and SCHWARTZ, 1970; AYHAN and RANDRUP, 1972; BUXBAUM et al., 1973) in the morphine-dependent rat. Reserpine's actions are somewhat controversial in that GUNNE (1963) found that reserpine did not alter behavior in the tolerant rat. L-DOPA reverses the antagonistic effects of α-MT (EIDELBERG and SCHWARTZ, 1970). The role of DA in stereotypy seen in the dependent rat cannot be discounted since spiramide and pimozide also antagonize stereotypy in the morphine-dependent rat (AYHAN and RANDRUP, 1972, 1973a).

V. Cat

The observation of VOGT (1954) that morphine decreased brain levels of NE in the cat has stimulated the interest of several investigators in the neurohumoral basis of cat mania. LOEWE (1956) found that chlorpromazine and desoxycorticosterone decreased, whereas dibenzyline and reserpine increased mania. STURTEVANT and DRILL (1957) found that large doses of reserpine and chlorpromazine decreased cat mania. DHASMANA et al. (1972) found that reserpine, tetrabenazine, haloperidol, and chlorpromazine decreased mania induced by morphine (20 mg/kg), while phenoxybenzamine, propranolol, atropine, mepyramine, and LSD-25 did not. Neither α-MT nor 5-HTP altered morphine inhibitory effects on sleep (ECHOLS and JEWETT, 1972). Morphine ritualistic and stereotypic behavior in the cat was inhibited by injection of haloperidol into the rostral-medial caudate and NE into the raphe nucleus (COOLS et al., 1974).

VI. Protracted Abstinence

The neurochemical basis of the negative, hypophoric feeling states associated with protracted abstinence is unknown. In the rat, which shows a slight increase in "wet dog" shakes during protracted abstinence and perhaps more hostility on handling and aggression (MARTIN et al., 1963; WIKLER and PESCOR, 1970; GIANUTSOS et al., 1973), adrenal glands are somewhat larger and contain increased quantities of norepinephrine (SLOAN and EISENMAN, 1968). Brain levels of NE are somewhat increased. Rats that have received repeated doses of morphine or methadone and have been abstinent for several weeks show primarily behavioral and EEG arousal responses to morphine and methadone (KHAZAN and COLASANTI, 1971c; NASH et al., 1973; KHAZAN and ROEHRS, 1973). Patients in protracted abstinence excrete more epinephrine (EISENMAN et al., 1969). Neither naloxone nor naltrexone alter signs of protracted abstinence in the dog (MARTIN et al., 1974).

VII. EEG Effects

The effects of narcotic analgesics on the EEG in man and animals have been recently reviewed (MARTIN and KAY, 1977). In general, morphine, as well as other narcotic analgesics such as methadone, levorphanol, codeine, and meperidine, slow the frequency of the EEG, increase high-voltage delta and slow-wave activity, and cause high-voltage spike and spike-and-dome activity. These types of changes have been observed in the mouse for methadone (LEIMDORFER, 1948), in the rat for morphine, methadone, meperidine, and codeine (CAHEN and WIKLER, 1944; VERDEAUX and MARTY, 1954; SAWYER et al., 1955; KHAZAN and COLASANTI, 1971a; COLASANTI and KHAZAN, 1973b), in the cat for morphine, levorphanol, and levallorphan (LEIMDORFER, 1948; MARTIN and EADES, unpublished observations; CREPAX and INFANTELLINA, 1956; SCHALLEK and KUEHN, 1959), in the dog for morphine and methadone (WIKLER and ALTSCHUL, 1950; CREPAX and INFANTELLINA, 1958), and in the rabbit for morphine, heroin, morphinone, levorphanol, levallorphan, methadone, meperidine, codeine, phenazocine, dextromoramide, and thebaine (GANGLOFF and MONNIER, 1955, 1957; GOLDSTEIN and ALDUNATE, 1960; BRADLEY, 1968; CARRUYO et al., 1968; CORRADO and LONGO, 1961; LONGO, 1962; NAVARRO and ELLIOTT, 1971).

In the cat small doses of narcotic analgesics produce EEG activation (LEIMDORFER, 1948; BUXBAUM, 1967; BRADLEY, 1968). During feline mania, the EEG may be activated and hippocampal spiking observed (TUTTLE and ELLIOTT, 1969; NAVARRO and ELLIOTT, 1971). In human subjects, the most frequently observed EEG changes in nontolerant, nondependent subjects have been a delay in blocking of the EEG, slowing of alpha frequency, production of delta activity, and increased theta activity. Some or all of these changes have been seen with morphine (GIBBS et al., 1937; ANDREWS, 1941, 1943; WIKLER, 1954), codeine (HIMMELSBACH et al., 1940), meperidine (ANDREWS, 1942), ketobemidone (ISBELL, 1949; ALTSCHUL and WIKLER, 1951), methadone (ISBELL et al., 1947, 1948a, b; MARTIN et al., 1973b), and heroin (ZAKS et al., 1969; VOLAVKA et al., 1970).

The effects of a variety of narcotic analgesics on activation of the EEG by sensory nerve stimulation and by direct stimulation of the reticular formation have been studied. SILVESTRINI and LONGO (1956) found that activation of the EEG of the rabbit by nociceptive stimulation was depressed by morphine, whereas activation by tactile and auditory stimulation was not. Similarly, activation of the EEG of the guinea pig by tooth pulp stimulation was depressed by morphine, whereas activation produced by auditory stimulation and stimulation of the reticular formation was not (RADOUCO-THOMAS et al., 1962). In the dog, EEG activation by both nociceptive and visual stimuli was depressed by morphine, whereas activation produced by auditory, tactile, and proprioceptive stimuli was not markedly affected (DOMINO, 1968). Electrical stimulation of the rat tail evoked high-voltage theta activity in the hippocampus and parietal cortex, and this activity was suppressed by morphine, meperidine, and dextromoramide (CHARPENTIER, 1965; SOULAIRAC et al., 1967). Nociceptive stimuli cause catatonia in morphinized mice and rats which is associated with increased cortical and striatal high-voltage slow-wave activity in the rat (STILLE, 1971).

Although the narcotic analgesics selectively antagonize the EEG activating effects of nociceptive stimuli, they also appear to have a direct depressant effect upon reticular formation. Morphine (FUJITA et al., 1953; GANGLOFF and MONNIER, 1955, 1957; BRADLEY, 1968; MARTIN and KAY, 1977), levorphanol (GANGLOFF and MONNIER, 1955, 1957), heroin (BRADLEY, 1968), and meperidine (MONNIER et al., 1962; BRADLEY, 1968) elevate the threshold of the mesencephalic reticular formation and the hypothalamus for producing EEG activation. Although these effects of the narcotic analgesics provide a neurophysiologic basis for their sedative actions, the effects of these drugs on sleep are not easily reconciled with this hypothesis. Thus, morphine suppresses REMS in the rat (KHAZAN et al., 1967), rabbit (KHAZAN and

SAWYER, 1964), cat (ECHOLS and JEWETT, 1969), and man (KAY et al., 1969). Further, morphine has been shown to suppress slow-wave sleep in the rat (KHAZAN et al., 1967), cat (ECHOLS and JEWETT, 1969), and man (KAY et al., 1969). LEWIS et al. (1970) found that heroin decreased REMS and delayed the onset of sleep spindles but did not alter slow-wave sleep in man.

Chronic administration of morphine to the rat is associated with a development of tolerance to morphine's slow-wave–inducing activity and enhancement of the desynchronized activity and behavioral arousal (KHAZAN and COLASANTI, 1971b; NASH et al., 1973; NAKAMURA and WINTERS, 1973). In man, tolerance develops to morphine's suppressant effects on REMS and slow-wave sleep (KAY and KELLY, 1970). Results obtained on patients stabilized on methadone are somewhat conflicting. MARTIN et al. (1973b) found that the quantity of REMS and slow-wave sleep were slightly but not significantly greater during methadone stabilization than during the control period. HENDERSON et al. (1970) found that chronic methadone administration decreased REMS and delta sleep.

VIII. Discussion and Conclusions

Single doses of narcotic analgesics are associated with feelings of sedation and euphoria. The sedative effects may well be related to the catatonic effect of narcotic analgesics, to elevation of the threshold for EEG activation, and to the production of EEG slow-wave activity. An argument can be made that catatonia seen in the rat is a consequence of morphine antagonizing the effects of dopamine. Further, many of the subjective effects produced by L-DOPA in Parkinsonian patients, such as agitation, restlessness, irritability, anxiety, anger, aggressiveness, and hostility, appear to be polarly opposite to the effects of narcotic analgesics (cf. MURPHY, 1973). COSTALL and NAYLOR (1974) have suggested that there may be two antagonistic dopaminergic systems, an excitatory one involving the striatum and a cataleptic one involving the amygdala.

Enhanced serotonergic influence may also play a role in the sedative and catatonic actions of morphine. The relationship between morphine-induced hyperactivity in the mouse (circling), cat mania, and euphoria is more tenuous. Both dopaminergic and noradrenergic mechanisms may be involved in circling in the mouse and in cat mania. Serotonin does not appear to be involved in these phenomena. It is tempting to relate the similarities of the subjective effects of the narcotic analgesics and amphetamine to this hypothesis and the fact that amphetamine is an indirectly acting sympathomimetic amine as well as being dopaminergic (see Part II, Sect. I, Chap. 1).

This hypothesis is not consistent with data obtained in tolerant and dependent subjects and animals. Chronic administration of narcotic analgesics in man is associated with a hypophoric subjective state. The hyperactivity and stereotypy seen in the intoxicated tolerant dependent rat appears to involve noradrenergic and possibly dopaminergic mechanisms.

OLDS and TRAVIS (1960) found that morphine and chlorpromazine depressed the rate of hypothalamic self-stimulation, whereas ADAMS et al. (1972) found that it increased hypothalamic self-stimulation. LORENS and MITCHELL (1973) found that the depressant effect of morphine appeared first and was followed by its enhancing effect, and that with chronic administration tolerance developed to the depressant effect but not to the stimulant effect. Further, the inhibitory effect of morphine, but not haloperidol, can be antagonized by naloxone (WAUQUIER et al., 1974).

Self-stimulation can be interpreted in several ways (OLDS, 1958). These data would be consistent with the paradigm that the depressant effects of morphine are need-reducing and euphoria-producing and thus decrease wants and desires. The stimulant actions, on the other hand, increase need states and hypophoria.

The facts that drugs of abuse produce euphoria and that hypophoria is a salient characteristic of the psychopath and an important aspect of the acute and protracted abstinence syndrome make the understanding of these phenomena one of the potentially most fruitful areas of research, which will lead to the definitive diagnosis and treatment of the drug abuser. A considerable body of evidence has evolved implicating noradrenergic, tryptaminergic, dopaminergic, serotonergic, and possibly phenethylaminergic mechanisms in the actions of the narcotic analgesics, the amphetamines, and the LSD-like hallucinogens (see Part II, Sect. I). A unified hypothesis has yet to be evolved, and it is possible that there may be many mechanisms by and pathologic states in which euphoria and hypophoria can become manifest.

C. Analgesia and Pain

I. Introduction

The affinity of organisms for nutritious and favorable environments and their aversion to hostile destructive environments and stimuli are properties that can be identified even in certain one-celled organisms. As organisms have become more highly organized not only has the nature of unpleasantness and noxious stimuli as well as their perception become more involved, but mechanisms for reacting to hostile states have become more subtle and complex. Not only must the organism have the capacity to identify a variety of noxious environmental elements such as extremes of temperature, radiation, pH, salinity, osmolality but also the absence of food, water, oxygen, and other nutrients. A number of complex reflex responses as well as more highly integrated and learned responses and patterns of behavior have developed to cope with adverse environments.

In addition to the evolution of protective reflexes and coping behavior, other mechanisms and devices have evolved that are used for defense against as well as for attack and immobilization of other organisms. These mechanisms also must have evolved quite early since they are present in coelenterates. Both the defensive-fighting as well as the food-acquiring immobilizing nematocyst are present in this phylum. Further, chemicals that serve as neurotransmitters, such as serotonin in species with nervous systems (WELSH, 1960; WELSH and MOORHEAD, 1960), are present in coelenterates. It is possible that certain neurotransmitters may have had their origin in defensive and immobilizing chemical devices.

In the pattern of things, it is equally important that animals perceive the dart of the aggressor, the sting of the defender, and the intent of the predator, as it is to develop defensive mechanisms. Thus, it is equally important that the development of pain-inflicting and perceiving mechanisms develop in parallel. The response to danger and injury takes many forms with differing levels of complexity such as reflex responses, avoidance, flight or escape responses, and aggressive action. In addition, there is the generation of associated affective states such as anxiety, apprehension, insecurity, fear, anger, suffering, discomfort, and pain.

In the study of pain and its relief, a variety of techniques have been devised to evoke anxiety and suffering and to measure responses to stimuli and circumstances that produce associated affective states. The reader is referred to the following authoritative articles, reviews, volumes, and monographs (HILL et al., 1952, 1957; BEECHER, 1959; SWEET, 1959; LIM, 1960; KEELE and ARMSTRONG, 1964; LIM and GUZMAN, 1968; SOULAIRAC et al., 1968; HOFFMEISTER, 1968).

II. Neuropharmacology of Pain and the Narcotic Analgesics

1. Peripheral Nerve and the Myoneural Junction

KRIVOY (1960) studied the effects of several agents on the response of the frog sciatic nerve to tetanic stimulation and found that morphine, dihydromorphine, codeine, and acetylsalicylic acid depressed the response to tetanic stimulation and produced a prolonged enhancement of the positive afterpotential. Methadone produces a much shorter enhancement of the positive afterpotential than morphine. CAIRNIE and KOSTERLITZ (1962) found that morphine did not affect threshold, conduction velocity, or the recovery cycle of the rabbit vagus or the cat saphenous nerve. In subsequent studies, KOSTERLITZ and WALLIS (1964) were unable to observe any effects of morphine in the amplitude of the C-fiber action potentials of the cat hypogastric nerve or upon its conduction velocity or its recovery cycle. Further, morphine did not alter posttetanic hyperpolarization of the cat hypogastric or the rabbit vagus nerve. In the desheathed rabbit vagus and cat hypogastric nerves, morphine and nalorphine, as well as other nonanalgesic drugs, antagonized the depolarizing effects of ACh (RITCHIE and ARMETT, 1963). Levorphanol, levallorphan, and dextrophanol block conduction in the giant squid axon (SIMON and ROSENBERG, 1970). Further, levorphanol antagonizes spontaneous firing in the squid axon induced by lowering C^{++} and Mg^{++}; however, morphine was ineffective in relatively high concentrations. On the other hand, FRAZIER et al. (1972) found that morphine, when perfused internally in the giant squid axon, depressed the action potential, as well as Na^{+} and K^{+} conductance in high concentrations (10^{-2} to 10^{-3}M). Morphine and naloxone depress in a dose-related manner posttetanic potentiation of the soleus muscle of the cat, and naloxone antagonizes the depression produced by morphine (SOTERPOULOS and STANDAERT, 1973).

2. Spinal Cord

a) Segmental and Spinal Cord Reflexes

α) Cat

The reported effects of morphine and related narcotic analgesics on spinal reflex activity in the cat are conflicting. In the chronic spinal cat, morphine depresses nociceptive reflexes. BODO and BROOKS (1937) studied the effects of morphine (8 mg/kg) on hindlimb reflex activity of cats that had been spinalized (C-6 or 7) a week or less prior to experimentation and found that morphine clearly depressed the flexor, crossed extensor, and Phillipson reflexes and depressed the clonic aspect of the knee jerk. WIKLER (1944) observed that morphine (2–15 mg/kg) depressed the flexor and crossed extensor reflexes, but left the patellar reflex unchanged in the acute and chronic spinal cat. In the acute spinal cat, morphine also depresses nociceptive and polysynaptic reflexes but not the patellar reflex. COOK and BONNYCASTLE (1953) found that a variety of narcotics depressed the flexor reflex of the acute spinal cat but this effect was variable. Morphine either depressed or facilitated the flexor reflex; codeine facilitated the flexor reflex; while Dilaudid and hydrocodone produced an initial facilitation which was followed by depression. Although methadone enhanced the patellar reflex, most other narcotics studied did not affect it. KRUGLOV (1955, 1959) found that morphine did not affect the patellar reflex but decreased inhibition evoked by ipsilateral peroneal nerve stimulation. DESALVA and OESTER (1960) found that the patellar reflex amplitude was increased in the acute spinal cat by doses of morphine up to 6 mg/kg and decreased by doses of 8 mg/kg. These doses decreased or abolished inhibition produced by stimulating the ipsilateral and contralateral sciatic nerves and facilitation produced by stimulating the contralateral sciatic nerve.

JURNA (1966) found that both morphine (1 and 2 mg/kg) and meperidine (1 and 2 mg/kg) suppressed α-motoneuron discharges evoked by muscle stretch. Morphine and meperidine did not alter spontaneous γ-efferent discharge rate or the rates of discharges evoked by muscle stretch in the spinal cat. In the intact and decerebrate cat, morphine, but not meperidine, suppressed γ-efferent discharge rate. The effects of morphine were the greatest in the intact cat (JURNA, 1965); however, morphine did suppress γ-efferent activity evoked by cutaneous nerve stimulation in the spinal cat (JURNA, 1966).

The electrophysiologic correlates of spinal cord reflex activity have not been clearly defined nor have the effects of narcotic analgesics on spinal cord electrophysiologic signs. WIKLER (1945) found that morphine (5 mg/kg) enhanced the monosynaptic and depressed the polysynaptic reflex of the high spinal cat. Fifteen mg/kg of morphine depressed the monosynaptic reflex. TAKAGI et al. (1955) found that morphine (7 mg/kg) depressed the mono- and polysynaptic potentials of the intact, midbrain, thalamic, and high spinal cat, but not of the low spinal cat. Following electrolytic destruction of the bulbar inhibitory center, morphine facilitated the mono- and polysynaptic potentials, whereas the depressant effect of morphine was seen following destruction of the facilitatory centers. Ohton, in contrast to morphine, depressed the mono- and polysynaptic reflexes in the low spinal cat. The depressant effects of morphine were enhanced by adrenaline and ephedrine. KRIVOY et al. (1973) found that 0.5, 2.5, and 12.5 mg/kg of morphine in the low spinal cat depressed both the mono- and polysynaptic potentials in a dose-related manner and presented evidence that the depressant actions of morphine, particularly with regard to monosynaptic potentials, were greater with submaximal high-frequency stimulation. These depressant effects of morphine were antagonized by naloxone. JURNA et al. (1973) found that morphine did not alter antidromically evoked potentials in the motor neurons or EPSPs evoked monosynaptically but did depress EPSPs evoked polysynaptically as well as evoked activity in interneurons. They observed further that facilitation of motor neuron EPSPs produced by repetitive stimulation of Ia afferents was decreased by morphine. NEUMAN et al. (1974) have shown that iontophoretically applied morphine will depress the firing of cat spinal cord neurons activated by nociceptive stimuli.

Narcotic analgesics depress other electrophysiologic signs that are related to polysynaptic activity and presynaptic inhibition. Morphine (1.5–10 mg/kg), methadone, and meperidine were found to depress the dorsal root V (DRV) potential in the decerebrate or spinal cat (KRIVOY and HUGGINS, 1961). SATOH and TAKAGI (1971b) reported that morphine (5 mg/kg) depressed the contralateral but not the ipsilateral dorsal root potential. In the intact cat, both morphine and meperidine decreased DRV and this effect was diminished by spinal cord transection (CHIN et al., 1974). These observations argue against the hypothesis that narcotic analgesics enhance spinal cord presynaptic inhibition. In contrast to these findings are those of REPKIN et al. (1974) who found that both morphine and stimulation of the nucleus raphe magnus increased the frequency of dorsal root antidromic potentials which result from depolarization of primary afferent fibers.

Although the monosynaptic segmental reflex potential is closely related to the patellar reflex, the polysynaptic potential has a much shorter duration than does the flexor reflex and does not exhibit the phenomenon of afterdischarge. KOLL et al. (1963) have studied the effects of small doses of morphine (0.3–0.4 mg/kg) on a C-fiber reflex evoked by tetanic stimulation of the superficial peroneal nerve. The evoked ventral root potential, which is characterized by a long latency, long-persisting discharge, is markedly depressed by morphine, and the depressant effect is antagonized by nalorphine. Persisting ventral root discharges evoked by stimulation of post δ-fibers were also depressed by small doses of morphine and this effect was antagonized by nalorphine. Morphine (1 and 5 mg/kg) and meperidine (1 and 3 mg/kg) diminished posttetanic potentiation of the monosynaptic potentials (JURNA and SCHAFER, 1965).

β) Dog

WIKLER and his collaborators studied the effects of narcotic analgesics and other drugs on the flexor, crossed extensor, Phillipson, extensor thrust, and patellar reflexes of the chronic spinal dog and found that morphine (5–100 mg/kg) and methadone (3 and 10 mg/kg) depressed the flexor, crossed extensor, and Phillipson reflexes and enhanced the ipsilateral extensor thrust. The effect of morphine and methadone on the poorly developed patellar reflex was usually depressant; however, when the reflex was well developed, neither drug had a consistent effect (HOUDE et al., 1951). MARTIN and EADES (1964) confirmed these observations, showing that a single dose of morphine (10 mg/kg) as well as an 8-h infusion of morphine (3 mg/kg/h) depressed the flexor and crossed extensor reflexes but did not alter the amplitude of the patellar reflex. Subsequent studies have demonstrated that a variety of analgesics including codeine, an isoquinoline analgesic (I-K-1) (MARTIN et al., 1964), nalorphine, and cyclazocine (MCCLANE and MARTIN, 1967) depressed the flexor reflex, and there was relatively good agreement between their potency in suppressing the flexor reflex and their analgetic activity in man.

γ) Rat

GROSSMAN et al. (1973) found that morphine, meperidine, and aminophenazone depressed facilitation of evoked spinal cord unit activity in the lightly anesthetized intact rat and to a lesser degree in the spinal preparation. Morphine did not affect the monosynaptic and facilitated the polysynaptic potential of the segmental reflex in the rat pretreated with reserpine. The interactions between these analgesics and amine depletion in spinal cord reflex activity were complex.

b) Spinal Cord Inhibitory Processes

Kruglov (1964) found that morphine (10 mg/kg) decreased facilitation of polysynaptic potentials, enhanced direct inhibition, and decreased recurrent inhibition in the decerebrate cat. In the chloralose-anesthetized spinal cat, morphine (1–12 mg/kg) and codeine (1 and 10 mg/kg), although depressing the monosynaptic potential, did not alter direct inhibition of this reflex. Thebaine (1–3 mg/kg), on the other hand, decreases direct inhibition (Corrado and Longo, 1961), and antagonizes the depressant actions of glycine but not GABA on interneurons and Renshaw cells (Curtis et al., 1968). Morphine and codeine, but not etorphine and nalorphine, also antagonize the depressant effect of iontophoretically applied glycine but not GABA on cat spinal cord interneurons (Curtis and Duggan, 1969; Lodge et al., 1974). Curtis and Duggan (1969) reported that morphine decreased both direct and recurrent inhibition in the cat spinal cord. Felpel et al. (1970) found that morphine (5–10 mg/kg) decreased recurrent but not direct inhibition in the decerebrate cat. The depression of recurrent inhibition was antagonized by nalorphine. The firing rate of Renshaw cells was not altered by morphine. Satoh and Takagi (1971 b) found that morphine (5 and 10 mg/kg) enhanced direct inhibition and that its effects were greater in the decerebrate than in the spinal cat. Presynaptic inhibition was only slightly depressed (10 mg/kg). Iontophoretically administered morphine increases the latency of the early Renshaw potentials (Duggan and Curtis, 1972), whereas systemically administered morphine was without effect. Lodge et al. (1974) found that iontophoretically applied morphine enhanced the stimulatory action of ACh but not homocysteate on the Renshaw cell, while etorphine depressed the stimulatory effects of both. Davies and Duggan (1974) found that morphine administered iontophoretically stimulated Renshaw cells, and this effect was decreased by both naloxone and dihydro-β-erythroidine and to a lesser degree by atropine. Naloxone also antagonized the excitant effect of ACh and nicotine but not that of acetyl-β-methacholine and DL-homocysteate. It did not alter the depressant actions of glycine and GABA on the Renshaw cell. The effects of morphine on presynaptic inhibition are conflicting (see Subsect. C.II.2.a.α). In this regard, however, it is important to distinguish between the direct and supraspinal influences of morphine.

3. Supraspinal Influences

The skin twitch reflex evoked by both thermal and mechanical stimulation (Houde and Wikler, 1951) in the dog and the tail flick reflex evoked by thermal stimulation (Irwin et al., 1951) in the rat are mediated at the spinal cord level; however, the amplitudes of both the skin twitch and the flexor reflex are greater after spinal cord transection (Irwin et al., 1951) which suggests that supraspinal influences modulate nociceptive spinal cord reflexes. Irwin et al. (1951) and subsequent investigators have demonstrated that the ability of narcotic analgesics to depress spinal cord nociceptive reflexes resides, at least in part, at a supraspinal level. Irwin et al. (1951) found that morphine (1–10 mg/kg) and methadone increased the latency of the tail flick reflex of the rat in a dose-related manner, and the dose response curves were less steep in the spinal than the intact preparation. In this regard, it is important to note that spinal cord transection did not affect the latency of the tail flick reflex (IRWIN et al., 1951).

In the intact dog, the latency of the skin twitch reflex was only slightly increased by morphine (5 mg/kg) but its amplitude was markedly depressed (HOUDE and WIKLER, 1951). The action of morphine in depressing the skin twitch reflex was greater in the intact than in the spinal animal. SATOH and TAKAGI (1971a) found that whereas morphine had a marked depressant effect on potentials evoked in the ventrolateral funiculus of the spinal cord by splanchnic nerve stimulation in the intact pentobarbital-anesthetized cat, its effects were much less in the spinal preparation and after spinal cord transection. IRWIN et al. (1951) suggested that the relative ineffectiveness of morphine in suppressing the tail flick response in the spinal rat was in part a consequence of the hyperreflexia induced by spinal cord transection. The efficacy of morphine in suppressing spinal nociceptive reflex activity is determined in part by the reflex and the strength of stimulation. Thus, in the chronic spinal dog the flexor reflex evoked by modest strengths of stimuli can be reproducibly depressed by 0.1 mg/kg of morphine (MCCLANE and MARTIN, 1967; MARTIN, 1967), a degree of analgesic sensitivity comparable to the most sensitive methods employed in the intact animal. Unfortunately, this reflex cannot be studied in the intact dog for the response to nociceptive stimuli applied to the hind foot is a high variable. To attempt to resolve the role of supraspinal influences on spinal cord reflex activity, the direct effects of narcotic analgesics on bulbar facilitatory and inhibitory centers have been studied. TAKAGI et al. (1955) concluded that morphine depressed spinal reflex activity by stimulating bulbar inhibitory areas; however, VAL'DMAN (1958) and SINCLAIR (1973) found that morphine depressed the inhibitory action of bulbar stimulation on the monosynaptic segmental reflex. SINCLAIR (1973) also found that meperidine depressed bulbar inhibition when administered systemically or close arterially, and suggested that meperidine's depressant action was on the spinal cord. The depressant effect of morphine but not that of meperidine was antagonized by nalorphine. JURNA (1966) found that both enhancement and inhibition of spindle discharges produced by stimulation of the bulbar reticular formation were depressed by morphine and meperidine. Closely related to these observations are the findings of RIBLET and MITCHELL (1972) who found that the depressant effect of morphine on the jaw opening reflex evoked by tooth pulp stimulation in the normal cat was greater than in the encephale isolé preparation, suggesting that actions of morphine on this reflex were indirect and a consequence of activating inhibitory pathways located in the caudal portion of the spinal nucleus of the trigeminal nerve.

Localized injection of narcotic analgesic agents into the brain has been used to determine their site of action. LOCKETT and DAVIS (1958), as well as HORLINGTON and LOCKETT (1959), while testing the hypothesis of BECKETT et al. (1956) that the N-dealkylated metabolites of narcotic analgesics were the active form, demonstrated that morphine and normorphine administered intracisternally elevated the threshold of electrical stimulation of the tail to evoke squeaking and that nalorphine antagonized these effects. Subsequent studies demonstrated that intracisternally administered morphine and meperidine respectively antagonized the analgesic effect of intracisternally administered normorphine and normeperidine. ADLER (1963) found that intraventricularly administered morphine, codeine, and norcodeine produced analgesia and hyperexcitability in the mouse. Codeine and morphine produced a Straub tail reaction, and codeine, clonic convulsions. The analgesic action of morphine but not codeine was antagonized by nalorphine. Both morphine and N-methyl-morphine locally administered into the hypothalamus of the rat produce analgesia, using the hot plate technique (FOSTER et al., 1967; TSOU and JANG, 1964).

HERZ et al. (1970) localized the site of action of morphine and fentanyl in suppressing licking evoked by tooth pulp stimulation in the rabbit to the aqueduct and fourth ventricle. Injection of morphine into the ventromedial hypothalamus, subthalamus, and medial mesencephalon also depressed the nociceptive reflex. Further studies (ALBUS et al., 1970) demonstrated that levallorphan and nalorphine administered intraventricularly or into the aqueduct and fourth ventricle antagonized the depressant effect of morphine on the licking response and respiratory depression. Of interest was the fact that the effects of morphine on behavioral depression and EEG activation were not always antagonized by the intraventricular administration of levallorphan. VIGOURET et al. (1973) found that intraventricularly administered morphine suppressed the flexor reflex in the rabbit and localized the site of action to the fourth ventricle. Further, levallorphan when placed in the fourth ventricle antagonized the depressant effects of morphine on the flexor reflex. Using radioautography for determining the distribution of morphine, TESCHEMACHER et al. (1973) concluded that the receptors for this action of morphine must be located within a millimeter of the surface of the fourth ventricle. Stimulation of the dorsal raphé but not the lateral dorsal raphé nucleus facilitated the analgesic action of morphine in the rat but did not produce analgesia in its own right (SAMANIN and VALZELLI, 1971). Electrical stimulation of the nucleus raphé magnus, on the other hand, produced analgesia (REPKIN et al., 1974) in the rat using the tail flick test. Morphine also increased spontaneous activity in dorsal root filaments, a consequence of depolarization of afferent terminals, and this action was antagonized by cinanserin. Lesions of caudal raphé nuclei decreased morphine's analgesic activity in the rat as measured by both the tail flick and writhing procedures (PROUDFIT and ANDERSON, 1974). Morphine depressed EEG activation evoked by both nociceptive and nonnociceptive stimuli in the rabbit when placed in the fourth ventricle but not in the third (ALBUS and HERZ, 1972). GROSSMAN et al. (1973) found that lesions in the substantia nigra of the rat antagonized morphine analgesic effects. JACQUET and LAJTHA (1973) found that morphine placed in the third ventricle or the posterior hypothalamus produced analgesia, while producing hyperalgesia when placed in the medial septum of the rat. PERT and YAKSH (1974) have identified two brain stem loci in the monkey where locally placed morphine and etorphine produced analgesia using the shock titration procedure: (1) A medial periventricular-periaqueductal region which extended from the fourth ventricle to the thalamus and hypothalamus, and (2) a lateral region which extended from the pons to the red nucleus and ventral thalamus.

4. Sensory Pathways

Although a great deal of experimental work has been done on the effects of narcotic analgesics on spinal cord reflexes, it is highly probable that these agents alter perception of and responsivity to pain and that this action is at higher brain levels. Attempts to determine the effects of narcotic analgesics on pain pathways have yielded equivocal results. Several analgesics including morphine, methadone, and ohton depress spinal cord potentials evoked by stimulating visceral afferents (splanchnic and phrenic nerves) and recorded from the ipsilateral fasciculus proprius in the cat (FUJITA et al., 1953, 1954). Morphine did not depress cortical responses evoked by tactile stimulation, but both morphine and methadone did depress responses in the contralateral medial lemniscus evoked by sciatic nerve stimulation following the placement of a lesion in the ipsilateral dorsolateral funiculus at the T-12–L-1 level. These findings suggested that impulses produced by nociceptive stimuli were attenuated shortly after they entered the brain or spinal cord. Potentials evoked in the cortex, thalamus, and midbrain, but not the medulla, by vagal stimulation were depressed by morphine. OGIU et al. (1958) also found that morphine depressed spinal cord potentials evoked by splanchnic nerve stimulation and that these effects were antagonized by nalorphine.

In an extensive series of experiments conducted in the unanesthetized dog, CHIN and DOMINO (1961) studied the effects of morphine (2–10 mg/kg) on cortical and brain-stem potentials evoked by single shocks to the tooth pulp. Previous studies (HENG and DOMINO, 1960) had shown that morphine elevated the threshold for mouth movements associated with tooth pulp stimulation. Potentials recorded in the medulla, midbrain, and some diffusely projecting and associational thalamic nuclei were enhanced, not depressed by morphine. The interpretation of these data favored by the investigators was that morphine decreased occlusion. In contrast, MCKENZIE and BEECHEY (1962) found that morphine (1–6 mg/kg) and meperidine (2–7 mg/kg) suppressed potentials evoked by stimulation of the tibial nerve in the unanesthetized curarized cat at a variety of brain-stem sites, including the ventral tegmentum, which was most sensitive, the paralemniscal areas, deep layer of superior colliculus, spinobulbar thalamic tract, dorsal tegmentum, and the central grey. MIZOGUCHI (1964) found that morphine depressed potentials evoked in the spinal tract of the trigeminal nerve of the dog by tooth pulp stimulation. In a subsequent study, MCKENZIE (1964) found that morphine and meperidine depressed hippocampal potentials evoked by stimulating the tibial nerve but not potentials evoked by septal stimulation. Morphine does not effect hippocampal or entorhinal excitability in the cat when potentials in these areas are evoked by stimulating the septum, hippocampus, entorhinal cortex, and piriform cortex, but does depress activity evoked by stimulating the radial nerve and the mesencephalic reticular formation (NAKAMURA and MITCHELL, 1972).

Observations concerning the effects of narcotic analgesics on potentials evoked in the neocortex are conflicting. Primary cortical potentials evoked by stimulation of the tooth pulp, sciatic nerve, auditory nerve, and eye in the unanesthetized dog, cat, and man are not altered by morphine (CHIN and DOMINO, 1961; CORSSEN and DOMINO, 1964; SINITSIN, 1964; DOMINO, 1968); however, DOMINO (1968) found that long latency cortical potentials evoked by tooth pulp stimulation in the dog were depressed by morphine. SINITSIN (1964) found cortical responses evoked by sensory stimulation in associational areas were also depressed by morphine. In contrast, morphine depressed potentials evoked by splanchnic, vagus, phrenic, and inferior cardiac nerves in the somaesthetic cortex of the barbiturate-anesthetized cat (FUJITA et al., 1953, 1954; OGIU et al., 1958).

BISCOE et al. (1972) found that etorphine administered intravenously prolonged the latency of primary potential and depressed repetitive afterdischarge of responses evoked from rat cerebral cortex by peripheral nerve stimulation. The effects were antagonized by the opioid antagonist, diprenorphine. Similar effects were observed in unit activity recorded from pyramidal cells of layer V. Morphine suppressed repetitive afterdischarge, but did not increase latency of the evoked unit potentials. Etorphine applied iontophoretically to cortical units also depressed repetitive afterdischarge. JURNA et al. (1972) found that morphine enhanced both the positive and negative primary cortical potentials evoked by stimulation of the superficial radial nerve as well as the direct cortical response in the rat and that this enhancement was partially antagonized by levallorphan.

5. Discussion and Conclusions

The relevance of changes produced by narcotic analgesics in spinal cord reflexes to the specific problem of the analgesic action of these agents and to the general problem of dependence is not clear. The issue of selectivity of both the site and mode of action of narcotic analgesics has not been clarified. Large doses of narcotic analgesics depress peripheral nerve and monosynaptic reflexes, suggesting that they may have actions on the presynaptic terminal of large afferent fibers or on the motor neuron nerves; however, the fact that only small doses of narcotic analgesics are needed to suppress the flexor reflex as well as ventral root discharges following delta and C-fiber stimulation, while leaving the patellar, extensor thrust, and short latency ventral root potentials either unaffected or facilitated, suggests that analgesics do have some selectivity in depressing spinal nociceptive reflex pathways. It is quite clear from these data that narcotic analgesics have multiple sites of action and probably modes of action (see Subsect. III.8. Discussion). In this regard, MAYER et al. (1971) have identified many mesencephalic and diencephalic sites in the rat brain which when electrically stimulated induce analgesia. Narcotic analgesics decrease spinal and supraspinal facilitatory mechanisms and both enhance and depress inhibitory mechanisms. They appear to selectively depress neurophysiologic phenomena characterized by rapid firing rate, afterdischarge, and long latency in peripheral nerve, spinal cord, and supraspinal structures. It must be remembered that nonanalgesic drugs such as the anticonvulsants and interneurone depressants share these types of activity. The question of where in the nervous system discomfort, pain, and anxiety are perceived and whether there are neurophysiologic phenomena that are valid indices of these emotions and affective states remains to be answered. Certainly, study of neuronal activity and the effects of narcotic analgesics on pathways that project to the limbic system and associational cortices hold promise.

III. Neurochemical and Neurohumoral Changes Associated with Analgesia

1. Introduction

A large literature has evolved concerning the role of neurotransmitters in the perception of and responsivity to pain and the effects of narcotic analgesics on the actions and metabolism of neurotransmitters.

2. Acetylcholine

a) Analgesia

Relationships between the analgesic actions of narcotics and cholinergic processes have been recognized for some time. SLAUGHTER and GROSS (1940), reasoning from the fact that some of the actions of morphine in the dog were parasympathomimetic, found that physostigmine enhanced gut tonus and the fall in blood pressure produced by morphine. Subsequent studies demonstrated that prostigmine enhanced the analgesic effects of morphine in the cat and that this enhancement was partially antagonized by atropine (0.04 mg/kg) (SLAUGHTER and MUNSELL, 1940). Results obtained on the effects of prostigmine on the analgesic effects of morphine in the rat have been conflicting. SAXENA and GUPTA (1957) found that prostigmine enhanced the analgesic action of morphine, whereas HERKEN et al. (1957) found that neither physostigmine nor prostigmine enhanced the analgesic action of morphine in the rat. PLEUVRY and TOBIAS (1971) found that physostigmine and oxotremorine enhanced morphine analgesia in the mouse, using the hot plate method.

MARTIN et al. (1958) and BUSCH et al. (1958) found that several β-halogenated pyruvic acid derivatives including dibromopyruvic acid, which produced a variety of muscarinic effects, induced analgesia in the mouse. HERZ (1962) found that arecoline produced analgesia in the mouse and the rat and that its analgesic actions were antagonized by atropinelike agents but not by nicotonic blockers such as mecamylamine. In contrast, atropinelike agents either did not affect morphine analgesia, a finding confirmed by ZETLER et al. (1963), or antagonized it in the mouse. Tremorine and its active metabolite oxotremorine, also have been shown to be analgesics in the mouse (CHEN, 1958; KERANEN et al., 1961; LESLIE, 1969; IRESON, 1969; HARRIS et al., 1969; HOWES et al., 1969; PLEUVRY and TOBIAS, 1971). Although atropine antagonized the analgesic effects of oxotremorine and physostigmine, it did not antagonize the analgesia produced by either nalorphine or morphine (IRESON, 1969). Although IRESON (1969) and PLEUVRY and TOBIAS (1971) were unable to demonstrate antagonism of physostigmine and oxotremorine analgesia by naloxone (2.5 mg/kg), HOWES et al. (1969) found antagonism with 10 mg/kg. METYS et al. (1969) found that oxotremorine and arecoline were effective in antagonizing some but not all responses to nociceptive stimuli in the rat. Carbachol intraventricularly, but not systemically, produced analgesia. Oxotremorine, arecoline, and intraventricularly administered carbachol also produced analgesia in the rabbit. Microinjection of carbachol into the reticular formation and the septum produced analgesia in the rat. Much higher quantities of carbachol were needed to produce analgesia when injected into the dorsal hippocampus and the caudate. The analgesic activity of nalorphinelike agonists can be demonstrated in mice pretreated with physostigmine (HOWES et al., 1969). PLEUVRY and TOBIAS (1971) found that the analgesic activity of morphine was decreased by reserpine, diethyldithiocarbamate, L-DOPA, and pCPA and enhanced by α-MT, whereas oxotremorine analgesia was enhanced by reserpine, α-MPT, and diethyl-dithiocarbamate, unchanged by pCPA, and decreased by L-DOPA.

In the mouse, intracerebrally injected physostigmine, and to a lesser extent DFP, enhanced morphine-induced analgesia (BHARGAVA and WAY, 1972), whereas intracerebrally injected hemicholinium antagonized the analgesic action of morphine (BHARGAVA et al., 1972a). CHARPENTIER (1961, 1965) studied the effects of atropine and physostigmine on the affective responses in the rat to electrical stimulation of the tail. Physostigmine lowered the threshold for stimulation and increased the responsivity to the second of a pair of electrical stimuli (excitability). The effects of atropine were more complicated; threshold was increased and the reaction to the stimulus diminished; however, the response to a second stimulus (excitability) was increased. In a series of experiments, Martin and Eades (1967) and Vaupel and Martin (1973) found that tremorine, oxotremorine, and physostigmine enhanced the flexor reflex of the chronic spinal dog and evoked fragmentary stepping movements, and these changes were antagonized by atropine. The narcotic analgesics and nalorphinelike agents depress the flexor reflex (McClane and Martin, 1967). These findings suggest that the mode of action of cholinergic agents in producing analgesia is different from that of the narcotic analgesics.

The inhibitory effects of narcotic analgesics on cholinesterases and acetylcholine synthesis have been reviewed by Martin (1963), Weinstock (1971), and more recently by Christian (1972), and for this reason will not be reviewed in this monograph. Although several narcotic analgesics are potent inhibitors of cholinesterases, there are several reasons to believe that this is not a significant mode of action of these drugs in producing analgesia: (1) Inactive congeners of potent analgesics are also potent inhibitors of cholinesterases and acetylcholine synthesis; (2) there are marked discrepancies between the analgesic and AChE-inhibitory activities among various narcotic analgesics; (3) doses of narcotic analgesics that produce marked analgesia do not inhibit brain AChE in vivo.

b) Guinea Pig Ileum and Other Isolated Tissues

During the course of studies of the preparatory and emptying phases of the guinea pig ileum independently conducted by Trendelenberg (1917), Schaumann (1955), and Kosterlitz and Robinson (1957), it was found that morphine depressed these responses. Subsequently, Paton (1957) and Schaumann (1957) demonstrated that morphine decreased acetylcholine release from the guinea pig ileum. Paton (1957) also showed that both morphine and nalorphine, as well as other narcotic analgesics, depressed the response of the guinea pig ileum to electrical stimulation. Morphine has little effect upon ACh-induced contraction of the guinea pig ileum (Kosterlitz and Robinson, 1958). In an extensive series of continuing experiments, Kosterlitz and his collaborators have shown that the action of morphine on peripheral cholinergic neurones is widespread. The slowing of the heart induced by vagal stimulation is reduced by morphine in the rabbit and rat, to a lesser degree in the cat, and not at all in the guinea pig (Kosterlitz and Taylor, 1959). Nalorphine also produces a modest depression of vagal inhibition and partially antagonizes the actions of morphine. Trendelenberg (1957) first reported that morphine reduced the response of the cat nictitating membrane to both pre- and postganglionic stimulation and speculated that morphine inhibited the release of a transmitter. Cairnie et al. (1961) found that morphine diminished the contraction of the nicitating membrane to postganglionic stimulation, but not the effects of stimulation of the cardiac accelerator nerve on the heart or the splenic nerve on the contraction of the spleen. The effects of morphine on contraction of the nicitating membrane were greater at low frequencies of stimulation than at high frequencies. Kosterlitz and Wallis (1966) observed that morphine reduced the synaptic potential in the superior cervical ganglion in which hexamethonium had been added to partially block transmission. PELIKAN (1960) had previously shown that morphine depressed the contraction of the nicitating membrane to preganglionic stimulation more for low-frequency stimulation than for high-frequency stimulation, indicating that the release of acetylcholine for each volley was decreased. GYANG et al. (1964), GYANG and KOSTERLITZ (1966), KOSTERLITZ and WATT (1968), KOSTERLITZ et al. (1972), and COX and WEINSTOCK (1966) have shown a good correlation between analgesic activity of the morphine- and nalorphinelike drugs and their ability to decrease the responsiveness of the guinea pig ileum to coaxial electrical stimulation. Morphine also decreases the release of ACh from the frog myoneural junction but increases the responsiveness of the muscle to ACh (FREDERICKSON and PINSKY, 1971).

c) *Brain ACh Release*

Morphine may well also decrease the release of acetylcholine from brain neurones of the anesthetized cat. BELESLIN and POLAK (1965) and BELESLIN et al. (1965) found that morphine depressed the outflow of ACh from the lateral ventricle (caudate nucleus, olfactory grey, or septum) and from the pons, medulla, and the interpeduncular fossa. Morphine also has been shown to decrease the release of ACh from the rat cerebral cortex (MATTHEWS et al., 1973), from potassium-stimulated rat cortex slices (SHARKAWI and SCHULMAN, 1969), and inhibit the uptake of ACh in mouse cortex slices (SCHUBERTH and SUNDWALL, 1967). JHAMANDAS et al. (1971) found that not only did several narcotic analgesics (morphine, meperidine, methadone, codeine, and propoxyphene) inhibit the release of ACh from the sensorimotor cortex of the anesthetized cat but that this effect was also shared by several antagonists with agonistic activity (levallorphan and pentazocine). Naloxone and, surprisingly, nalorphine, which has agonistic activity, did not alter ACh release rate. Naloxone antagonizes the depressant actions of morphine and levallorphan on ACh release (JHAMANDAS et al., 1970). Morphine also inhibits the release of ACh from the caudate (YAKSH and YAMAMURA, 1973). Morphine enhances the release of ACh in the unanesthetized cat, and the enhancement is antagonized by naloxone (MULLIN et al., 1973; PHILLIS et al., 1973).

LABRECQUE and DOMINO (1974) found doses of morphine less than 5.6 mg/kg decreased the release of ACh from the neocortex of the unanesthetized midpontine pretrigeminal transected cat, whereas 10 mg/kg increased release. LARGE and MILTON (1971) found that morphine, levorphanol, and naloxone decreased the K^+ evoked release of ACh from rat cerebral cortical slices, while nalorphine increased release. Morphine increased release of ACh from hippocampal slices while levorphanol, naloxone, and nalorphine decreased release. DOMINO and WILSON (1973a) have extended these findings by showing that morphine and related analgesics antagonized the brain ACh-depleting effect of hemicholinium and that this effect was antagonized by nalorphine and naloxone in the rat.

Morphine, when iontophoretically applied to brain stem neurones, may have either a facilitatory or inhibitory effect (BRADLEY and DRAY, 1973). Further, morphine antagonizes as well as occasionally potentiates the excitatory effects of iontophoretically applied ACh but not its depressant effects. It does not antagonize the excitatory effects of either glutamate or homocysteic acid.

d) Brain ACh Metabolism

In rats, single doses of morphine have been reported to either increase (HERKEN et al., 1957; GIARMAN and PEPEU, 1962; MAYNERT, 1967; CROSSLAND and SLATER, 1968; LARGE and MILTON, 1970; WAJDA and DATTA, 1971) or not affect total brain ACh levels (LAVIKAINEN and MATTILA, 1959). CROSSLAND and SLATER (1968) found that morphine decreased free but increased bound rat brain ACh, whereas RICHTER and GOLDSTEIN (1970) found both free and total ACh levels increased. In the mouse, brain levels of bound and free ACh are increased by single doses of morphine, nalorphine, pentazocine, cyclazocine, and naloxone (HANO et al., 1964; HOWES et al., 1969; RICHTER and GOLDSTEIN, 1970). Nalorphine antagonizes the increase seen with morphine. Very high concentrations of morphine, nalorphine, and naloxone decrease synthesis and release of ACh in mouse cerebral cortical slices, whereas pentazocine and cyclazocine increase synthesis. Cyclazocine increases release, whereas pentazocine decreases release (HOWES et al., 1970). The effects of morphine on synthesis but not release are antagonized by nalorphine. TORDA and WOLFF (1947) found that morphine decreased the synthesis of ACh in the frog.

3. Serotonin

Study of the relationships between pain, the analgesic actions of narcotics, and the neurochemistry of 5-HT is complicated by the fact that many of the agents used to manipulate brain 5-HT are not selective in action. For this reason and to prevent unnecessary duplication, some of these agents that affect serotonin metabolism such as reserpine will be discussed in the following section on catecholamines.

Analgesia

There are several lines of evidence that indicate that serotonergic neurons may in some way modulate the perception of pain and in some way be related to the analgesic actions of narcotic analgesics. HARVEY and LINTS (1965) demonstrated that lesions of the medial forebrain bundle decreased the flinch-jump threshold, as well as brain levels of serotonin, in the rat. Further studies (LINTS and HARVEY, 1969) showed that lesions in the septum and the dorsomedial tegmentum, areas to which the medial forebrain bundle projects, also lowered jump threshold and brain serotonin. Median dorsal raphe lesions facilitate the development of the conditioned avoidance response in the rat (LORENS and YUNGER, 1974). The decrease in jump threshold was related to a lowering of telencephalic but not brain stem serotonin, and the effect of medial forebrain bundle lesions on both jump threshold and serotonin levels could be antagonized by DL-5-HTP (HARVEY and LINTS, 1971). SAMANIN et al. (1970) found that lesions placed in the midbrain raphe of rats increased the sensitivity to pain as measured by tail compression and stimulation of the tail but not by the hot plate test, whereas YORK and MAYNERT (1973) and LORENS and YUNGER (1974) found either no effect or an increase in threshold or reaction time. The analgesic effects of morphine have been reported to be decreased by midbrain raphe lesions (SAMANIN et al., 1970) or unchanged (BLÄSIG et al., 1973a). YUNGER and HARVEY (1973) observed that medial forebrain bundle lesions lowered threshold and increased responsivity to nociceptive responses but not to the startle response in the rat.

DL-5-HTP, which in its own right in the intact mouse has no analgesic activity, enhances morphine-, meperidine- and methadone-induced analgesia in the mouse (SCHAUMANN, 1958; HUIDOBRO, 1963; DEWEY et al., 1970; HARRIS, 1970; GARDINER and EBERHART, 1970; CONTRERAS et al., 1973a, b) and in the rabbit (SAARNIVAARA, 1969a). Intraventricularly administered 5-HT produces analgesia in the mouse (CALCUTT and SPENCER, 1971; SEWELL and SPENCER, 1974) but not in the rat (SPARKES and SPENCER, 1969). Intracerebrally administered 5-HT enhances the analgesic activity of morphine, heroin, etorphine, meperidine, nalorphine, pentazocine, and cyclazocine in the rat (SEWELL and SPENCER, 1974). Intravenously administered 5-HT produces analgesia in the rabbit (SAARNIVAARA, 1969a). Systemically administered 5-HT enhances the analgesic action of morphine in the rat (NICAK, 1965; SPARKES and SPENCER, 1969), mouse (SIGG et al., 1958; CALCUTT and SPENCER, 1971), and rabbit (MATTILA and SAARNIVAARA, 1967a, b; SAARNIVAARA, 1969a).

TENEN (1967) found that pCPA lowered the threshold for the flinch-jump response in the rat, and that it decreased the analgesic effect of morphine, methadone, propoxyphene, possibly codeine, but not meperidine in the rat. SETHY et al. (1970), FENNESSY and LEE (1970), MAJOR and PLEUVRY (1971), and CONTRERAS et al. (1973a, b) also found that pCPA antagonized the analgesic action of morphine and also meperidine in the mouse. Further, MAJOR and PLEUVRY (1971) found that pCPA antagonized the analgesic effect of amphetamine and that 5-HTP enhanced the analgesic actions of both morphine and amphetamine. Similar observations have been made by GORLITZ and FREY (1972) in the rat. GENOVESE et al. (1973) found that intraventricularly administered 5,6-dihydroxytryptamine selectively decreased brain serotonin and morphine analgesia in the rat, whereas BLASIG et al. (1973a) found morphine analgesia was not altered by 5,6-dihydroxytryptamine. Related to these observations are the findings of FLÓREZ et al. (1972) that reserpine and pCPA but not 6-OH dopamine antagonized the respiratory depressant effect of morphine in the cat and that pargyline and tranylcypramine enhanced it. It is known that morphine can release 5-HT from the small intestine of the dog (BURKS and LONG, 1967). Although morphine inhibits the uptake of 5-HT by rabbit brain synaptosomes, naloxone shares this property, indicating that it is not an important mechanism of action (CIOFALO, 1973).

There are some data that are not readily reconciled with the hypothesis that the activation of serotonergic synapses is a necessary condition for the analgesic effects of morphine. 5-HT does not affect morphine analgesia in the rat (VEDERNIKOV and AFRIKANOV, 1969) and decreases it in the mouse (HUIDOBRO, 1963; HO et al., 1972a). Further, 5-HT and 5-HTP facilitate the flexor reflex of the chronic spinal dog, a nociceptive reflex depressed by morphine (MARTIN and EADES, 1970). Cyproheptadine, a tryptamine and serotonin antagonist, enhances morphine analgesia in the rabbit (SAARNIVAARA, 1969a) and depresses the flexor reflex of the chronic spinal dog (VAUPEL and MARTIN, 1973). HO et al. (1971, 1972c, 1973b, c) found that cAMP (or dibutyryl-cAMP) injected either intracerebrally or systemically antagonized the analgesic effects of morphine in the mouse and have related this finding to the ability of cAMP to increase serotonin synthesis and turnover. Further, GEBHART and MITCHELL (1973) found that morphine was more potent as an analgesic in the CFI strain of mice in which the turnover and brain levels of 5-HT were lower than in CFW strain. In the rat, the relationship between analgesia and 5-HT utilization is more complex

(RECH and TILSON, 1973). SAARNIVAARA (1969a) found an enhancement of the analgesic effect of morphine by pCPA in the rabbit. On the other hand, MARUYAMA et al. (1971) and HO et al. (1972b) were unable to observe any effect of pCPA on analgesia produced by morphine, methadone, pentazocine, cyclazocine, and nalorphine in the mouse. BUXBAUM et al. (1973) could not antagonize the analgesic action of morphine in the rat by either pCPA or raphe nuclei lesion. They further observed a prolongation of morphine analgesia by reserpine. Although in the mouse some narcotic analgesics may block (meperidine) the uptake of 5-HT by neurons, others do not (morphine) (CARLSSON and LINDQVIST, 1969). SAMANIN et al. (1973) found that lesions of the midbrain raphe of the rat decreased the analgesic activity of morphine, but not of meperidine, codeine, methadone, and *d*-propoxyphene. As previously mentioned, PROUDFIT and ANDERSON (1974) found that lesions of the medullary raphe nucleus did decrease morphine analgesia.

4. NE and E

The role of catecholamines in pain perception and in the analgesic activity of the narcotic analgesics is not well understood. Morphine has been shown to inhibit the spontaneous firing of noradrenergic neurons in the locus ceruleus but not to alter the firing of raphe serotonergic neurons (KORF et al., 1974). Naloxone antagonized the inhibitory effect of morphine.

a) Analgesia

LEIMDORFER et al. (1947) first demonstrated that E administered into the cisterna magna in the dog produced sedation, sleep, and analgesia. Subsequent studies confirmed these findings and demonstrated that NE, isoproterenol, and butanephrine also produced analgesia, whereas ephedrine, amphetamine, propradrine, synephrine, and tuamine did not (LEIMDORFER and METZNER, 1949; LEIMDORFER, 1950). FELDBERG and SHERWOOD (1954) found that intraventricularly administered E and NE produced analgesia and sleep in the cat. Intraventricularly administered NE produces analgesia in the rabbit (GARDELLA and IZQUIERDO, 1970) but not in the rat (SPARKES and SPENCER, 1969) or mouse (CALCUTT and SPENCER, 1971).

Systemically administered NE produced analgesia in the mouse (MILLER et al., 1955; EMILE et al., 1961) and rabbit (SAARNIVAARA, 1969b), but not the rat (CONTRERAS and TAMAYO, 1966). Most investigators have found that systemically administered E produces analgesia in the rat (CONTRERAS and TAMAYO, 1966), mouse (MILLER et al., 1955; EMILE et al., 1961), dog (IVY et al., 1944; LEIMDORFER and METZNER, 1949), cat (WEBER, 1904), guinea pig (LEIMDORFER and METZNER, 1949), and man (IVY et al., 1944; GROSS et al., 1948). MILOSEVIC (1955), however, was unable to see an analgesic effect in the mouse. Methamphetamine, amphetamine, methylphenidate, ephedrine, and metaraminol also produce analgesia in the mouse (MAJOR and PLEUVRY, 1971). Methoxamine, a directly acting α-adrenergic agonist, produces analgesia in the rabbit (SAARNIVAARA, 1969a) but facilitates the flexor reflex and produces the stepping reflex in the chronic spinal dog (MARTIN and EADES, 1968). Clonidine produces analgesia in mice and rats, as well as sedation, piloerection, tremors, and exophthalmus in the rat (PAALZOW, 1974). Clonidine, xylazine, and 2-(2,6-dimethylphenylamino)-2Δ_2-oxazoline have analgesic activity in the rat when administered systemically but other sympathomimetic amines such as naphazoline, tetryzoline, and oxymetazoline do not. All of these sympathomimetic amines had analgesic activity when administered intracerebrally (SCHMITT et al., 1974). Methoxamine produces analgesia in the dog using the skin twitch reflex (MARTIN and EADES, unpublished observations).

b) Interactions with Narcotic Analgesics

Most investigators have found that systemically administered NE either does not affect morphine analgesia or decreases it in the rat (CONTRERAS and TAMAYO, 1966; SPARKES and SPENCER, 1969), mouse (CALCUTT and SPENCER, 1971), and rabbit (MUNOZ, 1968; SAARNIVAARA, 1969b). SEWELL and SPENCER (1974) found that intracerebrally administered NE decreased the analgesic effect of morphine, heroin, meperidine, and etorphine in the mouse. Data concerning the interaction of E and morphine on analgesia are conflicting. In the mouse, E has been reported to increase (SIGG et al., 1958) and decrease (MILOSEVIC, 1955) the analgesic effects of morphine, while in man it produces a decrease (WOLFF et al., 1940). MILOSEVIC (1955) also found that E decreased the analgesic effects of methadone and meperidine in the mouse. E either increases morphine analgesia or does not alter it in the rat (PUHARICH and GOETZL, 1947; CONTRERAS and TAMAYO, 1966) and the rabbit (SAARNIVAARA, 1969b). HELLER et al. (1968) found that intraperitoneally administered E, NE, and isoproterenol produced analgesia in the mouse, and E and NE enhanced the analgesic effects of morphine. Propranolol antagonized the analgesic actions of morphine, E, NE, and isoproterenol, while dihydroergotamine did not. FENNESSY and LEE (1970) found that phentolamine, propranolol, and methysergide did not alter morphine analgesia in the mouse. Intraventricularly administered E and NE antagonize the analgesic effect of morphine in the rat (MUDGILL et al., 1974). FRIEND and HARRIS (1948) found that adrenalectomy decreased morphine analgesia in the rat. MILLER et al. (1955) found that adrenalectomy and adrenal demedulation were without effect in the rat, while WEI (1973c) observed that adrenalectomy increased morphine analgesia. Phenoxybenzamine and 1-phenyl-3-(2-thiazolyl)-2-thiourea, a dopamine-β-hydroxylase inhibitor, enhanced morphine analgesia in the rat while propranolol did not (CICERO, 1974). GORLITZ and FREY (1972) found that neither phentolamine nor propranolol alter morphine analgesia in the rat. Phenoxybenzamine and phentolamine displaced naloxone from rat-brain homogenates (CICERO et al., 1974b). Phenoxybenzamine produced analgesia in the rat.

c) Metabolism

Morphine, pentazocine, and nalorphine increased rat spinal cord concentrations of normetanephrine. This change was antagonized by naloxone and by spinal cord but not intercollicular transection (SHIOMI and TAKAGI, 1974).

α) Dopamine β-Hydroxylase

Dopamine β-oxidase inhibitors disulfiram and diethyldithiocarbamate do not consistently affect morphine analgesia. In the rat, disulfiram has been reported to decrease morphine analgesia (VEDERNIKOV and AFRIKANOV, 1969; VEDERNIKOV, 1969) or have no effect (GORLITZ and FREY, 1972), whereas diethyldithiocarbamate increases it

(WATANABE et al., 1969). Ouabain has been reported to increase morphine analgesia in the mouse (CALCUTT et al., 1971). The dopamine β-hydroxylase inhibitor 1-phenyl-3-(2-thiozalyl)-2-thiourea potentiates morphine analgesia in the mouse (BHARGAVA et al., 1972b).

β) Tyrosine Hydroxylase

In the mouse, α-MT has an analgesic effect (MAJOR and PLEUVRY, 1971) and has been reported to increase (MAJOR and PLEUVRY, 1971), have no effect on (RUDZIK and MENNEAR, 1965; DEWEY et al., 1970; FENNESSY and LEE, 1970), and antagonize (MEDAKOVIC and BANIC, 1964; VERRI et al., 1968; CONTRERAS et al., 1973a) morphine-induced analgesia. In the rat and rabbit, α-MT does not produce analgesia (MEDAKOVIC and BANIC, 1964; SAARNIVAARA, 1969b), and in the rat has been reported to have no effect on morphine and meperidine analgesia (MEDAKOVIC and BANIC, 1964), to decrease (CONTRERAS and TAMAYO, 1966; GORLITZ and FREY, 1973), or to slightly enhance and prolong it (BUXBAUM et al., 1973). α-MT antagonizes morphine analgesia in the rabbit, and the antagonism is reversed by L-DOPA (SAARNIVAARA, 1969b).

γ) Reserpine

Although some investigators have found that reserpine enhances the analgesic effect of narcotic analgesics in the mouse (TRIPOD and GROSS, 1957; LEME and ROCHA E SILVA, 1961; DANDIYA and MENON, 1963; ROSS and ASHFORD, 1967; JOUNELA and MATTILA, 1968; CONTRERAS et al., 1973a, b) especially within the first hours after its administration at a time when it is liberating large quantities of monoamines (MATTILA and LAVIKAINEN, 1960; HOWES et al., 1969; DEWEY et al., 1970; FENNESSY and LEE, 1970), the preponderance of data indicates that reserpine and reserpinelike agents antagonize the analgesic effects of narcotic analgesics in the rat (MEDAKOVIC and BANIC, 1964; CONTRERAS and TAMAYO, 1966; SPARKES and SPENCER, 1969; VEDERNIKOV and AFRIKANOV, 1969; GROSSMAN et al., 1973), mouse (SCHNEIDER, 1954; SIGG et al., 1958; SCHAUMANN, 1958; ZETLER et al., 1963; TAKAGI et al., 1964; RUDZIK and MENNEAR, 1965; ROSS and ASHFORD, 1967; TAKAGI and NAKAMA, 1968; SHEN et al., 1969; HOWES et al., 1969; SETHY et al., 1970; HARRIS, 1970; DEWEY et al., 1970; FENNESSY and LEE, 1970; MAJOR and PLEUVRY, 1971), rabbit (MŨNOZ and PAEILE, 1967; MUNOZ, 1968; MATTILA and SAARNIVAARA, 1968; VERRI et al., 1968), and guinea pig (RADOUCO-THOMAS et al., 1967). Reserpine alone does not appear to markedly alter responsivity to nociceptive stimuli.

Reserpine antagonism of morphine analgesia in the mouse, rat, and guinea pig is reversed by MAO inhibitors (SCHAUMANN, 1958; SIGG et al., 1958; MEDAKOVIC and BANIC, 1964; RADOUCO-THOMAS et al., 1967; JOUNELA and MATTILA, 1968; SHEN et al., 1969). Both 5-HTP and L-DOPA enhance analgesia in guinea pigs treated with reserpine and meperidine in the presence and absence of tranylcypramine (RADOUCO-THOMAS et al., 1967), either enhance analgesia in mice treated with reserpine or tetrabenazine and morphine (MATTILA and LAVIKAINEN, 1960; TAKAGI et al., 1964; TAKAGI and NAKAMA, 1968), or have no effect (CONTRERAS et al., 1973a). On the other hand, L-DOPA does not increase analgesia in rabbits treated with morphine and reserpine in the presence and absence of nialimide (MŨNOZ and PAEILE,

1967). Neither does pyragallol in the presence or absence of nialimide reverse reserpine's antagonism of morphine analgesia in the rabbit; however, L-DOPA in the presence of pyragallol did antagonize reserpine's effect on morphine analgesia both in the presence and absence of nialimide (MŨNOZ and PAEILE, 1967).

δ) 6-OHDA

SAMANIN and BERNASCONI (1972) and AYHAN (1972) found that intraventricularly administered 6-OHDA did not alter the pain threshold using either a hot plate method or tail compression in the rat, whereas BLÄSIG et al. (1973a) found that it produced hyperalgesia using electric tail stimulation. SAMANIN and BERNASCONI (1972) found that 6-OHDA enhanced the analgesic effect of morphine, whereas AYHAN (1972), BLASIG et al. (1973a), BHARGAVA et al. (1973), ELCHISAK and ROSECRANS (1973), and YORK and MAYNERT (1973) found that it antagonized the analgesic action of morphine and that the antagonism was partially reversed by L-DOPA. NAKAMURA et al. (1973a) further demonstrated that in the rat 6-OHDA-induced depletion of NE content in the hypothalamus potentiates morphine analgesia, whereas depletion of DA in the caudate decreases morphine analgesia. In rats treated with 6-OHDA and desmethylimipramine that showed a marked decrease in brain DA but not NE, morphine's analgesic activity was markedly reduced (BLASIG et al., 1973a). 6-OHDA did not alter the analgesic activity of buprenorphine in the rat (COWAN et al., 1974). FRIEDLER et al. (1972) found that 6-OHDA decreased the analgesic action of morphine in the mouse.

ε) Pyragallol

Pyragallol, which inhibits COMT and which would presumably increase central NE and DA, does not produce analgesia in the rat (CONTRERAS and TAMAYO, 1966) or the rabbit (MŨNOZ and PAEILE, 1967) and does not have a consistent effect on morphine analgesia (CONTRERAS and TAMAYO, 1966; MŨNOZ and PAEILE, 1967; VEDERNIKOV and AFRIKANOV, 1969).

ζ) Isolated Tissue

Morphine inhibits contraction of the electrically stimulated mouse vas deferens and the release of NE. This inhibition is antagonized by naloxone (HENDERSON et al., 1972). Electrically induced contractions are also antagonized by phentolamine and bretylium but not by scopolamine or hexamethonium.

5. Dopamine (DA)

Systemically administered DA produces analgesia in the rat (CONTRERAS and TAMAYO, 1966) but not in the rabbit (MŨNOZ and PAEILE, 1967), whereas intraventricularly administered DA produces analgesia in the rabbit (GARDELLA and IZQUIERDO, 1970) but not in the mouse (CALCUTT and SPENCER, 1971; CALCUTT et al., 1971). In addition, apomorphine, which is reputed to be a dopaminergic agonist, produces analgesia in the rat (AMSLER, 1923; VEDERNIKOV and AFRIKANOV, 1969), rabbit (SAARNIVAARA, 1969b), and guinea pig (AMSLER, 1923), although in the rabbit, data are conflicting (SILVER, 1930). Apomorphine produces a modest depression of the flexor reflex, does not increase the latency of the skin twitch reflex, dilates pupils, and increases pulse rate in the chronic spinal dog (MARTIN et al., 1976). Its profile of activity is not morphinelike. Although DA appears to either antagonize the effects of morphine or have no effect in the rat (CONTRERAS and TAMAYO, 1966), mouse (CALCUTT and SPENCER, 1971; CALCUTT et al., 1971), and rabbit (MŨNOZ and PAEILE, 1967; MŨNOZ, 1968), CALCUTT et al. (1971) have shown that intracerebrally injected DA and ouabain, which increase brain DA levels, enhance the analgesic activity of morphine in the mouse. MŨNOZ and PAEILE (1967) were unable to produce analgesia in the rabbit with DL-DOPA; whereas, SAARNIVAARA (1969b) produced analgesia in the rabbit with L-DOPA and RADOUCO-THOMAS et al. (1967) with DL-DOPA. Both DL-DOPA and L-DOPA enhance morphine analgesia in the mouse (DEWEY et al., 1970; CONTRERAS et al., 1973b), the rabbit (MŨNOZ and PAEILE, 1967; MŨNOZ, 1968; SAARNIVAARA, 1969b), and meperidine analgesia in the guninea pig (RADOUCO-THOMAS et al., 1967). DEWEY et al. (1970) found that tyrosine increased and tyramine decreased morphine analgesia in the mouse. Chlorpromazine and haloperidol antagonize the analgesic effects of *d*-amphetamine, *d*-parachloroamphetamine, and morphine in the rat (GORLITZ and FREY, 1972).

6. Other Drugs Which Affect Indole and Catecholamine Metabolism

The effects of a variety of agents which alter the metabolism of E, NE, DA, and 5-HT have been studied on analgesia. MAO inhibitors which decrease the destruction of these monoamines but also have other actions have been found by some investigators to produce analgesia in some species such as the mouse (EMILE et al., 1961; JOUNELA and MATTILA, 1968), rabbit (MŨNOZ and PAEILE, 1967), guinea pig (RADOUCO-THOMAS et al., 1967), and man (BOSWORTH, 1959; SHERBEL and HARRISON, 1959; CESARMAN, 1959; COSSIO, 1959; SCHWEIZER, 1959; MASTER and DONOSO, 1959); however, others have been unable to find analgesic activity among these types of agents in the rat (GUPTA and KULKARNI, 1966) and rabbit (DEFALQUE, 1965). The findings concerning interactions between MAO inhibitors and morphine are not consistent. Thus, it has been found by some investigators that MAO inhibitors enhance analgesia produced by narcotic analgesics in the rat (GUPTA and KULKARNI, 1966; YEH and MITCHELL, 1971), mouse (DEWEY et al., 1970; IWAMOTO et al., 1971), and rabbit (DEFALQUE, 1965), while others have found a decrease or no change in the mouse and rat (JOUNELA and MATTILA, 1968; VEDERNIKOV and AFRIKANOV, 1969; DEWEY et al., 1970). DL-DOPA had no effect on analgesia in rabbits treated with morphine and nialimide (MŨNOZ and PAEILE, 1967), whereas DOPA enhanced morphine analgesia in mice treated with iproniazid (SCHAUMANN, 1958). Although many investigators have attributed the effects of MAO inhibitors to changes in brain amine metabolism, YEH and MITCHELL (1971) have found that pargyline acutely inhibits the formation of morphine glucuronide.

α-Methyl DOPA inhibits decarboxylation of precursors of 5-HT, E, NE, and DA and is metabolized to the false transmitters α-methyl dopamine and α-methyl norepinephrine, and thus presumably decreases the availability of the above-mentioned neurohumors. α-Methyl DOPA is reported to produce analgesia in the rat (CON-

TRERAS and TAMAYO, 1966) but not in the mouse (RUDZIK and MENNEAR, 1965) and to enhance morphine analgesia in the rat (CONTRERAS and TAMAYO, 1966; VEDERNIKOV and AFRIKANOV, 1969) and the rabbit (MATTILA and SAARNIVAARA, 1967a) but not in the mouse (RUDZIK and MENNEAR, 1965).

7. Cyclic AMP, Prostaglandins, and Polypeptides

Intracerebrally or systemically administered cAMP antagonizes the analgesic effects of morphine in the mouse (Ho et al., 1971, 1972c, 1973b, c). CONTRERAS et al. (1972) found that agents which lower cAMP increase analgesia in the mouse.

Intracerebrally administered prostanglandin E_1 antagonized the analgesic activity of morphine in the rat (FERRI et al., 1974). Prostaglandins E_1 and E_2's stimulatory actions on cAMP are antagonized by morphine, heroin, and methadone but not by dextrorphan, and this effect of the narcotic analgesics is antagonized by naloxone (COLLIER and ROY, 1974). Apomorphine and morphine stimulate prostaglandin synthetase (COLLIER et al., 1974).

β-melanocyte stimulating hormone and tetracosactin ($ACTH_{1-24}$) antagonize morphine's depressant actions on the segmental reflex of the acute spinal cat (KRIVOY et al., 1974b). ZIMMERMANN and KRIVOY (1973) found that tetracosactin antagonized the depressant effects of morphine on the segmental reflex of the isolated rat hemicord, while tetracosactin alone was without activity.

TERENIUS and WAHLSTRÖM (1974, 1975a) first identified a polypeptide extracted from rat brain that would reversibly displace dihydromorphine from a rat brain receptor. Subsequently, a polypeptide was isolated from human cerebrospinal fluid which would displace naltrexone from rat brain synaptic membranes (TERENIUS and WAHLSTRÖM, 1975b). Increasing Na^+ decreased the binding affinity of this polypeptide. HUGHES (1975) and HUGHES et al. (1975) independently isolated a substance from the brains of pigs, rats, mice, rabbits, and guinea pigs which produced a naloxone antagonizable inhibition of electrically evoked twitches of the mouse vas deferens and the guinea pig ileum. This substance, which appears to be a polypeptide, is more active on the mouse vas deferens than on the guinea pig ileum. It is present in highest concentrations in the rabbit corpus striatum, in lower concentrations in the midbrain, pons, and medulla, in low concentrations in the hippocampus and cerebral cortex, and absent in the cerebellum. PASTERNAK et al. (1975) have also isolated what appears to be a polypeptide from calf and rat brain which also has properties similar to morphine in displacing both naloxone and dihydromorphine from receptors isolated from rat brain. TESCHEMACHER et al. (1975) and COX et al. (1975) have also isolated a polypeptide from cow and pig pituitary gland which has morphinelike properties on the mouse vas deferens and guinea pig ileum and which displaces etorphine and naloxone from synaptic membranes of guinea pig brains.

TERENIUS (1975) and TERENIUS et al. (1975) have studied the ability of a variety of polypeptides and other substances to antagonize the binding of dihydromorphine to rat brain synaptic membranes. $ACTH_{1-28}$, $ACTH_{1-24}$, and the subsequences $_{1-10}$, $_{4-10}$, $_{5-14}$, $_{7-16}$, and $_{11-24}$ were active. MSH, oxytocin, SP, bradykinin, GABA, desglycinamide lysine vasopressin, and the subsequences of ACTH $_{4-9}$, $_{4-7}$, $_{5-10}$, and $_{7-10}$ were inactive.

8. Discussion and Conclusions

It seems clear that there is a cholinergic analgesic mechanism of the muscarinic type operating in the brain. This analgesic mechanism must be different from the mechanism mediating morphine and nalorphine analgesia, but probably interacts with it in a synergistic manner. Analgesia produced by muscarinic agents is antagonized by atropine, whereas analgesia produced by narcotic analgesics is not. The locus of this cholinergic analgesic mechanism must be supraspinal, for muscarinic agents facilitate the flexor reflex in the chronic spinal dog, while narcotic analgesics suppress it. Further, muscarinic agents administered into the brain produce analgesia. Narcotic analgesics do inhibit the release of ACh; however, cholinergic mechanisms do not seem to be of importance in mediating spinal cord nociceptive reflexes. Whether the inhibition of ACh release is of importance in the production of analgesia at higher centers cannot be definitively resolved at this time. Arguing for this hypothesis are the following facts: (1) The inhibition of ACh release is an agonistic action which is antagonized selectively by narcotic antagonists. (2) There is a good correlation between the potency and efficacy of narcotic analgesics to inhibit the release of ACh and to produce analgesia. Against this hypothesis are the following facts: (1) Neither muscarinic nor nicotinic antagonists share the pharmacologic properties of narcotic analgesics. (2) Cholinergic agonists have analgesic activity. There are possible ways of reconciling these apparently contradictory types of evidence. Thus, it is possible that there may be cholinergic mechanisms in the brain that are neither muscarinic nor nicotinic in nature (Martin et al., 1958). Alternatively, narcotic analgesics may have multiple modes of action involving a variety of receptor-effector mechanisms in which the receptors have similar affinities for narcotic analgesics. It is conceivable that analgesia induced by narcotic analgesics may be a consequence of the interaction of multiple sites and modes of action.

Several lines of evidence suggest that there is a serotonergic mechanism involved in analgesia. Thus, intraventricularly injected 5-HT produces analgesia, and lesions of the medial forebrain bundle and raphe nuclei produce hyperalgesia. Although there is not complete concordance among existing data, the preponderance of evidence would suggest that the analgesic actions of narcotic analgesics are reduced by agents such as pCPA and reserpine which lower brain 5-HT levels, and enhanced by 5-HTP.

Intraventricularly and systemically administered NE and E produce analgesia in several species. The mechanism whereby this analgesia is produced is not known. The fact that intraventricularly administered amphetamine and ephedrine do not produce analgesia argues against the hypothesis that analgesia can be produced by the activation of central noradrenergic systems. Further, inhibition of dopamine β-hydroxylase, tyrosine hydroxylase, as well as the administration of 6-OHDA and reserpine, all of which reduce brain NE, do not consistently produce hyperalgesia, which further argues against a central noradrenergic analgesic system. On the other hand, there is a substantive body of data that there is an interaction between noradrenergic systems in the brain and narcotic analgesics. Evidence suggests that inhibition of dopamine β-hydroxylase or tyrosine hydroxylase and the administration of reserpine and 6-OHDA decrease the analgesic actions of narcotic analgesics, and that reserpine's antagonism may be reversible by DOPA and MAO inhibitors. The importance of morphine's ability to inhibit the release of NE is not clear.

There is suggestive evidence that there is a dopaminergic analgesic mechanism in the brain, but its relationship to morphine analgesia is not known. As indicated earlier, there is evidence that morphine may have DA antagonistic activity.

It has been known for sometime that there are interactions between polypeptides and narcotic analgesics. Recently, evidence for the presence of brain polypeptides which can interact with morphine receptors has been accumulating. Further, there is evidence that some polypeptides mimic the actions of morphine, while others antagonize it. The functional role of these ligands in the brain is unknown. Narcotic antagonists do not produce antimorphinelike (μ) effects in either the dog or man (MARTIN, 1967). Recently pharmacologic evidence has been obtained for three related receptors (μ, $\varkappa$, and σ) in the dog that must be stereochemically similar (MARTIN et al., 1976; GILBERT and MARTIN, 1976) and that naloxone and naltrexone produce anti-$\varkappa$ effects (MARTIN et al., 1976).

D. Tolerance and Physical Dependence

I. Introduction

Tolerance and physical dependence have, as has been previously mentioned, occupied a central position in formulating concepts of narcotic addiction. The commonly held pharmacologic formulation of narcotic addiction is that with continued use of narcotics, tolerance develops to their euphorigenic effects. This forces the addict to use large doses of narcotics more frequently to attain the desired effect. The increase in the quantity of narcotics consumed is associated with the development of physical dependence which is manifest by an abstinence syndrome when use of the narcotic is discontinued. The abstinence syndrome is discomforting to the addict, and when he learns that narcotics will relieve his discomfort and suffering, a true dependence has been established which reinforces drug-seeking behavior.

Certain aspects of tolerance and physical dependence can be quantitated and this has led to a systematic investigation of some aspects of these phenomena. Underlying many of these investigations is the assumption that tolerance to and physical dependence on narcotic analgesics are manifestations of a single phenomenon. It is important to emphasize that the terms tolerance and physical dependence designate a constellation of phenomena that involve a variety of mechanisms. The failure to recognize the complexities of tolerance and physical dependence has given rise to apparent conflicts between theories where no conflict exists.

The purpose of this section will be to describe tolerance and dependence, to identify the conceptual problems involved in their definition, to review the mechanistic explanations of these phenomena, and to relate them to the problems of addiction.

II. Definitions

The definitions used in the area of drug addiction are ad hoc which have had to serve in legal, medical, and pathologic contexts. Efforts have been made to develop a terminology that will be generally applicable and internally consistent (World Health Organization, 1950; EDDY et al., 1965), but in so doing the resulting terminology ignores conceptual issues. Part of the commonly used vocabulary refers to the appetitive properties of drugs.

Habituation refers to the pattern of repeated taking of a drug where the need and gratification are small, the dose is not increased, and no physical dependence on the drug is produced. The implication is that learning rather than the appetitive characteristics of the drug is mostly responsible for recurrent use. *Psychologic dependence* generally means that the effect of the drug satisfies some psychologic need and is gratifying to the individual. Although habituation has frequently been equated to psychologic dependence, the latter term alludes to a mechanism underlying drug-taking behavior, while the former designates the pattern of drug-taking behavior. *Addiction* refers to a state of repeated drug-taking behavior where the need or compulsion is much stronger than it is in habituation. Thus definition of addiction has been formalized and extended to include: (1) a strong compulsion, (2) the tendency to increase dose, and (3) the production of physical dependence (World Health Organization, 1950).

The term *tolerance* is used in a general sense to indicate that an organism is less susceptible to the effect of drug as a consequence of its prior administration. Thus, in this general sense any reduction of any effect of a drug following its prior administration can be considered tolerance. As will be subsequently pointed out, there are many mechanisms whereby tolerance develops to narcotic analgesics. Although these mechanisms are of general theoretical interest, their relationship and relevance to narcotic dependence and addiction are also of importance. From a more limited but relevant point of view, *tolerance* to narcotic analgesics is the alteration in response (change in character or intensity) induced by prior administration of the drug which either directly or indirectly contributes to or sustains drug-seeking behavior. When tolerance develops very rapidly either following a single dose or a few doses given over a short period of time, it is called *acute tolerance*, or *tachyphylaxis*. When the drug must be administered over a longer period of time to induce tolerance, it is called *chronic tolerance*. *Cross-tolerance* exists when tolerance to one drug confers tolerance to another.

The term *physical dependence* has several conceptual and operational definitions. Dependence implies a need. The concept of dependence, need, or want itself has several meanings which include desire and the sustaining of health and life. As previously indicated, those aspects of dependence that are related to desire are called *psychic dependence*, while those that are related to body health are called *physical dependence*. The evidence that a physical need exists is the emergence of signs of abstinence *(abstinence syndrome)*. Although the physiology of the body is markedly disturbed during the abstinence syndrome, many of the signs of abstinence are not overtly harmful. Some signs of abstinence are associated with discomforting symptoms which in turn give rise to a psychic need state. When physical dependence develops at a very rapid rate over a matter of hours, it is called *acute physical dependence*. When it is produced by prolonged treatment with a drug, it is called *chronic physical dependence*. The abstinence syndrome that results from termination of the drug is termed *withdrawal abstinence*. In contradistinction, abstinence that is produced by administering a competitive antagonist such as nalorphine to a subject physically dependent on morphine is called *precipitated abstinence* (sometimes referred to as acute abstinence). The signs of abstinence that first emerge and which may be present for from 2 days to 1–2 months are called *early* or *primary*, while those that follow and may persist for over 6 months are called *secondary* or *protracted abstinence.*

III. Description of Tolerance and Dependence

Much of the earlier literature on tolerance and dependence has been reviewed by KRUEGER et al. (1941) and by EDDY (1955) (also see Sect. II, Chap. 2).

1. Mouse

Most of the studies of tolerance and physical dependence in the mouse have been recent and were stimulated by the work of HUIDOBRO and collaborators. MAGGIOLO and HUIDOBRO (1961) demonstrated that implanted pellets of compressed free base of morphine released morphine slowly and induced tolerance to morphine's catatonic and analgesic effects. Tolerance does not appear to develop to morphine's lethal effect (FICHTENBERG, 1951). Acute tolerance develops to the increase in motor activity (DEWEY and HARRIS, 1973). When nalorphine was administered to the dependent mouse, an abstinence syndrome was precipitated which consisted of increased spontaneous activity, jumping, tremors and convulsions, increased respiratory rate, urination and defecation, and tearing. HUIDOBRO (1967) described a scoring system based on changes in motor activity. WAY et al. (1969) subsequently developed a method in the mouse for quantitatively measuring tolerance and physical dependence produced by pellet implantation. To measure tolerance, WAY et al. (1969) determined the AD_{50} of morphine before and following pellet implantation. Two measures of precipitated abstinence were used to assess physical dependence: (1) The number of jumps following the administration of naloxone to the dependent mouse, and (2) the determination of the ED_{50} of naloxone in precipitating jumping. These investigators showed that maximum tolerance and physical dependence developed by the third day following pellet implantation, that narcotic analgesics were relatively selective in suppressing jumping, and that there was an excellent correlation between the development of tolerance and physical dependence. Subsequently, Ho et al. (1973c) demonstrated that weight loss was also a useful measure of precipitated and withdrawal abstinence. Naloxone does not precipitate jumping in mice that have received pentazocine chronically (KAMEI et al., 1973).

2. Rat

HIMMELSBACH et al. (1935) studied the effects of morphine on struggling of the restrained rats through a cycle of dependence. Initially, morphine, heroin, and codeine markedly suppressed struggling. With chronic administration of morphine, the amount of struggling prior to injection increased, and although morphine did not reduce the amount of struggling to the level seen prior to chronic administration, the decrease in struggling using the preinjection level as a base line was as great or greater in the tolerant and dependent animal as in the nontolerant animal. Undoubtedly, this effect was the reduction of abstinence; nevertheless, the effect of morphine did persist. Similar observations have been made in the spinal dog (vide infra) and in man. When morphine, heroin, and codeine were permanently withdrawn, the level of struggling increased and then decreased approaching control level by the 7th to 11th days of abstinence.

Subsequent studies have shown that with chronic administration of morphine, tolerance develops to analgesia (LEWIS, 1949; FICHTENBERG, 1951; HANNA, 1960), sedation (FICHTENBERG, 1951; KAYMAKCALAN and WOODS, 1956; HANNA, 1960; MARTIN et al., 1963), hypothermia (GUNNE, 1960; MARTIN et al., 1963; DABROWSKI, 1966), and the depressant effect on motor activity (MARTIN et al., 1963). In the tolerant rat, morphine increases activity (JOEL and ETTINGER, 1926; FICHTENBERG, 1951; KAYMAKCALAN and WOODS, 1956; MARTIN et al., 1963) and metabolic rate (MARTIN et al., 1963). Single doses of morphine induce a measurable degree of tolerance to its analgesic effect in the rat (ERCOLI and LEWIS, 1945; GREEN and YOUNG, 1951). COCHIN and KORNETSKY (1964) studied the time course of the development of acute tolerance to the analgesic effects of morphine and its ability to diminish swimming. Tolerance to its analgesic effects developed after both single and repeated doses of morphine and persisted even after single doses for more than a year. Loss of tolerance to depression of swimming occurred more rapidly. KAYAN et al. (1969) have shown that rats receiving repeated doses of morphine show a higher degree of tolerance to the analgesic effect of morphine using the hot plate procedure if they had previously been placed on the hot plate while under the influence of morphine than if they had not. It is the interaction between the testing procedure and the analgesic that is important, for neither the procedure nor the drug experience alone will account for all of the tolerance (ADAMS et al., 1969). Acute tolerance to morphine can be demonstrated in very young rats (HUIDOBRO and HUIDOBRO, 1973). FICHTENBERG (1951) found that chronic administration of morphine inhibited growth and caused atrophy of genital organs in both males and females.

When morphine is abruptly withdrawn from or narcotic antagonists administered to dependent rats, a variety of signs emerge which include weight loss (FICHTENBERG, 1951; HOSOYA, 1959; MARTIN et al., 1963; GOODE, 1971), decreased consumption of food (MARTIN and SLOAN, 1968; GOODE, 1971) and water (MARTIN et al., 1963; MARTIN and SLOAN, 1968; GOODE, 1971), decreased body temperature (MAYNERT and KLINGMAN, 1962; MARTIN et al., 1963; MARTIN and SLOAN, 1968), urinary output (MARTIN and SLOAN, 1968), and an increased number of "head and wet dog" shakes (WIKLER et al., 1963; MARTIN et al., 1963). Other changes are seen including increased irritability and squealing on handling, hyperalgesia (TILSON et al., 1973), ptosis, writhing, jumping, teeth chattering, escape attempts, exploring, and ear blanching (KAYMAKCALAN and WOODS, 1956; HANNA, 1960; WIKLER et al., 1960; MAYNERT and KLINGMAN, 1962; MARTIN et al., 1963; BUCKETT, 1964; LORENZETTI and SANCILIO, 1970; WEI, 1973a; RISNER and KHAVARI, 1973; BLÄSIG et al., 1973b; CICERO and MEYER, 1973). Morphine-dependent rats show withdrawal aggression which is enhanced by amphetamine; whereas, rats dependent on alcohol or phenobarbital do not (PURI and LAL, 1974b). Withdrawal and precipitated abstinence syndromes may differ qualitatively. BLÄSIG et al. (1973b) have shown that the qualitative characteristics of the precipitated abstinence syndrome change with the level of dependence and the intensity of abstinence. Fighting and aggression are also seen in the abstinent rat (DAVIS and KHALSA, 1971; GIANUTSOS et al., 1974). A variety of narcotic analgesics including morphine, heroin, codeine (KUHN and FRIEBEL, 1962; BUCKETT, 1964), methadone, meperidine, apomorphine, codeine, and *d*-propoxyphene (LORENZETTI and SANCILIO, 1970) suppress abstinence in the rat.

Several investigators have felt that the rat was sedated while in early and precipitated abstinence (KAYMAKCALAN and WOODS, 1956; MAYNERT and KLINGMAN, 1962), basing their interpretations on the observations that the abstinent rat was immobile and exhibited ptosis. It has been difficult to reconcile the immobility with the many signs of irritability seen during early abstinence as described above. The immobile abstinent rat is most commonly in a prone position with his head up and eyes closed and does not assume sleep postures (MARTIN et al., 1963). Further, during early abstinence slow wave and REM sleep are decreased (KHAZAN and COLASANTI, 1972).

Signs of early abstinence persist for several days in the rat and are followed by a long-persisting group of signs—protracted abstinence. During protracted abstinence, rats that were dependent on morphine consume more water (MARTIN et al., 1963; MARTIN and SLOAN, 1968; WIKLER and PESCOR, 1970) and food (MARTIN and SLOAN, 1968), gain weight more rapidly than controls (MARTIN et al., 1963; WIKLER and PESCOR, 1970), have increased body temperature and metabolic rate (MARTIN et al., 1963; MARTIN and SLOAN, 1968), and exhibit aggression (GIANUTSOS et al., 1974) as well as more slow wave and REMS (KHAZAN and COLASANTI, 1972). WIKLER et al. (1963) found that rats dependent on etonitazene exhibited an early and protracted abstinence syndrome and that rats in protracted abstinence exhibited increased water consumption and a rapid weight gain.

3. Guinea Pig

Tolerance develops to the analgesic effect of morphine in the guinea pig when it is administered chronically (MULÉ et al., 1968; GOLDSTEIN and SCHULZ, 1973). Signs of abstinence in the dependent guinea pig include stretching, chewing, tooth grinding, squealing, motor excitability, and scratching (FRIEBEL and KUHN, 1965). Precipitated abstinence is characterized by urination, defecation, tooth grinding, hyperactivity, jumping, hindlimb stretching, and a decrease in body temperature (GOLDSTEIN and SCHULZ, 1973).

4. Rabbit

Tolerance develops to the antinociceptive actions of morphine when it is administered intramuscularly and intraventricularly (HERZ et al., 1972). HERZ et al. (1972) have characterized the precipitated abstinence syndrome which includes hypothermia, increased motor activity, and escape attempts.

5. Dog

PLANT and PIERCE (1928) characterized in some detail the effects of chronic administration of morphine and its withdrawal in the dog. They found that morphine initially produced vomiting associated with an initial loss of weight, salivation, sedation, constipation, and convulsions. With continued administration, marked tolerance developed to its sedative action. Of especial importance was the observation that these effects were less when animals were readdicted to morphine. When morphine was withdrawn from animals stabilized on 40–260 mg/kg/day administered once or twice a day, tremors, twitching, groaning, howling, whining, restlessness, gnawing, salivation, vomiting, weakness, sleepiness, diarrhea, panting and jerky respiration, and increased body temperature were observed.

TATUM et al. (1929) studied the acute effects of morphine in the dog, rabbit, cat, and monkey. In the dog, morphine produced a slowing of heart and respiratory rate, vomiting, defecation, depression, convulsions, and death. With chronic administration of morphine, tolerance developed to the depressant effects as well as to its emetic action and defecation, but not to depression of heart and respiratory rate. WIKLER and FRANK (1948) found that tolerance developed to the depressant effects of morphine and methadone on the flexor and crossed extensor reflexes of the chronic spinal dog when administered chronically but not to their stimulant effect on the extensor thrust.

These findings have been partially confirmed and extended. MARTIN et al. (1964) found that with chronic administration of morphine, tolerance developed to its depressant effects on the flexor and crossed extensor reflex, its excitant effects on the ipsilateral extensor thrust, to its hypothermic effect and associated panting, its ability to produce bradycardia and its miotic effect (Fig. 1). Although GOETZL et al. (1944) reported that chronically administered morphine produced hyperalgesia in the dog, neither MARTIN and EADES (1964) nor KAYAN and MITCHELL (1968) could confirm this observation.

Abrupt withdrawal in the dependent dog is associated with yawning, lacrimation, rhinorrhea, salivation, tremor, restlessness, vocalization, gnawing, head tossing, piloerection, retching, emesis, urination, panting, as well as increased temperature, respiratory rate, pulse rate, and pupillary diameter (PLANT and PIERCE, 1928; TATUM et al., 1929; MARTIN et al., 1974). In the chronic spinal dog the flexor and crossed extensor reflex are enhanced, the hind limb stepping reflex becomes manifest and the ipsilateral extensor thrust is depressed (WIKLER and FRANK, 1948; MARTIN and EADES, 1964). The decorticate dog also shows signs of abstinence which include irritability, yawning, rhinorrhea, salivation, circling, gnawing, rooting, vomiting, and an increase in respiratory rate, pulse rate, and body temperature (WIKLER, 1952). Convulsions is a very rare sign of abstinence from narcotic analgesics in the dog. We have seen it in only one dog but in this animal it was a reproducible sign of abstinence.

WIKLER and CARTER (1953) first demonstrated that nalorphine precipitated abstinence in the chronic spinal dog. The precipitated abstinence is characterized by many of the signs seen in withdrawal abstinence. The relative contribution of the various signs of abstinence to the total syndrome differs for various levels of dependence and is different from the withdrawal abstinence syndrome (MARTIN et al., 1974). In this regard, the qualitative characteristics of the withdrawal syndrome do not seem to differ for different levels of dependence.

The early abstinence syndrome becomes clearly manifest within 12–18 h after the last dose of morphine, persists for 3 or 4 days and is followed by the protracted abstinence syndrome. The protracted abstinence syndrome in the chronic spinal dog

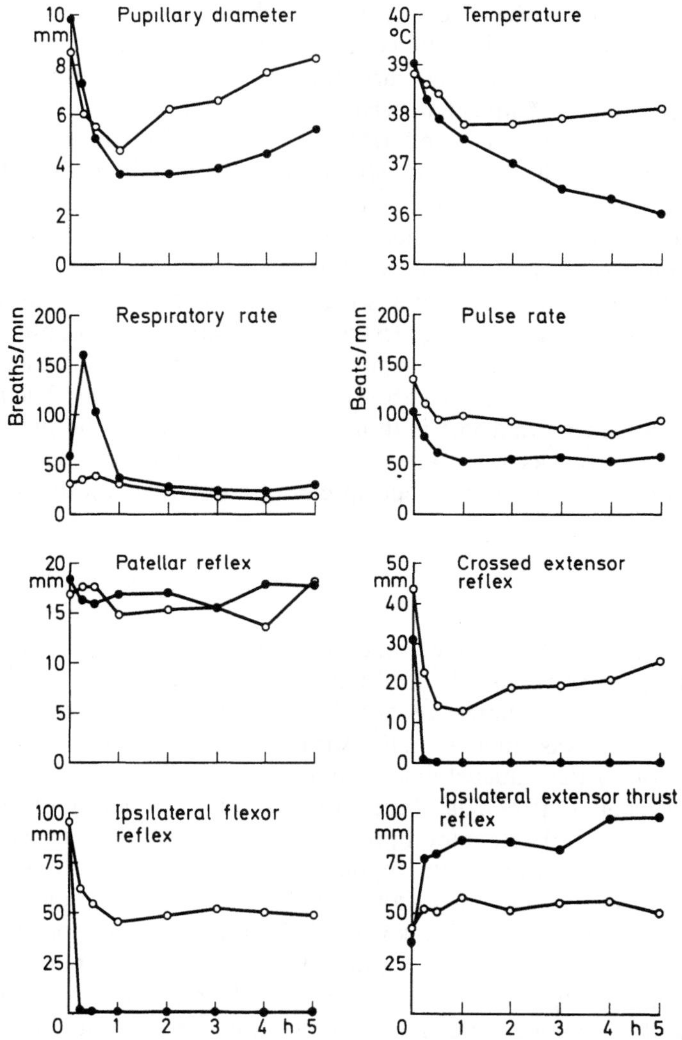

Fig. 1. Time action curves for effects of 10 mg/kg of morphine on various physiologic variables in nontolerant low spinal dogs (•) and in chronically tolerant low spinal dogs (○). (Reproduced from MARTIN and EADES, 1964, with the permission of the publisher)

is characterized by a decrease in temperature, pulse rate, and respiratory rate, an increase in the responsivity of the flexor reflex to strongly nociceptive stimuli but a decrease in responsivity to a low intensity stimulus. The latency of the skin twitch reflex evoked by radiant heat was shortened.

6. Monkey

TATUM et al. (1929) and KOLB and DUMEZ (1931) conducted the first extensive studies of morphine in the monkey. TATUM et al. (1929) found that morphine produced behavioral depression, pupillary dilatation, convulsions, and respiratory depression. When monkeys that had received morphine chronically were withdrawn, they exhibited nervousness, irritability, muscle tremors, priapism, and vomiting. KOLB and DUMEZ (1931) found that chronic administration of morphine, heroin, or codeine induced tolerance and that heroin and morphine but not codeine would cross suppress abstinence. KOLB and DUMEZ (1931) identified the crouching posture, grimacing, hypersensitivity to sensory stimulation and handling, and a fall in body temperature as signs of withdrawal abstinence. HOLTZMAN and VILLARREAL (1969, 1971) found that withdrawal hypothermia had a more rapid onset and was more intense if the monkey was restrained. SEEVERS (1936) identified many signs of abstinence in the rhesus monkey, including yawning, rhinorrhea, lacrimation, shivering, hiccups, perspiration, irritability, restlessness, vocalization, tremor, muscle twitches and rigidity, piloerection, anorexia, vomiting, diarrhea, priapism and masturbation, and rarely convulsions. Miosis is an abstinence sign in the monkey. The nalorphine-precipitated abstinence syndrome is thought to be similar to the morphine withdrawal syndrome (SEEVERS and DENEAU, 1963; HOLTZMAN and VILLARREAL, 1969). Signs of abstinence may vary from one species of monkey to another (KILLAM and DENEAU, 1973).

7. Man

The nature and characteristics of tolerance to and dependence on narcotic analgesics have been more intensively studied in man than in other species. Although these phenomena had received clinical recognition during the 19th century, systematic investigation of them was initiated on the ward of the Philadelphia General Hospital at the suggestion of the Philadelphia Committee for the Clinical Study of Opium Addiction (LIGHT et al., 1929–30). These investigators found that in the addict stabilized on morphine, heart rate, blood pressure, and respiration were within normal limits (LIGHT and TORRANCE, 1929a, b); they exhibited an anemia, leukocytosis, and a high blood lactic acid content (KARR et al., 1929), and had an abnormal glucose tolerance response (LIGHT and TORRANCE, 1929c). LIGHT and TORRANCE (1929d) also systematically studied patients during withdrawal and reported yawning, lacrimation, sneezing, stuffy nose, mydriasis, chills, piloerection, perspiration, muscle twitching, diarrhea, nausea, restlessness, feeling of weakness, and an increase in body temperature and metabolic rate as signs and symptoms of abstinence. Subsequently, HIMMELSBACH (1937), KOLB and HIMMELSBACH (1938), and HIMMELSBACH (1939) devised several scoring systems for quantitatively assessing the severity of abstinence by giving points for nonmeasurable signs such as yawning, lacrimation, rhinorrhea, perspiration, piloerection, mydriasis, tremors, restlessness, vomiting, and diarrhea as well as for the measurable signs weight loss, hyperpnea, fever, decrease in food intake, and increase in blood pressure. This scoring system allowed an objective assessment of the efficacy of agents for the detoxification of morphine-dependent subjects and subsequently for the ability of new analgesic agents to substitute for morphine in dependent subjects (see Sect. II, Chap. 3). Both the investigators at Philadelphia General Hospital and HIMMELSBACH of the U.S. Public Health Service Hospital at Lexington studied subjects who were dependent on narcotics at the time they were admitted to the hospital. For this reason, abstinence points were assessed using the difference between the state of the patient when stabilized and when abstinent. FRASER and ISBELL (1960) subsequently computed abstinence points in patients who were made dependent on morphine using both preaddiction and addiction values.

Because dependent patients had been used, the patients in the "nontolerant" and "tolerant" state were not definitively compared. MARTIN and JASINSKI (1969) reinvestigated the phenomena of tolerance and dependence in a group of subjects who had not received any narcotic from 1 to 7 months prior to being admitted to the research ward. Control observations were obtained over a 2-month period. Subjects were then made dependent on 240/mg/day of morphine over a 5-week period, were stabilized on morphine for 29 weeks, were gradually withdrawn over a 3-week period and observed for a 30-week postwithdrawal period. Subjects stabilized on morphine exhibited an elevation of blood pressure, pulse rate, and rectal temperature (MARTIN and JASINSKI, 1969) and decreased respiratory rate, sensitivity of the respiratory center to carbon dioxide (MARTIN et al., 1968) and pupillary diameter (Fig. 2). Com-

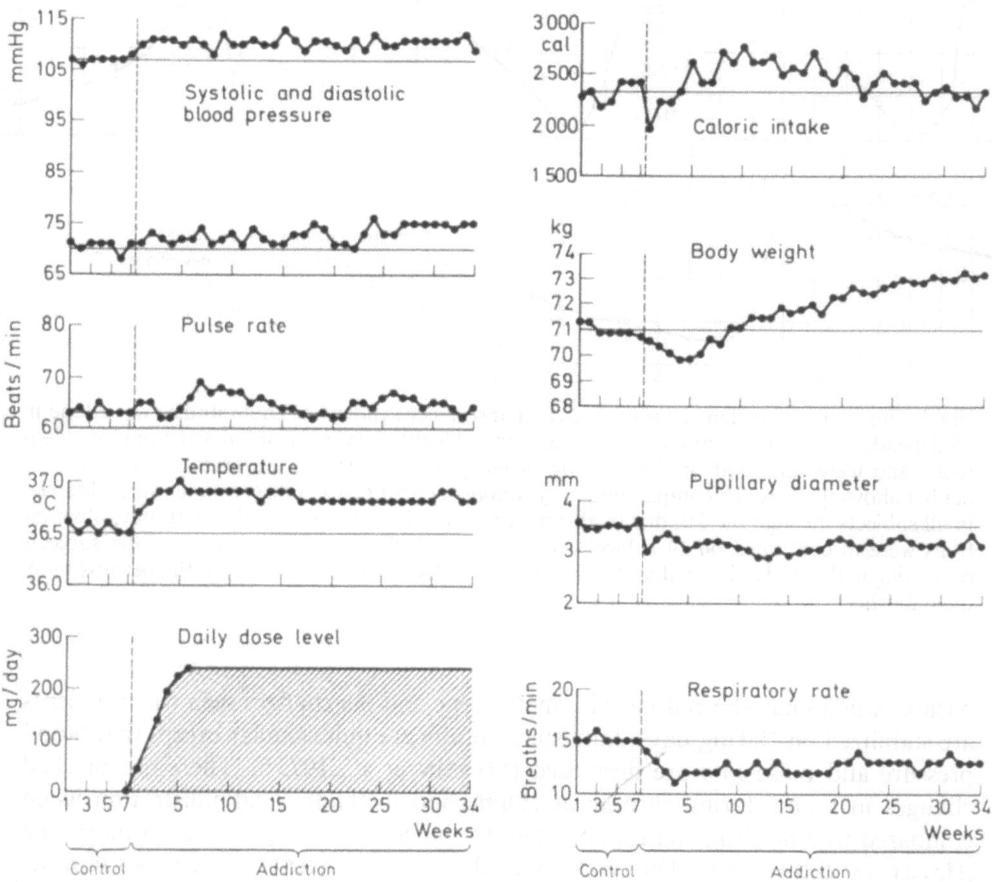

Fig. 2. Effects of chronic administration of morphine on clinical variables. Each point represents weekly mean a.m. observations for 7 subjects. Horizontal line represents mean of control determinations for 7 subjects. (Reproduced from MARTIN and JASINSKI, 1969, with the permission of the publisher)

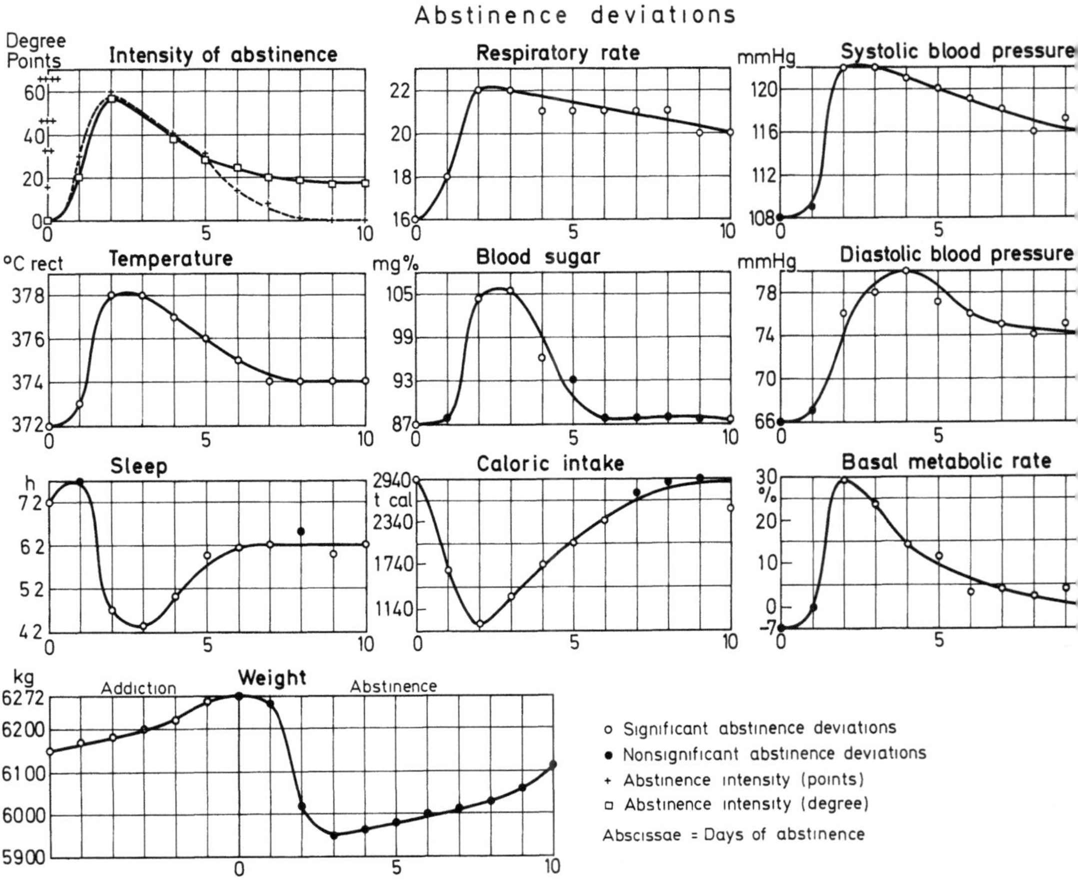

Fig. 3. Time course for signs of abstinence and abstinence syndrome. These studies were done in 65 dependent subjects, admitted to the U.S. Public Health Service Hospital at Lexington, Kentucky and were stabilized on 240–340 mg of morphine per day to the extent that they were neither showing signs nor complaining of symptoms of abstinence. Observations were obtained in all subjects through the 5th day of abstinence. From the 6th through the 10th day, observations were discontinued on 14 subjects. Values at each point represent mean of the subjects remaining in the study. (Reproduced from KOLB and HIMMELSBACH, 1938, with the permission of the publisher)

parable studies have been done with methadone and the changes seen when patients are stabilized on 100 mg/day administered orally are quite similar except that blood pressure and pulse rate are decreased (MARTIN et al., 1973b). There are marked changes in mood during chronic morphine and methadone administration when compared to a no drug control state and to the effects of single doses of morphine (HAERTZEN and HOOKS, 1969; MARTIN et al., 1973b). Whereas single doses of morphine increase MBG and Ef scale scores, which measure feelings of well-being and efficiency in a dose-related manner, chronic administration of morphine is associated with feelings of lethargy, weakness, tiredness, resentment, withdrawal, boredom, and hypochondriasis.

Following abrupt withdrawal from morphine, the intensity of abstinence reaches a maximum within 2 days and thereafter subsides. As can be seen from Figure 3, major signs of abstinence as assessed using a clinical grading system (degree) had returned to control level by the eighth day at a time when abstinence points were still elevated. KOLB and HIMMELSBACH (1938) followed patients through the tenth day of abstinence and noted that "... we believe that recovery is not complete in 10 days." HIMMELSBACH (1942) subsequently studied the changes associated with withdrawal from narcotic analgesics over a period of 270 days following withdrawal and his data are presented in Figure 4. Because of the many difficulties associated with the execution of a study of protracted abstinence, several comments concerning HIMMELSBACH'S experiments may be pertinent to problems of experimental design. (1) HIMMELSBACH'S patients were dependent when they were admitted for study and were stabilized on morphine or other narcotic analgesics. As has been previously discussed, the preaddiction and stabilization base-lines differ with regard to several important physiologic variables. (2) During the course of the study there was attrition in the number of subjects. HIMMELSBACH started with 21 subjects who participated through the 30th day; however, at the 270th day only nine of these patients were available for study. This is an important consideration in the interpretation of results for as has been shown (MARTIN and JASINSKI, 1969) the between-subjects variance is large and is comparable in magnitude to the treatment effects in studies of this sort. Thus, it is difficult to know whether some of the changes that HIMMELSBACH observed were attributable to change in the character of the patient population or to the recovery from the chronic administration of narcotic analgesics. (3) Eleven of the patients that HIMMELSBACH studied received only morphine, while the remaining ten received either dihydrocodeinone, tetrahydrothebaine, dihydroalphaisomorphine, alphaisomorphine, dihydroisocodeine, or isocodeine. Although narcotic analgesics do show a common constellation of effects, there are differences not only in their acute effects but in their effects during chronic administration and withdrawal. HIMMELSBACH (1942) found that the different manifestations of abstinence did not return to an equilibrium value at the same rate. Thus, blood levels of glucose, inorganic phosphate, and lactic acid returned to an equilibrium level within 1 month; body temperature, caloric intake, sleep and respiratory rate within 2–3 months and body weight, metabolic rate, blood pressure, hematocrit, sedimentation rate and blood specific gravity within 4–6 months (see Fig. 4). Thus, HIMMELSBACH saw the abstinence syndrome (early) as a physiologic perturbation which required up to 6 months to dissipate itself.

MARTIN and JASINSKI (1969) observed the long-term abstinence changes in patients stabilized on 240 mg of morphine per day following 3 weeks of dose reduction (Fig. 5). The primary abstinence syndrome persisted for approximately 5 weeks following the last dose of morphine and was followed by the protracted abstinence syndrome, which was characterized by a decrease in blood pressure, pulse rate, pupillary diameter, and responsivity of the respiratory center to carbon dioxide (MARTIN et al., 1968; MARTIN and JASINSKI, 1969). In addition, skin conductance was decreased and the responsivity of skin conductance to hypercapnia was increased (JONES et al., 1969). There was also a significant increase in excretion of E seen through the 17th week of abstinence (EISENMAN et al., 1969). Protracted abstinence was also seen in patients who had been dependent on methadone (MARTIN et al.,

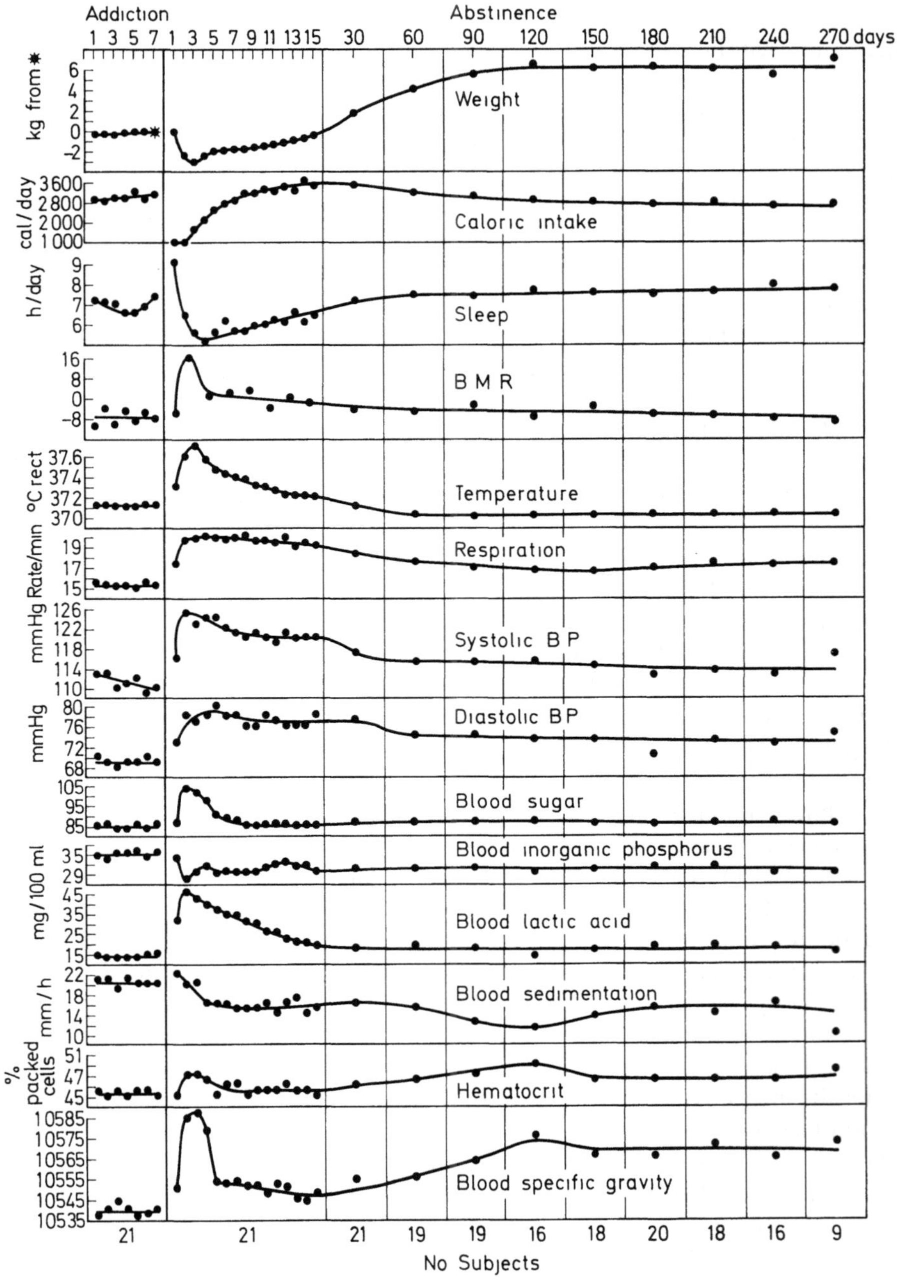

Fig. 4. These observations were made on 27 patients admitted to the U.S. Public Health Service Hospital at Lexington, Kentucky for treatment of their narcotic dependence. Eleven of these dependent patients were stabilized on morphine and the remaining 10 on morphine congeners which substituted for morphine. As can be seen, during the 270 days over which the study was conducted, 12 patients did not complete all observations. When patients were stabilized on narcotics and were abruptly withdrawn, observations were made at various intervals throughout course of the study. (Reproduced from HIMMELSBACH, 1942, with the permission of the publisher)

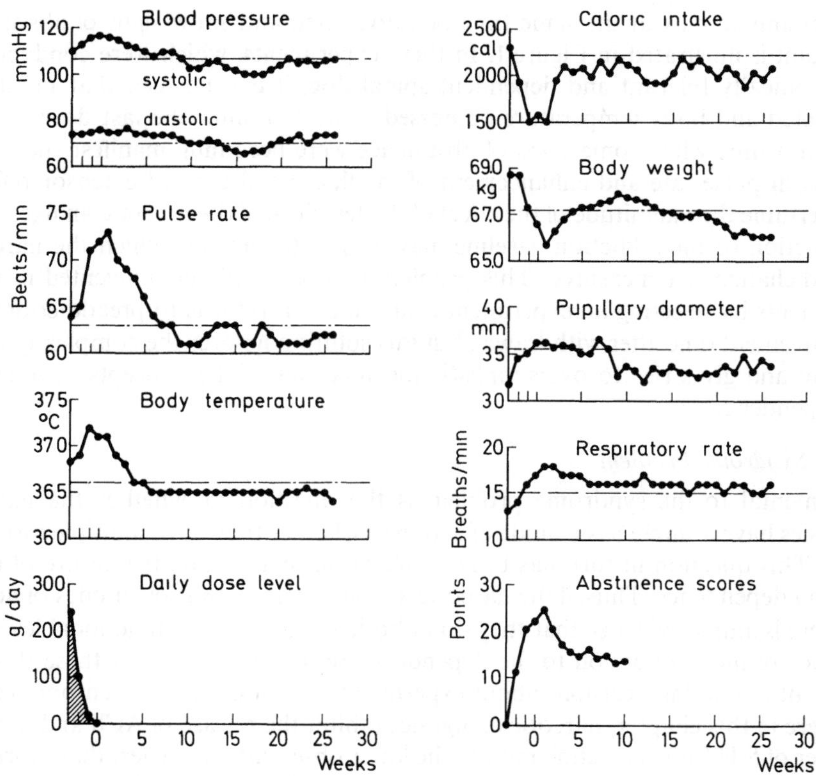

Fig. 5. Changes seen during early and protracted abstinence. Each point represents mean weekly a.m. values for 6 subjects. First point of each curve represents mean value for last 7 weeks of addiction. Horizontal line represents mean control value for the 6 subjects. One subject was withdrawn from the study near the end of the chronic intoxication phase because of episodes of acute cholecystitis. (Reproduced from MARTIN and JASINSKI, 1969, with the permission of the publisher)

1973 b). It emerged approximately 6 weeks after the last dose of methadone and was characterized by a trend for blood pressure, pulse rate, body temperature, and pupillary diameter to be decreased below preaddiction levels, for REM and slow wave sleep to be increased and for the patients to be hypophoric.

IV. Problems of Quantitating Tolerance and Physical Dependence

1. Tolerance

There are a number of technical problems associated with the measurement of tolerance.

a) Baseline Problem

Because some actions of narcotics persist longer than others in the dependent animal, the time that the test dose is administered following the last stabilization dose is arbitrary but critical to assessing tolerance. Thus, paradoxically the withdrawn dependent animal may at the same time be intoxicated and show signs of abstinence. This point is illustrated in Figure 1. In these experiments, which were conducted in the chronically tolerant and dependent spinal dog, it can be seen that pupils are constricted and body temperature depressed some 17 h after the last dose of morphine at a time when some signs of abstinence were becoming manifest such as an increase in pulse rate and enhancement of the flexor and crossed extensor reflexes. To determine the magnitude of the effect of the test dose of the narcotic analgesic, the preaddiction or the addiction baseline may be selected against which the narcotic-induced changes are measured. This problem is superficially circumvented in many experiments by selecting an experimental variable that returns to preaddiction level at a convenient time after withdrawal, but this solution ignores the complexity of the problem and gives rise to oversimplistic formulations of the concepts of tolerance and dependence.

b) *The Syndrome Problem*

Fundamental to the syndrome problem is the question or whether the narcotic analgesics have a single basic mode of action or whether they have multiple modes of action. This question in turn has basic implications concerning the nature of tolerance and dependence. Thus, if the narcotic analgesics have multiple modes of action, and there is ample evidence that they do at both a neuronal and functional level, the relevance of mode of action to the dependence-producing actions of these drugs is important in making decisions about experimental tactics. As has been pointed out elsewhere in this chapter, narcotic analgesics inhibit the release of ACh and increase turnover of 5-HT and catecholamines. The latter observations suggest that morphine increases activity of serotonergic, noradrenergic, and dopaminergic neurones. At a functional level it is well known that morphine constricts pupils in man, mouse, and dog, while it dilates them in the cat and monkey; that it is a potent respiratory depressant in man but may stimulate respiration in the dog; that in man it produces, paradoxically, both euphoria and sedation. There are several possible explanations of the different syndromes that single and repeated doses of narcotic analgesics can produce. Thus, narcotic analgesics may simultaneously stimulate functional systems that are antagonistic to each other such as the parasympathetic and sympathetic nervous system, and the dominant influence may depend upon the number of neurones activated in each system and this balance may vary from species to species and individual to individual. An interaction of this type has been demonstrated for morphine's respiratory effects. Morphine depresses the responsivity of the respiratory center to CO_2, and in man, where morphine is a potent respiratory depressant, this is its major mode of action. In the dog, morphine also lowers the set point of the temperature homeostat which in turn initiates panting as a means of dissipating heat (MARTIN and EADES, 1961; MARTIN, 1968a), and this facilitatory influence counteracts morphine's direct depressant effect on the respiratory center. As a consequence of the multiple modes of action of narcotic analgesics and the complex interactions that may occur, it is difficult to identify appropriate causal relationships with a high degree of certainty.

c) *Change in Effect*

During the course of chronic administration of narcotics, their effects change. Thus, in the nontolerant rat large doses of morphine depress body temperature, while in the tolerant rat body temperature is increased. As has been previously discussed, in the nontolerant man single doses of narcotic analgesics produce euphoria and sedation, while in the tolerant patient they result in apathy, anhedonia, withdrawal from social interactions, and hypochondriasis (see Subsect. B.II.2). These changes in effect also represent a dimension of the tolerance-dependence phenomena, which are not readily or simply reconciled with the commonly held view that tolerance is a diminution of effect as a consequence of chronic administration and that the effect can be recaptured in the tolerant animal if the quantity of drug administered is increased sufficiently.

d) Quantitation of Tolerance

Tolerance can be quantitated when the aforementioned difficulties do not interfere, by comparing the responsivity of narcotic analgesics prior to and during chronic administration. This was first done in man (MARTIN and FRASER, 1961) by obtaining dose-effect relationships for several parameters (e.g., pupils, subjective effects) of morphine and heroin prior to chronic admistration of heroin or morphine, and subsequently ascertaining the effects of doses of heroin and morphine when the subjects were receiving these drugs chronically. The magnitude of the effect of a given dose of narcotic in the tolerant patients was equated to the same magnitude of effects when the subjects were nontolerant, using the previously determined dose response relationship, and a ratio of the doses required to produce the effect in nontolerant and tolerant subjects was calculated. This ratio was called a "tolerance index." This technique was subsequently used to study tolerance in the dog (MARTIN and EADES, 1964) and was extended by ASTON (1965) in his study of barbiturate tolerance by determining a dose response relationship for sleeping time in rats prior to and while receiving pentobarbital chronically and calculating a potency ratio between these two circumstances. In a similar manner, WAY et al. (1969) have determined analgesic dose 50s (AD_{50}) prior to and during chronic administration of narcotics.

2. Physical Dependence

The baseline and syndrome problems that confound measurement and conceptualization of tolerance are similarly inherent in measuring physical dependence.

a) Baseline Problem

As previously pointed out, not all signs of abstinence evolve pari passu, and signs of abstinence and intoxication are present at the same time. One of the practicalities of measuring physical dependence is determining when a deviation is a sign of abstinence (e.g., when the value deviates from the stabilization or the preaddiction baseline). When the concept of physical dependence was being formulated, the physical sickness that resulted from abstinence or the discontinuation of drug administration was in a general way termed "abstinence". When HIMMELSBACH first attempted to quantitate this sickness for the practical purpose of determining whether the course of the illness could be modified by chemotherapeutic intervention, this problem was not attended to; however, when theoretical constructs were made to explain physical dependence, in terms of hypotheses concerning the number of receptors occupied, changes in the physiologic response to a drug receptor interaction and changes in contra-adaptive mechanisms, the issue of baseline became of critical importance.

b) Syndrome Problem

The abstinence syndrome consists of a large number of signs and symptoms. HIMMELSBACH in his original papers (HIMMELSBACH, 1937; KOLB and HIMMELSBACH, 1938) describes a method of measuring and weighting a variety of abstinence signs to obtain a number representing the intensity of abstinence (see Sect. II, Chap. 3). Analysis of an extensive series of unpublished experiments, which were conducted by Dr. E.G. WILLIAMS to determine the minimum chronically administered dose of morphine which would produce a clinically significant degree of physical dependence, as well as other data (cf. MARTIN, 1966), indicated that the relative contribution of the various sources of abstinence points did not change with the level of dependence. Further, it has been shown that morphine and other narcotic analgesics suppress abstinence in man in a dose-dependent manner (FRASER and ISBELL, 1960) and that narcotic antagonists produce a dose-dependent increase in intensity of precipitated abstinence (JASINSKI et al., 1967).

In the rat several scoring systems have been evolved in which it has been shown that the intensity of withdrawal abstinence can be suppressed in a dose-related manner by narcotic analgesics (BUCKETT, 1964; LORENZETTI and SANCILIO, 1970) or that the intensity of precipitated abstinence is increased by increasing doses of antagonists (WEI, 1973a; WEI et al., 1973a) (see Sect. II, Chap. 2). BLÄSIG et al. (1973b) have investigated the role of the degree of dependence in the relationship of signs of precipitated abstinence. Some signs of precipitated abstinence (e.g., exploring) are seen in animals with low levels of dependence, while others are manifest only in animals with high levels of dependence. With some signs of precipitated abstinence intensity increases with increasing doses of the antagonist (jumping) or plateaus (e.g., writhing), while with other signs intensity decreases (e.g., "wet shakes"). The fact that all signs of precipitated abstinence are not related linearly to the degree of dependence complicates interpretation of data as BLÄSIG et al. (1973b) have pointed out. WEI et al. (1973a) found that not all signs of precipitated abstinence were equally sensitive to naloxone.

In an attempt to circumvent this difficulty, MARTIN et al. (1974) used as the criterion for incorporating items into an abstinence scoring system for the chronic spinal dog an antagonist dose-related increase of the intensity, degree, or frequency of signs of precipitated abstinence. Using this scoring system, antagonists produced a dose-related increase in the score for precipitated abstinence, and morphine produced a dose-related decrease in the intensity of withdrawal abstinence. It was observed that the character of the precipitated abstinence syndrome changed with level of dependence, but the character of the withdrawal abstinence syndrome did not. A similar observation concerning withdrawal abstinence in man has been made (MARTIN, 1966).

It is thus apparent that all signs of tolerance and physical dependence do not evolve or become manifest in parallel. The basic question, which cannot be answered at this time, is which signs are related to symptoms that lead to continued drug-seeking behavior, relapse, and the exacerbation of psychopathy. Because this question cannot be answered and because different signs and symptoms of abstinence have different neuronal substrates, quantifying the abstinence syndrome allows the acquisition of data that may prove useful in the future. In this regard, it is important

Table 2. The relative percentage and rank of various sources of "Himmelsbach" points for morphine, cyclazocine, and nalorphine abstinence syndromes. The values for morphine and cyclazocine have been previously reported (MARTIN et al., 1965). The values for nalorphine were calculated from data obtained from the 7 experimental subjects reported in the study by MARTIN and GORODETZKY (1965). Spearman correlation coefficients (r_s) for the various conditions are also indicated. The values in parentheses indicate the level of significance for the various correlations

Source of points	Morphine		Cyclazocine		Nalorphine	
	% of total points	Rank	% of total points	Rank	% of total points	Rank
+ Signs	4.4	6	12.8	4	11.0	3
++ Signs	9.3	5	16.7	2	3.8	7
Caloric intake	1.9	8	5.5	6	6.7	6
Restlessness	0.8	9	0	9	1.1	8
Emesis	2.8	7	0.7	8	0	9
Fever	12.3	3	33.9	1	35.8	1
Hyperpnea	31.1	1	11.1	5	10.8	4
Systolic BP	25.5	2	3.1	7	9.9	5
Weight loss	11.5	4	15.8	3	20.9	2

$r_{s M \times C} = 0.47$ (>0.05)
$r_{s M \times N} = 0.60$ (>0.05)
$r_{s C \times N} = 0.72$ (<0.05)
M = morphine, C = cyclazocine, N = nalorphine

to not only compute a total abstinence score but to determine the relative contribution of each source of abstinence points to the total score. This permits an assessment of the qualitative characteristics of the abstinence syndrome. This type of analysis has been done to distinguish the nalorphine (MARTIN and GORODETZKY, 1965) and the cyclazocine (MARTIN et al., 1965) abstinence syndromes from the morphine abstinence syndrome (see Table 2) and has been used to distinguish types of dependence produced by other drugs from those produced by morphine and cyclazocine (JASINSKI et al., 1971).

V. Neurophysiology

The thrust of research concerning the neurophysiology of tolerance and dependence has had two purposes, the finding of the neurophysiologic substrate of signs of tolerance and dependence, and the elucidation of those sites which mediate suffering associated with the abstinence syndrome.

1. Isolated Tissues

PATON (1957) showed that tolerance or tachyphylaxis developed to morphine's depressant action on electrically-induced twitch response of the guinea pig ileum. When morphine was washed out, the twitch height was further reduced and thereafter slowly recovered, and further addition of morphine was necessary for the contraction to obtain its control height. GYANG and KOSTERLITZ (1966) viewed both narcotic analgesics and narcotic antagonists of the nalorphine type as partial agonists and tachyphylaxis or tolerance to the depressant effects of these agents on the guinea pig ileum as a manifestation of their blocking properties. They have observed that depressant or agonistic effects have a rapid onset, while the antagonistic (tolerance) effect has a slower onset and that the recovery from the antagonistic effects (tolerance) is more rapid for morphinelike agents than for nalorphinelike agents. Both PATON (1957) and GYANG and KOSTERLITZ (1966) have related tachyphylaxis to the central actions of both the narcotic analgesics and the antagonists to tolerance and physical dependence in the central nervous system.

Ilea obtained from guinea pigs that have been chronically exposed to morphine show tolerance to morphine's depressant effects on electrically-evoked twitch responses (EHRENPREIS et al., 1972; GOLDSTEIN and SCHULZ, 1973). TAKAGI et al. (1965) found that ilea from guinea pigs that had been treated chronically with morphine and withdrawn had increased spontaneous activity and an apparent increase in the affinity of 5-HT and morphine for the 5-HT receptor. The apparent affinities of nicotine, hexamethonium, and ACh were not different in the tolerant guinea pig from their affinities in control animals. GOLDSTEIN and SCHULZ (1973) also observed no change in responsivity of ilea obtained from guinea pigs tolerant to and dependent on morphine to ACh. These findings appear to differ from those of EHRENPREIS et al. (1972) who found increased responsivity to ACh in ilea obtained from chronically morphinized guinea pigs. GOLDSTEIN and SCHULZ (1973) further found that the depressant effects of NE, isoproterenol, and DA were decreased in ilea from tolerant dependent guinea pigs, while the effects of 5-HT were enhanced (SCHULZ and GOLDSTEIN, 1973). Morphine does not appear to alter the release of 5-HT in either ilea from tolerant or nontolerant guinea pigs, but antagonizes the effects of 5-HT (SCHULZ and CARTWRIGHT, 1974). Tolerance to the depressant effects of morphine and E and the associated hypersensitivity to 5-HT in ilea taken from pellet-implanted guinea pigs is lost in about 7 days after removal of the pellet (SCHULZ et al., 1974).

BURKS et al. (1974) observed that the stimulatory action of morphine but not 5-HT on the bowel was decreased in morphine-tolerant dogs. BURKS and GRUBB (1974) also observed perfusion of the small intestine with morphine or levorphanol produced acute tolerance to their stimulatory actions. The responsivity of the acutely tolerant small intestine to bethanechol, DMPP, and 5-HT was not altered. Perfusion of the small intestine with naloxone antagonized the stimulatory actions of morphine but not of bethanechol, DMPP, or 5-HT. When tachyphylaxis to the stimulatory effects of 5-HT was produced by perfusion, the responsivity to morphine but not bethanechol and DMPP was reduced. These data were interpreted as indicating that narcotic analgesics released 5-HT from intestinal tryptaminergic neurons and that tolerance involved the morphine receptor or releasing mechanism.

Colon and vas deferens from the morphine-tolerant rat show increased sensitivity to cholinergic agonists, 5-HT and NE and an increased maximal induced tension; however, when the dose response curve is plotted as percent of maximal contraction, the curve is not changed (POLLOCK et al., 1972).

According to EHRENPREIS et al. (1972), naloxone induces a contracture in ilea from chronically morphinized guinea pigs. GOLDSTEIN and SCHULZ (1973) report no difference in the effects of naloxone in antagonizing the effects of morphine in tolerant ilea.

2. Spinal Cord and Brain

WIKLER and FRANK (1948) demonstrated that morphine depressed the flexor and crossed extensor and enhanced the ipsilateral extensor thrust reflexes of the chronic spinal dog and that with chronic administration tolerance developed to these effects. These findings were confirmed by MARTIN and EADES (1964) who further found that tolerance to morphine's depressant actions on spinal cord reflex activity developed more rapidly in the intact dog in which supraspinal influences modulate reflex activity than in the chronic spinal dog. With prolonged chronic administration of morphine, a high degree of tolerance developed to its effects on spinal cord reflexes in the chronic spinal dog. Conservatively, 7–40 times as much morphine is required to produce a comparable degree of depression in the tolerant as in the nontolerant dog (MARTIN and EADES, 1964; also see Fig. 6).

The dependent chronic spinal dog also exhibits both precipitated and withdrawal abstinence signs consisting of exaggerated flexor and crossed extensor reflexes, a depressed ipsilateral extensor thrust reflex, and spontaneous fragmentary and continuous stepping movements (WIKLER and FRANK, 1948; WIKLER and CARTER, 1953; MARTIN and EADES, 1964). Spinal cord signs of precipitated abstinence can be erratically seen following the administration of a single large dose (WIKLER and CARTER, 1953) or reproducibly seen after a single 8-h infusion (3 mg/kg/h) of morphine (MARTIN and EADES, 1964). WIKLER and RAYPORT (1954) observed a patient with an almost complete upper thoracic spinal cord lesion as a sequelae of a syphilitic meningomyelitis during a cycle of morphine dependence. During chronic administration of morphine, a modest degree of tolerance developed to morphine's depressant effects on lower limb nociceptive reflexes. Following withdrawal or the administration of nalorphine, lower limb response to nociceptive stimuli was enhanced and spontaneous movements emerged. Spinal cord signs of abstinence developed in parallel with supraspinal autonomic signs.

HERZ and TESCHEMACHER (1973) have shown that tolerance develops to the depressant effects of morphine on the licking response to tooth pulp stimulation and to the flexor reflex to radiant heat stimulation of the sole of the hindfeet in the rabbit when it is administered systemically and intraventricularly. They further found that the degree of tolerance to the depressant effects of intramuscular test doses of morphine on the flexor reflex was greater when chronic morphine was administered intraventricularly than when administered intramuscularly. The spinal cord levels of morphine were much lower when it was administered intraventricularly than when it was administered intramuscularly. These experiments support the view that the site of action of morphine responsible for some of its nociceptive actions are near the bulbar ventricular system (see Analgesia). Further, HERZ et al. (1972) have been able to precipitate abstinence in the rabbit by both intramuscularly and intraventricularly administered nalorphine. They have localized the site of action of intraventricularly administered nalorphine in precipitating abstinence to the fourth ventricle.

WEI et al. (1972, 1973b) were able to precipitate a variety of signs of abstinence including diarrhea, ear blanching, teeth chattering, escape behavior, and "wet shakes" by applying naloxone crystals to the brain stem of the dependent rat. The site that showed greatest responsivity and reproducibility to naloxone was the posterior medial thalamus; however, abstinence was occasionally precipitated when naloxone was applied to the corpus striatum, neocortex, hippocampus, and subthalamus.

EIDELBERG and BARSTOW (1971) produced tolerance and physical dependence in the stump-tailed macaque by injecting morphine intraventricularly. Although the withdrawal abstinence syndrome was mild, nalorphine administered either intraventricularly or subcutaneously precipitated an abstinence syndrome consisting of hyperactivity, hyperirritability, salivation, retching, vomiting, diarrhea, and a contorted posture suggesting pain.

KERR and POZUELO (1971) explored the effects of hypothalamic, medial forebrain bundle, and septal lesions on the effects of chronically administered morphine and naloxone precipitated abstinence in the rat. Lesions of the ventromedial hypothalamus (VMH), and to a lesser degree rostral septum, enhanced many of the effects of large doses of chronically administered morphine such as rigidity, Straub tail, immobility, darting, head bobbing, catatonia, stupor, respiratory depression, gnawing, exophthalmus, and convulsions. Further, lesions of the VMH also markedly decreased the intensity of the naloxone-precipitated abstinence syndrome. Lesions of the medial forebrain bundle and dorsomedial hypothalamus did not affect the precipitated abstinence syndrome. KERR and POZUELO emphasize the role of the VMH in satiety as well as its role in emotional states characterized by fear, anxiety, and aggression. In the nondependent rat, morphine decreases the firing rate of units in the lateral hypothalamic area (LHA), which is concerned with hunger, while increasing the firing rate of units in the VMH. These effects were antagonized by nalorphine and naloxone (KERR et al., 1974). The effects of morphine and naloxone on the activity of VMH and LHA units were similar in the nondependent and abstinent morphine-dependent rats. The firing rate of LHA was greater in the abstinent than the nondependent rats. EIDELBERG and BOND (1972) found that in the naive rat morphine but not naloxone depressed firing of hypothalamic units; whereas, in the dependent rat both morphine and naloxone produced a marked acceleration of unit activity. AMIT et al. (1973) found that lesions in the ventral,and to a lesser degree dorsal, hypothalamus markedly decreased consumption of a morphine-containing solution by dependent rats. Transection of the dependent rat brain stem at the level of the mesencephalon decreased naloxone-precipitated wet dog shakes, but transection anterior to the mesencephalon did not (WEI, 1973b).

WIKLER (1952) studied two decorticate dogs through a cycle of morphine and methadone dependence. In the decorticate dog, morphine and methadone decreased body temperature, pulse rate, and respiratory rate, produced analgesia as assessed by the tooth pulp method and decreased circling and hyperactivity. With chronic administration of morphine and methadone, tolerance developed to these effects, and following withdrawal the animals exhibited salivation, rhinorrhea, vomiting, hyperthermia, tachypnea, tachycardia, circling, irritability, rooting behavior, and paradoxically an elevation of tooth pulp threshold. These signs of abstinence were felt to be as or more intense than those seen in comparably dependent intact dogs. WIKLER et al. (1952) also studied several patients before and after a therapeutic bilateral frontal lobotomy through a cycle of morphine dependence. Lobotomy did not affect signs of abstinence. Three of the patients were schizophrenic, who could not in a meaningful way communicate their feelings during withdrawal. One patient suffering intractable phantom limb pain did not complain during withdrawal following bilateral lobotomy.

FOLTZ and WHITE (1957) studied the effects of bilateral frontal lobectomy, cingulumotomy, and resection of the cingulate gyrus on signs of precipitated and withdrawal abstinence in the morphine-dependent rhesus monkey. Each of these three lesions decreased withdrawal signs such as irritability, aggression, grimacing, vocalization and anguished facies indicating discomfort, as well as autonomic and somatic signs such as piloerection, conjunctival injection, and diarrhea. TEITELBAUM et al. (1974) have reported that lesions of the medial thalamus but not the caudate and hippocampus restore morphine's ability to cause EEG slow wave activity in the morphine-tolerant rat. TRAFTON and MARQUES (1971) found that lesions placed in the anterior cortex decrease the consumption of morphine-containing solutions in rats dependent on modest doses of morphine. Lesion of the septum did not. WIKLER et al. (1972) studied the effects of lesion of the dorsal medial thalamic nucleus, the anterior temporal pole and the septum in the rat on signs of abstinence, fluid intake activity, and drug-seeking behavior during a control period, as well as during early and protracted abstinence. Lesion of the dorsal medial thalamus decreased activity and increased "wet dog" shakes, did not alter signs of early abstinence, except that the lesioned animals consumed less water than controls, or drug-seeking behavior during early abstinence and attenuated the increase in fluid intake seen during protracted abstinence. Lesions of the anterior temporal lobe decreased activity, and except for an additional decrease in the consumption of water did not alter signs of early abstinence. Anterior temporal lobe lesions did not alter drug-seeking behavior or signs of protracted abstinence. Septal lesions made the animals more vicious and decreased activity while lesions of the cingulum decreased colonic temperature. Neither septal nor cingulum lesions altered signs of early or protracted abstinence or drug-seeking behavior. In studies of relapse, using the quantity of etonitazene consumed as a measure, it was found that all post-dependent lesioned rats consumed more etonitazene solution than nondependent lesioned controls. The fact that formerly dependent rats consumed more water during protracted abstinence confounded, to a degree, interpretation of the effects of lesion of the dorsal medial thalamus and the cingulum, and it could not be determined if these lesions affected relapse. Lesions of the anterior temporal lobe and the septum were without effect on relapse.

VI. Neurochemical and Neurohumoral Changes

1. ACh

As indicated in the preceding section, changes in the responsivity of isolated tissues obtained from morphine-dependent animals to ACh is conflicting with investigators finding either no change (TAKAGI et al., 1965; GOLDSTEIN and SCHULZ, 1973) or an enhancement (POLLOCK et al., 1972; EHRENPREIS et al., 1972).

WIKLER and FRANK (1948) found that physostigmine enhanced the flexor reflex and evoked the stepping reflex. These findings were confirmed by MARTIN and EADES (1967) who further demonstrated that atropine antagonized these effects.

Subsequently, VAUPEL and MARTIN (1973) demonstrated that tremorine and oxotremorine facilitated the flexor reflex and evoked the stepping reflex and that these changes were selectively antagonized by atropine. Superficially, the actions of these cholinergic agents resembled signs of precipitated and withdrawal abstinence in the chronic spinal dog. MARTIN and EADES (1967) found that atropine produced a modest antagonism of spinal cord signs of abstinence; however, the interpretation of these findings is complicated by the fact that atropine itself produces a modest depression of the flexor reflex in the nondependent chronic spinal dog (VAUPEL and MARTIN, 1973).

In the morphine-dependent rat, physostigmine decreases and atropine increases hyperexcitability seen during abstinence (GRUMBACH, 1969). Interpretation of these results is also complicated by the fact that atropine alone causes excitability. Atropine did delay death by convulsions in the reserpinized abstinent rat. HOFFMEISTER and SCHLICHTING (1972) observed little effect of physostigmine or scopolamine on withdrawal abstinence in the morphine-dependent rat. PINSKY et al. (1973) also studied the effect of cholinergic and anticholinergic drugs in the dependent rat. Mecamylamine and atropine enhanced and later (44–45 h of abstinence) suppressed abstinence in rats made dependent on morphine. Choline also suppressed abstinence in the rat. JHAMANDAS and DICKINSON (1973) have reported that both atropine and mecamylamine suppress precipitated jumping in morphine and methadone-dependent mice. COLLIER et al. (1972) found that atropine (40 mg/kg) suppressed some signs (jumping, diarrhea, and chewing) of precipitated abstinence in the acutely dependent rat, while enhancing others (irritability and paw tremor). When the rats were pretreated with atropine prior to being made dependent, some signs of precipitated abstinence were exaggerated (jumping) and irritability to handling depressed. COLLIER et al. (1972) cites Fuentes' thesis and findings that atropine lessens and physostigmine enhances abstinence in the rat.

BHARGAVA and WAY (1972) found that intracerebrally injected physostigmine and DFP enhanced the analgesic actions of morphine to approximately the same degree in tolerant and nontolerant mice. Both DFP and physostigmine administered acutely but not chronically antagonized the increase in jumping associated with naloxone-precipitated abstinence but did not alter weight loss associated with withdrawal abstinence. These authors concluded that cholinergic influences were not of primary importance in the genesis of tolerance and physical dependence. Intracerebrally administered hemicholinium decreased the analgesic activity of morphine in the tolerant and dependent mouse, increased weight loss following withdrawal, and decreased the amount of naloxone necessary to precipitate abstinence (BHARGAVA et al., 1972a).

Findings concerning the effects of chronic morphine administration on brain ACh metabolism are conflicting. No changes in brain ACh levels as a consequence of either acute or chronic administration of morphine to the rat (JOHANNESSON and LONG, 1964) or chronic administration of levorphanol to the mouse (RICHTER and GOLDSTEIN, 1970) have been reported. On the other hand, MAYNERT (1967) observed an increase in brain levels of ACh in chronically morphinized rats. MERALI et al. (1974) similarly found that striatal ACh, DA, and AChE were increased in the chronically morphinized rat. LARGE and MILTON (1970) observed an increase in ACh turnover. DATTA et al. (1971) found that chronic morphine initially depressed choline acetylase activity in the caudate, but not the thalamus and motor cortex, but no change was seen when the dose was escalated. Choline acetylase activity was also depressed in the caudate after 48 h of abstinence. Reports of changes of ACh brain levels during abstinence in morphine-dependent rats are conflicting. LARGE and MILTON (1970) have reported that brain levels of ACh are increased during withdrawal. DOMINO and WILSON (1973b) and MERALI et al. (1974), on the other hand, have reported a decrease during morphine withdrawal and precipitated abstinence. Further, these changes are enhanced by pretreatment with HC-3. JHAMANDAS (1973) found that tolerance developed to morphine's depressant action on the release of ACh from the cortex of the lightly anesthetized rat with chronic administration, and that narcotic antagonists produced a large increase in the release of ACh in the dependent rat. LABRECQUE and DOMINO (1974) also found that with chronic administration partial tolerance developed to morphine's ability to inhibit the release of ACh from the neocortex of the brainstem transected cat.

In the tolerant mouse, brain levels of ACh are not different from control levels (HANO et al., 1964); however, HO and LOH (1970) found a significant reduction in choline acetyltransferase activity. Reserpine alone had no effect on choline acetyltransferase activity, but in combination with morphine partially antagonized the effects of morphine (HO and LOH, 1970). No changes in brain ACh levels have been observed in the abstinent mouse (HANO et al., 1964). Intracerebrally administered HC-3 did not alter the development of morphine tolerance or dependence in the rat (BHARGAVA et al., 1974).

One strain of mice which exhibited a high level of spontaneous and morphine-induced activity, a lower maze-learning ability, lower brain AChE and ACh acetylase activity and higher NE levels showed the most tolerance to morphine-induced hyperactivity. Yet another strain characterized by less spontaneous activity, no morphine-induced spontaneous activity, a higher maze-learning ability, higher AChE and ACh acetylase activity and a lower level of brain NE showed a higher degree of tolerance to morphine analgesic activity (OLIVERIO and CASTELLANO, 1974).

Naloxone has a biphasic effect on the release of cortical ACh in the morphine-dependent cat, initially decreasing and then increasing release (LABRECQUE and DOMINO, 1974).

2. E and NE

a) Adrenal Medulla

There is an extensive literature on the ability of morphine to increase blood sugar through the release of E and NE from the adrenal medulla. Reviews of the older literature on this phenomenon may be found in the monographs of KRUEGER et al. (1941, 1943), SATAKE (1954), and of REYNOLDS and RANDALL (1957). A more recent exposition and review may be found in the work of BORISON and his collaborators (BORISON et al., 1962, 1964; MOORE et al., 1965) who have described a chemoreceptor in the anterior brainstem which when activated by morphine causes the release of E and NE from the adrenal medulla, which is responsible for the increase in blood sugar. FELDBERG and SHALIGRAM (1972) and FELDBERG and GUPTA (1972) have more recently obtained evidence that these chemoreceptors may be located more caudally, perhaps in the fourth ventricle.

The effects of morphine and other narcotics on the adrenal medulla have been studied using levels, turnover and secretion of E, NE, and DA.

α) Single Doses (Cat, Dog, Rat, Mouse, Rabbit)

In the cat, single doses of morphine decrease medullary catecholamine levels (ELLIOTT, 1912; STEWART and ROGOFF, 1916; EMMELIN and STRÖMBLAD, 1951; VOGT, 1954; GUNNE, 1963; MAYNERT and LEVI, 1964) and increase the quantity of adrenaline in adrenal venous blood (STEWART and ROGOFF, 1922). Morphine increases adrenal and systemic blood levels of E and NE in the dog (SATO and OHMI, 1933; WADA et al., 1938; SIBUTA et al., 1949; RICHARDSON et al., 1958) and urinary excretion of E and NE in the dog (GUNNE, 1962b, 1963). Modest doses of morphine do not markedly alter dog adrenal E and NE levels (STEWART and ROGOFF, 1916; MAYNERT and KLINGMAN, 1962), while large dose levels (60 or 125 mg/kg) decrease E but not NE levels (MAYNERT and KLINGMAN, 1962). In the rat, single doses of morphine clearly decrease adrenal E levels (VOGT, 1954; MAYNERT and KLINGMAN, 1962; GUNNE, 1963). Repeated doses of morphine are more effective in lowering adrenal E levels (OUTSCHOORN, 1952) and also decrease NE levels (GUNNE et al., 1969). Pretreatment with αMT enhances morphine's adrenal depleting effect in the rat (GUNNE et al., 1969). In this regard, morphine also increases CA turnover in the mouse (BEDNARCZYK et al., 1970). Morphine increases urinary excretion of E and NE in the rat (CRAWFORD and LAW, 1958; GUNNE, 1963; AKERA and BRODY, 1968). STEWART and ROGOFF (1916) found that in the rabbit the adrenaline content of adrenals was reduced by morphine. MAYNERT and KLINGMAN (1962) found a decrease in E and an increase in NE levels with 150–3000 mg/kg but not with 30 or 60 mg/kg of morphine.

β) Chronic Administration

Several investigators have shown that the chronic administration of morphine causes adrenal hypertrophy in the intact rat (MACKAY and MACKAY, 1926; MACKAY, 1931; SUNG et al., 1953; TANABE and CAFRUNY, 1958; SLOAN et al., 1963) but not in hypophysectomized rats (TANABE and CAFRUNY, 1958). Further, SLOAN and EISENMAN (1968) showed that this hypertrophy persisted for over 5 months following withdrawal. AKERA and BRODY (1968) using smaller doses of morphine (120 mg/kg/day) and levorphanol (6 mg/kg/day) administered 3 times daily did not observe an increase in adrenal weight nor did WEI (1973c) using pellet implantation. DEANSLEY (1931) found that mice treated for 12–22 days with about 20 mg of morphine acetate per day and sacrificed 24 h after the last injection exhibited hypertrophy of the adrenal cortex and medulla.

Reports on the effects of chronic morphine administration on adrenal CA levels in the rat are conflicting. Although MAYNERT and KLINGMAN (1962) did not find any changes in adrenal NE after the chronic administration of morphine in rats treated with a MAO inhibitor 5 h before sacrifice, MAYNERT (1968) found that with chronic administration of morphine, rat adrenal levels of NE increased. GUNNE (1963) found no change in adrenal E and NE levels following chronic administration of morphine, nor did pretreatment with α-MT reveal any difference in E or NE levels between control and morphine-dependent rats (GUNNE et al., 1969). Repeated doses of morphine increase the incorporation of ^{14}C-tyramine into adrenal CA in the mouse (BEDNARCZYK et al., 1970). Chronic administration of morphine to the dog did not alter adrenal E and NE levels (MAYNERT and KLINGMAN, 1962; GUNNE, 1963). In the rabbit, chronic administration of morphine was associated with an increase in adrenal NE but not E levels (MAYNERT and KLINGMAN, 1962).

Chronic administration of morphine initially increases excretion of E and NE in rat urine (GUNNE, 1961, 1962a, 1963; SLOAN and EISENMAN, 1968; AKERA and BRODY, 1968); however, with continued administration partial tolerance develops to this effect. In the dog, a similar initial increase in the excretion of E and NE is observed, but a much higher degree of tolerance develops to the increased excretion of NE than to E (GUNNE, 1963). In man, excretion of E, NE, DA, VMA, metanephrine, and normetanephrine was increased after about 3 weeks of administration of increasing dose levels of morphine (WEIL-MALHERBE et al., 1965). In a subsequent study in which morphine was administered for a much longer time, EISENMAN et al. (1969) observed an early increase in the excretion of E and NE; however, when subjects had been stabilized on 240 mg/kg/day for 29 weeks, E excretion had returned to control levels but NE levels remained slightly but significantly elevated.

γ) Abstinence (Rat, Dog, Rabbit, Man)

Adrenal levels of E and NE either do not change, or a small but significant decrease in E but not NE has been seen following the administration of nalorphine in morphine-dependent rats (MAYNERT and KLINGMAN, 1962; GUNNE, 1963; GUNNE et al., 1969). Following abrupt withdrawal of morphine in the dependent rat, E and NE excretion is increased, with the increase in E being greater than the increase in NE (GUNNE, 1961, 1962a, 1963; SLOAN and EISENMAN, 1968; AKERA and BRODY, 1968). Adrenal gland weight is increased during precipitated abstinence (WEI, 1973c). During protracted abstinence (143 days after withdrawal) NE but not E levels were decreased (SLOAN and EISENMAN, 1968). In the dependent dog, MAYNERT and KLINGMAN (1962) and GUNNE (1963) found that adrenal levels of E and NE were decreased following abrupt withdrawal or the administration of nalorphine. The decrease in adrenal E following nalorphine was seen in the innervated but not the denervated adrenal gland (GUNNE, 1963). Blood levels of adrenaline were elevated in the abstinent but not the nonabstinent dog (TACHIKAWA, 1932). A delayed and modest increase in NE excretion was seen on the third to sixth day following abrupt withdrawal of morphine in the dependent dog (GUNNE, 1962b). MAYNERT and KLINGMAN (1962) found no change in rabbit adrenal levels of E or NE following administration of nalorphine. No increase in E or NE excretion was seen during abrupt withdrawal in man (WEIL-MALHERBE et al., 1965; EISENMAN et al., 1969). These observations are complicated by the fact that urinary production during abstinence is much decreased; however, HOELDTKE and MARTIN (1970) found that differences in urine production obtained by varying water intake did not affect CA excretion. GUNNE (personal communication) found that there was no change in the excretion of either E or NE with chronic administration of methadone in man. During withdrawal abstinence there was no increase in E or NE excretion. During precipitated abstinence no change was seen following 0.5 and 1.5 mg of nalorphine, but an increase in E and NE excretion was seen following 3 mg.

Neither tolerance nor dependence are affected by adrenalectomy in the rat (WAY et al., 1954; GEBHART and MITCHELL, 1972a; WEI, 1973c).

b) Brain

α) Single Doses (Cat, Dog, Rat, Mouse)

VOGT (1954) first observed that morphine decreased brain NE levels in the cat. Apomorphine also produced a decrease. These findings concerning the effects of morphine in the cat have been confirmed by QUINN et al. (1958), GUNNE (1963), MAYNERT and LEVI (1964), LAVERTY and SHARMAN (1965), and REIS et al. (1969). Most investigators have associated these changes in brain NE levels with increased autonomic and behavioral excitation. BORISON et al. (1962) found that morphine injected into the lateral ventricle produced hyperglycemia and reduced hypothalamic NE levels (MOORE et al., 1965), and as previously discussed, these authors (BORISON et al., 1964) believed that hyperglycemic response is due in part to morphine's interaction with a chemoreceptor in the subfornical organ and have pointed out that hyperglycemia may be produced in the cat with morphine without behavioral excitation.

Neither VOGT (1954) nor MAYNERT and KLINGMAN (1962) found changes in brain NE levels in the dog with doses of 40–125 mg/kg of morphine; but with larger doses (200 mg/kg) a significant decrease was observed. SEGAL and DENEAU (1962) and SEGAL et al. (1972) found that 3 mg/kg of morphine in the monkey decreased brain NE levels, while 30 mg/kg produced an increase. Morphine either has no effect (MAYNERT and KLINGMAN, 1962; LAVERTY and SHARMAN, 1965) or decreases rabbit brain NE (LAVERTY and SHARMAN, 1965).

GUNNE (1959) first reported that 20 mg/kg of morphine did not affect brain NE; whereas, repeated 30 mg/kg doses decreased and 60–90 mg/kg increased levels in the rat (GUNNE, 1959, 1963). MAYNERT and KLINGMAN (1962) and MAYNERT and LEVI (1964) confirmed and extended these observations showing that single 100 and 200 mg/kg doses decreased rat brain stem NE levels. FREEDMAN et al. (1961) reported that single doses of morphine (20–60 mg/kg) caused an initial decrease in brain NE levels which was followed 24 and 48 h later by an increase. SLOAN et al. (1962) found that 15 and 30 mg/kg of morphine did not affect rat brain NE levels, but 60 mg/kg produced a transient increase, while a convulsant dose of thebaine produced a nonsignificant decrease in brain NE levels. AKERA and BRODY (1968) found that morphine (10 mg/kg), levorphanol (2 mg/kg), and a nonconvulsant dose of thebaine did not alter rat brain NE levels. CLOUET and RATNER (1970) found that morphine (60 mg/kg) did not alter the quantity of radiolabeled NE formed from labeled tyrosine. SUGRUE (1974) also observed that morphine (20 mg/kg), methadone (10 mg/kg), and pentazocine did not alter NE whole brain levels, rate of NE synthesis or rate of NE loss in the rat. Morphine, but not methadone or pentazocine, increased the rate of pontine-medullary synthesis and loss of NE and the increased rate of loss was antagonized by naloxone. Histochemical studies by HEINRICH et al. (1971) indicate that morphine (20 mg/kg) does not alter the fluorescence of the ventromedial tegmentum or the midbrain reticular formation.

TAKAGI and NAKAMA (1966) found no change in mouse brain NE levels following 10 and 15 mg/kg of morphine. LEE and FENNESSY (1970) and FENNESSY and LEE

(1972) found that dose levels of morphine below 2 mg/kg had no effect, while dose levels of 2–20 mg/kg decreased mouse brain NE levels. With larger doses (10–300 mg/kg), RETHY et al. (1971) found a decrease in mouse brain CA levels; whereas, FENNESSY and LEE (1972) found an increase with 50 and 100 mg/kg. A wide range of dose levels of morphine increase brain ^{14}C CA levels in mice that received ^{14}C-tyrosine (SMITH et al., 1970; BEDNARCZYK et al., 1970; SHELDON and SMITH, 1971; FUKUI et al., 1972). Midbrain fluorescence attributable to NE was increased by morphine (HEINRICH et al., 1971). Large and in some instances convulsant doses of morphine, meperidine, ketobemidone, levorphanol, dextromethorphan, *l* but not *d*-methadone and thebaine also decrease brain CA levels (RETHY et al., 1971) Nalorphine does not alter NE levels (TAKAGI and NAKAMA, 1966), and naloxone antagonizes the depleting effect of morphine (RETHY et al., 1971).

β) Chronic Administration (Monkey, Cat, Dog, Rat, Mouse)

Chronic administration of morphine decreases brain NE levels in the monkey (SEGAL and DENEAU, 1962) and does not alter NE levels in the cat or dog (MAYNERT and KLINGMAN, 1962; GUNNE, 1963). Chronic morphine administration does not alter rat spinal cord levels of normetanephrine and tolerance developed to the increase in brain NE seen in the nontolerant rat (SHIOMI and TAKAGI, 1974). α-MT delayed the development of tolerance to morphine's respiratory depressant effects and decreased the respiratory sign of precipitated abstinence in the decerebrate cat (FLÓREZ et al., 1973).

Chronic administration of morphine and levorphanol does not alter brain CA levels in the mouse (RETHY et al., 1971). Small doses (10 and 30 mg/kg/day) of morphine do not affect the incorporation of ^{14}C-tyrosine in CA; whereas, large doses do (100–300 mg/kg/day) (SMITH et al., 1970, 1972) indicating that tolerance develops to the increased incorporation of ^{14}C-tyrosine into NE and DA caused by morphine (SMITH et al., 1970, 1972; ROSENMAN and SMITH, 1972). The increase in incorporation is antagonized by naloxone. MARSHALL and SMITH (1974) found that single doses of morphine increased caudate nucleus tyrosine hydroxylase activity and that during chronic administration of morphine tolerance developed to this effect, while CICERO et al. (1973) found that neither single doses nor chronically administered morphine alter rat brain tyrosine hydroxylase activity.

Chronically administered morphine increases brain levels of NE in the rat (FREEDMAN et al., 1961; MAYNERT and KLINGMAN, 1962; GUNNE, 1963; SLOAN et al., 1963; AKERA and BRODY, 1968). Chronically administered levorphanol and methadone do not increase rat brain NE levels nor does thebaine. The administration of MAO inhibitors to morphine-dependent rats causes a further increase in brain NE levels (MAYNERT and KLINGMAN, 1962; GUNNE, 1963; AKERA and BRODY, 1968). In morphine-dependent rats, the NE and DA depleting action of reserpine administered intraperitoneally and α-methyldopamine is diminished (GUNNE, 1963; GUNNE et al., 1970); whereas, the effect of α-MT is unchanged (GUNNE et al., 1969). Studies by BLOSSER and CATRAVAS (1974) indicate that acutely and chronically administered morphine does not affect the uptake of NE by the brain nor does chronic morphine administration alter the NE depleting effect of reserpine. Chronic morphine administration does decrease the absorption of intraperitoneally administered reserpine.

CLOUET and RATNER (1970) presented evidence that NE turnover was increased in morphine-dependent rats; whereas, NEAL (1968) and ALGERI and COSTA (1971) were unable to see any change.

γ) Abstinence

In the morphine-dependent dog, rabbit and monkey, both abrupt withdrawal and nalorphine-precipitated abstinence cause a decrease in brain NE (GUNNE, 1962b, 1963; MAYNERT and KLINGMAN, 1962; SEGAL and DENEAU, 1962; SEGAL et al., 1972). In keeping with the hypothesis that enhanced adrenergic activity is associated with abstinence is the observation that the α-adrenergic agonist methoxamine facilitates the flexor reflex and evokes the stepping reflex in the chronic spinal dog, signs of abstinence in the morphine-dependent spinal dog. Phenoxybenzamine, which antagonizes the facilitatory effect of methoxamine, does not, however, suppress abstinence in the spinal dog. Abrupt withdrawal of morphine in the rat has been shown to decrease NE levels as compared to preaddiction levels (GUNNE, 1959); have no effect as compared to saline controls (SLOAN et al., 1963; GUNNE, 1963; AKERA and BRODY, 1968); but decreased levels compared to chronic morphine levels (SLOAN et al., 1963; AKERA and BRODY, 1968). Nalorphine-precipitated abstinence does not alter brain NE levels in the rat (MAYNERT and KLINGMAN, 1962; GUNNE, 1963). The decrease in brain NE levels is prevented by treatment with MAO inhibitors (AKERA and BRODY, 1968). HUIDOBRO et al. (1963a), MARUYAMA and TAKEMORI (1973) and MARSHALL and SMITH (1973) found that α-MT suppressed nalorphine- and naloxone-induced jumping in the dependent mouse. DOPA but not dihyroxyphenylserine reversed these effects (MARUYAMA and TAKEMORI, 1973). GUNNE et al. (1969) found that brain levels of NE were decreased as a consequence of both withdrawal and precipitated abstinence in dependent rats treated with α-MT. Further, there was histochemical evidence of depletion of NE in the cerebral cortex, hypothalamus, and medulla in dependent rats pretreated with α-MT that had received nalorphine. The DA β-hydroxylase inhibitor diethylthiocarbamate reduced the intensity of abstinence in morphine-dependent rats (SCHWARTZ and EIDELBERG, 1970; HERZ et al., 1974). SCHWARTZ and EIDELBERG (1970) found that imipramine increased the intensity of abstinence while HERZ et al. (1974) found that desipramine increased jumping, eye twitches, and rhinorrhea, and decreased teeth chattering, "wet dog" shakes, and diarrhea.

CICERO et al. (1974a) found that phenoxybenzamine, phentolamine, haloperidol, and chlorpromazine but not propranolol and promethazine suppressed the precipitated abstinence signs diarrhea and "wet dog" shakes in the morphine-dependent rat. The dopamine β-hydroxylase inhibitor 1-phenyl-3-12-thiazolyl-2-thiourea reduced naloxone-precipitated abstinence in the mouse (BHARGAVA et al., 1972b).

There are some findings that suggest that the abstinence syndrome is associated with a decrease in brain noradrenergic activity. With abrupt withdrawal of morphine in tolerant and dependent mice there is a decrease in the incorporation of ^{14}C-tyrosine into NE and DA without any change in the endogenous brain levels of NE or DA (ROSENMAN and SMITH, 1972). HOFFMEISTER and SCHLICHTING (1972) found that methamphetamine and cocaine suppress abstinence in the rat, while dihydroergotamine enhances it. There are other observations that are not consistent with the hypothesis that abstinence is associated with increased utilization of NE. Rats chronically dependent on methadone do not show an increase in brain NE levels, nor is a decrease seen during withdrawal (AKERA and BRODY, 1968). NEAL (1968) found no difference between control and morphine abstinent rats (60 h) in the quantity of brain-deaminated metabolites and normetanephrine following the injection of DL [^{3}H]-NE intracisternally.

MAGGIOLO and HUIDOBRO (1962, 1965) found that reserpine enhanced abstinence in the mouse. FRIEDLER et al. (1972) found that the intracerebral injection of 6-OHDA increased the analgesic activity of morphine in pellet-implanted mice over appropriate control animals; however, the increase was proportionately the same as the increase in analgesia seen following 6-OHDA in naive animals (no pellet implantation). These investigators concluded that 6-OHDA did not interfere with the process responsible for tolerance. The 6-OHDA-treated mice showed a greater loss of weight when the pellets were removed and increased naloxone-precipitated jumping. BHARGAVA et al. (1973) have made similar observations in the rat. HUIDOBRO et al. (1963a, b) have also reported that DOPA, DOPA plus 5-HTP, α-methyl DOPA, serine, alanine, nicotinamide, and dioxypyridoxine suppress precipitated abstinence in the dependent mouse. Imipramine which has diverse actions on catechol and indole amines enhances nalorphine-induced abstinence in the mouse (CHIOSA et al., 1968).

The evidence of the effect of MAO inhibitors on abstinence in the mouse is conflicting. MAGGIOLO and HUIDOBRO (1962) found that iproniazid, nialamide, phenelzine, and pheniprazine suppressed nalorphine-precipitated abstinence, while pargyline (MARUYAMA et al., 1971; IWAMOTO et al., 1971) and marplan (MAGGIOLO and HUIDOBRO, 1962) enhanced the precipitated abstinence syndrome.

3. DA

a) Single Doses (Cat, Monkey, Rabbit, Mouse, Rat)

Morphine does not appear to change DA levels in cat brain (MOORE et al., 1965; LAVERTY and SHARMAN, 1965), although it increases HVA levels, suggesting that it increases DA turnover. In the monkey, both morphine and thebaine increase DA levels in the caudate (SEGAL and DENEAU, 1962; SEGAL et al., 1972). In the rabbit, morphine does not alter DA levels in the caudate (LAVERTY and SHARMAN, 1965).

In the mouse, morphine (2–20 mg/kg) transiently decreases brain levels of DA and NE (TAKAGI and NAKAMA, 1966; FENNESSY and LEE, 1972) and increases HVA and DOPAC levels (FUKUI and TAKAGI, 1972). Higher doses of morphine do not affect NE levels but increase DA levels (FENNESSY and LEE, 1972). HITZEMANN and LOH (1973) suggest that morphine may inhibit transport of DA into neuronal tissue. Etorphine increases brain levels of HVA significantly, and morphine nonsignificantly, and the effect of etorphine was blocked by the opioid antagonist cyprenorphine (M285) and α-MT (SHARMAN, 1966). Morphine also increases the incorporation of ^{14}C-tyrosine administered intravenously to DA and its metabolite 3-methoxy-tyramine in the cerebral and cerebellar cortex, diencephalon, midbrain, and brain stem, and these effects are antagonized in some areas of the brain, but not others, by naloxone. Naloxone itself increased levels of 3-methoxy-tyramine but not

DA (HITZEMANN et al., 1971). The fluorescence of the substantia nigra but not the ventromedial tegmentum is increased by morphine in the mouse, and this increase is blocked by α-MT (HEINRICH et al., 1971).

In rats treated with α-MT, brain and adrenal DA levels were decreased markedly by single doses of morphine. Nalorphine had no effect (GUNNE et al., 1969). GAUCHY et al. (1973) have shown that rats pretreated with morphine exhibit an increase in the synthesis and release of DA. In the rat, methadone increases corpus striatum HVA and the conversion of tyrosine into DA, but does not increase brain levels of DA (PEREZ-CRUET et al., 1972). Morphine, methadone, and pentazocine do not affect rat striatal levels of DA but increase its rate of synthesis and loss (SUGRUE, 1974). Naloxone antagonizes these effects. Morphine increases fluorescence in substantia nigra (HEINRICH et al., 1971) and increases the incorporation of ^{14}C-tyrosine into rat brain DA.

b) Chronic Administration (Rat, Mouse, Dog, Monkey, Man)

Chronic administration of morphine produces a small but not a statistically significant decrease in brain and adrenal DA in rats pretreated with α-MT (GUNNE et al., 1969) and increases incorporation of ^{14}C-tyrosine into DA (CLOUET and RATNER, 1970). Urinary excretion of DA is increased in the rat (SLOAN and EISENMAN, 1968). Although both morphine and haloperidol increase DA in the nontolerant rat brain, neither morphine nor haloperidol increase it in the morphine-tolerant rat (PURI and LAL, 1974a). Chronic and single dose administration of morphine antagonizes the early accumulation of brain DA, but only chronic administration of morphine antagonizes the irritability associated with the intraventricular administration of 6-OHDA (NAKAMURA et al., 1972). The effects of 6-OHDA in the morphine-tolerant rat are controversial. NAKAMURA et al. (1973b) found that 6-OHDA administered intraventricularly and into the hypothalamus did not affect morphine's analgesic activity in the tolerant rat but reduced it in the nontolerant rat. BHARGAVA et al. (1973) found that 6-OHDA decreased morphine analgesic activity in both the tolerant and nontolerant rat.

PLEUVRY (1971) has shown that mice tolerant to the analgesic effect of morphine are cross tolerant to the analgesic effect of methamphetamine. LITTLE and REES (1974) have shown that mice tolerant to the analgesic effect of morphine are cross tolerant to amphetamine and that mice who are tolerant to the analgesic activity of amphetamine are cross tolerant to morphine.

Chronic administration of morphine does not alter brain DA levels in the dog (GUNNE, 1963) or monkey (SEGAL and DENEAU, 1962; SEGAL et al., 1972). A trend toward increased excretion of DA is seen in man during the ascending dose phase; however, the data are conflicting during stabilization where increases are seen in some subjects but not in others (WEIL-MALHERBE et al., 1965; EISENMAN et al., 1969). Dihydroxyphenyl-acetic acid excretion is also increased by chronic morphine administration (WEIL-MALHERBE et al., 1965).

c) Abstinence (Rat, Dog, Mouse)

Brain levels of NE and DA tend to be somewhat elevated 24 h after withdrawal in the dependent rat (NAKAMURA et al., 1972). The incorporation of ^{3}H-tyrosine into rat striatal DA is increased during naloxone-precipitated abstinence (TSENG et al., 1974). GUNNE et al. (1969) also found DA levels elevated in dependent rats pretreated with α-MT during withdrawal or precipitated abstinence. Adrenal levels of DA were not affected. IWAMOTO et al. (1973) found a rapid and transient increase in striatal DA following the administration of naloxone to morphine-dependent rats. There was a tendency for DA excretion to be increased during early abstinence and unchanged during protracted abstinence in the rat (SLOAN and EISENMAN, 1968).

Apomorphine, *d*- and *l*-amphetamine, and methylphenidate, which have dopaminergic actions, enhance withdrawal aggression in rats, and these effects are antagonized by α-MT, haloperidol, morphine, and methadone. L-DOPA, DL-DOPA, and DL-dihydroxyphenylalanine also enhance morphine withdrawal aggression (LAL et al., 1971a and b; LAL and PURI, 1972; PURI and LAL, 1973b). In contrast, BHARGAVA et al. (1973) found that 6-OHDA increased withdrawal abstinence in the rat. Abstinent rats are tolerant to both the cataleptic actions and the increase in striatal DA turnover produced by morphine and haloperidol (LAL and PURI, 1973). Rats that have been chronically medicated with large doses of haloperidol and then withdrawn are not tolerant to these effects. Rats in protracted abstinence (30 days) show withdrawal aggression which is enhanced by apomorphine and antagonized by nigrostriatal lesions, but not by medial forebrain bundle lesions (GIANUTSOS et al., 1973). HERZ et al. (1974) found that *d*-amphetamine, cocaine and apomorphine, which were presumed to have a dopaminergic mechanism of action as well as the elevation of brain DA levels with L-DOPA and Ro 4-4602 increased precipitated abstinence signs, jumping, and possibly rhinorrhea while decreasing "wet dog" shakes, ptosis, and diarrhea in the morphine-dependent rat. α-MT decreased jumping and increased "wet dog" shakes and convulsions.

Apomorphine enhances some but not all signs of withdrawal abstinence in the morphine-dependent dog (MARTIN et al., 1976). Thus temperature, pupillary diameter, and pulse rate were increased, while rhinorrhea, tremors, and respiratory rate were decreased. The interpretation of some changes is ambiguous since apomorphine increases pupillary diameter and pulse rate in the nondependent dog.

In the mouse, repeated administration of morphine is not associated with an increase in brain levels of HVA and DOPAC (FUKUI and TAKAGI, 1972). A rapid transient increase in brain DA, particularly striatal, following the administration of naloxone was seen in the dependent mouse which was related to jumping (IWAMOTO et al., 1973). This increase was seen in reserpine- and pargyline-treated animals. Physostigmine but not atropine antagonized this increase and precipitated jumping. During withdrawal or precipitated abstinence, brain DA levels tend to decrease in the dog (GUNNE, 1962b, 1963) and to increase in the monkey (SEGAL and DENEAU, 1962; SEGAL et al., 1972). Haloperidol, a DA antagonist, suppresses some but not all signs of physical dependence in the dog (MARTIN et al., 1974).

In the abstinent man, DA excretion is not markedly different from control conditions (WEIL-MALHERBE et al., 1965; EISENMAN et al., 1969). Chlorpromazine and reserpine did not depress abstinence in man (FRASER and ISBELL, 1956); however, haloperidol has been reported to suppress some abstinence signs but not others in man, rat (LAL et al., 1971b), and dogs (MARTIN et al., 1974).

The older literature on apomorphine (cf., KRUEGER et al., 1943) as well as current work suggests that there are both similarities and differences between morphine and apomorphine. Thus, both drugs produce sedation (cf., KRUEGER et al., 1941, 1943), analgesia, apomorphine to a limited degree (cf., KRUEGER et al., 1943), emesis, through an action of the chemoreceptive trigger zone (BORISON, 1958, 1959) and pupillary dilation in the cat. It differs from morphine in that it dilates pupils in the dog (cf., KRUEGER et al., 1943). DENEAU and SEEVERS (1964) have reported that apomorphine in doses to 1 mg/kg does not suppress abstinence in the monkey.

4. Serotonin

a) Single Doses (Cat, Dog, Rat, Mouse)

The effects of single doses of narcotic analgesics on brain 5-HT metabolism are conflicting. In the cat there are reports of both increases (LAVERTY and SHARMAN, 1965) and decreases (TÜRKER and AKCASU, 1962) of brain 5-HT following the administration of morphine. LAVERTY and SHARMAN (1965) also reported increased brain levels of 5 HIAA. Neither morphine nor nalorphine produce changes in brain levels of 5-HT in the dog (MAYNERT et al., 1962), monkey (SEGAL and DENEAU, 1962), or rabbit (BRODIE et al., 1956; MAYNERT et al., 1962). Most investigators have found no changes in brain 5-HT levels in the rat (MAYNERT et al., 1962; SLOAN et al., 1962; YARBROUGH et al., 1971; GOODLET and SUGRUE, 1974) or mouse (SHEN et al., 1970) following morphine administration; however, BONNYCASTLE et al. (1962) reported that both morphine and meperidine elevated brain levels of 5-HT in the rat, and YARBROUGH et al. (1970, 1973) observed an increased incorporation of ^{14}C-L-tryptophan and ^{3}H-L-tryptophan into 5-HT following morphine. YARBROUGH et al. (1971) also found increased brain levels of 5-HT following morphine in rats treated with pargyline. HAUBRICH and BLAKE (1969, 1973) and GOODLET and SUGRUE (1974) observed increased brain levels of 5-HIAA in rats treated with morphine while GOODLET and SUGRUE (1974) found a decrease following meperidine. HOLTZMAN and JEWETT (1972) found that both pentazocine and morphine lowered total brain 5-HT in the rat. Naloxone antagonized this effect of morphine but not of pentazocine. SHEIN et al. (1970) found that the addition of morphine to rat pineal organ cultures had no effect on the synthesis of 5-HT. Using both MAO inhibitors and probenecid to measure 5-HT turnover in rat brain, GOODLET and SUGRUE (1974) found that morphine but not meperidine, methadone, or pentazocine increased it. SHEN et al. (1970) found that morphine had no effect on brain 5-HT in mice treated with pargyline. Similarly, YARBROUGH et al. (1970) did not observe an increase in incorporation of ^{14}C-5-HTP into 5-HT following morphine, but in subsequent experiments found that morphine increased both mouse brain 5-HIAA levels and the incorporation of radio tryptophan in 5-HT (YARBROUGH et al., 1972). FENNESSY and LEE (1972) found that morphine (1–20 mg/kg) decreased mouse brain 5-HT levels; however, 5-HIAA levels were increased by the 5 mg/kg dose level but were lowered by 10–50 mg/kg. BOWERS and KLEBER (1971) found that methadone increased mouse brain 5-HIAA levels, and although this increase was prevented by pargyline, there was no difference in brain levels of 5-HIAA between control and methadone-treated mice that had received pargyline. The mechanism whereby morphine exerts its effect on 5-HT metabolism and utilization is not known. GADDUM and VOGT (1956) found that morphine antagonized the depressant effects of reserpine and intraventricularly administered 5-HT in the cat.

b) Chronic Administration (Dog, Rat, Mouse, Rabbit, Man)

Chronic administration of morphine does not affect the brain levels of 5-HT in the dog (MAYNERT et al., 1962; GUNNE, 1963) and decreases 5HIAA excretion (GUNNE, 1963). In the rat, chronic administration of morphine does not alter brain levels of 5-HT but increases 5HIAA levels (COCHIN and AXELROD, 1959; MAYNERT et al., 1962; GUNNE, 1963; SLOAN et al., 1963). Tolerance develops to increases in brain 5HIAA levels produced by morphine; however, 5HIAA levels are increased in rats pretreated with probenecid, suggesting an increase in 5-HT turnover in the tolerant animal (HAUBRICH and BLAKE, 1969). YARBROUGH et al. (1970, 1973) however failed to see an increase in the incorporation of ^{14}C-L-tryptophan and ^{3}H-L-tryptophan in rat brain into 5-HT in rats receiving morphine chronically. Similarly, ALGERI and COSTA (1971) and BHARGAVA et al. (personal communication) failed to observe any change in brain 5-HT of rats receiving morphine chronically and pargyline. HAUBRICH and BLAKE (1973) observed an initial increase in rat brain 5HIAA and analgesia following morphine pellet implantation in the rat and as tolerance developed to the analgesic effect, there was a parallel decrease in brain 5HIAA concentration.

Chronic administration of morphine either by repeated injection or by pellet implantation does not alter brain levels of 5-HT in the mouse (BARTLET, 1960; SHEN et al., 1970; MARSHALL and GRAHAME-SMITH, 1970; MARUYAMA et al., 1971; CHENEY et al., 1971). There is a conflict between different investigators on the effects of chronic morphine on 5-HT turnover. In mice treated with pargyline, chronic administration of morphine increased brain 5-HT (WAY et al., 1968; LOH et al., 1969; SHEN et al., 1970; MARUYAMA et al., 1970, 1971). On the other hand, YARBROUGH et al. (1970) and CHENEY et al. (1971) found that chronic morphine administration did not alter 5-HT turnover. No increase in brain 5-HT was seen in dependent mice treated with tranylcypromine (MARSHALL and GRAHAME-SMITH, 1970). Cyclohexamide lessened the increase in brain 5-HT seen in the pargyline-treated dependent mouse (LOH et al., 1969). Further, chronic morphine and methadone administration with or without probenecid treatment increases brain 5HIAA in the mouse (SHEN et al., 1970; BOWERS and KLEBER, 1971). Treatment with pCPA, on the other hand, decreases brain 5-HT in morphine-dependent mice (WAY et al., 1968; MARSHALL and GRAHAME-SMITH, 1971; CHENEY and GOLDSTEIN, 1971a).

Rabbits receiving methadone chronically had decreased platelet levels of 5-HT but unaltered brain levels of 5-HT and 5HIAA (AHTEE and KÄÄRIÄINEN, 1973a). Chronically administered morphine on the other hand did not alter platelet levels of 5-HT and decreased brain 5-HT levels. Chronic administration of morphine, meperidine, and nalorphine in man (FRASER et al., 1957) tended to increase urinary excretion of 5HIAA. Patients receiving methadone chronically exhibit significantly lower levels of 5HIAA in CSF (TAMARKIN et al., 1970; BOWERS et al., 1971).

c) Abstinence

During either precipitated or withdrawal abstinence in morphine-dependent mice (MARUYAMA et al., 1971), rats (MAYNERT et al., 1962; SLOAN et al., 1963; GUNNE, 1963), dogs (MAYNERT et al., 1962; GUNNE, 1963), monkeys (SEGAL and DENEAU, 1962), and rabbits (MAYNERT et al., 1962), brain 5-HT levels do not change. Further, no change in turnover of 5-HT was seen during precipitated abstinence in the mouse

(SHEN et al., 1970; MARUYAMA et al., 1971). AZMITIA (1970) has found that brain hydroxylase activity in the abstinent rat is not different than saline-treated controls, but less than dependent rats. A decrease in 5HIAA excretion is seen during withdrawal and precipitated abstinence in the rat and an increase is seen in the dog (GUNNE, 1963).

KNAPP and MANDELL (1972) found that short-term administration of morphine depressed tryptophan hydroxylase activity in the particulate fraction obtained from septal nerve ending and to a lesser degree in that obtained from the supernatant fraction of the midbrain. Long-term administration of morphine significantly enhanced tryptophan hydroxylase activity in the particulate fraction but produced only a slight and not statistically significant enhancement of activity in the soluble fraction.

To further elucidate the role of serotonergic influences on tolerance and dependence, the effects of agents which alter its metabolism have been studied with conflicting results. WAY et al. (1968) found that pCPA decreased naloxone-induced jumping in the dependent pellet-implanted mouse. HO et al. (1972a) found that L-tryptophan increased abstinence intensity and that pCPA antagonized this effect. pCPA also inhibited the development of tolerance and physical dependence as measured by jumping as a sign of precipitated abstinence and weight loss as a sign of withdrawal abstinence (SHEN et al., 1970; HO et al., 1972b). Further, HO et al. (1973a) have demonstrated that 5,6 DHT, which produced a modest but selective decrease in brain 5-HT, also decreased tolerance to and dependence on morphine in the mouse. 5-HTP, which enhances morphine's analgesic effect in the nontolerant mouse produces hyperalgesia in the tolerant mouse and this effect is antagonized by pCPA (CONTRERAS et al., 1973b). OPITZ and REIMANN (1973) found that cyproheptadine and fenfluramine suppressed abstinence in the rat.

Other findings are not readily reconciled with the hypothesis that there is a serotonergic mechanism responsible for signs of abstinence in the tolerant mouse. HUIDOBRO et al. (1963a) reported that DL-tryptophan, DL-5HTP, and 5-HT combined with iproniazid decreased the intensity of nalorphine-induced abstinence. In seeming agreement with these findings, CHENEY et al. (1970) and CHENEY and GOLDSTEIN (1971a) found that pCPA enhanced levorphanol-induced running, but did not alter tolerance or physical dependence. MARSHALL and GRAHAME-SMITH (1970, 1971) and MARUYAMA et al. (1970, 1971) found that pCPA did not alter naloxone-induced jumping. Reserpine was also without effect (MARUYAMA et al., 1970).

SCHWARTZ and EIDELBERG (1970) were unable to observe an effect of pCPA on the rat abstinence syndrome. COLLIER et al. (1972) found that pCPA antagonized some but not all signs of precipitated abstinence in the acutely dependent rat. FLÓREZ et al. (1973) found that pCPA facilitated the development of acute tolerance to morphine's respiratory depressant effect and that it did not alter or enhance the increase in minute volume associated with the administration of naloxone in the decerebrate cat. pCPA does not increase the effectiveness of morphine in the dependent chronic spinal dog. As can be seen in Figure 6, the effects of morphine on the skin twitch, flexor, and crossed extensor reflexes are either not altered or are decreased following pCPA pretreatment. Similarly, the effects of morphine on pupillary diameter, temperature, and pulse are either decreased or unchanged by pCPA pretreatment. The respiratory depressant effect of morphine in the dependent spinal dog

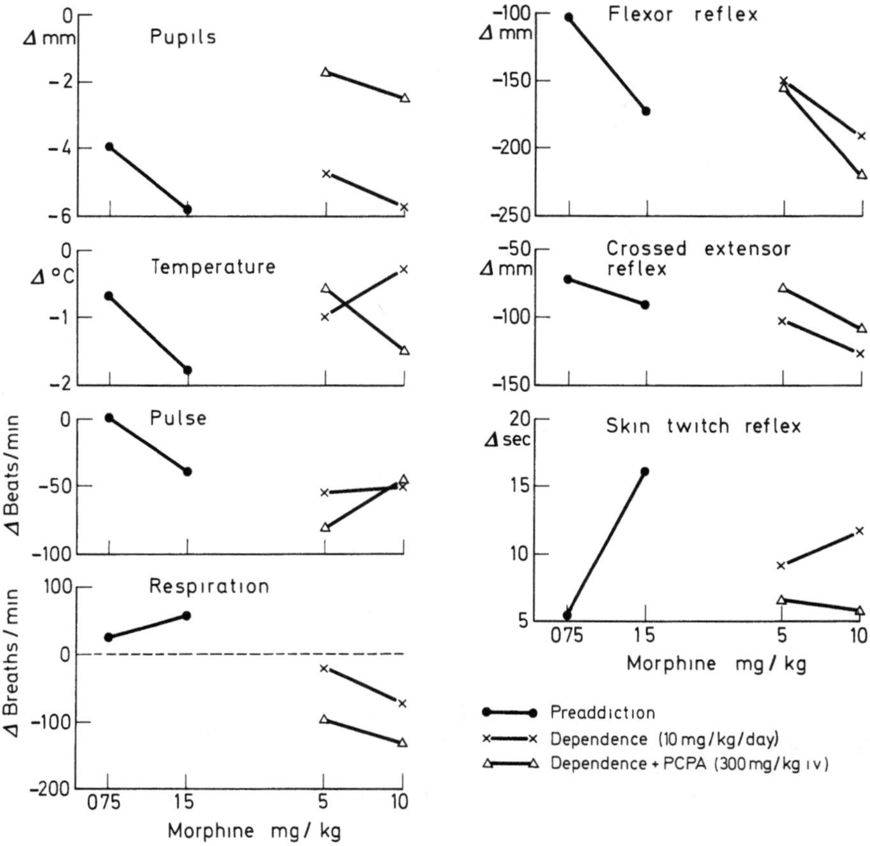

Fig. 6. Effects of pCPA (300 mg/kg) on tolerance to morphine in chronic spinal dogs that were stabilized on 10 mg/kg/day of morphine sulfate administered in 2 equally divided doses s.c. During preaddiction phase, the effects of 0.75 and 1.5 mg/kg of morphine administered intravenously were assessed on the above-indicated parameters. Flexor reflex evoked using a clamp (800 grams) and skin twitch using a radiant heat method. Test doses of morphine during preaddiction and addiction phases were administered at weekly intervals. Approximately 24 h following the intravenous administration of pCPA to the dependent dog either a 5 or 10 mg/kg test dose of morphine was administered intravenously and 24 h later either a 10 or 5 mg/kg test dose was administered. Three control observations were obtained prior to the administration of the test dose and observations were repeated at hourly intervals for 3 h. Mean of 3 control observations calculated and difference between this mean and observations obtained after test dose was determined. Sum of these differences for each dog and each parameter calculated. Each point represents mean of these sums for the 3 dogs studied. A more complete description of methods employed can be found in papers of MARTIN and EADES (1964), MCCLANE and MARTIN (1967), and MARTIN et al. (1974)

may be somewhat enhanced by pCPA pretreatment. Further, pCPA pretreatment does not alter the nalorphine-precipitated abstinence syndrome. Thus, nalorphine facilitated the flexor and crossed extensor reflexes, evoked the stepping reflex, and decreased the ipsilateral extensor thrust reflex to an equal degree in precipitation studies conducted before and following the administration of pCPA. The degree of mydriasis, hyperthermia, tachycardia, and tachypnea evoked by nalorphine in the

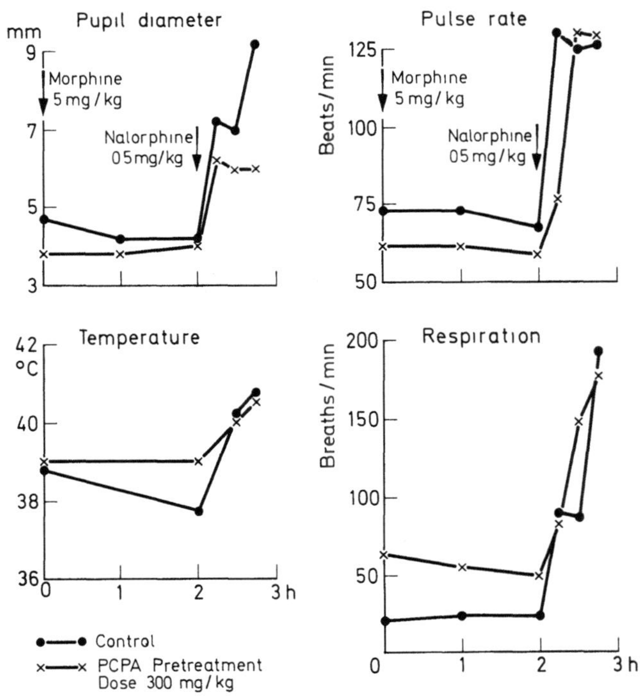

Fig. 7. Effects of pCPA on precipitated abstinence in chronic spinal dogs dependent on morphine (10 mg/kg/day). A morning stabilizing dose (5 mg/kg) of morphine was administered at 0800. Abstinence was precipitated with nalorphine (0.5 mg/kg/i.v.) administered at 1100. Prior to the administration of nalorphine, 3 control values were obtained at hourly intervals. Observations were repeated 15, 30, and 45 min after administration of nalorphine. Each value is the mean of observations obtained in 3 dogs. Interval between control and pCPA pretreatment-precipitated abstinence studies was approximately 1 week. Stepping reflex evoked by nalorphine in morphine-dependent spinal dog was more vigorous following treatment with pCPA

morphine-dependent chronic spinal dog was also not altered by pCPA (Fig. 7). Thus, pCPA does not appear to decrease tolerance or alter the precipitated abstinence syndrome in the morphine-dependent chronic spinal dog.

GEBHART and MITCHELL (1973) found no difference in the rate of tolerance development to morphine in two strains of mice, one of which had low levels and a low rate of utilization of 5-HT. On the other hand, pCPA attenuated the development of tolerance and the manifestation of physical dependence more in Sprague-Dawley rats which have a higher 5-HT turnover rate than in Fisher rats (RECH and TILSON, 1973).

5. Proteins, Polypeptides, Cyclic AMP, Prostaglandins, and GABA

COHEN et al. (1965) first demonstrated that actinomycin-D, an inhibitor of protein synthesis, could retard the development of tolerance to morphine in the mouse and rat. SMITH et al. (1966, 1967) showed that both actinomycin and puromycin prevented the development of long-term but not acute tolerance to the lenticular effects of levorphanol in the mouse. These studies have been extended to other inhibitors of protein synthesis and to other species. Thus, 8-azaguanine (SPOERLEIN and SCRAFANI, 1967) and cycloheximide (WAY et al., 1968; LOH et al., 1969; HUIDOBRO, 1971) have been shown to inhibit the development of tolerance in mice and actinomycin-D, cycloheximide, puromycin, 6-mercaptopurine, and 5-fluorouracil in the rat (COX et al., 1968; COX and OSMAN, 1969, 1970; FEINBERG and COCHIN, 1972). WAY et al. (1968) and LOH et al. (1969) demonstrated that cycloheximide retarded the development of physical dependence to morphine in mice. HUIDOBRO (1971) found that in mice (13 days after a morphine pellet was completely adsorbed) a single dose injection of cycloheximide 30 min before morphine retarded the development of acute tolerance. TAKEMORI and TULUNAY (1973), on the other hand, found that in the mouse, single doses of cycloheximide 12 h before morphine did not alter the ED_{50} of morphine or increase the antagonistic potency of naloxone; whereas, daily treatment with cycloheximide inhibited the development of tolerance and increased the antagonistic potency of naloxone. LOH et al. (1971) found that actinomycin-D increased the uptake of morphine by the brain. GRECKSCH et al. (1974) showed that orotic acid enhanced the development of tolerance in the rat. The nature of the changes that are caused by these agents that change RNA activity and protein synthesis is not known. HODGSON et al. (1974) found that chronic but not acute morphine treatment in the rat reduced the ability of chromatin to synthesize RNA.

Substance P (SP) has been reported to suppress abstinence signs in the rat (STERN, 1963) and the mouse (STERN and HADZOVIC, 1973). The studies of STERN and HADZOVIC (1973) were conducted with synthetic SP which suppressed nalorphine-precipitated jumping of the dependent mouse. These findings are not easily reconciled with the observation that SP antagonizes the analgesic actions of morphine (STERN and HUKOVIC, 1960; ZETLER, 1956a, b) and its presumed role as the neurotransmitter responsible for transmission of sensory input in the spinal cord. MARTIN and EADES (unpublished observations) found that 0.5 mg/kg of synthetic SP did not alter pupillary diameter, pulse rate, body temperature, or respiratory rate in the abstinent spinal dog. It produced some depression of the flexor reflex, and a transient reduction of the stepping reflex and sedation. Although there was a modest depression of the abstinence score (MARTIN et al., 1974) by SP, it was not significantly different from that obtained with the acetic acid vehicle. KRIVOY et al. (1974a) showed that desglycinamide^{9}-lysine vasopressin facilitates the development of acute tolerance to morphine in the mouse.

In the morphine-dependent mouse, cAMP increased the intensity of abstinence (HO et al., 1971, 1972c) and this effect was antagonized by cycloheximide (HO et al., 1973c). Theophylline, which inhibits phosphodiesterase, has been reported by COLLIER (1974) to produce a quasi-abstinence syndrome. Indomethacin, which blocks the production of prostaglandins, inhibits some and exaggerates other signs of precipitated abstinence in the acutely dependent rat (COLLIER et al., 1972).

GABA and aminooxyacetic acid apparently increase tolerance and physical dependence in the mouse (HO et al., 1973d). CLOUET and NEIDLE (1970) found no effect of single doses of morphine on brain GABA in the rat, while LIN et al. (1973) found an increase in brain stem GABA in the dependent rat. MAYNERT et al. (1962) found no change in brain levels of GABA in the dependent dog. SHERMAN and MITCHELL (1974) found that morphine in the nontolerant but not in the tolerant rat increased brain levels of gammahydroxybutyrate.

VII. Theories of Tolerance and Physical Dependence

1. Neuronal and Neurohumoral Factors

Tolerance and dependence to narcotic analgesics have been clearly demonstrated in the mouse, rat, guinea pig, dog, monkey, and man. The syndromes that have been demonstrated have to a degree represented the opportunistic application of available observational techniques. A variety of autonomic, somatomotor, and behavioral changes have been described following withdrawal of narcotics from dependent animals. In the rat and dog, early abstinence persists for several days and in man for approximately a month. Protracted abstinence has been studied and demonstrated in the rat, dog, and man. It persists for at least 6 months in these species.

A great deal of effort has been expended in an effort to understand the neuronal substrate and the neurohumoral and neurochemical mechanisms underlying tolerance, dependence, and abstinence. It is quite clear that tolerance and dependence can develop at several levels of the nervous system including peripheral neurones (guinea pig ileum), the spinal cord, the brain stem, and higher structures. Studies involving both the local application of narcotic antagonists, the direct application of narcotics to the brain, and lesion studies indicate that the diencephalon is a site necessary for the manifestation of many of the behavioral and motoric signs of abstinence. At the present time data on the sites necessary for the autonomic signs of abstinence have not been elucidated. There is considerable disagreement as to the role of the limbic system and the neocortex in the manifestation of behavioral and autonomic symptoms and signs of abstinence. The frontal lobe, cingulate gyrus, and perhaps other parts of the limbic system and corpus striatum may be involved in the suffering and hypophoria associated with early and protracted abstinence.

The importance of ACh in morphine tolerance and dependence cannot be assessed. Morphine depresses ACh release; tolerance may develop to the mechanism underlying this effect or apparently develop because ACh may accumulate in neurons and thus increase release. When morphine is withdrawn, ACh release may be increased. Which sign of abstinence this increase in release may account for is problematic, for data on the efficacy of cholinergic antagonists in suppressing signs of abstinence are conflicting.

There is ample evidence that single doses of morphine stimulate NE- and E-containing neurons both in the adrenal medulla and the brain, probably indirectly (i.e., by activating chemoreceptive neurons which project through interneurons to adrenergic and noradrenergic neurons), in a variety of species including the cat, dog, rabbit, rat, and mouse, resulting in an increase in turnover of these CA. It is less certain that other narcotic analgesics share all of the central actions of morphine on central noradrenergic neurons. With chronic administration, tolerance develops to the NE- and E-releasing action of morphine. When narcotics are withdrawn the preponderance of data suggests that E and NE are released from both brain and adrenal gland neurons although the magnitude of this increase in release varies greatly from species to species, probably being of more importance in the rat than in the mouse or man. A role for noradrenergic processes in facilitation of somatomotor reflexes, production of behavioral arousal, and enhancement of the sympathetic nervous system activity has been suggested by many lines of evidence and investigators, and these types of activity which are manifest during the early abstinence could be explained in part by increased release of NE by noradrenergic neurons. Noradrenergic processes do not easily fit into the commonly held concept that neurons which are depressed by narcotics show hyperactivity during withdrawal. However, it is possible that there may be several functional groups of noradrenergic neurons in the CNS, some of which are depressed by single doses of narcotic analgesics, while others are stimulated either directly or indirectly, and it is the neurons that are depressed that exhibit increased activity during abstinence.

Closely related are observations that narcotics increase the turnover of DA in the cat, mouse, and rat. It would appear that perhaps some signs of acute and protracted abstinence such as aggression could be mediated through a dopaminergic mechanism. LAL and his collaborators proposed that narcotic analgesics block the actions of DA in the corpus striatum and reduce negative feedback to dopaminergic cells, thus increasing the liberation and turnover of DA. When morphine is withdrawn there is an excess of dopaminergic tone. Several explanations for this can be postulated: (1) The postsynaptic receptor is hypersensitive to DA perhaps as a consequence of a chemical denervation; (2) the capacity of the neuron to synthesize and liberate DA is increased; and (3) a redundant pathway in parallel with the presynaptic elements that innervate the dopaminergic neuron has hypertrophied. All signs of abstinence cannot be explained by a dopaminergic mechanism.

Most studies indicate that single doses of narcotic analgesics increase the turnover of 5-HT in the rat and the mouse. This could be related to morphine's ability to antagonize the actions of serotonin (GADDUM and VOGT, 1956; GADDUM and PICARELLI, 1957; GYERMEK and BINDLER, 1962) and a consequent decrease in negative feedback. The fact that morphine antagonizes the stimulatory actions of serotonin on the intestine of the guinea pig but not the dog, cat, or monkey (PRUITT et al., 1974) cautions against overgeneralizing about serotonin-blocking hypotheses. Partial tolerance appears to develop to the increase in turnover in the rat. In the tolerant mouse, it is not clear whether the turnover of 5-HT is increased or not. In man, evidence suggests that it may be decreased. There is a body of evidence, controversial though it is, that suggests that 5-HT is necessary for the manifestation of some signs of tolerance and dependence. The mechanism whereby narcotics increase 5-HT turnover and facilitate the manifestation of signs to tolerance and dependence has not been elucidated.

2. General Theories

a) Homeostasis

Morphine has many sites of action in the central nervous system and affects many processes either directly or indirectly. The central nervous system reacts to both the direct and indirect actions of the narcotic analgesics using many adaptive mechanisms. Further, dependence, as manifest by the withdrawal or precipitated abstinence syndrome, is also a consequence and the result of the complex interaction of the adaptive mechanisms in the absence of the narcotic.

HIMMELSBACH (1943) first postulated that in chronic physical dependence, contra-adaptive changes in homeostatic mechanisms that are depressed by narcotics hypertrophy and that in the absence of the narcotic the contra-adaptive forces in conjunction with released homeostats gave rise to the abstinence syndrome. Regulation of respiration (MARTIN et al., 1968), gonadal hormones (MARTIN et al., 1973b), possibly adrenal cortical function (EISENMAN et al., 1958, 1961) in man, respiration in the cat (MARTIN and EISENMAN, 1962), and temperature in the dog (MARTIN, 1968a) may fit this model. Part of acute tolerance and the precipitated abstinence syndrome in both the acutely and chronically dependent animal can be explained in terms of changes in homeostats. The effects of morphine on the respiratory homeostat in man and the cat can be used to illustrate this mechanism. When the onset of action of a narcotic analgesic is very rapid, a profound transient depression of respiration is seen which reaches a peak in minutes and is followed again in minutes by partial recovery. This early lessening of effect cannot be explained on the basis of drug distribution for neurologic functions which do not involve homeostats remain or become further depressed when respiratory depression has lessened. The depression is associated with a rapid lowering of the set point of the respiratory homeostat and the partial recovery or acute tolerance with the accumulation of CO_2 to a level sufficient to again drive the depressed respiratory center.A similar type of phenomena explains some signs of acute dependence and precipitated abstinence. When the set point is rapidly normalized by administering a narcotic antagonist, the internal environment which is obtained in the depressed state acts as a strong driving force for the restorative mechanisms. This mechanism has been shown to operate in the cat (MARTIN and EISENMAN, 1962) and man (LANDMESSER et al., 1953). Figure 8 (first column) shows in a general way the interrelationship between the effect of the narcotic on the homeostatic set point and the physiologic functioning of the restorative mechanism. In chronic physical dependence a similar mechanism operates except that contra-adaptive forces have been recruited which alter the set point in the abstinent dependent subject (Fig. 8, column 2). MARTIN et al. (1968) found that the respiratory centers of chronically dependent subjects in early abstinence were hyperresponsive to CO_2 and their set point was lowered. This formulation leaves unanswered the question of the basic changes that are involved in altering the set point through contra-adaptive influences.

b) Learning and Adaptation

Learning and experience may enable animals to adapt to the actions of narcotics as well as other drugs. MITCHELL and his collaborators have paid particular attention to the interaction between the narcotic and the test situation. They have observed that the analgesic action of morphine is diminished after the first dose (KAYAN et al., 1969; ADAMS et al., 1969; GEBHART and MITCHELL, 1971) when the animal is placed in a situation that has some of the attributes of the original test situation while under the influence of morphine. Neither previous experience with the environmental factor nor the narcotic alone will induce a degree of tolerance comparable to the interaction test situation (GEBHART and MITCHELL, 1972b). Subjecting rats to several types of stress prior to testing did not increase acute tolerance (GEBHART et al., 1972). GEBHART et al. (1972) have suggested that certain aspects of the testing procedure as well as morphine may repress transmitter synthesis and suggest that tolerance is a consequence, in part, of negative feedback de-repression. KAYAN et al. (1973) also have seen an interaction between the test procedure and the analgesic when both were administered chronically either on a daily or weekly basis. Tolerance developed

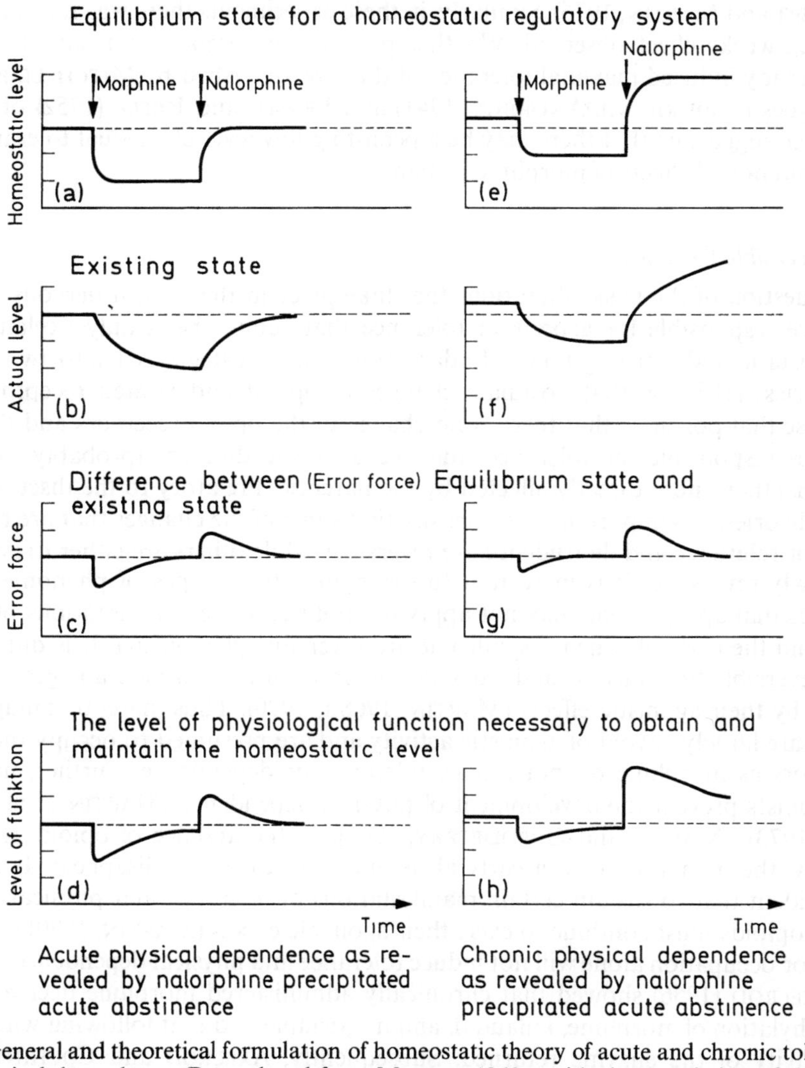

Fig. 8. General and theoretical formulation of homeostatic theory of acute and chronic tolerance and physical dependence. (Reproduced from MARTIN, 1968a, with the permission of the publisher)

most rapidly when morphine was administered and the animals tested daily. Regardless of the dosing schedule, near maximal tolerance developed by the third or fourth injection. The significance of this type of tolerance has not been assessed.

In a large number of factorially designed experiments measuring the effects of narcotic analgesics on pupillary diameter and subjective state conducted on the wards of the Addiction Research Center, in patients participating at weekly intervals (see Sect. II, Chap. 3), no significant between-weeks effect has been seen. More recently, similar studies have been conducted in chronic spinal dogs in which the graded doses of analgesics have been compared for their ability to depress the flexor and skin twitch reflexes, constrict pupils, slow pulse rate, and lower body temperature

(GILBERT and MARTIN, 1976). Similarly, in these experiments there was no significant between-weeks effect observed. Whether previous experience with narcotic drugs had already induced maximal tolerance of the type described by MITCHELL and his colleagues is not known. ANDREWS (1941) and FRASER and ISBELL (1952) obtained evidence suggesting that there may be a persisting low level of residual tolerance for some but not all effects of morphine in man.

c) Reversible Changes

The question of the basic alterations that take place in the central nervous system that are responsible for aspects of tolerance that cannot be readily explained by homeostatic and learning types of adaptation can be categorized into two major categories: (1) Those that postulate changes in opioid and related receptors and (2) those that postulate that there is no change in the opioid receptors and that the changes responsible for tolerance must be at a site different (probably different neurons) than those directly affected by the narcotic. Prefatory to the discussion of these theories, it is important to emphasize that some of the changes that are responsible for tolerance and dependence are reversible, while others are either irreversible or slowly reversible. It is important to distinguish these types of phenomena, for theories that apply to one may not apply to another. Some manifestations of tolerance and the early abstinence syndrome are reversible phenomena. It is quite clear that reversible tolerance to and early dependence on the narcotic analgesics is induced by their agonistic effects (MARTIN, 1968a, 1970). Thus, narcotic antagonists which are largely devoid of agonistic activity and are presumed to occupy the same receptors as morphine do not induce tolerance or dependence. Further, narcotic antagonists prevent the development of physical dependence (MARTIN et al., 1965, 1966, 1973c; MARTIN and GORODETZKY, 1967). In the absence of opioid agonistic activity, the manifestations of early abstinence and tolerance disappear. It is thus self-evident that to sustain certain manifestations of tolerance and physical dependence opioids must continue to exert their agonistic effects (MARTIN, 1970) and that receptor occupation alone will not induce tolerance and physical dependence.

AXELROD (1956) showed that chronically administered morphine decreased N-demethylation of morphine, Dilaudid, and meperidine and that following withdrawal, activity of the enzyme returned. Subsequently, AXELROD and COCHIN (1957) showed that nalorphine inhibited the N-demethylating enzyme and that chronic administration of nalorphine and normorphine would decrease the activity of this enzyme (COCHIN and AXELROD, 1959). Both of these latter agents produce tolerance and physical dependence (FRASER et al., 1958; MARTIN and GORODETZKY, 1965). AXELROD (1968) postulated that chronic administration of a drug could decrease the number of active receptors and felt that the N-demethylating enzyme might be a model of the narcotic receptor in explaining tolerance. The reduction of the number of receptors in itself does not explain physical dependence. COLLIER (1965) proposed that the number of silent and active narcotic receptors could increase or decrease as a consequence of chronic drug administration. Thus, tolerance can be explained by the induction of silent receptors. Some aspects of tolerance and dependence can be explained by the induction of active receptors. In COLLIER's formulation of tolerance to and dependence on narcotics, neurotransmitter receptors would be induced as a consequence of the narcotics acting as antagonists at the receptor site (e.g., DA) or presynaptically by decreasing the release of the neurotransmitter (e.g., ACh). TAKEMORI et al. (1973) found that pretreatment of mice with morphine but not pentazocine enhanced the potency of naloxone in antagonizing morphine's analgesic activity and increased its apparent affinity for the morphine receptor by a factor of 2. These studies have been extended (TULUNAY and TAKEMORI, 1974a and b) to show that mice acutely or chronically pretreated with morphine show both tolerance and an increased susceptibility to the antagonistic effects of naloxone. These effects were prevented by cycloheximide. These data have been interpreted as indicating that pretreatment with, and tolerance to, morphine are associated with a conformational change in the morphine receptor. An alternative explanation is that the apparent increase in potency of naloxone could be due to the induction of acute physical dependence.

De-repression theories of tolerance and dependence have been proposed at both an enzymatic and cellular level. GOLDSTEIN and GOLDSTEIN (1961, 1968) and SHUSTER (1961) have proposed that narcotics may inhibit enzymes whose product represses their synthesis, which in turn gives rise to increased quantities of the enzymes and tolerance. Upon withdrawal, the enzyme becomes uninhibited and because of increased quantities functions at an increased rate until substrate repression reduces its quantity.

EMMELIN (1961) alluded to and GRUMBACH (1961) as well as JAFFE and SHARPLESS (1962) specifically suggested that a chemical or pharmacologic denervation supersensitivity might be involved in the genesis of tolerance and physical dependence. There are probably several mechanisms that give rise to the phenomenon of "denervation supersensitivity." In the context of this discussion and the discussion of de-repression, it is possible that neurotransmitters repress the synthesis of their receptors and that drugs which suppress neurotransmitter release may increase receptor synthesis. As discussed earlier, it is difficult at this time to fit the interaction of morphine and its actions on ACh, NE, DA, and 5-HT into this formulation. Recently, however, SHOHAM and WEINSTOCK (1974) have shown that guinea pig ilea that are acutely tolerant to morphine's depressant effect on the electrically evoked twitch have exaggerated responsivity to ACh.

Compensatory changes may also occur at a cellular and functional level. Thus MARTIN and EADES (1960) have proposed that functional pathways in the central nervous system may be composed of parallel pathways which employ different neurohumoral processes and called this arrangement *pharmacologic redundancy* (Fig. 9-I). MARTIN (1968a, 1970) postulated that tolerance to a drug which depressed one of the parallel pathways but not others could be produced by the hypertrophy of a redundant pathway (Fig. 9-II). When the drug was withdrawn, the increased activity which resulted from the summation of the activity of the hypertrophied redundant pathway and the restoration of activity of the depressed pathway gave rise to the abstinence syndrome. This formulation offered an explanation of a number of perplexing dimensions of tolerance and the abstinence syndrome (MARTIN, 1968a) such as partial tolerance and tolerance to phasic actions of each stabilization dose in the presence of tonic alteration in function such as is seen with persistent miosis and respiratory depression in the dependent subject. Thus, when the redundant pathway has a limited ability to hypertrophy, partial tolerance is seen with a tonic depression.

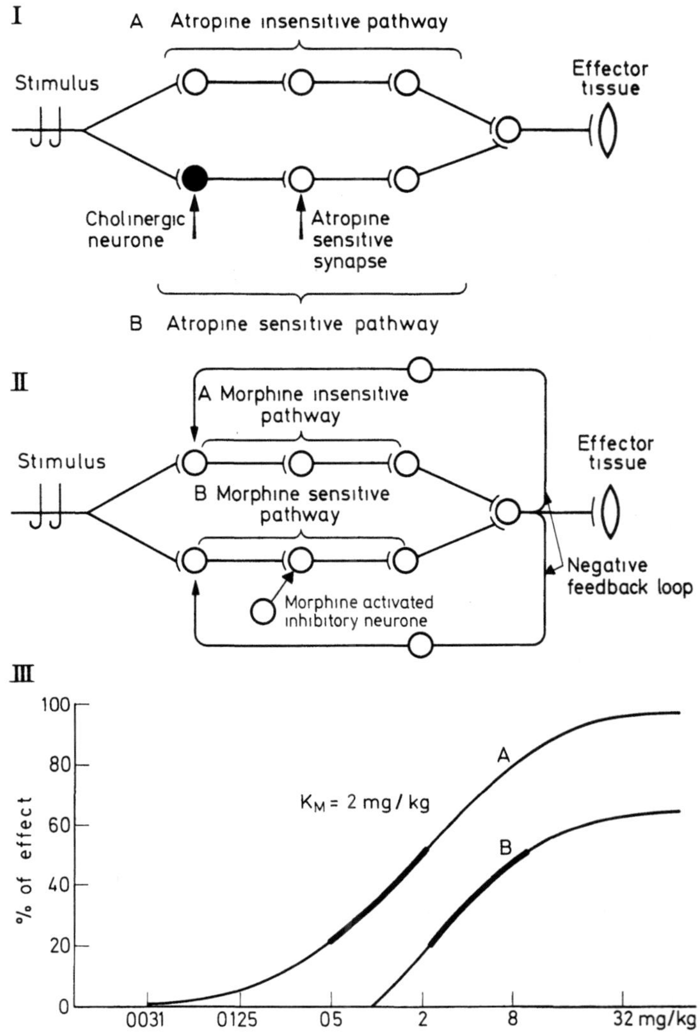

Fig. 9. Schematic drawing indicating principles of pharmacologic redundancy in hypothetical conditions necessary for development of tolerance and dependence (I and II). Panel III shows theoretical morphine dose-response curve (A) in which morphine has an affinity constant of 2 mg/kg. Heavy portion of line indicates range of doses and range of effects that can be studied in nontolerant animal. Maximal depression is seen with a dose of 2 mg/kg. Curve B represents the theoretical dose-response curve obtained with hypertrophy of the redundant pathway (see text)

This formulation also explains the parallel shift of the morphine dose response curve to the right in the tolerant animal (see Fig. 6 and WAY et al., 1969). As has been previously stated, there is reason to believe that narcotic analgesics continue to exert their full agonistic activity in the tolerant and dependent subject. Further, maximal measurable effects of narcotic analgesics may well be produced when only a small portion of the active receptors are occupied by the narcotic agonist in the nontolerant animal. In the tolerant subject, a much greater portion of the receptors would be occupied. On the basis of data of ANDREWS and HIMMELSBACH (1944), it has been calculated that the affinity of morphine for the morphine receptor may be of the order of magnitude of 1 mg/kg (MARTIN et al., 1973a). In Fig. 9-III a hypothetical dose response curve (A) is drawn in which it is assumed that morphine has an intrinsic activity of 1 and an affinity constant (Km) of 2 mg/kg. Further, it is assumed that the redundant pathway has negligible spontaneous tone and participates to a negligible extent in the mediation of the nociceptive reflex. The heavy part of curve A is the portion of the dose response curve that can be studied experimentally, and the upper end of this segment represents that dose of morphine (2 mg/kg) that will completely abolish the nociceptive reflex in the nontolerant animal. Curve B represents the dose response curve that can be obtained in the tolerant animal in which the morphine-resistant redundant pathway has hypertrophied to the extent that 33% of the nociceptive response is mediated over it. As can be seen 1 mg/kg of morphine is necessary to prevent hyperalgesia (abstinence) from becoming manifest and that over the range of detectable and observable analgesic effects of morphine (heavy part of curve B) approximately 5 times as much morphine is needed to produce a comparable degree of analgesia.

There are other factors that may play a role in phenomena that can be and are generally interpreted as signs of tolerance and dependence. Thus TATUM et al. (1929) proposed the well-known "dual action" hypothesis of physical dependence in which it was postulated that morphine had two mutually physiologically antagonistic modes of action. One mode, called "stimulant," was less intense but had a longer persistence than the "depressant" mode. Cumulation of stimulant effects was responsible for manifestations of tolerance and dependence. SEEVERS and DENEAU (1962, 1963) subsequently suggested that the "dual action" theory could not explain either chronic precipitated or withdrawal abstinence; whereas, it could explain precipitated abstinence in the acutely dependent animal. MARTIN (1968a) and more recently MICHAUD and JACOB (1973) have pursued this problem and have obtained for the most part concordant results and reached similar conclusions; namely, that the homeostatic and dual action theories provide an explanation for some of the signs of precipitated abstinence syndrome in the acutely dependent dog. It is presumed that stimulant actions of morphine are not antagonized by narcotic antagonists (SEEVERS and DENEAU, 1963) while the depressant are. Thus pretreatment with an antagonist is presumed to unmask and reveal morphine's stimulant properties. Table 3 compares the effects of nalorphine and naloxone in the acutely dependent dog to the effects of nalorphine and naloxone in the chronically dependent dog and the effects of morphine in dogs pretreated with naloxone and nalorphine. As can be seen, these three syndromes resemble each other in several regards but can be distinguished by others. In this regard, the precipitated abstinence syndrome can be distinguished qualitatively from the withdrawal abstinence syndrome (MARTIN et al., 1974). Acute tolerance and dependence in the mouse cannot be antagonized by naloxone (EIDELBERG and ERSPAMER, 1974); whereas, chronic dependence can be antagonized by cyclazocine and naltrexone in man (MARTIN et al., 1966; MARTIN et al., 1973c). These findings add additional weight to the argument that tolerance to and dependence on narcotic analgesics probably involves several processes.

Table 3. A summary and comparison of the actions and interactions of morphine, nalorphine, and naloxone in the nondependent, acutely dependent and chronically dependent chronic spinal dog

Effects of treatment	Morphine	Nalorphine	Nalorphine followed by morphine	Naloxone followed by morphine (MICHAUD and JACOB 1973)	Morphine infusion followed by nalorphine	Morphine infusion followed by naloxone (MICHAUD and JACOB 1973)	Nalorphine or naloxone in the chronically dependent dog (WIKLER and FRANK, 1948; MARTIN and EADES, 1964; MARTIN et al., 1974)
Lacrimation	+	0	+	0	+	+	+
Rhinorrhea	+	0	+	+	+	+	+
Salivation	+	0	+	+	+	+	+
Urination	0	0	0	+	+	+	+
Tremors	0	0	0	Occ	+	+	+
Emesis	+	0	+	+	Occ	0	Occ
Restlessness	+→−	−	+	+	+	+	+
Pupillary diameter	↓	↓	0	0	↑	↑	↑
Heart rate	↓	0	↑	↑	↑	↑	↑
Body temperature	↓	0	↓	↑	↑	↑	↑
Respiratory rate	↑→↓	0	↑	↓	0	0	↑
Flexor reflex	↓	↓	0		↑		↑↑
Crossed extensor reflex	↓	↓	0		↑		↑↑
Ipsilateral extensor thrust	↑	↓	0		↓		↓
Running movements	Occ		Occ early		Occ		Almost Always
No. signs similar to precipitated abstinence in chronically dependent dog	7	1	8 (7)	(7)	14 (9)	(9)	15 (11)

\+ the sign is commonly produced by the treatment
− the sign is decreased by the treatment.
↑ the level of function is increased by the treatment.
↓ the level of function is decreased by the treatment condition.
Occ Occasionally.
The studies of MICHAUD and JACOB were performed in the intact dog. The values given in parenthesis are the number of signs seen in the intact dog.

d) Persisting Changes

In 1964 COCHIN and KORNETSKY published the first systematic long-term study of the development and persistence of tolerance in the rat and found that tolerance developed after a single dose of morphine to its analgesic activity and that tolerance persisted for approximately a year following the termination of either acute or chronic administration of morphine. Acute and chronic tolerance to the slowing of swimming by morphine was less complete and persisted for a shorter time. MARTIN et al. (1963) demonstrated the phenomenon of secondary or protracted abstinence in the rat and showed that it persists for at least 6 months. Subsequent studies have demonstrated the phenomenon of protracted abstinence in the dog (MARTIN et al., 1974) and in man (MARTIN and JASINSKI, 1969; MARTIN et al., 1973b). These two lines of evidence strongly suggest that there are several types of changes that are designated as *tolerance* and *dependence* that persist for many months following termination of narcotic analgesics. There are many experimental findings that may bear on these phenomena.

Single Cells

There is extensive literature on the effects of narcotic analgesics and related drugs on the growth of cells and microorganisms which has recently been reviewed by SIMON (1971). Narcotic analgesics inhibit growth of viruses, bacteria, and cultured cells and cause histologic changes in cells. On prolonged exposure, cells become tolerant to the inhibitory effects of morphine and other narcotic analgesics on growth and when morphine is omitted from the nutrient media, growth is inhibited, histologic changes are seen, and cells die. In general, the tolerance and dependence seen in single reproducing cells differ from that seen in the intact animal in several important respects: (1) Much larger doses of narcotic analgesics are employed. Measurable pharmacologic effects are seen in most species with concentrations of less than 1–20×10^{-6} M; whereas, concentrations employed in single cell studies range from 10^{-5} to 10^{-3} M. (2) The end points are quite different. Cell death, failure to reproduce, or changes in cytoarchitecture are signs that emerge following withdrawal of narcotics. In contrast, dependence of the central nervous system to narcotics is presumed to occur in cells that do not reproduce. Further, because many of the changes seen during abstinence are reversible, it has been presumed that neurons in the central nervous system retain their viability. Critical data do not exist on whether there are neurohistologic changes seen in mammalian brain during abstinence (SEEVERS, 1955). (3) Those cells that do reproduce in the presence of narcotics may represent the emergence of resistance mutants, and this adaptive mechanism would not be present in the central nervous system.

The use of a homogeneous population of reproducing cells or organisms does provide a convenient way of studying biochemical changes that are associated with tolerance and physical dependence. Work in the area of nucleic acid metabolism, membrane phenomena, intermediary metabolism, and protein metabolism is outside our competence, and the reader is referred to several recent reviews and volumes in which these subjects are dealt with in depth (DOLE, 1970; CLOUET, 1971; MULÉ and BRILL, 1972).

Of particular interest are those changes that result in the creation of new macromolecules. KORNETSKY and KIPLINGER (1963) treated nontolerant rats with sera from tolerant dogs and found that the depressant action of morphine in swimming was enhanced. At about the same time D.E.ROSENBERG at the Addiction Research Center (unpublished observations) replaced the blood of nontolerant chronic spinal dogs with blood from morphine tolerant dogs. The transfusion procedure itself produced a modest depression of the flexor reflex but did not alter the depressant effects of a test dose of morphine. Subsequently, KIPLINGER and CLIFT (1964) showed that sera from morphine-tolerant dogs also enhanced the analgesic action of morphine in the mouse, using the hot plate technique. They further demonstrated that the potentiating action of the sera persisted for over 96 h, was not reduced by dialysis of the sera, developed in parallel with tolerance, and exhibited selectivity of action. In contrast to these results, KORNETSKY and COCHIN (1964) found that sera from dependent rabbits did antagonize the analgesic action of morphine in the mouse. SPECTOR and PARKER (1970) and SPECTOR (1971) demonstrated that an antigen formed of carboxymethylmorphine and bovine serum albumin induced antibodies in the rabbit that would bind morphine and chemically related molecules. BERKOWITZ and SPECTOR (1972) decreased the analgesic activity of morphine by perhaps a factor of 2 by pretreating mice with the carboxymethylmorphine—bovine serum albumin antigen. Although it is unlikely that newly synthesized sera proteins which bind morphine contribute to the phenomenon called tolerance, these data clearly indicate that narcotic analgesics can induce the synthesis of macromolecules.

FRIEDLER (1968) and FRIEDLER and COCHIN (1972) found that offspring of post-addict female rats who were mated 5 days after their last dose of morphine showed a less rapid growth rate and tolerance to morphine when compared to appropriate controls born of nonaddict mothers. The influence of nursing was eliminated as a possible cause for tolerance and slowed growth. Offspring born of nonaddict mothers but postaddict fathers did not differ in responsivity to morphine or growth rate from appropriate control rats (Dr. JOSEPH COCHIN, personal communication). FRIEDLER (1974a), however, has shown that offspring through the F_2 generation born of postaddict male and nonaddict female mice were significantly smaller than appropriate control mice. Further, FRIEDLER (1974b) has shown that offspring of postaddict male and nonaddict female mice show a modestly decreased response to morphine's analgesic effect as assessed by the shock attenuation procedure, while offspring of postaddict females and nonaddict males and of postaddict males and females show a marked decrement. The cause of these transmitted changes is not known. FRIEDLER and COCHIN (1972) speculated that they could be endocrinal in nature, possibly being mediated through the pituitary. UNGAR and COHEN (1966) reported that brain extracts from morphine-tolerant rats and dogs decreased the analgesic activity of morphine and characterized the substance as a dialysable substance which was destroyed by chymotrypsin but not by pancreatic ribonuclease. TIRRI (1967) was unable to observe any difference between control mice and mice treated with an extract from the brain of tolerant rats. SMITS and TAKEMORI (1968) were unable to observe significant differences in the analgesic effect of morphine in mice treated with a brain extract of dependent rats and mice treated with an extract from nondependent rats, although there was a consistent trend for morphine to have a lesser analgesic effect in mice treated with the extract from tolerant animals. Subsequent studies by UNGAR and GALVIN (1969) confirmed their earlier findings.

VIII. Summary and Conclusions

It is highly probable that narcotic analgesics have several primary sites and modes of action mediating their agonistic activity. Several lines of evidence indicate that narcotic analgesics act at a spinal cord, medullary, mesencephalic, diencephalic, and cortical level. Further, each of the primary sites and modes of actions alter the working of several functional systems. A variety of adaptive mechanisms are recruited to restore the function of these altered systems and some, perhaps all, of these adaptive changes give rise to a disordered physiology when the narcotic is withheld. It is also clear that changes in neuronal function associated with tolerance and dependence occur in peripheral tissues such as the ilea, colon, and vas deferens as well as the spinal cord, medulla, and diencephalon in several species of mammals.

The role of certain well-established neurotransmitters in tolerance and dependence has been extensively investigated. Although cholinomimetic drugs simulate some of the signs of acute abstinence, there are several reasons for believing that cholinergic mechanisms do not, either directly or indirectly, cause or mediate most signs of abstinence: (1) Cholinergic antagonists reduce, do not affect, and enhance various signs of abstinence. (2) HC-3 does not alter the development of tolerance or dependence. (3) Anticholinesterases do not enhance and may even suppress some signs of abstinence. Since acutely and chronically administered narcotic analgesics do alter the metabolism and release of ACh, a role of ACh in some actions of narcotic analgesics as well as tolerance to and dependence on them cannot be excluded.

Morphine increases the release and turnover of NE in the cat and mouse but not in the dog and rat. With chronic administration tolerance develops to these effects. During abstinence NE utilization is probably increased in the rat and mouse, and although some data suggest that increased activity of central noradrenergic neurons may be responsible for some signs of precipitated and withdrawal abstinence, there are other data that are not easily reconciled with this hypothesis. The mechanism whereby narcotic analgesics alter NE metabolism is not known, but studies of its effects on the adrenal medulla suggest that its actions may be indirect. Narcotic analgesics also increase and in higher doses may decrease the release of brain DA and a degree of tolerance develops to these effects. Whether there is increased or decreased utilization of DA during early abstinence cannot be definitely stated at this time. Pharmacologic evidence suggests that a dopaminergic mechanism could mediate or modulate some signs of both early and protracted abstinence such as aggression, jumping, and hyperthermia, but not others.

Morphine increases 5-HT utilization in the mouse but whether this action is shared by other narcotic analgesics and is of importance in other species is not clear. Tolerance may develop to this effect. The evidence as to whether there is increased utilization of 5-HT during abstinence and whether this increased utilization is responsible for certain abstinence signs is controversial.

The enormous endeavor in synthesizing a variety of drugs that are both pharmacologically and structurally related to morphine has given rise to studies which will give us insight into not only the evolution and function of the nervous system but about its pathology. We have been increasingly involved in speculations about the evolution and role of receptors in the central nervous system. The observation that drugs which mimic neurotransmitters produce pharmacologic effects while their antagonists in themselves are inactive (MARTIN and EADES, 1967, 1968; VAUPEL and MARTIN, 1973) can be interpreted as indicating that there are neurotransmitter receptors in certain brain neurones which have no appropriate presynaptic elements and which in the normal course of events are not activated. Presumptive evidence has been obtained for three receptors, two of which are closely related to but different from the morphine receptor which when activated by the appropriate agonist produce three distinct types of syndromes and physical dependence (MARTIN et al., 1976; GILBERT and MARTIN, 1976). These have been designated the μ, $\varkappa$, and σ receptors. Tolerance develops to the three types of agonistic action when administered chronically and naloxone and naltrexone antagonize the three syndromes. Yet in untreated cats and dogs only some anti-$\varkappa$ or μ effects have been observed when naloxone or naltrexone are administered (MARTIN et al., 1976; BELL and MARTIN, 1977). There is difficulty in reconciling these observations with identification of naturally occurring polypeptides which bind to morphine receptors and have morphinelike or related agonistic actions. Although a large amount of experimentation has been conducted based on the hypothesis that narcotic analgesics mimic a naturally occurring neurotransmitter, to date this neurotransmitter has not been convincingly demonstrated. The fact that naloxone has so few pharmacologic actions in animals who have not received a narcotic analgesic led MARTIN (1967) to conclude that there was no such transmitter or that its role in the normal conduct of brain function was of little importance [1]. There are other possibilities. It may be that many neurons in the brain have undifferentiated receptors which interact with and respond to a variety of drugs. The specificity of the narcotic antagonists as well as other antagonists argues against the undifferentiated receptor hypothesis. Some observations can be reconciled by assuming that there may be natural agonists for $\varkappa$ and possibly μ receptors in some functional systems in the nervous system but not in other systems that have μ and $\varkappa$ receptors. Thus the fact that antagonists do not produce pharmacologic effects in functional systems that are activated by μ and $\varkappa$ agonists may indicate a total absence of a natural agonist or the absence of a natural agonist in conjunction with its appropriate receptor. It is possible that in the evolution of receptors, variants are formed some of which may become functionally associated with a endogenous transmitter, others not. These latter types of receptors may be of special interest to pharmacologists for they may be activated by drugs and produce effects for which there is physiologic counterpart.

Substantive progress has been made in the identification of in vitro macromolecules that interact selectively with narcotic analgesics. GOLDSTEIN et al. (1971) studied a population of receptors that exhibited a saturable binding of levorphanol but not dextrorphanol. This receptor population accounted for only 2% of the binding of levorphanol, was present in nuclear, synaptosomal and mitochondrial and membrane fractions and was present in the cerebrum, thalamus, pons, medulla, and

[1] Since this chapter was written, a large body of data has been published indicating that there are brain and hypophyseal polypeptides that mimic many of the actions of morphine and related drugs on peripheral tissue, on in vitro receptors, and in the brain. These data have been recently reviewed by GOLDSTEIN [Science **193**, 1081—1085 (1976)] and SNYDER [Sci. Amer. **236**, 44—56 (1977)].

cerebellum. Further, binding of levorphanol was enhanced by EDTA and decreased by Ca^{++} and Mg^{++}. An AD_{50} of levorphanol occupied only a portion of these receptors. As has been previously suggested by MARTIN et al. (1973a), only a portion of the morphine receptors are occupied when pharmacologically active doses are administered to the nontolerant animal (see Fig. 9-III). Subsequent studies (LOWNEY et al., 1974) indicated that the receptor was a heterogeneous lipoprotein with several dissociation constants, and was present in the brain stem, cerebrum, but not the cerebellum. They further observed that although naloxone could antagonize the binding of levorphanol to the receptor, naloxone was preferentially bound to a different fraction than levorphanol. CHO et al. (1974) have found that cerebroside sulfate has many of the same chemical characteristics of the morphine receptor identified by LOWNEY et al. (1974).

PERT and SNYDER (1973a, b) defined a naloxone saturable receptor, whose binding could be decreased by levorphanol. In this regard, dextrorphan was 1/4000 as effective. The receptor was present in the corpus striatum, cerebral cortex, mesencephalon, and brain stem but not in the cerebellum. It was found to be present in a decreasing proportion in the mitochondrial-synaptosomal, microsomal, and nuclear fractions. Binding was decreased by Ca^{++} and Mg^{++} and increased by EDTA. The number of receptors increased with chronic administration of morphine (pellet implantation) and decreased rapidly following excision of the pellet. There was a poor correlation between the binding affinity of narcotics and narcotic antagonists for the receptor and their potencies (see Table 4) either as agonists or antagonists. SIMON et al. (1973) have identified a saturable binding site for etorphine and TERENIUS (1973) has identified a binding site for dihydromorphine for which binding was antagonized

Table 4. The relative potency of opioid agonists and antagonists as assessed in man and by mouse brain receptor (PERT and SNYDER, 1973)

Drug	Agonistic potency Morphine = 1		Antagonistic potency Naloxone = 1	
	Man	Mouse receptor	Man	Mouse receptor
Etorphine	300.0	23.0		
Cyclazocine	50.0	0.7	0.5	1.0
Hydromorphone	8.0	0.35		
Levorphanol	4.0	3.5		
Morphine	1.0	1.0		
Oxycodone	1.0	0.0002		
Methadone	1.0	0.23		
Nalorphine	1.0	2.3	0.1	3.0
Pentazocine	0.25	0.14		
Meperidine	0.14	0.007		
Codeine	0.1	0.0004		
d-Propoxyphene	0.04	0.007		
Dextrorphanol	0	0.001		
Naloxone	0	0.7	1.0	1.0
Levallorphan	?	7.0	0.3	10.0

Each value is expressed as the number of mg of morphine or naloxone equivalent to 1 mg of the compound.

by *l*-methadone but not *d*-methadone. LEE et al. (1973) found a small (5%) but not a statistically significant increase in saturable binding of dihydromorphine in tolerant rats. KLEE and STREATY (1974) found no evidence for a difference in the number of binding sites between tolerant and nontolerant rats.

Although receptors which bind narcotics and their antagonists share many properties in common with a hypothetical morphine receptor, there are also substantive differences. Further, there are apparent differences between receptors identified using different agonists and antagonists. The relationship between these saturable receptors which have a degree of stereospecificity and the pharmacologic effects of narcotics is tenuous. Nevertheless the identification of binding sites for narcotics in the central nervous system may allow the critical testing of some hypotheses about tolerance and physical dependence.

The relationship between tolerance, dependence, personality, and criminality remains the major psychiatric and medical problem of addiction. There is an increasing body of evidence that suggests that dependence on narcotic analgesics may produce long-persisting physiologic and behavioral changes following withdrawal which appear to be associated with the exacerbation of certain subjective states common to a variety of psychopaths.

Information has been obtained in the rat and man which indicates that the effects of morphine are different in subjects who have been previously tolerant to and dependent on narcotics. Further, there may be long persisting disorders in postaddicts that may be maternally transmitted to offspring. The role of hereditary factors in narcotic dependence has not been rigorously investigated as it has been for schizophrenia and depression. There is suggestive evidence that there may be hereditary influences in alcoholism and criminality (Sect. I, Chap. 1). The possibility that there may be drug inducible factors which may predispose to relapse and which may be transmittible to offspring may facilitate the search for chemotherapeutic measures for definitively treating narcotic addiction. DOLE and NYSWANDER (1968) have suggested that the narcotic addict may have a metabolic defect which may be responsible for continued drug-seeking behavior. This position is not greatly different from that of MARTIN (1968b) except for differences in opinion about the characteristics and etiology of the defect.

A fundamental action of most addictive drugs is to alter subjective states in a manner (euphoria) that is polarly opposite to some of the feeling states of the psychopath (hypophoria). The neuronal substrate for these feeling states (euphoria and hypophoria) is not known; however, there is suggestive evidence that catecholamines and related compounds may play a role. The narcotic analgesics have significant pharmacologic effects on the spinal cord, medulla, midbrain, hippocampus, septum, and cerebral cortex. No demonstrable actions of the narcotic analgesic have been reported on the cerebellum. These areas are all involved in feeling states or perception.

Abbreviations

ACh	Acetylcholine
AChE	Acetylcholine esterase
AD_{50}	Analgesic dose 50

ARCI	Addiction Research Center Inventory
CA	Catecholamine
cAMP	cyclic Adenosine-3′, 5′-monophosphate
COMT	Catechol O-methyl transferase
DA	Dopamine
DFP	Di-isopropyl phosphoroflouridate
DMPP	Dimethylphenylpiperazinium
DOPA	Dihydroxyphenylalanine
DOPAC	3,4-Dihydroxyphenyl acetic acid
E	Epinephrine
ED_{50}	Effective dose-50
EEG	Electroencephalogram
EPSP	Excitatory postsynaptic potential
FLA-63	Bis(4-methyl-1-homopiperazinyl-thiocarbonyl/) disulphide
GABA	γ-Aminobutyric acid
HC-3	Hemicholinium
5-HIAA	5-Hydroxyindole acetic acid
5-HT	5-Hydroxytryptamine
5-HTP	5-Hydroxytryptophan
HVA	Homovanillic acid
LSD	Lysergic acid diethylamide
MAO	Monoamine oxidase
α-MPT	Alpha-methylparatyrosine
α-MT	Alpha-methyltyrosine
NE	Norepinephrine
6-OHDA	6-Hydroxydopamine
pCPA	Parachlorophenylalanine
REMS	Rapid eye movement sleep
RNA	Ribonucleic acid
SP	Substance P
VMA	4-Hydroxy-3-methoxymandelic acid

References

Adams, W. J., Lorens, S. A., Mitchell, C. L.: Morphine enhances lateral hypothalamic self-stimulation in the rat. Proc. Soc. exp. Biol. (N.Y.) **140**, 770—771 (1972)

Adams, W. J., Yeh, S. Y., Woods, L. A., Mitchell, C. L.: Drug-test interaction as a factor in the development of tolerance to the analgesic effect of morphine. J. Pharmacol. exp. Ther. **168**, 251—257 (1969)

Adler, T. K.: The comparative potencies of codeine and its demethylated metabolites after intraventricular injection in the mouse. J. Pharmacol. exp. Ther. **140**, 155—161 (1963)

Ahtee, L., Kääriäinen, I.: 5-Hydroxytryptamine in platelets and brain of rabbits treated chronically with imipramine, morphine or methadone. Naunyn-Schmiedebergs Arch. exp. Path. Pharmacok. **277**, 429—436 (1973a)

Ahtee, L., Kääriäinen, I.: The effect of narcotic analgesics on the homovanillic acid content of rat nucleus caudatus. Europ. J. Pharmacol. **22**, 206—208 (1973b)

Ahtee, L., Kääriäinen, I., Paasonen, M. K.: Effect of nalorphine and antiparkinsonian drugs on methadone-induced rigidity; relation to homovanillic acid content of nucleus caudatus. Ann. Med. exp. Fenn. **50**, 180—185 (1972)

Akera,T., Brody,T.M.: The addiction cycle to narcotics in the rat and its relation to catecholamines. Biochem. Pharmacol. **17**, 675—688 (1968)
Albus,K., Herz,A.: Inhibition of behavioural and EEG activation induced by morphine acting on lower brain-stem structures. Electroenceph. clin. Neurophysiol. **33**, 537—545 (1972)
Albus,K., Schott,M., Herz,A.: Interaction between morphine and morphine antagonists after systemic and intraventricular application. Europ. J. Pharmacol. **12**, 53—64 (1970)
Algeri,S., Costa,E.: Physical dependence on morphine fails to increase serotonin turnover in rat brain. Biochem. Pharmacol. **20**, 877—884 (1971)
Altschul,S., Wikler,A.: Electroencephalogram during a cycle of addiction to keto-bemidone hydrochloride. Electroenceph. clin. Neurophysiol. **3**, 149—153 (1951)
Amit,Z., Corcoran,M.E., Amir,S., Urca,G.: Ventral hypothalamic lesions block the consumption of morphine in rats. Life Sci. **13**, 805—816 (1973)
Amsler,C.: Beiträge zur Pharmakologie des Gehirns. Naunyn-Schmiedebergs Arch. exp. Path. Pharmak. **97**, 1—14 (1923)
Andrews,H.L.: Brain potentials and morphine addiction. Psychosom. Med. **3**, 399—409 (1941)
Andrews,H.L.: Cortical effects of demerol. J. Pharmacol. exp. Ther. **76**, 89—94 (1942)
Andrews,H.L.: Changes in the electroencephalogram during a cycle of morphine addiction. Psychosom. Med. **5**, 143—147 (1943)
Andrews,H.L., Himmelsbach,C.K.: Relation of the intensity of the morphine abstinence syndrome to dosage. J. Pharmacol. exp. Ther. **81**, 288—293 (1944)
Aston,R.: Quantitative aspects of tolerance and posttolerance hypersensitivity to pentobarbital in the rat. J. Pharmacol. exp. Ther. **150**, 253—258 (1965)
Axelrod,J.: Possible mechanism of tolerance to narcotic drugs. Science **124**, 263—264 (1956)
Axelrod,J.: Cellular adaptation in the development of tolerance to drugs. In: Wikler,A. (Ed.): The addictive states. Ass. Res. nerv. ment. Dis., Vol.46, pp.247—264. Baltimore: Williams and Wilkins 1968
Axelrod,J., Cochin,J.: The inhibitory action of nalorphine on the enzymatic N-demethylation of narcotic drugs. J. Pharmacol. exp. Ther. **121**, 107—112 (1957)
Ayhan,I.H.: Effect of 6-hydroxydopamine on morphine analgesia. Psychopharmacologia (Berl.) **25**, 183—188 (1972)
Ayhan,I.H., Randrup,A.: Role of brain noradrenaline in morphine-induced stereotyped behaviour. Psychopharmacologia (Berl.) **27**, 203—212 (1972)
Ayhan,I.H., Randrup,A.: Behavioural and pharmacological studies on morphine-induced excitation of rats. Possible relation to brain catecholamines. Psychopharmacologia (Berl.) **29**, 317—328 (1973a)
Ayhan,I.H., Randrup,A.: Inhibitory effects of amphetamine, L-DOPA and apomorphine on morphine-induced behavioural excitation of rats. Arch. int. Pharmacodyn. **204**, 283—292 (1973b)
Azmitia,E.C., Jr.: Tryptophan hydroxylase changes in midbrain of the rat after chronic morphine administration. Life Sci. **9**, 633—637 (1970)
Bartlet,A.L.: The 5-hydroxytryptamine content of mouse brain and whole mice after treatment with some drugs affecting the central nervous system. Brit. J. Pharmacol. **15**, 140—146 (1960)
Beckett,A.H., Casy,A.F., Harper,N.J.: Analgesics and their antagonists: Some steric and chemical considerations. Part III. The influence of the basic group on the biological response. J. Pharm. Pharmacol. **8**, 874—883 (1956)
Bednarczyk,J.H., Smith,C.B., Sheldon,M.I., Villarreal,J.E.: Tolerance to the effects of morphine upon the incorporation of ^{14}C-tyrosine (^{14}C-TYR) into ^{14}C-catecholamines (^{14}C-CA) in mouse brain and adrenals. Pharmacologist **12**, 230 (1970)
Beecher,H.K.: Measurement of subjective responses; quantitative effects of drugs. New York: Oxford University Press 1959
Beleslin,D., Polak,R.L.: Depression by morphine and chloralose of acetylcholine release from the cat's brain. J. Physiol. (Lond.) **177**, 411—419 (1965)
Beleslin,D., Polak,R.L., Sproull,D.H.: The release of acetylcholine into the cerebral subarachnoid space of anesthetized cats. J. Physiol. (Lond.) **177**, 420—428 (1965)
Bell,J.A., Martin,W.R.: The effect of the narcotic antagonists naloxone, naltrexone and nalorphine on spinal cord C-fiber reflexes evoked by electrical stimulation or radiant heat. Europ. J. Pharmacol. **42**, 147—154 (1977)

Berkowitz, B., Spector, S.: Evidence for active immunity to morphine in mice. Science **178**, 1290—1292 (1972)

Berney, S. A., Buxbaum, D. M.: The effect of morphine on catecholamine turnover and its relationship to morphine-induced motor activity. Pharmacologist **15**, 202 (1973)

Bhargava, H. N., Afifi, A. H., Way, E. L.: Effect of chemical sympathectomy on morphine antinociception and tolerance development in the rat. Biochem. Pharmacol. **22**, 2769—2772 (1973)

Bhargava, H. N., Chan, S. L., Way, E. L.: Morphine analgesia, tolerance and physical dependence: Effect of intracerebral hemicholinium-3 (HC-3). Fed. Proc. **31**, 527 (1972a)

Bhargava, H. N., Chan, S. L., Way, E. L.: Influence of hemicholinum (HC-3) on morphine analgesia, tolerance, physical dependence and on brain acetylcholine. Europ. J. Pharmacol. **29**, 253—261 (1974)

Bhargava, H. N., Ho, I. K., Way, E. L.: Effect of dopamine beta-hydroxylase (DBH) inhibition on morphine analgesia, tolerance and physical dependence in mice. Fifth International Congress on Pharmacology, Abstracts of Volunteer Papers, p. 21, San Francisco, Calif., 1972b

Bhargava, H. N., Way, E. L.: Acetylcholinesterase inhibition and morphine effects in morphine tolerant and dependent mice. J. Pharmacol. exp. Ther. **183**, 31—40 (1972)

Biscoe, T. J., Duggan, A. W., Lodge, D.: Effect of etorphine, morphine and diprenorphine on neurones of the cerebral cortex and spinal cord of the rat. Brit. J. Pharmacol. **46**, 201—212 (1972)

Bläsig, J., Herz, A., Reinhold, K., Zieglgänsberger, S.: Development of physical dependence on morphine in respect to time and dosage and quantification of the precipitated withdrawal syndrome in rats. Psychopharmacologia (Berl.) **33**, 19—38 (1973b)

Bläsig, J., Reinhold, K., Herz, A.: Effect of 6-hydroxydopamine, 5,6-dihydroxytryptamine and raphe lesions on the antinociceptive actions of morphine in rats. Psychopharmacologia (Berl.) **31**, 111—119 (1973a)

Blosser, J. C., Catravas, G. N.: Action of reserpine in morphine-tolerant rats: Absence of an antagonism of catecholamine depletion. J. Pharmacol. exp. Ther. **191**, 284—289 (1974)

Bodo, R. C., Brooks, C. McC.: Effects of morphine on blood sugar and reflex activity in the chronic spinal cat. J. Pharmacol. exp. Ther. **61**, 82—88 (1937)

Bonnycastle, D. D., Bonnycastle, M. F., Anderson, E. G.: The effect of a number of central depressant drugs upon brain 5-hydroxytryptamine levels in the rat. J. Pharmacol. exp. Ther. **135**, 17—20 (1962)

Borison, H. L.: Influence of area postrema ablation on the emetic effect of adrenaline, apomorphine and pilocarpine administered by cerebral intraventricular injection in the cat. J. Physiol. (Lond.) **143**, 14P (1958)

Borison, H. L.: Effect of ablation of medullary emetic chemoreceptor trigger zone on vomiting responses to cerebral intraventricular injection of adrenaline, apomorphine and pilocarpine in the cat. J. Physiol. (Lond.) **147**, 172—177 (1959)

Borison, H. L., Fishburn, B. R., Bhide, N. K., McCarthy, L. E.: Morphine-induced hyperglycemia in the cat. J. Pharmacol. exp. Ther. **138**, 229—235 (1962)

Borison, H. L., Fishburn, B. R., McCarthy, L. E.: A possible receptor role of the subfornical organ in morphine-induced hyperglycemia. Neurology (Minneap.) **14**, 1049—1053 (1964)

Bosworth, D. M.: Iproniazid: A brief review of its introduction and clinical use. Ann. N.Y. Acad. Sci. **80**, 809—815 (1959)

Bowers, M. B., Jr., Kleber, H. D.: Methadone increases mouse brain 5-hydroxyindoleacetic acid. Nature (Lond.) **229**, 134—135 (1971)

Bowers, M. B., Jr., Kleber, H. D., Davis, L.: Acid monoamine metabolites in cerebrospinal fluid during methadone maintenance. Nature (Lond.) **232**, 581—582 (1971)

Bradley, P. B.: The effects of analgesic drugs on the electrical activity of the brain. In: Soulairac, A., Cahn, J., Charpentier, J. (Eds.): Pain (Proceedings of the International Symposium on Pain, Paris, April 11—13, 1967), pp. 411—423. New York: Academic Press 1968

Bradley, P. B., Dray, A.: Actions and interactions of microiontophoretically applied morphine with transmitter substances on brain stem neurones. Brit. J. Pharmacol. **47**, 642P (1973)

Brodie, B. B., Shore, P. A., Pletscher, A.: Serotonin-releasing activity limited to *Rauwolfia* alkaloids with tranquilizing action. Science **123**, 992—993 (1956)

Brownlee, G., Williams, G. W.: Potentiation of amphetamine and pethidine by monoamine oxidase inhibitors. Lancet **1963 I**, 669

Buckett, W. R.: A new test for morphine-like physical dependence (addiction liability) in rats. Psychopharmacologia (Berl.) **6**, 410—416 (1964)
Burks, T. F., Grubb, M. N.: Sites of acute morphine tolerance in intestine. J. Pharmacol. exp. Ther. **191**, 518—526 (1974)
Burks, T. F., Jaquette, D. L., Grubb, M. N.: Development of tolerance to the stimulatory effect of morphine in dog intestine. Europ. J. Pharmacol. **25**, 302—307 (1974)
Burks, T. F., Long, J. P.: Release of intestinal 5-hydroxytryptamine by morphine and related agents. J. Pharmacol. exp. Ther. **156**, 267—276 (1967)
Busch, H., Nair, P. V., Frank, M., Martin, W. R.: Comparative pharmacological effects of a number of β-halogenated pyruvic acids. J. Pharmacol. exp. Ther. **123**, 48—53 (1958)
Buxbaum, D.: Some behavioral and electrophysiological changes produced by low doses of morphine in the cat. Arch. Biol. Med. Exp. (Santiago) **4**, 154 (1967)
Buxbaum, D. M., Yarbrough, G. G., Carter, M. E.: Biogenic amines and narcotic effects. I. Modification of morphine-induced analgesia and motor activity after alterations of cerebral amine levels. J. Pharmacol. exp. Ther. **185**, 317—327 (1973)
Cabanac, M.: Physiological role of pleasure. Science **173**, 1103—1107 (1971)
Cahen, R. L., Wikler, A.: Effects of morphine on cortical electrical activity of the rat. Yale J. Biol. Med. **16**, 239—243 (1944)
Cairnie, A. B., Kosterlitz, H. W.: The action of morphine on impulse propagation in nerve fibres. Int. J. Neuropharmacol. **1**, 133—136 (1962)
Cairnie, A. B., Kosterlitz, H. W., Taylor, D. W.: Effect of morphine on some sympathetically innervated effectors. Brit. J. Pharmacol. **17**, 539—551 (1961)
Calcutt, C. R., Doggett, N. S., Spencer, P. S. J.: Modification of the antinociceptive activity of morphine by centrally administered ouabain and dopamine. Psychopharmacologia (Berl.) **21**, 111—117 (1971)
Calcutt, C. R., Spencer, P. S. J.: Activities of narcotic and narcotic-antagonist analgesics following the intraventricular injection of various substances. Brit. J. Pharmacol. **41**, 401P—402P (1971)
Carlsson, A., Lindqvist, M.: Central and peripheral monoaminergic membrane-pump blockade by some addictive analgesics and antihistamines. J. Pharm. Pharmacol. **21**, 460—464 (1969).
Carruyo, L., Florio, V., Longo, V. G., Scotti de Carolis, A.: The effect of narcotics and narcotic-antagonists on the electrical activity of the brain: Its relationship with their pain-obtunding activity. In: Soulairac, A., Cahn, J., Charpentier, J. (Eds.): Pain (Proceedings of the International Symposium on Pain, Paris, April 11—13, 1967), pp. 425—439. New York: Academic Press 1968
Cesarman, T.: Iproniazid in cardiovascular therapy, with special reference to its action in angina pectoris and blood pressure. Ann. N.Y. Acad. Sci. **80**, 988—1008 (1959)
Charpentier, J.: Variations des réponses à la douleur des rats traités par differéntes substances de type adrénergique et cholinergique. C. R. Soc. Biol. (Paris) **155**, 1490 (1961)
Charpentier, J.: Etude neuro-pharmacologique et électrophysiologique du comportement a la douleur chez le rat. Paris: Librairie Arnette 1965
Chen, G.: The anti-tremorine effect of some drugs as determined by Haffner's method of testing analgesia in mice. J. Pharmacol. exp. Ther. **124**, 73—76 (1958)
Cheney, D. L., Goldstein, A.: The effect of *p*-chlorophenylalanine on opiate-induced running, analgesia, tolerance and physical dependence in mice. J. Pharmacol. exp. Ther. **177**, 309—315 (1971a)
Cheney, D. L., Goldstein, A.: Tolerance to opioid narcotics; Time course and reversibility of physical dependence in mice. Nature (Lond.) **232**, 477—478 (1971b)
Cheney, D. L., Goldstein, A., Algeri, S., Costa, E.: Narcotic tolerance and dependence: Lack of relationship with serotonin turnover in the brain. Science **171**, 1169—1170 (1971)
Cheney, D. L., Goldstein, A., Sheehan, P.: Rate of development and reversibility of brain tolerance and physical dependence in mice treated with opiates. Fed. Proc. **29**, 685 (1970)
Chin, J. H., Crankshaw, D. P., Kendig, J. J.: Changes in the dorsal root potential with diazepam and with the analgesics aspirin, nitrous oxide, morphine and meperidine. Neuropharmacology **13**, 305—315 (1974)
Chin, J. H., Domino, E. F.: Effects of morphine on brain potentials evoked by stimulation of the tooth pulp of the dog. J. Pharmacol. exp. Ther. **132**, 74—86 (1961)

Chiosa, L., Dumitrescu, S., Banaru, A.: The influence of imipramine on the "abstinence syndrome" to morphine in mice. Int. J. Neuropharmacol. **7**, 161—164 (1968)

Cho, T. M., Wu, Y. C., Loh, H. H., Way, E. L.: Stereospecific binding of narcotics to cerebroside sulfates. Pharmacologist **16**, 248 (1974)

Christian, S. T.: Enzymes. In: Mulé, S. J., Brill, H. (Eds.): Chemical and biological aspects of drug dependence, pp. 449—463. Cleveland, Ohio: Chemical Rubber Co. Press 1972

Cicero, T. J.: Effects of α-adrenergic blocking agents on narcotic-induced analgesia. Arch. int. Pharmacodyn. **208**, 5—13 (1974)

Cicero, T. J., Meyer, E. R.: Morphine pellet implantation in rats: Quantitative assessment of tolerance and dependence. J. Pharmacol. exp. Ther. **184**, 404—408 (1973)

Cicero, T. J., Meyer, E. R., Bell, R. D.: Effects of phenoxybenzamine on the narcotic withdrawal syndrome in the rat. Neuropharmacology **13**, 601—607 (1974a)

Cicero, T. J., Wilcox, C. E., Meyer, E. R.: Effect of α-adrenergic blockers on naloxone-binding in brain. Biochem. Pharmacol. **23**, 2349—2352 (1974b)

Cicero, T. J., Wilcox, C. E., Smithloff, B. R., Meyer, E. R., Sharpe, L. G.: Effects of morphine, in vitro and in vivo, on tryosine hydroxylase activity in rat brain. Biochem. Pharmacol. **22**, 3237—3246 (1973)

Ciofalo, F. R.: Inhibition of synaptosomal 5-HT uptake by narcotics. Proc. West. Pharmacol. Soc. **16**, 38 (1973)

Clouet, D. H. (ed.): Narcotic drugs; biochemical pharmacology. New York-London: Plenum Press 1971

Clouet, D. H., Neidle, A.: The effect of morphine on the transport and metabolism of intracisternally-injected leucine in the rat. J. Neurochem. **17**, 1069—1074 (1970)

Clouet, D. H., Ratner, M.: Catecholamine biosynthesis in brains of rats treated with morphine. Science **168**, 854—856 (1970)

Cochin, J., Axelrod, J.: Biochemical and pharmacological changes in the rat following chronic administration of morphine, nalorphine and normorphine. J. Pharmacol. exp. Ther. **125**, 105—110 (1959)

Cochin, J., Kornetsky, C.: Development and loss of tolerance to morphine in the rat after single and multiple injections. J. Pharmacol. exp. Ther. **145**, 1—10 (1964)

Cohen, M., Keats, A. S., Krivoy, W., Ungar, G.: Effect of actinomycin D on morphine tolerance. Proc. Soc. exp. Biol. (N.Y.) **119**, 381—384 (1965)

Colasanti, B., Khazan, N.: Antagonism of the acute electroencephalographic and behavioral effects of morphine in the rat by depletion of brain biogenic amines. Neuropharmacology **12**, 463—469 (1973a)

Colasanti, B., Khazan, N.: Agonistic properties of narcotic analgesics and antagonists on the electroencephalogram and behavior in the rat and their reversal by naloxone. Neuropharmacology **12**, 619—627 (1973b)

Collier, H. O. J.: A general theory of the genesis of drug dependence by induction of receptors. Nature (Lond.) **205**, 181—182 (1965)

Collier, H. O. J.: The concept of the quasi-abstinence effect and its use in the investigation of dependence mechanisms. Pharmacology **11**, 58—61 (1974)

Collier, H. O. J., Francis, D. L., Schneider, C.: Modification of morphine withdrawal by drugs interacting with humoral mechanisms: Some contradictions and their interpretation. Nature (Lond.) **237**, 220—222 (1972)

Collier, H. O. J., McDonald-Gibson, W. J., Saeed, S. A.: Apomorphine and morphine stimulate prostaglandin biosynthesis. Nature (Lond.) **252**, 56—58 (1974)

Collier, H. O. J., Roy, A. C.: Morphine-like drugs inhibit the stimulation of E prostaglandins of cyclic AMP formation by rat brain homogenate. Nature (Lond.) **248**, 24—27 (1974)

Contreras, E., Castillo, S., Quijada, L.: Effect of drugs that modify 3′,5′-AMP concentrations on morphine analgesia. J. Pharm. Pharmacol. **24**, 65—66 (1972)

Contreras, E., Quijada, L., Tamayo, L.: A comparative study of the effects of reserpine and *p*-chlorophenylalanine on morphine analgesia in mice. Psychopharmacologia (Berl.) **28**, 319—324 (1973a)

Contreras, E., Tamayo, L.: Effects of drugs acting in relation to sympathetic functions on the analgesic action of morphine. Arch. int. Pharmacodyn. **160**, 312—320 (1966)

Contreras, E., Tamayo, L., Quijada, L., Silva, E.: Decrease of tolerance development to morphine by 5-hydroxytryptophan and some related drugs. Europ. J. Pharmacol. **22**, 339—343 (1973b)
Cook, L., Bonnycastle, D.D.: An examination of some spinal and ganglionic actions of analgetic materials. J. Pharmacol. exp. Ther. **109**, 35—44 (1953)
Cools, A.R., Janssen, H.J., Broekkamp, C.L.E.: The differential role of the caudate nucleus and the linear raphé nucleus in the initiation and the maintenance of morphine-induced behaviour in cats. Arch. int. Pharmacodyn. **210**, 163—174 (1974)
Corrado, A.P., Longo, V.G.: An electrophysiological analysis of the convulsant action of morphine, codeine and thebaine. Arch. int. Pharmacodyn. **132**, 255—269 (1961)
Corssen, G., Domino, E.F.: Visually evoked responses in man: A method for measuring cerebral effects of preanesthetic medication. Anesthesiology **25**, 330—341 (1964)
Cossio, P.: Contribution of iproniazid to the treatment of angina pectoris. Ann. N.Y. Acad. Sci. **80**, 1009—1015 (1959)
Costall, B., Naylor, R.J.: A role for the amygdala in the development of the cateleptic and stereotypic actions of the narcotic agonists and antagonists in the rat. Psychopharmacologia (Berl.) **35**, 203—213 (1974)
Cowan, A., Ghezzi, D., Samanin, R.: Effect of midbrain raphé lesion and of 6-hydroxydopamine on the antinociceptive action of buprenorphine in rats. Arch. int. Pharmacodyn. **208**, 302—305 (1974)
Cox, B.M., Ginsburg, M., Osman, O.H.: Acute tolerance to narcotic analgesic drugs in rats. Brit. J. Pharmacol. **33**, 245—256 (1968)
Cox, B.M., Opheim, K.E., Teschemacher, H., Goldstein, A.: A peptide-like substance from pituitary that acts like morphine. 2. Purification and properties. Life Sci. **16**, 1777—1782 (1975)
Cox, B.M., Osman, O.H.: The role of protein synthesis inhibition in the prevention of morphine tolerance. Brit. J. Pharmacol. **35**, 373—374P (1969)
Cox, B.M., Osman, O.H.: Inhibition of the development of tolerance to morphine in rats by drugs which inhibit ribonucleic acid or protein synthesis. Brit. J. Pharmacol. **38**, 157 (1970)
Cox, B.M., Weinstock, M.: The effect of analgesic drugs on the release of acetylcholine from electrically stimulated guinea-pig ileum. Brit. J. Pharmacol. **27**, 81—92 (1966)
Cox, C., Smart, R.G.: Social and psychological aspects of speed use—A study of types of speed users in Toronto. Int. J. Addict. **7**, 201—217 (1972)
Crawford, T.B.B., Law, W.: The urinary excretion of adrenaline and noradrenaline by rats under various experimental conditions. Brit. J. Pharmacol. **13**, 35—43 (1958)
Crepax, P., Infantellina, F.: Azione della morfina sul lembo isolato di corteccia cerebrale di gatto. Arch. Sci. biol. (Bologna) **40**, 147—162 (1956)
Crepax, P., Infantellina, F.: Effetti dell' applicazione locale di morfina sull'attivita elettrica del lembo isolato di corteccia cerebrale di cane predisposto ovvero non predisposto alla epilessia reflessa. Arch. Sci. biol. (Bologna) **42**, 415—441 (1958)
Crossland, J., Slater, P.: The effect of some drugs on the "free" and "bound" acetylcholine content of rat brain. Brit. J. Pharmacol. **33**, 42—47 (1968)
Curtis, D.R., Duggan, A.W.: The depression of spinal inhibition by morphine. Agents Actions **1**, 14—19 (1969)
Curtis, D.R., Hösli, L., Johnston, G.A.R.: A pharmacological study of the depression of spinal neurones by glycine and related amino acids. Exp. Brain Res. **6**, 1—18 (1968)
Dabrowski, R.: Adaptation of rats to morphine. Bull. Acad. pol. Sci. Cl. 2 **14**, 663—666 (1966)
Dandiya, P.C., Menon, M.K.: Studies on central nervous system depressants (III). Influence of some tranquillizing agents on morphine analgesia. Arch. int. Pharmacodyn. **141**, 223—232 (1963)
Datta, K., Thal, L., Wajda, I.J.: Effects of morphine on choline acetyltransferase levels in the caudate nucleus of the rat. Brit. J. Pharmacol. **41**, 84—93 (1971)
Davies, J., Duggan, A.W.: Opiate agonist-antagonist effects on Renshaw cells and spinal interneurones. Nature (Lond.) **250**, 70—71 (1974)
Davis, W.M., Brister, C.C.: Acute effects of narcotic analgesics on behavioral arousal in the rat. J. pharm. Sci. **62**, 974—979 (1973)
Davis, W.M., Khalsa, J.H.: Increased shock induced aggression during morphine withdrawal. Life Sci. **10**, 1321—1327 (1971)

Deanesly, R.: The histology of adrenal enlargement under experimental conditions. Amer. J. Anat. **47**, 475—509 (1931)

Defalque, R. J.: Potentiation of morphine by intraperitoneal injections of iproniazid in rabbits. Anesthet. et Analg. **44**, 190—193 (1965)

Deneau, G. A., Seevers, M. H.: Evaluation of new compounds for morphine-like physical dependence in the Rhesus monkey. Addendum 1, Minutes of 26th meeting of Committee on Drug Addiction and Narcotics, National Research Council, Washington, D.C. 1964

De Salva, S. J., Oester, Y. T.: The effects of central depressants on certain spinal reflexes in the acute high cervical cat. Arch. int. Pharmacodyn. **124**, 255—262 (1960)

Dewey, W. L., Harris, L. S.: Antagonistic activity of morphine and other narcotics in the mouse locomotor activity test. Pharmacologist **15**, 167 (1973)

Dewey, W. L., Harris, L. S., Howes, J. F., Nuite, J. A.: The effect of various neurohumoral modulators on the activity of morphine and the narcotic antagonists in the tail-flick and phenylquinone tests. J. Pharmacol. exp. Ther. **175**, 435—442 (1970)

Dhasmana, K. M., Dixit, K. S., Jaju, B. P., Gupta, M. L.: Role of central dopaminergic receptors in manic response of cats to morphine. Psychopharmacologia (Berl.) **24**, 380—383 (1972)

Dole, V. P.: Biochemistry of addiction. Ann. Rev. Biochem. **39**, 821—840 (1970)

Dole, V. P., Nyswander, M. E.: Methadone maintenance and its implication for theories of narcotic addiction. In: Wikler, A. (Ed.): The addictive states. Ass. Res. nerv. ment. Dis., Vol. 46, pp. 359—366. Baltimore: Williams and Wilkins 1968

Domino, E. F.: Effects of narcotic analgesics on sensory input, activating system and motor output. In: Wikler, A. (Ed.): The addictive states. Ass. Res. nerv. ment. Dis., Vol. 46, pp. 117—147. Baltimore: Williams and Wilkins 1968

Domino, E. F., Wilson, A.: Effects of narcotic analgesic agonists and antagonists on rat brain acetylcholine. J. Pharmacol. exp. Ther. **184**, 18—32 (1973a)

Domino, E. F., Wilson, A. E.: Enhanced utilization of brain acetylcholine during morphine withdrawal in the rat. Nature (Lond.) **243**, 285—286 (1973b)

Dressler, W. E., D'Alonzo, G., Rossi, G. V., Orzechowski, R. F.: Modification of certain vascular responses to dopamine by morphine. Europ. J. Pharmacol. **28**, 108—113 (1974)

Duggan, A. W., Curtis, D. R.: Morphine and the synaptic activation of Renshaw cells. Neuropharmacology **11**, 189—196 (1972)

Echols, S. D., Jewett, R. E.: Effects of morphine on the sleep of cats. Pharmacologist **11**, 254 (1969)

Echols, S. D., Jewett, R. E.: Effects of morphine on sleep in the cat. Psychopharmacologia (Berl.) **24**, 435—448 (1972)

Eddy, N. B.: The phenomena of tolerance. In: Sevag, M. G., Reid, R. D., Reynolds, O. E. (Eds.): Origins of resistance to toxic agents, pp. 223—243. New York: Academic Press 1955

Eddy, N. B., Halbach, H., Isbell, H., Seevers, M. H.: Drug dependence: Its significance and characteristics. Bull. Wld Hlth Org. **32**, 721—733 (1965)

Ehrenpreis, S., Light, I., Schonbuch, G. H.: Use of the electrically stimulated guinea pig ileum to study potent analgesics. In: Singh, J. Mm., Miller, L., Lal, H. (Eds.): Drug addiction: Experimental pharmacology. Vol. 1, pp. 319—342. Mount Kisco, New York: Futura Publishing Co. Inc. 1972

Eidelberg, E., Barstow, C. A.: Morphine tolerance and dependence induced by intraventricular injection. Science **174**, 74—76 (1971)

Eidelberg, E., Bond, M. L.: Effects of morphine and antagonists on hypothalamic cell activity. Arch. int. Pharmacodyn. **196**, 16—24 (1972)

Eidelberg, E., Erspamer, R.: Failure of naloxone to prevent acute morphine tolerance and dependence. Arch. int. Pharmacodyn. **211**, 58—63 (1974)

Eidelberg, E., Schwartz, A. S.: Possible mechanism of action of morphine on brain. Nature (Lond.) **225**, 1152—1153 (1970)

Eisenman, A. J., Fraser, H. F., Brooks, J. W.: Urinary excretion and plasma levels of 17-hydroxycorticosteroids during a cycle of addiction to morphine. J. Pharmacol. exp. Ther. **132**, 226—231 (1961)

Eisenman, A. J., Fraser, H. F., Sloan, J. W., Isbell, H.: Urinary 17-ketosteroid excretion during a cycle of addiction to morphine. J. Pharmacol. exp. Ther. **124**, 305—311 (1958)

Eisenman, A. J., Sloan, J. W., Martin, W. R., Jasinski, D. R., Brooks, J. W.: Catecholamine and 17-hydroxycorticosteroid excretion during a cycle of morphine dependence in man. J. psychiat. Res. **7**, 19—28 (1969)
Elchisak, M. A., Rosecrans, J. A.: Effect of central catecholamine depletions by 6-hydroxydopamine on morphine antinociception in rats: Involvement of brain dopamine. Pharmacologist **15**, 167 (1973)
Ellinwood, E. H.: Amphetamine psychosis: I. Description of the individuals and process. J. nerv. ment. Dis. **144**, 273—283 (1967)
Elliott, T. R.: The control of the suprarenal glands by the splanchnic nerves. J. Physiol. (Lond.) **44**, 374—409 (1912)
Emile, J. F., Shanaman, J., Warren, M. R.: The analgesic activity of phenelzine and other compounds. J. Pharmacol. exp. Ther. **134**, 206—209 (1961)
Emmelin, N.: Supersensitivity following "pharmacological denervation". Pharmacol. Rev. **13**, 17—37 (1961)
Emmelin, N., Strömblad, R.: Adrenaline and noradrenaline content of the suprarenals of cats in chloralose and morphine-ether anesthesia. Acta. physiol. scand. **24**, 261—266 (1951)
Ercoli, N., Lewis, M. N.: Studies on analgesics. I. The time-action curves of morphine, codeine, dilaudid and demerol by various methods of administration. II. Analgesic activity of acetylsalicyclic acid and aninopyrine. J. Pharmacol. exp. Ther. **84**, 301—317 (1945)
Estler, C. J.: Effect of α- and β-adrenergic blocking agents and para-chlorophenylalanine on morphine- and caffeine-stimulated locomotor activity of mice. Psychopharmacologia (Berl.) **28**, 261—268 (1973)
Feinberg, M. P., Cochin, J.: Inhibition of development of tolerance to morphine by cycloheximide. Biochem. Pharmacol. **22**, 3082—3085 (1972)
Feldberg, W., Gupta, K. P.: Morphine hyperglycaemia; the site of action. J. Physiol. (Lond.) **224**, 85P—86P (1972)
Feldberg, W., Shaligram, S. V.: The hyperglycaemic effect of morphine. Brit. J. Pharmacol. **46**, 602—618 (1972)
Feldberg, W., Sherwood, S. L.: Injections of drugs into the lateral ventricle of the cat. J. Physiol. (Lond.) **123**, 148—167 (1954)
Felpel, L. P., Sinclair, J. G., Yim, G. K. W.: Effects of morphine on Renshaw cell activity. Neuropharmacology **9**, 203—210 (1970)
Fennessy, M. R., Lee, J. R.: Modification of morphine analgesia by drugs affecting adrenergic and tryptaminergic mechanisms. J. Pharm. Pharmacol. **22**, 930—935 (1970)
Fennessy, M. R., Lee, J. R.: Comparison of the dose-response effects of morphine on brain amines, analgesia and activity in mice. Brit. J. Pharmacol. **45**, 240—248 (1972)
Ferri, S., Santagostino, A., Braga, P. C., Galatulas, I.: Decreased antinociceptive effect of morphine in rats treated intraventricularly with prostaglandin E_1. Psychopharmacologia (Berl.) **39**, 231—235 (1974)
Fichtenberg, D. G.: Study of experimental habituation to morphine. Bull. Narcot. **3**, 19—42 (1951)
Flórez, J., Delgado, G., Armijo, J. A.: Adrenergic and serotonergic mechanisms in morphine-induced respiratory depression. Psychopharmacoligia (Berl.) **24**, 258—274 (1972)
Flórez, J., Delgado, G., Armijo, J. A.: Brain amines and development of acute tolerance to and dependence on morphine in the respiratory center of decerebrate cats. Neuropharmacology **12**, 355—362 (1973)
Fog, R.: Behavioural effects in rats of morphine and amphetamine and of a combination of the two drugs. Psychopharmacologia (Berl.) **16**, 305—312 (1970)
Foltz, E. L., White, L. E., Jr.: Experimental cingulumotomy and modification of morphine withdrawal. J. Neurosurg. **14**, 655—673 (1957)
Foster, R. S., Jenden, D. J., Lomax, P.: A comparison of the pharmacologic effects of morphine and N-methyl morphine. J. Pharmacol. exp. Ther. **157**, 185—195 (1967)
Fraser, H. F., Eisenman, A. J., Brooks, J. W.: Urinary excretion of 5HIAA and corticoids after morphine, meperidine, nalorphine, reserpine and chlorpromazine. Fed. Proc. **16**, 298 (1957)
Fraser, H. F., Isbell, H.: Comparative effects of 20 mgm. of morphine sulfate on non-addicts and former morphine addicts. J. Pharmacol. exp. Ther. **105**, 498—502 (1952)

Fraser, H. F., Isbell, H.: Chlorpromazine and reserpine. (A) Effects of each and of combinations of each with morphine. (B) Failure of each in treatment of abstinence from morphine. Arch. Neurol. Psychiat. (Chic.) **76**, 257—262 (1956)

Fraser, H. F., Isbell, H.: Human pharmacology and addiction liabilities of phenazocine and levophenacylmorphan. Bull. Narcot. **12**, 15—23 (1960)

Fraser, H. F., Wikler, A., Van Horn, G. D., Eisenman, A. J., Isbell, H.: Human pharmacology and addiction liability of normorphine. J. Pharmacol. exp. Ther. **122**, 359—369 (1958)

Fraser, H. F., Van Horn, G. D., Martin, W. R., Wolbach, A. B., Isbell, H.: Methods for evaluating addiction liability. (A) "Attitude" of opiate addicts toward opiate-like drugs. (B) A short-term "direct" addiction test. J. Pharmacol. exp. Ther. **133**, 371—387 (1961)

Frazier, D. T., Murayama, K., Abbott, N. J., Narahashi, T.: Effects of morphine on internally perfused squid giant axons. Proc. Soc. exp. Biol. (N.Y.) **139**, 434—438 (1972)

Frederickson, R. C. A., Pinsky, C.: Morphine impairs acetylcholine release but facilitates acetylcholine action at a skeletal neuromuscular junction. Nature (Lond.) New Biol. **231**, 93—94 (1971)

Freedman, D. X., Fram, D. H., Giarman, N. J.: The effect of morphine on the regeneration of brain norepinephrine after reserpine. Fed. Proc. **20**, 321 (1961)

Friebel, H., Kuhn, H. F.: Über den Nachweis von "physical dependence" bei Codein-behandelten Meerschweinchen. Med. Pharmacol. exp. **12**, 92—96 (1965)

Friedler, G.: The role of immune factors in the development of tolerance to morphine. Dissertation for degree of Doctor of Philosophy, Boston University Graduate School, Boston, Massachusetts 1968

Friedler, G.: Morphine administration to male mice: Effects on subsequent progeny. Fed. Proc. **33**, 515 (1974a)

Friedler, G.: Effect of pregestational morphine administration to mice on behavior of their offspring. Pharmacologist **16**, 203 (1974b)

Friedler, G., Bhargava, H. N., Quock, R.; Way, E. L.: The effect of 6-hydroxy-dopamine on morphine tolerance and physical dependence. J. Pharmacol. exp. Ther. **183**, 49—55 (1972)

Friedler, G., Cochin, J.: Growth retardation in offspring of female rats treated with morphine prior to conception. Science **175**, 654—656 (1972)

Friend, F. J., Harris, S. C.: The effect of adrenalectomy on morphine analgesia in rats. J. Pharmacol. exp. Ther. **93**, 161—167 (1948)

Fujita, S., Yasuhara, M., Ogiu, K.: Studies on sites of action of analgesics. 1. The effect of analgesics on afferent pathways of several nerves. Jap. J. Pharmacol. **3**, 27—38 (1953)

Fujita, S., Yasuhara, M., Yamamoto, S., Ogiu, K.: Studies on sites of action of analgesics. 2. The effect of analgesics on afferent pathways of pain. Jap. J. Pharmacol. **4**, 41—51 (1954)

Fukui, K., Shiomi, H., Takagi, H.: Effect of morphine on tyrosine hydroxylase activity in mouse brain. Europ. J. Pharmacol. **19**, 123—125 (1972)

Fukui, K., Takagi, H.: Effect of morphine on the cerebral contents of metabolites of dopamine in normal and tolerant mice: its possible relation to analgesic action. Brit. J. Pharmacol. **44**, 45—51 (1972)

Gaddum, J. H., Picarelli, Z. P.: Two kinds of tryptamine receptor. Brit. J. Pharmacol. **12**, 323—328 (1957)

Gaddum, J. H., Vogt, M.: Some central actions of 5-hydroxytryptamine and various antagonists. Brit. J. Pharmacol. **11**, 175—179 (1956)

Gangloff, H., Monnier, M.: Démonstration de l'impact des analgésiques morphiniques et d'une substance antimorphinique sur le cortex et le subcortex du lapin (thalamus, rhinencéphale et substance réticulée). Helv. physiol. pharmacol. Acta **13**, C47—C48 (1955)

Gangloff, H., Monnier, M.: The topical action of morphine, levorphanol (levorphan) and the morphine antagonist levallorphan on the unanesthetized rabbit's brain. J. Pharmacol. exp. Ther. **121**, 78—95 (1957)

Gardella, J. L., Izquierdo, J. A.: The analgesic action of catecholamines and of pyrogallol. Europ. J. Pharmacol. **10**, 87—90 (1970)

Gardiner, T. H., Eberhart, G.: Effects of L-trypotophan and DL-hydroxytryptophan on narcotic analgesia in mice. Fed. Proc. **29**, 685 (1970)

Gauchy, C., Agid, Y., Glowinski, J., Cheramy, A.: Acute effects of morphine on dopamine synthesis and release and tyrosine metabolism in the rat striatum. Europ. J. Pharmacol. **22**, 311—319 (1973)
Gebhart, G. F., Mitchell, C. L.: Further studies on the development of tolerance to the analgesic effect of morphine: The role played by the cylinder in the hot plate testing procedure. Arch. int. Pharmacodyn. **191**, 96—103 (1971)
Gebhart, G. F., Mitchell, C. L.: The effect of adrenalectomy on morphine analgesia and tolerance development in rats. Europ. J. Pharmacol. **18**, 37—42 (1972a)
Gebhart, G. F., Mitchell, C. L.: The relative contributions of the testing cylinder and the heated plate in the hot plate procedure to the development of tolerance to morphine in rats. Europ. J. Pharmacol. **18**, 56—62 (1972b)
Gebhart, G. F., Mitchell, C. L.: Strain differences in the analgesic response to morphine as measured on the hot plate. Arch. int. Pharmacodyn. **201**, 128—140 (1973)
Gebhart, G. F., Sherman, A. D., Mitchell, C. L.: The influence of stress on tolerance development to morphine in rats tested on the hot plate. Arch. int. Pharmacodyn. **197**, 328—337 (1972)
Genovese, E., Zonta, N., Mantegazza, P.: Decreased antinociceptive activity of morphine in rats pretreated intraventricularly with 5,6-dihydroxytryptamine, a long-lasting selective depletor of brain serotonin. Psychopharmacologia (Berl.) **32**, 359—364 (1973)
Gianutsos, G., Hynes, M. D., Drawbaugh, R., Lal, H.: Morphine-withdrawal aggression during protracted abstinence: Role of latent dopaminergic supersensitivity. Pharmacologist **15**, 218 (1973)
Gianutsos, G., Hynes, M. D., Puri, S. K., Drawbaugh, R. B., Lal, H.: Effect of apomorphine and nigrostriatal lesions on aggression and striatal dopamine turnover during morphine withdrawal.: Evidence for dopaminergic supersensitivity in protracted abstinence. Psychopharmacologia (Berl.) **34**, 37—44 (1974)
Giarman, N. J., Pepeu, G.: Drug-induced changes in brain acetylcholine. Brit. J. Pharmacol. **19**, 226—234 (1962)
Gibbs, F. A., Gibbs, E. L., Lennox, W. G.: Effect on the electroencephalogram of certain drugs which influence nervous activity. Arch. intern. Med. **60**, 154—166 (1937)
Gilbert, J. G., Lombardi, D. N.: Personality characteristics of young male narcotic addicts. J. cons. Psychol. **31**, 536—538 (1967)
Gilbert, P. E., Martin, W. R.: The effects of morphine- and nalorphine-like drugs in the nondependent, morphine-dependent and cyclazocine-dependent chronic spinal dog. J. Pharmacol. exp. Ther. **198**, 66—82 (1976)
Goetzl, F. R., Burrill, D. Y., Ivy, A. C.: Observations on the analgesic effect of morphine during continued daily administration of small and uniform doses to dogs. J. Pharmacol. exp. Ther. **82**, 110—119 (1944)
Goldstein, A., Goldstein, D. B.: Enyzme expansion theory of drug tolerance and physical dependence. In: Wikler, A. (Ed.): The addictive states. Ass. Res. nerv. ment. Dis., Vol. 46, pp. 265—267. Baltimore: Williams and Wilkins Co. 1968
Goldstein, A., Lowney, I., Pal, B. K.: Stereospecific and nonspecific interactions of the morphine congener levorphanol in subcellular fractions of mouse brain. Proc. nat. Acad. Sci. (Wash.) **68**, 1742—1747 (1971)
Goldstein, A., Schulz, R.: Morphine-tolerant longitudinal muscle strip from guinea pig ileum. Brit. J. Pharmacol. **48**, 655—666 (1973)
Goldstein, A., Sheehan, P.: Tolerance to opioid narcotics. I. Tolerance to the "running fit" caused by levorphanol in the mouse. J. Pharmacol. exp. Ther. **169**, 175—184 (1969)
Goldstein, D. B., Goldstein, A.: Possible role of enzyme inhibition and repression in drug tolerance and addiction. Biochem. Pharmacol. **8**, 48 (1961)
Goldstein, L., Aldunate, J.: Quantitative electroencephalographic studies on the effects of morphine and nalorphine on rabbit brain. J. Pharmacol. exp. Ther. **130**, 204—211 (1960)
Goode, P. G.: An implanted reservoir of morphine solution for rapid induction of physical dependence in rats. Brit. J. Pharmacol. **41**, 558—566 (1971)
Goodlet, I., Sugrue, M. F.: Effect of acutely administered analgesic drugs on rat brain serotonin turnover. Europ. J. Pharmacol. **29**, 241—248 (1974)
Gorlitz, B.-D., Frey, H.-H.: Central monoamines and antinociceptive drug action. Europ. J. Pharmacol. **20**, 171—180 (1972)

Grecksch, G., Ott, T., Matthies, H.: Effects of orotic acid on the development of morphine tolerance. Psychopharmacologia (Berl.) **36**, 337—346 (1974)

Green, A. F., Young, P. A.: A comparison of heat and pressure analgesiometric methods in rats. Brit. J. Pharmacol. **6**, 572—587 (1951)

Gross, E. G., Holland, H., Carter, H. R., Christensen, E. M.: The role of epinephrine in analgesia. Anesthesiology **9**, 459—471 (1948)

Grossman, W., Jurna, I., Nell, T., Theres, C.: The dependence of the antinociceptive effect of morphine and other analgesic agents on spinal motor activity after central monoamine depletion. Europ. J. Pharmacol. **24**, 67—77 (1973)

Grumbach, L.: A hypothesis of opiate action. Presented at 23rd meeting, Committee on Drug Addiction and Narcotics, National Research Council, New York, New York 1961

Grumbach, L.: The effect of cholinergic and cholinergic-blocking drugs on the abstinence syndrome in the rat. Fed. Proc. **28**, 262 (1969)

Gunne, L.-M.: Noradrenaline and adrenaline in the rat brain during acute and chronic morphine administration and during withdrawal. Nature (Lond.) **184**, 1950—1951 (1959)

Gunne, L.-M.: The temperature response in rats during acute and chronic morphine administration. A study of morphine tolerance. Arch. int. Pharmacodyn. **129**, 416—428 (1960)

Gunne, L.-M.: The excretion of noradrenaline and adrenaline in the urine of rats during chronic morphine administration and during abstinence. Psychopharmacologia (Berl.) **2**, 214—220 (1961)

Gunne, L.-M.: Catecholamine metabolism and morphine abstinence. Ann. N.Y. Acad. Sci. **96**, 205—210 (1962a)

Gunne, L.-M.: Catecholamine metabolism in morphine withdrawal in the dog. Nature (Lond.) **195**, 815—816 (1962b)

Gunne, L.-M.: Catecholamines and 5-hydroxytryptamine in morphine tolerance and withdrawal. Acta physiol. scand. **58**, Suppl. 204, pp. 1—91 (1963)

Gunne, L.-M., Jonsson, J., Fuxe, K.: Effects of morphine intoxication on brain catecholamine neurons. Europ. J. Pharmacol. **5**, 338—342 (1969)

Gunne, L.-M., Jonsson, J., Fuxe, K.: Effects of chronic morphine administration on the catecholamine depletion induced by reserpine. J. Pharm. Pharmacol. **22**, 550—552 (1970)

Gupta, S. K., Kulkarni, H. J.: Modification of morphine analgesia in rats by MAO inhibitors. J. Indian med. Ass. **46**, 197—198 (1966)

Gyang, E. A., Kosterlitz, H. W.: Agonist and antagonist actions of morphine-like drugs on the guinea-pig isolated ileum. Brit. J. Pharmacol. **27**, 514—527 (1966)

Gyang, E. A., Kosterlitz, H. W., Lees, G. M.: The inhibition of autonomic neuroeffector transmission by morphine-like drugs and its use as a screening test for narcotic analgesic drugs. Naunyn-Schmiedebergs Arch. exp. Path. Pharmak. **248**, 231—246 (1964)

Gyermek, L., Bindler, E.: Blockade of the ganglionic stimulant action of 5-hydroxytryptamine. J. Pharmacol. exp. Ther. **135**, 344—348 (1962)

Haertzen, C. A.: Development of scales based on patterns of drug effects, using the Addiction Research Center Inventory (ARCI). Psychol. Rep. **18**, 163—194 (1966)

Haertzen, C. A., Hill, H. E., Belleville, R. E.: Development of the Addiction Research Center Inventory (ARCI): Selection of items that are sensitive to the effects of various drugs. Psychopharmacologia (Berl.) **4**, 155—166 (1963)

Haertzen, C. A., Hooks, N. T., Jr.: Changes in personality and subjective experience associated with the chronic administration and withdrawal of opiates. J. nerv. ment. Dis. **148**, 606—614 (1969)

Haertzen, C. A., Panton, J. H.: Development of a "psychopathic" scale for the Addiction Research Center Inventory (ARCI). Int. J. Addict. **2**, 115—127 (1967)

Hanna, C.: A demonstration of morphine tolerance and physical dependence in the rat. Arch. int. Pharmacodyn. **124**, 326—329 (1960)

Hano, K., Kaneto, H., Kakunaga, T., Moribayashi, N.: Pharmacological studies of analgesics. VI. The administration of morphine and changes in acetylcholine metabolism mouse brain. Biochem. Pharmacol. **13**, 441—447 (1964)

Harris, L. S.: Central neurohumoral systems involved with narcotic agonists and antagonists. Fed. Proc. **29**, 28—32 (1970)

Harris, L. S., Dewey, W. L., Howes, J. F., Kennedy, J. S., Pars, H.: Narcotic-antagonist analgesics: Interactions with cholinergic substances. J. Pharmacol. exp. Ther. **169**, 17—22 (1969)
Harvey, J. A., Lints, C. E.: Lesions in the medial forebrain bundle: Delayed effects on sensitivity to electric shock. Science **148**, 250—252 (1965)
Harvey, J. A., Lints, C. E.: Lesions in the medial forebrain bundle: Relationship between pain sensitivity and telencephalic content of serotonin. J. comp. physiol. Psychol. **74**, 28—36 (1971)
Haubrich, D. R., Blake, D. E.: Effect of acute and chronic administration of morphine on the metabolism of brain serotonin in rats. Fed. Proc. **28**, 793 (1969)
Haubrich, D. R., Blake, D. E.: Modification of serotonin metabolism in rat brain after acute or chronic administration of morphine. Biochem. Pharmacol. **22**, 2753—2759 (1973)
Heinrich, U., Lichtensteiger, W., Langemann, H.: Effect of morphine on the catecholamine content of midbrain nerve cell groups in rat and mouse. J. Pharmacol. exp. Ther. **179**, 259—267 (1971)
Heller, B., Saavedra, J. M., Fischer, E.: Influence of adrenergic blocking agents upon morphine and catecholamine analgesic effect. Experientia (Basel) **24**, 804—805 (1968)
Heller, M. E., Mordkoff, A. M.: Personality attributes of the young, nonaddicted drug abuser. Int. J. Addict. **7**, 65—72 (1972)
Henderson, A., Nemes, G., Gordon, N. B., Roos, L.: Sleep and narcotic tolerance. Psychophysiology **7**, 346—347 (1970)
Henderson, G., Hughes, J., Kosterlitz, H. W.: A new example of a morphine-sensitive neuro-effector junction: adrenergic transmission in the mouse vas deferens. Brit. J. Pharmacol. **46**, 764—766 (1972)
Heng, J. E., Domino, E. F.: Effects of morphine and nalorphine upon tooth pulp thresholds of dogs in the alert and drowsy state. Psychopharmacologia (Berl.) **1**, 433—436 (1960)
Herken, H., Maibauer, D., Müller, S.: Acetylcholingehalt des Gehirns und Analgesie nach Einwirkung von Morphin und einigen 3-Oxymorphinanen. Naunyn-Schmiedebergs Arch. exp. Path. Pharmak. **230**, 313—324 (1957)
Herz, A.: Wirkungen des Arecolins auf das Zentralnervensystem. Naunyn-Schmiedebergs Arch. exp. Path. Pharmak. **242**, 414—429 (1962)
Herz, A., Albus, K., Metys, J., Schubert, P., Teschemacher, Hj.: On the central sites for the antinociceptive action of morphine and fentanyl. Neuropharmacology. **9**, 539—551 (1970)
Herz, A., Bläsig, J., Papeschi, R.: Role of catecholaminergic mechanisms in the expression of the morphine abstinence syndrome in rats. Psychopharmacologia (Berl.) **39**, 121—143 (1974)
Herz, A., Teschemacher, Hj.: Development of tolerance to the antinociceptive effect of morphine after intraventricular injection. Experientia (Basel) **29**, 64—65 (1973)
Herz, A., Teschemacher, Hj., Albus, K., Zieglgänsberger, S.: Morphine abstinence syndrome in rabbits precipitated by injection of morphine antagonists into the ventricular system and restricted parts of it. Psychopharmacologia (Berl.) **26**, 219—235 (1972)
Hill, H. E.: The social deviant and initial addiction to narcotics and alcohol. Quart. J. Stud. Alcohol **23**, 562—582 (1962)
Hill, H. E., Haertzen, C. A., Davis, H.: An MMPI factor analytic study of alcoholics, narcotic addicts and criminals. Quart. J. Stud. Alcohol **23**, 411—431 (1962)
Hill, H. E., Haertzen, C. A., Wolbach, A. B., Jr., Miner, E. J.: The Addiction Research Center Inventory: Standardization of scales which evaluate subjective effects of morphine, amphetamine, pentobarbital, alcohol, LSD-25, pyrahexyl and chlorpromazine. Psychopharmacologia (Berl.) **4**, 167—183 (1963a)
Hill, H. E., Haertzen, C. A., Wolbach, A. B., Jr., Miner, E. J.: The Addiction Research Center Inventory: Appendix. I. Items comprising empirical scales for seven drugs. II. Items which do not differentiate placebo from any drug condition. Psychopharmacologia (Berl.) **4**, 184—205 (1963b)
Hill, H. E., Haertzen, C. A., Yamahiro, R. S.: The addict physician. A Minnesota Multiphasic Personality Inventory study of the interaction of personality characteristics and availability of narcotics. In: Wikler, A. (Ed.): The addictive states. Ass. Res. nerv. Ment. Dis., Vol. 46, pp. 321—332. Baltimore: Williams & Wilkins 1968
Hill, H. E., Kornetsky, C. H., Flanary, H. G., Wikler, A.: Studies on anxiety associated with anticipation of pain. I. Effects of morphine. Arch. Neurol. Psychiat. (Chic.) **67**, 612—619 (1952)

Hill, H. E., Pescor, F. T., Belleville, R. E., Wikler, A.: Use of differential bar-pressing rates of rats for screening analgesic drugs. I. Techniques and effects of morphine. J. Pharmacol. exp. Ther. **120**, 388—397 (1957)
Himmelsbach, C. K.: Clinical studies of drug addiction. II. "Rossium" treatment of drug addiction. Publ. Hlth Rep. (Wash.), Suppl. **125** (1937)
Himmelsbach, C. K.: Studies of certain addiction characteristics of: (a) Dihydromorphine ("paramorphan"), (b) dihydrodesoxymorphine-D ("desomorphine"), (c) dihydrodesoxycodeine-D ("desocodeine"), (d) methyldihydromorphinone ("metopon"). J. Pharmacol. exp. Ther. **67**, 239—249 (1939)
Himmelsbach, C. K.: Clinical studies of drug addiction; physical dependence, withdrawal and recovery. Arch. intern. Med. **69**, 766—772 (1942)
Himmelsbach, C. K.: Morphine, with reference to physical dependence. Fed. Proc. **2**, 201—203 (1943)
Himmelsbach, C. K., Andrews, H. L., Felix, R. H., Oberst, F. W., Davenport, L. F.: Studies on codeine addiction. Publ. Hlth Rep. (Wash.), Supp. 158, 1940
Himmelsbach, C. K., Gerlach, G. H., Stanton, E. J.: A method for testing addiction, tolerance and abstinence in the rat. Results of its application to several morphine alkaloids. J. Pharmacol. exp. Ther. **53**, 179—188 (1935)
Hitzemann, R. J., Loh, H. H.: Effect of morphine on the transport of dopamine into mouse brain slices. Europ. J. Pharmacol. **21**, 121—129 (1973)
Hitzemann, R. J., Loh, H. H., Ho, A. K. S., Way, E. L.: The effects of acute morphinization on the metabolism of ^{14}C-tyrosine II. Proc. West. Pharmacol. Soc. **14**, 91—94 (1971)
Ho, A. K. S., Loh, H. H.: Effect of morphine and amphetamine on brain cholineacetylase and esterase activity in mice and rats. Pharmacologist **12**, 231 (1970)
Ho, I. K., Loh, H. H., Way, E. L.: Effect of L-tryptophan on morphine analgesia, tolerance and physical dependence. Reported to 34th meeting, Committee on Problems of Drug Dependence, National Research Council, Ann Arbor, Mich., 1972a
Ho, I. K., Loh, H. H., Way, E. L.: Effect of cyclic AMP on morphine analgesia tolerance and physical dependence. Nature (Lond.) **238**, 397—398 (1972c)
Ho, I. K., Loh, H. H., Way, E. L.: Influence of 5,6-dihydroxytryptamine on morphine tolerance and physical dependence. Europ. J. Pharmacol. **21**, 331—336 (1973a)
Ho, I. K., Loh, H. H., Way, E. L.: Cyclic adenosine monophosphate antagonism of morphine analgesia. J. Pharmacol. exp. Ther. **185**, 336—346 (1973b)
Ho, I. K., Loh, H. H., Way, E. L.: Effects of cyclic 3',5'-adenosine monophosphate on morphine tolerance and physical dependence. J. Pharmacol. exp. Ther. **185**, 347—357 (1973c)
Ho, I. K., Loh, H. H., Way, E. L.: Influence of GABA on morphine analgesia, tolerance and physical dependence. Proc. West. Pharmacol. Soc. **16**, 4—7 (1973d)
Ho, I. K., Lu, S. E., Loh, H. H., Way, E. L.: Effects of c-AMP on morphine tolerant and physically dependent mice. Pharmacologist **13**, 314 (1971)
Ho, I. K., Lu, S. E., Stolman, S., Loh, H. H., Way, E. L.: Influence of *p*-chlorophenylalanine on morphine tolerance and physical dependence and regional brain serotonin turnover studies in morphine tolerant-dependent mice. J. Pharmacol. exp. Ther. **182**, 155—165 (1972b)
Hodgson, J. R., Bristow, R. L., Castles, T. R.: Repression of RNA transcription during the development of analgesic tolerance to morphine. Nature (Lond.) **248**, 671—673 (1974)
Hoeldtke, R. D., Martin, W. R.: Urine volume and catecholamine excretion. J. Lab. clin. Med. **75**, 166—174 (1970)
Hoffmeister, F.: Tierexperimentelle Untersuchungen über den Schmerz und seine pharmakologische Beeinflussung. Arzneimittel-Forsch. 16. Beiheft (1968)
Hoffmeister, F., Schlichting, U.: Einfluß von Sympathikomimetika, Sympathikolytika, Cholinomimetika und Anticholinergika auf das Abstinenzsyndrom von morphinabhängigen Ratten. In: Janzen, R., Keidel, W. D., Herz, A., Steichele, C.: Schmerz: Grundlagen — Pharmakologie — Therapie. Stuttgart: Georg Thieme Verlag 1972
Hollinger, M.: Effect of reserpine, α-methyl-p-tyrosine, p-chlorophenylalanine and pargyline on levorphanol-induced running activity in mice. Arch. int. Pharmacodyn. **179**, 419—424 (1969)
Holtzman, S. G., Jewett, R. E.: Some actions of pentazocine on behavior and brain monoamines in the rat. J. Pharmacol. exp. Ther. **181**, 346—356 (1972)

Holtzman,S.G., Villarreal,J.E.: Morphine dependence and body temperature in Rhesus monkeys. J. Pharmacol. exp. Ther. **166**, 125—133 (1969)
Holtzman,S.G., Villarreal,J.E.: Pharmacologic analysis of the hypothermic responses of the morphine-dependent Rhesus monkey. J. Pharmacol. exp. Ther. **177**, 317—325 (1971)
Horlington,M., Lockett,M.F.: Antagonism between alkylated and norcompounds of the morphine group injected intracisternally into mice. J. Pharm. Pharmacol. **11**, 415—420 (1959)
Hosoya,E.: Some withdrawal symptoms of rats to morphine. Pharmacologist **1**, 77 (1959)
Houde,R.W., Wikler,A.: Delineation of the skin-twitch response in dogs and the effects thereon of morphine, thiopental and mephenesin. J. Pharmacol. exp. Ther. **103**, 236—242 (1951)
Houde,R.W., Wikler,A., Irwin,S.: Comparative actions of analgesic, hypnotic and paralytic agents on hindlimb reflexes in chronic spinal dogs. J. Pharmacol. exp. Ther. **103**, 243—248 (1951)
Howes,J.F., Harris,L.S., Dewey,W.L.: The effect of morphine, nalorphine, naloxone, pentazocine, cyclazocine and oxotremorine on the synthesis and release of acetylcholine by mouse cerebral cortex slices in vitro. Arch. int. Pharmacodyn. **184**, 267—276 (1970)
Howes,J.F., Harris,L.S., Dewey,W.L., Voyda,C.A.: Brain acetylcholine levels and inhibition of the tail-flick reflex in mice. J. Pharmacol. exp. Ther. **169**, 23—28 (1969)
Hughes,J.: Isolation of an endogenous compound from the brain with pharmacological properties similar to morphine. Brain Res. **88**, 295—308 (1975)
Hughes,J., Smith,T., Morgan,B., Fothergill,L.: Purification and properties of encephalin—the possible endogenous ligand for the morphine receptor. Life Sci. **16**, 1753—1758 (1975)
Huidobro,F.: Effects of some indoles and acridine derivatives on the analgesic effect of morphine. Acta physiol. lat.-amer. **13**, 71—72 (1963)
Huidobro,F.: Morphine dependence in mice. Arch. Biol. Med. Exp. (Santiago) **4**, 155—161 (1967)
Huidobro,F.: Some relations between tolerance and physical dependence to morphine in mice. Europ. J. Pharmacol. **15**, 79—84 (1971)
Huidobro,F., Contreras,E., Croxatto,R.: Studies on morphine.—III. Action of metabolic precursors to serotonin and noradrenaline and related substances on the "abstinence syndrome" to morphine on white mice. Arch. int. Pharmacodyn. **146**, 444—454 (1963a)
Huidobro,F., Contreras,E., Croxatto,R.: Studies on morphine.—IV. Effects of some metabolites on the "abstinence syndrome" to morphine on white mice. Arch. int. Pharmacodyn. **146**, 455—462 (1963b)
Huidobro,F., Huidobro,J.P., Larrain,G.: Tolerance to morphine in the white mice. Acta physiol. lat.-amer. **18**, 59—67 (1968)
Huidobro,J.P., Huidobro,F.: Acute morphine tolerance in newborn and young rats. Psychopharmacologia (Berl.) **28**, 27—34 (1973)
Ireson,J.D.: Opioid and muscarinic anti-nociception. Brit. J. Pharmacol. **37**, 504P—505P (1969)
Irwin,S., Houde,R.W., Bennett,D.R., Hendershot,L.C., Seevers,M.H.: The effects of morphine, methadone and meperidine on some reflex responses of spinal animals to nociceptive stimulation. J. Pharmacol. exp. Ther. **101**, 132—143 (1951)
Isbell,H.: The addiction liability of some derivatives of meperidine. J. Pharmacol. exp. Ther. **97**, 182—189 (1949)
Isbell,H., Eisenman,A.J., Wikler,A., Frank,K.: The effects of single doses of 6-dimethylamino-4-4-diphenyl-3-heptanone (amidone, methadon or "10820") on human subjects. J. Pharmacol. exp. Ther. **92**, 83—89 (1948a)
Isbell,H., Wikler,A., Eisenman,A.J., Daingerfield,M., Frank,K.: Liability of addiction to 6-dimethylamino-4,4-diphenyl-3-heptanone (methadon, "amidone" or "10820") in man. Arch. intern. Med. **82**, 362—392 (1948b)
Isbell,H., Wikler,A., Eisenman,A.J., Frank,K.: Effect of single doses of 10820 (4,4-diphenyl-6-dimethylamino-heptanone-3) in man. Fed. Proc. **6**, 341 (1947)
Ivy,A.C., Goetzl,F.R., Harris,S.C., Burrill,D.Y.: The analgesic effect of intracarotid and intravenous injection of epinephrine in dogs and of subcutaneous injection in man. Quart. Bull. Northw. Univ. med. Sch. **18**, 298—306 (1944)
Iwamoto,E.T., Ho,I.K., Way,E.L.: Elevation of brain dopamine during naloxone-precipitated withdrawal in morphine-dependent mice and rats. J. Pharmacol. exp. Ther. **187**, 558—567 (1973)

Iwamoto, E.T., Shen, F., Loh, H.H., Way, E.L.: The effects of pargyline on morphine tolerant-dependent mice. Fed. Proc. **30**, 278 (1971)

Jacquet, Y.F., Lajtha, A.: Morphine action at central nervous system sites in rat: Analgesia or hyperalgesia depending on site and dose. Science **182**, 490—492 (1973)

Jaffe, J.H., Sharpless, S.K.: The rapid development of physical dependence on barbiturates. J. Pharmacol. exp. Ther. **150**, 140—145 (1965)

Jasinski, D.R., Martin, W.R., Haertzen, C.A.: The human pharmacology and abuse potential of N-allylnoroxymorphone (naloxone). J. Pharmacol. exp. Ther. **157**, 420—426 (1967)

Jasinski, D.R., Martin, W.R., Hoeldtke, R.: Studies of the dependence-producing properties of GPA-1657, profadol, and propiram in man. Clin. Pharmacol. Ther. **12**, 613—649 (1971)

Jasinski, D.R., Nutt, J.G., Bunney, W.E.: Lithium: Lack of effect on morphine euphoria. Pharmacologist **15**, 168 (1973)

Jhamandas, K.: Effect of morphine and its antagonists on the release of cortical acetylcholine (ACh) in the normal and morphine dependent rats. Pharmacologist **15**, 202 (1973)

Jhamandas, K., Dickinson, G.: Modification of precipitated morphine and methadone abstinence in mice by acetylcholine antagonists. Nature (Lond.) New Biol. **245**, 219—221 (1973)

Jhamandas, K., Phillis, J.W., Pinsky, C.: Effects of narcotic analgesics and antagonists on the in vivo release of acetylcholine from the cerebral cortex of the cat. Brit. J. Pharmacol. **43**, 53—66 (1971)

Jhamandas, K., Pinsky, C., Phillis, J.W.: Effects of morphine and its antagonists on release of cerebral cortical acetylcholine. Nature (Lond.) **228**, 176—177 (1970)

Joël, E., Ettinger, A.: Zur Pathologie der Gewöhnung. III. Mitteilung: Experimentelle Studien über Morphingewöhnung. Naunyn-Schmiedebergs Arch. exp. Path. Pharmak. **115**, 334—350 (1926)

Johannesson, T., Long, J.P.: Acetylcholine in the brain of morphine tolerant and non-tolerant rats. Acta pharmacol. (Kbh.) **21**, 192—196 (1964)

Jones, B.E., Martin, W.R., Jasinski, D.R.: Skin conductance responses to hypercapnia in man during a cycle of addiction to morphine. Psychopharmacologia (Berl.) **14**, 394—403 (1969)

Jounela, A.J., Mattila, M.J.: Modification by phenelzine of morphine and pethidine analgesia in mice. Ann. Med. exp. Fenn. **46**, 66—71 (1968)

Jurna, I.: The effect of morphine on the discharge of muscle spindles with intact motor innervation. Int. J. Neuropharmacol. **4**, 177—183 (1965)

Jurna, I.: Inhibition of the effect of repetitive stimulation on spinal motoneurones of the cat by morphine and pethidine. Int. J. Neuropharmacol. **5**, 117—123 (1966)

Jurna, I., Grossmann, W., Theres, C.: Inhibition by morphine of repetitive activation of cat spinal motoneurones. Neuropharmacology **12**, 983—993 (1973)

Jurna, I., Schafer, H.: Depression of post-tetanic potentiation in the spinal cord by morphine and pethidine. Experientia (Basel) **21**, 226—227 (1965)

Jurna, I., Schlue, W.R., Tamm, U.: The effect of morphine on primary somatosensory evoked responses in the rat cerebral cortex. Neuropharmacology **11**, 409—415 (1972)

Kamei, C., Shimomura, K., Ueki, S.: Significance of withdrawal jumping response in predicting physical dependence in mice. Jap. J. Pharmacol. **23**, 421—426 (1973)

Karr, W.G., Light, A.B., Torrance, E.G.: Opium addiction. IV. The blood of the human addict during the administration of morphine. In: Light, A.B., Torrance, E.G., Karr, W.G., Fry, E.G., Wolff, W.A.: Opium addiction, pp. 33—39. Chicago, Illinois: American Medical Assn. 1929—30. (Reprinted from Arch. intern. Med., Vols. 43—44)

Kay, D.C., Eisenstein, R.B., Jasinski, D.R.: Morphine effects on human REM state, waking state and NREM sleep. Psychopharmacologia (Berl.) **14**, 404—416 (1969)

Kay, D.C., Kelly, O.A.: Some changes in human sleep during chronic intoxication with morphine. Psychophysiology **7**, 346 (1970)

Kayan, S., Ferguson, R.K., Mitchell, C.L.: An investigation of pharmacologic and behavioral tolerance to morphine in rats. J. Pharmacol. exp. Ther. **185**, 300—306 (1973)

Kayan, S., Mitchell, C.L.: The effects of chronic morphine administration on tooth pulp thresholds in dogs and cats. Proc. Soc. exp. Biol. (N.Y.) **128**, 755—760 (1968)

Kayan, S., Woods, L.A., Mitchell, C.L.: Experience as a factor in the development of tolerance to the analgesic effect of morphine. Europ. J. Pharmacol. **6**, 333—339 (1969)

Kaymakcalan, S., Woods, L. A.: Nalorphine-induced "abstinence syndrome" in morphine-tolerant albino rats. J. Pharmacol. exp. Ther. **117**, 112—116 (1956)
Keele, C. A., Armstrong, D.: Substances producing pain and itch. Baltimore: Williams & Wilkins, 1964
Keranen, G. M., Zaratzian, V. L., Coleman, R.: Studies on 1,4-dipyrrolidino-2-butyne (Tremorine) in mice. Toxicol. appl. Pharmacol. **3**, 481—492 (1961)
Kerr, F. W. L., Pozuelo, J.: Suppression of physical dependence and induction of hypersensitivity to morphine by stereotaxic hypothalamic lesions in addicted rats. Mayo Clin. Proc. **46**, 653—665 (1971)
Kerr, F. W. L., Triplett, J. N., Jr., Beeler, G. W.: Reciprocal (push-pull) effects of morphine on single units in the ventromedian and lateral hypothalamus and influences on other nuclei: With a comment on methadone effects during withdrawal from morphine. Brain Res. **74**, 81—103 (1974)
Khalsa, J. H., Davis, W. M.: Effects of chronic alpha-methyltyrosine on the locomotor activity response to morphine and amphetamine in rats. Pharmacologist **15**, 219 (1973)
Khazan, N., Colasanti, B.: Differential pharmacodynamics of morphine injection in naive and post-addict rats. Fed. Proc. **30**, 277 (1971 a)
Khazan, N., Colasanti, B.: Decline in the mean integrated electroencephalogram voltage during morphine abstinence in the rat. J. Pharmacol. exp. Ther. **177**, 491—499 (1971 b)
Khazan, N., Colasanti, B.: EEG correlates of morphine challenge in postaddict rats. Psychopharmacologia (Berl.) **22**, 56—63 (1971 c)
Khazan, N., Colasanti, B.: Protracted rebound in rapid eye movement sleep time and electroencephalogram voltage output in morphine-dependent rats upon withdrawal. J. Pharmacol. exp. Ther. **183**, 23—30 (1972)
Khazan, N., Roehrs, T.: EEG responses to morphine test dose in morphine- and methadone-treated rats. Pharmacologist **15**, 168 (1973)
Khazan, N., Sawyer, C. H.: Mechanisms of paradoxical sleep as revealed by neurophysiologic and pharmacologic approaches in the rabbit. Psychopharmacologia (Berl.) **5**, 457—466 (1964)
Khazan, N., Weeks, J. R., Schroeder, L. A.: Electroencephalographic, electromyographic and behavioral correlates during a cycle of self-maintained morphine addiction in the rat. J. Pharmacol. exp. Ther. **155**, 521—531 (1967)
Killam, K. F., Deneau, G. A.: A study of morphine dependence in four species of subhuman primates. Proc. West. Pharmacol. Soc. **16**, 1—3 (1973)
Kiplinger, G. F., Clift, J. W.: Pharmacological properties of morphine-potentiating serum obtained from morphine-tolerant dogs and men. J. Pharmacol. exp. Ther. **146**, 139—146 (1964)
Klee, W., Streaty, R. A.: Narcotic receptor sites in morphine-dependent rats. Nature (Lond.) **248**, 61—63 (1974)
Knapp, S., Mandell, A. J.: Narcotic drugs: Effects on the serotonin biosynthetic systems of the brain. Science **177**, 1209—1211 (1972)
Kolb, L.: Pleasure and deterioration from narcotic addiction. Ment. Hyg. (N.Y.) **9**, 699—724 (1925)
Kolb, L., DuMez, A. G.: Experimental addiction of animals to opiates. Publ. Hlth Rep. (Wash.) **46**, 698—726 (1931)
Kolb, L., Himmelsbach, C. K.: Clinical studies of drug addiction. III. A critical review of the withdrawal treatments with method of evaluating abstinence syndromes. Publ. Hlth Rep. (Wash.), Suppl. 128 (1938)
Koll, W., Haase, J., Block, G., Mühlberg, B.: The predilective action of small doses of morphine on nociceptive spinal reflexes of low spinal cats. Int. J. Pharmacol. **2**, 57—65 (1963)
Korf, J., Bunney, B. S., Aghajanian, G. K.: Noradrenergic neurons: Morphine inhibition of spontaneous activity. Europ. J. Pharmacol. **25**, 165—169 (1974)
Kornetsky, C., Cochin, J.: Evidence of tolerance to morphine in the mouse after injection of serum from morphine-tolerant rabbits. Fed. Proc. **23**, 283 (1964)
Kornetsky, C., Kiplinger, G. F.: Potentiation of an effect of morphine in the rat by sera from morphine-tolerant and abstinent dogs and monkeys. Psychopharmacologia (Berl.) **4**, 66—71 (1963)

Kosterlitz, H. W., Lord, J. A. H., Watt, A. J.: Morphine receptor in the myenteric plexus of the guinea pig ileum. In: Kosterlitz, H. W., Collier, H. O. J., Villarreal, J. E. (Eds.): Agonist and antagonist actions of narcotic analgesic drugs (Proceedings of the symposium of the British Pharmacological Society, Aberdeen, July 1971). pp. 45—61. London: Macmillan Press, 1972

Kosterlitz, H. W., Robinson, J. A.: Inhibition of the peristaltic reflex of the isolated guinea pig ileum. J. Physiol. (Lond.) **136**, 249—262 (1957)

Kosterlitz, H. W., Robinson, J. A.: The inhibitory action of morphine on the contraction of the longitudinal muscle coat of the isolated guinea pig ileum. Brit. J. Pharmacol. **13**, 296—303 (1958)

Kosterlitz, H. W., Taylor, D. W.: The effect of morphine on vagal inhibition of the heart. Brit. J. Pharmacol. **14**, 209—214 (1959)

Kosterlitz, H. W., Wallis, D. I.: The action of morphine-like drugs on impulse transmission in mammalian nerve fibres. Brit. J. Pharmacol. **22**, 499—510 (1964)

Kosterlitz, H. W., Wallis, D. I.: The effects of hexamethonium and morphine on transmission in the superior cervical ganglion of the rabbit. Brit. J. Pharmacol. **26**, 334—344 (1966)

Kosterlitz, H. W., Watt, A. J.: Kinetic parameters of narcotic agonists and antagonists, with particular reference to N-allylnoroxymorphone (naloxone). Brit. J. Pharmacol. **33**, 266—276 (1968)

Krivoy, W. A.: The action of analgetic agents on positive afterpotentials of frog sciatic nerve. J. Pharmacol. exp. Ther. **129**, 186—190 (1960)

Krivoy, W. A., Huggins, R. A.: The action of morphine, methadone, meperidine and nalorphine on dorsal root potentials of cat spinal cord. J. Pharmacol. exp. Ther. **134**, 210—213 (1961)

Krivoy, W., Kroeger, D., Zimmermann, E.: Actions of morphine on the segmental reflex of the decerebrate spinal cat. Brit. J. Pharmacol. **47**, 457—464 (1973)

Krivoy, W., Kroeger, D., Taylor, A. N., Zimmermann, E.: Antagonism of morphine by β-melanocyte-stimulating hormone and by tetracosactin. Europ. J. Pharmacol. **27**, 339—345 (1974b)

Krivoy, W., Zimmermann, E., Lande, S.: Facilitation of development of resistance to morphine analgesia by desglycinamide[9]-lysine vasopressin (DGVP). Proc. nat. Acad. Sci. (Wash.), **71** 1852—1856 (1974a)

Krueger, H., Eddy, N. B., Sumwalt, M.: The pharmacology of the opium alkaloids, Part 1. Publ. Hlth Rep. (Wash.), Suppl. 165 (1941)

Krueger, H., Eddy, N. B., Sumwalt, M.: The pharmacology of the opium alkaloids, Part 2. Publ. Hlth Rep. (Wash.), Suppl. 165 (1943)

Kruglov, N. A.: Influence of analgesic substances on the lability and some other functional characteristics of the nerve center [in Russian]. Doctor's Thesis, Leningrad, 1955, p. 210

Kruglov, N. A.: Influence of some analgesic and narcotic substances on reciprocal inhibition [in Russian]. Pharmacol. Toxicol. **22**, 488—493 (1959)

Kruglov, N. A.: Effect of the morphine-group analgesics on the central inhibitory mechanisms. Int. J. Neuropharmacol. **3**, 197—203 (1964)

Kuhn, H.-F., Friebel, H.: Über den Nachweis von "physical dependence" bei Codein-behandelten Ratten. Med. exp. **6**, 301—306 (1962)

Kumar, R., Mitchell, E., Stolerman, I. P.: Disturbed patterns of behaviour in morphine tolerant and abstinent rats. Brit. J. Pharmacol. **42**, 473—484 (1971)

Kuschinsky, K., Hornykiewicz, O.: Morphine catalepsy in the rat: Relation to striatal dopamine metabolism. Europ. J. Pharmacol. **19**, 119—122 (1972)

Labrecque, G., Domino, E. F.: Tolerance to and physical dependence on morphine: Relation to neocortical acetylcholine release in the cat. J. Pharmacol. exp. Ther. **191**, 189—200 (1974)

Lal, H., O'Brien, J., Puri, S. K.: Morphine-withdrawal aggression: Sensitization by amphetamines. Psychopharmacologia (Berl.) **22**, 217—223 (1971a)

Lal, H., Puri, S. K.: Morphine-withdrawal aggression: Role of dopaminergic stimulation. In: Singh, J. M., Miller, L., Lal, H. (Eds.): Drug addiction: Experimental pharmacology, Vol. 1, pp. 301—310. Mount Kisco, N. Y.: Futura Publishing Co. 1972

Lal, H., Puri, S. K.: Effect of acute morphine or haloperidol administration on catalepsy and striatal dopamine turnover in rats chronically treated with morphine or haloperidol. Pharmacologist **15**, 259 (1973)

Lal, H., Puri, S. K., Karkalas, Y.: Blockade of opioid-withdrawal symptoms by haloperidol in rats and humans. Pharmacologist **13**, 263 (1971b)

Landmesser,C.M., Cobb,S., Converse,J.G.: Effects of N-allylnormorphine upon the respiratory depression due to morphine in anesthetized man with studies on the respiratory response to carbon dioxide. Anesthesiology **14**, 535—549 (1953)
Large,W.A., Milton,A.S.: The effect of acute and chronic morphine administration on brain acetylcholine levels in the rat. Brit. J. Pharmacol. **38**, 451P—452P (1970)
Large,W.A., Milton,A.S.: Effects of morphine, levorphanol, nalorphine and naloxone on the release of acetylcholine from slices of rat cerebral cortex and hippocampus. Brit. J. Pharmacol. **41**, 398P (1971)
Lasagna,L., von Felsinger,J.M., Beecher,H.K.: Drug-induced mood changes in man. 1. Observations on healthy subjects, chronically ill patients, and "postaddicts." J. Amer. med. Ass. **157**, 1006—1020 (1955)
Laverty,R., Sharman,D.F.: Modification by drugs of the metabolism of 3,4-dihydroxyphenylethylamine, noradrenaline and 5-hydroxytryptamine in the brain. Brit. J. Pharmacol. **24**, 759—772 (1965)
Lavikainen,P., Mattila,M.: The convulsions and the brain acetylcholine metabolism after morphine and nalorphine. Ann. Med. exp. Fenn. **37**, 133—140 (1959)
Lee,C.Y., Stolman,S., Akera,T., Brody,T.M.: Saturable binding of (^{3}H)-dihydromorphine to rat brain tissue in vitro: characterization and effect of morphine pretreatment. Pharmacologist **15**, 202 (1973)
Lee,J.R., Fennessy,M.R.: The relationship between morphine analgesia and the levels of biogenic amines in the mouse brain. Europ. J. Pharmacol. **12**, 65—70 (1970)
Leimdorfer,A.: An electroencephalographic analysis of the action of amidone, morphine, and strychnine on the central nervous system. Arch. int. Pharmacodyn. **76**, 153—162 (1948)
Leimdorfer,A.: The action of sympathomimetic amines on the central nervous system and the blood sugar: Relation of chemical structure to mechanism of action. J. Pharmacol. exp. Ther. **98**, 62—71 (1950)
Leimdorfer,A., Arana,R., Hack,M.H.: Hyperglycemia induced by the action of adrenalin on the central nervous system. Amer. J. Physiol. **150**, 588—595 (1947)
Leimdorfer,A., Metzner,W.R.T.: Analgesia and anesthesia induced by epinephrine. Amer. J. Physiol. **157**, 116—121 (1949)
Leme,J.G., Rocha E Silva,M.: Analgesic action of chlorpromazine and reserpine in relation to that of morphine. J. Pharm. Pharmacol. **13**, 734—742 (1961)
Leslie,G.B.: The effect of anti-parkinsonian drugs on oxotremorine-induced analgesia in mice. J. Pharm. Pharmacol. **21**, 248—250 (1969)
Lewis,J.R.: Tolerance in rats to new synthetic analgesic drugs. Fed. Proc. **8**, 315 (1949)
Lewis,S.A., Oswald,I., Evans,J.I., Akindele,M.O., Tompsett,S.L.: Heroin and human sleep. Electroenceph. clin. Neurophysiol. **28**, 374—381 (1970)
Light,A.B., Torrance,E.G.: Opium addiction. II. Physical characteristics and physical fitness of addicts during administration of morphine. In: Light,A.B., Torrance,E.G., Karr,W.G., Fry,E.G., Wolff,W.A.: Opium addiction, pp.13—20. Chicago: American Medical Assn. 1929—30a. (Reprinted from Arch. intern. Med., Vols.43—44)
Light,A.B., Torrance,E.G.: Opium addiction. III. The circulation and respiration of human addicts during the administration of morphine. In: Light,A.B., Torrance,E.G., Karr,W.G., Fry,E.G., Wolff,W.A.: Opium addiction, pp.21—32. Chicago: American Medical Assn. 1929—30b. (Reprinted from Arch. intern. Med., Vols.43—44)
Light,A.B., Torrance,E.G.: Opium addiction. V. Miscellaneous observations on human addicts during the administration of morphine. In: Light,A.B., Torrance,E.G., Karr,W.G., Fry,E.G., Wolff,W.A.: Opium addiction, pp.40—50. Chicago: American Medical Assn., 1929—30c. (Reprinted from Arch. intern. Med., Vols.43—44)
Light,A.B., Torrance,E.G.: Opium addiction. VI. The effects of abrupt withdrawal followed by readministration of morphine in human addicts, with special reference to the composition of the blood, the circulation and the metabolism. In: Light,A.B., Torrance,E.G., Karr,W.G., Fry,E.G., Wolff,W.A.: Opium addiction. pp.51—66. Chicago, American Medical Assn., 1929—30d. (Reprinted from Arch. intern. Med., Vols.43—44)
Light,A.B., Torrance,E.G., Karr,W.G., Fry,E.G., Wolff,W.A.: Opium addiction. Chicago: American Medical Assn., 1929—30. (Reprinted from Arch. intern. Med., Vols.43—44)
Lim,R.K.S.: Visceral receptors and viscera pain. Ann. N.Y. Acad. Sci. **86**, 73—89 (1960)

Lim,R.K.S., Guzman,F.: Manifestations of pain in analgesic evaluation in animals and man. In: Soulairac,A., Cahn,J., Charpentier,J. (Eds.): Pain (Proceedings of the International Symposium on Pain, Paris, April 11—13, 1967), pp.119—152, New York: Academic Press 1968
Lin,S.C., Sutherland,V.C., Way,E.L.: Brain amino acids in morphine tolerant and non-tolerant rats. Proc. West. Pharmacol. Soc. **16**, 8—13 (1973)
Lints,C.E., Harvey,J.A.: Altered sensitivity to foot-shock and decreased brain content of serotonin following brain lesions in the rat. J. comp. physiol. Psychol. **67**, 23—31 (1969)
Little,H.J., Rees,J.M.H.: Tolerance development to the antinociceptive actions of morphine, amphetamine, physostigmine and 2-aminoindane in the mouse. Experientia (Basel) **30**, 930—932 (1974)
Lockett,M.F., Davis,M.M.: The analgesic action of normorphine administered intracisternally to mice. J. Pharm. Pharmacol. **10**, 80—85 (1958)
Lodge,D., Headley,P.M., Duggan,A.W., Biscoe,T.J.: The effects of morphine, etorphine, and sinomenine on the chemical sensitivity and synaptic responses of Renshaw cells and other spinal neurones in the rat. Europ. J. Pharmacol. **26**, 277—284 (1974)
Loewe,S.: Influence of chlorpromazine, reserpine, dibenzyline, and desoxycorticosterone upon morphine-induced feline mania. Arch. int. Pharmacodyn. **108**, 453—456 (1956)
Loh,H.H., Shen,F.-H., Way,E.L.: Inhibition of morphine tolerance and physical dependence development and brain serotonin synthesis by cycloheximide. Biochem. Pharmacol. **18**, 2711—2721 (1969)
Loh,H.H., Shen,F.-H., Way,E.L.: Effect of d-actinomycin on the acute toxicity and brain uptake of morphine. J. Pharmacol. exp. Ther. **177**, 326—331 (1971)
Longo,V.G.: Electroencephalographic atlas for pharmacological research. Rabbit Brain Research, Vol.II. Amsterdam: Elsevier Publishing Co. 1962
Lorens,S.A., Mitchell,C.L.: Influence of morphine on lateral hypothalamic self-stimulation in the rat. Psychopharmacologia (Berl.) **32**, 271—277 (1973)
Lorens,S.A., Yunger,L.M.: Morphine analgesia, two-way avoidance, and consummatory behavior following lesions in the midbrain raphé nuclei of the rat. Pharmacol. Biochem. Behav. **2**, 215—221 (1974)
Lorenzetti,O.J., Sancilio,L.F.: Morphine dependent rats as a model for evaluating potential addiction liability of analgesic compounds. Arch. int. Pharmacodyn. **183**, 391—402 (1970)
Lowney,L.I., Schulz,K., Lowery,P.J., Goldstein,A.: Partial purification of an opiate receptor from mouse brain. Science **183**, 749—753 (1974)
MacKay,E.M.: The relation of acquired morphine tolerance to the adrenal cortex. J. Pharmacol. exp. Ther. **43**, 51—60 (1931)
MacKay,E.M., MacKay,L.L.: Resistance to morphine in experimental uremia. Proc. Soc. exp. Biol. (N.Y.) **24**, 129 (1926)
McClane,T.K., Martin,W.R.: Effects of morphine, nalorphine, cyclazocine, and naloxone on the flexor reflex. Int. J. Neuropharmacol. **6**, 89—98 (1967)
McKenzie,G.M., Sadof,M.: Effects of morphine and chlorpromazine on apomorphine-induced stereotyped behaviour. J. Pharm. Pharmacol. **26**, 280—282 (1974)
McKenzie,J.S.: The influence of morphine and pethidine on somatic evoked responses in the hippocampal formation of the cat. Electroenceph. clin. Neurophysiol. **17**, 428—431 (1964)
McKenzie,J.S., Beechey,N.R.: The effects of morphine and pethidine on somatic evoked responses in the midbrain of the cat, and the relevance to analgesia. Electroenceph. clin. Neurophysiol. **14**, 501—519 (1962)
Maggiolo,C., Huidobro,F.: Administration of pellets of morphine to mice: Abstinence syndrome. Acta physiol. lat.-amer. **11**, 70—78 (1961)
Maggiolo,C., Huidobro,F.: The influence of some drugs on the abstinence syndrome to morphine in mice. Arch. int. Pharmacodyn. **138**, 157—168 (1962)
Maggiolo,C., Huidobro,F.: Studies on morphine. VIII. Action of drugs that mobilize aromatic alkyl-amines on the intensity of the abstinence syndrome to morphine in white mice. Acta physiol. lat.-amer. **15**, 292—299 (1965)
Major,C.T., Pleuvry,B.J.: Effects of α-methyl-p-tyrosine, p-chlorophenylalanine, L-β-(3,4-dihydroxyphenyl)alanine, 5-hydroxytryptophan and diethyldithiocarbamate on the analgesic activity of morphine and methylamphetamine in the mouse. Brit. J. Pharmacol. **42**, 512—521 (1971)

Marshall, I., Grahame-Smith, D.G.: Unchanged rate of brain serotonin synthesis during chronic morphine treatment and failure of parachlorophenylalanine to attenuate withdrawal syndrome in mice. Nature (Lond.) **228**, 1206—1208 (1970)
Marshall, I., Grahame-Smith, D.G.: Evidence against a role of brain 5-hydroxytryptamine in the development of physical dependence upon morphine in mice. J. Pharmacol. exp. Ther. **173**, 634—641 (1971)
Marshall, I., Smith, C.B.: Blockade of development of physical dependence on morphine by chronic inhibition of tyrosine hydroxylase. Pharmacologist **15**, 243 (1973)
Marshall, I., Smith, C.B.: Acute and chronic morphine treatment and the hydroxylation of (1-^{14}C)-L-tyrosine in the mouse brain. Brit. J. Pharmacol. **50**, 428—430 (1974)
Martin, W.R.: Strong analgesics. In: Root, W.S., Hofman, F.G.: Physiological pharmacology, Vol. 1, pp. 275—312. New York-London: Academic Press, 1963
Martin, W.R.: Assessment of the dependence producing potentiality of narcotic analgesics. In: Radouco-Thomas, C., Lasagna, L. (Eds.): International Encyclopedia of Pharmacology and Therapeutics, Sec. 6, Vol. 1, pp. 155—180. Glasgow: Pergamon Press 1966
Martin, W.R.: Opioid antagonists. Pharmacol. Rev. **19**, 463—521 (1967)
Martin, W.R.: A homeostatic and redundancy theory of tolerance to and dependence on narcotic analgesics. In: Wikler, A. (Ed.): The addictive states, Ass. Res. nerv. ment. Dis., Vol. 46, pp. 206—225. Baltimore: Williams and Wilkins Co. 1968a
Martin, W.R.: The basis and possible utility of the use of opioid antagonists in the ambulatory treatment of the addict. In: Wikler, A. (Ed.): The addictive states, Ass. Res. nerv. ment. Dis., Vol. 46, pp. 367—371. Baltimore: Williams and Wilkins 1968b
Martin, W.R.: Pharmacological redundancy as an adaptive mechanism in the central nervous system. Fed. Proc. **29**, 13—18 (1970)
Martin, W.R.: Assessment of the abuse potentiality of amphetamines and LSD-like hallucinogens in man and its relationship to basic animal assessment programs. In: Goldberg, L., Hoffmeister, F.: Psychic dependence, Bayer Symposium IV, pp. 146—155. Berlin-Heidelberg-New York: Springer 1973
Martin, W.R., Abdulian, D.H., Unna, K.R., Busch, H.: A study of the peripheral and central actions of dibromopyruvic acid. J. Pharmacol. exp. Ther. **124**, 64—72 (1958)
Martin, W.R., Eades, C.G.: A comparative study of the effect of drugs on activating and vasomotor responses evoked by midbrain stimulation: Atropine, pentobarbital, chlorpromazine, and chlorpromazine sulfoxide. Psychopharmacologia (Berl.) **1**, 303—335 (1960)
Martin, W.R., Eades, C.G.: Demonstration of tolerance and physical dependence in the dog following a short-term infusion of morphine. J. Pharmacol. exp. Ther. **133**, 262—270 (1961)
Martin, W.R., Eades, C.G.: A comparison between acute and chronic physical dependence in the chronic spinal dog. J. Pharmacol. exp. Ther. **146**, 385—394 (1964)
Martin, W.R., Eades, C.G.: Pharmacological studies of spinal cord adrenergic and cholinergic mechanisms and their relation to physical dependence on morphine. Psychopharmacologia (Berl.) **11**, 195—223 (1967)
Martin, W.R., Eades, C.G.: Interactions between norepinephrine antagonists and potentiators (chlorpromazine, chlorpromazine sulfoxide, and imipramine) and sympathetic amines (amphetamine and methoxamine) on the flexor reflex of the chronic spinal dog. Int. J. Neuropharmacol. **7**, 493—501 (1968)
Martin, W.R., Eades, C.G.: The action of tryptamine on the dog spinal cord and its relationship to the agonistic actions of LSD-like psychotogens. Psychopharmacologia (Berl.) **17**, 242—257 (1970)
Martin, W.R., Eades, C.G.: Effects of phenethylamine (PEA) in the chronic spinal dog. Pharmacologist **16**, 205 (1974)
Martin, W.R., Eades, C.G., Fraser, H.F., Wikler, A.: Use of hindlimb reflexes of the chronic spinal dog for comparing analgesics. J. Pharmacol. exp. Ther. **144**, 8—11 (1964)
Martin, W.R., Eades, C.G., Thompson, J.A., Huppler, R.E., Gilbert, P.E.: The effects of morphine- and nalorphine-like drugs in the nondependent and morphine-dependent chronic spinal dog. J. Pharmacol. exp. Ther. **197**, 517—532 (1976)
Martin, W.R., Eades, C.G., Thompson, W.O., Thompson, J.A., Flanary, H.G.: Morphine physical dependence in the dog. J. Pharmacol. exp. Ther. **189**, 759—771 (1974)
Martin, W.R., Eisenman, A.J.: Interactions between nalorphine and morphine in the decerebrate cat. J. Pharmacol. exp. Ther. **138**, 113—119 (1962)

Martin, W.R., Fraser, H.F.: A comparative study of physiological and subjective effects of heroin and morphine administered intravenously in postaddicts. J. Pharmacol. exp. Ther. **133**, 388—399 (1961)
Martin, W.R., Fraser, H.F., Gorodetzky, C.W., Rosenberg, D.E.: Studies of the dependence-producing potential of the narcotic antagonist 2-cyclopropylmethyl-2′-hydroxy-5,9-dimethyl-6,7-benzomorphan (cyclazocine, WIN-20, 740, ARC II-C-3). J. Pharmacol. exp. Ther. **150**, 426—436 (1965)
Martin, W.R., Gorodetzky, C.W.: Demonstration of tolerance to and physical dependence on N-allylnormorphine (nalorphine). J. Pharmacol. exp. Ther. **150**, 437—442 (1965)
Martin, W.R., Gorodetzky, C.W.: Cyclazocine, an adjunct in the treatment of narcotic addiction. Int. J. Addict. **2**, 85—93 (1967)
Martin, W.R., Gorodetzky, C.W., McClane, T.K.: An experimental study in the treatment of narcotic addicts with cyclazocine. Clin. Pharmacol. Ther. **7**, 455—465 (1966)
Martin, W.R., Gorodetzky, C.W., Thompson, W.O.: Receptor dualism: Some kinetic implications. In: Kosterlitz, H.W., Collier, H.O.J., Villarreal, J.E.: Agonist and antagonist actions of narcotic analgesic drugs (Proceedings of the symposium of the British Pharmacological Society, Aberdeen, July 1971), pp. 30—44. London: Macmillan Press 1973a
Martin, W.R., Jasinski, D.R.: Physiological parameters of morphine dependence in man—tolerance, early abstinence, protracted abstinence. J. psychiat. Res. **7**, 9—17 (1969)
Martin, W.R., Jasinski, D.R., Haertzen, C.A., Kay, D.C., Jones, B.E., Mansky, P.A., Carpenter, R.W.: Methadone—A reevaluation. Arch. gen. Psychiat. **28**, 286—295 (1973b)
Martin, W.R., Jasinski, D.R., Mansky, P.A.: Naltrexone, an antagonist for the treatment of heroin dependence. Arch. gen. Psychiat. **28**, 784—791 (1973c)
Martin, W.R., Jasinski, D.R., Sapira, J.D., Flanary, H.G., Kelly, O.A., Thompson, A.K., Logan, C.R.: The respiratory effects of morphine during a cycle of dependence. J. Pharmacol. exp. Ther. **162**, 182—189 (1968)
Martin, W.R., Kay, D.C.: Effects of opioid analgesics and antagonists on the EEG. In: Longo, V.G. (Ed.): International Handbook of EEG and Clinical Neurophysiology, Vol. 7, Part C, pp. 97—109. Amsterdam: Elsevier Publishing Co. 1977
Martin, W.R., Sloan, J.W.: The pathophysiology of morphine dependence and its treatment with opioid antagonists. Pharmakopsychiat. Neuropsychopharmakol. **1**, 260—270 (1968)
Martin, W.R., Wikler, A., Eades, C.G., Pescor, F.T.: Tolerance to and physical dependence on morphine in rats. Psychopharmacologia (Berl.) **4**, 247—260 (1963)
Maruyama, Y., Hayashi, G., Smits, S.E., Takemori, A.E.: Studies on the relationship between 5-hydroxytryptamine turnover in brain and tolerance and physical dependence in mice. J. Pharmacol. exp. Ther. **178**, 20—29 (1971)
Maruyama, Y., Hayashi, G., Takemori, A.E.: Relation between 5-hydroxytryptamine (5-HT) and tolerance and physical dependence in mice. Pharmacologist **12**, 231 (1970)
Maruyama, Y., Takemori, A.E.: The role of dopamine and norepinephrine in the naloxone-induced abstinence of morphine-dependent mice. J. Pharmacol. exp. Ther. **185**, 602—608 (1973)
Master, A.M., Donoso, E.: Iproniazid in the anginal syndrome due to coronary disease. Ann. N.Y. Acad. Sci. **80**, 1020—1038 (1959)
Matthews, J.D., Labrecque, G., Domino, E.F.: Effects of morphine, nalorphine and naloxone on neocortical release of acetylcholine in the rat. Psychopharmacologia (Berl.) **29**, 113—120 (1973)
Mattila, M., Lavikainen, P.: The mouse tail reaction induced by morphine and the sedative action after reserpine and nalorphine. Ann. Med. exp. Fenn. **38**, 115—120 (1960)
Mattila, M., Saarnivaara, L.: Potentiation by indomethacin and 5-hydroxytryptamine of opiate analgesia in rabbits. Scand. J. clin. Lab. Invest. **19**, 62—63 (1967a)
Mattila, M.J., Saarnivaara, L.: Potentiation with indomethacin of the morphine analgesia in mice and rabbits. Ann. Med. exp. Fenn. **45**, 360—363 (1967b)
Mattila, M.J., Saarnivaara, L.: Modification by antihistaminic drugs of the morphine analgesia in rabbits. Ann. Med. exp. Fenn. **46**, 72—77 (1968)
Mayer, D.J., Wolfle, T.L., Akil, H., Carder, B., Liebeskind, J.C.: Analgesia from electrical stimulation in the brainstem of the rat. Science **174**, 1351—1354 (1971)
Maynert, E.W.: Effects of morphine on acetylcholine and certain other neurotransmitters. Arch. Biol. Med. exp. (Santiago) **4**, 36—41 (1967)

Maynert, E. W.: Catecholamine metabolism in the brain and adrenal medulla during addiction to morphine and in the early abstinence period. In: Wikler, A. (Ed.): The addictive states. Ass. Res. nerv. ment. Dis., Vol. 46, pp. 89—95. Baltimore: Williams and Wilkins 1968
Maynert, E. W., Klingman, G. I.: Tolerance to morphine. I. Effects on catecholamines in the brain and adrenal glands. J. Pharmacol. exp. Ther. **135**, 285—295 (1962)
Maynert, E. W., Klingman, G. I., Kaji, H. K.: Tolerance to morphine. II. Lack of effects on brain 5-hydroxytryptamine and γ-aminobutyric acid. J. Pharmacol. exp. Ther. **135**, 296—299 (1962)
Maynert, E. W., Levi, R.: Stress-induced release of brain norepinephrine and its inhibition by drugs. J. Pharmacol. exp. Ther. **143**, 90—95 (1964)
Medaković, M., Banić, B.: The action of reserpine and α-methyl-*m*-tyrosine on the analgesic effect of morphine in rats and mice. J. Pharm. Pharmacol. **16**, 198—206 (1964)
Menon, M. K., Dandiya, P. C., Bapna, J. S.: Modification of the effect of some central stimulants in mice pretreated with α-methyl-*l*-tyrosine. Psychopharmacologia (Berl.) **10**, 437—444 (1967)
Merali, Z., Ghosh, P. K., Hrdina, P. D., Singhal, R. L., Ling, G. M.: Alterations in striatal acetylcholine, acetylcholine esterase, and dopamine after methadone replacement in morphine-dependent rats. Europ. J. Pharmacol. **26**, 375—378 (1974)
Metÿs, J., Wagner, N., Metyšová, J., Herz, A.: Studies on the central antinociceptive action of cholinomimetic agents. Int. J. Neuropharmacol. **8**, 413—425 (1969)
Michaud, G. M., Jacob, J. J.: Interactions between morphine and naloxone in the intact dog: A contribution to the problem of acute dependence. Presented at 35th Meeting, Committee on Problems of Drug Dependence, National Research Council, Chapel Hill, N.C. 1973
Miller, J. W., George, R., Elliott, H. W., Sung, C. Y., Way, E. L.: The influence of the adrenal medulla in morphine analgesia. J. Pharmacol. exp. Ther. **114**, 43—50 (1955)
Milošević, M. P.: Effect of adrenaline on the analgesic response of mice to morphine and related drugs. Arch. int. Pharmacodyn. **104**, 50—56 (1955)
Mizoguchi, K.: The sites of action of morphine and the antagonistic action of levallorphan on the central nervous system of the dog. Folia pharmacol. jap. **60**, 326—346 (1964)
Monnier, M., Nosal, Gl., Radouco-Thomas, C.: Central mechanisms of pain analysed by the action of analgesics. In: Keele, C. A., Smith, R. (Eds.): The assessment of pain in man and animals, pp. 144—155. London: Universities Federation for Animal Welfare 1962
Moore, K. E., McCarthy, L. E., Borison, H. L.: Blood glucose and brain catecholamine levels in the cat following the injection of morphine into the cerebrospinal fluid. J. Pharmacol. exp. Ther. **148**, 169—175 (1965)
Mosnaim, A. D., Inwang, E. E.: A spectrophotometric method for the quantification of 2-phenylethylamine in biological specimens. Analyt. Biochem. **54**, 561—577 (1973)
Mudgill, L., Friedhoff, A. J., Tobey, J.: Effect of intraventricular administrations of epinephrine, norepinephrine, dopamine, acetylcholine, and physostigmine on morphine analgesia in mice. Arch. int. Pharmacodyn. **210**, 85—91 (1974)
Mulé, S. J., Brill, H. (Eds.): Chemical and biological aspects of drug dependence. Cleveland, Ohio: Chemical Rubber Co. Press 1972
Mulé, S. J., Clements, T. H., Layson, R. C., Haertzen, C. A.: Analgesia in guinea pigs: A measure of tolerance development. Arch. int. Pharmacodyn. **173**, 201—212 (1968)
Mullin, W. J., Phillis, J. W., Pinsky, C.: Morphine enhancement of acetylcholine release from the brain in unanaesthetized cats. Europ. J. Pharmacol. **22**, 117—119 (1973)
Mũnoz, C.: Role of catecholamines in morphine analgesia. Acta physiol. lat.-amer. **18**, 277 (1968)
Mũnoz, C., Paeile, C.: Changes in morphine analgesia induced by drugs which modify catecholamine content of the brain. Arch. Biol. Med. exp. (Santiago) **4**, 63—68 (1967)
Murphy, D. L.: Mental effects of L-DOPA. Ann. Rev. Med. **24**, 209—216 (1973)
Nakamura, J., Mitchell, C. L.: The effects of morphine, pentobarbital and saline on bioelectrical potentials recorded in limbic structures of the cat evoked by radial nerve and direct brain stimulation. Arch. int. Pharmacodyn. **200**, 70—87 (1972)
Nakamura, J., Winters, W. D.: Attenuation of the morphine EEG continuum following a repeat dose within 16 days: Delayed tolerance in the rat. Neuropharmacology **12**, 607—617 (1973)
Nakamura, K., Kuntzman, R., Maggio, A. C., Augulis, V.: Influence of 6-hydroxydopamine on the effect of morphine on the tail-flick latency. Psychopharmacologia (Berl.) **31**, 177—189 (1973a)
Nakamura, K., Kuntzman, R., Maggio, A., Conney, A. H.: Effect of 6-hydroxy-dopamine on catecholamine concentrations and behaviour in the morphine-tolerant rat. J. Pharm. Pharmacol. **24**, 484—487 (1972)

Nakamura,K., Kuntzman,R., Maggio,A., Conney,A.H.: Restoration of morphine analgesia in morphine-tolerant rats after the intraventricular administration of 6-hydroxydopamine. J. Pharm. Pharmacol. **25**, 584—587 (1973b)
Nash,P., Colasanti,B., Khazan,N.: Long-term effects of morphine on the electroencephalogram and behavior of the rat. Psychopharmacologia (Berl.) **29**, 271—276 (1973)
Navarro,G., Elliott,H.W.: The effects of morphine, morphinone, and thebaine on the EEG and behavior of rabbits and cats. Neuropharmacology **10**, 367—377 (1971)
Neal,M.J.: Failure of morphine dependence in rats to influence brain noradrenaline turnover. J. Pharm. Pharmacol. **20**, 950—953 (1968)
Neuman,R.S., Calvillo,O., Henry,J.L.: Blockade by morphine of nociceptive neurones in the dorsal horn of the cat. Pharmacologist **16**, 203 (1974)
Nicák,A.: The influence of serotonine and amphetamine on analgesic effect of morphine after reserpine premedication in rats and mice. Med. Pharmacol. exp. **13**, 43—48 (1965)
Nutt,J.G., Jasinski,D.R.: Comparison of intravenously administered methadone, morphine, heroin, and placebo. Fed. Proc. **32**, 694 (1973)
Ogiu,K., Takagi,H., Yamamoto,S.: Studies on the site of antagonistic action of nalorphine on morphine. Folia pharmacol. jap. **54**, 1—6 (1958)
Olds,J.: Selfs-stimulation of the brain. Its use to study local effects of hunger, sex, and drugs. Science **127**, 315—324 (1958)
Olds,J., Travis,R.P.: Effects of chlorpromazine, meprobamate, pentobarbital, and morphine on self-stimulation. J. Pharmacol. exp. Ther. **128**, 397—404 (1960)
Oliverio,A., Castellano,C.: Genotype-dependet sensitivity and tolerance to morphine and heroin: Dissociation between opiate-induced running and analgesia in the mouse. Psychopharmacologia (Berl.) **39**, 13—22 (1974)
Olson,R.W.: MMPI sex differences in narcotic addicts. J. gen. Psychol. **71**, 257—266 (1964)
Opitz,K., Reimann,I.: Suppression of the drug-induced morphine withdrawal syndrome by cyproheptadine. Psychopharmacologia (Berl.) **28**, 165—170 (1973)
Outschoorn,A.S.: The hormones of the adrenal medulla and their release. Brit. J. Pharmacol. **7**, 605—615 (1952)
Paalzow,L.: Analgesia produced by clonidine in mice and rats. J. Pharm. Pharmacol. **26**, 361—363 (1974)
Parker,R.B.: Mouse locomotor activity: Effect of morphine, narcotic antagonists, and the interaction of morphine and narcotic antagonists. Psychopharmacologia (Berl.) **38**, 15—23 (1974)
Pasternak,G.W., Goodman,R., Snyder,S.H.: An endogenous morphine-like factor in mammalian brain. Life Sci. **16**, 1765—1769 (1975)
Paton,W.D.M.: The action of morphine and related substances on contraction and on acetylcholine output of coaxially stimulated guinea pig ileum. Brit. J. Pharmacol. **12**, 119—127 (1957)
Pelikan,E.W.: The mechanism of ganglionic blockade produced by nicotone. Ann. N.Y. Acad. Sci. **90**, 52—69 (1960)
Perez-Cruet,J., DiChiara,G., Gessa,G.L.: Accelerated synthesis of dopamine in the rat brain after methadone. Experientia (Basel) **28**, 926 (1972)
Pert,A., Yaksh,T.: Sites of morphine induced analgesia in the primate brain: relation to pain pathways. Brain Res. **80**, 135—140 (1974)
Pert,C.B., Snyder,S.H.: Opiate receptor: Demonstration in nervous tissue. Science **179**, 1011—1014 (1973a)
Pert,C.B., Snyder,S.H.: Properties of opiate-receptor binding in rat brain. Proc. nat. Acad. Sci. (Wash.) **70**, 2243—2247 (1973b)
Phillis,J.W., Mullin,W.J., Pinsky,C.: Morphine enhancement of acetycholine release into the lateral ventricle and from the cerebral cortex of unanesthetized cats. Comp. gen. Pharmacol. **4**, 189—200 (1973)
Pinsky,C., Frederickson,R.C.A., Vazquez,A.J.: Morphine withdrawal snydrome responses to cholinergic antagonists and to a partial cholinergic agonist. Nature (Lond.) **242**, 59—60 (1973)
Plant,O.H., Pierce,I.H.: Studies of chronic morphine poisoning in dogs. I. General symptoms and behavior during addiction and withdrawal. J. Pharmacol. exp. Ther. **33**, 329—357 (1928)
Pleuvry,B.J.: Cross tolerance between methylamphetamine and morphine in the mouse. J. Pharm. Pharmacol. **23**, 969—970 (1971)

Pleuvry, B.J., Tobias, M.A.: Comparison of the antinociceptive activities of physostigmine, oxotremorine and morphine in the mouse. Brit. J. Pharmacol. **43**, 706—714 (1971)
Pollock, D., Muir, T.C., MacDonald, A., Henderson, G.: Morphine-induced changes in the sensitivity of the isolated colon and vas deferens of the rat. Europ. J. Pharmacol. **20**, 321—328 (1972)
Proudfit, H.K., Anderson, E.G.: Blockade of morphine analgesia by destruction of a bulbospinal serotonergic pathway. Pharmacologist **16**, 203 (1974)
Pruitt, D.B., Grubb, M.N., Jaquette, D.L., Burks, T.F.: Intestinal effects of 5-hydroxytryptamine and morphine in guinea pigs, dogs, cats, and monkeys. Europ. J. Pharmacol. **26**, 298—305 (1974)
Puharich, V., Goetzl, F.R.: The influence of adrenalectomy upon analgesic effectiveness of morphine in rats. Permanente Fdn. med. Bull. **5**, 19—22 (1947)
Puri, S.K., Lal, H.: Effect of apomorphine, benztropine or morphine on striatal dopamine turnover: Evidence for latent supersensitivity of dopaminergic receptors in morphine dependent rats. Pharmacologist **15**, 247 (1973a)
Puri, S.K., Lal, H.: Effect of dopaminergic stimulation or blockade on morphine-withdrawal aggression. Psychopharmacologia (Berl.) **32**, 113—120 (1973b)
Puri, S.K., Lal, H.: Tolerance to the behavioral and neurochemical effects of haloperidol and morphine in rats chronically treated with morphine or haloperidol. Naunyn-Schmiedebergs Arch. exp. Path. Pharmakol. **282**, 155—170 (1974a)
Puri, S.K., Lal, H.: Reduced threshold to pain induced aggression specifically related to morphine dependence. Psychopharmacologia (Berl.) **35**, 237—241 (1974b)
Puri, S.K., Reddy, C., Lal, H.: Blockade of central dopaminergic receptors by morphine: Effect of haloperidol, apomorphine or benztropine. Res. Comm. chem. Pathol. Pharmacol. **5**, 389—401 (1973)
Quinn, G.P., Brodie, B.B., Shore, P.A.: Drug-induced release of norepinephrine in cat brain. J. Pharmacol exp. Ther. **122**, 63A (1958)
Radouco-Thomas, C., Nosal, Gl., Radouco-Thomas, S. (in collaboration with Chaumontet, J.M. and Capt, M.): On the experimental pain threshold in animals. In: Keele, C.A., Smith, R. (Eds.): The assessment of pain in man and animals. pp. 271—289. London: Universities Federation for Animal Welfare 1962
Radouco-Thomas, S., Singh, P., Garcin, F., Radouco-Thomas, C.: Relationship between experimental analgesia and brain monoamines: Catecholamines and 5-hydroxytryptamine. Effects of precursors (DOPA, 5HTP) and monoamines modifying drugs. Arch. Biol. med. exp. (Santiago) **4**, 42—62 (1967)
Rech, R.H., Tilson, H.A.: Effects of p-chlorophenylalanine (p-CPA) on morphine analgesia and development of tolerance and dependence in two strains of rats. Pharmacologist **15**, 202 (1973)
Reis, D.J., Rifkin, M., Corvelli, A.: Effects of morphine on cat brain norepinephrine in regions with daily monoamine rhythms. Europ. J. Pharmacol. **9**, 149—152 (1969)
Repkin, A.H., Proudfit, H.K., Anderson, E.G.: Primary afferent depolarization as a mechanism of morphine analgesia. Pharmacologist **16**, 203 (1974)
Rethy, C.R., Smith, C.B., Villarreal, J.E.: Effects of narcotic analgesics upon the locomotor activity and brain catecholamine content of the mouse. J. Pharmacol. exp. Ther. **176**, 472—479 (1971)
Reynolds, A.K., Randall, L.O.: Morphine and allied drugs. Toronto: University of Toronto Press 1957
Riblet, L.A., Mitchell, C.L.: The effect of cervical spinal section on the ability of morphine to elevate the jaw jerk threshold to electrical stimulation of the tooth pulp in cats. J. Pharmacol. exp. Ther. **180**, 610—615 (1972)
Richardson, J.A., Woods, E.F., Richardson, A.K.: Elevations in plasma catecholamines following morphine. J. Pharmacol. exp. Ther. **122**, 64A (1958)
Richter, J.A., Goldstein, A.: Effects of morphine and levorphanol on brain acetylcholine content in mice. J. Pharmacol. exp. Ther. **175**, 685—691 (1970)
Risner, M.E.: The possible lack of noradrenergic mechanisms in *d*-amphetamine (A) self-administration. Fed. Proc. **34**, 768 (1975)

Risner, M. E., Khavari, K. A.: Morphine dependence in rats produced after five days of ingestion. Psychopharmacologia (Berl.) **28**, 51—62 (1973)

Ritchie, J. M., Armett, C. J.: On the role of acetylcholine in conduction in mammalian nonmyelinated nerve fibers. J. Pharmacol. exp. Ther. **139**, 201—207 (1963)

Rosenman, S. J., Smith, C. B.: ^{14}C-Catecholamine synthesis in mouse brain during morphine withdrawal. Nature (Lond.) **240**, 153—156 (1972)

Ross, J. W., Ashford, A.: The effect of reserpine and α-methyldopa on the analgesic action of morphine in the mouse. J. Pharm. Pharmacol. **19**, 709—713 (1967)

Rudzik, A. D., Mennear, J. H.: Antagonism of analgesics by amine-depleting agents. J. Pharm. Pharmacol. **17**, 326—327 (1965)

Saarnivaara, L.: Effect of 5-hydroxytryptamine on morphine analgesia in rabbits. Ann. Med. exp. Fenn. **47**, 113—123 (1969a)

Saarnivaara, L.: Analgesic activity of some sympathetic drugs and their effect on morphine analgesia in rabbits. Ann. Med. exp. Fenn. **47**, 180—190 (1969b)

Saavedra, J. M.: Enzymatic isotopic assay for and presence of β-phenylethylamine in brain. J. Neurochem. **22**, 211—216 (1974)

Samanin, R., Bernasconi, S.: Effects of intraventricularly injected 6-OH dopamine or midbrain raphe lesion on morphine analgesia in rats. Psychopharmacologia (Berl.) **25**, 175—182 (1972)

Samanin, R., Ghezzi, D., Mauron, C., Valzelli, L.: Effect of midbrain raphé lesion on the antinociceptive action of morphine and other analgesics in rats. Psychopharmacologia (Berl.) **33**, 365—368 (1973)

Samanin, R., Gumulka, W., Valzelli, L.: Reduced effect of morphine in midbrain raphé lesioned rats. Europ. J. Pharmacol. **10**, 339—343 (1970)

Samanin, R., Valzelli, L.: Increase of morphine-induced analgesia by stimulation of the nucleus raphé dorsalis. Europ. J. Pharmacol. **16**, 298—302 (1971)

Sasame, H. A., Perez-Cruet, J., DiChiara, G., Tagliamonte, A., Tagliamonte, P., Gessa, G. L.: Evidence that methadone blocks dopamine receptors in the brain. J. Neurochem. **19**, 1953—1957 (1972)

Satake, Y.: Secretion of adrenaline and sympathins. Tôhoku, J. exp. Med. **60**, Suppl. II, 1954

Sato, H., Ohmi, F.: Action of morphine on the epinephrine output, blood sugar content and blood pressure in dogs. Tôhoku J. exp. Med. **21**, 411—432 (1933)

Satoh, M., Takagi, H.: Enhancement by morphine of the central descending inhibitory influence on spinal sensory transmission. Europ. J. Pharmacol. **14**, 60—65 (1971a)

Satoh, M., Takagi, H.: Effect of morphine on the pre- and postsynaptic inhibitions in the spinal cord. Europ. J. Pharmacol. **14**, 150—154 (1971b)

Saxena, P. N., Gupta, G. P.: Potentiating effect of prostigmine on morphine-induced analgesia. Indian J. med. Res. **45**, 319—325 (1957)

Sawyer, C. H., Critchlow, B. V., Barraclough, C. A.: Mechanism of blockade of pituitary activation in the rat by morphine, atropine, and barbiturates. Endocrinology **57**, 345—354 (1955)

Schallek, W., Kühn, A.: Effects of drugs on spontaneous and activated EEG of cat. Arch. int. Pharmacodyn. **120**, 319—333 (1959)

Schaumann, W.: The paralysing action of morphine on the guinea pig ileum. Brit. J. Pharmacol. **10**, 456—461 (1955)

Schaumann, W.: Inhibition by morphine of the release of acetylcholine from the intestine of the guinea pig. Brit. J. Pharmacol. **12**, 115—117 (1957)

Schaumann, W.: Beeinflussung der analgetischen Wirkung des Morphins durch Reserpin. Naunyn Schmiedebergs Arch. exp. Path. Pharmak. **235**, 1—9 (1958)

Schmitt, H., Le Douarec, J.-C., Petillot, N.: Antinociceptive effects of some α-sympathomimetic agents. Neuropharmacology **13**, 289—294 (1974)

Schneider, J. A.: Reserpine antagonism of morphine analgesia in mice. Proc. Soc. exp. Biol. (N.Y.) **87**, 614—615 (1954)

Schuberth, J., Sundwall, A.: Effects of some drugs on the uptake of acetylcholine in cortex slices of mouse brain. J. Neurochem. **14**, 807—812 (1967)

Schulz, R., Cartwright, C.: Effect of morphine on serotonin release from myenteric plexus of the guinea pig. J. Pharmacol. exp. Ther. **190**, 420—430 (1974)

Schulz, R., Cartwright, C., Goldstein, A.: Reversibility of morphine tolerance and dependence in guinea pig brain and myenteric plexus. Nature (Lond.) **251**, 329—331 (1974)

Schulz, R., Goldstein, A.: Morphine tolerance and supersensitivity to 5-hydroxytryptamine in the myenteric plexus of the guinea pig. Nature (Lond.) **244**, 168—170 (1973)
Schwartz, A. S., Eidelberg, E.: Role of biogenic amines in morphine dependence. Life Sci. **9**, 613—624 (1970)
Schweizer, W.: Isopropythydrazides in angina pectoris. Ann. N. Y. Acad. Sci. **80**, 1016—1019 (1959)
Seevers, M. H.: Opiate addiction in the monkey. I. Methods of study. J. Pharmacol. exp. Ther. **56**, 147—156 (1936)
Seevers, M. H.: Possible mechanisms of physical dependence to narcotics. In: Sevag, M. G., Reid, R. D., Reynolds, O. E. (Eds.): Origins of resistance to toxic agents, pp. 244—263. New York: Academic Press 1955
Seevers, M. H., Deneau, G. A.: A critique of the "dual action" hypothesis of morphine physical dependence. Arch. int. Pharmacodyn. **140**, 514—520 (1962)
Seevers, M. H., Deneau, G. A.: Physiological aspects of tolerance and physical dependence. In: Root, W. S., Hofmann, F. G. (Eds.): Physiological pharmacology. Vol. 1, pp. 565—640. New York: Academic Press 1963
Segal, M., Deneau, G. A.: Brain levels of epinephrine, norepinephrine, dopamine and 5-HT during administration and withdrawal of morphine in monkeys. Fed. Proc. **21**, 327 (1962)
Segal, M., Deneau, G. A., Seevers, M. H.: Levels and distribution of central nervous system amines in normal and morphine-dependent monkeys. Neuropharmacology **11**, 211—222 (1972)
Sethy, V. H., Prahdan, R. J., Mandrekar, S. S., Sheth, U. K.: Role of brain amines in the analgesic action of meperidine hydrochloride. Psychopharmacologia (Berl.) **17**, 320—326 (1970)
Sewell, R. D. E., Spencer, P. S. J.: Modification of the antinociceptive activity of narcotic agonists and antagonists by intraventricular injection of biogenic amines in mice. Brit. J. Pharmacol. **51**, 140—141 P (1974)
Sharkawi, M., Goldstein, A.: Antagonism by physostigmine of the "running fit" caused by levorphanol, a morphine congener, in mice. Brit. J. Pharmacol. **37**, 123—128 (1969)
Sharkawi, M., Schulman, M. P.: Inhibition by morphine of the release of $\{^{14}C\}$ acetylcholine from rat brain cortex slices. J. Pharm. Pharmacol. **21**, 546—547 (1969)
Sharman, D. F.: Changes in the metabolism of 3,4-dihydroxyphenylethylamine (dopamine) in the striatum of the mouse induced by drugs. Brit. J. Pharmacol. **28**, 153—163 (1966)
Shein, H. M., Larin, F., Wurtman, R. J.: Lack of a direct effect of morphine on the synthesis of pineal ^{14}C-indoles in organ culture. Life Sci. **9**, 29—33 (1970)
Sheldon, M. I., Smith, C. B.: Effects of morphine and of forced running upon ^{14}C-catecholamine synthesis in various tissues of the mouse. Pharmacologist **13**, 313 (1971)
Shen, F., Loh, H. H., Way, E. L.: Reserpine antagonism of morphine analgesia in tolerant mice. Fed. Proc. **28**, 793 (1969)
Shen, F. H., Loh, H. H., Way, E. L.: Brain serotonin turnover in morphine tolerant and dependent mice. J. Pharmacol. exp. Ther. **175**, 427—434 (1970)
Sherbel, A. L., Harrison, J. W.: The effects of iproniazid and other amine oxidase inhibitors in rheumatoid arthritis. Ann. N.Y. Acad. Sci. **80**, 820—834 (1959)
Sherman, A. D., Mitchell, C. L.: Effect of morphine on regional levels of gamma-hydroxybutyrate in mouse brain. Neuropharmacology **13**, 239—243 (1974)
Shiomi, H., Takagi, H.: Morphine analgesia and the bulbospinal noradrenergic system: Increase in the concentration of normetanephrine in the spinal cord of the rat caused by analgesics. Brit. J. Pharmacol. **52**, 519—526 (1974)
Shoham, S., Weinstock, M.: The role of supersensitivity to acetylcholine in the production of tolerance to morphine in stimulated guinea pig ileum. Brit. J. Pharmacol. **52**, 597—603 (1974)
Shuster, L.: Repression and de-repression of enzyme synthesis as a possible explanation of some aspects of drug action. Nature (Lond.) **189**, 314—315 (1961)
Shuster, L., Hannam, R. V., Boyle, W. E., Jr.: A simple method for producing tolerance to dihydromorphinone in mice. J. Pharmacol. exp. Ther. **140**, 149—154 (1963)
Sibuta, H., Endo, K., Nagakura, G.: The increase of the epinephrine output under morphine depends upon the integrity of the splanchnic nerves. Tôhoku J. exp. Med. **50**, 1—6 (1949)
Sigg, E. B., Caprio, G., Schneider, J. A.: Synergism of amines and antagonism of reserpine to morphine analgesia. Proc. Soc. exp. Biol. (N.Y.) **97**, 97—100 (1958)

Silver, S.: Über die Schmerzüberempfindlichkeit durch Schlafmittel und ihre Beeinflussung. Ein Beitrag zum Mechanismus der Morphiumwirkung. Naunyn-Schmiedebergs Arch. exp. Path. Pharmak. **158**, 219—232 (1930)
Silvestrini, B., Longo, V. G.: Selective activity of morphine on the "EEG arousal reaction" to painful stimuli. Experientia (Basel) **12**, 436—437 (1956)
Simon, E. J.: Sites of action of narcotic analgesic drugs: Single cells. In: Clouet, D. H. (Ed.): Narcotic drugs; biochemical pharmacology, pp. 310—341. New York-London: Plenum Press 1971
Simon, E. J., Hiller, J. M., Edelman, I.: Stereospecific binding of the potent narcotic analgesic (^{3}H) etorphine to rat-brain homogenate. Proc. nat. Acad. Sci. (Wash.) **70**, 1947—1949 (1973)
Simon, E. J., Rosenberg, P.: Effects of narcotics on the giant axon of the squid. J. Neurochem. **17**, 881—887 (1970)
Sinclair, J. G.: Morphine and meperidine on bulbospinal inhibition of the monosynaptic reflex. Europ. J. Pharmacol. **21**, 111—114 (1973)
Sinitsin, L. N.: Effect of morphine and other analgesics on brain evoked potentials. Int. J. Neuropharmacol. **3**, 321—326 (1964)
Slaughter, D., Gross, E. G.: Some new aspects of morphine action. Effect on intestine and blood pressure; toxicity studies. J. Pharmacol. exp. Ther. **68**, 96—103 (1940)
Slaughter, D., Munsell, D. W.: Some new aspects of morphine action. Effects on pain. J. Pharmacol. exp. Ther. **68**, 104—112 (1940)
Sloan, J. W., Brooks, J. W., Eisenman, A. J., Martin, W. R.: Comparison of the effects of single doses of morphine and thebaine on body temperature, activity and brain and heart levels of catecholamines and serotonin. Psychopharmacologia (Berl.) **3**, 291—301 (1962)
Sloan, J. W., Brooks, J. W., Eisenman, A. J., Martin, W. R.: The effect of addiction to and abstinence from morphine on rat tissue catecholamine and serotonin levels. Psychopharmacologia (Berl.) **4**, 261—270 (1963)
Sloan, J. W., Eisenman, A. J.: Long-persisting changes in catecholamine metabolism following addiction to and withdrawal from morphine. In: Wikler, A. (Ed.): The addictive states. Ass. Res. nerv. ment. Dis., Vol. 46, pp. 96—105. Baltimore: Williams & Wilkins 1968
Smith, A. A., Karmin, M., Gavitt, J.: Blocking effect of puromycin, ethanol, and chloroform on the development of tolerance to an opiate. Biochem. Pharmacol. **15**, 1877—1879 (1966)
Smith, A. A., Karmin, M., Gavitt, J.: Tolerance to the lenticular effects of opiates. J. Pharmacol. exp. Ther. **156**, 85—91 (1967)
Smith, C. B., Sheldon, M. I., Bednarczyk, J. H., Villarreal, J. E.: Morphine-induced increases in the incorporation of ^{14}C-tyrosine into ^{14}C-dopamine and ^{14}C-norepinephrine in the mouse brain: Antagonism by naloxone and tolerance. J. Pharmacol. exp. Ther. **180**, 547—557 (1972)
Smith, C. B., Villarreal, J. E., Bednarczyk, J. H., Sheldon, M. I.: Tolerance to morphine-induced increases in $\{^{14}C\}$ catecholamine synthesis in mouse brain. Science **170**, 1106—1107 (1970)
Smith, G. M., Beecher, H. K.: Measurement of "mental clouding" and other subjective effects of morphine. J. Pharmacol. exp. Ther. **126**, 50—62 (1959)
Smith, G. M., Beecher, H. K.: Subjective effects of heroin and morphine in normal subjects. J. Pharmacol. exp. Ther. **136**, 47—52 (1962)
Smith, G. M., Semke, C. W., Beecher, H. K.: Objective evidence of mental effects of heroin, morphine and placebo in normal subjects. J. Pharmacol. exp. Ther. **136**, 53—58 (1962)
Smits, S. E., Takemori, A. E.: Lack of transfer of morphine tolerance by administration of rat cerebral homogenates. Proc. Soc. exp. Biol. (N.Y.) **127**, 1167—1171 (1968)
Soteropoulos, G. G., Standaert, F. G.: Neuromuscular effects of morphine and naloxone. J. Pharmacol. exp. Ther. **184**, 136—142 (1973)
Soulairac, A., Cahn, J., Charpentier, J.: Pain (Proceedings of the International Symposium on Pain, Paris, April 11—13, 1967). New York: Academic Press 1968
Soulairac, A., Gottesmann, Cl., Charpentier, J.: Effects of pain and of several central analgesics on cortex, hippocampus and reticular formation of brain stem. Int. J. Neuropharmacol. **6**, 71—81 (1967)
Sparkes, G. G., Spencer, P. S. J.: Modification of morphine analgesia in the rat by biogenic amines administered intraventricularly. Brit. J. Pharmacol. **35**, 362P—363P (1969)
Spector, S.: Quantitative determination of morphine in serum by radioimmunoassay. J. Pharmacol. exp. Ther. **178**, 253—258 (1971)

Spector, S., Parker, C. W.: Morphine: Radioimmunoassay. Science **168**, 1347—1348 (1970)
Spoerlein, M. T., Scrafini, J.: Effects of time and 8-azaquanine on the development of morphine tolerance. Life Sci. **6**, 1549—1564 (1967)
Steinert, H. R., Holtzman, S. G., Jewett, R. E.: Some agonistic actions of the morphine antagonist levallorphan on behavior and brain monoamines in the rat. Psychopharmacologia (Berl.) **31**, 35—48 (1973)
Stern, P.: Substance P as a sensory transmitter and its other central effects. Ann. N.Y. Acad. Sci. **104**, 403—415 (1963)
Stern, P., Hadžović, J.: Pharmacological analysis of central actions of synthetic Substance P. Arch. int. Pharmacodyn. **202**, 259—262 (1973)
Stern, P., Hukovíc, S.: Beziehungen zwischen zentraler und peripherer Wirkung der Substanz P. Med exp. (Berl.) **2**, 1—7 (1960)
Stewart, G. N., Rogoff, J. M.: The influence of certain factors, especially emotional disturbances, on the epinephrin content of the adrenals. J. exp. Med. **24**, 709—738 (1916)
Stewart, G. N., Rogoff, J. M.: The action of drugs on the output of epinephrin from the adrenals. VIII. Morphine. J. Pharmacol. exp. Ther. **19**, 59—85 (1922)
Stille, G.: Zur Pharmakologie katatonigener Stoffe. 4. Mitteilung: Die Wirkung von Morphin. Arzneimittel-Forsch. (Drug Res.) **21**, 650—654 (1971)
Sturtevant, F. M., Drill, V. A.: Tranquillizing drugs and morphine-mania in cats. Nature (Lond.) **179**, 1253 (1957)
Sugrue, M. F.: The effects of acutely administered analgesics on the turnover of noradrenaline and dopamine in various regions of the rat brain. Brit. J. Pharmacol. **52**, 159—165 (1974)
Sung, C. Y., Way, E. L., Scott, K. G.: Studies on the relationship of metabolic fate and hormonal effects of D,L-methadone to the development of drug tolerance. J. Pharmacol. exp. Ther. **107**, 12—23 (1953)
Sweet, W. H.: Pain. In: Handbook of Physiology, Section 1: Neurophysiology, Vol. 1, pp. 459—506. Washington, D.C.: American Physiological Society 1959
Tachikawa, Y.: Effect of morphine injections on blood-sugar and adrenalin in dogs. J. Orient. Med. **17**, 521—528 (1932)
Takagi, H., Nakama, M.: Effect of morphine and nalorphine on the content of dopamine in mouse brain. Jap. J. Pharmacol. **16**, 483—484 (1966)
Takagi, H., Nakama, M.: Studies on the mechanism of action of tetrabenazine as a morphine antagonist. II. A participation of catecholamine in the antagonism. Jap. J. Pharmacol. **18**, 54—58 (1968)
Takagi, H., Matsumura, M., Yanai, A., Ogiu, K.: The effect of analgesics on the spinal reflex activity of the cat. Jap. J. Pharmacol. **4**, 176—187 (1955)
Takagi, H., Takashima, T., Kimura, K.: Antagonism of the analgetic effect of morphine in mice by tetrabenazine and reserpine. Arch. int. Pharmacodyn. **149**, 484—492 (1964)
Takagi, K., Takayanagi, I., Irikura, T., Nishino, K., Ichinoseki, N., Shishido, K.: Responses of the isolated ileum of the morphine-tolerant guinea pig. Arch. int. Pharmacodyn. **158**, 39—44 (1965)
Takemori, A. E., Oka, T., Nishiyama, N.: Alteration of analgesic receptor-antagonist interaction induced by morphine. J. Pharmacol. exp. Ther. **186**, 261—265 (1973)
Takemori, A. E., Tulunay, F. C.: Further studies on the receptor alteration induced by morphine. Pharmacologist **15**, 242 (1973)
Tamarkin, N. R., Goodwin, F. K., Axelrod, J.: Rapid elevation of biogenic amine metabolites in human CSF following probenecid. Life Sci. **9**, 1397—1408 (1970)
Tanabe, T., Cafruny, E. J.: Adrenal hypertrophy in rats treated chronically with morphine. J. Pharmacol. exp. Ther. **122**, 148—153 (1958)
Tatum, A. L., Seevers, M. H., Collins, K. H.: Morphine addiction and its physiological interpretation based on experimental evidences. J. Pharmacol. exp. Ther. **36**, 447—475 (1929)
Teitelbaum, H., Catravas, G. N., McFarland, W. L.: Reversal of morphine tolerance after medial thalamic lesions in the rat. Science **185**, 449—451 (1974)
Tenen, S. S.: The effects of p-chlorophenylalanine, a serotonin depletor, on avoidance acquisition, pain sensitivity and related behavior in the rat. Psychopharmacologia (Berl.) **10**, 204—219 (1967)

Tenen,S.S.: Antagonism of the analgesic effect of morphine and other drugs by p-chlorophenylalanine, a serotonin depletor. Psychopharmacologia (Berl.) **12**, 278—285 (1968)

Terenius,L.: Stereospecific interaction between narcotic analgesics and a synaptic plasma membrane fraction of rat cerebral cortex. Acta pharmacol. toxicol. **32**, 317—320 (1973)

Terenius,L.: Effect of peptides and amino acids on dihydromorphine binding to the opiate receptor. J. Pharm. Pharmacol. **27**, 450—452 (1975)

Terenius,L., Gispen,W.H., De Wied,D.: ACTH-like peptides and opiate receptors in the rat brain: Structure-activity studies. Europ. J. Pharmacol. **33**, 395—399 (1975)

Terenius,L., Wahlström,A.: Inhibitor(s) of narcotic receptor binding in brain extracts and cerebrospinal fluid. Acta pharmacol. **35**, (Suppl.): 55, 1974

Terenius,L., Wahlström,A.: Search for an endogenous ligand for the opiate receptor. Acta physiol. scand. **94**, 74—81 (1975a)

Terenius,L., Wahlström,A.: Morphine-like ligand for opiate receptors in human CSF. Life Sci. **16**, 1759—1764 (1975b)

Teschemacher,H., Opheim,K.E., Cox,B.M., Goldstein,A.: A peptide-like substance from pituitary that acts like morphine. 1. Isolation. Life Sci. **16**, 1771—1776 (1975)

Teschemacher,Hj., Schubert,P., Herz,A.: Autoradiographic studies concerning the supraspinal site of the antinociceptive action of morphine when inhibiting the hindleg flexor reflex in rabbits. Neuropharmacology **12**, 123—131 (1973)

Tilson,H.A., Rech,R.H., Stolman,S.: Hyperalgesia during withdrawal as a means of measuring the degree of dependence in morphine dependent rats. Psychopharmacologia (Berl.) **28**, 287—300 (1973)

Tirri,R.: Transfer of induced tolerance to morphine and promazine by brain homogenate. Experientia (Basel) **23**, 278 (1967)

Torda,C., Wolff,H.G.: Effect of convulsant and anticonvulsant agents on acetylcholine metabolism (activity of choline acetylase, cholinesterase) and on sensitivity to acetylcholine of effector organs. Amer. J. Physiol. **151**, 345—354 (1947)

Trafton,C.L., Marques,P.R.: Effects of septal area and cingulate cortex lesions on opiate addiction behavior in rats. J. comp. physiol. Psychol. **75**, 277—285 (1971)

Trendelenberg,P.: Physiologische und pharmakologische Versuche über die Dünndarmperistaltik. Naunyn-Schmiedebergs Arch. exp. Path. Pharmak. **81**, 55—129 (1917)

Trendelenburg,U.: The action of morphine on the superior cervical ganglion and on the nictitating membrane of the cat. Brit. J. Pharmacol. **12**, 79—85 (1957)

Tripod,J., Gross,F.: Unterschiedliche Beeinflussung der analgetischen und der erregenden Wirkung von Morphin durch zentral dämpfende Pharmaka. Helv. physiol. pharmacol. Acta. **15**, 105—116 (1957)

Tseng,L.F., Loh,H.H., Ho,I.K., Way,E.L.: The role of brain catecholamines in naloxone induced withdrawal in morphine dependent rats. Proc. West. Pharmacol. Soc. **17**, 178—183 (1974)

Tsou,K., Jang,C.S.: Studies on the site of analgesic action of morphine by intracerebral microinjections. Scientia Sinica **8**, 1099—1109 (1964)

Tulunay,F.C., Takemori,A.E.: The increased efficacy of narcotic antagonists induced by various narcotic analgesics. J. Pharmacol. exp. Ther. **190**, 395—400 (1974a)

Tulunay,F.C., Takemori,A.E.: Further studies on the alteration of analgesic receptor-antagonist interaction induced by morphine. J. Pharmacol. exp. Ther. **190**, 401—407 (1974b)

Türker,K., Akcasu,A.: The effect of morphine on 5-HT content of cat's brain. New İstanbul Contr. clin. Sci. **5**, 89—97 (1962)

Tuttle,W.W., Elliott,H.W.: Electrographic and behavioral study of convulsants in the cat. Anesthesiology **30**, 48—64 (1969)

Ungar,G., Cohen,M.: Induction of morphine tolerance by material extracted from brain of tolerant animals. Int. J. Neuropharmacol. **5**, 183—192 (1966)

Ungar,G., Galvan,L.: Conditions of transfer of morphine tolerance by brain extracts. Proc. Soc. exp. Biol. (N.Y.) **130**, 287—291 (1969)

Val'dman,A.V.: The effect of analgesics on the descending reticular inhibitory system of the brainstem. New Data on the Pharmacology of the Reticular Formation of the Cerebrum, Leningrad Medical Institute, Leningrad, USSR 1958

Vaupel,D.B., Martin,W.R.: Interaction of oxotremorine with atropine, chlorpromazine, cyproheptadine, imipramine and phenoxybenzamine on the flexor reflex of the chronic spinal dog. Psychopharmacologia (Berl.) **30**, 13—26 (1973)

Vedernikov,Yu.P.: Interaction of amphetamine, apomorphine, disulfiram with morphine and the role played by catecholamines in morphine analgesic action. Arch. int. Pharmacodyn. **182**, 59—64 (1969)

Vedernikov,Yu.P.: The influence of single and chronic morphine administration on some central effects of amphetamine and apomorphine. Psychopharmacologia (Berl.) **17**, 283—288 (1970)

Vedernikov,Yu.P., Afrikanov,I.I.: On the role of a central adrenergic mechanism ın morphine analgesic action. J. Pharm. Pharmacol. **21**, 845—847 (1969)

Verdeaux,G., Marty,R.: Action sur l'électroencéphalogramme de substances pharmacodynamiques d'intérêt clinique. Revue neurol. **91**, 405—427 (1954)

Verri,R.A., Graeff,F.G., Corrado,A.P.: Effect of reserpine and alpha-methyl-tyrosine on morphine analgesia. Int. J. Neuropharmacol. **7**, 283—292 (1968)

Vigouret,J., Teschemacher,Hj., Albus,K., Herz,A.: Differentiation between spinal and supraspinal sites of action of morphine when inhibiting the hindleg flexor reflex in rabbits. Neuropharmacology **12**, 111—121 (1973)

Villarreal,J.E., Sheldon,M.I., Smith,C.B.: Separation of the effects of dependence producing analgesics upon locomotor activity of mice and catecholamine synthesis in the brain. Pharmacologist **13**, 313 (1971)

Vogt,M.: The concentration of sympathin in different parts of the central nervous system under normal conditions and after the administration of drugs. J. Physiol. (Lond.) **123**, 451—481 (1954)

Volavka,J., Zaks,A., Roubicek,J., Fink,M.: Electrographic effects of diacetylmorphine (heroin) and naloxone in man. Neuropharmacology **9**, 587—593 (1970)

Wada,M., Tanaka,H., Hirano,T., Taneiti,Y.: The effect of morphine admınistration upon the output rate of epinephrine, blood sugar level and blood pressure in normal and tolerant dogs. Tôhoku J. exp. Med. **34**, 52—71 (1938)

Wajda,I., Datta,K.: Effect of drugs on acetylcholine metabolism in the caudate nucleus of rats. Pharmacologist **13**, 313 (1971)

Wand,P., Kuschinsky,K., Sontag,K.-H.: Morphine-induced muscular rigidity in rats. Europ. J. Pharmacol. **24**, 189—193 (1973)

Watanabe,K., Matsui,Y., Iwata,H.: Enhancement of the analgesic effect of morphine by sodium diethylidithiocarbamate in rats. Experientia (Basel) **25**, 950—951 (1969)

Wauquier,A., Niemegeers,C.J.E., Lal,H.: Differential antagonism by naloxone of inhibitory effects of haloperidol and morphine on brain self-stimulatio. Psychopharmacologia (Berl.) **37**, 303—310 (1974)

Way,E.L., Loh,H.H., Shen,F.-H.: Morphine tolerance, physical dependence, and synthesis of brain 5-hydroxytryptamine. Science **162**, 1290—1292 (1968)

Way,E.L., Loh,H.H., Shen,F.-H.: Simultaneous quantitative assessment of morphine tolerance and physical dependence. J. Pharmacol. exp. Ther. **167**, 1—8 (1969)

Way,E.L., Sung,C.Y., Fujimoto,J.M.: The effect of adrenalectomy on the development of tolerance to morphine and methadone. J. Pharmacol. exp. Ther. **110**, 51—52 (1954)

Weber,H.: Ueber Anasthesie durch Adrenalin. Verh. dtsch. Kongr. inn. Med. **21**, 616—619 (1904)

Wei,E.: Assessment of precıpitated abstinence in morphine-dependent rats. Psychopharmacologia (Berl.) **28**, 35—44 (1973a)

Wei,E.: Brain lesions attenuating "wet shake" behavior in morphine-abstinent rats. Life Sci. **12**, 385—392 (1973b)

Wei,E.: Morphine analgesia, tolerance and physical dependence in the adrenalectomized rat. Brit. J. Pharmacol. **47**, 693—699 (1973c)

Wei,E., Loh,H.H., Way,E.L.: Neuroanatomical correlates of morphine dependence. Science **177**, 616—617 (1972)

Wei,E., Loh,H.H., Way,E.L.: Quantitative aspects of precipitated abstinence in morphine-dependent rats. J. Pharmacol. exp. Ther. **184**, 398—403 (1973a)

Wei,E., Loh,H.H., Way,E.L.: Brain sites of precipitated abstinence in morphine-dependent rats. J. Pharmacol. exp. Ther. **185**, 108—115 (1973b)

Weil-Malherbe, H., Smith, E. R. B., Eisenman, A. J., Fraser, H. F.: Plasma catecholamine levels and urinary excretion of catecholamines and metabolites in two human subjects during a cycle of morphine addiction and withdrawal. Biochem. Pharmacol. **14**, 1621—1633 (1965)

Weinstock, M.: Acetylcholine and cholinesterase. In: Clouet, D. H. (Ed.): Narcotic drugs; Biochemical pharmacology, pp. 254—261. New York: Plenum Press 1971

Weissman, A.: Cliff jumping in rats after intravenous treatment with apomorphine. Psychopharmacologia (Berl.) **21**, 60—65 (1971)

Welsh, J. H.: 5-Hydroxytryptamine in coelenterates. Nature (Lond.) **186**, 811—812 (1960)

Welsh, J. H., Moorhead, M.: The quantitative distribution of 5-hydroxytryptamine in the invertebrates, especially in their nervous systems. J. Neurochem. **6**, 146—169 (1960)

Wikler, A.: Studies on the action of morphine on the central nervous system of cat. J. Pharmacol. exp. Ther. **80**, 176—187 (1944)

Wikler, A.: Effects of morphine, nembutal, ether and eserine on two-neuron and multineuron reflexes in the cat. Proc. Soc. exp. Biol. (N.Y.) **58**, 193—196 (1945)

Wikler, A.: Reactions of dogs without neocortex during cycles of addiction to morphine and methadone. Arch. Neurol. Psychiat. (Chic.) **67**, 672—684 (1952)

Wikler, A.: Clinical and electroencephalographic studies on the effects of mescaline, N-allylnormorphine and morphine in man; A pharmacologic analysis of the functions of the spontaneous electrical activity of the cerebral cortex. J. nerv. ment. Dis. **120**, 157—175 (1954)

Wikler, A.: On the nature of addiction and habituation. Brit. J. Addict. **57**, 73—79 (1961)

Wikler, A.: Conditioning factors in opiate addiction and relapse. In: Wilner, D. M., Kassebaum, G. G.: Narcotics, pp. 85—100. New York: McGraw-Hill 1965

Wikler, A., Altschul, S.: Effects of methadone and morphine on the electroencephalogram of the dog. J. Pharmacol. exp. Ther. **98**, 437—446 (1950)

Wikler, A., Carter, R. L.: Effects of single doses of N-allylnormorphine on hindlimb reflexes of chronic spinal dogs during cycles of morphine addiction. J. Pharmacol. exp. Ther. **109**, 92—101 (1953)

Wikler, A., Frank, K.: Hindlimb reflexes of chronic spinal dogs during cycles of addiction to morphine and methadon. J. Pharmacol. exp. Ther. **94**, 382—400 (1948)

Wikler, A., Green, P. C., Smith, H. D., Pescor, F. T.: Use of a benzimidazole derivative with morphine-like properties orally as a presumptive reinforcer in conditioning of drug seeking behavior in rats. Fed. Proc. **19**, 22 (1960)

Wikler, A., Martin, W. R., Pescor, F. T., Eades, C. G.: Factors regulating oral consumption of an opioid (etonitazene) by morphine-addicted rats. Psychopharmacologia (Berl.) **5**, 55—76 (1963)

Wikler, A., Norrell, H., Miller, D.: Limbic system and opioid addiction in the rat. Exp. Neurol. **34**, 543—557 (1972)

Wikler, A., Pescor, F. T.: Persistence of "relapse-tendencies" of rats previously made physically dependent on morphine. Psychopharmacologia (Berl.) **16**, 375—384 (1970)

Wikler, A., Pescor, M. J., Kalbaugh, E. P., Angelucci, R. J.: Effects of frontal lobotomy on the morphine-abstinence syndrome in man; an experimental study. Arch. Neurol. Psychiat. (Chic.) **67**, 510—521 (1952)

Wikler, A., Rayport, M.: Lower limb reflexes of a "chronic spinal" man in cycles of morphine and methadone addiction. Arch. Neurol. Psychiat. (Chicago) **71**, 160—170 (1954)

Wolff, H. G., Hardy, J. D., Goodell, H.: Studies on pain. Measurement of the effect of morphine, codeine and other opiates on the pain threshold and an analysis of their relation to the pain experience. J. clin. Invest. **19**, 659—680 (1940)

World Health Organization: Expert Committee on Drugs Liable to Produce Addiction; Report on the Second Session. Wld Hlth Org. techn. Rep. Ser. **21**, 1950

Yaksh, T. L., Yamamura, H. I.: The effect of morphine *in vivo* on resting and evoked release of ^{3}H-acetylcholine from the cat caudate nucleus. Pharmacologist **15**, 203 (1973)

Yarbrough, G. G., Buxbaum, D. M., Bushing, J. A., Sanders-Bush, E.: Effect of acute and chronic morphine sulfate administration on brain serotonin (5HT) synthesis in rats and mice. Pharmacologist **12**, 211 (1970)

Yarbrough, G. G., Buxbaum, D. M., Sanders-Bush, E.: The role of serotonin (5-HT) in morphine locomotor effects in the rat. Pharmacologist **13**, 314 (1971)

Yarbrough, G.G., Buxbaum, D.M., Sanders-Bush, E.: Increased serotonin turnover in acutely morphine-treated mice. Biochem. Pharmacol. **21**, 2667—2669 (1972)

Yarbrough, G.G., Buxbaum, D.M., Sanders-Bush, E.: Biogenic amines and narcotic effects. II. Serotonin turnover in the rat after acute and chronic morphine administration. J. Pharmacol. exp. Ther. **185**, 328—335 (1973)

Yeh, S.Y., Mitchell, C.L.: Potentiation and reduction of the analgesia of morphine in the rat by pargyline. J. Pharmacol. exp. Ther. **179**, 642—651 (1971)

York, J.L., Maynert, E.W.: Role of analgesimetric procedure in assessing effects of chronic deficits of brain catecholamines or serotonin. Pharmacologist **15**, 242 (1973)

Yunger, L.M., Harvey, J.A.: Effect of lesions in the medial forebrain bundle on three measures of pain sensitivity and noise-elicited startle. J. comp. physiol. Pharmacol. **83**, 173—183 (1973)

Zaks, A.M., Bruner, A., Fink M., Freedman, A.M.: Intravenous diacetylmorphine (heroin) in studies of opiate dependence. Dis. nerv. Syst.-GWAN Suppl. **30**, 89—92 (1969)

Zetler, G.: Substanz P, ein Polypeptid aus Darm und Gehirn mit depressiven, hyperalgetischen und morphin-antagonistischen Wirkungen auf das Zentralnervensystem. Naunyn-Schmiedebergs Arch. exp. Path. Pharmak. **228**, 513—518 (1956a)

Zetler, G.: Wirkungsunterschied zwischen den Polypeptiden Bradykinin und Substanz P am Zentralnervensystem. Naunyn-Schmiedebergs Arch. exp. Path. Pharmak. **229**, 148—151 (1956b)

Zetler, G., Schafii-Laschteneschai, A.-H., Jacobsen, F.G.: Die Hemmung der Morphin-, Levorphan-, Methadon- und Pethidin-Analgesie der Maus durch „spezifische“ und „unspezifische“ Antagonisten. Naunyn-Schmiedebergs Arch. exp. Path. Pharmak. **246**, 191—202 (1963)

Zimmermann, E., Krivoy, W.: Antagonism between morphine and the polypeptides ACTH, $ACTH_{1-24}$, and β-MSH in the nervous system. In: Zimmermann, E., Gispen, W.H., Marks, B.H., De Wied, D.: Progress in Brain Research-vol. 39, Drug Effects on Neuroendocrine Regulation. Amsterdam: Elsevier Scientific Publishing Company 1973

CHAPTER 2

Assessment of the Abuse Potential of Narcotic Analgesics in Animals

W. R. MARTIN and D. R. JASINSKI

A. Introduction

The goal of assessment of the abuse potential of drugs is to minimize their abuse through the application of sound principles of preventive medicine. There are three major benefits that are a consequence of these efforts: (1) Physicians, who make most psychoactive drugs generally available, can be informed about the dangers associated with their abuse and prescribe them in a cautious and knowledgeable way. (2) Through the use of legal control mechanisms, manufacturers, distributors, and dispensers can be encouraged to employ proper precautions such that illicit diversion for the nonmedical use of psychoactive drugs is minimized. (3) New chemicals that have a high abuse potential but have no special therapeutic advantages can be identified before they are introduced into clinical medicine, and rational decisions can be made as to whether useful purposes would be served should they be made generally available to physicians.

The principles used in determining whether a drug should be controlled have not been formalized and are still actively evolving as more experience is gained with drug abuse and consequent problems. The history of drug abuse legislation in the USA has been reviewed in Section I, Chapter 1. From this history it is apparent that the problems with which drug abuse legislation has attempted to cope have changed markedly over the last 70 years and will undoubtedly continue to change. However, there appear to be several underlying concerns that will continue to be of importance for the foreseeable future. (1) Drug abuse is costly and harmful, is disruptive to social processes, and competes with other social priorities for resources. (2) The rapid development of chemistry, pharmacology, the brain sciences, and experimental therapeutics has allowed the discovery and development of psychoactive drugs at an unprecedented rate. To exemplify the magnitude of this problem, over 900 compounds have been assessed for their morphinelike abuse potentiality since 1953 at the University of Michigan Department of Pharmacology. Most of these compounds have properties similar to morphine and heroin and if made freely available would undoubtedly have been abused. Because of the great growth of knowledge about the relationship between chemical structure and neuropsychopharmacologic action of drugs, many chemicals have been synthesized that have a high abuse potential and many more will be synthesized in the future. Some abused drugs have been commonly used industrial chemicals whose psychoactivity has been discovered by lay persons who have become knowledgeable about the structures of psychotogens. It is imperative that the dangers of psychoactive drugs be recognized insofar as possible before they are introduced into clinical practice. It is equally important, however, not to stifle the discovery and development of new psychotherapeutic and other centrally acting drugs.

The two major characteristics of drugs of abuse are: (1) They have reinforcing properties that are responsible for at least certain individuals taking them chronically, and (2) they cause harm to the individual and to society. Both of these characteristics must be fulfilled before a drug can be considered to have an abuse liability or an abuse potential. The harm that abused drugs cause is discussed in relevant chapters of this volume. Various efforts have been made to formalize those precepts into workable laws and regulations.

The criteria established under the Comprehensive Drug Abuse Prevention and Control Act of 1970 for the scheduling of drugs are:

(1) Its actual or relative potential for abuse.
(2) Scientific evidence of its pharmacological effect, if known.
(3) The state of current scientific knowledge regarding the drug or other substance.
(4) Its history and current pattern of abuse.
(5) The scope, duration, and significance of abuse.
(6) What, if any, risk there is to the public health.
(7) Its psychic or physiological dependence liability.
(8) Whether the substance is an immediate precursor of a substance already controlled under this title.

A more general statement of principles for placing drugs that alter the functioning of the central nervous system under international control as proposed under the Convention on Psychotropic Substances (United Nations Commission on Narcotic Drugs, 1971) is:

(a) That the substance has the capacity to produce
  (i) (1) a state of dependence, and
    (2) central nervous system stimulation or depression, resulting in hallucinations or disturbances in motor function or thinking or behaviour or perception or mood, or
  (ii) similar abuse and similar ill effects as a substance in Schedule I, II, III, or IV,[1] and
(b) That there is sufficient evidence that the substance is being or is likely to be abused so as to constitute a public health and social problem warranting the placing of the substance under international control, the World Health Organization shall communicate to the Commission an assessment of the substance, including the extent or likelihood of abuse, the degree of seriousness of the public health and social problem, and the degree of usefulness of the substance in medical therapy, together with recommendations on control measures, if any, that would be appropriate in the light of its assessment.

Drugs in Schedule I were those requiring the highest level of control, while those in Schedule IV, the lowest.

The complicated conceptual ideas that guide the scheduling of drugs are not easily applied, particularly when practical considerations such as the impact of scheduling on medical practices and the societal benefits and detriments of control-

[1] The criteria for scheduling drugs in this convention were not stated; however, the following drugs were scheduled:

Schedule I - N,N-diethyltryptamine, DMHP, N,N-dimethyltryptamine, LSD, mescaline, parahexyl, psilocine, psilocybine, 2,5-dimethoxy-4-methyl amphetamine, and tetrahydrocannabinols
Schedule II - Amphetamine, dexamphetamine, methamphetamine, methylphenidate, phencyclidine, and phenmetrazine
Schedule III - Amobarbital, cyclobarbital, glutethimide, pentobarbital, and secobarbital
Schedule IV - Amfepramone, barbital, ethchlorvynol, ethinamate, meprobamate, methaqualone, methylphenobarbital, methyprylon, phenobarbital, pipradrol, and SPA.

ling and scheduling of drugs are actively debated. The final decisions are made as a consequence of the consideration of data by competent scientists and a review of their recommendations by the appropriate authorities. It is of great importance to recognize that once a drug has been introduced into clinical medicine and into drug-abusing subcultures, scheduling decisions are difficult to make. Control procedures have proved successful in preventing drug abuse but have had at best modest success in decreasing existing drug abuse problems. When the abuse potential of a drug is recognized early in its development, decisions about its abuse potential can be made on objective pharmacologic and therapeutic data unclouded by economic and political considerations.

Most major drugs of abuse have come under rigid legal control as a consequence of extensive street abuse. Epidemiologic and criminal investigations have been most commonly used to identify abuse problems and drugs of abuse. These techniques, however, are quite expensive, frequently provide evidence of uncertain validity, and identify problems after they have emerged. Consequently, scheduling of abused drugs on the basis of street abuse has had limited success in diminishing the extent of the problem. On the other hand, the identification of pharmacologic properties of new analgesics which could lead to their abuse has proved to be not only effective in preventing their abuse but has been economical. There is no reason why the pharmacologic principles that have evolved for narcotic analgesics and which will be described and discussed in this and the following chapter cannot be applied to all drugs of abuse.

B. Reinforcing Properties

There are two pharmacologic ways of identifying drugs with abuse potential: (1) the demonstration of reinforcing and harmful properties, and (2) demonstration of pharmacologic equivalence of an experimental drug to that of a drug of known abuse potentiality. The concept of a drug having positive reinforcing characteristics is a complex one and is related to some fundamental properties of the brain. Part of reinforcement is related to need reduction. Needs manifest themselves not only with appropriate acquisitive and avoidance behavior but with feeling states and associated affective behavior. The problem of whether the conscious experience of feelings of pleasure and displeasure is a fundamental property of the brain of mammals has concerned both the philosopher and the experimental biologist. The experiments of OLDS (1956, 1958) and others have shown that there are parts of the brain which when stimulated electrically give rise to self-stimulation, while there are other parts of the brain which when stimulated electrically are associated with avoidance behavior. An increasing body of data suggests that in animals as well as man drugs can be appetitive and aversive. The relationship between positive and negative reinforcing areas of the brain and those associated with needs has not been discovered.

The varying motivational and need states can be studied in animals with operant techniques or by behavioral observations (PLOOG, 1970) and in man by using rating questionnaires (HILL et al., 1963; HAERTZEN, 1966; CABANAC, 1971). The discovery that some drugs of abuse can have positive reinforcing properties in operant studies has led to the general belief that drugs that are positively reinforcing in animals are positively reinforcing in man. The narcotic analgesics do not appear to be reinforcing to normal subjects (von FELSINGER et al., 1955; SMITH and BEECHER, 1962) but are highly reinforcing to psychopaths (FRASER et al., 1961; HILL et al., 1963; HAERTZEN, 1966); whereas, amphetamines have been reinforcing in all populations studied (von FELSINGER et al., 1955). As discussed in Section I, Chapter 1, one of the characteristics of the psychopath is a feeling of hypophoria. Thus, narcotic analgesics appear to be reinforcing to psychopaths as a consequence of their ability to decrease hypophoria. This reinforcing action has been termed "negative euphoria" by KOLB (1925). The nature of the alterations in subjective states produced by drugs cannot be directly assessed in animals but using drugs as discriminative stimuli may provide indirect evidence (see Subsect. E.I.).

Another reinforcing property of certain drugs of abuse is their ability to induce tolerance and physical dependence. This phenomenon itself need not be reinforcing. Thus, it has been shown that cyclazocine (MARTIN et al., 1965) and nalorphine (MARTIN and GORODETZKY, 1965) induce both tolerance and physical dependence when administered chronically to man, yet these drugs have not been abused nor have patients who have exhibited signs and reported symptoms of abstinence exhibited any drug-seeking behavior. The two dimensions of tolerance and physical dependence to narcotic analgesics that give them reinforcing properties are: (1) the emergence of a discomforting abstinence syndrome which can be suppressed by narcotics, and (2) the exacerbation of hypophoric feelings when the addicts are stabilized, which are further worsened when they become abstinent (MARTIN et al., 1973). Subjective states existing when patients are chronically intoxicated with and dependent on sedative-hypnotics and alcohol have not been carefully characterized, but clinical reports and descriptions would suggest that they are predominantly negative. The abstinence syndrome associated with these dependences is unpleasant, and this unpleasantness is decreased by the sedative-hypnotics and alcohol. The chronic ingestion of large doses of amphetamine is also associated with negative feeling states (see Part II, Sect. I, Chap. 2). Thus, both tolerance to and dependence on drugs of abuse give rise to new need states of varying intensities which enhance the reinforcing properties of the drugs.

C. Pharmacologic Equivalence

From a practical point of view the best validated method of demonstrating abuse potentiality is by demonstrating pharmacologic equivalence. A variety of types of information help establish pharmacologic equivalence or the lack of it (Table 1).

I. Autonomic, Somatomotor, and Behavioral Effects in Nondependent and Dependent Animals (Table 2)

Physiologic and behavioral effects help establish equivalence. Thus, morphine and other narcotic analgesics reduce several types of responsivity to nociceptive stimuli, constrict pupils, depress body temperature, and are not commonly self-administered by naive animals. As can be seen from Table 1, this pattern of activity is different from that produced by amphetamines, LSD, and pentobarbital. Several types of

Table 1. Pharmacologic profiles of prototype drugs of abuse in dogs and rodents

	Morphine	Amphetamine	LSD	Pentobarbital
Single dose studies				
Pupils	↓	↑	↑	
Pulse	↓	↑–↓	↑	
Respiration	↑–↓	↑	↑	
Temperature	↓	↑	↑	
Skin twitch reflex	↓	↓	↓	
Flexor reflex	↓	↑	↑	↓
Stepping reflex	↓	↑	↑	↓
Self-administration	–, 0, +	+	?	+
Chronic dose studies				
Tolerance-cross tolerance	+	Some	+	Some
Suppression	+			+
Precipitation	+			
Direct addiction	+	?		+
Antagonists	Naloxone Naltrexone	Chlorpromazine	Cyproheptadine Chlorpromazine	

abused drugs including the morphinelike analgesics, amphetaminelike stimulants, and LSD-like hallucinogens act as agonists, and their effects can be antagonized selectively by specific antagonists. Thus, most of the effects of narcotic analgesics can be antagonized by the antagonists naloxone and naltrexone. Naloxone and naltrexone also antagonize the agonistic effects of cyclazocine- and nalorphinelike agonists; however, much larger dose levels are required.

II. Tolerance and Cross-Tolerance

The induction of tolerance and physical dependence provides several meaningful types of data concerning the mode of action of drugs of abuse. Tolerance to the narcotic analgesics is an extremely complicated phenomenon (see Sect. II, Chap. 1), and although tolerance does develop to many of their actions, it has not been widely employed in classifying these drugs. Tolerance and cross tolerance have, on the other hand, been quite useful in demonstrating the equivalence of LSD-like hallucinogens (see Part II, Sect. I, Chap. 3).

III. Suppression Studies

Studies of the ability of drugs to suppress signs of abstinence in dependent animals have been of great value in characterizing narcotic analgesics and barbituratelike depressants; however, as will be pointed out, they have limitations depending on the particular experimental parameters which may yield false positive and false negative results.

At the time of this writing, many dimensions of the power and specificity of suppression studies are not known. One of the important aspects of suppression studies is the complexity of the abstinence syndrome which has autonomic, somato-

Table 2. Effects of single doses of morphine and related narcotic analgesics

	Mouse	Rat	Dog	Monkey	Man
Autonomic					
Blood pressure					0–↓
Pulse rate			↓		0
Respiratory rate		↓	↑ panting		↓
Responsivity to CO_2			↓		↓
Pupil		↑	↓	↑	↓
Lenticular opacities	+				
Exophthalmus	+				
Temperature		↓↑	↓	↓	0–↓
Metabolic rate		↓↑			
Emesis			+	+	+
Constipation	+	+	+	+	+
Somatomotor					
Patellar reflex			0		
Flexor reflex			↓		
Crossed extensor reflex			↓		
Extensor thrust reflex			↑		
Skin twitch			↓		
Analgesia	+	+	+	+	+
Straub tail	+				
Convulsions	+	+	+	+	+
Behavioral					
Eating and drinking		↑	↓		↓
Self-administration					
Naive subjects	?	+	–	+	?
Conditioned subjects	?	+	+	+	+
Behavioral states	Depression and excitation, circling	Excitation and depression	Depression; sleep-conditioned salivation	Depression	Nodding; sedation

motor, behavioral and subjective characteristics which are summarized in Table 3. The pathophysiology of most of these signs of abstinence has not been either dissected or analyzed, and neither the kinds of changes in neuronal excitability nor the sites of changes in the central nervous system that occur in physical dependence are known. Thus, it is not known whether all of the signs of abstinence are the manifestation of a single or multiple types of changes that occur at one or several, possibly many, sites in the central nervous system. Since the abstinence syndrome may be a consequence of several distinct types of neurophysiologic changes, it is of great importance that multiple signs of abstinence be assessed. Some signs of abstinence appear only in certain species. Other signs are observed in several or many species but can only be conveniently measured in certain species.

IV. Direct Addiction and Precipitation Studies

Direct addiction studies in combination with precipitation studies not only provide evidence that an experimental drug produces physical dependence, but the characters of the withdrawal and precipitated syndromes help in classifying the drug.

Table 3. Signs of early withdrawal and precipitated abstinence

Autonomic		Somatomotor		Behavioral and subjective	
Blood pressure	(H↑)	Twitching	(Mo, H)	Tiredness	(H)
Pulse rate	(D↑, H↑)	Tremors	(M, D, Mo, H)	Weakness	(H)
Body temperature	(M – 0, R↓, D↑, Mo↓, H↑)	Shivering	(Mo, H)	Insomnia	(R, Mo, H)
Piloerection	(D, Mo, H)	Jumping	(M, R)	Irritability	(R, Mo, H)
Respiration	(M↑, D↑, H↑)	Head and wet shakes	(M, R)	Restlessness	(D, Mo, H)
Yawning	(D, Mo, H)	Hyperreflexia	(D)	Whining or crying	(D, Mo)
Rhinorrhea	(Mo, M, H)	Stepping reflex	(D, H)	Gnawing	(D)
Lacrimation	(M, R, D, Mo, H)	Hypermotility	(M)	Teeth chattering	(R, Mo)
Salivation	(D, Mo, H)	Convulsions	[M, D (rarely), Mo (rarely), H (infants)]	Aggression and fighting	(R, Mo)
Retching and emesis	(D, Mo, H)	Hiccups	[Mo (rarely), H (infants)]	Drug-seeking behavior	(R, D, Mo, H)
GI cramps	(Mo, H)	Writhing	(R)		
Defecation	(M, R, D, Mo, H)				
Pupillary diameter	(D↑, Mo↓, H↑)				
Ptosis	(R)				
Perspiration	(Mo, H)				
Urination	(M, R, D, H)				
Priapism and emission	(M, Mo, H)				

M = mouse; R = rat; D = dog; Mo = monkey; H = man.

Studies in the dog (MARTIN et al., 1974) and man (MARTIN, 1966) seem to indicate that the level of dependence to morphinelike drugs does not greatly alter the characteristics of the withdrawal abstinence syndrome. In contrast, the characteristics of the antagonist-precipitated abstinence syndrome are dependent on the level of dependence in the rat (BLASIG et al., 1973) and the dog (MARTIN et al., 1974). HIMMELSBACH (1937) was the first to quantitate the morphine abstinence syndrome in man (see Sect. II, Chap. 3) in which signs were weighted in accordance with the severity of the abstinence syndrome and the degree of change. Other rating systems have been developed to quantitate the degree of the abstinence syndrome in animals and will be discussed subsequently. It will suffice to say at this point that the values that result from scoring systems are complicated and ambiguous. To remove some of this ambiguity, it is necessary to examine the weighting of component signs of abstinence and to objectively compare syndromes. The morphine abstinence syndrome, at least in adults, can usually be easily distinguished from the barbiturate and alcohol abstinence syndromes. It is somewhat more difficult to distinguish the morphine abstinence syndrome from the abstinence syndrome of agonists of the cyclazocine- or nalorphine-type; however, this has been done in morphine-dependent man using correlation techniques (MARTIN et al., 1965; MARTIN and GORODETZKY, 1965; JASINSKI et al., 1970, 1971b) and more recently in morphine- and cyclazocine-dependent dogs (GILBERT and MARTIN, 1976). Precipitation of abstinence using narcotic antagonists such as naloxone and naltrexone is also useful in characterizing the type of dependence but even this technique lacks complete specificity for naloxone and naltrexone will precipitate abstinence in both the morphine- and cyclazocine-dependent dog (GILBERT and MARTIN, 1976). Naloxone and naltrexone are, however, less potent in precipitating abstinence in cyclazocine-dependent dogs than in morphine-dependent dogs.

D. Mouse

I. Pharmacologic Profile (Table 2)

There is very little well-validated data concerning the effects of narcotics on autonomic function in the mouse. Morphinelike drugs produce reversible lenticular opacities in mice which are antagonized by nalorphine. The potencies of narcotics for producing this effect parallel their analgesic potencies (WEINSTOCK, 1961). Lenticular opacities appear to be centrally mediated, involve an adrenergic component, and demonstrate tolerance (SMITH et al., 1966a, b, 1967). On the other hand, the somatomotor and behavioral effects of narcotics in the mouse have received wide attention. Small to moderate doses of morphine may produce excitement, compulsive circling, and a Straub tail. SHEMANO and WENDEL (1960) calculated the ratio of the LD_{50} to the ED_{50} for the Straub phenomenon for a number of narcotics (Straub tail index) which correlated with estimated addiction liability. ACETO et al. (1969) utilized the Straub tail phenomenon to study structure activity relationships among a number of opiates and opiate antagonists. Large doses of morphine produce a calming effect in which the mice may be immobile (catatonic) and do not attend stimuli as well as exophthalmos (see Sect. II, Chap. 1). Morphine analgesic action in the mouse can be demonstrated using thermal, electrical, chemical, and mechanical nociceptive stimuli (see WINTER, 1965).

II. Direct Addiction, Suppression and Precipitated Abstinence Studies (Table 3)

Knowledge of the effects of chronic morphine administration in the mouse and the associated development of tolerance and physical dependence owe much to HUIDOBRO and his colleagues (MAGGIOLO and HUIDOBRO, 1961). The administration of nalorphine to mice that had been previously implanted with pellets of morphine-free base precipitated an abstinence syndrome. The earliest signs of precipitated abstinence were increased, disorganized spontaneous activity and sniffing. Dragging of the perineum on the floor was also observed. When the animals were suspended by the tail, tremors were observed which increased in intensity until brief convulsive movements were seen. This response was called the "tail reaction." Mice also exhibited repeated and stereotypic jumping. Respiratory rate and depth was increased. Lacrimation, defecation, priapism associated with emission, and rhythmic pelvic movements were observed. HUIDOBRO and MAGGIOLO (1961) developed a grading system for the precipitated abstinence syndrome:

Grade 1. Restlessness and increased sniffing
Grade 2. Occasional running or jumping episodes
Grade 3. Frequent running and jumping
Grade 4. Increased running and jumping and a positive tail reaction
Grade 5. Intense running, jumping, and a tail reaction.

MARSHALL and WEINSTOCK (1971) have shown that morphine-dependent mice lose weight during withdrawal.

WAY et al. (1969) quantitated the precipitated abstinence syndrome by determining the proportion of mice which leaped off a raised platform in a 15-min period following the injection of naloxone. MARSHALL and WEINSTOCK (1971) quantitated the precipitated abstinence syndrome following subcutaneous administration of morphine by determining the frequency of jumps in an 8-min period following nalorphine administration and showed that the frequency was related to the level of morphine dependence.

WAY et al. (1969) showed that in mice made physically dependent with pellet implantations and abstinent by the surgical removal of the pellet that morphine, *l*-methadone, meperidine, levorphanol, dihydromorphinone, anileridine, and β-[-]-5-phenyl-9-methyl-2′-hydroxy-2-methyl-6,7-benzomorphan (GPA-1657) produce a marked suppression of jumping. Codeine, chlorpromazine, and pentobarbital, in that order, produced a lesser degree of suppression. *d*-Methadone, dextrorphan, and GPA-1658 (the *d*-isomer of GPA-1657) did not suppress abstinence. SHUSTER et al. (1963) showed that food-restricted mice would drink a milk solution containing dihydromorphinone and become tolerant to the analgesic and excitant effects of dihydromorphinone and cross-tolerant to morphine, codeine, and meperidine. Mice were also cross-tolerant to the excitant effects of amphetamine and cocaine. When withdrawn they showed irritability, piloerection, tremors, and weight loss. MCMILLAN et al. (1974) found that mice would consume 100–200 mg/kg of morphine daily when fluid intake was restricted to a morphine solution. When administered naloxone, they lost weight, exhibited jumping and diarrhea, and assumed a prone position.

SAELENS et al. (1971) produced withdrawal jumping with nalorphine and naloxone after 2 or 4 days of multiple intraperitoneal injections of morphinelike analgesics but not other central active agents and proposed this procedure as a screening method to estimate the physical dependence capacity of analgesics. KAMEI et al. (1973) showed that when morphine and morphine-6-glucuronide but not pentazocine were administered chronically by the subcutaneous route, jumping occurred following the administration of naloxone. Chlordiazepoxide, diazepam, amitriptyline, methamphetamine, and Δ^9-tetrahydrocannabinol produced a dose-related and marked suppression of naloxone-induced jumping. Diphenylhydantoin also suppressed jumping. Chlorpromazine, perphenazine, reserpine, pentobarbital, atropine, and mephenesin had little effect on naloxone-induced jumping in the morphine-dependent mouse.

E. Rat

I. Pharmacologic Profile (Table 2)

The pharmacologic effects of narcotics in the rat are discussed in detail in Section II, Chapter 1. In brief, morphine in the nontolerant animal produces both excitant or stimulatory and depressant effects. Which predominates, depends on the dose and the time after administration. Morphine depresses respiration, particularly in large doses, is said to dilate pupils, may both decrease and increase metabolic rate, and increase the release of ACTH (see KRUEGER et al., 1941). Morphine may increase motor activity, eating and drinking, cause jumping and gnawing, and in larger doses depression and sleep. It produces analgesia, decreasing the responsivity to a variety of nociceptive stimuli (WINTER, 1965).

The enteroceptive and perhaps subjective changes produced by narcotic analgesics and other drugs can act as discriminative stimuli. BELLEVILLE (1964) found that animals that acquired a lever pressing activity for food while under the influence of morphine extinguished less completely while under the influence of morphine. HILL et al. (1971) showed that morphine was less effective than pentobarbital as a discriminative stimulus for avoiding shock using a T-maze. BARRY (1974) also found that morphine was relatively weak as a discriminative stimulus; however, several investigators have shown that rats can discriminate between morphine and Δ^1-THC (BARRY and KUBENA, 1972), LSD (HIRSCHHORN and ROSECRANS, 1974), and alcohol (WINTER, 1975). SHANNON and HOLTZMAN (1975) have shown that rats trained to make an escape response with morphine as a discriminative stimulus exhibit a dose-related increase in accuracy when morphine, oxymorphone, levorphanol, methadone, and meperidine are administered. Their effectiveness and potency as discriminative stimuli agree well with their analgesic potency. Profadol, pentazocine, and cyclazocine were also effective as discriminative stimuli although the highest doses of cyclazocine were less effective than intermediate doses. COLPAERT et al. (1975a, b) have shown that parenterally administered morphine, fentanyl, phenoperidine, and piritramide but not haloperidol and orally administered morphine, codeine, fentanyl, and diphenoxylate but not loperamide act as equivalent discriminative stimuli in the rat.

II. Self-Administration

HEADLEE et al. (1955) first demonstrated, using an operant task, that rats with a modest degree of physical dependence on morphine exhibited a preference for intraperitoneally administered morphine and codeine. Subsequently, WEEKS (1962) showed that morphine-dependent rats with a polyethylene cannula in their jugular vein would bar-press for administration of morphine. COLLINS and WEEKS (1965) subsequently demonstrated that rats that had been made passively dependent on morphine would also bar-press for dihydromorphinone, methadone, and morphine. In these studies rats had an opportunity of self-administering two concentrations of morphine, codeine, dihydromorphinone, and methadone. Relative potencies were calculated on the basis of the number of injections taken in an hour of the different concentrations of the experimental drugs, and it was found that the potency of the drugs relative to morphine was: dihydromorphinone = 10, methadone = 3.4, and codeine = 0.67. COLLINS and WEEKS (1965) argued that relief of abstinence is the principle reason for the operant behavior but recognized that if a drug had aversive properties this would confound the interpretation of relative potency. Since the rat would bar-press less frequently for a narcotic with aversive properties, its potency in relieving abstinence would be overestimated. Another confounding variable is the duration of action of an intravenously administered dose. If the drug has a short duration of action, it will be self-administered more frequently and hence will appear to be less potent. Conversely, long-acting drugs will appear to be more potent. WEEKS and COLLINS (1964) and COLLINS and WEEKS (1967) also demonstrated that orally administered etonitazine or intravenously administered morphine, codeine, or meperidine decreased the rate of operant responding for morphine, while intravenously administered nalorphine or dexoxadrol increased it and intravenous dextromethorphan had no effect. WEEKS and COLLINS (1971) found that some untreated naive rats would bar-press for morphine. TROJNIAR et al. (1974) found that morphine-dependent rats would selfadminister morphine intragastrically.

III. **Suppression Studies** (Table 3)

Tolerance to and physical dependence on morphine have been extensively studied in the rat and have been reviewed in Section II, Chapter 1. When morphine is administered chronically, tolerance develops to its analgesic, sedative, hypothermic, and motor-depressant effects, and increases in metabolic rate and body temperature are seen when the stabilization dose is administered. When morphine is abruptly withdrawn or an antagonist administered, a variety of signs emerge including weight loss, decreased food and water consumption, diarrhea, head and body "wet dog" shakes, irritability, particularly on handling, fighting and aggression, ptosis, writhing, jumping and escape attempts, increased exploratory behavior, immobility, teeth chattering, hyperalgesia, hypothermia, ear blanching, penile erections and emissions, and chromadacryarrhea. A variety of techniques have been used to make rats physically dependent on morphine including subcutaneous or parenteral administration (HOSOYA, 1959; MARTIN et al., 1963; AKERA and BRODY, 1968), pellet implantation (WEI et al., 1973), reservoir implantation (GOODE, 1971), oral administration (STOLERMAN and KUMAR, 1970; STOLERMAN et al., 1971; RISNER and KHAVARI, 1973), intraperitoneal injection and perfusion (BUCKETT, 1964; LORENZETTI and SANCILIO, 1970; TEIGER, 1974), and programed intravenous injection (WEEKS, 1962). COUSSENS et al. (1973) have shown that physical dependence can be established and withdrawal abstinence observed within 48 h if large and frequent doses of morphine are administered.

BUCKETT (1964) devised an abstinence scoring system in the morphine-dependent rat in which writhing was given a score of 3, squealing and diarrhea 2, and teeth chattering, ptosis and "wet dog" shakes 1. He observed that heroin was 0.3 and codeine 0.03 times as potent as morphine in suppressing abstinence. Pyrilamine (1 and 10 mg/kg) also suppressed abstinence. LORENZETTI and SANCILIO (1970) using a similar technique found that meperidine was 0.7; methadone, 0.5; codeine, 0.03; and *d*-propoxyphene, 0.03 as potent as morphine in suppressing abstinence.

GOODE (1971) found that methadone, codeine, and meperidine, but not nalorphine, pentazocine, and cyclazocine, suppressed withdrawal weight loss and defecation in morphine-dependent rats. Further, naloxone, nalorphine, pentazocine, and cyclazocine precipitated abstinence in the dependent rat.

TEIGER (1974) showed that methadone (3 mg/kg/day), codeine (50 mg/kg/day), and meperidine (100 mg/kg/day) substituted for morphine in dependent rats. NOZAKI and HOSOYA (1972) observed that morphine, codeine, *d*-propoxyphene napsylate, meperidine, methadone, thienyl-aminobutene, nalorphine, and pentazocine would increase body weight in abstinent morphine-dependent rats. Barbital and meprobamate also increased body weight in abstinent rats.

IV. Direct Addiction and Precipitation Studies

AKERA and BRODY (1968) using weight loss as a measure of abstinence were able to make rats physically dependent on morphine and levorphanol but not methadone. TEIGER (1974) using an intraperitoneal infusion technique and weight loss and irritability as measures of abstinence made rats dependent on morphine and codeine. A lesser degree of physical dependence was produced by the intraperitoneal infusion of meperidine.

Administration of a narcotic antagonist to rats dependent on morphine precipitates an abstinence syndrome that is similar to the withdrawal syndrome (KAYMAKCALAN and WOODS, 1956; HANNA, 1960; MAYNERT and KLINGMAN, 1962; WEI et al., 1973; BLASIG et al., 1973). Nalorphine will precipitate an abstinence syndrome in rats chronically treated with codeine (KUHN and FRIEBEL, 1962). Dependence as measured by precipitated abstinence can be rapidly established in the rat. COLLIER et al. (1972) demonstrated that naloxone precipitated jumping 24 h after the administration of a sustained release preparation of morphine. FRUMKIN (1974) precipitated abstinence signs with naloxone in rats that had received three daily 10 mg/kg doses of morphine.

NOZAKI and HOSOYA (1972) observed that rats that received a variety of drugs chronically gained weight during the day, while control rats lost weight during the day as a consequence of their nocturnal feeding habits. These drugs included morphine, codeine, *d*-propoxyphene napsylate, azabicyclane, meperidine, methadone, thienyl-aminobutene, nalorphine, pentazocine, barbital, meprobamate, and aminopyrine. If these animals were challenged with naloxone, only those receiving morphine, codeine, *d*-propoxyphene napsylate, meperidine, methadone, thienyl-aminobutene, nalorphine, and pentazocine chronically lost weight. When abruptly withdrawn, only rats receiving morphine, codeine, *d*-propoxyphene napsylate, and azabicyclane chronically lost weight, while those receiving meperidine, methadone, thienyl-aminobutene, nalorphine, and pentazocine gained weight.

F. Dog

I. Pharmacologic Profile

The effects of narcotic analgesics on the central nervous system in the dog resemble their effects in man in some respects, differing in others (see Table 2). The striking features of modest single doses of morphine are to decrease pulse rate, responsivity to CO_2, pupillary diameter, body temperature, and reflex responsivity to nociceptive stimuli but not other reflexes (PLANT and PIERCE, 1928; TATUM et al., 1929; DRIPPS and DUMKE, 1943; WIKLER and FRANK, 1948; MARTIN and EADES, 1961, 1964; MARTIN et al., 1964). In addition, shortly after the administration of an intravenous dose of morphine, the dog exhibits emesis and panting, the latter a consequence of the lowering of the temperature homeostat (MARTIN, 1968).

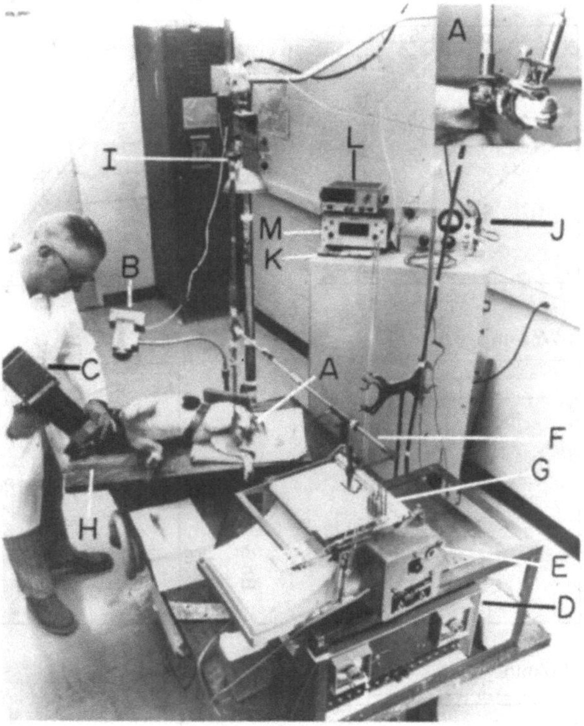

Fig. 1. Chronic spinal dog observation unit: *A* Pneumatic toe squeezer; *B* Thermal stimulator; *C* Polaroid close-up camera for photographing pupils; *D* Programmer and control unit for pneumatic toe squeezer; *E* Kymograph; *F* Isotonic recording system for flexor reflex; *G* Pens for marking stimulus strength; *H* Animal board. Note that the carrier of pneumatic programmer and kymograph can be rotated around an axis extending to the pivot of the ball socket (*I*) in order to obtain maximal reflex excursions; *J* Control for thermal stimulator; *K* Thermopile which in conjunction with digital voltmeter (*L*) is used for calibration of thermal stimulator; *M* Electronic timer for measuring the latency of skin twitch reflex

ANDREWS and WORKMAN (1941) showed that morphine, dilaudid, codeine, and aspirin but not pentobarbital or prostigmine elevated the threshold for the skin twitch reflex in the dog. HOUDE and WIKLER (1951) showed that morphine, as well as mephenesin and thiopental, decreased the amplitude of the skin twitch reflex. In additional studies, HOUDE and WIKLER (1951) and HOUDE et al. (1951) studied the effects of morphine, thiopental, and pentobarbital, as well as the interneuron depressants mephenesin and benzimidazole, on the patellar, extensor thrust, flexor, crossed extensor, and Phillipson's reflexes in the chronic spinal dog. Thiopental and pentobarbital depressed all reflexes; mephenesin and benzimidazole depressed all except the patellar reflex, which it enhanced; while morphine had little or no effect on the patellar reflex, enhanced the extensor thrust reflex and depressed the flexor, crossed extensor, and Phillipson's reflexes. Methadone's effects on these reflexes were similar to those of morphine (WIKLER and FRANK, 1948). MARTIN et al. (1964) first showed that depression of the flexor reflex, constriction of the pupil, and depression of body

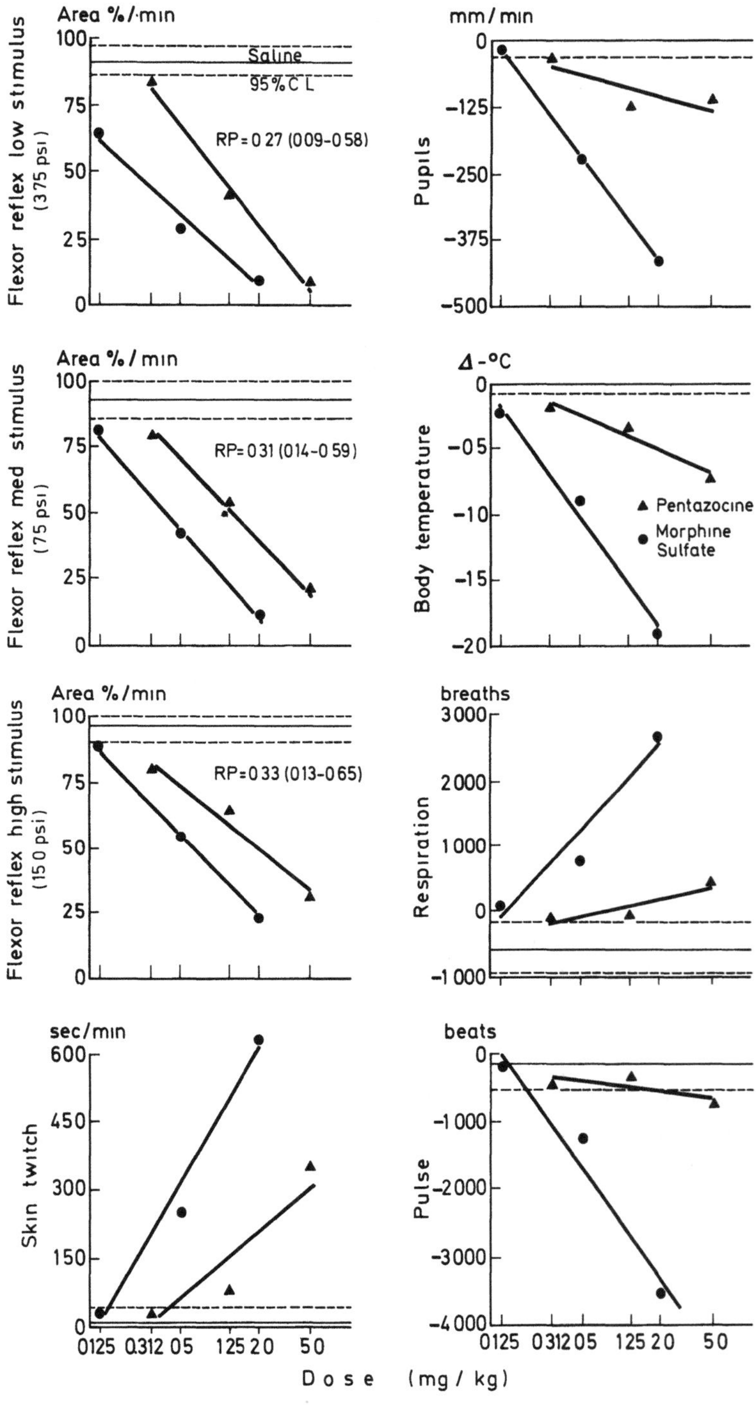

Fig. 2. Effects of graded doses of pentazocine and morphine on the flexor and skin twitch reflexes, pupillary diameter, body temperature, respiration, and pulse rate of chronic spinal dog. Each point represents the mean of area under time action curve for 6 dogs

temperature of the chronic spinal dog could be used for assaying the relative potency of narcotic analgesics. In these studies, it was shown that codeine and Versidyne [1-(chlorophenethyl)-2-methyl-6,7-dimethoxy-1,2,3,4-tetrahydroisoquinoline HCl] were $^1/_{10}$ to $^1/_{20}$ as potent as morphine. These techniques have been expanded not only to allow the calculation of the relative potencies of experimental drugs using a crossover design and their duration of action but a qualitative comparison of characteristics of several actions on the central nervous system of the chronic spinal dog (GILBERT and MARTIN, 1976; MARTIN et al., 1976). The apparatus used for making these observations is illustrated in Figure 1. The flexor reflex is evoked by either an electrical stimulus (MARTIN and EADES, 1967) or a programmed pneumatic toe squeezer (MARTIN et al., 1973) and pupillary diameter is determined photographically. The skin twitch is evoked using a radiant heat method. Body temperature, respiratory rate, and pulse rate, as well as the other observations, are determined periodically through the course of the experiment. The effects of graded doses of the experimental and test drugs are usually determined by calculating the area under the time action curves for each dog under each condition and for each parameter. When time action curves are disparate, peak effects may be used. Figure 2 illustrates a comparison that was made between morphine and pentazocine. In this comparison, it can be seen that pentazocine is about one-third as potent as morphine in suppressing the flexor reflex evoked by both mildly and strongly nociceptive stimuli. It should be noted that the dose response lines for morphine and pentazocine for all three strengths of stimuli are parallel, and all criteria for a valid assay were met. Further, the dose response lines produced by pentazocine on the skin twitch reflex, pupils, body temperature, respiration, and pulse rate had lower slopes than comparable curves produced by morphine. In accordance with the two-receptor theory (MARTIN, 1967), agonists of the morphine type depress the flexor reflex and skin twitch reflex, constrict pupils, slow heart rate, and lower body temperature. Agonists of the nalorphine type also depress the flexor reflex but are much less effective than morphinelike agonists in depressing the skin twitch reflex. They may dilate or constrict pupils, or have a biphasic pupillary dose-response curve. Further, they have either no effect on pulse rate or increase it. Body temperature is either not altered or decreased (GILBERT and MARTIN, 1976).

II. Self-Administration

JONES and PRADA (1973) studied the reinforcing properties of morphine in beagle dogs, using intravenous self-administration techniques, and found that it was aversive in 17 or 18 out of 22 dogs depending on whether the number of reinforcements (morphine administration) or the number of operant responses were used as a criterion. Approximately 20% of the dogs either exhibited no change in their bar-pressing rate or showed a modest increase. Subsequently, JONES and PRADA (unpublished observations) have shown that dogs that have bar-pressed for amphetamines will subsequently bar-press for morphine.

III. Suppression Studies (Table 3)

Both TATUM et al. (1929) and KOLB and DUMEZ (1931) felt that morphine dependence in the monkey was more similar to that in man than dependence in the dog. As can be seen from Tables 2 and 3, the effects of morphine and the characteristics of morphine abstinence are quite similar in all three species. The absence of data on a number of parameters, however, prevents making a critical judgment. Most investigators have found the dog somewhat more difficult to make dependent on morphine than the monkey. This is certainly true if the dog receives only one or two subcutaneous injections of morphine a day. Under these conditions dogs frequently lose their appetite, become debilitated, and die. The loss of appetite is in part confounded by decreased fluid intake, and if dogs are periodically hydrated by venoclysis, appetite improves and weight loss is minimal. The procedure commonly employed at the Addiction Research Center is to administer morphine at the beginning of the workday (ca. 08.00) and again, frequently using a larger dose, at the end of the workday (ca. 16.00). Using this dosing regime, dogs show signs of abstinence at 08.00 when they are dependent. Two economic procedures for the chronic administration off morphine in the dog have been developed. JONES and PRADA (unpublished observations), using techniques developed by WEEKS and COLLINS (1968) and DENEAU et al. (1969), have administered morphine through indwelling intravenous catheters with a programed injection schedule. MARTIN et al. (1973) have made dogs dependent by administering an 08.00 dose subcutaneously and a larger dose (usually 4 times larger) orally at 16.00. Using these techniques, dogs can be easily made dependent on large doses of morphine without the emergence of abstinence signs between doses. Morphine dependence has been extensively studied in the chronic spinal dog (WIKLER and FRANK, 1948; MARTIN and EADES, 1961; MARTIN et al., 1973). This preparation offers a number of advantages. The spinal dog can be easily trained to the observation apparatus, thus permitting the observation and measurement of a large number of autonomic, somatomotor, and behavioral signs. Some of the signs that have proven especially useful are presented in Table 4, as well as a weighting system used in determining the intensity of abstinence (MARTIN et al., 1974) in both substitution and precipitation studies. In developing scoring systems for quantitating the abstinence syndrome, several criteria have been used for determining the weight given the various signs of abstinence. A weighting system was developed by using only signs of precipitated abstinence whose intensity, frequency, or magnitude increased as the dose of the antagonist increased. Further, signs of abstinence were weighted such that their contribution to the total score would be approximately equal to that of all other signs. Subsequently, signs have been weighted such that the variance of the substitution or precipitation score was minimized. In experiments conducted to date, the choice of weighting factors does not seem to be critical. There is not only an excellent concordance between potency of antagonists in precipitating abstinence, but of agonists in suppressing abstinence using the different weighting systems. Further, there is good concordance between the relative potency of morphinelike agonists in suppressing abstinence and of antagonists in precipitating abstinence in the dog and potency estimates obtained in man using the HIMMELSBACH scoring system in which the clinical assessment of the intensity of abstinence provided the basis for weighting.

Since the morphine abstinence syndrome in dogs receiving morphine by both the oral and subcutaneous route does not become maximal until the 48th h of abstinence after the last oral dose of morphine, suppression studies are begun at 41.5 h of abstinence and carried through the 43rd h of abstinence. This procedure has advan-

			Precipitation abstinence						Suppression of withdrawal abstinence	
			54 mg/day			125 mg/day			54 mg/day	125 mg/day
			I	II	T_{54}	I	II	T_{125}	W_{54}	W_{125}
Graded signs										
Yawning		[1]	21.2 (13)	20.0 (12)	20.7 (11)	16.4 (15)	11.2 (9)	14.1 (12)	– 3.3 (14)	– 7.4 (9)
Lacrimation		[2]	16.6 (17)	20.8 (11)	18.7 (14)	16.8 (14)	4.0 (15)	11.1 (14)	–21.9 (5)	–13.6 (6)
Rhinorrhea		[1]	20.3 (14)	19.4 (14)	19.8 (12)	15.2 (16)	2.3 (16)	9.5 (16)	–13.4 (11)	–12.6 (7)
Salivation		[2]	1.4 (20)	11.6 (18)	6.5 (20)	10.8 (18)	–8.0 (21)	2.5 (21)	–37.8 (2)	–39.1 (1)
Tremor		[1]	24.5 (10)	13.4 (16)	18.9 (13)	13.9 (17)	1.6 (18)	8.9 (17)	–15.6 (8.5)	– 9.4 (8)
Restlessness		[1]	29.8 (7)	27.1 (9)	28.4 (7)	17.9 (13)	8.3 (13)	13.6 (13)	–13.7 (10)	–13.7 (5)
Whining		[2]	17.5 (15)	13.2 (17)	15.4 (16)	24.8 (10)	5.4 (14)	16.2 (11)	– 1.2 (15)	– 2.9 (13)
Present or not										
Gnawing		[2]	32.3 (6)	33.3 (6)	32.8 (6)	45.2 (6)	48.0 (4)	46.4 (4)	– 0.6 (17)	– 6.0 (10)
Head tossing		[2]	21.3 (12)	9.0 (19)	15.2 (17)	6.0 (20)	0.5 (19)	3.6 (19)	– 4.9 (13)	– 2.1 (15)
Barking		[10]	16.7 (16)	20.0 (13)	18.3 (15)	28.0 (8)	10.0 (10)	20.0 (9)	0 (19.5)	0 (18.5)
Piloerection		[10]	38.3 (5)	45.0 (5)	41.7 (5)	52.0 (4)	8.8 (11)	32.8 (6)	0 (19.5)	0 (18.5)
Retching		[50]	8.3 (18)	16.7 (15)	12.5 (18)	100.0 (1)	162.5 (1)	127.8 (1)	0 (19.5)	0 (18.5)
Emesis		[50]	0 (21)	0 (21)	0 (21)	30.0 (7)	12.5 (7)	22.2 (8)	0 (19.5)	0 (18.5)
Urination		[1]	22.8 (11)	21.0 (10)	21.9 (10)	21.1 (12)	12.1 (8)	17.1 (10)	– 0.7 (16)	– 0.4 (16)
Panting		[4]	27.3 (18)	27.5 (8)	27.4 (9)	8.0 (19)	8.3 (12)	8.1 (18)	–15.6 (8.5)	– 2.5 (14)
Stepping (Frag.)		[1]	5.8 (19)	8.5 (20)	7.2 (19)	4.2 (21)	1.5 (17)	3.0 (20)	– 7.7 (12)	– 4.6 (12)
Stepping (Cont.)		[3]	26.5 (9)	29.5 (7)	28.0 (8)	26.4 (9)	34.5 (5)	30.0 (7)	–17.4 (7)	– 5.8 (11)
Measurable signs										
Temperature	{1° C}	[10]	68.8 (2)	74.2 (3)	71.5 (2)	79.4 (3)	73.0 (3)	76.6 (3)	–45.3 (1)	–26.0 (3)
Respiration	{1 br/min}	[0.1]	61.2 (3)	67.6 (4)	64.4 (4)	24.1 (11)	–6.2 (20)	10.6 (15)	–29.9 (4)	–28.2 (2)
Pulse	{1 b/min}	[0.1]	85.9 (1)	112.0 (2)	98.9 (1)	80.6 (2)	76.2 (2)	78.6 (2)	–37.3 (3)	–22.7 (4)
Pupillary diameter	{1 mm}	[2]	55.0 (4)	74.8 (1)	64.9 (3)	51.3 (5)	13.0 (6)	34.3 (5)	–17.9 (6)	17.2 (21)

	r_s		r_s
$I \times II_{(54)}$	$=0.89\ (<0.01)$	$W_{54} \times W_{125}$	$=0.73\ (<0.01)$
$I \times II_{(125)}$	$=0.80\ (<0.01)$	$T_{54} \times W_{54}$	$=0.40$
$T_{54} \times T_{125}$	$=0.48\ (<0.05)$	$T_{125} \times W_{125}$	$= -0.25$

The values in each column are the mean of all animals for all experiments conducted under the particular condition. The values in parentheses are the ranks of the various signs and were used to calculate the Spearman Rank Correlation coefficients. The columns labeled I and II designate values obtained for various signs of precipitated abstinence for the two blocks of experiments conducted in dogs dependent on 54 and 125 mg/day. Columns labeled T designate the means of columns I and II. Columns labeled W are values obtained during withdrawal abstinence for the various signs. Values in brackets are weighting factors.

r_s are Spearman Rank Order Correlation coefficients. [Reproduced from MARTIN et al. (1974) with the permission of the publisher; some values have been recalculated.]

Table 5. The potency of morphine- and nalorphinelike agonists in suppressing abstinence in morphine-dependent dogs and men

Drug	Dog	Man
Fentanyl	70	
Levorphanol	12	Probably about 5
Phenazocine	8	8
Methadone	5	1–4
Morphine	1	1
d-Propoxyphene	0.16	ca. 0.1
Propiram	0.15	0.1
Meperidine	Will not suppress	Partial suppression
Normorphine	Will not suppress	Approximately as potent as morphine
Buprenorphine	Partial (37) suppression	
Pentazocine	Will not suppress	Will not or only marginally suppress

tages over the procedure used in man (FRASER et al., 1961) in which the subjects are only partially abstinent, since partial agonists will suppress abstinence when subjects are maximally abstinent but will either precipitate or not markedly affect abstinence when they are partially abstinent. Dogs are made abstinent at weekly intervals, and using a latin square design, the ability of two or three doses of morphine and two doses of an experimental drug to suppress abstinence are compared. The dose levels of morphine employed usually are 0.25, 1, and 4 mg/kg when dogs are dependent on approximately 10 mg/kg/day. Comparable doses of experimental drugs (based on their potency relative to morphine in suppressing the flexor and skin twitch reflexes and in constricting pupils in the nondependent chronic spinal dog) are administered using the previously mentioned fourfold increments in dose levels. Drugs are administered intravenously when possible. Only a limited experience has been obtained using the suppression technique in the dog. It is known that the results obtained with morphine are reproducible. In suppression studies conducted over a year apart in the same dogs maintained on a constant daily dose level of morphine, the effectiveness of morphine in suppressing signs of abstinence does not change. The results obtained with experimental drugs are presented in Table 5 and are compared with data obtained in man. As can be seen, the concordance is good. The slopes of suppression dose-response curves of partial agonists of the morphine type such as propiram and buprenorphine are less steep than stronger agonists such as morphine and *d*-propoxyphene (MARTIN et al., 1976).

IV. Direct Addiction and Precipitation Studies

Relatively little has been done in comparing the ability of narcotic analgesics and related drugs to produce physical dependence in the dog. WIKLER and FRANK (1948) demonstrated that chronic administration of morphine and methadone produced physical dependence in the chronic spinal dog. CARTER and WIKLER (1955) were unable to induce physical dependence with meperidine as indicated by signs of either precipitated or withdrawal abstinence in either the intact or chronic spinal dog when administered in dose levels of 10–20 mg/kg every 3 h. GILBERT and MARTIN (1976) have shown that physical dependence to cyclazocine can be induced in the chronic spinal dog; however, both the precipitated and withdrawal abstinence syndromes are qualitatively different from the morphine abstinence syndrome.

WIKLER and CARTER (1953) were the first to demonstrate that nalorphine precipitated an abstinence syndrome in the morphine-dependent dog which was qualitatively similar to the withdrawal abstinence syndrome. These observations were confirmed by MARTIN and EADES (1964); however, close analysis of the precipitated and withdrawal abstinence revealed that there were differences (MARTIN et al., 1974). Study of the ability of narcotic antagonists to precipitate abstinence in the morphine-dependent dog has proved to be useful in estimating their potency. Table 6 summarizes studies that have been conducted in the spinal dog and relates them to similar findings in man. The slope of the precipitation dose-response curve for antagonists with nalorphine- or cyclazocinelike agonist actions is similar to that of pure antagonists such as naloxone and naltrexone, while the slope of the precipitation dose-response curve for partial agonists of the morphine type such as buprenorphine is less steep and may have a ceiling.

G. Monkey

I. Pharmacologic Profile (Table 2)

The autonomic effects of morphine have not been studied to the same extent in the monkey that they have in the dog or man. Narcotic analgesics dilate pupils and produce respiratory depression. The constipating effects of narcotic analgesics are

Table 6. The potency of narcotic antagonists in precipitating abstinence in morphine-dependent dogs and human subjects

	Dog	Man
Pure antagonists		
Naloxone	1.0	1.0
Naltrexone	3.4	2.3
Antagonists with nalorphine or cyclazocinelike agonistic effects		
Pentazocine	0.002	0.002
Nalorphine	0.08	0.1
SKF 10,047	0.13	
Cyclazocine	0.47	0.52
Oxilorphan	0.66	0.64
Diprenorphine	4.6	
Partial agonists of the morphine type		
Propiram	0.0006	0.0005
Buprenorphine	0.15	

The potencies are expressed as the number of mg of naloxone that would produce the same effect as 1 mg of the experimental drug.

thought to be less in the monkey than in man and the dog. Appetite is depressed by narcotics, and they cause emesis. Small to moderate dose levels of morphine produce behavioral depression and reduce sexual activity, while large doses produce convulsions (TATUM et al., 1929; KOLB and DUMEZ, 1931; SEEVERS, 1936b). Morphine elevates body temperature and depresses spontaneous blinking rate (HOLTZMAN and VILLARREAL, 1969; VILLARREAL, 1970). The antinociceptive effects of morphine in the monkey have been demonstrated using an electric shock titration procedure (MALIS, 1972; WEISS and LATIES, 1964).

II. Self-Administration

THOMPSON and SCHUSTER (1964) first showed that the morphine-dependent rhesus monkey would self-administer morphine when abstinent. DENEAU et al. (1969) found that approximately 75% of naive rhesus monkeys would self-administer morphine and codeine to an extent that they would become physically dependent. Five naive monkeys self-administered codeine to the extent that they died, presumably of convulsions. This procedure in which rhesus monkeys without previous drug experience (naive) are given unlimited access to drugs such that each lever press results in a drug injection (FR 1) is termed *continuous self-administration.* Using this procedure, the rhesus monkey has been shown to self-administer propiram, *d*-propoxyphene, meperidine, methadone, dihydromorphinone, pentazocine, and dextromethorphan (DENEAU et al., 1969; YANAGITA, 1973b) but not nalorphine or saline (DENEAU et al., 1969). Morphine and morphinelike drugs appear to be weak reinforcers as measured by this procedure in the absence of previous conditioning history or prior to the development of physical dependence.

The development of physical dependence confounds the measurement of the primary reinforcing properties of morphine (WOODS and SCHUSTER, 1971; YANAGITA, 1973b). The rate of bar-pressing which serves as the measure of reinforcing properties of the drug has been shown to be related to unit dose. Morphine has rate-limiting effects on self-injection behavior to which tolerance develops (WOODS and SCHUSTER, 1971). A widely used technique which circumvents these confounding factors is the cross self-administration procedure in which rhesus monkeys are conditioned to respond to a standard drug. Access to the standard drug is limited and when a stable response rate is established, saline is substituted and the response extinguishes, showing that the monkey discriminates between the reinforcing drug and saline. The monkey is then given access to a test drug, and if response rates increase the test drug is presumed to be positively reinforcing. If response rates decrease, it is presumed to be aversive. YANAGITA et al. (1970) found that low doses of morphine and meperidine maintained responding rates above saline rates but below SPA rates; whereas, higher doses decreased responding rates below the saline rate in SPA-conditioned monkeys. HOFFMEISTER et al. (1970) found similar results with morphine using cocaine as a conditioning drug; however, in subsequent studies, HOFFMEISTER and SCHLICHTING (1972) studied morphine, codeine, *d*-propoxyphene, propiram, pentazocine, and nalorphine in cross-substitution tests in groups of monkeys conditioned to self-administer codeine or cocaine. Except for nalorphine, all drugs were self-administered by both groups of monkeys but response rates were higher in codeine than cocaine-conditioned monkeys. These findings indicate that prior drug history influences response rates. Further, morphine appeared to be less reinforcing than the other drugs. Similar findings have been reported by SCHUSTER and BALSTER (1972). YANAGITA (personal communication) found morphine and pentobarbital to be more reinforcing when substituted in pentazocine than in SPA-conditioned monkeys.

The rate of injections in the cross-substitution procedure, as well as in the continuous reinforcement procedure, is influenced by the duration of action of the drug, the behavioral disruptive effects of the drug and the development of physical dependence and tolerance (WOODS and SCHUSTER, 1971; HOFFMEISTER and SCHLICHTING, 1972; SCHUSTER and BALSTER, 1972; SCHUSTER and JOHANSON, 1974). Since the rate of self-administration is influenced by these various factors, other techniques have been developed for making less ambiguous measures of reinforcing properties of drugs. YANAGITA (1973b) developed a progressive ratio test in which monkeys are trained to self-administer a standard drug by pressing a lever 100 times (FR 100). The standard drug is replaced by saline until the monkey reduces the number of drug injections to less than half those for the standard drug. A test drug is then substituted for saline using a FR 100 schedule and self-administration is allowed. The lever press ratio is doubled 24 h later and again doubled repeatedly after a fixed number of injections. When the time interval between injections is greater than 48 h, the monkeys are said to have extinguished, and this extinction ratio is a measure of reinforcement. YANAGITA (1973b) found the extinction ratio for morphine is higher in dependent than nondependent monkeys. YANAGITA (personal communication) reports extinction ratios of 25600 for morphine 0.5 mg/kg i.v.; 12800 for cocaine 0.48 mg/kg i.v.; 6400 for codeine 1.0 mg/kg i.v.; pentazocine 0.25 mg/kg i.v., and pentobarbital 10 mg/kg intragastrically; 3200 for propiram 2.0 mg/kg i.v.; and 1600 for nicotine 200 ng/kg i.v. Morphine appears to be the most reinforcing drug using this procedure.

Procedures have also been developed to assess aversive properties of drugs. GOLDBERG et al. (1971) found that morphine-dependent monkeys would terminate bar-pressing if it resulted in the administration of nalorphine or naloxone, or if stimuli associated with their administration were presented. HOFFMEISTER and SCHLICHTING (1972) found nalorphine to be aversive in cross-substitution tests in codeine- and cocaine-conditioned monkeys. Subsequently, HOFFMEISTER and GOLDBERG (1973) found chlorpromazine but not imipramine to be aversive in cross-substitution tests in cocaine-conditioned monkeys. Thus, it was felt that monkeys could discriminate between drugs which were euphoriant and those which produced aversive effects in man, and a procedure was developed to measure the aversive properties of psychotropic drugs in which naive monkeys were trained to terminate electric shocks (escape) or an associate stimulus occurring 30 s before the shock (avoidance). When a saline infusion replaced the electric shocks, extinction of escape and avoidance behavior was observed. When nalorphine or cyclazocine was administered in place of saline, avoidance and escape behavior reappeared (HOFFMEISTER and WUTTKE, 1973). When naloxone, pentazocine, propiram, and codeine infusions replaced electric shocks, these led to extinction of avoidance and escape behavior. Thus, in the nondependent monkey, the narcotic antagonists nalorphine and cyclazocine have aversive properties, while naloxone, pentazocine, propiram, and codeine do not. Pentazocine and propiram, like nalorphine, were aversive in morphine-dependent monkeys (SCHLICHTING and HOFFMEISTER, 1971).

Procedures to measure the primary reinforcing properties of drugs are being developed to by-pass the rate-limiting effects of drugs in which animals are given a simultaneous choice between different drug solutions. These procedures appear not to have been used with opiates (SCHUSTER and JOHANSON, 1974).

On the basis of present studies of the injection procedures in monkeys, some conclusions may be drawn. Morphine and morphinelike drugs have primary reinforcing properties which are enhanced by the development of physical dependence. From the work of HOFFMEISTER and his colleagues, it appears that the naive rhesus monkey distinguishes morphinelike drugs which are euphoriants in man from the nalorphinelike agonists which produce dysphoric subjective effects and drugs which are devoid of agonistic effects.

III. Suppression Studies (Table 3)

The most widely used procedure for the assessment of morphinelike activity is the suppression technique in the morphine-dependent rhesus monkey. The development of this technique has been recounted by DENEAU (1970). In brief, following characterization of morphine dependence in the monkey by TATUM et al. (1929), KOLB and DUMEZ (1931), and SEEVERS (1936a, b), SEEVERS and his collaborators (IRWIN, DENEAU, MCCARTHY, and VILLARREAL) applied this basic knowledge, devising a clinical scoring system which had eight grades based on SEEVERS' (1936a) grading system, as well as a scoring system using a modification of the method of HIMMELSBACH (see Sect. II, Chap. 3). SEEVERS and IRWIN (1953) demonstrated an excellent correlation between these two systems of assessing the intensity of morphine abstinence in the monkey. The eight grade clinical scoring system has been used routinely in the study of the abstinence syndrome in the morphine-dependent rhesus monkey; however, observations have been validated from time to time using the abstinence point scoring system. Because of the importance of this work, the procedure used to measure the intensity of abstinence in the monkey as devised by SEEVERS (1936a) will be presented in detail.

Monkeys are made dependent on 12 mg/kg/day of morphine sulfate administered subcutaneously every 6 h in four equally divided doses. Suppression studies are conducted in monkeys that are 12–24 h abstinent. Signs of abstinence are assessed by trained observers into four clinical grades.

Mild abstinence consists of signs of apprehension, yawning, lacrimation, rhinorrhea, perspiration, hiccups, shivering, and irritability as evidenced by chattering, quarrelling, and fighting. These signs of abstinence are similar to the + and part of the + + signs of HIMMELSBACH (1937).

Intermediate signs of abstinence include intention tremor, anorexia, piloerection, muscle twitching, rigidity, and holding of the abdomen, which is interpreted as a sign of gastrointestinal cramps.

Severe signs of abstinence consist of extreme restlessness, assumptions of peculiar postures, vomiting, severe diarrhea, erections and continuing masturbation, conjunctival injection which is considered a sign of sleeplessness, continual calling and crying, spasticity, and lying on the side with eyes closed.

Very severe abstinence is characterized by docility and apathy, dyspnea, pallor, strabismus, dehydration and weight loss, prostration, circulatory collapse, and occasionally death.

DENEAU (1956) evolved a suppression test in the morphine-dependent rhesus monkey similar to that employed by HIMMELSBACH in man who were physically dependent on morphine. In this test morphine or an equivalent quantity of the experimental drug, as well as one-half and twice the equivalent quantity of the experimental drug, is administered to the abstinent monkey at weekly intervals and the decrease in intensity of abstinence observed. On the basis of these observations, the experimental physical dependence capacity of the experimental drug is graded high if it completely suppresses abstinence; intermediate if it suppresses only in dose levels that produce side effects; low if it only partially suppresses abstinence when maximally tolerated dose levels are administered; and none if it fails to suppress abstinence at any dose level.

This test has not only been of great value but has allowed the rapid and economic assessment of the physical dependence-producing properties of large numbers of synthetic analgesics. Further, this procedure has been better validated than any other procedures except those employed in man (see Sect. II, Chap. 3). DENEAU (1956) conducted an extensive series of studies in which he compared results obtained in the monkey using both suppression and direct addiction (see below) studies with those obtained in man in which comparable techniques were used. YANAGITA (1973b) has repeated many of the substitution and direct addiction experiments in the morphine-dependent rhesus monkey. Results obtained in substitution and direct addiction studies with morphine, methadone, meperidine, codeine, *d*-propoxyphene, and dextromethorphan were in agreement with those reported by DENEAU (1956). YANAGITA extended these procedures by making monkeys dependent on low doses of morphine (1.2 mg/kg/day) and found that propiram and pentazocine suppress abstinence in monkeys dependent on 1.2 mg/kg/day of morphine but precipitate abstinence in monkeys dependent on 12 mg/kg/day.

In the intervening period of approximately 20 years between DENEAU's original observations and this writing, a large number of additional analgesics have been studied both in the monkey and man, and a more definitive assessment of these techniques can be made. In evaluating and comparing suppression studies in the rhesus monkey with comparable techniques conducted in man, several important differences must be kept in mind: (1) A much higher level of dependence is produced in the monkey (12 mg/kg) than in man (0.5–4 mg/kg). When man is made dependent on 240 mg/day (ca. 3–4 mg/kg/day) "very severe" signs of abstinence such as dyspnea, circulatory collapse, and death are rarely or never seen. (2) The morphine abstinence syndrome may have a more rapid onset in the monkey than man, reaching maximal intensity by the 24th h of abstinence, while in man abstinence is maximal around the 48th h of abstinence (HIMMELSBACH, 1937; KOLB and HIMMELSBACH, 1938; FRASER et al., 1961). (3) Because of patients' intolerance to the suffering associated with abstinence, most suppression studies in man are conducted when patients are showing a lesser degree of abstinence than when similar studies are conducted in the monkey.

Since nearly 1000 drugs have been assessed in the Department of Pharmacology of the University of Michigan for their abuse potential, it is impossible within the confines of this volume to even cursorily summarize these data. Results obtained with several critical standard compounds are presented in Table 7 and by comparing these results with those obtained in man (see Sect. II, Chap. 3), it can be seen that the concordance is excellent. There have been a small number of drugs that have been assessed at the NIDA Addiction Research Center and at the University of Michigan

Table 7. A summary of the effects of single doses of opiate and opioid analgesics and nalorphine in morphine-dependent monkey as well as effects of chronic administration[a]

	Suppression studies	Direct addiction studies	Precipitated abstinence studies
Morphine	Complete	Severe	Severe
Codeine	Complete	Mild	Mild
Diacetylmorphine	Complete	Severe	Intermediate
Dilaudid	Complete	Intermediate-Severe	Severe
Dicodid	Complete	Intermediate	Intermediate
Oxymorphone	Complete	Intermediate	Intermediate-Severe
Metopon	Complete	Severe	Severe
Levorphanol	Complete	Intermediate-Severe	Intermediate
Dextrorphan	None	Mild	Mild
Dextromethorphan	None	Mild	Mild
Meperidine	Complete	Severe	Severe
Ketobemidone	Complete	Intermediate	Severe
Methadone	Complete	Severe	Severe
Propoxyphene	Incomplete	Mild	Mild

[a] These data are from the doctoral thesis of Dr. G. A. DENEAU, completed under the direction of Professor M. H. SEEVERS of the Department of Pharmacology of the University of Michigan, Ann Arbor, Michigan.

in which there have been substantive differences between results of the two laboratories. Some of these drugs are phenazocine, GPA-1657, GPA-2087 (*l*-etazocine), profadol, propiram, and tilidine. Some of these drugs and findings will be discussed subsequently in some detail in this chapter.

IV. Direct Addiction and Precipitation Studies (Table 3)

Direct Addiction

In the initial studies of the ability of morphinelike drugs to produce physical dependence, the drugs were administered at 12-h intervals (SEEVERS and IRWIN, 1953). SEEVERS et al. (1954), recognizing that shorter acting drugs would not completely sustain dependence on this schedule, decreased the interval between injections such that it was commensurate with the duration of action of the drug in the dependent monkey.

Referring to Table 7 and Section II, Chapter 3, it can be seen that by and large the agreement between results obtained in the monkey and in man is quite good although there are more discrepancies than have been found in suppression studies. The dependence-producing effects of dextromethorphan, meperidine, and ketobemidone seem greater in the monkey than in man while GPA-1657 and GPA-2087 are less. These differences will also be discussed in detail. Following the observations that antagonists such as nalorphine and cyclazocine which also have agonistic activity produced a type of physical dependence in man which was qualitatively different from morphine dependence (MARTIN et al., 1965; MARTIN and GORODETZKY, 1965), studies were conducted in monkeys to determine if agonists of the nalorphine type would produce physical dependence in this species. VILLARREAL (1972) reported that chronic administration of cyclazocine produced physical dependence in the rhesus monkey. In the cyclazocine-dependent monkey, naloxone, in doses 16–32 times higher than those required to precipitate abstinence in morphine-dependent monkeys, produced mild restlessness, frequent yawning and scratching accompanied by an increase in rectal temperature. Following withdrawal of cyclazocine, frequent yawning and scratching were seen. The fact that body temperature is elevated in the naloxone-precipitated abstinence syndrome in the cyclazocine-dependent monkey and is decreased in the morphine-dependent monkey suggests that cyclazocine dependence is also different from morphine dependence in the monkey. COWAN (1974) observed that chronically administered nalorphine, cyclazocine, and pentazocine produced physical dependence in the patas monkey and that both the precipitated and withdrawal abstinence syndromes were qualitatively different from the morphine abstinence syndrome.

IRWIN and SEEVERS (1952) reported that nalorphine (2 mg/kg) precipitated abstinence in monkeys dependent upon morphine, 6-methyldihydromorphine, racemorphan, methadone, and ketobemidone and compared the severity of the precipitated abstinence with withdrawal abstinence. BURNS et al. (1958) showed dose-response abstinence for nalorphine and levallorphan in morphine-dependent monkeys, and found levallorphan 3 times as potent as nalorphine. VILLARREAL and KARBOWSKI (1974) devised a scoring system based on 42 signs of abstinence in which 0, 1, 2, or 3 points were awarded for each sign depending upon intensity. Using bioassay techniques relative potency estimates were obtained for a number of antagonists and compared with potency estimates obtained in man, rodents and the guinea pig ileum. The potencies of levallorphan, naloxone, cyclazocine, profadol, and naltrexone relative to nalorphine but not nalbuphine were similar in the monkey and man.

H. Critique of Certain Critical Drugs

I. Meperidine and Ketobemidone

The ability of meperidine to create and sustain physical dependence has been studied in the rat, dog, monkey, and man. In the rat meperidine has a limited ability to suppress morphine abstinence and to produce physical dependence. Efforts to produce physical dependence on meperidine in the dog have been unsuccessful as determined by both nalorphine-precipitated and withdrawal abstinence. In contrast, in the monkey meperidine not only completely suppresses morphine abstinence but produces a high level of physical dependence. In man HIMMELSBACH (1942) found that meperidine only partially substituted for morphine in morphine-dependent subjects, and that upon abrupt withdrawal, the abstinence was mild in intensity, being less than that of codeine. ISBELL (1955) was able to precipitate abstinence with nalorphine only in patients who were stabilized on dose levels of meperidine in excess of 1600 mg/day. On the other hand, ISBELL (1949) found that ketobemidone, a congener of meperidine, was more effective than meperidine in suppressing abstinence and inducing physical dependence. Ketobemidone was also effective in suppressing abstinence and producing physical dependence in the monkey.

The reasons for the discrepancies between the physical dependence-producing ability of meperidine in monkey and man are not apparent. It is possible that the monkey may be more susceptible to the physical dependence-producing effects of phenylpiperidine analgesics than man. Several lines of evidence suggest that agonistic actions of the phenylpiperidines may be less in man than in some other species.

II. Dextrorphan and Dextromethorphan

Dextrorphan has been studied for its ability to suppress abstinence in morphine-dependent mice, monkeys, and man (ISBELL and FRASER, 1953) and is inactive. Further, dextrorphan does not produce either subjective or physiologic changes of the morphine type in man (ISBELL and FRASER, 1953). In the monkey, however, chronic administration of dextrorphan produces a type of physical dependence that is characterized by both a mild withdrawal and nalorphine-precipitated abstinence syndrome. Although dextromethorphan does not produce morphinelike subjective effects in man (ISBELL and FRASER, 1953), large doses do dilate pupils, increase blood pressure, and produce disturbing subjective feelings (JASINSKI et al., 1971 a); however, these changes are not antagonized by naloxone. Chronic administration of dextromethorphan, like dextrorphan, also produces a type of physical dependence characterized by a mild withdrawal and nalorphine-precipitated abstinence syndrome in the monkey. These provocative experiments in which dextrorphan and dextromethorphan were chronically administered in the monkey should be repeated in other species. It seems unlikely that the type of dependence that they produce could be of either the morphine or nalorphine type since morphine suppression studies are quite sensitive in measuring morphinelike activity and nalorphine will suppress abstinence in the cyclazocine dependent dog (GILBERT and MARTIN, 1976).

III. Phenazocine, GPA-1657 (Beta-[-]-5-phenyl-9-methyl-2′-hydroxy-2-methyl-6,7-benzomorphan) and GPA-2087 (*l*-alpha-5,9-diethyl-2′-hydroxy-2-methyl-6,7-benzomorphan; *l*-etazocine)

DENEAU et al. (1959) conducted three blind suppression studies with phenazocine in the morphine-dependent monkey. In dose levels up to 3 mg/kg, phenazocine did not suppress abstinence. In a subsequent study, 10 mg/kg suppressed abstinence to a degree equivalent to 3 mg/kg of morphine. Dose levels of 10 and 15 mg/kg produced convulsions. In further studies in which 6, 12, and 18 mg/kg were administered, it was estimated that 17 mg/kg of phenazocine was equipotent to 3 mg/kg of morphine in suppressing abstinence. Because phenazocine was approximately 10–20 times as potent as morphine as an analgesic in the mouse (MAY and EDDY, 1959) and only one-sixth as potent as morphine in suppressing abstinence in the monkey, it was felt that it might have a lower addiction liability than morphine.

When phenazocine was studied in man, FRASER and ISBELL (1960) found that it was 3–4 times more potent than morphine in producing morphinelike subjective effects and pupillary constriction and was 8 times more potent than morphine in suppressing abstinence in morphine-dependent subjects.

GPA-1657 did not suppress abstinence in abstinent morphine-dependent monkeys and precipitated abstinence in morphine-stabilized dependent monkeys (VILLARREAL, 1968, 1970); however, its agonistic action in depressing eye blinks in the monkey was antagonized by nalorphine. In man (JASINSKI et al., 1971 b) GPA-1657 produced typical morphinelike subjective and physiologic effects and was somewhat more potent than morphine in this regard. In subjects dependent on 240 mg/day of morphine, GPA-1657 was somewhat less potent than morphine in suppressing abstinence. In patients in whom GPA-1657 was administered chronically, nalorphine precipitated an abstinence syndrome. Further, when these patients were abruptly withdrawn, an abstinence syndrome emerged which most nearly resembled the morphine abstinence syndrome. In preliminary experiments, two additional observations were made concerning GPA-1657 (MARTIN, unpublished observations). GPA-1657 did not precipitate abstinence early in the addiction cycle when patients were receiving small doses of morphine or when they had reached their stabilization dose. GPA-1657 suppressed the stepping reflex and the hyperactive flexor reflex in the abstinent chronic spinal dog. Supraspinal signs of abstinence were not obviously suppressed.

GPA-2087 precipitates a modest abstinence syndrome in the morphine-dependent monkey. In man GPA-2087 was found to be equipotent to morphine in producing pupillary constriction and morphinelike subjective effects (JASINSKI and MANSKY, 1970) as well as suppressing the morphine abstinence syndrome (JASINSKI et al., 1971 a). In this regard, *d*-etazocine suppresses abstinence in the monkey.

It is clear that the monkey differs markedly from man in its responsivity to the benzomorphans in that this series of compounds have greater agonistic activity in man than in the monkey. Because of the small number of reactive sites on some of the compounds of this series, it is highly unlikely that the differences between the two species can be explained by differences in metabolism. A more likely explanation is that there is a difference in the configuration of the morphine receptor between the two species.

IV. Profadol and Propiram

Profadol and propiram were studied at the ARC at a time when two hypotheses were being seriously considered and tested: (1) It was hypothesized (MARTIN, 1967) that drugs that had antagonistic properties could be (a) partial agonists of the morphine type, or (b) a competitive antagonist at the morphine receptor and either a partial or strong agonist at the nalorphine or cyclazocine receptor (MCCLANE and MARTIN, 1967) or a competitive antagonist at both the morphine and nalorphine receptor (MCCLANE and MARTIN, 1967; JASINSKI et al., 1968). (2) A second hypothesis was related to the nature of morphine tolerance in which it was postulated that tolerance did not occur at the receptor, but rather as a consequence of hypertrophy of redundant functional pathways (MARTIN, 1970). One corollary of this hypothesis was that suppression of abstinence was a manifestation of the agonistic action of narcotic analgesics and that partial agonists should be able to suppress abstinence in subjects with low levels of dependence and precipitate abstinence in subjects with high levels of dependence.

Profadol produced analgesia in rodents that was antagonized by nalorphine and thus appeared to be morphinelike (WINDER et al., 1966). In the monkey, profadol failed to suppress abstinence (DENEAU and SEEVERS, 1965). In man single doses of profadol produced morphinelike subjective effects and pupillary constriction, being approximately one-half to one-third as potent as morphine. In patients dependent and stabilized on 240 mg/day of morphine, profadol was $^{1}/_{40}$ as potent as nalorphine in precipitating abstinence. In abstinent subjects who were dependent on 60 mg/day of morphine, profadol was one-third as potent as morphine in suppressing abstinence (Jasinski et al., 1971b). In subsequent studies, Villarreal (1970) found that chronic administration of profadol in the monkey produced physical dependence which was manifested by an abstinence syndrome of low to intermediate intensity.

Propiram has analgesic action in animals and man which can be antagonized by narcotic antagonists (Hoffmeister et al., 1974). In the morphine-dependent monkey, propiram precipitates but does not suppress abstinence (Villarreal, 1970). When propiram was administered chronically to monkeys, a severe abstinence syndrome was precipitated with naloxone but the withdrawal abstinence syndrome was mild. Propiram in single doses is $^1/_8$ as potent as morphine in producing subjective effects and miosis in man. It is $^1/_{10}$ as potent as morphine in suppressing abstinence in patients dependent on 60 mg/day of morphine and $^1/_{200}$ as potent as nalorphine in precipitating abstinence in patients dependent on 240 mg/day of morphine (Jasinski et al., 1971b). Subsequently, Yanagita (1973a, b) suppressed abstinence with propiram in monkeys dependent on $^1/_{10}$ of the standard dose of morphine.

With intensive investigations, the actions of profadol and propiram proved to be very similar in monkey and man. The overwhelming body of evidence supports the conclusions of Jasinski et al. (1971b), that these drugs are partial agonists of the morphine type.

J. Discussion and Conclusions

I. Species

It is apparent that a number of agonistic actions and dependence-producing properties of narcotic analgesics and closely related agonists can be studied in several species. Practical considerations related to the ease, convenience and cost of making observations rather than pharmacologic actions of narcotics frequently determine the desirability of a given species. Studies can be done most economically in small rodents. Rodents have proved useful for the measurement of certain behavioral and somatomotor manifestations of both narcotic intoxication and abstinence; however, the measurement of autonomic changes is more difficult. The fact that the dog is not only a large animal but one that can be readily trained to an experimental setting and procedure makes it a species especially useful for making autonomic and reflex measurements. The monkey has proved to be very useful for the measurement of behavioral, particularly affective, signs of abstinence. Man is unique in that he alone can communicate his feeling states verbally. Further, many measures of autonomic and somatomotor signs can be easily measured in man.

There are both similarities and differences in the effects of morphine on different species (see Tables 2 and 3), and because the types of observations made in various species are not the same, it is difficult to determine if any species can best serve as a prototype for man. It is quite clear that there are major differences in the responsivity of different species to various chemical classes of narcotic analgesics. Thus, it would appear that the monkey is more susceptible to the agonistic actions of the phenylpiperidines and less responsive to the benzomorphans than man. Further, the *d*-isomers of the morphinans and benzomorphans have agonistic activity in the monkey.

The *l*-isomers of the benzomorphans have lesser activity in the monkey than man. VILLARREAL (1972) correlated the potencies of drugs to suppress abstinence in morphine-dependent monkeys with their ability to produce analgesia in the mouse, and found an excellent correlation for morphine congeners ($r=0.87$), the phenylpiperidines ($r=0.93$), and the bridged oripavine and thebaines ($r=0.96$). The correlation was less good for the morphinans ($r=0.49$), the methadones ($r=0.54$), and the benzomorphans ($r=0.32$). Although some of the difference between the responsivity of the monkey and man to the phenylpiperidines, morphinans, and benzomorphans may be accounted for by differences in metabolism, this is an unlikely explanation for the benzomorphan differences. A more likely explanation is that morphine and related receptors may show subtle structural differences from one species to another. This conclusion has profound implications in the strategy of drug assessment for it is unlikely that any two species have identical morphine receptors; hence, no one species can be considered as a model for another for all morphinelike drugs. It is thus important that the dependence-producing capabilities of an experimental drug be assessed in several species before it is finally studied in man. The choice of species for any drug class can be decided only when adequate validating information has been generated.

Because of species difference in the responsivity to opioids, it is important that these differences not be mistaken for a dissociation in pharmacologic activity. Thus, the high analgesic activity of phenazocine in the mouse and its low potency in suppressing abstinence in the monkey did not in all probability represent any dissociation between analgesic and physical dependence-producing activities but only species difference. The confounding effect of species difference can be avoided by studying the agonistic action (e.g., analgesia) of the experimental drug in the same species that dependence and self-administration studies are done.

II. Sensitivity of Measures and Experimental Design

It is important to employ sensitive measures of the agonistic actions of narcotic analgesics (MARTIN, 1967). Thus, the analgesic actions of partial agonists of both the morphine and nalorphine type may go undetected if strong nociceptive stimuli are employed, yet may be easily seen if mild nociceptive stimuli are used. A similar principle applies to the measurement of abstinence signs in physically dependent animals.

In morphine-dependent men, liminal doses of antagonists will dilate pupils without eliciting other signs or symptoms of abstinence. Frequently feelings of weakness and restlessness are mentioned as symptoms of withdrawal abstinence without other signs of abstinence being manifest. On the other hand, signs of abstinence such as retching and emesis are seen infrequently when the abstinence syndrome is of mild intensity but commonly during severe abstinence. Thus, if subjects who have only a liminal degree of physical dependence or are mildy abstinent are observed only for signs characteristic of severe abstinence, the presence of dependence and abstinence may go undetected.

When the magnitude of drug-induced changes are of comparable magnitude to spontaneous changes and normal variability, it is necessary to make systematic observations using precise measurements and to conduct appropriate control experiments to detect effects. A crossover design is preferable to a group comparison.

III. Validity of Tests

1. Pharmacologic Profile

The pharmacologic profile of analgesics can frequently allow serendipitous guesses about their dependence and abuse potentiality. In this regard, it is important to reiterate that studies of the dependence-producing capabilities should be done in the same species that quantitative assessment of the pharmacologic profile is done for the data to provide even a first estimate of abuse potentiality in man. It is also important that in showing concordance or differences in profiles quantitative estimates be made of the potency of the experimental drug relative to a standard drug for various effects. Although the characterization of pharmacologic profiles has proved of great value in man, it has not been extensively employed using animals and for this reason only a modest amount of data has been generated in any species which bears on this issue of validity.

2. Self-Administration

The use and limitation of self-administration have been critically reviewed by SCHUSTER and THOMPSON (1969). From a pharmacologic perspective much more experimentation needs to be done to establish the validity of these methods in measuring primary reinforcing properties in animals that are relevant to the properties of abused drugs that predispose to their abuse in man. This aspect of the problem is further confounded by the fact that the reinforcing properties of drugs in man have been studied only in select patient populations such as addicts and habitual drug users. Thus, there are insufficient clinical data against which to validate self-administration techniques. At first glance, the correlation between the reinforcing properties of abused drugs in animals and the perniciousness of their abuse in man is poor. Thus, cocaine and amphetamine are the most reinforcing in animals and barbiturates next. The LSD-like hallucinogens and marihuana have liminal, if any, reinforcing properties. Narcotic analgesics have modest reinforcing properties in naive monkeys and are aversive in naive dogs. In man cocaine and amphetamine also are euphorigenic to most people, while narcotics are euphorigenic to a much more restricted population (Sect. I, Chap. 1; Sect. II, Chaps. 1 and 3; Part II, Sect. I, Chaps. 1 and 2). On the other hand, the magnitude of amphetamine and cocaine abuse has waxed and waned, being ameliorated frequently by modest legal control measures while narcotic and alcohol abuse have been more intractable problems. There is also a lack of concordance between techniques. Thus, in monkeys that have been conditioned to bar-press for codeine or cocaine, morphine appears to be less reinforcing than either; however, if the progressive ratio test is used, morphine appears to be more reinforcing than cocaine and codeine.

Animals conditioned with cocaine and amphetaminelike drugs are used most commonly for self-administration. This procedure has a profound influence on drug-taking behavior. Conditioned animals will bar-press for drugs that are aversive or devoid of reinforcing properties in naive animals and which are rarely if ever subject to clinical abuse. On the other hand, this pattern of behavior (conditioning) is commonly seen also among drug abusers. Thus, most heroin abusers have a history of experimenting with other drugs prior to using narcotics. The possibility that a prior history of drug use may cause either a type of psychologic conditioning or a pharmacologic physiologic conditioning that increases the need for or the euphorigenic properties of drugs of abuse cannot be discounted.

3. Suppression Studies

Suppression tests have proved to be one of the most economic and selective methods for identifying drugs that have dependence-producing capacity of the morphine type. There have been two circumstances, however, where suppression tests have provided misleading information.

a) Partial Agonists

The fact that there are partial agonists of both the morphine and nalorphine type has implications when using the dependent animal for assessing the dependence-producing capacity of these agents. When the dependent animal is stabilized (is not abstinent), a partial agonist will reduce the quantity of agonistic activity and precipitate abstinence. On the other hand, when no agonist receptors are occupied and the dependent animal is maximally abstinent, partial agonists will suppress abstinence. When animals are partially withdrawn, the effects of a partial agonist are less easy to predict. Depending on whether they reduce, leave unchanged, or increase agonistic activity, they may exacerbate, not affect, or suppress abstinence. When partially abstinent animals with high levels of dependence have been administered partial agonists, the trend has been for abstinence signs to be worsened (YANAGITA, 1973a, b, and personal communication); whereas, when partially abstinent human subjects with low levels of dependence have received partial agonists, abstinence has been suppressed (JASINSKI et al., 1971b). The demonstration that some drugs may be partial agonists of the morphine type may have important therapeutic implications, for it is possible for a partial agonist to have enough agonistic activity to produce a clinically significant degree of analgesia but not enough to produce a clinically significant degree of physical dependence.

b) Species

As discussed above (Subsect. J.I.) there are marked differences between species in their responses to certain chemical classes of narcotic analgesics which may be a consequence of dissimilarities in drug metabolism or to differences in configuration of receptors.

The overinterpretation of data generated in several species which suggests selectivity of action and a dissociation of therapeutic effects and toxicity or dependence-producing capability can be avoided if the relevant pharmacologic parameters are studied in the same species.

4. Direct Addiction and Precipitation Studies

Direct addiction studies provide another body of information which assists in characterizing a drug as belonging to a particular pharmacologic class. Because of subtle but important differences in pharmacologic actions among drugs closely related structurally to narcotic analgesics, it is of great importance to qualitatively characterize both the withdrawal and precipitated abstinence syndrome. This has assisted in distinguishing nalorphine and morphinelike agonists. Direct addiction studies, however, transcend the goal of obtaining another pharmacologic attribute for they provide insight into the relationship between physical dependence and drug need. Identification of the protracted abstinence syndrome and the demonstration that the nalorphine-like abstinence syndrome is not associated with drug-seeking behavior have shown that the presence or absence of overt dramatic signs of abstinence are not in themselves evidence of abuse potentiality. These observations have further indicated the importance of observations that are indicative of the affective consequences of abstinence.

Precipitation of abstinence by narcotic antagonists also provides important evidence about physical dependence-producing characteristics. In general, precipitated abstinence is a more sensitive indicator of physical dependence than withdrawal abstinence, particularly when abstinence is precipitated with a pure competitive antagonist such as naloxone. For the qualitative assessment of the abstinence syndrome, precipitated abstinence is less reliable than withdrawal abstinence for the characteristics of precipitated abstinence change with the level of dependence.

IV. Conclusions

No single technique used in a single species will allow a valid assessment of the abuse potentiality of psychoactive drugs. With the enormous growth of the chemical and pharmaceutical industries and their impact upon health, it is to society's self interest to identify the abuse potentiality of new drugs and chemicals prior to their introduction into the open market or medical practice. Broad-based studies of new and useful drugs and chemicals in several species using a variety of techniques are not only justified but indicated. There is every reason to believe that pharmacologic techniques exist to detect those properties of new analgesics which will give rise to abuse. To the present, all drugs that have morphinelike properties in animals have been found to be morphinelike in man. There have been some narcotic analgesics whose morphinelike properties were not initially identified in animal studies that were found to be morphinelike in man. When further studies of the appropriate nature were conducted in animals, the human studies were confirmed.

References

Aceto, M.D., McKean, D.B., Pearl, J.: Effects of opiates and opiate antagonists on the Straub tail reaction in mice. Brit. J. Pharmacol. **36**, 225—239 (1969)

Akera, T., Brody, T.M.: The addiction cycle to narcotics in the rat and its relation to catecholamines. Biochem. Pharmacol. **17**, 675—688 (1968)

Andrews, H.L., Workman, W.: Pain threshold measurements in the dog. J. Pharmacol. exp. Ther. **73**, 99—103 (1941)

Barry, H., III: Classification of drugs according to their discriminable effects in rats. Fed. Proc. **33**, 1814—1824 (1974)

Barry, H., III, Kubena, R.K.: Discriminative stimulus characteristics of alcohol, marihuana and atropine. In: Singh, J.M., Miller, L., Lal, H. (Eds.): Drug Addiction: Experimental Pharmacology, Vol. 1, pp. 3—16. Mount Kisco, N.Y.: Futura Publishing Company, Inc. 1972

Belleville, R. E.: Control of behavior by drug-produced internal stimuli. Psychopharmacologia (Berl.) **5**, 95—105 (1964)

Blasig, J., Herz, A., Reinhold, K., Zieglgansberger, S.: Development of physical dependence on morphine in respect to time and dosage and quantification of the precipitated withdrawal syndrome in rats. Psychopharmacologia (Berl.) **33**, 19—38 (1973)

Buckett, W. R.: A new test for morphine-like physical dependence (addiction liability) in rats. Psychopharmacologia (Berl.) **6**, 410—416 (1964)

Burns, R. H., McCarthy, D. A., Deneau, G. A., Seevers, M. H.: Comparison of dose dependent effects of nalorphine with those of levallorphan in production of graded abstinence in morphine dependent monkeys. Fed. Proc. **17**, 355 (1958)

Cabanac, M.: Physiological role of pleasure. Science **173**, 1103—1107 (1971)

Carter, R. L., Wikler, A.: Chronic meperidine intoxication in intact and chronic spinal dogs. Fed. Proc. **14**, 955 (1955)

Collier, H. O. J., Francis, D. L., Schneider, C.: Modification of morphine withdrawal by drugs interacting with humoral mechanism: Some contraindications and their interpretation. Nature (Lond.) **237**, 220—222 (1972)

Collins, R. J., Weeks, J. R.: Relative potency of codeine, methadone and dihydromorphinone to morphine in self-maintained addict rats. Naunyn-Schmiedebergs Arch. exp. Path. Pharmak. **249**, 509—514 (1965)

Collins, R. J., Weeks, J. R.: Lack of effect of dexoxadrol in selfmaintained morphine dependence in rats. Psychopharmacologia (Berl.) **11**, 287—292 (1967)

Colpaert, F. C., Lal, H., Niemegeers, C. J. E., Janssen, P. A. J.: Investigations on drug produced and subjectively experienced discriminative stimuli. 1. The fentanyl cue, a tool to investigate subjectively experienced narcotic drug actions. Life Sci. **16**, 705—716 (1975a)

Colpaert, F. C., Niemegeers, C. J. E., Lal, H., Janssen, P. A. J.: Investigations on drug produced and subjectively experienced discriminative stimuli. 2. Loperamide, an antidiarrheal devoid of narcotic cue producing actions. Life Sci. **16**, 717—728 (1975b)

Coussens, W. R., Crowder, W. F., Smith, S. G.: Acute physical dependence upon morphine in rats. Behav. Biol. **8**, 533—543 (1973)

Cowan, A.: Evaluation in nonhuman primates: Evaluation of the physical dependence capacities of oripavine-thebaine partial agonists in patas monkeys. In: Braude, M. C., Harris, L. S., May, E. L., Smith, J. P., Villarreal, J. E. (Eds.): Narcotic antagonists. Advances in Biochemical Psychopharmacology, Vol. 8. New York: Raven Press 1974

Deneau, G. A.: An analysis of the factors influencing the development of physical dependence to narcotic analgesics in the rhesus monkey with methods for predicting physical dependence liability in man. Dissertation for degree of Doctor of Philosophy, University of Michigan, Ann Arbor, Mich., 1956

Deneau, G. A.: The monkey colony in studies of tolerance and dependence. Univ. Mich. Med. Ctr. J. **36**, 212—214 (1970)

Deneau, G. A., McCarthy, D. A., Seevers, M. H.: Physical dependence liability studies in the monkey. Presented at 20th Meeting, Committee on Drug Addiction and Narcotics, National Research Council, Washington, D.C., 1959

Deneau, G. A., Seevers, M. H.: Evaluation of new compounds for morphinelike physical dependence in the Rhesus monkey. Presented at 27th Meeting, Committee on Drug Addiction and Narcotics, National Research Council, Houston, Tex., 1965

Deneau, G., Yanagita, T., Seevers, M. H.: Self-administration of psychoactive substances by the monkey. Psychopharmacologia (Berl.) **16**, 30—48 (1969)

Dripps, R. D., Dumke, P. R.: The effect of narcotics on the balance between central and chemoreceptor control of respiration. J. Pharmacol. exp. Ther. **77**, 290—300 (1943)

Fraser, H. F., Isbell, H.: Human pharmacology and addiction liabilities of phenazocine and levophenacylmorphan. Bull. Narcot. **12**, 15—23 (1960)

Fraser, H. F., Van Horn, G. D., Martin, W. R., Wolbach, A. B., Isbell, H.: Methods for evaluating addiction liability. (A) "Attitude" of opiate addicts toward opiate-like drugs, (B) A short-term "direct" addiction test. J. Pharmacol. exp. Ther. **133**, 371—387 (1961)

Frumkin, K.: Physical dependence in rats after low morphine doses. Life Sci. **15**, 455—462 (1974)

Gilbert, P. E., Martin, W. R.: The effects of morphine- and nalorphine-like drugs in the nondependent, morphine-dependent and cyclazocine-dependent chronic spinal dog. J. Pharmacol. exp. Ther. **198**, 66—82 (1976)

Goldberg, S.R., Hoffmeister, F., Schlichting, U., Wuttke, W.: Aversive properties of nalorphine and naloxone in morphine-dependent rhesus monkeys. J. Pharmacol. exp. Ther. **179**, 268—276 (1971)
Goode, P.G.: An implanted reservoir of morphine solution for rapid induction of physical dependence in rats. Brit. J. Pharmacol. **41**, 558—566 (1971)
Haertzen, C.A.: Development of scales based on patterns of drug effects, using the Addiction Research Center Inventory (ARCI). Psychol. Rep. **18**, 163—194 (1966)
Hanna, C.: A demonstration of morphine tolerance and physical dependence in the rat. Arch. int. Pharmacodyn. **124**, 326—329 (1960)
Headlee, C.P., Coppock, H.W., Nichols, J.R.: Apparatus and technique involved in a laboratory method of detecting the addictiveness of drugs. J. Amer. pharm. Ass. **44**, 229—231 (1955)
Hill, H.E., Haertzen, C.A., Wolbach, A.B., Jr., Miner, E.J.: The Addiction Research Center Inventory: Standardization of scales which evaluate subjective effects of morphine, amphetamine, pentobarbital, alcohol, LSD-25, pyrahexyl and chlorpromazine. Psychopharmacologia (Berl.) **4**, 167—183 (1963)
Hill, H.E., Jones, B.E., Bell, E.C.: State dependent control of discrimination by morphine and pentobarbital. Psychopharmacologia (Berl.) **22**, 305—313 (1971)
Himmelsbach, C.K.: Clinical studies of drug addiction. II. "Rossium" treatment of drug addiction. Publ. Hlth Rep. (Wash.), Suppl. 125, 1937
Himmelsbach, C.K.: Studies of the addiction liability of "Demerol" (D-140). J. Pharmacol. exp. Ther. **75**, 64—68 (1942)
Hirschhorn, I.D., Rosecrans, J.A.: A comparison of the stimulus effects of morphine and lysergic acid diethylamide (LSD). Pharmacol. biochem. Behav. **2**, 361—366 (1974)
Hoffmeister, F., Goldberg, S.R.: A comparison of chlorpromazine, imipramine, morphine and *d*-amphetamine self-administration in cocaine-dependent rhesus monkeys. J. Pharmacol. exp. Ther. **187**, 8—14 (1973)
Hoffmeister, F., Goldberg, S.R., Schlichting, U., Wuttke, W.: Self-administration of *d*-amphetamine, morphine and chlorpromazine by cocaine "dependent" rhesus monkeys. Naunyn-Schmiedebergs Arch. exp. Path. Pharmakol. **266**, 359—360 (1970)
Hoffmeister, F., Kroneberg, G., Schlichting, U., Wuttke, W.: Zur Pharmakologie des Analgetikums Propiramfumarat (N-(1-Methyl-2-piperidino-äthyl)-N-(2-pyridyl)-propionamid-fumarat). Arzneimittel-Forsch. (Drug Res.) **24**, Nr. 4a, 600—624 (1974)
Hoffmeister, F., Schlichting, U.U.: Reinforcing properties of some opiates and opioids in rhesus monkeys with histories of cocaine and codeine self-administration. Psychopharmacologia (Berl.) **23**, 55—74 (1972)
Hoffmeister, F., Wuttke, W.: Negative reinforcing properties of morphine-antagonists in naive rhesus monkeys. Psychopharmacologia (Berl.) **33**, 247—258 (1973)
Holtzman, S.G., Villarreal, J.E.: Morphine dependence and body temperature in rhesus monkeys. J. Pharmacol. exp. Ther. **166**, 125—133 (1969)
Hosoya, E.: Some withdrawal symptoms of rats to morphine. Pharmacologist **1**, 77 (1959)
Houde, R.W., Wikler, A.: Delineation of the skin-twitch response in dogs and the effects thereon of morphine, thiopental and mephenesin. J. Pharmacol. exp. Ther. **103**, 236—242 (1951)
Houde, R.W., Wikler, A., Irwin, S.: Comparative actions of analgesic, hypnotic and paralytic agents of hindlimb reflexes in chronic spinal dogs. J. Pharmacol. exp. Ther. **103**, 243—248 (1951)
Huidobro, F., Maggiolo, C.: Some features of the abstinence syndrome to morphine in mice. Acta physiol. lat.-amer. **11**, 201—209 (1961)
Irwin, S., Seevers, M.H.: Comparative study of regular and N-allylnormorphine induced withdrawal in monkeys addicted to morphine, 6-methyldihydromorphine, Dromoran, methadone, and ketobemidone. J. Pharmacol. exp. Ther. **106**, 397 (1952)
Isbell, H.: The addiction liability of some derivatives of meperidine. J. Pharmacol. exp. Ther. **97**, 182—189 (1949)
Isbell, H.: Withdrawal symptoms in 'primary' meperidine addicts. Fed. Proc. **14**, 354 (1955)
Isbell, H., Fraser, H.F.: Actions and addiction liabilities of dromoran derivatives in man. J. Pharmacol. exp. Ther. **107**, 524—530 (1953)
Jasinski, D.R., Mansky, P.: The subjective effects of GPA-2087 and nalbuphine (EN-2234A). Presented at 32nd Meeting, Committee on Problems of Drug Dependence, National Research Council, Washington, D.C., 1970

Jasinski, D. R., Martin, W. R., Hoeldtke, R. D.: Effects of short- and long-term administration of pentazocine in man. Clin. Pharmacol. Ther. **11**, 385—403 (1970)

Jasinski, D. R., Martin, W. R., Hoeldtke, R.: Studies of the dependence-producing properties of GPA-1657, profadol, and propiram in man. Clin. Pharmacol. Ther. **12**, 613—649 (1971 b)

Jasinski, D. R., Martin, W. R., Mansky, P. A.: Progress report on the assessment of the antagonists nalbuphine and GPA-2087 for abuse potential and studies of the effects of dextromethorphan in man. Presented at 33rd Meeting, Committee on Problems of Drug Dependence, National Research Council, Toronto, Canada, 1971 a

Jasinski, D. R., Martin, W. R., Sapira, J. D.: Antagonism of the subjective, behavioral, pupillary, and respiratory depressant effects of cyclazocine by naloxone. Clin. Pharmacol. Ther. **9**, 215—222 (1968)

Jones, B. E., Prada, J. A.: Relapse to morphine use in dog. Psychopharmacologia (Berl.) **30**, 1—12 (1973)

Kamei, C., Shimomura, K., Ueki, S.: Significance of withdrawal jumping response in predicting physical dependence in mice. Jap. J. Pharmacol. **23**, 421—426 (1973)

Kaymakcalan, S., Woods, L. A.: Nalorphine-induced "abstinence syndrome" in morphine tolerant albino rats. J. Pharmacol. exp. Ther. **117**, 112—116 (1956)

Kolb, L.: Pleasure and deterioration from narcotic addiction. Ment. Hyg. (N.Y.) **9**, 699—724 (1925)

Kolb, L., DuMez, A. G.: Experimental addiction of animals to opiates. Publ. Hlth Rep. (Wash.) **46**, 698—726 (1931)

Kolb, L., Himmelsbach, C. K.: Clinical studies of drug addiction. III. A critical review of the withdrawal treatments with method of evaluating abstinence syndromes. Amer. J. Psychiat. **94**, 759—797 (1938)

Krueger, H., Eddy, N. B., Sumwalt, M.: The pharmacology of the opium alkaloids, Part 1. Publ. Hlth Rep. (Wash.), Suppl. 165 (1941)

Kuhn, H. F., Friebel, H.: Über den Nachweis von "physical dependence" bei Codein-behandelten Ratten. Med. exp. (Basel) **6**, 301—306 (1962)

Lorenzetti, O. J., Sancilio, L. F.: Morphine dependent rats as a model for evaluating potential addiction liability of analgesic compounds. Arch. int. Pharmacodyn. **183**, 391—402 (1970)

Maggiolo, C., Huidobro, F.: Administration of pellets of morphine to mice: abstinence syndrome. Acta physiol. lat.-amer. **11**, 70—78 (1961)

Malis, J. L.: Analgesic testing in primates. In: Kosterlitz, H. W., Collier, H. O. J., Villarreal, J. E. (Eds.): Agonist and antagonist actions of narcotic analgesic drugs (Proceedings of the symposium of the British Pharmacological Society, Aberdeen, July 1971). London: Macmillan Press 1972

Marshall, I., Weinstock, M.: Quantitative method for assessing one symptom of the withdrawal syndrome in mice after chronic morphine administration. Nature (Lond.) **234**, 223—224 (1971)

Martin, W. R.: Assessment of the dependence-producing potentiality of narcotic analgesics. In: Radouco-Thomas, C., Lasagna, L. (Eds.): International Encyclopedia of Pharmacology and Therapeutics, Sec. 6, Vol. I. Glasgow: Pergamon Press 1966

Martin, W. R.: Opioid antagonists. Pharmacol. Rev. **19**, 463—521 (1967)

Martin, W. R.: A homeostatic and redundancy theory of tolerance to and dependence on narcotic analgesics. In: Wikler, A. (Ed.): The addictive states. Ass. Res. nerv. ment. Dis., Vol. 46. Baltimore: Williams and Wilkins 1968

Martin, W. R.: Pharmacological redundancy as an adaptive mechanism in the central nervous system. Fed. Proc. **29**, 13—18 (1970)

Martin, W. R., Eades, C. G.: Demonstration of tolerance and physical dependence in the dog following a short-term infusion of morphine. J. Pharmacol. exp. Ther. **133**, 262—270 (1961)

Martin, W. R., Eades, C. G.: A comparison between acute and chronic physical dependence in the chronic spinal dog. J. Pharmacol. exp. Ther. **146**, 385—394 (1964)

Martin, W. R., Eades, C. G.: Pharmacological studies of spinal cord adrenergic and cholinergic mechanisms and their relation to physical dependence on morphine. Psychopharmacologia (Berl.) **11** 195—223 (1967)

Martin, W. R., Eades, C. G., Fraser, H. F., Wikler, A.: Use of hindlimb reflexes of the chronic spinal dog for comparing analgesics. J. Pharmacol. exp. Ther. **144**, 8—11 (1964)

Martin,W.R., Eades,C.G., Thompson,J.A., Gilbert,P.E., Sandquist,V.L.: Progress report on the use of the dog for assessing morphine-like and nalorphine-like agonists as well as depot preparations of antagonists. Presented at 35th Meeting, Committee on Problems of Drug Dependence, National Research Council, Chapel Hill, N.C., 1973
Martin,W.R., Eades,C.G., Thompson,J.A., Huppler,R.E., Gilbert,P.E.: The effects of morphine- and nalorphine-like drugs in the nondependent and morphine-dependent chronic spinal dog. J. Pharmacol. exp. Ther. **197**, 517—532 (1976)
Martin,W.R., Eades,C.G., Thompson,W.O., Thompson,J.A., Flanary,H.G.: Morphine physical dependence in the dog. J. Pharmacol. exp. Ther. **189**, 759—771 (1974)
Martin,W.R., Fraser,H.F., Gorodetzky,C.W., Rosenberg,D.E.: Studies of the dependence-producing potential of the narcotic antagonist 2-cyclopropylmethyl-2′-hydroxy-5,9-dimethyl-6,7-benzomorphan (cyclazocine; WIN 20,740; ARC II-C-3). J. Pharmacol. exp. Ther. **150**, 426—436 (1965)
Martin,W.R., Gorodetzky,C.W.: Demonstration of tolerance to and physical dependence on N-allylnormorphine (nalorphine). J. Pharmacol. exp. Ther. **150**, 437—442 (1965)
Martin,W.R., Wikler,A., Eades,C.G., Pescor,F.T.: Tolerance to and physical dependence on morphine in rats. Psychopharmacologia (Berl.) **4**, 247—260 (1963)
May,E.L., Eddy,N.B.: A new potent synthetic analgesic. J. Org. Chem. **24**, 294 (1959)
Maynert,E.W., Klingman,G.I.: Tolerance to morphine. I. Effects on catecholamines in the brain and adrenal glands. J. Pharmacol. exp. Ther. **135**, 285—295 (1962)
McClane,T.K., Martin,W.R.: Effects of morphine, nalorphine, cyclazocine, and naloxone on the flexor reflex. Int. J. Neuropharmacol. **6**, 89—98 (1967)
McMillan,D.E., Waddell,F.B., Cathcart,C.F.: Establishment of physical dependence in mice by oral ingestion of morphine. J. Pharmacol. exp. Ther. **190**, 416—419 (1974)
Nozaki,M., Hosoya,E.: Screening of morphine type physical dependence liability using rats. Fifth International Congress on Pharmacology, Abstracts of Volunteer Papers, p.170, San Francisco, Calif. 1972
Olds,J.: A preliminary mapping of electrical reinforcing effects in the rat brain. J. comp. physiol. Psychol. **49**, 281—285 (1956)
Olds,J.: Self-stimulation of the brain. Its use to study local effects of hunger, sex, and drugs. Science **127**, 315—324 (1958)
Plant,O.H., Pierce,I.H.: Studies of chronic morphine poisoning in dogs. I. General symptoms and behavior during addiction and withdrawal. J. Pharmacol. exp. Ther. **33**, 329—357 (1928)
Ploog,D.: Social communication among primates. In: Schmitt,F.O. (Ed.): The neurosciences. New York: Rockefeller University Press 1970
Risner,M.E., Khavari,K.A.: Morphine dependence in rats produced after five days of ingestion. Psychopharmacologia (Berl.) **28**, 51—62 (1973)
Saelens,J.K., Granat,F.R., Sawyer,W.K.: The mouse jumping test—A simple screening method to estimate the physical dependence capacity of analgesics. Arch. int. Pharmacodyn. **190**, 213—218 (1971)
Schlichting,U.U., Hoffmeister,F.: Positive and negative reinforcing properties of weak morphine antagonists in physical dependent and non-dependent rhesus monkeys. Naunyn-Schmiedebergs Arch. exp. Path. Pharmakol., Suppl. to Vol. **270**, R 123 (1971)
Schuster,C.R., Balster,R.L.: Self-administration of agonists. In: Kosterlitz,H.W., Collier,H.O.J., Villarreal,J.E. (Eds.): Agonist and antagonist actions of narcotic analgesic drugs (Proceedings of the symposium of the British Pharmacological Society, Aberdeen, July 1971). London: Macmillan Press 1972
Schuster,C.R., Johanson,C.E.: The use of animal models for the study of drug abuse. In: Gibbins,R.J., Israel,Y., Kalant,H., Popham,R.E., Schmidt,W., Smart,R.G. (Eds.): Research advances in alcohol and drug problems, Vol.1. New York: John Wiley and Sons 1974
Schuster,C.R., Thompson,T.: Self-administration of and behavioral dependence on drugs. Ann. Rev. Pharmacol. **9**, 483—502 (1969)
Seevers,M.H.: Opiate addiction in the monkey. I. Methods of study. J. Pharmacol. exp. Ther. **56**, 147—156 (1936a)
Seevers,M.H.: Opiate addiction in the monkey. II. Dilaudid in comparison with morphine, heroin and codeine. J. Pharmacol. exp. Ther. **56**, 157—165 (1936b)

Seevers, M. H., Deneau, G., Kissel, J.: I. Annual report on "studies in the monkey (Macacca Mulatta) designed to determine the value of this animal for predicting addiction liability to the newer synthetic analgesics." Presented at 13th Meeting, Committee on Drug Addiction and Narcotics, Rahway, N. J. 1954

Seevers, M. H., Irwin, S.: Studies in the monkey (Macaca mulatta) designed to determine the value of this animal for predicting addiction liability to the newer synthetic analgesics. Presented at 11th Meeting, Committee on Drug Addiction and Narcotics, National Research Council, Lexington, Ky., 1953

Shannon, H. E., Holtzman, S. G.: A pharmacologic analysis of the discriminative effects of morphine in the rat. Presented at 37th Meeting, Committee on Problems of Drug Dependence, National Research Council, Washington, D. C. 1975

Shemano, I., Wendel, H.: Pharmacological indices of addiction liability. Pharmacologist **2**, 97 (1960)

Shuster, L., Hannam, R. V., Boyle, W. E., Jr.: A simple method for producing tolerance to dihydromorphinone in mice. J. Pharmacol. exp. Ther. **140**, 149—154 (1963)

Smith, A. A., Karmin, M., Gavitt, J.: Interaction of catecholamines with levorphanol and morphine in the mouse eye. J. Pharmacol. exp. Ther. **151**, 103—109 (1966a)

Smith, A., Karmin, M., Gavitt, J.: Central origin of the lenticular opacities induced in mice by opiates. J. Pharm. (Lond.) **18**, 545—546 (1966b)

Smith, A. A., Karmin, M., Gavitt, J.: Tolerance to the lenticular effects of opiates. J. Pharmacol. exp. Ther. **156**, 85—91 (1967)

Smith, G. M., Beecher, H. K.: Subjective effects of heroin and morphine in normal subjects. J. Pharmacol. exp. Ther. **136**, 47—52 (1962)

Stolerman, I. P., Kumar, R.: Preferences for morphine in rats: Validation of an experimental model of dependence. Psychopharmacologia (Berl.) **17**, 137—150 (1970)

Stolerman, I. P., Kumar, R., Steinberg, H.: Development of morphine dependence in rats: Lack of effect of previous ingestion of other drugs. Psychopharmacologia (Berl.) **20**, 321—336 (1971)

Tatum, A. L., Seevers, M. H., Collins, K. H.: Morphine addiction and its physiological interpretation based on experimental evidences. J. Pharmacol. exp. Ther. **36**, 447—475 (1929)

Teiger, D. G.: Induction of physical dependence on morphine, codeine and meperidine in the rat by continuous infusion. J. Pharmacol. exp. Ther. **190**, 408—415 (1974)

Thompson, T., Schuster, C. R.: Morphine self-administration, food-reinforced, and avoidance behaviors in rhesus monkeys. Psychopharmacologia (Berl.) **5**, 87—94 (1964)

Trojniar, W., Cytawa, J., Frydrychowski, A., Luszawska, D.: Intragastric self-administration of morphine as a measure of addiction. Psychopharmacologia (Berl.) **37**, 359—364 (1974)

United Nations Commission on Narcotic Drugs: The Convention on Psychotropic Substances: Full text as adopted. Bull. Narcot. **23(3)**, 5—14 (1971)

Villarreal, J. E.: Some antagonist analgesics with unusual pharmacological spectra uncovered in the Michigan program. Presented at 30th Meeting, Committee on Problems of Drug Dependence, National Research Council, Indianapolis, Ind. 1968

Villarreal, J. E.: Recent advances in the pharmacology of morphine-like drugs. In: Harris, R. T., McIsaac, W. M., Schuster, C. R., Jr. (Eds.): Drug dependence. Advances in mental science II. Austin: University of Texas Press 1970

Villarreal, J. E.: The effects of morphine agonists and antagonists on morphine-dependent rhesus monkeys. In: Kosterlitz, H. W., Collier, H. O. J., Villarreal, J. E. (Eds.): Agonist and antagonist actions of narcotic analgesic drugs (Proceedings of the symposium of the British Pharmacological Society, Aberdeen, July 1971). London: Macmillan Press 1972

Villarreal, J. E., Karbowski, M. G.: The actions of narcotic antagonists in morphine-dependent rhesus monkeys. In: Braude, M. C., Harris, L. S., May, E. L., Smith, J. P., Villarreal, J. E. (Eds.): Narcotic antagonists. Advances in Biochemical Psychopharmacology, Vol. 8. New York: Raven Press 1974

Von Felsinger, J. M., Lasagna, L., Beecher, H. K.: Drug-induced mood changes in man. 2. Personality and reactions to drugs. J. Amer. med. Ass. **157**, 1113—1119 (1955)

Way, E. L., Loh, H. H., Shen, F.-H.: Simultaneous quantitative assessment of morphine tolerance and physical dependence. J. Pharmacol. exp. Ther. **167**, 1—8 (1969)

Weeks, J. R.: Experimental morphine addiction: Method for automatic intravenous injections in unrestrained rats. Science **138**, 143—144 (1962)

Weeks,J.R., Collins,R.J.: Factors affecting voluntary morphine intake in self-maintained addicted rats. Psychopharmacologia (Berl.) **6**, 267—279 (1964)
Weeks,J.R., Collins,R.J.: Patterns of intravenous self-injection by morphine-addicted rats. In: Wikler,A. (Ed.): The addictive states. Ass. Res. nerv. ment. Dis., Vol. 46. Baltimore: Williams and Wilkins 1968
Weeks,J.R., Collins,R.J.: Primary addiction to morphine in rats. Fed. Proc. **30**, 277 (1971)
Wei,E., Loh,H.H., Way,E.L.: Quantitative aspects of precipitated abstinence in morphine-dependent rats. J. Pharmacol. exp. Ther. **184**, 398—403 (1973)
Weinstock,M.: Similarity between receptors responsible for the production of analgesia and lenticular opacity. Brit. J. Pharmacol. **17**, 433—441 (1961)
Weiss,B., Laties,V.G.: Analgesic effects in monkeys of morphine, nalorphine, and a benzomorphan narcotic antagonist. J. Pharmacol. exp. Ther. **143**, 169—173 (1964)
Wikler,A., Carter,R.L.: Effects of single doses of N-allylnormorphine on hindlimb reflexes of chronic spinal dogs during cycles of morphine addiction. J. Pharmacol. exp. Ther. **109**, 92—101 (1953)
Wikler,A., Frank,K.: Hindlimb reflexes of chronic spinal dogs during cycles of addiction to morphine and methadon. J. Pharmacol. exp. Ther. **94**, 382—400 (1948)
Winder,C.V., Welford,M., Max,J., Kaump,D.H.: Pharmacologic and toxicologic studies of *m*-(1-methyl-3-propyl-3-pyrrolidinyl)phenol (CI-572), an analgetic and antitussive agent. J. Pharmacol. exp. Ther. **154**, 161—175 (1966)
Winter,C.A.: The physiology and pharmacology of pain and its relief. In: De Stevens,G. (Ed.): Analgetics. Medicinal Chemistry, Vol. 5. New York: Academic Press 1965
Winter,J.C.: The stimulus properties of morphine and ethanol. Psychopharmacologia (Berl.) **44**, 209—214 (1975)
Woods,J.H., Schuster,C.R.: Opiates as reinforcing stimuli. In: Thompson,T., Pickens,R. (Eds.): Stimulus properties of drugs. New York: Appleton-Century-Crofts 1971
Yanagita,T.: Drug-interaction on physical dependence liability of some analgesics. Presented at 35th Meeting, Committee on Problems of Drug Dependence, National Research Council, Chapel Hill, N.C. 1973a
Yanagita,T.: An experimental framework for evaluation of dependence liability of various types of drugs in monkeys. Bull. Narcot. **25(4)**, 57—64 (1973b)
Yanagita,T., Ando,K., Takahashi,S.: A testing method for psychological dependence liability of drugs in monkeys. Presented at 32nd Meeting, Committee on Problems of Drug Dependence, National Research Council, Washington, D.C. 1970

CHAPTER 3

Assessment of the Abuse Potentiality of Morphinelike Drugs (Methods Used in Man)

D. R. JASINSKI

A. Introduction

I. Rationale for Assessment

As therapeutic agents, morphine and morphinelike drugs are indispensable, but these same agents are liable to nontherapeutic self-ingestion by segments of the population. Society has condemned this self-ingestion and its associated behaviors and has invoked stringent legal mechanisms controlling the manufacture and distribution of morphinelike drugs. In an attempt to develop analgesics, antitussives, and antidiarrheals that were devoid of those properties of morphine leading to abuse, a large number of drugs have been synthesized and studied. Few drugs have been discovered which exhibit this selectivity, but many have been found that are pharmacologically equivalent to morphine with a potential similar to morphine for abuse.

Paralleling this effort to obtain more selective agents has been the development of methods in man and animals to assess the abuse potential of morphinelike agents. This assessment is quite obviously an integral part of the development of selective agents but more importantly, is directed toward preventing the public health problems which have attended introduction of morphinelike agents into medical practice without recognition of the abuse potential of such drugs.

II. Origin of Assessment

1. Development of Morphine Substitutes

The initiation of the assessment of the abuse potentiality of morphinelike agents can be attributed to the research strategy formulated by the National Academy of Sciences, National Research Council, Committee on Drug Addiction, formed in 1929 to carry out research in the area of narcotics. The research strategy and conclusions that were arrived at by the Committee on Drug Addiction has been quoted by EDDY (1973):

> That further sociological studies are not likely, at this point, to help in the solution of drug addiction;
>
> that research at the beginning should be confined to the study of one drug (after consultation with many men, including those in charge of the narcotic divisions of the U.S. Public Health Service and of the Treasury Department, morphine was selected as the most important drug of the group);
>
> that since there are many specific uses of morphine in therapeutic practice and since no one drug can function for all of these uses, it is necessary to replace the legitimate uses of morphine with a number of substitutes;

that, if at all possible to substitute for all legitimate uses of morphine other chemical compounds without addiction properties, it should render morphine an unnecessary commodity in international commerce, and a definite step forward will have been taken.

That, if any success from the researches planned develops, it will come:

a) From some base line chemical and biological study such as is described hereafter,

b) From the provision of new implements for the use of those charged with control of drug addiction, (and)

c) From the provision of substitutes for every legitimate use of morphine first, and thereafter of other drugs, in order that the use of habit-forming drugs may be outlawed.

According to EDDY (1973) this strategy had two bases: (1) oral codeine was widely used for pain relief but was generally held to have little addiction liability. Although implied, quantitative proof of a dissociation of analgesia and addiction liability in codeine was lacking; this belief suggested the possibility of dissociation of analgesia and addiction through chemical modifications of morphine; (2) the decrease in the abuse of cocaine following the introduction of the synthetic local anesthesic, procaine, suggested that a synthetic substitute for morphine might also decrease its abuse.

To implement its research strategy, the Committee on Drug Addiction formed two research units. These were a chemical unit under LYNDON SMALL at the University of Virginia and a Pharmacological Unit at the University of Michigan under NATHAN EDDY. It was recognized that clinical testing would be necessary. This was accomplished by a cooperative effort with the U.S. Public Health Service. In 1929, the United States Public Health Service (then a bureau in the Department of the Treasury) was given the responsibility for care of federal prisoner opiate addicts with facilities at Lexington, Kentucky, and Fort Worth, Texas. An integral activity of the Lexington installation was research on problems of narcotic addiction.

Initially, a clinical research unit under HIMMELSBACH was established at the Federal Penitentiary at Leavenworth, Kansas. In 1935, this unit relocated at Lexington. The liaison formed between the Committee on Drug Addiction and the Lexington facility was for the clinical assessment at the Lexington facility of selected agents synthesized by Small's laboratory and characterized pharmacologically by Eddy's laboratory. With respect to the development program conducted by the Committee on Drug Addiction, HIMMELSBACH evaluated some 19 compounds all of which substituted for morphine and hence had addiction liability with indications that chemical changes in the morphine molecule affected equally analgesic action and addiction liability (HIMMELSBACH et al., 1938; HIMMELSBACH, 1939; HIMMELSBACH, 1942). This development program of the Committee on Drug Addiction was terminated in 1939 and the laboratories of SMALL and EDDY were transferred to the National Institutes of Health. In the following years the Committee on Drug Addiction maintained an advisory role but was largely inactive. In 1947, a number of events led to a prominent role of the committee directed toward protecting public health (EDDY, 1973).

2. Protection of the Public Health

In addition to the morphine substitutes produced by the research program of the National Academy of Sciences, additional efforts to obtain morphine substitutes preceded and paralleled the program of the Committee (EDDY, 1957). Although morphine derivatives had been produced in the late 1800's, the first morphine derivative introduced (in 1898) as a morphine substitute of increased effectiveness and decreased addiction liability was diacetylmorphine (heroin) (EDDY, 1957). In the period 1909–1922, laws were passed in the United States to limit the use of opium and its derivatives to legitimate medical and scientific uses. Heroin, codeine, and morphine, as derivatives of opium, were controlled. Concern over the addiction problems led to a ban of legal manufacture of heroin in 1924.

In the 30-year period following the introduction of heroin, a number of semisynthetic derivatives of morphine were produced. The most important was Dilaudid (dihydromorphinone). As a derivative of opium, Dilaudid was controlled under the United States narcotic laws; however, its introduction to the United States in 1932 was accompanied by claims of greater therapeutic effectiveness than morphine and less or no addiction liability. EDDY (1933) and KING et al. (1935) concluded that Dilaudid had "addiction liability" based on studies in animals and in morphine-dependent addicts. These were the first evaluations of an agent primarily for the purpose of providing information important to the public health.

Of additional consequence were the studies of the morphine derivative desomorphine (dihydrodesoxymorphine-D), a compound prepared by SMALL under the National Academy of Sciences' program. Direct addiction studies in monkeys failed to demonstrate physical dependence with desomorphine probably because of insufficient dose and insufficient frequency of dosing (EDDY and HIMMELSBACH, 1936; EDDY, 1973). Subsequently, HIMMELSBACH demonstrated the morphinelike properties of desomorphine in morphine-dependent addicts (EDDY and HIMMELSBACH, 1936). The inconsistency between the animal and human studies led to concern and criticism that the results obtained in addict patients were not truly indicative of addiction liability (HIMMELSBACH, personal communication). To resolve this issue, HIMMELSBACH conducted trials with desomorphine at the Pondville Cancer Hospital in Massachusetts in nonaddict patients with chronic pain. Desomorphine produced physical dependence in these nonaddict patients (EDDY and HIMMELSBACH, 1936) confirming the findings in addict patients. In 1935, the Committee on Drug Addiction recommended that production of desomorphine not be licensed in the interest of the public and in 1936 the Surgeon General of the Public Health Service recommended to the Secretary of the Treasury that the importation, sale, or manufacture of desomorphine in the United States be prohibited (EDDY and HIMMELSBACH, 1936). This action was the first where the experimental demonstration of "addiction liability" was the factor influencing a decision for narcotics control (EDDY, 1973) and apparently established a precedent for making such decisions.

A subsequent study by HIMMELSBACH was indicative of a greater role of assessment studies in making public health decisions. In the late 1930s, meperidine (pethidine), developed in Germany as a synthetic substitute for atropine, was discovered to have analgesic properties which led to its introduction as a morphine substitute lacking the addiction liability of morphine. HIMMELSBACH (1942, 1943) concluded that meperidine had addiction liability similar to morphine. Since meperidine was not a derivative of opium but was a totally synthetic compound chemically dissimilar to the opium alkaloids, it could not be controlled under legal mechanisms which recognized only derivatives of opium. In 1944, legal controls in the United States were extended to include "isonipecaine" and its derivatives thereby subjecting meperidine to control.

The successful development of meperidine, even though it had addiction liability, stimulated organic chemists to systematically modify the meperidine molecule attempting to dissociate analgesia from addiction liability and to similarly explore other chemical series (EDDY, 1957). Methadone, synthesized in Germany, was introduced into the United States in 1945. ISBELL et al. (1947) found that this synthetic drug, structurally different from morphine or meperidine, had the pharmacologic properties of morphine and hence had "addiction liability." These findings of ISBELL and his coworkers led to a further modification of United States laws extending controls for morphinelike drugs to any substance that had "addiction forming" or "addiction sustaining" properties. Thus, decisions for control of compounds required pharmacologic evaluation for morphinelike properties. As a result, the Committee on Drug Addiction was activated in 1947 to play the prime advisory role with respect to narcotics control. This advisory role persisted until 1970, when the Comprehensive Drug Abuse Prevention and Control Act provided that the Secretary of the Department of Health, Education, and Welfare give such advice.

From 1935, the research unit at Lexington (currently the National Institute on Drug Abuse Addiction Research Center) has maintained its interest in the assessment of narcotics in man and has continued to assist the Committee of Drug Addiction and Narcotics (currently the Committee on Problems of Drug Dependence) in its advisory role. More importantly, for the last forty years, there has been a virtually continuous sustained effort to develop, improve, and validate measures of those pharmacologic effects of morphine held to contribute to its abuse potential.

B. Origin of Methods

I. Physical Dependence

It is quite clear that at the initiation of his studies HIMMELSBACH (1934) equated the habit-forming properties of opiates with ability to produce physical dependence. He considered that physical dependence or addiction liability of an agent could be demonstrated through regular administration over a long period of time (direct addiction) or by substitution of the agent to maintain previously established physical dependence on morphine.

1. Substitution Hypothesis

HIMMELSBACH chose the substitution method as his primary experimental technique to demonstrate physical dependence and in his first publication (HIMMELSBACH, 1934) stated the hypothesis:

> Given valid addiction to morphine, a definite syndrome of abstinence phenomena will set in shortly after its abrupt and complete withdrawal. Hence a substance that can be completely substituted for morphine without permitting the appearance of that syndrome may be addicting in itself even though it is an adequate substitute. If stability can be maintained by the substituted product over a period sufficiently long to rule out abstinence resulting from withdrawal of morphine, and then can be withdrawn without permitting the appearance of that syndrome, it is probably an adequate non-addicting substitute for morphine; but if abstinence phenomena do set in, that substance is addicting. A substance that cannot be substituted has nothing to offer.

2. Experimental Procedures

The initial experimental procedure was developed to assess the addiction liability of codeine (HIMMELSBACH, 1934). Subjects, physically dependent on narcotics at the time of admission, were stabilized on four hypodermic injections of morphine per day with dosages adjusted to the minimum necessary to prevent withdrawal signs and symptoms and to minimize discomfort. With subjects unaware, codeine injections were gradually substituted for morphine injections such that after 3 days subjects were receiving only codeine injections. The subjects were maintained on codeine for 1–2 weeks. Then the codeine was abruptly withdrawn. The resulting abstinence syndrome was taken to indicate the addiction liability of codeine.

In these and subsequent experiments, HIMMELSBACH emphasized and developed several important characteristics of his experimental methods (HIMMELSBACH, 1937a and 1937b; KOLB and HIMMELSBACH, 1938). All experiments were conducted under controlled conditions. Newly admitted patients were isolated and the presence of physical dependence was established by abrupt withdrawal from narcotics. For substitution tests, only patients with marked withdrawal were subsequently stabilized on subcutaneously administered morphine with the dose adjusted to prevent objective signs of abstinence from emerging (usually 240–360 mg per day in four equally divided doses). Observations were made using a systematic standardized procedure. Control experiments for substitution studies were conducted in patients similarly stabilized on morphine and then abruptly withdrawn.

Initially, HIMMELSBACH (1934) characterized the abstinence syndrome into four grades to reflect the clinical severity. Grades were determined by the presence or absence of certain selected signs (Degree Method, Table 1). Subsequently, HIMMELSBACH developed a point system based on the composite signs of abstinence to more quantitatively reflect the abstinence syndrome (HIMMELSBACH, 1937b; KOLB and

Table 1. Degree method for scoring abstinence intensity (HIMMELSBACH, 1934)

Intensity	Signs
Mild (+)	Yawning Lacrimation Rhinorrhea Perspiration
Moderate (+ +)	Muscle tremor Dilated pupils Goose flesh Anorexia
Marked (+ + +)	Hyperpnea Restlessness Insomnia Elevated blood pressure
Severe (+ + + +)	Emesis Diarrhea Weight loss

HIMMELSBACH, 1938). The subjective aspects of abstinence were ignored and only selected signs of abstinence were measured and recorded. Abstinence signs were classified as "measurable signs" or those representing physical deviations from baselines obtained during stabilization on morphine (i.e., systolic blood pressure increases, pyrexia, hyperpnea, weight loss, decrease in caloric intake) and "nonmeasurable" signs or those which could not be expressed as physical deviations from baseline but merely as occurrence data (i.e., yawning, lacrimation, rhinorrhea, perspiration, piloerection, mydriasis, tremor, restlessness, and emesis). To obtain a measure of the intensity of the abstinence syndrome a numerical value or weighting system was assigned to each abstinence sign to reflect both the severity of abstinence syndrome and the appearance of the sign in the developing abstinence syndrome (Table 2).

Utilizing this scoring system (Table 2) as well as the Degree Method (Table 1), the abstinence syndrome was studied following the first 10 days of abrupt withdrawal in 65 subjects in whom strong physical dependence on morphine had been established (KOLB and HIMMELSBACH, 1938). HIMMELSBACH acknowledged that the point assignments to the various phenomena were arbitrary but felt the method was valid be-

Table 2. Point systems for measuring abstinence syndrome intensities[a]

	Himmelsbach system for withdrawal[b]				Modified scores for precipitation[d] and substitution[e] tests	
	By day		By hour		Points	Limit
	Points	Limit	Points	Limit		
Mydriasis	3	3	3	3	1/0.1 mm increase	—
Yawning	1	1	1	1	1	1
Lacrimation	1	1	1	1	1	1
Rhinorrhea	1	1	1	1	1	1
Perspiration	1	1	1	1	1	1
Tremor	3	3	3	3	3	3
Gooseflesh	3	3	3	3	3	3
Anorexia (40% decrease in caloric intake)	3 3	3	—	—	—	—
Restlessness	5	5	5	5	5	5
Emesis (each spell)	5	—	5	5	5	5
Fever (each 0.1° C rise over mean control)	1	—	1	10	1	—
Hyperpnea (each resp./min increase over control)	1	—	1	10	1	—
Systolic blood pressure (each 2 mm Hg increase over mean control)	1[c]	15	1	10	1	—
Weight loss (each lb. from control)	1	—	—	—	—	—

[a] Intensity is the sum of the points scored in each column.
[b] KOLB and HIMMELSBACH, 1938. HIMMELSBACH, 1939.
[c] A.m. systolic blood pressure.
[d] JASINSKI et al., 1967b.
[e] JASINSKI et al., 1971a.

cause of the experience with large numbers of addicts and the close agreement between the point system and the degree methods for determining abstinence intensity. A method for obtaining hourly scores (Table 2) was developed by deleting the scores for weight loss and anorexia and limiting scores on blood pressure, temperature, and respiration to 10 points each (HIMMELSBACH, 1939).

ANDREWS and HIMMELSBACH (1944) demonstrated that the abstinence syndrome intensity as measured by the point system following abrupt withdrawal of morphine was directly related to the stabilization dose of morphine. Further, this was a curvilinear rather than a linear function with maximum abstinence intensity occurring with 500 mg of morphine per day.

During substitution tests, the dose of the substituted drug would be adjusted to stabilize the patients and the single doses of that drug equivalent to stabilization dose of morphine would be used to estimate potency (HIMMELSBACH, 1939). Single doses rather than daily equivalents were utilized because of differences in duration of action of agents.

HIMMELSBACH and ANDREWS (1943) contributed one additional technique for assessing addiction liability—the suppression of morphine abstinence in dependent subjects with single doses of morphinelike drugs. Nineteen subjects were stabilized on morphine and then withdrawn. These subjects received either no medication, saline subcutaneously, or thiamine intravenously. The intensity of abstinence was assessed using the hourly point system from the 24th to the 40th h of abstinence. A three constant equation was fit to these data and they concluded that the intensity of abstinence from the 31st to the 40th could be predicted from the point scores obtained from the 24th to the 30 th h. Thus, the ameliorating effects of drugs on the morphine abstinence syndrome administered at the 30th h could be determined by comparison of the point scores obtained following drug administration with those calculated as the expected course of the untreated abstinence syndrome. Among the drugs studied, HIMMELSBACH and ANDREWS (1943) demonstrated that morphine, codeine, and meperidine clearly ameliorated the expected course of abstinence while prostigmin, pentobarbital, and atropine produced no amelioration. They concluded that such single dose suppression tests offered a quick method for detecting addiction liability.

Substitution and suppression studies to determine dependence liability as described by HIMMELSBACH were initially conducted in patients who had been dependent upon narcotics at the time of admission and subsequently stabilized on morphine. Changes in sentencing procedures so that patients were no longer physically dependent at admission and the need to study physical dependence under standard conditions required that subjects be made physically dependent on morphine. Such studies are conducted in volunteer prisoner subjects with documented histories of long-term narcotic abuse characterized by relapse after periods of enforced abstinence.

In the mid-1950s, the 24-h substitution test replaced those tests in which abstinence was suppressed when the patient was 30 h abstinent. In the 24-h substitution test, three doses of an unknown drug or placebo were substituted for three maintenance doses of morphine with no intervening period of abstinence. Multiple observations for abstinence were made during the period of substitution and patients were restabilized on morphine 24 h after beginning the test. FRASER (personal communication) introduced this technique because of the ability to utilize multiple doses of the agent being tested, the lesser discomfort of the patients during tests, the ability to conduct multiple tests in the same patient at weekly intervals, and the ability to compare abstinence scores from controls (placebo) and unknown drugs in the same patient (crossover). FRASER and ISBELL (1960a) demonstrated dose response curves for suppression of abstinence with partial maintenance doses of morphine in the 24-h substitution tests and further demonstrated that such dose response curves could be compared with similarly obtained dose response curves for graded doses of an unknown drug utilizing statistical techniques for parallel line bioassays to determine the relative potency of drugs in suppressing abstinence.

II. Euphoria and Subjective Effects

1. Initial Definition

Using substitution, suppression, and direct addiction procedures, ISBELL et al. (1947) found that methadone was a morphinelike drug. In these studies, ISBELL recognized that the characteristic pharmacologic profile produced by single doses of morphine-like drugs in nondependent subjects could be used to classify agents and first attempted to define and assess the subjective changes produced by morphine and related drugs that contributed to their abuse potential. ISBELL (1948a) writes:

> Since most persons begin the use of drugs and become addicted because the drugs produce effects which they regard as pleasurable, the detection of euphoria is a very important procedure in evaluating addiction liability. The method used is simple: Single doses of the drug under test are administered to former morphine addicts, and the subjects are unobtrusively watched for a period of 6 h or more by specially trained observers. For our purposes, euphoria is defined as a series of effects similar to those produced by morphine. These effects are: increased talkativeness, boasting, greater ease in the experimental situation, expression of satisfaction with the effects of the drug, requests for increased doses of the drug, increased motor activity, and, with larger doses, slurring of speech, motor ataxia, and evidence of marked sedation.As many experiments are done as are necessary to reach a clear-cut conclusion. The observations are controlled by administering 30 mg of morphine to the same subjects on other occasions. Initially, small subcutaneous doses of the drug under test are used, and if no untoward toxic effects are observed, the dosage is increased progressively in subsequent experiments until evidence of euphoria, roughly equivalent to that produced by 30 mg of morphine, is detected, or, if no evidence of euphoria is detected, the dosage is elevated until further increases would be regarded as dangerous. If euphoria is detected, blind experiments are arranged in which neither the subject nor the observer are aware whether the drug given was morphine or the compound under test. Finally, various doses of the drug are administered intravenously.

2. Experimental Procedures

The introduction of the measurement of subjective effects and euphoria to the assessment of addiction liability led to the development of methods to quantitatively measure the subjective effects not only of morphine but also of other psychoactive drugs. The first systematic studies to classify drugs at the Addiction Research Center were by ISBELL et al. (1956) who characterized the pharmacologic profile of graded doses of lysergic acid diethylamide (LSD-25) in morphine addicts and quantitated drug-induced changes in thought, mood, and perception with questionnaires completed by the subjects (subjective response) as well as drug-induced physiologic changes (objective response). Later, potencies of LSD-like agents were calculated using standard statistical techniques for parallel line assays on the subjective responses (ISBELL, 1959; ISBELL et al., 1959). FRASER et al. (1954) introduced the miotic effect as an objective measure of morphinelike effects in man.

A major inflence on the development of methods at the Addiction Research Center for measuring subjective effects was the work of BEECHER and his colleagues who quantitatively measured the effects of drugs on subjective responses with special attention to pain and its relief by morphine and other analgesic drugs (BEECHER, 1959). Methodologically, BEECHER and his colleagues established the importance and utility of the crossover design with inclusion of standard and placebo drug controls, the double-blind technique, and the randomized allocation of treatments in experiments measuring subjective effects. As a result, these principles of experimental design were introduced into the clinical studies at the Addiction Research Center (ISBELL, personal communication).

FRASER and ISBELL (1960a) introduced the first questionnaires into studies of drugs for morphinelike addiction liability. Responses indicating subjective effects similar to those of morphine and heroin were scored as "positive for opiates." Mean scores were utilized to construct dose response curves. Relative potencies calculated from these dose response curves for a subjective measure were observed to be similar to the relative potency for constricting pupils obtained from concurrent measures of changes in pupillary diameter.

FRASER (personal communication) concluded that the measurement of physical dependence alone was insufficient to assess the abuse liability of morphinelike drugs. He felt that measurement of the attitude of the patient toward the drug effects would contribute to the assessment of abuse liability. As a result, FRASER and ISBELL developed the "single dose and chronic dose opiate questionnaires" to determine the attitude of subjects toward the effects of various opiates (FRASER et al., 1961d). MARTIN and FRASER (1961) compared the effects of heroin and morphine with the single dose opiate questionnaires and introduced the standard mode of analysis of the single dose opiate questionnaires utilized in the assessment of morphinelike drugs.

The development of the other instrument utilized to measure subjective effects—the Addiction Research Center Inventory (ARCI)—was initiated because of the recognized need for valid specific scales of drug-induced subjective effects (ISBELL, personal communication). The ARCI was designed to measure not only drug-induced subjective changes but also those associated with personality and psychiatric disorders (HAERTZEN et al., 1963, and HAERTZEN, 1974). The original scales distinguished the subjective states induced by morphine, pentobarbital, chlorpromazine, LSD, pyrahexyl, amphetamine, and alcohol from those of placebo and no drug conditions (HILL et al., 1963a, b). The commonality of a large number of similar items in these empirical scales led to the development of the group variability and drug correction scales to better differentiate drugs (HAERTZEN, 1966). HAERTZEN (1970) demonstrated that certain of these more specific scales distinguished the subjective states of nalorphine and cyclazocine from that of morphine. MCCLANE and MARTIN (1976) were the first to utilize shortened lists of items from these scales to distinguish morphine from pentobarbital. Subsequently, similar questionnaires based on these ARCI scales were utilized to classify agents as morphinelike or nalorphinelike for the purposes of assessing abuse potential (JASINSKI et al., 1968a; JASINSKI et al., 1971a).

C. Current Methods

I. Physical Dependence

1. Substitution Tests

In the original procedures for substitution studies described by HIMMELSBACH (Subsect. B.I.2) only subjects demonstrating marked (+ + +) abstinence were chosen. Empirically, 240–360 mg of morphine daily were required to stabilize such patients. Consequently, when it became necessary to make addict patients dependent upon morphine for substitution studies, the initial dose levels were usually 240 mg of morphine per day or greater.

The procedure for experimental morphine dependence consists of admitting patients to a ward where observations of blood pressure, pulse rate, respiratory rate, and rectal temperature are made 3 times daily (6 a.m., 12 p.m., 6 p.m.). In addition, at each observation time, pupil size is determined and subjects are observed for signs of abstinence utilized in calculating HIMMELSBACH scores (lacrimation, rhinorrhea, perspiration, yawning, muscle tremor, piloerection, restlessness, emesis). Body weight is obtained daily and total daily caloric intake is estimated from weights of food portions and standard caloric tables. After a suitable period of baseline observations, morphine is administered subcutaneously 4 times daily (6 a.m., 10 a.m., 4 p.m., 10 p.m.) in gradually increasing doses until subjects are receiving 60 mg of morphine at each dose (240 mg morphine sulfate daily) and are felt to have sufficient physical dependence to allow substitution and suppression studies. During stabilization on morphine all observations are continued.

The standard procedure for substitution tests is the 24 h substitution procedure (FRASER and ISBELL, 1960a). In the 24 h substitution technique subjects dependent upon morphine, 240 mg daily, receive their last maintenance dose of 60 mg of morphine at 4 p.m. (hour 0). At 10 p.m. (hour 6), 6 a.m. (hour 14), and 10 a.m. (hour 18) a medication is administered subcutaneously under blind conditions instead of the scheduled 60 mg dose of morphine. Hourly, from 6 a.m. through 4 p.m. (hour 14 through hour 24), blood pressure, rectal temperature, pulse rate, respiratory rate, and pupil size are measured and the presence of lacrimation, rhinorrhea, perspiration, yawning, muscle tremor, piloerection, restlessness, and emesis are recorded. Utilizing mean averages of the three daily observations during stabilization on morphine (usually 7–14 days), baselines are obtained for changes in systolic blood pressure, pulse rate, respiratory rate, rectal temperature, and pupil size during the substitution test. Thus, for each hour from the 14th through 24th h of substitution an abstinence score is calculated (HIMMELSBACH hourly method, Table 2). A total abstinence score (point hours) for the substitution period is obtained by calculating the area under the curve or, more simply, by adding the eleven hourly scores. With this technique, FRASER and ISBELL (1960a) demonstrated that substitution of placebo and graded doses of morphine representing 10%, 20%, and 50% of the morphine maintenance dose to subjects dependent upon morphine 240 mg daily yielded total abstinence point scores which were inversely proportional to the dose of morphine. The regression line obtained from these responses could be compared with a similar regression line obtained with similarly conducted 24-h substitution tests in the same subjects with graded doses of another drug having morphinelike activity and relative potencies could be calculated for the ability of the drug to suppress morphine abstinence using standard statistical techniques for parallel line bioassays.

With the introduction of the Polaroid close-up camera for determining pupil size (MARQUARDT et al., 1967; JASINSKI and MARTIN, 1967b) the method of scoring mydriasis during abstinence has been changed. In the original procedure (FRASER and ISBELL, 1960a), pupil size was estimated by comparison with circles of known diameter. For assessing abstinence during substitution tests, the scoring method of HIMMELSBACH for hourly observations has been modified to give 1 point for each 0.1 mm of dilation over mean control instead of an invariant 3 points (Table 2). In addition, no limits are placed on the points as is done in the HIMMELSBACH scores.

In studies of partial agonists of morphine, 24-h substitution tests were conducted on subjects dependent upon 60 mg of morphine sulfate daily, administered subcutaneously in four 15 mg doses. It was demonstrated that sufficient physical dependence was present so that substitution tests and calculation of relative potencies for suppression of abstinence are obtained (JASINSKI et al., 1971a). Substitution studies have also been conducted in subjects dependent on 30 mg and 120 mg daily of morphine. Subjects dependent upon 30 mg of morphine daily administered subcutaneously in four 7.5 mg doses show abstinence signs of sufficient magnitude for the conduct of suppression tests.

Abstinence scores obtained with placebo and partial maintenance doses of morphine in groups of subjects dependent upon 30, 60, and 120 mg of morphine sulfate daily are shown in Figure 1, a, b, c. These data show certain characteristics of the 24-h substitution procedure. Within each group, there is homogeneity of variance for the abstinence scores. At all levels of dependence, morphine decreases the intensity of the abstinence syndrome in a dose-related manner. In this regard, the standard errors shown in Figure 1 represent the variability of the group response but the use of crossover design allows segregation of the between subjects variability when using standard techniques for parametric statistics. In one instance (Fig. 1b) a test-retest circumstance separated by a 6-week interval (subjects maintained on morphine) indicated the reproducibility of the total abstinence scores obtained in the same group of subjects and, by inference, the stability of the dependence level across time. There is significant variability between groups and as a consequence total abstinence scores in the 24-h substitution tests cannot be strictly related to the stabilization dose level. Thus, it is both necessary and advantageous to conduct substitution tests utilizing a crossover design.

The demonstration that withdrawal signs occurred without patient discomfort following abrupt withdrawal of cyclazocine and nalorphine (MARTIN et al., 1965; MARTIN and GORODETZKY, 1965c) led to the measurement of the discomfort or sickness experienced by the subject during withdrawal (MARTIN et al., 1973a). One consequence of these studies was to include a subjective measure of sickness in the 24-h substitution test (JASINSKI et al., 1970; JASINSKI et al., 1971a). When observations are made for abstinence scores, subjects are asked to rate the intensity of their withdrawal sickness on a 4 grade ordinal scale: 0 = no sickness; 1 = slight sickness; 2 = moderate sickness; and 3 = severe sickness. As with the abstinence scores, the sickness scores are summed to give a total sickness score for the substitution procedure (Fig. 1, lower half). Figure 1 indicates that significant sickness scores are obtained in substitution tests conducted in 30, 60, and 120 mg levels of dependence and

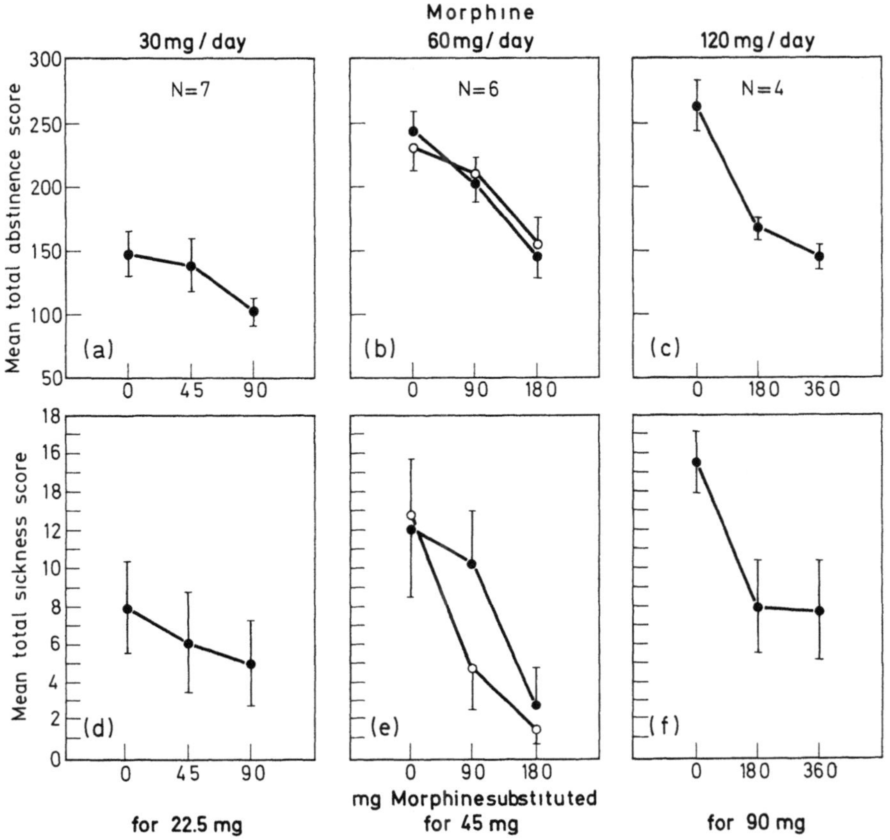

Fig. 1a–f. Abstinence scores and corresponding sickness scores from 24-h substitution tests in subjects dependent upon 30 mg morphine per day (a and d), 60 mg morphine per day (b and e), and 120 mg of morphine per day (c and f) (JASINSKI, unpublished). N = number of subjects in each group. Each point represents mean total response for the 11 hourly observations from the 14th through the 24th h of substitution. Doses of morphine substituted represent total of three doses of placebo or morphine substituted for three consecutive maintenance doses of morphine. In b and e, a test-retest result of an initial series of substitution tests (closed circles) followed by a replication of studies 6 weeks later.

that significantly lower sickness scores are obtained with partial maintenance doses of morphine, however, the variability of this measure is greater than that obtained for abstinence scores and consequently valid bioassays are not consistently obtained.

2. Direct Addiction

The chronic administration of an agent is generally the definitive test for determining the ability of an agent to produce physical dependence. The occurrence of a morphinelike abstinence syndrome upon abrupt withdrawal of the drug or following the administration of a narcotic antagonist (nalorphine or naloxone) are valid indicators of the production of physical dependence. However, execution of direct addiction experiments is difficult and hampered by the availability of suitable volunteer subjects willing and able to participate in such experiments as well as concern over actual or potential toxicity which may occur through long-term administration of agents since the dosages and the period of administration required for such experiments exceed those for therapeutic use. For these reasons, direct addiction experiments to demonstrate physical dependence are limited in number and are usually conducted with agents which are not clearly morphinelike from single dose studies and substitution tests.

Numerous problems attend the execution of direct addiction experiments which include the choice of the initial dose, frequency and rate of dose increases and the stabilization dose level to be achieved. In this regard there are not only differences among agents in their potencies to produce morphinelike effects but differences in their toxicity. There are also differences in sensitivity of individual subjects for both these effects.

The usual procedure is to employ doses equally effective to dose levels of morphine and to attain maximum stabilization doses equivalent to 240 mg of morphine per day. FRASER et al. (1961 d) have given the possible basis for the choice of doses for the potency estimates to be utilized in choosing doses for direct addiction studies. Thus potencies relative to morphine based on analgesia, suppression of abstinence in 24-h substitution tests, scale scores for subjective effects and miosis can be used to pick equivalent doses. In actual practice, however, the dose levels are usually limited by the occurrence of toxic effects such that maximal doses achieved are usually those which are just subtoxic. The occurrence of toxic effects prevents increasing the dose and may cause subjects to object and withdraw from studies. Some of the toxic effects that have limited the rate of dose increase and maximum doses achieved in direct addiction studies are tissue irritation with associated pain and sterile abscess formation at the injection sites; respiratory depression; toxic psychosis; convulsive behavior; stimulant effects such as hyperreflexia and tremor; nervousness; irritability; nausea and vomiting; urinary retention; and excessive sedation.

With regard to the occurrence of toxic effects in direct addiction experiments two factors are of importance. First, there is individual susceptibility to many of these toxic effects. In a group of subjects undergoing direct addiction studies, only one or two subjects may exhibit toxic effects such as nausea and vomiting or sedation which would preclude further dose increases. Subjects differ in their acceptance and tolerance of many of these effects. The second factor of importance is the development of tolerance to toxic effects. If patients are maintained on subtoxic dose levels, tolerance develops and higher dose levels can be attained. For example, in one set of direct addiction studies with pentazocine two of three subjects discontinued pentazocine after 7 and 12 days because of a lack of opiate effect and irritation at the injection site; one subject continued to take pentazocine for 25 days reaching a maximum daily dose of 385 mg (FRASER and ROSENBERG, 1964). In a subsequent study pentazocine was again administered subcutaneously to six subjects but dose levels were increased at a slower rate and all six subjects continued to participate and achieved stabilization doses ranging from 522 to 684 mg/day for a period of 28 days (JASINSKI et al., 1970).

The most extensive series of direct addiction studies was conducted by FRASER and his colleagues (1961 d). In these "short direct addiction tests," eight subjects received subcutaneously administered morphine to a maximum of 240 mg daily, orally administered morphine to a maximum of 240 mg daily, orally administered codeine to a dose of 1500 mg daily, subcutaneously administered phenazocine to a maximum dose of 12–36 mg daily, subcutaneously administered levophenacylmorphan to a dose of 24–36 mg daily, diphenoxylate administered orally to doses of 160–480 mg daily, and D-3-methoxy-N-phenethylmorphinan orally to a dose of 1200 mg daily. Patients received drugs in random order for a period of 18–20 days followed by a period of 10 days of substitution of placebo and subsequent administration of another drug in the series until seven of the eight subjects completed all drug conditions. Under the conditions of these experiments the subjects recognized the substitution of placebo for active drug in all instances; in the case of D-3-methoxy-N-phenethylmorphinan observers did not immediately recognize the change. Under conditions of the experiment, significant dependence was developed as evidenced by the occurrence of significant Himmelsbach scores following the substitution of placebo for an active drug. An additional study conducted by the same experimenters utilizing similar techniques compared subcutaneously administered morphine 207 mg daily, heroin 86.8 mg daily, and placebo as well as orally administered etonitazene 2.95 mg daily. Again significant physical dependence could be demonstrated through 18–20 days of administration. MARTIN and FRASER (1961) demonstrated physical dependence with intravenous heroin and morphine utilizing similar procedures.

The effects of chronically administered agents in direct addiction studies and the abstinence syndrome which occurs upon abrupt withdrawal are related to the effects of morphine, the standard agent. Certain characteristics of the chronic administration of morphine (KOLB and HIMMELSBACH, 1938; HIMMELSBACH, 1939; FRASER et al., 1961d; MARTIN and JASINSKI, 1969; MARTIN et al., 1968; and HAERTZEN and HOOKS, 1969) need to be emphasized from the viewpoint of interpreting direct addiction studies with other morphinelike agents. During chronic administration of morphine when subjects are stabilized there are persistent changes in mood and feeling state which are not euphoric but are characterized by apathetic sedation and hypochondriasis. During chronic administration of morphine pupils remained constricted, respiratory rate is depressed and blood pressure, pulse rate, and body temperature are elevated. The response of the respiratory center to CO_2 is depressed after administration of stabilization doses of morphine and large additional doses produce only minimal degrees of additional depression indicating development of tolerance. Abrupt withdrawal of morphine is accompanied by the development of the characteristic abstinence syndrome which begins at about 8 h after the last administration of morphine, increases in intensity until the 2nd or 3rd day and then slowly decreases through a period of 4–10 weeks after complete withdrawal. There is a secondary phase of withdrawal (a secondary abstinence syndrome) which persists for at least 6 months after abrupt withdrawal of morphine.

In direct addiction studies, attempts have been made to characterize the attitude of the subjects toward the drug during the period of chronic administration, stabilization, and withdrawal with emphasis on the aversive qualities of the withdrawal syndrome. In this regard three questionnaires have been utilized. FRASER et al. (1961d) developed the chronic opiate questionnaires for subjects and observers which emphasized the identification of a drug as an opiate or a nonopiate, the acceptability of the effects (liking), the strength of the drug, and the perception or recognition of abstinence phenomena. This questionnaire was utilized in the short direct addiction procedures described above (FRASER et al., 1961d) where it was found that subjects distinguished among drugs and that their preference rating for morphine, heroin, and codeine corresponded with actual abuse rates of these drugs. MARTIN and FRASER (1961) demonstrated that responses on the estimate of strength item in the chronic opiate questionnaires was dose-related to single doses of intravenous heroin and morphine and that the relative potency on this measure was similar to that for potencies for miosis and scales from the single-dose opiate questionnaires. MARTIN et al. (1965) modified the chronic questionnaire with the addition of items to determine on a graded scale (1) how "bad" or "good" the drug effects were and (2) the degree of "sickness' during abstinence. This questionnaire was utilized during direct addiction studies with cyclazocine (MARTIN et al., 1965), nalorphine (MARTIN and GORODETZKY, 1965c), pentazocine (JASINSKI et al., 1970), profadol, propiram, GPA-1657 (JASINSKI et al., 1971a), and nalbuphine (JASINSKI and MANSKY, 1972). MARTIN et al. (1973a) utilized a new chronic opiate questionnaire during a direct addiction study with methadone. At the present time, the use of subjective effects during chronic administration of morphinelike agents and withdrawal has not been sufficiently validated as a method for classifying drugs for the purpose of assessing abuse potential.

Although it is recognized that tolerance develops to the effects of morphine during chronic administration and cross tolerance exists to other narcotic analgesics, this phenomena has not been routinely utilized as a means of classifying agents for morphinelike properties. MARTIN and FRASER (1961) studied the rate of development of tolerance and cross tolerance with morphine and heroin and developed a tolerance index. MARTIN et al. (1965) and MARTIN and GORODETZKY (1965c) showed cross tolerance between cyclazocine and nalorphine.

The abrupt withdrawal of morphine or drugs capable of producing morphinelike physical dependence is accompanied by an abstinence syndrome which is assessed with the Himmelsbach score (Table 2). In general, the Himmelsbach scores are usually calculated from average values for the physiologic parameters during the last 7 days of addiction. However, FRASER and ISBELL (1960a) recognized that the persistent effects of morphine on some of the physiologic measures used in calculating Himmelsbach scores might lead to an overestimation of the intensity and duration of abstinence when controls were taken during addiction. They felt that inappropriate points were given in the Himmelsbach score with certain values merely representing a return to preaddiction control levels. As a consequence they developed a modified point score in which preaddiction values for caloric intake, temperature, respiratory rate, and systolic blood pressure are used instead of the addiction values.

Nalorphine intensifies the signs and symptoms of abstinence from morphine (ISBELL and FRASER, 1950). WIKLER et al. (1953) found that the administration of nalorphine to subjects after 2 or 3 days of administration of heroin, morphine, or methadone resulted in the precipitation of characteristic withdrawal phenomena. They concluded: (1) that the precipitated abstinence syndromes resembled the syndrome produced by abrupt withdrawal but was of a shorter duration, (2) that the intensity of precipitated abstinence was related to dose of nalorphine, and (3) that nalorphine could be used for the rapid diagnosis of addiction to morphine. Subsequently, the nalorphine-induced abstinence syndrome was taken as a specific indicator of morphinelike physical dependence (ISBELL, 1953a).

The procedure for assessing precipitated abstinence during direct addiction studies is that utilized to measure the relative potency of narcotic antagonists and is described in Section D.1.

Naloxone is 7–10 times more potent than nalorphine as an antagonist but lacks the agonistic effects of nalorphine and is regarded as a pure antagonist (JASINSKI et al., 1967b). Consequently, naloxone has proved more effective than nalorphine in precipitating abstinence, particularly in subjects with liminal levels of dependence. The relative ability of nalorphine and naloxone to precipitate abstinence have been compared for various levels of morphine dependence, 240 mg/day (JASINSKI et al., 1967b); 60 mg/day (JASINSKI et al., 1971a); and 30 mg/day (JASINSKI and NUTT, 1972). Nalorphine is $^1/_{14}$ as potent as naloxone in precipitating abstinence in subjects dependent on 30 mg of morphine a day while it is one-seventh as potent in subjects dependent on 240 mg/day (see Table 5).

II. Euphoria and Subjective Effects

The instruments, the experimental procedures and design, and the data analysis utilized to assess the effects of single doses of morphinelike drugs have been formalized as a result of demonstration of both the validity and utility of the various aspects of this method.

Basically, drugs are administered according to a crossover design in which each subject receives two or three doses of the standard drug (usually morphine), and appropriate blank medication (placebo), and two or three doses of the test drug under double-blind conditions. The drugs are administered in random order with minimum 7-day intervals between each drug administration. Effects of the drugs are measured with structured questionnaires completed by the subjects, a structured questionnaire completed by observers, and by change in pupil size determined photographically. The responses on scales and the questionnaires are quantitated using a standard scoring system; dose response curves are constructed using the mean responses for the standard drug and for morphine from each of the scales, and relative potencies are calculated from these dose response curves with statistical methods for parallel line bioassays.

The initial and still basic instruments utilized to measure the subjective effects of morphinelike drugs are the "single dose opiate questionnaires" introduced by FRASER et al. (1961d). The single dose opiate questionnaires (Table 3) consist of four questions. For the purpose of assessing subjective effects, most emphasis has been placed on questions 2, 3, and 4 of the single dose questionnaires which are seemingly diverse in content but reflect the subjective state induced by morphinelike agents in the postaddict population. Question 2 is concerned with drug identification; question 3 lists commonly observed symptoms and signs of morphine seen in postaddicts; and question 4 is an ordinal scale which measures the degree of liking. Greater importance is placed upon the subjects' responses rather than the observers' responses; however, the observers' responses have equal reliability in measuring morphinelike subjective effects and thus can importantly serve as a concurrent validating measure of the subjects' responses. As measured by single dose opiate questionnaires, observers agree closely with subjects but in general are more sensitive such that observers consistently identify minimal effects more frequently and generally estimate drug effects to be more intense.

The drug identification section (question 2) is characterized in relation to the response to graded doses of morphine (MARTIN, 1966). About half of the subjects will recognize and identify 10 mg of morphine administered subcutaneously or intramuscularly with the other half reporting they have received a blank medication. This dose for 50% response is less than 6 mg when morphine is administered intravenously. With increasing doses of morphine the subjects more correctly identify the drug such that approximately 90% of the patients can identify 20 mg of morphine as a narcotic when administered intramuscularly or subcutaneously. The drug identification section has been expanded recently to include additional agents (Table 3). The population of subjects sophisticated in the use of narcotics and other drugs is able to differentiate between morphine and certain other psychoactive drugs as measured by the drug identification question of the single dose questionnaire. In this regard morphine has been directly compared in a crossover experiment under double-blind conditions with pentobarbital (MARTIN et al., 1974b), nalorphine (MARTIN et al., 1965; JASINSKI et al., 1971a), cyclazocine (MARTIN et al., 1965), and *d*-amphetamine (JASINSKI and NUTT, 1972). In general, subjects and observers will confuse lower doses of these agents with morphine and vice versa, however, with larger doses there are very few confusions with subjects correctly identifying the drugs. An additional characteristic is that the subjects rarely identify placebo as an active medication.

The symptom and signs section of the single-dose questionnaires (question 3) was analyzed and scoring system devised in a study in which graded intravenous doses of morphine and heroin were compared (MARTIN and FRASER, 1961). Analysis of responses in this study indicated that almost all the symptoms and signs listed were reported in response to heroin and morphine. However, only responses to the symptoms: "itchy skin," "relaxed," "coasting," and "drive;" and the signs "scratching," "coasting," and "nodding" were shown to increase with dose. On this basis a weighted scoring system was introduced by MARTIN and FRASER (1961) in which those items which showed dose-related responses were weighted as 2, while other items were weighted as 1 (Table 3).

The fourth question in the single dose questionnaire is the "liking" question which is a quantitative measure of euphoria-producing properties of morphinelike drugs. The responses to these questions are scored as indicated in Table 3 (MARTIN and FRASER, 1961).

The other instrument utilized to assay subjective effects is a multiple scale questionnaire containing items from the Morphine-Benzedrine Group Scale (MBG), the Pentobarbital, Chlorpromazine, Alcohol Group Scale (PCAG), and the LSD Specific Scale (LSD) of the Addiction Research Center Inventory (JASINSKI et al., 1968a). Such questionnaires were introduced into assessment studies to distinguish between drugs of different classes. Since the greatest use has been in classifying narcotic antagonists, the origin and use of this questionnaire will be discussed in detail in the section on antagonists (Subsect. D.1). In this context, morphine, but not nalorphine, produces dose-related scores on the MBG Scale which can be used as a bioassay measure for morphinelike effects. On the other hand, nalorphine, but not morphine, produces dose-related scores on the PCAG and LSD Scales (JASINSKI et al., 1971a).

Table 3. Items of single-dose opiate questionnaires (FRASER et al., 1961d)

Subject rating	Observer rating
1. Do you feel the medicine? Yes / No	Any evidence of drug effect? Yes / No
2. Drug is most like what drug listed below (check 1):	Behavior is like that seen after:
a) Blank	a) Blank
b) "Dope" (opiate or potent synthetic analgesics)	b) "Dope" (opiate or potent synthetic analgesics)
c) Cocaine	c) Cocaine
d) Marihuana (pot)	d) Marihuana (pot)
e) Barbiturate ("goofballs")	e) Barbiturate ("goofballs")
f) Alcohol*	f) Alcohol*
g) Benzedrine (amphetamine, speed, bennies, methedrine)	g) Benzedrine (amphetamine, speed, bennies, methedrine)
h) LSD (acid)*	h) LSD (acid)*
i) Thorazine*	i) Thorazine*
j) Miltown or Librium*	j) Miltown or Librium*
k) Other	k) Other
3. Check each of the sensations which you feel:	Check each item which you think the patient shows:
a) Normal (no change)	a) Normal
b) Turning of stomach (1)	b) Scratching (2)
c) Skin itchy (2)	c) Red eyes (1)
d) Relaxed (1)	d) Relaxed (1)
e) "Coasting" (2)	e) "Coasting" (2)
f) "Soapbox" (1)	f) "Soapbox" (1)
g) Pleasant sick (1)	g) Vomiting (1)
h) Drive (2)	h) Nodding (2)
i) Sleepy (2)	i) Sleepy (1)
j) Nervous (1)	j) Nervous (1)
k) Drunken (1)	k) Drunken (1)
l) Other	l) Other
4. My liking for this drug is most nearly described by which of the following:	How much do you think patient liked the effects of this drug?
a) Not at all (0)	a) Not at all (0)
b) Slight (1)	b) Slight (1)
c) Moderate (2)	c) Moderate (2)
d) A lot (3)	d) A lot (3)
e) An awful lot (4)	e) An awful lot (4)
f) Other	f) Other

The figures given in parentheses following items of questions 3 and 4 are the weights given to the particular responses for the purpose of calculating the opiate symptom, opiate sign, subject's "liking" and observer's "liking" scores (MARTIN and FRASER, 1961).
Those items marked with an asterisk in question 2 were not in the original single-dose opiate questionnaire but have been added subsequently.

The desirability and need had been recognized for an independent objective quantifiable index of morphinelike effects as a validating measure for the measurement of subjective effects. FRASER and his colleagues (1954) photographed pupils of subjects in complete darkness after morphine administration and demonstrated that miosis produced by morphine was dose related. Doses to 30 mg of subcutaneously administered morphine produced a dose-related response suitable for assay purposes. ISBELL et al. (1956) demonstrated that drug-induced changes in pupillary diameter could be assessed with the subjects in a room with constant ambient light and their eyes fixed on a point to maintain constant accommodation. Pupil size was measured by comparison of the pupil with black circles of known diameter on a card held alongside the patient's eyes. The large variability between observers' estimates of pupil size with this technique led to the reintroduction of photographic methods. The procedures and validation were described by JONES et al. (1962) and GORODETZKY and MARTIN (1965b). The basic procedure involves photographing the pupil and a ruler in the same plane as the pupil while maintaining constant accommodation and constant light intensity striking the retina. The pupillary size can then be measured by relating the diameter of the pupil to the ruler.

In assaying for morphinelike subjective effects and miosis the following experimental procedure is utilized. The subject's pupils are photographed at 1 and again at one-half h prior to drug administration. The pupillary diameters for these two observations are then averaged to give an estimate of control pupillary diameter. At fixed intervals after administration of the drug (usually $^1/_2$, 1, 2, 3, 4, and 5 h) pupils are again photographed; subjects complete the single dose questionnaire and the multiple scale questionnaires; and observers complete the observers' single dose questionnaire. For each subject after each drug administration, scores are calculated for the opiate symptoms scale, the opiate sign scales and subjects' liking scale, the observers' liking scale, and the MBG Scale as well as change in pupillary diameter from the mean of the predrug control pupillary diameters. Utilizing mean responses at each postdrug interval for each drug, time action curves are calculated. For the various scales and the change in pupillary diameter, total response scores are calculated by summing the postdrug response for each subject. Dose response curves are then constructed by utilizing mean total response scores for each measure. From these dose response curves, relative potencies can be calculated with the statistical methods for parallel line bioassays utilizing a crossover design (FINNEY, 1964).

D. Narcotic Antagonists

ISBELL and his collaborators studied the human pharmacology of nalorphine in the late 1940s and early 1950s. They concluded that nalorphine did not have addiction liability and suspected it might relieve clinical pain (ISBELL and FRASER, 1950). Single doses as well as chronic administration of nalorphine produced dysphoric effects (WIKLER et al., 1953; ISBELL, 1956). In the abstinent morphine-dependent subject, nalorphine increased the intensity of the abstinence syndrome (ISBELL and FRASER, 1950) and in nonabstinent morphine-dependent subjects, nalorphine precipitated abstinence (WIKLER et al., 1953). Direct addiction studies with nalorphine gave no indication of physical dependence since no abstinence signs were observed upon abrupt withdrawal (ISBELL, 1956). Subsequently, nalorphine was shown to be equipotent to morphine in clinical pain relief but was unsuitable for clinical use because of the dysphoric effects (LASAGNA and BEECHER, 1954; KEATS and TELFORD, 1956). As a result, the search for a nonaddicting analgesic led to the synthesis of compounds having morphine antagonist activity as potential analgesics of low abuse potential.

The introduction of such compounds necessitated systematic evaluations for abuse liability.

The development of quantitative and specific methods to characterize the effects of narcotic antagonists for the purpose of abuse potential began with the study of Martin et al. (1965) who evaluated the subjective effects and physical dependence producing properties of cyclazocine. Subsequent studies of a number of agents having morphine antagonistic properties indicated that these agents, even though they share the common feature of antagonizing the effects of morphine, differed in their spectrum of agonist activity when studied for their ability to produce subjective effects and to produce physical dependence. On the basis of these studies, the compounds are operationally classified as those having nalorphinelike agonist effects, those lacking agonist effects, and those having morphinelike agonist effects.

Since the assessment of dependence liability depends to a great extent upon classification of the agonistic effects of the antagonists as possessing one of these types of characteristic agonistic activities, each classification will be discussed separately.

I. Nalorphine and Cyclazocinelike Antagonists

Martin et al. (1965) and Martin and Gorodetzky (1965b, c) evaluated the subjective effects of nalorphine and cyclazocine utilizing the single dose opiate questionnaires and pupillography. Cyclazocine (0.4–2 mg) and nalorphine (8–32 mg) administered subcutaneously were compared with subcutaneously administered morphine. A clear-cut pattern of subjective effects of nalorphine and cyclazocine were distinguishable from the subjective effects of morphine. In lower doses, both nalorphine and cyclazocine were identified as an opiate and produced symptoms which were similar to those produced by low doses of morphine. Both constricted pupils to a degree equivalent to low doses of morphine. Larger doses of both cyclazocine and nalorphine produced psychotomimetic and sedative effects. On the single dose opiate questionnaires, subjects most frequently identified large doses of cyclazocine and nalorphine as a barbiturate while morphine was correctly identified as an opiatelike compound. A clear distinction was demonstrated between dose-related symptom and sign complexes for morphine on the one hand and nalorphine and cyclazocine on the other hand. In contrast to morphine, the symptoms and signs, sleepy and drunken, were demonstrated to be dose related for cyclazocine and nalorphine. An additional characteristic of nalorphine and cyclazocine was that plateaus or maximal effects were reached on miosis and scores on the subjects' liking scale at a level of 1.0 mg of cyclazocine and 16 mg of nalorphine since larger doses did not produce significantly greater responses. The miotic response and the subjects' liking scores were significantly less at these doses of cyclazocine and nalorphine than were produced by larger doses of morphine.

The study of the subjective effects of nalorphine and cyclazocine were continued by Haertzen (1970) using the Addiction Research Center Inventory (ARCI) to develop additional subjective effects questionnaires which would help in distinguishing nalorphinelike drugs from morphine. Using the ARCI the subjective effects of cyclazocine (0.6 and 1.2 mg/70 kg) and nalorphine (16 and 32 mg/kg) were compared with no drug and placebo conditions. Cyclazocine and nalorphine produced similar

Table 4. The forty items from the Addiction Research Center Inventory utilized in multiple-scale questionnaires to distinguish the subjective effects of nalorphine and cyclazocine from those of morphine (JASINSKI et al., 1968a, 1971a)

Morphine-Benzedrine group (MBG)	Pentobarbital-Chlorpromazine-Alcohol Group (PCAG)	LSD-Specific
I would be happy all the time if I felt as I do now.	*My speech is slurred.	*I have a weird feeling.
I feel as if I would be more popular with people today.	I am not as active as usual.	I have a disturbance in my stomach.
Today I say things in the easiest possible way.	I have a feeling of just dragging along rather than coasting.	I feel an increasing awareness of bodily sensations.
*I feel more clear-headed than dreamy	*I feel more clear-headed than dreamy (answered negatively).	*I would be happy all the time if I felt as I do now (answered negatively).
Things around me seem more pleasing than usual.	I feel sluggish.	I feel anxious and upset.
I have a pleasant feeling in my stomach.	*A thrill has gone through me one or more times since I started the test (answered negatively).	*A thrill has gone through me one or more times since I started the test.
I feel a very pleasant emptiness.	My head feels heavy.	My movements are free, relaxed, and pleasurable (answered negatively).
I fear that I will lose the contentment I now have.	I feel like avoiding people although I usually do not feel this way.	I feel very patient (answered negatively).
I feel in complete harmony with the world and those about me.	I feel dizzy.	I have unusual weakness of my muscles.
I feel less discouraged than usual.	*I am full of energy (answered negatively).	Some parts of my body are tingling.
I can completely appreciate what others are saying when I am in this mood.	People might say that I am a little dull today.	It seems I'm spending longer than I should on each of these questions.
I would be happy all the time if I felt as I feel now.	It seems harder than usual to move around.	My hands feel clumsy.
*I am full of energy.	I am moody.	I notice my hand shakes when I try to write.
I am in the mood to talk about the feeling I have.	*I feel drowsy.	*I feel drowsy (answered negatively).
I feel so good that I know other people can tell it.	I feel more excited than dreamy (answered negatively).	
I feel as if something pleasant had just heppened to me.		

Items marked with an asterisk are contained in more than one scale but are scored for the appropriate scale depending upon the positive or negative answers.

changes on ARCI scales on items. These changes, however, were different from those produced by morphine. Cyclazocine and nalorphine produced no significant changes on euphoria items but produced significant effects on the items measuring psychotomimetic phenomena and those associated with the sedative, drunken, and disorienting effects of barbiturates and alcohol. Previous studies by MCCLANE and MARTIN (1976) had demonstrated the usefulness of short questionnaires derived from the

ARCI in distinguishing the subjective effects of pentobarbital from those of morphine. This technique was extended by constructing a short questionnaire (see Table 4) utilizing the Morphine-Benzedrine Group Scale (MBG) which measures opiatelike euphoria, the LSD Specific Scale (LSD) which measures certain psychotomimetic effects, and the Pentobarbital, Chlorpromazine, Alcohol Group Scale (PCAG) which measures sedative effects (MARTIN, 1967a; JASINSKI et al., 1968a). Comparisons of subcutaneously administered morphine and nalorphine and intravenously administered morphine and nalorphine demonstrated the ability of this questionnaire, as well as the single dose questionnaire and pupillography to distinguish the subjective effects of morphine and nalorphine (JASINSKI et al., 1971a). Morphine produced significant dose-related scores on the MBG Scales but not on the LSD or PCAG Scales while nalorphine produced no significant scores on the MBG items, but produced significant dose-related scores on the LSD and PCAG Scales. Neither nalorphine (ISBELL and FRASER, 1950) nor cyclazocine (FRASER, personal communication) suppress abstinence in subjects dependent on large doses of morphine (240 mg per day or greater). Both will precipitate abstinence in morphine-dependent subjects (WIKLER et al., 1953; MARTIN et al., 1965).

ISBELL (1956) found no physical dependence in direct addiction studies with nalorphine (80–120 mg/day for 30 days). SCHRAPPE (1959), however, observed typical morphine withdrawal signs, but mild, following abrupt withdrawal of 150 mg/day of nalorphine (40 days) in a psychiatric patient.

In the assessment of cyclazocine for dependence producing properties, it was concluded that 13.2 mg/day of cyclazocine (70–104 days) (MARTIN et al., 1965) and 240 mg/day of nalorphine (26–32 days) (MARTIN and GORODETZKY, 1965c) produced physical dependence as evidenced by an abstinence syndrome upon abrupt withdrawal. It was further shown that chronic administration of cyclazocine (MARTIN et al., 1965) and nalorphine (MARTIN and GORODETZKY, 1965c) induced direct tolerance as well as cross tolerance. Neither agent precipitated abstinence in either cyclazocine or nalorphine-dependent subjects. Withdrawal of nalorphine and cyclazocine was followed by abstinence characterized by signs similar to those seen in morphine withdrawal such as rhinorrhea, yawning, lacrimation, chills, diarrhea, fever, loss of body weight, and loss of appetite. However, blood pressure and respiratory rate were only slightly increased. Mean peak Himmelsbach scores of 15.5 for nalorphine and 18.8 for cyclazocine were obtained. In morphine-dependent subjects, scores of this magnitude would indicate a mild to moderate degree of dependence. MARTIN et al. (1965) and MARTIN and GORODETZKY (1965c) showed that the abstinence syndromes for nalorphine and cyclazocine were qualitatively similar to each other but were qualitatively distinct from the morphine abstinence syndrome. These were the first studies in which qualitative comparisons between abstinence syndromes were done objectively using correlation coefficients. MARTIN (1966) subsequently demonstrated the similarity between various signs of abstinence at various levels of morphine dependence and intensities of abstinence. These observations demonstrated that differences between the nalorphine and cyclazocine abstinence syndrome could not be accounted for by the low intensities of the nalorphine, cyclazocine, and morphine abstinence syndrome. An early symptom of abstinence observed with nalorphine and cyclazocine but not in abstinence with morphine and morphinelike drugs is a sensation in the head which was described as electric shocks or lightheadedness. Another difference between the morphine and nalorphine abstinence syndrome was itching which was a symptom of nalorphine withdrawal but not of cyclazocine or morphine withdrawal. The abstinence syndrome with nalorphine had a more rapid onset than that of cyclazocine.

An important characteristic differentiating the nalorphine and cyclazocine abstinence syndrome from the morphine abstinence syndrome was the lack of concern of the abstinent subjects over their symptoms and the absence of drug seeking behavior.

The assessment of compounds having morphine antagonistic activity for abuse potential require that a quantitative method of estimating morphine antagonistic activity be developed. WIKLER et al. (1953) demonstrated that increasing the dose of nalorphine administered to subjects dependent on morphine, methadone, or heroin produced an increase in the intensity of the abstinence syndrome. JASINSKI et al. (1967b) quantitatively defined the relationship between the dose of antagonist and the intensity of abstinence and showed that conventional bioassay procedures could be used to determine the relative potency of antagonists in precipitating abstinence. Subjects maintained on 240 mg of morphine daily received their normal 60 mg stabilization dose of morphine at 6 a.m. At 8 a.m. and 8:30 a.m. systolic and diastolic blood pressure, respiratory rate, pulse rate, rectal temperature, and pupil size (JASINSKI et al., 1967b) were measured. At 9 a.m. an antagonist was administered subcutaneously under double-blind conditions. At 9:15 a.m., 10:00 a.m., 11:00 a.m., and 12:00 p.m. systolic and diastolic blood pressure, rectal temperature, pulse rate, respiratory rate and pupil size were again measured. At these times subjects were also observed for yawning, lacrimation, rhinorrhea, perspiration, piloerection, restlessness, emesis, and tremor. At each postantagonist observation time an abstinence score for precipitated abstinence was obtained by utilizing the scoring system shown in Table 2 which is a modified version of the Himmelsbach scoring system. In this

Table 5. Relative potencies of various agents in precipitating abstinence in morphine dependent subjects. Potencies expressed as mg of drug equivalent to 1 mg of nalorphine with 95% confidence limits in parentheses

Level of morphine dependence	Drug	Relative potency	Reference
240 mg daily	naltrexone	0.06 (0.04–0.17)	MARTIN et al., 1973b
	l-cyclazocine	0.12 (0.10–0.16)	JASINSKI and NUTT, 1973
	naloxone	0.14 (0.12–0.20)	JASINSKI et al., 1967b
	oxilorphan	0.16 (0.13–0.20)	JASINSKI and NUTT, 1973
	d, l-cyclazocine	0.19 (0.14–0.26)	JASINSKI and NUTT, 1973
	levallorphan	0.52 (0.38–0.73)	JASINSKI et al., 1967b
	nalorphine	1.00	
	profadol	43.5 (32.8–52.6)	JASINSKI et al., 1971a
	pentazocine	51.2 (34.3–96.3)	JASINSKI et al., 1970
	propiram	192 (127–503)	JASINSKI et al., 1971a
60 mg daily	naloxone	0.10 (0.06–0.16)	JASINSKI et al., 1970
	nalorphine	1.00	
	nalbuphine	3.7 (2.4–5.5)	JASINSKI and MANSKY, 1972
30 mg daily	naloxone	0.04 (0.02–0.05)	JASINSKI and NUTT, 1972
	nalorphine	1.00	

modified scoring system, 1 point is awarded for every 0.1 mm increase in pupil size over the mean predrug control.

Table 5 lists those agents assayed by this procedure and their potency relative to nalorphine in precipitating abstinence. It should be noted that there is a high concordance between potency estimates for morphine antagonist activity by this method in man with those from precipitated abstinence in the dog (MARTIN et al., 1974a) and monkey (VILLAREAL and KARBOWSKI, 1974), from antagonism of acute morphine effects in rodents (BLUMBERG and DAYTON, 1974; PACHTER, 1974; PIERSON, 1974) and from antagonism of morphine in the guinea pig ileum (KOSTERLITZ et al., 1974).

II. Antagonists Lacking Agonist Effects (Naloxone)

Reports of possible analgesic effects led to the study of the narcotic antagonist naloxone for abuse potential (JASINSKI et al., 1967b). In these studies, single doses of naloxone produced no miotic or subjective effects; chronic administration and withdrawal of naloxone produced no signs or symptoms of abstinence, indicating that naloxone did not produce physical dependence. When administered to morphine-dependent subjects (240 mg/day) naloxone was approximately 7 times more potent than nalorphine in precipitating abstinence. Thus, naloxone was considered an antagonist lacking agonist activity.

III. Partial Agonists of the Morphine Type

The compounds profadol and propiram were known to be analgesics and morphine antagonists. Studies of their subjective effects and dependence-producing properties in man indicated that narcotic antagonists could have agonist activity which resembled that of morphine rather than that of nalorphine (JASINSKI et al., 1971a). In single doses profadol produced dose-related miosis, scale scores on the symptom, sign, "liking," and MBG Scales with no significant effects on the LSD and PCAG Scales. Subjects identified profadol as an opiate. Potencies for the responses on these measures of morphinelike effects indicated that profadol produced miosis and morphinelike subjective effects with a potency similar to that for analgesic effects in man. Direct addiction studies with profadol indicated morphinelike physical dependence rather than nalorphinelike physical dependence. Nalorphine precipitated abstinence during the period of chronic administration of profadol. This precipitated abstinence was identified by the subjects as opiatelike withdrawal. Abrupt withdrawal of profadol produced an abstinence syndrome qualitatively similar to the morphine abstinence syndrome and distinct from the nalorphine abstinence syndrome as evidenced by analysis of the sources of points in the Himmelsbach scores. During profadol abstinence, subjects reported marked discomfort and exhibited drug-seeking behavior.

Like profadol, single doses of propiram also produced dose-related miosis, and elevated scores on the symptom, sign, liking, and MBG Scales. Further, propiram was consistently identified as an opiate. With large subcutaneous doses, however, there was a slight but significant elevation of PCAG and LSD scores but characteristic symptoms of nalorphine and cyclazocine psychotomimetic effects were not reported. Direct addiction studies with propiram also indicated a capability to produce morphinelike physical dependence. Nalorphine precipitated abstinence during chronic administration of propiram and abrupt withdrawal produced an abstinence syndrome associated with drug-seeking behavior. This abstinence syndrome was mild as judged by peak Himmelsbach scores and was neither typically morphine or nalorphinelike.

When administered to subjects dependent upon 240 mg of morphine, profadol and propiram did not substitute for morphine and suppress abstinence with the 24-h substitution procedure. When administered using the procedure for precipitated abstinence, profadol and propiram precipitated abstinence with potencies less than nalorphine.

These observations supported previous speculations that there could be partial agonists of the morphine type (MARTIN, 1963, 1967 b).

To demonstrate the lesser intrinsic activity, profadol and propiram were substituted in subjects dependent upon 60 mg of morphine per day (15 mg 4 times daily) using the 24-h substitution procedure. The rationale underlying this procedure was the hypothesis that the agonist activities of morphine were undiminished in the tolerant dependent subject (MARTIN, 1970) and that 240 mg of morphine per day would produce near maximal dependence with near maximal receptor occupation and activity (HIMMELSBACH and ANDREWS, 1943). Thus at this level of morphine dependence partial agonists such as profadol and propiram would displace morphine from the receptor, reduce the level of agonist activity, and precipitate abstinence. At a lower level of morphine dependence, such as 60 mg per day, a smaller proportion of receptors would be occupied such that the activity produced by this level of morphine could be equaled by a sufficiently strong partial agonist. Thus, the agonist would now suppress abstinence.

The results of the 24-h substitution tests in subjects dependent upon 60 mg of morphine per day with profadol and propiram were in keeping with this formulation. Both suppressed abstinence with a potency relative to morphine and similar to that for miosis, subjective effects, and analgesia (JASINSKI et al., 1971 a). Profadol and propiram were concluded to be partial agonists of the morphine type. As a result of these experiments, 24-h substitution tests at various levels of morphine dependence were introduced as a technique for assessing intrinsic activity.

E. Validity

Two issues are involved in the validity of the assessment methods. First is the validity of the experimental procedures and instruments to demonstrate equivalence to morphine and second, is the validity of the assertion that the demonstration of such equivalence in an institutionalized addict population is indicative of an abuse potential.

I. Subjective Effects and Euphoria

The population in which the instruments for measuring morphinelike subjective effects were developed and in which assays are conducted essentially have the common characteristics of documented histories of abuse of heroin or other narcotic analgesics as well as varying degrees of experience with other psychoactive agents.

Table 6. Relative potencies of morphinelike drugs in man as assessed by miosis, single dose questionnair equivalent to 1 mg of the first. Values in parentheses are 95% confidence limits

Drugs compared	Pupils (Miosis)	Questions	Signs	Symptoms
H (i.v.) × MS (i.v.)[a]	2.6 (2.1–3.2)		2.4 (1.9–3.0)	2.2 (1.9 –3.0)
H (i.v.) × MS (i.v.)[b]	2.1 (1.6–2.9)		2.0 (1.4 –3.0)	2.1 (1.4 –3.0)
C (o) × MS (s.c.)[c]	0.06 (0.01–0.12)		0.05 (0.0 –7.1)	0.05 (0.01–0.10)
C (i.m.) × MS (i.m.)[a]	0.07 (0.05–0.09)		0.12 (0.10–0.15)	0.10 (0.08–0.13)
C (i.v.) × MS (i.v.)[e]	0.07 (0.03–0.11)		0.12 (0.09–0.15)	0.12 (0.08–0.16)
MS (o) × MS (s.c.)				
MS (i.v.) × MS (s.c.)[e]	1.1 (0.9 –1.4)		1.2 (0.9 –1.4)	1.2 (1.0 –1.4)
d-PH (o) × MS (i.v.)[c]	0.03 (0.00–0.07)		0.03 (0.00–0.12)	0.04 (0.00–0.11)
d-PH (o) × MS (s.c.)[f]	0.02 (0.01–0.06)		0.03 (0.00–0.12)	0.03 (0.00–1.10)
d-PN (o) × *d*-PH (o)[f]	0.67 (0.53–0.83)		0.45 (0.06–0.77)	0.38 (0.04–0.67)
d-PH (i.v.) × MS (i.v.)[e]	0.06 (0.05–0.08)		0.09 (0.07–0.11)	0.09 (0.07–0.13)
M (i.m.) × MS (i.m.)[c]	0.07 (—)		0.10 (0.00–0.27)	0.14 (0.04–0.27)
Me (o) × Me (i.m.)[j]	0.4 (—)		0.5 (—)	0.7 (0.2 –1.0)
Me (s.c.) × MS (s.c.)[j]	1.1 (0.8–1.6)		1.00 (0.8 –1.6)	— (—)
Me (i.v.) × MS (i.v.)[b]	0.9 (0.6 –1.2)		0.7 (0.5 –1.0)	0.8 (0.5 –1.2)
d-Me (o) × MS (s.c.)				
O (s.c.) × MS (s.c.)[a]	0.6 (0.4 –1.0)		1.3 (—)	1.0 (—)
I-A-40 (s.c.) × MS (s.c.)[a]	0.7 (0.4 –1.1)		1.7 (1.2 –2.0)	1.7 (1.4 –2.0)
I-C-27 (s.c.) × MS (s.c.)[a]	0.21 (0.19–0.23)		0.3 (0.2 –0.4)	0.4 (0.3 –0.4)
L (s.c.) × MS (s.c.)[a]	5.2 (2.7 –8.0)	6.1 (5.0–7.5)		
P (s.c.) × MS (s.c.)[a]	3.8 (1.3 –5.6)	3.2 (2.3–5.0)		
F (s.c.) × MS (s.c.)[a]	25.0 (14–33)			50.0
1-K-1 (i.v.) × C (i.v.)[a]	1.2 (1.0 –1.4)			1.0 (0.9 –1.0)
1-K-1 (i.m.) × C (i.m.)[a]				
GPA 1657 (s.c.) × MS (s.c.)[e]	3.1 (2.1 –4.5)		1.6 (1.5 –3.6)	1.6 (1.4 –2.2)
GPA 1657 (i.v.) × MS (i.v.)[e]	1.2 (0.02–50.0)		1.7 (1.0 –2.0)	1.4 (1.0 –2.0)
GPA 2087 (s.c.) × MS (s.c.)[u,v]	1.4 (0.8 –2.8)		0.8 (0.4 –1.6)	0.9 (0.5 –1.4)
E (s.c.) × MS (s.c.)[n]	333		500	1000
T (i.m.) × MS (s.c.)[h]	0.10 (0.03–0.18)		0.04 (0.00–0.14)	
T (o) × MX (s.c.)[c]	0.13 (0.04–0.24)		0.12 (0.01–0.26)	0.11 (0.03–0.21)
Cx (i.m.) × MS (i.m.)[o]	0.06 (0.03–0.10)		0.14 (0.10–0.19)	0.14 (0.02–0.45)
D (i.m.) × MS (i.m.)[o]	0.8 (0.3 –1.2)		1.0 (0.6 –1.7)	1.2 (0.8 –2.5)
A (s.c.) × MS (s.c.)	11.1 (—)		25.0 (10.9–111.1)	20.0 (9.9 –142.8
Pm (s.c.) × MS (s.c.)[e]	0.12 (—)		0.14 (0.12–0.15)	0.13 (0.10–0.17)
Pm (i.v.) × MS (i.v.)[e]	0.12 (0.07–0.17)		0.11 (0.07–26)	0.14 (0.06–0.45)
Pl (i.m.) × MS (i.m.)[e]	0.35 (0.17–0.83)		0.45 (0.05– 1.0)	0.50 (0.16–1.42)
Pl (i.v.) × MS (i.v.)[e]	0.38 (0.27–0.50)		0.43 (0.34–0.62)	0.47 (0.35–0.52)
Dp (o) × MS (s.c.)				

H = Heroin; MS = Morphine; C = codeine; *d*-PH = *d*-propoxyphene HCl; *d*-PN = *d*-propoxyphene napsyla M = Meperidine; Me = methadone; O = Oxycodone; I-A-40 = 6-methylene-6-desoxy-14-hydroxydihydr morphine HCl (Abbott A-25443); I-C-27 = d-2-acetoxy-1,]-diphenyl-3-methyl-4-pyrollidine butane H Lilly 31518); L = levophenacylmorphan; P = Phenazocine; F = Fentanyl; I-K-1 = 1-(p-chlor-phenethyl)- methyl-6,7-dimethoxy-1,2,3,4-tetrahydro-isoquinoline HCl; GPA 1657 = β[–]-5-phenyl-9-methyl-2′-hydroxy- methyl-6,7-benzomorphan; GPA 2087 = l-alpha-5,9-diethyl-2′-hydroxy-2-methyl-6,7-benzomorphan; E = Et phine (M-99); T = tilidine; Cx = Codoxime; D = dihydrocodeinone; Pm = propiram; Pl = profadol; A = azid morphine; *d*-Me = *d*-methadone; Dp = diphenoxylate; i.v. = intravenous; o = orally; s.c. = subcutaneous i.m. = intramuscularly.

BG Scale, 24-h substitution tests, and relief of pain. Potencies are expressed as mg of the second drug

king (Aides)	Liking (Patients)	MBG	Substitution test	Analgesia
2.7 (2.3 -3.1)	2.1 (1.8 -2.6)			2.6[a]
2.0 (1.2 -3.0)	1.8 (0,9 -2.7)	2.0 (1.3 -3.1)		
0.04 (—)	0.02 (0.00–0.13)	0.04 (0.00–0.20)[d]	0.07[a]	0.05[d]
0.10 (0.09–0.12)	0.09 (0.07–0.12)			0.08[d]
0.10 (0.08–0.12)	0.10 (0.06–0.14)	0.05 (0.09–0.15)		
			0.34[t]	0.16[d]
1.4 (1.1 -1.7)	1.7 (1.2 -2.0)	1.0 (0.6 -1.4)		
0.03 (0.00–0.07)	0.03 (0.00–0.12)	0.02 (0.00–0.11)		Ca. 0.03[c]
0.02 (—)	0.03 (0.00–0.09)	0.03 (0.00–0.10)	0.04[c]	
0.48 (—)	0.43 (0.14–0.71)	0.34 (0.03–0.62)	0.50[f]	0.67[g]
0.10 (0.08–0.12)	0.09 (0.07–0.12)	0.07 (0.04–0.10)		
0.10 (0.00–0.29)	0.11 (0.02–0.23)	0.08 (—)		0.09[i]
0.4 (—)	0.5 (—)	0.6 (—)		0.4[k]
1.0 (0.6 -1.9)	— (—)	— (—)		1.1[k]
1.0 (0.6 -1.6)	0.9 (0.4 -1.8)	0.8 (0.5 -1.4)		
			0.06 (0.03–0.09)[s]	
1.6 (—)	1.3 (—)	1.1 (0.8 -1.6)		
1.7 (1.2 -2.0)	2.0 (1.4 -2.5)	0.9 (0.8 -1.1)		
0.4 (0.3 -0.4)	0.5 (0.4 -0.6)	0.3 (0.2 -0.7)		0.25[a]
			9.1 (4.8 -19.5)	
			9.2 (4.2 -17.2)	3.0[a]
0.0	50.0			
			0.14 (0.09–0.83)	
1.9 (1.8 -2.1)	1.5 (1.2 -7.4)	1.2 (0.04-3.2)	0.74 (0.49–1.12)	1.2[l]
1.4 (1.1 -2.0)	1.7 (1.2 -2.5)	1.4 (0.8 -2.0)		
1.1 (0.7 -1.7)	1.0 (0.6 -1.6)	1.1 (0.4 -3.1)	1.1 (0.5 -2.5)	1.2[m]
0	1000	500		
				0.045[p]
).13 (0.05–0.21)	0.13 (0.01–0.26)	0.12 (0.01–0.28)	0.04 (0.01–0.09)	
).10 (0.02–0.18)	0.14 (0.04–0.50)			
1.1 (0.6 -1.7)	1.2 (0.8 -3.3)			
).0 (10.2–66.7)	16.7 (7.6 -90.9)	25.0 (9.2 -25.0)	0.04 (0.02–0.06)	
).10 (0.05–0.16)	0.11 (0.05–0.19)	0.13 (0.07–0.16)	0.6 (0.5—0.8)	0.10[q]
).12 (0.08–0.17)	0.21 (0.11–5.0)	0.11 (0.04–0.18)	0.10 (0.02–0.23)[e]	
).55 (0.38–0.76)	0.58 (0.34–0.83)	0.52 (0.13–1.11)		0.25[r]
).47 (0.40–0.58)	0.50 (0.37–0.62)	0.43 (0.30–0.62)	0.31 (0.16–0.47)[e]	
			0.55 (0.30–1.72)[t]	

ᴍARTIN, 1966, for original reference.
ASINSKI and NUTT, 1972.
ASINSKI et al., 1974.
VALLENSTEIN et al., 1967.
ASINSKI et al., 1971a.
ASINSKI et al., 1975b.
UNSHINE et al., 1971.
ASINSKI and NUTT, 1973.

[i] FORREST et al., 1974.
[j] MARTIN et al., 1973a.
[k] BEAVER et al., 1967.
[l] FORREST, 1968.
[m] FORREST, 1970.
[n] JASINSKI et al., 1975a.
[o] JASINSKI and MARTIN, 1967a.
[p] HOUDE et al., 1974.

[q] FORREST et al., 1972.
[r] BEAVER et al., 1969.
[s] FRASER and ISBELL, 1962.
[t] FRASER and ISBELL, 1961.
[u] JASINSKI et al., 1971b.
[v] JASINSKI and MANSKY, 1970.

This drug-sophisticated population as well as the use of trained and experienced observers makes it difficult if not impossible to disguise the nature of the drug under investigation. FRASER et al. (1961d) evaluated the responses to the single dose questionnaire following three successive weekly doses of codeine phosphate, 140 mg orally, and 3 weekly doses of diphenoxylate, 70 mg orally, and demonstrated with analysis of variance that the largest source of variability for all measures whether evaluated by patients or aides, was the difference between subjects. There was no significant variability between weeks indicating that subjects and observers answered the questions consistently when the same dose of given drug is repeated on successive weeks. Subsequently, the use of the latin square design where subjects are considered as rows and weeks as columns has consistently indicated a lack of significant variability between weeks (columns) (For example, JASINSKI et al., 1971a).

The best validated instrument has been the single dose opiate questionnaires. Morphine has been distinguished from pentobarbital (MARTIN et al., 1974b), nalorphine and cyclazocine (MARTIN et al., 1965), and *d*-amphetamine (JASINSKI and NUTT, 1972). In addition, subjects rarely identify placebo as an active drug. FRASER and ISBELL (1960a) pointed out the concordance between the potencies calculated from miosis and questionnaire responses in nondependent subjects as well as the concordance with potency estimates for the suppression of abstinence. MARTIN and FRASER (1961) demonstrated in a comparison of graded doses of intravenously administered heroin and morphine that total response scores on the opiate symptom and sign scales, subjects' and observers' liking scales, and miosis were linearly related to dose and covaried among themselves. Further, from the dose response curve for heroin and morphine on each measure a relative potency could be calculated which was similar across each measure. MARTIN (1966) pointed out the advantages of this procedure for assessing subjective effects since the concordance between potency estimates indicated the internal validity of the method. To demonstrate external validity, MARTIN (1966) compared potency estimates calculated from these measures with estimates of potency for morphine and morphinelike drugs obtained in analgesic studies with nonaddict patients (HOUDE et al., 1965) and the potency of narcotic analgesics in suppressing abstinence. In the interim many other drugs have been compared using these techniques and the results of these studies are summarized in Table 6.

Examination of Table 6 indicates the general concordance between potency estimates from miosis, scores from the single dose opiate questionnaires, MBG Scale, suppression of abstinence and pain relief. In addition, the reproducibility of the methods are indicated in the similar potency estimates obtained from two separate comparisons of intravenous heroin and morphine conducted 11 years apart and two separate comparisons of oral *d*-propoxyphene hydrochloride to subcutaneous morphine conducted 1 year apart (Table 6). A further indicator of reproducibility are similar potency estimates for the intravenous route as against the intramuscular or subcutaneous route. This is true for codeine, methadone, GPA-1657, profadol, and propiram (Table 6).

The present multiple scale questionnaire containing selected items from the scales of the Addiction Research Center Inventory clearly distinguish morphine from nalorphine (JASINSKI et al., 1971a) and morphine from pentobarbital (MCCLANE and MARTIN, 1976) such that it appears to be a valid means of separating morphinelike drugs from these two classes of agents. This questionnaire, however, will not distinguish morphine from *d*-amphetamine (JASINSKI and NUTT, 1972).

The concordance of potency estimates for subjective effects with those for miosis, suppression of abstinence and pain relief, the reproducibility of potency estimates for subjective effects and euphoria, and the demonstrated ability to distinguish morphine from other psychoactive drugs are strong evidence that the measurement of subjective effects utilizing current methods in an addict population are a valid means for demonstrating pharmacologic equivalence to morphine.

II. Physical Dependence

Originally, the addiction liability of morphinelike drugs was equated to their ability to produce physical dependence as evidenced by a characteristic withdrawal syndrome. To the addict patient abstinence symptoms are discomforting. When he learns to relieve these symptoms with the administration of a morphinelike drug, the abstinence symptoms are generally agreed to be a drive which is responsible for drug-seeking behavior. Operationally, the Himmelsbach scores based upon signs of abstinence served as the measure of the capacity of a drug to produce physical dependence and presumably drug-seeking behavior. However, the basic assumption that certain signs of abstinence are specific indicators of the capacity of a drug to induce drug seeking seems to be unwarranted. MARTIN et al. (1965) and MARTIN and GORODETZKY (1965c) demonstrated that abrupt withdrawal of chronically administered cyclazocine and nalorphine produced signs of abstinence. These abstinence syndromes were similar to each other but differed from the morphine abstinence syndrome. In addition to qualitative differences between the nalorphine and morphine abstinence syndromes, symptoms occurred in the nalorphine and cyclazocine abstinence syndromes which did not occur in the morphine abstinence syndrome and most importantly, abrupt withdrawal of nalorphine and cyclazocine was not associated with reports of discomfort nor a strong drive to obtain the drug.

Direct addiction experiments with a number of drugs have shown physical dependence resembling that of morphine as indicated by Himmelsbach scores upon abrupt withdrawal or precipitated by nalorphine. Agents such as the naturally occurring opiate codeine; semisynthetic opiates such as heroin; and synthetic opioids such as methadone, meperidine, diphenoxylate, and *d*-propoxyphene have the capacity to produce morphinelike physical dependence. It must be recognized, however, that there are inherent difficulties in conducting direct addiction experiments. The variability of response to the morphinelike properties and toxic effects, differences in toxic effects among the agents, the exclusion of the subjective effects of abstinence from the Himmelsbach score, and the lack of a validated quantitative measure of the aversiveness of withdrawal limit to some extent the comparison of physical dependence characteristics among agents. Nevertheless, it is recognized that there are differences in the time course and intensity of abstinence among agents as measured by the Himmelsbach score. There are probably also differences among drugs in the character of the autonomic signs contributing to the withdrawal phenomena. Most importantly, the lack of a systematic validated measure of aversiveness and the exclusion of the subjective aspects of abstinence from the Himmelsbach score give no indication of the relationship between the aversiveness of the abstinence syndrome and the intensity of abstinence as measured by the Himmelsbach score. Further, three difficulties arise with the interpretation of precipitation tests in direct addiction studies: (1) Large doses of nalorphine and especially naloxone will precipitate abstinence signs in patients who have received only a few doses of a narcotic and at a time when no detectable abstinence can be demonstrated following abrupt withdrawal (WIKLER et al., 1953; NUTT and JASINSKI, 1974). (2) The effectiveness of nalorphine, but not naloxone, in precipitating morphine abstinence is related to the level of dependence (see Table 5). Nalorphine is less effective in precipitating abstinence following long-term administration of 30 mg of morphine daily than 60 or 240 mg daily. Thus, the observation of seemingly lesser effectiveness of nalorphine in precipitating abstinence in a direct addiction experiment must be interpreted cautiously. This is especially true in the case of less potent morphinelike agents where toxicity limits the maximum dose. (3) Naloxone will antagonize the effects of single doses of cyclazocine in man (JASINSKI et al., 1968a). It is suspected, but not demonstrated, that naloxone would precipitate abstinence in man after long term cyclazocine or nalorphine administration. Thus, naloxone-precipitated abstinence may not be specific for dependence of the morphine type.

It is important to emphasize that the intensity of physical dependence measured with Himmelsbach scores may not be a prime indicator of an agent to produce compulsive drug-seeking behavior. Both HIMMELSBACH (1942, 1943) and ISBELL (1955a) had difficulty demonstrating intense abstinence with meperidine, yet there is ample evidence that meperidine is an opioid which has been widely abused (RASOR and CRECRAFT, 1955).

For a number of years, significant suppression of abstinence as measured by decreases in Himmelsbach scores was felt to indicate specificity for morphinelike agents since only morphinelike agents would specifically suppress all parameters of abstinence and decrease the abstinence score based on composite signs of abstinence. Conversely, agents which did not suppress the Himmelsbach score were concluded to lack morphinelike activity. In this regard, a number of agents seemingly distinguished from morphine in their pharmacologic profile did not suppress abstinence when administered in clinically tolerated doses (Table 7). However, there are reservations about the specificity of the suppression tests.

In the 24-h substitution tests, only mild abstinence occurs since maximal Himmelsbach scores after abrupt withdrawal of morphine do not occur until the second day of abstinence. Consequently, signs of severe abstinence such as emesis, restlessness, and tremor occur usually only at the last two or three observations and thus have little contribution to the total abstinence scores. The failure to suppress abstinence cannot be taken as a lack of morphinelike activity since the demonstration that partial agonists of morphine suppress at 60 mg of morphine daily but not at 240 mg daily (JASINSKI et al., 1971a). On the other hand, the demonstration that nalorphine is less effective than naloxone in precipitating abstinence in subjects receiving 30 mg of morphine per day than at higher dose levels (JASINSKI and NUTT, 1972) indicates that the agonist effects of nalorphine may modify the ability of nalorphine to precipitate abstinence. Although not demonstrated, this observation suggests that at low levels of morphine dependence suppression tests may not be specific. An additional reservation concerning the specificity of the suppression tests

Table 7. Agents which in clinical doses have failed to suppress morphine abstinence in man as measured by the HIMMELSBACH abstinence score

Drug	Reference
1. preparin (6,7-diethoxyl-1[3'z4'-diethoxyl-benzyl]-isoquinoline)	HIMMELSBACH, 1937a
2. rossium (bis-1-phenyl-3-methyl-5-pyrozolone)	HIMMELSBACH, 1937b
3. thiamine	HIMMELSBACH, 1940
4. prostigmin	HIMMELSBACH et al., 1942
5. atropine	HIMMELSBACH and ANDREWS, 1943
6. pentobarbital	HIMMELSBACH and ANDREWS, 1943 FRASER et al., 1961a
7. pyridoxine	HIMMELSBACH and ANDREWS, 1943
8. pyrahexyl	HIMMELSBACH, 1944
9. cortisone acetate	FRASER and ISBELL, 1953
10. adrenocorticotropic hormone	FRASER and ISBELL, 1953
11. methotrimeprazine	FRASER and ROSENBERG, 1963a
12. phenyramidol	FRASER et al., 1961a
13. dextrorphan	ISBELL and FRASER, 1953b
14. dextromethorphan	ISBELL and FRASER, 1953b
15. chlorpromazine	FRASER and ISBELL, 1956c
16. reserpine	FRASER and ISBELL, 1956c
17. narcotine	FRASER, 1954
18. papaverine	FRASER and ISBELL, 1956a
19. trifluoperazine	FRASER and ISBELL, 1959b
20. mephenesin	ISBELL, 1951
21. ethyl alcohol (intravenous)	ISBELL, 1951
22. pyribenzamine	ISBELL, 1951

is indicated by the study of FRASER and ISBELL (1962) who found that D-3-methoxy-N-phenethylmorphinan orally suppressed abstinence in the 24-h substitution test being $^1/_{24}$ as potent as subcutaneous morphine. Substitution was incomplete because of toxicity. Single doses produced predominantly barbituratelike subjective effects with sedation behaviorally and pupillary constriction. In a short direct addiction test (18–20 days), abrupt withdrawal of D-3-methoxy-N-phenethylmorphinan (1200 mg orally) was not associated with a detectable abstinence syndrome.

The continuing use of the Himmelsbach score in suppression tests is based upon the observation that relative potencies and confidence limits calculated from 24-h substitution tests are similar to relative potencies and confidence limits obtained for miosis, scales for subjective effects and relief of pain (Table 6) indicating that suppression tests can be a valid bioassay measure for morphinelike activity. The suppression test at various levels of dependence is also used to detect differences in intrinsic activity.

III. Indication of Abuse Potential

In the last 40 years a large number of agents have been evaluated in man for morphinelike dependence liability (Table 8). The demonstration of pharmacologic equivalence to morphine has resulted in many instances in legal restrictions on the drugs to protect public health. Some would argue that the only true test of abuse

Table 8. A selected list of drugs evaluated in man at the Addiction Research Center for morphineli properties and/or for morphine antagonist properties. The numbering system and classifications a internal to the ARC. Drugs in which data are unpublished are excluded from this list. Not all agents this list are morphinelike

Drug No.	Name	Subjective effects	Direct addiction	Morphine abstinence, suppression, or precipitati
Morphine and congeners				
I-A-2	benzylmorphine myristyl ester (myrophen)	Fraser and Isbell, 1955a Isbell, 1955b		Fraser and Isbell, 1955 Isbell, 1955b
I-A-3	β-chloro-morphide			Himmelsbach et al., 193
I-A-4	codeine	Fraser et al., 1960a Fraser et al., 1961d Jasinski et al., 1971b Kay et al., 1967 Fraser et al., 1961c Fraser et al., 1961b Martin et al., 1966 Jasinski et al., 1974 Fraser and Isbell, 1960c Fraser and Isbell, 1961a	Fraser et al., 1960a Fraser et al., 1961d Fraser et al., 1961c Fraser et al., 1961b Fraser and Isbell, 1961a	Himmelsbach, 1934 Himmelsbach et al., 193 Himmelsbach, 1941 Himmelsbach et al., 194 Himmelsbach and Andrews, 1943 Fraser and Isbell, 1961 Himmelsbach et al., 193 Fraser et al, 1961c Fraser and Isbell, 1960 Fraser and Isbell, 1960
I-A-5	diacetyldi-hydromorphine			Himmelsbach et al., 193
I-A-6	diacetylmorphine (heroin)	Martin and Fraser, 1961 Nutt and Jasinski, 1973 Jasinski and Nutt, 1972	Martin and Fraser, 1961 Fraser et al., 1961d	
I-A-7	dihydro-α-isomorphine			Himmelsbach et al., 193 Himmelsbach, 1941
I-A-8	dihydrocodeine			Himmelsbach, 1941
I-A-9	dihydrocodeine methyl ether (tetrahydro-thebaine)			Himmelsbach, 1938
I-A-10	dihydrocodeinone (hydrocodone, dicodide, dicodid)	Fraser and Isbell, 1950 Jasinski and Martin, 1967b Isbell, 1949a	Isbell, 1949a	Fraser and Isbell, 1950 Jasinski and Martin, 1967b Isbell, 1949a
I-A-11	dihydro-codeinone enol acetate; acetyldihydro-codinone (Acedicon)			Himmelsbach et al., 193
I-A-12	dihydrodesoxy-codeine-D			Himmelsbach, 1941 Himmelsbach, 1939
I-A-13	dihydrodesoxy-morphine-D (desomorphine)			Himmelsbach, 1938 Eddy and Himmelsbach, 1936 Himmelsbach, 1941 Himmelsbach, 1939
I-A-14	dihydrohetero-codeine			Himmelsbach et al., 193

ble 8 (continued)

rug o.	Name	Subjective effects	Direct addiction	Morphine abstinence, suppression, or precipitation
A-15	dihydrohydroxy codeinone (oxycodone, Eucodal, Percodan)	Martin et al., 1966		Martin et al., 1966
A-16	dihydrohydroxy-morphinone (oxymorphone)	Fraser and Isbell, 1955a Isbell, 1954a	Fraser and Isbell, 1955a Isbell, 1954a	Fraser and Isbell, 1955a Himmelsbach et al., 1938 Isbell, 1954a
A-17	dihydroiso-codeine			Himmelsbach et al., 1938 Himmelsbach, 1941
A-18	dihydromorphine (paramorphan)			Himmelsbach, 1941 Himmelsbach, 1939
A-19	dihydro-morphinone (hydromorphone, Dilaudid)			Himmelsbach et al., 1938 Himmelsbach, 1941 King et al., 1935
A-21	isocodeine			Himmelsbach, 1941 Himmelsbach et al., 1938
A-22	α-isomorphine			Himmelsbach et al., 1938 Himmelsbach, 1941
A-23	norcodeine	Fraser et al., 1960a Fraser and Isbell, 1958a	Fraser et al., 1960a Fraser and Isbell, 1958a	Fraser et al., 1960a Fraser and Isbell, 1958a
A-24	6-methyl-Δ^6-desoxymorphine			Isbell and Wikler, 1953
A-25	6-methyldihydro-morphine	Fraser and Isbell, 1950 Isbell and Eisenman, 1948a		Fraser and Isbell, 1950 Isbell and Eisenman, 1948a
A-26	methyldihydro-morphinone (metopon, methyl-dilaudid)			Himmelsbach et al., 1938 Himmelsbach, 1939
A-29	morphine sulphuric ether			Himmelsbach and Andrews, 1940
A-30	morpholinyl-ethylmorphine (pholcodeine, homocodeine, ethnine)	Isbell, 1952a		Isbell, 1952a
A-31	normorphine	Fraser et al., 1958 Fraser and Isbell, 1958a	Fraser et al., 1958 Fraser and Isbell, 1958a	Fraser et al., 1958 Fraser and Isbell, 1958a
A-33	N-(3-oxo-3-phenyl-propyl) normorphine	Fraser and Isbell, 1958a		Fraser and Isbell, 1958a
A-34	N-phenethyl-normorphine bitartrate	Fraser and Isbell, 1958a		Fraser and Isbell, 1958a
A-35	pseudocodeine			Himmelsbach et al., 1938
A-38	6-acetyl-3-ethoxyldihydro-morphine	Fraser and Wolbach, 1961		Fraser and Wolbach, 1961

Table 8 (continued)

Drug No.	Name	Subjective effects	Direct addiction	Morphine abstinence, suppression, or precipitation
I-A-40	6-methylene-6-desoxy-14-hydro-dihydromorphine	MARTIN et al., 1966		MARTIN et al., 1966
I-A-41	dihydro-codeinone-0-(carboxymethyl)-oxime dihydrate (codoxime)	JASINSKI and MARTIN, 1967a		JASINSKI and MARTIN, 1967a
I-A-42	6-azidomorphine	JASINSKI et al., 1975b		JASINSKI et al., 1975b
Morphinans				
I-B-2	*d*-3-hydroxy-2,N-dimethyl-morphinan	ISBELL, 1954b ISBELL, 1955b		ISBELL, 1954b
I-B-3	*l*-3-hydroxy-2-N-dimethyl-morphinan	ISBELL, 1954b		ISBELL, 1954b
I-B-4	dextrorphan (*d*-3-hydroxy-N-methyl morphinan, D-dromoran)	ISBELL and FRASER, 1953b ISBELL, 1952a		ISBELL and FRASER, 1953b ISBELL, 1952a
I-B-5	*d*, *l*-3-hydroxy-N-methyl-morphinan (dromoran)	FRASER and ISBELL, 1950 ISBELL, 1949b	ISBELL, 1949b	FRASER and ISBELL, 1950
I-B-6	*l*-3-hydroxy-N-methyl-morphinan (levorphanol, levophan, levo-dromoran)	ISBELL and FRASER, 1953b ISBELL, 1952a		ISBELL and FRASER, 1953b ISBELL, 1952a
I-B-8	*l*-3-hydroxy-morphinan (norlevorphanol)	FRASER and ISBELL, 1958b	FRASER and ISBELL, 1958b	FRASER and ISBELL, 1958b
I-B-9	*d*-3-hydroxy-N-phenethyl morphinan	FRASER and ISBELL, 1962 FRASER and ISBELL, 1958b FRASER and ISBELL, 1961	FRASER and ISBELL, 1962	FRASER and ISBELL, 1962 FRASER and ISBELL, 1958b FRASER and ISBELL, 1961
I-B-10	*l*-3-hydroxy-N-phenethyl morphinan	FRASER and ISBELL, 1962 FRASER and ISBELL, 1956a		FRASER and ISBELL, 1962 FRASER and ISBELL, 1956a
I-B-11	dextro-methorphan (*d*-3-methoxy-N-methyl-morphinan, D-methyl dromoran)	ISBELL and FRASER, 1953b ISBELL, 1953b JASINSKI et al., 1971b MANSKY and JASINSKI, 1970		ISBELL and FRASER, 1953b ISBELL, 1953b

ble 8 (continued)

ug).	Name	Subjective effects	Direct addiction	Morphine abstinence, suppression, or precipitation
-12	Racemethorphan (*d*, *l*-3-methoxy-N-methyl-morphinan methorphan)	ISBELL and FRASER, 1953b ISBELL, 1952a		ISBELL and FRASER, 1953b ISBELL, 1952a
-13	*l*-3-methoxy-N-methyl-morphinan (levomethorphan)	ISBELL and FRASER, 1953b		ISBELL and Fraser, 1953b
-14	*d*-3-methoxy-N-phenethyl-morphinan (NIH 7296A)	FRASER and ISBELL, 1962 FRASER and ISBELL, 1958b FRASER and ISBELL, 1961	FRASER and ISBELL, 1962 FRASER et al., 1961d	FRASER and ISBELL, 1962 FRASER and ISBELL, 1958b FRASER and ISBELL, 1961
-15	*l*-3-methoxy-N-phenethyl-morphinan	FRASER and ISBELL, 1957		FRASER and ISBELL, 1957
-17	*l*-3-hydroxy-N-phenacyl-morphinan methane sulphonate (levophenacyl-morphan, Präparat)	FRASER and ISBELL, 1960a FRASER et al., 1961d FRASER and ISBELL, 1959a	FRASER and ISBELL, 1960a FRASER et al., 1961d	FRASER and ISBELL, 1960a FRASER and ISBELL, 1959a
-19	*l*-3-hydroxy-N-(3,3-dimethylallyl) morphinan	FRASER et al., 1960b		FRASER et al., 1960b
	thadone and congeners			
-1	*d*-α-acetyl-methadol	ISBELL, 1951 FRASER and ISBELL, 1952 ISBELL, 1952a		ISBELL, 1951 FRASER and ISBELL, 1952
-2	*d*, *l*-α acetyl-methadol (racemic acetylmethadol)	FRASER and ISBELL, 1952 ISBELL, 1951 ISBELL, 1952a		FRASER and ISBELL, 1952 ISBELL, 1951
-3	*l*-α-acetyl-methadol	FRASER and ISBELL, 1952 ISBELL, 1951 ISBELL, 1952a		FRASER and ISBELL, 1952 ISBELL, 1951
-4	methadol (α-*dl*-methadol)	ISBELL and EISENMAN, 1948a ISBELL and EISENMAN, 1948b		ISBELL and EISENMAN, 1948a ISBELL and EISENMAN, 1948b
-5	α-*l*-methadol	ISBELL and FRASER, 1954 ISBELL, 1953b		ISBELL and FRASER, 1954 ISBELL, 1953b
-6	*β*-acetyl-*d*-methadol	ISBELL and FRASER, 1954 ISBELL, 1953b		ISBELL and FRASER, 1954 ISBELL, 1953b
-7	*β*-*d*,*l*-methadol	ISBELL, 1955b		ISBELL, 1955b

Table 8 (continued)

Drug No.	Name	Subjective effects	Direct addiction	Morphine abstinence, suppression, or precipitation
I-C-8	*d*-methadon (dextro-methadon)	Fraser and Isbell, 1962 Isbell and Eisenman, 1948a Isbell and Eisenman, 1948b Fraser and Isbell, 1958b Fraser and Isbell, 1961	Fraser and Isbell, 1962	Fraser and Isbell, 1962 Isbell and Eisenman, 1948a Isbell and Eisenman, 1948b Fraser and Isbell, 1958 Fraser and Isbell, 1961
I-C-9	*d,l*-methadon (Amidone, Dolophine, Palomidon)	Isbell et al., 1948a Isbell et al., 1947 Isbell et al., 1948b Isbell and Vogel, 1949 Martin et al., 1973a Jasinski and Nutt, 1972 Nutt and Jasinski, 1973	Isbell et al., 1947 Isbell et al., 1948b Isbell and Vogel, 1949 Martin et al., 1973a	Isbell et al., 1947 Isbell et al., 1948b Isbell and Vogel, 1949
I-C-10	*l*-methadon (levomethadon)	Isbell and Eisenman, 1948a Isbell and Eisenman, 1948b		Isbell and Eisenman, 1948a Isbell and Eisenman, 1948b
I-C-11	4,4-diphenyl-6-dimethylamino-hexanone-3 (Ticardia)	Isbell and Fraser, 1955 Isbell, 1954b Isbell, 1955b		Isbell and Fraser, 1955 Isbell, 1954b Isbell, 1955b
I-C-12	*d*-propoxyphene hydrochloride	Jasinski et al., 1971a Fraser et al., 1961d Fraser et al., 1961c Fraser and Isbell, 1960c Fraser and Rosenberg, 1964 Fraser and Isbell, 1956a Jasinski et al., 1974 Jasinski et al., 1975b	Fraser et al., 1961c Fraser and Isbell, 1960c	Fraser et al., 1961c Fraser and Isbell, 1956 Fraser and Isbell, 1960 Jasinski et al., 1975b
I-C-13	*d, l*-propoxy-phene hydrochloride	Fraser and Isbell, 1960c Fraser and Isbell, 1956a	Fraser and Isbell, 1960c	Fraser and Isbell, 1960
I-C-15	*d, l*-4,4-diphenyl-5-methyl-6-dimethyl-amino-3-hexanone (isomethadon)	Isbell and Eisenman, 1948a Isbell and Eisenman, 1948b		Isbell and Eisenman, 1948a Isbell and Eisenman, 1948b
I-C-19	piperidyl methadon (dipi-panone, *d, l*-4, 4-diphenyl-6-piperi-dino-3-heptanone)	Fraser and Isbell, 1956a		Fraser and Isbell, 1956
I-C-20	*d, l*-ethyl-2,2-diphenyl-4-dimethylamino-butyrate	Isbell, 1954b		Isbell, 1954b

ble 8 (continued)

ug).	Name	Subjective effects	Direct addiction	Morphine abstinence, suppression, or precipitation
-21	*d*-2,2-diphenyl-4-dimethylamino ethyl valerate	ISBELL, 1954b		ISBELL, 1954b
-22	*dl*-2,2-diphenyl-4-dimethylamino ethyl valerate	ISBELL, 1954b		ISBELL, 1954b
-23	*l*-2,2-diphenyl-4-dimethylamino ethyl valerate	ISBELL, 1954b		ISBELL, 1954b
-24	*d, l*-ethyl-2,2-diphenyl-4-morpholino-butyrate	ISBELL, 1954b FRASER and ISBELL, 1956b		ISBELL, 1954b FRASER and ISBELL, 1956b
-25	*α-d, l*-3-acetoxy-4, 4-diphenyl-6-methyl-amino-heptane	FRASER and WOLBACH, 1961		FRASER and WOLBACH, 1961
-26	*d*-3-dimethyl-amino-1,1-diphenyl butyl-ethyl sulfone hydrochloride	WOLBACH and FRASER, 1963 FRASER et al., 1962		WOLBACH and FRASER, 1963 FRASER et al., 1962
-27	*d*-2-acetoxy-1,2-diphenyl-3-methyl-4-pyrolidino-butane	MARTIN and GORODETZKY, 1965a MARTIN et al., 1966		MARTIN and GORODETZKY, 1965a MARTIN et al., 1966
-28	*d*-propoxyphene napsylate	JASINSKI, et al., 1975b		JASINSKI et al., 1975b
peridine and congeners				
-1	alphaprodine (Nisentil) NU1196	ISBELL, 1949c		ISBELL, 1949c
-2	alphameprodine (NU-1932)	ISBELL, 1949c		ISBELL, 1949c
-3	anileridine (Leritine)	FRASER and ISBELL, 1956a		FRASER and ISBELL, 1956a
-4	*d,l-β*-1,3-dimethyl-4-phenyl-4-proprionoxy-piperidine (NU1779)	ISBELL, 1949c		ISBELL, 1949c
-5	oxypheneridine 1-(2-hydroxy-2-phenethyl)-4-phenyl-4-carbethoxy-piperidine	FRASER and ISBELL, 1957		FRASER and ISBELL, 1957

Table 8 (continued)

Drug No.	Name	Subjective effects	Direct addiction	Morphine abstinence, suppression, or precipitation
I-D-6	diphenoxylate	FRASER and ISBELL, 1961a FRASER et al., 1961d FRASER and ISBELL, 1960b	FRASER and ISBELL, 1961a FRASER et al., 1961a FRASER and ISBELL, 1960b	FRASER and ISBELL, 1961a FRASER and ISBELL, 1960b
I-D-11	bemidone	ISBELL, 1949c		ISBELL, 1949c
I-D-12	ketobemidone	ISBELL, 1949c ISBELL, 1948b	ISBELL, 1949c ISBELL, 1948b	ISBELL, 1949c ISBELL, 1948b
I-D-13	meperidine (pethidine, Demerol Dolantin)	JASINSKI and NUTT, 1973	HIMMELSBACH, 1942 HIMMELSBACH, 1943 ISBELL, 1955a	HIMMELSBACH and ANDREWS, 1943 HIMMELSBACH, 1942
I-D-15	morpholinoethyl-norpethidine; 1-(2-morpholino-ethyl)-4-carbethoxy-piperidine	FRASER and ISBELL, 1957		FRASER and ISBELL, 1957
I-D-18	alvodine {ethyl-4-phenyl-1[3(phenyl-amino)-propyl]-4-piperidine-carboxylate ethane sulfonate}	ISBELL and FRASER, 1959		ISBELL and FRASER, 1959
I-D-19	1-proprio-phenone-4-carbethoxy-4-phenyl-piperidine	ISBELL and FRASER, 1959		ISBELL and FRASER, 1959
I-D-20	ethyl-1-(2-carbamlethyl)-4-phenyl-piperidine-4-carboxylate hydrochloride	WOLBACH and FRASER, 1963 WOLBACH and FRASER, 1962		WOLBACH and FRASER, 19 WOLBACH and FRASER, 19
I-D-21	2,2-diphenyl-4 (1-[4-(N-piperidine)-4-carboxamide]-piperidine)-butyronitrite	FRASER et al., 1962	FRASER et al., 1962 FRASER et al., 1963	FRASER et al., 1962
I-D-22	1-hydroxy-ethoxy-ethyl-4-propionyl-piperidine	FRASER et al., 1962	FRASER et al., 1963	FRASER et al., 1962
Hexamethyleneimines				
I-E-1	*d*, *l*-alpha-1,3, dimethyl-4-phenyl-4-propionoxy-hexamethy-leneimine	FRASER, 1956		FRASER, 1956 FRASER and ISBELL, 1956

ble 8 (continued)

ug).	Name	Subjective effects	Direct addiction	Morphine abstinence, suppression, or precipitation
:-2	1,3-dimethyl-4-phenyl-4-carbethoxy hexamethyleneimine	FRASER, 1956 FRASER and ISBELL, 1956b		FRASER, 1956 FRASER and ISBELL, 1956b
:-3	1,2-dimethyl-4-phenyl-4-carbmethoxy-hexyleneimine	FRASER, 1956 FRASER and ISBELL, 1956b		FRASER, 1956 FRASER and ISBELL, 1956b
:-4	1-methyl-4-phenyl-4-carbethoxy-hexamethyleneimine (ethoheptazine, NIH5835)	ISBELL, 1954b ISBELL, 1955b FRASER, 1956 FRASER and ISBELL, 1956b		ISBELL, 1954b ISBELL, 1955b FRASER, 1956 FRASER and ISBELL, 1956b
hienylbutenylamines				
-1	3-diethylamino-1,1-dı-(2′-thienyl)-*l*-but-1-ene (NIH4185, diethyl-thiambutene, C-49)	ISBELL and FRASER, 1953a ISBELL, 1951		ISBELL and FRASER, 1953a ISBELL, 1951
-3	3-ethylmethyl-amino-1,1-di-(2′-thienyl) but-1-ene (NIH5145, ethyl-methyl-thıambutene)	ISBELL and FRASER, 1953a ISBELL, 1953b	ISBELL and FRASER, 1953a ISBELL et al., 1953	ISBELL and FRASER, 1953a ISBELL, 1953b
zimidazoles				
-1	clonitazine [1-(β-diethyl-amınoethyl)-2-(benzyl-4-chloro-)-5-nitro-benzimıdazole]	FRASER et al., 1960b		FRASER et al., 1960b
-2	etonitazene [1-(β-diethyl-amınoethyl)-2-(p-ethoxybenzyl)-5-nitro-benzimidazole, NIH7607]	FRASER et al., 1960b	FRASER et al., 1961d FRASER et al., 1960b	FRASER et al., 1960b
zomorphans				
-1	phenazocine	FRASER and ISBELL, 1960a FRASER and ISBELL, 1959a FRASER et al., 1962	FRASER and ISBELL, 1960a	FRASER and ISBELL, 1960a FRASER and ISBELL, 1959a FRASER et al., 1962

Table 8 (continued)

Drug No.	Name	Subjective effects	Direct addiction	Morphine abstinence, suppression, or precipitation
I-H-2	*l*-2′-hydroxy-2, 5,9-trimethyl-6, 7-benzomorphan (levomethazocine)	FRASER et al., 1962	FRASER et al., 1963	FRASER et al., 1962
I-I-5	etorphine	JASINSKI et al., 1975a JASINSKI et al., 1974		JASINSKI et al., 1975a JASINSKI et al., 1974
Propionanilides				
I-J-1	diampromide	FRASER et al., 1960b		FRASER et al., 1960b
I-J-2	phenampromide	FRASER et al., 1960b		FRASER et al., 1960b
I-J-4	fentanyl	GORODETZKY and MARTIN, 1965a GORODETZKY and MARTIN, 1965b		
Miscellaneous				
I-K-1	1-(p-chloro-phenethyl)-2-methyl-6,7-di-methoxy-1,2,3,4-tetrahydro-isoquinoline	FRASER et al., 1961b FRASER et al., 1961c FRASER et al., 1961d	FRASER et al., 1961c	FRASER et al., 1961b FRASER et al., 1961c
I-L-1	phenyramidol	FRASER et al., 1961a	FRASER et al., 1961a	FRASER et al., 1961a
I-M-1	carisoprodol	FRASER et al., 1961a	FRASER et al., 1961a	FRASER et al., 1961a
I-N-1	1-dimethyl-amino-3-phenylidane	FRASER et al., 1962	FRASER et al., 1963	FRASER et al., 1962
I-N-2	2-amino-indane	FRASER et al., 1962	FRASER et al., 1963	FRASER et al., 1962
I-O-1	1,2-dimethyl-3-phenyl, 3-propionoxy pyrollidine	FRASER, 1964 FRASER et al., 1962	FRASER, 1964 FRASER et al., 1962	FRASER, 1964 FRASER et al., 1962
I-O-2	*d, l*-profadol	JASINSKI et al., 1968b JASINSKI et al., 1971a JASINSKI et al., 1969	JASINSKI et al., 1971a JASINSKI et al., 1969	JASINSKI et al., 1968b JASINSKI et al., 1971a JASINSKI et al., 1969
I-O-3	*l*-profadol	JASINSKI and NUTT, 1973		
I-O-4	*d*-profadol	JASINSKI and NUTT, 1973		
I-Q-1	dexoxadrol	JASINSKI et al., 1968b		JASINSKI et al., 1968b
I-R-1	propiram	JASINSKI et al., 1968b JASINSKI et al., 1971a JASINSKI et al., 1969	JASINSKI et al., 1971a JASINSKI et al., 1969	JASINSKI et al., 1968b JASINSKI et al., 1971a JASINSKI et al., 1969
I-T-1	tilidine	JASINSKI and NUTT, 1973 JASINSKI et al., 1974		JASINSKI and NUTT, 1973
Nalorphine congeners				
II-A-1	N-allyl-norheroin	FRASER and ISBELL, 1955b ISBELL, 1953b	ISBELL, 1954b	FRASER and ISBELL, 1955b ISBELL, 1953b

ble 8 (continued)

rug o.	Name	Subjective effects	Direct addiction	Morphine abstinence, suppression, or precipitation
A-4	nalorphine	Haertzen, 1970 Isbell, 1952b Isbell, 1953a Martin et al., 1965 Jasinski et al., 1971a Jasinski et al., 1967b Wikler et al., 1953	Isbell, 1956 Isbell, 1953a Martin and Gorodetzky, 1965b Isbell, 1952b Martin and Gorodetzky, 1965c	Isbell, 1952b Isbell, 1953a Wikler et al., 1953 Jasinski et al., 1967b
A-5	N-propyl-dihydro-morphine	Fraser and Isbell, 1955b Isbell, 1954b		Fraser and Isbell, 1955b Isbell, 1954b
A-6	naloxone	Jasinski et al., 1967b Jasinski et al., 1967a	Jasinski et al., 1967b Jasinski et al., 1967a	Jasinski et al., 1967b Jasinski et al., 1967a
A-7	nalbuphine	Jasinski and Mansky, 1972 Jasinski and Mansky, 1970	Jasinski and Mansky, 1972 Jasinski et al., 1971b	Jasinski and Mansky, 1972 Jasinski et al., 1971b
A-8	naltrexone	Martin et al., 1971 Martin et al., 1973b		Martin et al., 1971 Martin et al., 1973b
vallorphan congeners				
B-1	L-3-acetoxy-N-allyl morphinan	Fraser and Isbell, 1955b		Fraser and Isbell, 1955b
B-2	D-3-hydroxy-N-allylmorphinan	Isbell, 1954b Isbell and Fraser, 1955		Isbell, 1954b Isbell and Fraser, 1955
B-3	levallorphan	Jasinski et al., 1967b Fraser and Isbell, 1955b Isbell, 1954b		Jasinski et al., 1967b Fraser and Isbell, 1955b Isbell, 1954b
B-4	L-3-hydroxy-N-propargyl-morphinan	Isbell and Fraser, 1955 Isbell, 1955b		Isbell and Fraser, 1955 Isbell, 1955b
B-5	*l*-3-methoxy-N-allylmorphinan	Fraser and Isbell, 1955b Isbell, 1955b		Fraser and Isbell, 1955b Isbell, 1955b
B-6	oxilorphan *l*-BC-2605	Jasinski and Nutt, 1973		Jasinski and Nutt, 1973
B-7	butorphanol	Jasinski et al., 1975b		Jasinski et al., 1975b
nzomorphan antagonists				
C-2	pentazocine	Fraser and Rosenberg, 1963c Fraser and Rosenberg, 1964 Jasinski et al., 1970 Jasinski et al., 1971a Jasinski et al., 1975b	Fraser and Rosenberg, 1963c Fraser and Rosenberg, 1964 Jasinski et al., 1970	Fraser and Rosenberg, 1963c Fraser and Rosenberg, 1964 Jasinski et al., 1970
C-3	*d*, *l*-cyclazocine	Haertzen, 1970 Martin et al., 1964 Jasinski et al., 1968 Martin et al., 1965 Fraser and Rosenberg, 1963b	Martin et al., 1964 Martin et al., 1965 Fraser and Rosenberg, 1963b	Martin et al., 1964 Martin et al., 1965

Table 8 (continued)

Drug No.	Name	Subjective effects	Direct addiction	Morphine abstinence, suppression, or precipitation
II-C-4	β-(−)-5-phenyl-9-methyl-2′-hydroxy-2-methyl-6,7-benzomorphan (GPA1657)	JASINSKI et al., 1971a MARTIN et al., 1967 JASINSKI et al., 1968b JASINSKI et al., 1969	JASINSKI et al., 1971a JASINSKI et al., 1968b	JASINSKI et al., 1971a JASINSKI et al., 1968b
II-C-5	*l*-α-5,9-diethyl-2′-hydroxy-2-methyl-6,7-benzomorphan (GPA 2087)	JASINSKI and MANSKY, 1970		JASINSKI et al., 1971b
II-C-7	*l*-cyclazocine	JASINSKI and NUTT, 1973		JASINSKI and NUTT, 1973
Miscellaneous				
	morphine-nalorphine mixtures	ISBELL, 1954b FRASER et al., 1956	ISBELL, 1954b ISBELL, 1955b	ISBELL, 1954b
	phenazocine-N-allylnorphenazocine mixtures	FRASER et al., 1963		FRASER et al., 1963
	methadone-naloxone mixtures	NUTT and JASINSKI, 1974		NUTT and JASINSKI, 1974
	morphine-amphetamine mixtures	JASINSKI et al., 1972 JASINSKI and NUTT, 1972		
	morphine-daptazole mixtures	FRASER et al., 1957a FRASER et al., 1957b		FRASER et al., 1957a FRASER et al., 1957b

potential is the extent that a drug is abused in an uncontrolled situation. A number of considerations refute the practicality of this index as the absolute indicator of abuse potential. The scientific validity of the epidemiologic methods to provide precise indices of cases of abuse even with such drugs as heroin is uncertain. The pharmacologic similarities of a drug to morphine indicates its potential for abuse. The actual frequency of the abuse of a drug at a point in time depends upon controls, the overall availability of the drug, knowledge of the drug's actions, customs and fads, the available preparation of the drug, as well as attitudes, laws, and customs of society. Important is the recognition that agents demonstrated pharmacologically equivalent to morphine are in most instances not marketed. If marketed, they are subject to the same legal controls as morphine. As a consequence, the potentiality of the majority of these compounds to induce drug-seeking behavior is most commonly not realized. Recently, the majority of those morphinelike agents introduced into therapy are those felt to have a lesser liability for abuse. Nevertheless, existing data of incidence of abuse support the validity of the methods in man to predict drug-seeking behavior.

MARTIN (1966) tabulated the frequency of abuse of narcotic analgesics obtained from three studies of the clinical records of patients admitted to the U.S. Public Health Service Hospital, Lexington, and from statistical tables prepared by the Bureau of Narcotics. These data have been incorporated into Table 9 (columns A, B, C, D) and expanded to reflect more recent surveys of abuse frequency of various narcotics. The ARC numbers have been included in this listing. With the exception of pentazocine, all drugs on this list have been judged to have a morphinelike mode of action from studies in man (cross reference Table 9 to Table 8).

To interpret Table 9, Spearman's rank order correlation coefficients (SNEDECOR and COCHRAN, 1973) were calculated for the relative frequency of abuse of narcotics as judged by different surveys. Comparison of the data (columns B, C, D, Table 9) tabulated by MARTIN (1966) shows similar patterns for frequency of abuse (Table 10, comparison 1 and 2).

In the late 1960s, the Lexington Hospital was devoted to care of narcotic addicts civilly committed under the Narcotic Addicts Rehabilitation Act (NARA). From July 21, 1967, to May 20, 1969, 1096 patients completed upon admission a social data form under the direction of Dr. JOHN A. O'DONNELL. One section lists 102 drugs broken down into classes. For the narcotic analgesics, the patient was asked to check which he had used and to which he had been "hooked" (physically dependent). These data are tabulated in Table 9, column E (O'DONNELL and VOSS, unpublished). The pattern of drug use in this group of patients ("hooked") is similar to previous surveys (Table 10, comparison 3). Most interesting was the high incidence of use of cough syrups and codeine. WEPPNER (1971) investigated the subsample of addict patients who had reported codeine as their first addiction. From these data, WEPPNER (1971) concluded that codeine cough syrups were the second most frequently abused drug after heroin and were close to morphine and dilaudid in the number of first addictions reported, accounting for 12% of the first narcotic addictions.

In 1973, the Drug Enforcement Administration instituted the Drug Abuse Warning Network (DAWN) to monitor drug abuse trends in the United States. This system tabulates nonmedical use of substances and drug-related deaths from four sources. These are (1) emergency rooms, (2) in-patient units in nonfederal, short-term, general hospitals, (3) county medical examiners or county coroners, and (4) crisis intervention centers. Table 9, column F, lists total mentions of narcotic drugs reported by the entire DAWN System for the period of July 1, 1973, through April 30, 1975 (unpublished data kindly furnished by Mr. ERNEST A. CARABILLO, JR. of the Drug Enforcement Agency and Dr. PHILLIP H. PERSON, JR. of the National Institute of Drug Abuse). Comparison of the rankings of frequency of mentions for the DAWN System with those of the data of O'DONNELL in column E indicates a lack of correlation (Table 10, comparison 4). Comparison of the relative rankings for individual drugs indicates that the differences are predominantly accounted for by the relatively lower ranking of paregoric and the relatively higher ranking of *d*-propoxyphene, pentazocine, and methadone in the DAWN data. To a surprising extent, these changes are consistent with other observations and probably reflect changes in the pattern of narcotic use. In 1972, concern over numerous reports of intravenous abuse of paregoric led to its restriction to prescription sale (Federal Register, Vol. 37, No. 65, April 4, 1972, p. 6734) suggesting that the decrease in relative

Table 9. Frequency of abuse of morphinelike drugs

	ARC No.	*A* Drug preferred 1936–1937	*B* Drug of choice 1962–1963	*C* Drug used 1965	*D* Active addicts 1963	*E* Civilly committed addicts 1968–1969	
						Ever Used	Ever Hooked
Morphine		697	291	112	1036	690	252
Opium derivatives (total)		123	266	75	674		
Paregoric			229	65	553	449	153
Opium			8	4	121	395	55
Laudanum			1	—	—	66	11
Pantopon			28	6	—	208	33
Codeine	I-A-4		101	26	341	482	66
Cough Syrups						756	208
Heroin	I-A-6	240	1870	670	44428	964	866
Dihydromorphinone	I-A-19		214	60	568	526	230
Oxycodone	I-A-15		—	9	59	121	22
Oxymorphone	I-A-16		—	1	—	132	66
Meperidine	I-D-13		118	55	584	504	121
Methadone	I-C-9		61	9	100	723	110
d-Propoxyphene	I-C-12 I-C-28					460	44
Dihydrocodeinone	I-A-10					66	11
Ethylmorphine						33	—
Dromoran	I-B-5					11	—
Levodromoran	I-B-6					11	—
Metopon	I-A-26					22	—
Alphaprodine	I-D-1					11	—
Pentazocine	II-C-2					77	1
Anileridine	I-D-3						
Diphenoxylate	I-D-6						
Other Narcotics		62	12		115		

Data in columns A, B, C, D, are from MARTIN (1966).

Data in column A are from a study of clinical records of patients admitted to the U.S. Public Health Service Hospital, Lexington, Ky. (PESCOR, 1943).

Data given in columns B and C are unpublished data of Dr. J. A. O'DONNELL obtained from records of the PHS Hospital, Lexington, Ky.

Data in column D were obtained from statistical tables prepared by the Bureau of Narcotics.

Data in column E represents data from surveys of civilly committed narcotic addicts at the Lexington Hospital for the years 1968 and 1969 (O'DONNELL and VOSS, unpublished).

Data in column F are from the Drug Abuse Warning System (DAWN).

F DAWN System mentions 1973–1975	*G* General population of men 20–30 1974–1975	*H* Mayo Clinic admissions 1966–1972	Estimated relative availability		
			I Aggregate U.S. production (kg) 1973	*J* Parenteral potency	*K* Morphine equivalence (kg)
3760	51	11	535	1	535
			1146	0.1	115
77	56				
341	169	2			
	3				
4343	398	17	32564	0.1	3256
31808	103				
1079	13	4	58	10	580
2029	29	12	1240	1	1240
28		1	60	10	600
1689	77	37	16292	0.1	1629
9864	48	2	3339	1	3339
9692	298	14	88120	.07	6168
4	2		485	1	485
4			22	ca .07	1.5
1					
10			8	5	40
			52	0.2	10.4
1832	22	32	10087	0.3	3026
16			276	0.4	110
			660	0.5 (oral)	330
	5	1			

Data in column G are from a survey of psychoactive drug use in 2508 men representative of the general population of age 20–30-year-old men in the United States for the period of October 1974 to May 1975 (O'DONNELL et al., unpublished).

Data in column H are from admissions to Mayo Clinic for prescription drug abuse (SWANSON et al., 1973).

Data in column I are from COCHIN and HARRIS (1975). Pentazocine estimates kindly provided by Dr. MONROE TROUT of Sterling Winthrop, Inc. *d*-Propoxyphene estimates kindly provided by Dr. GLENN KIPLINGER of Eli Lilly Company.

Potencies in column J are from Table 6 or estimated from data on relative analgesic effectiveness.

Data in column K represent the relative availability of the various drugs in terms of morphine equivalence.

Table 10. Spearman's rank order correlation coefficients for relative frequency of abuse of various narcotics from the listings in Table 9

Comparison No.	Columns	Number of drugs compared	r_s	P
1	C vs B	10	+1.000	0.00
2	D vs C	10	+0.906	< .01
3	E[a] vs C	12	+0.885	< .01
4	F vs E[a]	12	+0.126	n.s.[b]

[a] Responses to the question "hooked?"
[b] Non-significant.

frequency was in response to increasing control. Pentazocine was introduced into general prescription sales in August 1967.

As will be discussed below, the increased prominence of methadone as a drug of abuse from 1968 to 1973–1975 in the DAWN data is supported by other studies. Of special interest is the high ranking of *d*-propoxyphene in the DAWN data. It is also noted that codeine and codeine preparations were highly ranked in the DAWN data supporting the observations in NARA population. Three other studies support the observation that *d*-propoxyphene and codeine are drugs of abuse and of public health concern. O'DONNELL et al. (unpublished) surveyed 2508 men of age 20–30 with respect to the nonmedical use of psychoactive drugs. This sample was designed to be representative of all men in this age bracket in the United States who were 20–30 years old in 1974. In this sample, 642 reported use of opiates other than as medically directed. The specific drugs and their reported frequency of use of specific drugs is shown in Table 9, column G, and indicates that codeine and *d*-propoxyphene were the most frequently reported drugs. SWANSON et al. (1973) investigated 225 patients requiring inpatient psychiatric treatment at the Mayo Clinic between July 1, 1966, and July 1, 1972, for abuse of prescription medications. Of these patients, 87 had abused various analgesics. The types of analgesics abused by these patients are listed in Table 9 and indicates that pentazocine, codeine, and *d*-propoxyphene were the second, third, and fourth most frequently abused drugs. CHAMBERS et al. (1971) also found *d*-propoxyphene to be a drug of abuse in both addict and nonaddict populations.

Since 1967 at least there appears to have been a relative increase in the abuse of *d*-propoxyphene and codeine. Of interest is the parallel with the increase in heroin addiction which began in 1966 and peaked in 1971 (GREENE and DUPONT, 1974). Recently, SEITNER et al. (1975) surveyed a sample of physicians with regard to their prescribing habits with respect to opiate drugs and their alternates. The most frequently prescribed analgesic for acute and chronic conditions with mild pain were *d*-propoxyphene (42.4% for acute and 48.5% for chronic pain) and codeine combinations (30.8% for acute and 19.6% for chronic conditions). For acute severe pain, the major drugs prescribed were meperidine (60.3%), codeine and combinations (25.6%), morphine (24.1%), and pentazocine (9.0%). For chronic severe pain, the major drugs prescribed were codeine and combinations (34.4%), meperidine (22.5%), pentazocine (16.8%), and morphine (8.6%). For cough, codeine and combinations were used 52%

of the time. GREENE and DUPONT (1974) observed the illicit use of methadone as an alternative to heroin in Washington, D.C. and attributed this to the availability of methadone. Thus, the foregoing considerations suggest that the increased incidence of nonmedical use of *d*-propoxyphene and codeine parallels the increased incidence of heroin addiction and methadone addiction and reflects the nonmedical use of available morphinelike drugs.

Although the morphinelike properties of codeine and *d*-propoxyphene are well recognized (Table 8), both have been held to have a lesser addiction liability than morphine. Consequently, a discussion of the concept of relative abuse liability is in order.

When legal mechanisms first limited opium and its derivatives to legitimate medical and scientific use, oral preparations containing limited amounts of these agents (including heroin, morphine, and codeine) were exempt from legal controls. Pharmacologic equivalence among these drugs as well as potency differences were not recognized. This is especially true with respect to codeine which was utilized mainly as an oral preparation and was thought to have little or no addiction liability. The experimental demonstration that codeine had morphinelike addiction liability (HIMMELSBACH, 1934; HIMMELSBACH et al., 1940) and subsequently the recognition that a group of agents in the morphine-codeine series differed for the most part only in potency and duration of action with no dissociation in ability to produce or sustain physical dependence (HIMMELSBACH, 1941) clearly demonstrated the pharmacologic equivalence of codeine to morphine. From a public health viewpoint, the addiction liability of codeine, however, was felt to be less than that of morphine. The incidence of abuse as reported in the literature was relatively low and was predominantly by the oral route although parenteral abuse was known (HIMMELSBACH et al., 1940; DAVENPORT, 1938). Also codeine was not a choice in patients admitted to the hospital at Lexington (HIMMELSBACH et al., 1940). Reasons for the lesser addiction liability were sought in other properties of codeine. The characteristics of codeine which limited its addiction liability relative to morphine were thought to be greater expense, lesser potency, lesser solubility, tissue irritant properties, side-effects which limited ability to attain strong morphinelike effects, and availability predominantly as an oral preparation in low dosage units (HIMMELSBACH, 1934; HIMMELSBACH et al., 1940). Although the addiction potentiality of codeine was recognized, its low abuse incidence and its extensive use as antitussive and analgesic indicated that it should not be controlled to the same degree as morphine (WOLFF, 1938; HIMMELSBACH et al., 1940). Codeine thus came to serve as a criterion drug for control purposes with those agents having a greater abuse potential than codeine being controlled while those with a lesser abuse potential being exempted from control.

In recent years, additional studies have confirmed the morphinelike properties of codeine and provided more precise quantitative estimates of its potency. Codeine produces morphinelike subjective effects by the intramuscular route (KAY et al., 1967), and by the intravenous route (JASINSKI et al., 1971a) where it is estimated $^1/_{10}$ to $^1/_{15}$ as potent as morphine. As an analgesic codeine is $^1/_{13}$ as potent as morphine when both are administered intramuscularly (WALLENSTEIN et al., 1967). Orally codeine is $^1/_{20}$ as potent as subcutaneously administered morphine in producing miosis and subjective effects (JASINSKI et al., 1974) which is similar to the potency of $^1/_{21}$ for oral codeine to intramuscular morphine in relieving pain (WALLENSTEIN et al., 1967). Codeine suppresses morphine abstinence (HIMMELSBACH, 1934; HIMMELSBACH et al., 1940) and it has been estimated that oral codeine is $^1/_{15}$ as potent as subcutaneous morphine in suppressing abstinence (FRASER and ISBELL, 1961a). Direct addiction studies with codeine showed morphinelike physical dependence as evidenced by a withdrawal syndrome with nalorphine precipitation and abrupt withdrawal (FRASER et al., 1960a; FRASER et al., 1961d). Thus there is no quantitative evidence of a dissociation of therapeutic effect from the ability to produce morphinelike subjective effects or physical dependence. As can be seen from Table 9, column K, codeine is used about 5 times as much as morphine (including opium derivatives), yet their abuse incidences are comparable (Table 9, column F). These data argue that although codeine has a significant abuse liability it is less than that of morphine.

FRASER et al. (1961c) and FRASER and ISBELL (1960c) demonstrated that *d*-propoxyphene produced morphinelike euphoria and physical dependence but concluded because of certain characteristics, *d*-propoxyphene has lesser abuse potential than codeine. When compared to codeine, *d*-propoxyphene was less potent. The degree of morphinelike effects in single dose studies and substitution studies was limited by toxic effects such as drug-induced nervousness, convulsions, and psychoses. Similar toxic effects limited chronic administration and in addition marked tissue irritation was felt to prevent sustained parenteral self-administration of *d*-propoxyphene. Further, the lesser liability of abuse of *d*-propoxyphene was related to the availability only as an oral preparation in formulations difficult to dissolve. JASINSKI et al. (1971a) found intravenously administered *d*-propoxyphene hydrochloride $^1/_{10}$ to $^1/_{15}$ as potent as intravenously administered morphine in producing morphinelike subjective effects and miosis. Orally, *d*-propoxyphene hydrochloride was $^1/_{25}$ to $^1/_{40}$ as potent as subcutaneously administered morphine in producing miosis, subjective effects, and in suppressing morphine abstinence (JASINSKI et al., 1974, 1975b). The napsylate salt of *d*-propoxyphene orally was 1.5 to 2.5 times less potent than the hydrochloride salt orally (JASINSKI et al., 1975b). As can be seen from Table 9 the incidence of abuse and the use of *d*-propoxyphene (column K) are both about twice that of codeine. These data as well as the above cited data suggest that the abuse liability of *d*-propoxyphene is comparable to that of codeine.

FRASER and ISBELL (1961a) demonstrated diphenoxylate produced morphinelike euphoria and physical dependence but concluded that diphenoxylate had advantages from the viewpoint of abuse relative to codeine and morphine since diphenoxylate was insoluble in water and could not be easily extracted from oral preparations for intravenous use. It was not suitable for subcutaneous use and the onset of euphoria was markedly slower by all routes of administration than that of codeine or morphine. Even though diphenoxylate (Lomotil) is the most frequently prescribed opiate for diarrhea (59.2%) (SEITNER et al., 1975), there is no evidence of abuse of diphenoxylate despite its widespread use and availability (Table 9).

Dextromethorphan is an antitussive devoid of morphinelike properties (ISBELL and FRASER, 1953b; JASINSKI et al., 1971b; MANSKY and JASINSKI, 1970) and judged to lack addiction liability. Even though it is the second most widely prescribed antitussive (30%) (SEITNER et al., 1975), there is little evidence of abuse (see Project DAWN Reports).

It is important to emphasize that the decisions regarding the potential for public health problems with respect to codeine and related drugs were based on social experiences of a number of decades ago. Unfortunately, these decisions were formulated into legal mechanisms which influence the availability and choice of various morphinelike drugs. Even more unfortunate is the tendency for these precedents to obscure the essential pharmacologic nature of these agents and their potential for abuse.

The introduction of dihydromorphinone (Dilaudid) was accompanied by reports and opinions that it was a nonaddicting substitute for morphine; however, KING et al. (1935) demonstrated that dihydromorphinone was capable of producing and sustaining physical dependence and hence had addiction liability. Approximately equal amounts of dihydromorphinone and morphine are used (Table 9, column K) with the incidence of abuse of dihydromorphinone approaching the same order of magnitude as that of morphine (Table 9).

HIMMELSBACH (1942) found that meperidine produced morphinelike physical dependence and concluded that meperidine had an addiction liability. As a consequence of these studies, meperidine was subject to legal controls. This conclusion of addiction liability was supported by the significant incidence of abuse which followed the introduction of meperidine (RASOR and CRECRAFT, 1955). However, HIMMELSBACH (1942) could only partially suppress morphine abstinence during 10 days of meperidine substitution and observed that withdrawal of the substituted meperidine produced only mild abstinence signs of more abrupt onset but with less subjective complaints than morphine withdrawal. In a direct addiction study, HIMMELSBACH (1942) observed muscular twitches and tremors, hyperactive reflexes, startle responses, seizures and a toxic psychosis during chronic meperidine administration (1–3 g daily). Abrupt withdrawal of meperidine for 24 h after 1 or 2 months of administration produced only mild signs of abstinence with no subjective complaints. Abrupt withdrawal after 10 or 11 weeks of meperidine administration produced mild typical morphine abstinence signs of more abrupt onset than morphine abstinence and with typical complaints the second day of withdrawal. ISBELL (1955) stabilized primary meperidine addicts on meperidine and found that nalorphine 15 mg precipitated abstinence only when the stabilization dose of meperidine was 1600 mg daily. When the stabilization dose was increased to 2800 mg daily, nalorphine 15 mg precipitated definite abstinence (ISBELL, 1952). Abrupt withdrawal of meperidine (2800 mg daily) produced an abstinence syndrome reaching peak intensity in 7–12 h and characterized by lesser signs of autonomic dysfunction but greater restlessness and muscular twitching than in morphine abstinence. JASINSKI and NUTT (1973) showed that in nondependent subjects single intramuscular doses of meperidine were $^1/_7$–$^1/_{14}$ as potent as morphine in producing morphinelike subjective effects and miosis but with a more rapid onset and shorter duration. In these subjects meperidine 300 mg produced nervousness, a fine tremor at rest, hyperactive deep tendon reflexes, and ankle clonus. In addition, meperidine in equally euphoriant doses produced significantly greater sedation than morphine as measured by the PCAG Scale. From the viewpoint of the codeine-morphine paradigm, these observations taken together suggest that meperidine has an abuse liability less than that of morphine and more similar to that of codeine and *d*-propoxyphene. Supporting this conclusion is the data in Table 9 which indicates that meperidine is used about 3 times as much as morphine (column K) yet the incidence of abuse is one-half that of morphine. Further, the incidence of abuse when adjusted for relative availability is similar to that of codeine (Table 9).

ISBELL et al. (1947) demonstrated that methadone was highly euphoric, suppressed morphine abstinence, and produced morphinelike physical dependence and concluded that methadone had the same addiction liability as morphine. On the basis of these studies, methadone was subject to the same legal restrictions as morphine. Over the next 20 years there was only a relatively low incidence of abuse [see tabulation by MARTIN (1966) in Table 9, column B, C, D, and SAPIRA et al., 1968]. The introduction of methadone as a chemotherapeutic agent for opiate addiction with its widespread accompanying availability, the increasing knowledge of its properties, and the changes in societal attitudes appear to contribute to its increasing appearance as a drug of abuse as has been documented by a number of investigators (GREENE and DUPONT, 1974; GREENE et al., 1974; CHABALKO et al., 1973; BAZELL, 1973; STEPHENS and WEPPNER, 1973; CHAMBERS and INCIARDI, 1972). Supporting these observations is the increased frequency of abuse of methadone (Table 9). Also surveys of addict patients entering federally funded treatment programs through the Drug Abuse Reporting System (DARP) for the period 1971–1973 (SIMPSON, 1974) and the Client Oriented Data Acquisition Process (CODAP) for the period 1973–1974 (National Institute on Drug Abuse, 1974) indicated that among these patients methadone was the most frequently abused opioid after heroin.

The available epidemiologic evidence supports the validity of the conclusion that demonstration of pharmacologic equivalence to morphine is also indicative of a similar potential for public health problems.

F. Pentazocine and Related Compounds

Pentazocine produces effects which resemble both those of morphine and those of nalorphine. As a result, the classification of this agent as either morphine or nalorphinelike has been difficult and is still uncertain.

FRASER and ROSENBERG (1964) found that pentazocine in low doses produced morphinelike subjective effects which were limited by side-effects as the dose was increased. Further, they found that pentazocine did not suppress abstinence in subjects dependent upon 240 mg of morphine per day using the 24-h substitution technique and produced only mild abstinence in direct addiction studies. As a result, FRASER and ROSENBERG (1964) concluded that pentazocine had less abuse liability than codeine. JASINSKI et al. (1970) reassessed the abuse potentiality of pentazocine because of reports of abuse following its introduction into clinical medicine. In studies of subjective effects, pentazocine in doses up to 40 mg subcutaneously produced significant MBG responses equivalent to 10 mg of morphine with no significant PCAG or LSD responses. These responses were distinguished from those of nalorphine 10 mg which produced no MBG scores but did produce LSD and PCAG scores. A similar pattern of significant MBG scores with no LSD or PCAG scores was obtained when 40 mg of pentazocine was administered intravenously (JASINSKI et al., 1971 a). Increasing the dose of pentazocine to 60 mg/70 kg produced an abrupt change in the pattern of effects from morphinelike to a nalorphinelike syndrome (JASINSKI et al., 1970). There was a decrease in the MBG response and an increase in the PCAG and LSD Scale scores. Thus when related to morphine and nalorphine, pentazocine appeared to be capable of producing morphinelike subjective effects in lower doses and nalorphinelike effects in larger doses. Pentazocine is $^1/_{50}$ as potent as nalorphine in precipitating abstinence in subjects dependent upon 240 mg of morphine per day. Through slow elevation of dose, physical dependence on pentazocine was produced in a direct addiction study. During stabilization on pentazocine, administration of nalorphine (12 mg) did not precipitate abstinence. However, naloxone in doses larger than those necessary to precipitate abstinence in morphine-dependent subjects precipitated an abstinence syndrome which the subjects identified as opiatelike withdrawal. Abrupt withdrawal of pentazocine was accompanied by an abstinence syndrome. Patients sought relief from this abstinence syndrome by requesting drugs. Analysis of the sources of points in the Himmelsbach score revealed that qualitatively the pentazocine abstinence syndrome had characteristics of both the morphine and nalorphine abstinence syndrome. In 24-h substitution studies, pentazocine did not suppress abstinence in subjects dependent upon 60 mg of morphine per day.

Overall, pentazocine appeared to have both morphinelike and nalorphinelike agonist activity. An explanation for these mixed properties was sought (JASINSKI et al., 1970) in the concepts of receptor dualism, competitive dualism, and competitive antagonism which had been proposed by MARTIN (1967b) to explain the relationship of the agonist and the antagonist activities of the narcotic antagonists and morphine. Pentazocine was thought to have greater affinity and lesser intrinsic activity than morphine at the morphine receptor accounting for its ability to produce elements of both morphinelike subjective effects and morphinelike physical dependence and a lesser affinity but greater intrinsic activity at the nalorphine receptor to explain its ability to produce nalorphinelike subjective effects and elements of nalorphinelike physical dependence. To test this hypothesis, it was felt necessary to demonstrate that pentazocine could substitute for morphine and suppress abstinence, if indeed pentazocine was a partial morphine agonist. Although pentazocine suppresses abstinence in subjects dependent upon 30 mg/day of morphine (JASINSKI, unpublished) there are reservations that pentazocine has morphinelike activity. These are based on: (1) the possibility that 24 h substitution tests at 30 mg of morphine daily are not specific for morphinelike activity (Subsect. E.2), (2) the lack of a validated measure of aversiveness of abstinence (Subsect.E.2), (3) single-dose studies indicating that pentazocine should have greater agonistic activity than observed in suppression studies, and (4) the observation that nalorphine and cyclazocine also have the ability to produce a type of euphoria. With regard to the latter, MARTIN et al. (1965) demonstrated that low doses of cyclazocine and nalorphine produced a significant elevation of the liking scales of the single dose opiate questionnaire. JASINSKI et al. (1968a) found that cyclazocine elevated scores on the MBG Scale. JASINSKI et al. (1975b) compared the subjective and miotic effects of cyclazocine 0.5 and 1 mg and pentazocine 40 and 80 mg with the effects of butorphanol and morphine. The pattern of effects of pentazocine and cyclazocine could not be differentiated; however, statistically valid potencies could not be obtained on all measures because of significant "F" ratios for mean squares in the analysis of variance. These observations suggest differences in the profile of subjective effects between pentazocine and cyclazocine. Thus, it is possible that within this class of drugs there may be inherent differences in their ability to produce euphoria, sedation, and psychotomimetic effects such that pentazocine as well as other agents in this class may be more reliably euphorigenic. Also supporting this observation are the studies of JASINSKI and MANSKY (1972) who found nalbuphine to have a pharmacologic profile most similar to pentazocine. More recently JASINSKI (unpublished observations) found nalbuphine suppressed abstinence in subjects dependent upon 30 mg of morphine per day. Similar reservations about the interpretation of mixed morphine-nalorphine agonist activity apply to nalbuphine as they do to pentazocine. More importantly, nalbuphine in these studies (JASINSKI and MANSKY, 1972) appeared to be significantly less psychotomimetic than pentazocine indicating that the analgesic activity in this class of compounds can be dissociated from psychotomimetic activity.

In this regard, studies of the reinforcing properties of pentazocine in monkeys also have not been useful in classifying pentazocine (see Sect. II, Chap. 2, for discussion of procedures in the monkey for measuring the effects of morphinelike drugs). One set of studies of the reinforcing properties of pentazocine were those of YANAGITA (1973 and personal communication). He found that pentazocine suppressed morphine abstinence in monkeys dependent on 0.3 mg/kg per day but not at 3.0 mg/kg per day. Physical dependence was produced in direct addiction studies with an abstinence syndrome of mild intensity, however, this abstinence syndrome did not intensify self-injection behavior. Pentazocine was highly reinforcing in the continuous self-administration (naive monkeys) and the cross self-administration procedures (SPA).

HOFFMEISTER and WUTTKE (1974) found pentazocine and propiram to be more reinforcing than nalorphine and cyclazocine with self-injection procedures in the monkey. All four drugs had negative reinforcing properties in the morphine-dependent monkey. In naive monkeys, pentazocine and propiram, but not nalorphine or cyclazocine, had positive reinforcing properties. On the other hand, nalorphine and cyclazocine, but not pentazocine or propiram, had negative reinforcing properties in naive monkeys since both initiated and maintained stable avoidance-escape responses.

Thus from both the human and the monkey studies pentazocine would be judged more reinforcing than nalorphine or cyclazocine but less than morphine, codeine, or *d*-propoxyphene. Supporting this are the data in Table 9 which indicates that the availability of pentazocine approaches that of codeine (column K) yet the incidence of abuse is less than one-half that of codeine (column F). Further, the mode of action of the reinforcing effects of pentazocine is not known at the present time; however, it is likely that it is not morphinelike.

VII. Summary and Conclusions

The concordance of potency estimates from the measures of miosis, subjective effects, suppression of abstinence, and analgesia is strong evidence that these measures of subjective effects and physical evidence are valid indicators of a morphinelike mode of action. The actual abuse incidences of morphine, codeine, dihydromorphinone, meperidine, methadone, *d*-propoxyphene, dextromethorphan, diphenoxylate, and pentazocine clearly indicate that the liability of an agent to be abused is validly and accurately judged from the methods developed to assess such abuse liability and further that the abuse of morphinelike drugs is related to their ability to produce euphoria and physical dependence.

The systematic assessment of analgesics, antitussives, and antidiarrheals for morphinelike activity has been an effective preventive medicine measure. Physicians have been accurately informed about the abuse potential of opioids introduced into therapeutics and have most frequently prescribed those of a lesser abuse potential. Many agents pharmacologically equivalent to morphine and heroin have not been introduced into therapeutics. As a consequence, there has been no repetition of the public health problems of the magnitude which followed the introduction of heroin into therapeutics.

In this regard, there have been three groups of compounds demonstrated to have a lesser abuse potential than morphine. First, codeine, *d*-propoxyphene, and diphenoxylate have morphinelike modes of action. However, factors such as greater toxicity and lesser aqueous solubility than with heroin and morphine, as well as availability predominantly or only as oral preparations of low activity, have significantly lowered the abuse incidence. Even with lesser controls, codeine and *d*-propoxyphene have been only modestly abused; diphenoxylate not at all. Second, dextromethorphan is an antitussive which does not produce morphinelike euphoria and physical dependence. It has not been significantly abused. Third, pentazocine produces euphoria and physical dependence which resemble that of morphine; however, the mode of action of pentazocine is most likely not morphinelike but similar to that of nalorphine and cyclazocine. The abuse liability of pentazocine has been judged to be less than that of codeine or *d*-propoxyphene. The incidence of abuse of pentazocine supports this conclusion.

More recently, partial agonists of morphine have been recognized. Because of a markedly lesser ability to produce morphinelike effects, propiram was judged to have a lesser abuse potential than codeine or *d*-propoxyphene. Since propiram has not yet been introduced into therapeutics there is no evidence to support the validity or accuracy of this judgment of lesser abuse potential.

At issue is the necessity of assessment studies in man as animal methods are progressively developed. The evidence of species differences in responsivities to morphinelike drugs clearly indicates the need for final assessment of these agents in man.

In the face of a persistent and overwhelming abuse of heroin, few would agree with the original hypothesis advanced 45 years ago that the introduction of nonaddicting morphine substitutes would solve the addiction problem. What must not be overlooked, however, is the possible preventive medicine benefits through the replacement of morphinelike drugs in therapeutics with effective and more selective substitutes lacking the abuse potential of morphine.

References

Andrews, H. L., Himmelsbach, C. K.: Relation of the intensity of the morphine abstinence syndrome to dosage. J. Pharmacol. exp. Ther. **81**, 288—293 (1944)

Bazell, R. J.: Drug abuse: Methadone becomes the solution and the problem. Science **179**, 772—775 (1973)

Beaver, W. T., Wallenstein, S. L., Houde, R. W., Rogers, A.: A clinical comparison of the analgesic effects of methadone and morphine administered intramuscularly, and of orally and parenterally administered methadone. Clin. Pharmacol. Ther. **8**, 415—426 (1967)

Beaver, W. T., Wallenstein, S. L., Houde, R. W., Rogers, A.: A comparison of the analgesic effects of profadol and morphine in patients with cancer. Clin. Pharmacol. Ther. **10**, 314—319 (1969)
Beecher, H. K.: Measurement of Subjective Responses. New York: Oxford University Press 1959
Blumberg, H., Dayton, H. B.: Naloxone, naltrexone, and related noroxymorphones. In: Braude, M. C., Harris, L. S., May, E. L., Smith, J. P., Villarreal, J. E. (Eds.): Narcotic Antagonists. Advances in Biochemical Psychopharmacology, Vol. 8. New York: Raven Press 1974
Chabalko, J., LaRosa, J. C., DuPont, R. L.: Death of methadone users in the District of Columbia. Int. J. Addict. **8**, 897—908 (1973)
Chambers, C. D., Inciardi, J. A.: An empirical assessment of the availability of illicit methadone. Proceedings of the Fourth National Conference on Methadone Treatment, San Francisco, California, pp. 149—151. New York: National Association for the Prevention of Addiction to Narcotics 1972
Chambers, C. D., Moffett, A. D., Cuskey, W. R.: Five patterns of Darvon abuse. Int. J. Addict. **6**, 173—189 (1971)
Cochin, J., Harris, L.: Synthetic Substitutes for Opiate Alkaloids. Report prepared for Committee on Problems of Drug Dependence, National Research Council, Washington, D.C., under contract No. BNDD 72-9 between the National Academy of Sciences and Bureau of Narcotics and Dangerous Drugs, 1975
Davenport, L. F.: The abuse of codeine. Publ. Hlth. Rep. (Wash.), Suppl. 145 (1938)
Eddy, N. B.: Dilaudid (dihydromorphinone hydrochloride). J. Amer. med. Ass. **100**, 1031—1035 (1933)
Eddy, N. B.: The history of the development of narcotics. Law Contemp. Prob. **22**, 3—8 (1957)
Eddy, N. B.: The National Research Council involvement in the opiate problem, 1928—1971. Washington, D.C.: National Academy of Sciences 1973
Eddy, N. B., Himmelsbach, C. K.: Experiments on the tolerance and addiction potentialities of dihydrodesoxymorphine-D ("Desomorphine"). Publ. Hlth. Rep. (Wash.), Suppl. 118 (1936)
Finney, D. J.: Statistical method in biological assay. New York: Hafner Publishing Company 1964
Forrest, W. H., Jr.: Report of the Veterans Administration Cooperative Analgesic Study. Presented at 30th Meeting, Committee on Problems of Drug Dependence, National Research Council, Indianapolis, Indiana, 1968
Forrest, W. H., Jr.: Report of the Veterans Administration cooperative analgesic study. Presented at 32nd meeting, Committee on Problems of Drug Dependence, National Research Council, Washington, D.C. 1970
Forrest, W. H., Brown, C. R., Gold, M., Mahler, D. L., Schettini, A., Teutsch, G., James, K. E.: Report of the Veterans Administration cooperative analgesic study. Presented at 35th meeting, Committee on Problems of Drug Dependence, National Research Council, Mexico City, Mexico 1974
Fraser, H. F.: Addiction liability of narcotine. Presented at 14th meeting, Committee on Drug Addiction and Narcotics, National Research Council, Rensselaer, N.Y. 1954
Fraser, H. F.: Addictive potentialities of hexamethyleneimines. Fed. Proc. **15**, 423 (1956)
Fraser, H. F.: Addictiveness of 1,2-dimethyl-3-phenyl-3-propionoxy-pyrrolidine hydrochloride (ARC-I-0-1). Bull. Narcot. **16**, 37—43 (1964)
Fraser, H. F., Essig, C. F., Wolbach, A. B.: Evaluation of carisoprodol and phenyramidol for addictiveness. Bull. Narcot. **13**, 1—5 (1961a)
Fraser, H. F., Isbell, H.: Addiction liabilities of morphinan, 6-methyldihydromorphine and dihydrocodeinone. J. Pharmacol. exp. Ther. **100**, 128—134 (1950)
Fraser, H. F., Isbell, H.: Actions and addiction liabilities of alphaacetylmethadols in man. J. Pharmacol. exp. Ther. **105**, 458—465 (1952)
Fraser, H. F., Isbell, H.: Failure of cortisone and ACTH in treatment of the morphine abstinence syndrome. Ann. intern. Med. **38**, 234—238 (1953)
Fraser, H. F., Isbell, H.: Addictive properties of morphine derivatives. J. Pharmacol. exp. Ther. **113**, 21—22 (1955a)
Fraser, H. F., Isbell, H.: Morphine antagonists. Fed. Proc. **14**, 340 (1955b)
Fraser, H. F., Isbell, H.: Reports on addiction liability tests of new substances. Presented at 17th meeting, Committee on Drug Addiction and Narcotics, National Research Council, Washington, D.C., 1956a

Fraser, H. F., Isbell, H.: Work of the NIMH Addiction Research Center, PHS Hospital, Lexington, Kentucky. Calendar Year 1955. Presented at the 17th meeting, Committee on Drug Addiction and Narcotics, National Research Council, Washington, D.C., 1956b

Fraser, H. F., Isbell, H.: Chlorpromazine and reserpine. (A) Effects of each and of combinations of each with morphine. (B) Failure of each in treatment of abstinence from morphine. Arch. Neurol. Psychiat. (Chic.) **76**, 257—262 (1956c)

Fraser, H. F., Isbell, H.: Addiction liability of new analgesics. Presented at 18th meeting, Committee on Drug Addiction and Narcotics, National Research Council, Indianapolis, Ind., 1957

Fraser, H. F., Isbell, H.: Human pharmacology and addiction liability of certain compounds related to morphine or codeine. Presented at 19th meeting, Committee on Drug Addiction and Narcotics, National Research Council, Washington, D.C., 1958a

Fraser, H. F., Isbell, H.: Progress report. NIMH Addiction Research Center, PHS Hospital, Lexington, Kentucky. Presented at 19th meeting, Committee on Drug Addiction and Narcotics, National Research Council, Washington, D.C., 1958b

Fraser, H. F., Isbell, H.: Addiction liabilities of (a) *dl*-2'-hydroxy-5,9-dimethyl-2-(phenethyl)-6,7-benzomorphan HBr (NIH-7519) and (b) *l*-3-hydroxy-N-phenacyl-morphinan methane sulfonate (NIH-7525). Presented at 20th meeting, Committee on Drug Addiction and Narcotics, National Research Council, Washington, D.C., 1959a

Fraser, H. F., Isbell, H.: Addictiveness of trifluoperazine (SKF-5019). Presented at 20th meeting, Committee on Drug Addiction and Narcotics, National Research Council, Washington, D.C., 1959b

Fraser, H. F., Isbell, H.: Human pharmacology and addiction liabilities of phenazocine and levophenacylmorphan. Bull. Narcot. **12**, 15—23 (1960a)

Fraser, H. F., Isbell, H.: Human pharmacology and addictiveness of ethyl 1-(3-cyano-3,3-diphenylpropyl)-4-piperidine carboxylate hydrochloride (R-1132, diphenoxylate). Presented at 21st meeting, Committee on Drug Addiction and Narcotics, National Research Council, Philadelphia, Pennsylvania 1960b

Fraser, H. F., Isbell, H.: Pharmacology and addiction liability of *dl*- and *d*-propoxyphene. Bull. Narcot. **12**, 9—14 (1960c)

Fraser, H. F., Isbell, H.: Human pharmacology and addictiveness of ethyl 1-(3-cyano-3,3-diphenylpropyl)-4-piperidine carboxylate (R-1132, diphenoxylate). Bull. Narcot. **13**, 29—43 (1961a)

Fraser, H. F., Isbell, H.: II. Human pharmacology and addictiveness of certain dextroisomers of opioids: 1. *d*-3-hydroxy-N-phenethylmorphinan, 2. *d*-3-methoxy-N-phenethylmorphinan, and 3. *d*-methadone. Presented at 23rd meeting, Committee on Drug Addiction and Narcotics, National Research Council, New York, New York 1961b

Fraser, H. F., Isbell, H.: Human pharmacology and addictiveness of certain dextroisomers of synthetic analgesics: I. *d*-3-hydroxy-N-phenethylmorphinan, II. *d*-3-methoxy-N-phenethylmorphinan, III. *d*-methadone. Bull. Narcot. **14**, 25—35 (1962)

Fraser, H. F., Isbell, H., Nash, T. L., van Horn, G. D.: Use of miotic effects in evaluating analgesic drugs in man. Arch. int. Pharmacodyn. **98**, 443—451 (1954)

Fraser, H. F., Isbell, H., Rosenberg, D. E., Wolbach, A. B.: Annual report of the National Institute of Mental Health Addiction Research Center. Presented at 24th meeting, Committee on Drug Addiction and Narcotics, National Research Council, New York, New York 1962

Fraser, H. F., Isbell, H., Van Horn, G. D.: Effects of morphine as compared with a mixture of morphine and diaminophenylthiazole (Daptazole). Presented at 18th meeting, Committee on Drug Addiction and Narcotics, National Research Council, Indianapolis, Indiana 1957a

Fraser, H. F., Isbell, H., Van Horn, G. D.: Effects of morphine as compared with a mixture of morphine and diaminophenylthiazole (Daptazole). Anesthesiology **18**, 531—535 (1957b)

Fraser, H. F., Isbell, H., Van Horn, G. D.: Human pharmacology and addiction liability of norcodeine. J. Pharmacol. exp. Ther. **129**, 172—177 (1960a)

Fraser, H. F., Isbell, H., Wolbach, A. B.: Addictiveness of new synthetic analgesics. I. Benzimidazole derivatives: (a) 2-(p-chlorobenzyl)-1-diethylaminoethyl-5-nitrobenzimidazole methane sulfonate (NIH-7586, ARC I-G-1), and (b) 2-(p-ethoxybenzyl)-1-diethylaminoethyl-5-nitrobenzimidazole hydrochloride (NIH-7607, ARC I-G-2). II. (-) 3-hydroxy-N-(3,3-dimethylallyl)-morphinan hydrobromide (NIH-7446). III. (a) N-(methyl-2-piperidinoethyl-propionanilide hydrochloride) (Phenampromid), and (b) N-(2-(methyl)-phenethylamino)-propyl)-propionanilide sulfate (Diampromid). Presented at 21st meeting, Committee on Drug Addiction and Narcotics, National Research Council, Philadelphia, Pennsylvania 1960b

Fraser,H.F., Martin,W.R., Wolbach,A.B., Isbell,H.: Addiction liability of 1-(p-chlorophenethyl)-6,7-dimethoxy-2-methyl-1,2,3,4-tetrahydroisoquinoline (I-K-1, No.4-1778/1). Presented at 23rd meeting, Committee on Drug Addiction and Narcotics, National Research Council, New York, New York 1961b

Fraser,H.F., Martin,W.R., Wolbach,A.B., Isbell,H.: Addiction liability of an isoquinoline analgesic, 1-(p-chlorophenethyl)-2-methyl-6,7-dimethoxy-1,2,3,4-tetrahydroisoquinoline. Clin. Pharmacol. Ther. **2**, 287—299 (1961c)

Fraser,H.F., Rosenberg,D.E.: Observations on the human pharmacology and addictiveness of methotrimeprazine. Clin. Pharmacol. Ther. **4**, 596—601 (1963a)

Fraser,H.F., Rosenberg,D.E.: Preliminary report on the human pharmacology and addiction liability of 2-cyclopropylmethyl-2′-hydroxy-5,9-dimethyl-6,7-benzomorphan (ARC II-C-3); (WIN 20,740). Presented at 25th meeting, Committee on Drug Addiction and Narcotics, National Research Council, Ann Arbor, Michigan 1963b

Fraser,H.F., Rosenberg,D.E.: Studies on addiction liability of 2′-hydroxy-5,9-dimethyl-2-(3,3-dimethylallyl)-6,7-benzomorphan (II-C-2): a narcotic antagonist. Presented at 25th meeting, Committee on Drug Addiction and Narcotics, National Research Council, Ann Arbor, Michigan 1963c

Fraser,H.F., Rosenberg,D.E.: Studies on the human addiction liability of 2-hydroxy-5,9-dimethyl-2-(3,3-dimethylallyl)-6,7-benzomorphan (WIN 20,228); a weak narcotic antagonist. (II-C-2). J. Pharmacol. exp. Ther. **143**, 149—156 (1964)

Fraser,H.F., Rosenberg,D.E., Isbell,H.: Progress report of the NIMH Addiction Research Center on certain analgesic drugs. Presented at 25th meeting, Committee on Drug Addiction and Narcotics, National Research Council, Ann Arbor, Michigan 1963

Fraser,H.F., Van Horn,G.D., Isbell,H.: Studies on N-allylnormorphine in man: Antagonism to morphine and heroin and effects of mixtures of N-allylnormorphine and morphine. Amer. J. med. Sci. **231**, 1—8 (1956)

Fraser,H.F., Van Horn,G.D., Martin,W.R., Wolbach,A.B., Isbell,H.: Methods for evaluating addiction liability. (A) "Attitude" of opiate addicts toward opiate-like drugs. (B) A short-term "direct" addiction test. J. Pharmacol. exp. Ther. **133**, 371—387 (1961d)

Fraser,H.F., Wolbach,A.B.: The addiction liability of alpha-*dl*-3-acetoxy-4,4-diphenyl-6-methylamino-heptane hydrochloride (NIH-7667, ARC I-C-25) and 6-acetyl-3-ethoxy-dihydromorphine (NIH-7623, ARC-I-A-38). Presented at 23rd meeting, Committee on Drug Addiction and Narcotics, National Research Council, New York, New York 1961

Gorodetzky,C.W., Martin,W.R.: A comparative study of fentanyl, droperidol and morphine. Presented at 27th meeting, Committee on Problems of Drug Dependence, National Research Council, Houston, Texas 1965a

Gorodetzky,C.W., Martin,W.R.: A comparison of fentanyl, droperidol and morphine. Clin. Pharmacol. Ther. **6**, 731—739 (1965b)

Greene,M.H., DuPont,R.L.: Heroin addiction trends. Amer. J. Psychiat. **131**, 545—550 (1974)

Greene,M.H., Brown,B.S., DuPont,R.L.: Illicit methadone abuse in Washington, D.C.: In: Senay,E., Shorty,V., Alksne,H. (Eds.): Developments in the field of drug abuse. Cambridge, Mass.: Schenkman Publishing Co., Inc. 1974

Haertzen,C.A.: Development of scales based on patterns of drug effects, using the Addiction Research Center Inventory (ARCI). Psychol. Rep. **18**, 163—194 (1966)

Haertzen,C.A.: Subjective effects of narcotic antagonists cyclazocine and nalorphine on the Addiction Research Center Inventory (ARCI). Psychopharmacologia (Berl.) **18**, 366—377 (1970)

Haertzen,C.A.: An overview of Addiction Research Center Inventory Scales (ARCI): an appendix and manual of scales. DHEW Publication No. (ADM) 74—92. Rockville, Maryland: National Institute on Drug Abuse 1974

Haertzen,C.A., Hill,H.E., Belleville,R.E.: Development of the Addiction Research Center Inventory (ARCI): Selection of items that are sensitive to the effects of various drugs. Psychopharmacologia (Berl.) **4**, 155—166 (1963)

Haertzen,C.A., Hooks,N.T., Jr.: Changes in personality and subjective experiences associated with the chronic administration and withdrawal of opiates. J. nerv. ment. Dis. **148**, 606—614 (1969)

Hill, H. E., Haertzen, C. A., Wolbach, A. B., Miner, E. J.: The Addiction Research Center Inventory. Standardization of scales which evaluate subjective effects of morphine, amphetamime, pentobarbital, alcohol, LSD-25, pyrahexyl and chlorpromazine. Psychopharmacologia (Berl.) **4**, 167—183 (1963a)

Hill, H. E., Haertzen, C. A., Wolbach, A. B., Miner, E. J.: The Addiction Research Center Inventory: I. Items comprising empirical scales for seven drugs. II. Items which do not differentiate placebo from any drug condition. Psychopharmacologia (Berl.) **4**, 184—205 (1963b)

Himmelsbach, C. K.: Addiction liability of codeine. J. Amer. med. Ass. **103**, 1420 (1934)

Himmelsbach, C. K.: Clinical studies of drug addiction. I. The absence of addiction liability in "Perparin". Publ. Hlth Rep. (Wash.), Suppl. **122** (1937a)

Himmelsbach, C. K.: Clinical studies of drug addiction. II. "Rossium" treatment of drug addiction. Publ. Hlth Rep. (Wash.), Suppl. **125** (1937b)

Himmelsbach, C. K.: Studies of certain addiction characteristics of (a) dihydromorphine ("paramorphan"), (b) dihydrodesoxymorphine-D ("desomorphine"), (c) dihydrodesoxycodeine-D ("desocodeine") and (d) methyldihydromorphinone ("metopon"). J. Pharmacol. exp. Ther. **67**, 239—249 (1939)

Himmelsbach, C. K.: Thiamine in the treatment of the morphine abstinence syndrome in man. J. Pharmacol. exp. Ther. **70**, 293—296 (1940)

Himmelsbach, C. K.: The effects of certain chemical changes on the addiction characteristics of drugs of the morphine, codeine series. J. Pharmacol. exp. Ther. **71**, 42—48 (1941)

Himmelsbach, C. K.: Studies of the addiction liability of "Demerol" (D-140). J. Pharmacol. exp. Ther. **75**, 64—68 (1942)

Himmelsbach, C. K.: Further studies of the addiction liability of Demerol (1-methyl-4-phenylpiperidine-4-carboxylic acid ethyl ester hydrochloride). J. Pharmacol. exp. Ther. **79**, 5—9 (1943)

Himmelsbach, C. K.: Treatment of the morphine abstinence syndrome with a synthetic cannabis-like compound. Sth. med. J. (Bgham, Ala.) **37**, 26—29 (1944)

Himmelsbach, C. K., Andrews, H. L.: Studies on the modification of the morphine abstinence syndrome by drugs. J. Pharmacol. exp. Ther. **77**, 17—23 (1943)

Himmelsbach, C. K., Andrews, H. L., Felix, R. H., Oberst, F. W., Davenport, L. F.: Studies on codeine addiction. Publ. Hlth Rep. (Wash.), Suppl. **158** (1940)

Himmelsbach, C. K., Eddy, N. B., Davenport, L. F.: Studies on drug addiction. Part III. Clinical studies of drug addiction, with special reference to opium derivatives and allied synthetic substances. Publ. Hlth Rep. (Wash.), Suppl. **138** (1938)

Himmelsbach, C. K., Oberst, F. W., Brown, R. R., Williams, E. G.: Studies of the influence of prostigmine on morphine addiction. J. Pharmacol. exp. Ther. **76**, 50—56 (1942)

Hoffmeister, F., Wuttke, W.: Self administration: Positive and negative reinforcing properties of morphine antagonists in Rhesus monkeys. In: Braude, M. C., Harris, L. S., May, E. L., Smith, J. P., Villarreal, J. E. (Eds.): Narcotic antagonists. Advances in biochemical psychopharmacology, Vol. 8. New York: Raven Press 1974

Houde, R. W., Wallenstein, S. L., Beaver, W. T.: Clinical measurement of pain. In: deStevens, G. (Ed.): Analgetics. New York: Academic Press 1965

Houde, R. W., Wallenstein, S. L., Rogers, A.: Analgesic studies in cancer patients: Tilidine, propiram, SU-19713B, levorphanol and nalbuphine. Presented at 36th meeting, Committee on Problems of Drug Dependence, National Research Council, Mexico City, Mexico 1974

Isbell, H.: Methods and results of studying experimental human addiction to the newer synthetic analgesics. Ann. N.Y. Acad. Sci. **51**, 108—122 (1948a)

Isbell, H.: Report on K-4710 (keto-bemidone, 10720). Presented at 3rd meeting, Committee on Drug Addiction and Narcotics, National Research Council, Washington, D.C. 1948b

Isbell, H.: Addiction liability of dihydrocodeinone (No. 154). Presented at 5th meeting, Committee on Drug Addiction and Narcotics, National Research Council, Washington, D.C. 1949a

Isbell, H.: Addiction liability of morphinan (NU-2206). Presented at 5th meeting, Committee on Drug Addiction and Narcotics, National Research Council, Washington, D.C. 1949b

Isbell, H.: The addiction liability of some derivatives of meperidine. J. Pharmacol. exp. Ther. **97**, 182—189 (1949c)

Isbell, H.: Addiction liability of the acetylmethadols. Presented at 7th meeting, Committee on Drug Addiction and Narcotics, National Research Council, Washington, D.C. 1951

Isbell, H.: Activities of the NIMH Addiction Research Center, USPHS Hospital, Lexington, Kentucky. I. Clinical studies of addiction; II. Clinical investigations of barbiturate addiction; III. Addiction liabilities of *l*- and *d*-Dromoran (3-hydroxy-N-methylmorphinan); IV. Addiction liability of *dl*-3-methocy-N-methylmorphinan (methyl ether of Dromoran, Ro-1-5470). Presented at 9th meeting, Committee on Drug Addiction and Narcotics, National Research Council, Washington, D.C. 1952a
Isbell, H.: Studies on N-allylnormorphine. Presented at 9th meeting, Committee on Drug Addiction and Narcotics, National Research Council, Washington, D.C., 1952b
Isbell, H.: Nalline—a specific narcotic antagonist; clinical and pharmacological observations. Merck Rep. **62**, 23—26 (1953a)
Isbell, H.: Studies on addiction liabilities of new agents and related work of the Research Division. Presented at 11th meeting, Committee on Drug Addiction and Narcotics, National Research Council, Lexington, Kentucky 1953b
Isbell, H.: Addiction liability of dihydrohydroxymorphinone (Numorphan or No. 5501). Presented at 14th meeting, Committee on Drug Addiction and Narcotics, National Research Council, Rensselaer, New York 1954a
Isbell, H.: Work of the NIMH Addiction Research Center, Public Health Service Hospital, Lexington, Kentucky. Presented at 13th meeting, Committee on Drug Addiction and Narcotics, National Research Council, Nutley, New Jersey 1954b
Isbell, H.: Withdrawal symptoms in "primary" meperidine addicts. Fed. Proc. **14**, 354 (1955a)
Isbell, H.: Work of the NIMH Addiction Research Center, U.S. Public Health Service Hospital, Lexington, Kentucky, for the calendar year 1954. Presented at 15th meeting, Committee on Drug Addiction and Narcotics, National Research Council, Lexington, Kentucky 1955b
Isbell, H.: Attempted addiction to nalorphine. Fed. Proc. **15**, 442 (1956)
Isbell, H.: Comparison of the reactions induced by psilocybin and LSD-25 in man. Psychopharmacologia (Berl.) **1**, 29—38 (1959)
Isbell, H., Belleville, R. E., Fraser, H. F., Wikler, A., Logan, C. R.: Studies on lysergic acid diethylamide (LSD-25). Effects in former morphine addicts and development of tolerance during chronic intoxication. Arch. Neurol. Psychiat. (Chic.) **76**, 468—478 (1956)
Isbell, H., Eisenman, A. J.: The addiction liability of some drugs of the methadone series and 6-methyldihydromorphine. Presented at 2nd meeting, Committee on Drug Addiction and Narcotics, Washington, D.C. 1948a
Isbell, H., Eisenman, A. J.: The addiction liability of some drugs of the methadone series. J. Pharmacol. exp. Ther. **93**, 305—312 (1948b)
Isbell, H., Eisenman, A. J., Wikler, A., Frank, K.: The effects of single doses of 6-dimethylamino-4-4-diphenyl-3-heptanone (amidone, methadon or 10820) on human subjects. J. Pharmacol. exp. Ther. **92**, 83—89 (1948a)
Isbell, H., Fraser, H. F.: Addiction to analgesics and barbiturates. J. Pharmacol. exp. Ther., Part II, **90**, 355—397 (1950)
Isbell, H., Fraser, H. F.: Actions and addiction liability of dithienylbutenylamines in man. J. Pharmacol. exp. Ther. **109**, 417—421 (1953a)
Isbell, H., Fraser, H. F.: Action and addiction liabilities of Dromoran derivatives in man. J. Pharmacol. exp. Ther. **107**, 524—530 (1953b)
Isbell, H., Fraser, H. F.: Addictive properties of methadone derivatives. Fed. Proc. **13**, 369 (1954)
Isbell, H., Fraser, H. F.: Addiction liability of 4,4-diphenyl-6-dimethylamino-hexanone-3. J. Pharmacol. exp. Ther. **113**, 29—30 (1955)
Isbell, H., Fraser, H. F.: Report on the work of the National Institute of Mental Health Addiction Research Center. Presented at 20th meeting, Committee on Drug Addiction and Narcotics, National Research Council, Washington, D.C. 1959
Isbell, H., Fraser, H. F., Wikler, A.: Work of the Research Division of the National Institute of Mental Health. Demonstrations. Presented at 11th meeting, Committee on Drug Addiction and Narcotics, National Research Council, Lexington, Kentucky 1953
Isbell, H., Miner, E. J., Logan, C. R.: Relationship of psychotomimetic to anti-serotonin potencies of congeners of lysergic acid diethylamide (LSD-25). Psychopharmacologia (Berl.) **1**, 20—28 (1959)
Isbell, H., Vogel, V. H.: The addiction liability of methadon (amidone, Dolophine, 10820) and its use in the treatment of the morphine abstinence syndrome. Amer. J. Psychiat. **105**, 909—914 (1949)

Isbell,H., Wikler,A., Eddy,N.B., Wilson,J.H., Moran,C.F.: Tolerance and addiction liability of 6-dimethylamino-4-4-diphenyl-heptanone-3 (methadon). J. Amer. med. Ass. **135**, 888—894 (1947)

Isbell,H., Wikler,A., Eisenman,A.J., Frank,K., Daingerfield,M.: Liability of addiction to 6-dimethylamino-4,4-diphenyl-3-heptanone (methadon, amidone or 1082) in man. Arch. intern. Med. **82**, 362—392 (1948b)

Jasinski,D.R., Griffith,J.D., Carr,C.B.: Etorphine in man. I. Subjective effects and suppression of morphine abstinence. Clin. Pharmacol. Ther. **17**, 267—272 (1975a)

Jasinski,D.R., Griffith,J.D., Carr,C.B., Gorodetzky,C.W., Kullberg,M.P.: Progress report from the Clinical Pharmacology Section of the Addiction Research Center. Presented at 36th Meeting, Committee on Problems of Drug Dependence, National Research Council, Mexico City, Mexico 1974

Jasinski,D.R., Griffith,J.D., Pevnick,J.S., Clark,S.C.: Progress report on studies from the Clinical Pharmacology Section of the Addiction Research Center. Presented at 37th Meeting, Committee on Problems of Drug Dependence, National Research Council, Washington, D.C. 1975b

Jasinski,D.R., Mansky,P.: The subjective effects of GPA-2087 and nalbuphine (EN-2234A). Presented at 32nd meeting, Committee on Problems of Drug Dependence, National Research Council, Washington, D.C. 1970

Jasinski,D.R., Mansky,P.A.: Evaluation of nalbuphine for abuse potential. Clin. Pharmacol. Ther. **13**, 78—90 (1972)

Jasinski,D.R., Martin,W.R.: Assessment of the dependence-producing properties of dihydrocodeinone and codoxime. Clin. Pharmacol. Ther. **8**, 266—270 (1967a)

Jasinski,D.R., Martin,W.R.: Evaluation of a new photographic method for assessing pupil diameters. Clin. Pharmacol. Ther. **8**, 271—272 (1967b)

Jasinski,D.R., Martin,W.R., Haertzen,C.A.: The human pharmacology and abuse potential of N-allynoroxymorphone (naloxone). Presented at 29th meeting, Committee on Problems of Drug Dependence, National Research Council, Lexington, Ky. 1967a

Jasinski,D.R., Martin,W.R., Haertzen,C.A.: The human pharmacology and abuse potential of N-allylnoroxymorphone (naloxone). J. Pharmacol. exp. Ther. **157**, 420—426 (1967b)

Jasinski,D.R., Martin,W.R., Hoeldtke,R.: Progress report on the abuse potential of weak narcotic antagonists. Presented at 31st meeting, Committee on Problems of Drug Dependence, National Research Council, Palo Alto, Calif. 1969

Jasinski,D.R., Martin,W.R., Hoeldtke,R.D.: Effects of short- and long-term administration of pentazocine in man. Clin. Pharmacol. Ther. **11**, 385—403 (1970)

Jasinski,D.R., Martin,W.R., Hoeldtke,R.D.: Studies of the dependence-producing properties of GPA-1657, profadol and propiram in man. Clin. Pharmacol. Ther. **12**, 613—649 (1971a)

Jasinski,D.R., Martin,W.R., Mansky,P.A.: Progress report on the assessment of the antagonists nalbuphine and GPA-2087 for abuse potential and studies of the effects of dextromethorphan in man. Presented at 33rd Meeting, Committee on Problems of Drug Dependence, National Research Council, Toronto, Ontario, Canada 1971b

Jasinski,D.R., Martin,W.R., Sapira,J.D.: Antagonism of the subjective, behavioral, pupillary and respiratory depressant effects of cyclazocine by naloxone. Clin. Pharmacol. Ther. **9**, 215—222 (1968a)

Jasinski,D.R., Martin,W.R., Sapira,J.D.: Progress report on the dependence-producing properties of GPA-1657, profadol hydrochloride (CI-572), propiram fumarate (BAY-4503) and dexoxadrol. Presented at 30th meeting, Committee on Problems of Drug Dependence, National Research Council, Indianapolis, Ind. 1968b

Jasinski,D.R., Nutt,J.G.: Progress report on the assessment program of the NIMH Addiction Research Center. Presented at 34th meeting, Committee on Problems of Drug Dependence, National Research Council, Washington, D.C. 1972

Jasinski,D.R., Nutt,J.G.: Progress report on the clinical assessment program of the Addiction Research Center. Presented at 35th meeting, Committee on Problems of Drug Dependence, National Research Council, Chapel Hill, North Carolina 1973

Jasinski,D.R., Nutt,J.G., Carr,C.B.: Evaluation in man of the effects of a mixture of morphine and *d*-amphetamine (MA). Fed. Proc. **31**, 530 (1972)

Jones,B.E., Martin,W.R., Isbell,H., Fraser,H.F.: Evaluation of a photographic method of estimating pupil diameter in man. Fed. Proc. **21**, 326 (1962)

Kay, D. C., Gorodetzky, C. W., Martin, W. R.: Comparative effects of codeine and morphine in man. J. Pharmacol. exp. Ther. **156**, 101—106 (1967)
Keats, A. S., Telford, J.: Nalorphine, a potent analgesic in man. J. Pharmacol. exp. Ther. **117**, 190—199 (1956)
King, M. R., Himmelsbach, C. K., Sanders, R. S.: Dilaudid (dihydromorphinone). A review of the literature and a study of its addictive properties. Publ. Hlth Rep., Suppl. 113 (1935)
Kolb, L., Himmelsbach, C. K.: Clinical studies of drug addiction. III. A critical review of the withdrawal treatments with method of evaluating abstinence syndromes. Amer. J. Psychiat. **94**, 759—797 (1938)
Kosterlitz, H. W., Waterfield, A. A., Berthoud, V.: Assessment of the agonist and antagonist properties of narcotic analgesic drugs by their actions on the morphine receptor in the guinea pig ileum. In: Braude, L. S., Harris, L. S., May, E. L., Smith, J. P., Villarreal, J. E. (Eds.): Narcotic antagonists. Advances in biochemical psychopharmacology, Vol. 8. New York: Raven Press 1974
Lasagna, L., Beecher, H. K.: The analgesic effectiveness of nalorphine and nalorphine-morphine combinations in man. J. Pharmacol. exp. Ther. **112**, 356—363 (1954)
Mansky, P. A., Jasinski, D. R.: Effects of dextromethorphan (D) in man. Pharmacologist **12**, 231 (1970)
Marquardt, W. G., Martin, W. R., Jasinski, D. R.: The use of the Polaroid CU camera in pupillography. Int. J. Addict. **2**, 301—304 (1967)
Martin, W. R.: Strong analgesics. In: Root, W. S., Hofman, E. G. (Eds.): Physiological pharmacology, Vol. 1. New York: Academic Press 1963
Martin, W. R.: Assessment of the dependence-producing potentiality of narcotic analgesics. In: Radouco-Thomas, C., Lasagna, L. (Eds.): International encyclopedia of pharmacology and therapeutics. Glasgow: Pergamon Press 1966
Martin, W. R.: Clinical evaluation for narcotic dependence. In: Way, E. L. (Ed.): New concepts in pain and its clinical management. Philadelphia: F. A. Davis Co. 1967a
Martin, W. R.: Opioid antagonists. Pharmacol. Rev. **19**, 463—521 (1967b)
Martin, W. R.: Pharmacological redundancy as an adaptive mechanism in the central nervous system. Fed. Proc. **29**, 13—18 (1970)
Martin, W. R., Eades, C. G., Thompson, W. O., Thompson, J. A., Flanary, H. G.: Morphine physical dependence in the dog. J. Pharmacol. exp. Ther. **189**, 759—771 (1974a)
Martin, W. R., Fraser, H. F.: A comparative study of physiological and subjective effects of heroin and morphine administered intravenously in postaddicts. J. Pharmacol. exp. Ther. **133**, 388—399 (1961)
Martin, W. R., Fraser, H. F., Gorodetzky, C. W., Rosenberg, D. E.: Additional studies on the addictiveness of 2-cyclopropyl methyl-2′-hydroxy-5,9-dimethyl-6,7-benzomorphan (Win 20,740, ARC-II-C-3). Presented at 26th meeting, Committee on Drug Addiction and Narcotics, National Research Council, Washington, D.C. 1964
Martin, W. R., Fraser, H. F., Gorodetzky, C. W., Rosenberg, D. E.: Studies of the dependence-producing potential of the narcotic antagonist 2-cyclopropylmethyl-2′-hydroxy-5,9-dimethyl-6,7-benzomorphan (cyclazocine, Win 20,740; ARC-II-C-3). J. Pharmacol. exp. Ther. **150**, 426—436 (1965)
Martin, W. R., Gorodetzky, C. W.: A preliminary report on the morphine-like properties and physical dependence liability of alpha-*d*-acetoxy-l, 2-diphenyl-3-methyl-4-pyrrolidinobutane hydrochloride (Lilly -31,518, NIH-7662, ARC-I-C-27). Presented at 27th meeting, Committee on Problems of Drug Dependence, National Research Council, Houston, Texas 1965a
Martin, W. R., Gorodetzky, C. W.: Demonstration of tolerance to and physical dependence on N-allylnormorphine (nalorphine). Presented at 27th meeting, Committee on Problems of Drug Dependence, National Research Council, Houston, Texas 1965b
Martin, W. R., Gorodetzky, C. W.: Demonstration of tolerance to and physical dependence on N-allynormorphine (nalorphine). J. Pharmacol. exp. Ther. **150**, 437—442 (1965c)
Martin, W. R., Gorodetzky, C. W., Kay, D. C., McClane, T. K., Jasinski, D. R.: Activities of the Addiction Research Center during 1965. Presented at 28th meeting, Committee on Problems of Drug Dependence, National Research Council, New York, N.Y. 1966
Martin, W. R., Jasinski, D. R.: Physiological parameters of morphine dependence in man—tolerance, early abstinence, protracted abstinence. J. psychiat. Res. **7**, 9—17 (1969)

Martin, W.R., Jasinski, D.R., Haertzen, C.A., Kay, D.C., Jones, B.E., Mansky, P.A., Carpenter, R.W.: Methadone—a reevaluation. Arch. gen. Psychiat. **28**, 286—295 (1973a)

Martin, W.R., Jasinski, D.R., Mansky, P.A.: Characteristics of the blocking effects of EN-1639A (N-cyclopropylmethyl-7,8-dihydro-14-hydroxynormorphinone HCl). Presented at 33rd meeting, Committee on Problems of Drug Dependence, National Research Council, Toronto, Ontario, Canada 1971

Martin, W.R., Jasinski, D.R., Mansky, P.A.: Naltrexone, an antagonist for the treatment of heroin dependence. Arch. gen. Psychiat. **28**, 784—791 (1973b)

Martin, W.R., Jasinski, D.R., Sapira, J.D.: Progress report on the assessment of the ability of GPA-1657 to produce drug dependence of morphine type in man. Presented at 29th meeting, Committee on Problems of Drug Dependence, National Research Council, Lexington, Ky. 1967

Martin, W.R., Jasinski, D.R., Sapira, J.D., Flanary, H.G., Kelly, O.A., Thompson, A.K., Logan, C.R.: The respiratory effects of morphine during a cycle of dependence. J. Pharmacol. exp. Ther. **162**, 182—189 (1968)

Martin, W.R., Thompson, W.O., Fraser, H.F.: Comparison of graded single intramuscular doses of morphine and pentobarbital in man. Clin. Pharmacol. Ther. **15**, 623—630 (1974b)

McClane, T.K., Martin, W.R.: Subjective and physiologic effects of morphine, pentobarbital, and meprobamate. Clin. Pharmacol. Ther. **20**, 192—198 (1976)

National Institute on Drug Abuse: Report on drug abuse treatment and rehabilitation services. For the CODAP National Management Central Administrative Unit. Rockville, Maryland: NIDA, August 1974

Nutt, J.G., Jasinski, D.R.: Comparison of intravenously administered methadone (ME), morphine (MO), heroin (H) and placebo (P). Fed. Proc. **32**, 694 (1973)

Nutt, J.G., Jasinski, D.R.: Methadone-naloxone mixtures for use in methadone maintenance programs, I. An evaluation in man of their pharmacological feasibility. II. Demonstration of acute physical dependence. Clin. Pharmacol. Ther. **15**, 156—166 (1974)

Pachter, I.J.: Synthetic 14-hydroxymorphinan narcotic antagonists. In: Braude, M.C., Harris, L.S., May, E.L., Smith, J.P., Villarreal, J.E. (Eds.): Narcotic antagonists. Advances in biochemical psychopharmacology, Vol. 8. New York: Raven Press 1974

Pierson, A.K.: Assays for narcotic antagonist activity in rodents. In: Braude, M.C., Harris, L.S., May, E.L., Smith, J.P., Villarreal, J.E. (Eds.): Narcotic antagonists. Advances in biochemical psychopharmacology, Vol. 8. New York: Raven Press 1974

Rasor, R.W., Crecraft, H.J.: Addiction to meperidine (Demerol) hydrochloride. J. Amer. med. Ass. **157**, 654—657 (1955)

Sapira, J.D., Ball, J.C., Cottrell, E.S.: Addiction to methadone among patients at Lexington and Fort Worth. Publ. Hlth Rep. (Wash.) **83**, 691—694 (1968)

Schrappe, O.: "Physical dependence" nach chronischer Verabreichung von n-Allylnormorphin. Drug Res. **9**, 130—132 (1959)

Seitner, P.G., Martin, B.C., Cochin, J., Harris, L.: Survey of analgesic drug prescribing patterns. Washington, D.C.: Drug Abuse Council, Inc. 1975

Simpson, D.: Pretreatment drug use by patients entering drug treatment programs during 1971—1973. Texas Christian University Institute of Behavioral Research Report No. 74-5, 1974

Snedecor, G.W., Cochran, W.G.: Statistical Methods. Ames, Iowa: Iowa State University Press 1973

Stephens, R.C., Weppner, R.S.: Legal and illegal use of methadone: One year later. Amer. J. Psychiat. **130**, 1391—1394 (1973)

Sunshine, L.E., Slafta, J., Fleischman, E.: A comparative analgesia study of propoxyphene hydrochloride, propoxyphene napsylate and placebo. Toxicol. appl. Pharmacol. **19**, 512—518 (1971)

Swanson, D.W., Weddige, R.L., Morse, R.M.: Abuse of prescription drugs. Mayo Clin. Proc. **48**, 359—367 (1973)

U.S. Department of Justice, Drug Enforcement Administration: Drug Abuse Warning Network, Phase II Report, July 1973—March 1974. Washington, D.C.: U.S. Government Printing Office 1974

Villarreal, J. E., Karbowski, M. G.: The actions of narcotic antagonists in morphine-dependent Rhesus monkeys. In: Braude, M. D., Harris, L. S., May, E. L., Smith, J. P., Villarreal, J. E. (Eds.): Narcotic antagonists. Advances in biochemical psychopharmacology, Vol. 8. New York: Raven Press 1974

Wallenstein, S. L., Houde, R. W., Beaver, W. T.: Analgesic studies of orally and parenterally administered morphine and codeine in patients with cancer. Fed. Proc. **26**, 742 (1967)

Weppner, R. S.: "Cheap Kicks": Codeine cough syrup abusers and some of their social characteristics. Int. J. Addict. **6**, 647—660 (1971)

Wikler, A., Fraser, H. F., Isbell, H.: N-allylnormorphine: Effects of single doses and precipitation of acute "abstinence syndromes" during addiction to morphine, methadone or heroin in man (post addicts). J. Pharm. exp. Ther. **109**, 8—20 (1953)

Wolbach, A. B., Fraser, H. F.: Addiction liability of ethyl 1-(2 Carbamyl-ethyl)-4-phenylpiperidine-4-carboxylate hydrochloride. Presented at the 24th meeting, Committee on Drug Addiction and Narcotics, National Research Council, New York, New York 1962

Wolbach, A. B., Fraser, H. F.: Addiction liability of I-C-26 (dextro-3-dimethylamino-1,1-diphenylbutyl ethyl sulfone hydrochloride) and I-D-20 (ethyl 1-(2-carbamethyl)-4-phenylpiperidine-4-carboxylate hydrochloride. Bull. Narcot. **14**, 25—28 (1963)

Wolff, P.: The significance of codeine as a habit-forming drug. Bull. Hlth. Org. L.o.N. **7**, 546—580 (1938)

Yanagita, T.: An experimental framework for evaluation of dependence liability of various types of drugs in monkeys. Bull. Narcot. **25**, 57—64 (1973)

CHAPTER 4

Psychiatric Treatment of Narcotic Addiction

H. T. CONRAD

A. Introduction

The pendulum of drug addiction treatment has yet to come to rest. It has swung from the punitive position of imprisonment to more humanistic forms. Civil court commitment now competes with prison sentences. If we can imagine a spectrum of different treatment modalities, we can better deal with the reality of different diagnostic classifications of drug addicts (AUSUBEL, 1948). We cannot dismiss the argument of different etiology and cultural background in individual addicts, which serves to militate against fitting all into the Procrustean bed of psychoanalysis.

In recent years, some professionals have observed a sense of despair and retreat from life's stresses and problem-solving under *affluence*, as well as under poverty, among the young drug seekers today (COHEN, 1969). In view of the epidemic nature of drug addiction among the noncompeting young, many have felt a sense of urgency to develop treatment methods applicable to large numbers of persons.

Psychiatric treatment of the narcotic addict seems to have evolved from a unilateral psychoanalytic approach to a multimodal one. The traditional psychoanalytic approach has failed to demonstrate its efficacy or its applicability to typical drug abusers. Some who abuse drugs respond positively to each of several different treatment models. Therefore, it appears at this time that the availability in a comprehensive program of multiple approaches, a spectrum of overlapping modalities, provides the greatest likelihood of effectiveness with the range of personality variables seen among drug-dependent persons.

Lay persons tend to expect successful treatment of the drug abuser to be absolute. If the addict recurrently relapses, he is regarded as having failed. Instead of counting productive, drug-free periods as partial successes, intermittent periods of relapse are considered as tangible evidence of failure.

Professional persons, on the other hand, tend to define successful treatment in terms of the achievement of socially oriented goals. But there is lack of agreement either as to what goals should serve as criteria, or as to how soon, or after how long, or how often, or how many times, the goals may be attained with success still being acknowledged. Some professional workers are rigid in their requirements; others are extremely flexible and liberal. Since complete abstinence from the use of narcotics is rare at this time, some researchers define "success" as reduced frequency of use, use under medical control or supervision, decreased criminality, or increased employment.

With many forms of illness, the physician determines what constitutes appropriate medical treatment. He may also decide to establish lesser goals than trying to attempt a cure. This has not always been so in the treatment of narcotic addiction. In this field of mental health, legal restrictions are imposed concerning treatment so that, for example, if the physician decides that a feasible goal would entail less than complete abstinence, he may be in difficulty with legal authorities. FREEDMAN and SHAROFF (1965) contend that rethinking is needed, so that relative results may be acceptable:

> The first step is to revise our own thinking in terms of goals. We must think as realistically about goals in the treatment of narcotic addiction as we do in any other chronic, relapsing illness. Cure or total abstinence will always be our ultimate goal; but until we are really in a position to achieve it, we must set lesser goals for ourselves and our patients.

B. Psychoanalytic Treatment

In the 1930s, a period that many regard as the "Golden Age" of psychoanalysis, there were a number of papers written by the great pioneers in the field elucidating the dynamics and psychopathology of addiction. Addicts were regarded as impulse-ridden individuals who attempted to use drugs in order to achieve the satisfaction of primitive inner needs. There was some difference of opinion as to the nature of these primitive inner needs but many felt that they were archaic longings of an oral nature, and combined elements of sexuality, security, and self-esteem, simultaneously (FENICHEL, 1945).

It was felt that there was a factor in the premorbid personality, which some regarded as genetically transmitted, which made the individual attribute such overwhelming significance to the effects achieved from the use of drugs. This premorbid personality was usually described as a passive one with large elements of orality and narcissism. The typical inability of the addict to delay gratification or to tolerate physical or psychic pain was observed.

There was disagreement as to whether modifications of standard psychoanalytic technique were necessary in order to treat persons who were addicted. It was generally agreed, however, that these patients were as difficult to treat as those suffering from perversions, because the "illness" was basically one which gave the patient pleasure rather than pain. The observation was also made that it was relatively easy to effect withdrawal from the drug, but difficult to prevent relapse. FENICHEL (1945) summarized this brilliantly in his single statement, "it is not the chemical effect of the drug that must be combatted but the morbid wish to be drunkenly euphoric."

FENICHEL advocated psychoanalysis only in an institutional setting, feeling that ambulatory treatment of an addict would be pointless. He also made the interesting observation that there were probably some cases of addiction where the use of drugs serves as a retreat from environmental or sociologic conditions which where overwhelming or unbearable. He wisely observed that treatment in such cases would be pointless as long as these environmental conditions continued unchanged, and might become unnecessary if they were corrected. Interestingly enough, this same point of view is often expressed by contemporary psychotherapists, who have come to realize that not only the internal psychological problems of the addict have to be dealt with, but that his entire lifestyle has to be modified, including at times very direct environmental manipulation.

In a 1933 article RADO described addiction as a narcissistic disorder involving an artificial and autoerotic sexual organization modeled on infantile masturbation. The phenomenon of addiction was based on this conceptual scheme:

> The ingestion of drugs, it is well known, in infantile archaic thinking represents an oral insemination; planning to die from poisoning is a cover for the wish to become pregnant in this fashion ... The wish to be pregnant is a mute appeal to the function of reproduction, to "divine Eros" to testify to the immortality of the ego. [By permission of Grune & Stratton, Inc. Publishers (see references).]

As stated by ROSENFELD (1960), most of the detailed case studies of addicts treated by psychoanalysis appeared in the literature before 1939. The reason is presumably that this method of treatment is usually impracticable with addicts, or rather that addicts are usually neither willing nor able to meet the constraints imposed by analysis. ROSENFELD (1960) states frankly that his own psychoanalytic investigation of addicts was limited to a few patients, yet he was of the opinion as late as 1960 that psychoanalysis could be effective with this population. Moreover, he stated, he found it unnecessary to modify his customary psychoanalytic approach. He felt that orality and very early sadistic impulses formed a psychopathologic basis for drug abuse. Addiction, he believed, was compounded by manic and depressive mechanisms which were reinforced and consequently altered pharmacologically. He added, "drugging often has a depressive meaning also, the drug symbolizing a dead or ill object which the patient feels compelled to incorporate out of guilt" (ROSENFELD, 1960).

Some who claim success with the analytic approach do not describe cases of the young, group-inducted addicts with signs of character disorder; they refer instead to isolated individuals trying to cope with marital, educational, or success hang-ups: people who abuse drugs and still try to maintain some responsibility and a facade of respectability.

The author feels that the cases of addicts treated analytically and reported in the literature were atypical of addicts generally. Regardless of whether or not one accepts analytic findings and attested therapeutic dynamics, the process of psychoanalysis is simply too time-consuming, too expensive, and too limited in terms of the proportion of addicts amenable to this genre of treatment to be practicable today.

C. Institutional Treatment

From 1935 until 1967, the United States Public Health Service operated two hospitals for the treatment of narcotic addicts. For many years the hospitals in Lexington, Kentucky and Fort Worth, Texas were the only institutions attempting to deal therapeutically with the problems of addiction. It was in these centers that institutional treatment for narcotic addicts was primarily developed. These hospitals treated prisoner patients, who were usually sentenced to serve relatively long periods of time, and voluntary patients who had no direct legal constraints on them at all. There were often indirect constraints on the voluntary patients in the sense that local authorities had requested them to "voluntarily" seek treatment or face criminal charges.

The experience with the voluntary patients was largely unsatisfactory, inasmuch as three-fourths of them left the hospital against medical advice prior to the completion of the recommended treatment period, which was usually from 4 to 6 months. In the case of both prisoners and voluntary patients there was no statutory authority for follow-up care after leaving the institution. Some prisoners were released on parole, where a follow-up of sorts was available.

Because the hospitals were chronically underfunded and understaffed, and because the patient census was usually high, "treatment" was often no more than the passage of time in an environment that was relatively drug-free. The environment was maintained in a drug-free condition, however, by architectural and physical constraints which made the hospital a prison for all practical purposes. Many dedicated staff members made real efforts at providing individual and group psychotherapy, along with vocational counseling and opportunities for continuing education, but the patients tended to band together in adherence to the "street code" and there was a distinct dichotomy between the patients and the staff.

A large security force was maintained, and in many day-to-day activities they exercised more control over the addict patients' program of rehabilitation than did the professional treatment staff. This, coupled with an extensive system of locks, bars, grilles, and all of the other trappings of a penal institution, made the hospital much more of a prison than most hospitals, although more of a hospital than most prisons.

Despite the lack of postinstitutional care and evaluation of formal follow-up of patients released from Lexington and Fort Worth prior to 1967, a number of studies were done in an effort to provide follow-up data (PESCOR, 1943; BALL and PABON, 1965; VAILLANT, 1966a, b, c, d; O'DONNELL, 1969). Many of these studies had real methodologic problems, but the three that were done most carefully involved follow-up of patients from Puerto Rico (BALL and PABON, 1965), Kentucky (O'DONNELL, 1969), and New York City (VAILLANT, 1966a, b, c, d). Although all of these studies showed that most patients tended to relapse quickly to the use of narcotics soon after leaving the hospital, they also showed that the abstinence rate was roughly proportional to the passage of time from discharge. That is, it would appear that somewhere between 20 and 40% of the patients who had been to Lexington in the old days became permanently abstinent by the age of 40.

VAILLANT conducted a 12-year follow-up of 100 New York City addicts first admitted to the Public Health Service Hospital in Lexington between August 1952 and January 1953. By 1962, 41% were known to be off drugs and living in the community. Moreover, at least 30% had achieved good social adjustment. Though punishment in itself does not seem to benefit the addict, it was found that periods of abstinence correlated with enforced periods of compulsory supervision, e.g., conditions of parole following a court-imposed sentence. However, addicts were less successful after voluntary treatment. Thus though 67% of addicts treated under long sentences with parole succeeded, 96% of addicts treated by voluntary hospitalization failed (VAILLANT, 1966a).

As a psychiatrist, VAILLANT stresses the need for consistent concern with authority. As for psychotherapy, he feels that it should be directed toward alleviating or correcting unrewarding behavior. The therapist may represent an ally who cares when the addict is honest and independent, and who serves as an external superego to back up shaky impulse control (VAILLANT, 1966a).

A professional person with whom the addict identifies is in a favored position to serve *in loco parentis* because he may have the objectivity and emotional detachment that the real parents were unable to muster. A common characteristic among VAILLANT's population was lack of adequate identification with the adult role. It is interesting to note that all of the 30 addicts abstinent 12 years after hospitalization (and who had been abstinent for at least 3 years at the time of the study) had been living apart from their parents when they achieved abstinence. When questioned, none felt that his parents had helped him to achieve this goal (VAILLANT, 1966c).

Addicts often assume a facade of unconcerned cheerfulness. However, this may mask marked underlying depression. The therapist should be aware that depression may exist, even when it does not show. Even though an additional 2% of the addicts studied by VAILLANT achieved extended abstinence annually, another 1% was lost by attrition, either by death or by being institutionalized. Since the death rate of the sample of addicts studied was roughly 3 times that expected of males of similar age, they may be regarded by some as either consciously or unconsciously suicidal. VAILLANT states: "By the age of forty, a quarter of the addicts in this study had died directly or indirectly as a result of their own acts, and what mental hospitalization they received was often for attempted suicide" (VAILLANT, 1966b).

The most accurate predictor of chronic recidivism among the New York City addicts followed by VAILLANT was a poor prior work history. New and meaningful relationships, on the other hand, seemed of importance among those addicts with sustained abstinence. Meaningful work and an accepting and appreciative employer were commonly found among those who gave up addiction. It would seem then that job counseling, placement services, and reinforcing support of steady vocational efforts might be helpful to the addict during his recovery period. It seems reasonable to expect that improved methods and services both in the initial treatment facility and in aftercare should raise the abstinence figures cited by VAILLANT (1966d).

VAILLANT and RASOR (1966) contend that "initial abstinence must be coerced" for the reason that addicts are commonly motivated to remain addicted. These coauthors incisively commented that "addicts with imprisonment and parole did five times as well as the addict with flat sentences."

Recently, VAILLANT (1971) has reported that today 44% of his New York addicts previously treated at Lexington have achieved total abstinence for 5 years or more; 10% are on methadone, so do not meet the criterion of being drug-free, or their status is unknown; 25% are institutionalized; and 21% are dead.

If an addict remained abstinent for 1 year, the chances of relapsing or remaining drug-free were essentially equal. It was found, however, that the addict who remained abstinent for 3 or 4 years usually remained drug-free (VAILLANT, 1971).

Historically, then, institutional treatment for narcotic addiction was primarily developed at the Lexington and Fort Worth Public Health Service Hospitals. Continuity of hospital and posthospital care was urged by the professional staff at Lexington and Fort Worth but legislative authority was absent until the passage of the Narcotic Addict Rehabilitation Act of 1966.

Efforts were made to provide continuity of hospital and posthospital care in 1952 at Riverside Hospital and in 1959 at Metropolitan Hospital, both in New York City. Also, in 1966 the Fort Worth Public Health Service Hospital developed with 14 community agencies in San Antonio a carefully designed array of hospital and community services for San Antonio addicts. This was done on a research basis only. The community services were only intermittently utilized during crises by the addicts to whom they were made available. However, in neither the New York programs nor the Fort Worth-San Antonio continua were results significantly better than institutional care alone (MADDUX et al., 1971). It is to Maddux' credit that he conducted one of the few carefully designed and controlled studies in the field of treatment of narcotic addiction. Most studies have been little more than anecdotal or program description studies, and in very few of them have there been any efforts made to provide a control group for comparative purposes.

In retrospect, it can be seen that the Lexington and Fort Worth Hospitals were struggling alone for many years with a problem of great magnitude. Beginning in the 1950s these hospitals made efforts to implement group psychotherapy in treating the drug addict (OSBERG and BERLINER, 1956). At that time professionals tended to be perceived by addict patients as individuals who could be fooled into believing the addicts' verbalizations, no matter where they led him or by what digressive maneuvers. AUSUBEL (1948) felt that group psychotherapy with narcotic addicts was impractical. Because of loyalty among group members to one another, excuses for abusing drugs would tend to receive support from the other members, AUSUBEL contended, with the result that sessions would be "usually transformed into group rationalization sessions."

However, FORT (1955) and other skillful therapists were able to provide leadership through several stages of group existence until the group members took over the leadership in progressive ways. Defenses of the members gradually broke down through pressures from their peers. This was one harbinger of the use of peer groups as an additional weapon in the therapeutic armamentarium.

D. Community Treatment Efforts

I. Psychiatrically Oriented Programs

In New York City, Riverside Hospital, a coeducational rehabilitation center for adolescents was established, and the assumption was made that once a patient was cared for in this hospital he would be socially rehabilitated and free of his addiction to drugs. An evaluation of the program at Riverside Hospital in the mid-1950s showed this assumption to be incorrect: "Of 248 patients whose progress was followed for 3 years after their first admission, 11 died (1.5% per annum), a rate not greatly different from other treatment programs for this age group; however, only 8 had not gone back to heroin, a low rate. A retrospective review of the medical records of the 8 who were 'cured' substantiated these patients' claims that they never had been addicted in the first place" (TRUSSELL, 1971).

In 1959 there was established at the New York Medical College-Metropolitan Hospital Center in New York City a voluntary treatment facility for the study of narcotic addiction. An integral aspect of the therapeutic program was continued involvement in the addict's own community (FREEDMAN and SHAROFF, 1965). This practice anticipated the financial encouragement that was to be provided during the years following 1966 by the National Institute of Mental Health for the development of state and local agencies designed to treat addicts. The adjunctive, supportive services offered through the treatment facility of the Metropolitan Hospital Center included vocational rehabilitation and evaluation of education needs. Patients were in many instances encouraged to return to or to continue in school.

Addicts generally are reluctant to postpone what they feel they immediately require. Thus when they requested treatment and beds were not immediately available to them at the Metropolitan Hospital Center it was found that as many as 50% of the applicants disappeared. The staff at this center found that their greatest problem was in getting the addict to accept what it was they had to offer. They found that attendance by the patients was highly irregular and that after discharge from the period of hospitalization only about 40% of the patients kept their outpatient appointments. Even those who kept their appointments kept them irregularly and often at times other than the scheduled ones. The overall impression of the staff of this center was that their results were not satisfactory.

The Washington Heights Rehabilitation Center conducted a 5-year program in the treatment of narcotic addiction. Sponsored jointly by the National Institute of Mental Health, the New York City Department of Health, and the New York City Community Mental Health Board, the program anticipated some of the features which now characterize civil commitment programs. Component elements included the following: (1) The 95 cases treated at the center were committed to participating in the program by reason of their being probationers who had been convicted by the court of either a criminal or a drug-related act. (2) At a meeting of the principals involved in his case, the violator was reminded of the conditions of probation, and of the consequences of failure to adhere to these conditions. (3) The addict was further informed that he was expected to conform to standards of behavior, and that failure to do so would not be tolerated. The standards included abstinence from drug use and from antisocial or criminal behavior (LIEBERMAN and BRILL, 1968). Personal and professional authority were employed to back a graduated series of sanctions. Built into the program were provisions for evaluation and research. It was felt clinically that the structuring of relationships, the setting of limits, and the direct confrontation with realities and societal requirements could be of specific benefit in treating narcotic addicts.

Unfortunately, sustained data were never produced to substantiate the following conclusions: "Preliminary analysis of our data indicates that the number of those who reverted back to their old negative adaptations is very low. The majority of patients are working or in school, or both, at the present time. We feel quite certain that our early impressions will be sustained through the final evaluation of the programme" (LIEBERMAN and BRILL, 1968).

II. Synanon

The first community effort in treating narcotic addiction that was developed and structured entirely without professional staff was the Synanon program in California in 1958. Synanon was created by CHARLES E. DEDERICH, a former business executive who founded the organization after working through problems which had led to the symptom of alcoholism. After the arrest of his alcoholism, he was capable and motivated to transmit to others the forces which had led to his becoming a teetotaler. The operation began with a small cadre of alcoholics and addicts who decided to join forces with him in starting what has developed into a powerful social movement.

Synanon now has major residential facilities in Santa Monica, Oakland, and Tomales Bay, California. The narcotic addict on entering Synanon is informed that he is not sick but simply stupid, and that he must give up all elements of status and prestige which he possessed on the outside and accept the fact that he is an irresponsible baby who must learn all over again how to live his life. Out of this process total involvement develops in a new social role where there is great emphasis on growth, the formation of a new identity, and the seeking and assuming of responsibilities. In addition to the total community which Synanon represents for those who live within it, Synanon uses a technique known as the "game." This is a form of leaderless group confrontation, which has been referred to as "human sport." A good deal of verbal hostility is vented during these games but there are strict prohibitions against violence or threats of violence. The Synanon game is interesting when thought of in conjunction with the ethologic concern with human intraspecies aggression. A pressing need exists for man to find more effective sublimations of his irrational aggressions, or as the ethologists would state it, aggression directed to the point of a nonfatal outcome.

A typical narcotic addict handles aggression poorly; he fears his own capacity to destroy and thus will avoid and deny even tolerable aggression. This frequently promotes the chronic low grade depression observed in addicts. In the Synanon games the addict learns, through the protection of rules to titrate sadistic aggression; to aggress in a way that says to another addict "I care enough to pull your covers." Thus in Synanon and in all other self-help movements modeled after it, the street code or the criminal code (that one never reveals the deficiencies or illegal activities of one's peers) is rejected vehemently. Thus, exposing the inadequacies or illegal activities of a fellow addict in a Synanon game is regarded as an expression of concern, and is highly praised as such.

The content of the games deals with present reality and the addict's degree of responsible behavior within the social system of Synanon. The objective is to achieve emotional engagement by concentrating on each individual and drawing him into a maelstrom of animated discussion by whatever means will serve the purpose; thus there is the use of "engrossments" which means exaggeration of petty or imagined grievances employed to provoke emotion. The question of the validity of this approach seems to be irrelevant in view of the very wide target presented by "dope fiend behavior" which is under attack. The sum total effect of this gamed aggression is to modulate depression so the residents can continue to function within the system without the use of drugs. The critical factor that seems to produce positive results in this self-help model, correctly pursued, lies not simply in the various group sessions but in the territorial awareness for the residents that this operation is their own. The thrust of this territoriality defeats the typical state of failure, despair, and impotency that enmeshes the addict personality. Even more than the professionally directed therapeutic community which utilizes self-help principles, Synanon eschews the medical model that reinforces passive dependency on a benevolent representative of authority. The success of any program based on the purely medical model is often capriciously related to the degree of cultural compatibility between the individual addict and the "benevolent representative."

Claims have been made that Synanon has had unusual success, and that its success rate was far higher than that of other institutions (YABLONSKY, 1965).

Little has been published concerning retention rates for even long established self-help communities. The mystique surrounding Synanon may have arisen in part because of seeming reluctance to subject this progenitor of all self-help communities to study and research by disinterested investigators.

CRESSEY and VOLKMAN (1963) report that of the 263 persons either admitted or readmitted to Synanon between 1958 and 1961, 190 left against advice. The 71 residents who remained in the program represented 27.2% of the total.

Their data collected for the former's master's thesis, summarized dropout rates at Synanon during the first 3 years of its existence as follows:

> Between May, 1958 (when Synanon started), and May, 1961, 263 persons were admitted or readmitted to Synanon. Of these, 190 (72%) left Synanon against the advice of the Board of Directors and the older members. Significantly, 59% of all dropouts occurred within the first month of residence, 90% within the first three months. Synanon is not adverse to giving a person a second chance, or even a third or fourth chance: of the 190 persons dropping out, eighty-three (44%) were persons who had been readmitted. The dropout rate behavior of persons who were readmitted was, in general, similar to the first admission; 64% of their dropouts occurred within the first month, 93% within the first three months of readmission (CRESSEY and VOLKMAN, 1963).

How successful is Synanon, the most famous of all selfhelp communities for drug addicts? Actually, this is most difficult to determine. It was once Synanon's aim to assist members to reenter the outside world, but now it attempts to hold them for the remainder of their lives. CHARLES DEDERICH explains: "We once had the idea of 'graduates.' This was a sop to social workers and professionals who wanted me to say that we were producing 'graduates.' I always wanted to say to them: 'A person with this fatal disease will have to live here all his life.'

"I know damn well if they go out of Synanon they are dead. A few, but very few, have gone out and made it. When they ask me, 'If an addict goes to Synanon, how long will it take?' my answer is, 'If he's lucky, it will take forever."

"We have had 10000–12000 persons go through Synanon. Only a small handful who left became ex-drug addicts. Roughly one in ten has stayed clean outside for as much as two years" (GRAFTON, 1971).

We do not know how many persons applied to Synanon and were turned away. The current number of Synanon residents is approximately 1500. Let us assume then that a round figure of 11000 would approximate the number of Synanon dropouts, defined in terms of Dederich's view that the successful Synanon member remains indefinitely. In view of the fact that 1500 (or 12%) remain of the 12500 admitted to Synanon since May 1958, the holding power of this movement appears to be quite low.

It should also be kept in mind that an unspecified number of the 1500 present Synanon residents are nonaddicts, or 'lifestylers' who have chosen to affiliate themselves with the organization for the purpose of developing a more satisfying style of life.

Since the founding of Synanon, there have been a number of "Synanon-like" treatment and rehabilitation programs for addicts that have developed. Many of these have been organized or staffed by either former narcotic addicts or professionals who have had some experience and exposure to the Synanon movement. This has been true for such programs as the Daytop program and the Phoenix House program in New York City. Other programs have developed as spin-offs from these programs. Most of these other programs have professional staff involved in them in either a supervisory or advisory capacity. Many of them differ from Synanon in that their philosophy holds that members should be prepared to leave and reenter the greater society after a period of from 12–18 months.

This is true, for example, of Daytop in New York, which was initially funded through a grant from the National Institute of Mental Health. At the point in its existence where 300 individuals had entered Daytop, it was estimated that 10% had completed its program (LOURIA, 1968).

Phoenix House is another rehabilitation center for addicts which has gained national reputation. In 4 $^1/_2$ years of operation with over 4000 admissions, Phoenix is currently running a drop-out or "split" rate of approximately 60%. However, even those who drop out of the 18-month Phoenix program prior to the completion of their treatment seem to derive some benefit from the experience. For example, a follow-up study has shown that among those persons who stayed in Phoenix from 3 to 11 months before leaving against advice, there was a one-third decrease in criminal activity compared to their previous life styles. For those persons who stayed 12 months or more before "splitting" there was approximately a two-thirds decrease in criminal activities (DELEON et al., 1971).

While these programs do not appear to be eminently successful in terms of the quantity of their output compared to the total addict population, there is no question that the qualitative change effected by programs such as Synanon, Daytop, and Phoenix is a remarkable one, and one that deserves much closer study in the future.

Professionally staffed and directed multiple modality treatment centers based in the addict's home community, and utilizing the services of qualified and trained ex-addicts, are now being developed throughout the United States as a result of the efforts of the National Institute of Mental Health. Twenty-nine such centers have already been funded, and a systematic data collection method has been developed that will eventually provide an accurate reflection of the relative effectiveness of the different approaches to the treatment of addiction. These centers all provide for not only hospitalization, but also for partial hospitalization, emergency services, consultation, and education and prevention services, as well as additional (e.g., chemical) treatment modalities depending upon local needs and expertise.

E. Civil Commitment Programs

The professional staff of the Public Health Service Hospitals at Lexington, Kentucky and Fort Worth, Texas had long felt two major areas of frustration in their attempts to treat narcotic addicts: (1) There was no way to hold voluntary patients until they achieved the full benefits of the period of hospitalization recommended by the staff, and (2) there was no provision for any form of aftercare services or supervision following release from the hospital and discharge to the addict's home community. Based upon the Lexington experience, civil commitment programs were suggested and eventually enacted in California and New York. The California Civil Commitment Program was enacted in 1961 and provided for compulsory treatment and supervised aftercare of narcotic addicts. The program was administratively placed in the California Department of Corrections. This has resulted in a program that has been basically a prison program, both in institutional settings and in internal operations. The program is one which is rather carefully and strictly enforced. The addict is committed to a minimum and maximum period of time and after a period of inpatient treatment in a rehabilitation facility can be released under supervised parole. If he violates his parole he can be returned to the inpatient facility. The rehabilitation facility is one which is kept locked, with guards, fences, and barbed wire, and escape is regarded as a criminal offense.

In a study of the first 1209 addicts released to parole under the California Commitment Program it was found that within the first year half of the addicts had been returned to the institution for parole violation while only one-third remained in good standing. The others had violated various conditions of their parole but for one reason or another had not been returned to the institution. The number of persons able to remain continuously on supervised parole dropped to about one in six by the time 3 years had elapsed. As later cohorts were examined the success rate dropped even lower and by 1967 only 25% of those released on parole were able to make it in good standing for a year, as opposed to the previous $33\,^1/_3\%$.

By the end of 3 years, only one addict out of six was able to continue to function on parole. This meant that he had avoided significant entanglement with legal authorities, and he had not been caught using illegal drugs. Subsequent examination of the records of the 16% or so of the addicts who had successfully completed 3 years on parole revealed that some were subsequently recommitted following new criminal charges or the evidence of relapse to drug use (KRAMER, 1970).

The United States federal government passed a civil commitment law in 1966 known as the Narcotic Addict Rehabilitation Act. This law represented a breakthrough in the sense that it provided that narcotic addicts could apply through a civil court procedure for hospital care, followed by up to 3 years of continuing supervised treatment and rehabilitation services in their home community. Eventually all these services will be provided from start to finish in the home community of the addict, but initially the hospital services have been provided for most of the addicts at the federal facilities in Lexington, Kentucky and Fort Worth, Texas.

In 1967 the Public Health Service Hospitals in Lexington and Fort Worth were transferred to the National Institute of Mental Health and redesignated as Clinical Research Centers. The Fort Worth Clinical Research Center and subsequently the Lexington Clinical Research Center were transferred to another agency of the government and are no longer part of the National Institute of Mental Health.

The Narcotic Addict Rehabilitation Act (known as NARA) has three main titles.

Title I authorizes federal courts to commit for treatment certain narcotic addicts who are charged with a federal offense and who desire to be treated for their addiction instead of being prosecuted for the criminal charge. Under this title the total period of treatment may not exceed 36 months, and the in-hospital phase is of indeterminate length depending on the patient's response to treatment. When the patient responds sufficiently he may be conditionally released from the in-hospital phase and placed under supervised outpatient care in his home community. This supervised outpatient care is individually arranged for the addict by more than 150 local contract aftercare agencies, and may include vocational training, continued education, group or individual psychotherapy, half-way houses, financial assistance, and regular urine testing to determine if the addict has relapsed to the use of illicit drugs. If the patient completes the treatment program successfully he may be discharged and the criminal charges against him will be dismissed. If the patient fails to complete his program successfully or relapses to the use of narcotics and refuses to cooperate in the treatment program, his commitment may be ended and the prosecution on the original criminal charge may be resumed.

Title II provides for a sentencing procedure to commit for treatment narcotic addicts who have already been convicted of a crime in a federal court. These patients are treated in special facilities of the Bureau of Prisons, and then provided with aftercare when they are conditionally released.

Title III provides for the civil commitment of addicts who are not charged with a federal offense. The majority of the patients in the civil commitment program fall under this Title. These patients may be provided with hospital care up to 6 months, then are released to a period of supervised aftercare which may last as long as 36 months. The same comprehensive services are provided, along with regular urine testing to detect relapse to the use of illicit drugs. The in-hospital phase for these patients is currently conducted by local community agencies under contract.

Since Title II patients are treated within the Federal Bureau of Prisons, this title will not be discussed further in this text. Before Title I or Title III patients are accepted in the treatment program, they must undergo a period of examination and evaluation which may last up to 30 days. During this time the patients are examined by two physicians, one of whom is a psychiatrist, and two questions are answered for the court: (1) Is the patient in fact a narcotic addict? and (2) Is the patient likely to be rehabilitated through treatment under the act? If these questions are answered affirmatively by the physicians, the patient is committed to the treatment program. Experience under the act thus far has indicated that approximately 49 out of every 100 patients referred for examination and evaluation are found to be addicts who are likely to be rehabilitated under the act. The other 51 are usually found to be narcotic addicts but are found to be not likely to be rehabilitated because of very low motivation levels or because of disruptive and antagonistic behavior in the hospital setting.

At Lexington, the total admissions during 1969 were 2008, of which 732 (36.5%) were found to be likely to be rehabilitated; of 2083 admissions in 1970, 1090 (52.3%) were considered likely to be rehabilitated; of 949 admissions in 1971, 555 (58.5%) were considered likely to be rehabilitated by the Lexington facility staff.

Of 5859 total admissions since the first admission in 1967 through August 1971, 2880 (49.2%) were found likely to be rehabilitated.

The professional staff at the Clinical Research Center spent a great deal of time trying to persuade the narcotic addicts who were sent for examination and evaluation to stay and undertake the full treatment program. The addicts were told in the most blunt terms possible that if they did not remain in treatment they would be leaving the institution to return to a life of continued degradation and a good chance of either death or a long-term prison sentence at some time in the near future.

However, the typical experience was that the addict may have had a state or local charge facing him, and he was in difficulty with the local law enforcement people, or his wife was angry with him, or his habit had gotten out of size and he could no longer financially continue to support his habit, or he was physically ill. He then sought civil commitment and was sent to the Clinical Research Center at Lexington. There he was withdrawn and after 4 or 5 days he was usually free of opiates and relatively comfortable, and he began to feel better. He then discovered that the local or state charges against him had been dropped because he entered the civil commitment treatment program. He then began to indicate to the professional staff that he was not interested in further treatment, that he felt his problem had been solved, and that he wished to be released immediately. The professional staff of course realized what had happened. The patient had withdrawn from his habit; he was "clean"; his legal pressure had disappeared, and he wanted to go back out on the street and get a "fix" again. He realized that he could now resume his addiction at a much lower cost and with a much greater "high" than he had before entering the center. Both the professional staff and those patients who had been in treatment for several months and had matured somewhat customarily spent a great deal of time trying to convince the addict that he was not yet ready to leave and that he should remain in treatment. Yet the center staff found itself faced with the problem of an addict who was basically hostile. This hostility expressed itself either directly, through threats or actual acts of violence directed against center staff or other patients or indirectly, through passive obstructionism. That is, the addict simply refused to get out of bed; he would not come to an interview; he would not talk in his group therapy meetings; he refused to take a shower. Faced with this situation, the professional staff usually found that this type of patient was not likely to be rehabilitated, and he was returned to the court at the end of his period of examination and evaluation and the proceedings were dismissed.

It is interesting to note that under Title III of the act, approximately 50% of the narcotic addicts who were sent for examination and evaluation were found to be not likely to be rehabilitated because of their disruptive and antagonistic or uncooperative behavior during the period of examination and evaluation. By contrast, only approximately 15% of those addicts sent under Title I of the act were found not likely to be rehabilitated. Apparently, the immediate threat of impending prosecution on a federal charge had an effect on the addict's level of motivation or degree of cooperation.

For those patients who were found likely to be rehabilitated, the treatment program at the Lexington Clinical Research Center was an energetic and open one. Most of the architectural features of the institution which were prisonlike had been removed, including virtually all of the bars, grilles, and locks. Physical seclusion was no longer used and the ultimate disciplinary sanction was expulsion from the program. The doors of the center were unlocked, and rules and regulations were modified in order to encourage and reward desirable behavior rather than to focus on punishment of undesirable behavior. The previously monolithic treatment program was broken down into a unit system, with five separate treatment units, each of which had its own autonomous staff and unit chief. Each unit was provided with its own budget, and decided itself how that money would be spent. The development of individual life styles and treatment programs was encouraged among the units, and measurement instruments were developed to record the differences among them. The patients were encouraged to be very active participants in the day-to-day operations of the units, and as they demonstrated increasing responsibility in their behavior and attitudes, they were afforded increasing degrees of involvement in the decision-making processes.

The previously large security force was reduced in numbers and maintained only a perimeter-type control. That is, they attempted to prevent the introduction of contraband drugs, and they attempted to identify those patients who decided to leave the institution. Although there are penalties for escape provided in the Narcotic Addict Rehabilitation Act, those penalties have not been enforced because of legal technicalities in the wording and interpretation of the act. Interestingly enough, only 592 (10.1%) of the first 5859 admissions under the Narcotic Addict Rehabilitation Act walked away without authorization, some of those during the examination and evaluation phase and a lesser number during the actual treatment phase.

In order to maintain a drug-free environment without using rigid physical security measures, the center initiated a program of routine random urine testing. Every patient was assigned a three-digit number on admission and every morning a random numbers table provided a list of patients who were to submit a urine specimen that day for examination. The patient was given 3 h within which to produce the urine under direct observation of the staff, and the urine was tested for the presence of opiates, quinine, amphetamines, barbiturates, antihistamines, tranquilizers, or other unidentifiable substances. In addition, every patient returning from pass or leave was asked to submit a urine specimen, and any patient whom the staff suspected of using contraband drugs was asked to submit a urine specimen. The overall number of positives usually ran around 8%. Out of that 8%, $3^{1}/_{2}$ were "stalls," i.e., the patient claimed that he was unable to produce any urine during the 3 h given him. All such "stalls" were regarded by the treatment staff as positives. An additional 3% were actual positives for drugs such as tranquilizers, barbiturates, amphetamines, or antihistamines. Only 1.5% of the urines tested were positive for opiates. When a patient turned up with a positive, the matter was referred to his treatment group and his therapist for discussion, confrontation, and disposition.

The patients were thus acquainted with the urine testing system procedure which they would continue in the aftercare phase of the program when they returned to their home community. If their general adjustment in aftercare and their urine tests indicated that they were relapsing to the use of narcotics, they could be recommitted for another phase of inpatient treatment. Of the first 1537 narcotic addicts discharged to aftercare, 354 had been recommitted for another period of inpatient treatment at the center. This was a recommitment rate of approximately 23%.

The actual treatment program within the center varied from unit to unit, but there were certain characteristics most units had in common. Each unit hired one or more rehabilitated ex-addicts to work in conjunction with the treatment staff.

On all units the treatment staff felt that a great deal of personal involvement with the addict was mandatory to induce him to enter treatment and to keep him in treatment. The major technique used after personal involvement had been gained was that of confrontation, rather than introspection or reflection. There was a great deal of emphasis placed upon the present, although consideration of past events was not excluded in helping to understand the present. Much time was spent on discussing behavior and stressing to the addict that ultimately he is responsible for his own behavior and that ultimately society will hold him accountable for his behavior. While the focus was on behavior, thoughts and feelings were also examined and dealt with, especially the widespread dysphoric feelings which emerge as the addict gets involved in treatment. The management of the depression which becomes evident in the addict as he gets engaged in treatment is a very difficult clinical problem, and one that seems to become more manifest the longer the addict remains in treatment.

At some point early in his 6 months' stay at the center, after being confronted with his past and present irresponsible behavior, the addict was called upon to make a commitment to change. In getting him to make this commitment, his personal involvement with the therapist was utilized as leverage along with the influence of the ex-addict therapy assistants and the more mature patients who are at more advanced stages of treatment. It is the influence of a positive peer group that seems to have the greatest effect in eliciting a commitment to change from the addict, and without this, most professional staff intervention seems doomed to failure. The commitment to change was sought in very concrete terms, and a detailed description of the types of behavior that were to be changed was outlined for the addict, by both the treatment staff and his peer group. The addict was then expected to make a demonstration of the change that had been called for. That is, he was asked to "walk the walk as well as talk the talk." From that point on, the awarding of status and prestige and material comforts within the treatment unit was dependent upon the individual's seeking and assuming responsibility. He was called upon, over and over again in group confrontation sessions, to not only be honest but to be responsible and to demonstrate *both* by his behavior.

During regular staffing, the patient's progress was evaluated and plans were made for his eventual release to aftercare. The eventual plans for aftercare began the first day the patient was accepted for treatment, and were made final during the course of the 6 months' inpatient phase.

One treatment unit was much different from the others in that it was run for research and demonstration for 2 years, during which time it was also designated as an aftercare agency under the Narcotic Addict Rehabilitation Act, and patients resided there for approximately 18 months before returning to their home community. The last 12 months of this 18-month period involved considerable freedom to come and go, and was not really what would usually be termed institutional treatment. This community was operated exclusively by ex-addicts. Two professional staff, both psychiatrists, were available as advisors and consultants.

Members had their own quarters, located in a building separated geographically from the complex of interconnected buildings that constituted the Clinical Research Center, and which contained the other four treatment units.

The unit was very similar to Synanon in some respects but had developed unique characteristics of its own. Patients entered by applying for an interview from one of the other units. After being evaluated by the ex-addict staff, the applicant was either accepted, accepted conditionally, or rejected. Almost all applicants were accepted conditionally, which meant they were asked to come over daily to demonstrate the sincerity of their interest by performing menial tasks such as mopping the floor or cleaning out latrines. If an applicant did this regularly and faithfully for a week, he was accepted as a resident member. The self-help unit members not only entered the adjoining community of Lexington for purposes such as shopping, attending movies, and socializing, but they also invited the community to visit them, with regular Saturday night open houses and a Sunday evening "square game club"—group confrontation sessions in which members of the local community joined the members of the house. These "games" were similar if not identical to those at Synanon, and occurred within the regular House activities on Monday, Wednesday, and Friday evenings from 7:30 to 9:00 p.m.

Interestingly enough, the internal operation of this ex-addict run unit was extremely authoritarian in its style, yet succeeded in creating a living atmosphere of considerable warmth and friendliness. The unit was coeducational; that is, there were residents of both sexes. Sexual interaction among unmarried members was strictly forbidden, and even among married members was sometimes prohibited on the grounds that they had not yet demonstrated sufficient responsibility in their life style to warrant such a privilege.

There was heavy emphasis on the achievement of individual responsibility, and material rewards and status were awarded to individuals on the basis of their demonstrated accomplishments in this regard. Such rewards then were regarded as a deserved consequence of responsible behavior, rather than as merely ends in themselves. There was a great deal of emphasis on hard work, thrift, cleanliness, and self-education and self-reliance. A great deal of time was spent reading Emerson and Thoreau and discussing their philosophical concepts. The members' stated goal was to become self-actualizing and self-fulfilling persons, and they sought to achieve the "good life" through hard work and reason, accompanied by fervent disavowal of any form of chemical intoxication.

These addicts who mutually selected and were selected by the self-help unit were younger by about 3 years than the population of the rest of the Clinical Research Center. They also stayed in school longer. Analysis of the Lexington Personality Inventory data revealed that, compared with other center residents, the self-help residents tended to be "more intellectually oriented, more verbal, more middle-class in attitudes, more actively coping than defensive in regard to personal problems, less like an alcoholic in attitudes, less positive in attitude toward drugs, less inclined to devalue self; but on the negative side, a bit more impulsive, more morally deviant in past behavior, and a bit more anti-society in behavioral and attitudinal history. Our self-care group had personality profiles that were a bit less pathological, but tended to be similarly high on psychopathic deviate and schizophrenic scales, with somewhat greater ego strength and the suggestion of less depression" (JONES, 1971).

When an experimental revision of the *Rotter I-E Scale* was administered to self-help residents and to center residents, it was discovered that the former had greater internal control. Accordingly, personal and peer group controls were sufficient to discourage drug abuse among these residents. Drug use was practically nonexistent in this unit, as measured by the urine surveillance program (which was applied equally to the ex-addict employees as well as to the patient-residents in treatment).

Later measurements of self-help and center member traits revealed less marked differences. This trend may be explained in part by somewhat less selectivity later than earlier in the screening of self-help candidates, and by progressive efforts by Center residents to emulate in their respective living units the self-help life style. This latter tendency was very evident.

Twenty months after its inception a total of 115 candidates had been accepted for residence in the house. Of this number 68 (59%) had left the house against advice, and 37 (32%) remained active in the 18-month program, 5 (4%) had been discharged as rehabilitated under the Narcotic Addict Rehabilitation Act, and 5 (4%) had been transferred to complete the last months of their aftercare in facilities in other communities.

Of the 115 members, 75 were civilly committed under the Narcotic Addict Rehabilitation Act and the remainder were the previously mentioned "special study" cases. Of the 75 patients committed under the Narcotic Addict Rehabilitation Act, 17 remained within the unit continuing in their treatment, 2 former NARA patients resided at the House as staff members employed by the Clinical Research Center, 5 had been discharged as completely rehabilitated, and 5 had been transferred to complete their aftercare in other communities.

Toward the end of the 2nd year, the ex-addict staff of this unit no longer lived up to their positions of trust, and behaved in an irresponsible manner which required termination of the unit.

How effective has the federal civil commitment program been? Since the first patients were accepted for treatment under the Narcotic Addict Rehabilitation Act in 1967, and the patient census did not build up significantly until 1968, only preliminary data are yet available on the relative effectiveness of the total NARA program. Long-term, comprehensive evaluative procedures must wait for the mere passage of time. In the meantime, gross effectiveness criteria are being used. Such indicators include the number of days worked, money earned, kinds and quantities of illicit drugs consumed, the number of arrests, and kinds of legal charges.

It can be argued that it is economical to treat patients under the civil commitment program. The cost for $3\,^1/_2$ years is approximately \$ 21000. Incarceration in an average correctional facility for the same length of time would cost approximately \$ 26000. By contrast, the nontreated addict costs the public approximately \$ 43000 for a $3\,^1/_2$ year period. This cost is calculated in terms of stolen property, welfare payments, emergency treatment, and medical care (PERSON et al., 1971).

A study of 1200 patients who were in aftercare in the civil commitment program in 1970 showed that approximately 85% were employed, 70% were not arrested and spent no time in jail during that period, 35% were in self-help therapy, and 33% were pursuing their education. During the first year of the aftercare phase, 60% do not become readdicted, whereas 25% do become sufficiently readdicted as to require further hospital treatment either in the community or at Lexington. The remaining 15% drop out of the program. Of all the days the patients spend in the civil commitment program during aftercare for 3 months or more, 80% were free of heroin use, and 60% were free of abuse of any type of illicit drugs (BROWN, 1971).

It appears that most patients after leaving the hospital phase of the treatment program do experiment to some degree with drugs during their initial months in aftercare, for only 13% of the persons in aftercare have been completely drug-free throughout. However, drug experimentation and use tends to decrease with the length of time spent in aftercare (PERSON et al., 1971).

On the whole, when one considers the relatively large numbers of narcotic addicts who are still found not likely to be rehabilitated during the initial examination and evaluation phase, it appears that the overall effectiveness of the federal civil commitment program yet remains to be demonstrated. However, it does appear that for some of the narcotic addict population it may represent an approach that, along with all others, may be useful. It is important to emphasize that civil commitment is only a means of entry into a treatment program, and does not represent a treatment modality per se.

F. Conclusions

Many psychiatrists and other professionals who are intimately involved in the treatment of narcotic addicts on a day-to-day basis are now in agreement with certain philosophical principles. Most would feel that the use of the narcotic drug is an epiphenomenon, and that most of the patients whom they see have had grossly disordered lives long before they ever stuck a needle in their arm. However, the use of narcotic drugs is a severely complicating epiphenomenon which makes the patient very resistant to treatment. This is because this particular form of illness initially affords the patient pleasure rather than pain, and pleasure of a peculiarly intense and psychologically archaic kind. In most of the nonchemical treatment programs, as well as in many of the chemical treatment programs, growth through responsible *behavior* (not verbal declamation, however impressive) is the objective. It must be stressed to the patient that the responsibility for this growth, while initially shared by treatment staff, is ultimately his in the end.

There should be no secret power-trust meetings in the treatment setting; policy and decisions about patients should be openly discussed and arrived at in an air of mutual honesty. Since treatment programs for addicts that have been run on a completely democratic basis have sometimes become corrupt, most of the currently successful programs are either outright autocracies or guided democracies where monumental decisions are not left to a vote. Here staff and patient leaders alike must make the hard, edifying decisions to demonstrate responsibility together. A need to be firm and consistent is imperative with regard to the addict, and when a patient breaks one of the essential rules of any program (such as no use of illicit drugs, no violence, no threats of violence) he must be dealt with in confrontation within his therapy group, and he may find himself quickly expelled from the treatment program. This has a positive feedback to the other patients in terms of real, predictable consequences to behavior. Most addicts have learned to disregard consequences of behavior because of insouciant authority and their own silver-tongued ability to verbally defend their behavior.

We have learned that it is vitally important in treatment of addicts, particularly young addicts, to distrust insight gathering as a catalyst for behavior change. Responsible behavior begins with the first act of positive caring about oneself and others (GLASSER, 1965); it is this behavior that must be nurtured and constantly examined, as it is in the encounter group situation. The so-called dope fiend has a double-think view of verbalization. He both treasures it as his magical front for the hustle, and at the same time, he is able to put it down as a vehicle of insincerity practiced by all dope fiends. He therefore marvels at the naivete with which professionals will often accept him simply on his verbal productions. This duplicitous behavior allows the addict to both invite failure in a treatment program and then to sanctimoniously criticize the same program, a typical alloplastic defense (narcissistic projection).

The group model in which the ex-addict is exclusively, or almost exclusively, in charge, has greater appeal to the younger addict. There is undoubtedly a group induction process in the introduction to the use of illicit drugs, probably accurately reflected in the street addict's phrase "the best way to become a dope fiend is to know one." The adolescent's adherence to his group's values becomes a part of his magical thinking in a technotronic culture. Where chemicals are widely used to alter outer appearances, he too can use them and make himself look "cool." Through the use of opiates he can be a "cat" who gets above the frustrating social scene. This type of animism flourishes in the adolescent world, and in this world the enchanting elixir has become the most holy of grails. To channel this type of group-approved search into a search where the addicts as a group are seeking and assuming greater responsibility for their own regeneration has become one of the most exciting yet difficult chapters of drug abuse treatment.

Interested readers may wish to review the published articles by GLOVER (1932), SAVITT (1954), LOWRY (1956), NYSWANDER (1956), and OSNOS (1963) which have been included in the bibliography.

References

Ausubel, D. P.: The psychopathology and treatment of drug addiction in relation to the mental hygiene movement. Psychiat. Quart. Suppl. **22**, part 2, 219—250 (1948)

Ball, J. C., Pabon, D. O.: Locating and interviewing narcotic addicts in Puerto Rico. Sociol. Social Res. **49**, 401—411 (1965)

Brown, B. S.: Statement before the Subcommittee on Alcoholism and Narcotics of the Senate Labor and Public Welfare Committee. Drug dependence-extent of problem and treatment modalities. Hearings before the Subcommittee on Alcoholism and Narcotics. May (1971) GPO, Wash.

Cohen, S.: The drug dilemma. New York: McGraw Hill 1969

Cressey, D. R., Volkman, R.: Differential association and the rehabilitation of drug addicts. Amer. J. Sociol. **49**(2), 129—142 (1963)

DeLeon, G., Holland, S., Rosenthal, M. S.: The Phoenix therapeutic community: changes in criminal activity of resident drug addicts. In press J. Amer. med. Ass.

Fenichel, O.: The psychoanalytic theory of neurosis. New York: Norton 1945

Fort, J. P.: The psychodynamics of drug addiction and group psychotherapy. Int. J. Gr. Psychother. **5**, 150—156 (1955)

Freedman, A. M., Sharoff, R. L.: Crucial factors in the treatment of narcotic addiction. Amer. J. Psychother. **19**, 397—411 (1965)

Glasser, W.: Reality therapy; a new approach to psychiatry. New York: Harper and Row 1965

Glover, E.: Common problems in psycho-analysis and anthropology; drug ritual and addiction. Brit. J. med. Psychol. **12**, 112 (1932)

Grafton, S.: Synanon asks a question: can the ex-addict walk alone? Addiction and Drug Abuse Report, **2**, 1—3 (June 1971)

Jones, R. E.: Characteristics of addicts opting for a selfcare program. Paper read at 79th Annual Convention of the Amer. psychol. Ass., Wash. Sept. 3, 1971

Kramer, J. C.: The state versus the addict: uncivil commitment. Boston Univ. Law Review **50**, No. 1, 1—22 (1970)

Lieberman, L., Brill, L.: Rational authority and the treatment of narcotic offenders. Bull. Narcot. **20**, 33—37 (1968)

Louria, D. B.: The drug scene. Toronto: McGraw-Hill 1968

Lowry, J. V.: Hospital treatment of the narcotic addict. Fed. Probation **20**, 42—51 (1956)

Maddux, J. F., Berliner, A. K., Bates, W. M.: Engaging opioid addicts in a continuum of services. A community-based study in the San Antonio area. Fort Worth, Texas: Texas Christian University Press 1971. Behavioral Sciences Monographs, No. 71-1

Nyswander, M.: The drug addict as a patient. New York: Grune and Stratton 1956

O'Donnell, J. A.: Narcotic addicts in Kentucky. Public Health Service Pub. 1881. National Institute of Mental Health, Wash.: GPO. 1969

Osberg, J. W., Berliner, A. K.: The developmental stages in group psychotherapy with hospitalized narcotic addicts. Int. J. Gr. Psychother. **6**, 436—446 (1956)

Osnos, R.: The treatment of narcotics addiction. N.Y. State J. Med. **63**, 1182—1188 (1963)
Person, P. H., Chatham, L. R., Doran, R. F.: Program evaluation in NIMH funded treatment and rehabilitation programs. Paper read at the 99th Annual meeting of the Amer. publ. Hlth. Ass. Minneapolis, Minn. October 13, 1971
Pescor, M. J.: A statistical analysis of the clinical records of hospitalized drug addicts. Publ. Hlth Rep. Suppl. 143 (Wash.) (1943)
Rado, S.: The psychoanalysis of pharmacothymia (drug addiction). Psychoanalysis of behavior: The collected papers of Sandor Rado. New York: Grune and Stratton 1956
Rosenfeld, H. A.: (On drug addiction). Int. J. Psycho-Anal. **41**, 467—475 (1960)
Savitt, R. A.: Extramural psychoanalytic treatment of a case of narcotic addiction. J. Amer. psychoanal. Ass. **II**, No. 3, 494—502 (1954)
Trussell, R. E.: Treatment now. Hospitals **45**, pt. 1, 47—48 (1971)
Vaillant, G. E.: A twelve-year follow-up of New York narcotic addicts: I. The relation of treatment to outcome. Amer. J. Psychiat. **122**, 727—737 (1966a)
Vaillant, G. E.: A twelve-year follow-up of New York narcotic addicts: II. The natural history of a chronic disease. New Engl. J. Med. **275**, 1282—1288 (1966b)
Vaillant, G. E.: A twelve-year follow-up of New York narcotic addicts: III. Some social and psychiatric characteristics. Arch. gen. Psychiat. **15**, 599—609 (1966c)
Vaillant, G. E.: A twelve-year follow-up of New York narcotic addicts: IV. Some characteristics and determinants of abstinence. Amer. J. Psychiat. **123**, 573—584 (1966d)
Vaillant, G. E.: What has happened to 100 Lexington addicts 20 years later? Paper read at in-service training meeting of NIMH Clinical Research Center staff, Lexington, Ky. Sept. 15, 1971
Vaillant, G. E., Rasor, R. W.: The role of compulsory supervision in the treatment of addiction. Fed. Probation, **30**, 1—7 (June 1966)
Yablonsky, L.: The tunnel back—Synanon. New York: MacMillan 1965

CHAPTER 5

Chemotherapy of Narcotic Addiction

W. R. MARTIN

A. Introduction

Chemotherapy of narcotic addiction began in the 1920s with efforts to detoxify patients without suffering or discomfort. Procedures were devised which allowed patients not only to be withdrawn with minimal discomfort but to be restored to apparently good body health after several months of treatment. Many patients who were successfully detoxified and restored to good health, however, not only relapsed to drug use but continued to demonstrate sociopathic behavior following treatment. The major thrust of the treatment of narcotic addiction has since shifted from detoxification to therapies that would facilitate social adjustment and decrease sociopathic behavior. Chemotherapy has come to play a role in achieving these ends.

With the use of drugs to treat behavior has come a realistic fear that drugs may be used for political as well as therapeutic ends. It is important that the physician restricts his role to the treatment of pathologies of the body and brain with the end of restoring normal function. Whether or not a given type of behavior is sociopathic depends to a degree on prevailing cultural values which may not be adaptive or useful. The use of chemotherapeutic agents to treat narcotic dependence should be restricted to physicians who can make a diagnosis of the disease processes and are knowledgeable about the indications as well as the dangers of the drugs that are employed. It is important to emphasize that not all of the basic pathologies associated with psychopathy and drug addiction have been identified nor have specific drug therapies been evolved to deal with them.

B. Diagnosis of Narcotic Addiction

The pathologies of narcotic addiction have been discussed in Section I, Chapter 1, and Section II, Chapter 1. In brief, most narcotic addicts exhibit a high level of psychopathy which is characterized by impulsivity, egocentricity, hypophoria, increased need states, and sociopathic behavior. Narcotic analgesics when first used reduce hypophoric feelings and need states; however, with continued use tolerance and physical dependence develop which are associated with an exacerbation of hypophoric feeling states. When the patient becomes abstinent, hypophoric feeling states and needs are further increased. Early signs and symptoms of abstinence are followed by a protracted abstinence syndrome which is a syndrome characterized by both physiologic deviations and feelings of hypophoria. In making a diagnosis of narcotic addiction each of these three pathologic processes plays a role.

I. Diagnosis of Psychopathy

A variety of types of evidence assist in the diagnosis of psychopathy. An elevated psychopathic deviate scale score on the MMPI alone or in the presence of other pathologic elevation is important. A history of both impulsive and sociopathic behavior is commonly obtained. The patient will commonly have had attendance and disciplinary problems in school, exhibited precocious sexual behavior, and been involved in delinquent and criminal activity. Police and court records are helpful in establishing sociopathic behavior. In initial interactions with psychopathic patients one may be impressed with their charm, loquacity, and fluency; however, psychopaths will become demanding and when their wishes are not met, become overtly hostile and plaintive, blaming most of their misfortunes on others. Patients frequently give a history of poly-drug use, of being arraigned for possession or sale and possession of narcotics, of having been dependent on and withdrawn from narcotics, and of having been institutionalized for the treatment of narcotic dependence.

II. Diagnosis of Tolerance and Physical Dependence

Before initiating treatment, it is necessary and important to determine whether or not the patient is dependent on narcotics or other drugs. If the patient is physically dependent on narcotics, detoxification is necessary before a narcotic antagonist is given therapeutically. On the other hand, if the patient is not known to be dependent, considerable thought should be given before initiating maintenance therapy. There may be few clinical signs in the dependent patient. He may have needle marks and pigmented needle tracks over veins that are being or have been used for administering narcotics intravenously. Pupils may be constricted, however, the large individual variability in pupil's size makes this sign of little diagnostic value. Some patients are thin and even emaciated because of a neglect of eating. Irritant drugs such as barbiturates, pentazocine, propoxyphene, and methadone may cause induration, phlebitis, and ulcers in regions where they have been injected.

To develop a clinically significant and detectable degree of physical dependence, a patient must use approximately 10 mg of heroin or 30 mg of morphine or more a day (JASINSKI and NUTT, 1972). Patients using dependence-producing quantities of morphine or heroin will have urines that are positive for morphine using standard analytic techniques (see Sect. II, Chap. 6) for at least 24 h following their last dose. By this time a clinically significant degree of abstinence should become manifest and detectable by a knowledgeable observer. The signs and symptoms of morphine abstinence in man are discussed in detail in Chapters 1 and 3 of this section. Urine testing methods will detect most of the commonly abused narcotics in the urine of addicts who are dependent upon them. The presence of a narcotic in the urine is, however, not evidence of dependence, only of use.

ISBELL (1954) proposed the use of nalorphine for the rapid diagnosis of morphine and methadone dependence. TERRY and BRAUMOELLER (1956) first used the technique for the diagnosis of street narcotic addiction. The technique most commonly employed is to administer 3 mg of nalorphine subcutaneously and observe for signs of abstinence. Several systems of grading abstinence signs have been proposed (see Sect. II, Chap. 3; State of California, 1958). Measurement of pupillary diameter has proved to be a sensitive method for detecting precipitated abstinence. In patients who have not received a narcotic, nalorphine may produce pupillary constriction; whereas, in patients who have received a narcotic it will tend to produce mydriasis. ELLIOTT et al. (1964) found that 3 mg of nalorphine produced an average pupillary constriction of 0.62 mm following the administration of a placebo; however, in patients who had received a single 15 mg dose of morphine, the same dose of nalorphine dilated pupils for 4 h and after 8 h produced pupillary constriction. The pupillary constricting effects of nalorphine are a consequence of its agonistic actions; whereas, its ability to dilate pupils is a consequence of its ability to competitively antagonize the agonistic effects of morphine and precipitate abstinence in either the acutely or chronically dependent subject. Even under conditions of fixed illumination and accommodation, pupillary diameter varies up to 1.5 mm on replicate photographic determinations with an average difference of about 0.25 ($\pm$0.03 SEM) mm between replications (JONES et al., 1962). A simple and convenient method for photographing pupils using a modified Polaroid camera has been described and validated for the detection of precipitated abstinence (JASINSKI and MARTIN, 1967; JASINSKI et al., 1967; MARQUARDT et al., 1967) in which replicate measurements of pupillary diameter are made at 30-min intervals before the administration of antagonist and 15, 30, 60, and 90 min thereafter. In addition, other signs of abstinence are determined and the intensity of precipitated abstinence assessed using the scoring system of HIMMELSBACH (Sect. II, Chap. 3). Although the cost of doing a precipitation test in this way is more expensive than that used in California, it is more reliable. The cost and implications of treatment of narcotic addiction are such that extensive diagnostic workups are entirely justified. In the chronically dependent patient, nalorphine is very useful in precipitating abstinence especially for the first 8–16 h of withdrawal (WAY et al., 1966); however, as withdrawal abstinence becomes manifest, nalorphine becomes less effective in precipitating additional signs of abstinence.

Naloxone, a relatively pure antagonist, has no effect on pupillary diameter in its own right and has also been used for the diagnosis of narcotic dependence (JASINSKI et al., 1967; ELLIOTT, 1971; BLACHLY, 1973). Although the naloxone precipitation test has not been as well validated as the nalorphine test, it probably has several advantages: (1) In the absence of a narcotic, naloxone has no effects at all in therapeutic or diagnostic doses and will not therefore add to the depressant effects of sedative-hypnotic agents as nalorphine may. (2) It is a more effective antagonist of codeine, meperidine, propoxyphene, and pentazocine than nalorphine. (3) It is more effective than nalorphine in precipitating abstinence in patients with low levels of physical dependence.

C. Natural History of Narcotic Addiction

In order to determine whether a treatment helps a chronic disease process such as narcotic addiction, it is necessary to have some estimate of what the mental health status of the patient would have been had he not received therapy and compare this estimate with his mental health status while in therapy. There are three ways that this estimate can be made, each with its limitations: (1) Through history of the patient and examination of the patient's social and criminal record, an estimate of the incidence of sociopathic behavior before treatment can be obtained and used as a basis for comparing his behavior while in treatment. Because many recollections of patients, their family, and their associates cannot be validated and because hard data on many important outcome variables cannot be obtained, this retrospective method has serious limitations. (2) A superficially stronger method is to randomly assign patients or alternatively have matched samples of patients assigned to a treatment group and a control group. Under ideal conditions these two groups of patients would receive comparable care except for the experimental treatment modality. There are several practical difficulties in executing this type of experiment. Usually patient populations are small, and for this reason, sampling differences may be large compared to treatment effects. Further, because of the availability of treatment programs, control and experimental patients may leave the experiment and join other treatment groups. (3) Longitudinal studies or reviews of patients' records of social adjustment and drug use have provided the best data about the natural history of drug addiction. This method also has limitations. Because data are obtained opportunistically, relevant experimental facts may not be acquired. Further, the groups studied may be atypical. Longitudinal studies are very expensive and time-consuming to execute.

VAILLANT (1966, 1973) has studied 100 New York City male heroin addicts for 20 years following their admission to the USPHS Hospital at Lexington. Figure 1 (VAILLANT, 1966) illustrates several important points: (1) The death rate among addicts is high. This and other studies indicate that it is 1–2% per year. (2) There was a trend for the addiction incidence to decrease across time and for the number of

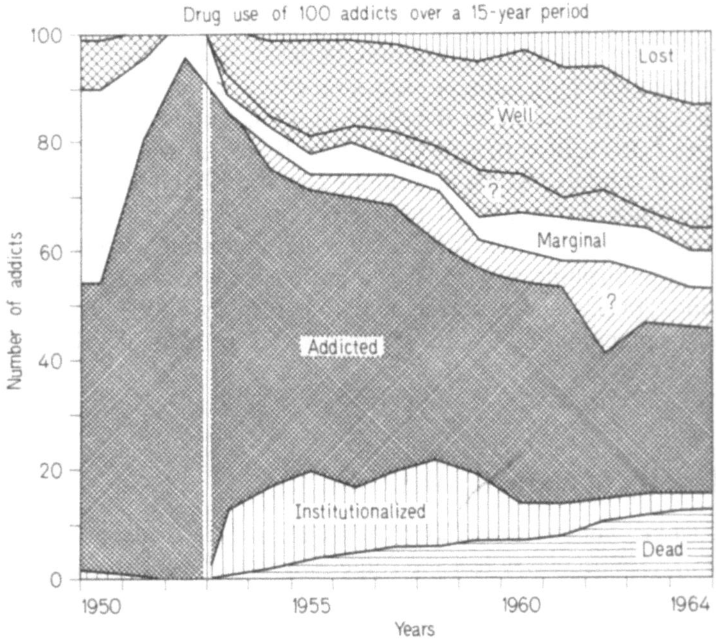

Fig. 1. Reproduced from VAILLANT (1966) with the permission of the publisher (copyright 1966, the American Psychiatric Association)

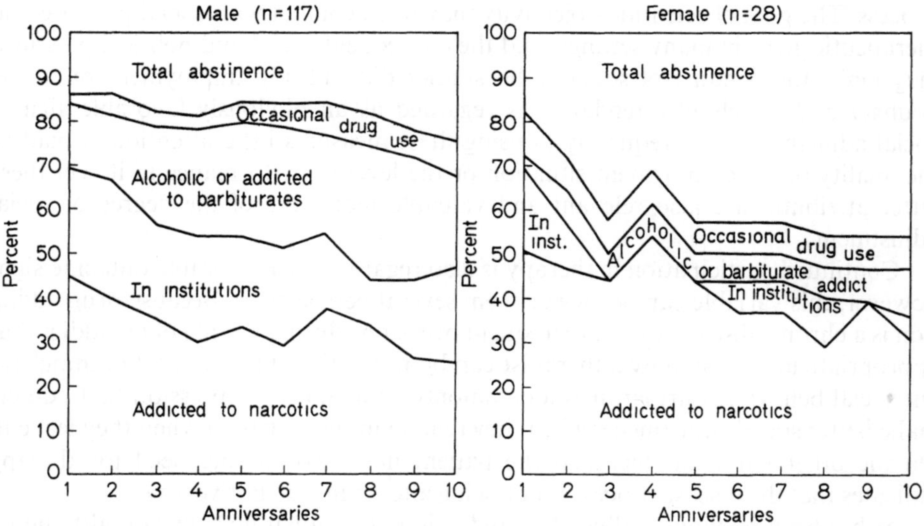

Fig. 2. Percent in various drug statuses on first ten anniversaries of first admission, by sex, based on subjects who were alive and followed for at least ten years after first admission. Reproduced from O'DONNELL (1969) with the permission of the author

patients who were doing well to increase, with 20–25%, depending on the base, doing well. Figure 2 (O'DONNELL, 1969) shows data on 145 Kentucky addicts that also illustrate this latter point. Further, less than 50% of the subjects were actively using narcotics, but many were using other drugs. These as well as other studies have shown that addiction is a chronic disease characterized by remissions and exacerbations. In the absence of control data, all treatment data must be assessed keeping these facts in mind (MADDUX et al., 1971; BOWDEN and MADDUX, 1972; MADDUX and BOWDEN, 1972).

D. Assessment of Efficacy

I. Relevant Outcome Variables and Goals of Therapy

It would greatly simplify the study of drug addiction if the fundamental pathology of the disease was known and valid methods for assessing the degree of the disease process were available. Unfortunately, if drug addiction or psychopathy has pathognomonic signs, their validity has not been demonstrated. Until such time as such signs are identified, it will be necessary to use indices of sociopathic behavior to assess the disease process. Many of these signs are listed in Table 5 of Section I, Chapter 1. Drug use, criminal activity, and lack of employment are the most commonly used indicators of sociopathic behavior: (1) Drug use as assessed by urine testing procedures is the most objective, the best validated, and the most reliable indicator (see Sect. II, Chap. 6). (2) Criminality is difficult to detect since it is usually a surreptitious activity, and only a modest portion of it is identified and becomes a matter of record. For criminality to be validly measured by law enforcement records, it is essential that they not be distorted by the intercessions of therapists in the legal process. The goal of scientific objectivity may be in conflict with social purpose and therapeutic goals in many settings, and the consequent social and political pressures may make the conduct of sound valid studies difficult. (3) Employment or in the younger addict, school attendance, is regarded as an obviously favorable sign of social adjustment, but frequently in using these measures little attention is paid to the quality of the employment situation or the level of performance in it. Yet these latter attributes are also relevant and valuable indicators of the degree of social adjustment.

Continued participation in therapy is also regarded as a favorable outcome sign; however, this variable can be viewed from several perspectives. Because drug addiction is a chronic disease, there are frequent occasions during the life of an addict that appropriate intercession by a therapist can blunt the thrust and impact of impulsive antisocial behavior. Further, it is a commonly held clinical impression that addicts make better social adjustments when they remain in therapy than when they leave it. On the other hand, the fact that the patient has a continuing need for therapy indicates that the disease process is not being altered in a definitive way.

As has been previously alluded to, reduction of sociopathic behavior, although a socially desirable consequence of medical and psychiatric therapy, is not the purpose of chemotherapy. The purpose of medical therapy is to treat the pathologic abnormalities that are responsible for sociopathic behavior. The facilitation of social adjustment by chemotherapeutic agents should not be regarded as an end in itself, but a consequence of the appropriate alteration of pathophysiologic processes.

II. Experimental Design

Prefatory to the discussion of individual experimental problems associated with the assessment of the efficacy of therapies in drug addiction, it is important to emphasize that definitive guidelines cannot be given. It has long been recognized by thoughtful experimenters studying drug addiction that serious attempts have not been made to assess the effectiveness of therapies; however, many have felt that methods do exist for studying the impact of therapy on addictive states, but have not been applied. There are, however, a number of problems associated with the assessment of therapies in a clinical setting which at this time seem insurmountable.

1. Controls

The function of a control group in any experiment is to obtain an estimate of the impact of uncontrolled variables on outcome variables. In the study of chronic disease processes the control group provides a first estimate of the natural history of the disease. In the study of therapeutic measures in addiction, it is difficult to obtain an adequate control group for the experimenter cannot prevent subjects from leaving the experiment and participating in other therapeutic programs. If it could be assumed that newly elected treatment programs by the patient did not alter the course of the disease, this election by the patient would not change his value as a control subject. Unfortunately, the efficacy of alternative treatment programs has not been assessed, and this destroys the value of eloped control subjects. This is a serious problem because a large number of patients leave all treatment programs. Another problem created by patients leaving the treatment program is the bias created by patient selection. Thus, those who remain in treatment may well be self-selected subjects who are not representative of the addict population at large. For these reasons, the generalization that could be drawn from such an experiment would of necessity be quite limited.

2. Dropouts

It is important to continue to study and follow-up dropouts from both control and treatment groups for two reasons: (1) If it is not done, it is impossible to determine which changes are a consequence of patient selection and which are a consequence of treatment. (2) If the treatment itself has long-term adverse effects, these may be manifest in the dropouts of the treatment group and can be best detected by comparing the treatment dropouts with the control dropouts.

3. Use of a Blind Design

The use of a blind design is to prevent the expectations of the experimenters from biasing the outcome. Unfortunately, it is nearly impossible to eliminate charismatic influences of experimenters from affecting treatment outcome. Further, the fact that most chemotherapeutic agents produce effects (tastes, changes in mood, autonomic function, etc.) makes it easy for both the experimental subjects and observers to distinguish between treatment conditions. Thus, a blind design is difficult to maintain.

4. Random Selection of Patients

The principle of random selection, to prevent sample bias from confounding interpretation of treatment results, is not easily applied in any therapeutic setting but is particularly difficult to apply in the treatment of addicts. Well over 50% of all addicts will refuse either initially or during the course of treatment to participate in therapeutic endeavors. As has been previously emphasized, the self-selection process confounds the interpretation of treatment data.

Our knowledge of the natural history of the addictions will allow its definition within only the roughest limits and, as a consequence, small changes in outcome variables produced by therapeutic intervention cannot be detected with any degree of certainty. These difficulties perhaps can be partially surmounted through the use of large patient populations; however, the great expense of conducting such studies and the variability that is introduced by differences in procedures between investigators raise the questions of both their feasibility and value. For these reasons, efforts to determine the basic pathology of psychopathy and the development of meaningful methods of measuring it should have the highest priority in research commitments to this disease.

E. Detoxification

Early endeavors to treat narcotic addicts were directed to helping them "break their habits" by alleviating withdrawal symptoms and accompanying signs (cf., KOLB and HIMMELSBACH, 1938). To this end a number of treatments have been proposed. It was not until Dr. CHARLES SHULTZ' systematic study of the ability of treatments to relieve abstinence signs that the efficacy of detoxification procedures was first rigorously assessed. SCHULTZ (Mayor's Committee on Drug Addiction, 1930) used a 4-point ordinal scale (0-absent; +-moderate; ++-marked; +++-excessive) to measure the intensity of four abstinence sign complexes: (1) gastrointestinal, (2) muscular discomfort, (3) restlessness, and (4) prostration or weakness. SHULTZ studied a number of treatment regimes using a blind design and compared the outcome with control addicts who were abruptly withdrawn using an inactive placebo: (1) Narcosan, a suspension of lipids, proteins, and carbohydrates, was found to be inactive, perhaps even worsening the abstinence syndrome. It was theorized on the basis of the Meyer-Overton hypothesis concerning the relationship between narcosis and lipid solubility that Narcosan would neutralize toxic body substances produced to counteract all the effects of the narcotic. (2) Atropine and scopolamine were studied and found to be inactive in relieving any signs of abstinence, including gastrointestinal ones, and worsened restlessness. Some patients became irrational and delirious. (3) Sedative-hypnotics including the barbiturates amobarbital, aprobarbital, barbital and phenobarbital, paraldehyde, chloral hydrate, alcohol, and bromide were found to be ineffective in suppressing abstinence signs and frequently produced their own toxic effects which were manifest as irrational behavior. Magnesium sulfate had a quieting effect in some patients but not in others. (4) Only morphine and codeine relieved abstinence signs, and gradual dose reduction of these narcotics was found to be the most efficacious way of withdrawing patients.

HIMMELSBACH (1937) (see Sect. II, Chap. 3) developed a sophisticated way of clinically and objectively assessing the intensity of the narcotic abstinence syndrome in man and studied the ability of a number of substances to relieve abstinence (see Sect. II, Chap. 3). Rossium (bis-1-phenyl-3-methyl-5-pyrazolone) and thiamine were found to be ineffective (HIMMELSBACH, 1937, 1940). Prostigmine when administered with morphine chronically did not alter the intensity of abstinence (HIMMELSBACH et al., 1942). In an extensive study, HIMMELSBACH and ANDREWS (1943) found that thiamine, pyridoxine, atropine, prostigmine, and pentobarbital did not alter the morphine abstinence syndrome while codeine, morphine, and meperidine, in a dose-related manner, did.

ISBELL et al. (1948a, b, c) first observed that the abstinence syndrome of methadone-dependent patients had a latency of several days, was milder, and persisted longer than the morphine-abstinence syndrome. Because of the relative mildness of the abstinence syndrome, he proposed it as a useful agent for detoxifying addicts (ISBELL, 1950). Until quite recently, it has been felt that detoxification should be conducted in a drug-free institutional setting using a reduction schedule determined by the physician. Recently RAYNES and PATCH (1973) detoxified patients using either a schedule where methadone was reduced 5 mg daily (control) or where the patient could determine when the methadone dose would be reduced. In the latter group, however, if the patient did not request a decrease in dose it was automatically reduced 5 mg every 3 days. The number of patients who received day care and were successfully detoxified using the semivoluntary procedure was twice that of the control group; however, for the total patient population, the two procedures differed only slightly.

If narcotic-dependent patients are gradually withdrawn using either an intermediate (morphine) or a long-acting (methadone) narcotic, only minor discomforts will be experienced. Tranquilizers do not markedly alter the morphine- or heroin-abstinence syndrome. Although several investigators have reported that chlorpromazine and reserpine reduce the intensity of abstinence of narcotic-dependent subjects, FRASER and ISBELL (1956) were unable to confirm these findings. In patients stabilized on 240 mg of morphine a day, neither chlorpromazine nor reserpine alter the intensity of abstinence as assessed using the Himmelsbach technique.

F. Maintenance Therapy

I. Introduction

The value and appropriateness of maintenance therapy for the treatment of narcotic addiction have been debated by the medical community for over 50 years. With the Supreme Court decision (1922) that narcotic addicts should not be given narcotics for self-administration and the interpretation that the administration of narcotics for the "gratification" of addiction was not proper professional practice and therefore illegal, narcotic dispensaries (clinics) were established in several cities. These clinics were closed by the Secretary of the Treasury on the recommendation of the AMA.

The history of these clinics was subsequently reexamined by the AMA's Council on Mental Health in response to a request by a delegate for the endorsement by the AMA for the legal distribution of narcotics to addicts. The report of the AMA was published in 1957 (Council on Mental Health, American Medical Association) and was against the reestablishment of narcotic clinics for the following reasons: (1) They could see no feasible way for the distribution of narcotics to addicts that would prevent their diversion into illicit traffic. (2) There was the possibility that certain physicians might become "script doctors" providing addicts with narcotic prescriptions for profit. These reasons probably were also responsible for the AMA's 1924 recommendation that Federal and State government "narcotic clinics" and ambulatory treatment of narcotic addicts by private physicians be discontinued. (3) It was felt that the underlying pathology of the addict would not be altered by maintenance and that maintenance would not be effective.

A different position was taken by the New York Academy of Medicine (1955, 1963) who felt that there were several legitimate reasons why maintenance would be effective and why they questioned the wisdom of prohibiting the giving of narcotics to addicts: (1) It was felt to be unjust to treat a sign of a disease, drug use, as a crime. (2) They felt by taking the profit of narcotics by making narcotics easily available and cheap, the addict pusher would be eliminated, and hence the spread of the use of narcotics. (3) They further felt that the control of the administration of narcotics by health professionals would provide a better circumstance for treatment and rehabilitation than street use.

An attempt to experimentally resolve these issues started when DOLE initiated a methadone maintenance program at Rockefeller University. In their first report, DOLE and NYSWANDER (1965) reported on 22 patients who were stabilized on 50–150 mg of methadone. These patients reported a disappearance of their "narcotic hunger," only rarely had urines that were positive for morphine, and 14 of the patients were employed. The only medical problem associated with methadone maintenance was constipation. As Dole's program expanded, the results continued to be encouraging. DOLE et al. (1966) reported on 128 heroin addicts who had been admitted to their methadone maintenance program. The retention rate was 89% (114 patients), and of these 83% were employed as opposed to 17% prior to maintenance therapy.

These successes of methadone maintenance in the hands of DOLE and his collaborators encouraged many other therapists to employ the treatment modality and in less than a decade after its initiation over 80000 patients are currently being maintained on methadone. The data that have been obtained from this experience suggest that both the protagonists and the antagonists of maintenance therapy were correct in some of their predictions and wrong in others. As has been previously discussed, there are probably no currently available methods for rigorously assessing the efficacy of narcotic addiction treatment modalities. Analysis of the data bearing on the efficacy of methadone maintenance clearly demonstrates the difficulties that are encountered in trying to objectively evaluate both the usefulness and the disadvantages of maintenance therapy. The Proceedings of the National Conferences on Methadone Treatment have been published (1968, 1969, 1970, 1972, 1973). These proceedings not only record much of the ongoing debate between protagonists and antagonists of maintenance therapy but most of the problems, successes, and aspirations of professionals who are earnestly trying to effectively treat narcotic-dependent patients. Many of the clinical reports at these conferences do not deal with the problem of efficacy in a critical manner. The following discussion will reiterate the problems of assessing efficacy of therapeutic endeavors as they relate to methadone maintenance therapy as well as review and attempt to evaluate reported data from several large methadone maintenance programs.

There have been a number of criticisms of the statistical analysis and interpretation of data concerning the outcome of patients in methadone maintenance programs which include: (1) Participants are not typical of the addict population and represent a subpopulation with a better than average prognosis. (2) The quality of ancillary care given methadone maintenance subjects in some treatment programs is superior to that available in other types of treatment clinics. (3) The performances of the dropouts have not been included in the estimates of efficacy, only that of the successes. (4) Some of the measures of efficacy such as urine testing were not relevant since the subjects were being sustained on high levels of narcotics. (5) Patients in treatment programs receive preferential treatment by law enforcement agencies and courts (MARTIN, 1970; EPSTEIN, 1974).

II. Acceptance and Retention Rates

The overall acceptance of methadone maintenance as treatment is probably between 20–25% for the total addict population. In 1973 when the number of addicts may have decreased to 360000 from 500000 in 1971, there were approximately 80000 patients on methadone maintenance (22%) in the United States. In New York City where the total addict population is approximately 150000, 30000 patients were in methadone maintenance programs (20%). It is unlikely that the percentage of patients who accept methadone maintenance will increase greatly with time since the rate of induction of new patients into maintenance programs has stabilized or decreased in many regions. The attrition rate from methadone maintenance programs appears to be about 10–15% per year in the most successful programs. The retention rate in the largest, best established and most rigorously assessed programs varies greatly. Thus, the methadone maintenance program of New York City under the direction of VINCENT DOLE had a retention rate of 80% (GEARING, 1970b). The patient population in this program had more whites, fewer Puerto Ricans and Negroes, and were older than the subjects comprising the 1967 New York City Narcotics Register (GEARING, 1970a). The Santa Clara County Methadone Program which is comprised chiefly of older (mean age 30 years) whites and Mexican-Americans has had a retention rate of 65 or 77%, depending on how it was calculated (WILMARTH and GOLDSTEIN, 1974). On the other hand, the Illinois Drug Abuse Program whose narcotic addict patient population is comprised primarily of young black addicts has had a retention rate of about 50% of their methadone ambulatory subjects and less than 10% of their methadone residential subjects (SENAY et al., 1973). It is thus obvious that a variety of types of parameters related to patient selection (age, race, etc.) may influence retention rate. On the basis of current evidence, it is unlikely that over 25% of narcotic addicts will accept maintenance therapy and that no more than 75% of those who initially accept it will remain in therapy.

III. Efficacy

Several of the large methadone maintenance programs have attempted to assess the effectiveness of their programs in decreasing illicit drug use and criminality and increasing socially desirable behavior, such as employment, in an objective and systematic manner. Since it is difficult to interpret the therapeutic significance of decreased illicit drug use in patients who are being given much larger doses of narcotics by treatment programs than they have ever used on the street, this summary of results will be primarily addressed to criminality as measured by arrests and social adjustment as measured by employment as relevant outcome variables. Further, although an enormous effort has gone into the testing of urines of methadone-maintained patients, little of these data has been acquired, analyzed, and published in a systematic manner. The studies of DOBBS (1971) and BROWN et al. (1973) indicate that at least 40% of the urines of patients being maintained on methadone are positive for drugs. JONES and PRADA (1977) have found that the administration of methadone to dogs that maintain a dependence by self-administering morphine only transiently decreases the quantity of self-administered morphine. Thus, it may be of importance to conduct definitive experiments determining the effects of methadone maintenance on drug-seeking behavior of a general population of addicts. This is not to discount the social and economic significance of maintenance therapy on decreased illicit drug use for it undoubtedly is closely related to decreased criminality. Methadone maintenance does not, however, in any sense cause a decrease in need for narcotics or other drugs.

1. New York City Methadone Maintenance Program (NYCMMP)

As previously described, the patient population (2205) studied by GEARING (1970a) was predominantly white and somewhat older than the average New York City heroin addict. Further, they exhibited somewhat less criminality (10–15%) prior to participating in the NYCMMP than that of a contrast group of patients from a Detoxification Unit. The patients in the NYCMMP showed markedly less criminality (94%) than the contrast group. The percentage of time that the patients were employed prior to being admitted to the NYCMMP for different periods of time ranged from 15–25%. While in treatment the percentage ranged from 35–50%. There was a clear trend for employment to increase the longer patients remained in treatment. GEARING'S report (1970b) to the Third National Conference on Methadone Maintenance on 3485 patients gave similar results. JOSEPH and DOLE (1970) found that parolees and probationers had a somewhat lesser retention rate but exhibited less criminal activity and more involvement in constructive activities.

2. Illinois Drug Abuse Program (IDAP)

Data on criminality and employment for those patients in the IDAP on methadone maintenance have not been separately reported (SENAY et al., 1973); however, since approximately 85% of the patients are in methadone maintenance, the values for outcome variables for the program probably represent for the most part the impact of methadone maintenance. The patients in this program were predominantly black and were younger than the patient population studied by GEARING (1970a, b, 1972). In a subgroup of 218 (ca. 25% of admission) patients whose arrest records were studied during 2 years prior to treatment and up to 134 weeks of treatment, criminality was decreased 62%. Ninety-six percent of the crimes committed were for vice, narcotic offenses, and theft. Crimes against persons, however, increased 46% (10–15) while patients were in treatment. The mean percent of patients employed at time of admission to treatment was 28% as compared to 47% at the time the study was conducted. In contrast to Gearing's findings, the probability of a patient being employed was not related to the length of stay in treatment.

3. Santa Clara County Methadone Program (SCCMP) (WILMARTH and GOLDSTEIN, 1974)

The incidence of arrest was also markedly decreased in SCCMP, a predominantly white male patient population with a mean age of 30.3 years. When calculated on the basis of active patients and dropouts, there was, respectively, a 70% reduction and an 82% reduction in criminality for those patients on methadone maintenance. Only 26% of the patients who were on methadone maintenance were employed full-time as opposed to 21% of patients who were not receiving methadone maintenance.

4. California Department of Corrections Methadone Maintenance Program (CDCMMP)

The patient population of the CDCMMP (JONES and BERECHOCHEA, 1973) was made up of predominantly white male addicts of which 38% were paroled felons and 62%

were civilly committed addicts. The percentage of patients in treatment who were arrested was approximately 35% less than both a retrospective and an independent control group. This decrease was statistically significant. Forty-seven percent of the patients were employed full-time while in active treatment; whereas, only 28% were employed at time of entry into the program.

5. Pittsburgh Black Action Methadone Program (PBAMP) (Cleveland et al., 1974)

The PBAMP is a predominantly black innercity methadone maintenance program in which 62% of the patients were under 30. Data obtained from nurse and hospital records of 300 patients admitted to the program from June 1969 to September 1972 were analyzed. The attrition in this program was slightly over 20% a year. About 6–7% of the urines of patients in maintenance therapy were positive for narcotics. The incidence of criminality was reduced by about 50% in those patients who remained in therapy but was unchanged in dropouts. Methadone maintenance did not change employment histories.

6. Drug Abuse Reporting Program (DARP)

DARP has analyzed data obtained from federally supported treatment programs. Sells (1975) has described the 36 treatment programs and the 15831 patients admitted from June 1971 to June 1972 to these programs. Gorsuch (1975) reported on the outcome of over 3500 patients on methadone maintenance and compared their outcome with approximately 5000 patients treated without drugs. The retention rate of the methadone-maintained patients (60%) was higher than the no-drug treatment programs (14–42%) (Joe and Simpson, 1975). Criminality and drug use were decreased and to a lesser extent employment and productive activities increased in the methadone-maintained patients.

7. Summary

From these varied treatment programs with experiences with approximately 7500 patients, the following conclusions can be drawn: (1) Criminal behavior of patients treated with methadone maintenance is reduced by one-third to two-thirds, indicating that a substantive part of their illicit activity is associated with drug acquisition. In support of this conclusion are the observations of Dole et al. (1969) in 12 methadone-maintained prisoner-addicts and 16 nonmaintained prisoner-addicts. Fifteen of 16 nonmaintained prisoner-addicts were reincarcerated within a year of release while only 3 of 12 methadone-maintained prisoner-addicts were. (2) Methadone-maintained patients have shown a modest increase in involvement in social productive activity such as employment but the magnitude of this increase varies greatly from one program to another. In some programs the rate of employment of methadone-maintained patients nearly doubled while in other programs there was no significant change. The characteristics of the populations studied indicate that a certain amount of patient selection had occurred in all programs and that the prognosis of the patients in these programs is probably better than that of the general addict population. Further, there were other social forces that helped decrease sociopathic behavior. The programs described are among the best administered and most closely supervised. For these reasons the results herein summarized probably are not typical and are superior to those obtained in most maintenance treatment programs.

Another very important measure of the success of treatment programs is the addict mortality rate. Here the data do allow a clear and definitive judgment to be made as to whether patients on methadone maintenance have a lower death rate than untreated addicts for several reasons. Although a death rate among the addict population at large has been assumed to be approximately 1% per year, the population base upon which this estimate is made is uncertain. Further, the base upon which the methadone maintenance mortality rate is calculated has a large impact on the value. If the calculation includes both patients who are in active treatment as well as those who have been discharged, the death rate among methadone-maintained patients is approximately 1% per year (GEARING, 1970a; BADEN, 1970). On the other hand, if death rate is calculated on the base of patients while they are in methadone treatment, the rate is 0.76% per year (GEARING, 1972). CHAPPEL (1974) has reported that the death rate for the Illinois Drug Abuse Program was 0.6% for all patients entering treatment. In patients who left methadone maintenance programs, the death rate was 2.8% per year. CHAPPEL (1974) cited BARR who found that the annual death rate among addicts in Philadelphia was 2%, among methadone-maintained patients 4%, and among methadone maintenance dropouts 3.6%. It thus appears that although methadone maintenance has decreased certain types of sociopathic behavior, it has not markedly altered the high death rate of addicts.

IV. Toxicity

Methadone is a narcotic analgesic which is approximately equipotent to morphine in producing analgesia (BEAVER et al., 1967), euphoria, and feelings of well-being as well as miosis when administered subcutaneously or intravenously (NUTT and JASINSKI, 1973; MARTIN et al., 1973b). When administered orally it is one-half as potent as it is when administered parenterally (BEAVER et al., 1967; MARTIN et al., 1973b). When administered chronically, methadone is approximately four times as potent as morphine in producing and maintaining physical dependence. The physical dependence produced by methadone is qualitatively similar to that produced by morphine and heroin (ISBELL et al., 1948a, b, c,; MARTIN et al., 1973b). When methadone is abruptly withdrawn from patients who are dependent upon it, a typical morphine abstinence syndrome becomes manifest which has a slower onset, is less intense, and persists for a longer period of time than morphine (ISBELL et al., 1948a, b c,; JAFFE et al., 1970; MARTIN et al., 1973b). Reports which have appeared in the literature stating that patients dependent on methadone either do not have an abstinence syndrome or a very mild one when abstinent are not correct. In studies in which methadone was administered in known dose levels, in which the administration of every dose was verified under carefully controlled conditions and in which the obtaining of additional medications was precluded, a typical morphine- or heroinlike abstinence syndrome has been seen by experienced observers (ISBELL et al., 1948a, b, c; MARTIN et al., 1973b).

Withdrawal signs have also been reported in newborn infants born of mothers maintained on methadone. The abstinence syndrome seen in these newborns

(REDDY et al., 1971; RAJEGOWDA et al., 1972; ZELSON et al., 1973; ZELSON, 1973) is similar to that seen in infants born of heroin-dependent mothers (COBRINIK et al., 1959; ZELSON et al., 1971) and consists of tremors, hypertonicity, hyperirritability, high pitched cry, vomiting, fever, respiratory distress, diarrhea, increased mucous secretions, sweating and, in some infants, convulsions. Although mild signs of abstinence may be seen during the first few days of life, the most severe signs of methadone abstinence are seen from the 7th to the 14th post-partum day (ZELSON, 1973). Sudden deaths in babies born of methadone-maintained mothers have been reported (PIERSON et al., 1972); however, the significance of this observation cannot be assessed at this time.

The prevalence of methadone-related deaths is increasing. Thus, in Washington, D.C. in 1971, of the 82 reported narcotic overdose deaths, 60 were attributed to heroin while 17 were attributed to methadone; however, for the period from March 1972 through February 1973, of the 55 opiate overdose deaths 29 were attributable to methadone and 11 to heroin (DUPONT, 1971). In 1971 there were 1268 deaths of narcotic users in New York City and of these 320 were associated with methadone use (25%). Of these, 81 were not in methadone maintenance programs and were presumed to have taken methadone in place of heroin (BADEN and TUROFF, 1973). In the first 6 months of 1973 in New York City there were 143 deaths in which methadone was implicated and an additional 104 in which both methadone and heroin were found on autopsy (FARBER, 1974). Methadone overdose deaths present a pathologic picture similar to that seen in heroin overdose cases characterized by pulmonary edema and congestion (KJELDGAARD et al., 1971; FRASER, 1971; THORNTON and THORNTON, 1974).

V. Types of Maintenance Therapy

1. High Dose Maintenance

DOLE and NYSWANDER (1965) in their initial studies of methadone maintenance administered methadone in dose levels of 50–150 mg/day in a hospital setting. During induction, methadone was administered twice daily in increasing doses; however, patients were stabilized on a single daily dose of methadone of 80–120 mg/day (DOLE and NYSWANDER, 1968). When patients achieved their stabilization dose level, they were discharged to outpatient community units where methadone was dispensed daily, urine samples obtained, and certain social services provided. DOLE felt that these high doses of methadone were necessary to completely abolish drug hunger and produce a high level of blockade against the effects of street heroin. Many have quarrelled with the use of the word "blockade" to redesignate the phenomena of "tolerance and cross tolerance" for methadone does not block the effects of other narcotic analgesics any more than it can block its own effects. In studies by MARTIN et al. (1973b) and JONES and PRADA (1975) it was observed that 4 mg of Dilaudid administered intravenously could still produce discernible changes in subjective state and pupillary diameter until the maintenance dose level of methadone was in excess of 50 mg/day, supporting Dole's position that high dose levels of methadone are necessary for the production of maximal degrees of cross tolerance. With the expansion of methadone maintenance programs, three innovations were introduced: ambulatory induction and stabilization, the issuance of take-home medication, and the prescribing of methadone for maintenance. Ambulatory induction and stabilization introduced by JAFFE and ZAKS (1968) proved to be much more economical than hospitalizing addicts and its popularity and use have increased; however, it minimizes contacts between the patient and the therapeutic team, and as a consequence, less is known about the patient, his narcotic use pattern, and his behavior while under the influence of methadone.

Because many patients objected to the inconvenience of daily visits to maintenance units to obtain their daily dose of methadone many clinics issued take-home methadone. The issuance of take-home methadone also decreased the operating costs of maintenance clinics which further increased the popularity of take-home medication. Most well-run clinics have reserved the take-home privilege of medication until a certain degree of reliability has been demonstrated by patients. Many methadone-maintained patients found that they had only to take a fraction of their maintenance dose (less than 50%) to prevent the emergence of abstinence signs and symptoms and were thus able to sell large portions of their medication. Further, many physicians who initiated their own methadone treatment programs prescribed large quantities of methadone which increased the quantity illegally diverted (HOOGERBEETS, 1970). As previously mentioned, the rapid increase in the incidence of methadone-related deaths is a consequence of this diversion.

A number of patients on methadone maintenance who receive a single daily dose complain of symptoms and signs of abstinence prior to taking their medication. In the studies of ISBELL et al. (1948c) and MARTIN et al. (1973b), no patient exhibited identifiable signs of abstinence for at least 24 h after their last dose of methadone. GOLDSTEIN (1972) compared on a blind basis the performance of patients who received a single oral dose of methadone daily and a placebo with a group of patients who received their daily methadone in two divided doses. There was no evidence that the once-a-day group had more symptoms or problems that were related to abstinence than the twice-a-day group.

2. Low Dose Maintenance

JAFFE (1970) was the first to compare low dose with high dose maintenance by studying these dose regimes in two groups of nearly matched patients. The mean dose of methadone for the low dose group was 36 mg/day; the mean dose for the high dose group was 100 mg/day. These groups were compared for retention rate in the program, employment, arrest rate, and narcotic use. There was no significant difference in retention rate for the two groups and by the 14th week of treatment approximately 50% of the patients from both groups still remained in continuous treatment. There was no significant difference between the incidence of drug use for the two groups; however, there was a trend for the low dose maintenance group to have more morphine-positive urines. This finding would be consistent with the hypothesis that high level maintenance is necessary for maximal cross tolerance. There were also no significant differences between the employment rates for the two groups although more of the high dose patients were employed than the low dose patients. Three patients in the high dose group and one patient in the low dose group were arrested while in treatment. JAFFE (1970) concluded that the use of low dose methadone maintenance does not significantly decrease the effectiveness of methadone maintenance. Although low dose methadone maintenance may not produce as high a degree of cross tolerance to heroin and other narcotic analgesics as high dose maintenance, it may decrease the quantity of methadone that is illegally diverted.

Knowledge of the dose of methadone by patients does not appear to increase complaining about the dose of medication or the incidence of imagined withdrawal symptoms (RENAULT, 1973). GOLDSTEIN et al. (1975) allowed a group of methadone-maintained patients latitude in adjusting their dose. For the group there was a trend to modestly increase their dose; however, there were some subjects (15%) who decreased their dose while others (30%) increased it. The decreasers tended to have increasing numbers of dirty urines while the increasers had progressively fewer dirty urines. The average for all patients was about 1 in 5 urines being dirty. Most of the patients and all of the staff preferred having the patients adjust their dose. It decreased unpleasant negotiations between the staff and patients. For this reason and because it may be a practical and simple way of adjusting for differences between patients in their metabolism and sensitivity to methadone, GOLDSTEIN et al. (1975) recommended its adoption with the following safeguards: (1) No take-home methadone at high dosages, (2) dose increments and decrements should be small, and (3) changed no more frequently than at weekly intervals.

3. Levomethadyl (LAAM, *dl*-α-acetylmethadol, l-α-acetylmethadol)

FRASER and ISBELL (1952) found that when LAAM was administered parenterally it had a latency of onset of action of 4–6 h, but when administered orally had a more rapid onset of action. Further, it had to be given orally only once every 72 h to morphine-dependent patients to effectively suppress abstinence. When dependent patients were stabilized on LAAM and abruptly withdrawn, signs of abstinence emerged slowly. JAFFE et al. (1970) compared the emergence of abstinence in methadone-maintained patients (20–55 mg/day) with patients maintained on *dl*-α-acetylmethadol (24–66 mg/day). Methadone was administered daily, while *dl*-α-acetylmethadol patients received their medications on Monday, Wednesday, and Friday. Abstinence symptoms became manifest between the 36th and 48th hour of withdrawal from methadone but did not become manifest until the 96th hour of abstinence following *dl*-α-acetylmethadol withdrawal. Subsequently, JAFFE and SENAY (1971) showed that LAAM could also prevent the appearance of signs of abstinence for 72 h. ZAKS et al. (1972) compared a group of patients on methadone maintenance (100 mg/day) with a group of patients receiving low doses (30–50 mg Monday, Wednesday, and Friday) of LAAM and another group receiving high doses (80 mg Monday, Wednesday, and Friday). The degree of cross tolerance to heroin was greater and the incidence of positive urines was lower in the methadone and high dose LAAM patients than in the low dose LAAM patients. JAFFE et al. (1972) compared a group of patients stabilized on methadone (30–80 mg/day) with a group stabilized on LAAM (36–80 mg Monday, Wednesday, and Friday) and found that employment was higher, arrests the same, and drug use more in the LAAM than the methadone-maintained patients; however, none of the differences were statistically significant. LEVINE et al. (1973) found that doses of 70–80 mg of LAAM administered on Monday, Wednesday, and Friday were necessary to create a sufficient degree of cross tolerance such that the subjective and pupillary effects of 25 mg of intravenously administered heroin could not be detected by either subjects or observers.

JAFFE and SENAY (1971) feel that the long duration of action of LAAM may have two advantages over methadone in maintenance therapy: (1) The fact that it need be administered only 3 times a week will decrease the inconvenience of the dispensation of daily doses of maintenance narcotics from a clinic. (2) As a consequence, it should be possible to nearly abolish the policy of take-home medication and in this way decrease diversion of methadone into the illicit market. It would appear that somewhat better results are obtained with large triweekly doses of LAAM than with smaller doses. Since there should be little reason to issue take-home doses of LAAM and since all doses could be administered under direct supervision, illicit diversion should not be a concern in arriving at an appropriate stabilization dose. Because of the long latency of onset of LAAM, particularly by the parenteral route, it has been presumed that it is converted into an active metabolite or metabolites. The metabolism of LAAM has been shown to be quite complicated and as yet the metabolite which confers its long duration of action in man has not been definitely established but may be the normetabolite. In the studies conducted in man, the only differences that have been observed between LAAM and methadone have been reports of irritability by some patients receiving LAAM (JAFFE et al., 1970; ZAKS et al., 1972). These side effects as well as difficulties in adjusting the dose of LAAM may limit its acceptability.

4. Heroin

Heroin has been proposed for maintenance because its advocates feel that a larger number of patients will be attracted into maintenance therapy as a consequence of the belief that heroin is more euphorigenic than methadone. Evidence that is available, however, lends little support to this belief. The subjective effects produced by heroin are qualitatively similar to those produced by morphine and methadone (MARTIN and FRASER, 1961; MARTIN et al., 1973b; NUTT and JASINSKI, 1973) and experienced addicts cannot distinguish intravenously administered heroin from methadone (NUTT and JASINSKI, 1973). Further, the subjective states that chronically administered heroin and morphine produce are similar (MARTIN and FRASER, 1961), are not euphoric in nature, and are not in themselves reinforcing to the addict (HAERTZEN and HOOKS, 1969; MARTIN et al., 1973b). To be adequately stabilized patients must take heroin approximately every 6 h and for a level of cross tolerance to be established which would be equivalent to that produced by 50–100 mg of methadone would necessitate a daily dose of 100–200 mg/day. No ready solution has been offered to the potential diversion problems that would be associated with providing large quantities of take-home heroin to patients. Heroin as well as methadone maintenance is used in Great Britain (cf., BEWLEY et al., 1972); however, its use for maintenance purposes is restricted to a small number of well-qualified physicians and special treatment clinics.

VI. Rationale and Critique

It is apparent that both the protagonists and critics of maintenance therapy have been correct in many of their positions and predictions. Much of the criminal activity of the narcotic addict is associated with his efforts to acquire drugs. The data obtained from patients on methadone maintenance clearly indicate that the incidence of crimes related to stealing, vice, and narcotics is decreased when addicts are in maintenance therapy. There have been problems in delivering methadone to patients which have resulted in illicit diversion which may contribute to the total supply of illicit narcotics. The question of whether maintenance alters the underlying disease process responsible for narcotic addiction is debatable. There have been two proposed mechanisms whereby methadone may ameliorate disease processes responsible for narcotic-seeking behavior: (1) DOLE and NYSWANDER (1967) have proposed that the narcotic addict has a metabolic defect that is rectified or ameliorated by methadone and other narcotics; however, the defect has not been identified nor has any narcotic been shown to alter the defect. Further, neither the nature, cause, nor the manifestations of this metabolic defect have been demonstrated. The concept, however, should not be discounted for several reasons that have been argued in Section I, Chapter 1 and Section II, Chapter 1. Many narcotic addicts as well as other drug abusers have a psychopathic personality, hypophoric feelings, and perhaps increased needs and wants. There is some evidence that hereditary factors may play a part in drug abuse. Finally, there is evidence that chronic use of narcotics may produce long-lasting physiologic and personality alterations, some of which may be transmitted to offspring. There is no evidence that chronically administered narcotic helps any of these defects. Rather there is evidence that some of the personality defects of the addicts are made worse when narcotics are administered chronically (see below). (2) As a consequence of the development of high levels of cross tolerance to narcotics in methadone-maintained patients, they no longer exert any effect when used and thus conditioned drug-seeking behavior and abstinence may be extinguished. There are no data available as to whether extinction of either conditioned abstinence or conditioned drug-seeking behavior occurs as a consequence of methadone maintenance. The fact that drug experimentation is greater and more frequent early in methadone maintenance therapy than later could, however, be a consequence of extinction processes.

The relatively low acceptance rate (ca. 25%) and the high attrition rate (10–15%) of methadone maintenance per year has come as a surprise to many therapists. Part of the reason for their surprise is a misconception of the nature of the subjective effects produced by narcotics in the tolerant subject. It is commonly held that the stabilized narcotic addict has normal feeling states; however, this is not so. The subjective effects produced by narcotics in nontolerant addicts are ones of enhanced self-image, feelings of energy and efficiency and a decrease in hypophoric feelings (see Sect. II, Chap. 1). When narcotic analgesics are administered chronically, the predominant mood state is one of negative feeling states, hypochondriasis, and social withdrawal (HAERTZEN and HOOKS, 1969; MARTIN et al., 1973b). Whereas single doses of narcotics in the nontolerant addict relieve feelings of hypophoria and are thus reinforcing, chronically administered narcotics not only do not alleviate hypophoric feeling states but make them worse. Thus, this aspect of the effects of chronically administered narcotics is not only not positively but is negatively reinforcing. When signs of abstinence emerge, however, there is a further exacerbation of negative feeling states (MARTIN et al., 1973b) and the emergence of discomforting feelings which are relieved by a stabilizing dose of narcotic. Thus, the positively reinforcing property of narcotics in the tolerant and dependent subject is relief of symptoms of abstinence. The worsening of feelings of hypophoria in the maintained addict probably accounts for the increased use of other euphoriants such as amphetamines, sedative-hypnotics, and alcohol. Thus, there are substantive pharmacologic reasons for the ambivalence of many addicts toward methadone. The hope that heroin will prove more seductive or reinforcing to addicts is probably unrealistic for chronically administered heroin is neither more nor less reinforcing than methadone.

The narcotic-maintained subjects differ from normal subjects in other ways. Some are markedly sedated, many become socially withdrawn, others are impotent. Further, when they are unable to obtain or are withdrawn from methadone they not only exhibit a primary but a secondary abstinence syndrome (see Sect. I, Chap. 1 and Sect. II, Chap. 1). The prognosis of patients who leave methadone maintenance programs remains uncertain. There are anecdotal accounts of patients who have discontinued methadone maintenance and have remained abstinent. CUSHMAN and DOLE (1973) studied 53 methadone-maintained patients who had been detoxified. Five were not interviewed, 38 were still abstinent 9.4 (range 2–29) months following their last dose of methadone and 10 (21%) reentered methadone maintenance programs. PERKINS and BLOCH (1971) identified a subgroup of 72 patients who had been discharged from the methadone maintenance program of the Morris J. Bernstein Institute. These authors (PERKINS and BLOCH, 1970) had previously reported on the 512 patients who constituted the treatment group. Of the 72 patients, 6 were not located and 6 had returned to maintenance programs. Of the remaining 60, 6 had died (10%). Of the 54 living patients, 40 had been arrested (74%) and 39 were currently using heroin or other narcotics (72%). The differences in the outcome variables of these two studies are undoubtedly related to the reasons for the termination of methadone administration. For those patients who are discharged from methadone maintenance, the prognosis is quite poor and the possibility that it was worsened as a consequence of having been stabilized on methadone cannot be excluded. It has been shown experimentally that stabilization on and withdrawal from methadone will result in protracted abstinence and a worsening in personality characteristics that are associated with psychopathy (MARTIN et al., 1973b).

The diversion and ill-advised prescribing of methadone has created problems. To limit ill-advised use of methadone, the Food and Drug Administration (FDA) essentially restricted methadone use to approved treatment programs and hospital pharmacies. The American Pharmaceutical Association (APA) challenged the right of the FDA to administratively restrict the distribution of methadone or any other drug, and at this writing the court has upheld the APA position (HOLDEN, 1974). The logistics of administering maintenance programs continue to present problems to which proven solutions have yet to be found. Whether the gains achieved with maintenance therapy of narcotic addiction outweigh its disadvantages is still not clear although much evidence suggests that this may be so.

G. Narcotic Antagonists

I. Pharmacology

There are several excellent reviews and published symposia on narcotic antagonists which summarize what is known of their chemistry and pharmacology (WOODS, 1956; ARCHER and HARRIS, 1965; FRASER and HARRIS, 1967; MARTIN, 1967; LEWIS et al., 1971; BRAUDE et al., 1974). Only the salient features of their chemistry and pharmacology as they relate to their use as therapeutic agents for the treatment of narcotic addiction and dependence will be discussed. These issues are: (1) the relationship between their agonistic and antagonistic actions, (2) their potency, (3) duration of action, and (4) acceptability for depot administration. Cyclazocine was the first narcotic antagonist proposed as an agent for the treatment of narcotic addiction (MARTIN et al., 1965, 1966) following a series of studies delineating some of its pharmacologic properties in man. In these studies the agonistic actions of cyclazocine were characterized and differentiated from its antagonistic effects by showing that tolerance developed to its agonistic effects but not to its antagonistic effects (MARTIN et al., 1965, 1966). These studies gave rise to theoretical speculations concerning the agonistic and antagonistic actions of the narcotic antagonists (MARTIN, 1967). Figure 3 shows the structural formulae of narcotic antagonists that have been studied in man for their utility in treating narcotic addicts.

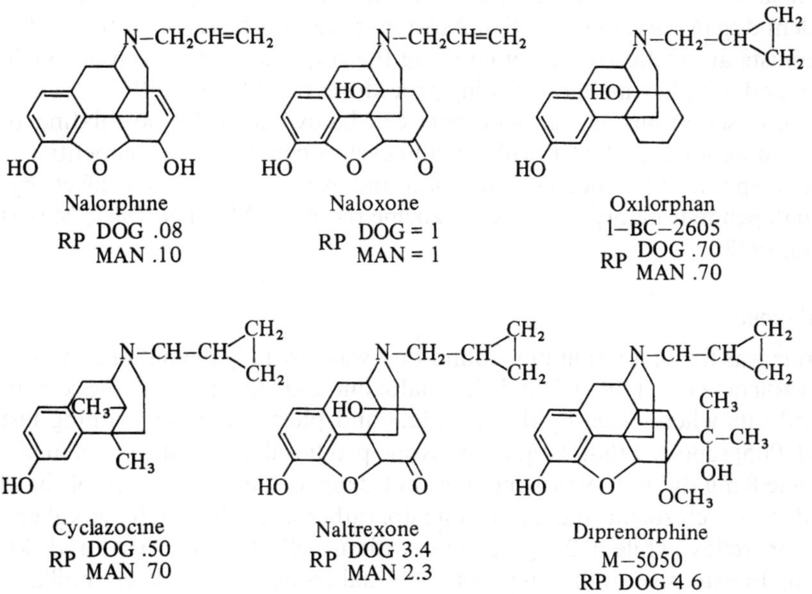

Fig. 3. Structures and relative potencies (*RP*) of narcotic antagonists

Agonistic and Antagonistic Actions of Narcotic Antagonists

a) Cyclazocine and Nalorphine

The agonistic actions of nalorphine, cyclazocine, and related antagonists have been carefully studied in man and the spinal dog. Their agonistic actions are similar in some respects to morphine and differ in others. Drugs with both morphinelike and cyclazocinelike agonistic actions produce analgesia, respiratory depression, and pupillary constriction in man (cf., MARTIN, 1967). There are, however, marked differences in their effects. Cyclazocine and nalorphine produce feelings of sleepiness, tiredness, apathy, drunkenness, irritability, racing thoughts, delusions and hallucinations in postaddicts and nonaddict subjects; whereas, morphinelike agents produce feelings of well-being, energy, and efficiency in postaddicts. Further, cyclazocine and nalorphine induce tolerance and a type of physical dependence in man which can be distinguished from that produced by morphine (Martin et al., 1965; Martin and Gorodetzky, 1965).

In the chronic spinal dog the cyclazocine-type agonists are effective depressants of the flexor reflex but have either a modest depressant or no effect on the skin twitch reflex and a modest depressant effect on body temperature. In contrast, morphinelike agonists are not only effective depressants of the flexor reflex but also of the skin twitch reflex and body temperature (Martin et al., 1973a). Nalorphine depresses the ipsilateral extensor thrust reflex in the chronic spinal dog; whereas, morphine increases it (Wikler and Frank, 1948; Wikler and Carter, 1953; Martin and Eades, 1964; Martin, 1967). Morphinelike agonists produce a dose-related miosis; whereas, cyclazocinelike agonists may produce either a marked miosis, little effect, a dose-related biphasic effect, or mydriasis (Martin et al., 1973a). They are effective antagonists of morphine in the chronic spinal dog and precipitate abstinence in the dependent dog (Martin et al., 1974). Most investigators have found that nalorphine-like agonists are ineffective in diminishing the response in the conventionally used tail flick and hot plate analgesic testing procedures (cf., Martin, 1967).

These observations as well as others can be explained by postulating that the narcotic antagonists of the nalorphine- and cyclazocine-type are competitive antagonists of morphine at the morphine receptor and exert their agonistic effects by acting on an independent nalorphine or cyclazocine receptor (Martin, 1967; Martin and Jasinski, 1972).

b) Naloxone

Naloxone was the first antagonist found that was virtually devoid of agonistic activity. Blumberg et al. (1961) found that naloxone did not have analgesic activity in mice and rats when tested by the hot plate and phenylquinone writhing tests. Lasagna (1965) reported that 2 mg of naloxone produced a mild degree of analgesia in man while 8 mg did not. Naloxone does not depress the flexor reflex of the chronic spinal dog as cyclazocine and nalorphine do, rather large doses (20 mg/kg) enhanced the flexor reflex while 0.2 mg/kg was without effect (McClane and Martin, 1967a, b). In extensive studies in postaddicts, naloxone did not affect pupillary diameter, produce subjective effects or physical dependence (Jasinski et al., 1967). In man naloxone was found to be over 7 times as potent as nalorphine as an antagonist but was not effective orally and had a short duration of action.

c) Naltrexone

Naltrexone which shares structural similarities with both cyclazocine and naloxone was selected for study in man because it was hoped that the 14-OH group would decrease agonistic properties and the methylcyclopropyl substitution on the nitrogen would confer a long duration of action. Blumberg et al. (1967) had already shown that naltrexone was a more potent antagonist than naloxone and that it was virtually devoid of analgesic activity. Indeed the first studies in man, which could not have been conducted without the interest and help of Drs. Blumberg, Jacobsen, and Pachter of Endo Laboratories, Inc., revealed that it was several times more potent

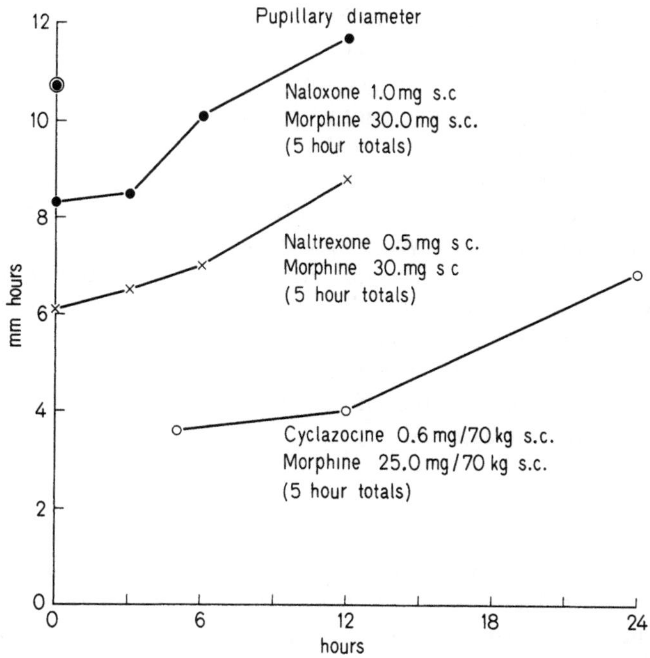

Fig. 4. Comparison of time action courses of naloxone, naltrexone, and cyclazocine in antagonizing pupillary constrictive effects of morphine. Experiment with cyclazocine differed from those with naloxone and naltrexone in that the dose level of morphine was 25 mg/70 kg. Results with cyclazocine were obtained from studies previously reported (MARTIN et al., 1966). Uppermost dot and open circle on ordinate are a measure of constrictive effects of morphine when administered with placebo. Lower dot and cross on ordinate are effects of morphine when administered simultaneously with naloxone and naltrexone. Remaining symbols are effects of naloxone, naltrexone, and cyclazocine when administered indicated number of h prior to administration of morphine. [Reproduced from MARTIN et al. (1971)]

than naloxone, that it had longer duration of action and was effective orally (MARTIN et al., 1971, 1973c). Further studies in the mouse, guinea pig, and rabbit showed that naltrexone was also several times as potent as naloxone and had a half-life of approximately 3 h as opposed to $1\,^1/_2$ h for naloxone (BLUMBERG and DAYTON, 1972, 1974). In man the half-life of subcutaneously administered naltrexone was from 10–12 h as compared to 4–6 h for subcutaneously administered naloxone and 24 h for subcutaneously administered cyclazocine (Fig. 4; MARTIN et al., 1973c). A single 15 mg orally administered dose of naltrexone produced a significant and sustained antagonism of the effect of morphine for 24 h. In the morphine-dependent chronic spinal dog naltrexone is 3.5 times more potent than naloxone in precipitating abstinence and has no significant agonistic actions (MARTIN et al., 1974). It does not have significant agonistic effects in man.

d) Oxilorphan (1-BC-2605)

Oxilorphan, the levorphanol congener of naltrexone, was found to be one-half as potent as pentazocine as an analgesic in the mouse writhing test and approximately equipotent to naloxone as an antagonist in the rat (PACHTER, 1974). In the morphine-dependent chronic spinal dog, oxilorphan was 0.7 times as potent as naloxone and 1.4 times as potent as cyclazocine in precipitating abstinence. In the nondependent spinal dog, oxilorphan is a partial agonist of the nalorphine or cyclazocine type whose intrinsic activity or efficacy is less than that of pentazocine and equal to that of nalorphine. It is more than 6 times as potent as pentazocine and equipotent to cyclazocine as an agonist (MARTIN et al., 1973a, 1974). In man oxilorphan was one-third to one-fourth as potent as cyclazocine in producing subjective effects and 1.2 times more potent than cyclazocine as an antagonist (JASINSKI and NUTT, 1973). The duration of oxilorphan's miotic and subjective effects was equal to or slightly less than those of cyclazocine. Oxilorphan produced cyclazocine-type subjective effects in man. To date neither patients' acceptance of chronically administered oxilorphan nor its efficacy in antagonizing the effects of acutely or chronically administered narcotics in man have been assessed.

e) Diprenorphine

Diprenorphine is a narcotic antagonist which was found by COWAN (1974) to be 20 times more potent than naloxone in antagonizing the analgesic effects of morphine and to be devoid of analgesic activity in the rat and mouse. It did not produce direct dependence when administered chronically to monkeys. BLUMBERG and DAYTON (1972) found diprenorphine to be approximately equipotent to naltrexone, which in turn was twice as active as naloxone in antagonizing oxymorphone-induced Straub tail reaction in mice and narcosis in rats. It was devoid of analgesic activity in mice but had limited analgesic activity in rats. In the chronic spinal dog, diprenorphine was about 10 times as potent as morphine in depressing the flexor reflex but had a ceiling effect equivalent to about 0.5 mg/kg of morphine. It did not alter the latency of the skin twitch reflex. It produced miosis. It was somewhat more potent than naltrexone in precipitating abstinence in the dependent chronic spinal dog (MARTIN et al., 1974). Although diprenorphine has not as yet been studied in man, its pharmacologic properties in animals would suggest that it would exhibit properties in man similar to those of nalorphine, being a partial agonist of the nalorphine or cyclazocine type, however, being approximately 30 times more potent than nalorphine as a competitive antagonist of morphine.

Naloxone and naltrexone are the only potent narcotic antagonists that have been shown to be devoid of agonistic activity. Diprenorphine and oxilorphan are also potent antagonists which have a lesser degree of agonistic activity than cyclazocine. Although naltrexone would appear to be the antagonist of choice in treating narcotic addiction, cyclazocine and, in all probability, oxilorphan and diprenorphine are or could be useful agents.

II. Pharmacologic and Therapeutic Rationales

1. Pharmacologic

SEEVERS and DENEAU (1963), as well as ISBELL and WIKLER (see MARTIN, 1967), in the quest for a nonaddicting analgesic, had recognized that chronic administration of mixtures of the narcotic antagonist nalorphine or levallorphan and morphine would produce a lesser degree of physical dependence than the same dose of morphine alone; however, neither the theoretical nor therapeutic implication of these findings was recognized. One of the factors that obscured interpretation of these data was the failure to recognize the fact that nalorphine and later cyclazocine were agonists and that this property was distinct from their morphine antagonistic property. When the effects of chronically administered cyclazocine (MARTIN et al., 1965) and nalorphine (MARTIN and GORODETZKY, 1965) were studied in man, it was shown that they were associated with the presence of both direct and cross tolerance to drugs of the nalorphine type as well as the development of physical dependence. It was apparent that the mode of action of nalorphinelike agonists (see above) was different than that of morphine but that they also acted as competitive antagonists of morphine (MARTIN, 1967). It was also observed that the cyclazocine abstinence syndrome had a latency of several days. These findings suggested that cyclazocine had a long duration of action in man. This attribute remains unexplained for in the rat, dog, and mouse the duration of action of cyclazocine is short compared to man.

The recent demonstration that naltrexone is metabolized to 6-β-OH-naltrexone in man and guinea pig but not in the dog or rat suggests that a metabolite of naltrexone and possibly cyclazocine may be responsible for their long duration of action (CONE, 1973; CONE et al., 1974, 1975). To demonstrate the long duration of action of cyclazocine in man, MARTIN et al. (1966) studied the time action course of single doses of both its agonistic and antagonistic effects in man and found that both cyclazocine's agonistic and antagonistic effects persisted for over 24 h. It was further demonstrated that when cyclazocine was administered chronically in dose levels of 4 mg per day protection was afforded against very large single doses of morphine (up to 120 mg) and heroin. In one patient 100 mg of heroin was administered intravenously with only modest effects being observed and felt. Further, when morphine (240 mg/day) was administered chronically to patients who were receiving cyclazocine (4 mg) daily, only a modest degree of physical dependence was produced. These findings were interpreted as indicating that tolerance develops only to the agonistic but not the antagonistic effects of nalorphine- and cyclazocinelike drugs when they are administered chronically. It was not until studies of the effects of chronically administered naloxone were complete that this hypothesis was rigorously demonstrated (MARTIN, 1967).

2. Therapeutic

When cyclazocine was first proposed for the treatment of abstinent narcotic addicts, it was thought that several therapeutic goals might be obtained that would decrease the probability of relapse as well as prevent other adverse consequences of narcotic use. These beneficial goals were: (1) It would prevent the induction of physical dependence as a consequence of stress-initiated spree use of narcotics. (2) It would antagonize the pharmacologic actions of narcotic analgesics that are reinforcing (euphorigenic and physical dependence-producing effects and the associated abstinence syndrome) and in so doing allow the extinction of conditioned abstinence and conditioned drug-seeking behavior. (3) By preventing the development of physical dependence, cyclazocine treatment might allow the physiologic or pathologic extinction of the protracted abstinence syndrome with a consequent decrease in hypophoric feeling states and psychopathy. (4) When patients are taking therapeutic doses of cyclazocine or other narcotic antagonists, it is almost impossible for them to ingest an overdose. (5) It was also thought that the knowledge by the patient that he could not experience the effects or the consequences of opiate use might diminish anxieties that would be a consequence of his ambivalence toward narcotic use (MARTIN et al., 1966; MARTIN and GORODETZKY, 1967; MARTIN, 1968; MARTIN and SLOAN, 1968). It was further recognized that cyclazocine when administered either acutely or chronically had little or no reinforcing properties (MARTIN et al., 1965) and it was felt that only addicts who had a high level of motivation for remaining abstinent would voluntarily accept and continue with this treatment modality (MARTIN et al., 1966).

The complete scope of the utility of narcotic antagonists has yet to be explored and demonstrated. Only recently DAVIS and SMITH (1974) have shown that conditioned and unconditioned bar-pressing behavior for morphine by rats can be extinguished by pretreatment with naloxone. As previously mentioned, it is possible that they may in a substantive and definitive way intercede in psychopathologic processes. In addition, there may be special populations of addicts where antagonists may have particular utility (JAFFE, 1974). Thus, for both the young and the narcotic initiate, the antagonist may have special value by preventing the pharmacologic actions of narcotics from making them physically dependent. They could serve as adjuncts to other treatment modalities and processes such as civil commitment. Finally, they may be of help to methadone-maintained patients who have been voluntarily withdrawn by diminishing the impact of spree use of narcotics while they are still experiencing early and protracted withdrawal symptoms.

III. Therapeutic Trials

1. Cyclazocine

Because of its dysphoric agonistic properties, special precautions must be undertaken in stabilizing patients on a blocking dose of cyclazocine. Most investigators use a method originally described by MARTIN et al. (1966) in which patients are started on an oral dose of 0.1 mg twice daily or 0.2 mg daily, which only rarely produces disturbing side effects. This dose is then slowly and opportunistically increased, allowing tolerance to develop to cyclazocine's agonistic effects until a dose level of 4–6 mg per day is being administered. This dose level is more than enough to block the effects of most street doses of heroin, although very large doses (50–100 mg) can surmount the competitive antagonism. The dysphoric effects of cyclazocine, as well as its other agonistic effects, can be antagonized by naloxone and naltrexone (BLUMBERG et al., 1966; MCCLANE and MARTIN, 1967a, b; MARTIN et al., 1967; JASINSKI et al., 1968), and this property has been used by RESNICK et al. (1971) to accelerate the induction of patients on cyclazocine by administering large doses of naloxone orally with cyclazocine when needed and requested.

JAFFE and BRILL (1966) first reported on their clinical experiences with cyclazocine in the treatment of 11 predominantly white middle class narcotic addicts. At the time that the report was written, some patients had been on cyclazocine up to 3 months. Three patients were discontinued, one because of an emergent hepatitis, probably unrelated to cyclazocine, and another for continued heroin use. The third patient, a physician, stopped cyclazocine after 5 weeks, but remained in treatment. The remaining patients seemed to have found the drug helpful and appeared to make better adjustments than they did prior to treatment. JAFFE and BRILL (1966) concluded that it "may be helpful to well-motivated narcotic users." Patients who experimented with narcotics found that they were without effect. Subsequently, LASKOWITZ et al. (1972) reported on a total of 35 patients treated with cyclazocine. Of these, 10 discontinued the program, 5 because of side effects, and 5 so that they could continue to use narcotics. Three patients died of causes unrelated to cyclazocine, 7 patients were placed in methadone maintenance programs, and one in a therapeutic community. Of the 14 successful patients, 7 were off cyclazocine and were abstinent. Seven continued to take cyclazocine and were opiate-free. Excluding the patients who died, 45% were drug-free. CHAPPEL et al. (1971) found that of 186 patients who requested abstinence treatment, 33 (18%) accepted and were placed on cyclazocine. There were no marked differences with regard to age, sex, ethnic origin, drug use, or criminality in the patients who were placed on cyclazocine and those who were treated with withdrawal alone. Of these 33 patients, 13 remained in aftercare for 14 months and 11 for 20 months. The attrition rate of these patients was high, with about one-half leaving the treatment program within a year. About 85% of the patients who remained in treatment were employed.

FREEDMAN et al. (1967) studied the effects of cyclazocine treatment in two groups of addicts, those being inducted into a therapeutic community and a second group for which no intensive rehabilitation measures were undertaken. The therapeutic community did not provide an adequate circumstance for the study of cyclazocine because of its stringent regulations and the high dropout rate. Nevertheless, 24 of 51 patients of this group who volunteered for cyclazocine therapy remained in treatment. Fifty-three patients comprising the second group volunteered for cyclazocine therapy. In these studies FREEDMAN et al. (1967) introduced a heroin challenge similar to the morphine challenge employed by MARTIN et al. (1966) and found that a stabilization dose of cyclazocine antagonized the subjective and pupillary effects of large doses of heroin. Although most patients were stabilized on a single 4 mg a day dose level, when side effects persisted the dose was divided and administered twice daily. Of the 53 patients in the second group, 30 remained in treatment, 22 dropped out, and one patient signed out against medical advice at the time the report was written. FREEDMAN et al. (1968) subsequently reported on 58 patients who were stabilized on cyclazocine. Of these, 27 continued in treatment for 2 months or more and 31 had dropped out. Anecdotically, the patients treated with cyclazocine had adequate sexual activity, reported less criminality, and had more interest in employment. RESNICK et al. (1970) compared a group of 14 patients who had accepted cyclazocine with 17 patients who had dropped out of the cyclazocine program. Of the cyclazocine-treated patients, 86% were married and 91% employed, while 18% of the dropouts were married and 29% employed. In addition, through the use of a questionnaire, they determined a "Q" score which essentially assessed the remembrance of the patient of the subjective effects of heroin. If patients remembered heroin as producing feelings of euphoria (see Sect. II, Chap. 1) and well-being, they were scored low. If they did not, they scored high. The patients who dropped out of the cyclazocine treatment program had somewhat lower "Q" scores (mean of 29.3) than did the cyclazocine-treated patients (mean of 39.7). Some patients have remained in cyclazocine treatment from 1–4 years and have shown a low rate of drug use (9%), a high rate of employment (86%), and have had a stable marital adjustment (71%) (RESNICK et al., 1971). RESNICK et al. (1971) believe that stable heterosexual relationships and a lack of need for heroin are good predictors for success in cyclazocine therapy. RESNICK et al. (1974a) have treated patients with doses of cyclazocine (10–30 mg) and have found that the duration of blockade can be thus extended to 48–72 h. Of a total of 140 patients who were placed on cyclazocine by FREEDMAN and his collaborators, 10% were drug-free for from 6–18 months, and 10–15% would drop out and then return (FINK, personal communication).

BANAY (1968) reported on 50 patients who started in a cyclazocine treatment program. At the time of his report, 30 (60%) were in treatment and doing well (drug-free, no criminal activity and working). A conference held in 1970 (FINK, 1970) revealed that at that time approximately 450 had been treated with narcotic antagonists (almost exclusively cyclazocine), and it was estimated that 40% of the patients accept antagonists as a treatment modality. KISSIN et al. (1973) found that of 842 patients who exhibited an interest in cyclazocine, 192 (23%) were stabilized on cyclazocine. In a study of 92 of these patients, criminality was reduced 76%, social adjustment was improved (79%), and heroin use decreased by 97%. There was only a modest decrease (20%) in secondary drug abuse. Although adequate comparative data were not available, the holding power of cyclazocine was less than that of methadone. KLEBER et al. (1974) found that 42% (12/27) of patients who were not eligible for or did not want methadone accepted cyclazocine therapy and had a somewhat lower incidence of drug use than a comparable control group.

SCHOOFF et al. (1973, 1974) compared the outcome of 20 suburban and 20 innercity addicts treated with cyclazocine with 5 suburban and 9 innercity control patients. The different groups of patients were approximately of the same age, had similar histories of heroin use, and had completed approximately the same number of years of schooling. Patients were hospitalized for 6 weeks, were detoxified, and were placed on cyclazocine after having been drug-free for 1 week. Attempts were made to involve both the patient and his family in therapy and vocational counseling and to place the patient in a job. Patients were maintained on cyclazocine from 4–6 months. Follow-up data were obtained on 37 of the 54 patients. Fifteen of the 25 cyclazocine subjects were regarded as successes (60%) while 10 were regarded as failures. Three of the 12 control subjects were regarded as successes (25%) while 75% were failures. Sixty-nine percent of urines of treated patients were negative while 12.5% of urines of control patients were negative.

During these clinical trials of cyclazocine several difficulties in its use were encountered: (1) Even in the hands of skilled therapists the dysphoric subjective effects of cyclazocine could not be completely avoided in all patients during the induction and stabilization procedure. (2) Stabilized patients would deliberately miss doses of cyclazocine and when its antagonistic effects had been partially dissipated would then engage in spree use of heroin. This type of behavior had two adverse consequences: (a) It allowed the patient to experience the reinforcing effects of heroin, which thus precluded either the psychologic extinction of conditioned drug-seeking behavior or abstinence or the physiologic extinction of protracted abstinence. (b) As a consequence of dropping one or two doses of cyclazocine, patients would lose part of their tolerance to the agonistic dysphoric effects of cyclazocine, and at the end of a heroin spree when cyclazocine was reinstituted, dysphoric and unpleasant subjective effects were experienced which discouraged its continued use. Most therapists using cyclazocine felt that its dysphoric side effects were partly responsible for its low acceptance by patients. It was felt that a longer acting antagonist or antagonist preparation that was devoid of dysphoric side effects would have greater therapeutic efficacy and patient acceptance.

2. Naloxone

In our initial studies of naloxone, when it was found that it was devoid of cyclazocine- or nalorphinelike agonistic effects (McClane and Martin, 1967a; Jasinski et al., 1967), studies were undertaken to determine both its duration of action and its oral efficacy. It was found that the antagonistic effects of a large dose (15 mg subcutaneously) of naloxone had markedly diminished 4–6 h after administration (Jasinski et al., 1967). Subsequently, a more precise estimate of its duration of action in man was made which indicated that it had a half-life of 4–6 h. Further, it was not as effective orally as cyclazocine. When naloxone was administered subcutaneously every 4 h in a dose of 15 mg (90 mg/day), it blocked the pupillary and subjective effects of 90 mg/70 kg of subcutaneously administered morphine. Fink et al. (1968) administered naloxone orally and chronically in dose levels ranging from 100–200 mg/day. It was given twice daily in divided doses. Only a slight reaction to 20 mg of heroin administered intravenously 4–6 h after the morning dose was observed in patients receiving 100–140 mg/day and no reaction in patients receiving 160 or 200 mg/day. The blockade produced by naloxone was demonstrable 10 h after an oral dose of 160 or 200 mg/day. Zaks et al. (1971) administered naloxone in single doses of 400–3000 mg. Doses of 800–3000 mg produced antagonism of the effects of 25 and 50 mg of heroin for 18 h and 3000 mg for 24 h. No toxicity was associated with these doses of naloxone although one patient complained of apathy, depression, and decreased appetite. Kurland et al. (1973a) has used naloxone in the treatment of parolee narcotic addicts. Fifty-two of these patients were placed on naloxone immediately after being paroled. Another 23 parolees, transferred from an abstinence program because of excessive drug abuse, were also treated with naloxone. Patients reported to the treatment facility daily between the hours of 6 and 9 p.m., at which time a urine sample was obtained and naloxone administered orally as a tablet. If patients were using narcotics, the dose level was progressively increased from 200 mg to 800 mg. If patients exhibited continued abstinence, naloxone was discontinued. Approximately 50% of the 75 patients who were placed in treatment remained in treatment for 6 months and 10% for 12 months. About twice as many of the direct admissions remained in treatment as relapsing patients who were transferred from abstinence programs. Kurland et al. (1973b) compared 39 naloxone-treated patients with 39 patients who received a placebo and 41 other control patients. There was no difference in the percentage of patients who remained in treatment among the three groups. It is important to emphasize two aspects of Kurland's studies. (1) Patients were not protected from the effects of heroin during the day and were subjected to episodes of precipitated abstinence when they returned to the clinic in the evening. This regime precluded extinction of conditioned abstinence and drug-seeking behavior, probably had aversive characteristics, and created anxieties in patients. (2) No outcome variables other than continuation in treatment were measured.

3. Naltrexone

Naltrexone may circumvent some of the difficulties associated with the use of both cyclazocine and naloxone (MARTIN et al., 1973c). If its acceptance by patients is greater than that of cyclazocine, this would support the view that the failure of patients to accept cyclazocine is due in part to cyclazocine's dysphoric side effects. Because of its potency and relatively long duration of action, a single daily oral dose should produce an adequate degree of blockade. At the time of this writing naltrexone has been chronically administered in dose levels of 50 mg/day or more to over 500 patients and appears to be a safe drug. A study was conducted under the auspices of the National Academy of Sciences in which addicts who received naltrexone in a dose level of 50 mg/day were compared with addicts who did not receive naltrexone (placebo). Although the study is not complete, the data to date indicate that the incidence of side effects and physiologic abnormalities has not been different between the two groups. In the initial studies of naltrexone in man, MARTIN et al. (1973c) demonstrated that in dose levels of 30 mg orally it did not produce detectable subjective effects, that it produced a modest increase in blood pressure and decrease in pupillary diameter and body temperature. When administered subcutaneously, it was several times more potent than naloxone and had a half-life of about 10 h. When administered orally, its blocking effects persisted for over 24 h. When administered in daily oral doses of 30–50 mg, it decreased the effects of subcutaneously administered morphine by over 20 times and markedly attenuated the development of physical dependence when morphine was also administered chronically. RESNICK et al. (1974b) in initial clinical trials with naltrexone found that with 20 and 30 mg oral dose levels about $^1/_3$ of the patients had complaints including feelings of tiredness, sluggishness, irritability, problems falling asleep, abdominal pains, nausea, and vomiting. These changes seen during the first 2 days of naltrexone administration may have been related to precipitated abstinence. When oral doses of 120–200 mg a day were administered, some patients complained of tiredness, sluggishness, irritability, and difficulty sleeping. No blood pressure changes or changes in body temperature were noted. Using doses of 120–200 mg, a significant antagonism of the effects of 25 mg of heroin administered intravenously was seen for 72 h.

SCHECTER et al. (1974) have studied that effects of naltrexone in dose levels up to 125 mg a day in two normal and 30 addict subjects. The effects most commonly reported by subjects receiving naltrexone were irritability, fatigue, difficulty in sleeping, dyspepsia, a decrease in appetite, and feelings of being sleepy and tired. Many of these symptoms were also reported by subjects prior to receiving naltrexone. No psychotomimetic or dysphoric effects such as those reported by patients receiving nalorphine or cyclazocine were recounted by these subjects.

Preliminary results of the National Academy of Science study indicate that patients receiving naltrexone stay in treatment longer than patients receiving a placebo.

IV. Discussion

The acceptance rate of narcotic antagonists by addicts is low. Probably 10–20% of addicts would accept antagonist therapy and of these probably 10–30% would remain in therapy for 6 months to a year. Existing data would suggest that this 1–6% of the addict population that would accept antagonist therapy would show substantive improvements in all aspects of their social performance including drug use, criminality, and employment performance. It is possible that the acceptance rate could be higher in other circumstances. At the present, federally funded treatment programs have more treatment openings available than there are addicts who will accept treatment. Since treatment programs are funded and staffed on the basis of the number of patients in active treatment, there is the natural tendency for program administrators to keep their census as high as possible by offering treatment modalities that are attractive to addicts, and rejecting treatment modalities that do not have a high patient acceptance. Although this attitude has an apparent face validity, the long-term outcome of patients who are treated with different treatment modalities is not known. For this reason no hard assessment of medical efficacy or total social cost can be made of various treatment modalities.

The narcotic antagonists do not alter the hypophoria, impulsiveness, or the egocentricity that are salient features of the psychopathic disease process. They, thus, lack a reinforcing and a therapeutic property that is important to most addicts. The studies of RESNICK et al. (1970, 1971) would suggest that addicts who do accept antagonist therapy are less psychopathic than those who reject it. The psychopathology of the narcotic addict is probably quite complex and involves not only the psychopathology associated with psychopathy, but conditioning processes and the learning of deviant patterns of behavior (FREEDMAN, 1968). The primary defect in most addicts is their psychopathy and extinction of conditioned behavior and rehabilitation will not be possible in most addicts until the psychopathic process is definitively treated. The treatment of the psychopathic process alone will also be inadequate and as many clinicians have felt, multiple types of treatment, possibly administered concurrently, will be necessary.

The extinction of conditioned drug-seeking behavior may have different manifestations in man than it does in animals. KLEBER et al. (1974) have reported that patients receiving antagonists frequently do not test its blocking effects by using narcotics. ROGER MEYER and his collaborators (personal communication) have allowed patients to perform a simple operant task to earn heroin in a controlled experimental setting. When patients were not treated with an antagonist, they would work to obtain nearly the maximum amount of heroin that was offered to them. However, when they were under the influence of an antagonist (naloxone, 500 mg orally 4 times a day, or naltrexone, 15 mg orally once a day) patients only infrequently worked for heroin. JONES and PRADA (1975) have studied narcotic-seeking behavior using an operant procedure during a cycle of methadone dependence (MARTIN et al., 1973c). They observed that most patients would complete an arduous task to obtain 4 mg of Dilaudid as long as they could perceive the effect of the drug, and coincidentally, as long as Dilaudid produced pupillary constriction. However, when a high level of tolerance was established, drug-acquiring work decreased. These observations raise some questions as to whether extinction in man is associated with the same pattern of extinction behavior as is seen in animals. Alternatively, extinction may not be the principal reason for patients receiving narcotic antagonists abstaining from narcotic use.

In the street setting it has been impossible to assess the role of conditioned abstinence and drug-seeking behavior on the clinical course of narcotic addiction because patients have voluntarily discontinued the use of the antagonist and thus reinforced conditioned processes. One possible solution to this particular problem would be to develop a narcotic antagonist depot that would provide a therapeutic level of antagonist for some days after its administration. The characteristics of the antagonist that might be suitable for such a depot are: (1) The antagonist must be very potent and have a long duration of action. Although it is possible that a pure antagonist which has the potency and duration of action of cyclazocine may be discovered where a release rate of 5–10 mg/day would produce a clinically effective level of blockade, a release rate of 25–50 mg/day for naltrexone will probably be necessary. To attain a blockade for 1 month, the depot must therefore contain a gram or more of naltrexone. Since most depot vehicles cannot carry more than 35% antagonist by weight and retain their integrity, the smallest depot device will probably have to weigh about 5 grams. (2) The antagonist must be virtually devoid of nalorphinelike agonistic actions. Because of variations of release rate of antagonists from depots, the emergence of psychotomimetic effects could not be precluded if the antagonist has agonistic actions. At the present time only naltrexone fulfills these characteristics; however, diprenorphine and oxilorphan may have sufficiently low agonistic properties to have utility. Other characteristics of the depot are that it is: (a) nonirritating, (b) nontoxic, (c) compatible with the antagonist, (d) biodegradable, and (e) can be implanted in a way that is acceptable to patients.

WOODLAND et al. (1973) suspended naltrexone in polylactide plastic (35% by weight) chips. This preparation was studied in the chronic spinal dog. The preparation was suspended in methyl cellulose and injected intramuscularly such that a dose level of approximately 17 mg/kg of naltrexone monohydrate was achieved. In the dog naltrexone has a blocking half-life of approximately 2 h. The naltrexone polylactide preparation produced a significant level of antagonism of morphine analgesic and pupillary effects for 29 days (MARTIN and SANDQUIST, 1974). GRAY and ROBINSON (1974) have prepared a zinc tannate salt of naltrexone and suspended it in aluminum monostearate gel. When this preparation was administered intramuscularly to rats in a dose level of 40 mg/kg, a significant level of antagonism to morphine's analgesic effects was achieved for over 12 days, and some antagonism was seen through the 21st day. Naltrexone has a very short half-life in the rat and the dog. Thus, both of these preparations can in all probability provide a clinically significant level of antagonism to street heroin in addict patients for about a month. The prospect of attaining a long-acting antagonist depot preparation is technically feasible and may permit a more definitive assessment of the role of conditioning in relapse. They may also provide the therapist with a treatment which will minimize the effects of capricious drug-seeking behavior.

H. Summary

Depending on one's perspective, the development of the chemotherapy of narcotic addiction can be said to have made rapid progress and at the same time be considered inadequate. Serious efforts to understand narcotic abuse, its diagnosis, and treatment date from the latter part of the 1920s, and in that time (1) the concept of physical dependence and its diagnosis and treatment have been discovered, (2) the importance of psychopathy in the disease process has been elucidated, (3) methods of detoxification which are both safe and spare the patient much suffering have been developed, and (4) two classes of drugs have been discovered which may have a significant therapeutic effect in some addicts. The largest portion of addicts will not accept treatment and of those who do less than half, although receiving some benefits, are treated definitively.

The major therapeutic benefits that have been observed with maintenance therapy with methadone or LAAM are that there is relatively high acceptance by addict patients, and a decrease in drug-seeking behavior and associated sociopathic behavior. Patients remain in therapy in part because they have a high level of physical dependence. There have been some disadvantages associated with methadone maintenance which include: (1) A diversion of methadone into the illicit market with a consequent increase in the availability of licit and illicit narcotics and associated problems such as addiction, illicit drug traffic, and overdose deaths. (2) Patients maintained on a high level of methadone are highly dependent not only on methadone but on the clinic or therapist from which methadone is obtained. With this circumstance is the ever-present danger that the therapist or clinic is in turn dependent on the addict for its livelihood. The script doctor is one of the most pernicious results of this relationship. (3) It is difficult to withdraw patients from maintenance therapy, and both the early and protracted abstinence syndromes are associated with an exacerbation of hypophoric feelings. (4) Maintenance therapy produces a number of side effects such as sedation and impotence that reach serious proportions in some patients.

Several difficulties have been encountered in the use of narcotic antagonists in treatment. They have no reinforcing properties in their own right and thus are not accepted by most addicts. Of lesser importance is the fact that patients on narcotic antagonists are markedly resistant to not only the reinforcing properties of narcotic analgesics but their therapeutic actions such as analgesia. This is not a serious difficulty; however, it will take a major educational effort to convince therapists who wish to give narcotics to produce analgesia that the antagonist blockade can be surmounted if large doses of narcotics are employed. Those narcotics that have toxic effects when large doses are administered cannot be used; however, potent analgesics such as morphine, methadone, and levorphan can be administered in dose levels that will surmount the blockade. Thus, some degree of analgesia may be produced by 50–100 mg of morphine or its equivalent in patients receiving 4 mg/day of cyclazocine or 50 mg/day of naltrexone. There are a number of advantages to the narcotic antagonists. (1) Naloxone and naltrexone are virtually devoid of side effects. (2) They do not produce physical dependence associated with drug-seeking behavior, and patients may be withdrawn safely without consequence. (3) It is virtually impossible for patients who are on a maintenance dose of an antagonist to overdose themselves with narcotics. (4) There appears to be a marked decrease in sociopathic behavior and a marked increase in socially acceptable behavior in patients who are treated with antagonists. (5) It is not anticipated that these agents will create any serious diversion problem.

The ultimate potential of presently available chemotherapeutic agents can only be roughly approximated. Maintenance therapy (methadone or LAAM) may be accepted by 25% of the addict population, 20% will remain in therapy and of these one-third to two-thirds will have a significant reduction in their sociopathic behavior depending on the criteria used and the population in treatment. Treatment with narcotic antagonists has had and will have in all probability, if used alone, a much lower acceptance, probably around 5%. Approximately one-half of these will remain in treatment and about two-thirds of these will have a significant reduction in sociopathic behavior. When viewed as percentages, the impact of these treatments seems small but when considered in terms of the number of patients who may benefit from treatment the impact is more substantive. Thus, if the number of narcotic addicts is between 300 and 500 thousand, then between 21 and 75 thousand will benefit substantively from maintenance therapy and 5–10 thousand from antagonist therapy.

A major challenge in the chemotherapy of narcotic addiction will be the treatment of psychopathy and its associated signs and symptoms of hypophoria, impulsivity, egocentricity, and exaggerated needs and wants. With the rapid growth of knowledge concerning the way drugs affect moods, feeling states, needs and appetites, this should be an attainable goal. Unfortunately, many drugs that do alter mood and need states also induce tolerance, and in some instances, physical dependence. It may be that the underlying mechanisms of these phenomena will have to be demonstrated to allow a rational approach in the search for definitive chemotherapeutic agents.

Abbreviations

AMA	American Medical Association
LAAM	(*dl*-α-acetylmethadol, *l*-α-acetylmethadol)
MMPI	Minnesota Multiphasic Personality Inventory

References

Archer, S., Harris, L. S.: Narcotic antagonists. Progr. Drug Res. **8**, 261—320 (1965)

Baden, M. M.: Methadone related deaths in New York City. Proceedings of the Second National Methadone Maintenance Conference, New York, N.Y., 1969. Int. J. Addict. **5**, 489—498 (1970)

Baden, M. M., Turoff, R. S.: Deaths of persons using methadone in New York City—1971. Presented at 35th Meeting, Committee on Problems of Drug Dependence, National Research Council, Chapel Hill, North Carolina, 1973

Banay, R. S.: Progress in ‘cyclazocine plus’. Correct. Psychiat. Soc. Ther. **14**, 187—195 (1968)

Beaver, W. T., Wallenstein, S. L., Houde, R. W., Rogers, A. A.: A clinical comparison of the analgesic effects of methadone and morphine administered intramuscularly, and of orally and parenterally administered methadone. Clin. Pharmacol. Ther. **8**, 415—426 (1967)

Bewley, T. H., James, I. P., Mahon, T.: Evaluation of the effectiveness of prescribing clinics for narcotic addicts in the United Kingdom (1968—1970). In: Zarafonetis, C. J. D. (Ed.) Drug Abuse; Proceedings of the International Conference, pp. 73—92. Philadelphia: Lea & Febiger 1972

Blachly, P. H.: Naloxone for diagnosis in methadone programs. J. Amer. med. Ass. **224**, 334—335 (1973)

Blumberg, H., Dayton, H. B.: Narcotic antagonist studies with EN-1639A (N-cyclopropylmethylnoroxymorphone hydrochloride). Fifth International Congress on Pharmacology, Abstracts of Volunteer Papers, p. 23, San Francisco, California 1972

Blumberg, H., Dayton, H. B.: Naloxone, naltrexone, and related noroxymorphones. In: Braude, M. C., Harris, L. S., May, E. L., Smith, J. P., Villarreal, J. E. (Eds.): Narcotic antagonists. Advances in Biochemical Psychopharmacology, Vol. 8. New York: Raven Press 1974

Blumberg,H., Dayton,H.B., George,M., Rapaport,D.N.: N-allylnoroxymorphone: A potent narcotic antagonist. Fed. Proc. **20**, 311 (1961)
Blumberg,H., Dayton,H.B., Wolf,P.S.: Counteraction of narcotic antagonist analgesics by the narcotic antagonist naloxone. Proc. Soc. exp. Biol. (N.Y.) **123**, 755—758 (1966)
Blumberg,H., Dayton,H.B., Wolf,P.S.: Analgesic and narcotic antagonist properties of noroxymorphone derivatives. Toxicol. appl. Pharmacol. **10**, 406 (1967)
Bowden,C.L., Maddux,J.F.: Methadone maintenance: Myth and reality. Amer. J. Psychiat. **129**, 435—440 (1972)
Braude,M.C., Harris,L.S., May,E.L., Smith,J.P., Villarreal,J.E. (Eds.): Narcotic antagonists. Advances in Biochemical Psychopharmacology, Vol. 8. New York: Raven Press 1974
Brown,B.S., DuPont,R.L., Bass,U.F., III, Brewster,G.W., Glendinning,S.T., Kozel,N.J., Meyers,M.B.: Impact of a large-scale narcotics treatment program: A six month experience. Int. J. Addict. **8**, 49—57 (1973)
Chappel,J.N.: Methadone and chemotherapy in drug addiction. Genocidal or lifesaving? J. Amer. med. Ass. **228**, 725—728 (1974)
Chappel,J.N., Jaffe,J.H., Senay,E.C.: Cyclazocine in a multimodality treatment program: Comparative results. Int. J. Addict. **6**, 509—523 (1971)
Cleveland,W.H., Bowles,B., Hicks,W., Burks,C., Rogers,K.D.: Outcomes of methadone treatment of 300 innercity addicts. Publ. Hlth Rep. (Wash.) **89**, 563—568 (1974)
Cobrinik,R.W., Hood,R.T. Jr., Chusid,E.: The effect of maternal narcotic addiction on the newborn infant. Review of literature and report of 22 cases. Pediatrics **24**, 288—304 (1959)
Cone,E.J.: Human metabolite of naltrexone (N-cyclopropylmethylnoroxymorphone) with a novel C-6 isomorphine configuration. Tetrahedron Letters **28**, 2607—2610 (1973)
Cone,E.J., Gorodetzky,C.W., Yeh,S.Y.: Biosynthesis, isolation and identification of β-hydroxynaltrexone. Pharmacologist **16**, 225 (1974)
Cone,E.J., Gorodetzky,C.W., Yeh,S.Y.: Biosynthesis, isolation, and identification of 6β-hydroxynaltrexone, a major human metabolite of naltrexone. J. pharm. Sci. **64**, 618—621 (1975)
Council on Mental Health, American Medical Association: Report on narcotic addiction. J. Amer. med. Ass. **165**, 1968—1974 (1957)
Cowan,A.: Evaluation in nonhuman primates: Evaluation of the physical dependence capacities of oripavine-thebaine partial agonists in patas monkeys. In: Braude,M.C., Harris,L.S., May,E.L., Smith,J.P., Villarreal,J.E. (Eds.): Narcotic antagonists. Advances in Biochemical Psychopharmacology, Vol. 8. New York: Raven Press 1974
Cushman,P., Dole,V.P.: Detoxification of rehabilitated methadone-maintained patients. J. Amer. med. Ass. **226**, 747—752 (1973)
Davis,W.M., Smith,S.G.: Naloxone use to eliminate opiate-seeking behavior: Need for extinction of conditioned reinforcement. Biol. Psychiatry **9**, 181—189 (1974)
Dobbs,W.H.: Methadone treatment of heroin addicts. Early results provide more questions than answers. J. Amer. med. Ass. **218**, 1536—1541 (1971)
Dole,V.P., Nyswander,M.: A medical treatment for diacetylmorphine (heroin) addiction. J. Amer. med. Ass. **193**, 646—650 (1965)
Dole,V.P., Nyswander,M.E.: Heroin addiction—A metabolic disease. Arch. intern. Med. **120**, 19—24 (1967)
Dole,V.P., Nyswander,M.E.: Methadone maintenance and its implication for theories of narcotic addiction. In: Wikler,A. (Ed.): The addictive states. Ass. Res. nerv. ment. Dis., Vol. 46. Baltimore: Williams and Wilkins 1968
Dole,V.P., Nyswander,M.E., Kreek,M.J.: Narcotic blockade. Arch. intern. Med. **118**, 304—309 (1966)
Dole,V.P., Robinson,J.W., Orraca,J., Towns,E., Searcy,P., Caine,E.: Methadone treatment of randomly selected criminal addicts. New Engl. J. Med. **280**, 1372—1375 (1969)
DuPont,R.L.: Profile of a heroin-addiction epidemic. New Engl. J. Med. **285**, 320—324 (1971)
Elliott,H.W.: The nalorphine (pupil) test in the detection of narcotic use. In: Clouet,D.H. (Ed.): Narcotic drugs; biochemical pharmacology. New York-London: Plenum Press 1971
Elliott,H.W., Nomof,N., Parker,K., Dewey,M.L., Way,E.L.: Comparison of the nalorphine test and urinary analysis in the detection of narcotic use. Clin. Pharmacol. Ther. **5**, 405—413 (1964)
Epstein,E.J.: Methadone: The forlorn hope. Public Interest No. **36**, 3—24 (1974)

Farber, M. A.: Methadone deaths exceed those from heroin here. The New York Times, pages 1 and 23, March 20, 1974
Fink, M.: Narcotic antagonists in opiate dependence. Science **169**, 1005—1006 (1970)
Fink, M., Zaks, A., Sharoff, R., Mora, A., Bruner, A., Levit, S., Freedman, A. M.: Naloxone in heroin dependence. Clin. Pharmacol. Ther. **9**, 568—577 (1968)
Fraser, D. W.: Methadone overdose. Illicit use of pharmaceutically prepared parenteral narcotics. J. Amer. med. Ass. **217**, 1387—1389 (1971)
Fraser, H. F., Harris, L. S.: Narcotic and narcotic antagonist analgesics. Ann. Rev. Pharmacol. **7**, 277—300 (1967)
Fraser, H. F., Isbell, H.: Actions and addiction liabilities of alphaacetylmethadols in man. J. Pharmacol. exp. Ther. **105**, 458—465 (1952)
Fraser, H. F., Isbell, H.: Chlorpromazine and reserpine. (A) Effects of each and of combinations of each with morphine. (B) Failure of each in treatment of abstinence from morphine. Arch. Neurol. Psychiat. (Chic.) **76**, 257—262 (1956)
Freedman, A. M.: Toward a rational approach to the treatment of narcotic addiction: Basic notions. In: Wikler, A. (Ed.): The addictive states. Ass. Res. nerv. ment. Dis., Vol. 46, pp. 378—386. Baltimore: Williams and Wilkins 1968
Freedman, A. M., Fink, M., Sharoff, R., Zaks, A.: Cyclazocine and methadone in narcotic addiction. J. Amer. med. Ass. **202**, 191—194 (1967)
Freedman, A. M., Fink, M., Sharoff, R., Zaks, A.: Clinical studies of cyclazocine in the treatment of narcotic addiction. Amer. J. Psychiat. **124**, 1499—1504 (1968)
Gearing, F. R.: Evaluation of methadone maintenance treatment program. Proceedings of the Second National Methadone Maintenance Conference, New York, N.Y., 1969. Int. J. Addict. **5**, 517—543 (1970a)
Gearing, F. R.: Successes and failures in methadone maintenance treatment of heroin addiction in New York City. Proceedings of the Third National Conference on Methadone Treatment, San Francisco, California, pp. 2—16. PHS Publication No. 2172. Washington, D.C.: U.S. Government Printing Office 1970b
Gearing, F. R.: Deaths before, during, and after methadone maintenance treatment in New York City. Proceedings of the Fourth National Conference on Methadone Treatment, San Francisco, California, pp. 493—494. New York: National Association for the Prevention of Addiction to Narcotics 1972
Goldstein, A.: Blind comparison of once-daily and twice-daily dosage schedules in a methadone program. Clin. Pharmacol. Ther. **13**, 59—63 (1972)
Goldstein, A., Hansteen, R. W., Horns, W. H.: Control of methadone dosage by patients. J. Amer. med. Ass. **234**, 734—737 (1975)
Gorsuch, R. L.: The impact of drug treatments on during-treatment criteria: 1971—1972 DARP admissions. Amer. J. Drug Alcohol Abuse **2**, 73—98 (1975)
Gray, A. P., Robinson, D. S.: Naltrexone zinc tannate: A prolonged-action narcotic antagonist complex. J. pharm. Sci. **63**, 159—161 (1974)
Haertzen, C. A., Hooks, N. T.: Changes in personality and subjective experience associated with the chronic administration and withdrawal of opiates. J. nerv. ment. Dis. **148**, 606—614 (1969)
Himmelsbach, C. K.: Clinical studies of drug addiction. II. "Rossium" treatment of drug addiction. Publ. Hlth Rep. (Wash.), Suppl. **125** (1937)
Himmelsbach, C. K.: Thiamine in the treatment of the morphine abstinence syndrome in man. J. Pharmacol. exp. Ther. **70**, 293—296 (1940)
Himmelsbach, C. K., Andrews, H. L.: Studies on modification of the morphine abstinence syndrome by drugs. J. Pharmacol. exp. Ther. **77**, 17—23 (1943)
Himmelsbach, C. K., Oberst, F. W., Brown, R. R., Williams, E. G.: Studies of the influence of prostigmine on morphine addiction. J. Pharmacol. exp. Ther. **76**, 50—56 (1942)
Holden, C.: Methadone: Court ruling threatens FDA regulations. Science **185**, 46—47 (1974)
Hoogerbeets, J.: Methadone in Miami. Proceedings of the Second National Methadone Maintenance Conference, New York, N.Y., 1969. Int. J. Addict. **5**, 499—506 (1970)
Isbell, H.: Manifestations and treatment of addiction to narcotic drugs and barbiturates. Med. Clin. N. Amer. **34**, 425—438 (1950)
Isbell, H.: Rapid diagnosis of addiction to morphine. J. Amer. med. Ass. **154**, 414 (1954)

Isbell, H., Eisenman, A. J., Wikler, A., Frank, K.: The effects of single doses of 6-dimethylamino-4-4-diphenyl-3-heptanone (Amidone, methadon, or 10820) on human subjects. J. Pharmacol. exp. Ther. **92**, 83—89 (1948a)

Isbell, H., Vogel, V. H., Chapman, K. W.: Present status of narcotic addiction, with particular reference to medical indications and comparative addiction liability of the newer and older analgesic drugs. J. Amer. med. Ass. **138**, 1019—1026 (1948b)

Isbell, H., Wikler, A., Eisenman, A. J., Frank, K., Daingerfield, M.: Liability of addiction to 6-dimethylamino-4-4-diphenyl-3-heptanone (methadon, amidone or 10820) in man. Arch. intern. Med. **82**, 362—392 (1948c)

Jaffe, J. H.: Further experience with methadone in the treatment of narcotics users. Proceedings of the Second National Methadone Maintenance Conference, New York, N.Y., 1969. Int. J. Addict. **5**, 375—389 (1970)

Jaffe, J.: Pharmacological and societal factors in the management of drug dependence. In: Braude, M. C., Harris, L. S., May, E. L., Smith, J. P., Villarreal, J. E. (Eds.): Narcotic antagonists. Advances in Biochemical Psychopharmacology, Vol. 8. New York: Raven Press 1974

Jaffe, J. H., Brill, L.: Cyclazocine, a long acting narcotic antagonist: Its voluntary acceptance as a treatment modality by narcotics abusers. Int. J. Addict. **1**, 99—123 (1966)

Jaffe, J. H., Schuster, C. R., Smith, B. B., Blachley, P. H.: Comparison of acetylmethadol and methadone in the treatment of long-term heroin users. A pilot study. J. Amer. med. Ass. **211**, 1834—1836 (1970)

Jaffe, J. H., Senay, E. C.: Methadone and *l*-methadyl acetate. J. Amer. med. Ass. **216**, 1303—1305 (1971)

Jaffe, J. H., Senay, E. C., Schuster, C. R., Renault, P. R., Smith, B., DiMenza, S.: Methadyl acetate *vs* methadone. A double-blind study in heroin users. J. Amer. med. Ass. **222**, 437—442 (1972)

Jaffe, J. H., Zaks, M. S.: Experience with the use of methadone. Proceedings of the First National Conference on Methadone Treatment, New York, N.Y., pp. 23—29. New York: National Association for the Prevention of Addiction to Narcotics 1968

Jasinski, D. R., Martin, W. R.: Evaluation of a new photographic method for assessing pupil diameters. Clin. Pharmacol. Ther. **8**, 271—272 (1967)

Jasinski, D. R., Martin, W. R., Haertzen, C. A.: The human pharmacology and abuse potential of N-allylnoroxymorphone (naloxone). J. Pharmacol. exp. Ther. **157**, 420—426 (1967)

Jasinski, D. R., Martin, W. R., Sapira, J. D.: Antagonism of the subjective, behavioral, pupillary, and respiratory depressant effects of cyclazocine by naloxone. Clin. Pharmacol. Ther. **9**, 215—222 (1968)

Jasinski, D. R., Nutt, J. G.: Progress report on the assessment program of the NIMH Addiction Research Center. Presented at 34th Meeting, Committee on Problems of Drug Dependence, National Research Council, Ann Arbor, Michigan 1972

Jasinski, D. R., Nutt, J. G.: Progress report on the clinical assessment program of the Addiction Research Center. Presented at 35th Meeting, Committee on Problems of Drug Dependence, National Research Council, Chapel Hill, North Carolina 1973

Joe, G. W., Simpson, D. D.: Retention in treatment of drug users: 1971—1972 DARP admissions. Amer. J. Drug Alcohol Abuse **2**, 63—71 (1975)

Jones, B. E., Martin, W. R., Isbell, H., Fraser, H. F.: Evaluation of a photographic method of estimating pupil diameter in man. Fed. Proc. **21**, 326 (1962)

Jones, B. E., Prada, J. A.: Drug-seeking behavior during methadone maintenance. Psychopharmacologia (Berl.) **41**, 7—10 (1975)

Jones, B. E., Prada, J. A.: Effects of methadone and morphine maintenance on drug-seeking behavior in dog. Psychopharmacology (Berl.), in press

Jones, W. A., Berechochea, J. E.: California Department of Corrections Methadone Maintenance Program: An evaluation. Research Report No. 50, Research Division, Department of Corrections, State of California, December 1973

Joseph, H., Dole, V. P.: Methadone patients on probation and parole. Fed. Prob. **34**, 42—48 (1970)

Kissin, B., Ottomanelli, G., Sang, E., Halloran, G.: Cyclazocine treatment for heroin addicts. Proceedings of the Fifth National Conference on Methadone Treatment, Washington, D.C., Vol. 1, pp. 658—666. New York: National Association for the Prevention of Addiction to Narcotics 1973

Kjeldgaard, J. M., Halm, G. W., Heckenlively, J. R.: Methadone-induced pulmonary edema. J. Amer. med. Ass. **218**, 882—883 (1971)

Kleber, H., Kinsella, J. K., Riordan, C., Greaves, S., Sweeney, D.: The use of cyclazocine in treating narcotic addicts in a low-intervention setting. Arch. gen. Psychiat. **30**, 37—42 (1974)

Kolb, L., Himmelsbach, C. K.: Clinical studies of drug addiction. III. A critical review of the withdrawal treatments with method of evaluating abstinence syndromes. Amer. J. Psychiat. **94**, 759—797 (1938)

Kurland, A. A., Hanlon, T. E., McCabe, L.: Naloxone and the narcotic abuser: A controlled study of partial blockade. Presented at 35th Meeting, Committee on Problems of Drug Dependence, National Research Council, Chapel Hill, North Carolina 1973b

Kurland, A. A., Krantz, J. C. Jr., Henderson, J. M., Kerman, F.: Naloxone and the narcotic abuser: A low-dose maintenance program. Int. J. Addict. **8**, 127—141 (1973a)

Lasagna, L.: Drug interaction in the field of analgesic drugs. Proc. roy. Soc. Med. **58**, 978—983 (1965)

Laskowitz, D., Brill, L., Jaffe, J. H.: Cyclazocine intervention in the treatment of narcotics addiction: Another look. In: Brill, L., Lieberman, L. (Eds.): Major modalities in the treatment of drug abuse. New York: Behavioral Publications 1972

Levine, R., Zaks, A., Fink, M., Freedman, A. M.: Levomethadyl acetate. Prolonged duration of opioid effects, including cross tolerance to heroin, in man. J. Amer. med. Ass. **226**, 316—318 (1973)

Lewis, J. W., Bentley, K. W., Cowan, A.: Narcotic analgesics and antagonists. Ann. Rev. Pharmacol. **11**, 241—270 (1971)

Maddux, J. F., Berliner, A., Bates, W. F.: Engaging opioid addicts in a continuum of services. A community-based study in the San Antonio area. Behavioral Science Monograph No. 71-1. Fort Worth: Texas Christian University Press 1971

Maddux, J. F., Bowden, C. L.: Critique of success with methadone maintenance. Amer. J. Psychiat. **129**, 440—446 (1972)

Marquardt, W. G., Martin, W. R., Jasinski, D. R.: The use of the Polaroid CU camera in pupillography. Int. J. Addict. **2**, 301—304 (1967)

Martin, W. R.: Opioid antagonists. Pharmacol. Rev. **19**, 463—521 (1967)

Martin, W. R.: The basis and possible utility of the use of opioid antagonists in the ambulatory treatment of the addict. In: Wikler, A. (Ed.): The addictive states. Ass. Res. nerv. ment. Dis., Vol. 46, pp. 367—377. Baltimore: Williams and Wilkins 1968

Martin, W. R.: Commentary on the Second National Conference on Methadone Treatment. Int. J. Addict. **5**, 545—552 (1970)

Martin, W. R., Eades, C. G.: A comparison between acute and chronic physical dependence in the chronic spinal dog. J. Pharmacol. exp. Ther. **146**, 385—394 (1964)

Martin, W. R., Eades, C. G., Thompson, J. A., Gilbert, P. E., Sandquist, V. L., Workman, M.: Progress report on the use of the dog for assessing morphine-like and nalorphine-like agonists as well as depot preparations of antagonists. Presented at 35th Meeting, Committee on Problems of Drug Dependence, National Research Council, Chapel Hill, North Carolina 1973a

Martin, W. R., Eades, C. G., Thompson, W. O., Thompson, J. A., Flanary, H. G.: Morphine physical dependence in the dog. J. Pharmacol. exp. Ther. **189**, 759—771 (1974)

Martin, W. R., Fraser, H. F.: A comparative study of physiological and subjective effects of heroin and morphine administered intravenously in postaddicts. J. Pharmacol. exp. Ther. **133**, 388—399 (1961)

Martin, W. R., Fraser, H. F., Gorodetzky, C. W., Rosenberg, D. E.: Studies of the dependence-producing potential of the narcotic antagonist 2-cyclopropylmethyl-2′-hydroxy-5,9-dimethyl-6,7-benzomorphan (cyclazocine, WIN-20,740, ARC II-C-3). J. Pharmacol. exp. Ther. **150**, 426—436 (1965)

Martin, W. R., Gorodetzky, C. W.: Demonstration of tolerance to and physical dependence on N-allylnormorphine (nalorphine). J. Pharmacol. exp. Ther. **150**, 437—442 (1965)

Martin, W. R., Gorodetzky, C. W.: Cyclazocine, an adjunct in the treatment of narcotic addiction. Int. J. Addict. **2**, 85—93 (1967)

Martin, W. R., Gorodetzky, C. W., McClane, T. K.: An experimental study in the treatment of narcotic addicts with cyclazocine. Clin. Pharmacol. Ther. **7**, 455—465 (1966)

Martin, W.R., Jasinski, D.R.: The mode of action and abuse potentiality of narcotic antagonists. In: Janzen, R., Keidel, W.D., Herz, A., Steichele, C. (Eds.); Payne, J.P., Burt, R.A.P. (Eds., English Edition): Pain: Basic principles—pharmacology—therapy, pp. 225—234. Stuttgart: Georg Thieme Publishers 1972
Martin, W.R., Jasinski, D.R., Haertzen, C.A., Kay, D.C., Jones, B.E., Mansky, P.A., Carpenter, R.W.: Methadone-A reevaluation. Arch. gen. Psychiat. **28**, 286—295 (1973b)
Martin, W.R., Jasinski, D.R., Mansky, P.A.: Characteristics of the blocking effects on EN-1639A (N-cyclopropylmethyl-7,8-dihydro-14-hydroxynormorphinone HCl). Presented at 33rd Meeting, Committee on Problems of Drug Dependence, National Research Council, Toronto, Ontario, Canada 1971
Martin, W.R., Jasinski, D.R., Mansky, P.A.: Naltrexone, an antagonist for the treatment of heroin dependence. Arch. gen. Psychiat. **28**, 784—791 (1973c)
Martin, W.R., Jasinski, D.R., Sapira, J.D.: Antagonism of the subjective, pupillary and respiratory depressant effects of cyclazocine by naloxone in man. Pharmacologist **9**, 230 (1967)
Martin, W.R., Sandquist, V.L.: A sustained release depot for narcotic antagonists. Arch. gen. Psychiat. **30**, 31—33 (1974)
Martin, W.R., Sloan, J.W.: The pathophysiology of morphine dependence and its treatment with opioid antagonists. Pharmakopsychiat. Neuro-Psychopharmakol. **1**, 260—270 (1968)
Mayor's Committee on Drug Addiction: Report of the Mayor's Committee on Drug Addiction to the Hon. Richard C. Patterson, Jr., Commissioner of Correction, New York City. Amer. J. Psychiat. **10**, 433—538 (1930)
McClane, T.K., Martin, W.R.: Effects of morphine, nalorphine, cyclazocine, and naloxone on the flexor reflex. Int. J. Neuropharmacol. **6**, 89—98 (1967a)
McClane, T.K., Martin, W.R.: Antagonism of the spinal cord effects of morphine and cyclazocine by naloxone and thebaine. Int. J. Neuropharmacol. **6**, 325—327 (1967b)
New York Academy of Medicine: Report on drug addiction. Bull. N.Y. Acad. Med. **31**, 592—607 (1955)
New York Academy of Medicine: Report on drug addiction-II. Bull. N.Y. Acad. Med. **39**, 417—473 (1963)
Nutt, J.G., Jasinski, D.R.: Comparison of intravenously administered methadone, morphine, heroin and placebo. Fed. Proc. **32**, 694 (1973)
O'Donnell, J.A.: Narcotic addicts in Kentucky. PHS Publication No. 1881. Washington, D.C.: U.S. Government Printing Office 1969
Pachter, I.J.: Synthetic 14-hydroxymorphinan narcotic antagonists. In: Braude, M.C., Harris, L.S., May, E.L., Smith, J.P., Villarreal, J.E. (Eds.): Narcotic antagonists. Advances in Biochemical Psychopharmacology, Vol. 8. New York: Raven Press 1974
Perkins, M.E., Bloch, H.I.: Survey of a methadone maintenance program. Amer. J. Psychiat. **126**, 1389 (1970)
Perkins, M.E., Bloch, H.I.: A study of some failures in methadone treatment. Amer. J. Psychiat. **128**, 47—51 (1971)
Pierson, P.S., Howard, P., Kleber, H.D.: Sudden deaths in infants born to methadone-maintained addicts. J. Amer. med. Ass. **220**, 1733—1734 (1972)
Proceedings of the National Conference on Methadone Treatment: First Conference, New York, N.Y., June 21—22, 1968. New York: National Association for the Prevention of Addiction to Narcotics, 1968. Second Conference, New York, N.Y., Oct. 26—27, 1969. Int. J. Addict., Vol. 5, No. 3, 1970. Third Conference, New York, N.Y., Nov. 14—16, 1970. PHS Publication No. 2172. Washington: U.S. Government Printing Office 1970. Fourth Conference, San Francisco, Calif., Jan. 8—10, 1972. New York: National Association for the Prevention of Addiction to Narcotics 1972. Fifth Conference, Washington, D.C., March 17—19, 1973. New York: National Association for the Prevention of Addiction to Narcotics 1973
Rajegowda, B.K., Glass, L., Evans, H.E., Maso, G., Swartz, D.P., Leblanc, W.: Methadone withdrawal in newborn infants. J. Pediat. **81**, 532—534 (1972)
Raynes, A.E., Patch, V.D.: An improved detoxification technique for heroin addicts. Arch. gen. Psychiat. **29**, 417—419 (1973)
Reddy, A.M., Harper, R.G., Stern, G.: Observations on heroin and methadone withdrawal in the newborn. Pediatrics **48**, 353—358 (1971)
Renault, P.F.: Methadone maintenance: The effect of knowledge of dosage. Int. J. Addict. **8**, 41—47 (1973)

Resnick, R. B., Fink, M., Freedman, A. M.: A cyclazocine typology in opiate dependence. Amer. J. Psychiat. **126**, 1256—1260 (1970)
Resnick, R. B., Fink, M., Freedman, A. M.: Cyclazocine treatment of opiate dependence: A progress report. Compr. Psychiat. **12**, 491—502 (1971)
Resnick, R., Fink, M., Freedman, A. M.: High-dose cyclazocine therapy of opiate dependence. Amer. J. Psychiat. **131**, 595—597 (1974a)
Resnick, R. B., Volavka, J., Freedman, A. M., Thomas, M.: Studies of EN-1639A (naltrexone): A new narcotic antagonist. Amer. J. Psychiat. **131**, 646—650 (1974b)
Schecter, A. J., Friedman, J. G., Grossman, D. J.: Clinical use of naltrexone (EN-1639A): Part I: Safety and efficacy in pilot studies. Amer. J. Drug Alcohol Abuse **1**, 253—269 (1974)
Schooff, K. G., Hersch, R. G., Lowy, D. G.: Cyclazocine—A reasonable alternative to methadone. Presented at 35th Meeting, Committee on Problems of Drug Dependence, National Research Council, Chapel Hill, North Carolina 1973
Schooff, K. G., Keegan, J. F., Lowy, D. G.: Cyclazocine—A prerequisite in the treatment of heroin addiction. Presented at 36th Meeting, Committee on Problems of Drug Dependence, National Research Council, Mexico City, Mexico 1974
Seevers, M. H., Deneau, G. A.: Physiological aspects of tolerance and physical dependence. In: Root, W. S., Hofmann, F. G. (Eds.): Physiological pharmacology, Vol. 1, pp. 565—640. New York: Academic Press 1963
Sells, S. B.: The DARP research program and data system. Amer. J. Drug Alcohol Abuse **2**, 1—14 (1975)
Senay, E. C., Jaffe, J. H., Chappel, J. N., Renault, P., Wright, M., Lawson, C., Charnett, C., DiMenza, S.: IDAP—Five-year results. Proceedings of the Fifth National Conference on Methadone Treatment, Washington, D.C., Vol. 2, pp. 1437—1450. New York: National Association for the Prevention of Addiction to Narcotics 1973
State of California, Department of Public Health: Recommended procedure for the nalorphine test. A guide to physicians. Berkeley, Calif.: California State Printing Office 1958
Terry, J. G., Braumoeller, F. L.: Nalline: An aid in detecting narcotic users. Calif. Med. **85**, 299—301 (1956)
Thornton, W. E., Thornton, B. P.: Narcotic poisoning: A review of the literature. Amer. J. Psychiat. **131**, 867—869 (1974)
Vaillant, G. E.: A twelve-year follow-up of New York narcotic addicts: I. The relation of treatment to outcome. Amer. J. Psychiat. **122**, 727—737 (1966)
Vaillant, G. E.: A 20-year follow-up of New York narcotic addicts. Arch. gen. Psychiat. **29**, 237—241 (1973)
Way, E. L., Mo, B. P. N., Quock, C. P.: Evaluation of the nalorphine pupil diagnostic test for narcotic usage in long-term heroin and opium addicts. Clin. Pharmacol. Ther. **7**, 300—311 (1966)
Wikler, A., Carter, R. L.: Effects of single doses of N-allylnormorphine on hindlimb reflexes of chronic spinal dogs during cycles of morphine addiction. J. Pharmacol. exp. Ther. **109**, 92—101 (1953)
Wikler, A., Frank, K.: Hindlimb reflexes of chronic spinal dogs during cycles of addiction to morphine and methadone. J. Pharmacol. exp. Ther. **94**, 382—400 (1948)
Wilmarth, S. S., Goldstein, A.: Therapeutic effectiveness of methadone maintenance programs in the USA. WHO Publication No. 3. Geneva: World Health Organization 1974
Woodland, J. H. R., Yolles, S., Blake, D. A., Helrich, M., Meyer, F. J.: Long-acting delivery systems for narcotic antagonists. J. Med. Chem. **16**, 897—901 (1973)
Woods, L. A.: The pharmacology of nalorphine (N-allylnormorphine). Pharmacol. Rev. **8**, 175—198 (1956)
Zaks, A., Fink, M., Freedman, A. M.: Levomethadyl in maintenance treatment of opiate dependence. J. Amer. med. Ass. **220**, 811—813 (1972)
Zaks, A., Jones, T., Fink, M., Freedman, A. M.: Naloxone treatment of opiate dependence. A progress report. J. Amer. med. Ass. **215**, 2108—2110 (1971)
Zelson, C.: Infant of the addicted mother. New Engl. J. Med. **288**, 1393—1395 (1973)
Zelson, C., Lee, S. J., Casalino, M.: Neonatal narcotic addiction. Comparative effects of maternal intake of heroin and methadone. New Engl. J. Med. **289**, 1216—1220 (1973)
Zelson, C., Rubio, E., Wasserman, E.: Neonatal narcotic addiction: 10 year observation. Pediatrics **48**, 178—189 (1971)

CHAPTER 6

Detection of Drugs of Abuse in Biological Fluids

CH. W. GORODETZKY

A. Introduction

Detection of drugs of abuse and their metabolites in body fluids plays a role in various aspects of drug dependence. Qualitative screening usually of urine specimens is an integral part of most treatment and prevention programs for drug abusers; and, it was estimated in 1972 that 8 million urine samples were being analyzed a year for methadone programs alone in the United States (CATLIN, 1973a). To assist in the diagnosis of medical emergencies involving possible drug overdose and to aid in the determination of cause of death, toxicology laboratories frequently perform qualitative and sometimes quantitative analyses of body fluids and tissues for drugs. Research in metabolism and pharmacokinetics of drugs of abuse usually requires quantitative determination of drugs and known metabolites and simple and sometimes very complex qualitative methods for detection and identification of new metabolites. In a related area, using many similar methods, law enforcement and forensic laboratories and investigators are also often faced with the problem of the chemical identification of unknown drugs and poisons. The major emphasis in this review is on qualitative screening methods for the detection of drugs of abuse. It considers the general principles and concepts of their use and interpretation, provides a comprehensive review of most commonly used methods, and evaluates the validity of these methods to detect drugs or their metabolites following ingestion by humans.

B. General Principles

I. Uses of Biological Fluid Screening

The purpose of screening biological fluids for drugs of abuse is to provide objective laboratory evidence to aid in the diagnosis of drug use or ingestion. Such information can be useful in several clinical and research situations. In making the initial diagnosis of drug abuse, chemical analysis of urine (or other biological fluid) is a useful adjunct to a history and physical examination. The identification in the urine of a drug or its metabolites indicates the presence of drug intake. However, it will not give information about such important clinical parameters as drug dose, frequency of administration, or the presence or absence of physical dependence. For the diagnosis of some clinical conditions, such as heroin dependence prior to the onset of overt signs of abstinence, presence of drug or metabolite in the urine is necessary but not sufficient evidence. Urine screening as an aid in initial diagnosis is most frequently used in patients presenting themselves for treatment of drug addiction and in patients for whom the diagnosis of drug addiction is suspected. Mass screening has also been used in large populations with a probable low incidence of drug abuse, such as the military (BAKER, 1972) and job applicants (SOHN, 1972), to provide initial identification of drug abusers. Although the intent of such testing is usually to allow early treatment intervention, the results can be used as a basis for legal action or as a condition precluding employment, as pointed out by GOLDBERG (1975).

Probably the most frequent use of urine screening is to monitor the progress of treatment of drug abusers with regard to the parameter of illicit drug use. The results of such tests can aid the clinician in evaluating a patient's progress toward discontinuance of illicit drug use and indicating when additional treatment modalities may be necessary. Urine screens are one objective measure of the efficacy of treatment; that is, a measure of the influence of the program on the incidence of illicit drug use among its patients. It is not either the only or the most relevant measure of treatment effectiveness. Illicit drug use is only one possible measure of treatment outcome, as has been emphasized by GEARING (1972) (also see Sect. II, Chap. 5). Also, especially for comparison among different treatment programs, one must consider such factors as the details of the screening procedures (e.g., logistics of sample collection, frequency of testing), the capabilities of the test methods used, and the laboratories performing them, and, as pointed out in a study of NIGHTINGALE et al. (1972), the attitude of the program toward the test results.

Whether urine screening in a treatment program is in itself of any therapeutic value is currently a matter of debate. GOLDSTEIN and JUDSON (1974) reported a study, replicated in three methadone maintenance treatment clinics, in which one patient group had their urine monitored according to the usual clinic protocol and a second group was not monitored. After 3 months a random, surprise urine sample was obtained from patients in both groups. No differences between groups were found in clinic attendance or retention rate in the program. With regard to the incidence of positive findings for morphine in the surprise urine sample, in one clinic the monitored group had statistically significantly fewer positives than in the unmonitored group, and in a second clinic there was a trend in the same direction. However, in the third clinic there was a trend toward a higher incidence of positives in the monitored group. The authors concluded that urine screening may have a modest impact on decreasing drug use. A study of the effect of feedback of urine test results to clinic staff and patients was conducted by GREVERT and WEINBERG (1973). They found no difference between feedback groups in the incidence of urine samples positive for illicit drugs. However, they did feel that in some cases in the feedback group a positive test result had allowed beneficial early treatment intervention. BLUMBERG and HEATON (1973) also felt that urine screening was beneficial in control of drug abuse in a psychiatric hospital. The need for close collaboration between the laboratory and the clinic to make screening tests most useful was emphasized by BOWEN and GURR (1970). Clearly more studies are needed of the influence of urine screening on treatment outcome. In epidemiologic studies urine screening can provide objective data pertaining to the prevalence of drug use and the availability of particular drugs to the population under study.

II. Test Parameters

There are many test characteristics which are considered important in the description, evaluation, and selection of biological fluid screening methods. For purposes of discussion these parameters are categorized into the following groups: socioeconomic, chemical, and pharmacologic.

1. Socioeconomic Parameters

The socioeconomic parameters of screening tests are those which pertain to setting up and performing the tests in an ongoing operational laboratory; and they include such factors as cost, complexity of test performance, and speed of analysis. These parameters will be discussed specifically for each screening method in later sections. Tabular summaries of speed and cost of some commonly used screening procedures have been reported by CATLIN (1973a), World Health Organization (1974), MULÉ (1974), and KAISTHA and TADRUS (1975).

In estimating cost one must consider initial, nonrecurring start-up costs (primarily for instruments and equipment, nonexpendable supplies, and laboratory space), daily expendable materials (such as reagents), personnel costs, and miscellaneous recurring expenses (such as equipment and instrument maintenance). For some methods these expenses vary with the volume of tests run (e.g., cost of specific test reagents for immunoassays). Based on the number of samples per day to be analyzed, operating expenses can be combined to estimate a single cost per sample figure. Analysis of cost factors for establishing and operating a laboratory to perform drug screening tests (with emphasis on thin-layer chromatographic procedures) was reported by KAISTHA and JAFFE (1972a), KAISTHA (1972), and KAISTHA and TADRUS (1975).

The speed of performing a screening test can be considered in terms of the turn around time for analysis of a single sample (CATLIN, 1973a); that is, the time from collection of the sample to reporting the results of the analysis. However, since many samples can be processed simultaneously by some screening procedures, the number of samples which can be analyzed in a given time period (usually one 8-h working day) should also be considered.

The complexity of carrying out a screening test involves not only the technical capability of performing the procedure and operating the instruments, but also the judgment necessary to evaluate the final result. In some screening methods the end point is objective (e.g., the homogenous enzyme immunoassay, EMIT), while in others a subjective judgment is required (e.g., thin-layer chromatography). In general, the simpler tests can be performed by less experienced and technically trained personnel and frequently have fewer sources of potential error compared to the more complex procedures.

2. Chemical Parameters

a) Sensitivity

Sensitivity is defined as the minimal concentration of drug in the original undiluted biological fluid which is detectable with a high probability. This definition allows comparisons among different screening methods and includes such variables as volume of fluid used, extracting solvent, pH of sample, and losses due to taking aliquots, splitting samples, etc. The statistical nature of the definition (i.e., detectable with high probability) takes into account the variability encountered in the procedures. For qualitative screening methods, such as thin-layer chromatography, GORODETZKY

(1972a, b) proposed defining sensitivity as the drug concentration detectable in urine 99% of the time, a value easily determined statistically with 95% confidence limits, by simple experiments. Such sensitivity values have been determined for several thin-layer chromatographic methods for morphine detection (GORODETZKY, 1973a; KULLBERG and GORODETZKY, 1974) and for a colorimetric screening method for detection of barbiturates (PEHR, 1975). Unfortunately the term "sensitivity" has been used in various and sometimes confusing ways in the literature on drug screening methods. It has been used to indicate total amount of drug detectable (without specifying volume) and also to represent the absolute sensitivity of the detection system for pure drug.

For some screening methods which give continuous quantitative results (such as some immunoassays), minimal "cutoff concentrations" have been selected representing the operational sensitivity of the methods for use as screening tests. Assay values above the values given by the cutoff concentration are considered "positive" and those below it "negative." Selection of the appropriate cutoff concentration is usually based on assay values of normal, drug-free biological fluid and values given by compounds which might be found in the biological fluid but which are not desired to be detected. Because of the distribution of values at any specified concentration (the magnitude of which depends on the variability of the assay), the concentration of drug in biological fluid detectable 95% (or higher) of the time will be greater than the cutoff concentration. SCHNEIDER et al. (1974) defined the "detection concentration" (or "detection level") as the concentration of drug detectable in 95% of samples, and reported such data for the free radical assay technique (FRAT) and the homogeneous enzyme immunoassay (EMIT).

b) Specificity

The specificity of a test is defined as its ability to distinguish among compounds. It may apply to distinguishing between various drugs of interest, between a drug of interest and one not of primary concern (but which may be closely chemically related), or between a drug of interest and a natural constituent of the biological fluid. Lack of specificity can lead to false positive results; and, therefore, a high degree of specificity is generally desirable. However, it has been pointed out (CATLIN, 1973a) that the inability of a test to distinguish between a drug and its metabolites (e.g., reactivity of some immunoassays with both morphine and morphine glucuronide) can be advantageous and increase the capability for detection of use of that drug.

3. Pharmacologic Parameters

Validity

The validity of a method is defined as its ability to detect a drug or its metabolites in biological fluid following human drug administration (GORODETZKY, 1972a, b). As applied to screening methods it is a broad concept, including a method's sensitivity, some aspects of specificity, and a large number of drug metabolic variables, such as drug dose, route of administration, concentration and pH of the biological fluid, and between subject variability in absorption, metabolism, and excretion of drug. To accurately define a method's validity one must specify the drug administered (including dose, route, and time of administration), the compound detected, the time of sample collection, and the biological fluid analyzed. For example, validity pertains to answering such questions as: For how long after a 5 mg intravenous dose of heroin is a given method likely to continue to detect morphine in the urine? Knowledge of a method's validity is most important to the clinician to aid in interpretation of the test results and to select a method for testing which is most likely to meet his clinical needs.

4. Other Terms

The terms "accuracy," "reproducibility," and "reliability" have been frequently applied to the evaluation and description of screening methods. However, they are usually used in a general sense, referring to one or more of the chemical or pharmacologic parameters described above. Accuracy often refers to the precision of quantitative determination, but can also apply to the ability to correctly identify a drug or metabolite in biological fluid or detect specific human drug use. Reproducibility and reliability generally refer to a method's variability; that is, its ability to produce the same results on multiple analyses of the same or similar samples.

III. Interpretation of Screening Tests

In qualitative screening tests each sample is reported as either positive or negative for a particular drug or drug group (depending on the test). As pointed out by CATLIN (1973a), since each sample is obtained from a person who either did or did not take the drug (or one of the drugs) in question, there are four possible interpretations of the test result, as follows:

	Person has taken drug	*Person has not taken drug*
Test result: positive	True positive	False positive
Test result: negative	False negative	True negative

The two true test results accurately reflect the clinical situation. However, a true positive on a screening test will not in itself indicate the dose or time of drug administration nor distinguish between chronic or single dose use. A true negative result indicates that the compound determined is absent from the biological fluid, or, in clinical terms, no drug was taken within the time desired for its use to be detected.

A false positive is a positive result from analysis of a biological fluid from a person who has not taken the drug (or drugs) in question; that is, an incorrect identification of the presence of the determined compound. It is an important category, since a false diagnosis of drug taking can have far-reaching consequences for the individual. Such incorrect positive test results can occur because of errors in judgment (e.g., reading a thin-layer plate), mistakes in performance of the test procedures, instrument malfunction, clerical errors, or inherent limitations in the specificity of the test. Because of differences in interpretation and possibly in remedies, it is useful to identify three classes of false positives: physiologic, pharmacologic, and procedural. A physiologic false positive is a positive result in a normal, drug-free sample;

that is, a sample from a subject who has not taken any drugs. Such a result may be caused by a naturally occurring constituent of the biological fluid and may indicate that insufficient account has been taken of this background activity in the test (e.g., the cutoff level has been set too low). Because of the ubiquity of the substances, positive results caused by smoking tobacco products (primarily due to nicotine and its metabolites) and by drinking coffee, tea, cocoa, or cola drinks (principally from caffeine, other xanthines, and their metabolites) are usually considered in this class. A pharmacologic false positive is a positive result from a sample of a subject who has taken a drug other than the drug in question. From a chemical viewpoint some tests (e.g., some of the immunoassays) are appropriately used only to identify drug groups and a positive result should not be interpreted to represent only one drug in that group. For example, many morphine radioimmunoassays strongly cross react with codeine; therefore, a positive result in a sample containing codeine would not be considered a false positive by this test. If it is desired to identify these drugs with greater specificity, a test other than, or in addition to, the radioimmunoassay should be used. From a clinical point of view, whether taking a particular drug produces a true or false positive depends on what is defined by the tester as desired to be diagnosed. For example, if the tester wished to identify opiate use, a positive result from a subject taking codeine would be a true positive. On the other hand, if he wished to specifically identify heroin use, the positive result caused by codeine use would be a false positive. A procedural false positive is an incorrectly positive result which can be traced to an error somewhere in the overall testing procedures. This could include such factors as clerical errors, mistakes in collection or sample identification, and instrument malfunctions.

A false negative is a negative finding in a sample from a person who has taken the drug in question. Clinically the false negative should be defined in terms of the parameters of validity. The tester must specify the drug, dose, route of administration, and time after administration which is desired to be detected. Any negative finding in that specified time period would be a false negative. Chemically a false negative may be defined as a negative finding in a sample known to contain the compound to be determind in a concentration greater than the sensitivity of the test (or the concentration defined as necessary to be detected). The presence of the compound in the sample may be known from analysis by another method or because the compound had been added to a drug-free sample (i.e., a "spiked" sample).

IV. Other Factors

1. Confirmation of Test Results

Because of the implications of a positive finding, especially in urine screening for initial diagnosis of drug abuse or monitoring drug treatment, it has been advocated that all positive test results be confirmed using a second test procedure (e.g., World Health Organization, 1974). To lessen the possibility of similar errors, the second test method should be qualitatively different than the first; for example, to confirm a radioimmunoassay result, thin-layer chromatography would be better than homogeneous enzyme immunoassay. It may be useful, especially when a large number of samples are routinely screened and many negatives are expected, to use the first test primarily as a negative exclusion test; that is, principally to identify true negatives, which will then require no further analysis. Ideally such a method would be rapid, simple (possibly automated), inexpensive, and very sensitive (to minimize false negatives, within the validity specifications). Poor specificity and a moderate number of false positives would be acceptable at this stage of the analysis. The immunoassays, particularly radioimmunoassay, have been recommended for use as initial tests. The confirmatory test would be principally to eliminate false positives from the first test, and, therefore, high specificity would be essential. Sensitivity must be sufficient to meet the clinical needs with regard to validity; cost, speed and simplicity are less important than in the primary test. Methods such as thin-layer and gas chromatography have been recommended for such use.

2. Frequency of Testing

In routine monitoring of urine during treatment for drug abuse to detect use of illicit drugs, the frequency of testing is an important consideration. The desire to diagnose return to drug taking as quickly as possible must be balanced against available resources of money, facilities, and personnel. GOLDSTEIN and BROWN (1970), assuming 24-h drug detectability, calculated that testing once every 5 days on a random schedule would be sufficient to allow a median of 4 days of undetected drug use and a maximum of 14 days. Graphs were provided so that other estimates could be made to meet particular clinical needs. MULÉ and KRAMER (1973) provided formulas for making similar calculations, taking some additional factors into consideration and allowing estimation of cost of testing. KLEBER and GOULD (1971) briefly described a schedule using varying probability of random testing with a maximum number of days between tests.

3. Proficiency Testing of Laboratories

Testing proficiency of a laboratory's performance by submission of samples of known composition can be beneficial to both the laboratory and the client. Identified deficiencies can be remedied and performance to minimal requirements of such parameters as sensitivity and specificity can be evaluated and encouraged. Ideally test samples would be obtained from subjects at a time following known drug administration to meet desired validity requirements and would be submitted to the laboratory under blind conditions (i.e., just as a usual clinical sample). Unfortunately such test samples are seldom available for any large-scale testing. Normal, drug-free samples to which drugs and appropriate relative amounts of metabolites (if available) have been added in concentrations at or above a defined minimum sensitivity requirement can also serve as test samples. A procedure for preparation of such samples for use as either internal controls in the laboratory or possibly as proficiency testing samples has been described by FRINGS and QUEEN (1972b). The splitting of clinical samples to submit to more than one laboratory or as separate samples to the same laboratory can also give some information about capability of laboratory performance. In the United States the Center for Disease Control has run a proficiency testing program for Toxicology-Drug Abuse since 1972, in which 10 samples are submitted on a quarterly basis to several hundred laboratories under their jurisdiction or who volunteer to participate. Because of logistic problems these samples have been submitted on an open rather than blind basis. SOHN (1973) has described the program run in the city and state of New York since 1970.

4. Urine Sampling

Since biological fluid screening entails considerable time, effort, and expense, it is important to make every effort to insure that the sample is authentic; that is, that the sample is obtained from the subject of record at the time requested. This is particularly pertinent in urine screening, where various subterfuges have been used to substitute a known negative sample for the requested one. Therefore, although it may be inconvenient and require additional personnel and facilities, it is necessary that all urine samples be obtained under direct observation by an experienced and trustworthy observer, as emphasized, for example, by GOLDSTEIN and BROWN (1970). In the logistics of sample handling care must be taken to avoid clerical mix-up. Also it should be kept in mind that as a urine sample ages and bacterial and fungal growth occurs, the pH and ionic strength of the sample may change; and these parameters must be checked if they are critical variables in the test procedure being used. In this regard FRINGS and QUEEN (1972c) reported that urine samples containing a large variety of drugs of abuse were stable for 14 days without refrigeration or added preservatives.

C. Methods

Many analytical chemical techniques have been applied to the detection, identification, and quantitation of drugs. This review concentrates on those procedures which have been used in screening methods for drugs of abuse in biological fluids, especially urine. They are classified for discussion according to the detection technique utilized. Earlier reviews for additional reference include those by BRANDT (1973), CATLIN (1973a), KAISTHA (1972), MULÉ (1972b), SIEK (1973), TAYLOR (1971), and the World Health Organization (1974). Alcohol detection will not be discussed; recent reviews for alcohol analysis in blood (CRAVEY and JAIN, 1974a) and breath (JAIN and CRAVEY, 1974a) have been published.

I. Thin-Layer Chromatography (TLC)

Thin-layer chromatography was one of the earliest and probably is still the most commonly used analytic technique in screening urine for the presence of drugs of abuse. The basic elements of the great majority of TLC systems described are: extraction of drugs from urine; application of the concentrated and reconstituted extract to a TLC medium (e.g., plate, microfiber sheet); development of the chromatogram; and, detection of drugs by chromogenic spray reagents and/or physical methods (such as exposure to heat and ultraviolet light). Drugs are identified by their migration distance relative to the solvent front (i.e., R_f) and by specific color reactions to the detection reagents compared to known standards. There are a great many variations among the screening methods reported in the literature with regard to these basic elements.

1. Extraction Methods

There are three techniques which have been most commonly reported for use in screening methods for extraction of drugs of abuse from biological fluids, principally urine. These are organic solvent extraction (frequently referred to as liquid-liquid extraction), ion exchange resin impregnated paper extraction (using the cation exchange resin paper, Reeve Angel SA-2), and nonionic resin column extraction (using Amberlite XAD-2).

a) Organic Solvent Extraction

The major papers describing TLC urine screening procedures using organic solvent extraction are summarized in Table 1. Direct urine extraction with organic solvents was the most frequently reported technique used in screening procedures, and was the first commonly used extraction procedure, as described in the early papers of Cochin and Daly (1962, 1963) and in the more general toxicologic paper of Sunshine (1963). As shown in the first column of the table, the majority of authors described systems which screened for several classes of commonly abused drugs, most often narcotic analgesics (NA), sedative-hypnotics (SH), and stimulants (ST); and they also frequently included other commonly prescribed drugs which were extracted by the described procedures, such as phenothiazines and antidepressants (PH), and antihistamines (AH). Major hallucinogens, such as LSD, and cannabinoids were not included in most papers, although Mulé (1969) described several psychotomimetics and Bastos et al. (1970) provided data on a wide range of organic basis, including LSD and also autonomic drugs and local anesthetics.

The three columns in Table 1 under the major heading of "extraction" summarize several of the principal parameters of the organic solvent extraction procedures used. The most common volume of urine used was 10–20 ml, although some authors recommended aliquots of 25 or even 50 ml and several described systems using small volumes of 1–6 ml. Cochin and Daly (1962, 1963) and Sunshine (1963) described the use of variable urine volumes, adjusting the volume of extracting solvent according to the amount of urine to be extracted. Most authors did not note any pretreatment of the urine; however, Blass et al. (1974) recommended centrifugation of urine prior to extraction, apparently to remove debris and precipitates.

Procedures utilizing a single extraction or two or three separate extractions at different pH's have been described. Those procedures utilizing multiple extractions may use a separate urine aliquot for each extraction (e.g., Mulé, 1971) or may extract the same aliquot sequentially (e.g., Kaistha and Jaffe, 1972c). The most recent and complete triple extraction procedure was that described by Mulé (1971). An acid extraction at lower than pH 1 (using sulfuric acid) was used to detect barbiturates, glutethimide, and diphenylhydantoin. A second extraction was carried out at pH 9 using a potassium phosphate buffer after prior acid hydrolysis of the urine; this extract was primarily for detection of narcotic analgesics and quinine. A pH of 10–11, also using a potassium phosphate buffer, was used for the third extraction, which extracts a large group of opioids, stimulants, phenothiazines, antidepressants, antihistamines, and some minor tranquilizers such as diazepam, chlordiazepoxide, and meprobamate. Double extraction procedures used an acid pH (usually about 1) for detection of sedative-hypnotics and an extraction at pH 9–10 for the remaining

Table 1 Thin-layer chromatography urine screening procedures Organic solvent extraction

Drug or[a] Drug groups tested	Extraction			TLC			Other comments	Reference
	Urine vol used	pH-buffers[b]	Solvents[b,c]	Plates	Primary developing solvents	Detection		
NA, ST, PH, AH, HL	10 ml	pH 8 5 — NH_4Cl/NH_4OH	EtOH, ether	3 plates run· Silica Gel G	$CHCl_3$:MeOH (90 10) MeOH NH_3 (100:1.5)	Ninhydrin, carbon disulfide-ammonia, cupric chloride, iodoplatinate, p-dimethylaminobenzaldehyde, p-nitroaniline	Describes acid hydrolysis—R_f and color data for 133 compounds	BASTOS et al (1970)
ST	1–5 ml	Alkaline with NaOH	Diethyl ether	Alumina G Silica Gel G	10 systems given	Iodoplatinate, p-nitroaniline, ninhydrin	Primary emphasis on GC with TLC for confirmation, describes GC/MS—R_f data for 12 compounds	BECKETT et al. (1967)
NA, SH, ST, PH	5–10 ml	3 extractions HCl(SH) NaOH(ST) sodium bicarbonate (NA, PH)	$CHCl_3$(SH, ST) $CHCl_3$ Isopro (9:1) (NA, PH)	3 plates run. Silica Gel G	$CHCl_3$:acetone (9·1) benzene dioxane· EtOH NH_3(50.40 5 5) MeOH.NH_3(100 1 5)	Mercuric chloride—diphenylcarbazone, iodoplatinate, FPN reagent	Describes stimulants by GC only, confirmatory GC and fluorescence methods—R_f data for 11 compounds	BERRY and GROVE (1971)
NA, SH, PH	about 15 ml	2 separate extractions acid-H_2SO_4(SH) base-NaOH(NA, PH)	$CHCl_3$-aqueous—$CHCl_3$	Silica Gel G	$CHCl_3$.acetone (9:1) (SH) MeOH.NH_3 (100 1 5) (NA, PH)	Mercuric chloride—diphenylcarbazone, mercurous nitrate, Dragendorff's reagent, furfural, iodoplatinate, Mandelin's reagent	Oriented for hospital toxicology labs; describes confirmatory color tests, separate extraction for benzodiazepines	BERRY and GROVE (1973)
NA, SH, ST, PH	10 ml	pH 9.5 — NH_4Cl/NH_4OH	Not given	Toxi-Gram ITLC or Gelman ITLC sheets (type SAF)	EtAc MeOH NH_4OH (85·10·5)	Ninhydrin, mercuric sulfate, diphenylcarbazone, iodoplatinate, Dragendorff's reagent	Sample application to TLC plates using a disc—R_f and color data for 25 compounds	BLASS et al (1974)
NA, SH, ST, PH	2 ml	2 extractions H_2SO_4(SH) pH 8 5-9, NH_4OH, sodium bicarbonate (NA, ST, PH)	$CHCl_3$(SH) $CHCl_3$ Isopro (4:1) (NA, ST, PH)	Eastman Chromatogram Sheets with fluorescent indicator	Hexane EtOH (93 7) EtAc·MeOH NH_4OH (85.10.5)	Ninhydrin, iodoplatinate, Dragendorff's reagent, N,2,6-trichlorobenzoquinoneimine, mercurous nitrate, vanillin, furfural	Sample application to TLC plates using microdot techniques; describes confirmation by microcrystal procedure—R_f and color data for 57 compounds	BROICH et al (1971a)

NA	$^1/_4$ volume of extracting solvent	pH 9 — NaOH	Ethylene chloride with 10% isoamyl alcohol	Silica Gel G or Aluminum oxide G	6 systems are given	Iodoplatinate	Early paper, describes acid hydrolysis—R_f data for 16 compounds	COCHIN and DALY (1962)
SH	$^1/_3$ volume of extracting solvent	pH 5 — HCl	Methylene chloride	Silica Gel G	$CHCl_3$ acetone (9 1), benzene·HAc (9·1) dioxane benzene NH_3 (20 75 5)	Mercurous nitrate, potassium permanganate	Early paper—R_f and color data for 20 compounds	COCHIN and DALY (1963)
NA, PH, AL	10 ml	conc. NH_4OH	$CHCl_3$ EtAc EtOH (3 1 1)	Silica Gel G	EtAc. MeOH NH_4OH (85.10.5)	H_2SO_4, iodoplatinate, Dragendorff's reagent	Separation of morphine from other basic and neutral drugs—R_f and color data for 48 compounds	DAVIDOW et al (1966)
NA, SH, ST, PH, AH	10 ml	pH 9.5 — NH_4Cl/NH_4OH	$CHCl_3$·Isopro (24 1)	Silica Gel G	EtAc MeOH.NH_4OH (85 10 5)	Ninhydrin, diphenylcarbazone, mercuric sulfate, iodoplatinate, Dragendorff's reagent	R_f and color data for 89 compounds	DAVIDOW et al. (1968)
NA, SH, ST, PH	10 ml	pH 9 5 — NH_4Cl/NH_4OH	$CHCl_3$	EM Silica Gel G with and without fluorescent indicator	EtAc MeOH NH_4OH (85 10·5)	Ninhydrin, diphenylcarbazone, mercuric sulfate, iodoplatinate, Dragendorff's reagent, potassium permanganate, H_2SO_4	Describes acid hydrolysis, ion exchange resin impregnated paper, and XAD-2 extractions, "one plate" and "two plate" TLC system — R_f and color data for 17 compounds	EM Lab Product Brochure
NA, SH, ST, PH	15 ml	pH 9 5 — NH_4Cl/NH_4OH	$CHCl_3$:Isopro (≅44 1)	3 sheets run Gelman ITLC Sheets (SAF-D)	EtAc cyclohexane MeOH NH_4OH (≅67·32 0 6·0 4) and (≅80 4 12 4 6 7 0 5) EtAc·MeOH.NH_4OH (87 5.8 4 5)	Silver acetate, diphenylcarbazone, potassium permanganate, ninhydrin, iodoplatinate	Describes a mini-TLC system, packaged as a kit—R_f and color data for 36 compounds	Gelman Drug Ident Systems
Morphine	15 ml	pH 9 3-borate	$CHCl_3$ Isopro (3 1)	Silica Gel G	EtAc:MeOH NH_4OH (85 10 5)	Iodoplatinate	Recovery and sensitivity study, compared to ion exchange resin impregnated paper extraction	GORODETZKY (1973a)
SH	5 ml	pH 6-potassium phosphate	Diethyl ether	Analtech Silica Gel G	$CHCl_3$.acetone	Diphenylcarbazone, mercuric sulfate	Gives method for amphetamine and opiates using ion exchange resin impregnated paper—R_f and color data for 7 compounds	HEATON and BLUMBERG (1969)

Table 1 (continued)

Drug or[a] Drug groups tested	Extraction			TLC			Other comments	Reference
	Urine vol. used	pH-buffers[b]	Solvents[b,c]	Plates	Primary developing solvents	Detection		
NA	10 ml	pH 9-10-Na borate/KOH	$CHCl_3$ Isopro (3 1)	Quantum LQD	EtAc. MeOH: NH_3 (80·20.0 5)	H_2SO_4, iodoplatinate, Dragendorff's reagent	Describes ammonolysis of urine, acid and enzymatic hydrolysis, GC confirmation—R_f and color data for 11 compounds	HIGGINS and TAYLOR (1974)
NA, SH, ST, PH	15 ml	2 separate extractions: pH 1-HCl (SH) pH 10 1 NH_4Cl/ NH_4OH (NA, ST, PH)	Benzene CHI_3 (8.2) (SH) $CHCl_3$ Isopro (9 1) (NA, ST, PH)	Gelman ITLC Sheets	Not given	Diphenylcarbazone, silver acetate, mercuric sulfate, ninhydrin, H_2SO_4, iodoplatinate, ammoniacal silver nitrate, iodine-potassium iodide	Describes acid hydrolysis, ion exchange resin impregnated paper extraction, primarily concerned with extraction	KAISTHA and JAFFE (1971)
NA, SH, ST, PH, AH	15 ml	2 extractions: pH 1-HCl (SH) pH 9-10-NH_4Cl/ NH_4OH(NA, ST, PH, AH)	Benzene $CHCl_3$ (8:2) (SH) $CHCl_3$ Isopro (9 1) (NA, ST, PH, AH)	Gelman ITLC Sheets (Type SA)	6 systems with EtAc cyclohexane. NH_4OH (some with MeOH, H_2O and/or dioxane)	Ninhydrin, H_2SO_4, iodoplatinate, ammoniacal silver nitrate, potassium permanganate, bromcresol green, sodium bicarbonate, diphenylcarbazone, silver acetate, mercuric sulfate, iodine-potassium iodide	Describes also ion exchange resin impregnated paper extraction—R_f and color data for 46 compounds	KAISTHA and JAFFE (1972c)
NA, SH, ST, PH	10 ml	2 separate extractions pH 9-NH_4OH (NA, ST, PH) pH 5-NaOH or H_2SO_4 (SH)	$CHCl_3$ Isopro (9.1) (NA, ST, PH) $CHCl_3$ (SH)	Silica Gel G	EtAc MeOH NH_4OH (85 10 5) or (85 10 0 25) EtOH HAc H_2O (60 30 10)	Ninhydrin, H_2SO_4, iodoplatinate, mercurous nitrate, p-dimethylamino-benzaldehyde, antimony trichloride-acetic anhydride	Describes acid hydrolysis—R_f data for 7 compounds	KOKOSKI (1970)

NA, SH, ST, PH	25 ml (SH) 50 ml (NA, ST, PH)	pH 5-HCl(SH) pH 9-NaOH(NA, ST, PH)	Methylene chloride (SH) ethylene dichloride with 10% isoamyl alcohol (NA, ST, PH)	Silica Gel G	$CHCl_3$ acetone (9 1) (SH) BuOH HAc H_2O (4 1 5) (ST) EtOH dioxane benzene NH_3 (5 40 50 5) (NA) EtAc HAc (5 2 2) (PH)	Mercurous nitrate, ninhydrin, iodoplatinate, H_2SO_4	Also reports TLC of some hallucinogens, describes acid hydrolysis, UV confirmation of morphine—R_f and color data for 21 compounds (mainly PH and hallucinogens)	McIsaac (1966)
NA	6 ml	pH 10-potassium phosphate	Ethylene dichloride with 25% isobutanol	Silica Gel G	EtOH dioxane benzene NH_4OH (5 40 50 5) also 6 other systems	Iodoplatinate	Early paper, extensive UV and GC data, describes acid hydrolysis—R_f data for 21 narcotics	Mulé (1964)
NA, SH, ST, PH, HL	15 ml	3 extractions pH 2.2-citrate (SH) pH 9 3-borate (NA, PH, HL) pH 11-carbonate (ST)	$CHCl_3$ (SH, ST) $CHCl_3$ Isopro (3 1) (NA, PH, HL)	Silica Gel G	EtOH dioxane benzene NH_4OH (5 40 50 5) EtAc MeOH NH_4OH (85 10 5) also 12 other systems	Silver acetate, potassium permanganate, mercurous nitrate, iodoplatinate, ammoniacal silver nitrate, Dragendorff's reagent, ninhydrin, H_2SO_4	Describes also ion exchange resin impregnated paper extraction—R_f and color data for 29 compounds	Mulé (1969)
NA, SH, ST, PH	15 ml (2) 25 ml	3 extractions pH <1-H_2SO_4 (SH) pH 9-potassium phosphate (NA) pH 10-11-potassium phosphate (NA, ST, PH)	$CHCl_3$ Isopro (3 1) (NA) $CHCl_3$ (SH) $CHCl_3$ with 25% EtOH (NA, ST, PH)	Brinkmann plates, Silica Gel G, and Silica Gel F-254 Eastman Chromatogram Sheets (No 6060 with fluorescent indicator)	EtAc MeOH NH_3 (85.10 10) (85 5 2 5) $CHCl_3 \cdot$MeOH NH_3 (90 10.1)	H_2SO_4, iodoplatinate, ammoniacal silver nitrate, ammonium hydroxide, potassium permanganate, diphenylcarbazone, ninhydrin, iodine, furfural	Describes fluorometric screen for morphine and quinine, GC confirmation, acid hydrolysis before pH 9 extraction—R_f and color data for 41 compounds	Mulé (1971)
NA, SH, ST	50 ml	3 extractions acidic (SH) pH 13 (ST) pH 8 5-$NaHCO_3$ (NA)	Ether (SH) $CHCl_3$ (ST) $CHCl_3$ Isopro (3 1) (NA)	Silica Gel G	$CHCl_3$ MeOH (4 1)	Iodoplatinate	Describes acid hydrolysis, confirmation by microcrystal tests, emphasis on NA, ST—gives order of R_f values for 44 compounds	Ono et al (1969)

Table 1 (continued)

Drug or[a] Drug groups tested	Extraction			TLC			Other comments	Reference
	Urine vol used	pH-buffers[b]	Solvents[b,c]	Plates	Primary developing solvents	Detection		
NA, ST	20 ml	2 separate extractions: pH 8 6 KOH, NaHCO (NA) pH 10-potassium carbonate (ST)	butanol-H_2SO_4-$CHCl_3$ with 10% EtOH(NA) $CHCl_3$-H_2SO_4-$CHCl_3$(NA, ST)	Silica Gel G	EtOH NH_4OH dioxane benzene (10.11:80·100) (NA) NH_4OH MeOH (3 300) (NA, ST)	Iodoplatinate, phenol reagent, Dragendorff's reagent	Gives R_f values for 24 compounds	PARKER and HINE (1967)
NA, SH, ST, PH	15 ml	pH 6-K-Na phosphate	$CHCl_3$ Isopro (3:1)	2 plates, Eastman Chromatogram Sheets (No 6061)	$CHCl_3$ acetone (9 1) (SH) EtAc·MeOH NH_4OH (85 10 5) (NA, ST, PH)	Mercuric sulfate, diphenylcarbazone, ninhydrin, H_2SO_4, iodoplatinate, Dragendorff's reagent, ammoniacal silver nitrate	Uses bromcresol purple to increase extraction of basic drugs at pH 6—R_f data for 22 compounds	STONER and PARKER (1974)
NA, SH, PH	About 1/10 vol of extracting solvent	2 separate extractions: acid (SH) base (NA, PH)	$CHCl_3$—NaOH—$CHCl_3$(SH) $CHCl_3$—H_2SO_4—$CHCl_3$(NA, PH)	Silica Gel G	$CHCl_3$ acetone (9 1) (SH) MeOH NH_3(100 1 5) (NA, PH) also describes 4 other systems	Mercuric sulphate, diphenylcarbazone, potassium permanganate, FPN reagent, furfural, iodoplatinate, chlorine-starch iodide	General toxicologic procedure for urine, blood, tissue, stomach contents, early paper on use of TLC in toxicology—R_f and color data for 61 compounds	SUNSHINE (1963)
NA, PH	10 ml	pH 8 5-9-potassium carbonate, or borate	$CHCl_3$.t-butanol (9.1)	Analtech, Silica Gel G	BuOH:H_2O:HAc (4 2 1) BuOH.conc HCl (9 1) MeOH H_2O.HAc benzene (80:15 2.5)	Iodoplatinate	Describes acid hydrolysis—R_f and color data for 19 compounds	WALLACE et al (1972a)

[a] NA = Narcotic analgesics and antagonists, SH = Sedative-hypnotics (including barbiturates and minor tranquilizers), ST = Stimulants, PH = Phenothiazines and antidepressants; AH = Antihistamines; HL = Hallucinogens

[b] Letters in ( ) indicate drug or drug group

[c] EtOH = ethanol, MeOH = methanol, $CHCl_3$ = chloroform, Isopro = isopropanol, EtAc = ethyl acetate, HAc = acetic acid, BuOH = butanol

neutral and basic drugs, including the opioids and stimulants (e.g., KAISTHA and JAFFE, 1972c). Based on the observation of STEVENSON (1961) that significant quantities of barbiturate could be extracted at alkaline pH's, DAVIDOW et al. (1968) described a full screening procedure utilizing a single extraction at pH 9.5 using ammonium chloride/ammonium hydroxide buffer. Most subsequently described single extraction procedures have used similar pH and buffer conditions. STONER and PARKER (1974) described a single extraction at pH 6. However, in this procedure bromcresol purple was included in the buffer, forming salts with the basic drugs and increasing their solubility in the extracting solvent. The salts were subsequently split by ammonia in the TLC developing solvent.

As can be seen in the fourth column of Table 1, many different extraction solvents and solvent mixtures have been reported. Chloroform: isopropanol in combination, most commonly varying from 3:1–9:1 in composition but reaching as high as 44:1, appears to be the most common extracting solvent for single pH extractions (e.g., DAVIDOW et al., 1968), for the neutral-basic extraction in a 2-pH procedure, and for the pH 9–10 (i.e., opioid) extraction in a triple extraction system. Chloroform alone was the most frequently used extractant for the strongly acidic (i.e., barbiturate) and strongly basic (i.e., stimulant) extractions when they were done separately. Chloroform also has been used in combination with benzene (e.g., KAISTHA and JAFFE, 1972c), butanol (e.g., WALLACE et al., 1972a), and ethanol (e.g., MULÉ, 1971); and the use of ethanol and ether has been described (e.g., BASTOS et al., 1970). Some of the earlier methods used methylene chloride and ethylene dichloride (e.g., MCISAAC, 1966). The majority of screening methods used a single solvent extraction. Although the extract may be "dirtier" (that is, it may contain more interfering substances, such as urinary pigments) than could be achieved by multiple extractions, there was a gain in time, simplicity, and economy. However, several authors described triple extraction procedures, extracting first into organic solvent, then into an aqueous phase, and finally back into solvent (PARKER and HINE, 1967; SUNSHINE, 1963). Organic and aqueous phases were shaken by hand or machine and were separated, usually after centrifugation, by use of separatory funnels, aspiration, or, recently phase separation filter paper (e.g., EM Lab. Product Brochure). Emulsions can be a problem, as discussed by KAISTHA and JAFFE (1971), who broke them by adding a small volume of absolute ethanol. The separated organic phase was then evaporated to dryness using moderate heat (under 100° C), usually using a water bath, heating block, oven, or by a stream of air or nitrogen. Several special manifolds have been described to facilitate the process (LEDERER and GERSTBREIN, 1973; YEH et al., 1973), which can be time consuming if large volumes of solvent are used or many samples are run. If amphetamines were to be determined, the organic extract was made acidic (usually with a few drops of methanolic sulfuric or hydrochloric acid) to convert these compounds to their less volatile salts. The extraction procedure was completed by redissolving the residue in a small volume of solvent for application to the TLC medium.

In addition to the major elements of the extraction procedure summarized in Table 1, there are many modifications and variations used by different laboratories. For example, "salting out" procedures to increase drug recovery have been described using sodium chloride (GORODETZKY, 1973a) and potassium carbonate (BASTOS et al., 1970). HIGGINS and TAYLOR (1974) described the use of ammonium hydroxide, which they felt increased the detection of methadone and optimized conditions for extracting all opioids containing a phenolic hydroxyl group. Procedures for "cleaning up" the organic extract have been discussed by several authors. Most commonly a wash was used for removal of interfering substances, such as lead acetate (BERRY and GROVE, 1971), sodium bicarbonate (BROICH et al., 1971a), and potassium phosphate (WALLACE et al., 1972a). HIGGINS and TAYLOR (1974) advocated filtering the organic extract. BASTOS et al. (1970) extracted first into ethanol from urine saturated by potassium carbonate and then purified by extraction of drugs from ethanol into ether at pH 8.5.

It is difficult to estimate a single extraction efficiency for the organic solvent extraction procedures described because of the great variability between the methods. Recovery figures have been reported by several authors, usually for radioisotopically labeled drugs added to normal urine. However, it is not always clear at what point in the extraction procedure the recovery was evaluated, nor whether corrections were made for such variables as aliquots taken. MULÉ et al. (1971) reported comparative recoveries in the organic phase of C^{14} and H^3 labeled drugs added to normal urine by the 3-pH extraction procedure of MULÉ (1971) and the single pH methods of BASTOS et al. (1970) and DAVIDOW et al. (1968). Extraction efficiency of stimulants and opiates was similar for the three methods, with recoveries of approximately 65–85% for morphine (in the pH 9 extract) and methadone, 91–99% for meperidine, 74–93% for cocaine, and 59–65% for amphetamine. Comparative recoveries of meprobamate, pentobarbital, and phenobarbital in the pH 1 extraction of the 3-pH method with the single pH procedures showed comparable figures for meprobamate (77–89%). However, lower recoveries were found for pentobarbital and phenobarbital in the pH 9.5 extraction of DAVIDOW et al. (1968) (63 and 39%, respectively) compared to the other two methods (82–92% for pentobarbital and 72–90% for phenobarbital). BERRY and GROVE (1971) reported similarly low recoveries for phenobarbital (49%) and barbital (31%) using a pH 9.6 extraction. By another 3-pH extraction procedure, MULÉ (1969) reported recoveries of 61% for morphine and amphetamine and 86% for pentobarbital. In an earlier paper (MULÉ, 1964) recoveries of 81–93% were reported for a wide range of opiates using UV spectrophotometry of unlabeled drugs added to normal urine in high concentrations (25–200 µg/ml). GORODETZKY (1973a) defined extraction efficiency in terms of the amount of drug reaching the thin-layer plate versus the amount in the total urine aliquot analyzed and reported approximately 61% recovery of C^{14}-morphine added to normal urine. An extraction efficiency of 63% for C^{14}-morphine at pH 9 was reported by KOKOSKI et al. (1968) and KOKOSKI (1970). By optimizing morphine extraction conditions WALLACE et al. (1972a) achieved 92% recovery of morphine into chloroform:butanol at pH 8.5–9.0. Using UV spectrophotometry and gas chromatography BASTOS et al. (1970) reported recoveries of large concentrations of 11 different drugs in both the ethanol and ether extracts in their procedure. Most drugs were recovered in amounts exceeding 60% in the final ether extract, although amphetamine and methamphetamine gave 46 and 52% recoveries, respectively; and chlordiazepoxide and propoxyphene resulted in only 34 and 38% recoveries, respectively. No barbiturate recoveries were reported. COCHIN and DALY (1963) reported 90–100% recovery of barbiturates in an early paper using a pH 5 extraction. In summary, from the data available it appears that extraction efficiencies of greater than 60% are obtainable for most drugs by most of the described organic solvent extraction procedures.

Morphine and amphetamine are generally at the lower end of the recovery range and recovery of greater than 80% for many drugs is not uncommon. Single pH extractions at pH greater than 9 may result in low recoveries of barbiturates, in the 40–60% range.

b) Ion Exchange Resin Impregnated Paper Extraction

Table 2 summarizes papers describing TLC screening systems utilizing ion exchange resin impregnated paper extraction of urine. This extraction procedure was first proposed by DOLE et al. (1966), who subsequently published a revised method (DOLE et al., 1972). As in the direct organic solvent extraction methods, the majority of the described procedures are full screening methods for most of the abused drugs of interest, with the exception of the hallucinogens.

The aliquot of urine used for analysis in the ion exchange paper methods was usually greater than that used for direct organic solvent extraction. A 50 ml aliquot was the most frequently recommended volume, with a range of 20–100 ml. Most authors advocated rough pH adjustment of urine prior to extraction, in most cases to pH 5–6. However, DOLE et al. (1972) recommended an initial urine pH of 4–5 and the EM Lab Product Brochure suggested pH 6–7. Urine was extracted undiluted in most methods, although DOLE et al. (1966; 1972) as well as HEATON and BLUMBERG (1969) described a 1:1 dilution with water. Urine was extracted in the majority of procedures with one piece of Reeve Angel SA-2 cation exchange paper (6 × 6 cm square), which was placed with the urine in a wide-mouthed jar and shaken gently by machine or hand or allowed to soak (usually for 12–24 h). JUSELIUS and BARNHART (1973) have shown evidence that soaking or intermittent hand shaking for only a 1-h extraction period resulted in decreased extraction efficiency compared to continuous machine shaking. The use of two squares of SA-2 paper has been recommended by GORODETZKY (1973a) and extraction in a test tube using a 1 $^{1}/_{4}$ × 4″ accordion-folded strip of SA-2 paper was described in the EM Lab Product Brochure. JAFFE and KIRKPATRICK (1966) have also described a procedure using a "sandwich" of cation (SA-2) and anion (SB-2) exchange paper for extraction, with the SB-2 paper used specifically for determination of barbiturates. Following extraction, the paper (on which are adsorbed the drugs of interest) was removed and the urine was discarded. The paper may be analyzed immediately or dried and analysis completed at a later date. In the initial description of this extraction procedure DOLE et al. (1966) proposed remote extraction of urine (for example, at a treatment clinic), and mailing the dried paper to the laboratory for analysis. In a later publication DOLE et al. (1972) reported that dried paper can be stored for up to 4 years without deterioration.

The extraction procedure was continued by elution of drugs from the exchange resin paper into organic solvents at one pH or using two or three sequential elutions at different pH's. Papers may be analyzed singly or pooled for elution. This latter procedure has been recommended in several papers (JUSELIUS and BARNHART, 1973; KAISTHA and JAFFE, 1971; KAISTHA et al., 1975), pooling papers from urine specimens of the same patient. This procedure does not compromise the sensitivity of the analysis and resulted in considerable savings in expense; however, it is not possible to determine from which specimen a positive result originated. Three sequential elutions were originally advocated by DOLE et al. (1966) and subsequently described with some modifications by MULÉ (1969) and DOLE et al. (1972). Analogous to the

Table 2 Thin-layer chromatography urine screening procedures Ion exchange resin impregnated paper extraction, Reeve Angel SA-2

Drug or[a] drug groups tested	Extraction			TLC			Other comments	Reference
	Urine vol used	Paper elution		Plates[a]	Primary developing solvents	Detection		
		pH-buffers[a]	Solvents[a]					
NA, PH	50–100 ml	pH 9 3-borate	$CHCl_3$ Isopro (3.1)	Silica Gel G	BuOH.HAc H_2O (top layer) (4 1 5)	Iodoplatinate, ammoniacal silver nitrate	Describes GC methods for barbiturates and amphetamine with organic solvent extraction—R_f and color data for 11 compounds	BASELT and CASARETT (1971)
NA, SH, ST, PH	50–100 ml (diluted 1:1 with H_2O) (at pH 5–6)	3 sequential elutions pH 2.2-citrate (SH) pH 9.3-borate (NA, PH) pH 11-carbonate (SH)	$CHCl_3$ (SH, ST) $CHCl_3$ Isopro (3:1) (NA, PH)	Silica Gel G	EtAc MeOH·NH_4OH (85.10.5) $CHCl_3$ acetone (9 1) (SH)	H_2SO_4, silver acetate, potassium permanganate, iodoplatinate, ammoniacal silver nitrate, ninhydrin	Suggest use of 6 cm × 6 cm TLC plates—R_f and color data for 14 compounds	DOLE et al. (1966)
NA, SH, ST, PH	30–50 ml (diluted 1·1 with H_2O) (at pH 4–5)	3 sequential elutions pH 2.2-citrate (SH) pH 9 3-borate (NA, PH, ST) pH 11-carbonate (ST)	$CHCl_3$ (SH, ST) $CHCl_3$·Isopro (3:1) (NA, PH, ST)	Analtech Silica Gel G (SH) Brinkmann Silica Gel F-254 (NA, PH) Quantum S/G Q 1 Silica Gel (ST)	EtAc.MeOH.NH_4OH (85:10:5) EtAc EtOH BuOH. con NH_4OH (28 14 2.0 4)	Silver acetate, diphenylcarbazone, mercuric sulfate, iodoplatinate, ammoniacal silver nitrate, ninhydrin, Marquis reagent	R_f and color data for 30 compounds	DOLE et al. (1972)
NA, SH, ST, PH	25 ml (pH 6–7)	Single elution pH 9 5-NH_4Cl/NH_4OH	$CHCl_3$ Isopro (9:1)	Same as in Table 1, EM Lab Product Brochure			Describes organic solvent and XAD-2 extractions, "one plate" and "two plate" TLC systems—R_f and color data for 17 compounds	EM Lab. Product Brochure
Morphine	50 ml (at pH 5–6)	pH 9 3-borate	$CHCl_3$.Isopro (3:1)	Same as in Table 1, GORODETZKY (1973a)			Recovery and sensitivity study, compared to organic solvent extraction, uses 2 papers	GORODETZKY (1973a)

NA, ST, PH	50–100 ml (diluted 1.1 with H_2O) (at pH 5–6)	pH 9 3-borate	$CHCl_3$.Isopro (3 1)	Analtech Silica Gel G	EtAc MeOH NH_3 (85·10.5)	Bromcresol green, H_2SO_4, iodoplatinate, ammoniacal silver nitrate	Washes paper at pH 2 2 (citrate buffer) with $CHCl_3$ and discards, describes SH method using organic solvent extraction—R_f and color data for 7 compounds	HEATON and BLUMBERG (1969)
NA, SH, ST, PH	40–50 ml	2 sequential elutions: pH 1-citrate (SH) pH 10 1-NH_4Cl/ NH_4OH (NA, ST, PH)	$CHCl_3$ (SH) $CHCl_3$.Isopro (3 1) (NA, ST, PH)	Same as in Table 1, KAISTHA and JAFFE (1971)			Describes also organic solvent extraction, primarily concerned with extraction, suggests pooling papers for elution	KAISTHA and JAFFE (1971)
NA, SH, ST, PH, AH	40–50 ml (at pH 5–6)	2 sequential elutions· pH 1-citrate (SH) pH 10.1-NH_4Cl/ NH_4OH (NA, ST, PH, AH)	$CHCl_3$ (SH) $CHCl_3$ Isopro (3 1) (NA, ST, PH, AH)	Same as in Table 1, KAISTHA and JAFFE (1972c)			Describes also organic solvent extraction—R_f and color data for 46 compounds	KAISTHA and JAFFE (1972c)
NA, SH, ST, PH	20–50 ml (pH 5–6)	Single elution pH 10 1-NH_4Cl/NH_3	$CHCl_3$ Isopro (3 1)	Gelman ITLC Sheets (Type SA)	4 systems with EtAc, cyclohexane, NH_3 (3 with MeOH, 1 with MeOH and H_2O)	Ninhydrin, diphenylcarbazone, silver acetate, mercuric sulfate, H_2SO_4, iodoplatinate, iodine-potassium iodide, ammoniacal silver nitrate, potassium permanganate	Suggests pooling papers for elution, describes two stage development, complex spraying sequence—color data for 74 compounds—R_f data for 122 compounds alone and in various combinations	KAISTHA et al (1975)
NA, SH, ST, PH, HL	50 ml (pH 5–6)	3 sequential elutions. pH 2 2-citrate (SH) pH 9 3-borate (NA, PH, HL) pH 11-carbonate (ST)	$CHCl_3$ (SH, ST) $CHCl_3$ Isopro (3 1) (NA, PH, HL)	Same as in Table 1, MULÈ (1969)			Describes also organic solvent extraction—R_f and color data for 29 compounds	MULÉ (1969)

[a] Same as Table 1

direct solvent extraction methods, barbiturates were eluted at an acid pH (pH 2.2 using a citrate buffer), narcotic analgesics and phenothiazines were found in the pH 9.3 eluate (using a borate buffer), and stimulants were determined after eluting at pH 11 (with a carbonate buffer). DOLE et al. (1972) and JUSELIUS and BARNHART (1973) noted that a significant fraction of amphetamines also elute at pH 9.3 and approximately one-third of the eluant was included in the pH 11 elution. A method using two sequential elutions has been advocated by KAISTHA and JAFFE (1971; 1972c), using pH 1 for elution of barbiturates and pH 10.1 (with ammonium chloride/ammonium hydroxide buffer) for the remaining drugs. Recently, KAISTHA et al. (1975) have described a single elution at pH 10.1, and the EM Lab Product Brochure recommended a single elution at pH 9.5.

Chloroform:isopropanol (3:1) (and in one case 9:1) was the most common eluting solvent described, as seen in the fourth column of Table 2. It was described for use in the pH 9.3 elution in the 3-pH procedures, the pH 10.1 elution in the 2-pH methods, and as the one eluting solvent when a single elution was used. Chloroform alone was used in the low pH elution of 2- or 3-pH methods and also for the pH 11 elution for stimulants. Solvent, buffer and paper (or pooled papers) were mixed, usually by shaking by machine, and the organic phase was separated (by pipette or in a separatory funnel) and evaporated to dryness, using the same techniques as described above for completion of the organic solvent extraction.

The minimal extraction efficiency data which have been reported using ion exchange resin impregnated paper procedures indicate that this method is less efficient than direct organic solvent extraction. Using C^{14} and H^3 labeled drugs added to normal urine, MULÉ (1969) reported recoveries from urine of approximately 22, 2.5, and 2% for morphine, pentobarbital, and d-amphetamine. GORODETZKY (1973a) found 48% recovery of C^{14}-morphine, comparing amount of drug reaching the TLC plate to the amount in the urine aliquot analyzed. The very low recoveries of pentobarbital and amphetamine found by MULÉ have been widely quoted. However, DOLE et al. (1972) have noted that some poor batches of SA-2 paper were released by the manufacturer between 1966 and 1969 and could account for the low recoveries reported. Consistent with this possibility is the overall sensitivity of the ion exchange paper methods for amphetamine and pentobarbital (discussed in detail below), which would be unlikely to be achieved with extraction efficiencies as low as those reported by MULÉ (1969).

c) *Nonionic Resin Extraction, Amberlite XAD-2*

In Table 3 are summarized the major papers describing use of the nonionic resin, Amberlite XAD-2, for urine extraction in TLC screening systems for drugs of abuse. First proposed by FUJIMOTO and WANG (1970), this is the most recent of the three commonly used extraction procedures. As for the other extraction procedures, the majority of authors described full screening systems for drugs of abuse in urine; also IBRAHIM et al. (1975) have reported a method using XAD-2 resin extraction for general toxicologic analysis of stomach contents and large tissue samples.

The volume of urine used in analysis tends to be smaller than that used in ion exchange paper extraction and larger than that used for direct organic solvent extraction. An aliquot of 20–30 ml was most commonly used, although WEISSMAN et al. (1971) recommended a small volume of 5 ml (to reduce the amount of urinary pig-

Table 3 Thin-layer chromatography urine screening procedures Nonionic resin extraction, Amberlite XAD-2

Drug or[a] drug groups tested	Extraction			TLC			Other comments	Reference
	Urine vol used	Resin	Eluting solvents[a]	Plates[a]	Primary developing solvents[a]	Detection		
NA, SH, ST, PH	25 ml	Self-packed plastic column, 45 × 10 mm resin bed	2 sequential elutions isopropyl ether (SH) $CHCl_3$ Isopro (3 1) (SH, ST, PH)	Quantum Silica Gel F-254 Q5WF (SH) Brinkmann Polygram Silica Gel Sheets (SH) Brinkmann Silica Gel (NA, ST, PH)	$CHCl_3$ MeOH NH_3 (90 10. 1) (SH) EtAc: MeOH H_2O. NH_4OH (85 10 3 1) (NA, ST, PH)	Mercuric sulfate, diphenylcarbazone, iodine acid, furfural, ninhydrin, H_2SO_4, iodoplatinate, p-nitroaniline, alcoholic sodium hydroxide	Adsorption with special hydraulic flow control apparatus, eluate washed with aqueous acid or base	BASTOS et al. (1973)
NA, SH, ST	20 ml (buffered at pH 9.5)	Brinkmann prepacked columns	1,2-dichlor-ethane. EtAc (4 6) or $CHCl_3$ Isopro (6 1)	Brinkmann Silica Gel G	EtAc MeOH. NH_4OH (85 10:5)	Ninhydrin, diphenyl-carbazone, mercuric sulfate, iodoplati-nate, Dragendorff's reagent	No flow control during adsorption, elution through phase separation filter paper in filter cart-ridge, describes acid hydrolysis of dried extract	Brinkmann Product Brochure
NA, SH, ST, PH	20 ml	Brinkmann prepacked columns	$CHCl_3$ Isopro (6·1)	Silica Gel G (may use com-mercial plates-not specified)	EtAc MeOH NH_4OH (85 10 5)	Ninhydrin, diphenyl-carbazone, mercuric sulfate, iodoplatinate, Dragendorff's reagent, ammoniacal silver nitrate	No flow control during adsorption, elution through filter cartridge, describes confirmation of morphine by Technicon Autoanalyzer	DAVIDOW et al (1973)
NA, SH, ST, PH	20 ml (buffered at pH 9.0)	Prepacked or self-packed columns	$CHCl_3$ Isopro (3 1)	Same as in Table 1, EM Lab Product Brochure			No flow control during adsorption, describes organic solvent and ion exchange resin impregnated paper extraction, "one plate" and "2 plate" TLC system—R_f and color data for 17 compounds	EM Lab Product Brochure

Table 3 (continued)

Drug or[a] drug groups tested	Extraction: Urine vol used	Extraction: Resin	Extraction: Eluting solvents[a]	TLC: Plates[a]	TLC: Primary developing solvents[a]	TLC: Detection	Other comments	Reference
NA, SH, ST, PH	20 ml (buffered at pH 9.5)	Kodak Chromat/0/ Screen prepacked columns	1,2-dichloro-ethane: EtAc (4.6)	Eastman Chromatogram Sheet (No 6061)	EtAc. MeOH: NH_4OH (85.10:5) BuOH HAc: H_2O (4.1 1) (ST) $CHCl_3$.acetone (9·1) (SH) EtAc. MeOH. H_2O (7.2·1) (NA, PH)	Ninhydrin, mercuric sulfate, H_2SO_4, di-phenylcarbazone, iodoplatinate, ammoniacal silver nitrate, potassium ferricyanide—ferric chloride	Describes acid hy-drolysis, no flow control during ad-sorption, elution through filter car-tridge—R_f and color data for 24 com-pounds—gives param-eters for photographing TLC plates	FISHER et al (1972)
NA, SH, ST, PH	30 ml	Self-packed columns, 100 × 5 mm resin bed	Methanol (discard 1st 8 ml of eluate)	4 plates run 1—Brinkmann Silica Gel F-254 on alu-minum (ST) 3—Gelman ITLC Sheets (Type SA); (NA, PH, SH, benzodiazepines)	EtAc. MeOH NH_4OH (85 10:5)	Iodoplatinate, Dragenorff's reagent, ninhydrin, mercurous nitrate, bleach, phenol, starch-iodine	No flow control during adsorption—gives R_f and color data for 16 compounds	HETLAND et al. (1972)
NA, SH, ST	20 ml (buffered at pH 8 5–9)	Brinkmann prepacked columns	$CHCl_3$ MeOH (3 1)	Quantum LQD	EtAc MeOH NH_4OH (180 17:7)	Iodoplatinate	Flow control during adsorption with Tech-nicon proportional pumps, TLC described for morphine only, gives 3 methods of acid hydrolysis	KULLBERG and GORODETZKY (1974)

NA, SH, ST, PH, AH	25 ml	Self-packed plastic columns, 45 × 10 mm resin bed	$CHCl_3$ Isopro (3 1)	2 plates run· Brinkmann Silica Gel F-254, Brinkmann Polygram Sheets	EtAc MeOH H_2O. NH_3 (85 10.3:1) (F-254 plates) $CHCl_3$.MeOH NH_3 (90 10·1) (Polygram Sheets)	Ninhydrin, iodoplatinate, sodium iodide, p-nitroaniline, alcoholic sodium hydroxide, H_2SO_4, diphenylcarbazone, mercuric sulfate	Adsorption with special hydraulic flow control apparatus, study of factors effecting adsorption and elution—R_f and color data for 40 compounds	MULÉ et al (1971)
NA, SH, ST, PH	35 ml	Self-packed in cone-shaped polyethylene bag, 12 cm high resin bed	MeOH	3 plates run Gelman ITLC Sheets (Type SG)	Benzene hexane: diethylamine (25 10.1) (NA, PH) $CHCl_3$ NH_4OH (50.0 1) (ST) $CHCl_3$.HAc (50 0 1) (SH)	Diphenylcarbazone, mercurous nitrate, Fluoram (fluorescamine), iodoplatinate	No flow control during adsorption, describes confirmation by RIA and GC—R_f and color data for 27 compounds	ROERIG et al (1975a)
SH	50 ml	Self-packed in cone-shaped polyethylene bag, 7 cm high resin bed	MeOH	Gelman ITLC Sheets	$CHCl_3$ HAc (50.0 1)	Diphenylcarbazone, mercurous nitrate	No flow control during adsorption—R_f and color data for 12 compounds	WANG and MUELLER (1973)
NA, SH, ST, PH	5 ml	Self-packed in columns; 60 × 10 mm resin bed	MeOH	Brinkmann Silica Gel on aluminum	EtAc.MeOH·NH_4OH (85.10.5)	Ninhydrin, diphenylcarbazone, mercuric sulfate, iodoplatinate, Dragendorff's reagent	No flow control during adsorption—R_f and color data for 24 compounds	WEISSMAN et al (1971)
NA, SH, ST, PH	15 ml (buffered at app pH 8)	Self-packed plastic columns; approx. 5 cm high resin bed	MeOH	2 plates run Bakeflex Sheets (Type IB 2-F)	EtAc.MeOH:NH_4OH (85 10 5) $CHCl_3$ acetone: NH_4OH (90·9·1)	Ninhydrin, diphenylcarbazone, mercuric sulfate, iodoplatinate	No flow control during adsorption	WISLOCKI et al. (1974)

[a] Same as Table 1.

ments in the extract) and WANG and MUELLER (1973) used 50 ml in a method for sedative-hypnotic detection. MILLER et al. (1973) and KULLBERG et al. (1973) reported that for a fixed amount of resin (2 g) adsorption efficiency of drug onto resin with increasing urine volume was not changed for amphetamine, decreased slightly for morphine, and decreased markedly for phenobarbital. These effects were exaggerated by high flow rates. On the basis of these studies a volume of 20 ml of urine at a controlled flow rate was recommended.

Adjustment of urinary pH to about 8–9.5 prior to extraction has been recommended by some authors. HETLAND et al. (1972) reported no effect of urine pH in the range of 3–9.5. However, in detailed studies MILLER et al. (1973), KULLBERG et al. (1973), and BASTOS et al. (1972) did find significant effects of urinary pH on drug adsorption. Morphine was most efficiently adsorbed at about pH 8–10; amphetamine adsorption was optimal over a wide range, about pH 6–12; and phenobarbital was best adsorbed at a lower range, about pH 2–7. All studies agreed on an optimal compromise pH of approximately 8–9, which should result in 80–95% adsorption efficiency for all drugs.

XAD-2 resin was most frequently packed in columns for urine extraction; these may be purchased prepacked or the resin can be purchased in bulk. When obtained in bulk for self-packing, the resin must first be prepared by multiple washes; several preparation procedures have been described (e.g., MULÉ et al., 1971). Small plastic or glass columns were most frequently used; the size of the resin bed varied considerably in the described procedures, as seen in the third column of Table 3. WANG and MUELLER (1973) and ROERIG et al. (1975a) described a column fabricated from a polyethylene bag by heat-sealing a wedge-shaped channel; this allowed easy compression of the column when it was desired to put pressure on the resin. FISHER et al. (1972) briefly described a bulk XAD-2 resin extraction, in which 2 g of resin were added to 10 ml of pH 9.5 buffered urine in a beaker. After 5 min of swirling, the urine was decanted and discarded and the resin was washed with 5% methanol in water and then eluted by swirling in the beaker with 10 ml of 1,2-dichlorethylene:ethyl acetate (4:6). Finally the organic phase was decanted and concentrated for TLC. BASTOS et al. (1972) investigated four other resins and concluded that, all factors considered, none of these was superior to the XAD-2.

The effect of the rate of flow of urine through the XAD-2 resin column on adsorption efficiency has been studied by several authors. MULÉ et al. (1971) described a special hydraulic flow control apparatus to control flow of urine during adsorption simultaneously for a batch of 130–150 samples. Although the actual flow rate was not reported, from the time given for flow through part of the column and, in a later paper (BASTOS et al., 1973), for the total extraction, the flow rate appears to be approximately 1–2 ml/min. Compared to gravitational flow, use of the hydraulic flow control apparatus resulted in 10–31% increase in recovery of morphine, quinine, phenobarbital, and meprobamate from the resin. MILLER et al. (1973) and KULLBERG et al. (1973) used a Technicon Proportioning Pump to control urine flow rates and found a decrease in adsorption efficiency with increasing flow rates. This was most severe with phenobarbital (with 90% adsorption only at flow rates less than 0.5 ml/min), less marked with morphine (90% adsorption at less than 2 ml/min), and least for amphetamine (greater than 90% adsorption between 0.75 and 4 ml/min). In one study (MILLER et al., 1973) gravitational flow rate was found to be highly variable, with most columns giving 2–6 ml/min flows, and resulting in a range of 65–85% adsorption efficiency for morphine. KULLBERG et al. (1973) recommended controlled flow at an intermediate rate of 2.1 ml/min. In a later paper KULLBERG and GORODETZKY (1974) described simultaneous control of 23 columns with the Technicon Proportioning Pump at a flow rate of 2.5 ml/min. From these studies it appears that flow control during adsorption at 1.5–2.5 ml/min is desirable.

Most reported methods used a single elution of all drugs from the resin. Methanol and chloroform:isopropanol (in ratios of 3:1–6:1) were the most commonly described eluting solvents; and ethyl acetate:dichloroethane and chloroform:methanol have also been recommended. BASTOS et al. (1973) described two sequential elutions, first with isopropyl ether primarily for acidic drugs and then with chloroform:isopropanol. Ethyl acetate:dichloroethane (3:2) was studied in detail by MILLER et al. (1973) and KULLBERG et al. (1973); they found that addition of 25% isopropanol to the mixture gave greater than 90% recoveries of morphine, phenobarbital, and amphetamine from the resin. MULÉ et al. (1971) also studied several solvents and solvent mixtures before deciding on chloroform:isopropanol.

Because air and water were frequently trapped in the resin after adsorption, techniques to improve flow characteristics during elution have been developed. KULLBERG and GORODETZKY (1974) added 1 ml of acetone to the column prior to elution and initiated flow of eluting solvent by application of positive air pressure to the column; BASTOS et al. (1973) made similar use of air pressure. ROERIG et al. (1975a) and WANG and MUELLER (1973), using a column fabricated from a polyethylene bag, applied manual pressure to the resin to remove as much urine as possible prior to elution. Addition of eluting solvent in two or three aliquots was recommended by DAVIDOW et al. (1973) and MULÉ et al. (1971) to improve recovery of drugs into the eluate. Additional procedures have also been described to decrease the amount of urinary pigments and other interfering substances in the eluate. Elution into a small volume of buffered aqueous wash, which was subsequently separated and discarded, was the most commonly used technique (BASTOS et al., 1973; KULLBERG and GORODETZKY, 1974). BASTOS et al. (1973) described a novel phase separation, freezing the acid aqueous wash following the first of two sequential elutions and pouring off the liquid organic phase. The tube containing the frozen acidic wash was then used to receive the second eluate and the mixture was then buffered to pH 8–9.5. The upper aqueous phase was finally removed by aspiration. HETLAND et al. (1972) discarded the first 8 ml of eluate, which they reported contains no detectable drug; and WISLOCKI et al. (1974) washed the column with dilute ethanol in water prior to elution. Following elution the separated organic eluate was evaporated to dryness in preparation for TLC, as described above for the other extraction methods.

Extraction efficiency data have been reported in several papers, using mostly radioisotopically labeled drugs added to normal urine and examining recovery at different steps in the extraction procedure. MULÉ et al. (1971) reported recoveries into the organic eluate of 75–90% for several barbiturates, meprobamate, cocaine, and meperidine. Amphetamine recovery was 49%; and methadone and morphine recoveries were 56 and 64%, respectively. In the lower recoveries the major loss was in the urine effluent (that is, failure of the drugs to adsorb to the resin); only small amounts were lost on elution or into the bicarbonate wash. Recoveries were generally comparable to those achieved by the organic solvent extraction methods of

MULÉ (1971), BASTOS et al. (1970), and DAVIDOW et al. (1968), although the amphetamine recovery by XAD-2 resin was 10–15% lower than in the other methods. In a later report, using minor modifications of the XAD-2 method of MULÉ et al. (1971), BASTOS et al. (1972) found recoveries of 84–96% for amphetamine, morphine, methadone, and phenobarbital. Using two sequential elutions of the resin, BASTOS et al. (1973) reported similar good recoveries of greater than 80% for several barbiturates, narcotics, cocaine, and amphetamine. KULLBERG and GORODETZKY (1974) reported recoveries of 84–93% into the organic eluate for four narcotics (including morphine and methadone), four barbiturates, amphetamine, and cocaine. In this study final recoveries at the thin-layer plate were 63–78%. Recoveries in the same range in the organic phase have been reported by ROERIG et al. (1975a) and WEISSMAN et al. (1971). In summary, it appears that recoveries into the organic eluate of greater than 80% for the majority of drugs of interest are achievable using the XAD-2 resin column extraction, especially if care is taken to adjust urine pH to 8–9.5 and control the flow during adsorption at 1.5–2.5 ml/min. These recoveries are generally slightly greater than those reported using organic solvent extraction methods.

d) Other

In addition to the commonly used extraction procedures described above, several new methods have been proposed for use in TLC screening systems. BROICH et al. (1971b) described lyophilization of urine by freeze-drying and then direct organic solvent extraction of the lyophilized sample. The authors felt this method offered convenience in concentration and storage of urine and the liquid-solid extraction should result in high drug recoveries. However, no extraction efficiencies were reported. Adsorption of drugs from urine onto charcoal was described by MEOLA and VANKO (1974). A 10 ml urine aliquot was mixed with 0.5 ml of prepared charcoal slurry and 5 ml of pH 11 carbonate buffer. After centrifuging, the liquid layer was aspirated and discarded. The charcoal was then sequentially eluted, first with ethyl ether for barbiturates, glutethimide, and cocaine, and then with chloroform: isopropanol (5:1) for narcotics, amphetamines, and other drugs. The method was reported to be simple and rapid and used only a small volume (2.5 ml) of eluting solvents. Extraction efficiencies were not reported; however, overall sensitivity of detection was in the same range as reported for TLC methods using other extraction techniques.

2. Thin-Layer Chromatography (TLC)

After evaporation to dryness, the residue of the urine extract was redissolved in a small volume of solvent and applied to a TLC medium. All of the extract may be applied to a single TLC plate or sheet, the extract may be split for application to more than one plate and/or sheet, or some of the extract may be saved for possible later use in a confirmatory procedure such as gas chromatography. Use of both TLC plates and sheets has been described. Silica Gel G, sometimes including a fluorescent indicator, was by far the most common TLC coating material used, as seen in Tables 1–3. In early papers, homemade plates were described; however, commercially available precoated plates are now in common usage. Plates marketed by Analtech,

Brinkmann, and Quantum were most frequently described. DOLE et al. (1972) evaluated several precoated plates and recommended specific plates for each of three ion exchange resin paper extracts (see Table 2). WEISSMAN et al. (1971) chose Brinkmann plates after a comparative evaluation because of consistent and even solvent front migration. Some authors have recommended the use of TLC sheets, such as Gelman ITLC (usually type SA), Eastman Chromatogram Sheets (usually No. 6061), and Brinkmann Polygram Sheets. Plates and sheets of various sizes have been advocated, from 6 cm × 6 cm (as described by DOLE et al., 1966) to the more common 20 cm × 20 cm. The smaller plates have the advantage of speed of development, but the disadvantage of a shorter distance to effect compound separation. The redissolved residue was usually applied to the TLC medium using disposable capillary pipettes or microsyringes. BROICH et al. (1971a) used a microdot application technique, in which the entire extract in 5 μl was applied in a spot not exceeding 1 mm in diameter; they reported application of up to 35 spots on one side of a single TLC sheet.

A large variety of solvent mixtures have been described for development of the TLC plate, as seen in Tables 1–3. The most commonly described developing solvent was ethyl acetate:methanol:ammonium hydroxide (usually 85:10:5), first proposed by DAVIDOW et al. (1966). Many variants of this mixture have also been used, such as changing the proportion of the constituents, addition of water, and omission of ethyl acetate. Chloroform:acetone (9:1) was another common mixture, used especially for separation of barbiturates and other acidic drugs. Most screening systems used one-dimensional development, with the solvent front allowed to travel 8–12 cm from the origin. Usually no special precautions were taken to control temperature, humidity, or saturation of the tank. However, HETLAND et al. (1972) recommended lining the developing tank with filter paper to achieve saturation. KOKOSKI et al. (1968 and 1970) emphasized the use of two-dimensional TLC for confirmation of positive tests. KAISTHA et al. (1975) described a two-stage developing procedure, in which a plate was developed first in one solvent mixture (until the solvent front reached 9 cm) and was then dried and developed again in the same direction in a second solvent mixture (until the front reached 14.5 cm). The authors felt this afforded optimum separation of a wide variety of compounds.

The chromogenic sprays used for detection of compounds on the developed TLC plate are listed for each paper in Tables 1–3. Those in most common use are further summarized by the drug, or group of drugs, detected in Table 4. Although many described screening systems used the same spray reagents, the details of the detection procedures were quite variable. As discussed above, frequently more than one extraction was performed, resulting in more than one TLC plate; or the final residue from a one-extraction procedure may be split prior to TLC. In general, those screening systems resulting in a single plate used simpler schemes for each drug (or drug group), with less emphasis on confirmatory sprays than multiple plate systems. Compounds were detected by the formation of colored products following application of the detection reagent. The colors formed after each spray within a given group of drugs were usually similar, but may differ slightly in shade, intensity, or stability. The sequence in which the different reagents were applied may also effect the color of the products formed. Extensive and detailed tables of color reactions are found in the papers of MULÉ (1971), MULÉ et al. (1971), KAISTHA and JAFFE (1972c), and KAISTHA et al. (1975).

Table 4. Common TLC detection reagents by drug or drug groups detected

Drug (or drug group)	Detection reagent
Narcotic analgesics and antagonists Cocaine	Primary: iodoplatinate Confirmatory: Dragendorff's reagent; ammoniacal silver nitrate; potassium permanganate
Amphetamine Methamphetamine Other stimulants (excluding cocaine)	Primary: ninhydrin-heat-UV; Bromcresol Green; fluorescamine (Fluram) Confirmatory: p-nitroaniline (PNA)
Barbiturates Glutethimide Ethchlorvynol	Primary: diphenylcarbazone (DPC); mercuric sulfate; mercurous nitrate; silver acetate; N,2,6-trichlorobenzoquinoneimine Confirmatory: potassium permangnate
Phenothiazines Antidepressants Antihistamines	Primary: sulfuric acid; FPN reagent Corfirmatory: iodoplatinate
Quinine	Primary: UV[a] Confirmatory: iodoplatinate; Dragendorff's reagent
Meprobamate	Primary: furfural

[a] UV = exposure to ultraviolet light.

For detection of alkaloids, including narcotic analgesics and antagonists and cocaine, iodoplatinate was the primary spray reagent of choice. For confirmation of narcotic analgesics iodoplatinate may be followed by either Dragendorff's reagent or ammoniacal silver nitrate, and the latter sequence may be further followed by potassium permanganate for additional confirmation.

Ninhydrin (made up in acetone or various alcohols) followed by heat and exposure to UV light was by far the most common detection procedure for amphetamine, other primary amines, and some secondary amines. A distinction between acid- and acetone-ninhydrin was made by BASTOS et al. (1970) and MULÉ et al. (1971); acetone-ninhydrin was used for primary amines and acid-ninhydrin for secondary amines such as methamphetamine. Also BASTOS et al. (1970) described a spraying sequence aimed specifically at different functional groups of basic drugs. They used freshly diazotized p-nitroaniline for confirmation of primary amines and dithiocarbonate formation (by exposure to carbon disulfide and ammonia) made visible with cupric chloride for confirming secondary amines. In this system tertiary amines were detected with iodoplatinate and overspraying with p-dimethylaminobenzaldehyde revealed primary aromatic amines, indoles, hydrazides, and phenothiazines. Use of alcoholic sodium hydroxide after p-nitroaniline has been reported to increase the intensity of the confirmatory spots produced by primary amines (MULÉ et al., 1971). Also the possible use of p-nitroaniline for detection of narcotics has been noted by NEESBY (1973). Bromcresol Green has been recommended as an alternate amphetamine spray and also reported to detect large amounts of opiates (HEATON and BLUMBERG, 1969; KAISTHA and JAFFE, 1972c). Fluram (fluorescamine) was first described by KLEIN et al. (1974) and detected amphetamine as a bright greenish fluorescence under long wavelength UV light (360 nm). It has been described for use in screening systems by MEOLA and VANKO (1974) and ROERIG et al. (1975a). Large amounts of amphetamine may also be visualized with iodoplatinate (DOLE et al., 1966; MULÉ, 1971).

For detection of barbiturates either diphenylcarbazone or one of the three inorganic salts listed in Table 4 (usually mercuric sulfate) may be used as the primary reagent with the other serving as a confirmatory spray. BERRY and GROVE (1971) described a combination mercuric chloride-diphenylcarbazone reagent for barbiturates. N,2,6-trichlorobenzoquinoneimine was described by BROICH et al. (1971a) for detection of cyclic imide type compounds, such as barbiturates, glutethimide, and diphenylhydantoin. VINSON and HOOYMAN (1975) combined this reagent with bicarbonate and dimethyl sulfoxide and described its use as a universal spray reagent for stimulants, phenothiazines, and narcotics, as well as barbiturates.

Sulfuric acid was the most commonly used reagent for phenothiazines, antidepressants, and antihistamines, with confirmation after spraying with iodoplatinate for narcotic analgesic detection. These compounds may also be detected by heating following spraying for barbiturates with mercuric sulfate, which was made up in sulfuric acid. Also Mandelin's reagent has been recommended for this group of drugs by BERRY and GROVE (1973).

The fluorescence of quinine and its metabolites was observed under UV light, in some procedures preceded by spraying with sulfuric acid. As with phenothiazines, confirmation was usually made after narcotic analgesic sprays. For meprobamate, in addition to furfural (which was the most commonly used detection reagent), the use of vanillin has been described (BROICH et al., 1971a).

In addition to the reagents aimed at particular drugs or drug groups, MULÉ (1971) noted that iodine (1% in methanol) may be used a universal reagent, giving yellow-brown spots for most narcotic analgesics, stimulants, phenothiazines, and antidepressants. Also most of these drugs quenched fluorescent indicator plates when exposed to short wavelength (254 nm) UV light.

A simple spraying sequence has been described with minor procedural variations in several papers (e.g., DAVIDOW et al., 1968) for detection of all groups of drugs on a single plate. In general outline it is as follows: (1) spray with ninhydrin, apply heat, expose to UV light (observe for amphetamine and other primary amines); (2) spray with diphenylcarbazone and mercuric sulfate (observe for barbiturates and glutethimide); (3) apply heat (observe for phenothiazines, antidepressants, antihistamines); (4) observe under UV light for quinine fluorescence; (5) spray with iodoplatinate (observe for narcotic analgesics and cocaine; confirm quinine, phenothiazines, antidepressants, antihistamines); (6) spray with Dragendorff's reagent to confirm narcotic analgesics. A highly complex and complete detection sequence for all drugs on a single plate was described by KAISTHA et al. (1975).

3. Sensitivity

Because of the great variation among TLC screening methods described, it is difficult to determine single values for the sensitivity of TLC methods to detect different drugs or groups of drugs. Also, little systematic sensitivity data have been reported, although many authors gave approximate urine concentrations, above which the described method was felt to be able to consistently detect the drugs of interest.

Table 5. Sensitivity of TLC screening procedures

Drug (or drug group)	[a]Sensitivity range (μg/ml)
Morphine	0.2–1
Quinine	0.1–0.5
Other narcotic analgesics and antagonists	0.5–2
Sedative-hypnotics	0.5–2
Stimulants	0.5–2

[a] Reported as the minimal detectable concentration of drug in urine.

Table 5 summarizes the range of frequently reported sensitivities, in terms of urine concentration in μg/ml. Morphine and quinine are listed separately since they were frequently reported individually in the screening literature because of their importance in the detection of heroin abusers.

The best available sensitivity data are for the detection of morphine. GORODETZKY (1973a) and KULLBERG and GORODETZKY (1974) determined sensitivity by adding morphine to normal urine and analyzing multiple samples at several concentrations in random order under blind conditions. Using a method described earlier for TLC of pure standards (GORODETZKY, 1972), they determined sensitivity as the concentration of morphine statistically detectable 99% of the time, with 95% confidence limits of that value. Sensitivities reported (with 95% confidence limits) for detection of morphine with iodoplatinate as the spray reagent were 0.19 (0.14–0.25) μg/ml using organic solvent extraction, 0.16 (0.07–0.35) μg/ml with extraction by ion exchange resin impregnated paper, and 0.08 (0.06–0.10) μg/ml with XAD-2 resin column extraction.

Although the above-noted studies showed a greater sensitivity for detection of morphine using an XAD-2 resin extraction compared to use of organic solvents or ion exchange paper, the majority of reported sensitivities are generally in the same range for screening methods using all three common extraction procedures. Lesser extraction efficiencies were frequently compensated for by increasing the volume of urine analyzed. MULÉ (1969) and MONTALVO et al. (1970) reported lesser sensitivities using ion exchange paper compared to organic solvent extraction for barbiturates, methadone, and amphetamine. However, in more recent papers of DOLE et al. (1972) and KAISTHA et al. (1975) sensitivities were reported using ion exchange paper techniques at the lower end of the ranges shown in Table 5, comparable to those reported using other extraction procedures. KAISTHA et al. (1975) reported routine detection of morphine at a concentration of 0.15 μg/ml; and, by increasing the volume of urine analyzed from 20 ml to 35 ml and 50 ml, they noted sensitivities of 0.10 μg/ml and 0.07 μg/ml, respectively. Using an XAD-2 resin screening procedure to analyze normal urine to which drugs were added, ROERIG et al. (1975a) reported high sensitivities for detection of morphine, methadone, barbiturates, amphetamine, and methamphetamine, also at the lower end of the ranges given in Table 5.

Several authors have noted methods for increasing sensitivity. HIGGINS and TAYLOR (1974) reported a general increase in sensitivity by washing TLC plates by developing them overnight in the chromatography solvent prior to activation and use. WALLACE et al. (1972a) noted the need to wait 3–4 h for maximal color development of morphine spots after iodoplatinate spray; they reported an increase in sensitivity for detection of morphine from 0.5 μg/ml immediately after spraying to 0.15 μg/ml after 3–4 h. An increased sensitivity for detection of pure standards of 11 opioids by following iodoplatinate with the confirmatory stains, ammonical silver nitrate, and potassium permanganate was reported by GORODETZKY (1972c). BASTOS et al. (1973) emphasized the decrease in sensitivity which occurred when the extract from a single extraction was split for application to more than one TLC plate.

a) Hydrolysis

Since many opioids are excreted in human urine as conjugated metabolites (WAY and ADLER, 1962), mainly as the glucuronide, hydrolysis has been recommended in many screening procedures to split the conjugates and make available additional free drug for extraction and detection. The highly water-soluble conjugates themselves are not extracted to a significant extent by the extraction procedures described above. The major concern is heroin, which is excreted in the urine mainly as morphine and morphine glucuronide in a ratio of 1:8–9 (ELLIOTT et al., 1971; OBERST, 1943; YEH and GORODETZKY, 1974). The primary purpose of hydrolysis in urine screening is to increase the probability of detection of morphine in the urine of a subject using heroin by increasing the concentration of free morphine.

Both acid and enzymatic methods of urine hydrolysis have been described; those papers including these methods are so indicated in the last column of Tables 1–3. Acid hydrolysis was most commonly carried out by adding to the urine 10–20% by volume of concentrated (12 N) hydrochloric acid and heating in a boiling water bath for 1 h (e.g., KAISTHA and JAFFE, 1971), or, more commonly, placing the sample in an autoclave at 120° C and 15–20 pounds/sq. in pressure for 15–30 min (for example, KOKOSKI, 1970). Following hydrolysis the acid was neutralized and the urine was then extracted. Use of 6 N hydrochloric acid (WALLACE et al., 1972) and sulfuric acid (FISHER et al., 1972) has also been described and HIGGINS and TAYLOR (1974) recommended a mild alkaline hydrolysis with ammonium hydroxide. Enzymatic hydrolysis can be accomplished with β-glucuronidase (DOLE et al., 1966; YEH, 1975) or Glucalase (HIGGINS and TAYLOR, 1974; YEH, 1975), a mixture of β-glucuronidase and aryl sulfatase, by incubating the urine with enzyme for 24 h at 37° C and appropriate pH. Although the conditions of enzymatic hydrolysis are milder than acid hydrolysis, it is less favored because of the long incubation time.

Hydrolysis procedures were most commonly described with methods in which organic solvent extraction was used. The urine was usually hydrolyzed prior to extraction, although BASTOS et al. (1970) hydrolyzed the water-soluble fraction after extraction of free drugs first with ethanol and then ether. GORODETZKY et al. (1974) described use of acid hydrolysis prior to ion exchange resin impregnated paper extraction and DOLE et al. (1966) suggested hydrolysis by incubating the ion exchange paper with β-glucuronidase after adsorption of drugs from the urine. KULLBERG and GORODETZKY (1974) described three methods of acid hydrolysis using an XAD-2 resin column extraction. Hydrolysis of urine prior to extraction resulted in approximately 75% recovery of morphine from morphine glucuronide initially added to the urine; hydrolysis on the resin after adsorption gave 40% recovery, but cleaner TLC plates than the first method; and, hydrolysis of the concentrated organic eluate from the column gave only 10% morphine recovery. MULÉ et al. (1971) used a slightly different XAD-2 extraction and found little adsorption of morphine glucuronide on the resin and no recovery of the conjugate in the organic eluate. The Brinkman Product Brochure recommended hydrolysis of the dried methanol eluate from the XAD-2 column; however, no data were given. KULLBERG et al. (1973) and MILLER et al. (1973) found slightly decreased adsorption efficiency onto XAD-2 resin of morphine, phenobarbital, and amphetamine added to hydrolyzed urine compared to nonhydrolyzed urine. From these studies, if hydrolysis is used with an XAD-2 resin extraction, hydrolysis prior to application of the urine to the resin appears to be the procedure of choice.

There is some disagreement in the literature concerning possible destruction of some drugs of interest during acid hydrolysis. DOLE et al. (1966) stated that acid hydrolysis destroys much of the free morphine. However, other authors (for example, COCHIN and DALY, 1962; YEH, 1975) have not substantiated this finding; and, with the wide use of acid hydrolysis in screening procedures for morphine detection, it seems unlikely that there is extensive destruction of free morphine with the commonly used procedures. Depending on the severity of the hydrolysis procedure, there is likely to be some destruction of drugs with acid-labile moieties (such as esters and amides), which include barbiturates, cocaine, and meperidine; and there may be some loss of low molecular weight compounds, such as amphetamine, by volatilization. Also HIGGINS and TAYLOR (1974) reported destruction of methadone during hydrolysis. Most authors describing methods using more than one extraction advocated hydrolysis only of that aliquot used for narcotic analgesic detection (e.g., MULÉ, 1971) or recommended extraction of both a hydrolized and nonhydrolyzed sample.

4. Specificity

Specificity in TLC screening procedures is achieved by a combination of differential pH extraction, different mobilities of compounds on TLC plates (i.e., different R_f values) and differential staining reactions with chromogenic detection sprays.

Those procedures in which multiple extractions were used at different pH's were meant to optimize extraction efficiencies and achieve initial separation of the major chemical groups of drugs, acid, basic, and neutral groups. However, specificity is not great at this stage of the analysis, even for drug groups; and, as seen in Tables 1–3, the trend is toward simpler extraction techniques, using one or two extractions. Several authors have described procedures to increase specificity at the extraction step. Washing the organic extract (for example, BROICH et al., 1971a) or the eluate from an XAD-2 column (for example, KULLBERG and GORODETZKY, 1974) was used to remove interferring substances such as urinary pigments which could mask or be confused with a drug of interest. BASTOS et al. (1973) washed the first eluate (for acidic drugs) from a differential XAD-2 column elution with acid to remove basic drugs and then added the wash to the second eluate (for organic bases); this increased the specificity of extraction into the two eluates. Similarly, DOLE et al. (1972) and JUSELIUS and BARNHART (1973) added a portion of the pH 9.3 ion exchange resin paper eluate to the pH 11 eluate, to enrich the latter in amphetamine, which was partially eluted at pH 9.3.

The separability of compounds on a TLC plate is influenced by the size of the plate (i.e., the absolute distance of travel of the solvent front), the relative solubility of the compounds of interest in the developing solvent system, and the affinity of the compounds to the adsorbent layer. Silica gel was almost universally used in screening procedures as the adsorbent stationary phase. Plate or sheet size was most commonly 20 cm × 20 cm, with the solvent front allowed to travel 8–12 cm. Use of smaller plates has been suggested (e.g., DOLE et al., 1966); however, although time of development may be shortened, the ability to separate compounds is compromised. Many different developing solvent systems have been described, each with different separation capabilities, as discussed above and detailed in Tables 1–3; the number of compounds for which R_f data are tabulated in each paper is noted in the last column of these tables. In procedures in which multiple pH extractions were used, a different solvent system may be used to develop the chromatogram from each extract, increasing specificity within each drug group. Further specificity may be achieved by making more than one chromatogram from the same extract, developing each in a different solvent system. For optimal specificity multiple aliquots of the same urine sample would be extracted and each extract chromatographed and developed in a different solvent system. However, since this can be time consuming and expensive, the extract from one urine aliquot may be split to make several chromatograms, sacrificing some sensitivity for an increase in specificity (for example, ROERIG et al., 1975a). Two-dimensional TLC allows development of the same chromatogram in two solvent systems, the second system developing in a direction at right angles to the first. This technique allows an increase in specificity without compromising sensitivity; however, to get maximum separability from each solvent system, only a single extract is applied to each plate. Also, as noted above, KAISTHA et al. (1975) have described a sequential development in two solvent systems in the same direction, which they felt optimized separability for a single urine extract for all drugs on one plate. Choice of the best solvent system, or systems, to achieve the desired separations is often difficult. MOFFAT et al. (1974a) and MOFFAT and SMALLDON (1974) described a method for evaluating and comparing the discriminating power of various chromatographic systems and its general application to TLC. In a later study (MOFFAT and CLARE, 1974), the method was applied to TLC (and paper chromatography) separation of basic drugs, including many drugs of abuse. They found that using an adsorbent of silica gel sprayed with 0.1 N NaOH the following systems had optimal discriminating power: chloroform:methanol (90:10); cyclohexane:toluene:diethylamine (75:15:10); and acetone. This paper included a table of R_f values for 100 basic drugs, using six different solvent systems.

The R_f value of a compound may be influenced by the presence of other compounds in the same urine extract. FISHER et al. (1972) reported that urea affects the shape and R_f of the morphine spot; DAVIDOW et al. (1966) noted a higher R_f of morphine in extracted urine than of pure standard on the plate, an effect they felt was due to displacement by urea. BERRY and GROVE (1971) reported that co-extracted material (not specified) retarded the movement of morphine and methadone. MULÉ (1971) also noted that R_f values of drugs in urine extracts did not always agree with pure reference standards on the plate. Large amounts of chlorpromazine metabolites were found by HEATON and BLUMBERG (1969) to alter the R_f of amphetamine. They suggested splitting the urine sample and adding amphetamine to one aliquot; the spiked aliquot then served to determine the appropriate comparison R_f for amphetamine.

The presence of drug metabolites can also aid in the identification and differentiation of drugs on TLC plates. For example, metabolites have been reported to help in identification of methadone, cocaine, methamphetamine and codeine (FISHER et al., 1972; MULÉ, 1971).

Differential staining reactions with the chromogenic detection sprays have also been used to distinguish many drugs. Differentiation between major drug groups, even on the same plate, is generally not difficult. Also, BASTOS et al. (1970) have described a spraying sequence aimed at differentiating certain chemical groups (primary, secondary, tertiary, and aromatic amines; and indoles). However, within groups, especially for compounds of similar chemical structure, the colors developed are frequently quite similar, and may differ only slightly in shade, intensity, speed of development, or stability of color. The colors listed in the extensive tables referenced earlier (see discussion of detection reagents) may be of some use in differentiating drugs to the experienced technician, who has used the same spraying sequence for a long time. Several specific differential staining reactions have been described. Secobarbital may be differentiated from other barbiturates by formation of a yellow product after spraying with potassium permanganate, a reaction which does not occur with other barbiturates (MULÉ, 1971; and others). The black color developed by morphine after treatment with ammoniacal silver nitrate and heat has been used as a confirmatory procedure to help differentiate morphine from some other opioids (MULÉ, 1971). For differentiating cocaine and methadone, KAISTHA and JAFFE (1972c) sprayed with iodoplatinate followed by ammoniacal silver nitrate; both drugs were detectable after iodoplatinate and after silver nitrate the methadone spot bleached, while cocaine turned yellow. In addition, many authors have presented techniques to solve differentiation problems encountered in their particular screening procedures.

False positives can be a significant problem with TLC procedures, especially since the final determination of a positive result is a subjective judgment (i.e., visual detection of a spot on a TLC plate with the same R_f and color reaction as a known standard on the same plate or as published reference data). Spots with similar R_f and staining characteristics can be confused with a drug of interest. Commonly used TLC developing solvents and spraying sequences have been chosen to minimize this possibility. Nevertheless, problems still remain, especially with the simpler systems (for example, all drugs on a single plate) and inexperienced technicians. Nicotine (e.g., GOENECHEA and BERNHARD, 1969), caffeine and their metabolites (from smoking tobacco products and drinking coffee) are the most frequently experienced problems, particularly on the TLC plates intended for detection of narcotic analgesics. These compounds are partially extracted by most of the single pH procedures described above and in the alkaloid extract in those methods in which more than one extraction is used. They stain darkly with iodoplatinate, producing colors which may be similar to some opiates (MULÉ, 1971); and, depending on the developing solvent used, may be confused with drugs such as morphine, codeine, and methadone (BERRY and GROVE, 1971; FISHER et al., 1972). Use of multiple solvent systems and confirmatory sprays can aid in the differentiation. TLC spots in extracts of urine of subjects taking phenothiazines, which could be confused with morphine, have been noted by several authors (BERRY and GROVE, 1971; DAVIDOW et al., 1968; FRINGS and QUEEN, 1972b). Ninhydrin positive spots in extracts of normal urine have been described, however, they can be differentiated from amphetamine by R_f values (DAVIDOW et al., 1968; FISHER et al., 1972). Other ninhydrin positive spots noted by FISHER et al. (1972) are phenylpropanolamine, vitamin C, methadone, amino acids, and estrogen compounds from some oral contraceptives. In a discussion of false positives using an ion exchange paper extraction with elution at 3 pH's, DOLE et al. (1972) noted that diphenylhydantoin was extracted in the pH 2.2 eluate. It produced similar color reactions with silver nitrate spray as barbiturates; but it was distinguishable by a slightly different R_f value from amobarbital, if amobarbital was kept in mind and included as a reference drug. These authors also pointed out that oxymetazoline (contained in some nasal decongestant sprays and solutions) had the same R_f and color reactions as morphine in the common developing solvent mixture of ethyl acetate:methanol:NH_4OH (85:10:5), and it must be differentiated with a second solvent system.

TLC screening procedures are capable of great specificity if sufficient time is taken for multiple differential pH extraction, use of several solvent systems for TLC development (or two-dimensional TLC), and several primary and confirmatory sprays on separate plates, if necessary. The more these elements are simplified (for example, by use of a single pH extraction, one small plate for all drugs, and a simple spraying sequence) for the sake of speed, simplicity, or economy, the more specificity is compromised.

5. Socioeconomic Parameters

The complexity of TLC screening procedures can vary considerably; and, therefore, their speed and cost. In carrying out even the simpler procedures, experienced, careful, and well-trained technicians are desirable, especially for accurate reading of chromatograms. The turn around time for analysis of a single sample has been estimated to be generally 1–2 h, with a single technician able to analyze 60–100 samples per day by processing multiple samples simultaneously (CATLIN, 1973a; KAISTHA and TADRUS, 1975; World Health Organization, 1974). Establishing a laboratory to do TLC screening analysis requires little major equipment; a $ 500 (U.S.) cost was estimated by CATLIN (1973a). KAISTHA and JAFFE (1972a) and KAISTHA and TADRUS (1975) calculated start-up costs of approximately $ 2200 (U.S.), including a reciprocating shaker, UV viewer, water bath, oven, and analytical balance. Approximate cost per sample has been estimated in the range of $ 0.80–1.50 (U.S.) (CATLIN, 1973a; KAISTHA and TADRUS, 1975).

6. Special TLC Procedures

a) *Mini-TLC*

Several authors have described TLC screening procedures in which very small plates or sheets were used. These generally had very short chromatogram development times (5 min or less) and used only small volumes of developing solvents, saving time and money. Usually one sample (or at most a few samples) was placed on each plate, and these procedures were aimed primarily at laboratories analyzing a small number of samples. Reported sensitivities were generally in the same range as those described for TLC methods in which larger plates were used. The major disadvantage of the small plates was the short distance available for separation of drugs, their metabolites, and interfering substances. COPENHAVER et al. (1972) described an organic solvent extraction at pH 9.3 for narcotic analgesics and quinine, followed by TLC on a silica gel G plate made on a microscope slide and developed in a spray jar. Polyamide TLC plates, 3 cm × 3 cm, were used for detection or narcotics (HO et al., 1971; LOH et al., 1973), amphetamine, and methamphetamine (LOH et al., 1972) after reacting the organic solvent extract of urine or tissue with dansyl chloride. HO et al. (1972) also described a method in which 3 cm × 3 cm Eastman chromatogram sheets (No. 6061) were used for narcotics detection with iodoplatinate spray reagent.

Kits for urine extraction and mini-TLC have been described and marketed by Gelman Instrument Co. (Gelman Drug Ident. Systems) and Eastman Kodak Co. (FISHER et al., 1972; Kodak Chromat/o/Screen, 1971 a, b, c, d). In the Gelman system (Seprachrom) organic solvent extraction of urine is used with TLC in small plastic chambers on Gelman ITLC sheets; separate developing solvents are provided for acidic and basic drug groups. Amberlite XAD-2 resin extraction is used in the Kodak procedure; the concentrated extract is applied to Eastman chromatogram sheets, which are developed in plastic chambers using one of three developing solvent gels (supplied for amphetamines, alkaloids, and barbiturates).

b) Specific Drug Problems

Detection of cocaine and its metabolites in urine has received special attention. Because cocaine is not well separated from methadone, propoxyphene and pentazocine on TLC plates developed in the common solvent system of ethyl acetate:methanol:ammonium hydroxide (85:10:5) and also has a similar color reaction with iodoplatinate as these drugs, special developing solvent systems have been described to accomplish the desired separation. The use of two dimensional TLC has also been advised (EM Lab Product Brochure; KOKOSKI, 1970). Since cocaine is rapidly metabolized (FISH and WILSON, 1969a; WOODS et al., 1951), several methods have been described for detection of the major urinary metabolites of cocaine, the metabolites benzoylecgonine and ecgonine. Noting the highly water-soluble nature of the metabolites, VALANJU et al. (1973) described an organic solvent extraction at pH 8.5 with chloroform:isopropanol:1,2-dichloroethane (8:1:3). The TLC plate was developed sequentially in two solvent systems in the same direction and cocaine and its metabolites were detected after spraying with a modified Dragendorff's reagent (to give a red spot) and then iodoplatinate. A microcrystal test on the eluted spots was recommended for confirmation. In the procedure described by BASTOS et al. (1974), chloroform:ethanol (3:2) was used for extraction of cocaine and its metabolites into the ethanolic aqueous phase and then into an ethanolic extract. The compounds were then butylated, re-extracted, concentrated, and applied to a silica gel Polygram TLC sheet. Sequential or two-dimensional development of the sheet was recommended, with detection using iodoplatinate; a sensitivity of 3–5 μg/ml was reported.

Separation and differentiation of amphetamine from other amines has been the subject of several papers. BUSSEY and BACKER (1974) described a XAD-2 resin extraction of urine (Brinkmann Product Brochure) followed by TLC with sequential development in two solvent systems (or one of two confirmatory systems) for separation of amphetamine and other primary amine drugs. Fluorescence following spraying with ethanolic phenylacetaldehydeninhydrin and aqueous sodium phosphate was used for detection; chromotropic acid was used to selectively react with 3,4-methylenedioxyamphetamine (MDA), distinguishing this drug from amphetamine. Detection sensitivity for all drugs of interest was reported as 1 μg/ml. SHAW and PEEL (1975) recommended use of fluorescamine and gallic acid to achieve the desired differentiation of phenethylamine derivatives. Both GUPTA et al. (1974) and VAN HOOF and HEYNDRICKX (1974) described formation of fluorescent derivatives of extracted amphetamine and other amines by coupling with 4-chloro-7-nitrobenzofurazan (NBD-Cl) and observing fluorescence under longwave UV light. GUPTA et al. (1974) noted that observation of spots for both amphetamine and methamphetamine in the same sample indicated the likelihood of methamphetamine use.

Separation of methadone, its metabolites, and congeners has become increasingly important with the use of methadone in the treatment of heroin dependence. JAIN et al. (1975) described a method in which organic solvent extraction of urine was used, followed by TLC development in one of three solvent systems, all of which separated methadone and its primary metabolite, 2-ethylidene-1,5-dimethyl-3,3-diphenylpyrrolidene, and were not interfered with by other drugs of abuse in urine. Sensitivity for detection of methadone and its metabolite using iodoplatinate spray was reported as 0.25 μg/ml. GUPTA et al. (1974) reported detection of the primary methadone metabolite by formation of colored compounds (seen on TLC) on reaction with 4-chloro-7-nitrobenzofurazan; methadone, its other metabolites, and other drugs of abuse did not form such compounds. MCINTYRE et al. (1975) described detection of methadone and three metabolites and l-alphaacetylmethadol and four metabolites, which could be distinguished by R_f values and differential color reactions with ninhydrin, ninhydrin plus UV light, and iodoplatinate followed by Dragendorff's reagent. Also MISRA et al. (1972) reported several TLC and paper chromatography developing systems for separating methadone and some of its metabolites and congeners. Narcotic antagonists have also been used in the treatment of narcotic addicts, and DIGREGARIO and O'BRIEN (1974) have reported a screening method for extraction from urine and TLC identification of cyclazocine, naloxone, and naltrexone.

Specific procedures for the detection of the benzodiazepine minor tranquilizers, diazepam (Valium), chlordiazepoxide (Librium), and oxazepam (Serax), have been described by KOKOSKI et al. (1974). Hydrolysis of urine with acid and heat produced 2-amino-5-chlorobenzophenone (ACB) as a hydrolytic product formed from metabolites of all three drugs. ACB was then extracted with chloroform and detected on TLC as a blue spot with appropriate R_f value after spraying with sodium nitrate and N-1-naphthylethylenediamine dihydrochloride. Sensitivity of 0.3 μg/ml was reported for detection of oxazepam. BERRY and GROVE (1973) described a similar benzodiazepine detection method as part of a general toxicologic screening procedure. SLEEMAN et al. (1975) reported a method for detecting methaqualone and four of its metabolites after urine hydrolysis, chloroform extraction, and TLC with detection by iodoplatinate or quenching of fluorescence on a fluorescent indicator plate.

Although TLC screening procedures have not been routinely used for the detection of cannabis users, compounds from cannabis and their metabolites have been determined in drug metabolism studies by TLC after organic solvent extraction of urine from subjects who have ingested cannabis preparations or synthetic cannabinoids (for example, ANDERSEN et al., 1971; CHRISTIANSEN and RAFAELSON, 1969; KANTER et al., 1975).

II. Gas Chromatography (GC)

In recent years gas chromatography has been increasingly used for detection and, especially, quantitative determination of drugs of abuse in biologic fluids. Some general screening methods have been advocated for use in high volume urine screening (such as in a drug treatment program), both as primary screening methods and as confirmatory procedures for specimens found positive by another test (e.g., TLC). However, the greater use of GC is for screening and quantitative determination in hospital toxicology laboratories, to aid in the diagnosis of overdose and in the evaluation of autopsy material (BASELT et al., 1975; SUNSHINE, 1969, 1971). Here biological material other than urine is frequently used (for example, blood or tissue) and emphasis is not placed on high volume output, simplicity, or low cost. Gas chromatographic methods are also used in a clinical pharmacology setting to monitor therapeutic blood levels of drugs and in drug metabolism studies.

In general, most gas chromatographic procedures involve extraction of drugs from the biological material, injection of the concentrated extract into the gas chromatograph, and detection of separated compounds as they emerge from the GC column. Within the chromatograph the extract passes through a column packed with a solid support coated with a stationary phase under specified temperature and carrier gas flow conditions. Separation of drugs occurs from their different partitioning between the gas and stationary phases. Drugs are identified primarily by the time they (or their derivatives) take to traverse the column under specified conditions (that is, their retention times) compared to known standards. Within these basic elements there is great variability among the published GC methods. Areas of variability include: type and complexity of extraction; possible synthesis of derivatives, either prior to injection into the chromatograph or formed on the GC column; material, length, diameter, shape, and treatment of the column; solid support; stationary phase; pretreatment of packed columns; temperature conditions; carrier gas; and detector.

1. Methods

a) General Screening Methods

Some of the important parameters of 10 gas chromatography general screening procedures are summarized in Table 6. For additional methods and references the reader is referred to earlier reviews (noted above) and to SUNSHINE (1969, 1971). In five of the papers noted in Table 6, use of the procedure for mass urine screening for the major drugs of abuse was emphasized (ADAMS, 1971; ADAMS et al., 1973; GOLDBAUM et al., 1972; SANTINGA, 1971; and SOLON and MIKKELSEN, 1971). Organic solvent extraction into two or three fractions for gas chromatographic analysis was used in three of the methods and ADAMS et al. (1973) advocated adsorption of drugs onto charcoal, as described by MEOLA and VANKO (1974). For simplicity and speed,

Table 6. Gas chromatography general screening procedures

Extraction	Gas chromatography						Comments	Reference
	Drug form	Column	Solid support	Stationary phase[a]	Temperature	Detector[b]		
4 organic solvent extractions combined into 2 extracts	Free	Glass	Gas Chrom Q	OV-17	Programmed	FID	Emphasis on mass urine screening—uses solvent-free injection, automatic data processing	ADAMS (1971)
Charcoal and organic solvents (MEOLA and VANKO, 1974)	Free	Glass	Gas Chrom Q	OV-17 OV-1	Programmed	FID	Emphasis on mass urine screening—uses solvent-free injection, automatic data processing—injection split for two columns	ADAMS et al. (1973)
4 separate organic solvent extractions	Free and derivatized (methyl, acetyl, benzyl, trimethylsilyl)	Glass	Gas Chrom Q	OV-17 SE-30	Various isothermal and programmed	FID	Emphasis on toxicologic use—describes primary and confirmatory GC analyses	BARRETT (1971)
Organic solvent extraction to give 5 fractions	Free (mostly) and derivatived (methyl, acetyl, trimethylsilyl)	Glass	Chromosorb G Teflon 6 Chromosorb W	SE-30 Hallcomid M-18 plus Carbowax 600 Carbowax 6000 plus KOH	Various isothermal	FID	Emphasis on toxicologic use—retention time data for 600 compounds	FINKLE et al. (1971)
Organic solvent extraction to give 3 fractions	Free	Not given	Not given	UCON W98 OV-1 OV-210 OV-17 Apiezon-L plus KOH	Various isothermal	FID	Emphasis on mass urine screening—morphine by fluorometry	GOLDBAUM et al. (1972)

Table 6 (continued)

Extraction	Gas chromatography						Comments	Reference
	Drug form	Column	Solid support	Stationary phase[a]	Temperature	Detector[b]		
Organic solvent extraction to give 4 fractions	Free	Glass	Anakrom ABS	SE-30 QF-1	Various isothermal	Sr^{90} argon ionization	Emphasis on toxicologic use—early paper—retention time data for 59 compounds	KAZYAK and KNOBLOCK (1963)
Organic solvent extraction to give 5 fractions	Free and derivatized (trimethylsilyl)	Stainless steel	Chromosorb W	SE-30 QF-1	Various isothermal and programmed	FID	Emphasis on toxicologic use—described in several sequential papers	MCMARTIN and STREET (1966b) STREET (1967) STREET (1969)
Not given	Free	Stainless steel	Chromosorb W	SE-30 Carbowax 20M plus KOH	Various isothermal	FID	Emphasis on toxicologic use—retention time data for 41 compounds	PARKER et al. (1963)
Organic solvent extraction to give 3 fractions	Free	Glass	Not given	OV-1 OV-17 SE-30	Various isothermal	FID	Emphasis on mass urine screening—morphine and quinine by fluorometry	SANTINGA (1971)
Not given	Free	Glass	Chromosorb W HP	OV-17 OV-1	Multilevel programmed	FID	Emphasis on mass urine screening—uses 2 GC's, autosamplers, integrators, computer	SOLON and MIKKELSEN (1971)

[a] KOH = Potassium hydroxide.
[b] FID = Flame ionization detector.

drugs were analyzed underivatized. OV-1 and OV-17 were the most commonly recommended stationary phases and all authors, except ADAMS (1971), used more than one column to achieve the desired separations. Both temperature programming and various isothermal temperatures were described; and drugs were detected by a flame ionization detector in all procedures. ADAMS (1971) and ADAMS et al. (1973) advocated the use of Perkin-Elmer instrumentation for rapid mass urine screening, using a solvent-free capsule injection system and on-line automated data processing equipment. The injection technique eliminated the solvent front, allowing detection of compounds with very short retention times; the analysis system compared GC data of unknowns to previously injected standards and printed out drug identity and amount for each sample. In one method (ADAMS et al., 1973) the injected extract was split, part entering each of two columns to produce simultaneous chromatograms. For positive identification agreement with known standards on both columns for both retention time and also quantitative estimate was required. In the screening methods described by GOLDBAUM et al. (1972) and SANTINGA (1971) morphine was determined by fluorometry, because of greater sensitivity of this technique compared to the gas chromatographic procedures used. Also SANTINGA (1971) emphasized the use of a Shimadzu gas chromatograph, which he felt was best suited for mass urine screening with high sensitivity because of the designs of the on-column injector and flame ionization detector. Two separate gas chromatographs, each with a different column, were used by SOLON and MIKKELSEN (1971) with automated sample injection and data analysis to achieve a rapid, objective screening analysis. In a manual edited by SCHWARTZ (1971), several general analytic schemes were described for specific use with the Bendix Toxichron gas chromatographic system for screening for drugs of abuse, with emphasis on both toxicologic use and mass urine screening.

In the remaining papers summarized in Table 6, a general toxicologic analysis was emphasized, which included drugs of abuse. Organic solvent extraction into 4 or 5 fractions for analysis was used, and extracts were analyzed with the compounds both in free form and after derivatization. Use of multiple columns of both glass and stainless steel have been described, with OV-17 and SE-30 the most commonly recommended stationary phases, and Chromosorb W the most frequent solid support. Isothermal temperature conditions, as well as temperature programming were used, generally with flame ionization detectors. BARRETT (1971) described both primary and confirmatory gas chromatographic procedures, using derivatized compounds for the initial screening tests, and both free and derivatized compounds in the confirmation. FINKLE et al. (1971) used four different columns under seven different isothermal conditions and provided extensive retention time data on approximately 600 compounds, mostly in the free form. In an early paper KAZYAK and KNOBLOCK (1963) suggested SE-30 as a primary column, with injection of each extract at three different isothermal temperature conditions for separation and identification of all drugs of interest. PARKER et al. (1963) also found SE-30 to be the most generally useful column. In a series of papers MCMARTIN and STREET (1966b) and STREET (1967, 1969) described screening procedures for sedative-hypnotics and alkaloids (including stimulants and some narcotic analgesics and antagonists). The organic solvent extraction procedure was described by STREET and MCMARTIN (1963) and details of column preparation and treatment were as presented by MCMARTIN and STREET (1966a). MOFFAT (1975) reviewed the use of SE-30 as a stationary phase for analysis of drugs and tabulated retention time data (as retention indices) for 480 compounds, including many drugs of abuse. In general, the methods in which toxicologic use is emphasized involve more extensive sample preparation, more frequent use of derivatives, a larger number of different columns and multiple chromatographic conditions as well as greater emphasis on quantitative estimation than those screening procedures in which emphasis is placed on high volume urine screening.

Several authors have described general urine screening procedures in which gas chromatography was used as part of the primary screening test and thin-layer chromatography was used for the remainder. BASELT and CASARETT (1971) described a TLC method for opiates and gas chromatography on a 1% SE-30 column for screening for barbiturates (as methylated derivatives) and amphetamine and methamphetamine (as acetyl derivatives). Opiates and barbiturates were determined by TLC by BERRY and GROVE (1971), who used GC on a 10% Apiezon-L plus 10% KOH column to detect amphetamines.

b) Confirmation Methods

It has been advocated that all positive results from a primary urine screening test be confirmed by an independent method (World Health Organization, 1974); and the use of gas chromatography as a confirmatory test (especially for primary TLC screening procedures) has been recommended (RUBIN, 1973). Several authors have described gas chromatographic methods specifically for this purpose. MULÉ (1971) described a GC confirmatory procedure for all major drugs of abuse following a 3-pH differential organic solvent extraction; primary screening was accomplished with TLC and fluorometry. He used two columns at various isothermal temperatures, 3% SE-30 for barbiturates and narcotics, and 10% Apiezon-L plus 10% KOH for stimulants. Drugs were analyzed underivatized; retention time data were given for 25 compounds. The method of ROERIG et al. (1975a) was also meant for general confirmation of all drugs of interest, except morphine. The drugs were analyzed underivatized on an OV-1 or OV-17 column at various isothermal temperatures. Data on retention times of 18 drugs were tabulated. AGGARWAL et al. (1974) described a method for confirmation of organic bases other than morphine and codeine. A 10 ml urine sample was extracted into 100 µl of solvent; a small volume of the extract was reacted with acetic anhydride (to make acetyl derivatives) and analyzed on an OV-17 column with temperature programming and using a solvent-free injection technique. Procedures for confirmation of narcotics have been described by FENIMORE and DAVIS (1970), HIGGINS and TAYLOR (1974), and SINE et al. (1973). All converted the drugs into trimethylsilyl derivatives prior to analysis. FENIMORE and DAVIS (1970) extracted urine using the ion exchange paper method of DOLE et al. (1966), and analyzed the pH 9.3 eluate on an OV-17 column to confirm narcotics found in a color reaction screening test. HIGGINS and TAYLOR (1974) and SINE et al. (1973) both used organic solvent extraction, the former using an OV-1 column and the latter using SE-30 as the stationary phase. Columns packed with Apiezon-L and Carbowax 20 were recommended by CARTONI and CAVALLI (1968) to confirm stimulants found on TLC, after scraping the spots off the plate and extracting them with ether. BERRY and GROVE (1971) used an OV-225 column and organic solvent extraction of urine to confirm underivatized cocaine suspected from a TLC screening test. HAMMER et al. (1974) suggested converting cocaine to its methyl derivative on column, by injecting it along with trimethylanilinium hydroxide. All the gas chromatographic confirmatory methods noted have used flame ionization detectors.

c) Methods for Specific Drugs or Drug Groups

Since the early 1960's, gas chromatography has become increasingly sophisticated in technique and instrumentation and has gained greatly in popularity. There have been a large number of papers published describing GC methods for use in specific areas of toxicology and for quantitative determination of drug levels in body fluids to aid in drug treatment and to carry out studies in drug metabolism and pharmacokinetics. It is beyond the scope of this paper to comprehensively review this extensive literature. However, methods pertaining to the major classes of drugs of abuse will be briefly discussed and some of the pertinent references cited.

In an early paper MULÉ (1964) described a gas chromatographic method for separation and identification of 31 narcotic analgesics in organic solvent extracts of urine, blood, and tissue. He used an SE-30 column and a radium-226 argon ionization detector. To aid in identification, both free and derivatized drugs were determined, using the techniques of ANDERS and MANNERING (1962) for forming derivatives on column by injecting onto the column acetic or propionic anhydride immediately following the injection of the extracted drug. Data were tabulated for free compounds as well as acetyl and propionyl derivatives. Using an OV-1 column and flame ionization detection MEDZIHRADSKY and DAHLSTROM (1975) determined eight opioids and cocaine, some underivatized and some as trimethylsilyl derivatives.

Methods have been described for both the qualitative and quantitative determination of morphine and heroin and their metabolites in urine, blood, and various tissues. The majority of these methods have used OV-17, SE-30, or OV-1 columns, flame ionization detectors, and trimethylsilyl derivatization (ELLIOTT et al., 1971; FISH and WILSON, 1969b; FRY et al., 1974; IKEKAWA et al., 1969; NAKAMURA and WAY, 1975; WILKINSON and WAY, 1969; YEH, 1973, 1975; YEH and MCQUINN, 1975). Use of an electron capture detector was suggested by DAHLSTROM and PAALZOW (1975), who prepared pentafluoropropionyl derivatives for analysis with a tritium detector, and by WALLACE et al. (1974a), who used a Ni^{63} detector for analysis of trifluoroacetyl derivatives. SMITH and COLE (1975) recommended a nitrogen (or alkali) flame ionization detector for determination of heroin and monoacetylmorphine in blood after derivatization with trifluoroacetic anhydride.

DIGREGARIO and O'BRIEN (1974) have described a screening method for detection of the narcotic antagonists, cyclazocine, naloxone, and naltrexone in urine using trimethylsilyl derivatives, a flame ionization detector, and an OV-17 column. Quantitative GC methods used in metabolic studies of naloxone and naltrexone have also been described. CONE et al. (1974) and VEREBELY et al. (1975) used OV-17 columns and flame ionization detectors for determination of naltrexone and metabolites. The former used pentafluoropropionyl derivatives and the latter trimethylsilyl derivatives. MALSPEIS et al. (1975) recommended a Ni^{63} electron capture detector for analysis of naloxone and naltrexone and their metabolites after derivatization with pentafluoropropionic anhydride. Methadone and metabolites were determined by INTURRISI and VEREBELY (1972) in free form on an SE-30 column, using a flame ionization detector. Several methods have been described for determination of propoxyphene and its N-demethylated metabolites as free compounds with flame ionization detectors and on various columns, including OV-17, OV-7, SE-30, and W 98 (EVENSON and KOELLNEL, 1973; NASH et al., 1975; VEREBELY and INTURRISI, 1973; WOLEN and GRUBER, 1968). A GC method for detection of meperidine in blood using an OV-17 column and flame ionization detector was described by GOEHL and DAVISON (1973).

Many procedures have been described for identification and quantitative estimation of barbiturates and other sedative-hypnotics, in addition to the general screening methods already noted above (for example, COOPER et al., 1972; DVORCHIK, 1975; FLANAGAN and WITHERS, 1972; KANANEN et al., 1972; PARKER et al., 1962; SOLON and WISNIEWSKI, 1970a; WATSON and KALMAN, 1972; WILLIAMS et al., 1973). Most have used flame ionization detectors with various stationary phases (including OV-17, OV-1, SE-30, and UCW-98), usually under isothermal conditions. Tailing and adsorption with loss of sensitivity have been problems with barbiturates (MCMARTIN and STREET, 1966b) and remedies such as special treatment and preparation of columns and inclusion of tristearin in the stationary phase (MCMARTIN and STREET, 1966a) have been used. Use of a nitrogen flame ionization detector was recommended by DVORCHIK (1975) to improve sensitivity and allow use of a very small biological sample; also BREIMER and VAN ROSSUM (1974) used a nitrogen flame ionization detector in a method for determination of hexobarbital in plasma. Barbiturates have been analyzed both as free compounds and after derivatization, frequently as methyl derivatives, formed by procedures such as that of BROCHMANN-HANSSEN and OKE (1969) using trimethylanilinium hydroxide or using trimethylphenylammonium hydroxide (as, for example, KANANEN et al., 1972). STREET (1969) formed trimethylsilyl derivatives on column. Recently improved barbiturate derivatization techniques were described by GREELEY (1974) and by EHRSSON (1974), who used an extractive methylation procedure.

Among the other sedative-hypnotics, a method for determination of 1,4-benzodiazepines and-diazepin-2-ones was described by DESILVA et al. (1976), who used a Ni^{63} electron capture detector, with some compounds converted to methyl or silyl derivatives. Methods for individual tranquilizers of this group have also appeared, such as that of ZINGALES (1971) for chlordiazepoxide and DESILVA et al. (1964) for diazepam (both of which used an electron capture detector). Methaqualone determination has been described by DOUGLAS and SHAHINIAN (1973) and by EVENSON and LENSMEYER (1974); and GC determinations of glutethimide (HANSEN and FISHER, 1974; KADAR and KALOW, 1972) and ethchlorvynol (EVENSON and POQUETTE, 1974) have also been reported. MAES et al. (1969) described methods for ethchlorvynol, paraldehyde, meprobamate, and carisoprodol.

In an early paper FALES and PISANO (1962) described the application of gas chromatography to the separation of biologically important amines (including amphetamine and ephedrine) using an SE-30 stationary phase and an argon ionization detector. BECKETT and ROWLAND (1965) described a method for determination of amphetamine in urine using a column of Carbowax 6000 plus KOH, and confirmation as an acetone derivative. A routine screening procedure for stimulants in urine was described by BECKETT et al. (1967), using both GC and TLC. In the first stage of the analysis, following diethyl ether extraction, the drugs were analyzed underivatized; four columns were used (Carbowax 6000 plus KOH, Carbowax 20 M plus KOH, SE-30, and Apiezon-L plus KOH) at various isothermal temperatures and using a flame ionization detector. Derivatives for confirmation of identification were made in the second stage of the analysis, including acetone and ketone Shiff's bases, oxazolidines, carbinolamines, acetates, propionates, heptafluorobutyrates, and trimethylsilyl derivatives. Retention time data were tabulated for 40 compounds and later BECKETT and MOFFAT (1968) added 74 more compounds. LEBISH et al. (1970) also described a screening method for primarily toxicologic use for determination of amphetamine and related amines in blood. They used a flame ionization detector, acetamide derivatives (with confirmation as free bases), and columns of SE-30 and Carbowax 6000 plus KOH; retention time data were presented for 50 compounds.

Methods for amphetamine and methamphetamine analyzed as free bases on Apiezon-L plus KOH columns were given by SOLON and WISNIEWSKI (1970b) and JAIN et al. (1974); the latter confirmed the drugs as trifluoroacetamide derivatives on an OV-17 column. The trichloroacetamide derivative of methamphetamine was determined by DRISCOLL et al. (1971) using an OV-1 column and tritium electron capture detector. A rapid method for ephedrine determination was described by PICKUP and PATERSON (1974) using a flame ionization detector and Carbowax 20 M plus KOH column; and WELLS et al. (1974) have presented a method for measurement of methylphenidate and a metabolite in urine. Methods for determination of cocaine and/or its metabolite, benzoylecgonine, have been described in several papers. FISH and WILSON (1969b) determined underivatized cocaine with an OV-17 column and flame ionization detector; WALLACE et al. (1976) used a similar column and detector but methylated the extract to reconvert benzoylecgonine to cocaine. KOONTZ et al. (1973) removed the benzoylecgonine from a TLC plate, then methylated it and chromatographed it on an SE-30 column. BLAKE et al. (1974) used an electron capture detector to determine cocaine after conversion to a heptafluorobutyryl or pentafluoropropionyl derivative.

Although GC methods are not in common clinical use for detection of hallucinogens, GARRETT and HUNT (1973) have described a method for determination of tetrahydrocannabinol in biological fluids. Both underivatized compounds (with flame ionization detector) and pentafluorobenzoate derivatives (with Ni^{63} electron capture detector) were used.

2. Sensitivity

As with TLC, it is difficult to generalize about the sensitivity of gas chromatographic methods, because of both the lack of systematic data and, especially, the variability of the reported methods. Estimates of the sensitivity ranges of GC screening procedures for various classes of drugs of abuse are given in Table 7. These are primarily for methods which used a flame ionization detector and underivatized drugs; they are in general agreement with the estimates given by CATLIN (1973a), KAISTHA and TADRUS (1975), and the World Health Organization (1974).

The overall sensitivity of gas chromatographic procedures can be influenced by the volume of sample used, the extraction procedure, volume of extract chromatographed, preparation of derivatives, pretreatment of the column, solid support and stationary phase, and detector used. Since many of the methods described were for the analysis of blood, the sample volume was generally smaller than in TLC methods. Methods using very small blood volumes (50–100 µl) obtained by finger-prick have been reported (e.g., DVORCHIK, 1975; FLANAGAN and WITHERS, 1972). Howev-

Table 7. Sensitivity of GC screening procedures

Drug group	[a]Sensitivity range (μg/ml)
Narcotic analgesics and antagonists	0.5—2
Sedative-hypnotics	0.5—2
Stimulants	0.1—2

[a] Minimal detectable concentration of drug in biological fluid.

er, IKEKAWA et al. (1969) described methods for extracting 500–800 ml of urine for subsequent determination of morphine, and reported detection of 5–10 μg in 1000 ml of urine using trimethylsilyl derivatization. Organic solvent extraction was most commonly used in the GC methods discussed and many authors recommended triple extractions to reduce background interference in the final extract and, therefore, increase effective sensitivity. Column preparation and treatment was discussed in detail by MCMARTIN and STREET (1966a) who described some methods to reduce adsorption of compounds on the column and its packing. Occasional injections of a silylating reagent into the column to block adsorption sites were noted in many papers. Preparation of derivatives generally increased sensitivity, but was sometimes employed primarily as an additional aid in compound identification or to facilitate passage of compounds through the column (for example, to reduce peak tailing). Those screening methods which used acetyl and trimethylsilyl derivatives of narcotic analgesics and antagonists usually reported sensitivities in the range of 0.05–0.5 μg/ml.

The large majority of screening methods discussed have used flame ionization detectors; and several papers in the early 1960's described use of an argon ionization detector. However, the electron capture detector (with both tritium and Ni^{63}) and the nitrogen flame ionization detector have been increasing in use, especially for drug metabolic studies and for determination of low drug levels following therapeutic doses. BLAKE et al. (1973) generally recommended the use of the electron capture detector to achieve sensitivities in the range of 25–100 ng/ml; and methods with sensitivities of 1–30 ng/ml have been described for morphine (WALLACE et al., 1974a), cocaine (BLAKE et al., 1974), methamphetamine (DRISCOLL et al., 1971), and benzodiazepines (DESILVA et al., 1976). DAHLSTROM and PAALZOW (1975) reported detection of morphine at a level of 0.5 ng/ml in plasma and 0.1 ng/30 mg of brain tissue using a pentafluoropropionyl derivative and a tritium electron capture detector. Use of the nitrogen flame ionization detector was advocated by RIEDMANN (1974) for determination of low nanogram quantities and use of small biological samples, as later reported by DVORCHIK (1975). SMITH and COLE (1975) reported detection of heroin and monoacetylmorphine in blood at levels of 100 ng/ml with the nitrogen flame ionization detector. All three detectors have been used in methods for quantitative determination.

3. Specificity

Identification of a compound by gas chromatography depends primarily on the retention time of the unknown compound or its derivatives compared to known standards. To compensate for differences between individual chromatographs, columns, and conditions, retention times are frequently divided by the retention time of a selected standard compound and are reported as relative retention times. As in TLC, some specificity can be obtained in the extraction procedure. In the methods summarized in Table 6, three to five fractions were most frequently obtained after extraction; these separate fractions can then each be chromatographed on a different column or under different conditions to give optimal separation of the particular drug group. Analogous to choice of developing solvent in TLC, choice of column stationary phase in GC can be difficult. MOFFAT et al. (1974b) applied a discriminating power analysis (MOFFAT et al., 1974a) to the evaluation of the ability of eight stationary phases to separate 62 basic drugs. SE-30, Apiezon-L, OV-17, Carbowax 20M, GDMS, DEGS, and Carbowax 20M and DEGS plus KOH were evaluated, each coated on acid-washed, DMCS-treated, Chromosorb G. SE-30 and OV-17 were found to be the best discriminators; these stationary phases were also commonly used in the methods noted in Table 6. In some methods each extract was chromatographed on more than one column and agreement was required on both columns for retention time and quantitative estimate for positive identification (ADAMS et al., 1973).

Preparation of derivatives is frequently used to increase specificity, since the formation of different derivatives depends on the presence of specific functional groups, and the retention time of these derivatives can also be compared to known standards. Derivatives have been prepared prior to chromatography and also have been formed on-column. Injecting the appropriate anhydride immediately after the injection of the sample extract has been used to form on-column acetyl and propionyl derivatives (for example, ANDERS and MANNERING, 1962; MULÉ, 1964). Also methyl and trimethylsilyl derivatives of barbiturates have been formed by injecting the derivatizing agent along with the extract (for example, BROCHMANN-HANSSEN and OKE, 1969; STREET, 1969).

HATCH (1972) used methaqualone as an internal standard in a barbiturate gas chromatographic analysis and proposed using the slope of the curve of peak height ratio vs. weight ratio to distinguish among the different barbiturates.

As with thin-layer chromatography, gas chromatographic procedures are capable of great specificity if sufficient time is taken for multiple extractions, preparation of derivatives, and, if necessary, use of several columns under different conditions. In simple systems some of the specificity is likely to be lost.

4. Socioeconomic Parameters

Except for a few semiautomated urine screening procedures, the majority of GC methods require a well-trained, experienced, and careful technician to carry out the often complex procedures of extraction, derivative preparation, operation of the instrument, and interpretation of the record. For analysis of a single sample the turn around time for most procedures has been estimated at 1–2 h (CATLIN, 1973a; World Health Organization, 1974), with probably 10–25 samples screened per work day on a single instrument, depending on the completeness of the analysis (KAISTHA and TADRUS, 1975; SCHWARTZ, 1971). ADAMS (1971) reported that using a semiautomated system running 24 h per day with automatic sample injector and data processor, and computer print-out of results, 65 samples could be screened per day for narcotics and amphetamines and 32 samples a day could be analyzed for all the major drugs of abuse. Start-up costs for a laboratory to perform gas chromatographic analyses have been estimated at $ 4000–10000 (U.S.), with an operating cost of $ 1–2.40 (U.S.) per sample (CATLIN, 1973a; KAISTHA and TADRUS, 1975; SCHWARTZ, 1971; World Health Organization, 1974). The initial cost of a semiautomated procedure (such as described by ADAMS, 1971; SCHWARTZ, 1971; SOLON and MIKKELSEN, 1971) would be considerably higher.

III. Fluorometry

Fluorometry has been used in mass screening procedures in several specific areas, for toxicologic use or determination of therapeutic blood levels, and in drug metabolism studies. Most commonly the biological sample is extracted and the extract is analyzed in a fluorometer for fluorescence either of the drugs themselves or of derivatives formed chemically prior to measurement. The degree of fluorescence can be determined using specific excitation and emission wavelengths (which frequently allows quantitative determination) or full or partial excitation and/or emission spectra can be recorded. Compounds may be identified by the wavelengths of maximum or minimum excitation and emission or by their ability to form specific fluorescent derivatives.

1. Methods

The major use of fluorometry in screening methods has been for the detection of morphine. KUPFERBERG et al. (1964a) described a fluorometric method for quantitative determination of morphine in plasma and brain, in which the extracted drug was converted to the highly fluorescent pseudomorphine by reaction with potassium ferro-ferricyanide in weakly alkaline solution. Fluorescence of the derivative was measured at an emission wavelength of 440 nm with excitation at 250 nm. TAKEMORI (1968) modified this method to increase its sensitivity by using very clean and siliconized glassware, reducing the reaction volume, using only ferricyanide, and precipitating plasma proteins. The method has been further modified by SANTINGA (1971) and by GOLDBAUM et al. (1972) for use in a general urine screening procedure. Fluorescence was measured in the morphine containing fraction from a multiple pH differential organic solvent extraction procedure; then ferricyanide solution was added and the fluorescence measured again. Appreciable increase in fluorescence at the appropriate wavelengths for excitation and emission was evidence for the presence of morphine. SANTINGA (1971) also described determination of quinine by its native fluorescence, with emission maximum at 448 nm using an excitation wavelength of 350 nm, as part of the screening procedure. The quinine was removed prior to morphine determination. KUPFERBERG et al. (1964b) also adapted this method for use as a detection technique on a TLC plate and it was recommended also by COCHIN (1966). The plate was sprayed with ferro-ferricyanide reagent and fluorescence of the resulting compounds was visualized under UV light. BERRY and GROVE (1971) converted morphine to pseudomorphine on the TLC plate as a confirmation test by exposure of the plate to concentrated ammonia after spraying with iodoplatinate. Scraping the iodoplatinate positive morphine spot off the TLC plate, eluting it with methanol, and then converting it to pseudomorphine by reaction with ferricyanide was recommended by DOEDENS and FORNEY (1974) to confirm the presence of morphine. Ho et al. (1971) and LOH et al. (1973) advocated preparation of a dansyl derivative from extracted morphine, which was then determined on a mini-TLC plate by visualization under UV light.

Treatment with sulfuric acid, then ammonium hydroxide and heat was first used to produce a fluorescent compound from morphine by FULTON (1937), who described a fluorescence spot test for morphine. Similar methods have been applied for the determination of morphine in raw opium by NADEAU and SOBOLEWSKI (1958), for codeine and codèthyline (ethylmorphine) by BALATRE et al. (1961), and to the estimation of codeine and morphine by BRANDT et al. (1961). Using this same technique for the production of the fluorophore, DAL CORTIVO et al. (1970b) and MULÉ and HUSHIN (1971) described very similar routine screening procedures utilizing an automated turret spectrofluorometer (ATS) to rapidly scan manually prepared samples. MULÉ (1971) incorporated this procedure into a general urine screening method, in which a small urine aliquot was first screened unhydrolyzed fluorometrically for the presence of morphine or quinine. If a positive result was found for either drug, another urine aliquot was hydrolyzed and then analyzed by TLC. In this ATS fluorometric procedure a 2 ml urine sample was first extracted into $CHCl_3$:isopropanol (3:1) at pH 9–10; two-thirds of the organic phase was used for morphine analysis and one-third for quinine. The quinine aliquot was back extracted into acid and fluorescence measured with excitation at 365 nm and emission scanned from 410 nm to 510 nm to determine the quinine peak at 450 nm. The morphine aliquot was evaporated to dryness, treated with acid-base-heat to produce the fluorophore, and then analyzed in the ATS at 392 nm excitation and the emission scanned as for quinine to determine the morphine peak emission at 425 nm.

A fully automated fluorometric method for screening urine for morphine was first described by BLACKMORE et al. (1971), who used Technicon Autoanalyzer equipment for extraction and measured fluorescence of pseudomorphine after addition of ferricyanide. The method was modified and improved by SANSUR et al. (1972) and was subsequently marketed by Technicon Instruments Corp. In this procedure the sample was split after extraction, one-half serving as a blank and one-half being treated with ferricyanide to produce pseudomorphine. Both halves were read simultaneously in two fluorometers, producing simultaneous records; presence of pseudomorphine fluorescence was indicated by difference in the two readings, thereby controlling individually for the background fluorescence of each sample. The urine was analyzed unhydrolyzed. DAVIDOW et al. (1973) reported on routine use of this procedure for confirmation of positive morphine findings.

Methods for determination of meperidine (DAL CORTIVO et al., 1970b) and methadone (MCGONIGLE, 1971) fluorometrically have also been described. Both fluorophores were produced by reaction of the extracted drug with paraformaldehyde in concentrated sulfuric acid. VALENTOUR et al. (1974) described a fluorometric method for determination of propoxyphene, using a fluorescent derivative formed by reaction with 4-chloro-7-nitrobenzo-2,1,3-oxadiazole.

A fully automated fluorometric method for urine screening for amphetamine has been described (HAYES, 1973) using Technicon Autoanalyzer equipment. Amphetamine was extracted into chloroform at basic pH, washed with buffer, reextracted into an aqueous phase, then reacted with formaldehyde and acetylacetone to form a fluorescent dihydrolutidine derivative. Fluorescence was read in a single fluorometer with excitation at 410 nm and emission at 476 nm. A general method for detection of primary and secondary amines in blood and urine was reported by MONTFORTE et al. (1972). After a chloroform extraction the amines were derivatized with 4-chloro-7-nitrobenzo-2,1,3-oxadiazole to form fluorescent compounds, which gave different excitation spectra for primary and secondary amines at a constant emission wavelength of 510 nm. Further compound identification was accomplished by TLC. VAN HOOF and HEYNDRICKX (1974) advocated primary use of TLC with the same derivatives with visualization under UV light at 350 nm and identification by R_f and color of spots; quantitative fluorescence measurements could be made on the plate using a fluorometer with a thin film scanner. A different derivatization technique to make a fluorescent coumarinamine salt was described by STEWART and LOTTI (1971) to determine amphetamine in pharmaceutical preparations. The use of fluorescamine as a TLC detection reagent for amphetamine, producing fluorescent spots on the plate under UV light, was noted above (see TLC Section); the spots can also be scraped from the plate, eluted, and fluorescence quantitatively determined in a fluorometer (KLEIN et al., 1974). Preparation of dansyl derivatives of amphetamine and methamphetamine with detection on mini-TLC plates under UV light was described by LOH et al. (1972).

Although not in common clinical use, fluorometric methods for determination of LSD in biological fluids have been described and used in drug metabolism studies. AXELROD et al. (1957) determined LSD in animal plasma and tissues by extraction from salt-saturated alkaline solution with heptane containing 2% isoamyl alcohol, reextraction into acid, and determination of native fluorescence with excitation wavelength of 325 nm and emission at 445 nm. This method was slightly modified by AGHAJANIAN and BING (1964), who studied plasma levels of LSD in man. A similar method, including use of TLC, for determination of LSD in illicit preparations was described by DAL CORTIVO et al. (1966). FAED and MCLEOD (1973) described a urine screening test for LSD in which drug was extracted from urine with heptane, separated, and purified by paper chromatography, then eluted and determined by decrease in fluorescence after irradiation for 30 min under shortwave (254 nm) UV light. Fluorometric methods for use primarily in metabolic studies and for determination of therapeutic blood levels have also been described for methaqualone (BROWN and SMART, 1969), amobarbital (SWAGDIS and FLANAGAN, 1964), pentobarbital (HOLLISTER et al., 1963), and chlordiazepoxide (KOECHLIN and D'ARCONTE, 1963). Additional drug references can be found in the bibliography published by K.K. TURNER et al. (1971). MELIKIAN and FORREST (1973) have described preparation of dansyl derivatives of Δ^9- and Δ^8-tetrahydrocannabinols, which they felt might be used for development of a fluorometric method for determination of these compounds.

2. Sensitivity

Sensitivites of the fluorometric screening procedures described are summarized in Table 8. TAKEMORI (1968) used careful modifications of the method of KUPFERBERG

Table 8. Sensitivity and specificity of fluorometric screening procedures

Drug	Sensitivity [a](in μg/ml)	Reported interfering compounds
Morphine		
Manual method	0.1–0.2	Nalorphine, apomorphine, heroin, morphine glucuronide, dihydromorphine, 6-acetylmorphine
ATS method	0.2–0.5	
Autoanalyzer method	0.1–0.2	
Meperidine	0.3	Quinine, quinidine, methapyrilene
Methadone	1	Amphetamine, meperidine, cocaine
Propoxyphene	0.02	None reported
Quinine	0.1	None reported
LSD	0.02	Ergotamine, methysergide
Amphetamine		
Manual method	0.5–1	None reported
Autoanalyzer method	0.05–0.15	Phenethylamine, tuaminoheptane

[a] Minimal detectable concentration in biological fluid.

et al. (1964a) and reported sensitivity to detect 20 ng/ml morphine in plasma, approximately 5–10 times greater sensitivity than obtained in the urine screening methods using the same fluorophore. MULÉ and HUSHIN (1971) determined morphine sensitivity to be 0.22 μg/ml in their ATS method and DAL CORTIVO et al. (1970b) reported approximately 0.5 μg/ml sensitivity. AXELROD et al. (1957) and AGHAJANIAN and BING (1964) reported sensitivity of 1 ng/ml for LSD in plasma and tissue. For the other methods noted, the general range of sensitivities reported was 0.25–1 μg/ml.

3. Specificity

The fluorometric screening procedures in most common use are meant to determine a particular drug, rather than to serve as a general screen to identify and distinguish a large number of drugs. Therefore, to evaluate specificity it is necessary to determine which compounds will interfere with the analysis, producing either a false positive or false negative result. The interfering compounds reported for the various procedures are listed in Table 8. Those compounds noted for meperidine and methadone determination were reported to fluoresce within the same wavelength region as the drugs of interest; however, it was not stated whether they would result in false negative or false positive readings. The other listed compounds could produce false positive results in the procedures noted. Those compounds listed to interfere with morphine determination were reported by one or more authors. Both KUPFERBERG et al. (1964b), for the ferricyanide reaction, and MULÉ and HUSHIN (1971), for the production of a fluorophore by treatment with acid-base-heat, concluded that the phenanthrene nucleus and free phenolic hydroxy were necessary to produce a fluorescent compound. DAL CORTIVO et al. (1970b) reported fluorescent compounds formed by heroin and, to a lesser extent, by morphine glucuronide; however, it is possible that the phenolic hydroxyl substituents in these compounds were cleaved during sample preparation. In the majority of papers describing urine screening procedures the authors reported lack of interference from a wide variety of drugs of various chemical and pharmacologic classes, both added to normal drug-free urine or in urine obtained from subjects who had received drugs. For those fluorometric methods used as part of a general screening procedure (such as GOLDBAUM et al., 1972; and SANTINGA, 1971) added specificity was obtained from the extraction procedure used prior to fluorometric analysis. In the manual method for detection of primary and secondary amines (MONTFORTE et al., 1972), the two general drug classes were distinguished by their different excitation spectra and further identification was achieved by TLC of the derivatives.

4. Socioeconomic Parameters

For the manual fluorometric procedures, well-trained technical personnel would be required to carry out extraction procedures, derivative preparation, operation of the fluorometer and interpretation of the records. The semiautomated and automated procedures involve less sample preparation, but still require the expertise to operate, standardize, and interpret the records from instruments which can be complex. Analysis of a single sample by a manual method would probably require 30 min to 1 h, depending on the complexity of the sample preparation. Both described ATS methods for morphine analysis used a turret holding 16 manually prepared samples, which were automatically analyzed in 5.3 min. Using a spare turret, MULÉ and HUSHIN (1971) estimated that 150 samples could be scanned and recorded in 1 h. It has been estimated that 400–500 samples could be analyzed in a single day by one technician for morphine by the ATS method (CATLIN, 1973a; KAISTHA and TADRUS, 1975). The Technicon Autoanalyzer was reported to complete morphine analyses at a rate of 40/h (SANSUR et al., 1972), with an estimated 200–300 samples completed per day (CATLIN, 1973a; KAISTHA and TADRUS, 1975). The automated amphetamine method of HAYES (1973) completed 15–30 samples/h. Start-up costs involve primarily the purchase of instruments which have been estimated at $ 6000 (U.S.) for an ATS and $ 25000 (U.S.) for a Technicon Autoanalyzer; operating costs per sample were estimated at approximately $ 0.80–1.10 (U.S.) for the ATS method and $ 0.15–0.30 (U.S.) for the Autoanalyzer (CATLIN, 1973a; World Health Organization, 1974; KAISTHA and TADRUS, 1975).

IV. Immunoassays

Procedures based on immunoassays are the most recent of the commonly used methods for screening biological fluids for drugs of abuse. In addition to the general references given earlier, immunoassays for drugs of abuse have been recently reviewed by BIDANSET (1974), BRATTIN and SUNSHINE (1973), and in a book edited by MULÉ et al. (1974b). All of the immunoassays in common use are based on the competitive displacement of labeled drug from antibody binding sites by unlabeled drug in the biological sample. The labeled drug either remaining bound or, more commonly, displaced and free is measured as an indication of the presence of unlabeled drug in the sample. Since the amount of labeled drug displaced is related to the amount of unlabeled drug in the sample, immunoassays may be used for quantitative estimation. Since the relationship between concentration of drug and quantity of labeled drug displaced is not linear, appropriate controls and standard curves are important. The specificity of the assay depends primarily on the specificity of the antibodies. Antibodies to the drugs of interest have been obtained by coupling the drug or a derivative (as a hapten) to a protein or polypeptide to form an antigen, which is injected into an animal as in other immunization procedures. Principles and methods for preparing drug-protein complexes for possible use as antigens have been reviewed by ERLANGER (1973). Antibodies are obtained from the serum of the injected animal. Within these general principles immunoassays may differ with regard to synthesis and nature of the antigen, animal used to produce antibodies, label on the drug, need to separate bound and free drug for measurement, and technique used to measure bound or free labeled drug. Because of these differences in technique, immunoassays differ in their sensitivity, specificity, speed, complexity, and cost. Four immunoassay techniques are currently in common use for determination of drugs of abuse: (1) radioimmunoassay (RIA); (2) homogeneous enzyme immunoassay (EMIT); (3) hemagglutination-inhibition (HI); and, (4) free radical assay technique (FRAT).

1. Methods

a) Radioimmunoassay (RIA)

Since its introduction for the measurement of hormones in the early 1960's, the use of RIA has expanded rapidly for detection and quantitative estimation of a large number of compounds. This technique is coming into routine use in the clinical laboratory (LAPOINTE, 1973) and automation is being introduced (MARSHALL, 1974), as well as computerized data processing (DUDDLESON et al., 1972). Recent general reviews of RIA include those of SKELLEY et al. (1973) and BUTLER (1973); also HARWOOD (1974) has briefly reviewed application of RIA to detection of drugs of abuse.

The general procedure for RIA involves incubation of the biological sample (usually with no pretreatment) with radioisotopically labeled drug and antidrug antibodies (usually as antiserum), separation of the bound and free labeled drug, and determination of free or bound radioactivity. The important parameters of radioimmunoassays described for drugs of abuse are summarized in Table 9. The number of drugs for which radioimmunoassays are available is rapidly increasing and commercial kits and reagents are available for most major drugs of abuse (e.g., from Roche Diagnostics). The first morphine RIA was described by SPECTOR and PARKER (1970). SPECTOR (1971) then described its use for quantitative determination of morphine in serum (with the incubation time shortened to one hour at room temperature) and SPECTOR and VESELL (1971) used the technique to measure serum drug levels after morphine administration in man. Subsequent modification of the procedure, described by CATLIN et al. (1973b) and SPECTOR and SEIDNER (1974), resulted in a test which was suitable to use for mass urine screening and which was made commercially available as a kit by Roche Diagnostics (as Abuscreen). This morphine RIA was the first easily available for general urine screening and is currently the most commonly used of the RIA's for this purpose. RIA's with emphasis on use for urine screening have been described also for barbiturates (CLEELAND et al., 1975) and for morphine and barbiturates in combination in a single test (USATEGUI-GOMEZ et al., 1975). Also clinical screening trials have been reported using RIA kits for morphine, barbiturates, amphetamine, and the morphine-barbiturate combination (MULÉ et al., 1974a, 1975; ROERIG et al., 1975a, b); and a trial with a methaqualone kit has been described (BERMAN et al., 1975). COUMBIS and KAUL (1974) described use of a mor-

Table 9. Radioimmunoassays for drugs of abuse

Drug	Antigen[a]	Anti-serum source	Labeled drug	Assay conditions	Separation of bound and free[b]	Deter-mination	Comments	Reference
Morphine	3-carboxymethyl-morphine-BSA	Goat	H^3-dihydromorphine	Incubate 10 min to 1 h at room temp.	ppt. bound with $(NH_4)_2SO_4$	Count free	Emphasizes urine screening	CATLIN et al. (1973b)
Morphine	Morphine-3-glucuronide-BSA	Rabbit	C^{14}-morphine	Incubate overnight at 4° C	ppt. bound with $(NH_4)_2SO_4$	Count bound		KOIDA et al. (1974b)
Morphine	3-carboxymethyl-morphine-BSA	Rabbit	H^3-dihydromorphine	Incubate overnight at 4 °C	ppt. bound with $(NH_4)_2SO_4$	Count bound	First morphine RIA	SPECTOR and PARKER (1970)
Morphine	3-carboxymethyl-morphine-BSA	Goat	I^{125}-morphine	Incubate 10 min to 1 h at room temp.	ppt. bound with $(NH_4)_2SO_4$	Count free	Emphasizes urine screening	SPECTOR and SEIDNER (1974)
Morphine	3-carboxymethyl-morphine-poly-L-lysine (complexed to succinylated hemocyanin)	Rabbit Guinea pig	Copolymer of carboxymethyl-morphine and a polypeptide containing I^{125}	Incubate 1 h at 37° C	ppt. bound with anti-γ-globulin antibodies (overnight at 2–4° C)	Count bound		VAN VUNAKIS et al. (1972)
Barbiturates	5-allyl-5-(β-carboxyl-α-methyl-ethyl) barbituric acid-BGG	Goat	I^{125}-secobarbital	Incubate 10 min to 1 h at room temp.	ppt. bound with $(NH_4)_2SO_4$	Count free	Emphasizes urine screening	CLEELAND et al. (1975)
Phenobarbital	5-phenyl-5-(4-aminobutyl) barbituric acid-BSA	Rabbit	H^3-phenobarbital	Incubate overnight at 4° C	ppt. bound with $(NH_4)_2SO_4$	Count free		CHUNG et al. (1973)

Barbiturates	5-allyl-5-(β-carboxyl-α-methyl-ethyl) barbituric acid-BGG	Rabbit	C^{14}-barbital	Incubate overnight at 4° C	ppt. bound with $(NH_4)_2SO_4$	Count bound		FLYNN and SPECTOR (1972)
Barbiturates	5-allyl-5-(β-carboxyl-α-methyl-ethyl) barbituric acid-BGG	Rabbit	C^{14}-pentobarbital	Incubate overnight at 4° C	ppt. bound with $(NH_4)_2SO_4$	Count bound		SPECTOR and FLYNN (1971)
Phenobarbital	p-azophenobarbital-BSA	Rabbit	C^{14}-phenobarbital or H^3-phenobarbital	Incubate overnight at 4° C	ppt. bound with $(NH_4)_2SO_4$	Count bound		SATOH et al. (1974)
Morphine-barbiturates	Morphine as per SPECTOR and PARKER (1970) Barbiturate as per SPECTOR and FLYNN (1971)		I^{125}-morphine I^{125}-secobarbital	Incubate 1 h at room temp.	ppt. bound with $(NH_4)_2SO_4$	Count free	Emphasizes urine screening with combined antibodies	USATEGUI-GOMEZ et al. (1975)
Amphetamines	N-(4-aminobutyl) methamphetamine-BSA	Rabbit	H^3-amphetamine	Incubate 30 min at room temp.	ppt. bound with $(NH_4)_2SO_4$	Count free		CHENG et al. (1973)
LSD	LSD-BSA	Rabbit	H^3-LSD	Incubate overnight at room temp.	ppt. free with dextran-coated charcoal	Count bound		CASTRO et al. (1973)
LSD	LSD-HSA (absorbed on polyacrylamide gel)	Sheep	H^3-LSD	Incubate 2 h at room temp.	ppt. free with dextran-coated charcoal	Count bound		LOEFFLER and PIERCE (1973)
LSD	LSD-HSA	Rabbit	Copolymer of LSD and polypeptide containing I^{125}	Incubate for 4 h at 2–4° C	ppt. bound with anti-γ-globulin antibodies (20 hrs. at 2–4° C)	Count bound		TAUNTON-RIGBY et al. (1973)

Table 9 (continued)

Drug	Antigen[a]	Antiserum source	Labeled drug	Assay conditions	Separation of bound and free[b]	Determination	Comments	Reference
Cannabinoids	Δ^9THC-hemisuccinate-BSA	Sheep	H^3-Δ^9THC	Incubate 4 h at 4° C	ppt. free with dextran-coated charcoal	Count bound	Uses Triton K-405 in assay mixture—requires preextraction to assay plasma	TEALE et al. (1974a, 1974b, 1975)
Cannabinoids	Δ^9THC-azobenzoyl-KLH	Goat	H^3-Δ^8THC	Incubate 2 h at 4° C	ppt. free with dextran-coated charcoal	Count bound	Requires preextraction to assay plasma	GRANT et al. (1972) GROSS et al. (1974)
Methaqualone	Not given (commercial kit)		I^{125}-methaqualone	Incubate 1 h at room temp.	ppt. bound with $(NH_4)_2SO_4$	Count free	Emphasizes clinical screening	BERMAN et al. (1975)
Fentanyl	Carboxyfentanyl-BGG	Rabbit	H^3-fentanyl	Incubate 30 min at 37° C	ppt. bound with $(NH_4)_2SO_4$	Count bound		HENDERSON et al. (1975)
Etorphine	Etorphine-3-hemisuccinate-BSA	Rabbit	H^3-etorphine	Incubate 1 h at room temp.	ppt. free with dextran-coated charcoal	Count bound		ROBINSON et al. (1975)
Pentazocine	Azobenzoyl-pentazocine poly-L-lysine or carboxymethyl-pentazocine-poly-L-lysine	Rabbit	H^3-pentazocine	Incubate 1 h at room temp.	ppt. bound with $(NH_4)_2SO_4$	Count free		WILLIAMS and PITTMAN (1974)

[a] BSA = bovine serum albumin. BGG = bovine gamma globulin. HSA = human serum albumin. KLH = keyhole limpet hemocyanin.
[b] $(NH_4)_2SO_4$ = ammonium sulfate.

phine RIA for quantitative determination of drug in autopsy specimens. Production of antibodies to mescaline and related compounds has been described by SCHNOLL et al. (1973) and VAN VUNAKIS et al. (1969).

Antigens used for the production of antibodies for use in RIA's have been formed by coupling the drug of interest or, more commonly, one of its derivatives to a protein or polypeptide. The proteins most frequently used have been bovine serum albumin and gamma globulin; also use of human serum albumin, keyhole limpet hemocyanin, and the polypeptide poly-L-lysine has been described. As seen in Table 9, 3-carboxymethylmorphine has been the most common derivative of morphine used for antigen formation; however, in studies concerning primarily antibody specificity, antibodies have been produced to antigens formed from morphine-3-hemisuccinate, morphine-6-hemisuccinate, and 2(p-aminophenylazo) morphine (KOIDA et al., 1974a; SPECTOR et al., 1973; WAINER et al., 1972). Goat, guinea pig, rabbit, and sheep have been used as sources of antisera, with rabbit the most frequently described. Radioisotopes used as drug labels include H^3, C^{14}, and I^{125}; liquid scintillation and gamma spectrometry have been used to evaluate radioactivity. These isotopes were usually incorporated into the primary drug to be assayed, although the first morphine RIA and the clinical screening test for morphine used H^3-dihydromorphine as the labeled drug. In several methods (VAN VUNAKIS et al., 1972; TAUNTON-RIGBY et al., 1973) the labeled drug was a copolymer of the drug of interest or a derivative and a synthetic polypeptide containing amino acids labeled with I^{125}.

Assay conditions have varied in the described tests, with incubation time from 10 min to overnight at temperatures of 2–4° C, room temperature, or 37° C. Those assays described for mass urine screening used incubation times of 10 min to 1 h at room temperature. RIA's are heterogenous assays, requiring physical separation of free and bound radioactivity for evaluation. Precipitation of bound radioactivity with saturated ammonium sulfate has been by far the most common method used to effect this separation, and was recommended in all of the screening assays. Use of dextran-coated charcoal was also described, as well as immunoprecipitation with antibodies directed against the antibodies used in the assay. VAN VUNAKIS and LEVINE (1974) have recently reviewed the use of double antibody techniques and nitrocellulose membranes for separation of bound and free drug. LINDQUIST and SPRATT (1973) have described solid phase radioimmunoassays for digoxin and for morphine in which antiserum is coated on a thin-walled polystyrene cup. After incubation the free radioactivity is poured off and the bound, coated on the cup, is solubilized and counted. Both free and bound radioactivity have been used in final evaluation of the assay, with the tests recommended for urine screening counting free drug (i.e., the supernatant after ammonium sulfate precipitation). Although the great majority of RIA's described used a biological sample without pretreatment (usually urine, serum, or plasma), the RIA's for cannabinoids in plasma require extraction prior to assay to reduce nonspecific binding.

b) Homogeneous Enzyme Immunoassay (EMIT)

The homogeneous enzyme immunoassay technique was first reported by RUBENSTEIN et al. (1972); modifications described by SCHNEIDER et al. (1973) produced a test suitable for use in mass urine screening. Based on this procedure screening tests for morphine, methadone, barbiturates, amphetamine, and the cocaine metabolite, benzoylecgonine, have been developed and marketed by Syva Company under the trade name EMIT. Additional details and summaries of these procedures have been given by BASTIANI et al. (1973) and SCHNEIDER et al. (1974).

In this technique the drug is labeled with an enzyme, lysozyme, in such a way that the drug-enzyme complex retains enzymatic activity. However, when bound with previously prepared antidrug antibodies, the complex is inactive, presumably due to steric hindrance from the large antibody molecule. Competitive displacement of the drug-enzyme complex from the antibody by drug in the biological sample results in restoration of enzyme activity, which can be measured by clearing of a bacterial solution. The greater the amount of drug in the sample, the greater the enzymatic activity. Since the bound enzyme-labeled drug is inactive, physical separation of bound and free is not necessary to measure the enzymatic activity of the mixture, due to free enzyme-labeled drug. For this reason the assay is referred to as homogeneous. In practice, to a 200 μl suspension of the nonpathogenic bacteria M. luteus are added 50 μl each of urine, antidrug antibody, and lysozyme-labeled drug. Each is washed in with 250 μl of buffer using an automatic diluter/pipette. The assay mixture is aspirated into a temperature-controlled cell of a UV spectrophotometer set at 436 nm and the difference in absorbance between the 7th and 47th second after sample aspiration is automatically calculated and printed out as a measure of enzymatic activity. For high volume output the procedure can be automated. A sample is considered "positive" if the change in absorbance exceeds a previously selected magnitude, representing the operating sensitivity of the assay. BROUGHTON and ROSS (1975) have described an application of this technique to use with the GEMSAEC centrifugal analyzer, using half the volume of reagents, providing more complete automation, and increasing the speed for a large number of samples compared to the standard procedure.

c) Hemagglutination-Inhibition (HI)

Use of an immunoassay based on hemagglutination-inhibition for detection of morphine was reported by ADLER and LIU (1971); and its application as a qualitative and semiquantitative screening test for morphine in urine was described by ADLER et al. (1972). Assays for detection of methadone (LIU and ADLER, 1973) and glutethimide (VALENTOUR et al., 1973) have been reported; and reagents and kits for most major drugs of abuse are commercially available (e.g., from R.D. Products and Technam, Inc.).

In the HI technique a drug (or one of its derivatives) is labeled with tanned sheep red blood cells. When bound to antidrug antibody these indicator cells agglutinate, forming a diffuse filmlike pattern. However, when the red cell-labeled drug is competitively displaced from the antibody by drug in the biological sample, agglutination is inhibited and the indicator cells settle in a sharp pellet or button. Visual observation of this nonagglutinated settling pattern indicates a "positive" test. The operating sensitivity of the test can be varied by diluting the antibody (i.e., antiserum) so that agglutination is inhibited by a selected minimal drug concentration in a standard urine sample. Semiquantitative estimation is achieved using a precalibrated antiserum with serial dilutions of the urine sample. As with EMIT, no separation of free and bound labeled drug is necessary in HI. In practice, the assay may be carried out on a microtiter plate using several dilutions of antiserum, as well as a negative control. Reagent and sample volumes of 25–30 µl are used and an incubation time of 1.5–2 h is necessary before visual reading of the end point.

Ross et al. (1975) have described a closely related assay, based on inhibition of latex agglutination. This test was carried out in the same manner as HI, except that the indicator cells were replaced by latex-labeled morphine. The end point in this assay was also visual. The assay was carried out in a test tube which, after 2 h incubation at 37° C, was observed against a fluorescent light source. A "positive" test was indicated by a clear or translucent specimen (i.e., no latex agglutination).

d) Free Radical Assay Technique (FRAT)

Leute et al. (1972a) reported detection of morphine using an immunoassay with determination of free labeled drug by electron spin resonance (ESR) spectrometry, and its application to a clinical screening test for morphine in urine (Leute et al., 1972b). Assays under the trade name FRAT are now commercially available from Syva Company for morphine, methadone, barbiturates, amphetamine, and benzoylecgonine, a metabolite of cocaine. These tests were recently reviewed by Bastiani et al. (1973) and Schneider et al. (1974).

Like EMIT and HI, the free radical assay technique is a homogeneous immunoassay, not requiring separation of free and bound labeled drug. Drug is labeled with nitroxide, a stable free radical. The spin-labeled drug tumbles freely in solution producing a sharp three-lined ESR spectrum. However, when bound to antidrug antibody its mobility is decreased and the ESR spectrum is broad. Identification of the sharp ESR spectrum peaks is an indication of competitive displacement of spin-labeled drug from antidrug antibody by unlabeled drug in the biological sample. The height of the peaks can be used for quantitative estimation, with appropriate controls and standard curves. In practice, using a calibrated capillary pipette and plastic microbeaker, a 50 µl urine sample is first oxidized with 5 µl of dichromate; a 20 µl aliquot is then mixed with 10 µl of mixture of antibody and spin-labeled drug. The combined solution is drawn into the capillary tube, the end is sealed with putty, and the tube is incubated at room temperature for 15–30 min. The spectrum is then recorded by a specially modified ESR spectrometer.

2. Sensitivity

The sensitivities of immunoassays vary not only between the different immunoassay methods but also between the assays within any one method. Sensitivity can be influenced by the affinity of the labeled and unlabeled drug for the antibody, nonspecific binding, the background measurement from normal drug-free biological fluid, and the sensitivity of the analytical technique used for measuring labeled drug. RIA sensitivity is partially dependent on the specific activity of the labeled drug. In an analogous manner the sensitivity of EMIT can be changed by alteration of the enzyme label. Gorodetzky and Kullberg (1974a) described use of a homogeneous enzyme immunoassay developed by Syva Company for determination of morphine in saliva. Morphine coupled to malic dehydrogenase was the labeled drug; and the conversion of oxaloacetic acid to malic acid with the consequent oxidation of NADH to NAD was the monitoring enzymatic reaction. The assay was approximately 20 times more sensitive than the EMIT urine screening assay.

For use of the immunoassays as qualitative screening tests, "cutoff" concentrations have been chosen, representing the operational sensitivity of the assays. An assay measurement less than that given by the selected cutoff concentration of drug added to normal, drug-free urine is considered a "negative" test; and a measurement greater than that of the cutoff concentration is "positive." Cutoff concentrations have been selected on the basis of minimal overlap with the distribution of assay measurements from normal, drug-free urine and also from urine containing drugs (or drug metabolites) which are not desired to be detected (i.e., possible pharmacologic false positives). BRATTIN and SUNSHINE (1974) examined four morphine immunoassays by comparing "spiked" urine samples (i.e., normal, drug-free urine to which drug is added) with normal, drug-free urine. On the basis of this comparison they defined maximal sensitivity of RIA, FRAT, EMIT, and HI of 10, 100, 400, and 30 ng/ml for morphine, respectively; however, because of the possibility of pharmacologic false positives in urine from addicts in treatment programs they recommended cutoff concentrations of 100, 500, 600, and 100–200 ng/ml for the four assays. Similarly, MULÉ et al. (1975) studied several RIA's and reported the following concentrations statistically distinguishable from normal, drug-free urine: morphine, 5 ng/ml; barbiturates, 10 ng/ml; amphetamine, 500 ng/ml; and for the morphine-barbiturate combination 50 ng/ml for morphine and 100 ng/ml for secobarbital. These authors also recommend higher cutoff concentrations for general use as screening tests. SCHNEIDER et al. (1974) presented histograms showing distribution of measurements for normal, drug-free urines and spiked samples for the FRAT and EMIT assays. In addition to cutoff concentrations they defined detection levels, concentrations which would be found positive at least 95% of the time. These detection concentrations were higher than the cutoff concentrations because of the distribution of the measurements at the cutoff concentration.

In Table 10 are summarized the sensitivities of the urine screening immunoassays, expressed as cutoff and detection concentrations. RIA and HI are, in general, more sensitive than EMIT and FRAT. Within the limits of the maximal attainable sensitivity, cutoff levels of the immunoassays can be changed depending on the acceptable level of physiologic and pharmacologic false positives. For example, the EMIT morphine assay has been described for use with an 0.300 μg/ml cutoff (detection concentration equals 0.500 μg/ml), although the probability of pharmacologic false positives following dextromethorphan ingestion increased (GORODETZKY, 1975). Similarly, especially if confirmatory testing is not done, it has been advised to raise the cutoff concentration to at least 0.100 μg/ml for morphine RIA and HI (BRATTIN and SUNSHINE, 1974).

RIA is capable of considerably greater sensitivity than the cutoff levels generally recommended in the urine screening assays. Many of the assays summarized in Table 9 were reported to have sensitivities ranging from 0.5–10 ng/ml of biological sample, and absolute sensitivities of 20–100 pg in the assay tube. The HI glutethimide assay (VALENTOUR et al., 1973) was reported to detect 500 ng/ml; and ROSS et al. (1975) recommended a 0.200 μg/ml cutoff for the morphine latex agglutination inhibition test.

Table 10. Sensitivity of available urine screening immunoassays. Sensitivities are expressed as cutoff and detection concentrations

Assay	Cutoff[a] (detection[b]) concentration in μg/ml			
	RIA	EMIT	HI	FRAT
Morphine	0.040–0.100	0.500 (0.700)	0.030–0.060	0.250 (0.500)
Barbiturate[c]	0.100	1.000 (2.000)		1.000 (2.000)
Morphine-barbiturate combination	0.100			
Amphetamine[d]	1.000	1.000 (3.000)		1.000 (3.000)
Benzoylecgonine[e]		1.000 (1.600)	0.030–0.060	0.500 (1.000)
Methadone		0.500 (0.700)	0.030–0.060	0.250 (0.500)
Methaqualone	0.100			

[a] Cutoff = concentration in urine giving measurement below which a sample is considered negative and above which a sample is considered positive.
[b] Detection = concentration in urine at which at least 95% of the samples will be read positive.
[c] Sensitivity expressed as concentration of secobarbital.
[d] Sensitivity expressed as concentration of d-amphetamine.
[e] Major metabolite of cocaine.

3. Specificity

The specificity of an immunoassay is dependent primarily on the specificity of the antibodies, that is, their ability to bind selectively one compound strongly and others only weakly or not at all. Antibody specificity can be influenced by the antigen used for its production (including the drug or drug derivative used as hapten, the protein or polypeptide to which the hapten is coupled, and the configuration of the hapten on the protein), the animal used, immunization procedure and schedule, and the time of bleeding. Even with these factors held constant, there can be variability between individual immunized animals; CATLIN (1973b) concluded that every solution containing antibodies should be considered an unique reagent. Because of this variability, precise specificity generalizations are not possible; however, some common cross-reactivity characteristics have been described for the commonly used screening tests. Most of the papers cited earlier in this section describing immunoassays contained some specificity data; extensive cross reactivity tables were presented by BRATTIN and SUNSHINE (1973), MULÉ et al. (1974a), MULÉ and BASTOS (1974), MULÉ et al. (1975), SCHNEIDER et al. (1974), and SPECTOR et al. (1973).

The morphine screening immunoassays have all been reported to cross react with codeine, usually with equal or slightly greater reactivity than with morphine. Reactivity with morphine glucuronide has been reported consistently, usually at 30–80% of the reactivity with morphine. Cross reactivity was also usually reported with heroin and to a lesser degree (usually 10–50% as great as with morphine) to normorphine, dihydromorphinone (Dilaudid), dihydromorphine, and sometimes levorphan. There was generally reported negligible or very low cross reactivity to other opioids (including methadone, propoxyphene, meperidine, and naloxone), barbiturates, major and minor tranquilizers, antihistamines, and diphenoxylate (Lomotil). However, ADLER et al. (1972) did report pharmacologic false positives with the morphine HI assay (using a 30 ng/ml cutoff) in urine samples of hospitalized patients administered meperidine. Dextromethorphan has been the subject of several studies. Cross reactivity in dextromethorphan spiked normal drug-free urine has generally been reported to be low; however, in clinical studies some pharmacologic false positives have been found, especially if low cutoff levels were used. CATLIN et al. (1973b) administered 30 mg of dextromethorphan orally every 6 h for 4 days; of 75 urine samples collected and analyzed using a morphine RIA, 52 had morphine equivalent concentrations below 25 ng/ml and 23 had concentrations between 25 and 46 ng/ml morphine equivalent. By 12 h after the last dose all urine samples were below the 25 ng/ml level. ADLER et al. (1972) reported positives using the morphine HI assay with a 30 ng/ml cutoff concentration for 9 h following a single oral dose of 30 mg dextromethorphan. GORODETZKY (1975) gave a single oral dose of 60 mg dextromethorphan to 10 subjects and collected urine in approximately 8-h-aliquots. All samples were negative using the morphine RIA with cutoff concentration as low as 40 ng/ml and by EMIT with a 0.5 μg/ml cutoff; however, with EMIT with a 0.3 μg/ml cutoff, 22% of the samples in the first 8-h period following drug administration were positive. Pharmacologic false positives in urine have also been reported for 5–12 h following ingestion of poppy seeds in pastry using both the RIA and HI for morphine (CATLIN et al., 1973b; ADLER et al., 1972). Although in vitro studies have shown very low cross reactivity with naloxone, KOKOSKI and JAIN (1975) reported positives using the morphine RIA with a 60 ng/ml cutoff for 12 h following a single oral dose of 2000 mg of naloxone.

The commonly used clinical screening morphine immunoassays use antibodies produced in response to antigens formed from morphine derivatives with moieties attached to the 3-position (generally 3-0-carboxymethylmorphine). Antibodies with different specificities have been produced using different morphine derivatives in the antigen. For example, using morphine derivatized in the 6-position in the antigen produced antibodies relatively more sensitive to morphine glucuronide; and coupling through the 2-position gave increased cross reactivity to nalorphine (SPECTOR et al., 1973). KOIDA et al. (1974a) also produced antibodies using an oxymorphone derivative as the hapten; these had a different pattern of specificity than the antibodies produced using an antigen containing morphine. It is likely that highly specific antibodies to many different compounds could be produced by judicious choice of antigen, animal, and immunization schedule.

The immunoassays used for barbiturate screening have generally been designed to cross react with several barbiturates. Antibodies for these assays have been produced with an antigen incorporating a derivative of secobarbital; and this barbiturate has generally been used as the reactivity standard. Cross reactivity varying from approximately 10–200% compared to secobarbital has been frequently reported for pentobarbital, phenobarbital, butabarbital, amobarbital, barbital, mephobarbital, and talbutal. The RIA is relatively nonreactive to thiopental; however, FRAT and EMIT do cross react significantly with this barbiturate. Lack of cross reactivity has been reported with diphenylhydantoin (Dilantin), caffeine, a large number of major and minor tranquilizers, amphetamine, cocaine, urea, and pyrimidines. Although glutethimide in vitro showed a low degree of cross reactivity, ROERIG et al. (1975b) reported positive urine specimens for approximately 24 h following a single oral dose of 500 mg of glutethimide using, for analysis, the RIA with a 100 ng/ml cutoff. From their specificity studies, FLYNN and SPECTOR (1972) and SPECTOR et al. (1973) concluded that reactivity with barbiturate antibodies used in the RIA required a barbiturate 6-membered ring with an intact urea portion and carbonyl groups on the 4 and 6 positions. Many different substituents could be present on the 5 position; and maximal reactivity was achieved when there were two substituents at this position. CHUNG et al. (1973) described production of highly specific phenobarbital antibodies, with little cross reactivity with other barbiturates. The combination morphine-barbiturate assay has been reported to have approximately the same specificity as each test alone (USATEGUI-GOMEZ et al., 1975; MULÉ et al., 1975).

MULÉ et al. (1975) reported slight cross reactivity with methamphetamine and phenylpropanolamine in the amphetamine RIA. For FRAT and EMIT cross reactivity in the amphetamine assays ranging from approximately 20–200% compared to amphetamine has been reported for phenethylamine, phenylpropanolamine, methamphetamine, phenmetrazine, ephedrine, mephentermine, phentermine, benzphetamine, cyclopentamine, nylidrin, isoxsuprine, and methoxyphenamine. No cross reactivity was described with barbiturates and opiates. In the benzoylecgonine assay there has been slight cross reactivity (on the order of 10% or less) reported with ecgonine and cocaine; lack of reactivity was described with amphetamine, barbiturates, and opiates. Reported assays for methadone have shown some cross reactivity with alpha-acetylmethadol, but very low or no reactivity with other opiates (including propoxyphene), methadone metabolites, or barbiturates. The methaqualone RIA showed cross reactivity with the monohydroxylated metabolites of methaqualone and did not react with morphine, barbiturates, methadone, or amphetamine (BERMAN et al., 1975).

The LSD RIA's summarized in Table 9 reported good specificity (especially TAUNTON-RIGBY et al., 1973), with cross reactivity in some cases to other ergot alkaloids but not to simple indoles (such as serotonin and tryptophan) nor to a large number of other tested compounds. The cannabinoid RIA's were directed primarily toward detection of tetrahydrocannabinol (THC); some cross reactivity with THC metabolites and other 3-ringed cannabinoids was reported. The RIA's noted for fentanyl, etorphine, and pentazocine were reported to have good specificity, with little reaction with other opioids or metabolites of the primary drug of interest. The HI for glutethimide described by VALENTOUR et al. (1973) was reported by the authors to show no significant cross reactivity with several barbiturates and opiates, amphetamine, quinine, and diphenylhydantoin.

All EMIT assays are subject to a small proportion of physiologic false positives due to the presence of endogenous lysozyme in some urine specimens. The incidence has been estimated at 1% (SCHNEIDER et al., 1974). This may be controlled for by running a background blank for each positive sample (i.e., containing only the test urine sample, bacterial suspension, and buffer).

Because of their relative lack of specificity, immunoassays have been recommended for use as negative exclusion tests, with operating sensitivity set at the highest practical level (i.e., lowest cutoff concentration) (e.g., BRATTIN and SUNSHINE, 1974). No further analysis would be performed on negative samples and all positive findings would be confirmed by methods of greater specificity (such as TLC or GLC) to achieve more definitive drug identification and to eliminate physiologic and pharmacologic false positives. Use of the immunoassays as confirmatory tests for positives determined by other primary screening procedures has also been recommended (e.g., ROERIG et al., 1975a).

4. Socioeconomic Parameters

Estimates of the cost and speed of immunoassays have been reviewed by BRATTIN and SUNSHINE (1973, 1974), CATLIN (1973a), KAISTHA and TADRUS (1975), MULÉ (1974), MULÉ et al. (1974a), and the World Health Organization (1974). A comparative summary of these estimates and other socioeconomic parameters is shown in Table 11. All of the immunoassays have some relative advantages and disadvantages. All share the disadvantage of relative lack of specificity compared to some other screening methods (as discussed in detail above). The operating cost per sample (representing mainly reagent costs) is comparable between the screening immunoassays, with a slightly lower range for HI than the others. Also, the speed for running a large number of samples is in the same range for all of the assays, as reflected in the number of samples which can be analyzed per day.

RIA is potentially the most sensitive of the immunoassays. Also it is amenable to automation (which simplifies the procedure technically and increases somewhat the speed of processing) and is suitable for quantitative evaluation. The time required to run a single sample is relatively long because of incubation time and because the assay is heterogeneous and requires physical separation of the free and bound labeled drug. Although the amount of radioactivity involved in the screening RIA's does not present a significant radiation hazard, licensing is required and radioisotope working precautions must be taken. Moderately expensive radioactivity counting instrumentation is needed for RIA; however, these instruments are also of general use in many other research and clinical assays involving radioisotopes.

The EMIT assay is simple to perform and even the manual system is semiautomated; greater automation with further simplification of the procedure and increase in speed of processing large numbers of samples is also available. The EMIT is not as sensitive as the RIA and HI and is able to provide some quantitative estimation;

Table 11. Socioeconomic parameters of screening immunoassays

Parameter	RIA	EMIT	HI	FRAT
1. Complexity, skill required	Pipetting, separating, radioisotope caution, instrument operation	Semiautomated pipetting and instrument operation	Tedious micropipetting	Simple pipetting, instrument operation
a) Endpoint	Objective	Objective	Subjective	Objective
2. Speed (one assay)				
a) for single sample	1–2 h	2–4 min	1.5–2 h	2–4 min
b) no. samples/day	200–400	200–300	200–400	200–300
3. Cost (U.S. $)				
a) start-up	8000–15000	6000–8000	200–500	25000–27000
b) operating cost/test	0.50–1.50	0.50–1.00	0.25–0.55	0.50–1.00

however, it has the advantage of a very short turn around time for a single sample. Once the instrument is operating, standards run, and the equipment and supplies set up, a single sample can be run for one assay and the answer obtained in approximately 1 min. Instrumentation cost is somewhat less than RIA and the UV spectrophotometer, automatic diluter/pipette, and printing calculator are generally useful for other analytical procedures.

The HI is also simple, but requires pipetting into microtiter plate wells (usually using a calibrated dropper pipette) which has been reported to be tedious and error prone (BRATTIN and SUNSHINE, 1974). The subjective endpoint of the assay is also a disadvantage. This assay is used for screening in the same sensitivity range as the RIA and is capable of semiquantitative estimation. The time for analysis of a single sample is relatively long because of the long incubation time. A definite advantage of the HI is the very low start-up cost, requiring no major instrumentation. Some problems with quality control of reagents have been reported (BRATTIN and SUNSHINE, 1974; MULÉ and BASTOS, 1974).

The FRAT assay is similar to EMIT in its sensitivity range, capability to provide some quantitative estimation, and ability to run a single sample in a very short time once the instrument has been set up and calibrated. The pipetting and mixing procedures are simple (using calibrated capillary tubes) although they are not as readily automated as they are in the EMIT system. The greatest disadvantage of this assay is the high cost of a specialized instrument modified to perform only this assay.

V. Other Techniques

Most analytical methods applicable to organic compounds have been used to some degree in detection or identification of drugs in biological fluids. With regard to drugs of abuse, methods other than those already discussed have been used as adjuncts to other screening procedures, for confirmation of positive samples, and for additional aids in positive identification of unknown compounds (especially in forensic toxicology). In this section the following methods will be briefly discussed: paper chromatography, colorimetry, ultraviolet (UV) spectrophotometry, microcrystallography, mass spectrometry (MS), integrated gas chromatography/mass spectrometry (GC/MS), infrared (IR) spectrophotometry, and high pressure liquid chromatography (HPLC).

1. Paper Chromatography

For general screening of urine and other biological fluids the use of paper chromatography has been largely superseded by thin-layer and gas chromatography. MANNERING et al. (1954) described a method for detection of morphine in urine and tissues and for separation of nine opium alkaloids and meperidine using paper chromatography; and LIN and WAY (1963) described a system for separation of 20 narcotic drugs. WADDELL (1965) gave a rapid method for barbiturate identification by extraction of biological fluids with ether and analysis with paper chromatography. A general toxicologic scheme using high temperature reversed phase paper chromatography was described by STREET (1962). Recent reviews of paper chromatographic methods have been presented by TAYLOR (1971), with emphasis on narcotic analgesic analysis, and by JAIN and CRAVEY (1974b), with particular reference to barbiturates. Additional paper chromatographic R_f data on drugs as well as summaries of papers, solvent systems, and detection reagents used have been described by SUNSHINE (1969). Also MOFFAT and SMALLDON (1974) and MOFFAT and CLARE (1974) applied the discrimination power analysis of MOFFAT et al. (1974a) to several commonly used paper chromatographic systems for analysis of basic drugs.

2. Colorimetry

Colorimetric methods have been used for the rapid identification of drugs in solid form (e.g., in seized material) and in biological fluids or their extracts. Although these methods are usually simple and inexpensive, they are generally of limited utility because of relative lack of sensitivity and specificity. Colorimetric tests for drugs of abuse have been reviewed by TAYLOR (1971) and MULÉ (1972b); a tabular summary of spot tests for drugs in general was given by SUNSHINE (1969). MASOUD (1975) has described a flowsheet for systematic spot testing for more than 40 commonly encountered street drugs. Commercial test kits are available, such as Screendex, evaluated by DECKER and LEWIS (1973). They reported the tests useful for drug formulations, urine, serum, and stomach contents in the sensitivity range of 5–50 µg/ml for a large number of drugs. However, some difficulties with interpretation were described, resulting in a high proportion of false positives; and, it was emphasized that all positives should be confirmed by another method. Colorimetric methods have also been used for quantitative estimation of drugs of abuse in biological fluids (MULÉ, 1972b), although more sensitive and specific methods are now more commonly used. A novel method combining continuous drug-dye complex extraction and CM-cellulose chromatography for analysis of basic drugs was described by MCMARTIN et al. (1969). Bromcresol Green was used as the dye and a sensitivity of 1–2 µg/ml in biological fluids was achieved. RUTTER (1972) used an autoanalyzer to estimate amphetamine and other related primary amines in urine by colorimetric determination of a chromophore formed by reaction of the drugs with 2,4,6-trinitrobenzene sulfonic acid. A total of 20 samples per hour could be analyzed. PEHR (1975) reported a new, simple colorimetric screening method for barbiturates in urine. An ether extract of the urine was reacted with a divalent mercury/dithizone chelate and diluted pyridine to form a visually determined pinkish-violet color. Sensitivity to detect 5–10 µg/ml of phenobarbital was reported.

3. Ultraviolet (UV) Spectrophotometry

Identification and quantitative estimation of drugs by their pattern of UV absorbance and the degree of absorbance at a selected wavelength has been used extensively. However, from the point of view of screening methods, this technique is limited by general lack of sensitivity and specificity and the necessity for relatively clean extracts for analysis. JATLOW (1973) discussed the general use of UV methods and concluded that they could be useful in hospital toxicologic analysis if plasma levels could be expected above 2 µg/ml. This paper emphasized use in analysis of barbiturates, chlordiazepoxide, glutethimide, and meprobamate. Extensive UV absorption reference data can be found in the book of SUNSHINE (1969). Coleman

Instruments Division of Perkin-Elmer Corporation (1970) published a booklet describing UV methods for several drugs of abuse. Also SIEK and OSIEWICZ (1975) discussed UV absorption properties of a large number of drugs (including drugs of abuse) divided into 13 structural groups.

FARMILO (1954) reviewed the use of UV methods for analysis of narcotics and OESTREICHER et al. (1954) presented spectral data for 90 compounds. MARTIN et al. (1963) provided additional UV data as did MULÉ (1964), who described a UV method for narcotic analysis in biological fluids. WALLACE et al. (1972b) presented a method for UV analysis of methadone with sensitivity increased to 1 μg/ml by oxidizing the extracted drug to benzophenone, which has a much greater absorbtivity than methadone. The most commonly used methods for UV analysis of barbiturates are those based on papers by GOLDBAUM (1948 and 1952) and STEVENSON (1961). A differential UV absorbance spectrum was commonly run on extracted barbiturate at approximately pH 10 (reference) and pH 13 (sample). Absorbance at 260 nm in the differential spectrum or the difference in absorbance at 240 nm and 260 nm in the spectrum could be related to barbiturate concentration for quantitative estimation. GOLDBAUM (1952) advocated use of difference ratios (i.e., difference in absorbance at the two pH's) at various wavelengths and 260 nm to distinguish between specific barbiturates. BLACKMORE et al. (1971) used differential UV absorbance of barbiturates at two pH's for their determination in an autoanalyzer. UV methods have also been described for ethchlorvynol (WALLACE et al., 1974) and methaqualone in postmortem tissues (CHRISTOPOULOS et al., 1974); and a general review of analysis of nonbarbiturate sedative-hypnotics, including UV methods, was presented by CRAVEY and JAIN (1974b). WALLACE (1967) and WALLACE et al. (1968) described UV methods for determination of ephedrine and amphetamine.

4. Microcrystallography

Microcrystal tests have been used as additional or confirmatory procedures in the identification of drugs in both solid form and in biological fluids. FULTON (1969) presented a detailed review of the application of microcrystallography to drug identification; and an extensive tabular summary of microcrystal tests can be found in SUNSHINE (1969). The hanging microdrop technique has been commonly used for carrying out this test. An approximately 0.1 μl microdrop of unknown solution was mixed on a cover slip with a microdrop of test reagent, the cover slip was overturned into a cavity slide, and crystal formation was observed under the microscope. Hundreds of test reagents have been described for use in microcrystal tests. The form and orientation of the crystals compared to standard descriptions, photographs or, most specifically, known standards treated in the same way as the unknown served as the basis for compound identification. This procedure may have sensitivity into the μg range and may be used with extracted and purified materials (e.g., eluted from a chromatogram). Also it can take up to 24 h for crystals to form. ONO et al. (1969) described microcrystal tests for morphine, dihydromorphinone, codeine, norcodeine, methadone, quinine, and methamphetamine; and TAYLOR (1971) reviewed the most commonly used microcrystal tests for identification of narcotics. HUANG and BADEN (1973) described a simple microcrystal test using a single reagent (a mercuric iodide) for confirmation of alkaloids eluted from a TLC plate.

5. Mass Spectrometry (MS) and Gas Chromatography/Mass Spectrometry (GC/MS)

Mass spectrometry alone and integrated gas chromatography/mass spectrometry have become important techniques in the analysis of drugs and drug metabolites (JENDEN and CHO, 1973; World Health Organization, 1974). Although the instrumentation is very expensive (generally $ 50000–150000 U.S., depending mainly on the degree of computerized data analysis), these methods are being increasingly used to aid in the identification of unknown drug metabolites and, especially in the very sensitive mass fragmentography mode, for quantitative estimation of drugs and metabolites. GC/MS has the capability for a high degree of sensitivity and specificity. However, it has not been widely used for mass urine screening because of the expense of the equipment, the technical skill necessary to operate it, and the extensive sample preparation needed for optimal use of the technique (and, therefore, the long turn around time and small number of samples run per day). MS and GC/MS have been recommended for use in forensic toxicology and in the diagnosis of drug overdose in the hospitalized patient (SKINNER et al., 1972; LAW, 1973). HAWKS (1974) has also discussed use of GC/MS in development and confirmation of immunoassays, especially those with high sensitivity. Extensive tables of GC/MS reference data, suitable for computerized storage and retrieval, were published by FINKLE et al. (1972) and further supplemented with additional data by FINKLE et al. (1974). Data were presented for both electron impact and chemical ionization spectra. SAFERSTEIN and CHAO (1973) provided isobutane chemical ionization spectral data for 62 commonly abused drugs. BOERNER et al. (1973) described the MS analysis of body fluids as an aid in the diagnosis of the overdosed patient. They introduced the organic solvent extract of the biological fluid directly into the mass spectrometer through a Llewellyn 3-stage membrane separator inlet system, referred to as chemical vapor analysis (CVA). The analysis was accomplished rapidly using computerized data handling and a general sensitivity of 5 μg/ml was reported. GREEN and LITTLEJOHN (1973) also briefly described a direct introduction MS system for use with biological fluids. BILLETS et al. (1973) used an MS analysis for drug detection in gastric contents of intoxicated patients. They analyzed the solid residue following evaporation of a chloroform extract by using the solid probe for sample introduction. DOWN and GWYN (1975) recommended scraping unsprayed spots (identified by visualization under UV light) from a TLC plate directly into a capillary tube for MS analysis, also using the solid probe.

6. Infrared (IR) Spectrophotometry

Infrared spectrophotometry is a useful analytical chemical technique to determine structural information about an unknown compound and aid in its identification. However, from the point of view of screening and clinical toxicology, relatively large amounts of purified (often crystalline) material are usually needed; also skill is needed to perform the analysis and interpretation of the spectra requires expertise and experience. HANNAH and PATTACINI (1972) presented a simplified flow diagram approach to interpretation of IR spectra aimed at identification of drugs of abuse. They looked sequentially for key bands at specific wave numbers, allowing classification of the compound into 1 of 10 major groups; a search for a matching spectrum could then be limited to that group. Representative spectra for each group were included in the paper. TAYLOR (1971) has reviewed the application of IR techniques to narcotic analysis; and JAIN and CRAVEY (1974b) noted micro infrared spectrophotometric methods for analysis of barbiturates, using drugs recovered from GC and TLC analyses.

7. High Pressure Liquid Chromatography (HPLC)

High pressure liquid chromatography using both absorption and ion exchange columns is a new separation technique, which is being increasingly applied to the analysis of drugs. Currently available detectors use primarily UV, fluorescence, or refractive index measurements and lack great general sensitivity; some compounds (such as those which fluoresce or for which fluorescent derivatives can be easily made) are detectable in low concentrations. Some recent references have described separations with emphasis on solid forms (e.g., street drugs) for drugs of abuse in general (CHAN et al., 1974), alkaloids (VERPOORTE and SVENDSEN, 1974), narcotics (KNOX and JURAND, 1973a, b), barbiturates (JAIN and CRAVEY, 1974b), phenethylamines (CASHMAN et al., 1973), and LSD (JANG and WHEALS, 1973; WITTWER and KLUCKHOHN, 1973).

D. Validity

As with many other parameters of drug detection methods, it is difficult to make generalizations concerning their validity to detect drugs or their metabolites in body fluids following human drug administration. In addition to significant methodologic variability (as described above), validity can also be influenced by a large number of pharmacologic variables, such as drug dose, route and time of administration, drug metabolism, and concentration and pH of the biological fluid (especially for urine). Also there have been only a few systematic studies reported concerned principally with the evaluation of the validity of screening methods; some additional data have been given in papers which were primarily methodologic. Quantitative metabolism and disposition studies of drugs of abuse have been reported, many of which were recently reviewed (World Health Organization, 1974). However, these studies are of limited value in predicting the time course of detectability by screening tests, primarily because drug concentration in urine at various time intervals after drug administration is seldom reported. Also the time course studied is frequently short (e.g., 24 h) and the analytic methods used are considerably more complex than the screening methods in common use. These metabolic studies, as well as studies of hospitalized drug-intoxicated patients, have provided plasma level data of use in interpretation of toxicologic analyses, especially in drug overdose patients. Expected plasma, serum, and blood levels of over 100 drugs in both therapeutic and toxic conditions were recently compiled and reviewed by BASELT et al. (1975).

I. Narcotic Analgesics and Antagonists

1. Heroin and Morphine

The most extensive and systematic data on the validity of screening methods have been gathered for heroin and morphine. In a series of validity studies single intravenous doses of heroin or morphine were administered to drug-free former narcotic addict volunteers. Using a crossover design, in each study approximately 10 subjects each received two or three minimally to highly euphorigenic doses of opiate at weekly intervals in random order. All urine was collected ad lib, and, in addition, each subject was asked to urinate every 8 h for a 1 week period following each drug administration. Urine was also collected prior to drug administration to serve as negative control samples. At the end of the collection period each subject's urines were combined into approximately 4–8 h aliquots to give sufficient volume in each sample to do all the desired chemical tests. In one heroin study doses of 2.5 and 5 mg/70 kg were used (GORODETZKY et al., 1974); in a second study (to evaluate additional new screening methods) the same doses plus a 10 mg/70 kg heroin dose were administered (GORODETZKY and KULLBERG, 1974b; GORODETZKY, 1975; YEH et al., 1976). A morphine validity study used doses of 6 and 12 mg/70 kg (GORODETZKY, 1973b), equieuphorigenic to the two lower doses of heroin. All urine samples, including negative controls, were randomized, coded, and analyzed under blind conditions by well-specified and defined, commonly used screening methods. Data were compiled as the percent of urine samples positive for morphine by 8-h periods following drug administration.

The following methods for detection of morphine were used in these studies: (1) TLC preceded by organic solvent, ion exchange resin impregnated paper, and XAD-2 resin column extraction (GORODETZKY, 1973a; KULLBERG and GORODETZKY, 1974) of unhydrolyzed urine (OS, IE, XAD-2) and acid-hydrolyzed urine (OSH, IEH, XAD-2H) with sensitivities of 0.19 (OS), 0.16 (IE), and 0.08 (XAD-2) μg/ml; (2) GC using trimethylsilyl derivatization (YEH and MCQUINN, 1975) preceded by organic solvent extraction of unhydrolyzed (GC) and hydrolyzed (GC-H) urine, with sensitivity of 0.05 μg/ml; (3) the free radical assay technique (FRAT) with cutoff sensitivity of 0.5 μg/ml; (4) the Technicon Autoanalyzer fluorometric method (TECH) without hydrolysis, with sensitivity of 0.2 μg/ml; (5) homogeneous enzyme immunoassay with cutoff sensitivities of both 0.3 and 0.5 μg/ml (EMIT-.3, EMIT-.5); and, (6) the morphine Abuscreen RIA using both ^{3}H and a cutoff sensitivity of 0.05 μg/ml (RIA-H) and ^{125}I with cutoffs of 0.04, 0.1, and 0.2 μg/ml (RIA-0.04, RIA-0.1, RIA-0.2).

The results of these studies are summarized in Table 12. The methods with the shortest time course of detectability were those which detect only free morphine, that is, the three TLC methods, the GC technique, and the Technicon Autoanalyzer which used extraction of unhydrolyzed urine. Minimally to moderately euphorigenic doses of heroin were detectable with high probability for only 8 h following administration and a 10 mg dose was detectable for 16–24 h. Equieuphorigenic morphine doses were detectable approximately 8 h longer. These experimental results are in general agreement with clinical reports, although some of the latter have reported slightly longer time courses of detectability, using primarily urine samples from addicts in treatment or from a few hospitalized patients (BERRY and GROVE, 1971;

Table 12. Validity of screening methods for detection of morphine in urine following heroin and morphine administration. Each figure is the time (by 8-h periods) for detection of morphine in ≧50% of urine samples

Method[a]	Heroin dose mg/70 kg, IV			Morphine dose mg/70 kg, IV	
	2.5	5	10	6	12
OS[b,e], IE[b,e], XAD-2[c], TECH[b,e]	0–8	8	16	8	8–16
GC[d]			24		
OSH[b,e], IEH[b,e], FRAT[b,e], EMIT-0.5[c], RIA-0.2[c]	8–16	16–24	32	32–40	40–48
XAD-2H[c], EMIT-0.3[c], RIA-0.1[c]	24–32	32–40	48		
RIA-H[b,e], RIA-0.04[c]	32–40	48–56	56	72	80
GC-H[d]			80		

[a] See text for abbreviation.
[b] Heroin study (GORODETZKY et al., 1974).
[c] Heroin study (GORODETZKY and KULLBERG, 1974b; GORODETZKY, 1975).
[d] YEH et al., 1976.
[e] Morphine study (GORODETZKY, 1973b).

DOLE et al., 1966; DAVIDOW et al., 1966; MARKS et al., 1969; PARKER et al., 1966, KAISTHA and JAFFE, 1972b). Hydrolysis of urine prior to organic solvent or ion exchange resin paper extraction or use of the FRAT, EMIT or RIA (with highest cutoff levels) added approximately 8–16 h to the time course of detectability. In studies using clinical samples analyzed by several methods PAYTE et al. (1971) and SOHN et al. (1973) emphasized the need for urine hydrolysis to increase detectability of heroin use. SCHNEIDER et al. (1973) noted a longer detectability with EMIT than TLC without hydrolysis; and LEUTE et al. (1972b) reported that administration of a spoon of heroin (approx. 30 mg) was detected for 48 h using FRAT. The XAD-2 resin extraction with hydrolysis was the most sensitive TLC analysis, approximatly equal in time course of detectability to EMIT with the 0.3 μg/ml cutoff and RIA with the intermediate cutoff. Again, approximately 8–16 h were added to detectability compared to the next less sensitive methods. The longest detectability was seen with the RIA with 0.05 and 0.04 μg/ml cutoffs and GC with hydrolysis; a moderately to highly euphorogenic dose of heroin was detectable for 2–3 days. Equieuphorogenic doses of morphine were detected for 16–24 h longer. ROSS et al. (1975), using the morphine latex agglutination inhibition test, reported urine morphine levels greater than 0.04 μg/ml for approximately 72 h following a 6.5 mg morphine dose, in close agreement with the experimental results. USATEGUI-GOMEZ et al. (1975) detected morphine in urine for approximately 60 h after a 10 mg dose using the Mor-Barb combination RIA. Also IKEKAWA et al. (1969) found morphine detectable in urine for 72–96 h using a GC method with hydrolysis and trimethylsilyl derivatization. Physiologic false positives based on analysis of predrug control urines in the heroin and morphine validity studies were: 0% for OS, IE, XAD-2, IEH, all EMIT's, and all RIA's; 0.8% for XAD-2H; 1% for FRAT; 2.5% for OSH; and, 3% for TECH. False positives were not evaluated by GC.

Several extensive clinical studies of chronic heroin users have been reported by CATLIN (1973b), CATLIN et al. (1973a, b). A total of 75–110 known heroin users each gave one urine sample (and, in some cases, also a blood sample) and estimated the time since last heroin self-administration; no estimate of dose level was attempted. Urine samples were analyzed under blind conditions by RIA using ^3H (with an 0.025 μg/ml cutoff sensitivity), FRAT (with 0.5 μg/ml cutoff), hemagglutination inhibition (HI)(with 0.025 μg/ml cutoff), the Technicon Autoanalyzer (with sensitivity = 0.2 μg/ml), and TLC preceded by organic solvent extraction without hydrolysis (sensitivity = 0.5–1 μg/ml) (DAVIDOW et al., 1968). Using TLC and the Technicon Autoanalyzer there were greater than 50% positives for only the first 12–24 h following estimated time of last heroin administration. FRAT gave a high proportion of positives for 48 h. HI and RIA analyses showed 100% positives for 48 h after estimated time of last heroin use and greater than 50% positives for 72 h by RIA and 96 h by HI. Consistent with these studies PARKER et al. (1966) reported only 31% positive urines by TLC without hydrolysis 48 h after the last morphine dose in subjects given a total of 285 mg of morphine over a 5-day period with injection four times per day.

The possible use of biological fluids other than urine for screening for heroin use has been explored. GORODETZKY and KULLBERG (1974a) obtained finger-stick blood samples and saliva from five subjects at various time intervals after administration of single intravenous doses of 2.5, 5, and 10 mg/70 kg heroin. Samples were analyzed by modifications of RIA, FRAT, and EMIT screening methods with sensitivities in the range of 20–30 ng/ml. Following the moderate and high doses of heroin, morphine was detectable in a high proportion of samples for 2–4 h in the plasma and 1–2 h in saliva. During chronic morphine administration (30 mg subcutaneously four times per day for 3 months) morphine was detectable in plasma for at least 6 h and in saliva for 3–4 h after the last morphine dose. It was concluded that the time course of detectability was too short for routine use of plasma or saliva for screening for heroin use, although they could be of some utility in detection of high dose chronic abuse. These results are consistent with the plasma results reported by SPECTOR and VESELL (1971) following 10 mg of morphine, although these authors detected low concentrations of morphine in plasma (below the cutoff levels used in the screening study described above) for up to 48 h in four of five subjects, using a sensitive RIA procedure. CATLIN et al. (1973a, b) used RIA and HI methods with sensitivities of 10 ng/ml to analyze serum samples from known heroin users from whom urine was also obtained (as described above). They found fewer positives in serum than in urine and concluded the urine was the more useful biological fluid for screening for heroin use. LEUTE et al. (1972b) examined saliva and urine from addicts in treatment, but found only a weak correlation between urine and saliva morphine levels. OBERST (1942) reported detection of small amounts of morphine in perspiration of subjects taking large doses of morphine chronically; however, difficulty in obtaining samples would preclude use of this biological fluid in most routine screening.

2. Other Drugs

The majority of screening method validity data for drugs other than morphine and heroin comes primarily from qualitative reports in methodology papers and, in a few instances, from pilot studies in which one or two subjects were administered drug and urine collected and analyzed.

BĚRRY and GROVE (1971) reported detection of 10 mg of oral methadone for 24 h using a TLC method preceded by organic solvent extraction, and KAISTHA and JAFFE (1972b) found a 38 mg dose detectable for 56 h, also using a TLC method. Alpha acetylmethadol, 60 mg, was detected for 71–76 h (KAISTHA and JAFFE, 1972b). HI with a cutoff sensitivity of 30 ng/ml was used by LIU and ADLER (1973) to analyze 24-h urine specimens from an addict who had received a single 50 mg oral methadone dose. They reported positive urines for 12 days following drug administration; urine drug concentrations were below 0.5 µg/ml (a common sensitivity level for most other screening methods) by approximately 72 h. In a metabolic study, INTURRISI and VEREBELY (1972) followed urinary excretion of methadone and its primary metabolite for 96 h following a 10 mg oral dose using a GC method with sensitivity for detection of 15 ng/ml of drug in plasma. However, no urinary concentrations of drug were reported. HENDERSON and WILSON (1973) found methadone in sweat of subjects receiving 70 mg/day of oral methadone.

A 30 mg oral dose of codeine was reported detectable for 30 h by KAISTHA and JAFFE (1972b) using a TLC method; and, using FRAT, LEUTE et al. (1972b) found positive urines for 4 days following three doses of codeine totalling 65 mg. Codeine doses of 8–32 mg were detected for 48–72 h by Ross et al. (1975) with the morphine latex agglutination inhibition test. SOLOMON (1974) analyzed urine samples from five subjects receiving single oral doses of codeine (amount not given) by the TLC method of DAVIDOW et al. (1968). He reported detection of codeine for 72 h and morphine from 48–96 h following drug administration.

Detection of 5 mg/day oral cyclazocine or 50 mg/day oral naltrexone for 24 h by TLC or GLC screening methods was reported by DIGREGARIO and O'BRIEN (1974). A 50 mg oral dose of pentazocine was detected for 60–70 h by KAISTHA and JAFFE (1972b). GORODETZKY and KULLBERG (1975) analyzed urine samples from seven subjects after administration of a highly euphorigenic dose of etorphine (100 µg) using TLC preceded by XAD-2 resin extraction (sensitivity = 0.2 µg/ml), GC with trimethylsilyl derivatization (sensitivity = 0.1 µg/ml), and the morphine immunoassays RIA (Abuscreen) and EMIT. No positive urine specimens were found and it was concluded that it was unlikely that etorphine abuse could be diagnosed using these screening methods.

Using TLC sensitivities, published drug metabolism data, and making assumptions concerning extraction efficiency, GORODETZKY (1972c) predicted the relative detectability of 15 opioids administered in approximately equieuphorigenic doses. The detectability data noted above have thus far been generally consistent with these predictions (e.g., relative detectability of codeine, 65 mg, morphine, 6 mg, and heroin, 2.5 mg). Several comparative studies have been reported in which clinical samples from addicts in treatment programs have been analyzed by several screening methods (MULÉ et al., 1974a; SINE et al., 1974; SOHN et al., 1973). These studies have generally found that the number of positive samples directly correlated with the sensitivity of the method, the greater sensitivity resulting in the larger number of positives. MONTALVO et al. (1972) emphasized the need for cautious interpretation of highly diluted urines with low specific gravity; and EGNER and CAMPBELL (1972), in a theoretical paper, noted the need to consider the time elapsed since last urination in evaluation of probability of detection.

II. Sedative/Hypnotics

The most extensive study of the validity of screening methods to detect barbiturate administration was reported by ROERIG et al. (1975b). Each of six drug-free volunteers received single 100 mg oral doses of each of two barbiturates at a 2-week interval and urine was collected at 4, 8, 12, and 16 h following drug administration, and then at 12-h intervals for 112 h. Secobarbital, pentobarbital, and amobarbital were used in the study. Samples were coded and analyzed under blind conditions by TLC preceded by XAD-2 resin extraction (sensitivity = 0.5 μg/ml), GC (sensitivity = 0.2 μg/ml), and RIA (cutoff sensitivity = 0.1 μg/ml). Using the TLC method 90% of the samples were positive for 30 h following drug administration and there were greater than 50% positives for 88 h. GC and RIA gave 100% positives for 52 and 76 h postdrug, respectively; and both methods showed greater than 50% positives for 112 h.

Clinical reports of barbiturate detection have been generally consistent with the experimental results. Barbiturate doses in the dose range of 100–200 mg have usually been reported to be detectable with high probability for 3–4 days by several screening methods, such as UV (BLACKMORE et al., 1971), RIA (CLEELAND et al., 1975), the Mor-Barb combination RIA (USATEGUI-GOMEZ et al., 1975), TLC using organic solvent extraction (HEATON and BLUMBERG, 1969), and XAD-2 resin extraction (ROERIG et al., 1975a). A 30 mg phenobarbital dose was reported detectable for 72 h by RIA (CLEELAND et al., 1975) and for 24 h by TLC (KAISTHA and JAFFE, 1972b); in the latter paper 60 mg of secobarbital was reported detected for 36–40 h. PEHR (1975) used a colorimetric screening test and reported positive urines at concentrations greater than 2 μg/ml for 7 days following administration of a 400 mg dose of phenobarbital. KOKOSKI et al. (1974) reported detection of positive urines for 8–9 days following a single 5 mg dose of diazepam (Valium) using a TLC method preceded by conversion of the drug metabolites to a detectable hydrolytic product.

WALBERG (1974) analyzed urine by the EMIT barbiturate assay and serum by a UV method in patients suspected of barbiturate overdose. He found that a barbiturate level in the urine greater than 1.5 μg/ml was as good an indicator as a serum level greater than 3 μg/ml of the involvement of a barbiturate as an etiologic agent in the comatose patient. He concluded that the EMIT assay could be a useful adjunct for rapid toxicologic analysis for barbiturates.

III. Stimulants

Several studies have reported urine concentrations of amphetamine and methamphetamine following drug administration to a small number of subjects. CARTONI and CAVALLI (1968), using a combination of TLC and GC, reported urine concentrations of amphetamine in one subject after administration of 10 mg of amphetamine. They found greater than 1 μg/ml for 24 h, between 0.1 and 1 μg/ml from 24–32 h, and less than 0.1 μg/ml (but detectable by their method) to 72 h. RUTTER (1972) studied the same amphetamine dose using an automated colorimetric method and also found greater than 1 μg/ml urine concentration for 24 h and greater than 0.4 μg/ml for 48 h. Doses of 10 mg of methamphetamine were administered to three subjects by LEBISH et al. (1970) and urine concentrations greater than 1 μg/ml were found for 24–36 h. The clinical reports of validity data have generally been consistent with these urine concentrations. Doses of 5–20 mg of amphetamine and methamphetamine and 8 mg of phenmetrazine have been reported to be detectable for approximately 24 h using TLC (KAISTHA and JAFFE, 1971 and 1972b), GC (BERRY and GROVE, 1971), and fluorometry (MONTFORTE et al., 1972). BECKETT et al. (1967) reported detection of therapeutic doses of amphetamines and ephedrines for 48 h using GC methods. BECKETT et al. (1969) also pointed out the great variability in amphetamine excretion in the urine with alteration in urinary pH; a low pH favored urinary excretion. Excretion of amphetaminelike compounds in sweat was reported by VREE et al. (1972), who suggested possible use of this biological fluid to detect amphetamine use after athletic performance.

FISH and WILSON (1969a) analyzed urine from a subject during chronic cocaine use (120 mg/day) and reported very low concentrations of free cocaine (at times not detectable) varying with urinary pH; the major urinary excretion product was benzoylecgonine. Clinical reports have been in agreement with this study. BERRY and GROVE (1971) reported detection of free cocaine in urine for only 2 h following a 20 mg oral dose, using a GC method with sensitivity of 0.1 μg/ml. VALANJU et al. (1973) examined 1000 clinical urine samples in which 15% were found positive for benzoylecgonine while only 1% were positive for cocaine. BASTOS et al. (1974) found 9% benzoylecgonine positives in 3000 urine samples using EMIT, with the majority confirmed by TLC. SCHNEIDER et al. (1974) analyzed the urine of five subjects following self-administration of an undetermined cocaine dose using FRAT, with a detection sensitivity of 1 μg/ml. They reported benzoylecgonine detectable for 24–48 h after cocaine administration.

IV. Other Drugs

TAUNTON-RIGBY et al. (1973) analyzed urine for 24 h after LSD administration in doses of 200–400 μg to eight subjects, using an RIA method. All but one postdrug sample were positive for LSD at concentrations of 1.5–55 μg/ml. FAED and MCLEOD (1973) detected LSD in urine using paper chromatography and fluorescence 4 days after ingestion of 200 μg of the drug by one subject. Urine and plasma were analyzed by a THC RIA in four subjects following smoking of a cigarette containing 5 mg THC by TEALE et al. (1974b). Plasma levels were detectable for 2–3 h in three or four subjects and urines were positive at concentrations above 5 ng/ml for 48 h following THC administration. CHRISTIANSEN and RAFAELSON (1969), using TLC preceded by organic solvent extraction of enzyme-hydrolyzed urine detected cannabis metabolites for 7 h following ingestion of 750 mg of cannabis resin. ANDERSEN et al. (1971) reported similar results, detecting oral use of 300 mg of cannabis resin containing 14 mg of THC for 6–7 h in the urine of one volunteer subject using organic solvent extraction and TLC.

Validity for quinine detection has been reported in several primarily methodologic papers. DOLE et al. (1966) reported that the amount of quinine usually found in a bag of heroin could be detected in urine for 5–14 days using TLC preceded by ion exchange paper extraction. The same detection time was noted by FISHER et al. (1972). MULÉ and HUSHIN (1971) used an ATS fluorometric method to analyze morning urines for 12 days following ingestion of 325 mg of quinine by three volunteers. Quinine was detected for 10–11 days following drug administration.

E. Summary

Biological fluid screening for drugs of abuse is an important and useful adjunct in the initial detection of drug abuse, monitoring patients for illicit drug use during long-term drug abuse treatment regimens, and diagnosis of drug overdose. Depending on the primary purpose of the testing and the particular desire to the tester, requirements may differ with regard to such test parameters as cost, speed, complexity, sensitivity, specificity, and validity. For example, treatment programs frequently require inexpensive tests, simple to perform, and capable of analyzing a large number of samples with relatively rapid results (sometimes requesting at least provisional results within a few minutes); at least moderate sensitivity, specificity, and validity are usually necessary, although the particular clinical desires of the program may show great variability. For hospital toxicologic analysis, cost and complexity are usually less important, and volume of testing is lower; rapid results are important; great sensitivity is usually not critical to diagnosis of overdose; and good specificity is desirable.

Thin-layer chromatography has low start-up costs, moderate to low cost per sample, and provides sufficiently rapid results for many needs. It has moderate sensitivity and validity for most drugs of abuse, and good specificity. Experienced personnel are required, especially in final reading and interpretation of the TLC plates. Gas chromatography is generally moderate in cost, both for initial instrument purchase and in operating cost per sample; it is capable of high sensitivity and good specificity, but is complex and not useful for achieving rapid results or high volume output when used for maximal sensitivity and specificity. Some rapid, high volume gas chromatographic methods have been described; however, they generally require special and expensive instrumentation and achieve only moderate sensitivity and specificity. Fluorometric methods are somewhat limited in scope for analysis of drugs of abuse, but have some selective use, especially in automated or semiautomated testing for morphine and, perhaps, quinine. Instrumentation is moderate to expensive and cost per sample is moderate; rapid results can be achieved as well as high volume output (especially with fully automated procedures). Sensitivity and specificity are moderate, and some experience is necessary to operate the instrumentation and interpret the records. In general, the immunoassays have similar and relatively poor specificity, compared to the other screening methods. Results are usually interpreted as positive or negative for a group of drugs with greater or lesser cross reactivity depending on the drug group and, to a degree, on the assay. Cost per sample is relatively high due to the expense of the special test reagents and the need for separate reagents for each drug group tested. As a group, the immunoassays are simple to perform, amenable to automation, and, with one exception (hemagglutination inhibition) have objective, easily interpretable end points. Radioimmunoassay and hemagglutination inhibition are both capable of high sensitivity. RIA instrumentation is moderate to expensive, while HI has very low start-up costs; however, HI requires subjective judgment to determine the final test result, which requires experienced personnel and can be a source of error and test variability. The free radical assay technique and the homogeneous enzyme immunoassay have moderate sensitivity (in the same range as TLC); however, they are capable of very rapid analysis of a single sample (only several minutes when the assay is set up and calibrated), more rapid than any of the other screening tests. FRAT requires a very expensive and specialized instrument; EMIT has only moderate start-up costs.

In the last 5 years much progress has been made in the development of both screening and quantitative analytical methods for drugs of abuse in biological fluids; and a wide range of capabilities is now available. Continued methodologic development is likely, with increases in sensitivity and specificity, decreasing cost, and greater automation and simplicity of performance. A broader range of more specific, sensitive, and quantitative immunoassays are likely to be available. Additional systematic clinical and experimental human drug metabolic and excretion data is needed to improve interpretation of test results. Also possible use of other biological samples, such as breath, and more extensive use of quantitative analyses for drugs and their metabolites to aid in more precise diagnosis of drug-related conditions needs to be further explored.

References

Adams, R. F.: A drug screening procedure utilizing gas liquid chromatography. Clin. Chem. Appl., Study No. 41 (Perkin-Elmer Corp.) 1971

Adams, R. F., Purcell, J. E., Ettre, L. S.: Rapid drug analysis in biological samples by gas chromatography. Amer. Lab. **5**, 51—60 (1973)

Adler, F. L., Liu, C.-T.: Detection of morphine by hemagglutination-inhibition. J. Immunol. **106**, 1684—1685 (1971)

Adler, F. L., Liu, C.-T., Catlin, D. H.: Immunological studies on heroin addiction. I. Methodology and application of a hemagglutination-inhibition test for detection of morphine. Clin. Immunol. Immunopath. **1**, 53—68 (1972)

Aggarwal, V., Bath, R., Sunshine, I.: Technique for rapidly separating drugs from biological samples. Clin. Chem. **20**, 307—309 (1974)

Aghajanian, G. K., Bing, O. H. L.: Persistence of lysergic acid diethylamide in the plasma of human subjects. Clin. Pharmacol. Ther. **5**, 611—614 (1974)

Anders, M. W., Mannering, G. J.: New peak-shift technique for gas-liquid chromatography. Preparation of derivatives on the column. Analyt. Chem. **34**, 730—733 (1962)

Andersen, M. J., Nielsen, E., Schou, J., Steentoft, A., Worm, K.: A specific method for the demonstration of cannabis intake by TLC of urine. Acta pharmacol. (Kbh.) **29**, 111—112 (1971)

Axelrod, J., Brady, R. O., Witkop, G., Evarts, E.: The distribution and metabolism of lysergic acid diethylamide. Ann. N.Y. Acad. Sci. **66**, 435—444 (1957)

Baker, Jr., S. L.: U.S. Army heroin abuse identification program. Amer. J. publ. Hlth **62**, 857—860 (1972)

Balatre, P., Traisnel, M., Delcambre, J. P.: Microdetermination of codeine and codethyline by their fluorescence. Ann. pharm. franç. **19**, 171—174 (1961)

Barrett, M. J.: An integrated gas chromatographic program for drug screening in serum and urine. Clin. Chem. Newsletters **3**, 1—10 (1971)

Baselt, R. C., Casarett, L. J.: Detection of drugs in urine for methadone treatment programs. J. Chromatog. **57**, 139—141 (1971)

Baselt, R. C., Wright, J. A., Cravey, R. H.: Therapeutic and toxic concentrations of more than 100 toxicologically significant drugs in blood, plasma, or serum: a tabulation. Clin. Chem. **21**, 44—62 (1975)

Bastiani, R. J., Phillips, R. C., Schneider, R. S., Ullman, E. F.: Homogeneous immunochemical drug assays. Amer. J. med. Technol. **39**, 211—216 (1973)

Bastos, M. L., Jukofsky, D., Saffer, E., Chedekel, M., Mulé, S. J.: Modifications of the XAD-2 resin column method for the extraction of drugs of abuse from human urine. J. Chromatog. **71**, 549—553 (1972)

Bastos, M. L., Jukofsky, D., Mulé, S. J.: Routine identification of drugs of abuse in human urine. III. Differential elution of the XAD-2 resin. J. Chromatog. **81**, 93—98 (1973)

Bastos, M. L., Jukofsky, D., Mulé, S. J.: Routine identification of cocaine metabolites in human urine. J. Chromatog. **89**, 335—342 (1974)

Bastos, M. L., Kananen, G. E., Young, R. M., Monforte, J. R., Sunshine, I.: Detection of basic organic drugs and their metabolites in urine. Clin. Chem. **16**, 931—940 (1970)

Beckett, A. H., Moffat, A. C.: Routine detection and identification in urine of stimulants, analgesics, antihistamines, local anaesthetics and other drugs, some of which may be used to modify performance in sport. J. Pharm. (Lond.) **20** Suppl. 485—505 (1968)

Beckett, A. H., Rowland, M.: Determination and identification of amphetamine in urine. J. Pharm. (Lond.) **17**, 59—60 (1965)

Beckett, A. H., Salmon, J. A., Mitchard, M.: The relation between blood levels and urinary excretion of amphetamine under controlled acidic and under fluctuating urinary pH values using [^{14}C] amphetamine. J. Pharm. (Lond.) **21**, 251—258 (1969)

Beckett, A. H., Tucker, G. T., Moffat, A. C.: Routine detection and identification in urine of stimulants and other drugs, some of which may be used to modify performance in sport. J. Pharm. (Lond.) **19**, 273—294 (1967)

Berman, A. R., McGrath, J. P., Permisohn, R. C., Cella, J. A.: Radioimmunoassay of methaqualone and its monohydroxy metabolites in urine. Clin. Chem. **21**, 1878—1881 (1975)

Berry, D. J., Grove, J.: Improved chromatographic techniques and their interpretation for the screening of urine from drug-dependent subjects. J. Chromatog. **61**, 111—123 (1971)

Berry, D. J., Grove, J.: Emergency toxicological screening for drugs commonly taken in overdose. J. Chromatog. **80**, 205—219 (1973)

Bidanset, J. H.: Drug analysis by immunoassays. J. chromatog. Sci. **12**, 293—296 (1974)

Billets, S., Carruth, J., Einolf, N., Ward, R., Fenselau, C.: Rapid identification of acute drug intoxications. Johns Hopk. med. J. **133**, 148—155 (1973)

Blackmore, D. J., Curry, A. S., Hayes, T. S., Rutter, E. R.: Automated analysis for drugs in urine. Clin. Chem. **17**, 896—902 (1971)

Blake, J. W., Huffman, R., Noonan, J., Ray, R.: GLC and the electron capture detector, a screening procedure for drugs. Amer. Lab. **5**, 63—67 (1973)

Blake, J. W., Ray, R. S., Noonan, J. S., Murdick, P. W.: Rapid, sensitive, gas-liquid chromatographic screening procedure for cocaine. Analyt. Chem. **46**, 288—289 (1974)

Blass, K. G., Thibert, R. J., Draisey, T. F.: A simple, rapid thin-layer chromatographic drug screening procedure. J. Chromatog. **95**, 75—79 (1974)

Blumberg, A. G., Heaton, A. M.: Control of drug abuse in a psychiatric hospital by random urine analyses. Clin. Toxicol. **6**, 217—227 (1973)

Boerner, U., Abbott, S., Eidson, J. C., Becker, C. E., Horio, H. T., Loeffler, K.: Direct mass spectrometric analysis of body fluids from acutely poisoned patients. Clin. chim. Acta **49**, 445—454 (1973)

Bowen, D. A. L., Gurr, D. M.: Thin layer chromatographic laboratory analysis in cases from a drug addiction center. Clin. Toxicol. **3**, 89—95 (1970)

Brandt, M. K.: Thin-layer chromatography as a method of screening for narcotics usage. Amer. J. med. Technol. **39**, 217—222 (1973)

Brandt, R., Ehrlich-Rogozinsky, S., Cheronis, N. D.: Spectrophotofluorometric method for the microdetection and estimation of morphine and codeine. Microchem. J. **5**, 215—223 (1961)

Brattin, W. J., Sunshine, I.: Immunological assays for drugs in biological samples. Amer. J. med. Technol. **39**, 223—230 (1973)

Brattin, W. J., Sunshine, I.: A comparison of available immunoassays for drugs of abuse in urine. In: Mulé, S. J., Sunshine, I., Braude, M., Willette, R. E. (Eds.): Immunoassays for drugs subject to abuse, pp. 107—116. Cleveland: CRC Press 1974

Breimer, D. D., van Rossum, M.: Rapid and sensitive gas chromatographic determination of hexobarbital in plasma of man using a nitrogen detector. J. Chromatog. **88**, 235—243 (1974)

Brinkmann Product Brochure: Brinkmann drug-skreen. Drug screening system for the analysis of commonly abused drugs in urine

Brochmann-Hanssen, E., Oke, T. O.: Gas chromatography of barbiturates, phenolic alkaloids, and xanthine bases: flash-heater methylation by means of trimethylanilinium hydroxide. J. pharm. Sci. **58**, 370—371 (1969)

Broich, J. R., Hoffman, D. B., Andryauskas, S., Galante, N., Umberger, C. J.: An improved method for rapid, large-scale thin-layer chromatographic urine screening for drugs of abuse. J. Chromatog. **60**, 95—101 (1971a)

Broich, J. R., Hoffman, D. B., Goldner, S. J., Andryauskas, S., Umberger, C. J.: Liquid-solid extraction of lyophilized biological material for forensic analysis. I. Application to urine samples for detection of drugs of abuse. J. Chromatog. **63**, 309—312 (1971b)

Broughton, A., Ross, D. L.: Drug screening by enzymatic immunoassay with the centrifugal analyzer. Clin. Chem. **21**, 186—189 (1975)

Brown, S. S., Smart, C. A.: Fluorometric assay of methaqualone in plasma by reduction to 1,2,3,4-tetrahydro-2-methyl-4-oxo-3-0-tolylquinazoline. J. Pharm. (Lond.) **21**, 466—468 (1969)

Bussey, R. J., Backer, R. C.: Thin-layer chromatographic differentiation of amphetamine from other primary-amine drugs in urine. Clin. Chem. **20**, 302—304 (1974)

Butler, V. P.: Radioimmunoassay and competitive binding radioassay methods for the measurement of drugs. Metabolism **22**, 1145—1153 (1973)

Cartoni, G. P., Cavalli, A.: Detection of doping by thin-layer and gas chromatography. J. Chromatog. **73**, 158—161 (1968)

Cashman, P. J., Thornton, J. I., Shelman, D. L.: High pressure liquid chromatographic separation of phenethylamines of forensic interest. J. chromatog. Sci. **11**, 7—9 (1973)

Castro, A., Grettie, D. P., Bartos, F., Bartos, D.: LSD radioimmunoassay. Res. Commun. chem. Path. Pharmacol. **6**, 879—886 (1973)

Catlin, D. H.: A guide to urine testing for drugs of abuse, Special Action Office Monograph, Series B, No. 2. Washington: Special Action Office for Drug Abuse Prevention 1973a

Catlin, D. H.: A comparison of five current methods for detecting morphine. Amer. J. clin. Path. **60**, 719—728 (1973b)

Catlin, D. H., Adler, F. L., Liu, C.-T.: Immunological studies on heroin addiction. II. Applications of a sensitive hemagglutination-inhibition test for detecting morphine to diagnostic problems in chronic heroin addiction. Clin. Immunol. Immunopathol. **1**, 446—455 (1973a)

Catlin, D., Cleeland, R., Grunberg, E.: A sensitive, rapid radioimmunoassay for morphine and immunologically related substances in urine and serum. Clin. Chem. **19**, 216—220 (1973b)

Chan, M. L., Whetsell, C., McChesney, J. D.: Use of high pressure liquid chromatography for the separation of drugs of abuse. J. chromatog. Sci. **12**, 512—516 (1974)

Cheng, L. T., Kim, S. Y., Chung, A., Castro, A.: Amphetamines: new radioimmunoassay. FEBS Lett. **36**, 339—342 (1973)

Christiansen, J., Rafaelson, O. J.: Cannabis metabolites in urine after oral administration. Psychopharmacologia **15**, 60—63 (1969)

Christopoulos, G. N., Chen, N. W., Toman, A. J.: Isolation and identification of methaqualone from post-mortem tissues. J. chromatog. Sci. **12**, 267—268 (1974)

Chung, A., Kim, S. Y., Cheng, L. T., Castro, A.: Phenobarbital specific antisera and radioimmunoassay. Experientia (Basel) **29**, 820—821 (1973)

Cleeland, R., Davis, R., Heveran, J., Grunberg, E.: A simple, rapid ^{125}I radio-immunoassay for the detection of barbiturates in biological fluids. J. forens. Sci. **20**, 45—57 (1975)

Cochin, J.: Analysis for narcotic analgesics and barbiturates in urine by thin-layer chromatographic techniques without previous extraction and concentration. Psychopharmacol. Bull. **3**, 53—60 (1966)

Cochin, J., Daly, J. W.: Rapid identification of analgesic drugs in urine with thin-layer chromatography. Experientia (Basel) **18**, 294—298 (1962)

Cochin, J., Daly, J. W.: The use of thin-layer chromatography for the analysis of drugs. Isolation and identification of barbiturates and nonbarbiturate hypnotics from urine, blood, and tissues. J. Pharmacol. exp. Ther. **139**, 154—159 (1963)

Coleman Instruments Division of Perkin-Elmer Corporation: Characteristics of narcotic and dangerous drugs; their determination by ultraviolet spectrophotometry. Applications Data Sheet, CLIN-1A, (1970)

Cone, E. J., Gorodetzky, C. W., Yeh, S. Y.: The urinary excretion profile of naltrexone and metabolites in man. Drug Metab. Dispos. **2**, 506—512 (1974)

Cooper, R. G., Greaves, M. S., Owen, G.: Gas-liquid chromatographic isolation, identification, and quantitation of some barbiturates, glutethimide, and diphenylhydantoin in whole blood. Clin. Chem. **18**, 1343—1349 (1972)

Copenhaver, J. H., Blose, I. L., Carver, M. J.: A micro-thin layer chromatographic technique for the detection of morphine in urine, pp. 329—331. Proc. of the Fourth Nat. Conf. on Methadone Treatment, San Francisco, Calif., Jan., 1972
Coumbis, R. J., Kaul, B.: Distribution of morphine and related compounds in human tissues and biological fluids using radioimmunoassay techniques. J. forens. Sci. **19**, 307—312 (1974)
Cravey, R. H., Jain, N. C.: Current status of blood alcohol levels. J. chromatog. Sci. **12**, 209—213 (1974a)
Cravey, R. H., Jain, N. C.: The identification of non-barbiturate hypnotics from biological specimens. J. chromatog. Sci. **12**, 237—245 (1974b)
Dahlstrom, B., Paalzow, L.: Quantitative determination of morphine in biological samples by gas-liquid chromatography and electron-capture detection. J. Pharm. (Lond.) **27**, 172—176 (1975)
Dal Cortivo, L. A., Broich, J. R., Dihrberg, A., Newman, B.: Identification and estimation of lysergic acid diethylamide by thin-layer chromatography and flourometry. Analyt. Chem. **38**, 1959—1960 (1966)
Dal Cortivo, L. A., De Mayo, M. M., Weinberg, S. B.: Fluorometric determination of microgram amounts of meperidine. Analyt. Chem. **42**, 941—942 (1970a)
Dal Cortivo, L. A., Kallet, E., Matusiak, W.: Semiautomated fluorometric system for detecting morphine in urine, pp. 6391—6402. Proc. of the NAS-NRC Committee on Problems of Drug Dependence, Washington, D.C., May, 1970b
Davidow, B., Petri, L. N., Quame, B.: A thin-layer chromatographic screening procedure for detecting drug abuse. Amer. J. clin. Path. **50**, 714—719 (1968)
Davidow, B., Petri, L. N., Quame, B., Searle, B., Fastlich, E., Savitzky, J.: A thin-layer chromatographic screening test for the detection of users of morphine or heroin. Amer. J. clin. Path. **46**, 58—62 (1966)
Davidow, B., Quame, B., Abell, L. L., Lim, B.: Screening for drug abuse. Hlth Lab. Sci. **10**, 329—334 (1973)
Decker, W. J., Lewis, L. R.: Evaluation of Screendex—a multireagent system for detection of drug abuse and drug overdose. Clin. Toxicol. **6**, 201—209 (1973)
DeSilva, J. A. F., Bekersky, I., Puglisi, C. V., Brooks, M. A., Weinfeld, R. E.: Determination of 1,4-benzodiazepines and -diazepin-2-ones in blood by electron-capture gas-liquid chromatography. Analyt. Chem. **48**, 10—19 (1976)
DeSilva, J. A. F., Schwartz, M. A., Stefanovic, V., Kaplan, J., D'Arconte, L.: Determination of diazepam (Valium) in blood by gas chromatography. Analyt. Chem. **36**, 2099—2104 (1964)
Digregario, G. J., O'Brien, C.: Chromatographic detection of narcotic antagonists in human urine. J. Chromatog. **101**, 424—427 (1974)
Doedens, D. J., Forney, R. B.: Confirmation of morphine on thin-layer plates by fluorometry. J. Chromatog. **100**, 225—226 (1974)
Dole, V. P., Crowther, A., Johnson, J., Monsalvatge, M., Biller, B., Nelson, S. S.: Detection of narcotic, sedative, and amphetamine drugs in urine. N.Y. St. J. Med. **72**, 471—476 (1972)
Dole, V. P., Kim, W. K., Eglitis, I.: Detection of narcotic drugs, tranquilizers, amphetamines, and barbiturates in urine. J. Amer. med. Ass. **198**, 349—352 (1966)
Douglas, J. F., Shahinian, S.: GLC determination of methaqualone in plasma. J. pharm. Sci. **62**, 835—836 (1973)
Down, G. J., Gwyn, S. A.: Investigation of direct thin-layer chromatography—mass spectrometry as a drug analysis technique. J. Chromatog. **103**, 208—210 (1975)
Driscoll, R. C., Barr, F. S., Gragg, B. J., Moore, G. W.: Determination of therapeutic blood levels of methamphetamine and pentobarbital by GC. J. pharm. Sci. **60**, 1492—1495 (1971)
Duddleson, W. G., Midgley, A. R., Niswender, G. D.: Computer program sequence for analysis and summary of radioimmunoassay data. Comp. biomed. Res. **5**, 205—217 (1972)
Dvorchik, B. H.: Gas chromatographic method for microdetermination of barbiturates in blood using a nitrogen-selective flame ionization detector. J. Chromatog. **105**, 49—56 (1975)
Egner, D. O., Campbell, D.: Significance of detector sensitivity in detection of drug abusers. Technical Report No. 72—07. U.S. Army Land Warfare Laboratory, Aberdeen Proving Ground, Maryland 21005 (1972)
Ehrsson, H.: Gas chromatographic determination of barbiturates after extractive methylation in carbon disulfide. Analyt. Chem. **46**, 922—924 (1974)

Elliott, H. W., Parker, K. D., Wright, J. A., Nomof, N.: Actions and metabolism of heroin administered by continuous intravenous infusion to man. Clin. Pharmacol. Ther. **12**, 806—814 (1971)
EM Laboratories Product Brochure: Drug abuse detection using pre-coated TLC plates.
Erlanger, B. F.: Principles and methods for the preparation of drug protein conjugates for immunological studies. Pharmacol. Rev. **25**, 271—280 (1973)
Evenson, M. A., Koellnel, S.: Rapid method for quantitative determination of propoxyphene in serum by gas-liquid chromatography. Clin. Chem. **19**, 492—495 (1973)
Evenson, M. A., Lensmeyer, G. L.: Qualitative and quantitative determination of methaqualone in serum by gas chromatography. Clin. Chem. **20**, 249—254 (1974)
Evenson, M. A., Poquette, M. A.: Rapid gas chromatographic method for quantitation of ethchlorvynol ("Placidyl") in serum. Clin. Chem. **20**, 212—216 (1974)
Faed, E. M., McLeod, W. R.: A urine screening test for lysergide (LSD-25). J. chromatog. Sci. **11**, 4—6 (1973)
Fales, H. M., Pisano, J. J.: Gas chromatography of biologically important amines. Analyt. Biochem. **3**, 337—342 (1962)
Farmilo, C. G.: The physical methods for the identification of narcotics (cont.) Part IIIA. The ultraviolet spectrophotometric method. Bull. Narcot. **6 (3—4)**, 18—41 (1954)
Fenimore, D. C., Davis, C. M.: 18. Rapid screening of urine for detection of narcotic drugs. In: Harris, R. T., McIsaac, W. M., Schuster, C. R. (Eds.): Drug Dependence, pp. 242—250. Austin: University of Texas Press 1970
Finkle, B. S., Cherry, E. J., Taylor, D. M.: A GLC based system for the detection of poisons, drugs, and human metabolites encountered in forensic toxicology. J. chromatogr. Sci. **9**, 393—419 (1971)
Finkle, B. S., Foltz, R. L., Taylor, D. M.: A comprehensive GC-MS reference data system for toxicological and biomedical purposes. J. chromatog. Sci. **12**, 304—328 (1974)
Finkle, B. S., Taylor, D. M., Bonelli, E. J.: A GC/MS reference data system for the identification of drugs of abuse. J. chromatog. Sci. **10**, 312—332 (1972)
Fish, F., Wilson, W. D. C.: Excretion of cocaine and its metabolites in man. J. Pharm. (Lond.) **21**, 135S—138S (1969a)
Fish, F., Wilson, W. D. C.: Gas chromatographic determination of morphine and cocaine in urine. J. Chromatog. **40**, 164—168 (1969b)
Fisher, W. T., Baitsholts, A. D., Grau, G. S.: Use of precoated, flexible thin-layer sheets in drug screening. J. chromatog. Sci. **10**, 303—311 (1972)
Flanagan, R. J., Withers, G.: A rapid micro-method for the screening and measurement of barbiturates and related compounds in plasma by gas-liquid chromatography. J. clin. Path. **25**, 899—904 (1972)
Flynn, E. J., Spector, S.: Determination of barbiturate derivatives by radioimmunoassay. J. Pharmacol. exp. Ther. **181**, 547—554 (1972)
Frings, C. S., Queen, C. A.: Interpretation of drug abuse screening tests on urine from persons receiving various medications. Clin. Chem. **18**, 713 (1972a)
Frings, C. S., Queen, C. A.: Preparation and use of a urine control for certain drugs of abuse. Clin. Chem. **18**, 1440 (1972b)
Frings, C. S., Queen, C. A.: Stability of certain drugs of abuse in urine specimens. Clin. Chem. **18**, 1442 (1972c)
Fry, D. E., Wills, P. D., Twycross, R. G.: The quantitative determination of morphine in urine by gas-liquid chromatography and variations in excretion. Clin. chim. Acta **51**, 183—190 (1974)
Fujimoto, J. M., Wang, R. I. H.: A method of identifying narcotic analgesics in human urine after therapeutic doses. Toxicol. appl. Pharmacol. **16**, 186—193 (1970)
Fulton, C. C.: Sulfomorphid; and the purple fluorescence test, a new derivative test for morphine. J. Amer. med. Ass. **26**, 726—729 (1937)
Fulton, C. C.: Modern microcrystal tests for drugs. The identification of organic compounds by microcrystallographic chemistry. New York-London: Wiley Interscience 1969
Garrett, E. R., Hunt, C. A.: Picogram analysis of tetrahydrocannabinol and application to biological fluids. J. pharm. Sci. **62**, 1211—1214 (1973)
Gearing, F. R.: People versus urines, pp. 325—326. Proc. of the Fourth Nat. Conf. on Methadone Treatment, San Francisco, Calif., Jan., 1972

Gelman Drug Identification Systems: Procedure manual. Gelman Instrument Co., Publication M2001-173
G.K. Turner Associates: Drugs. A bibliography of fluorometric methods of analysis. Flourometry Review, Acc. No. 11429 (1971)
Goehl, T.J., Davison, C.: GLC determination of meperidine in blood plasma. J. pharm. Sci. **62**, 907—909 (1973)
Goenechea, S., Bernhard, W.: Über die Störung des Morphiumnachweises bei der dünnschichtchromatographischen Untersuchung von Raucherurinextrakten durch Nicotin. A. Analyt. Chem. **246**, 130—132 (1969)
Goldbaum, L.: An ultraviolet spectrophotometric procedure for the determination of barbiturates. J. Pharmacol. exp. Ther. **94**, 68—75 (1948)
Goldbaum, L.: Determination of barbiturates. Ultraviolet spectrophotometric method with differentiation of several barbiturates. Analyt. Chem. **24**, 1604—1607 (1952)
Goldbaum, L.R., Santinga, P., Dominguez, A.M.: A procedure for the rapid analysis of large numbers of urine samples for drugs. Clin. Toxicol. **5**, 369—379 (1972)
Goldberg, P.: The uses and abuses of urinalysis. In: Senay, E., Shorty, V., Alkane, H. (Eds.): Developments in the field of drug abuse, pp. 931—938. Cambridge: Schenkman Publishing Comp. 1975
Goldstein, A., Brown, B.W.: Urine testing schedules in methadone maintenance treatment of heroin addiction. J. Amer. med. Ass. **214**, 311—315 (1970)
Goldstein, A., Judson, B.A.: Three critical issues in the management of methadone programs. In: Bourne, P.G. (Ed.): Addiction, pp. 130—148. New York: Academic Press 1974
Gorodetzky, C.W.: Urinalysis: practical and theoretical consideration, pp. 155—156. Proc. of the Fourth Nat. Conf. on Methadone Treatment, San Francisco, Calif., Jan., 1972a
Gorodetzky, C.W.: Validity of urine tests in monitoring drug abuse, pp. 53—62. Proc. of the NAS-NRC Committee on Problems of Drug Dependence, Ann Arbor, Mich., May, 1972b
Gorodetzky, C.W.: Sensitivity of thin-layer chromatography for detection of 16 opioids, cocaine, and quinine. Toxicol. appl. Pharmacol. **23**, 511—518 (1972c)
Gorodetzky, C.W.: Efficiency and sensitivity of two common screening methods for detecting morphine in urine. Clin. Chem. **19**, 753—755 (1973a)
Gorodetzky, C.W.: Time course of morphine (M) detection in human urine after IV morphine. Fed. Proc. **32**, 764 (1973b)
Gorodetzky, C.W.: Time course of morphine (M) detection in human urine after IV heroin (H) by EMIT and RIA-I^{125}. Fed. Proc. **34**, 814 (1975)
Gorodetzky, C.W., Angel, C.R., Beach, D.J., Catlin, D.H., Yeh, S.Y.: Validity of screening methods for drugs of abuse in biological fluids. I. Heroin in urine. Clin. Pharmacol. Ther. **15**, 461—472 (1974)
Gorodetzky, C.W., Kullberg, M.P.: Validity of screening methods for drugs of abuse in biological fluids. II. Heroin in plasma and saliva. Clin. Pharmacol. Ther. **15**, 579—587 (1974a)
Gorodetzky, C.W., Kullberg, M.P.: Time course of morphine (M) detection in human urine after IV heroin (H) by XAD-2 resin and TLC. Pharmacologist **16**, 193 (1974b)
Gorodetzky, C.W., Kullberg, M.P.: Etorphine in man. II. Detectability in urine by common screening methods. Clin. Pharmacol. Ther. **17**, 273—276 (1975)
Grant, J.D., Gross, S.J., Lomax, P., Wong, R.: Antibody detection of marihuana. Nature (Lond.) New Biol. **236**, 216—217 (1972)
Greeley, R.H.: New approach to derivatization on gas-chromatographic analysis of barbiturates. Clin. Chem. **20**, 192—194 (1974)
Green, D.E., Littlejohn, D.P.: Automated analysis of biological specimens using a compound-specific detector. Proc. West. Pharmacol. Soc. **16**, 226—230 (1973)
Grevert, P., Weinberg, A.: A controlled study of the clinical effectiveness of urine test results in a methadone maintenance program, pp. 1052—1059. Proc. of the Fifth Nat. Conf. on Methadone Treatment, Washington, D.C., March, 1973
Gross, S.J., Soares, J.R., Wong, S.-L.R., Schuster, R.E.: Marijuana metabolites measured by a radioimmune technique. Nature (Lond.) **252**, 581—582 (1974)
Gupta, R.N., Chttim, B.G., Keane, P.M.: Screening for the major methadone metabolite and methamphetamine in the urine. J. chromatog. Sci. **12**, 67—70 (1974)

Hammer, R. H., Templeton, J. L., Panzik, H. L.: Definitive GLC method of identifying cocaine. J. pharm. Sci. **63**, 1963—1965 (1974)
Hannah, R. W., Pattacini, S. C.: The identification of drugs from their infrared spectra. Perkin Elmer Infrared Applications Study 11 (1972)
Hansen, A. R., Fisher, L. J.: Gas-chromatographic simultaneous analysis for glutethimide and an active hydroxylated metabolite in tissues, plasma, and urine. Clin. Chem. **20**, 236—242 (1974)
Harwood, C. T.: Radioimmunoassay: its application to drugs of abuse. Pharmacology (Basel) **11**, 52—57 (1974)
Hatch, R. C.: Simultaneous identification and quantitation of submicrogram amounts of barbiturates by gas chromatography. Amer. J. vet. Res. **33**, 203—207 (1972)
Hawks, R. L.: Gas chromatographic mass spectrometry in drug screening by immunoassay. In: Mulé, S. J., Sunshine, I., Braude, M., Willette, R. E. (Eds.): Immunoassays for drugs subject to abuse, pp. 73—86. Cleveland: CRC Press 1974
Hayes, T. S.: Automated fluorometric determination of amphetamine in urine. Clin. Chem. **19**, 390—394 (1973)
Heaton, A. M., Blumberg, A. G.: Thin-layer chromatographic detection of barbiturates, narcotics, and amphetamines in urine of patients receiving psychotropic drugs. J. Chromatog. **41**, 367—370 (1969)
Henderson, G. L., Frincke, J., Leung, C. Y., Torten, M., Benjamini, E.: Antibodies to fentanyl. J. Pharmacol. exp. Ther. **192**, 489—496 (1975)
Henderson, G. L., Wilson, B. K.: Excretion of methadone and metabolites in human sweat. Res. Commun. chem. Path. Pharmacol. **5**, 1—8 (1973)
Hetland, L. B., Knowlton, D. A., Couri, D.: A method for the detection of drugs at therapeutic dosages in human urine using adsorption column chromatography and thin-layer chromatography. Clin. chim. Acta **36**, 473—478 (1972)
Higgins, T. N., Taylor, J. D.: Optimization of extraction and separation of narcotics from urine. Clin. Biochem. **7**, 280—289 (1974)
Ho, I. K., Loh, H. H., Way, E. L.: Mini thin-layer chromatography in the detection of narcotics in the urine. Proc. West. Pharmacol. Soc. **14**, 183—186 (1971)
Ho, I. K., Loh, H. H., Way, E. L.: Mini thin-layer chromatography in the detection of narcotics in urine from subjects on a methadone maintenance program. J. Chromatog. **65**, 577—579 (1972)
Hollister, L. E., Kanter, S. L., Clyde, D. J.: Studies of prolonged-action medication. III. Pentobarbital sodium in prolonged-action form compared with conventional capsules: serum levels of drug and clinical effects following acute doses. Clin. Pharmacol. Ther. **4**, 612—618 (1963)
Huang, J. C., Baden, M. M.: Rapid methods of screening micro-quantities of abused drugs from urine samples for microcrystal tests. Clin. Toxicol. **6**, 325—350 (1973)
Ibrahim, G., Andryauskas, S., Bastos, M. L.: Application of amberlite XAD-2 resin for general toxicological analysis. J. Chromatog. **108**, 107—116 (1975)
Ikekawa, N., Takayama, K., Hosoya, E., Oka, T.: Determination of morphine in urine by gas chromatography. Analyt. Biochem. **28**, 156—163 (1969)
Inturrisi, C. E., Verebely, K.: A gas-liquid chromatographic method for the quantitative determination of methadone in human plasma and urine. J. Chromatog. **65**, 361—369 (1972)
Jaffe, J. H., Kirkpatrick, D.: The use of ion-exchange resin impregnated paper in the detection of opiate alkaloids, amphetamine, phenothiazines, and barbiturates in urine. Psychopharmacol. Bull. **3**, 49—52 (1966)
Jain, N. C., Cravey, R. H.: A review of breath alcohol methods. J. chromatog. Sci. **12**, 214—218 (1974a)
Jain, N. C., Cravey, R. H.: The identification of barbiturates from biological specimens. J. chromatog. Sci. **12**, 228—236 (1974b)
Jain, N. C., Leung, W. J., Budd, R. D., Sneath, T. C.: Thin-layer chromatographic separation of methadone and its primary metabolite in the presence of other drugs in urine specimens. J. Chromatog. **103**, 85—90 (1975)
Jain, N. C., Sneath, T. C., Budd, R. D.: Rapid gas-chromatographic determination of amphetamine and methamphetamine in urine. Clin. Chem. **20**, 1460—1462 (1974)
Jang, I., Wheals, B. B.: The characterization of LSD in illicit preparations by pressure-assisted liquid chromatography and gas chromatography. J. Chromatog. **84**, 181—186 (1973)

Jatlow, P.: Ultraviolet spectrophotometric analysis of drugs in biological fluids. Amer. J. med. Technol. **39**, 231—236 (1973)
Jenden, D. J., Cho, A. K.: Application of integrated gas chromatography/mass spectrometry in pharmacology and toxicology. Ann. Rev. Pharmacol. **13**, 371—390 (1973)
Juselius, R. E., Barnhard, F.: Detection of barbiturates, narcotics, and amphetamines in urine. Clin. Toxicol. **6**, 53—57 (1973)
Kadar, D., Kalow, W.: A method for measuring glutethimide (doriden) in human serum after intake of therapeutic doses. J. Chromatog. **72**, 21—27 (1972)
Kaistha, K. K.: Drug abuse screening programs: detection procedures, development costs, street-sample analysis, and field tests. J. pharm. Sci. **61**, 655—679 (1972)
Kaistha, K. K., Jaffe, J. H.: Extraction techniques for narcotics, barbiturates, and central nervous system stimulants in a drug abuse urine screening program. J. Chromatog. **60**, 83—94 (1971)
Kaistha, K. K., Jaffe, J. H.: Cost of a toxicology laboratory facility. Development expense and cost per urine test using thin-layer chromatographic techniques in a drug abuse urine screening program. Int. J. Addict. **7**, 585—592 (1972a)
Kaistha, K. K., Jaffe, J. H.: Reliability of identification techniques for drugs of abuse in a urine screening program and drug excretion data. J. pharm. Sci. **61**, 305—307 (1972b)
Kaistha, K. K., Jaffe, J. H.: TLC techniques for identification of narcotics, barbiturates, and CNS stimulants in a drug abuse urine screening program. J. pharm. Sci. **61**, 679—689 (1972c)
Kaistha, K. K., Tadrus, R.: Comparison of costs for testing a wide variety of drugs of abuse per urine specimen in a drug abuse urine screening program and frequent urine collections. J. Chromatog. **109**, 149—162 (1975)
Kaistha, K. K., Tadrus, R., Janda, R.: Simultaneous detection of a wide variety of commonly abused drugs in a urine screening program using thin-layer identification techniques. J. Chromatog. **107**, 359—379 (1975)
Kananen, G., Osiewicz, R., Sunshine, I.: Barbiturate analysis—a current assessment. J. chromatog. Sci. **10**, 283—286 (1972)
Kanter, S. L., Hollister, L. E., Moore, F.: Marihuana metabolites in urine of man. Res. Commun. chem. Path. Pharmacol. **10**, 215—219 (1975)
Kazyak, L., Knoblock, E. C.: Application of gas chromatography to analytical toxicology. Analyt. Chem. **35**, 1448—1452 (1963)
Kleber, H. D., Gould, L. C.: Urine testing schedules in methadone maintenance. J. Amer. med. Ass. **215**, 2115—2116 (1971)
Klein, B., Sheehan, J. E., Grunberg, E.: Use of fluorescamine ("Fluram") to detect amphetamine in urine by thin-layer chromatography. Clin. Chem. **20**, 272—274 (1974)
Knox, J. H., Jurand, J.: Application of high-speed liquid chromatography to the analysis of morphine, heroin, 6-(0-acetyl)morphine and methadone. J. Chromatog. **87**, 95—108 (1973a)
Knox, J. H., Jurand, J.: Separation of morphine alkaloids, heroin, methadone and other drugs by ion-exchange chromatography. J. Chromatog. **82**, 398—401 (1973b)
Kodak Chromat/O/Screen, Extraction kit, Kodak Publication No. JJ-179, 1971a
Kodak Chromat/O/Screen, Analysis kit for amphetamines, Kodak Publication No. JJ-175, 1971b
Kodak Chromat/O/Screen, Analysis kit for alkaloids, Kodak Publication No. JJ-176, 1971c
Kodak Chromat/O/Screen, Analysis kit for barbiturates, Kodak Publication No. JJ-177, 1971d
Koechlin, B. A., D'Arconte, L.: Determination of chlordiazepoxide (Librium) and of a metabolite of lactam character in plasma of humans, dogs, and rats by a specific spectrofluorometric micro method. Analyt. Biochem. **5**, 195—207 (1963)
Koida, M., Takahashi, M., Kaneto, H.: The morphine 3-glucuronide directed antibody: its immunological specificity and possible use for radioimmunoassay of morphine in urine. Jap. J. Pharmacol. **24**, 707—714 (1974b)
Koida, M., Takahashi, M., Muraoka, S., Kaneto, H.: Antibodies to BSA conjugates of morphine derivatives: strict dependency of the immunological specificity on the hapten structure. Jap. J. Pharmacol. **24**, 165—167 (1974a)
Kokoski, R. J.: Drug abuse detection by thin-layer chromatography in urine screening program, pp. 6406—6414. Proc. of NAS-NRC Committee on Problems of Drug Dependence, Wash., D. C., Feb., 1970

Kokoski,R.J., Hamner,S., Shiplet,M.: Benzodiazepine tranquilizer abuse in narcotic addict treatment programs: urinalysis as a detection and control measure, pp. 200—207. Proc. of the NAS-NRC Committee on Problems of Drug Dependence, Mexico City, Mexico, March, 1974

Kokoski,R.J., Jain,M.: Comparison of results for morphine urinalyses by radioimmunoassay and thin-layer chromatography in a narcotic clinic setting. Clin. Chem. **21**, 417—419 (1975)

Kokoski,R.J., Waitsman,E.S., Sands,F.L., Kurland,A.A.: Narcotic detection in thin-layer chromatography in a urine screening program, pp. 5433—5446. Proc. of NAS-NRC Committee on Problems of Drug Dependence, Indianapolis, Ind., Feb., 1968

Koontz,S., Besemer,D., Mackey,N., Phillips,R.: Detection of benzoylecgonine (cocaine metabolite) in urine by gas chromatography. J. Chromatog. **85**, 75—79 (1973)

Kullberg,M.P., Gorodetzky,C.W.: Studies on the use of XAD-2 resin for detection of abused drugs in urine. Clin. Chem. **20**, 177—183 (1974)

Kullberg,M.P., Miller,W.L., McGowan,F.J., Doctor,B.P.: Studies on the single extraction of amphetamine and phenobarbital from urine using XAD-2 resin. Biochem. Med. **7**, 323—335 (1973)

Kupferberg,H., Burkhalter,A., Way,E.L.: A sensitive fluorometric assay for morphine in plasma and brain. J. Pharmacol. exp. Ther. **145**, 247—251 (1964a)

Kupferberg,H.J., Burkhalter,A., Way,E.L.: Fluorometric identification of submicrogram amounts of morphine and related compounds on thin-layer chromatographs. J. Chromatog. **16**, 558—559 (1964b)

LaPointe,J.G.: Clinical implementation of radioimmunoassay. Lab. Management **11**, 20—48 (1973)

Law,N.C.: A modern approach for drug identification. Amer. J. med. Technol. **39**, 237—243 (1973)

Lebish,P., Finkle,B.S., Brackett,J.W.: Determination of amphetamine, methamphetamine and related amines in blood and urine by gas chromatography with hydrogen-flame ionization detector. Clin. Chem. **16**, 195—200 (1970)

Lederer,W.H., Gerstbrein,H.L.: A manifold for concentrating to dryness multiple organic solvent samples simultaneously and rapidly for thin-layer or gas chromatography. Analyt. Biochem. **55**, 326—327 (1973)

Leute,R., Ullman,E.F., Goldstein,A.: Spin immunoassay of opiate narcotics in urine and saliva. J. Amer. med. Ass. **221**, 1231—1234 (1972b)

Leute,R.K., Ullman,E.F., Goldstein,A., Herzenberg,L.A.: Spin immunoassay technique for determination of morphine. Nature (Lond.) **236**, 93—94 (1972a)

Lin,S.C., Way,E.L.: Use of paper chromatographic technics on urine for evaluating narcotic usage by the nalorphine pupil test. J. forens. Sci. **8**, 209—219 (1963)

Lindquist,C.A., Spratt,J.L.: Solid-phase radioimmunoassay of digoxin and morphine. Fed. Proc. **32**, 719 (1973)

Liu,C.-T., Adler,F.L.: Immunologic studies on drug addiction. I. Antibodies reactive with methadone and their use for detection of the drug. J. Immunol. **111**, 472—477 (1973)

Loeffler,L.J., Pierce,J.V.: Radioimmunoassay for lysergide (LSD) in illicit drugs and biological fluids. J. pharm. Sci. **62**, 1817—1820 (1973)

Loh,H.H., Ho,I.K., Cho,T.M., Lipscomb,W.: Estimation of morphine by polyamide mini thin-layer chromatography. J. Chromatog. **76**, 505—508 (1973)

Loh,H.H., Ho,I.K., Lipscomb,W.R., Cho,T.M., Selewski,C.: Mini thin-layer chromatography. III. A rapid and sensitive method for the estimation of amphetamine and methamphetamine. J. Chromatog. **68**, 289—293 (1972)

Maes,R., Hodnett,N., Landesman,H., Kananen,G., Finkle,B., Sunshine,I.: The gas chromatographic determination of selected sedatives (ethchlorvynol, paraldehyde, meprobamate, and carisoprodol) in biological material. J. forens. Sci. **14**, 235—254 (1969)

Malspeis,L., Bathala,M.S., Ludden,T.M., Bhat,H.B., Frank,S.G., Sokoloski,T.D., Morrison,B.E., Reuning,R.H.: Metabolic reduction of naltrexone I. Synthesis, separation and characterization of naloxone and naltrexone reduction products and qualitative assay of urine and bile following administration of naltrexone, α-naltrexol, or β-naltrexol. Res. Commun. chem. Path. Pharmacol. **12**, 43—65 (1975)

Mannering,G.J., Dixon,A.C., Carroll,N.V., Cope,O.B.: Paper chromatography applied to the detection of opium alkaloids in urine and tissues. J. Lab. clin. Med. **44**, 292—300 (1954).

Marks, V., Fry, D., Chapple, P. A. L., Gray, G.: Application of urine analysis to diagnosis and treatment of heroin addiction. Brit. med. J. **1969 II**, 153—155
Marshall, D. J.: Toward automating radioimmunoassay. Amer. Lab. **6**, 91—95 (1974)
Martin, L., Genest, K., Cloutier, J. A. R., Farmilo, C. G.: Physico-chemical methods for the identification of narcotics (contd.) Part VI—Common physical constants, UV, IR and X-ray data for 12 narcotics and related compounds. Bull. Narcot. **15(3—4)**, 17—38 (1963)
Masoud, A. N.: Systematic identification of drugs of abuse. I. Spot tests. J. pharm. Sci. **64**, 841—843 (1975)
McGonigle, E. J.: Determination of microgram quantities of methadone by fluorescence. Analyt. Chem. **43**, 966—967 (1971)
McIntyre, J. A., Armandi, A. E., Risen, L. P., Ling, W., Haberfelde, G. C.: Thin-layer chromatography and enzyme immunoassay of L-alpha-acetyl methadol and methadone metabolites in urine. Clin. Chem. **21**, 109—112 (1975)
McIsaac, W. M.: Establishment of a new drug addiction program. Psychopharmacol. Bull. **3**, 40—44 (1966)
McMartin, C., Simpson, P., Thorpe, N.: The specific assay of basic drugs in urine by CM-cellulose chromatography using continuous drug-dye complex extraction as a detection system. J. Chromatog. **43**, 72—83 (1969)
McMartin, C., Street, H. V.: Gas-liquid chromatography of submicrogram amounts of drugs. I. Preparation, scope and limitation of columns. J. Chromatog. **22**, 274—285 (1966a)
McMartin, C., Street, H. V.: Gas-liquid chromatography of submicrogram amounts of drugs. II. Analysis of barbiturates and related drugs in biological media. J. Chromatog. **23**, 232—241 (1966b)
Medzihradsky, F., Dahlstrom, P. J.: Concurrent determination of narcotic drugs in plasma by gas-liquid chromatography. Pharmacol. Res. Commun. **7**, 55—69 (1975)
Melikian, A. P., Forrest, I. S.: Dansyl derivatives of Δ^9- and Δ^8-tetrahydrocannabinols. J. pharm. Sci. **62**, 1025—1026 (1973)
Meola, J. M., Vanko, M.: Use of charcoal to concentrate drugs from urine before drug analysis. Clin. Chem. **20**, 184—187 (1974)
Miller, W. L., Kullberg, M. P., Banning, M. E., Brown, L. D., Doctor, B. P.: Studies on the quantitative extraction of morphine from urine using nonionic XAD-2 resin. Biochem. Med. **7**, 145—158 (1973)
Misra, A. L., Vadlamani, N. L., Mulé, S. J.: Chromatographic separation of methadone, some of its metabolites and congeners. J. Chromatog. **67**, 379—381 (1972)
Moffat, A. C.: Use of SE-30 as a stationary phase for the gas-liquid chromatography of drugs. J. Chromatog. **113**, 69—95 (1975)
Moffat, A. C., Clare, B.: The choice of paper and thin-layer chromatographic systems for the analysis of basic drugs. J. Pharm. (Lond.) **26**, 665—670 (1974)
Moffat, A. C., Smalldon, K. W.: Optimum use of paper, thin-layer and gas-liquid chromatography for the identification of basic drugs. II. Paper and thin-layer chromatography. J. Chromatog. **90**, 9—17 (1974)
Moffat, A. C., Smalldon, K. W., Brown, C.: Optimum use of paper, thin-layer and gas-liquid chromatography for the identification of basic drugs. I. Determination of effectiveness for a series of chromatographic systems. J. Chromatog. **90**, 1—7 (1974a)
Moffat, A. C., Stead, A. H., Smalldon, K. W.: Optimum use of paper, thin-layer and gas-liquid chromatography for the identification of basic drugs. III. Gas-liquid chromatography. J. Chromatog. **90**, 19—33 (1974b)
Monforte, J., Bath, R. J., Sunshine, I.: Fluorometric determination of primary and secondary amines in blood and urine after thin-layer chromatography. Clin. Chem. **18**, 1329—1333 (1972)
Montalvo, J. G., Klein, E., Eyer, D., Harper, B.: Identification of drugs of abuse in urine. I. A study of the Dole technique. J. Chromatog. **47**, 542—545 (1970)
Montalvo, J. G., Scrignar, C. B., Alderette, E., Harper, B., Eyer, D.: Flushing, pale-colored urines, and false negatives. Urinalysis of narcotic addicts. Int. J. Addict. **7**, 355—364 (1972)
Mulé, S. J.: Determination of narcotic analgesics in human biological materials. Application of ultraviolet spectrophotometry, thin-layer and gas-liquid chromatography. Analyt. Chem. **36**, 1907—1914 (1964)

Mulé, S. J.: Identification of narcotics, barbiturates, amphetamines, tranquilizers and psychotomimetics in human urine. J. Chromatog. **39**, 302—311 (1969)
Mulé, S. J.: Routine identification of drugs of abuse in human urine. I. Application of fluorometry, thin-layer and gas-liquid chromatography. J. Chromatog. **55**, 255—266 (1971)
Mulé, S. J.: Detection and identification of drugs of dependence. In: Mulé, S. J., Brill, H. (Eds.): Chemical and biological aspects of drug dependence, pp. 278—303. Cleveland: CRC Press 1972
Mulé, S. J.: Methods for the analysis of morphine and related surrogates: current status. J. chromatog. Sci. **12**, 245—253 (1974)
Mulé, S. J., Bastos, M. L.: A comparison of immunoassay methods for the detection of drugs subject to abuse. In: Mulé, S. J., Sunshine, I., Braude, M., Willette, R. E. (Eds.): Immunoassays for drugs subject to abuse, pp. 99—106. Cleveland: CRC Press 1974
Mulé, S. J., Bastos, M. L., Jukofsky, D.: Evaluation of immunoassay methods for detection, in urine, of drugs subject to abuse. Clin. Chem. **20**, 243—248 (1974a)
Mulé, S. J., Bastos, M. L., Jukofsky, D., Saffer, E.: Routine identification of drugs of abuse in human urine. II. Development and application of the XAD-2 resin column method. J. Chromatog. **63**, 289—301 (1971)
Mulé, S. J., Hushin, P. L.: Semiautomated fluorometric assay for submicrogram quantities of morphine and quinine in human biological material. Analyt. Chem. **43**, 708—711 (1971)
Mulé, S. J., Kramer, A. R.: Applied statistical techniques for use by the clinician in the urinalysis of drugs subject to abuse. Int. J. Addict. **8**, 939—957 (1973)
Mulé, S. J., Sunshine, I., Braude, M., Willette, R. E. (Eds.): Immunoassays for drugs subject to abuse. Cleveland: CRC Press 1974b
Mulé, S. J., Whitlock, E., Jukofsky, D.: Radioiummunoassay of drugs subject to abuse: critical evaluation of urinary morphine-barbiturate, morphine, barbiturate, and amphetamine assays. Clin. Chem. **21**, 81—86 (1975)
Nadeau, G., Sobolewski, G.: Une nouvelle méthode de dosage tres petites quantités de morphine. Canad. J. Biochem. **36**, 625—631 (1958)
Nakamura, G. R., Way, E. L.: Determination of morphine and codeine in post-mortem specimens. Analyt. Chem. **47**, 775—778 (1975)
Nash, J. F., Bennett, I. F., Bopp, R. J., Brunson, M. K., Sullivan, H. R.: Quantitation of propoxyphene and its major metabolites in heroin addict plasma after large dose administration of propoxyphene napsylate. J. pharm. Sci. **64**, 429—433 (1975)
Neesby, T.: New spray for thin-layer chromatograms of narcotics. Clin. Chem. **19**, 356—357 (1973)
Nightingale, S. L., Michaux, W. W., Platt, P. C.: Clinical implications of urine surveillance in a methadone maintenance program. Int. J. Addict. **7**, 403—414 (1972)
Oberst, F. W.: Studies on the fate of morphine. J. Pharmacol. exp. Ther. **74**, 37—41 (1942)
Oberst, F. W.: Studies on the fate of heroin. J. Pharmacol. exp. Ther. **79**, 266—270 (1943)
Oestreicher, P. M., Farmilo, C. G., Levi, L.: The physical methods for the identification of narcotics (cont.) Part IIIB. Ultraviolet spectral data for ninety narcotics and related compounds. Bull. Narcot. **6(3—4)**, 42—70 (1954)
Ono, M., Engelke, B. F., Fulton, C.: Procedures for assured identification of morphine, dihydromorphinone, codeine, norcodeine, methadone, quinine, methamphetamine, etc., in human urine. Bull. Narcot. **21**, 31—40 (1969)
Parker, K. D., Hine, C. H.: Manual for the determination of narcotics and dangerous drugs in the urine. Bull. Narcot. **19**, 51—57 (1967)
Parker, K. D., Fontan, C. R., Kirk, P. L.: Separation and identification of tranquilizers by gas chromatography. Analyt. Chem. **34**, 757—760 (1962)
Parker, K. D., Fontan, C. R., Kirk, P. L.: Rapid gas chromatographic method for screening of toxicological extracts for alkaloids, barbiturates, sympathomimetic amines, and tranquilizers. Analyt. Chem. **35**, 356—359 (1963)
Parker, K. D., Hine, C. H., Nomof, N., Elliott, H. W.: Urine screening techniques employed in the detection of users of narcotics and their correlation with the nalorphine test. J. forensic Sci. **11**, 152—166 (1966)
Payte, J. T., Wallace, J. E., Blum, K.: Hydrolysis: a requisite for morphine detection in urine. Curr. ther. Res. **13**, 412—416 (1971)

Pehr, F.: Simple, highly selective screening method for barbiturates in urine. Clin. Chem. **21**, 1609—1611 (1975)
Pickup, M. E., Paterson, J. W.: The determination of ephedrine plasma levels by a gas chromatographic method. J. Pharm. (Lond.) **26**, 561—562 (1974)
Riedmann, M.: Specific gas chromatographic determination of phenothiazines and barbiturate tranquilizers with the nitrogen flame ionization detector. J. Chromatog. **92**, 55—59 (1974)
Robinson, J. D., Morris, B. A., Marks, V.: Development of a radioimmunoassay for etorphine. Res. Commun. chem. Path. Pharmacol. **10**, 1—8 (1975)
Roerig, D. L., Lewand, D., Mueller, M., Wang, R. I. H.: Methods of identification and confirmation of abusive drugs in human urine. J. Chromatog. **110**, 349—359 (1975a)
Roerig, D. L., Lewand, D. L., Mueller, M. A., Wang, R. I. H.: Comparison of radioimmunoassay with thin-layer chromatographic and gas-liquid chromatographic methods of barbiturate detection in human urine. Clin. Chem. **21**, 672—675 (1975b)
Ross, R., Horwitz, C. A., Hager, H., Usateggui, M., Burke, M. D., Ward, P. C. J.: Preliminary evaluation of a latex agglutination-inhibition tube test for morphine. Clin. Chem. **21**, 139—143 (1975)
Rubenstein, K. E., Schneider, R. S., Ullman, E. F.: "Homogeneous" enzyme immunoassay. A new immunochemical technique. Biochem. biophys. Res. Commun. **47**, 846—851 (1972)
Rubin, M.: Requirements and problems in urinalysis for abuse drugs. The role of gas chromatography. Amer. J. med. Technol. **39**, 205—210 (1973)
Rutter, E. R.: Automated method for screening urine for amphetamine and some related primary amines. Clin. Chem. **18**, 616—620 (1972)
Saferstein, R., Chao, J.: Identification of drugs by chemical ionization mass spectroscopy. J. Ass. off. analyt. Chem. **56**, 1234—1238 (1973)
Sansur, M., Buccafuri, A., Morgenstern, S.: Automated fluorometric method for determination of morphine in urine. J. Ass. off. analyt. Chem. **55**, 880—887 (1972)
Santinga, P. H.: Application of fluorescence and gas chromatography to mass drug screening. Fluorescence News **6**, 1—7 (1971)
Satoh, H., Kuroiwa, Y., Hamada, A., Uematsu, T.: Radioimmunoassay for phenobarbital. J. Biochem. **75**, 1301—1306 (1974)
Schneider, R. S., Bastiani, R. J., Leute, R. K., Rubinstein, K. E., Ullman, E. F.: Use of enzyme and spin labeling ion homogeneous immunochemical detection methods. In: Immunoassays for drugs subject to abuse, Mulé, S. J., Sunshine, I., Braude, M., Willette, R. E. (Eds.): pp. 45—72. Cleveland: CRC Press 1974
Schneider, R. S., Lindquist, P., Wong, E. T., Rubenstein, K. E., Ullman, E. F.: Homogeneous enzyme immunoassay for opiates in urine. Clin. Chem. **19**, 821—825 (1973)
Schnoll, S. H., Vogel, W. H., Odstrchel, G.: The specificity of anti-mescaline antibody produced in rabbits. Fed. Proc. **32**, 719 (1973)
Schwartz, M., ed.: Clinical methods for drugs of abuse. The Bendix Corporation 1971
Shaw, M. A., Peel, H. W.: Thin-layer chromatography of 3,4-methylenedioxy-amphetamine, 3,4-methylenedioxymethamphetamine and other phenethylamine derivatives. J. Chromatog. **104**, 201—204 (1975)
Siek, T. J.: Thin-layer and gas-chromatography as identification aids in forensic science. Analabs Res. Notes **13**, 1—9 (1973)
Siek, T. J., Osiewicz, R. J.: Identification of drugs and other toxic compounds from their ultraviolet spectra. Part II. Ultraviolet absorption properties of thirteen structural groups. J. forens. Sci. **20**, 19—37 (1975)
Sine, H. E., Kubasik, N. P., Rejent, T. A.: Determination of the presence of morphine in urine. Clin. Biochem. **7**, 102—105 (1974)
Sine, H. E., Kubasik, N. P., Woytash, J.: Simple gas-liquid chromatographic method for confirming the presence of alkaloids in urine. Clin. Chem. **19**, 340—341 (1973)
Skelley, D. S., Brown, L. P., Besch, P. K.: Radioimmunoassay. Clin. Chem. **19**, 146—186 (1973)
Skinner, R. F., Gallaher, E. J., Knight, J. B., Bonelli, E. J.: The gas-chromatograph-mass spectrometer as a new and important tool in forensic toxicology. J. forens. Sci. **17**, 189—198 (1972)
Sleeman, H. K., Cella, J. A., Harvey, J. L., Beach, D. J.: Thin-layer chromatographic detection and identification of methaqualone metabolites in urine. Clin. Chem. **21**, 76—80 (1975)

Smith, D. A., Cole, W. J.: Rapid and sensitive gas chromatographic determination of diacetylmorphine and its metabolite monoacetylmorphine in blood using a nitrogen detector. J. Chromatog. **105**, 377—381 (1975)
Sohn, D.: Drug screening—a fact of life for the nineteen seventies. Industr. Med. **41**, 18—21 (1972)
Sohn, D.: Analysis for drugs of abuse: the validity of reported results in relation to performance testing. Int. J. Addict. **8**, 65—74 (1973)
Sohn, D., Simon, J., Hanna, M. A., Ghali, G. V., Tolba, R. A., Melkonian, V.: Screening for heroin—a comparison of current methods. Analyt. Chem. **45**, 1498—1502 (1973)
Solomon, M. D.: A study of codeine metabolism. Clin. Tox. **7**, 255—257 (1974)
Solon, J. M., Mikkelsen, L.: Semi-automated analysis of drugs. Hewlett Packard, Application Note ANC-27-71 (1971)
Solon, J. M., Wisniewski, J. V.: Qualitative screening for barbiturates. Hewlett Packard, Application Note, ANC-22-70 (1970a)
Solon, J. M., Wisniewski, J. V.: Determination of nanogram quantities of amphetamine in urine, Hewlett Packard, Application Note, ANC-18-70 (1970b)
Spector, S.: Quantitative determination of morphine in serum by radioimmunoassay. J. Pharmacol. exp. Ther. **178**, 253—258 (1971).
Spector, S., Berkowitz, B., Flynn, E. J., Peskar, B.: Antibodies to morphine, barbiturates, and serotonin. Pharmacol. Rev. **25**, 281—291 (1973)
Spector, S., Flynn, E. J.: Barbiturates: radioimmunoassay. Science **174**, 1036—1038 (1971)
Spector, S., Parker, C. W.: Morphine: radioimmunoassay. Science **168**, 1347—1348 (1970)
Spector, S., Seidner, A.: Radioimmunoassays for morphine. In: Mulé, S. J., Sunshine, I., Braude, M., Willette, R. E. (Eds.): Immunoassays for drugs subject to abuse, pp. 13—21. Cleveland: CRC Press 1974
Spector, S., Vesell, E. S.: Disposition of morphine in man. Science **174**, 421—422 (1971)
Stevenson, G. W.: Spectrophotometric determination of blood barbiturate. Analyt. Chem. **33**, 1374—1378 (1961)
Stewart, J. T., Lotti, D. M.: Fluorimetric determination of amphetamines with 3-carboxy-7-hydroxycoumarin. J. pharm. Sci. **60**, 461—463 (1971)
Stoner, R. E., Parker, C.: Single—pH extraction procedure for detecting drugs of abuse. Clin. Chem. **20**, 309—311 (1974)
Street, H. V.: The simultaneous detection of alkaloid, neutral and acidic poisons in human tissues by high temperature reversed phase paper chromatography. J. forens. Sci. **7**, 222—230 (1962)
Street, H. V.: Gas-liquid chromatography of submicrogram amounts of drugs. III. Analysis of alkaloids in biological media. J. Chromatog. **29**, 68—79 (1967)
Street, H. V.: Gas-liquid chromatography of submicrogram amounts of drugs. IV. Identification of barbiturates, hydantoins, amides, imides, carbamates, phenylbutazone, carboxylic acids, and hydrazine drivatives by direct derivative formation within the gas chromatograph. J. Chromatog. **41**, 358—366 (1969)
Street, H. V., McMartin, C.: Quantitative estimation and identification of barbiturates in blood in emergency cases. Nature (Lond.) **199**, 456—459 (1963)
Sunshine, I.: Use of thin-layer chromatography in the diagnosis of poisoning. Amer. J. clin. Path. **40**, 576—582 (1963)
Sunshine, I.: Handbook of analytical toxicology. Cleveland: Chemical Rubber Company 1969
Sunshine, I.: Manual of analytical toxicology. Cleveland: Chemical Rubber Company 1971
Swagdis, J. E., Flanagan, T. L.: Spectro-photofluorometric determination of low concentrations of amobarbital in plasma. Analyt. Biochem. **7**, 147—151 (1964)
Takemori, A. E.: An ultrasensitive method for the determination of morphine and its application in experiments in vitro and in vivo. Biochem. Pharmacol. **17**, 1627—1635 (1968)
Taunton-Rigby, A., Sher, S. E., Kelley, P. R.: Lysergic acid diethylamide: radioimmunoassay. Science **181**, 165—166 (1973)
Taylor, J. F.: Methods of chemical analysis. In: Clouet, D. H. (Ed.): Narcotic drugs, pp. 17—88. New York: Plenum 1971
Teale, J. D., Forman, E. J., King, L. J., Marks, V.: Production of antibodies to tetrahydrocannabinol as the basis for its radioimmunoassay. Nature (Lond.) **249**, 154—155 (1974a)
Teale, J. D., Forman, E. J., King, L. J., Marks, V.: Radioimmunoassay of cannabinoids in blood and urine. Lancet **1974IIb**, 553—555

Teale, J. D., Forman, E. J., King, L. J., Piall, E. M., Marks, V.: The development of a radioimmunoassay for canabinoids in blood and urine. J. Pharm. (Lond.) **27**, 465—472 (1975)
Usategui-Gomez, M., Heveran, J. E., Cleeland, R., McGhee, B., Telischak, Z., Awdziej, T., Grunberg, E.: Simultaneous detection of morphine and barbiturates in urine by radioimmunoassay. Clin. Chem. **21**, 1378—1382 (1975)
Valanju, N. N., Baden, M. M., Valanju, S. N., Mulligan, D., Verma, S. K.: Detection of biotransformed cocaine in urine from drug abusers. J. Chromatog. **81**, 170—173 (1973)
Valentour, J. C., Harold, W. W., Stavitsky, A. B., Kananen, G., Sunshine, I.: Detection of glutethimide (doriden) by hemagglutination inhibition. Clin. chim. Acta **43**, 65—67 (1973)
Valentour, J. C., Monforte, J. R., Sunshine, I.: Fluorometric determination of propoxyphene. Clin. Chem. **20**, 275—277 (1974)
Van Hoof, F., Heyndrickx, A.: Thin-layer chromatographic-spectrofluorometric analysis of amphetamine and amphetamine analogs after reaction with 4-chloro-7-nitrobenzo-2,1,3-oxadiazole. Analyt. Chem. **46**, 286—288 (1974)
Van Vunakis, H., Bradvica, H., Benda, P., Levine, L.: Production and specificity of antibodies directed toward 3,4,5-trimethoxyphenylethylamine, 3,4-dimethoxyphenylethylamine, and 2,5-dimethoxy-4-methylamphetamine. Biochem. Pharmacol. **18**, 393—404 (1969)
Van Vunakis, H., Levine, L.: Use of the double antibody and nitrocellulose membrane filtration techniques to separate free antigen from antibody bound antigen in radioimmunoassays. In: Mulé, S. J., Sunshine, I., Braude, M., Willette, R. E. (Eds.): Immunoassays for drugs subject to abuse, pp. 23—35. Cleveland: CRC Press 1974
Van Vunakis, H., Wasserman, E., Levine, L.: Specificities of antibodies to morphine. J. Pharmacol. exp. Ther. **180**, 514—521 (1972)
Verebely, K., Inturrisi, C. E.: The simultaneous determination of propoxyphene and norpropoxyphene in human biofluids using gas-liquid chromatography. J. Chromatog. **75**, 195—205 (1973)
Verebely, K., Mulé, S. J., Jukofsky, D.: A gas-liquid chromatographic method for the determination of naltrexone and beta-naltrexol in human urine. J. Chromatog. **111**, 141—148 (1975)
Verpoorte, R., Svendsen, A. B.: High-speed liquid chromatography of alkaloids. I. J. Chromatog. **100**, 227—230 (1974)
Vinson, J. A., Hooyman, J. E.: A universal thin-layer chromatographic visualization reagent for drugs. J. Chromatog. **105**, 415—417 (1975)
Vree, T. B., Muskens, A. T. J. M., van Rossum, J. M.: Excretion of amphetamines in sweat. Arch. int. Pharmacodyn. **199**, 311—317 (1972)
Waddell, W. J.: A simple, rapid method for paper chromatography of barbiturates. Clin. Chem. **11**, 37—39 (1965)
Wainer, B. H., Fitch, F. W., Rothberg, R. M., Fried, J.: Morphine-3-succinylbovine serum albumin: an immunogenic hapten-protein conjugate. Science **176**, 1143—1145 (1972)
Walberg, C. B.: Correlation of the "EMIT" urine barbiturate assay with a spectrophotometric serum barbiturate assay in a suspected overdose. Clin. Chem. **20**, 305—306 (1974)
Wallace, J. E.: Determination of ephedrine and certain related compounds by ultraviolet spectrophotometry. Analyt. Chem. **39**, 531—533 (1967)
Wallace, J. E., Biggs, J. D., Ladd, S. L.: Determination of amphetamine by ultraviolet spectrophotometry. Analyt. Chem. **40**, 2207—2210 (1968)
Wallace, J. E., Biggs, J. D., Merritt, J. H., Hamilton, H. F., Blum, K.: A sensitive thin-layer chromatographic technique for determining morphine in urine. J. Chromatog. **71**, 135—140 (1972a)
Wallace, J. E., Hamilton, H. E., Payte, J. T., Blum, K.: Sensitive spectrophotometric method for determining methadone in biological specimens. J. pharm. Sci. **61**, 1397—1400 (1972b)
Wallace, J. E., Hamilton, H. E., Blum, K., Petty, C.: Determination of morphine in biologic fluids by electron capture gas-liquid chromatography. Analyt. Chem. **46**, 2107—2111 (1974a)
Wallace, J. E., Hamilton, H. E., Riloff, J. A., Blum, K.: Spectrophotometric determination of ethchlorovynol in biologic specimens. Clin. Chem. **20**, 159—162 (1974b)
Wallace, J. E., Hamilton, H. E., King, D. E., Bason, D. J., Schwertner, H. A., Harris, S. C.: Gas-liquid chromatographic determination of cocaine and benzoylecgonine in urine. Analyt. Chem. **48**, 34—38 (1976)
Wang, R. I. H., Mueller, M. A.: Identification of barbiturates in urine. J. pharm. Sci. **62**, 2047—2049 (1973)

Watson, E., Kalman, S. M.: A thirty minute determination of sedatives in plasma by gas-liquid chromatography. Clin. chim. Acta **38**, 33—37 (1972)

Way, E. L., Adler, T. K.: The biological disposition of morphine and its surrogates. Geneva: World Health Organization 1962

Weissman, N., Lowe, M. L., Beattie, J. M., Demetrion, J. A.: Screening method for detection of drugs of abuse in human urine. Clin. Chem. **17**, 875—881 (1971)

Wells, R., Hammond, K. B., Rodgerson, D. O.: Gas-liquid chromatographic procedure for measurement of methylphenidate hydrochloride and its metabolite, ritalinic acid, in urine. Clin. Chem. **20**, 440—443 (1974)

Wilkinson, G. R., Way, E. L.: Sub-microgram estimation of morphine in biological fluids by gas-liquid chromatography. Biochem. Pharmacol. **18**, 1435—1439 (1969)

Williams, A. J., Jones, T. W. G., Cooper, J. D. H.: A rapid method for the determination of therapeutic barbiturate levels using gas-liquid chromatography. Clin. chim. Acta **43**, 327—332 (1973)

Williams, T. A., Pittman, K. A.: Pentazocine radioimmunoassay. Res. Commun. chem. Path. Pharmacol. **7**, 119—143 (1974)

Wislocki, A., Martel, P., Ito, R., Dunn, W. S., McGuire, C. D.: A method for the detection of drugs of abuse in urine. Hlth Lab. Sci. **11**, 13—19 (1974)

Wittwer, J. D., Kluckhohn, J. H.: Liquid chromatographic analysis of LSD. J. chromatogr. Sci. **11**, 1—3 (1973)

Wolen, R. L., Gruber, C. M.: Determination of propoxyphene in human plasma by gas chromatography. Analyt. Chem. **40**, 1243—1246 (1968)

Woods, L. A., McMahon, F. G., Seevers, M. H.: Distribution and metabolism of cocaine in the dog and rabbit. J. Pharmacol. exp. Ther. **101**, 200—204 (1951)

World Health Organization Meeting of Investigators: Detection of dependence-producing drugs in body fluids, World Health Organization Technical Report Series No. 556. Geneva: World Health Organization 1974

Yeh, S. Y.: Separation and identification of morphine and its metabolites and congeners. J. pharm. Sci. **62**, 1827—1829 (1973)

Yeh, S. Y.: Urinary excretion of morphine and its metabolites in morphine-dependent subjects. J. Pharmacol. exp. Ther. **192**, 201—210 (1975)

Yeh, S. Y., Flanary, H., Sloan, J.: Efficient, low-cost gas manifold for use in sample concentration. Clin. Chem. **19**, 687—688 (1973)

Yeh, S. Y., Gorodetzky, C. W.: Disposition of heroin in man. Pharmacologist **16**, 269 (1974)

Yeh, S. Y., Gorodetzky, C. W., McQuinn, R. L.: Urinary excretion of heroin and its metabolites in man. J. Pharmacol. exp. Ther. **196**, 249—256 (1976)

Yeh, S. Y., McQuinn, R. L.: GLC determination of heroin and its metabolites in human urine. J. pharm. Sci. **64**, 1237—1239 (1975)

Zingales, I. A.: Determination of chlordiazepoxide plasma concentrations by electron capture gas-liquid chromatography. J. Chromatog. **61**, 237—252 (1971)

SECTION III

Sedative/Hypnotics and Alcohol Dependence

CHAPTER 1

The Pharmacology of Sedative/Hypnotics, Alcohol, and Anesthetics: Sites and Mechanisms of Action

CEDRIC M. SMITH

A. Introduction: Sedative/Hypnotics of Interest

"Dependency," "habit," "addiction," "compulsion," "custom," are a few of the many terms denoting the propensity of some human beings to ingest or otherwise self-administer drugs, the major obvious action of which appears to be one of depression of nervous system activity. The taking of such drugs is implicitly associated with the fact that the agents produce a state objectively and subjectively defined as intoxication. All of the drugs discussed below alter nervous system functions and, concomitantly, the feeling state, self-perception, mood, behavior, and "consciousness."

The historical onset of human self-intoxication is lost in the ancient history of mankind; its cultural context is revealed in a number of ways, perhaps most directly in verbal communication. More specifically, the sequence of initiation into use, the nature of the habits that might ensue, the social circumstances of such use, and the tolerance and withdrawal signs and symptoms have been revealed not only in the relatively recent research treatises but also in the culture patterns and the language [1]. A prime point is that knowledge of the hedonistic and dependence-producing capacities of the sedative/hypnotic drugs and alcohol is and has been common knowledge to most who use the agent with some frequency and to their acquaintances.

The phenomenology and temporal sequence of dependency with habitual use of sedative/hypnotics and alcoholic beverages were, however, widely disputed and confused until the classic clinical studies of VICTOR and ADAMS (1953) and the experimental investigations of ISBELL and FRASER, and their co-workers (see Sect. III, Chaps. 2 and 3). These studies differentiated the signs and symptoms due the *primary* acute effects of the agent itself, such as alcohol or barbiturates, from the effects *derivative* of associated conditions such as vitamin deficiency, and from those due to *withdrawal* after repeated ingestion of the agent.

The topics of primary interest are the delineation of possible mechanisms and sites of action of the sedative/hypnotics in producing alterations in nervous system function, with particular reference to those mechanisms that might underlie habitual

[1] Such language frequently derives from the use of a specific drug or chemical agent, e.g., wino, alky, pill-head, pill-freak, pot-head, acid-head, goofed-up, drugged, rum-head, wine-head, rummy, rum-pot, hop-head, Martini alcoholic, on pills, pill baby, Sneaky Peter, Squirrel cage, stumble-bum, whiskey habit, to cite a few (see, for example, KELLER and MCCORMICK, 1968), and for the withdrawal syndrome sick, whammy, sweats, shakes, fits, goose-flesh, black-out, etc. (cf. e.g., LINGEMAN, 1969; HAERTZEN et al., 1970a, b, 1974; RUBINGTON, 1971, CULL and HARDY, 1974). Words and their accociations can be utilized to compare drug effects qualitatively and quantitatively (HILL et al., 1963a, b; MARTIN, 1966; HAERTZEN and HOOKS, 1973; HAERTZEN et al., 1963, 1970a, b, 1974; MARTIN et al., 1974b).

use and dependency (Subsect. C). Since other chapters present evidence on physical dependence on ethanol and barbiturates, I have chosen to devote a major subsection (D) to the phenomenology and possible mechanisms of functional or neural tolerance, since this may be an initiating event in dependence syndromes. Subsections A and B deal with definitions and identification of the drugs with sedative/hypnotic properties which are or have been used habitually by a sizable fraction of those exposed, i.e., which are agents of concern with respect to being associated with dependence or nonmedical self-administration. The patterns and characteristics of repeated use of the various sedative/hypnotic agents and alcohol—including the nature of continued intoxication and the sudden withdrawal therefrom—are discussed briefly (B). The final subsection (E) attempts a synthesis and projections for future studies.

A number of comprehensive and thorough reviews of these topics appeared while this analysis was being prepared. As will be apparent, the author is particularly indebted to the recent reviews of ISRAEL (1970), SHARPLESS (1970), WALLGREN and BARRY (1970), KALANT et al., (1971), MYERS and VEALE (1972), KISSIN and BEGLEITER (1971–1974); ASTON (1972); HUG (1972); MAYNERT (1972a, b, c); MULÉ and BRILL (1972); MELLO (1973b); CLARK and ROSNER (1973); ROSNER and CLARK (1973); SEIXAS and EGGLESTON (1973); KALANT (1974, 1975); NOBLE (1974); KRNJEVIĆ (1974a, b); HALSEY et al., (1974); NOBLE et al., (1975); FINK (1975); GOLDSTEIN (1975a and LEBLANC and CAPPELL (1975) for their authoritative compilations of previous work and for their provocative analyses of possible mechanisms of action and of dependence, tolerance, and withdrawal. Further, detailed considerations of these topics are presented in a number of books only recently published. In view of the ready availability of extensive and almost complete documentation in this area, the present review will focus on an analysis of some rather specific questions: The reader is urged to refer to the above citations for full referencing of previous work[2].

Dependence: For the purposes of this review, a broad definition will be used throughout to denote the propensity of human beings or laboratory animals to take (self-administer) a specific agent or class of agents repeatedly. It is, therefore, generally synonymous with the idea of compulsive use, repetitive self-administration, a habit of use, etc. The term *physical dependence* will apply only to situations in which sudden termination of chronic administration is followed by a specific set of withdrawal signs and symptoms in the organism, such as tremor or diarrhea. The expression *psychological dependence* will be eschewed since it is primarily a term used to denote repeated use without obvious physical dependence and yet implies various "psychological mechanisms."

It is understood that *dependence* may imply: behavior of repeated self-administering and the behavior associated with acquiring a supply and the taking; a subjective state of discomfort induced by relative withdrawal, and conversely, alleviation of the discomfort by taking the drug; or a subjective state of unease apparently unrelated to drug use or effects (discomfort, anxiety, tension, hunger, pain, etc.) abolished by actions of the drug and thus providing an apparent drive to continue administration as drug effects wane.

[2] The author humbly apologizes in advance for many pertinent references that will, undoubtedly, be found to have been overlooked or uncited because of the magnitude of the task and severe limitations in his memory and filing organization.

For our purposes, the most general definition would appear to be the behavioral one in combination with the verbalization of the subjective internal state occurring concomitantly with the behavior.

Tolerance: KALANT et al. (1971), KALANT (1973a, b), HUG (1972) and LEBLANC and CAPPELL (1975) appropriately differentiate the many connotations of "tolerance." *Initial* or spontaneous tolerance is taken to describe the fact that a given individual may exhibit less effect of a given dose of drug than others of a population of similar individuals, i.e., a tolerance not directly related to previous exposure to the specific or similar agents. The origins of such tolerance may be related to sex, species, age, and diet, as well as to genetic and environmental influences.

Acquired [increase in] tolerance, on the other hand, refers to changes in sensitivity and responsivity to drug effects due to a given event, usually the prior administration of the same or another drug. Acquired tolerance is generally measured either as a decrease in the effect of a given dose or as the increase in the dose required to produce a given effect. This type of tolerance is related only to the effect(s) under observation and the method of measurement; it provides no intrinsic inference as to mechanism(s) responsible for the development of tolerance. Possible mechanisms range from alterations in disposition (dispositional tolerance) (KALANT et al., 1971; HUG, 1972), which are due to changes in absorption, distribution, excretion, and metabolism, to a decrease in the response in the target (brain) tissue to the agent (functional or tissue tolerance). The latter type of tolerance will be considered at some length in Subsect. D.

The *agents* at issue which are commonly called sedative/hypnotics include the following. This classification and listing is derived from chapters in GOODMAN and GILMAN (1970, 1975).

Barbiturates:
(Among the very large number of barbiturates, the many studies cited below have mostly utilized barbital, phenobarbital, amobarbital, hexobarbital, pentobarbital, methohexital, secobarbital, or thiopental.)

Aliphatic Alcohols:
Ethanol
Methanol
Propanol
Butanol

Chloral Derivatives:
Trichloroethanol
Chloral hydrate
Trichloroethyl phosphate
Chlorobutanol

Carbamic Acid Esters of Glycols:
Meprobamate

Piperidinedione Derivatives:
Glutethimide
Methyprylon

2,3-Disubstituted Quinazalones:
Methaqualone

Benzodiazepine Compounds:
Chlordiazepoxide
Diazepam
Nitrazepam
Oxazepam

Others of perhaps lesser practical significance in either extent of use or information available on effects and mechanisms of action include the cyclic ether, paraldehyde. Many other agents have sedative or hypnotic activity. Of particular interest are: (1) the volatile and gaseous general anesthetics—diethylether, divinylether, chloroform, halothane, nitrous oxide, and cyclopropane; (2) the agents used and therefore classified as anti-convulsants or anti-epileptic, including the simple inorganic bromides; (3) the various glue solvents and petroleum products, including gasoline, lighter fluid, and kerosene vapors, benzene, toluene, acetone, amyl acetate, aliphatic ketones.

Throughout the preparation of this review an effort was made to ascertain what facts applied to given specific agents—avoiding generalizations to classes of substances, such as the general anestetics. In the later subsections, an attempt will be made in the conclusions to differentiate those situations in which generalizations appear to be warranted and to assist in conceptualizing problems which previous generalizations have tended to obfuscate. Thus, the topics of emphasis include:

— Molecular mechanisms underlying each drug or agent's actions,
— Sites of action involved in producing each specific effect, and
— Evidence relating the mechanism(s) of action *to* the action at the site(s) of interest.

The drug effects of interest include those actions evident on acute administration, including sedation, drowsiness, sleep, relief or prevention of anxiety, a relief of tension or worry, a liking for the drug, a sense of relaxation, a sense of well-being or euphoria, and other alterations of subjective feeling states.

Certainly, all of the agents cited above produce some degree of sedation or sleep; most have been self-administered by an appreciable number of people. It is important to the discussion to emphasize that in acting on the nervous systems, the alterations that are produced may or may not be directly perceived by the individual. Other drug effects that may be of relevance are amnesia, analgesia, and decreased ability to concentrate, as well as the phenomena and mechanisms causally associated with the changes in the nature and intensity in drug[3] effects that appear with repeated administration, i.e., tolerance, sensitization, and withdrawal syndromes.

Although some degrees of dependence and cross-tolerance and dependence have been unambiguously demonstrated among certain compounds (barbiturates, alcohol, chloral, benzodiazepines, methaqualone, glutethimide, meprobamate, paraldehyde, and a few general anesthetics), it does not follow directly that all act to produce

[3] "Drug" will be used in the general, neutral, pharmacologic context to refer to any substance with a biological effect.

general depression by the same or similar mechanisms or that general depressant agents will exhibit cross-tolerance and cross-dependence as well.

A special case of interest is the question of the degree to which there is "conditioning" or learning of the subjective responses and drug actions, including the possibility of "placebo" actions in anticipation prior to drug taking, as well as actions associated with and in response to prior experiences produced by the agent. Various aspects of this general problem are presented in "Drug Mystification" of LENNARD and associates (1971), "Ceremonial Chemistry" (SZASZ, 1974) and related to social factors (e.g., CHEIN, 1969a, b), and conditioning theory (see WIKLER, 1968a, b, 1973; OVERTON, 1973, 1974; WINTER, 1974; DEWS, 1973; STEINBERG, 1969, 1970; JOYCE, 1969; SCHUSTER, 1973).

It is recognized that the self-administration or dependence may range from low doses taken only occasionally (at intervals of days to years) to habitual, continual self-induced intoxication with overt physical dependence or chronic drug toxicity.

B. Patterns of Nonmedical Use of Sedative/Hypnotic Agents

The observation that at least some of the individuals exposed to sedatives, anesthetics, and similar agents choose to take them repeatedly is the basis for their inclusion in this chapter. A detailed examination of the natural frequency and the prevalence of such dependences is beyond the scope of this review. Further, prevalence figures are confounded by the many sources of obvious influence, such as fads, fashions, group identification, social conditions, accessibility, real and perceived risks, etc. Moreover, in any given population a variety of agents are used by the various members. How such data on use, even if available, should be expressed is also a complex issue.

It seems sufficient for the present purpose to state simply that an appreciable number of individuals have been known to become dependent on the sedative/hypnotic agents listed above. That certain individuals take small or large amounts of various sedative/hypnotics such as barbiturates or methaqualone, repeatedly and compulsively to the extent that intoxication, tolerance, and appearance of withdrawal symptoms are produced, is well established in a variety of sociocultural settings.

Barbiturates and minor tranquillizers are not uncommonly taken in conjunction with alcohol, or sequentially with amphetamine. What has not been well studied, however, is the incidence of dependence or habit formation of those exposed, of the strength of the habits, and of the tolerance and withdrawal characteristics, with the exception of sporadic studies and case reports (see, for example, GLATT, 1968, 1969; ESSIG, 1969; FINER, 1970; CHAMBERS et al., 1972; CROWLEY et al., 1974b; PATCH, 1974; GREENBLATT and SHADER, 1974a, b; WEISE et al., 1973; WEISE and PRICE, 1975; GREEN and MACDONALD, 1976; WESSON and SMITH, 1977).

The data and impressions that do bear on this question suggest that the incidence of sedative drug use in Western societies is less than that of alcohol, and that this difference appears to be, at least in part, related to the apparent circumstances and reasons behind initial usage, e.g., for legitimate medical reasons in contrast to use, for example, for "kicks" in a party setting. Moreover, a certain proportion of those who use sedatives repeatedly can be readily accounted for by "alcoholics" or narcotic users who shift to combined or sole use of barbiturates, some purposely to reduce apparent drinking (e.g., FINER, 1970; DEVENYI and WILSON, 1971a, b). It has been recently shown that nontolerant narcotic addicts rate pentobarbital as desirable and as producing euphoria as well as producing "sleepiness and drunkenness" (MARTIN et al., 1974b; MCCLANE and MARTIN, 1976). "Mixed habits" or "polydrug abuse" is a matter of current sociopolitical concern.

Similarly, the use of solvents, anesthetics and chloral hydrate has been known, largely anecdotally, for years; as an example the "chloral habit" has many similarities to alcohol or barbiturate abuse (see PRESS and DONE, 1967a, b; BRILLIANT, 1970; LYNN et al., 1971; SMITH, 1972; WEISE et al., 1973; Report Series, 1974). Likewise, paraldehyde dependency was a common clinical observation among those in whom it was being used to treat alcohol dependence; even bromides continue to be used (MCDANAL et al., 1974).

I. Characteristics of Dependency on Sedative/Hypnotic Agents

The variability in the nature and intensity of the compulsive drive for drug use among individuals and in the patterns of use appear to be as varied as the behaviors of individuals using the agents and the cultural and behavioral contexts in which they are used. These range from low frequency/low dose to low frequency/high dose (e.g., on binge, spree, party) to high frequency/high dose use. Further, the degree of "control" exercised by the individual or the group on the nature, amount, and frequency of use appears to vary widely. But it should be noted that the majority of those diagnosed as habitual users retain some measure of control over the situations and dosage, with the exception of intermittent periods of uncontrolled continuous intoxication to the degree of almost complete incapacitation (see MACANDREW and EDGERTON, 1969; FREEDMAN, 1971; ZINBERG, 1974; BALTER and LEVINE, 1971; BALTER, 1973; BALTER et al., 1974; WEISE and PRICE, 1975; WESSON and SMITH, 1977).

It is commonly inferred that repetitive self-use of alcohol or sedatives is a progressive phenomenon with an inexorable increase in intake; but examination of the population and patterns of individual use commonly reveals marked fluctuations over time in use by a given person, as well as apparently spontaneous reduction of intake in middle or late years of life (CAHALAN and ROOM, 1974). Even more marked movement might be detectable if such surveys were repeated more frequently. MADSEN (1974) in a uniquely personal review of the literature and individual cases, emphasizes, as few have, the importance of "going on a drunk" for the alcoholic. Interestingly, monkeys working for i.v. alcohol infusions exhibit marked fluctuations in amounts received per day; intake may be high for 3–5 days followed by spontaneous cessation of intake for up to 24 h to be followed again by a period of high levels of intake (WOODS et al., 1971; WOODS and WINGER, 1971; MELLO, 1973b; WINGER and WOODS, 1973).

Another point to consider is the dosage and effect reached and maintained. Most chronic users of any of the above agents appear to take the drugs to produce a certain level of intoxication. A few persistently take enough to produce unconsciousness, but most take, by very rough dosage regulation, only sufficient amounts to produce a given level of intoxication. When tolerant, this dose may be large, whereas when the individual is susceptible, it is appreciably smaller.

Some would argue that individuals who spontaneously reduce their intake were never truly alcoholics or barbiturate addicts, but this may be only a tautology if the definition of "alcoholic" requires or implies progressively greater levels of use. The point remains that the drive to take these drugs, as revealed by behavior examined over long time spans, fluctuates and varies with periods of abstinence.

These comments are made with full recognition of the variety of other factors influencing the amount taken at any one time—the setting, group pressure and activities, physical condition, etc. The primary point to be made, nevertheless, is the presence of some degree of intoxication and drug effect in such individuals. Looked at as free choice situation, the drug is not taken simply to alleviate impending withdrawal symptoms; rather, it *appears* to be usually taken to produce a given desirable subjective state.

Patterns of use of alcohol have been extensively described, but analogous data for the other agents under consideration, by and large, have not. Since it is felt by many that alcohol use is strongly patterned by cultural factors, it is hazardous to extrapolate these findings to other sedative/hypnotics. Nevertheless, one would be surprised if the usage of barbiturates did not parallel at least the categories of alcoholism delineated by JELLINEK (1960): JELLINEK differentiated four types of alcoholism in a very broad classification based on:

1. drinking patterns, amount, frequency and pattern, plus
2. presence of physical addiction (dependence)
3. apparent ability to abstain for long periods
4. presence of physical disabilities, such as "polyneuropathy, gastritis, and cirrhosis."

The types were characterized thus:

> ... *Beta alcoholism* is that species of alcoholism in which such alcoholic complications as polyneuropathy, gastritis, and cirrhosis of the liver may occur without either physical or psychological dependence upon alcohol ... *Gamma alcoholism* means that species of alcoholism in which (1) acquired increased tissue tolerance to alcohol, (2) adaptive cell metabolism, (3) withdrawal symptoms and "craving," i.e., physical dependence, and (4) loss of control are involved ... *Alpha alcoholism* represents *purely* psychological *continual* dependence or reliance upon the effect of alcohol to relieve bodily or emotional pain ... *but does not lead to "loss of control"* or "*inability to abstain*" ... *Delta alcoholism* shows the first three characteristics of gamma alcoholism as well as a less marked form of the fourth characteristic—that is, instead of loss of control there is inability to abstain ... There are, of course, many other species of alcoholism—if it is defined as any drinking that causes any damage ... Among these other species is periodic alcoholism, which in Europe and Latin America is still designated as dipsomania ... *Epsilon alcoholism* ...

Irrespective of whether or not these four types represent unique and distinct forms of drinking behavior, and many concur that they do (see SEIXAS, 1972; MADSEN, 1974), these types are characteristic patterns of use of a large group of alcohol and sedative abusers. It can, thus, serve as the initial step in describing and classifying use and abuse patterns for these agents. From case studies and anecdotal data, similar patterns appear to be identifiable with barbiturates and similar drugs with perhaps some special cases of mixed drug use and truly exploratory drug taking.

Barbiturates, ethanol and meprobamate have been generally classified on pharmacologic grounds in a single broad class of "general depressants" that have, by inference, a common syndrome and, possibly, common mechanisms of action. Usually included in this class are also the volatile and gaseous general anesthetics (see bibliography of WEISE et al., 1973; SMITH, 1972; BRILLIANT, 1970; LYNN et al.,

1971; NAGLE, 1968). Although not often stated explicitly, this holistic view more or less requires the added explanation that the differences in properties and usefulness of the various agents lie in their unique physical properties, convenient modes of administration, or differences in duration of action. For example, chloroform was used for a short time as an additive or active ingredient in a beverage. We can speculate that its use was not continued because of its irritating properties and because its action was just too brief to gain widespread use as an intoxicant.

The recreational use of nitrous oxide and ether to produce intoxication antedated by some years their application to the relief of pain in dentistry and surgery. SIMPSON, in searching for other anesthetics as a substitute for ether, introduced chloroform to his associates informally in his home; even his niece tried inhaling its vapors—and to their amusement she fell asleep crying: "I'm an angel! Oh, I'm an angel!" (HAGGARD, 1929; ROBINSON, 1946). Sir HUMPHREY DAVY made and tried nitrous oxide on convenient animals including a cat, kittens, dogs, rabbits, guinea pigs, mice, hens, and goldfinches. Most especially he examined its effects on man including his friends, associates, and himself. Most people found the effects of the gas were not only pleasant, but had a strong inclination to repeat the experience!

> From the strong inclination of those who have been pleasantly affected by the gas to respire it again, it is evident that the pleasure produced is not lost, but that it mingles with the mass of feelings, and becomes intellectual pleasure, or hope. The desire of some individuals acquainted with the pleasures of nitrous oxide for the gas has been so strong as to induce them to breathe with eagerness, the air remaining in the bags after the respiration of others.
>
> I have often felt very great pleasure when breathing it alone, in darkness and in silence, occupied only by ideal existence... Whenever I have breathed the gas after excitement from moral or physical causes, the delight has been often intense and sublime.
>
> The thrilling was very rapidly produced. The pleasurable sensation was at first local, and perceived in the lips and about the body, and in the middle of the experiment was so intense and pure as to absorb existence. At this moment, and not before, I lost consciousness; it was, however, quickly restored, and I endeavored to make a bystander acquainted with the pleasure I experienced by laughing and stamping.
>
> Cited in ROBINSON (1946)

Interestingly, some people did not find the experience especially pleasant. PRIESTLY, the discoverer of nitrous oxide, never enjoyed its effects because it caused severe throbbing of the arteries (ROBINSON, 1946).

The subjective phenomena of relaxation, insight, relief of anxiety and euphoria are described not only in these far past times; only recently has a curious book appeared by another SMITH (1972) who reexamines and extols the virtues and raptures to be achieved by inhalation of a number of "anesthetic" gases or volatile liquids.

It is obvious that the human being is subject to wish-fulfilling fantasies and the experiencing of suggested or sought-after sensations. To what degree this phenomenon explains the euphoria following ethanol or the anesthetics has not been ascertained experimentally. But there is little doubt that many of these agents are not only not aversive, but produce positive feelings, although some do find them unpleasant [cf., hypoxia and nitrogen narcosis, below, and the extensive studies of nitrous oxide (RUSSELL and STEINBERG, 1955; STEINBERG, 1954, 1955, 1956; SUMMERFIELD and STEINBERG, 1957; see review of PARBROOK, 1967a, b; an annotated bibliography of WEISE et al., 1973; BRILLIANT, 1970; LYNN et al., 1971; NAGLE, 1968), the old observations on general anesthetics (NAGLE, 1968; SMITH, 1972; DUNBAR, 1905), and the recent intense inquiries into the psychopharmacology of cannabis preparations]. These, together with the exploration of a variety of methods of altering "consciousness", continue to raise the question of whether some common denominator in neural sites and mechanisms of action might be responsible for actions perceived as pleasant or desirable.

Among the signs and symptoms of acute intoxication from all of the sedatives and hypnotics appear to be alterations in mood and feeling ranging from euphoria, reduction of felt anxiety, a sense of well-being, analgesia, a feeling of importance, to dysphoria and depression; the more pleasant feeling tones predominate in published reports. Compulsive laughter is commonly associated with cannabis and nitrous oxide, but it is also obviously occasionally experienced with ethanol and other agents.

Ideation and logical mental processes are generally perceived as dulled, cloudy, slowed, or as a difficulty in concentrating although fantasy and loose associations are frequently described as increased along with more dreaminess. Along the activity scale, most people and most sedative/hypnotic or anesthetic drugs tested would be rated as "downers," i.e., the subjects feel less drive and energy than in control states. Nevertheless, the sense of insight, revelation, wonder and tremendously important discovery accompanies many, if not all, intoxications with the above agents (cf. DUNBAR, 1905; SMITH, 1972; STEINBERG, 1956; HAERTZEN et al., 1974; MARTIN et al., 1974b; BRILLANT, 1970; LYNN et al., 1971; NAGLE, 1968).

Time sensation is altered, but in varied and complex fashions. The objective timing may be altered erratically, whereas the subjective sense of time duration may be increased or decreased depending on the time interval and the intervening activity (cf. STEINBERG, 1956) and method of expression employed.

Subjective phenomena also fairly consistently include mild alterations in sensory phenomena, as occurs certainly with nitrous oxide, ether or ethanol. There may be tingling, dizziness (vertigo), numbness, and some visual disturbances including diplopia.

The similarities of the syndromes produced by equivalent doses of nitrous oxide, ethanol, and barbiturates include motor unsteadiness and incoordination. The dose-response and potency relationships among these agents with respect to behavioral and mood alterations, on the one hand, and motor incoordination, on the other, have yet to be studied systematically. It would not be too surprising to find quantitative differences, and that such differences may underlie the apparent differences in spectra of action of many of the compounds such as the benzodiazepines and meprobamate, which appear to have little motoric effects compared to their subjective changes, while ethanol and mephenesin produce marked motor alterations. In this connection, although tolerance develops to a striking degree, the mild motor disturbances are usually thought to be a give-away to any heavy drinker. Even in some of the most experienced drinkers, a dose bordering on that producing motor incoordination is required to produce the desired subjective change. However generally true this is, W.C. FIELDS, for example, apparently kept himself rather precisely titrated and his motor system compensated sufficiently finely to be able to play expert golf and billiards; it appears that it was his cirrhotic liver "what done him in."

Thus, the search for the mechanisms of action of sedative drugs which as a class are habituating; may possibly be the same as the search for the mechanism of anesthetic action, possibly including that of the accompanying analgesia.

One thinks of certain effects as unique, but it is possible to consider that the overt behavior is the end result of a complex interplay of influences. For example, is sleep only one possible outcome of a drug with antianxiety or relaxing properties? "Hypnotics" in usual doses may not directly produce sleep; rather they can be viewed as facilitating or making more likely its occurrence. Thus, a distinction can be drawn between the amnesia and unconsciousness produced by anesthetics and the sleep facilitated by a single drink of alcohol or a night-time dose of a barbiturate hypnotic. This differentiation is consonant with the observation of the "stimulating" effects of alcoholic beverages and of barbiturates taken in settings associated with excitement and a party. Certainly, sleep and unconsciousness are perceived as the same or very similar phenomena, and largely as a period of amnesia except for some memories of dreams and partial awareness of sleep. The blackout of the alcoholic is a period of time of apparent consciousness but eventual amnesia. A very similar phenomenon is now being reproduced daily in dental practice with i.v. diazepam and barbiturates! Clinicians describe in positive terms the cooperativeness of the patient, absence of anxiety, the possible presence of analgesia, and—overall—a "beautiful" patchy amnesia for most events. Thus, we might redefine general anesthesia, Stage I, as involving a sense of well-being, altered consciousness, mild analgesia and patchy amnesia, and the possible occurrence of sleep if left alone; with increased concentrations, there appears increased analgesia, amnesia and unconsciousness (cf. SMITH, 1972).

In extrapolating this topic, the effects of hypoxia (and perhaps hypercapnia) come to mind. Mild hypoxia (from whatever causes, such as high altitude, carbon monoxide poisoning, or low oxygen tension) producing faintness, fantasy, disordered thought, and unconsciousness has been well described in the literatures of space- and aeromedicine, undersea physiology and of air pollution (see STEWART, 1975; FRISANCHO, 1975). The syndrome of mild hypoxia is thus quite similar to that evoked by ethanol or barbiturates—decreased mental effectiveness and visual acuity, emotional lability, faulty judgement, analgesia, and muscular incoordination progressively increasing until unconsciousness occurs. A similar acute picture is associated with poisoning with low levels of carbon monoxide and with a variety of other CNS depressants such as methane, ethane, and higher chain alcohols (see annotated bibliography of WEISE et al., 1973).

One reason for this digression away from sedative/hypnotic drugs is to address the possible common features of hangover following overindulgence of alcoholic beverages, anesthesia, or hypoxia. The hangover syndrome(s), defined as severe throbbing headache aggravated by straining and movement, nausea and vomiting, increase in anxiety, and fine tremor are obviously similar; however, the hangover does not seem to be solely the consequence of an episode of central nervous system (CNS) depression to a given level. If it were there, it would be anticipated routinely following surgical anesthesia. Although it is true that nausea and vomiting are relatively frequent after anesthesia, the characteristic throbbing headache is not. (However, the nausea and vomiting after an alcohol debauch may well be compounded by the vomiting due to the gastric irritation and/or injudicious eating and drinking. Further, the headache attributed to a hangover may well be actually a migraine or the weekend tension headache syndrome.) The tremor and anxiety of hangover are, as yet, hard to define accurately. The time course of the hangover is such that it characteristically may start some 3–4 h after drinking starts, i.e., even prior to going to sleep after an evening out on the town, whereas the peak of the headache intensity may appear any time from 6 to 10 h following drinking. In most instances, recovery is on its way before or by 6–12 h following the alcoholic excess (cf. CHAPMAN, 1970; WALLGREN and BARRY, 1970; YLIKAHRI et al., 1974). Although not studied systematically or reported consistently, alcohol itself is presumably one of the most effective treatments of hangover—"hair of the dog that bit one" and it seems to be the important ingredient in the majority of home or barroom remedies.

It is here postulated that the hangover is a mild withdrawal syndrome consistent with the discussion of WALLGREN and BARRY (1970) and the observations of MACDONNELL et al. (1975). If so, the treatment follows logically; on the other hand it may be that the reports of efficacy of alcohol for hangover alleviation are obtained only from heavy drinkers who are, indeed, going into mild withdrawal at the time the hangover appears. For them, a drink is precisely what will relieve their acute distress although this may not be the case for the occasional drinker.

A syndrome quite similar to the hangover or migraine is associated with toxicity caused by carbon monoxide (SALEN, 1946; BJERVER and GOLDBERG, 1948; PECORA, 1959; RÖSELER, 1961), cyanide (WARD, 1947; WARD and WHEATLEY, 1947) by natural gas, and perhaps by gasoline vapors (see bibliography of WEISE et al., 1973). With CO, there is a severe throbbing headache, weakness, nausea and vomiting. With CO, it is thought that the headache is due to cerebral edema and excessive transudation across capillary vessel walls consequent to local hypoxia.

It should be noted that gasoline vapors and a variety of organic solvents produce intoxication subjectively similar to that produced by ethanol, including habituation. The degree to which functional tolerance develops is not described precisely, although painters are said to be tolerant to ethanolic beverages. Whether or not there actually is cross-tolerance with ethanol or if there is a withdrawal syndrome is not known to this reviewer.

The syndrome associated with the chronic administration of barbiturates and ethanol has been elicited in a variety of species (MELLO, 1973b; PERRIN et al., 1974, 1975). Physical dependence, as evidenced by a withdrawal syndrome, is seen in animal studies with barbiturates (see works by ESSIG, 1962a, b, 1963, 1964, 1965, 1966a, b, 1967, 1968a, b, 1969, 1972; ESSIG and FLANARY, 1959, 1961; ESSIG and LAM, 1968; ESSIG et al., 1969; FRASER and ISBELL, 1954a; WATERS and OKAMOTO, 1972, 1973; ROSENBERG and OKAMOTO, 1974a, b; OKAMOTO et al., 1975; JONES et al., 1976) and with ether (GESSNER, 1974). Further, animals will self-administer a variety of sedative/hypnotic drugs including ethanol (WOODS and SCHUSTER, 1970; WOODS and WINGER, 1971; MYERS and VEALE, 1972; WOODS et al., 1971, 1973; WINGER and WOODS, 1973; SINCLAIR, 1974a; SMITH et al., 1976), barbiturates (DENEAU et al., 1965, 1968, 1969; SCHUSTER and THOMPSON, 1969; SCHUSTER and JONANSON, 1974; YANAGITA and TAKAHASHI, 1970, 1973; CICERO and SMITHLOFF, 1973; DAVIES and MILLER, 1963; WEEKS, 1971) and benzodiazepines (see Subsect. D.I.3). For reviews, see MYERS and VEALE (1972), SCHUSTER and JONANSON (1974), THOMPSON and PICKENS (1975), MELLO (1973b), and N.I.D.A. bibliography (1974).

It is generally concluded (cf. HARVEY, 1975; BOGGAN, 1974) that methaqualone has acute and chronic effects similar to barbiturates, although the data available in the literature are sparse and not comprehensive. An explicit test of tolerance and withdrawal hyperexcitability to audiogenic seizures has been carried out using rats—showing that methaqualone is quite similar to phenobarbital (KOHLI et al., 1974). Recent reviews also include those of HARVEY (1975), SCHWARTZBURG et al. (1973), INABA et al. (1973); JONES et al. (1976), and PASCARELLI (1973).

It is generally inferred that all of the agents of the sedative/hypnotic group, such as most barbiturates, exhibit a similar pattern of physical dependence. Although much less studied, the chronic ingestion of glutethimide, methaqualone, chlordiazepoxide, diazepam, chloralhydrate, and paraldehyde seem to produce a similar type of dependence (see also FRASER (Sect. III, Chap. 2); ESSIG, 1969; HOLLISTER, 1973; GREENBLATT and SHADER, 1974a, b; WEISE and PRICE, 1975; N.I.D.A. 1974; JONES et al., 1976).

Moreover, it is now quite clear that at least certain psychotropic agents may have, in addition to their intoxicating actions, negative reinforcing (aversive) properties in naive subjects or animals (HOFFMEISTER, 1975).

Physical dependence development generally requires doses that also produce tolerance. A detailed analysis of the factors, dose pattern of administration and duration, for other than a few agents, has never been accomplished in man or in analogous animal studies. For example, HOLLISTER (1973) claims that the dependence is obtained with chlordiazepoxide only by using doses far in excess of those encountered clinically. He claims that for this reason, and others perhaps, this substance has a very low dependence-producing liability in comparison with alcohol or barbiturates. This may be true, but the data presently available don't prove the point. Thus, the information is lacking on which to base a comprehensive comparison among sedative/hypnotic drugs with regard to their relative propensities to induce physical dependence. Even in animal studies, such comparisons present a number of technical difficulties (see Chap. 2). For example, the dependence-producing effects of long-acting depressants may be underestimated because addicted subjects would lose tolerance and dependence as the effect of the drug is slowly dissipated. The possible mechanisms for physical dependence and the relationship to tolerance are also briefly discussed below (see Subsect. D).

II. Mechanisms of Dependence

The following is an overview in outline form of the possible mechanisms and viewpoints regarding the development of a habit of repeated alcohol or drug use (cf. WIKLER, 1972). The topic can be posited as the explanation of the behavior termed dependence or drug habit.

1. Drug Effects Subjectively Perceived

Subjectively perceived drug effects (and perhaps those not directly perceived) may serve as primary reinforcing events. For example, euphoria or sense of well-being has been reported by many people for caffeine, amphetamine, cocaine, ethanol, barbiturates, morphine, marijuana (as well as related drugs and others including thyroid hormone). In a simplistic analysis, it might be suggested that such euphoria could result from any of the following:

— The drug and drug state (plus set and setting) could be the stimulus for a conditioned euphoric response. This category would include the phenomena of "drug mystification" (WIKLER, 1968a, b; LENNARD et al., 1971).

— The drug could act (in appropriate doses and schedules) on those neurons whose action (or inaction) is known to give rise to feeling tones of relaxation, relief of anxiety, and pleasure (the activation of so-called pleasure centers).

— The drug could act selectively to depress those neural systems mediating pain and anxiety, and thus relieve anxiety and reduce pain.

— The drug could act to reduce "judgement" and "inhibitions."

— The drug could act to depress and reduce most neural activity, tending to induce sleep and somnolence, and thus reduce "attention" to all internal/external stimuli.

— The drug could act to reduce all feeling tone states—excitement, as well as anger and anxiety, equanimity or well-being.

— The drug could act to exert its influence on feeling and behavior indirectly and by nonperceived neural actions. It is possible to postulate fanciful amplifying sequences; for example, a modest effect on muscle spindle afferents which in turn could cause cerebellar motor incoordinations which would cause ataxia and unsteadiness, which via unconscious muscle sensory input causes an increase in presynaptic inhibition and less pain input, resulting in less hurt and pain, less anxiety, etc.

— Some combination of the above.

Other possible drug actions of interest are: relief of anxiety and neurosis[4]; escape—a time out; relief of depression; increase in ego strength via daydream or "high;" relief of stress; sleep; antagonism of the effects of amphetamine, coffee or mania; drug effects as a substitute for exercise, sleep and relaxation.

Drug effects may thus be seen as positive and yet interactive with societal pressures, stressful situations, personal inadequacies, self-image, permitting or promoting fantasy, providing a sense of power, false courage, a decrease in the sense of guilt or simply as phantasy or recreation. In brief, the language is inadequate to cover the realm of subjective phenomena associated with the effects of drugs.

2. Conditioning or Learning Theories

(cf. DEWS, 1973)

— Habit-formation—What is "habit" and what is the habit being described? Is a habit but one class of learned behaviors, of which all of the associated activities and environments are the stimuli and, over time, the secondary reinforcer? Obviously, the more unique the sensory (afferent) input is and the more frequent the routine and associated activities, the stronger will be the habit.

---

[4] Masserman's landmark experiments (MASSERMAN et al., 1944; MASSERMAN and YUM, 1946) showed increased alcohol consumption with experimentally induced "neuroses," a classic demonstration of alcohol addiction secondary to a conflict situation; although this postulate is still attractive and the experiments of interest, the experiments are difficult to design and interpret (see e.q. CAPPELL and HERMAN, 1972; SMART, 1965; MYERS and VEALE, 1972; FALK and SAMSON, 1976; MELLO, 1973b).

— Drug actions can serve as stimuli for learned behavior and they also can give rise to the phenomenon of state-dependent learning (see Subsect. D.2.k.).

— "Vicarious" conditioning or creation of desire seems obvious and exploited extensively in advertising. It should be mentioned that a drug might also serve as a *labilizer* by altering subjective conditions or making mood more labile.

— Secondary conditioning and "Hustler Theories" (WIKLER, 1968a, b).

3. Drug Taking or Intoxication as Part of a Psychological Reaction

As just a few examples: The drug-taking behavior may be part of a behavior that serves as:

— Punishment of or revenge against significant others as a symbol of rebellion, to gain sympathy, to illustrate interpersonal dependence, to gain attention, to gain acceptance in a group or to become recognized as a leader.

— Alcohol behavior as a "Game Alcoholics Play"—the transactional analysis game theory of "alcoholism" includes a hangover component not readily applied to the abuse of barbiturates or similar agents. Although STEINER (1971) may well have alienated many by his style, the analysis and descriptions have a ring of accuracy with a validity equal to that of other psychologic/psychiatric theories. Whether or not the mechanisms completely describe the actual mechanisms involved in alcoholism or sedative drug dependencies, and whether these can be generalized to other individuals and situations, have not yet yielded to direct proof.

— Ceremonial use of drugs—Current society puts great weight on "drugs," "dope," "liquor," "pot," "pills," "drinking," (cf. SZASZ, 1974; WINICK, 1974; JOYCE, 1969) and the drugs, their use, and the social aspects include appreciable components of ceremonial and behavior-modification significance.

4. Drug-Induced Drug Dependence

Dependence can be correlated with tolerance and physical dependence with acute and protracted abstinence syndromes. These induced drive states are the factors most individuals focus on conceptually. Drug taking is viewed as being followed in turn by the actions of agent, subsidence of drug actions, perhaps some rebound anxiety, sleep disturbances, and tremulousness—all of which are relieved by the same drugs. Thus, a progressive increase in anxiety, sleep disturbance, behavioral mishaps and failures is viewed as leading to more and more drug taking. This classic description of the development of dependence is clearly obtained in man and animals (MELLO, 1973b) for most of the agents cited including anesthetics such as ether (GESSNER, 1974).

A central concern then becomes that of determining the nature of the "loss of control" with continued use. However well this physical addiction concept explains continued drug taking, it fails to address or explain the most troublesome aspect: the resumption of a habit after many months of abstinence.

More and more data are accumulating to suggest that a protracted abstinence syndrome not only exists but has appreciable influence on behavior. Put differently, a single cycle of addiction to narcotic agents is postulated to leave behind a sustained trace of altered physiology, including increased susceptibility (MARTIN and JASINSKI,

1969; MARTIN, 1972; cf. also COCHIN, 1973). It is here postulated that an analogous syndrome is present with sedatives/hypnotics and alcohol. The sparse evidence for such a protracted abstinence syndrome is presented below. If proved, it will constitute a new kind of confirmation of the general tenet of Alcoholics Anonymous—"once an alcoholic, always an alcoholic."

Such drug or alcohol habits result, sooner or later, in the assumption of a "new life" of the drug users or the boozers, drug hustlers, and drugs (alcohol) as a way of life—with a culture, a group, a common language and mores, and thus an alternative to squarer existences. These concepts couple a primary physiologic addiction with learning and adaptation in a social context.

The very fact that much of the craving for the psychoactive drug of choice, e.g., barbiturate, can be abolished by another drug, such as ethanol or chloral hydrate given by the same or another route, and in a milieu different from the customary one suggests that at the least a major component of the dependence in many situations of long-term chronic use of sedatives is related to the specific pharmacologic effects of the agent. Although dependent individuals commonly prefer the specific preparation and agent and route of administration to which they are accustomed, they will accept and employ a variety of alternative agents of the same pharmacologic class. Whether or not the specific nature of the pharmacologic effects is critical or if the alteration per se is more important in the early stages of repeated use is an open question about which much is speculated, directly and by implication.

But why does one person and not another develop problems related to habitual nonmedical drug use and why is one drug chosen, and not another? These questions are presumed to be the basic ones, and for the purposes of this review can, unfortunately, only be outlined. The following factors are plausibly and frequently identified:

1. Personality: Among alcohol and drug addicts, there are more personality problems; consistently over 20% of all patients admitted to mental hospitals could carry a primary diagnosis of alcoholism. Almost by definition, those individuals who repeatedly break societal norms by aggressive and assaultive acts, whether drunk or sober, are socio- or psychopaths. Of major significance is the resolution of the question whether the agent itself has anything intrinsic or biologically to do with such behavior (for one side of the question see SZASZ, 1974).

2. External: Privation, absence of love, heartbreak, broken homes, and hard luck are common themes among those patients seen clinically with alcohol or drug abuse problems. Certainly, drug use is more frequent in environments supporting the introduction into drug use by parents and associates.

3. Physiology and drug actions:

(a) For those at risk, perhaps the agents are less aversive, less punishing, or have less negative effects and after-effects, such as less hangover.

(b) Actions of the agents may interact with personality or problem states; for example, perhaps those who are not as goal-directed as others might find sedative drugs' effects less aversive (cf. HILL, 1962; MAYFIELD, 1968a, b; KLEIN, 1970).

KLEIN (1970) presents intriguing speculation regarding some specificity of a class of psychotropic drugs in their effects as functions of psychiatric, behavioral and affective states. For example, the contrasting effects of amphetamines or caffeine (cf. GOLDSTEIN et al., 1965) in different individuals and at different ages. KLEIN advances an attractive possibility that psychotropic drugs could have specific reparative ameliorative effects to correct mental/neurophysiologic disturbances.

By like reasoning, it is conceivable that individuals may exhibit specific vulnerability to altered states of neural functioning produced by drugs. It can be further speculated that these vulnerabilities may exhibit specificity to various drug classes (HILL, 1962). Certain individuals with manic/depressive disease have been reported to concentrate their alcoholic intake during the manic phase (REICH et al., 1974, 1975). Such behavior is reminiscent of the common clinical impression that depressants are used sequentially with stimulants, such as amphetamine, in order to induce sleep or wakefulness or alteration in mood.

Drug-seeking behavior or the habit-forming propensities may be related to the specific drug. Over the short time span, such influences are described by the terms of tolerance and physical dependence. Increasingly, attention is being drawn to effects that extend to weeks and months.

In addition, it is becoming recognized that a single cycle of physical dependence sets the stage for succeeding cycles in which the dependence appears to develop more rapidly than with the initial exposure (BRANCHEY et al., 1971; WALKER and ZORNETZER, 1974; MENDELSON et al., 1966; cf. WIKLER, 1972). STEINBERG and TOMKIEWICZ (1973) describe a shift in the dose-effect curves for dextroamphetamine and chlordiazeproxide due to a single exposure a week prior to the trial. WATERS (1975, personal communication) has described an increased sensitivity to barbiturates appearing 4–7 days or more after a single dose, similar to the earlier studies of ASTON (1966a, b).

MAYFIELD (1968a, b) emphasizes that the subjective states produced by ethanol are, to a large extent, functions of the pre-existing affective state. Patients receiving a given dose of alcohol by i.v. infusion when depressed underwent marked improvement in affect, whereas when normal they showed only a slight improvement in affect. Further, the moderate drinkers had significantly more improvement with alcohol than excessive drinkers.

(c) Perhaps there is a unique sensitivity of individual brains to the actions of agent(s) (e.g., on the limbic or frontal regions). Drug effects are known to be modulated by the state of neural system. Lesions of the nervous system may significantly alter the sensitivity to a number of agents [e.g., increased sensitivity to barbiturates after frontal lesions or partial decerebrations (SEGUIN and STAVRAKY, 1957), decreased effects after septal damage (HELLER et al., 1960), increased amphetamine effects after a variety of types of lesions (see ADLER, 1970; ADLER and GELLER, 1977)].

(d) Another mechanism of interest is the impairment of memory in that it would provide a mechanism for the inhibition of memories that are painful or anxiety producing and, thus, reduce a certain fraction of total inxiety.

(e) Genetic predisposition seems probable with alcohol use, but how such predisposition is mediated and manifested is not known (see reviews by ERIKSSON, 1975; GOODWIN, 1971; SCHUCKIT et al., 1972; ISRAEL and MARDONES, 1971; RODGERS, 1972; MCCLEARN, 1975).

Would ethanol, barbiturates, etc., be addictive if they did not exhibit tolerance or withdrawal syndromes? For some individuals, the prime habit appears to be periodic severe intoxication. For these the addiction would only consist of the weekly or monthly binge. On the other hand, for the six-a-day shots or pills, it would appear that repeated use of agents with rapid onset and of short duration would be more addictive than ones with a slow onset and slow decrement, since the magnitude of the withdrawal phenomena would be less. One consequence of this approach is that those individuals with rapid absorption and metabolism would be expected to exhibit greater proneness to addiction.

Physical dependence appears to be an irreversible phenomenon—like losing virginity, growing up, or getting old. Many workers would, moreover, sharply differentiate the symptomatic excessive drinker from the biologically addicted, even though the former may, over the passage of years, become the latter.

The prime questions remain:

What are the site(s) and nature of the positive reinforcing aspects of sedative drugs or, put differently, what makes drug and alcohol use pleasant, rewarding and reinforcing? Why do some people drink, on a given occasion, to excess—by their own or by other criteria? Why are drugs that result in unconsciousness, or close to it, taken in binges, particularly by those to whom such excess comes almost as a surprise; i.e., do they lose control or give control to another? The questions in abuse syndromes, then, concern the origins and mechanisms of control of consummatory behaviors.

The following section *is based on the assumption that the interactions of alcohol and related drugs with portions of the CNS underlie the habit-forming and dependence-producing properties of these agents*, and these *interactions are, in site and mechanism, not unique to the human species and not unique to those who are, or might become, drug addicts or alcoholics.*

(This is *not* to say that these assumptions are proven or that the susceptibility of individuals and groups to the development of various forms of drug dependence does not vary widely as the result of social, biological, and psychological differences.)

C. Mechanisms and Sites of Action of Barbiturates, Ethanol, and General Anesthetics

I. Introduction: Origin of Selectivity and Specificity of Action

The two aspects of mechanism(s) and of site(s) of action of sedative/hypnotics and anesthetics will be considered in separate sections, although they obviously cannot really be separated in a comprehensive explanation. The last part of this section will attempt a synthesis and summary of what is known of both, and, more importantly, what can be discovered.

In the consideration of any agent affecting the complex multisynaptic system of the nervous system, a comprehensive frame of reference is needed. The outline presented below has been derived in part from WALL (1967), PATON and SPEDEN (1965), and ESPLIN (1970).

The prime reference frame includes defining quantitatively the spectrum of a drug's central actions, i.e., the relative sensitivity of all of the relevant neuronal systems to the agent, the relative concentrations (as functions of dose, etc.) and the final change in behavior under examination. Several processes can be identified that, potentially, could provide selectivity of drug action as expressed quantitatively and qualitatively. Obviously, each of the following categories is not necessarily exclusive; many are probably interrelated:

1. *Selective localization/distribution* to organs, cell groups, or subcellular components.

2. *Differential mechanisms of action on different cell types* as categorized according to size, density of connection, basic membrane properties, blood supply, etc. (e.g., the postulate that ether and barbiturates selectively depress small neurons or interneurons).

3. *Differential action related to organization of neurons*—e.g., multineuronal sequential pathways vs. network vs. single synapse.

4. *Differential effects relating to different functional states of the neuron*—e.g., on repetitively active vs. inactive neurons—receptivity to neurohumoral transmitters; membrane permeability (resting and active); conduction at sites on dendrites, soma, axon, and terminals; membrane potential and excitability; propagation of local and action potentials; neurohumoral synthesis and release; metabolic status such as glucose uptake and utilization, oxygen uptake and utilization, protein synthesis and catabolism, and replacement of cellular constituents.

5. *Differential effects relating to synaptic and neurohumoral mechanisms*—differences relating to selective drug effects among any of the following aspects of neurotransmission: transmitter synthesis, storage, release, translocation to postsynaptic site, postsynaptic binding; postsynaptic receptor activation of and changes in membrane permeability; initiation of postsynaptic depolarization, hyperpolarization or propagated action potential; pre- and postsynaptic interactions among excitatory and inhibitory systems.

6. *Differential effects relating to persistence of* or repeated exposure to an agent—cellular or functional tolerance or adaptation, receptor induction.

7. *Differential effects resulting from metabolic biochemical differences*—e.g., action according to redox state, glucose level, state of oxygen transport system etc. (cf. also WOODBURY et al., 1975).

II. Mechanisms

1. Selectivity Due to Differential Distribution

The possibility exists that a selectivity of action on certain central nervous elements of anesthetics, sedatives, and hypnotics is the consequence of differential distribution, including limited access to specific sites. Certainly all of the agents under consideration distribute to brain in sufficient concentrations to produce an alteration in nerve cell function. In spite of the differences among these agents, all compounds must be soluble to some degree in both water and lipids. Interestingly, OLDENDORF (1974) points out that ethanol probably has the optimal partition coefficient of a CNS drug.

That barbiturates distribute first to gray matter and later to white matter has been known since the early days of autoradiographic studies of ROTH and BARLOW (1961). Their findings have been recently expanded on (SAUBERMANN et al., 1974; SAUBERMANN, 1975; CASSANO et al., 1967; and reviewed by MAYNERT, 1972a); however, these recent studies have not revealed cellular sites of distinctly high affinity, although the kinetics appear to be rather complex and not readily correlated directly with the behavioral response to the agent![5]

2. Alterations in Axonal Excitation and Conductance; ATPase as a Site of Action

This mechanism or derivations of it have been postulated, at one time or another, for all centrally acting depressants. By itself it doesn't provide any hint as to the origin of the selectivity of actions on restricted pathways nor an explanation of the differences in action among compounds, such as phenothiazines, local anesthetics, general anesthetics, or sedative/hypnotics.

Assuming that other factors can be found to explain selectivity, does assessment of drug actions on axonal excitation and conduction provide insights as to mechanisms of such drug action? Or put differently, can the axon serve as a useful isomorphic model for the more complex nervous system? THESLEFF (1956) used it to examine anesthetic effects on the nerve-muscle synaptic system; SEEMAN (1972b) has presented and reviewed extensive data that barbiturates (BLAUSTEIN, 1968) and volatile anesthetics alter NA^+ and K^+ conductances associated with action potential generation.

The thesis that elucidation of the peripheral actions of anesthetics may provide the answer to the mechanism of general anesthesia was addressed in an extensive review by ALPER and FLACKE (1969). The effects of various general depressants are complex. For example, although ethanol was found to depress the early transient currents for Na^+ and K^+, it was either without effect on a late inactivation of the K^+ current (ARMSTRONG and BINSTOCK, 1964) using squid axon, or it was enhanced, as observed by BERGMANN et al. (1974) in *Aplysia*.

It is frequently assumed that the homologous series of alcohols have qualitatively similar properties, but this is not actually the case. For example, in a lobster axon preparation, aliphatic alcohols have independent effects on both the spike-generating and resting potentials and thus the sodium and potassium conductances (HOUCK, 1969). In a similar vein, GAGE et al. (1975) report differences in the various alcohols' action on skeletal neuromuscular transmission.

The anesthetics and sedatives are well known to produce excitatory phenomena on behavior or neural reflexes. Ethanol is directly excitatory, at least for some sensory nerves and receptors (KUCERA and SMITH, 1971, 1972; INGLE, 1971; GREENHOUSE and SZUMSKI, 1972; LATHERS and SMITH, 1973, 1976; ANDERSON and RAINES, 1974; ANDERSON, 1975) and this action may be related to either modest sensory terminal depolarization or the decreased membrane conductance at the first node. Excitatory actions of the gas anesthetics have been commented on by a number of

[5] The import and consequences of the metabolism and metabolic products of the agents under discussion will not be mentioned except as they pertain to regulation of the time-course of action. By and large, these matters are unique to the specific agent, and have little direct consequence in terms of site or mechanism of action in the CNS.

investigators (cf. PATON and SPEDEN, 1965; RICHARDS et al., 1975), but the mechanism is not known. With respect to stimulation of sensory endings, such as carotid baroreceptors, the activity is limited to only a few of the volatile and gas anesthetics (reviewed by SMITH, 1967, 1973). It is possible that the excitatory action is separable from depressant activity in view of the selective antagonism by alpha methyl-p-tyrosine of ethanol-induced behavioral excitation in mice and rats (CARLSSON et al., 1972). The behavioral excitation is also consistent with depression of an inhibitory system, as has been reported in fish (FABER and KLEE, 1975, 1976, 1977) and spinal cord (MEYER-LOHMANN et al., 1972).

Within the range of concentrations producing intoxication, there is no evidence of blockade of large axons in vivo, although the possible influence on repetitively activated small diameter fibers has been little investigated. Since the safety margin for these smaller fibers is appreciably lower than in larger fibers, and since repetitively activated units are more susceptible to block of conduction, it is at least conceivable that certain sensory changes (such as numbness, mild paresthesias, cutaneous analgesia) are effects of anesthetics and ethanol on axons themselves or presynaptically in the cord. Moreover, repetitively active neurons may be much more susceptible to depression by a variety of depressants (cf. e.g., ESPLIN, 1970; SHAPOVALOV, 1963; KLEE et al., 1975a, b; FABER and KLEE, 1975, 1976, 1977). Nevertheless, DEJONG and NACE (1967) found no effect on C fibers of reasonable concentrations of the general anesthetics nitrous oxide, ether, methoxyflurane, or halothane.

Adenosine triphosphatase (ATPase) has been seriously considered as a site of action, at least of ethanol (ISRAEL et al., 1965, 1970, 1966; ISRAEL and SALAZAR, 1967; ISRAEL-JACARD and KALANT, 1965; ISRAEL and KURIYAMA, 1971; JÄRNEFELT, 1961, 1972; KALANT and ISRAEL, 1967; SUN and SAMORAJSKI, 1970; SUN, 1976; LIN, 1976; see ISRAEL et al., 1975a, b; WALLGREN et al., 1975). That ATPase is an unlikely site of action of the neuronal effects of barbiturates is amply demonstrated by data of WASER and SCHAUB (1971) that phenobarbital was essentially without effect on synaptosomal ATPase ($>10^{-2}$M). Also, ANDERSEN (1972), in reviewing the effects of anesthetics on Na^+ transport, concludes that the net effect on Na^+ transport appears to be unrelated to the anesthetic effect. SEEMAN (1972b) likewise concludes that it is unlikely that this enzyme is involved in the basic mechanism of anesthetic nerve blockade or anesthesia.

In related studies, ouabain administered intracerebroventricularly to mice does produce CNS and locomotor depression that might be due to ouabain's inhibition of Na^+, K^+ ATPase activity; directly applied ouabain depressed the cortical neuron responses to glutamate (GODFRAIND et al., 1971).

Inhalation anesthetics such as halothane may indeed, as a primary or secondary effect, inhibit ATPase activity including that from muscle, actomyosin triphosphatase (MERLIN et al., 1974, WILLIAMS et al., 1975); this action is intimately associated with the uptake and effect of Ca^{++}. In related studies, the effects of adenosine diphosphate to stimulate the interaction of actin and myosin and inhibit dissociation of the actomyosin complex are inhibited by reasonable concentrations of ethanol and acetaldehyde (PUSZKIN and RUBIN, 1975).

WALLGREN et al. (1972) have used tert. butanol to examine the mechanism of alcohol's action. The syndrome produced in animals by tert. butanol is similar to that of ethanol. In isolated slices of cerebral cortex, tert. butanol or ethanol depressed the rise in intracellular sodium associated with electrical stimulation of the slices, although there were only weak effects on K^+ loss. The major effect of the agents was to depress the rise in sodium conductance during the action potential, whereas the inhibition of ion transport was apparently of minor importance. They conclude that ethanol acts with special affinity to the K^+ binding site in the ion transport system.

Although blockade of axonal conduction has not been generally entertained as the site/mechanism of anesthetic or sedative drug action, the mechanisms involved in the maintenance of an excitable, polarized membrane continue to be a focus of interest for all general anesthetics and for such intoxicants as ethanol.

For example, the effects described above of ethanol on axonal membrane potential and ionic conductances are not directly consistent with a primary and selective effect of ethanol on membrane Na^+, K^+ ATPase. However, the finding that ethanol interacts more directly with K^+ concentration effects on ATPase activity presents more intriguing possibilities (ISRAEL et al., 1966, 1975a, b; ISRAEL and SALAZAR, 1967; LIN, 1976; SUN, 1976). A variety of data implicate anesthetic and ethanol actions with changes in K^+ permeability and with Ca^{++} interactions with K^+ permeability (LAZAREWICZ et al., 1974). However, as will be seen below, a convincing case has yet to be made for anesthetics or sedative/hypnotics altering specific axonal ionic conductance mechanisms. That such changes might relate to the central effects of ethanol is suggested by the finding that ethanol inhibits sodium transport in brain slices during electrical stimulation and active potassium ion transport in the recovery period following electrical stimulation in one strain of rat, whereas in cortical slices from the high-drinking AA rat strain this inhibition of K^+ transport was not observed (BOGUSLAWSKY and NIKANDER, 1972). In liver, chronic administration of ethanol appears to increase permeability to K^+ (BERNSTEIN et al., 1974).

The finding of an adaptive increase in brain Na^+, K^+ ATPase with chronic ethanol treatment raises the possibility of ATPase as a site of ethanol action, or at least tolerance development (ISRAEL et al., 1970, see also BERNSTEIN et al., 1974). AKERA et al., (1973), WALLGREN et al., 1975; and GOLDSTEIN and ISRAEL (1972) failed to confirm the findings of ISRAEL et al., (1970), whereas KNOX et al. (1972) found that chronic treatment of cats with alcohol resulted in increased brain Na^+, K^+ ATPase only in frontal association cortex and hippocampus, but not in the caudate, amygdala or reticular system measured days after withdrawal (see also the extensive review on ATPase function of SCHWARTZ et al., 1975, and the New York Academy of Science Symposium edited by ASKARI, 1974).

The fact that alcohol is an inhibitor of ATPase presents serious difficulties in its use as a solvent for experiments, e.g., on ionophore function (see, e.g., discussion of FORTE et al., 1974).

3. Alterations in Synaptic Transmission: Excitatory Transmission; Presynaptic Inhibition; Postsynpatic Depression

a) Depression of Central Excitatory Synaptic Transmission by Barbiturates and General Anesthetics

Most proponents of axonal conduction as a model would agree with the general conclusion that synaptic transmission is impeded by concentrations of anesthetics, ethanol, or sedative/hypnotic agents which are much lower than those required to alter conduction along axons[6]. The original study by LARRABEE and POSTERNAK (1952) (see also LARRABEE et al., 1952) concluded that pentobarbital, chloroform, and chloretone were about 10 times more effective as depressants of ganglionic synaptic transmission than of conduction; ether was 3 times more effective, whereas ethanol and urethane depressed both equally, a fact frequently overlooked. Thus, with urethane and ethanol as well as other short chain alcohols, the synaptic transmission was *not* more sensitive than axonal conduction, clearly indicating rather marked differences in selectivity of action among this group of general depressants. (Some of the reports are: BROOKS and ECCLES, 1947; SHAPOVALOV, 1963; CRAWFORD and CURTIS, 1966; SOMJEN and GILL, 1963; SOMJEN, 1963; SOMJEN et al., 1965; THESLEFF, 1956; CRAWFORD, 1970; WESTMORELAND et al., 1971; RICHARDS, 1971, 1972a, b, 1973, 1974; RICHARDS et al., 1975; NICOLL, 1972, 1974, 1975a, b; RICHENS, 1969; THOMSON and TURKANIS, 1973; GALINDO, 1969, 1972; DEJONG et al., 1970; ADAMS et al., 1970; ADAMS, 1974; BARKER, 1975; BARKER and GAINER, 1973; BARKER and NICOLL, 1973; RANSOM and BARKER, 1975; WEAKLY, 1969; SCHLOSSER, 1971; GAGE, 1965; GAGE et al., 1975; QUILLIAM, 1959; ZORYCHTA, 1974; FABER and KLEE, 1976, 1977.)

b) *Presynaptic Sites of Action to Reduce Transmitter Synthesis or Release*

Depression of excitatory postsynaptic potentials (EPSPs) can arise from decreased transmitter release consequent to a variety of mechanisms or from postsynaptic depression. The latter could be due either to general depression of neuronal excitability or to selective depression of responsitivity to transmitter or the receptor-coupled change in conductance. An accumulating body of evidence favors at least one mechanism and site of action for barbiturates and ethanol, i.e., presynaptic with depression of transmitter release and/or increase in presynaptic inhibition (ECCLES and MALCOLM, 1946; ECCLES et al., 1963; SCHMIDT, 1964; LØYNING et al., 1964; MIYAHARA et al., 1966; BANNA, 1969 (and chloral hydrate); BANNA and JABBUR, 1969; WEAKLY, 1969; DOWNES and WILLIAMS, 1969; DAVIDSON and RIX, 1972; QUASTEL et al., 1972; DAVIDOFF, 1973; ZORYCHTA, 1974; NICOLL, 1972, 1974, 1975a, b;

[6] Re-reading the old literature has been a revealing experience. LARRABEE and POSTERNAK (1952), in a delightfully complete study, showed that pentobarbital, chloretone, chloroform, ether, and n-octyl alcohol selectively depressed synaptic transmission more than axonal conduction through sympathetic ganglia. This depressant activity was accentuated by repetitive stimulation; special note was made of the phenomenon although little has been done to follow up on the observation (see KLEE et al., 1975a). Even by this date, the importance of frequency of excitation and anesthetic action had already been recognized by JARCHO (1949), MARSHALL (1941) and MARSHALL et al. (1941). The possibility that ethanol may have had some facilitatory action on the synaptic transmission, as well as a depressant one, was also recognized by LARRABEE and POSTERNAK (1952). With reference to the role of frequency of neural activation and selectivity of depressive action, see also: the general discussions by ESPLIN (1970) and SMITH (1965); LONGO (1961) with reference to depression of Renshaw cell discharge by mephenesin; WINTERS et al. (1967a) review; observations on cortical after-discharge; KLEE et al. (1975a) regarding ethanol actions on spinal reflexes; and the general phenomenon that tonic reflexes requiring sustained repetitive discharges are more susceptible to depression than phasic ones CHIN and SMITH (1962), SMITH and MURAYAMA (1964), and MATSUSHITA and SMITH (1970a, b). However, there are marked differences in relative sensitivity among compounds (CHIN and SMITH, 1962).

RANSOM and BARKER, 1975). This possibility has also been suggested by SOMJEN and GILL (1963) and PATON and SPEDEN (1965).

Bromide ion has been long known to act as a sedative and anticonvulsant, presumably by replacing a critical amount of chloride. The mechanism of subsequent neural events and the mechanism of bromide sedation are not known. MUCHNIK and GAGE (1968) reported that bromide reduced transmitter release at skeletal muscle fibers in a high Mg^{++}, low Ca^{++} medium, and suggested this as a mechanism underlying sedation. GINSBORG (1968) shortly thereafter showed that a decrease in transmitter release did not occur in sympathetic ganglia after replacement of 80% of the chloride by bromide. Thus, the matter remains open.

ZORYCHTA (1974) has completed a through comparison of ether, halothane, and procaine on the spinal monosynaptic EPSP in motoneurons. Ether and halothane were found to reduce the quantum content of the evoked release from a primary afferent fiber to a spinal motoneuron. In concentrations required for surgical anesthesia, neither general anesthetic altered presynaptic conduction or postsynaptic depolarization by the transmitter. In contrast, the depression of the monosynaptic reflex by procaine was the consequence of blockage in fine branches of the afferent fiber. ZORYCHTA's data are consistent with an action of the anesthetics resulting in either an increased level of existing presynaptic inhibition or altering more directly the transmitter release mechanism. It is, however, very difficult to delimit the degree to which the variable depression of electrical excitability of the postsynaptic neuron is significant to transmission block (SOMJEN, 1967; ZORYCHTA, 1974; RICHENS, 1969; EIDELBERG and WOOLEY, 1970; SHAPOVALOV, 1963).

NICOLL (1975a, b) has recently extended the understanding of anesthetics using the frog spinal cord preparation to demonstrate that barbiturates either mimic or potentiate the action of the transmitter from the interneuron whose axon terminates on the presynaptic ending. The action of this transmitter is to depolarize the presynaptic terminal and, thus, reduce the amount of transmitter released upon its orthodromic activation. [A promising candidate for the mediator of this presynaptic depolarization is gamma-aminobutyric acid (GABA)—cf. NICOLL, 1975a.]

Pentobarbital as well as amylobarbital, thiopental, and barbital prolonged the dorsal root potentials in an isolated frog spinal cord preparation; primary afferents were depolarized and their excitability increased. In higher concentrations of barbiturates, dorsal ganglion cells are also depolarized (NICOLL, 1975a). GABA has similar effects. Some 5–15 min after its application, the dorsal root ganglion cells are desensitized to the depolarizing action of GABA. In such a densensitized state, the cell is less sensitive to pentobarbital. NICOLL (1975a, b) concludes, firstly, that low concentrations of barbiturates mimic the action of GABA on receptors located presynaptically on excitatory afferents, and secondly, that barbiturates retard the reuptake of GABA following its release; both of these effects result in a decrease of transmitter release presynaptically. In addition, RICHENS (1969) reports block of presynaptic conduction in isolated frog cord preparations. Ethanol also augments presynaptic depolarization and potentiates the effects of GABA (DAVIDOFF, 1973).

In pursuing these studies further, NICOLL (1975a, b) found that barbiturates mimic the action of GABA on frog motoneurons—an inhibitory hyperpolarization. This hyperpolarization is coupled with an antagonism of the excitatory depolarization produced by either the natural transmitter or by applied glutamate. Why this hyperpolarization has not been seen in vivo is not clear; no change in resting membrane potential in vivo has been explicitly reported by a number of investigators—cf., e.g., SASAKI and OTANI (1962), LØYNING et al. (1964). RICHENS (1969) reports increased excitability of motoneurons in isolated frog spinal cord following methohexital. NICOLL (1975a, b) thus suggests that barbiturates act both pre- and postsynaptically, with the latter action consisting of at least two separable mechanisms.

RANSOM and BARKER (1975), using mouse spinal neurons in culture, reached almost identical conclusions following a comparative examination of the interactions of pentobarbital with glutamate or GABA applied iontophoretically [one cannot help but contrast this report with another by one of the same authors (BARKER, 1975) with strongly put—almost contrary—conclusions described below (Subsect. C.II.4)].

Interestingly, the monosynaptic depression and presynaptic actions of pentobarbital are antagonized not only by picrotoxin but also by catechol (BANNA, 1970), an agent having a prejunctional site of action at the neuromuscular junction (GALLAGHER and BLABER, 1973). The evocation and potentiation by anesthetics of what became known as presynaptic terminal depolarization and the associated presynaptic inhibition has a relatively long history (ECCLES and MALCOLM, 1946; ECCLES et al., 1963). Although these findings make a good story, there are a number of conflicting observations [7].

It seems unlikely that presynaptic inhibition enhancement is *the* basic neural mechanism of action of sedative/hypnotics in view of the different potencies among the agents and the fact that an agent such as mephenesin with subjective effects similar to alcohol and which relieves the tremulousness of alcohol withdrawal does not augment spinal presynaptic inhibition. In fact, mephenesin blocks such inhibition (LLINÁS, 1964; MIYAHARA et al., 1966), whereas trimethadione, an agent with little anesthetic action, produces augmentation of presynaptic inhibition.

Although there is much evidence of a presynaptic site of action, the conclusion that anesthetics, including barbiturates, gaseous general anesthetics, and ethanol act directly on excitable tissues to reduce excitability especially at the synapse, is also widely held. For example, SOMJEN (1963, 1967) and SOMJEN and GILL (1963), in definitive studies of spinal motoneuron reflexes, concluded that the anesthetics ether and thiopental act primarily postsynaptically to produce membrane stabilization; similar data are reported by SHAPOVALOV (1963) with convincing observations of the blockade of antidromic activation of motoneurons. However, SOMJEN points out that reduction in transmitter output would also be compatible with their results. And most investigators—even those addressing presynaptic sites—find postsynaptic depressant actions, especially with larger doses (e.g., SHAPOVALOV, 1963; LØYNING et al., 1964; WEAKLY, 1969; NICOLL, 1975a, b; ZORYCHTA, 1974).

But the more troublesome aspect is that for certain extensive studies and certain sites, the presynaptic actions have apparently not been detectable, in comparison with postsynaptic actions. In a series of recent studies, RICHARDS and his colleagues (RICHARDS, 1971, 1972a, 1972b, 1973; BLISS and RICHARDS, 1971; summarized by RICHARDS, 1974, and RICHARDS et al., 1975) have examined the actions of a series of

[7] One is reminded of the statement attributed to Linus Pauling, "But don't let a fact stand in the way of good theory."

anesthetics on synaptic transmission in an isolated preparation of olfactory cortex. All agents depressed excitatory synaptic transmission. Interestingly, there were significant differences among agents. Although KRNJEVIĆ, (1974a, b) (also see Subsect. C.II.6.c) reported that anesthetics had little, if any, effect on the responses of cortical cells to glutamate, RICHARDS (1974) found that three of the anesthetics tested (pentobarbital, methoxyflurane and tricholorethylene) caused a reversible depression of glutamate-induced activity of cortical cells; however, halothane failed to have any effect, although the concentration tested depressed synaptic transmission. RICHARDS points out that his results are compatible with the notion that halothane acts to interfere with transmitter release whereas the other anesthetic agents act, in part at least, to depress postsynaptic receptor functions.

A number of agents can be shown to increase or to evoke presynaptic inhibition in the spinal cord including general anesthetics, such as nitrous oxide, and the analgesic meperidine and benzodiazepines such as diazepam (SCHMIDT, 1965, 1971; SCHMIDT et al., 1967; SCHLOSSER, 1971; CHIN et al., 1974; POLZIN and BARNES, 1976). The presynaptic effects on the lumbar cord of diazepam and morphine are reduced but not abolished by spinal cord section (CHIN et al., 1974). Interestingly, nitrous oxide anesthesia blocked the presynaptic actions of meperidine, but not those of diazepam (see Subsect. C.IV).

Assessing the site of action of drugs increasing presynaptic inhibition poses another set of questions of possible sites from the primary afferent membrane, the proximal synapse, intercalated neurons and their segmental and central connections to the sensory nerves and terminals. For example, some of alcohol's apparently presynaptic actions could well derive in part from its stimulatory effects on muscle spindles (KUCERA and SMITH, 1971; LATHERS and SMITH, 1976).

EIDELBERG and WOOLEY (1970) were unable to find evidence of a spinal presynaptic action of ethanol, although a replication of those experiments by MURAYAMA (personal communication, 1974) did confirm the earlier findings of MIYAHARA et al. (1966) and indirectly LØYNING et al. (1964). Moreover, an action presynaptically does not preclude simultaneous effects directly on the motoneuron. The apparently analogous studies of SAUERLAND et al. (1970) with ethanol, using the trigeminal motor neurons, simply revealed no direct effect of ethanol on the trigeminal reflexes of presynaptic synapses on the primary afferent—essentially all of the actions of ethanol obtained in that study had their origin rostral to the site of the recording.

It should be noted that if ethanol—as well as other agents—had no other effect than to reduce and retard the spike potential, it would tend to have the end effect of reducing transmitter release from presynaptic terminals; thus, it might well be concluded to act selectively on presynaptic mechanisms. Recording from the dorsal root ganglia cells present a potentially useful procedure for examining in greater detail the transmitter and the ionic requirements of the putative transmitter, GABA, and agents influencing presynaptic systems (FELTZ and RASMINSKY, 1974).

SOMJEN and GILL (1963) found no alteration in post-tetanic potentiation (PTP) which suggested that certain presynaptic mechanisms per se are uninfluenced by the anesthetics; however, RICHARDS et al. (1975) depicts marked increases in PTP by methoxyflurane. Nevertheless, it seems generally true that anesthetics, barbiturates, alcohol or benzodiazepines may not selectively depress PTP (cf. WOODBURY and

ESPLIN, 1959; SCHLOSSER, 1971). Selectively localized effects of diazepam on hippocampal PTP has been recently described (MATTHEWS and CONNOR, 1976).

The complexity of analyzing the synaptic sites of drug action are readily illustrated by a recent study of FABER and KLEE (1975, 1976) on the effect of ethanol on the Mauthner cell of goldfish and its collateral inhibitory system. Intra- and extracellular recordings were obtained from the Mauthner cell before and after addition of 1–2% ethanol to the water respiring the fish. Low concentrations of ethanol (3–5 µg/mg brain weight) specifically blocked, over a period of 1 h, both the electrically and chemically mediated components of collateral inhibition of the Mauthner cell.

Only with appreciably higher concentrations was the excitability of the Mauthner cell itself altered; the safety factor for impulse transmission to the axon hillock was reduced and the action potential eventually failed. Common inhibitory interneurons mediate both collateral and afferent inhibitions of the Mauthner cell; the latter were not affected by ethanol. Furthermore, the same interneurons mediate the electrical and chemical transmission components of the collateral inhibition. Therefore, it was concluded that the block of collateral inhibition induced by ethanol occurred at the excitatory synapses between the Mauthner cell axon collaterals and those interneurons. On the basis of the changes in the frequency characteristics of this synapse during the development of the ethanol effect, these results have been interpreted as indicating that ethanol acts presynaptically to impair transmitter release from the Mauthner cell axona to the inhibitory interneuron (see FABER and KLEE, 1977).

There may well be multiple relevant mechanisms and sites of action of sedative and anesthetic agents (SALMOIRAGHI and WEIGHT, 1967). GALINDO (1969) compared the actions of procaine, pentobarbital, and halothane on cuneate neurons, including pre- and postsynaptic aspects. Pentobarbital appeared to affect most sensitively the postsynaptic excitation evoked trans-synaptically, and that due to acetylcholine (ACh) application (KRNJEVIĆ and PHILLIS, 1963a, b; BLOOM et al., 1965; KRNJEVIĆ, 1975), but it also acted at presynaptic sites. In contrast, halothane appeared to be rather inactive at presynaptic or postsynaptic sites, and GALINDO suggests that it acts by facilitating inhibition. CATCHLOVE et al. (1972) and KRNJEVIĆ (1975) emphasize that many anesthetic agents including methohexital, halothane, chloroform, methoxyflurane, ether, trichlorethylene, or nitrous oxide each depress spontaneous activity and the discharges of single cortical cells to iontophoretically applied ACh—whereas the activity evoked by glutamate is either increased or unaltered (but see RICHARDS, 1974 and above). In contrast, local anesthetics depress all activity irrespective of the initiating agent.

The possibility that there is a complex interaction between the mediator and the depressant anesthetic agent is emphasized by BLOOM et al. (1965) with the statement:

> the possibility of selective effects of anesthetics on the sensitivity to suspected transmitters of different nerve cell populations in the mammalian central nervous system should be borne in mind.

They point out that the facilitation produced on single units on the caudate nucleus induced by ACh was diminished or suppressed by small doses of barbiturates or ether, whereas the ACh depression and the responses to norepinephrine (NE), dopamine (DA), glutamate, and GABA were unaltered. Nevertheless, these findings support the suggestion of KRNJEVIĆ (1972, 1974a, b, 1975) that there exists a selective interaction between ACh-induced activation of certain cells and the actions of anesthetics, including barbiturates.

FRANK (1972) emphasizes the concept of "key" synaptic sites of action of anesthetics in the presence of the fact that many agents with different sites or modes of action may produce anesthesia-like states. In a series of experiments, he and his colleague [FRANK and JHAMANDAS (1969, 1970a, b), (reviewed in FRANK, 1972)] examined the interaction of agents with the dose-effect curve for phenobarbital and, in parallel experiments, the effects of the agents on cortical slabs. All of the agents with any depressant neural effects were synergistic with or potentiated the actions of phenobarbital, including agents generally known to produce convulsions in large doses such as local anesthetics and tetrodotoxin. All of the same agents reduced the electrical responses of the cortical slab to direct cortical stimulation. In contrast, the "convulsants" bemegride, picrotoxin and pentylenetetrazole had direct cortical excitant effects, and were antagonistic to phenobarbital in mice.

These experiments confirm that a variety of sedative and anesthetic agents *can* or may act on cortical neurons but they do not provide explicit evidence about the possible key sites for general anesthesia or the possible locus of action in synaptic transmission. Nevertheless, such studies do provide some indication of differential classification of such anesthetics. They also leave open the disturbing possibility that the site and mechanisms of action of sedative agents—as well as of the antagonistic convulsants—may have so far escaped attention and are not among the many mechanisms already discovered.

c) *Effects on Inhibitory Postsynaptic Potential (IPSP)*

The depressions of excitatory transmission by barbiturates or ethanol have generally been found to take place in the absence of any marked alteration of the IPSPs on the same neuron (WEAKLY, 1969; ECCLES et al., 1971 a, b; NICOLL, 1972). However, there may well be indirect activation or depression of IPSPs by actions on remote neurons, e.g., as observed with ethanol by FABER and KLEE (1975) with Mauthner cell collaterals or the excitation of Renshaw cells (MEYER-LOHMANN et al., 1972). Also, the subject is not closed in view of the unequivocal prolongation of IPSPs by hexobarbital reported briefly (LARSON and MAJOR, 1970); BARKER (1975) also noted a prolongation of the IPSP by alcohol and attributed it to a presynaptic action. GALINDO (1969) has also suggested that halothane may act by facilitating inhibition as a primary effect. The most interesting data in this connection are the findings of NICOLL et al. (1975) that pentobarbital markedly prolongs the inhibitory postsynaptic potentials in the hippocampus, presumably as a result of blockade of GABA uptake.

4. Evidence from Invertebrate Nervous Systems

The availability of simpler nervous systems in invertebrates has tempted many to use molluscan neurons as a model. BARKER (1975), using a variety of invertebrate preparations, has argued for "selective depression of postsynaptic receptor-coupled Na^+, K^+, Ca^{++} conductance" as *the* mechanism of the CNS depression observed with agents commonly used as general anesthetics (pentobarbital, chloroform, chloralose, urethane, and ethanol)[8]. In his view, selectivity of action such as an anticonvulsant effect is achieved as the result of varying degrees of hydrophobicity corresponding—in some undefined fashion—to sites of weaker or stronger Na^+-dependent postsynaptic excitation.

The report of BARKER (1975) is, in part, in conflict with that of BERGMANN et al. (1974) which addressed only the actions of ethanol—which produced a mixture of effects differing from cell type to cell type—membrane hyperpolarization in some, depolarization in others. These two reports are in agreement that the agents involved reduced Na^+ and Ca^{++} conductance changes. BERGMANN et al. (1974) found that ethanol enhanced delayed outward potassium currents, although a fast outward K^+-dependent current observable in some cells was reduced. These investigators also found a decrease in amplitude and rate of rise of the action potential following ethanol, consistent with a decrease in Na^+ conductance.

The effects of ethanol in reducing the membrane potential in one group of *Aplysia* neurons via an increase in resting Na^+ permeability (BERGMANN et al., 1974) is consonant with much earlier studies on frog nerve (GALLEGO, 1948), on frog muscle fibers (KNUTSSON, 1961; KNUTSSON and KATZ, 1967; INOUE and FRANK, 1967), and on squid giant axon (ARMSTRONG and BINSTOCK, 1964). In *Tritonia*, ethanol depressed excitatory synaptic transmission and reduced membrane potential and EPSP amplitudes; action potentials were decreased and prolonged (CHASE, 1975).

The effects of ethanol on *Aplysia* have recently been further elaborated (KLEE, FABER, BERGMANN, personal communication, 1975b; see recent review by FABER and KLEE, 1977) to show that concentrations of 0.5–4% produce graded, concentration-dependent depression of EPSPs and, in parallel, depression of the responses to ACh—whether the ACh action was hyperpolarizing (H cells) or depolarizing (D cells). The hyperpolarizing response is apparently Cl^- dependent, and in cells exhibiting a

---

[8] Barker's (1975) conclusion is based on a series of findings:
1. Depression by pentobarbital of EPSPs in lobster muscle fibers (EPSPs mediated by glutamate and coupled to increases in Na^+ and Ca^+ conductance); IPSPs, membrane potential or resistance were not altered in 0.2 mM.
2. Pentobarbital depression of the depolarizing phase of "biphasic postsynaptic potential" in a land snail neuron (DA is indicated as the probable transmitter and Na^+ and K^+ conductances are involved). The membrane potential and resistance were not altered; IPSPs in the same cells were not influenced; however, the duration and frequency of the IPSPs increased—and this was taken to indicate a presynaptic site of action as well.
3. The evocation by pentobarbital, chloroform, chloralose, urethane, and ethanol, etc. ("All of the agents used") of an increase in the number of spikes in the bursts of the snail cells exhibiting bursting pacemaker potentials. Also, the amplitudes of the potentials were increased. Thus, in this preparation, pentobarbital (and other anesthetics?) facilitated the release of an inhibitory transmitter and increased neuronal excitability.
4. Depression in the sea hare, *Aplysia*, by pentobarbital of the amplitude of the EPSP without altering its temporal facilitation.
5. Antagonism by pentobarbital of the glutamate effects on a lobster muscle preparation.
6. Antagonism by all of the anesthetic agents tested of the actions of ACh on the snail nerve cell, whereas they reportedly failed to influence a number of tests of responses presumably dependent on alterations of Cl^- or K^+ conductances.
Although BARKER (1975) claims that these observations are clear-cut and apply for all anesthetics tested, the figures presented do exhibit such effects as increased discharge frequency immediately following chloroform, an ACh-induced hyperpolarization in snail neurons and an increase by pentobarbital in the ACh-induced depolarization of a snail neuron.

mixed hyperpolarization consisting of two components, one K^+ and one Cl^--dependent, ethanol depressed both with the Cl^--dependent component more antagonized than the K^+ conductance component.

In *Aplysia* pentobarbital blocks ACh effects in both D and H cells (SATO et al., 1967; KLEE et al., 1975a, b); in addition, it produces hyperpolarization, eventually with irreversible hyperpolarization with large concentrations. Evidence was obtained that pentobarbital increased K^+ conductance; the reversal potential for the hyperpolarization was found to be -72 MV, concluded to be the K^+ equilibrium potential. The effects of pentobarbital parallel those just cited for ethanol.

The effects of ethanol on the electrical excitability of *Aplysia* neurons are similar to those found with mephenesin (KLEE and FABER, 1974). [Mephenesin is known to induce subjective feelings of drunkenness and to relieve the tremors and anxiety of early withdrawal from ethanol (SCHLAN and UNNA, 1949; SPRENG, 1953; SULLIVAN, 1954) or withdrawal from barbiturates in animal studies (CROSSLAND and TURNBULL, 1972). In addition, it has some local anesthetic activity in addition to its better known apparently selective actions on interneurons and polysynaptic reflexes (cf. review of SMITH, 1965).]

In invertebrate preparations one may see analogies of the more complex nervous system. Moreover, these nervous elements demonstrate, unequivocally, differentiation in sensitivity to volatile anesthetics among different cells and among different sites on the same cell (CHALAZONITIS, 1967). Whether these differences relate to neuronal cytostructural differences at sites of anesthetic molecule interaction, or whether they reflect differences in membrane and cell safety margins, is not known.

5. Evidence from the Skeletal Neuromuscular Junction

The skeletal neuromuscular junction has been extensively employed as a model synaptic system and certainly such studies have provided much of the knowledge and leads for investigations at other synaptic sites. Disruptions of skeletal neuromuscular transmission produced by sedatives, hypnotics, or alcohols are thought to be minimal when the agents are given systemically for their central effects of sleep or sedation (see review of ALPER and FLACKE, 1969). A sophisticated review of drug action on the skeletal muscle end-plate has recently appeared (COLQUHON, 1975).

SEYAMA and NARAHASHI (1975) have examined and reviewed the actions of pentobarbital on the isolated neuromuscular preparation of the frog. The primary and most obvious effect was a suppression and block of neuromuscular transmission due primarily to a decrease (0.5–1.5mM) in the sensitivity of the end-plate to ACh. In addition, and related to this action, was a reduction in the magnitude and time course of the end-plate currents, both electrically and ACh-induced. Both the Na^+ and K^+ end-plate currents were depressed in amplitude and the falling phase of the current appreciable shortened (see also ADAMS, 1974). Pentobarbital produced only a small degree of depolarization in the resting membrane potential; this decrease occurred in both the end-plate and in the non-end-plate regions of the muscle. The authors note, in discussing their results in the context of those of THOMSON and TURKANIS (1973) and the earlier investigations of THESLEFF (1956), that in low concentrations pentobarbital also has the effect of increasing the quantal content of the end-plate potential (e.p.p.). In higher concentrations, this effect decreases and with the highest concentrations, there was actually a decrease in the quantal content. These authors review briefly the synaptic effects of the barbiturates and point out that postsynaptic depression has been observed by a number of different investigators in vertebrate preparations (see Subsect. C.II.3). SEYAMA and NARAHASHI (1975) did not address the presynaptic effects in any detail. One can wonder if the modest depolarization observed with the barbiturates could be analogized to a modest degree of presynaptic depolarization which would be consonant with the observations of increase in presynaptic inhibition, even though they state that there is an increase in quantal content of the e.p.p.

Inhibitors of oxidative phosphorylation produce an increase in miniature e.p.p. (m.e.p.p.) frequency in isolated frog muscle preparations plus an augmentation of transmitter released per nerve impulse (ALNAES and RAHAMINOFF, 1975); the authors postulate that the mitochondria participate in the regulation of intracellular free Ca^{++}. Presumably, synaptic depressants could act via alteration in mitochondrial function (see Subsect. C.II.9). OKADA (1967) notes that the effects of Mg^{++} and Ca^{++} on m.e.p.p. frequency in frog muscle were facilitated by the presence of ethanol.

QUASTEL and LINDER (1975) have reviewed the results of a series of studies of anesthetics on the mammalian neuromuscular junction. They conclude that the actions of ethanol and pentobarbital and other depressants differ with respect to their underlying mechanisms. This conclusion is based on the observation that spontaneous frequency of m.e.p.p.s is increased by alcohols, chloral hydrate, pentobarbital, ether, chloroform, urethane, and chlorpromazine; these actions are apparently independent of the Ca^{++} level in the bathing medium. These agents also, in varying degrees, depress transmitter release, an action related to external Ca^{++} and nerve terminal polarization. Ethanol is cited as having little, if any, presynaptic inhibitory action in contrast with pentobarbital which has much, yet produces only a small augmentation of release. An initial portion of ethanol's action is due to potentiation of the twitch response as the result of repetitive action potentials initiated prejunctionally (COOPER and DRETCHEN, 1975).

Preliminary experiments with ether and halothane indicate that the major postsynaptic effects is a reduction in maximum miniature end-plate current (m.e.p.c. amplitude and rather complex and conflicting changes in the time course of the m.e.p.c. Halothane inhibits depolarization by cholinergic agents (GISSEN et al., 1966; WAUD et al., 1973).

QUASTEL and LINDER conclude definitively:

> The diversity and specificity of effects indicate that there is no single common action of these agents. The actions of ethanol and pentobarbital on the subsynaptic membrane are inconsistent with the concept of nonspecific membrane stabilization.

In agreement with these more definitive studies, GERGIS et al. (1975) noted in an isolated frog sciatic nerve gastrocnemius preparation that thiopental (10,100 mg/ml), ether (3, 6%), and methoxyflurane (0.1, 0.3%) depressed muscle contraction but did not alter ACh release as determined by actual measurement of ACh release upon tetanic stimulation; thus, they acted postsynaptically. Halothane or forane was without influence. KENNEDY and GALINDO (1974) conclude, using the rat diaphragm preparation, that ethers act selectively postjunctionally, whereas the nonethers act prejunctionally. Further evidence of differential actions is the finding that pentobarbital and thiopental inhibit the Ca^{++} uptake by rat brain synaptosomes stimulated by potassium (BLAUSTEIN and ECTOR, 1975); however, phenobarbital, ethanol, and chloroform were without effect. These data are consistent with the postulate that pentobarbital and thiopental act to decrease transmitter release.

The current associated with the m.e.p.p., in agreement with GAGE et al. (1975), was found by QUASTEL and LINDER (1975) to be appreciably prolonged by ethanol (0.2–0.8 M) or octanol (GAGE et al., 1974). They conclude that ethanol "acts to delay dissociation of the ACh-receptor complex," whereas pentobarbital was found to affect only a slow component of the m.e.p.c. to reduce and prolong it. In contrast, SEYAMA and NARAHASHI (1975) report a selective action of pentobarbital on the sodium component of the m.e.p.c., to decrease the rate of rise and accelerate the fall of ACh depolarization.

The actions of alcohol on the neuromuscular junction are complex; presynaptic effects are reflected in a marked increase in m.e.p.p. frequency (OKADA, 1967; GAGE, 1965). An increased e.p.p. amplitude is observed in low concentrations and a decrease in high.

The most recent in a series of innovative investigations by GAGE and co-workers (see last papers in the series in 1975) has resulted in a somewhat unique theory of alcohol action. GAGE et al. (1975) point out that the alcohols, and they place particular emphasis on ethanol, act postsynaptically to increase the amplitude and duration of m.e.p.p. and potentiate the depolarization evoked by ACh. They show, further, that these actions are the consequence of a prolongation of the decay phase of the endplate currents (without action on the "growth phase" of such currents). After examining a number of aliphatic alcohols, they suggest that the alcohols act in the lipid phase of the postsynaptic membrane to cause a change in its dielectric constant. The membrane so altered is viewed as the "environment of the rate-limiting reaction responsible for the decay of the endplate conductance."

GAGE and his colleagues appear to have their focus on the process of recovery of neuromuscular transmission and not so much on determining the mechanisms of alcohol action. As such, it is understandable that the actions of alcohol at other sites are not addressed in discussing the recent work, such as the actions presynaptically of increased frequency of m.e.p.p.s. and changes in the quantal content of the e.p.p.

GAGE et al. (1975) do present information valuable in ascertaining the site(s) of greatest susceptibility to alcohols. They found no evidence of effects on the passive electrical properties of the muscle membrane (other than the reflection of the action on m.e.p.c.). In contrast to other reports, the increase by ethanol of membrane resistance was slight and there was no alteration in the capacitance; however, small decreases in membrane resistance are reported by OKADA (1967). Further, they argue against alcohols acting as cholinesterase inhibitors; the results are in many ways consistent with that interpretation except for evidence that the membrane "noise" is altered in a fashion indicating that the alcohol alters the "elementary" process. These recent studies direct attention to the possibility that ethanol and other alcohols have a primary mechanism of prolonging postsynaptic junctional currents. The evidence is held to favor the view that this prolongation of junctional currents follows from a reduction in the rate of conformational change in a macromolecule possessing a dipole moment in the postsynaptic membrane.

It would be of particular interest to ascertain whether this type of action occurs at other synaptic sites and with other transmitters. One could speculate that such an action is consistent with known effects of ethanol to increase Renshaw cell activity, a neuron activated by cholinergic collaterals from motor neurons. It is also consonant with the rather old observation that ethanol and other alcohols potentiate ACh actions and antagonize neuromuscular block produced by tubocurarine (NELEMANS, 1962; SACHDEV et al., 1963), as do catechol and certain phenols (HOBBIGER, 1952; RUMMEL and SCHMITZ, 1954; MOGEY and YOUNG, 1949; BLABER and GALLAGHER, 1971).

If one speculates that the ethanol action will be seen in the recovery cycle in depolarizing synaptic mechanisms generally, then one would predict an increase and potentiation of spinal monosynaptic transmission, and an increase and prolongation of presynaptic inhibition, one consequence of which could be analgesia (cf. CHIN et al., 1974). Increased and prolonged presynaptic inhibition, of course, is known to be produced by ethanol and other depressants (see Subsect. C.II.3).

GAGE et al. (1975) make extremely important points regarding correlating in vitro findings on synaptic systems with those in vivo:

> Finally, two interesting points emerge from these results with the alcohols. First, because the environment of a molecular reaction occurring in a biological membrane may be an important determinant of the rate of that reaction, the rates of biologically important reactions occurring *in situ* may be quite different from the rates measured in a "test-tube environment," despite the fact that the molecules themselves are unchanged. Secondly, because the rate of the decay phase of a synaptic current can be an important determinant of the input-output function of a synapse (GAGE and MCBURNEY, 1973) and because it is possible that the environment of the important reacting molecules may play a large part in determining the duration of this phase of the current flow, it is possible that it is in fact the "environmental factors" which are modified in certain adaptive and pathological processes at synapses.

As noted, this publication avoids addressing previous studies which showed moderate depolarization (possibly consequent to inhibition of Na^+, K^+ ATPase), increased m.e.p.p. frequency (see OKADA, 1967), interaction of ethanol action with Ca^{++}, and increased membrane resistance. But also of interest is the long time course of onset of ethanol action reported by GAGE et al. (1975) in contrast to the rapid equilibration of ethanol with water assumed by many investigators; yet, to reach anything approaching steady-state effects in the nervous system may take of the order of 30 min (see GOSTOMZYK et al., 1969; LATHERS and SMITH, 1976).

Other peripheral synaptic systems have been used in the fashion of models for central synapses and the pioneer work of LARRABEE and POSTERNAK (1952) using autonomic ganglia was cited above. In a similar preparation QUILLIAM (1959) found that paraldehyde and methylpentynol also depress transmission, probably by altering ACh transmitter release. Using guinea pig ileum, SPEDEN (1965) found distinct differences among anesthetics with many having both pre- and postsynaptic actions. Chloroform and trichloroethylene stimulated ACh release; chloroform and ether depressed ACh release, whereas halothane and methoxyflurane were only depressant of induced activity. BISCOE and MILLAR (1966) also found evidence of impaired transmission and transmitter release with cyclopropane, halothane and ether. These results have some similarities with those cited above for the skeletal neuromuscular junction.

As a special case, at least some of the action of ethanol may be the consequence of its metabolism and the associated biosynthesis of tetrahydroisoquinoline alkaloids (TIQs). These alkaloids may be derived in vivo by the condensation of acetaldehyde with one of the CAs, DA, NE, or epinephrine (E). This subject has been reviewed authoritatively by COHEN (1973a, b, c) and DEITRICH and ERWIN (1975) with the general speculation that such isoquinolines *might* be responsible for some of the actions of ethanol, especially those of long duration, such as hangover or withdrawal symptomatology, but not the acute intoxication. That they are similar to, or act like, morphine as postulated by DAVIS and WALSH (1970a, b) has been amply shown to be fallacious, both factually and rationally (SEEVERS, 1970; GOLDSTEIN and JUDSON, 1971; SIMPSON, 1975) (see Subsect. C.II.11).

6. Alterations in Putative Central Neurotransmitters

KALANT (1974, 1975) and earlier WALLGREN and BARRY (1970) authoritatively reviewed an extensive literature on ethanol effects on putative neurotransmitter systems. It has been observed and postulated that certain of the actions of ethanol are due to increased NE release; however, the contrary is reported as well, and, similarly, with respect to DA and 5-hydroxytryptamine (5-HT). KALANT (1974, 1975) and NOBLE et al. (1975) conclude that interpretation of these data and resolution of conflicting findings are impossible at present. Ethanol directly depressed the responses of cortical neurons to ACh, NE, 5-HT, but not GABA (LAKE et al., 1973).

a) Serotonergic Systems

Recent studies of FRANKEL et al. (1974) have definitively confirmed the summaries of KALANT (1974), WALLGREN and BARRY (1970), NOBLE (1974), and NOBLE et al. (1975) to the effect that ethanol, administered acutely or chronically, fails to alter 5-HT turnover in rat brain. On the other hand, serotonergic systems have been strongly implicated in the regulation of ethanol ingestion (see AHTEE and ERIKSSON, 1973 and review of MYERS and MARTIN, 1973; MYERS et al., 1972) although this has been questioned recently (KIIANMAA, 1975).

b) Adrenergic Systems

POHORECKY (1974) utilized up-to-date techniques to examine central and peripheral noradrenergic functions after acute and chronic administration of ethanol. It has been known for some time that CAs are released peripherally following ethanol administration (e.g., KLINGMAN and GOODALL, 1957; PERMAN, 1958, 1960, 1961; GIACOBINI et al., 1960a, b; ANTON, 1965; SCHENKER et al., 1966; DAVIS et al., 1967). As POHORECKY (1974) notes, central alterations in noradrenergic systems following ethanol require clarification. Both central and peripheral noradrenergic neurons are affected by ethanol with a transient decrease in NE level and increase in its turnover. What role CAs may play in the effects of ethanol and during ethanol withdrawal remains to be established. It is possible that adrenergic neuronal systems may be involved in "reward systems" and alcohol consumption behavior; alcohol consumption increased concomitantly with degeneration of dorsal noradrenergic pathway in the midbrain induced by stereotaxic injections of 6-hydroxydopamine (KIIANMAA et al., 1975a, b). However, neither intoxication nor withdrawal appears to influence brain levels of 5-HT, NE or 5-hydroxy indole acetic acid (WALLGREN, 1973). The effects of β-blockade by propranolol on tremor of withdrawal are discussed below in Subsect. D.I.2).

c) Cholinergic Systems

There are extensive data implicating cholinergic systems in the actions of ethanol. The skeletal neuromuscular junction and the ganglia data have been reviewed in other sections on mechanisms of action. The excitation of Renshaw cells by ethanol (MEYER-LOHMANN et al., 1972) is presumably due to ACh release and/or its potentiation. A similar action has recently been explicitly demonstrated for collateral inhibition of the Mauthner cell in the goldfish (FABER and KLEE, 1975).

Ethanol is known to inhibit the release of ACh from cortical brain slices, presumably as a consequence of its reduction of passive Na^+ permeability (see reviews of KALANT, 1974, 1975). Systemic or topical administration of ethanol results in decreased ACh release from cortical sites (ERICKSON and GRAHAM, 1973; PHILLIS and JHAMANDAS, 1971; KALANT and GROSE, 1967; CARMICHAEL and ISRAEL, 1975; MORGAN and PHILLIS, 1975).

A number of agents alter such ACh release. The release of ACh from cortex induced by hemicholininum is blocked by tetrahydrocannabinol (THC), morphine, chlordiazepoxide, and pentobarbital (DOMINO, 1971). Reasoning by analogy from GAGE's studies on the neuromuscular junction, ethanol has a number of effects on cholinergic systems including increasing transmitter release associated with presynaptic depolarization, prolonging the EPSP, depressing the permeability responses post-synaptically, and in higher concentrations raising threshold for action potential generation and propagation pre- as well as postsynaptically. In contrast to ethanol's action to increase ACh release peripherally, it has the opposite action centrally for both the ACh-released spontaneously and that appearing upon stimulation (reviewed by KALANT, 1974), consonant with the original studies of MATTHEWS and QUILLIAM (1964) using the superior cervical ganglion preparation. Interestingly, in frog ganglia ethanol utterly fails to increase the frequency of miniature potentials or their amplitudes (R. J. MCISAAC, personal communication).

The effects of other depressants on ACh levels and turnover have been reviewed by KALANT (1974), SCHUBERTH and SUNDWALL (1973), KRNJEVIĆ (1974a, b), PEPEU (1973), and BRUNNER et al. (1975). Anesthetics, barbiturates, and narcotics tend to result in an increase in ACh levels in brain even though its release may be decreased (e.g., CROSSLAND and SLATER, 1968; CROSSLAND and MERRICK, 1954). [Note may be made of contrary findings with acute and chronic ethanol administration in mice by RAWAT (1974).] Barbiturates produce an increase in both labile and bound fractions as well as increase the rate of *de novo* synthesis from plasma choline (SCHUBERTH and SUNDWALL, 1973). The increases in levels vary appreciably according to the brain site, although it is not yet possible to define the roles of such changes in the drug-induced changes in nervous system function (cf. PEPEU, 1973).

The observations that prior administration of physostigmine shortened ethanol sleeping time and that atropine antagonized the effects of physostigmine are consistent with the view that ethanol interacts centrally with cholinergic systems (ERICKSON and BURNAM, 1971; ERICKSON and GRAHAM, 1973). However, physostigmine does not shorten barbiturate sleeping time (KAYAALP and NUMANOGLER, 1965 cited in ERICKSON and BURNAM, 1971) and anticholinergic agents fail to alter ethanol-induced behavioral depression (GRAHAM and ERICKSON, 1974).

KRNJEVIĆ (1974a, b; 1975) reviews a wealth of data and reports that anesthetics depress, perhaps selectively, the responses of cortical cells to ACh. One of the earliest of the depressant demonstrations on a synaptic system was the report of LARRABEE and BRONK (1952) using the cholinergic system in ganglia. However, effects of topical application to cortical neurons of ACh, NE, 5-HT, but not GABA, were antagonized by ethanol (LAKE et al., 1973).

Evidence from isolated preparations extends the conclusion that barbiturates and ethanol act to interfere with cortical cholinergic transmission. Release of various transmitters from chopped cortical tissues loaded with precursors can be induced by electrical stimulation. This stimulated release was sensitive to ethanol or barbiturates with the most sensitive being ACh, followed by 5-HT. DA, E, glutamate, and GABA were, in that order, less sensitive to ethanol's action (CARMICHAEL and ISRAEL, 1975). On the other hand, acute or chronic ethanol administration is without effect on striatal choline acetyltransferase (WAJDA et al., 1977).

d) GABA Systems

The active uptake of a variety of putative amino acid transmitters requires a low concentration of K^+ and the presence of Na^+ ions; moreover, inhibition of Na^+, K^+ ATPase leads to inhibition of uptake (IVERSEN, 1974) and alcohol is one such inhibitor (see Subsect. C.II.2).

Uptake mechanisms for amino acids may be altered by sedative/hypnotic agents. As an example, the ability of the choroid plexus to take up 5-HIAA against a concentration gradient is inhibited by 2, 4 dinitrophenol (2,4-DNP), iodoacetate, n-ethylmaleimide and pentobarbital (CZERR and VAN DYKE, 1971, cited by SHARMAN, 1974).

The possible roles of changes in GABA levels in the action and dependence-inducing properties of ethanol and barbiturates have been definitively reviewed by WALLGREN (1971, 1973), KALANT (1975), DAVIDSON and RIX (1975), RIX and DAVIDSON (1977), and in SUTTON and SIMMONDS (1973, 1974). Amino acid transmitters in the nervous sytem have been thoroughly discussed by CURTIS and JOHNSTON (1974) and KRNJEVIĆ (1974b). The attention that has been given to GABA is obviously related, at least in part, to the possibility that it is the probable transmitter for presynaptic inhibition (SCHMIDT, 1964; BANNA, 1969; DAVIDSON and SOUTHWICK, 1971; BELL and ANDERSON, 1972, 1974; DAVIDOFF, 1973; ROACH and REESE, 1971; ROACH et al., 1973; CHIN et al., 1974; RIX and DAVIDSON, 1977; BARKER and NICOLL, 1973; see Subsect. C.II.3). In various preparations and conditions, acute administration of ethanol (SUTTON and SIMMONDS, 1973) or pentobarbital (SUTTON and SIMMONDS, 1974) may fail to alter brain levels of GABA in the rat whereas chronic ethanol administration resulted in an increase in GABA levels.

In reviewing conflicting findings on GABA levels, RIX and DAVIDSON (1977) were unable to resolve the areas of conflict. They concluded that there is "no evidence to date which would implicate disturbance of the GABA system as having a causal role in the genesis or maintenance of alcoholism or other drug dependent states."

However, in Research Institute on Alcoholism laboratories, CHAN (1975) has recently shown a consistent 10–20% elevation of central GABA levels $^1/_2$–$1^1/_2$ h after ethanol administration to mice; base-line levels of GABA and the increases were almost the same in two strains of mice exhibiting marked differences in sensitivity and duration of sleep following ethanol—thus confirming, in general, the findings of WALLGREN and associates (see WALLGREN, 1971, 1973; also RAWAT, 1974).

The hypothalamus, of the various sites in the CNS, tended to exhibit the highest changes after ethanol, whereas the spinal cord was the lowest and least influenced by ethanol. SIEMENS and CHAN (1975) were also able to offer a possible explanation for varied results obtained by previous investigators. Injections of 0.9% saline alone resulted in an increase in GABA levels which declined over the next 2 h to reach the levels found in uninjected animals. This influence of the control could well account for much of the variability among previous studies.

SUTTON and SIMMONDS (1973, 1974) suggest that chronic ethanol may affect a metabolic pool of GABA as distinct from a transmitter pool. Ethanol (acute or chronic) did not alter the rate of disappearance of H^3 GABA injected intracisternally. [Nevertheless, PATEL and LAL (1973) found a decrease in brain GABA at the peak of ethanol withdrawal.] In contrast to ethanol, pentobarbital produced a small increase followed by a decrease in GABA level. No changes were found in L-glutamate decarboxylase (GAD) or 4-amino: 2-oxoglutarate aminotransferase (GABA-T). However, the initial rate of disappearance of H^3 GABA injected intracisternally (the "fast" component) was decreased following pentobarbital, whereas chronic pentobarbital decreased the rate constant of the slow component of H^3 GABA disappearance.

SUTTON and SIMMONDS (1973, 1974) suggest that the two components of GABA disappearance may have anatomic cellular representation as neurons and glia, respectively; and they argue for a specific barbiturate action. They are aware, though, of the fact that barbiturates also alter the turnover of DA, NE, and 5-HT. Moreover, there is little to indicate whether the effects on GABA levels or disappearance rate are direct or due to the drug-induced alterations in neuronal activities. The fact that the injection procedure alone results in elevation of GABA levels suggests that the changes in levels or turnover could well be the consequence, very indirectly, of changes in nerve cell activity.

CUTLER et al. (1974) present the intriguing possibility that the GABA-barbiturate interaction experiments may be subject to major reinterpretation. Membrane transport of GABA, measured in brain slices, is inhibited by pentobarbital and amobarbital, but not by phenobarbital and hexobarbital; they attribute the difference to the side chain constituent! If this is so, it suggests that only some depressant barbiturates act via alterations of GABA actions (see Subsect. C.II.9 regarding other amino acid transport studies).

Synergy between ethanol and GABA has been found both in man and a variety of animals. In addition, one metabolic product of GABA, gammahydroxy butyrate (as well as gamma butyrolactone), is mutually synergistic with ethanol (MCCABE et al., 1971), as are glycine, serine (BLUM et al., 1972), 5-HT, tryptamine, and DA (ROSENFELD, 1960); the synergism between ethanol and glycine could be due to additive effects or more complex changes in amino acid uptake, GABA synthesis and others (see e.g., DAVIDSON and SOUTHWICK, 1971).

However, the fact that amino-oxacetic acid (AOAA), an agent which increases in vivo GABA levels, can prevent withdrawal seizures from barbiturate (ESSIG, 1968b) or alcohol (GOLDSTEIN, 1973a, b) is fascinating although CROSSLAND and TURNBULL (1972) failed to find AOAA effective against withdrawal seizures that were antagonized by mephenesin. Also, n-dipropyl sodium acetate (n-DPA, "Epilim") has been reported to be effective against withdrawal seizures in man (BONFIGLIO et al.,

1972); it also elevates brain GABA levels and inhibits GABA-T. The possibility that AOAA acts via other mechanisms or that GABA systems may modify seizures without being the initial site of the disturbance is emphasized by RIX and DAVIDSON (1977) and by CROSSLAND and TURNBULL (1972).

More precise evidence that AOAA actions are not mediated by changes in GABA has recently been presented by BELL and ANDERSON (1974); their data led them to conclude that the GABA accumulating in the spinal cord after AOAA administration failed to gain access to GABA receptors (see discussion in Subsect. C.IV.b for effects of benzodiazepines on GABA systems.) In barbital tolerant animals, the cortical GABA levels may be increased slightly (BLAGOEVA et al., 1972) or not at all (CROSSLAND and TURNBULL, 1972) with no evidence that withdrawal hyperexcitability was related to a drop in GABA levels although no studies of turnover were undertaken (BLAGOEVA et al., 1972; CROSSLAND and TURNBULL, 1972). Anesthesia results in decreased GABA synthesis (CHENG and BRUNNER, 1975); GABA content and uptake are varied (cf. BRUNNER et al., 1975). As noted earlier, ethanol (DAVIDOFF, 1973) and barbiturates (NICOLL, 1975a, b) augment both presynaptic depolarization and the actions of GABA on presynaptic terminals (see Subsect. C.II.3).

e) Dopaminergic Systems[9]

It seems fairly well established that alcohols, barbiturates, and benzodiazepines do not achieve their effects by acting directly postsynaptically on dopaminergic systems or via an influence on DA-sensitive adenylate cyclase (KAROBATH and LEITICH, 1974); nevertheless, that DA systems may be involved directly or indirectly in states of CNS depression seems fairly obvious. As an example, a depressant action of ouabain revealed when it is given intracerebroventricularly is associated with increased brain DA levels but no changes in NE or 5-HT levels (DOGGETT and SPENCER, 1971); whether or not the inhibitory action of ouabain on Na^+, K^+ ATPase is involved in its central effects is problematical. Increased DA levels in the striatum have been demonstrated after acute and chronic administration of ethanol (WAJDA et al., 1977).

[9] It may be only a curiosity, but apomorphine has in times past had a position of questionable honor in relation to ingestion of excess alcohol. I have been told that the standard treatment in the 1930's in New York and Germany (Hamburg) where special physicians supervised individuals incarcerated for the night for being publicly intoxicated was "Apo in dem Po-Po"! Presumably, the apomorphine produced or facilitated emesis and precluded further absorption of alcohol. Anecdotally, the subjects were calmer and more relaxed than would be anticipated from the simple induction of emesis (see DENT, 1934; review by SCHLATTER and LAL, 1972). However, such treatment might also be expected to precipitate or accelerate the appearance of a withdrawal syndrome in chronic drinkers and the possible production of pneumonitis from the inspirated material usually accompanying vigorous vomiting. It is tempting to speculate on the possible relationship between apomorphine's known dopamine agonistic action and ethanol's acute and chronic effects (see reviews on dopamine by BARNETT, 1975; CALNE et al., 1975). The use of apomorphine was reviewed and studied relatively recently (SCHLATTER and LAL, 1972); evidence was presented for "decreased craving" for alcoholic beverages and a high incidence of spontaneous penile erections. These considerations are complicated by the probable presence of tetrahydroisoquinoline congeners appearing as a side consequence of alcohol metabolism to acetaldehyde and the latter's conjugation with DA (see Subsect. C.II.11.c). Clinical suggestions that apomorphine antagonizes or modifies ethanol's depressant effects deserve further study (cf. NOBES, 1953; MARTIN et al., 1960 cited by NOBLE, 1974, and NOBLE et al., 1975; EWING et al., 1976).

BLUM et al. (1973) report that l-dihydroxyphenylalanine (L-DOPA) pretreatment prolonged ethanol narcosis and concomitantly increased brain DA levels. [Although depletion of brain 5-HT did not alter ethanol sleeping time, administration of 5-HT enhanced ethanol effects. ROSENFELD (1960) observed prolongation of ethanol sleeping time with 5-HT, tryptamine and DA.] In a study of neuroleptics, SEEMAN and LEE (1974) noted that ethanol evokes the release of DA from synaptosomes with a threshold concentration of 0.05 M (0.23% wt/vol) but the possible implications of these observations were not addressed (cf. also SEEMAN, 1975).

f) Histamine

Anesthesia produced by barbiturates or gaseous anesthetics reduces brain histamine turnover in rats due to both a decreased synthesis and decreased disappearance (POLLARD et al., 1973; POLLARD et al., 1974); the authors suggest that the decrease in histamine turnover was secondary to changes in neuronal activity.

g) Tryptamine

Under behavioral conditions where responding is suppressed by punishment, pentobarbital and chlordiazepoxide increase it. Tryptamine antagonists have similar effects; whether the effects of such drugs as the sedative/tranquilizers are related to antagonism of tryptaminergic systems is not known (GRAEFF, 1974). MARLEY and VANE (1963) some years ago reported a selective depression by pentobarbital of the spinal reflex responses to tryptamine; halothane had much less effect on the responses to tryptamine than pentobarbital. The tryptamine-pentobarbital interaction was demonstrated at the spinal and, presumably, reticular formation levels. ROSENFELD (1960) reported prolongation of ethanol sleeping time by tryptamine, as well as 5-HT and DA; thus, prolongation was apparently not related to alterations in absorption, metabolism or distribution.

h) Cyclic AMP

The barbiturates or meprobamate do not appear to act via cyclic AMP (WEINRYB and CHASIN, 1972; BEER et al., 1972), whereas the benzodiazepines may (see also SCHULTZ, 1974a, b; SCHULTZ and HAMPRECHT, 1973). Evidence is accumulating that AMP (FABER and GREENBERG, 1975, and personal communication) and GMP (SURIA and COSTA, 1975a, b) are critically involved in slow potentials and subthreshold conductance changes in nervous system excitability and repetitive discharges that occur over rather longer time periods; their involvement in the short-lived processes, such as spike generation and propagation, may not be appreciable.

High concentrations of ethanol are generally required to stimulate adenylcyclase activity in various tissues (see definitive review of VOLICER and GOLD, 1975). Ethanol inhibits phosphodiesterase in some tissues but in brain only the "low-affinity" enzyme of the pons is affected. The alterations in ATP and ATPase levels following ethanol can lead to secondary changes in cAMP. Cyclic AMP levels are decreased in brain and liver by acute ethanol administration, whereas chronic ethanol administration results in increased basal adenylate cyclase activity and cAMP levels, and decreases the cyclase stimulation by NE. In withdrawing animals, the reverse occurs, an increase over controls in the stimulation by NE of cAMP formation (FRENCH and PALMER, 1973; VOLICER et al., 1977b). Interestingly, decrease in brain cAMP seems to be largely limited to the cerebellum; pentobarbital exhibited a similar effect (VOLICER and GOLD, 1973). These investigators also examined the rise in cAMP that follows decapitation; pentobarbital or low doses of ethanol could block the post-decapitation rise in cAMP that occurred in the pons and medulla. The question is raised whether this effect is related to the initial nervous system stimulation seen following ethanol administration.

Recently, GOLD and VOLICER (1976) discovered a new nucleotide associated with ethanol administration. They postulate that ethyladenylate is formed as a consequence of the observed increase in nucleotide synthesis.

Thus, acute ethanol administration, in sublethal doses and concentrations, fails to directly affect both brain adenylate cyclase activity and phosphodiesterase; the effects observed in various studies appear to be secondary to actions on neurotransmitter release or adenosine formation (VOLICER and GOLD, 1975). Moreover, ethanol and ethanol dependence seem to influence the cGMP system to a significantly greater degree than AMP, especially that of cerebellum (VOLICER et al., 1977a, b; VOLICER and HURTER, cited in REDOS et al., 1976).

7. Nonspecific Membrane Actions

This subject is extensively reviewed in the proceedings of recent symposia edited by FINK (1972, 1975) and HALSEY et al. (1974). WOODBURY et al. (1975), MULLINS (1975), E. B. SMITH (1974), and HALSEY et al. (1974) have discussed extensively various physiochemical model systems and presented cogent analyses of a number of different aspects. In spite of their attractiveness, holistic theories of anesthesia and/or CNS depression are limited to (1) the physical principles underlying the distribution in membranes and the differences in concentration reached at various cell sites at steady-state, and (2) the physical principles involved in the *kinetic* aspects of uptake, distribution, and excretion. Of particular interest is the report that ethanol increases fast axoplasmic flow in central neural systems (ISRAEL et al., 1975c).

SEEMAN (1972a, b, 1975) and others have recently expanded the *membrane theory* of anesthesia, if it can be so designated, to the *membrane expansion theory* of anesthesia. The question of specificity of drug action—such as, for example, why one agent produces analgesia and anesthesia whereas another produces more sedation and antianxiety effects than analgesia—raises persistent knotty problems for such theories. Further, theories of anesthesia that treat neuronal systems as uniform can encompass the wide variety of agents only by adding secondary, albeit important, hypotheses to explain chemical and syndrome specificity—e.g., hypoxia, barbiturates, benzodiazepines, and gaseous anesthetics all produce sedation and altered consciousness, but it seems highly unlikely that they have a common molecular site and mechanism of action.

Related to the concept of membrane expansion as a basic mechanism is the reversal of anesthetic action by an increase of hydrostatic pressure (JOHNSON and FLAGLER, 1950, 1951). This observation provides a potential procedure to test physical interaction theories of anesthetic and sedative drug action. Earlier enthusiasm, which stemmed from reports that pressure reversed all classes of anesthetics, from ethanol and barbiturates to gaseous and volatile anesthetics (MILLER, 1972, 1974), have now been tempered. An increase in pressure has only limited potency in reversing the signs of anesthesia in intact animals (frog, tadpole, newt) when such anesthesia is produced by gaseous or volatile general anesthetics. Anesthesia produced by pentobarbital is apparently not reversed by pressure (see review of ROTH, 1975) in contrast to previous reports (MILLER, 1972, 1974). Moreover, the reversal with pressure of the conduction blockade in isolated nerves is limited to gaseous anesthetics; the block by ethanol, butanol or procaine is not antagonized by increased pressure (ROTH et al., 1974, cited in ROTH, 1975).

KENDIG and COHEN (1975) and KENDIG et al. (1975), in a similar vein, report that although the partial block of preganglionic conduction (isolated rat superior cervical ganglion preparation) by halothane is reversed by increasing the pressure to 137 atmospheres, the block of synaptic transmission in the ganglion by halothane is actually enhanced by increased pressure.

Pressure does reverse, in a limited fashion, the signs in certain animals of anesthesia with gaseous agents; it does not antagonize the effects of anesthetics at all synapses. Thus, one or both of the following assumptions is probably false: (1) increase by pressure reverses anesthesia because it antagonizes the primary mechanism of the membrane effects of anesthetics; or (2) anesthetics act by a similar molecular mechanism on all excitable tissues, e.g., by causing membrane expansion.

ROTH and others do note that in isolated nerves the blocking concentration is some five times that required in intact animals, yet the same pressure is needed for reversal. Thus, a number of observations remain at question. The pioneer works on pressure reversal were the studies of JOHNSON and FLAGLER (1950, 1951) in which they reported pressure reversal of pentobarbital in the newt and ethylcarbamate and ethanol in the tadpole (see MILLER, 1972, and SMITH, 1974). E. B. SMITH (1974) summarizes experiments using mice and newts; pressure was reported to antagonize anesthesia induced by ether, halothane, butanol, and pentobarbital. But R. A. SMITH and ROTH (cited in ROTH, 1975) could not replicate the reversal of a barbiturate. Another potential problem is the fact that experiments on the very tissues that do not uniformly exhibit pressure antagonism are used to support these general theories of anesthesia.

Since increased pressure, by itself, causes some neuronal hyperactivity, the "high-pressure nervous syndrome," the limited antagonism in vivo of the anesthetic is not necessarily due to direct interactions between pressure, the anesthetic, and the membrane. The antagonism or enhancement by pressure of the effects of anesthetics on isolated tissues may reflect direct interactions, but the present data suggest a mixture of actions. In any event, general theories must account for the fact that the inert gases—xenon, krypton, argon, nitrogen, hydrogen, neon, and helium (in order of potency, with helium almost inactive)—do produce a state of anesthesia. The many similarities between the syndrome and the electrophysiologic findings with common anesthetics and the inert gases at high pressure are reviewed by BENNETT et al. (1975).

Gaseous anesthetics do appear to interact with large biological molecules in a rather discrete fashion as illustrated by examination of the interaction with proteins, hemoglobin, olive oils and enzymes (cf. ALLISON, 1974; HALSEY, 1974; WHITE, 1974; SMITH, 1974; METCALFE et al., 1974; BRUNNER et al., 1975; CHERKIN, 1969). It is, thus, plausible to envision anesthetics interacting with either lipids or proteins in neuronal or intracellular membranes, resulting in rather specific changes in conformation and functional properties, rather than a general unitary change.

That the anesthetics might have prime action on other sites than nerve cell membranes or have action within or around cells is only raised on occasion. But so long as such actions cannot be ruled out, they must continue to be included in discussions of mechanism.

ALLISON (1974), NUNN (1974) and HINKLEY and TELSER (1975) have recently summarized a number of these interactions which include:

—Reversible breakdown of microtubules.
—Arrest of mitosis.
—Decrease or increase in viscosity of cytoplasm.
—Decrease in motility or cytoplasmic streaming of cells, including mammalian lymphocytes.
—Alterations in the clustering of proteins associated with membrane fusion.
—Alterations in redox potential.
—Alterations in enzyme activity.
—Reduction of membrane-RNA synthesis.

After reviewing the evidence, ALLISON (1974) concludes that the loss of consciousness with hypoxia results from other mechanisms than does the loss of consciousness produced by anesthetics, although he does recognize the "escape clause" that if anesthetics act on only a small number of uniquely sensitive cells, then the assessment of biochemical changes in the entire brain would not detect the action.

A recent symposium (HALSEY et al., 1974) explored the variety of possible molecular mechanisms for general anesthesia. A number of the participants emphasized the fact that the state recognized as general anesthesia probably could result from a number of different primary drug actions inasmuch as a wide variety of substances were capable of causing such a state. Thus, one of the searches regarding mechanisms is for the final common mechanism of the reversible, quiet induction of unconsciousness and amnesia denoted general anesthesia.

As a corollary, the differences among syndromes thus relate both to different primary mechanisms and to the nature of the side effects. PATON (1974) in attempting to reconcile the different spectra of neural effects of different anesthetics, suggests revising "the unitary theory" to include membranes of different properties as a function of the neural pathway concerned. In so doing, he recognizes the potential for more selective anesthesia—e.g., "anesthesia" without depression of pathways not related to the anesthetic state, such as those involved in respiration. In animal studies (mice) the volatile anesthetics exhibit marked differences in their potency in blocking a nociceptive reflex, abolishing the righting reflex, and producing death presumably by depressing respiration. By these comparisons, ether is the most active analgesic, relatively speaking, with halothane and paraldehyde least effective relative to dose required to block the righting reflex (PATON, 1974). In addition, there may well be differences in the relative efficacies of anesthetics among various sensory modalities or types of pain (BURNS et al., 1960; ROBSON et al., 1965; see HALSEY et al., p. 59–60, 1974; ROSNER and CLARK, 1973 and Subsect. C.III.2).

8. Stereo- and Chemical Specificity

That general anesthetics may act rather nonspecifically is borne out by the fact that the two optical isomers of halothane had identical potencies in their effects on transmission through the rat superior cervical ganglion, in contrast to the difference in activity of optical isomers of some barbiturates and steroid anesthetics (see discussion by HALSEY, 1974). Barbiturate action is moderately sensitive to changes in stereochemical conformation (see review of DAVES et al., 1975; and reports of GIBSON et al., 1959; DOWNES et al., 1970; CHRISTENSEN and LEE, 1973; BÜCH et al., 1973; HALEY and GIDLEY, 1970; GORDIS, 1971; WAHLSTRÖM, 1966). Not only are there potency differences, one enantiomer may exhibit excitatory or convulsant activity, whereas the other produces only anesthesia (DOWNES et al., 1970; SITSEN and FRESEN, 1974) or selective anticonvulsant actions (GORDIS, 1971). One possible basis for the excitatory actions described above for barbiturates may be related to the presence of one enantiomer with excitatory action, as has been demonstrated for pentobarbital (WADDELL and BAGGETT, 1973). Potency differences for those pairs that have some similar properties range around a ratio of 2–4 (HALEY and GIDLEY, 1970; BÜCH et al., 1973). The selective effects of specific enantiomers of barbiturates are not due to differential distribution or transport into brain tissue, since the enantiomers of pentobarbital and secobarbital were distributed equally to various brain regions (FREUDENTHAL and MARTIN, 1975).

DAVES et al. (1975) present the conclusion that the receptor system for barbiturates that results in CNS depression and anesthesia "has minimal structural specificity." By contrast, excitatory receptor binding requires the meeting of very subtle structural parameters. Of especial interest in relation to possible use of the stereospecific actions is the observation that whereas the (−) isomers of 5-(1′,3′dimethylbutyl)-5-ethylbarbituric acid, which produce anesthesia, block the contracting effects of the (+) excitatory isomer on aortic strips (HUPKA et al., 1969). The utility and limitations of isolated tissue models for examining stereoisomers have been recently explored (EDNEY and DOWNES, 1975).

Moreover, the various barbiturate effects appear to exhibit distinct differential sensitivity to stereochemical factors; the anticonvulsant activities of two isomers of pentobarbital were equal, whereas one was almost twice as potent as the other as an anesthetic agent. It is tempting to speculate regarding what the differences in subjective effects of the various enantioners of the barbiturates would reveal. For example, is it possible that the euphoric effects detectable with pentobarbital (MARTIN et al., 1974b; MCCLANE and MARTIN, 1976) a property of one or the other or both of the optical isomers?

Judicious selection of optical isomers for study of site of action and antagonism by barbiturates may provide the tools to unravel the origins of drug dependence and anesthesia. For example, one can speculate that the excitatory form of the barbiturate serves as an agonist and is perhaps structurally similar to a naturally occurring substance. The analog blocking of this natural substance is the prime—the depressant—action of most of the barbiturates (this argument is obviously analogous to that utilized among the narcotics) (SHULMAN, 1970). In addition to the differences in central effects, there are differences among optical and other isomers of barbiturates with respect to rate and route of metabolism (see FREUDENTHAL and MARTIN, 1975;

HOLTZMAN and THOMPSON, 1975; BÜCH et al., 1970; MCCARTHY and STITZEL, 1971; FURNER et al., 1969; PALMER et al., 1969).

The structural similarity between barbiturates and pyrimidines has been long recognized, but a possible connection of their biological roles has been adduced only recently. PENN (1973, 1975b), in a search for a barbiturate antagonist based on structural analogies, developed a mixture of five substances that did exhibit antagonism; all four substances were required in addition to 5-hydroxy methyl cytosine (nicotinamide, thiamine, thymine and pyridoxal). Actually, only the cytosine compound and pyridoxal were reported as required to exhibit antagonism of ethanol (PENN, 1975a).

Following the lead of a possible naturally occurring sleep-inducing substance (FENCL et al., 1971; PAPPENHEIMER, 1976), KROOTH and MAY (1975) theorize that, since dihydrouracil has central stimulatory actions uracil is a prime candidate involved in the mediation of sleep states. This extensive theory could also encompass the limited increase in brain excitability produced by caffeine via its demethylation and its conversion to a ribotide that inhibits orotidine-5′-monophosphate decarboxylase. The end result of this sequence might be a deceleration of the synthesis of uracil, the postulated naturally occuring sleep catabolite. Alternatively, the barbiturates might act to inhibit an enzyme that converts endogenous barbituric acid to malonate and thus result in accumulation of uracil. These formulations distinguish the more or less natural sleep-inducing properties of barbiturates from their more generalized depressant properties.

This biochemical theory of action of barbiturates, although highly speculative, does provide a number of testable hypotheses including the determination of the existence of enzyme mechanisms only postulated at present, such as enzymes that catabolize uracil, via barbituric acid, to malonate and urea.

Assuming success of this speculative theory, one is left with integrating the mechanism of action of uracil (or any other natural "sleep agent") into neurophysiology and chemistry (cf. a recent review of pyrimidines, LEVINE et al., 1974). The theory is also consistent with the fact that thalidomide has only sleep-inducing properties and not general depressant activity as exhibited by the barbiturates. Such differences in action among structures serve to suggest strongly that the various actions such as sleep vs. anesthetic coma vs. relief of anxiety exhibited by the sedative/hypnotics represent different sites and, possibly, different mechanisms of action.

9. Mechanisms of Action Other than those Directly on Excitable Membranes and Synapses

a) Direct Action on Oxidation

Anesthetics and anesthesia decrease brain oxygen consumption and result in increased glucose and glycogen concentrations, decreased glucose utilization and suppressed glycolysis (see, e.g., summary of BRUNNER and PASSONNEAU, 1972; BRUNNER et al., 1975, QUASTEL, 1975, and reports of MICHENFELDER and THEYE, 1972; WOLLMAN, 1972; GEY et al., 1965). But in analogy with studies of neurotransmitter levels, changes in metabolism and metabolites centrally fail to reveal which is cause and which effect—anesthesia results in decreased oxygen and glucose utilization; but is anesthesia the consequence or the cause of such metabolic changes? Uncoupling of oxidative phosphorylation has long been suggested as a possible mechanism of barbiturate action (BRODY and BAIN, 1951, 1954; BRODY, 1955; see review by QUASTEL, 1975).

b) Free Intracellular Ca^{++} Changes and Mitochondrial Respiration

Most writers addressing mechanisms of depressant drug action focus on the neuronal membrane or directly on synaptic mechanisms involving transmitter synthesis, storage, release and postsynaptic events. However, the question of actions at other sites remains. KRNJEVIĆ (1972, 1975) has redirected attention to possible primary actions of anesthetics and hypnotics on cellular respiration. He postulates that the cellular mechanism primarily affected by anesthetics is the sequestration of free Ca^{++} by mitochondria, a process highly dependent on mitochondrial respiration but which does not necessarily require ATP utilization (see also the review of LEHNINGER, 1970, on mitochondria and Ca^{++} transport and ALNAES and RAHAMINOFF, 1975). The CNS requires a continuous supply of oxygen and glucose; unconsciousness supervenes within seconds or a minute without them. Early in hypoxic states or in anesthesia, ATP levels are unaffected and, therefore, some other factors are involved in the loss of consciousness.

It is postulated that hypoxia or anesthetics produce an intracellular increase in Ca^{++}, such an increase would be expected to reduce excitability "probably mainly by an increase in cell-membrane K^+ conductance," such as actually measured with *Aplysia* (MEECH, 1972). KRNJEVIĆ (1975) cites a wealth of data generally consistent with this postulate. He leans heavily on the observation that 2,4-DNP and uncouplers of oxidative phosphorylation, of which barbiturates are one of the classic agents (BRODY and BAIN, 1951, 1954; BRODY, 1955), depress excitability of cortical neurons (DOGGETT et al., 1970; DOGGETT and SPENCER, 1971; GODFRAIND et al., 1970, 1971). This depression is associated with an increase in K^+ permeability. Anesthetics such as amytal, by suppressing mitochondrial metabolism (CHAPPELL and CROFTS, 1965), are postulated to produce an increase in intracellular Ca^{++} which presumably results in changes in Na^+, K^+ permeabilities and in such process as anaerobic glycolysis (SHANKAR and QUASTEL, 1972).

Ca^{++} uptake into cells and into mitochondria has been found to be altered by a variety of anesthetic or sedative substances, such as barbiturates (see BLAUSTEIN and ECTOR, 1975) and halothane; a variety of inhalation anesthetics inhibit mitochondrial respiration (HALL et al., 1973; ROSENBERG and HAUGAARD, 1973). Phenobarbital, like phenytoin, inhibits Ca^{++} uptake into synaptosomes, as stimulated by raised K^+ (SOHN und FERRENDELLI, 1976). However, ethanol and chloroform failed to alter Ca^{++} uptake by nerve terminals in vitro (BLAUSTEIN and ECTOR, 1975). But Ca^{++} and Mg^{++} effects on m.e.p.p. release in frog muscle required the presence of ethanol (OKADA, 1967). Barbiturates also inhibit Ca^{++} uptake by frog muscle microsomes in the presence of K^+ and stimulate Ca^{++} uptake when K^+ concentration is low. However, some barbiturates are relatively inactive (DUGGAN, 1971).

KRNJEVIĆ (1972, 1975) connects anesthetic actions and hypoxia with a depression of muscarinic cholinergic excitation of cholinergic neurons (CATCHLOVE et al., 1972) and appropriately emphasizes the correlation of long trains of repetitive discharges and after-discharges with cortical functioning, possibly that of sensory awareness (LIBET, 1973). (See section on reticular system, cortical activation, and caudate neurons, Subsect. C.III.2). CATCHLOVE et al. (1972) found not only depression by anesthetics of the responsiveness of cortical cells to ACh but also no change or distinct enhancement of the responses of the same cell to glutamate iontophoresis.

However, the classic blocking agents for central muscarinic responses are anticholinergics such as atropine, an agent which fails to produce anesthesia, although sedation and mood depression are obtained with small doses and overt hallucinations with larger ones (see OSTFELD et al., 1960; and related reviews by ABOOD et al., 1959; OSTFELD et al., 1959; OSTFELD, 1960). The fact that cholinergic blocking agents, especially antimuscarinic ones such as atropine, fail to induce unconsciousness has always been a primary fact to be reckoned with in the theories of a "cholinergic reticular activating system."

The neuronal asphyxia/free intracellular Ca^{++} hypothesis may well also be a viable hypothesis for the presynaptic action of anesthetics. An intracellular increase in free Ca^{++} in nerve terminals would be expected to result in an increase in spontaneous release of transmitters but a decrease in the depolarization-induced release (GLAGOLEVA et al., 1970; KUSANO, 1969; ALNAES and RAHAMINOFF, 1975).

This exciting hypothesis deserves experimental tests: One of these would be to examine the interaction of anesthetics, sedatives, and hypnotics with hypoxic and asphyxic states—studies that are relatively infrequent. Surprisingly, CULLEN and EGER (1970) found no influence of hypoxia on the median anesthetic concentration of halothane in dogs, although uncouplers such as DNP markedly potentiate general anesthetics (KILLAM et al., 1958). In opposition, MILLER (1975) suggests that there is so much phosphate in mitochondria in vivo that an increase in intracellular free Ca^{++} is almost inconceivable. Thus, the issue remains unresolved to date.

Cerebral metabolic responses established for one anesthetic agent do not necessarily apply to another; generalizations about the interrelationships of anesthesia, cerebral metabolism, and function may well be premature (ALTENBURG et al., 1969).

In connection with the observations on mitochondrial inhibitors, it has been reported that haloperidol, in contrast to chlorpromazine, antagonizes the hyperpyrexic and lethal effects of 2,4-DNP (GATZ and JONES, 1972); the antagonism has been obtained with in vitro mitochondrial preparations and in vivo. It would be interesting to learn more about the interaction between ethanol and haloperidol, haloperidol and DA and mitochondrial function.

As just noted, hypoxia does not consistently synergize in any striking way with depressants. However, the literature is sparse and contradictory. STEWART (1975) has presented a useful up-to-date review of carbon monoxide, an agent which adds to the effects of ethanol (RÖSELER, 1961; BJERVER and GOLDBERG, 1948; MALLACH and RÖSELER, 1961; PEARSON and NEAL, 1970; PECORA, 1959), whereas SALEN (1946) claims marked synergy between ethanol and carbon monoxide.

That ethanol, at least, interacts with Ca^{++} has been long known. Recently, ROSS (1974, 1976), ROSS et al. (1974), and CARDENAS and ROSS (1975) have shown conclusively that after ethanol, the level of Ca^{++} in brain tissue significantly decreases. It is tempting to suggest that this decrease reflects a general movement from cells to extracellular fluid, whereas during withdrawal from chronic use, there is a movement in the reverse direction with resulting hypocalcemia (and possibly related—hypomagnesemia—see VICTOR, 1973 and Sect. III, Chap. 3).

The decrease in tissue Ca^{++} exhibits cross-tolerance to the calcium depleting effects of morphine; further, naloxone pretreatment prevented the Ca^{++}-depleting effects of ethanol, morphine, or salsolinol (CARDENAS and ROSS, 1975). Of special interest was the observation that only after near lethal doses of barbiturates did pentobarbital significantly lower brain Ca^{++}. [The antagonism of ethanol by naloxone is a fascinating observation and other reports of possible acute antagonism have appeared (Ho et al., 1976; Killam et al., 1976); however, experiments by SMITH and ABEL (personal observations) fail to reveal significant antagonism by naloxone of motor incoordination or sleep induced by ethanol.]

The area of study of the interactions between centrally active agents, intracellular and extracellular concentrations of Ca^{++} and other ions in specific cells is predictably one of significant advancement in the immediate future. For example, LAZAREWICZ et al. (1974) have demonstrated the feasibility of determining cellular movement of Ca^{++} and its increased cellular uptake in response to elevations of K^{+}.

Interestingly, ethanol was found to have no effects on mitochondrial oxygen consumption in concentrations less than 108 mM (HALL et al., 1973) implying that it was of influence in higher concentrations. This concentration is effective in altering the K^{+} binding sites in isolated rat brain cortex slices (WALLGREN et al., 1974).

Although a variety of membraneous energy-linked systems have been proposed as models for anesthetic action, most effects require concentrations of anesthetics appreciably higher than are thought to obtain in vivo. For example, MILLER et al. (1972) review that halothane concentrations greater than 20% or more are required to alter Ca^{++} uptake into mitochondria. They conclude that the effects of halothane on mitochondria could be the result of inhibition of electron transfer, partial uncoupling, an effect on the Ca^{++} carrier, or an alteration in mitochondrial membrane structure; the authors favor the last view which appears to be indistinguishable in general form from the membrane expansion theory expounded by SEEMAN (1972a, b) although the latter focuses on the plasma membrane.

SEEMAN, in his extensive reviews and studies (1972a, b, 1975), has striven after a complete theory of mechanism of action of all psychotropic agents. Some of the explanations of the dilemmas of conflicting evidence of selectivity of action, almost as a direct result, have had to be strained, and have led some to reject his proposals. Nevertheless, the conceptual models of the interaction between inhalation anesthetics or ethanol with membranes that are also involved with Ca^{++}, correlations involving transport and binding, do encompass many of the experimental observations.

Numerous mitochondrial functions are altered by inhalation anesthetics although at concentrations estimated to be some 2–2.5 times the minimal alveolar concentrations for anesthesia, the locus of action delineated by COHEN and MCINTYRE (1972) was NADH dehydrogenase. HALL et al. (1973) suggest, in analogy with GUTMAN et al. (1970), that anesthetics inhibit electron transfer between nonheme iron and ubiquinone. As these authors and others (cf. HALL et al., 1973) point out, it remains to be elucidated whether such mitochondrial changes result in anesthesia, or are implicated more in other effects of anesthetics such as the malignant hyperpyrexia syndrome.

Effects of anesthetics are seen not only at the chemical level but also on morphology of the mitochondrion as observed in electron micrographs (GREEN, 1972; TAYLOR et al., 1972). It is tempting to suggest that these changes are directly related to alterations in Ca^{++} (cf. LEHNINGER, 1970). FRENCH and MORIN (1969) have developed intriguing possibilities of mitochondrial changes (damage) with chronic ethanol administration. These changes were seen in liver mitochondria; whether or not they occur in cells of brain or other tissues has not been explored (FRENCH, 1967, 1968; FRENCH and MORIN, 1969; FRENCH et al., 1969; FRENCH and TODOROFF, 1971).

In contrast, HALL et al. (1973) and NAHRWOLD et al. (1974) present data consistent with the view that the potencies of anesthetic agents (as well as convulsants) are correlated fairly well with effectiveness as inhibitors of mitochondrial respiration. As NAHRWOLD et al. (1974) point out, the hypothesis is not sufficient to predict which agents will be anesthetic or to explain the phenomenon of anesthesia. However, FINK and HASCHKE (1973) have reviewed the bulk of the experimental evidence regarding the influence of general anesthetics on cerebral metabolism and conclude that there is little justification for basing the mechanism of the anesthetic state on alterations in neurotransmitter metabolism or in general metabolism.

Perhaps the most definitive review to date of the effects of anesthetics on intermediary metabolism is that of BRUNNER et al. (1975) and the book edited by FINK (1975). Briefly put, anesthetic agents to tend to inhibit mitochondrial respiration and to inhibit Ca^{++} uptake to varying degrees.

In recent studies, BANAY-SCHWARTZ et al. (1974) have demonstrated that ethanol and amytal decrease ATP levels in brain slices, with an associated increase in Na^+ and a decrease in K^+. Qualitatively similar effects were produced by DNP; all three substances inhibited amino acid uptake (see also IVERSEN, 1974; SHARMAN, 1974). BANAY-SCHWARTZ et al. (1974) conclude that the level of ATP is not the main limiting factor for metabolic transport. In addition, it should be noted that amino acid transport into brain may be inhibited by ethanol (FREUND, 1972; IVERSEN, 1974), as well as by some barbiturates (TELLER et al., 1974a, b) (see Subsect. C.II.13.c).

c) *Ethanol on Uptake of Other Drugs*

Ethanol administration clearly seems to alter the entry of other substances into brain (LESLIE et al., 1971; see OLDENDORF, 1974). Increases in uptake after ethanol have been reported for bovine albumin, trypan blue (LEE, 1962), pentobarbital (SEIDEL, 1967), pentobarbital and thiamine (LESLIE et al., 1971), and possibly amylobarbital (RATCLIFFE, 1969). COLDWELL et al. (1970, 1971, 1973) found, on the contrary, decreased pentobarbital levels after ethanol when both were administered i.p., whereas GRAHAM (1960) had found no differences due to ethanol. In contrast, no changes in uptake by ethanol are observed with GABA (LESLIE et al., 1971) or triiodothyronine (BLEEKER et al., 1969); however, barbiturates reduce GABA fluxes in cortex slices (CUTLER et al., 1974).

10. Alterations in Microcirculation

The discussion of self-use of sedative/hypnotics, ethanol, and the like is not complete without mention of some aspects commonly overlooked. The possibility has not been ruled out that ethanol's actions are, in part at least, associated with a reduction of microcirculation (MOSKOW et al., 1968; KNISELY et al., 1969; PENNINGTON and KNISELY, 1973). Data exist to support a direct relationship; there have been no confirming studies of which this writer is aware. The report of HADJI-DIMO et al. (1968) of alterations by ethanol of the cerebral blood flow and EEG in the cat is pertinent, especially so in the context of selective, regional changes in blood flow associated with various thought processes (RISBERG and INGVAR, 1971; INGVAR and LASSEN, 1975; review by KETY, 1975). Further, the suggestion of altered microcirculation takes on new significance if the hypotheses of KRNJEVIĆ (1975) and of ISRAEL et al. (1975a, b) receive further confirmation, thus creating a situation where a primary drug action is intracellular on mitochondrial function (and possibly Ca^{++} levels) with an action on circulation making oxygen availability even less than normal. From the cells' point of view, this would constitute an attack on two fronts: relative hypoxia and cellular impairment of oxygen transport.

The importance of such interactions for the acute effects of ethanol is not known. On the other hand, this interaction may have profound significance with repeated dosing or large doses or drug combinations in terms of inducing or exacerbating CNS toxicity (cf. FREUND, 1973b; VICTOR, 1973; VICTOR et al., 1971).

11. Interactions with Hormones, Antagonists, and Metabolic Products

a) *Hormonal Interactions*

A consideration of the interrelationships of hormonal function and the action of sedative/hypnotic agents presents some intriguing aspects (PRANGE, 1974). Of special interest are recent findings of antagonism of the sedative/hypnotic drugs—ethanol, pentobarbital—by thyrotropin-releasing hormone (TRH). The experiments lead to the conclusion that the action of TRH itself in the CNS is involved in the antagonism and it is not secondary to thyroid hormone release (BREESE et al., 1974a, b, c; 1975).

The subject is complex and is, perhaps, best summarized with a series of statements: Thyroid gland dysfunction is associated with alterations in affect and mood; hypothyroid states are generally associated with depression. Imipramine action in relieving depression is potentiated by the addition of liothyronine. (This effect is apparent in women but not in men.) Nevertheless, WHYBROW and FERRELL (1974) conclude that alterations of thyroid function are not specific to depression since they are found in other psychiatric states.

Not only is the action and toxicity of imipramine increased by thyroid treatment, so it is with a number of diverse types of drugs (see review of BREESE et al., 1975). In part, this potentiation can be related to alterations in metabolism, e.g., prolongation of hexobarbital action by thyroid is due to inhibition of metabolism (CONNEY and GARREN, 1961). Although the toxicity of barbiturates is enhanced by thyroid treatment, TRH antagonizes sleep induced by pentobarbital or ethanol (BREESE et al., 1975).

Intracisternally administered TRH produces a characteristic motor syndrome of mild hyperactivity; these effects are accentuated in animals receiving pentobarbital, although there was no change in the sleeping time (PRANGE et al., 1974a). TRH by itself has relatively little effect on several operant tasks and evokes only a modest increase in amphetamine-induced motor activity. Thus, it is difficult to conclude that TRH is acting either to interfere directly with depressant drug action or by pharmacologic antagonism via contravening CNS excitation.

In preliminary studies in euthyroid women with unipolar depression, a single injection of TRH caused a prompt, brief improvement of the depression; the authors suggest that this is a direct central action of TRH (PRANGE et al., 1972, 1974b).

Further, TRH appears to have mild stimulatory activity in normal female volunteers, and this activity was *not* correlated with thyrotropin release.

TRH has now been shown to have an antidepressant effect (cf. review of PRANGE et al., 1974b; WILSON et al., 1974a, b) but this action appears to be central and not mediated through the thyroid. Moreover, triiodothyronine has antidepressant (WILSON et al., 1974a) and, possibly, antipsychotic properties (cf. CAMPBELL and FISH, 1974). The nature of the interactions of acute and chronic administered triiodothyronine, thyroxine, thyroid stimulating hormone (TSH) and TSH releasing hormone (TRH) with a variety of depressant drugs would be of interest. Surprisingly, relatively little is known except for the generalizations that adrenergic function tends to be enhanced by thyroid hyperactivity and that anesthetic requirement is a function of metabolic rate. Thus, it is thought that hyperthyroid individuals require greater amounts of anesthetic to undergo anesthetization than do euthyroid patients. However, data to support these time-honored clinical maxims are sparse.

The purported efficacy of lithium therapy in reducing binge drinking in alcoholic patients (KLINE et al., 1974; WREN et al., 1973) and in reducing preference for alcohol in animals (SINCLAIR, 1974b; HO and TSAI, 1975) and an antagonism of some ethanol actions by lithium (LINNOILA et al., 1974b, c) raises further the possibility of common threads of action and interaction. For example, the following scheme is at least conceivable in terms of known facts: Lithium could act on the thyroid to reduce (transiently) the circulating thyroid level (CARLSON et al., 1974); this reduction, in turn, might result in increased TRH elaboration which could act to antagonize ethanol effects and also cause an increase in TSH and, ultimately, a normalization of thyroid function.[10] Whether or not chronic lithium administration results in sustained increases in TRH and, if so, if this would be sufficient to antagonize ethanol actions is unknown. This scheme is presented not so much to advance a mechanism but rather to illustrate the complexity of the problems.

Reasoning further by analogy, ISRAEL and colleagues (1973; see reviews of 1975a, b) have implicated altered thyroid hormone function in the acute and chronic liver changes accompanying alcohol ingestion. Their methodical investigations can be summarized by stating that ethanol induces an increased rate of cellular oxygen consumption which results from increased levels of active thyroid hormone plus increased "demand" associated with Na^+, K^+ ATPase inhibition and ethanol metabolism. This increased oxygen consumption in the face of a limited supply is thought to result in cellular hypoxia and death of those cells located farthest from the arterial circulation. Antagonism by propythiouracil, presumably acting peripherally to antagonize thyroid hormone, prevents the liver damage.

These experiments raise the important question of the mechanism(s) of these actions of ethanol. Ethanol does increase the level of free thyroid hormone presumably by causing an increase in free fatty acids which facilitates the dissociation of thyroid hormone from the protein bound inactive form. Ethanol could also be acting on the oxygen transport system as well as the Na^+, K^+ ATPase and other systems regulating cell permeability. In concentrations above 109 mM, it may alter mitochondrial respiration (HALL et al., 1973; see Subsect. C.II.9). The interrelationships

[10] However, triiodothyronine was ineffective in altering the acute alcohol intoxication in a well-designed investigation (KALANT et al., 1962).

with other agents such as barbiturates is speculative. There is a complex connection between depressant drug action and thyroid releasing and thyroid hormones which is summarized above. In addition, there are complex effects of sedatives on many hormonal systems (see e.g. GORSKI, 1974; STOKES, 1971).

b) Evidence Derived from Antagonism and Interaction Studies

The sedative/hypnotics and anesthetics have effects on the intact nervous system more or less opposed to those of various excitants and convulsants, such as strychnine, pentylenetetrazol, picrotoxin, doxaprim, bemegride, amphetamines, and caffeine (cf. WALLGREN and BARRY, 1970; annotated bibliography on alcohol and drug interactions of POLACSEK et al., 1972). Most observers agree that there is no specific competitive or molecular basis of the antagonism between these classes, and, in most instances, it is severely constrained in terms of the very limited dosage range over which antagonism can be demonstrated (see FORREST et al., 1972). Likewise, these stimulants can only, at optimal dosages, partially antagonize the sedative/hypnotics. This antagonism is so limited and so touchy to control that most clinicians have concluded that the stimulants have no place in the treatment of overdoses of the sedative/hypnotics. Conversely, cautious use of short-acting sedatives has been generally found useful in the management of stimulant overdose.

By and large, all sedative/hypnotics are additive or synergic, especially with ethanol (WALLGREN and BARRY, 1970) with the possible exception of chlordiazepoxide (see reviews of ethanol-drugs of abuse interactions by KISSIN in KISSIN and BEGLEITER, 1974 and SMITH, 1976).

The possibility that certain barbiturates—and other sedative/hypnotics—might exhibit agonist and analog antagonist properties has been of long-standing interest. SHULMAN(1970) makes the analogy between barbiturate derivatives and the narcotics thoroughly. Unfortunately for the general theory, bemegride and other related compounds are CNS stimulants that are rather nonspecific in their structural requirements. Although the stimulant property could originate from a conformational drug-receptor fit resulting in increased excitation or decreased inhibition, the sedative barbiturate analogs are viewed as binding with similar receptors but as blocking agents rather than as stimulants. However, data are consistent with either interpretation, a direct interaction with a common receptor, or an action on separate neural receptor systems. DOWNES et al. (1970) conclude from a study of barbiturate analogs and stereo-isomers that the convulsant and depressant actions of barbiturates represent separate actions at separate sites. Similarly, a comparison of *d*- and *l*-isomers of methobarbital reveal differences in sedative but not anticonvulsant potencies (GORDIS, 1971) (see Subsect. C.II.8). It would be particularly interesting to determine the interactions of depressant barbiturates with bemegride on spinal presynaptic and isolated cortex preparations. LEE-SON et al. (1975) conclude from the parallel dose-effect curves for a number of barbiturates in antagonizing the depolarizing action of carbachol on guinea pig lumbrical muscle, that barbiturates, including a convulsant analog, have a "common mechanism of action."

c) Actions Due to Metabolic and Condensation Products of Ethanol and Barbiturates

The possibility that ethanol actions are due, especially upon chronic administration, to the occurrence of condensation products of acetaldehyde and biogenic amines to form TIQs caused an intense flurry of interest (COHEN and COLLINS, 1970; DAVIS and WALSH, 1970a, b; DAVIS et al., 1970). That such condensation products can occur seems well demonstrated (reviewed by COHEN and COLLINS, 1970; COLLINS, 1973; WALSH, 1973; DAVIS, 1973; COHEN, 1973a, b) especially in the peripheral adrenergic system (COHEN, 1973a, b, c; SIMPSON, 1975). Such TIQs could serve as false transmitters (GREENBERG and COHEN, 1973) but whether they occur in sufficient quantities to produce major effects now seems doubtful. Certainly, ethanol's intoxicating and physical dependence properties are not causally related to the appearance of such alkaloids. Naloxone fails to induce a withdrawal syndrome in mice physically dependent on ethanol (GOLDSTEIN and JUDSON, 1971) although naloxone does interact with alcohol in some situations (LABELLA and PINSKY, 1976). The possibility that such alkaloids could function as false transmitters at least in peripheral nervous systems has been recently addressed by SIMPSON (1975); he concludes that as yet the weight of evidence fails to indicate substantial accumulation of alcohol-derived aldehydes or condensation products to be significance in the overall syndrome of ethanol intoxication.

A possible role of biogenic aldehydes formed in the metabolism of biogenic amines in the actions of barbiturates as well as a number of other centrally acting agents has been proposed and supported by experimental findings (VON WARTBURG et al., 1973; DAVIS, 1973; TABAKOFF et al., 1973, 1974; TABAKOFF and BOGGAN, 1974). Although a variety of barbiturates and anticonvulsants act as in vitro inhibitors of the isoenzymes of aldehyde reductase, correlation between in vitro and in vivo activity could be obtained only for some of the anticonvulsant activities (ERWIN and DEITRICH, 1973; RIS et al., 1975). A wide variety of drugs are *not* effective inhibitors of any of these isoenzymes including diazepam, meprobamate, promazine and haloperidol. Interestingly, the demethylated derivative of diazepam, which occurs in vivo, may be an active inhibitor. In view of the high inhibition constant found for 4-hydroxyphenylacetic acid, the authors doubt that the increase in biogenic acid levels after alcohol (TABAKOFF et al., 1974) would divert the metabolism of biogenic aldelydes from the reductive to the oxidative pathway (RIS et al., 1975).

12. Evidence from Genetic Differences

Interesting new data are appearing that tend to reinforce the view that aliphatic alcohols have distinctly different sites or mechanisms of action from those of the barbiturates or ether (RODGERS, 1972; RANDALL and LESTER, 1974; HESTON et al., 1974). MCCLEARN (1975) has inbred mice obtaining a strain of animals that consistently show long sleep (LS) after a given dose of ethanol and a strain of short sleep mice (SS) (see also reviews of ERIKSSON, 1973, 1974, 1975). A number of alcohol effects are demonstrably inheritable (REED, 1975); differences in mean metabolism rates in racial groups are also a factor.[11] These authors, and others (CHAN, 1975; SIEMENS and CHAN, 1975, 1976), have demonstrated that the varying duration of sleep reflects not differences in metabolism, but distinctly different sensitivities of the CNSs to ethanol. In fact, the strains that are more sensitive to ethanol are less

[11] Racial differences in effects in man have been long suggested and recently documented (WOLFF, 1972; EWING et al., 1974); whether or not such differences can be related to metabolism is disputed (FENNA et al., 1971; BENNION and LI, 1976).

sensitive to pentobarbital, whereas the two strains are equally sensitive to ether. The high sensitivity of the LS mice to ethanol is also seen with other aliphatic alcohols, methanol, propanol and butanol (McCLEARN, 1975).

These differential sensitivities are not the consequence of differences in blood or brain levels, but clearly seem to reflect different mechanisms of action (RANDALL and LESTER, 1974; SIEMENS and CHAN, 1975, 1976). Differences in levels of acetaldehyde (LIN, 1975) or of glutamate or GABA or the alterations of these putative neurotransmitters after alcohol are not likely to be responsible for the differences in sensitivity to alcohol (CHAN, 1975).

Earlier studies of KAKIHANA et al. (1966), SCHNEIDER et al. (1973), and RANDALL and LESTER (1974) clearly demonstrated that mouse strains with high preference for ethanol had less sensitivity to its neural effects than rodent strains with a low preference (confirmed also by LIN, 1975). The differences in preference for propylene glycol were in the same direction (SCHNEIDER et al., 1973).

Strain differences in sensitivity to other drugs have been noted by a number of investigators. For example, SANSONE and MESSERI (1974) have reported that BALB/c mice are more sensitive than SEC/IRe mice to both chlordiazepoxide and chlorpromazine depressant actions on avoidance behavior. Such studies require, for clear validity, a measure of drug level especially at the sites of action; without pharmacokinetic data, interpretation of strain differences is difficult.

Differences in drug sensitivity have been ascribed to levels of arousability (McLAREN and MICHIE, 1956; IRWIN, 1960), to variations in the sensitivity of target tissues (e.g. FULLER, 1970), or the presence of various anatomic or pathologic alterations in the CNS (see e.g. OLIVERIO et al., 1973; LIEB et al., 1974; LIEB and CRANDALL, 1974; ADLER, 1970; ADLER and GELLER, 1977) (see Subsect. C.II.13). These demonstrably different individual sensitivities to drugs and these differences can be detected as selective responsiveness of certain behaviors to drug influence. SOURBRIE et al. (1974) recently demonstrated that animals can be selected according to their "emotionality" level in the open-field test. The more emotional group of animals were more sensitive to benzodiazepines (diazepam or oxazepam in antagonizing immobilization-produced gastric ulcers and in improving a heated-floor maze performance). In comparison, chlorpromazine and dexamphetamine exhibited no differential effects. The article of SOURBRIE et al. (1974) discuss previous work with similar implications that there are definitive interactions between drugs, behavior, and individual differences!

These differences, even in subjective effects, are commonly recognized, but have been little studied. For example, are those in whom ethanol or the antianxiety drugs have subjectively perceived desirable (pleasant) effects more likely to use the agents habitually (see, for example, the discussion of EWING et al., 1976)? Rather long ago, NEWMAN (1935, 1941) studied alcohol given so that the subjects were not aware that they were receiving alcohol. In these circumstances, normal individuals responded with freedom of conversation, drowsiness and intensification of the previously existing mood. In strong contrast, psychoneurotic individuals showed exaggerated garrulity, emotional outbursts and further neurotic symptomatology. The schizophrenic group retired further from reality becoming antagonistic, mute and negativistic. These studies call for confirmation and extension in view of their profound significance (cf. MAYFIELD, 1968a, b).

13. Differences in Neural Organization

a) *Effects of Drugs Following Brain Damage*

Partial isolation of various regions of the nervous system unequivocally alters the sensitivity of the remaining systems to pharmacologic agents, such as barbiturates and pentylenetetrazole. For example, after frontal cortex or cerebral hemisphere removal in cats, the contralateral limbs exhibit more weakness and incoordination than the limbs of the other side after small doses of pentobarbital (SEGUIN and STAVRAKY, 1957). Operated cats were better protected by pentobarbital against seizures to pentylenetetrazol than intact animals. In 1960, HELLER et al. reported that rats with lesions in the septal forebrain slept longer after thiopental or barbital administration than normal rats or those with lesions in the cerebral cortex or caudate nucleus. Analogously, rats with lesions in the amygdalae have decreased motor activity and enhanced responsivity to amphetamine (ADLER, 1961, 1970; FURGIUELLE et al., 1964).

This entire subject of drug response following brain damage has been reviewed by a pioneer in the field (ADLER, 1970; ADLER and GELLER, 1977; see also SMITH, 1970a, b). The experimental assessment of tranquilizing drug action has frequently utilized animals with lesions designed to increase aggressiveness or emotionality, such as rats with septal lesions. A comparison across drug classes reveals a degree of differential effects on rats with lesions of the olfactory lobes, relative to normal animals and other drugs (NURIMOTO et al., 1974). Chlordiazepoxide and diazepam (along with haloperidol) were distinctly more selective in their depression of "hyperemotionality" than pentobarbital or meprobamate. It is tempting to speculate on the import of the recognized damage to the CNS sustained by alcoholics and by barbiturate users as a consequence both of trauma and of alcoholic neuropathies (cf. HAUG, 1968; reviews of WALLGREN and BARRY, 1970; VICTOR, 1968; VICTOR et al., 1971; FREUND, 1973b; NOBLE and TEWARI, 1973; DREYFUS, 1973; POSER, 1973). Does such damage predispose to further drug or alcohol use or alter the nature of the drug actions? A recent report that behavioral tolerance to repeated alcohol administration in rats fails to develop following lesions of frontal cortex or administration of cycloheximide (LEBLANC et al., 1976) emphasizes the importance of the neural organization for drug action.

b) *Patterns of Neuronal Discharge and Sensitivity of Drug Depression*

Many of the agents under consideration alter not only the transmission of a single impulse across a synaptic junction but also the pattern of repetitive activity. The fact is that most, if not all, neuronal systems discharge repetitively and the frequency and pattern of discharge functionally carry the information. This is significant in a consideration of both site and mechanism of drug effects since many agents appear to depress repetitive activity much more readily than single impulses [e.g. the effects of pentobarbital, ether, and cyclopropane on repetitive activity originating in motor nerve terminals (WERNER, 1961; RAINES and STANDAERT, 1969; RIKER and OKAMOTO, 1969; VAN POZNAK, 1967), mephenesin and parpanit on decerebrate rigidity (LONGO, 1961; CHIN and SMITH, 1962) and depression of neuronal after-discharge by benzodiazepines (see LONGO, 1972)].

The motoneuron and monosynaptic reflex is one of the most common model systems for examining drug actions, yet it may not be a very suitable model or test system for elucidating the actions of anesthetics, even though the preparation provides a technically feasible means to gain great insight as to sites and mechanisms. It is, of all of the commonly tested reflexes, one of the more resistant to the actions of all general anesthetics; the knee jerk/monosynaptic reflex is one of the last to disappear with gradual deepening of general anesthesia. In fact, its resistance is one of the pivotal observations underlying what might be called the statistical theory of anesthesia, i.e., the summation of effects in long neuronal chains of a drug effect of almost undetectable decrease in probability of transmission at each synapse in the chain (BARANY, 1947). Moreover, although the monosynaptic reflex is detectably affected by low doses of barbiturates and ethanol, the dose-effect curves are very flat indeed, and reflex abolition requires doses of ethanol that cause respiratory arrest (cf. KOLMODIN, 1953; CHIN and SMITH, 1962; LATHERS and SMITH, 1976). This point also raises the methodologic problem of comparing sensitivity of two or more synaptic systems in which dose-effect curves have different slopes. For example, at low concentrations one reflex system may be more depressed than another, whereas at a higher dose, the reverse may well obtain. In fact, this problem can be readily illustrated using two different measures in a given synaptic system—a phasic and tonic stretch reflex (CHIN and SMITH, 1962).

c) Interactions Among Sites, Mechanisms, and Agents. Other Possibilities

ZORYCHTA (1974), SOMJEN (1963), CHIN and SMITH (1962), EIDELBERG and WOOLEY (1970), CHANELET and LONCHAMPT (1972), as well as many others, have noticed excitatory alterations in spinal interneuron activity with anesthetics and ethanol, as reflected by the alterations in fluctuations of the membrane potential of the motoneuron. Some interneurons, such as Renshaw cells, are excited by ethanol (MEYER-LOHMANN, et al., 1972) although the same cells may be depressed by barbiturates or mephenesin (LONGO et al., 1960). Excitation of some neurons after gaseous anesthetics is not an uncommon observation (see e.g., RICHARDS et al., 1975; ROSNER and CLARK, 1973). Barbiturates stimulate respiration in some subjects (BROWN et al., 1973). The "paradoxical" stimulation by barbiturates in children is a well established clinical observation. In fact, rather large doses of barbiturates are needed to produce sedation in children. BROWN et al. (1973) cites a personal communication from R. TINKLENBERG that secobarbital was more frequently associated with assaultive crime in a group of 50 youthful offenders than any other drug, although marijuana and alcohol use were, of course, more frequent. The subjects reported that secobarbital produces a higher and greater irritability than do the other drugs.

The fact that the very low doses of barbiturates or ethanol used in man for self-administration produce apparent stimulation presents conceptual and semantic difficulties. Among the explanations has been the possibility that some degree of actual excitation occurs. Some would cite the EEG activation as objective evidence of excitation. Moreover, a variety of barbiturate analogs are known to have convulsant properties (cf. DOMINO, 1957; DOWNES and WILLIAMS, 1969; BÜCH et al., 1970; DAVES et al., 1975). Some of the convulsant barbiturates have been shown to increase the monosynaptic reflex, presumably by a direct action on the motoneuron. Many barbiturates appear to possess a mixture of depressant and stimulant activities; even pentobarbital may have transient stimulatory effects on spinal reflexes (CHIN and SMITH, 1962; DOWNES and WILLIAMS, 1969).

One of the reasons for these mixed effects is the presence of two enantiomers in a racemic mixture as obtains with pentobarbital; one of the enantiomers exhibits appreciable excitatory actions in addition to the depressant activity exhibited by both enantiomers. Whether this will serve to explain all of the excitatory action of barbiturates, remains to be seen (see Subsect. C.II.8). Such an explanation patently does not apply to gaseous anesthetics or ethanol—both of which exhibit in low concentrations a variety of excitatory effects. The stimulation by ethanol is evidenced in a wide variety of animals, including fish, as increased behavioral motor activity. This increased activity is similar to that observed with amphetamine, and like amphetamine, is antagonized by prior administration of alpha-methyl-p-tyrosine (CARLSSON et al., 1972; ENGLE et al., 1974).

Among other excitatory aspects, it has been suggested that sedatives/hypnotics, and alcohol facilitate acting out and aggression. That many of this group of compounds do promote aggression has been noted in both experimental animals and man (reviewed by KRSIAK, 1974); benzodiazepines have been likewise identified as increasing hostility in certain patients (see Subsect. C.IV and D.III).

One of the more curious, but possibly fruitful, observations relates to the actions of ethanol on muscle spasticity and rigidity. By and large, beverage alcohol's reputation as a skeletal muscle relaxant is anecdotal, but it has been found to be effective in alleviating essential tremor (WINKLER and YOUNG, 1971)[12] as is propranolol (SEVITT, 1971; WINKLER and YOUNG, 1971, 1974; MARSDEN et al., 1968). The mechanism of essential tremor is unknown, but it is interesting to speculate that withdrawal shakes and tremor, which are alleviated dramatically by both alcohol and propranolol, may have the same mechanisms as essential and physiologic tremor (ZILM et al., 1975; ZILM and SELLERS, 1976). Note also the similarity of thyrotoxicosis and the withdrawal state (see AUGUSTINE, 1967; MARSDEN et al., 1968; ISRAEL et al., 1973, 1975a, b; WINKLER and YOUNG, 1974; ZILM et al., 1975). The effects of ethanol on tremor are especially intriguing in view of its muscle spindle-stimulating property (KUCERA and SMITH, 1971, 1972) and the clinical report of the converse of relaxation, the elucidation of motor neuron discharges and muscle spasms in a unique patient (BLANK et al., 1974; see also BRIMBLECOMBE and PINDER, 1972). The effects of benzodiazepines and barbiturates on essential tremor would be interesting to study. Possibly related are propranolol's alleviation of withdrawal tremors (ZILM et al., 1975), its antianxiety effects (KELLNER et al., 1974), and its augmentation of ethanol effects (ALKANA et al., 1976). Propranolol also reportedly blocks the tremors induced by oxotremorine (BRIMBLECOMBE and PINDER, 1972; AGARWAL and BOSE, 1967; JACOBI, 1967; ACHARI and SINHA, 1967, 1968; COX and POTKONJAK, 1970; WATANABE et al., 1971).

Barbiturates alter amino acid uptake and transport (TELLER et al., 1974a, b). Some barbiturates were found to stimulate amino acid uptake by brain slices, whereas others were inhibitory. Although such changes might be related to toxicity, a primary site of neural action on amino acid transport seems unlikely because of the differences in qualitative effects on such transport of agents which share similar CNS effects.

[12] In discussing this paper FOLEY notes that "many alcoholics have started off as essential shakers..."

The effects of ethanol and minor tranquilizer drugs in decreasing an organism's response to stress can be detected and measured by the corticosteroid level (reviewed in the recent paper of LAHTI and BARSUHN, 1974); a relatively selective central action of benzodiazepines, meprobamate and phenobarbital in this test was found, although MERRY and MARKS (1972) found a significant decrease in plasma cortisol following barbiturates or ethanol, but not after diazepam. Phenothiazines, narcotic analgesics and a variety of other agents had little if any effect on corticosteroid level.

III. Sites of Action

An understanding of the nervous system based on correlations between the function and the behavior of the specific nerve cells underlying such function is yet far from complete. Such an understanding intrinsically involves solving the ultimate questions of the physiologic and biochemical bases of pain, unconsciousness, sleep-wakefulness, mood, perception, motor-sensory control, etc. However imperfect the understanding, a review of sedative/hypnotic drug effects on such functions and behaviors with the aim of identifying their site(s) and mechanism(s) can be addressed and summarized.

Among the neuronal systems and actions that will be reviewed are sites of action correlating to the neural mechanism underlying specific drug action(s) and the degree of selectivity/specificity of action of the agent:

1. Neural bases of drug-induced muscle incoordination (spinal cord studies reviewed in Subsect. C.II.3 above).
2. Origins and evidence of specificity of site and mechanism of action of sedative/hypnotic drugs on the cerebral cortex, reticular formation, and limbic system.
3. Evidence of, or mechanisms of: amnesia; change in affect, mood; analgesia; psychedelic effects.

1. Motor Systems and Incoordination

Overt, measurable effects of all sedative/hypnotic agents in relatively low doses are ataxia, motor incoordination, moderate muscle relaxation, ocular divergence; with larger doses these disruptions extend to dysarthria, dysmetria, and severe ataxia affecting the legs in particular. It is difficult to develop a comprehensive theory for motor disturbances based on interactions of the agent with neuronal structures, in part because of the mixture of effects that have been reported, the rather high levels that are generally reported in order to elicit such actions, the inability to make a coherent story out of the different sites and actions of such agents as ethanol, and, lastly, the presence of disparities between the effects of ethanol, for example, and the effects of presumably similar agents, such as barbiturates, on the other.

Early on, the suggestion was made that ethanol affected the cerebellum selectively to produce such ataxia (DUNBAR, 1905; FLOURENS, 1824, cited by DOW and MORUZZI, 1958), yet to this reviewer's knowledge only one study has explicitly begun to examine ethanol's action on the cerebellum. EIDELBERG et al. (1971), in a rather brief report, described a depression in discharge of Purkinje cells in decerebrate cats and a tendency for acceleration of discharge of cerebellar interneurons. The activity of vestibular neurons, by and large, was depressed following ethanol administration—whether or not the cerebellum was present; the doses used by EIDELBERG et al. (1971) were those known to be associated with mild to moderate ataxia in unanesthetized animals. The effects of ethanol are dramatically revealed by the failure of the development of cerebellar cells, including Purkinje cells, in newborns receiving daily doses of alcohol (BAUER-MOFFETT and ALTMAN, 1975).

Low concentrations of a wide variety of hypnotics and anesthetics alter the spontaneous electrical activity of the anterior lobe of the cerebellum and the red nucleus. GOGOLAK et al. (1972) differentiate three groups of agents on the basis of the qualitative differences in their effects. Barbiturates, a variety of depressant agents (paraldehyde, chloral hydrate, methaqualone, ethinamate, bromvalurea, clorazepam, diazepam, medazepam) and methoxyflurane produce a very regular rhythm. In contrast, diethyl ether, urethane, ethyl alcohol, and halothane caused desynchronization, whereas trichlorethylene induced in deep anesthesia, high frequency, high amplitude activity. The authors suggest that although these actions may arise from the cerebellum, other sites may be involved.

A marked sensitivity of cerebellar units to anesthetics in general has been observed frequently (cf., e.g., BLOEDEL and ROBERTS, 1969; ECCLES et al., 1971a, b; BUCHTEL et al., 1972; GORDON et al., 1971, 1972, 1973; KÖRLIN and LARSON, 1970). The effects of barbiturates on the cerebellum have been examined only recently. MURPHY and SABAH (1970) and LATHAM and PAUL (1971) have reported depression of Purkinje cell discharge by barbiturates similar to that seen with ethanol. GORDON et al. (1973) found that thiopental (0.5–8 mg/kg i.v. in cat preparations) depressed the axon discharge in granular cells and, thus, caused a failure of Purkinje cells to respond to peripheral stimulation. This dose of barbiturate did not modify the fast, mossy fiber pathway or the postsynaptic potentials of granular cells. In addition, the activity in climbing fibers was increased by an action outside the cerebellum. GORDON et al. (1973) conclude that pentothal in doses up to 8 mg/kg does not affect presynaptic activity in mossy fibers, or synaptic potentials in granule or of Golgi cells. It is suggested that the barbiturate may be acting on reticulocerebellar neurons or on granule cells; its action is not presynaptic on mossy fibers at the mossy fiber-granule cell synapse. ECCLES et al. (1971a, b) have shown that barbiturates depress Purkinje cell excitability and have a marked effect on the inhibition of Purkinje cells resulting from parallel fiber activation of inhibitory interneurons. MORTIMER (1973) also proposes that barbiturates reduce the background level of mossy fiber input as well as depress Purkinje cell excitability to parallel fiber inputs. (This paper also presents a powerful computer technique for understanding complex neuronal systems.)

The actions of the barbiturates on the cerebellum are complex, as might be anticipated, with excitation of some cells and depression in others. In an extensive recent report on cerebellar physiology, it was found that the spontaneous activity of nuclear cells of the cerebellum can be greatly increased by pentobarbital (ARMSTRONG et al., 1975); these authors presumably used commercial Nembutal containing 10% ethanol, i.e., for each 1 ml the animal receives 60 mg of pentobarbital plus 100 mg of ethanol. Many studies using pentobarbital are confounded by the simple oversight of the fact of the ethanol content.

With stimulation of peripheral nerves or the olive, interpositus neurons exhibited two phases of repetitive excitation; the second late phase was consistently more sensitive to depression by pentobarbital, and this dose was lower than that required to elicit acceleration of background discharge.

Although extensively studied, the pathophysiology of ataxia after ethanol or barbiturates has not been even accurately characterized. Tremor is a common event, yet its characteristics are in some dispute (BRIMBLECOMBE and PINDER, 1972). It is to be hoped that more elaborate and detailed human factors and engineering studies will facilitate the characterization of describing functions of human motor behavior and, thereby, permit the precise description of the nature of motor function disruption by drugs. Productive directions for such studies are presented in PORTER (1975) and STEIN et al. (1973b) among others.

To this reviewer's knowledge, the effects of other anesthetic agents, or sedative/hypnotics, and even the so-called centrally-acting muscle relaxants, have not been examined specifically for their effects on the cerebellum except for spontaneous activity (GOGOLAK et al., 1972). More importantly, their effects on motor coordination and on the cerebellum have not been compared in order to determine which sites are selectively affected. For example, it is not known if there are differences among sedative/hypnotic agents in regard to the degree of motor incoordination they produce relative to their sedative or hypnotic actions. The question is an old one. DUNBAR (1905) and PHILIP B. SMITH (1972) claim that alcohol produces more staggering than does ether—in presumably equally intoxicating doses.

Indirect cerebellar effects may have been involved in many studies, such as those cited below regarding benzodiazepines. The modulating influence of mid-brain reticular formation on segmental reflexes—such as facilitation—is generally more sensitive to depression by drugs, such as chlordiazepoxide or diazepam, than is the spinal reflex itself. Such observations are consistent with a supraspinal site of action, but do not prove it (cf. SMITH, 1965); moreover, the site(s) of primary effect could be almost anywhere below the level of the experimental section of the neural axis.

Caution is thus in order in interpreting all alterations of function involving cortical sites or spinal reflexes as originating from a site of drug action within the CNS. For example, a classic effect of ethanol is to produce nystagmus—so-called postalcohol nystagmus (PAN) (cf. FREGLY et al., 1967; ASCHAN et al., 1956; GOLDBERG, 1943) usually presumed to be central in origin. However, this PAN requires gravity and peripheral vestibular organs (OOSTERFELD, 1970) and is apparently due to a change in viscosity of endolymph (see MONEY and MYLES, 1974; KALANT, 1975). Further, we have shown that ethanol excites muscle spindle receptors (KUCERA and SMITH, 1971), receptors intimately involved in coordinated muscle activity. This explanation for the nystagmus produced by ethanol and other sedatives is not the whole or a general explanation, since barbiturates also induce nystagmus, and the site of this action is likely to lie in the vestibular and oculomotor neurons sensing and coordinating eye movements.

A related action of ethanol and barbiturates that has received almost no attention is the change in binocular convergence, an extension of the near-point of convergence; subjectively, the associated diplopia may or may not be suppressed. These two effects on eyeball movement—nystagmus and nonconvergence—seem to be among the signs that are subjectively perceived as indicating "under the influence."

Barbiturates appear to affect both in a fashion similar to ethanol, but whether or not either is more sensitive relative to the other actions has not been investigated.

Although the selectivity of site is generally assumed not to be due to differential distribution, there are at least suggestive autoradiographic data that ethanol is present in higher concentrations in the cerebellum and pituitary than at other central sites (FABRE et al., 1973). Differential distribution, temporally related, is known to obtain to some degree with certain barbiturates (ROTH and BARLOW, 1961). Such data are easily misinterpreted since they utilize a measure of total radioactivity at a given time after administration; for example, the study of FABRE et al. (1973) revealed relatively constant radioactivity in brain for "at least 12 hours" with the conclusion that the ethanol was incorporated into other molecules, such as amino acids or the isoquinoline condensation products (see Subsect. C.II.2).

The other possible sites of action at which motor incoordination could originate include spinal segmental reflexes, the reticular formation and basal ganglia. (The effects of the various agents on spinal motoneurons and presynaptic systems have been described in some detail in Subject. C.) None of these sites have received explicit comparative study or even a comparison of sensitivity to various agents; even the so-called muscle relaxants fail to reveal an unambiguous selectivity of effects on motor systems (SMITH, 1965). To illustrate, ethanol and barbiturates have never received an adequate clinical test to determine their efficacy, potency, and selectivity as skeletal muscle relaxants. By and large, the applicability to clinical situations of animal test models for skeletal muscle hyperactivity has not been critically analyzed.

In a similar vein, the protracted argument as to whether there were significant differences in the primary properties of meprobamate and phenobarbital hinged, in part, on meprobamate's touted muscle relaxant effects. No data in man or laboratory animals have been developed to distinguish between the agents on that ground (see references in SMITH, 1965).

Effects of ethanol at the spinal level have been summarized in a recent thesis by LATHERS and a paper by LATHERS and SMITH (1976) showing that there is depression of both mono- and polysynaptic reflexes and, in contrast to the repeated statements in the literature, no selectivity of action on polysynaptic reflexes. The individual investigation most frequently cited for the latter is KOLMODIN, yet KOLMODIN (1953) found that in about half of the preparations the monosynaptic reflex was more sensitive and, conversely, the polysynaptic was more depressed in the other half. Spinal interneurons are depressed by ethanol (EIDELBERG and WOOLEY, 1970), whereas Renshaw cells are activated (MEYER-LOHMANN et al., 1972).

EIDELBERG and WOOLEY (1970) are frequently cited as the most definitive study of the spinal motor neurons in which they showed there was a general neuronal membrane depressant effect on spinal motor neurons by doses of alcohol which produce moderate to severe ataxia. These authors failed to find an effect of ethanol on presynaptic inhibition in contrast to other investigators (see Subsect. C.II.3). But the question of site of action is indeed complex. For example, both presynaptic inhibition and monosynaptic reflex depression by ethanol are augmented by the sustained muscle spindle afferent activation (KUCERA and SMITH, 1971, 1972; LATHERS and SMITH, 1976), as well as by the Renshaw cell activation. The degree to which these systems and possibly other afferents contribute to the motor incoordination is not known.

2. Specificity of Site and Mechanism of Action on Cerebral Cortex, Reticular and Limbic Systems

a) *Cortical Activity and Evoked Potentials—Barbiturates, Anesthetics, Ethanol*

If a finite locus or representation of various moods, such as euphoria, anxiety, dysphorias, depression, etc., exists in the cerebral cortex, it seems possible that some combination of electrodes and stimulus parameters of cortex stimulation in humans would at least provide suggestive evidence of such loci. That both motor and sensory responses can be obtained by cortical stimulation has been long known; extensive investigations have been described and detailed by one of the prime investigators (PENFIELD, 1958) and brought up to date for a Handbook by LIBET (1973).

Although it is frequently assumed that there is a close correspondence between cortical electrical activity and motor, sensory, and thought process, a wide variety of spontaneous and stimulation-linked cortical activity has no directly corresponding subjective sensory experience. For example, the extensive cortical projections of group Ia afferents from muscle spindles are not associated with any subjective experience. It is thought, therefore, that these inputs serve in the integration and organization of movements mediated by the cortex without producing any subjective sensory experience.

A large portion of the cortex is said to be electrically inexcitable in that brief trains of stimuli applied to the cortical surface fail to elicit any responses. Vague psychic "memories" or hallucinations may be obtained in some cases with temporal lobe stimulation (PENFIELD, 1958). Nevertheless, the postulated emotional brain—the brain of motivation and rewards presumably activated with depth electrodes in reinforcing sites in animal studies—has not been studied in any detail in man although stimulation at various sites clearly can evoke emotional and affective responses (see, for example, HEATH, 1964, 1972; DELGADO, 1969; MYERS, 1974).

Assuming that the appropriate stimulus parameters can be generated, such studies would permit definitive determination of whether or not such functions as affective states, motivation, and reward system can be, as is generally assumed, localizable to anatomically defined sites. Although less likely, serious consideration should be given to the possibility that such descriptors as the various affective states, mood, drive, rewards, and emotions are strictly derivative abstracts that communicate the status of extremely large interactive neuronal networks extending from the spinal cord to the frontal cortex. Even if one takes the narrower view of the limbic system as composing the "emotional brain," affective states would appear to involve large, and presumably complex, nerve networks in contrast to the possibly more sharply defined sensory and motor pathways.

Of course, anesthetics and ethanol affect cortical cells, directly as well as indirectly; presumably, this applies to all of the agents under consideration. In summary, the spontaneous firing pattern of some neurons shifts to a rather irregular pattern; evoked discharges may be reduced or abolished; increases in neuronal activity may also be obtained, especially those of neurons exhibiting long latency and repetitive activity to remote stimulation.

The effects of barbiturates and general anesthetics on the EEG is too well-known and too large a topic for a comprehensive review here. Therefore, only some specific aspects that are germane to the questions of interpretation of experiments on site(s) and mechanism(s) of action of sedative/hypnotics will be reviewed. As SHARPLESS (1970) and HARVEY (1975) emphasize, the barbiturates and other similar agents clearly influence large numbers of neural systems and what selectivity can be found is, at best, relative and comparative. In this review, we are interested generally in low doses and in actions that occur at sites of appreciable sensitivity.

The changes produced in the EEG by the various sedative/hypnotic and anesthetic drugs vary widely, although large doses produce coma and slow waves of low amplitude. For example, the barbiturates and anesthetics readily and reproducibly produce in low doses a high frequency "activation" pattern in the EEG followed by a slowing of the predominant EEG frequencies; "sleep" spindles appear and characteristically the EEG activating response to cortical thalamic sensory stimulation or reticular formation stimulation is depressed.

Perhaps one of the most comprehensive reviews of this subject remains that of KILLAM (1962) (and recently, e.g., CLARK and ROSNER, 1973; ROSNER and CLARK, 1973; SALETU, 1974; SHAGASS, 1974). In landmark studies, FRENCH et al. (1953) and MAGOUN (1954) demonstrated that ether and pentobarbital blocked EEG arousal mediated by the reticular stimulation, and also depressed the potentials in the reticular formation evoked by sensory stimulation. The crucial experiments on selectivity have been those showing that the cortical responses evoked by sensory stimulation which are transmitted through thalamic relays are less sensitive to disruption by barbiturates than the responses mediated through the reticular formation.

Early evidence that anesthetics acted selectively to depress the reticular activating system includes initial studies of pentobarbital, thiopental, ether, nitrous oxide (HAUGEN and MELZACK, 1957), as well as chloroform, divinylether, and trichloroethylene (DAVIS et al., 1958). The striking differences in the EEG patterns obtained with the various gas/volatile general anesthetics need repeated emphasis; the origins of these differences have not been fully explained; yet they are of crucial significance in interpreting EEG data (see, e.g., FAULCONER and BICKFORD, 1960; CLARK and ROSNER, 1973; ROSNER and CLARK, 1973).

In an extensive review of anesthetics, ROSNER and CLARK (1973) and CLARK and ROSNER (1973) stress the need for the examination of general anesthetics on various sites as functions of dose; thus, they propose regional action sequence charts that depict "regional dose-response curves." With respect to regional specificity they conclude, tentatively, that the inhalational anesthetics can be classified into a spectrum with ether, nitrous oxide, and cyclopropane in one group with mixed excitatory/depressant effects on reticular function, coupled with progressive cortical, and eventually thalamic, depression. This depression is manifest in spontaneous, evoked unit and repetitive activity. The EEG effects of the agents methoxyflurane, halothane, trichlorethylene and chloroform appear similar, according to ROSNER and CLARK (1973), but differ from those of the previous group by less direct cortical depression and less depression, relatively, of nonspecific thalamic pathways. These authors appropriately emphasize the paucity of truly comparative data and the difficulty of establishing causal, rather than simple associative, relationships among the effects and sites. In addition, rigorous cardiovascular and respiratory controls are essential to such studies.

HORSEY and AKERT (1953) provide an accurate description of the effects of sedatives and hypnotics on the spontaneous electrical activity of the cortex and concluded that ether and barbiturates produce a pronounced activation state even greater in low doses than does alcohol, whereas the spindles after alcohol were not as "spikey" and have a lower frequency than those with barbiturates.

ROSNER and CLARK (1973) suggest that ether would be expected to disrupt psychological functioning more severely than corresponding doses of barbiturates, in view of ether's apparent greater effect on cortical activity than the barbiturates.

More specifically, ethanol has been described as having unique EEG effects (see reviews by HIMWICH and CALLISON, 1972; BEGLEITER and PLATZ, 1972; KALANT, 1975; PERRIN et al., 1974). In low doses, the EEG frequency spectrum shifts to higher frequencies ("activation") (see, e.g., HADJI-DIMO et al., 1968; SAUERLAND and HARPER, 1970). With increasing dosages and blood levels the frequency of the EEG declines and the amplitude of the waves is suppressed. HADJI-DIMO et al. (1968) examined not only the EEG frequency but found also that the cortical blood flow increased in parallel with the increase in EEG frequency; however, these stimulatory effects on the EEG and on blood flow could not be demonstrated in animals under pentobarbital anesthesia. That the reticular formation is not essential for the effects of ethanol on cortical electrogenesis has been the result of every test of the proposal; it was definitively examined by SAUERLAND and HARPER (1970) who found the same actions of ethanol on the cortex both before and after transection of the brain anterior to the reticular formation.

Recently, ROSADINI et al. (1974), in examining the antagonism of pyrithioxin and ethanol, found wide variability in the EEG effects of ethanol in human subjects—both within a given subject, across subjects, and with respect to blood alcohol levels. These variabilities make generalizations or correlations between behavior and cortical electrical activity questionable, at best. The "acute tolerance" described by PERRIN et al. (1974) is a further problem. HATCH et al. (1970) also emphasize the individual variability in EEG response to thiopental, as studied in cats.

Cerebral-evoked potentials extracted by averaging techniques applied to the EEG have become well-established techniques. Results with psychotropic drugs have been reviewed definitively by SHAGASS (1974). Barbiturates tend to increase latencies of evoked responses, small doses increase their amplitude while larger doses decrease them. The sites of action inferred are direct cortical and subcortical (see DOMINO et al., 1963; DOMINO, 1967; CIGÁNEK, 1961, 1967). A variety of doses of ethanol resulted in decreased auditory evoked responses in unanesthetized cats (PERRIN et al., 1974).

But such drug-induced changes in the cortical evoked responses are difficult to interpret with respect to the specific nature of the drug effect and its site. Often, the evoked response seems more to reflect the background state of modulating systems than direct influences on the pathway itself (see reviews of WINTERS et al., 1967a, b; ROSNER and CLARK, 1973; CLARK and ROSNER, 1973; SHAGASS, 1974). Cortical potentials evoked by sensory stimulation tend to be larger in the resting and drowsy states than in sleep or in the alert condition. Cortical responses to clicks (in unrestrained cats) were markedly increased by relatively low doses of pentobarbital, chloralose, and chlorpromazine; however, in comparable experiments, ethanol and urethane decreased the response (NAKAI et al., 1965, 1966; PERRIN et al., 1974). That sedative/hypnotics and ethanol alter cortical evoked potentials has been demonstrated in a number of animal studies and in man (e.g., MASSERMAN and JACOBSON, 1940; GRENELL, 1959; DRAVID et al., 1963; NAKAI and TAKAORI, 1965; NAKAI et al., 1966; GROSS et al., 1966, 1973; HIMWICH et al., 1966; DIPERRI et al., 1968; MORI et al., 1968; LEWIS et al., 1970; BOYD et al., 1971, 1974; BEGLEITER et al., 1972, 1974; GUHA and PRADHAN, 1974; reviewed by HIMWICH and CALLISON, 1972, ROSNER and CLARK, 1973, and SHAGASS, 1974).

The site of action of barbiturates, chloralose, and urethane in altering cortical evoked responses may well be subcortical; for example, NAKAI et al. (1965) demonstrated that the depression of the response to auditory stimuli could already be recorded in the inferior colliculus. In contrast to these agents, ethanol only in relatively large doses depressed a portion of the collicular response. Moreover, changes similar to those obtained with pentobarbital were seen after such diverse agents as tetrahydrocannabinol and mescaline (GUHA and PRADHAN, 1974).

On the basis of a series of studies, NAKAI and DOMINO (1969) concluded that pentobarbital has a neocortical effect somewhat greater than its effect on the midbrain reticular formation. In comparison, ethanol, relatively speaking, had an even greater cortical effect; but, of importance is the fact that both agents had cortical effects in the same doses that had an equal or lesser effect on their measures of reticular formation functions. The observations of GANGLOFF and MONNIER (1957, 1958) that phenobarbital had greater effects on thalamocortical circuits than the reticular system has always posed a difficulty in the development of the reticular activating system hypothesis, even though the significance of such differences was confounded by the use of different species and a different drug from those used by other investigators. Similarly, TAKAORI et al. (1966) showed that pentobarbital (and chloralose and chlorpromazine) abolished the depressant effects of reticular formation stimulation on cortical responses to clicks; however, ethanol was without influence on the response to reticular stimulation.

These recent studies confirm the old observations of MASSERMAN and JACOBSON (1940) that there was a rather striking decrease after ethanol in the threshold for the stimulation of hypothalamus in cats as measured in a response for the emotional responses to the hypothalamic stimulation. In general, the electrophysiologic response to stimulation on the cortex tended to be depressed by ethanol, although the authors did notice that in low doses there may be a decrease in the threshold for cortical stimulation. They quite appropriately pointed out that there may well be an interaction between cortex and hypothalamus and that the decrease in threshold of the hypothalamus may be the result of a decrease in cortical inhibition of the hypothalamus rather than of a direct action at that site. These observations are confirmed by those of DRAVID et al. (1963), HIMWICH et al. (1966), and DIPERRI et al. (1968) who concluded that the cortical association area was as sensitive or more sensitive to the depressant action of alcohol than was the reticular formation which was, in turn, more sensitive than the primary somatosensory cortex (on the basis of the cortical evoked responses to stimulation of peripheral nerves).

GRENELL (1959), in an overview presentation, reported that a number of aliphatic alcohols (methanol through butanol) produced in low doses an increase (excitation) in the cortical evoked response to a click stimulus, whereas in larger doses depression was observed. Ethanol doses ranged from those producing little behavioral effects in intact animals to those producing anesthesia. AMASSIAN (1954) stressed the susceptibility of the cortical association areas to small doses of barbiturates, although a quantitative study has not been undertaken for other agents.

Thus, there are little convincing data to show that ethanol acts selectively on the reticular system relative to the cerebellum or cerebral cortex. The disparities among drug actions strongly suggest that a single site/mechanism for the cortical effects of all these agents cannot be seriously entertained. Perhaps the differences are epiphenomena—but logical consistency demands either the inclusion of EEG data as evidence of CNS action, or the rejection of such data.

KALANT (1974) attempts to reconcile the apparent lack of effects of ethanol on the reticular stimulation as compared to barbiturates by citing the fact that NAKAI et al. (1965) used noncomparable doses of ethanol and pentobarbital. It is true that some of the conclusions of NAKAI et al. (1965) and NAKAI and DOMINO (1969) were based on low doses of ethanol relative to those of pentobarbital—yet had the ethanol dose been increased, it would have most likely altered the response via the reticular formation—but lowering the pentobarbital dose would not conceivably have resulted in a more selective action on the cortex.

PERRIN et al. (1974) interpreted their observations of a more rapid depression of the reticular formation evoked response during infusion of alcohol as indicating selective effects on the reticular system; depression of spontaneous frequency and amplitude occurred to lesser degrees in cerebral cortex, hypothalamus, and hippocampus, in that order. However, the figures and data presented fail to demonstrate convincingly a selective action on the reticular formation.

In correlating electrical activity and behavior, spontaneous electrical activity of cortex, limbic system, thalamus or tegmentum is really not markedly affected by intoxicating doses of ethanol (STORY et al., 1961) whereas the evoked responses in the somatosensory and in the visual cortex to the corresponding thalamic relay nucleus and optic tract are markedly depressed. Although the authors mistakenly quote KOLMODIN (1953) as showing greater effect of ethanol on polysynaptic cord reflexes, their conclusion that ethanol acts on cortical interneurons may be correct. (STORY et al., 1961, also present an interesting example of possible acute tolerance to the depression by ethanol of the augmenting and visual response in acute cat experiments.) Interestingly, cyclopropane abolished cortical-evoked responses in concentrations that did not markedly alter threshold or intensity perception in the same subject, thus revealing a marked dissociation between perception of a peripheral nerve stimulus and the cortical response (CLARK et al., 1969).

As mentioned above, interpretation of changes in cerebral-evoked potentials is complex. It seems well established that the evoked responses in frontal cortical areas to cortical or sensory stimulation is susceptible to depression by sedatives and hypnotics. Moreover, the more delayed components of the evoked response, the late responses, are more sensitive to depression by ethanol, pentobarbital, and ether than the early responses, with low doses altering the initial responses only variably. Among these three agents there is apparently little qualitative differences in effects or differential sensitivity among the components in squirrel monkeys (BOYD et al., 1974); slightly larger doses depressed both components. BOYD et al. (1974) failed to detect any stimulatory activity of these depressant agents. By contrast, Δ^9-THC markedly augmented both the early and the late evoked responses and facilitated the appearance of repetitive discharges. As has been repeatedly emphasized, repetitive activity or after-discharges are generally more susceptible to depression than the primary response (BERRY, 1965; STRAW and MITCHELL, 1966; WINTERS et al., 1967a, b; SMITH, 1965).

The changes in EEG and evoked responses reported by BOYD et al. (1971, 1974) could not readily be correlated to subjective state, mood, or alterations in sensory perception. In fact, Δ^9-THC produced EEG effects similar to those of picrotoxin and metrazol; LSD only had depressant effects on the late response in high doses, although mescaline effects on the late responses were similar to those of THC. Moreover, amphetamines appeared to be without effect; atropine was stated to be also without actions, but the dosage may have been too low.

Effects of meprobamate on the human EEG are generally considered to be similar to those of ethanol and barbiturates (BOKONJIC and TROJABORG, 1960); the EEG effects also exhibit tolerance and withdrawal hyperactivity. There is little evidence that meprobamate exerts reticular depression as selective as that produced by the barbiturates; perhaps the thalamus is affected selectively (KILLAM, 1962). Mephenesin, which has subjective effects similar to those of ethanol and was early shown to be effective in relieving the anxiety and tremor of the early withdrawal from alcohol (SCHLAN and UNNA, 1949; SPRENG, 1953; SULLIVAN, 1954; CROSSLAND and TURNBULL, 1972), has effects on the reticular system, on tonic reflexes (CHIN and SMITH, 1962), on spinal neurons, and also on peripheral nervous structures (cf. review of SMITH, 1965; GROSSIE and SMITH, 1966; SMITH, 1973).

In rhesus monkeys, pentobarbital and nitrazepam reproducibly produce slow-wave sleep but decrease wakefulness and REM sleep (DAVID et al., 1974). Of especial interest was the observation of enhanced REM levels for 7 days following a single dose of either drug. Similar sustained postdrug effects were also seen after imipramine. Hyperirritability and hyperarousability were evident 24 h after the single oral dose of pentobarbital or nitrazepam, even though the acute effects of somnolence were over within 3 h. Full recovery required 48–72 h. These well-documented findings in monkeys parallel and extend findings in human studies and reports (OSWALD and PRIEST, 1965; MALPAS et al., 1970; OSWALD, 1968, 1971; OSWALD et al., 1969; cf. also KAY et al., 1972).

The quantitative studies of DAVID et al. (1974) showed initial increased wakefulness and activity in monkeys following pentobarbital as discussed by BRAZIER (1963). Nitrazepam, in these studies, did not exhibit any stimulatory action. The fact that at least some benzodiazepines have effects that last long beyond the overt actions has been demonstrated unequivocally by DAVID et al. (1974); the rebound hyperexcitability was also observed with pentobarbital.

Direct tests have revealed that the sensitivities of cortical cells to excitant amino acids and to ACh are both reduced by systemically administered anesthetics but reportedly without a selective effect on the responses to either type of putative transmitter (CRAWFORD and CURTIS, 1966; CRAWFORD, 1970), [although there appears to be a selective action of ethanol on ACh release (CARMICHAEL and ISRAEL, 1975)]. The dose of anesthetic required may be appreciably higher than that required to alter EEG activation evoked by reticular formation stimulation. The agents tested with similar effects were diallyl-barbituric acid, sodium methylthioethyl-2-pentylthiobarbiturate, urethane (which required anesthetizing doses), and α-chloralose

(CRAWFORD and CURTIS, 1966). Halothane and nitrous oxide had only small depressant effects and the experimental results were variable, supporting the conclusions of ROSNER and CLARK (1973). Direct iontophoretic application of pentobarbital produced depression of the responses to D-L homocysteic acid and to acetylcholine. KRNJEVIĆ (1974b, 1975) claims universal lack of effect of anesthetics on the excitatory response to glutamate, whereas in an isolated preparation RICHARDS (1974) and RICHARDS et al. (1975) find only halothane inactive, whereas the excitation by glutamate was depressed by barbiturates, ether, and trichloroethylene. (The ACh effects on the same cells were selectively antagonized by atropine—and not by dihydrobetaerythroidine or gallamine; the latter agent produced excitation by itself.)

Isolated perfused brain exhibits effects of barbiturates and chloral hydrate similar to those in intact animals (see short review of KRIEGLSTEIN and STOCK, 1974).

With respect to specificity and selectivity of neural sites, the possibility remains that the sleep or unconsciousness induced by hypnotics and anesthetics is the result of drug action limited to a few sensitive sites. FOLKMAN et al. (1968) claim that teflurane and methoxyflurane produced light sleep, EEG spindles and analgesia when it was diffused from the tip of a cannula placed in the mesencephalon; cyclopropane was far less potent and nitrous oxide was without effect.

How specific such focal effects are remains in question since WILKINSON et al. (1971) found similar actions when the membrane-tipped cannula was in the thalamus or hypothalamus but not when it was in frontal white matter. In contrast, the caudate nucleus was selectively sensitive to microinjected phenobarbital, resulting in sedation and EEG changes (LEIGHTON and JENKINS, 1970, and see review in MYERS, 1974, p. 456); no such effects occurred when it was injected into the midbrain reticular system, amygdala, hippocampus, or thalamic nuclei. These studies deserve repetition in view of their profound importance regarding the possibility of selectively sensitive neural locations for sedation produced by barbiturates. Clearly these studies reveal different sites of action than those of FOLKMAN et al. (1968). Other evidence of sensitivity of the reticular system is the finding that the application of various agents to the mesencephalic reticular formation results in EEG alterations (GROSSMAN, 1968; MYERS, 1974).

Other evidence bearing on local drug sensitivity includes the in vitro studies cited above, and such evidence as increased presynaptic inhibition with local application of ethanol solution to the cuneate nucleus (DAVIDSON and RIX, 1972)[13] and diazepam to spinal interneurons (CHANELET and LONCHAMPT, 1972).

Regarding selectivity of action, there are various reports that may represent "flukes," but they may also provide significant clues for future studies. For example, ISHIKAWA et al. (1966) found that thiopental (2–20 mg/kg i.v. in cat preparations) completely blocked the after-discharge obtained by electrical stimulation of the

[13] WAYNER et al. (1975) claim that hypothalmic cells are more sensitive to the direct effects of ethanol than are those of the cortex. These studies compared "electrophoretic application of ethyl alcohol" through a capillary microelectrode with its effects upon i.v. administration. However, the absence of a charge on the molecule of ethanol precludes its "electrophoretic application." The data briefly presented imply that ethanol was ejected only by application of current, yet no correction for leakage is mentioned. In addition, the ethanol actions appear rather unimpressive since far more cells at all sites examined (thalamus, hypothalamus, cortex) were not influenced than were either excited or depressed.

hippocampus. In contrast, meprobamate and LSD were without effect, whereas chlordiazepoxide, chlorpromazine, atropine, and reserpine, in the lowest doses which had effects actually prolonged the after-discharge. Unfortunately, other barbiturates and anesthetics have not been explored. Note should be made of the absence of effects of meprobamate and the prolongation of the after-discharge by chlordiazepoxide; both of these effects are qualitatively different from those of thiopental.

A tentative summary of the sites of action of the barbiturates, anesthetics and ethanol can be outlined according to sensitivity, starting with the most sensitive.

Barbiturates:

— Variable, poorly understood, excitatory actions involving the reticular system (cf. ROSNER and CLARK, 1973; DUNDEE, 1974).
— Prolongation/potentiation of presynaptic inhibition.
— Depolarization of primary afferent terminals.
— Depression of association cortex.
— Depression of repetitive responses such as the muscarinic effects of ACh on cortical cells.
— Depression of cortical activation produced by reticular formation—action at reticular formation.
— Depression of Purkinje cells of the cerebellum.
— Depression of excitatory postsynaptic responses, such as the motoneuron response to glutamate.
— Depression of transmission by postjunctional block of the response to ACh.
— Depression of neuronal excitability, including block of small nerve fibers and repetitively discharging large fibers.

Gaseous/volatile anesthetics:

— Excitatory action on EEG; origin in reticular formation [agents can be differentiated on bases of degree of excitatory action (CLARK and ROSNER, 1973)].
— Depression of cortical activation induced by reticular formation.
— Increase in presynaptic inhibition in the cord.
— Facilitation of postsynaptic inhibition.
— Reduction of transmitter output from action presynaptically in cortex.

Ethanol:

— Suggestion of excitatory action on EEG cortex plus increased transmitter release, transient increase in reflexes.
— Depression of association cortex (depression of transmitter release to stimulation).
— Depression of Purkinje cells of cerebellum.
— Increase in presynaptic inhibition.
— Stimulation of Renshaw cells.
— Postsynaptic depression of EPSP in most neurons.
— Prolongation of e.p.p. at neuromuscular junction (little selectivity of effects on the reticular formation relative to cortex or spinal cord).

This consideration implies the conclusion that there are multiple sites and possibly a number of different mechanisms involved with even one sedative/hypnotic agent. Further, although a spectrum of sensitivities for each agent can be delineated, there are complex interactions among site, mechanism, agent, and behavior of interest.

It remains conceivable that more specific and widely applicable mechanisms of action are yet to be discovered, for example, for the barbiturates or inhalational anesthetics. Reading of current literature provides this reviewer with little optimism that there has yet been postulated the unique mechanism of action that will prove to be applicable to any of these agents at their many sites of action, such as the identification of a specific neurotransmitter or a blocking action of a given transmitter. Such specificity cannot be ruled out, however. It remains possible that general anesthesia or any one of the many behaviors comprising the syndrome of general anesthesia may be demonstrated to rest on the function of only a small number of neurons. This thesis receives support from the probability that electrically induced anesthesia derives from a narrowly defined locus of action in the midbrain (REYNOLDS, 1971; HOCHERMAN, 1972) and effectiveness of locally applied anesthetics in the same general region (GROSSMAN, 1968; FOLKMAN et al., 1968; WILKINSON et al., 1971; and MYERS, 1974 review—but see conflicting data of LEIGHTON and JENKINS, 1970).

b) Summary and Comment

This review of studies on EEG or cortical responses illustrates the complexity of the systems and perhaps the questionable value of extensive speculation based on present gross electrical recording techniques. The cortical responses to natural stimuli may be increased or decreased by stimulation of various sites in the reticular formation. Further, whether increased or decreased, the reticular influence may be exerted in the thalamus, colliculus, or cortex—and different pathways exist and are, in part, modality-specific. Thus, unless the drugs exert an exquisitely specific effect, the evoked response to the sensory stimulation may be increased or decreased—e.g., increased by depressing reticular inhibitory systems or decreased by cortical or other influence. Analogously, the effects of the agents on inhibition and facilitation induced by reticular formation stimulation may be difficult to interpret—in cases where effects actually occur. With ethanol, the situation appears controversial and relatively unstudied; whether or not the direct cortical effects appear with doses that are without appreciable influence on the responses to reticular formation stimulation remains to be defined unambiguously.

This review of recent studies, in the context of the extensive previous work, leads one to postulate that either (1) the EEG and evoked responses are too gross and mixed a response to be indicative of qualitative differences in site and mode of action of drugs with different effects on thought and behavior or (2) the cortical response signals drug effects at many sites beyond those of primary interest and the site of processes of the cortical system under test may not be of particular relevance to understanding the drug-induced alterations in thought processes and behavior.

This observer's bias is that the agents act on the cortical system and that such actions are of importance; however, our electrophysiologic techniques have been too gross and too indiscriminate, so that they pick up simultaneous actions at too many sites, each with a different functional role. What the evoked potential changes do represent, nevertheless, is unequivocal evidence that the drugs in question alter, directly or indirectly, the function of the neurons involved in the evoked response. The contrary proposition obviously does not hold; the failure to detect a change with such gross recording of neuronal activity does not denote an absence of an action at that brain locus. More sophisticated techniques for analyses of EEG effects may well provide precise data on selectivity of action (cf., e.g., GEHRMANN and KILLAM, 1975).

3. Consideration of a Variety of Purported Clinical Effects of Sedative/Hypnotic Agents

a) Antianxiety Action

The sedative/hypnotic class overlaps almost indistinguishably with antianxiety agents, partly as a reflection of the rationale for clinical prescribing. From the point of view of mechanism and site of action, an antianxiety agent might be one that: (1) antagonizes agents/mechanisms known to increase anxiety, including such agents as amphetamine, cocaine, and perhaps caffeine in susceptible individuals, or perhaps would have adrenergic β-blocking activity since such agents do have antianxiety properties (cf. propranolol and KELLNER et al., 1974); (2) acts at central sites suspected to be involved in mediating anxiety and its somatic components, such as the hypothalamus (such as MASSERMAN and JACOBSON, 1940, claimed for ethanol) or the limbic system as propounded for meprobamate and benzodiazepines.

Evidence that the associational neocortex exhibits as great a sensitivity to barbiturates and ethanol as any part of the nervous system studied to date is reviewed above. Even this observation should be tempered with the realization that all of the relevant comparisons have not been undertaken—neuronal discharge patterns, other CNS active agents, other possible sites.

b) Sedation vs. Antianxiety or Antiepileptic Effects

It is commonly claimed that the sedative effects of drowsiness and grogginess produced by pentobarbital or benzodiazepines will wear off over a few days with repeated ingestion, yet the antianxiety effects or antiepileptic actions will be retained. This implies, categorically, the selective development of tolerance to the sedative effects. It also imbues the term sedative with very narrow properties if it is, in fact, to be differentiated from antianxiety effects. Evidence to indicate that antiepileptic action is not related to sedation or hypnotic effects is mentioned above. Although no convincing evidence has come to my attention, I will subscribe to the postulate that tolerance can develop differentially to the various subtle mood- and consciousness-altering properties of these agents. But the degree to which this is feasible and the degree of differentiation actually possible among anxiety-reducing, sedative, and hypnotic properties has yet to be determined (see Subsect. D).

That sedative and anticonvulsant activity may be due to separate mechanisms or sites of action has long been postulated—among the actions of a given compound or among a series of analogues. The actions of the pair of optical isomers of mephobarbital are perhaps a good illustration; both *d*- and *l*-forms reportedly possess anticonvulsant properties but only the *l*-form was found to exhibit hypnotic activity (GOR-

DIS, 1971; see Subsect. C.II.8 regarding convulsant barbiturates). If these results can be confirmed and the action demonstrated to be due to these molecular entities, these data would demonstrate unequivocally that barbiturate action requires, firstly, stereospecificity and precise molecular configuration and, secondly, that the hypnotic/sedative actions (and perhaps other actions) of the barbiturates are selective for site and mechanism, rather than being the result of "nonspecific" neuronal depressant actions.

c) Amnesia

Amnesia may be produced by any of these agents, but it is usually of relevance as a toxic effect; however, amnesic effects of the benzodiazepines, especially diazepam, is becoming routinely exploited clinically—e.g., in oral surgery (KEILTY and BLACKWOOD, 1969). DUNDEE and PANDIT (1972) compared meperidine, hyoscine, and diazepam and showed that diazepam was most effective of the three in producing amnesia with a rapid onset and short duration of action of an hour or so. The authors claim that the incidence of amnesia is not related to the degree of drowsiness, but this is confused in the discussion because it is not clear whether they mean related across drugs or among patients receiving the same drug. They note that the use of the methohexitone-nitrous oxide anesthesia itself produced significant amnesia. Hence, the possibility exists of directly produced amnesia which is a synergistic effect with the anesthetic commonly administered. A discussion of memory mechanisms is beyond the scope of this chapter but note should be made of the possibility that amnesia may not be too far removed, behaviorally speaking, from relief of worry or tension, and might tend to preclude sustained anxiety reactions. Further studies in this area are indicated.

The blackouts known to be associated with marked alcohol intoxication and effects of alcohol on memory are reviewed by GOODWIN et al. (1969a, b), GOODWIN and HILL (1973), MELLO (1973a), MILLER and DOLAN (1974), RYBACK (1970, 1971, 1973), and TAMERIN et al. (1971) without definitive conclusion as to their mechanism or a precise delineation of the mechanisms of memory impairment or the specific conditions under which blackouts occur. It is interesting to speculate about whether the blackout of acute alcohol intoxication, the amnesia of barbiturates or anesthetics (e.g., MAZZIA and RANDT, 1966) and the retrograde amnesia with electroconvulsion therapy or concussion have similar underlying mechanisms. For example, does a syndrome equivalent to blackouts occur with barbiturates? The report of SQUIRE et al. (1975) describes retrograde amnesia following ECT, but what was due to ECT and what to the methohexital, atropine and succinylcholine used with each ECT session cannot be ascertained, although this report does present useful strategies for studying amnesic states (see also BARBIZET, 1970; ROSENZWEIG and BENNETT, 1975; JOHN, 1967; RUSSELL and NATHAN, 1946).

Convulsions produced by flurothyl are also associated with amnesia. CHERKIN (1970) reviews the evidence regarding the mechanism of the disturbances in memory processing produced by various experimental agents and procedures. He emphasizes the graded nature of the effect and the conclusion that its magnitude depends on the intensity of treatment and the time elapsing between learning and treatment. With flurothyl the evidence appears to favor disruption of memory consolidation more than its retrieval.

Barbiturates (pentobarbital) impair learning by an apparent interruption in consolidation—i.e., impairment is limited to the time period of up to 30–60 min following the training in goldfish (LIU and BRAUD, 1974); interestingly, picrotoxin facilitates learning in this test system analogous to previous facilitation reported for pentylenetetrazole. Inhibition by pentobarbital of consolidation of maze-learning in rats has also been reported (GARG and HOLLAND, 1968). As has been long known, material learned during sedation with thiopental and in anesthesia (MAZZIA and RANDT, 1966) is readily forgotten. Retention of such material is apparently not state-dependent since recall is not facilitated by reinstatement of sedation (OSBORN et al., 1967).

d) Impairment of Judgement, Increased Risk-taking Behavior, Release of Inhibitions

A wealth of observation and psychological studies attests to the conclusion that sedative drugs induce changes in judgement and an increase in risk-taking. Most descriptions and interpretations consist largely of verbal descriptions of behavior. In short, people and animals seem to do things under the drug they would not do without it. They take risks they would not normally take; they may "show off" and they may engage in otherwise unsanctioned behavior. Most of these behaviors appear to be learned, socially conditioned adaptations of more primary drug actions. The role of social influence in shaping effects of drugs has been a current popular theme—sometimes with considerable insight [see two recent polemical books (LENNARD et al., 1971; MACANDREW and EDGERTON, 1969)]. In fact, a recent investigation failed to find any actual improvement in mood and affective state following alcohol even though the subjects expected to experience an improvement and they described their tone as improved (MAYFIELD, 1968 b).

These actions, including those under the general heading such as the "release from inhibition" theories, are rather vaguely ascribed to actions in the brain and in the cerebral cortex—and well they might be. However, in experiments examining such behaviors there are very significant interactions among dose, compound, and fighting behavior in attacking and subordinating animals (HO and MCLEAN, 1974; cf. KRSIAK, 1974; MICZEK, 1974 a, b). The sites of these various drug actions have received little direct attention.

e) Analgesia

It is well known that ether, ethanol and cyclopropane have analgesic actions in subanesthetic doses. The analgesic and amnesic effects of ether are sufficiently profound to permit major surgical procedures. By contrast, it is frequently stated that barbiturates and related substances not only do not produce analgesia, they may even exaggerate pain sensibility (DUNDEE, 1960; SHARPLESS, 1970). The situation is not so simple, since some degree of analgesia may be obtained with barbiturates, equivalent to low doses of narcotic analgesics (KEATS and BEECHER, 1950). One explanation for this paradox is that the narcotics may produce a more profound and comprehensive calming effect than barbiturates. It is, of course, well known that intoxication with ethanol or barbiturates results in a quiet sleepy person primarily when the individual chooses, for other reasons, to sleep. During the daytime or in a party setting barbiturates may well appear to activate or release behaviors normally suppressed.

The analgesic action of all of these agents can be readily related to their known effects presynaptically on primary afferents. This action alone might be sufficient to result in analgesia (as well as the subjective alterations in sensory input). And such a mechanism has been postulated for the narcotics. However plausible this site/mechanism of action may be, supraspinal actions on the cortex and reticular system and at other sites may be more important. Thus, we can conclude that the accentuation of presynaptic inhibition by the sedative/hypnotics, barbiturates ethanol, as well as narcotics may cause analgesia, but might only be one contributing factor among many (CHIN et al., 1974). The actual sites of drug action probably resulting in such analgesic effects range from directly on the spinal presynaptic system for the sedative/hypnotics to sites in the midbrain for narcotics and diazepam (e.g., SATOH and TAKAGI, 1971; JACQUET and LAJTHA, 1975; CHIN et al., 1974; AKIL et al., 1976; HOCHERMAN, 1972; EIDELBERG and BARSTOW, 1971).

Of some interest is the claim that halothane lacks the analgesic actions obtainable with ether, ethanol, and cyclopropane. If this is so, a comparison with other agents of its actions at various sites, especially presynaptically, might provide valuable clues regarding the mechanisms of analgesia. However, ROBSON et al. (1965) have demonstrated differential analgesic effects of nitrous oxide, halothane, and thiopental on different types of pain.

It is conceivable that the sedative/hypnotics, to varying degrees, activate the endorphin system now known to operate with narcotics. In support of this view, naloxone has been reported to antagonize an "analgesic" effect of nitrous oxide in mice (BERKOWITZ et al., 1976). Naloxone antagonism has been taken to indicate a narcotic mechanism of action (MAYER and HAYES, 1975; AKIL et al., 1976), but caution regarding specificity is in order (HAYES et al., 1977). Naloxone is certainly not an antagonist of general depressant effects of ethanol (SMITH and ABEL, personal communication).

Further suggestive evidence for specific analgesic properties is the old laboratory observation of species differences—barbiturates produce sleep, respiratory depression and decreased responsiveness to a variety of sensory stimuli in most animals and man, to the degree that they can frequently be the sole anesthetic for modest surgical procedures. Such is not the case in rabbits since reflex jerks and struggling are the result of a skin incision with pentobarbital anesthesia in the rabbit using the same dose and with the same apparent degree of unconsciousness as would be obtained in the dog, or with ether in the same rabbit.

f) Mood

All of the sedative/hypnotics and anesthetics alter mood, as discussed in the introduction. One of the most important aspects appears to be the fact that the agents facilitate or evoke a *change* in mood state, i.e., sedative drugs appear to introduce a marked increase in lability of mood, whereas the direction of the change is strongly influenced by prior experience, expectation, surrounding stimuli ("set" and "setting"). As one example among many, in the studies of CHESSICK et al. (1966) the setting influence apparently overwhelmed any specific drug action of pentobarbital, as well as morphine or chlorpromazine. However, HOLLISTER (1976) discounts the potency of setting in altering THC actions. Although most of these agents are referred to as "downers," this appelation may well reflect the contrast in activity and drive with "uppers"—amphetamines and congeners. That sedative/hypnotics appear to decrease initiative and drive is obvious; knowledge about the site or neurophysiologic mechanism of such effects is almost nonexistent.

Perhaps more to the point are the studies of perception of internal states induced by drugs when given under blind conditions and the correlative studies of drugs serving as stimulus cues in animals. HAERTZEN (1965, 1966), HAERTZEN et al. (1963, 1969, 1970b, 1974), HAERTZEN and HILL (1963), HAERTZEN and HOOKS (1973), and HILL et al. (1963a, b) have been able to show that individuals can define the sedative/hypnotic (pentobarbital/secobarbital) drug state by adjective check lists. The utility of these descriptor lists has had extensive concurrent validation in the areas of narcotics and hallucinogens. With barbiturates not only are apparently trivial cues used, the agents clearly are perceived to be pleasurable and produce a "high" and to relieve anxiety, to promote relaxation and sleep (JASINSKI, 1973; MARTIN et al., 1974b).

It would appear that glutethimide, methaqualone, meprobamate, and some benzodiazepines share many of the subjective properties of the barbiturates, even though there may well be differences. Analogous animal data are being generated (e.g., OVERTON, 1968, 1974; WINTER, 1974); rats trained under barbital do not respond to the drug state provided by ethanol; however, animals trained under ethanol respond to barbital (YORK and WINTER, 1975b). These data suggest that although certain aspects of the drug state are similar across agents, they may not be identical (see Subsect. D.I.2).

g) *Psychedelic Actions of Hypnotics, Anesthetics, and Volatile Solvents*

In addition to many other substances, the general anesthetics have been classed along with alcohol in terms of a variety of their subjective effects. Among these are a large number of more fantastic, mystical, or heightened mood states. In addressing nonmedical, personal use of such agents it seems appropriate to consider all of the subjective effects in the search for clues regarding the bases of such use. A curious book recent entitled "The Chemical Glimpses of Paradise," by PHILIP B. SMITH (1972) is an essay and collection of accounts (mostly old) of the subjective effects of self-administered anesthetics, with particular emphasis on ether, chloroform and nitrous oxide. Ether and nitrous oxide intoxication have been known and described repeatedly over the years, for example, WILLIAM JAMES (1902):

> Nitrous oxide and ether, especially nitrous oxide, when sufficiently diluted with air, stimulate the mystical consciousness in an extraordinary degree. Depth beyond depth of truth seems revealed to the inhaler. This truth fades out, however, or escapes, at the moment of coming to; and if any words remain over in which it seemed to clothe itself, they prove to be the veriest nonsense. Nevertheless, the sense of a profound meaning having been there persists; and I know more than one person who is persuaded that in the nitrous oxide trance we have a genuine metaphysical revelation.
>
> Some years ago I myself made some observations on this aspect of nitrous oxide intoxication, and reported them in print. One conclusion was forced upon my mind at that time, and my impression of its truth has ever since remained unshaken. It is that our normal waking consciousness, rational consciousness as we call it, is but one special type of consciousness, whilst all about it, parted from it by the filmiest of screens, there lie potential forms of consciousness entirely different. We may go through life without suspecting their existence; but apply the requisite stimulus, and at a touch they are there in all their completeness, definite types of mentality which probably somewhere have their field of application and adaptation ...

DUNBAR (1905), and recently SMITH (1972), clearly distinguish between the effects of alcohol, on the one hand, and the effects of ether on the other, implying that although both are similar, the "fantastic" effects are more apparent with ether relative to disturbance of motor function than they are with alcohol, even though they occur with the latter. Staggering is said to be very apparent under alcohol but not under ether. DUNBAR (1905) concluded that the staggering after alcohol is due "to early affection of the cerebellum." The editing of the SMITH book leaves something to be desired and many of the striking contrasts among the descriptions from other people are not recognized or discussed. For example, some make a point of not having any nausea or any other effects of the agents; others mention that the taste or the feeling of nausea is unpleasant. But the description with ether of the tingling and vibration of the nerves by a number of different authors/observers is striking.

The editor of the book is quite right in pointing out that these old authors in fact describe vividly the psychedelic experience of revelation and insight which was to receive such great popularity many years later by T. LEARY and LSD proponents and to a lesser degree by solvent users (see WEISE et al., 1973). Some found that the psychedelic experience with ether was reproducible, repeatedly so, whereas others indicate that it was a one-time occurrence. Some writers obviously have used many of the drugs over and over again whereas others describe only a single experience. The book seems seriously biased since no mention is made of the—certainly large—number of individuals who had very unpleasant experiences, or the absence of any experience with these agents. It is not sufficient to point out that for some they produce unique fantasies and "new levels of consciousness;" the frequency of occurrence in the given individual and the frequency among different individuals of such experiences is needed.

Where the author of this book sits in the debate as to the use and methods of use of such anesthetics is not clear. The introduction contains an apparently tongue-in-cheek remark decrying any promiscuous use of anesthetics with the comment: "They are deadly." Certainly, it would appear that Dr. Smith had a mystic experience mediated by the effect of nitrous oxide as a young man. He subsequently found that other anesthetic agents produced similar effects on him and that although he introduces himself, his conclusions are not provided except to indicate that his experience was not a unique one and that God had not "loved me above all men and had given me alone this pure chemical grace."

The experiences he describes were undoubtedly pleasant; the laughing experience characteristic of nitrous oxide anesthesia is also pleasant and presumably would tend to make people retake the agent. A question arises as to whether these pleasurable and reinforcing subjective states induced by the drugs are common among many agents, such as alcohol, general anesthetics and barbiturates: are they a consequence of a common mode of action and are the differences among them minor aspects of their spectrums of actions on the other systems or are these differences indicative of totally different mechanisms of action which produce similar subjective states? Another point is whether or not these positive subjective experiences are in fact the prime reason for repeated use or are manifestations of the same mechanism which gives rise to the repeated and compulsive aspects of the repeated use.

There is abundant experience with anesthetics to the effect that they are subjectively intoxicating in a fashion similar to or almost identical to the intoxication induced by ethanol. Although there is much evidence to suggest that although the end states of anesthesia, sedation or drunken intoxication may be produced by a number of agents, the cellular mechanisms of these may differ (cf. e.g. QUASTEL, 1975, and conclusions below).

One of the major problems in trying to tie together electrophysiology, synaptic mechanisms, subjective states, and behavior is the general lack of comparability among species, specific drug dose, route, and neural system. For example, the excellent work on behavioral and subjective states with nitrous oxide by STEINBERG and associates (1954, 1955, 1956) has, unfortunately, no parallel in electrophysiological experimentation.

The suggestion keeps reappearing that the psychedelic and excitant properties of anesthetics and ethanol are due to excitatory actions and sites of action distinctly different from mechanisms and sites involved in the sedative or hypnotic actions. The topic was addressed in a different context above. Suffice it here to mention the antagonism of ethanol-induced excitation by α-methyl-p-tyrosine (CARLSSON et al., 1972), the presence of a convulsant enantiomer in pentobarbital (WADDELL and BAGGETT, 1973), the discussions of PATON (1974) and MARJOT (1974), and the clear separation of anticonvulsant activity from the sedative in homologous compounds.

(The abuse and illicit use of sedative/hypnotics has led to its literature and mystiques. As an illustration, MCLAUGHLIN (1974) recently presented an extensive review on the abuse of sedative/hypnotics with a focus on legal governmental reports and popular press literature; the recommendation and the conclusions reflect only a distillation of the popular press; the medical and biological literature was not even consulted!)

h) Conclusions

A consideration of the results of such direct studies on cortical neurons as RICHARDS (1974) and KRNJEVIĆ (1975)—after considering actions on spinal and other "model" neural systems—leads directly to the rather obvious conclusion that the critical studies of the anesthetic or sedative properties be done using the cells actually involved in the function. This generalization implies that spinal studies may well provide an understanding of some possible general properties of an agent, but will only be specifically relevant to the anesthetic effects on spinal neuron function and possibly relevant to its analgesic actions—under the assumption of discrete sites of action for the anesthetic impairment of consciousness. METCALFE et al. (1974) cogently summarize the basic problems regarding establishing the molecular mechanisms not only of anesthesia but also of other states of altered consciousness or mood:

> ... the problem of anaesthetic mechanism cannot be tackled rigorously at the molecular level until the relevant synapses and their subcellular structures which are most sensitive to anaesthetic action have been identified. We do not want to imply that general anaesthesia necessarily results from localized structural changes in a single type of neuronal structure from one region of the CNS, but simply to indicate that the target structures for anaesthetic action must be identified at the biochemical level before any molecular analysis of anaesthetic interactions can be assumed to be relevant. If it turns out that lipid bilayer perturbations are entirely irrelevant to the function of membrane proteins, which we have considered as one possibility, then any justification for the extensive use which has been made of the lipid bilayer as a model for anaesthetic interaction will be seriously undermined. In the same sense, firefly luminescence may only be able

to throw an uncertain light on the mechanism of anaesthesia. The plethora of effects which anaesthetics have on CNS preparations suggest that the outstanding physiological and biochemical problems of identifying the target structures for anaesthetic action are at least as difficult as the problems of molecular mechanism.

IV. Mechanisms and Sites of Action of Benzodiazepines

Benzodiazepines have a variety of CNS actions including sedation, amnesia, decrease in anxiety, sleep, antiepileptic and anticonvulsant effects, antagonism of conflict behavior, muscle relaxation, relief of trigeminal neuralgia, alleviation of delirium tremens, etc. Among the various therapeutic substances they are becoming some of the most widely and extensively prescribed. The benzodiazepines appear to present some unique actions and possible mechanisms of action. For example, general anesthesia and respiratory depression are not common effects of benzodiazepines currently in clinical use. In fact, the margin of safety with such agents as chlordiazepoxide or diazepam is especially large. In view of these differences, these benzodiazepines are being discussed separately, combining the materials on site and mechanism of action and finishing with a review of tolerance and dependence syndromes. Recent reviews and symposia include COOK and SEPINWALL, 1975; COSTA and GREENGARD, 1975; GARATTINI et al., 1973; GREENBLATT and SHADER, 1974a, b; PADJEN and BLOOM, 1975; MATTSON, 1972; WULFSOHN, 1973.

1. Sites of Detectable Effects

In view of their properties, it is not surprising that benzodiazepines have been found to have a wide spectrum of action as assessed using neurophysiologic techniques; among the sites where actions are found are cortex, cerebellum, limbic system, reticular formation, and spinal cord.

The alterations in cortical activity have been well described in man and animals (see LIEB et al., 1974; DAVID et al., 1974; GREENBLATT and SHADER, 1974a, b; LONGO, 1972; KILLAM et al., 1973; GUERRERO-FIGUEROA et al., 1969a, b, 1970a, b, 1973, 1974). In low doses, an increase in fast cortical activity appears; in large doses, slow waves become predominant. Concurrently, there is depression of cortical activity evoked by direct or peripheral nerve activity (see, e.g., SALETU, 1974).

Spontaneous and evoked activity in the reticular formation, septal region and hippocampus is likewise depressed by benzodiazepines (MORILLO, 1962; SCHALLEK and KUEHN, 1960, 1963; SCHALLEK et al., 1962, 1964; ARRIGO et al., 1965; EIDELBERG et al., 1965; GUERRERO-FIGUEROA et al., 1973; NGAI et al., 1966; ZBINDEN and RANDALL, 1967; TSENG and WANG, 1971a, b; NAKANISHI and NORRIS, 1971; MATTHEWS and CONNOR, 1976; UMEMOTO and OLDS, 1975).

Whether or not the amygdala and hippocampus are sites of especial sensitivity to benzodiazepines as emphasized by SCHALLEK and co-workers (1960, 1963, 1964) remains to be proven conclusively. MATTHEWS and CONNOR (1976) describe site specificity to depression of PTP within the hippocampus. GUERRERO-FIGUEROA et al. (1973) report qualitatively opposite effects when the benzodiazepines are administered orally from the effects, described above, after parenteral administration. The warning of VIETH et al. (1968) regarding solvent effects is well taken and might account for such differences. The vexing problem of undetectable interactions between solvents such as propylene glycol and ethanol with potential nervous system effects with compounds under question persists as a significant hazard of interpretation (cf. CHIN and SMITH, 1962, CRANKSHAW and RAPER, 1971).

Only a few studies, such as STEINER and HUMMEL (1968), have actually sought to compare two or more neural systems and two or more drugs in exactly comparable fashion. They concluded, tentatively, that nitrazepam was able to reduce polysynaptic sensory input to the hippocampus without disturbing the visual system, in contrast to phenobarbital. Such a decrease in input to the limbic system could conceivably modulate emotionality. Similarly, OISHI et al. (1972) suggest that chlordiazepoxide acts to alter behavioral outputs by an action on the hippocampus. Flurazepam exhibited similar actions with depression of hippocampus apparently to a greater extent than the depression of amygdala, hypothalamus and mid-brain functions (HASHIMOTO et al., 1973). HAEFELY et al. (1975) have conducted a series of studies that have led them to postulate that benzodiazepines act on a cortical-subcortical feedback system involving limbic forebrain, locus ceruleus, reticular system, and back to the occiptal cortex over a pontine/geniculate pathway.

The qualitative effects on cortical neuronal system resemble those of barbiturates; for example, HOCKMAN and LIVINGSTON (1971) argue that the benzodiazepines' prime site of action is the cerebral cortex, even for actions resulting in modification of autonomic reflexes.

It is of particular interest to point out that the specificity of neuronal actions of the benzodiazepines varies among the various molecular varieties; although chlordiazepoxide and diazepam have many similarities in their central actions, there are a number of differences. And one can look to the future of even more variations on the theme whereby one property is exaggerated over another (e.g., GARATTINI et al., 1973; RANDALL and KAPPELL, 1973; RANDALL and SCHALLEK, 1968; ITIL et al., 1971; GUERRERO-FIGUEROA et al., 1974).

The various benzodiazepines, especially diazepams, display potent anti-epileptic effects. Intravenous diazepam is now routine therapy for status epilepticus. In patients with intractable temporal lobe epilepsy, i.v. diazepam produced an increase in fast EEG frequencies with a decrease in low frequencies and "epileptic" activity. Subcortically, it reduces EEG activity over the entire frequency range (LIEB and CRANDALL, 1974; LIEB et al., 1974). Differences in drug effects among patients were marked; phenobarbital failed to have significant effects but the dose may have been insufficient. As might be anticipated, diazepam has been found effective in various animal models of epilepsy such as the epileptogenic focus induced by penicillin. In such preparations both diphenylhydantoin and diazepam exert brief anticonvulsant actions (STARK et al., 1974).

LONGO (1972) analyzed the actions of the tranquilizing drugs and concluded that some components of the EEG are similar with meprobamate, barbiturates, and benzodiazepines, but that there are also some differences. For example, the slow cortical EEG waves after benzodiazepines are not seen to be similar to those of sleep, and among a series of compounds, they are not correlated with the appearance of sedation. On the other hand, the after-discharge elicited by limbic stimulation is inhibited by all of the present group of agents, and may be considered correlated with anticonvulsant activity (LONGO, 1972) rather than indicative of anxiolytic activity as postulated by RANDALL and SCHALLEK (1968).

Ethanol and diazepam, when given in combination, produce at least additive effects and the clinical impression is one of distinct potentiation (HUGHES et al., 1965; FORNEY and HUGHES, 1968; MYRSTEN et al., 1971 cited in MØRLAND et al., 1974; LINNOILA and MATTILA, 1973; LINNOILA et al., 1974a, b, c; HAFFNER et al., 1973). The last authors present evidence that certain behaviors were affected in opposite fashion by low doses of ethanol and diazepam; generally diazepam slowed psychomotor test performance, whereas after ethanol there was increased speed and errors, and the combination resulted in even further slowing and greater errors. On the other hand, both agents had just detectable effects on flicker fusion frequency, and the combination markedly lowered the frequency that was detectable as fused. On clinical tests, ethanol clearly influenced balance more than diazepam, whereas diazepam had a more frequent effect on memorizing ability (MØRLAND et al., 1974).

It is claimed that the benzodiazepine derivative, medazepam, has no detrimental effects on driving and is *not* synergistic with ethanol (LANDAUER et al., 1974); the experimental data appear solid and the question of appropriate dosage arises. If this benzodiazepine has antianxiety effects in doses that do not have serious motor effects, and does not interact with ethanol, it will truly be a unique and valuable substance, although this is much to hope for.

Clearly, the benzodiazepines have one aspect that overwhelmingly differentiates them from the barbiturates; the respiratory system is relatively resistant, and death from overdose is difficult to accomplish either inadvertently or with suicidal intent—simply because of the large difference between effective and lethal doses (RANDALL and KAPPELL, 1973).

The muscle relaxant effect of the benzodiazepines is not only due to their general sedative properties but, more specifically, to the enhancement of presynaptic inhibition (SCHMIDT et al., 1967; CHIN et al., 1974) and supraspinal reduction of descending gamma bias (BRAUSCH et al., 1973) originating in reticular system and cerebellum, and without direct effects on motoneurons or muscle spindles (NGAI et al., 1966; PRZYBYLA and WANG, 1968; TSENG and WANG, 1971a, b). Nevertheless, the neural outcome of diazepam actions appears similar to that of other sedative muscle relaxants including meprobamate and chlorpromazine (HENATSCH and INGVAR, 1956; BUSCH et al., 1960; CHIN and SMITH, 1962).

Diazepam (as well as nitrous oxide, aspirin and meperidine) depresses the dorsal root potential, and the action is at least partly supraspinal, since spinal transection reduces the effects of both diazepam and morphine (SCHLOSSER, 1971; STRATTEN and BARNES, 1971; CHIN et al., 1974) without effect on spinal PTP, recurrent or direct inhibition, or motoneuron recovery (SCHLOSSER, 1971).

In spite of the apparently greater sensitivity of supraspinal structures to benzodiazepines, they clearly can and do have effects on spinal functions such as presynaptic inhibition (SCHMIDT et al., 1967; SCHLOSSER, 1971; CHIN et al., 1974; POLZIN and BARNES, 1974), accentuation of GABA action (POLC et al., 1974), binding with spinal glycine receptors (SNYDER and ENNA, 1975), effects on spinal interneurons (CHANELET and LONCHAMPT, 1972), as well as NE and 5-HT activity (LIDBRINK et al., 1973, 1974).

2. Mechanisms of Action

Many of the different actions of the various benzodiazepines can, apparently, be separated. The antianxiety actions observed clinically are better correlated, in a series of homologues, with "anticonflict" tests in animals than with sedative or muscle relaxant actions (see review by COOK and SEPINWALL, 1975; STEIN et al., 1975). These animal behavioral effects of benzodiazepines cannot be directly attributed to alterations in NE, DA or 5-HT turnover, phosphodiesterase inhibition, increased GABA action, or the mimicking of glycine, although reports of each of these actions and others have appeared (cf. BARTHOLINI et al., 1973; LIDBRINK et al., 1973, 1974; STEIN et al., 1973a; TAYLOR and LAVERTY, 1973). Some data suggestive of involvement of 5-HT have been obtained (COOK and SEPINWALL, 1975; LIDBRINK et al., 1974; STEIN et al., 1975) in that 5-HT antagonists have analogous behavioral effects to those of the benzodiazepines, methysergide, cinanserin, or bromolysergic acid and as also does p-chlorophenylalanine, a serotonin synthesis inhibitor, and lastly as does 5-HT nerve terminal damage by 5-6 dihydroxytryptamine. STEIN et al. (1975) muster evidence from diverse sources for a 5-HT hypothesis of benzodiazepine action; interestingly, intraventricular administration of 5-HT antagonizes the "punishment-lessening" effects of the benzodiazepines.

DA and NE turnovers are altered by benzodiazepines (see summary by FUXE et al., 1975) although the doses required are frequently large. STEIN et al. (1975) demonstrated that the NE turnover induced by oxazepam rapidly exhibited tolerance, whereas the decreased 5-HT turnover did not; these observations parallel the observation that the depressant actions purportedly undergo tolerance but the antianxiety action is sustained with repeated dosing (see also FUXE et al., 1975; GOLDBERG et al., 1967; MARGULES and STEIN, 1968).

The problem of deducing cause and effect in studies of such drug effects on neurotransmitters in intact brain systems remains; at best, the conclusions can only be couched in terms of strong or less strong sets of correlations. For example, STEIN et al. (1975) explicitly recognize that the involvement of 5-HT neurons in benzodiazepine action is quite possibly secondary to a primary action on GABA-mediated systems; the fact that picrotoxin antagonized the effects of benzodiazepines in their conflict test is consonant with this view.

Similarly, the changes in NE turnover following benzodiazepines could result from increased GABA receptor activity on locus ceruleus cells (FUXE et al., 1975). The localized effects of benzodiazepines to reduce DA turnover in portions of the limbic cortex are possibly important for understanding their unique spectra of action—whether the DA changes are primary or secondary to changes in neuronal discharge or GABA turnover cannot be decided at present.

Strong evidence for the involvement of GABA in the actions of benzodiazepines has recently been elucidated for a number of sites, including cerebellum (COSTA et al., 1975), spinal cord (POLC et al., 1974), striatal nigral pathways and cuneate nucleus (HAEFELY et al., 1975). GABA is likely involved as a transmitter in presynaptic inhibition (see Subsect. C.II.3, and BARKER and NICOLL, 1973; BARKER et al., 1975a, b; BELL and ANDERSON, 1972; CURTIS and JOHNSTON, 1974; DAVIDOFF, 1972; DAVIDOFF et al., 1973; KRNJEVIĆ, 1974b; LEVY, 1974; NICOLL, 1975a, b; NISHI et al., 1974). Benzodiazepines enhance presynaptic inhibition (CHANELET and LONCHAMPT, 1972; CHIN et al., 1974; POLC et al., 1974; SCHLOSSER, 1971; SCHMIDT et al., 1967; STRATTEN and BARNES, 1971; POLZIN and BARNES, 1974, 1976) as discussed previously with barbiturates and ethanol. The story is relatively consistent in view of the potent antipicrotoxin action of diazepam (BARNES and MOOLENAAR, 1971) and the recent report that pentylenetetrazol acts as an antagonist of GABA actions on primary afferents in the isolated frog spinal cord (NICOLL and PADJEN, 1976); however, HILL et al. (1974) report that it did not block postsynaptic inhibition induced by GABA, although it did depress presynaptic inhibition. Further, AOAA treatment, which results in elevated GABA levels, results in increased dorsal root potentials which can be further increased by benzodiazepines (POLC et al., 1974). The classic GABA receptor blockers, picrotoxin or bicuculline, suppress the dorsal root potentials in benzodiazepine-treated animals.

The failure of PADJEN and BLOOM (1975) to find effects of benzodiazepines on motoneurons or dorsal root potentials in isolated frog spinal cord preparations seriously detracts from a GABA hypothesis for the mode of action of benzodiazepines. Nevertheless, diazepam enhanced dorsal root potentials in the cat, and the action was antagonized by picrotoxin (POLZIN and BARNES, 1976; CURTIS et al., 1976). Also CHANELET and LONCHAMPT (1972) have noted that a number of spinal interneurons are sensitive to diazepam microinjected locally.

As noted by many of the participants of a recent symposium edited by COSTA and GREENGARD (1975), the mechanism of action of benzodiazepines in inducing antianxiety effects or changes in mood can, as yet, only be addressed by analogy with spinal, subcortical, or cerebellar studies (see short review by PADJEN and BLOOM, 1975). HAEFELY et al. (1975) summarize interesting studies implicating benzodiazepines and GABA on a pathway from the limbic forebrain, locus ceruleus, and reticular system, which in turn feeds back to alter pontogeniculo-occipital cortical electrographic waves.

With respect to their interaction with GABA systems, it has been suggested that the benzodiazepines do not act to mimic GABA at its receptor sites since they are very weak in competing with GABA for receptor binding (SNYDER and ENNA, 1975). These last authors reemphasize the possibility of separating muscle relaxant, sedative and antianxiety actions of drugs such as benzodiazepines.

Another area of appreciable interest centers around possible influence of benzodiazepines on the basic process of PTP. SURIA and COSTA (1973, 1974, 1975a) find that benzodiazepines selectively block PTP in bullfrog sympathetic ganglion in vitro, with diphenylhydantoin sharing a similar action. In contrast, little if any effect of benzodiazepines on spinal PTP has been detected (SCHLOSSER, 1971; SWINYARD and CASTELLION, 1966); however, PTP in the hippocampus is sensitive to diazepam (MATTHEWS and CONNOR, 1976). SURIA and COSTA (1975a, b) conclude that diazepam-induced increase in PTP is the consequence of increased prostaglandin synthesis and involves dibutyrl cGMP and GABA; it is postulated that GABA mediates a depolarizing action of benzodiazepines. It may well be that the preparation of SURIA and COSTA is relatively unique with its involvement of GABA in mediating PTP; their experiments suggest the need for comparative studies on the mammalian cord, sympathetic ganglia, and neuromuscular junction. The study and analysis of SURIA and COSTA, although intriguing, has some troublesome spots. In the figures presented in the 1974 paper, there are questions raised by the claim of no effect of the agent on the single shock response; actually diazepam and the prostaglandin appear to augment the depression associated with tetanic stimulation.

Nevertheless, it is intriguing to consider the structural relationships of benzodiazepines to prostaglandin receptor blockers and the effects of prostaglandins on presynaptic transmitter systems detailed by SURIA and COSTA. These authors cite relevant literature on the effects of diazepam on phosphodiesterase and the potentiation thereof by prostaglandins; they also cite preliminary observations of inhibition of PTP by 3′, 5′ GMP, which did not occur with the adenosine congener. Also, is this PTP depression—and the possible relationship to prostaglandins—of importance only in relation to its antiseizure action, or is this possibly the mechanism of most of the neural actions of benzodiazepines, diphenyhydantoin and of barbiturates? The complex modulatory effects of prostaglandins on autonomic transmission have been summarized (BRODY and KADOWITZ, 1974; WEEKS, 1974).

In somewhat related studies, it has been suggested that benzodiazepines act on the cAMP systems; at least they have weak phosphodiesterase inhibiting activity (SCHULTZ and HAMPRECHT, 1973; WEINRYB and CHASIN, 1972) and cAMP formation stimulated by prostaglandin, histamine, or NE is depressed (SCHULTZ and HAMPRECHT, 1973; SCHULTZ, 1974a, b). CROWLEY et al. (1974a) conclude that some diazepines inhibit both cAMP and cGMP phosphodiesterase but that absence of regional specificity of action makes the functional correlation difficult to draw.

A number of benzodiazepines are metabolized to N-desmethyl diazepam (nordiazepam). A clinical study of chlorazepate and diazepam led ROBIN et al. (1974) to conclude that clinical improvement was correlated with the level of N-desmethyl diazepam in plasma.

The question of the role of drug action on PTP deserves more than a passing concern in view of the probable actions of general anesthetics, barbiturates and ethanol on presynaptic inhibition. Following tetanic stimulation of afferent dorsal root fibers, there is hyperpolarization of presynaptic terminals of the fibers tetanized, and this is manifested by PTP. Its absence would, relatively speaking, be viewed as presynaptic depression and any presynaptic inhibition would be synergic with a block of PTP. It is conceivable that the two mechanisms (PTP and presynaptic inhibition) are even more closely coupled. Or put somewhat differently, the relative effects of the various agents on the two phenomena need to be assessed. As examples can be cited the report of RAINES and STANDAERT (1967) that diphenylhydantoin abolishes post-tetanic hyperpolarization, an action shared by thiopental but not by phenobarbital or diethylether. Whether all of the benzodiazepines depress PTP in a number of synergetic systems has yet to be ascertained. Moreover, the barbiturates, ethanol and other similar agents deserve reexamination for their actions on PTP.

It is also plausible that the anticonvulsant properties of the benzodiazepines are not related to their sedative and motor depressant activities in view of the claim that tolerance develops readily to the motor ataxia and hypnotic actions but not the anticonvulsant effects. (However, KILLAM et al., 1973, do observe tolerance in chronic studies in the baboon, *Papio papio*.) Similar observations have been made for phenobarbital and similar agents used as antiepileptic agents. (See later discussion of the mechanisms underlying functional tolerance.)

Especially striking is the finding that although benzodiazepines are potent antagonists of many convulsants such as picrotoxin and pentylenetetrazol, their ED_{50}s were some 7–10 times lower against isoniazid-induced seizures than against convulsions elicited by picrotoxin (SURIA and COSTA, 1975a); in parallel studies, barbiturates and diphenylhydantoin were only slightly more effective against isoniazid. (The effects of isoniazid were presumably the result of inhibition of GABA synthesis.)

Without doubt, the animal studies reveal dissociation of sedative and muscle-relaxant actions (ZBINDEN and RANDALL, 1967). Clinical studies appear to indicate a different relationship between antianxiety and sedative effects among the various benzodiazepines. It should be noted that the various benzodiazepines appear to have some selectivity of action within a congener series; this is especially evident in assays for muscle relaxant action in comparison with other properties (cf. WULFSOHN, 1973). Receptor binding might well be highly specific in view, for example, of the high stereospecificity of binding of benzodiazepines to human albumin (MÜLLER and WOLLERT, 1975). SNYDER and ENNA (1975) review the data on the potent binding of diazepam to a glycine receptor preparation derived from spinal cord (see also YOUNG et al., 1974). Potency for subjective effects in man is strongly correlated with muscle relaxation in cat, fighting test, and antipentylenetetrazol tests in mice, and with glycine receptor binding. CURTIS et al. (1976) present strong evidence against benzodiazepines acting on a glycine receptor.

The interactions of benzodiazepines with cholinergic systems are reviewed by CONSOLO et al. (1975). Diazepam increases central acetylcholine levels to varying degrees at different sites in doses that did not alter the CNS levels of 5-HT, NE, DA or their metabolites. Neither cholineacetyltransferase nor cholinesterase activities were affected by incubation in vitro with the high levels of diazepam of 10 µg/100 mg tissue/ml. These observations led to the conclusion that diazepam acted to block the release of ACh, an action which is conceivably secondary to its actions on a GABA system. Interestingly, pentylenetetrazol administration antagonized the changes in ACh levels produced by diazepam.

Carbamazepine (Tegretol, a benzodiazepine used selectively in the treatment of trigeminal neuralgia and as an anticonvulsant) has been investigated with respect to changes induced in ionic conductances of axons (SCHAUF et al., 1974). The expectation was that the changes would be similar to those of diphenylhydantoin; actually, there was a decrease in both Na^+ and K^+ conductances. In addition, the leakage current for hyperpolarizing steps was reduced with very modest membrane depolarization. Thus, the actions were different from those of diphenylhydantoin. Rather, the effects appear to be similar to those of alcohols whereas barbiturates in relatively large concentrations also have qualitatively similar actions (cf. BLAUSTEIN, 1968). Actually, the alcoholic vehicle that had to be employed for the benzodiazepines may well have complicated the results. The degree to which other benzodiazepines have actions similar to carbamazepine is not known.

3. Subjective Effects

The amnesia that appears after i.v. administered diazepam is striking in its reproducibility among different patients as well as its rapid onset and short duration. KORTTILA and LINNOILA (1975a, b) found the peak amnesia within 2–3 min of the injection in agreement with the earlier studies of DUNDEE and PANDIT (1972); the duration with the larger doses used (0.30 and 0.45 mg/kg) was 15–20 min (cf. also CLARKE et al., 1970; DUNDEE and PANDIT, 1972). Such amnesia has been found especially useful in dental practice (KEILTY and BLACKWOOD, 1969) as mentioned above.

Without question, i.v. diazepam produces feelings of well-being, elation, loquaciousness and slurring of speech (WYANT and STUDNEY, 1970; KORTTILA and LINNOILA, 1974, 1975a, b; WOODY et al., 1975; KRYSPIN-EXNER and DEMEL, 1975; annotated bibliography of WEISE and PRICE, 1975). Subjective effects of benzodiazepines are said to be minor, although a significant number of patients describe its tension-relieving properties and a sense of calmness and confidence. Such relief of anxiety is claimed to be less marked than that archieved with i.v. methamphetamine or amylobarbitone (e.g., BETHUNE et al., 1966).

The effects of benzodiazepines on thought and behavior are not all positive, however. A number of clinical observations attest to possible hostility with chlordiazepoxide therapy (e.g., SALZMAN et al., 1974) although others doubt its relevance in clinical management (e.g., RICKELS and DOWNING, 1974).

The probability that the rate of onset of subjective drug effects is important in influencing the degree to which an agent will be self-administered has been discussed for many years. More specifically, many have commented that the i.v. route and the selection of drugs or dosage formulations that facilitate the rapid translocation to the CNS—such as heroin or methamphetamine—are important reasons for these agents and this route being chosen. These concerns appear to apply also to the benzodiazepines. It is slowly appearing that the agent of greatest use for self-administration is diazepam, and to some degree, flurazepam. Among benzodiazepines, diazepam has one of the fastest rates of onset of effects, and a relatively short duration of action. This apparent short duration of action may be partially deceiving since 5–10% of the peak blood level may be detected a week after a 10-mg dose of diazepam (BLIDING, 1974).

Acute tolerance to the subjective effects of the benzodiazepines has been clearly shown recently by BLIDING'S (1974) thorough studies of diazepam. Analogous findings in animal studies of diazepam are described by GUERRERO-FIGUEROA et al. (1974). It was suggested that the subjective appreciation of drug effects is more related to the rate of onset of the effects, rather than the absolute drug level (this may be another way of describing the phenomenon of acute tolerance—see below).

The fact that diazepam exhibits a greater abuse than chlordiazepoxide is emerging clearly (KISSIN, 1974; IRWIN, 1973; KOKOSKI et al., 1974; ROSENBERG, 1974). This is consistent with the pharmacodynamics of the two agents; whether other factors obtain is an open question.

BLIDING (1974) postulates that the risk of abuse and habituation is related to subjectively experienced effects, and that these depend on the rate of onset of subjective effects which, in turn, is a function of the rate of increase in brain level. This speculation is plausible and generally consistent with the field observations of human behavior. Its implications are of general import, i.e., (1) the individuals take the drugs to achieve a given subjective endpoint, (2) that such drug taking is, of necessity, sporadic and repetitious, and (3) that the drug effect sought is not the antianxiety tranquilizing action but the sensations (euphoria?) associated with the change in central nervous state. Of course, these implications are consistent with a variety of the theories of drug use, including theories involving conditioning to both primary and secondary drug actions. It does raise again, however, the primary question of whether or not the change in the individual's subjective state induced by the drug, and the statements made about it, contain the elements of the prime motivations underlying a given drug's use. Alternatively, the elements of habit formation, and of a drug's reinforcing properties may well remain beyond (below?) the level of conscious appreciation or detection. If this is so, then all theories based on "feeling tone," consciousness expansion, relief of anxiety, etc., are functional only as rationalizations—or, at best, only as descriptors of corollaries and concomitant properties of the primary neural actions of the drug.

D. Tolerance and Dependence

I. Characterization of Acute and Chronic Acquired Functional Tolerance

1. Introduction

The detailed characterization of the physical dependence associated with chronic ethanol or barbiturate ingestion is presented in Section III, Chaps. 2 and 3. Thus, this subsection will address, firstly, those studies that provide insight into the mechanisms of tolerance with the sedative/hypnotic agents including ethanol and barbiturates, and, secondly, evidence for tolerance and physical dependence with the various sedative agents other than the alcohol-barbiturate group.

It is generally concluded since the classic studies of the Lexington group (ISBELL et al., 1950; ISBELL and FRASER, 1950; ISBELL et al., 1955; BELLEVILLE and FRASER, 1957; FRASER, 1957; FRASER and ISBELL, 1954a, b; FRASER, 1972; FRASER et al., 1953, 1954, 1956, 1957, 1958; WIKLER, 1968c, 1970, 1972, 1974; WIKLER et al., 1956; WIKLER and ESSIG, 1970) that all agents of the sedative/hypnotic class exhibit similar tolerance and dependence syndromes, and that the withdrawal syndrome for each agent is similar and alleviated, at least experimentally, by any other agent of the class. Nevertheless, it should be emphasized that this conclusion is an extrapolation based on only a few explicit studies with a very few compounds. For example, the volatile and gaseous anesthetics would presumably be effective in affording rapid relief of withdrawal symptomatology from ethanol or barbiturates; however, other than the casual clinical observations, the effectiveness of such anesthetics in treating the withdrawal syndrome has never been tested experimentally.

Comprehensive reviews on acquired tolerance to depressants and the possible mechanisms underlying it have appeared (WALLGREN and BARRY, 1970; KALANT et al., 1971; HUG, 1972; KALANT, 1973a, b; WALLGREN, 1973; GOLDSTEIN, 1975a, b; LEBLANC and CAPPELL, 1975). Tolerance to sedative and hypnotic agents was also definitively discussed some years back by SEEVERS and DENEAU (1963). The use of many, if not all, of the sedative/hypnotic agents have been observed to be associated with the development of acquired tolerance to the agent.[14] Although the term "tolerance" may connote a variety of rather different phenomena, it generally denotes either a *decrease in the response to a given dose or drug level*, or an *increase in the dose or level required to produce a given effect*. In the case of acquired tolerance, these changes in sensitivity are attributed to the previous exposure to the same substance; whereas, when it is due to exposure to another, similarly acting agent, cross-tolerance is generally specified. In the literature, other words are frequently used to describe the same or similar phenomena including, "getting used to," "adaptation,"

[14] KALANT et al. (1971) appropriately clarify terminology. Tolerance may be used to imply unique, individual, pre-existing or innate tolerance—or it may be used, probably more commonly, in reference to tolerance acquired consequent to previous exposure to the given drug or a similar agent (cross-tolerance). The marked variations in individual susceptibility, i.e. variations in "innate tolerance," of ethanol or barbiturates is a common observation. It was succinctly described in the classic papers of FRASER (1957) and FRASER et al. (1956).

"tachyphylaxis," "acclimation," "adjustment," and "habituation"—even though the last term has been ascribed a narrow meaning in certain psychological circles. All of these terms, including tolerance, have been given narrower or broader definitions by various writers and these very variations in definition and connotation make interpretation difficult. Moreover, since none of the definitions provide a comprehensive set of explanations or mechanisms for the phenomena observed, the terms remain, at best, simply descriptive.

A discussion of acquired tolerance requires defining those aspects requisite to characterization of the phenomena. The occurrence and magnitude of tolerance are functions of each of the following and their interactions:

1. Drug effect—what change(s) in function or biochemistry, and the degree of such change(s).
2. Dose and dose regimen—in relation to the agent and its effect(s) (site and magnitude). For example, the acquired tolerance to ethanol and sedative/hypnotics is generally held to be limited to a maximum increase of two- to fourfold in the dose required to produce a given effect.
3. Time course—of onset and of disappearance of a given level of tolerance.
4. Reversibility.

It should be possible to measure the magnitude of tolerance as a complex function as revealed by both the dose-response curve for a given drug effect (e.g., ataxia or euphoria) in the naive individual and following the repeated exposure to the agent. Further, tolerance should be measurable in terms of the dose and dosage frequency of the drug administration eliciting the tolerance (cf. KALANT et al., 1971). For example, acute tolerance has been used specifically for that tolerance which appears during the time course of a single administration of a drug (with ethanol this acute tolerance is commonly referred to as the Mellanby effect) (cf. MELLANBY, 1919; MIRSKY et al., 1941).

L. GOLDBERG (1943) used the word *habituation* where most now use the expression *acquired tolerance* to describe the decrease in effects of ingested alcohol. GOLDBERG distinguished three main concepts "increased rate of elimination, psychic compensation, and a changed reaction of the nerve cells to the penetration of alcohol." He recognized that a combination of the theories might result. In his pioneer experiments in man, he emphasized that acute compensation was marked for alcohol effects on psychological tests, considerable for motor functions, and essentially not detectable in its actions on sensory perception. Increased tolerance was also well demonstrated for most of the tests for heavy drinkers versus moderate drinkers and abstainers. The results unambiguously showed that such tolerance was due to "compensatory" mechanisms in the nervous system and/or alterations in the nerve cells' responses to ethanol. As can be readily appreciated, the present state of knowledge is little advanced over these studies and the subsequent investigations of FRASER, ISBELL, WIKLER, and colleagues (see citations above).

Changes associated with prior exposure in the *disposition* of the agent are termed dispositional tolerance and could arise from changes in absorption, distribution, metabolism, or excretion. Only alterations in metabolism seem to be frequently involved (cf. HUG, 1972), although it was recently reported (JOHNSON and PATEL, 1975) that thiopental distribution is markedly altered in animals receiving ethanol chronically. CICERO et al. (1971) found no differences in absorption or metabolism of ethanol in various strains of rats; in contrast, BELENKO and WOODS (1973) found higher peak blood levels in a nondrinker group, whereas the disappearance rate was essentially identical. Moreover, since metabolic tolerance and cross-interaction are both complex and result primarily in tolerance manifested because of lower tissue levels of an active agent, they will also not be addressed in detail, except insofar as they pertain to general considerations and to interpretation of individual experiments and agents.

The phenomenon of functional (or behavioral) acquired tolerance to a sedative agent does appear to occur with sedative/hypnotics generally, and is apparently a prime type of tolerance involved with chronic exposure. Possible mechanisms of such tolerance, such as alterations in drug receptor interactions subcellularly, cellular biochemical adaptations, homeostatic mechanisms involving small or large neural systems, fatigue depletion phenomena, learning or psychological mechanisms, will be the focus of the following discussion.

2. Ethanol

There are appreciable innate differences in the rate of *metabolic* disappearance of ethanol among naive individuals—human and animal (cf. reviews of KALANT et al., 1971; HUG, 1972; KATER et al., 1969). Nevertheless, it is clear that dispositional (metabolic) tolerance occurs with chronic exposure to ethanol beverages (ISBELL et al., 1955; MENDELSON et al., 1965; MENDELSON, 1968; HAWKINS et al., 1966; KHANNA et al., 1967; KATER et al., 1969; KHANNA and KALANT, 1970; cf. KALANT et al., 1971; RUBIN and LIEBER, 1971; HUG, 1972), although it is not obviously manifest in many experiments. For example, in MENDELSON and MELLO's series of studies with chronic administration of alcohol, metabolic tolerance would have been expected to be evidenced as a decline in measured blood alcohol levels, yet this frequently did not seem to occur.

The maximal increase in metabolic degradation is relatively small, approximately twice the control rate. As KALANT et al. (1971) appropriately emphasize, the rate of metabolism of ethanol is slow, relative to its absorption; thus, changes in rates of metabolism will have little, if any, influence on blood and brain levels over the time period of absorption of 30 min or an hour following oral administration. In other words, the peak effects of a single dose of ethanol, irrespective of route of administration, will not be appreciably affected by the presence or absence of metabolic tolerance. The tolerant individual, however, may be able to metabolize up to approximately twice as much ethanol over periods of hours and days as the nontolerant person.

Acute functional tolerance to the gross symptoms and signs of intoxication develops during the time period of 3–6 h following ingestion of ethanol. MELLANBY is attributed with the first citation of the effect in man (cf. KALANT et al., 1971). MIRSKY et al. (1941) have presented a vivid description thereof; but one can question the validity of the absolute values of high alcohol levels cited in MIRSKY et al. (1941). Usually, acute tolerance has been detected by the fact that the effects of ethanol are greater in the rising phase of the blood concentration curve than in the declining phase—and that the degree of intoxication is greater when the blood level is achieved rapidly than when it is achieved slowly (cf. recent studies of JONES and VEGA, 1972, which controls for practice effects). This phenomenon of acute tolerance has been repeatedly rediscovered and restudied (cf. e.g., ROSENBAUM, 1942; GOLDBERG, 1943; CASPERS and ABELE, 1956; MAYNERT and KLINGMAN, 1960; STORY et al., 1961; MACDONNELL and FESSOCK, 1972; JONES and VEGA, 1972; HURST and BAGLEY, 1972; GREIZERSTEIN and SMITH, 1972, 1973a, b, 1974; GREIZERSTEIN, 1975; SUGARMAN et al., 1973; PERRIN et al., 1974, 1975; LEBLANC et al., 1975), although it is not always evident (LOOMIS and WEST, 1958). Acute tolerance is readily demonstrated for both subjective effects of ethanol intoxication as well as psychomotor impairment; in fact, subjective phenomena appear to be sensitive indicators of ethanol action (EKMAN et al., 1963; EKMAN, 1964; GOLDBERG, 1966; BROWN et al., 1975). In these various studies the *rate* of acute tolerance development does not appear to be dose-concentration related; blood alcohol curves drop linearly after reaching their peak, whereas the intoxication scores decrease at a faster rate, but also linearly and at approximately the same rate for three different doses. Such a conclusion may be unwarranted in view of the complexity of scaling of subjective variables as functions of dose.

These conclusions regarding tolerance are based on an assumption that concentrations in tissues are proportional to the blood levels. In fact, theoretical analyses and acute measurements reveal that during the absorption and onset of ethanol effects, levels in brain and other tissues actually lag behind the values that will be attained under steady-state conditions (GOSTOMZYK et al., 1969). As metabolism occurs and blood levels decline, the brain levels are appreciably higher than in the steady-state. Thus, the degree of acute tolerance is actually even greater than that indicated by the simple comparison of the degree of impairment with blood levels early and late during the single exposure.

As MAYNERT and KLINGMAN (1960) discuss and as KALANT et al. (1971) subsequently suggest, for at least a limited number of drug effects the process(es) underlying chronic tolerance seen with repeated administration appears to be essentially the same as that evident in the later portion of a single intoxication. This conclusion has been recently reemphasized by GREIZERSTEIN and SMITH (1972, 1973a, b, 1974); GREIZERSTEIN (1975) and LEBLANC et al. (1975). In fact, tolerance to a challenge ethanol dose induced by continuous exposure of fish to given levels of ethanol appears within 3 h, is maximal in 3–6 h, and is sustained essentially unchanged in magnitude for days (GREIZERSTEIN and SMITH, 1973a, b, 1974; GREIZERSTEIN, 1975; see also RYBACK and INGLE, 1970; RICHARDSON, 1972). In fish at least, no distinction can be made between acute and chronic tolerance except on the basis of the focus of interest of the investigator. On the other hand, most studies of tolerance have used a daily dosage regimen that results in fluctuating blood and tissue levels, with a design which calls for testing at intervals of days to weeks. Under such circumstances, tolerance is seen in man and animals in a few days, e.g., for behavioral and EEG effects in monkeys (HOGANS et al., 1961; STORY et al., 1961), operant behavior in rats (LESTER, 1961), motor behavior in dogs (MAYNERT and KLINGMAN, 1960) and in rats (LEBLANC et al., 1975; see also EBERT et al., 1964; CICERO et al., 1971; RATCLIFFE, 1969); the behavioral effects reportedly exhibit greater tolerance than the effects on the EEG (HOGANS et al., 1961). Acute tolerance has also been reported for EEG actions of ethanol in unanesthetized cats (PERRIN et al., 1974), motor reflexes (MACDONNELL and FESSOCK, 1972) and EEG in monkeys (STORY et al., 1961).

Tolerance develops more rapidly when the subjects are tested or perform repeatedly under the influence of ethanol, as compared with those who receive the same tests in the nondrugged state yet receive the same amount of ethanol (see KALANT et al., 1971; CHEN, 1968). As will be described below, these findings are a critical aspect of one plausible theory of tolerance which involves the "adaptation" or "learning" to function while partially impaired by the agent.

The degree of tolerance exhibited by ethanol is limited both in man (ISBELL et al., 1955) and experimental animals. In man, only slight increases in the dose given to tolerant individuals result in gross impairment. Similar findings were obtained with barbiturates (FRASER, 1957). At best, tolerant individuals generally require up to three times as much to produce a given effect as in the nontolerant state—or put differently, after tolerance develops a given dose is about one-third as effective. In contrast, tolerance to narcotics and other agents may result in a 10- to 20-fold increase in the dose needed to exert detectable effects. MAYNERT and KLINGMAN (1960) present data showing clearly different degrees of tolerance developing with various sedative/hypnotics. Further, although there is only a quantitative shift in sensitivity to the motor incoordinating effects during chronic administration, the mood shifts from one of gaiety to morose belligerency.

NEWMAN and LEHMAN (1938), LEBLANC et al. (1969), and KALANT et al. (1971) describe a parallel shift in the dose-effect curve to ethanol in the tolerant animal. However, the shape of the curve and the relationship between the tolerance-producing dose, the time sequence of administration, and the behavioral endpoint have yet to be clearly defined (see discussion by LEBLANC and CAPPELL, 1975).

In their classic paper, MAYNERT and KLINGMAN (1960) clearly demonstrated that tolerance to ethanol and to barbiturates appeared quickly, that the magnitude of tolerance was dose-dependent up to a maximum beyond which no further tolerance could be elicited, and that the magnitude of tolerance was markedly different depending on the measure of drug effect used, even among different motor functions. These authors examined ethanol, paraldehyde, thiopental, pentobarbital and trichloroethanol.

SIDELL and PLESS (1971) have characterized the consistency and dose-response relationships of the tolerance to ethanol in healthy subjects, confirming much previous work. Interestingly, in the group data, there is no evidence of acute tolerance; for the most sensitive measure—a test of eye-hand coordination—there was no clear dose-effect relationship for the two lowest doses, even though the mean blood levels for these were consistently different by almost 50 mg-%. Further, euphoria was not experienced consistently, and when it did appear, it was frequently followed by distinctly uncomfortable feelings.

In his detailed review HUG (1972) puzzles over the apparent complete tolerance to the electro-shock seizure threshold-elevating property of ethanol (ALLAN and SWINYARD, 1949); however, it would appear that the full range of doses was not explored.

For some effects of ethanol, acute tolerance cannot be detected. Neither the depression of the monosynaptic reflex (LATHERS and SMITH, 1973, 1976) nor the stimulation of muscle spindle afferents (KUCERA and SMITH, 1971 and 1972; LATHERS and SMITH, 1973 and 1976) decline over the 2–6 h of sustained exposure to ethanol. Actually, both effects temporally lag 15–20 min behind arterial blood levels which were achieved in 5 min and declined over 30 min; the maximal effects on spinal reflexes and spindles tended to appear 20–30 min after administration. Such delays suggest that either the effects relate to penetration to sites of remote access to ethanol or that the end neural effect is the result of a chain of metabolic processes that require appreciable time be accomplished. In any event, in the spinal animal no acute tolerance to these actions can be detected.

In contrast, the depressant effect of ethanol on acetylcholine release from cortical slices observable in vitro is not evident in slices from animals tolerant to ethanol (KALANT and GROSE, 1967). In contrast, in a related study, the Na^+, K^+ ATPase of brain and the active transport of cations were higher in the brains of tolerant animals than in the nontolerant (ISRAEL et al., 1970, KNOX et al., 1972, but see GOLDSTEIN and ISRAEL, 1972; WALLGREN et al., 1975). In vitro acute tolerance or tachyphylaxis to the effects of ethanol on the isolated frog heart was reported many years ago (RANSOM, 1919).

PERRIN et al., 1974 (see also KALANT, 1973a, b, 1974) carried out extensive EEG studies in intact cats: the cortical effects of ethanol exhibit tolerance in parallel and slightly in advance of obvious behavioral tolerance. The tolerance was most marked in animals receiving the largest chronic doses (4.5 g/kg daily). Dependence was also evident by behavioral and EEG hyperactivity appearing upon withdrawal.

Although acute tolerance was not seen to ethanol's effects on the monosynaptic reflex, a study of spinal function after tolerance has been induced with ethanol has not been carried out (see studies below by OKAMOTO using barbiturates). Also, dependence on barbiturates, ethanol, or any of the sedative/hypnotics has not been explicitly examined in the chronic spinal dog preparation used so effectively in research on narcotic dependence (cf. WIKLER and FRANK, 1948; MARTIN et al., 1974a; JONES et al., 1976).

Functional tolerance appears to be sustained as long as the blood level persists. The rate of disappearance of tolerance and the rates of reinduction remain relatively unexamined. In fish, the tolerance to ethanol disappears upon placing the previously equilibrated fish in a large fresh water aquarium almost as fast as the brain level declines (over a period of 3 h).

The phenomenon of "recall tolerance," the more rapid development of tolerance after weeks of abstinence described by LEBLANC et al. (1969) and discussed by KALANT et al. (1971) and KALANT (1973a, b), clearly deserves further study. It could well be related to the long term effects of single hypnotic doses of barbiturate or benzodiazepine described by DAVID et al. (1974) and by OSWALD and PRIEST (1965); OSWALD (1968, 1971, 1973); OSWALD et al. (1969). Furthermore, BRANCHEY et al. (1971) have made the pregnant observation that a prior period of physical dependence predisposes the animals to experiencing a much more severe withdrawal period to a subsequent alcohol dependence than naive controls. WALKER and ZORNETZER (1974), in a related study, found that withdrawal signs in mice were significantly more severe during a second cycle of dependence than after the first. "Recall tolerance" was searched for in fish but could not be detected (GREIZERSTEIN, unpublished observations). GOLDSTEIN (1974) finds slow development of dependence but rapid decay (see GOLDSTEIN, 1975a, b; LEBLANC and CAPPELL, 1975).

Interestingly, in the classic chronic studies in man, the alcohol or barbiturate doses were increased over the first 4 days of administration (ISBELL and FRASER, 1950; ISBELL et al., 1955; FRASER, 1957, 1972; MENDELSON, 1964; MENDELSON et al., 1968; MELLO and MENDELSON, 1970; MENDELSON 1970a, b; FRASER et al., 1957, 1958). The reports appear to indicate that the dosage was, in fact, increased to sustain, in a very approximate fashion, a given level of intoxication. It is tempting to speculate that the tolerance was developed during a given day with some loss of tolerance as the tissue levels declined, especially over the night-time period. However, some degree of tolerance remained the next day and with resumption of intake of yet larger doses, tolerance was further enhanced. (In addition to functional tolerance, metabolic tolerance also developed with such schedules to ethanol or to agents, such as pentobarbital, over a period of 4 or more days.) (See Section III, Chaps. 2 and 3.)

3. Barbiturates

Studies of tolerance to the barbiturates parallel those of ethanol. However, the factors involved in barbiturate tolerance are frequently even more complex and relate primarily to the different pharmacokinetic and metabolic aspects of the various barbiturates. The bulk of the studies have utilized thiopental (rapid translocation, redistribution, rapid metabolism), pentobarbital (relatively rapid onset and extensive metabolism), or phenobarbital (slow onset, long duration of action, almost exclusively not metabolized, although an effective stimulus of drug-metabolizing enzymes).

Acute tolerance to thiopental occurs without doubt (to its anesthetic effects in man: MARK et al., 1949; BRODIE et al., 1951; DUNDEE et al., 1956; to its ataxia producing effects in dogs: MAYNERT and KLINGMAN, 1960; to its depression of cerebral oxygen consumption in dogs: ALTENBURG et al., 1969; in rats: SINGH, 1970). Tolerance to pentobarbital effects includes reduction of ataxia in dogs (SHIDEMAN et al., 1948; MAYNERT and KLINGMAN, 1960), of sleeping times in rats (ASTON, 1965, 1966a, b; WARBURTON, 1968; SINGH et al., 1970), and reduction of EEG effects in cats (HATCH et al., 1970). Amobarbital citations involve reduction of sleeping time in mice [NURMAND, 1969—Farmakol. i Toksikol. **32**, 41 (1969), cited by HUG (1972)] and of avoidance behavior in rats (WARBURTON, 1968). Hexobarbital tolerance can be measured using the parameters of anesthesia in rabbits (STUMPF and CHIARI, 1965) and EEG response (WAHLSTRÖM and WIDERLÖV, 1971). A tolerance similar to that with barbiturates is seen with glutethimide (CURRY and NORRIS, 1970; CURRY, 1974).

A wide variety of studies have used chronic, repeated exposure to barbiturates, but few permit an accurate determination of the latency or time course of the development of the tolerance. JAFFE and SHARPLESS (1965) detected tolerance and withdrawal hyperexcitability to pentobarbital after one dose of intermittent i.v. administration. Maximal tolerance for pentobarbital was apparently evident between 12 and 24 h for pentobarbital sleeping times (ASTON, 1965, 1966a, b; WARBURTON, 1968; SINGH et al., 1970; SINGH, 1971). No tolerance could be detected if the interval was 48 h or longer (ASTON, 1965, 1966a, b; SINGH et al., 1970; IRWIN et al., 1956). These findings are somewhat disconcerting, since clinical observations and the studies of FRASER et al. (1956, 1957) and FRASER (1957) suggest that little, if any, tolerance occurs with hypnotic doses of pentobarbital or seconal given nightly (24-h intervals); however, tolerance was grossly evident with large doses given daily to dogs (FRASER and ISBELL, 1954a); a number of studies of less frequent administration have found tolerance (e.g., CARMICHAEL and THOMPSON, 1941; see also OSWALD, 1968, 1971; OSWALD et al., 1969; OSWALD, 1973). Dependence, as evidenced by abstinence signs upon withdrawal, occurs with chronic barbital or pentobarbital administration in cats (ROSENBERG and OKAMOTO, 1974a, b; OKAMOTO et al., 1975), dogs (JONES et al., 1976), and mice and rats (WATERS and OKAMOTO, 1972, 1973); the dependence is graded in intensity and its magnitude is a function of both the dose and duration of chronic barbiturate administration.

ROSENBERG and OKAMOTO (1974a, b) have presented the results of detailed studies of the development of pentobarbital tolerance and dependence in cats. In some contrast to other studies, although both dispositional and functional tolerance appeared rapidly in the first 2 days, functional tolerance continued to increase even after more than a month of twice-daily dosing! Increasing tolerance with daily administration has been reported for man (FRASER et al., 1957) and a variety of animal studies, e.g., the cross-tolerance studies of WAHLSTRÖM (1968a).

KALANT et al. (1971) and HUG (1972) have summarized the bulk of the world literature on onset and degree of tolerance to barbiturates. At best, tolerance is stated to be severely limited with tolerant individuals showing approximately 60% less effect or requiring 60%–200% larger doses to produce a given effect (but see STEVENSON and TURNBULL, 1970). Further, the degree of tolerance is appropriately indicated by blood or tissue levels associated with a given effect. Using such criteria, tolerance to the cumulative dose of phenobarbital, for example, can be marked; BUTLER et al. (1954) cite no apparent effects of plasma levels of phenobarbital some five times those that initially produced marked symptoms (see also LOUS, 1952). Such tolerance was observed within 2 to 3 days and disappeared "coincidentally with the elimination of the drug." The question of functional tolerance to the antiepileptic effects remains; most observers conclude that it does not occur, but explicit proof is lacking (see, e.g., BUCHTHAL et al., 1968; BUCHTHAL and LENOX-BUCHTHAL, 1972; WOODBURY et al., 1972).

The fact that different functions develop tolerance to varying degrees and with different time courses has been stressed by many workers. Perhaps one of the clearest demonstrations of these aspects has been the study of MAYNERT and KLINGMAN (1960); these authors postulated that one factor involved in these differences was the degree of neurologic impairment represented by a given effect of the drug, such that tolerance would be less to those effects which constitute greater neural impairment.

FRASER (1957) concludes, in reviewing the studies in man with barbiturates, that tolerance appeared to develop more rapidly in situations where there was some motivation (reaction time, coordination) as compared with those with little motivation such as hours of sleep. KALANT (1973a) and LEBLANC et al. (1973) have described such tolerance as "behaviorally augmented"; confirmation has appeared recently (WOOD, 1977).

The relatively small change in the lethal dose in otherwise tolerant individuals has been frequently cited (cf. KALANT et al., 1971 and HUG, 1972 for references). However, human data suggest that significant tolerance to lethal doses does occur.

The necessity, in studies of tolerance, to determine concentration of active agent at sites of action is clear from studies in our laboratories with ethanol and fish; tolerance was difficult to demonstrate simply on the basis of effects of a given concentration in the water, but it could be readily shown by measuring brain levels associated with a given level of impairment. Following a period of exposure to low levels of ethanol, the brain level at the time of loss of equilibrium is increased almost three times over that of controls. Similarly, STEVENSON and TURNBULL (1970) demonstrated an increase in tolerant animals of the total barbital concentrations in brain at the time of awakening. In cross-tolerance experiments, after the chronic administration of barbital the sleeping time with an acute dose of pentobarbital was shortened and the total concentration of both barbiturates was, strikingly, some seven times greater in the tolerant rats.

The rate of recovery of tolerance from the barbiturates is difficult to assess since, for most studies, one is concerned with two different mechanisms—tolerance associated with increased rate of metabolism and functional tolerance. (As noted earlier, readers are referred to other sources for discussions of induced alterations in metabolism; there is no evidence that metabolic tolerance per se is connected in any direct way with functional tolerance or dependency syndromes.) Nevertheless, metabolic tolerance and cross-tolerance can be readily induced without necessarily producing functional tolerance, e.g., by the use of low doses of phenobarbital as the inducer of thiopental metabolism (SHARMA et al., 1970a, b). Moreover, the cross-tolerance to alcohol in persons who use barbiturates regularly may be not only of the functional type but also metabolic, due to induction of liver acetaldehyde dehydrogenase (REDMOND and COHEN, 1971; DEITRICH, 1970).

LEBLANC and CAPPELL (1975) and GOLDSTEIN (1975a, b) have examined the theoretical implications of the kinetics of tolerance and dependence. It seems probable that each neural system influenced exhibits different tolerance kinetics as well as different rates of cumulation. Such a possibility would be consistent with the observation that, although tolerant, the drug effect and the subjective responses in a chronic user may appear to be, to a degree, qualitatively different than in the naive user. At least certain aspects of functional tolerance to barbiturates may take weeks to disappear (see, e.g., WAHLSTRÖM, 1968a).

In extensive studies in dogs, propranolol and methysergide pretreatment significantly reduced both the dose of thiopental required for anesthesia and the duration of the anesthesia. The propranolol treatment also resulted in a slight reduction of the thiopental level determined upon awakening from anesthesia (HATCH, 1972). These results were interpreted as an interference by propranolol with development of acute tolerance; however, the situation is more complex since propranolol (as well as methysergide, atropine, and chloroethyl dibenzylamine) also reduced the dose and level required for anesthesia. Whether these effects are directly on the nervous system or due to altered distribution of thiopental is hard to ascertain. In contrast, propranolol may accentuate ethanol's actions (ALKANA et al., 1976).

4. Meprobamate

Tolerance occurs to the EEG and behavior effects of meprobamate and its close relative, phenaglycodol (CHIN and SWINYARD, 1958). It is suggested that both metabolic and functional tolerance occur (cf. summary by KALANT et al., 1971; e.g., XHENSEVAL and RICHELLE, 1965; PHILLIPS et al., 1962; HUG, 1972). Interestingly, meprobamate and phenaglycodal acutely elevate seizure threshold. This elevation exhibits tolerance, and withdrawal after chronic administration results in a lower threshold. The onset of the lowered threshold is within 8 h (SWINYARD et al., 1957; CHIN and SWINYARD, 1958, 1959).

5. Benzodiazepines

Acquired tolerance is produced by chlordiazepoxide (HOOGLAND et al., 1966; MATSUKI and IWAMOTO, 1966; GOLDBERG et al., 1967; BARNETT and FIORE, 1971a, b, 1973; QUENZER and FELDMAN, 1975), oxazepam (MARGULES and STEIN, 1968), lorazepam (STEIN and BERGER, 1971), flurazepam (CANNIZZARO et al., 1972) and diazepam (POIRE et al., 1967; see BARNETT and FIORE, 1971a, b, 1973; cf. reviews of GARATTINI et al., 1973; GREENBLATT and SHADER, 1974a, b). The degree and rate of onset of the tolerance have not been accurately quantified; daily treatment for weeks seems required in man. Acute tolerance occurs to the depression by diazepam of the linguomandibular reflex in cats (BARNETT and FIORE, 1971a, b, 1973). A portion of this acute tolerance may be related to an altered rate of appearance of a metabolite, and this metabolite may act antagonistically to diazepam itself. Interestingly, these authors did *not* obtain cross-tolerance between diazepam and phenobarbital or methocarbamol. However, cross-tolerance between diazepam and alcohol has been observed in man (BARNETT and FIORE, 1971a, b; 1973).

6. General Anesthetics

Although relatively unstudied, functional tolerance to inhalational anesthetics appears to occur (KRUG et al., 1965; LOURIA, 1968; HUG, 1972). The anesthesiologists' frequent observation of greater resistance to inhalational as well as barbiturate general anesthetics in alcohol and sedative drug users (SOEHRING and SCHÜPPEL, 1966; ADRIANI and MORTON, 1968; HAN, 1969; JAFFE, 1970) has led to the explicit demonstration of greater resistance to halothane in alcoholic patients (BOURNE, 1960; VAN HARREVELD et al., 1951; HARRFELDT, 1965; HAN, 1969), but it is not always detectable (SHAGASS and JONES, 1957).

A massive study of the tolerance to general anesthetics shown by animals made tolerant to ethanol was undertaken by SCHEININ (1971) following up the studies of LEE et al. (1964). Tolerance to alcohol was produced by giving alcohol every second day for months; tolerance to ethanol alone was detectable after 3 weeks of such treatment. Some degree of cross-tolerance was found to the anesthetics halothane and ether. Conversely, animals receiving halothane or thiopental five times weekly became markedly resistant to ethanol. These extensive studies are seriously flawed by the infrequent administration of agents; presumably similar results showing cross-tolerance could have been demonstrated within days or at the most 2 weeks using dosage schedules of once or twice daily (cf. ROSENBERG and OKAMOTO, 1974a, b; OKAMOTO et al., 1975). Animal studies confirm the cross tolerance; e.g., ABREU and EMERSON (1939) found increased time required for ether anesthesia in mice treated chronically with alcohol as AHLQUIST and DILLE found using rabbits (1940).

Cross-tolerance between isoflurane and ethanol reported recently (JOHNSTONE et al., 1975) presents unexpected findings. Animals receiving chronic ethanol administered in the drinking water exhibited a progressive increase in tolerance to isoflurane over a period of 20 days. Surprisingly, the tolerance persisted after the ethanol was discontinued for 20, 40, 55, and even 75 days. Such long-lasting tolerance would appear to be a different phenomenon from that more commonly observed in other studies; it could, however, present a model for examining a "protracted tolerance" syndrome.

As HUG (1972) emphasizes, the subject of cross-tolerance between different agents is especially complex, since two agents may interact to potentiate or antagonize each other's metabolism in the liver; with chronic administration, the natures and degrees of interactions are without question functions of the specific drugs, individual doses, duration and frequency of prior administration of each of the agents.

Paraldehyde and trichoroethanol exhibit acute and chronic tolerance similar qualitatively and quantitatively to that observed with ethanol, thiopental, and pentobarbital (MAYNERT and KLINGMAN, 1960).

II. Mechanisms of Functional Tolerance to Depressant Drugs

1. Introduction

A number of theories for tolerance to sedative/hypnotic agents have been advanced. Usually these (e.g., WALLGREN, 1971, 1973) are suggestions for a specific—usually a biologically plausible—mechanism. Little has been done in the way of direct theoretical analysis; the presentations of COLLIER (1966, 1969), GOLDSTEIN (1974, 1975a, b), AXELROD (1968); GOLDSTEIN and GOLDSTEIN (1961, 1968), HUG (1972), KALANT et al. (1971), WALLGREN (1973), MARTIN (1968), SHUSTER (1961), LEBLANC and CAPPELL (1975) cover the field. These formulations can now be extended by analytical classification to cover, albeit superficially, a large variety of potentially feasible mechanisms. It should be kept in mind that tolerance is a descriptive term, and that a number of different mechanisms could possibly be involved, depending on the specific behavior and test situation. It is of particular importance to draw the distinction between two classes of possible mechanisms. The first class of mechanisms is more directly concerned with the decline in drug effect due expressly to exposure to the agent which is the result in turn of a mechanism that is set in motion by the drug effect; the result is a returning of function towards the predrug state. This class of tolerance mechanisms can thus be viewed, using some common terms, as: compensatory (compensation), adaptive (adaptation), feedback controlled, self-regulatory, or homeostatic. This class of mechanisms directly implies a system that initially detects the drug-induced change and, subsequently, responds to alter the system such that the drug effect is reduced or abolished; finally, a detection system is needed to monitor and regulate the efficacy of this compensatory response mechanism itself[15].

[15] Mechanisms for tolerance phenomena can also be considered as analogous to the mechanisms developed to explain the recovery of movement after lesions of the CNS. GOLDBERGER (1974) notes the apparent contradiction between the belief that different neural components of the motor systems make unique contributions to movement control and with the fact of recovery (tolerance) of a motor function after the neural structure mediating it was destroyed. For the

2. Feedback Compensatory Systems

Although this discussion is couched in somewhat anthropomorphic terms, it is the frame of reference of servo- and cybernetic systems, and may be useful so long as the admonition is heeded that such detection systems no more imply an intelligent homunculus than the thermostatic control of a house heating/cooling system.

Possible mechanisms of acquired functional tolerance based on a feedback compensatory system include the theories of *enzyme induction, expansion, repression* or *de-repression*. The GOLDSTEIN's (see recent discussion of GOLDSTEIN, 1974, 1975a, b; GOLDSTEIN and GOLDSTEIN, 1961; GOLDSTEIN et al., 1968; AXELROD, 1968; WALLGREN, 1973) formulation is applicable to narcotics and sedatives as well.

Critical to this mechanism is (1) a receptor or enzyme site of drug action, (2) a detection of the altered enzyme activity coupled to 1 and (3) a feedback mechanism of altered rate and product appearance, such as increased synthesis of enzyme, decreased rate of catabolism of enzyme, or removal of its repression. A similar final result could appear by a feedback mechanism that either altered the level of a controlling enzyme co-factor or the level of activities of other enzymes in the pathway in question or a competing pathway.

As has been discussed extensively elsewhere, simply measuring enzyme activities alone is not sufficient to determine a tolerance mechanism or even detect if such an

---

analogy, substitute a unique effect of a drug on a neural system and the appearance of tolerance. A definitive enumeration of the mechanisms for neural recovery following lesions includes (HORN, 1970; GOLDBERGER, 1974; ROSNER, 1974):

1. *Functional reorganization* of the synaptic-neural system remaining:
   a) "Enzyme receptor induction."
   b) Denervation supersensitivity of the neurons remaining, including the mechanisms of increased reactivity to transmitters, spread of receptor sites, loss of receptor specificity [with reference to tolerance and dependence see SHARPLESS and JAFFE (1969)].
   c) Collateral sprouting or regeneration.
   d) Presynaptic mechanisms—e.g., alterations in PTP (see, e.g., WALL, 1970; SPENCER and APRIL, 1970).
   e) Alterations in postsynaptic cellular membranes—direct or indirect due, e.g., to changes in intracellular free Ca^{++}.
   f) Alterations in type or amount of RNA and protein synthesized (see, e.g., ROSE, 1970).
2. *Equipotentiality* of neural elements (in line with the Lashley conclusions).
   If some neurons are destroyed or depressed, the remainder have the potential to take over the functions of those destroyed as a general nervous system property, or at least for the cortical motor systems.
3. *Vicarious function.* The replacement of the function destroyed by another system. It implies a latent capacity of a system to mediate functions it does not normally subsume.
   Redundancy theories are related to this vicarious function, implying an overlap or interchangeability among nervous elements (cf. MARTIN, 1968). Whether or not one set of neurons is hierarchically or functionally dominant is a subset concern for any vicarious or redundancy formulation. Furthermore, one must account for the delays and slow time course of development of recovery or of tolerance.
4. *Substitution* or alternative strategies imply the utilization of other systems so that the same apparent end product behavior is elicited, even though there is little recovery in the usual sense—e.g., the compensatory hypertrophy of muscle groups not normally utilized in a given movement.

adaptive system is operating; the critical question is not the *potential* enzyme activity but the *actual* rate of turnover of substrate-product in vivo. This is perhaps most readily understood from the fact that many enzymes are present in concentrations that provide activities far exceeding those needed for cellular function; the control of metabolic processes in a given multi-enzyme pathway is generally limited to a few critical components and sites. MCILWAIN (1970) proposes and reviews small amounts of data indicative of adaptive alterations in such enzymes.

The general presence of marked cross-tolerance among most agents in the group of sedative/hypnotics has been used to argue rather convincingly that the tolerance mechanisms do not involve a specific molecular receptor upon which these agents act. In fact, most observers doubt if such a receptor exists, at least for ethanol or gas anesthetics. Some other examples of cross-tolerance are presented in KALANT et al. (1971), SCHEININ (1971), HUG (1972), STEVENSON and TURNBULL (1974). Cross-tolerance studies provide a powerful approach to defining mechanisms of action and mechanisms of tolerance in the context of the view that acquired functional tolerance to ethanol is an adaptation of the CNS to the functional deficiency (GREIZERSTEIN and SMITH, 1973b, 1974; KALANT et al., 1971; LEBLANC et al., 1975; see also WAHLSTRÖM and WIDERLÖV, 1971). Recent confirmation of this view (LEBLANC and KALANT, 1975) demonstrated that rats tolerant to ethanol on the moving belt test exhibit complete cross-tolerance to a number of aliphatic alcohols (n-propanol, isopropanol, n-butanol, t-butanol). Moreover, after correction by thermodynamic activity coefficients, almost identical dose-response curves are obtained for all of these alcohols.

Turning from general theoretical considerations to a few possible enzyme sites for actions of barbiturate and ethanol, ISRAEL et al. (1966), and ISRAEL and SALAZAR (1967), among others, have shown that ethanol inhibits the Na^+, K^+ ATPase isolated from brain tissue; after chronic ethanol administration, the activities of brain cortex ATPase are increased [ISRAEL et al., 1970; see also KNOX et al., 1972; WALLGREN et al., 1975; however, not with GOLDSTEIN and ISRAEL, 1972 (see Subsect. C.II.2).] These observations are consonant with the enzyme feedback mechanism for tolerance, given that the interaction of ATPase and ethanol is, in fact, related to ethanol's pharmacologic actions (see discussion above). However, this increase did not appear using the ethanol inhalation technique in the mouse (GOLDSTEIN and ISRAEL, 1972). Tolerance to ethanol is not reflected in tolerance to the depression in vitro of oxygen consumption of cerebral tissue (WALLGREN and LINDBOHM, 1961).

The mechanism underlying the increased ATPase enzyme activity is not known. Many factors involved in ATPase activity and the wide variety of possible mechanisms could underlie the above observations, including a shift in the relative concentration of various molecular forms of the ATPase(s) as actually observed in cultured cells (SYAPIN et al., 1976). Nevertheless, the observation might provide a clue to useful models for one tolerance mechanism. Presumably, ethanol affects enzyme activity by allosteric influence on the protein and/or lipid portion of the enzymes, possibly by competing with or acting allosterically at sites involved with K^+ sensitivity. But whether or not the changes in ATPases are a feedback, homeostatic mechanism or not is not known either, even though some results to date are consistent with such an interpretation. The results of SYAPIN et al. (1976) suggest selective actions on glial elements and differential effects on Mg^{++} and Na^+, K^+ ATPases.

ACh production from cortical slices subjected to stimulation is depressed by ethanol and, after chronic administration of ethanol, the productivity is altered as if a compensatory mechanism had been set in motion (KALANT and GROSE, 1967). Acute tolerance to this effect has been reported (MORGAN and PHILLIS, 1975). WAHLSTRÖM (1966, 1968a) also implicates a central cholinergic mechanism in the tolerance to hexobarbital, whereas HATCH (1972) implicates adrenergic and others based on the actions of physostigmine, methysergide and propranolol. Chronic ethanol also results in changes in cortical response and uptake of NE (FRENCH and PALMER, 1973; FRENCH, 1975). On the other hand, 5-HT turnover is not involved in the development of tolerance to ethanol (FRANKEL et al., 1974).

A portion of the tolerance with chronic ethanol exposure may be due to altered alcohol dehydrogenase (ADH) levels in brain, as demonstrated by RASKIN and SOKOLOFF (1972). Cross-tolerance to paraldehyde and chloral hydrate could also involve ADH; however, cross-tolerance to other agents such as inhalational anesthetics or barbiturates necessitate other explanations.

A compelling case that acetaldehyde does not play a significant role in tolerance or in the acute abstinence syndrome to ethanol has been made repeatedly (cf. WALLGREN, 1973). The dependence syndrome is obtained in pyrazole-treated animals and it is obtained with tert. butanol, an agent exhibiting full cross-tolerance with ethanol and cross-dependence (WALLGREN et al., 1974). These two situations alone indicate that dependence and tolerance *per se* following chronic ethanol exposure cannot be solely or primarily due to acetaldehyde or its condensation products. Nevertheless acetaldehyde, biogenic aldehydes, or TIQ derivatives could play a variety of roles—especially in acute and chronic toxicity of ethanol (cf. COHEN and COLLINS, 1970; COHEN, 1973a, b, c; TABAKOFF et al., 1973; COLLINS, 1973; see Subsect. C.II.11).

The uncoupling effect of barbiturates does not appear to have been examined with respect to possible changes with chronic administration.

3. Localization of Site of Tolerance

The tolerance to barbiturates can be detected locally in the CNS. The sleeping time in response to pentobarbital injected into the lateral ventricle is a useful measure of brain sensitivity (STEVENSON and TURNBULL, 1974). The chronic i.p. administration of barbital, methaqualone, chlordiazepoxide, nitrazepam, or ethanol resulted in tolerance to pentobarbital administered intracerebroventricularly. However, this tolerance was apparent and tested 48 h after the last chronic dose; although it is tolerance in the descriptive sense, it is probably more accurately described as withdrawal hyperexcitability. By 48 h one would expect much of any chronic functional tolerance that had occurred to have declined or disappeared and have been replaced with withdrawal hyperactivity. Thus, what STEVENSON and TURNBULL (1974) appear to be studying is not the tolerance mechanisms with chronic sedative drug administration, but decreased sensitivity of the brain mechanisms to pentobarbital as a component of the withdrawal syndrome, as is suggested by the authors. The method has great further potential; these authors state that they failed to find tolerance with chronic diazepam, and suggest that the dose may not have been adequate; the data in their table show effects to the contrary. It would be of great interest to correlate the time course and syndrome after intracerebroventricular administration of pentobarbital to rats with the microdistribution of pentobarbital in brain structures. This suggestion stems from the fact that the injected volume is only 25 µl with onset of sleep of a few minutes and a dose-dependent duration of 4 min or more at the lowest doses.

In the extensive study of STEVENSON and TURNBULL (1974) the use of the brain levels of pentobarbital provides a powerful measure, but in the data presented time and level are complexly interwoven.

In a somewhat similar study, tolerance to phenobarbital was readily produced when it is administered into the lateral ventricles of rats in doses far below those required with systemic administration (MYCEK and BREZENOFF, 1974). Tolerance was induced by four daily doses and was apparent on the 2nd day and progressed to almost complete tolerance to the given dose on day 5; recovery developed over a week following cessation of treatment—a rather long time considering the rate of development. Although this appears to be a good model for tolerance, it is confounded by apparent changes in liver metabolizing capability and by the fact that no central tolerance was found after chronic i.p. administration in parallel experiments.

4. Receptor Desensitization

A specific case of acute tolerance is the desensitization of cellular response to transmitter agonists. The phenomenon has been most intensively studied at the skeletal neuromuscular junction. A similar phenomenon, tachyphylaxis, has been long known with agents such as nicotine (cf. LARSON and SILVETTE, 1968).

The short 3–6 h time course of development of acute tolerance to depressants suggests that if enzymes are involved, the mechanisms are not directly involved with protein synthesis. On the other hand, desensitization mechanisms or tachyphylaxis appear to have a time course faster than tolerance development. The mechanisms underlying receptor desensitization to cholinergic substances are complex, involving a postulated three structural states of the receptor—resting, active, and desensitized, interacting with Ca^{++} and agonist (see COHEN et al., 1974; COHEN and CHANGEUX, 1975; KATZ and THESLEFF, 1957; MANTHEY, 1966; NASTUK, 1967; NASTUK and PARSONS, 1970; NASTUK et al., 1972). It seems to be implied that desensitization mechanisms are not feedback controlled in the conventional sense; perhaps these are appropriately included in a second class of tolerance mechanisms of "*rate-dependent drug action, relaxation-time*" (see below) (cf. PATON, 1961). Possibly related is the observation that ethanol's depleting effect on brain Ca^{++} exhibits tolerance after a single dose that lasts 24 h or more; such tolerance is blocked by pretreatment with naloxone or cycloheximide and exhibits cross-tolerance with morphine (ROSS, 1974; ROSS et al., 1974).

Another type of feedback cellular mechanism can be mentioned; changes in membrane potential are generally opposed by membrane/cellular mechanisms. Ethanol causes modest depolarization of many membranes, but compensatory repolarization would be anticipated with a time course of hours or days; but no evidence has been reported or looked for intensively.

5. Disuse Supersensitivity

JAFFE and SHARPLESS (1965, 1968) have tried to demonstrate disuse supersensitivity for pentobarbital, but not fully successfully. The phenomenon of supersensitivity seems to suffer from the analogous problem as the phenomenon of tolerance in that it could result from a variety of feasible mechanisms but none of them have been ruled in or out. For example, whether or not denervation supersensitivity occurs in single cells is not established. More likely, it requires at the least a functional synapse or nerve network involving many synapses. A synaptic, presynaptic feedback model of adaptation is succinctly presented by MANDELL (1975a) in addressing the question of tolerance to tricyclic antidepressant drugs in a book of papers addressed to the general concerns of neurobiological mechanisms of adaptation and behavior (MANDELL, 1975b).

6. Tolerance Involving Adaptation in Small and Large Neuronal Networks

The currently most popular theories of tolerance to depressant drugs focus on the adaptation of complex neuronal systems to the impaired function induced by the agent—an old idea emphasized by MAYNERT and KLINGMAN (1960), and more recently by CHEN (1968), GREIZERSTEIN and SMITH (1973a, b, 1974), GREIZERSTEIN (1975), KALANT (1973a, b), LEBLANC et al. (1973), KALANT et al. (1971), and LEBLANC et al. (1975). Most commonly put, it states that the tolerance develops as a function of the activity and "practice" of the neuronal (musculoskeletal sensorimotor systems) under the influence of the agent. Thus, the stimulus to the adapatation is the impaired function and the development of tolerance is related to the degree of impairment, the amount of functional demand (the "drive state"), and the amount of practice and experience with the drug state. Implied in this theory is an analogy with the learning of newer, compensated mechanisms of functioning—hence it is sometimes referred to as learned tolerance. For the most part, such studies have centered almost exclusively on motoric impairment. As KALANT et al. (1971) and MAYNERT and KLINGMAN (1960) emphasize, it is hazardous to extrapolate from one drug action end point to another.

The concept of learned adaptation is fully consonant with another prime conclusion of MAYNERT and KLINGMAN (1960) that the magnitude of apparent tolerance was inversely related to the degree of neuronal impairment. The greater the potential neuronal reserve, the greater the potential for compensatory tolerance.

Given that such a description of tolerance is accurate, what neuronal biological mechanisms produce such adaptive tolerance? It should be noted that this concept of tolerance is implicitly included in basic physiologic problem areas such as those pioneered by ADOLPH (1943, 1960, 1968) under the rubric of "Physiological regulation," habituation, neural plasticity and homeostasis (cf., e.g. KANDEL et al., 1970; HORN and HINDE, 1970; CAREW et al., 1972; CAREW and KANDEL, 1973; PEEKE and HERZ, 1973a, b; STEIN et al., 1973b; ZIPPEL, 1973; MANDELL, 1975a, b; BARACH, 1974) as well as the adaptation to exercise and altitude (cf., e.g. MARGARIA, 1967; FRISANCHO, 1975) and the area of learning of motor acts.

WAHLSTRÖM (1968b) attempted an ingenious approach to differentiate "learned" tolerance from other mechanisms by exploring two groups of rats—one during activity and one during the natural quiesence of the dark-light cycle. The results were confounded by a high percentage of deaths of unknown causes and by a design that used barbital, a nonmetabolized but long-acting agent, administered daily over weeks. Nevertheless, he did demonstrate more cross-tolerance to hexobarbital's EEG effects in the dark and more active group. Analogous experiments could well provide useful approaches to differentiating tolerance mechanisms (cf. LEBLANC and CAPPELL, 1975).

Temperature acclimation appears to involve alterations in the membrane and metabolic properties of central neurons (KONISHI and HICKMAN, 1964; ROOTS and PROSSER, 1962; SAARIKOSKI, 1970) including those of cerebellum (KOTCHABHAKDI and PROSSER, 1972; PROSSER, 1967). The time course of temperature acclimation may have a very rapid onset of 30 min and variations over weeks (SIDELL et al., 1973). Moreover, spinal and peripheral mechanisms may be sufficient for certain types of acclimation (ROOTS and PROSSER, 1962).

Adaptation to cold, such as that of the women divers of Korea, involves profound physiologic changes including shifts in basal metabolic rate, oxygen utilization, fat distribution, and circulation. The degree of adaptation fluctuates over the year in correlation with external temperatures to which they are exposed daily (see summary by HONG, 1973). Adaptive changes to many stresses involve central nervous mechanisms; moreover, these mechanisms may be blocked by anesthesia, such as those involved in temperature regulation in hibernation (HAMMEL et al., 1973).

The compensations in the disturbances of head position and nystagmus produced by labyrinthectomy have a time course similar to that of acute tolerance to motor impairment by ethanol—3–9 h. The compensatory mechanisms appear to involve many parts of the nervous system, including the contralateral cortex, cerebellum, and to some degree the spinal cord (see review and experiments of SCHAEFER and MEYER, 1973). Interestingly, the compensatory mechanisms involve neural activity, since the administration of drugs that impair motor functioning, such as phenobarbital or chlorpromazine given in single or repeated doses, delayed the appearance of compensation for 36 or more hours. Conversely, agents producing an increase in activity accelerated the compensatory process (SCHAEFER and MEYER, 1973). Although I infer that it is activity that is important, these drugs may be acting on more basic processes underlying adaptation and learning. As GESCHWIND (1974) notes in discussing plasticity of the nervous system, one is concerned with changes occurring over seconds, hours, days, weeks, months, and indeed, years.

Tolerance, habituation, and adaptation occur with many environmental conditions, such as hypoxia from altitude or carbon monoxide poisoning (see review of FRISANCHO,1975) or with chronic lung disease (RAMSEY, 1972). In the latter instance, a withdrawal of the stimulus of hypoxia may result in respiratory arrest, i.e., the control point for respiratory control has shifted from CO_2 to hypoxia.

Ethanol has complex interactions with respiratory gas exchange and hypoxia. Hypoxia and ethanol are synergic, at least additive, in their effects on higher brain functions of concentration, memory, etc. Also, tolerance appears to develop to augmented intoxication due to ethanol in newcomers at altitude (MCFARLAND and FORBES, 1936; HANSEN and CLAYBAUGH, 1975; MEERLOO, 1957). In addition, HANSEN and CLAYBAUGH (1975) have demonstrated that ethanol ingestion results in lower alveolar and arterial oxygen pressures in comparison to carbohydrate ingestion. The brain hypoxia that would be expected to result from carbon monoxide inhalation or hemodilution is appreciably compensated for, such that the oxygen tension in jugular blood is only slightly reduced. The compensatory mechanisms maintaining brain oxygen levels include vasodilation and increased cerebral blood flow; hemodilution could contribute to increased flow due to decreased viscosity (PAULSON et al., 1973).

These marked shifts in cerebral blood flow and oxygen availability pose a number of questions regarding differences in rate of onset and distribution of drugs and anesthetics given simultaneously or sequentially after carbon monoxide, hypoxia, or hemodilution.

Animals with frontal lesions show changes in responsivity to specific drugs. GLICK (1974), in a review of previous work, discusses the fact that matching behavior was depressed by scopolamine more in lesioned animals; the animals were less sensitive to amphetamine, chlorpromazine, alpha-methyl-dopa, but equally sensitive as normal animals to LSD, physostigmine, mecamylamine and pentobarbital. In fact, frontal rats exhibit hyper- or hyposensitivity to amphetamine depending upon the behavior situation (see review of ADLER, 1970 and Subsect. C.II.13 above). MEERLOO (1957) relates, anecdotally, a number of case studies in which sensitivity to alcohol was markedly enhanced by brain damage, concussion, or tumor; he also notes that the sensitivity might be decreased in some individuals. At least in a well-established animal model of tolerance, frontal cortical lesions prevent tolerance development to ethanol (LEBLANC et al., 1976), and atropine prevented tolerance development to intracerebroventricularly administered phenobarbital (BREZENHOFF and MYCEK, 1975).

Only a few studies have examined neurophysiological changes during chronic alcohol or barbiturate use and withdrawal. BEGLEITER et al. (1972, 1974) describe the decrease in somatosensory evoked potentials in man during acute alcohol intoxication, the exaggeration of the potentials reaching a peak increase at 24 h after withdrawal. Subsequently, the potentials return to control levels in a period of days. Similar findings have been recently reported for the spontaneous EEG activity in unanesthetized cats during oral administration of alcohol on a 8, 12 or 24 h schedule (PERRIN et al., 1975). The most marked changes were detected from hypothalamic electrodes. The EEG spiking during withdrawal was less in the cat experiments than in similar mouse studies (WALKER and ZORNETZER, 1974).

The only detailed studies of electrophysiological changes during chronic intoxication with barbiturates and subsequent withdrawal are the recent extensive and detailed observations of ROSENBERG and OKAMOTO (1974a, b) on experimentally induced dependence and withdrawal in cats. In cats dependent on barbital there are changes in spinal function paralleling the dependence and withdrawal as detected by changes in segmental reflexes in acute spinal cats. Whether reflex alterations are the consequences of changes in the spinal cord or in areas rostral to the section cannot be ascertained, since it is known that sustained supraspinal influences can alter spinal cord function measured after acute section (CHAMBERLAIN et al., 1963; GRIFFIN, 1970). The report of ROSENBERG and OKAMOTO (1974a, b) does make obvious the need for studies on chronic spinal animals, analogous to those conducted with narcotic dependence (WIKLER and FRANK, 1948; MARTIN et al., 1974a). That tolerance and perhaps dependence could involve alterations or synaptic connections receives appreciable support from the finding that ethanol reduces neuronal growth in the cerebellum in developing rats (BAUER-MOFFETT and ALTMAN, 1975).

Sustained effects on spinal function are discussed and demonstrated by GRIFFIN (1970), FITZGERALD and THOMPSON (1967), and CHAMBERLAIN et al. (1963). GRIFFIN (1970) reviews extensive studies on habituation of the flexor reflex in rats in which lesions of frontal cortex markedly reduce the degree of spinal habituation so long as the frontal lesions antedate the spinal cord section. He concludes that the habituation is localized to spinal internuncials which are subordinate to cortical centers that either accelerate or delay the habituating processes. WALL (1970) has defined this spinal habituation further in terms of the interneurones involved and the process of PTP. [It would be interesting to know whether anyone has examined the effects of diphenylhydantoin on spinal habituation. Nether picrotoxin nor strychnine had any effect on habituation (SPENCER et al., 1966).] General depressant drugs such as urethane, chloralose, or pentobarbital apparently do not block habituation completely (see SEGUNDO and BELL, 1970), although the drug actions have not received explicit examinations in comparison to untreated controls.

KANDEL et al. (1970) argue convincingly for the cellular-connection approach to studying learning and plasticity in the nervous system—and by this they mean the specification of neuronal systems as opposed to postulations of mechanisms that involve large undifferentiated aggregates of nervous elements. GROVES and THOMPSON (1973) have schematized a neuronal network that includes habituating synapses, sensitizing synapses as well as synapses that appear to lack plastic properties—such a network is based on extensive experimental studies and has significant potential for future analyses.

Tolerance could, thus, well utilize the same mechanisms as those that underlie other time-dependent CNS processes such as habituation (CAREW and KANDEL, 1973; MANDELL, 1975a, b) and learning (FRANKS, 1958; KANDEL, 1976). However, these mechanisms, as well as those of tolerance, have yet to be identified—even to the degree of specifying whether these processes are discretely or diffusely located. A potentially useful example of this is the report of tolerance and cross-tolerance to morphine of analgesia produced by focal electrical stimulation of the brain (AKIL et al., 1976; MAYER and HAYES, 1974, 1975). Consonant with these findings is the localization of certain actions and tolerance to the analgesic effects of morphine to the periaqueductal gray matter (JACQUET and LAJTHA, 1975) and with ventricular injections (EIDELBERG and BARSTOW, 1971). (See also YORK, 1972 and LINSEMAN, 1976). The complex interaction of lesions, tolerance and dependence has been demonstrated with morphine. Evidence suggesting a neural localization for analgesia, for tolerance and for dependence has been obtained; whether or not such evidence holds up or whether the apparent localization of function is the consequence of a complex interaction between drug effects and lesions remains to be ascertained (LINSEMAN, 1976). A possible role of the adrenal system was introduced in the recent paper reporting facilitation of tolerance to ethanol by dexamethasone (WOOD, 1977).

One of the mechanisms postulated to underlie both learning and tolerance is the utilization of previously inactive or *redundant pathways* (for narcotics, see MARTIN, 1968, and Sect. II, Chap. 1) and/or by other mechanisms resulting in neuronal hypertrophy. Not only are the stimulus and the rate of development of tolerance involving homeostatic feedback mechanisms of interest but also the degree of the retention of tolerance and its decline over time. Certain facets of tolerance to drugs appear to have a short time course and decline parallel with the brain level, but with some delay (e.g., GREIZERSTEIN and SMITH, 1973a, b, 1974). However, the retention may be over days, and a subsequent exposure may result in rapid resumption of tolerance (see recall tolerance—KALANT et al., 1971 and discussions of LEBLANC and CAPPELL, 1975 and MCILWAIN, 1970). On the basis of a simple learned response, one would anticipate fairly long retention of the recognition of the state and mechanisms of coping rather than a rapid decline of tolerance. On the other hand, in most of the experimental tests, adaptation may involve largely unconscious motor learning and, thus, recovery from the adaptation is a process that is, in essence, the reverse of the tolerance development and might be expected to exhibit a similar time course.

The process of functional tolerance, then, could be viewed as analogous with learning and memory. Or it can be postulated that tolerance *is* one aspect of the more general neural processes of learning and memory. As such, it can be studied phenomenologically and mechanistically from the same aspects as recently outlined for neural mechanisms of learning and memory (e.g. ROSENZWEIG and BENNETT, 1975; KANDEL, 1976) including such aspects as formal behavioral descriptions of the phenomenon, the identification and localization of the neural and synaptic processes involved, the exploitation of relatively simple biological systems for the study of cellular processes, and the exploration of the phenomena during development of the nervous system.

The most common conventional view of tolerance to sedatives, such as alcohol and barbiturates, involves adaptation, accommodation, adjustment and, thus, implies a feedback control—i.e., the correction of a drug-induced impairment after its detection. Such a feedback, adaptive correction could occur subcellularly or as part of a behavior of a neural network. However, the network involved is not obvious for the positive, desirable, rewarding, or anti-anxiety drug actions. Put differently, why would any nervous system drive to compensate for an impairment that results in an improved subjective or behavioral state? One is thus led to suggest that either the drug effect is perceived (or detected at some level) as being less than optimal (or standard) functioning, *or* that the compensation is at a site/mechanism that does not, in any way, sense the more final outcome detected by the observer.

It is indeed common knowledge that tolerance does appear to the antianxiety effects of alcohol, and apparently, to gaseous anesthetics as observed with chloroform (cf. MORGAN, 1974) and ether (VAN HARREVELD et al., 1951). Whether or not the subjective antianxiety actions exhibit tolerance at a different rate from the sedative/hypnotic actions has not been quantitatively characterized. But it is a common clinical comment that the antianxiety actions of meprobamate, chlordiazepoxide, and diazepam persist even though the sleepiness encountered with the initial doses wears off; this deserves explicit demonstration inasmuch as tolerance to the antianxiety effects may be one reason underlying the tendency toward chronic escalation of the dose. YORK and WINTER (1975a, c) have shown that daily administration of barbital for 8 days results in marked tolerance to the hypnotic effects, but no tolerance in the ability of barbital to serve as a discriminative stimulus was detectable.

The anticonvulsant or antiepileptic action of the sedative/hypnotics is apparently a property distinguishable from their sedative effects. Acquired tolerance to the anticonvulsant actions would not appear to be directly associated with a homeostatic mechanism. The degree to which such tolerance develops to the various agents is disputed; BOOKER (1972) concludes that there is tolerance to the anticonvulsant actions of phenobarbital, whereas BUCHTHAL and LENOX-BUCHTHAL (1972) conclude that there is not. Tolerance is apparent in the experimental studies of FREY and KAMPMANN (1965). Tolerance unquestionably compromises the utility of benzodiazepines such as nitrazepam (see MILLICHAP, 1972 and MATTSON, 1972). It is probable that there is some cross-tolerance in the intoxicating properties of marihuana and alcohol as suggested by such observations as reported by JONES and STONE (1970).

7. Neurotropic Processes

The changes associated with chronic drug treatment that could involve trophic processes has received little attention. Trophic interactions, at least at microdistances, are known to occur in the adult brain, and to a lesser degree, in the spinal cord (ECCLES, 1974). Although it is speculative, a depressant drug could conceivably alter not only neuronal activity but also the transport of trophic macromolecules. Such changes would result in alteration in synaptic structure and function. Moreover, this process is certainly consistent with the appearance of a withdrawal syndrome and the rather fixed time sequence of the appearance and disapperance of withdrawal signs.

Yet even more speculative are the possible interactions among ethanol and ACh (see Subsect. C.II.6) and the possibility of ACh serving as a neurotropic transmitter (see DRACHMAN, 1974). At the least, the possibility that the neuronal degeneration observed with chronic alcohol ingestion is due in part to ethanol's alteration of trophic maintenance functions deserves explicit attention; these processes can be viewed as the extended time domain from acute to chronic tolerance, dependence and sustained neuronal disruption (see BAUER-MOFFETT and ALTMAN, 1975).

Chronic use of alcohol has been long known to result in brain dysfunction, including learning difficulties. Such disruption has been recently reproduced in animal models (WALKER and FREUND, 1971; FEHR et al., 1976). It remains moot whether such brain dysfunction produced by chronic alcohol or marihuana administration is analogous to human syndromes or is related in a way, mechanistically, to tolerance, abuse or addiction syndromes.

8. Acquired Functional Tolerance: Not due to Homeostatic or Compensatory Functions

A decline in effect with repeated dosing *or* an increase in dose (tissue level) required for a given pharmacologic action need not necessarily require a compensatory or feedback controlled system. All that is required is that the original exposure to the drug set in motion series of actions which, although slightly delayed, will eventually antagonize or oppose the primary drug action. Obviously, if the two opposing actions occurred simultaneously and to exactly the same degree, there would be no overt drug effect. However, if the antagonistic effect is delayed over the primary effect, the result will be a gradual decline in drug effect over time (see figures in KALANT et al., 1971 and LEBLANC and CAPPELL, 1975).

The kinetics of this and some of the following types of tolerance may be identical in a formal sense; also, metabolic tolerances may exhibit a similar time courses.

9. Rate Theory of Drug Action

A form of tolerance can be readily envisioned for that situation in which a drug's action is related, molecularly, to the rate of change in concentration rather than the absolute concentration (PATON, 1961). However conceptually plausible such a mechanism is, it has little data to commend it in the case of ethanol and the sedatives. In fact, where acute tolerance has been directly examined (e.g., KUCERA and SMITH, 1971, 1972; LATHERS and SMITH, 1976) it does not hold. Actually, what have been described as possible examples of such rate-dependent actions may only be special illustrations of cellular feedback control that is manifested by kinetics of action in which the effect is more closely proportional to the rate of rise of concentration than to the concentration itself. Some analogies are perhaps worthy of mention. In sensory systems, adaptation to a constant stimulus is quite common, and drugs that act on such systems would be expected to—and some do—exhibit a response that is a function, in part, of the rate of change of concentration (see SMITH, 1963, 1967, 1973; MURTHY and DESHPANDE, 1974).

10. Depletion of Transmitter, Substrate or Mediators

For agents and actions that are the result of an evoked release of a neurotransmitter or a specific substrate, excessive action would result in depletion and, therefore, a subsidence of the drug effect—i.e., tolerance. Examples are found most readily in acute tolerance or tachyphylaxis such as to nicotine, capsaicin, the histamine-releasing actions of morphine or curare, reserpine's evoked release of CAs, ethanol's potentiation of neuromuscular transmission and release of ACh. For ethanol and the sedatives a depletion hypothesis for action and subsequent tolerance is plausible and has not been ruled out. On the other hand, there is relatively little to support it. There are a number of reports on the interactions of ethanol with putative neurotransmitters that provide vague intimations of support for such speculations, but to date, the subject has not been explicitly addressed. For example, FRENCH (1975) has recently described briefly a marked alteration following ethanol treatment of the uptake by the cortex of topically applied NE.

It is likely that tolerance will be detected in isolated preparations in which ethanol evokes release of transmitters. Following the decline in release with depletion, the subsequent exposure to ethanol may fail to elicit a comparable level of release. Tolerance would be, in such an instance, the plausible manifestation of a depletion of transmitter stores, or a depletion of the triggers for such release. It is certainly conceivable that the excitatory actions of ethanol—presynaptic ACh release, muscle spindle excitation, membrane depolarization, and Renshaw cell activation—could exhibit tolerance; however, none of these properties have been explicitly studied in animals made tolerant to ethanol, whereas chronic ethanol treatment does alter both cholinergic (KALANT, 1974, 1975; QUENZER and FELDMAN, 1975) and adrenergic systems (THADANI et al., 1974).

A special case has recently been made for a specific change in the PTP duration by ethanol in *Aplysia* that is interpreted as tolerance (TRAYNOR et al., 1975, 1976). The decay rate for PTP and ethanol's depressant effect have been linked theoretically with increased membrane fluidity (WOODSON et al., 1976). In a somewhat similar vein, but in an entirely different preparation, INGRAM (1976) obtained data indicating that *E. coli* have markedly different membrane fatty acid compositions after growing in a media containing an aliphatic alcohol; it is postulated that such changes in membrane lipids may constitute a biochemical basis of tolerance. The relevance of these findings rests in large measure on the degree to which these preparations are similar to mammalian neural systems.

The phenomenon of tolerance is seen with a wide variety of drugs including phenothiazines (e.g., BOYD, 1960; LAURENCE and WEBSTER, 1961; LAGERSPETZ, 1963;

TIRRI, 1966; MATSUKI and IWAMOTO, 1968; KUPFER et al., 1971), caffeine (COLTON et al., 1968), pepper, atropinic agents, and anticholinesterases. RUSSELL et al. (1975) have recently presented, in one paper in a series on the behavioral tolerance to irreversible anticholinesterases, very strong evidence that the mechanism of this tolerance is due either to a change in the acetylcholine release mechanism or, more likely, a decrease in the sensitivity of the muscarinic cholinergic receptors. They were able to demonstrate that the behavioral tolerance was not due to metabolic changes, to end product inhibition, to alterations in the inhibition of cholinesterase, or the development of an adrenergic system that had taken over the function of the impaired cholinergic system. The possibility that a neuronal feedback control system involving the utilization of a nonblocked, redundant neuronal pathway was not explored. The authors rule out adrenergic systems moderately well; however, they recognize the possibility that serotonergic systems could be involved in the tolerance development. Nevertheless, their experiments do not permit assuming a feedback system for the tolerance, a detector, or a mechanism whereby the tolerance adjustment occurs.

11. Learned Responses and Tolerance; Related Subjective Phenomena

As noted earlier, tolerance has frequently been likened to a learned response. The time course of functional tolerance, in relation to dose and task, also suggests, at least that functional tolerance is analogous to a learned response. GEBHART and MITCHELL (1971a, b, 1972) and co-workers (1971, 1972; KAYAN et al., 1969) have clearly shown that tolerance development to narcotics—as measured in animals—is a complex phenomenon depending on the state of the animal, the drug, route, and testing condition(s) stimuli; prior experience with drug and with the test situation are both important (see also WAHLSTRÖM, 1968b; PORSOLT et al., 1970). Tolerance to many agents is selective in its appearance and this is also true of the benzodiazepines; certain drug actions such as muricidal behavior require not only daily drug administration to maintain but also that the testing be done under the drugged condition (QUENZER and FELDMAN, 1975).

Another factor to be considered is the demonstration that the initial drug effect in naive animals or subjects may be appreciably different from that in subsequent tests, quite independent of the time interval separating them; this is well described for narcotic analgesics (cf. HUG, 1972).

As has also been observed with narcotics, certain protein synthesis inhibitors appear to antagonize the development of tolerance, at least to ethanol (LEBLANC et al., 1976; ROSS, 1976). Neither actinomysin nor cycloheximide, given during the induction of ethanol dependence, altered the head twiches seen in mice during ethanol withdrawal (COLLIER et al., 1976).

Excessive ethanol levels may not give rise to tolerance in part because ethanol inhibits protein synthesis (NOBLE and TEWARI, 1973). The difficult problems of the validity and impact of antagonism by protein synthesis inhibitors remain (HUG, 1972). (These difficulties center on ascertaining the degree to which the protein synthesis inhibitors' actions are limited to inhibition of protein synthesis. A second problem of interpretation relates to whether or not the inhibition of protein synthesis impacts on the actual systems involved in the development of tolerance; it is conceivable that tolerance may be interrupted as a secondary consequence to other metabolic alterations.) In addition to changes in protein synthesis, tolerance could also be linked with increased biosynthesis and turnover of membrane lipids (VIRTANEN and WALLGREN, 1975a, b).

Certainly functional tolerance or the behaviorally augmented tolerance (cf. KALANT, 1973a, b; LEBLANC et al., 1973) implies a learned response involving a memory system of some sort, and future studies of the mechanisms of tolerance will certainly involve determination of the recognition, storage and recall mechanisms of tolerance and cross-tolerance (cf., e.g., JOHN, 1967; ZIPPEL, 1973; ROSENZWEIG and BENNETT, 1975; KANDEL, 1976). Recent reports do, in fact, lend appreciable credence to learning and adaptive theories of tolerance development to ethanol. LEBLANC et al. (1976) have found that rats subjected to either polar frontal cortex lesions or dosing with cycloheximide fail to reacquire tolerance to ethanol in a test situation in which the controls readily exhibit tolerance. The fact that there is only rather limited tolerance and cross-tolerance among chemically diverse compounds of the sedative/hypnotic class weighs strongly in favor of the idea that the tolerance is an adaptive response involving a large number of complex neural systems. At the very least, the tolerance mechanisms involve neural or biochemical pathways and mechanisms separate from the primary molecular site(s) of drug action, since these latter are most likely different for the various agents and may be distinctly different depending on the specific drug effects, such as sleep or ataxia.

The internal states and stimuli produced by drugs are generally thought to be related to the way in which people take and respond to the agents. Animals can be trained to make a differential response on the basis of the presence or absence of a given drug; in such a situation, the drug and its actions constitute the discriminative stimulus and the measure of behavior depends on the drug effect. A symposium on this topic was presented in 1972, co-chaired by WINTER (1974) and by OVERTON (1974). Frequently, descriptions of the phenomenon state that the differentiated behavior depends on a "perceived" drug effect, as if the animal/man is consciously aware of some changes in internal state and makes a deliberate (rational) choice based on certain sensory input or perceptions. Although such a view is plausible, there are no data to indicate that direct perception of a given drug state is, of necessity, involved. To illustrate, the discriminating behavior may follow from the drug actions in the brain (and there is good evidence for a central site of action) in the same manner as posthypnotic behavior follows the appropriate trigger where conscious perception is not essential. Semantically, it is difficult since the internal state associated with drug administration clearly serves both as a stimulus to certain behaviors and the stimulus to certain perceptions. The perceived actions may or may not match objectively measured effects (cf., e.g., SMITH and BEECHER, 1959; SMITH et al., 1962; MALPAS, 1972).

The phenomenon of state-dependent learning may well be associated with, or isomorphic in mechanism with, certain aspects of functional tolerance. The discriminative stimulus cues provided by drugs for behavioral experiments do exhibit tolerance, but the degree of tolerance manifested varies with training condition (YORK and WINTER, 1975a, b, c); interpretation of such experiments is difficult, since the test may be superimposed on a relatively steady degree of tolerance imposed by a protocol that calls for drug administration every other day.

What is of special interest is the fact that discriminated learning occurs with all of the agents man frequently self-administers or abuses—and this learning is quite selective as regards pharmacologic class. The central sedatives constitute one rather distinct category (reviewed by BARRY, 1968, 1974; BARRY and KUBENA, 1972; OVERTON, 1966, 1974). In summary, each of the following agents elicit the pentobarbital response in pentobarbital trained animals (trained against saline control): alcohol, a number of barbiturates (pheno-, seco-, amo- and pentobarbital, barbital), meprobamate, chloral hydrate, paraldehyde and subanesthetic doses of ether. In alcohol-trained animals, the following evoked an alcohol-appropriate response: pentobarbital, phenobarbital, ethyl carbamate, chlordiazepoxide, chloral hydrate. The differentiation from other classes of drugs is generally quite sharp, e.g., against amphetamine, bemegride, LSD, nicotine, THC, atropine; similar results have recently been found with the failure to cross-react between morphine and ethanol (WINTER, 1975).

In spite of these rather precise dose and class of drug specificities, there are some complications, such as a chlorpromazine test eliciting a response in alcohol-trained animals, although this may reflect general depression of motor function (cf. BARRY, 1974). BARRY (1974) indicates that, in his hands, ethanol appears to interact with some qualitative differences from pentobarbital. The recent studies of YORK and WINTER (1975a, b, c) indicate that animals trained under barbital respond appropriately to barbital only when receiving it and not when they receive ethanol or saline. BARRY (1974) postulates that one of the actions underlying the discrimination—as well as the self-administration—is their anxiety-reducing properties; he classifies alcohol as being the weakest in this regard, pentobarbital intermediate, and chlordiazepoxide strongest of the above list. The studies of YORK and WINTER (1975a, b, c) are consistent. Whether or not such distinguishable properties actually exist (such as anxiety-reduction) and can be measured reliably by these techniques awaits further work.

Returning to an earlier issue, subjects may or may not be able subjectively to differentiate different levels of drug effects that are detectable readily by observation or behavioral tests; for example, subjects clearly differentiated subjectively between two doses of amylobarbital, whereas two doses of nitrazepam were not discriminated (MALPAS et al., 1970; MALPAS, 1972). Such complexities in the assessment and reporting of subjective effects is an area of great importance. It has yet to be established experimentally or theoretically whether or not important etiologies (causes) of individual drug taking can be identified by introspection or by analyses of the subjective phenomena. The historical approach suggests that such etiologies are not readily accessible via introspection and subjective analyses; for example, it has taken hundreds of years to identify and characterize physical dependence phenomena, tolerance and addictive behaviors (cf. also SCHUSTER and JONANSON, 1974).

On the other hand, the intrinsic validity of subjective scoring systems and the expansion of the evidence of their applicability for studying feeling and mood states (e.g., MARTIN et al., 1974b) suggest that many of clues regarding habitual drug taking and the motivational constructs surrounding this behavior will be found by exploiting these investigatory tools (see MALPAS, 1972; STEINBERG, 1954, 1955, 1956; and references to HAERTZEN and HILL and colleagues, in Subsect. C.III.3.g). Initial studies of patients previously dependent on sedative/hypnotics reveal marked variations in sensitivity and preference as functions of dose and regimen (GRIFFITHS et al., 1976, and PICKENS, R., personal communication).

Preferences among beverages and agents are also complex, and in part are related to rapidity of onset and to taste as well as to the quality of effect. Although there is a similarity in effects and addictions among the various alcohols, some people appear to prefer paraldehyde, or methanol, or isopropyl alcohol (MENDELSON et al., 1957). The users are reported to state that methanol or paraldehyde provides greater or faster effects and produces greater calming of the nerves. The case reports clearly indicate acquired tolerance to the nonethyl alcohols, and paraldehyde—and probable tolerance to ethanol as well. The authors succinctly pose the question: "Can it be that adaptation occurs through the establishment or reactive overgrowth of specific enzyme systems for the breakdown and metabolism of offending substances?"

CAPPELL and LEBLANC (1975) remark that little can be said of the unconditioned effects of psychoactive drugs that are responsible for their ability to act as positive reinforcers. The review of DOWNS et al. (1975) also concludes that tolerance to the reinforcing effects of drugs has never been demonstrated experimentally. The latter comment is at first astounding, but on reflection it seems clear that it is perhaps without much meaning as stated, since the very drug reinforcement experiments involve a complex interaction and manipulations of the performance, reinforcement, and dosage schedule—at best, a very artificial setting. What, in fact, would be taken as evidence of tolerance, a changed rate of responding after repeated administrations of drug? This certainly occurs with most sedatives and hypnotics. In most instances animals, if allowed free choice of dose and frequency, will self-administer most agents, such as barbiturates, morphine or cocaine, in amounts that are eventually life-threatening. Very few of the self-administration studies have examined the influence of other behaviors that might compete with drug-taking, drives or stimulation—with the exception of food or water (see SCHUSTER and JONANSON, 1974; EWING et al., 1976).

III. Benzodiazepine Tolerance and Dependence

1. Introduction

Tolerance to the benzodiazepines occurs, but it is a complex matter. Clinically, the antianxiety action is supposed to be resistant to tolerance, whereas tolerance is said to appear readily to the sedative or hypnotic actions (CANNIZZARO et al., 1972; GOLDBERG et al., 1967; MARGULES and STEIN, 1968). The tolerance to repeated doses of chlordiazepoxide appears to be both metabolic and functional (HOOGLAND et al., 1966). However such repeated doses result in more complex alterations in sensitivity to drug effects (see COOK and SEPINWALL, 1975; and BARNETT and FIORE, 1973, for animal studies, and KORTTILA and LINNOILA, 1974, 1975a, b, and LINNOILA et al., 1974a, b, c, for interesting studies in man).

Not only is tolerance observed with repeated dosing; acute tolerance is revealed by the observation that the behavioral alterations after a single dose are more marked on the rising phase of the blood concentration than on the falling phase (BLIDING, 1974); such acute tolerance phenomena were discussed above in the context of ethanol and barbiturates (Subsect. D.I.1).

Acute and chronic behavioral tolerance is evident with many of the effects of benzodiazepines as exemplified with diazepam—and the degree of such presumptive tolerance varies markedly among the performance tests and behaviors under examination. In the study of KORTTILA and LINNOILA (1975a) the blood levels reached after i.v. administered diazepam were a function of the dose; the levels achieved within 2–4 h were fairly well sustained for up to 10 h. The data reveal tolerance in that all of the behaviors examined at all doses had by the 10-h observations returned to control levels. For example, the mean level after 0.45 mg/kg i.v. was 389 at 4 h and 351 ± 91 mg/ml at 10 h after the same dose; note that this latter 10-h level exceeded the peak level of 339 obtained 4 h after a lower dose of 0.30 mg/kg. Moreover, the difference in effect over time is even more striking if one includes the lowest dose studied by KORTTILA and LINNOILA (1975a) of 0.15 mg/kg, the significant effects of which tended to be limited to less than 2 h duration. The Romberg sign returned to control levels after the largest doses in the short time of 36 min; in contrast, the increased flicker fusion frequency required 6–8 h to return to control levels. These differences in time course of drug action were evident in the means of blood levels in spite of the investigators' caution regarding the large inter-individual variations in serum concentration. (Also of note was the fact that the subjects found the effects of i.v. diazepam pleasant.)

Tolerance in man to the antiepileptic effects of benzodiazepines has been reviewed by MILLICHAP (1972) and MATTSON (1972). Tolerance to benzodiazepines undoubtedly appears with therapeutic or greater than therapeutic doses (see BLIDING, 1974; review of GREENBLATT and SHADER, 1974a, b; KORTTILA and LINNOILA, 1975a, b and HANNA, 1972).

The anticonvulsive effect of diazepam (against pentylenetetrazol) is more readily detectable with an ED_{50} of less than 1 mg/kg i.p. than is incoordination in the rotarod test with an ED_{50} of 4; a similar relationship was obtained for phenobarbital (FUXE et al., 1975). Yet, the tolerance to the two effects developed at different rates and to different degrees. Almost complete tolerance to the motor effect was evident on the third daily dose, whereas no tolerance to the anticonvulsant action was seen until the 5th day. Note should be made of the fact that partial tolerance to the anticonvulsant action did develop. Analogous studies with phenobarbital would certainly be of interest to ascertain if this is unique to benzodiazepines. Moreover, the apparent difference in percent tolerance between the two tests may be misleading without a comparison of the actual shifts in the dose-effect curves; the shifts could conceivably actually reflect a parallel alteration in the absolute sensitivities to given benzodiazepine concentrations.

STEIN et al. (1973a, b) postulate that benzodiazepines do not directly increase the tendency to act, but release behavior from suppression. One of the observations on which this theory rests is the tolerance which develops rapidly to the depressant actions whereas the antianxiety action purportedly fails to exhibit tolerance (GOLDBERG et al., 1967; MARGULES and STEIN, 1968). The biochemical and mediator data consistent with this postulate are extensive and will not be reviewed here except to note that with repeated dosing the oxazepam-induced decrease in NE turnover in the midbrain-hindbrain was no longer detectable after 6 days. However, 5-HT turnover remained substantially reduced. The fact that the 5-HT turnover reductions in midbrain-hindbrain regions were more marked than in the forebrain led the authors to emphasize that the site of action of the antianxiety effects of benzodiazepines may be in the central gray of the midbrain (STEIN et al., 1975).

2. Cross-tolerance and Dependence

The tolerance that develops with chronic use of benzodiazepines is complicated by their long duration of action and the presence of active and inactive metabolites (BARNETT and FIORE, 1971 a, b, 1973). Cross-tolerance does not, apparently, appear to phenobarbital or meprobamate (BARNETT and FIORE, 1971 b, 1973). Although cross-tolerance (on acute tests) between diazepam and meprobamate or phenobarbital is not readily apparent, both chlordiazepoxide and diazepam are becoming the drugs of choice in the treatment of alcohol withdrawal. The agents appear to relieve all the signs and symptoms of withdrawal readily and without hazard (cf. reviews and reports of KAIM, 1973; KAIM and KLETT, 1972; VICTOR, 1970; KISSIN, 1975). Of interest is the fact that all of the sedative/hypnotic agents and "minor tranquilizers" have been, as far as tested, effective in contravening the withdrawal syndrome following cessation of chronic ethanol ingestion.

In reverse fashion, many alcohol abusers are developing secondary habits to chlordiazepoxide (cf. FINER, 1970 for example) or diazepam in a fashion similar to earlier combined use of barbiturates. Diazepam, consonant with clinical evidence, effectively reduces the central excitatory actions characteristic of withdrawal from chronic alcohol administration in experimental animal preparations. [Of some interest is the study of GUERRERO-FIGUEROA et al. (1970 b) that whereas only three of the six normal animals developed seizures 22 h or more after withdrawal, all six "epileptic" cats exhibited seizures, starting 12 h after deletion of alcohol feeding. The most prominent EEG finding during withdrawal was said to be the increase in amplitude of the evoked response in the limbic system.]

Although overdose is unlikely to be sufficient to produce death, benzodiazepines cause—in low amounts—a hangover reportedly similar to that of other sleep medications, an impairment of intellectual and psychomotor performances, and drowsiness similar to, but more protracted than that after barbiturates.

Although tolerance and dependence may require repeated dosing for the appearance of full effects, a single dose can have prolonged actions, similar to those obtained with pentobarbital (DAVID et al., 1974; OSWALD, 1973). In very large doses, nitrazepam was found to induce EEG fast activity for up to 11 days afterwards with total whole night sleep reduced from more than 500 min to 200 min on the 12th night. After a large dose of chlordiazepoxide there was a rebound insomnia for more than a month (see OSWALD, 1973; KALES and KALES, 1974; KALES et al., 1974). Similar long-lasting effects of a single dose of nitrazepam (or pentobarbital) have been documented using rhesus monkeys (DAVID et al., 1974).

The benzodiazepines exhibit both tolerance and physical dependence and the magnitude of both are dose-dependent (see reviews of GARATTINI et al., 1973; GREENBLATT and SHADER, 1974 a, b; HOLLISTER et al., 1961); the effects at least on i.v. administration are reported to be pleasant and relaxing (DUNDEE and PANDIT, 1972; KORTTILA and LINNOILA, 1975 a, b). The subjective effects of diazepam, although similar to ethanol, are apparently sufficiently different in areas of greater feelings of relaxation and reduced concentration whereas equal depression of motivation and less impairment of motor function has been reported (HAFFNER et al., 1973).

The withdrawal syndrome following cessation of chronic use of large doses is stated to be qualitatively similar to that observed with barbiturates as originally studied experimentally with chlordiazepoxide by HOLLISTER et al. (1961); see bibliography of WEISE and PRICE (1975). GREENBLATT and SHADER (1974b) review the clinical data and confirm the dependence-producing capability of many of the diazepines and their relief by reinstituting the drug—including diazepam, chlordiazepoxide, nitrazepam, flurazepam or oxazepam. The withdrawal syndrome may well vary as a function of duration of drug action with the shorter-acting agents exhibiting more severe effects. The abuse liability may well be a function of the rapidity of onset of subjective drug effects as emphasized for other drugs and recently by BLIDING (1974) for benzodiazepines. Thus, diazepam would appear to have a much greater abuse liability than, for example, chlordiazepoxide (see also KISSIN, 1975; IRWIN, 1973). The withdrawal syndrome may have some unique features, including a protracted time course and a preponderance of hallucinatory phenomena (see FLOYD and MURPHY, 1976; DYSKEN and CHAN, 1977; PRESKORN and DENNER, 1977). Evidence for prenatal physical dependence has appeared (REMENTERIA and BHATT, 1977).

Animal studies provide confirmation of dependence potential for benzodiazepines. Physical dependence with apparent tolerance and drug self-administration has been demonstrated for chlordiazepoxide in monkeys similar to secobarbital (YANAGITA and TAKAHASHI, 1970; FINDLEY et al., 1972) and similarly with diazepam and oxazepam (YANAGITA and TAKAHASHI, 1973). Other studies include the demonstration that rats will lever press to administer medazepam intragastrically through an indwelling cannula (GOTESDAM, 1973).

Many observers conclude that the dependence risk of benzodiazepines is low, but no one knows. A simple calculation gives some cause for possible concern. If the risk of dependence were only 0.1% of those receiving a prescription (and excluding those in hospitals), the number of newly dependent individuals annually—based on 1972 data—would approximate 27,000 individuals a year. Assuming a stable population of 77,000,000 receiving benzodiazepines, (cf. GREENBLATT and SHADER, 1974b) this results in a figure of 77,000 (see also BALTER et al., 1974).

Such a postulated risk of 0.1% is quite low in view of estimates of alcohol problems of some 4–12% of the adult population and some 20–40% of all psychiatric hospital admissions for other than solely alcohol problems. BOWES (1965) reports overt dependence in 1% of 500 outpatients receiving diazepam; in contrast, REGGIANI et al. (1968) estimate dependence in less than one in a million users.

Data on use patterns of these agents and all sedative/hypnotics are direly needed. For example, it may well be that for some individuals, therapy with benzodiazepines could conceivably result in a reduction of the abusive use of alcohol (DITMAN, 1961; KISSIN, 1975), even though present impressions of most clinicians seem to be to the contrary (see SHAW et al., 1975). On the other hand, admissions for drug dependence on benzodiazepines are becoming of increasing importance (Medical Letter, 1975; KALES and KALES, 1974; KALES et al., 1974; SWANSON et al., 1973; EWING and BAKEWELL, 1967; FLOYD and MURPHY, 1976; DYSKEN and CHAN, 1977; PRESKORN and DENNER, 1977; REMENTERIA and BHATT, 1977) (see also BURKE and ANDERSON, 1962; SELIG, 1966; GLATT, 1967; GORDON, 1967; EDGLEY, 1970; CLARE, 1971; BRAMSON, 1973; FINK et al., 1974; MODELL, 1974; MOFENSON and GREENSHER, 1974; MILLER, 1974; WOODY et al., 1975; bibliography of WEISE and PRICE, 1975). Certainly, agents in this class are, in general, much more rational sleep medications than barbiturates, as emphasized by KOCH-WESER and GREENBLATT (1974).

In spite of many disclaimers, there are appreciable numbers of people who habitually ingest benzodiazepines; for the most part, these are obtained initially via one or more prescriptions (annotated bibliography of WEISE and PRICE, 1975; see, e.g. WOODY et al., 1975; BRAMSON, 1973). Anecdotal evidence and survey data indicate that diazepam, at least, enjoys an appreciable illicit traffic; certainly the agents are known to the alcoholic population as drugs with useful and appreciated effects; the New York Times Magazine devoted a major article to "Valiumania." Even subjects of driving test experiments reportedly found the effects of i.v. diazepam pleasant (KORTTILA and LINNOILA, 1975a, b).

At the present rate of increase for diazepam and new benzodiazepines, all other sedative/hypnotic agents will be, essentially, replaced by benzodiazepines. And, at this present rate of increased use, most of Western society will shortly be "benzodiazepined" most of the time!

The benzodiazepines—such as Librium and Valium—might have been better named Soma à la Aldous Huxley. The extent of current use in the United States suggests that they are approaching the extent of use envisioned for Soma in "Brave New World."

Little thought has apparently been given to the implications of the fact that almost a majority of the population are fairly continuously under the active influence of such agents—or if not them, then ethanol, marihuana, or barbiturates with which they may act synergically. Some of the broader implications were at least touched on in a small volume edited by EVANS and KLINE (1971)—"Psychotropic Drugs in the Year 2000;" in contrast, the essay by ZINBERG (1974) tended to confuse previous and present drug use with the questions of values, goals and desirable/useful drugs. To illustrate, the complex and special issues of operating vehicles and machinery while under the influence of ethanol or any of these potentially disruptive drugs are only beginning to be addressed.

It is suggested that a prime need remains for the explicit determination of such simple facts as the spectrum of strengths of habit entailed by the chronic use of benzodiazepines, the propensity and effects of a chronically incrementing dose, the consequences of discontinuation, and the possibility of i.v. self-administration. It may well be that chronic use of benzodiazepines will remain of minor personal and social consequence—but this has not been proven. In fact, the proposition that these "minor tranquilizers" will reduce interpersonal aggressive behavior is not generally true, and the reverse may in fact be a common consequence of use.

Of special interest in this regard is the exploration of combinations—such as amphetamine and chlordiazepoxide—analogous to the amphetamine-barbiturate mixtures or cocaine-heroin sequences of years past (see RUSHTON et al., 1973).

The problems of multiple interacting factors of environment and symptoms and agent in assessing clinically the potential for personal abuse of the antianxiety drugs are well illustrated by two papers in a benzodiazepine conference (HOLLISTER, 1973; RICKELS, 1973). HOLLISTER concludes that most of the agents of the class may be found to be "better" than placebo, but differences among the agents may be slight. He emphasizes that the relative lack of tolerance to the drug and long duration of action make chlordiazepoxide a poor candidate for production of physical dependence.

"Although I was able to show many years ago that physical dependence to chlordiazepoxide could occur, it has been generally overlooked that it took extremes of dose and duration of treatment to produce these signs (HOLLISTER et al., 1961)." The bland assurance of a "very low risk of abuse" of benzodiazepines of the authoritative panel cited in GARATTINI et al. (1973) is perhaps calming, but seriously misleading. Moreover, the literature reviews seem to this reviewer to play down any possible negative side effects of the benzodiazepines. For example, WULFSOHN (1973) allots less than one-third of a page to "side-effects" and one-third to tolerance and toxicity in a review text of 15 pages. In fact, what is reported seems to have been adapted from the manufacturer's package insert for one of the compounds.

GREENBLATT and SHADER (1974b) are appreciably more cautious in their assessment of the liability to abuse and dependence of the benzodiazepines including a table decribing case reports of abuse or dependence to date. A few of the more recent reports sustain this cautious view (FLOYD and MURPHY, 1976; PRESKORN and DENNER, 1977; DYSKEN and CHAN, 1977; REMENTERIA and BHATT, 1977). The reports of dependence or abuse with relatively low levels in the presence of other reports of relatively heavy use without dependence (DITMAN and BEJOR, 1966) make it difficult to reach a comprehensive conclusion. It is perhaps attributable to pharmacokinetic differences in as much as plasma levels of diazepam determined acutely, or chronically, are subject to wide variability and unknown sources of fluctuation. The statements regarding tolerance or interactions with other drugs are, therefore, subject to serious reservations in the absence of concurrent blood level determination (cf. BIANCHI et al., 1974).

Although the benzodiazepines have had extensive prescription use, the necessity for their inclusion in routine institutional pharmacopeia has been brought into question. KEELER and MCCURDY (1975) report that unavailability of the antianxiety drugs, benzodiazepines and hydroxyzine, as Medicaid prescription items statewide resulted in, essentially, no substitution by other agents in over two-thirds of the cases. The authors conclude that physicians consider these drugs less hazardous than other psychoactive agents, yet do not feel that the drug effects are sufficiently important to the patient or doctor to warrant substitution by other agents or worth the extra cost and effort. These findings suggest that the majority of prescriptions for anxiety were of marginal value to begin with. In the face of certain, but of unknown significance, risk of dependence on such, the Hippocratic injunction might well be applied routinely. The caveat regarding restricting prescriptions to those who might abuse them would, logically, rule out most instances of their use in individuals with psychoneurotic anxiety since these are the very individuals who are already at risk!

E. Conclusions and Postulates

I. Sites of Action

It is suggested that each of the substances in the general classes of sedative/hypnotic/minor tranquilizer, in producing given behavioral or mental states, acts on specific neuronal systems. This conclusion follows a corollary assumption that only a limited number of specifically identifiable neurons and neuronal pathways are involved in each behavior, thought and mental process.

II. Mechanisms of Action

After reviewing available evidence, it is tentatively concluded that each chemical class of the agents acts via different molecular mechanisms and cellular sites of action. Thus, ethyl alcohol is thought to have a different molecular mechanism(s) of action from the benzodiazepines, from the barbiturates, etc. Even though there may be some mechanisms that are common across drug classes, there are a significant number that are not.

Even for the general anesthetics, there is more than one such site and mechanism of action. For example, ethanol may increase presynaptic inhibition by potentiating GABA action; this presynaptic action presumably underlies some of the analgesia and paresthesias induced by ethanol, one action among many of the actions of ethanol. Evidence to date suggests that among these actions the following can be identified:

1. Ethanol acts to inhibit conduction in small or repetitively active neurons, to prolong and then block cholinergic transmission, and to increase presynaptic inhibition.

2. Sedative barbiturates act (a) as GABA agonists and/or prolong GABA-mediated presynaptic action, (b) to antagonize the actions of some excitatory transmitters by postsynaptic actions at some sites, (c) may act as well by blocking a naturally occurring barbiturate analog, (d) act, as anticonvulsants, on the mechanisms giving rise to repetitive, synchronous cortical discharges.

3. Inhalation anesthetics can be divided into 2 or 3 groups with apparently different spectra of nervous system activity—e.g., one group that includes nitrous oxide, cyclopropane, and halothane can be postulated to act primarily presynaptically, whereas that group including ether, methoxyflurane, and trichloroethylene act postsynaptically;

4. The benzodiazepines appear to act to produce their acute effects by actions distinguishable from barbiturates, ethanol, or inhalation anesthetics. The spectra of nervous system activity vary significantly among the various benzodiazepine derivatives and the possible mechanisms of action include interaction with the specific processes of prostaglandin modulation of transmitter release mechanisms involved in the phenomenon of PTP.

5. It is not unlikely that more specific delineation of site(s) and mechanism(s) of action will be discovered in the future. Such studies require the designation of the behavior end action under consideration (e.g., in relief of anxiety or increased ease of mental association, or analgesia) and the relationship of neural substrate to both the drug effect and the behavior.

III. Subjective Phenomena

1. Individuals perceive and value certain drug effects, and these actions are concluded to be desirable. These drug effects include those feelings of intrinsic well-being directly induced by the drugs or due to learned/conditioned/shaped phenomena associated with the drug taking. Included in such subjective sensations are those of revelation and insight (psychedelic phenomena) as well as relief of anxiety and tension. A change in mood/feeling state may also take on desirable or reinforcing properties.

2. The sedative/hypnotic drugs may well have primary positive reinforcing properties, independent of learning or conditioning; it is postulated that such primary reinforcing properties are not necessarily directly perceived by the individual—i.e., the reinforcing qualities of drug intoxication states may or may not be synonymous with subjective perceptions of well-being or they may occur in addition to other subjective effects. [To illustrate, the effects of heroin or ethanol on first use may or may not be perceived as pleasant (see WILLIS, 1969) yet it is possible that such exposure may significantly predispose the individual to subsequent selection and use].

3. The reinforcing properties of these drugs may be in part a function of prior exposure to the drug effects, i.e., the drug action itself may prove to be the initiating stimulus for the appearance of dependence-producing qualities, over and above the acute addiction and withdrawal syndromes. It is suggested that a period of high dose intake (of sedative/hypnotics/anesthetics) with tolerance and physical dependence results in long-lasting changes in the nature of drug effects in a given individual and the susceptibility for subsequent dependence; that is, it is postulated that sedative/hypnotics may give rise to a protracted abstinence syndrome analogous to, but different from, that demonstrated for opiates.

4. In addition, it is suggested that there are innate, possibly inherited, individual differences in the degrees of both positive and negative reinforcement a given agent can produce. (Moreover, it should be noted that the neural mechanisms underlying the development for the "simple" behaviors designated as "habits" have not been identified.)

IV. Tolerance

1. *Induced metabolic tolerance* occurs with many of the drugs under consideration, but the potential magnitude of such tolerance and the mechanisms responsible are different for each individual compound; these differences appear to be manifestations of the molecular uniqueness of each compound and the biochemical specificity of metabolic systems.

2. *Innate functional tolerance* varies over a two- to four-fold quantitative range; tolerance to the syndrome, to the subjective and aversive effects as well as to the reinforcing properties varies markedly, both among individuals and among compounds.

3. *Acquired functional tolerance* varies in magnitude and time course as functions of the different behaviors and neural systems involved, the drug, the dose, and the pattern of administration.

It is postulated that the mechanism(s) of acquired tolerance involves many cells and utilizes the same or analogous neural network mechanisms as other biological adaptive processes, such as adaptation to extreme temperature, learning a motor act, or becoming habituated to changed sensory input, etc.

The acquired tolerance for motor disruptions commonly exhibits a rapid onset and is sharply limited by the maximal dose that can be given; in contrast, tolerance to other actions such as to sleep-inducing or anti-epileptic actions may be less marked and more gradual in development.

Tolerance, with feedback adaptations as a mechanism, may utilize no common or general "neural tolerance system;" rather, the tolerance potential may be intrinsic to many neural systems.

The degree and the nature of acquired functional tolerance to the subjective effects of any of these agents remain to be elucidated; such assessment is intimately confounded by the presence of a changed internal state assiciated with incipient withdrawal. Thus, the measurement and detection of subjective tolerance is intrinsically uncertain. For example, the "euphorigenic" activity of alcohol in an individual chronically taking alcohol varies markedly with respect to time after the last of chronic doses as well as dose, frequency and duration of prior administration.

Considerations of tolerance (as well as physical dependence) must specify dose, route, frequency, and full history of drug ingestion and tissue levels as functions of time and of behaviors and significant life events. For the most part, drug use and abuse, including tolerance and dependence, are dynamic, continuing processes with waxing and waning tissue levels, dosage taken, functional, behavioral and metabolic tolerance, and the variations of human behaviors. These changes and the sporadic nature of events constitute the reality of drug use and abuse.

V. Dependence

Dependence may simply be defined as the repeated personal ingestion of intoxicating doses of a drug or drugs. The appearance of signs and symptoms upon abrupt withdrawal from chronic ingestion of a sedative/hypnotic agent denotes the presence of *physical dependence* and includes tremor, anxiety, delirium, hyperthermia, nausea/vomiting, seizures, and hallucinations. On the basis of incomplete data, it can be extrapolated that all of the current agents denoted as sedatives/hypnotics/inhalational anesthetics have the potential to produce physical dependence upon repeated administration of maximally tolerated doses.

The mechanisms of physical dependence underlying the signs and symptoms of physical dependence are not known. It is suggested that both tolerance and physical dependence begin to occur within the first hour of the initial administration of the first dose, and have a time course such that acute tolerance develops progressively over 6 h and blends, in time with chronic acquired behavioral (functional) tolerance.

Physical dependence on sedative/hypnotic agents appears to develop over hours, days and weeks and is *not* solely the overshoot of the adaptive processes of tolerance (see e.g., GOLDSTEIN, 1975a, b) although acquired behavioral tolerance appears to be basically a complex of adaptive processes. It is suggested that upon removal of the agent, the overshoot of adaptive processes is a readily applicable mechanism only for the tremor, "shakes", and anxiety. It is also plausible that the symptoms and signs of the alcoholic hangover represent a mild form of an "acute abstinence syndrome."

Further, it is concluded that the hallucinations, delirium, and generalized seizures occurring as the withdrawal syndrome have a much more complex origin. Among the speculations on their pathophysiology are:

1. The culmination of the sustained depletion of essential neurohormones
2. The accumulation of "toxic" materials either in the CNS or systemically
3. Secondary and tertiary consequences of other types of neural derangement such as long-lasting sleep derangements (GROSS et al., 1966, 1973; OSWALD et al.,

1969) or electrolyte disturbances such as hypocalcemia and the hypomagnesia studied by VICTOR and co-workers (VICTOR, 1973).

4. Manifestations of phenomena analogous to disuse supersensitivity (SHARPLESS and JAFFE, 1969), the utilization alternative or redundant neural pathways (MARTIN, 1966) or tissue growth, hypertrophy and atrophy (cf. MANDELL, 1975a).

VI. Incidence and Selection of Agent

The agent used for self-administration for incidental or regular chronic use is both a personal and a social matter, in addition to the constraints of pharmacokinetics of a drug's actions.

To date, there have been no comparative studies of the simple habit-forming, subjective desirability or dependence-producing liability among drugs of the sedative/hypnotic/anesthetic classes. Although this chapter reflects the reality of the fact that some people at some time and with some frequency have self-administered one or more of these agents, the degree of use varies from small doses infrequently to nearly immobilizing doses taken repeatedly during a given day and over days to months and years.

Whether or not animal studies of self-administration or preference analogously reflect human situations remains to be demonstrated, inasmuch as with this class of compounds, animals exhibit marked differences related to strain, species, and individual, with respect to preference and degree to which self-administration is undertaken (e.g. WINGER and WOODS, 1973; YANAGITA and TAKAHASHI, 1970; WEEKS, 1971; MELLO, 1973b). This wide variability among individuals in their more-or-less "natural" preference may well accurately depict the human situation, and as with alcohol use and abuse, the degree to which a given individual uses the agent to an extent that they or others describe it as "abuse" or dependence depends on a fairly large number of factors (cf. e.g., CHEIN, 1969a, b; MADSEN, 1974; CAHALAN and ROOM, 1974; MELLO, 1975; ROBINS, 1973; VAILLANT, 1969; CHAMBERS et al., 1975).

Among these many factors appear to be the positive subjective and reinforcing properties of the drug (as opposed to aversive properties), the presence of manic and anxiety states of endogenous origin or due to other drugs or events such as caffeine or amphetamines or the abstinence syndrome, the availability and social approbation or sanctions for use, as well as the "role" and "game" that drug taking and intoxication may play in interpersonal and group affairs.

The above list is obviously only a partial, sketchy outline but it does provide the contexts of the "natural" experiments, which, although they provide guides for immediate social and clinical intervention, do not provide substantial information regarding the relative potential for use/abuse by humans of a given drug of a class as compared with other drugs.

VII. A Classification of the Potential Hazards of Self-ingestion of Psychoactive Agents with Sedative and Hypnotic Actions

Underlying the scientific interest in the mechanisms of dependence are the individual and societal concerns for human well-being, as well as the avoidance and the treatment of a variety of hazards.

Inasmuch as these hazards, pernicious effects, or negative consequences constitute a basic stimulus for interest in such drugs, the delineation and classification of the various potential hazards can facilitate the explicit determination of the nature and magnitude of such risks.

Work directed at defining such a framework is presented below, recognizably incomplete and tentative, in outline format. As a sweeping generalization, it can be postulated that all of the direct and indirect actions of the drugs, as well as all behaviors and environment of the drug taking and action, have the potential of presenting hazards to the individuals, their environment or other people.

1. The Individual and the Drug

a) Hazard from Other than the Central/Primary, Predictable Drug Actions

(These are mentioned primarily for the sake of completeness).

1. Allergic, hypersensitivity reactions
2. Organ toxicity—other than brain/nervous tissue (e.g., liver damage with chronic use of alcoholic beverages, obesity, malnutrition, etc.)
3. Toxicity from vehicle, mode of administration (e.g., suffocation occasioned by inhalation from plastic bags used for solvent administration or thrombophlebitis from intravenous administration), or adulteration, or inadvertent poisoning due, e.g., to ignorance or to mislabeling qualitatively or quantitatively.

b) Hazards Associated with Intoxicant (Psychoactive) Drug Administration in an Acute or Single Dose

The primary hazards include quantitative and qualitative aspects of the drug dose and effects that are not "as intended," e.g., coma (instead of sleep, or relief of anxiety, as with ethanol or barbiturate or diazepam), ataxia, slurred speech, amnesia and blackouts, appearance of intoxication, unpredictable or uncertain reaction to ethanol or other drugs, altered (impaired?) judgement, impaired coordination, incapacitation and reduced ability to work, to read, to learn, to perform, to adjust, or to do anything else.

The assessment of the degree of hazard or incapacitation requires relating to the magnitude/nature and duration of drug effect (i.e., to dose and route and history of other use, etc.) and to interaction with "things" and other personal behaviors—e.g., secondary hazards to the person in the form of accidents and trauma, at home, on the street, in vehicles.

Among the hazards are the negative effects following the subsidence of an acute drug action, i.e., hangovers. Hangover is used to denote at least two distinguishable syndromes—sustained drug action, for example, the morning-after mental sluggishness from a sleeping pill, and the "acute withdrawal syndrome" as exemplified by the morning after a drunken debauch. Hangover incapacitation following ethanol includes nausea, vomiting, depression, anxiety, remorse, tremor, thirst.

The hazards can also include an altered response(s) to drug administration some days or months subsequently, due to the prior drug experience.

c) Hazards Associated with Repeated Administration

These include all of the above plus:

1. Organic alterations, such as chronic brain changes in the adult individual or in the fetus, and other physical alterations such as obesity, malnutrition.

2. The activity (habit?) itself of repeated self-intoxication in comparison with alternative behaviors (obviously a function of repetition frequency) [see below].

3. Tolerance and increased sensitivity (possibly) in partial withdrawal coupled with unpredictable action of subsequent doses—e.g., the apparent decrease in safety margin for lethality for acute toxicity with barbiturates in individuals with acquired tolerance. The degree to which this syndrome applies generally to barbiturates or to benzodiazepines is not known. The relationship between tolerance and physical dependence with ethanol and sedative/hypnotics remains problematic. At the least, tolerance can occur without appreciable physical dependence, and dosage regimens that induce physical dependence are uniformly accompanied by tolerance. Nevertheless, there are precise limits to acquired tolerance (MAYNERT and KLINGMAN, 1960).

4. Qualitative alteration in mood and behavior and dynamics of mood and behavior, such as severe augmentation of magnitude of mood. The increasing depression seen during a drinking bout is a case in point, coupled with the altered responses to a drink, such as temporary relief of the depression, relief of morning shakes or hangover "anxiety," or the induction of sleep in the insomnia and disturbed sleep of withdrawal.

5. "Adoption of the intoxicant option" as a "personal system of experience management" as a primary means of recreation, of adaptation, of experiencing life and environment, of solution to personal concerns, for revelation and inspiration, for escape, for treatment of behavioral/mood problems, exploration and development, or to treat "illness," pain, fatigue, and anxiety.

6. Ritual habits—the meaning, significance and hazards of drug use relate also to the perceived meaning and import of the *act* of taking a drug or alcohol plus the mode and ritual of administration. Ritual behavior may also be viewed as a form of habit. Habit, so-called psychic/psychological dependence, is manifested as the propensity or drive to take the drug, plus the consequences of disruption of any ingrained habit such as personal dissatisfaction, irritability, disruption, sense of frustration and anxiety that follows the failure to conform to the habit [such as tea for the English, lunch or dinner at a specific time, a cocktail before dinner, brushing the teeth, bathing, grooming, etc. The strength of a habit can be measured in comparison with respect to the degree to which it can compete with other drives and behaviors. The attention and frame of reference given to "habits" has varied markedly over the years (see e.g., John Dewey's "Human Nature and Conduct," 1922).]

7. Dependence—physical dependence is inexorably associated with "habit" or "psychological dependence;" the manifestation of physical dependence to sedative/hypnotics is the withdrawal syndrome which may include anxiety, depression, tremor and shakes, nausea and vomiting appearing initially as drug levels are falling (barbiturates, benzodiazepines, ethanol) followed by the sleep disturbances of insomnia and difficulty sleeping, hyperventilation (with hypocalcemia and hypomagnesemia), autonomic hyper-reactivity, delirium, hallucinations, and generalized seizures. The early withdrawal signs to physical dependence on sedative/hypnotics are readily

reversed by the same or similar drugs—a fact that presumably predisposes to chronic drug taking. However, the withdrawal signs that appear some days later are not so readily recognized as such. The relative efficacies of barbiturates. minor tranquillizers and other sedative/hypnotics in treating or terminating late-occurring withdrawal signs and symptoms have not been well established.

Primary concerns for understanding the hazards and implications of chronic sedative use focus on identifying the mechanisms and drives and stimuli controlling ingestion and escalation of the dose—i.e., the mechanisms used for individual personal prediction of drug effects and the provision for control or management of drug effects.

8. Interaction of drug action, personality and mental/psychological development—it is a truism that each intoxicant dose alters the learning and adaptive functioning of the individual, to some degree irreversibly, just as each experience and behavior puts a lasting mark on the biology and behavior. More important is the realization that the magnitude of drug effect, the frequency and the regularity of use are all important determination of the magnitude of the interactive effect of intoxicant psychoactive drug use, subjective effects and subsequent behavior. (The purposeful provision for "drug-free periods" from all kinds of drugs may well be worthy of serious consideration for alcohol, antianxiety, psychedelic and, for that matter, all kinds of drugs.)

An illustration is perhaps in order. The various intoxicants facilitate or evoke thought processes that might be characterized as nonlinear, as fantasies, fantastic, or magical, free-floating, associative, or emotional; at least some measure of their general appeal is just this alteration in mood and thought, i.e., intoxication. Although such modes of thought might well appeal, and perhaps be functional, to the painter, the creative writer or poet, they may seriously hamper the accountant, the surgeon, the auto-mechanic with a tough repair problem, or the bank manager. Moreover, the adoption of the more magical thought processes as a routine method of solving personal problems results inevitably in some increase in irrationality and unrealistic assessments, decisions and actions. In the sense of using drugs for ulterior purposes of changing thought ["the intoxicant option" (e.g., Low, 1976)], the successful use of one intoxicating drug most certainly predisposes toward the personal use of other drugs claimed to be desirable as intoxicants.

Recapitulating, the prime hazards of personal intoxicant use in probable order of importance in the context of just the individual and the agent are:

1. "Doing drugs" or drinking rather than something else, plus development of the *habit*;

2. The adoption of the intoxicated perspective, coupled with nonadaptive thought and behavior patterns;

3. The hazards of the various incapacitations and disruptions of human capabilities characterizing the intoxication on both the acute and chronic scales, including physical dependence, disruption of work or driving performance;

4. The various forms of potential organic damage and unpredictable adverse interactions with other drugs, medicaments, foods, and social activities.

2. Hazards of Intoxicant Drug Use in the Context of the Society and Social Setting, as Well as the Individual and the Drug

Individual behaviors and their consequences take place, for the most part, in the social setting. The social environment provides the history, the information, the introduction, and the context. It is frequently overlooked that people *learn* to use/administer drugs and they learn the attitudes, the mechanics, the settings, and the sources. Directly and indirectly, a very small number learn by innocent exploration, whereas most learn, primarily, directly from other people and secondarily from various sources of communication—newspapers, magazines, TV, movies, books, educational programs, etc.

1. Intoxication per se that damages/hurts others: Failure to meet responsibilities of dependents, a failure to accomplish needed work or expected activity, such as that related to job and income, work, personal service and interaction, family, communication—local and extended. The extreme case would be the individual who because of sustained chronic intoxication per se is noncontributory to society, yet is dependent on it for sustenance and protection.

2. Hazards of damage to the offspring: *Prenatal*—physical dependence and disturbances of neural development recognized as the "fetal alcohol syndrome;" to the degree that this syndrome is the consequence of derangements of neural development consequent to disordered neural activity, any prenatal exposure to sedatives would be expected to result in similar mental retardation, disordered facial and brain development. The degree to which sedative drug influences can be visited on the offspring and their offspring is not a trivial concern. Suggestive evidence exists not only for opiates but also for influences of effects of alcohol ingestion by the father resulting in alteration in the offspring and their reaction as adults to ethanol.

Postnatal—hazards of chronic intoxication of parents are well known and relate to neglect, setting a poor example, plus secondary consequences of inadequate nutrition, nurturing, love, and training.

3. Overt antisocial or anti-individual actions consequent to the intoxicated drug state include a wide spectrum of hazardous or undesirable consequences. Aggressive, violent, antisocial acts of personal violence, fraud and theft of property *may* be the consequence of, or associated with, ethanol, barbiturate or stimulant intoxication; faulty judgement is frequently said to be consequent to alcohol/drug use, e.g., in use of weapons, automobile, airplanes, boats, or by those in positions of authority or responsibility (e.g., the potential misjudgement of political or military leaders, of attorneys, of physicians, or family members in positions of responsibility, etc.). Certainly, this is a serious hazard during severe acute intoxication and with chronic mental deterioration, such as Korsakoff's psychosis or senile dementia.

Disruption of domestic tranquility by aberrant or aggressive behavior and by the intrinsic fact of failing to behave "as expected" in society, i.e., by being deviant, is a frequent presenting complaint about excessive alcohol use. Related are expenditures required of individuals and social institutions to care for those whose chronic or excessive use makes them of public concern, such as the public inebriate, petty crime by alcohol users, the psychiatric manifestations such as depression and suicide, as well as "accidental" suicides occasioned by drug use. Psychoses may also be aggravated, as well as marked in symptomatology, by excessive alcohol (and sedative/hypnotic) use.

Perhaps of more general social concern are the irritations, frustration and inappropriate responses (inappropriate to the majority) of the individual intoxicated in public, i.e., being a nuisance and irritating to others. (Note should be made of the likely secondary and tertiary consequences of each of these more primary actions, such as rejection of, or attack on, the offending individual. An individual's use clearly has hazards in terms of the drug or intoxication producing an adverse effect not only on children but on other members of the family group.)

4. Hazards related to patterns of drug and alcohol use as social phenomena: Drug use and intoxication of an individual present a pattern and example for others to emulate, whereas excessive use may result in unproductive social responses (e.g., exaggregated legal penalties) with their own repercussions. The act, the ritual and the habit in an individual and in groups provide extensive social reinforcement of all actions associated with the agent's use. The drug and the act serve as the signal—the mark—of the person has given traits or belongs to a certain group, e.g., those hold the view that the "in person drinks his whiskey," or "only the deviant or low classes use marihuana," etc.

The agent, the brand, the act, the ritual, or the intoxication carry with them tremendous potential hazards of being viewed by society as either (a) "immune from harm" (such as the well-to-do, the "beautiful people," the "jet-set," "the educated," the "well-adjusted," etc.) and, thus, vulnerable by their rejection of the reality of their humanity and biology, or (b) stereotyping drug- or alcohol-using individuals as being queer, deviant, an enemy, a sinner, as mentally ill, etc., resulting in their persecution, incarceration, banishment, forced hospitalization or conversion, etc.

Drug use leads to cultural/social/individual *reinforcement* of going to a site, participating in ritual, ingestion/administration, and social experiencing of effects. The behaviors, including intoxication, are truly infective and may be hazardous; at the least, the activity may be viewed as nonproductive or as wasted time.

The presence and use of various drugs lead to the economic factors of production, distribution, sales, capital and labor investment, advertising and escalating press for greater use, production, sales and social involvement. In turn, appreciable social/economic involvement leads to social and governmental regulation and control and taxation and bureaucracy, etc., etc.

In most societies the views and behaviors associated with drugs with the potential for intoxication are far more complex than this short outline might suggest. For example, the societal reaction and responses to the introduction of new agents and modes of behavior are extremely complex and are, at least, functions of the degree of "strangeness" for existing patterns, by whom introduced, and of the rate at which the changes occur, such that small changes introduced slowly encounter, in general, little opposition.

However, in times of sharp and marked changes associated with the appearance of a new drug with new actions and whose use is perceived to disrupt existing personal, family and social relationships, the societal response may well be intense, frustrated, and irrationally focused on passing rules, regulations and laws directed at those concluded to be responsible, such as the users, the promoters or dealers. In the presence of mores or laws providing persecution of such drug-using deviants, the possession, sale, or use of such drugs leads, expectedly, to the harm of such individuals.

On a milder note, the legal and social constraints on the use of alcoholic beverages by the youth of current United States society may result, where these are in effect, in young adults with little or no training or experience in how to drink safely such potentially toxic substances. The rather large number of inadvertent lethal poisonings by alcoholic beverages and/or sedative drugs in the United States is consonant with this view—a hazard, then, more directly, of the customs and laws surrounding intoxicant drug use in a society than of the drug(s) or their actions.

The habit- and dependence-forming property of drugs varies markedly with the individual, as revealed both in man and analogous animal studies. To what degree these differences relate to variations in general habit-forming or dependent physiology/psychology, to variations in reinforcing quality or degree, or to variations in the relative degree of aversive properties is not known and has been little studied.

The insights of Alcoholics Anonymous imply that the alcoholic person has a unique susceptibility to the addicting and abusive use of alcohol or of related sedatives. Their tenets are consonant with the postulate of an addiction-prone person, biology, or personality coupled with an ever-present potential for relapse in the "sober alcoholic." The implication of this assumption is that the hazards to both individual and to society with the intoxicating substances are intrinsic to the person with a unique predisposition or sensitivity.

An alternative view holds that the addictions and abuses have various motivations and origins with the unique addiction-prone individual constituting only a small percentage of the problems. For the many others, drug-related hazards of altered behavior can also be viewed in the context of behavioral manifestations, e.g.: suicide; schizophrenic withdrawal or acting out; manic/depressive affective disorders and acting out; personal hostility and aggression; family or job irresponsibility; antisocial behavior, such as driving while intoxicated in a fashion causing auto or plane accidents; social/medical dependence and irresponsibility; manipulatory interpersonal behavior and ideation ("Games People Play"); and self-medication for anxiety, depression, mania, feelings of guilt, worthlessness or powerlessness.

It can be seen that the conceptual considerations rest on distinguishing social pathology, from individual pathology, from the interaction between both individual and social pathology (cf. HILL, 1962).

Hazards in human and social terms relate directly to the prevalence, frequency, and magnitude of drug use and also the associated factors that operate to favor high prevalence of use, such as ready availability, profit, advertisement and promotion.

A primary hazard of personal drug use is their use in efforts to cope with personal problems such as anxiety, insomnia, mania, etc. Such use promotes dependence, a vicious cycle of drug/alcohol use. Further, their use tends to preclude more effective and safer alleviation of the problems. The classical dilemma centers on the degree to which it is beneficial for the organism to be pressured to adapt and learn to adjust to the trials of living such as anxiety and sleeplessness. This recalls us to the question of Soma of "Brave New World"—is a drug that relieves otherwise unavoidable pain, mild anxiety, and depression desirable, and should it be widely available and used? Does such use present the eventual hazard of individuals who have lost the capability of adapting to the unavoidable stresses of existence?

Social control systems regulate the availability and acceptability of experimentation with intoxicant drugs and also define the degree to which such use is viewed as deviant or accepted. If intoxicant drug use is widely accepted, personal use is likely to be high and controlled largely by the initial and habitual reinforcement contingencies as well as the degree to which the consequences of such use are perceived as desirable, undesirable or aversive. Paraphrasing MACANDREW and EDGERTON (1969) perhaps "Since societies, like individuals, get the sorts of [intoxicated] comportment that they allow, they deserve what they get."

A group of hazards frequently neglected are those associated with the promotion and perpetuation of what might be termed, the "Party Mentality" or as the "Search for the Cabaret" as a way of life. Among certain social groups in which intoxicant drugs or alcohol are used heavily a prime apparent activity is directed toward having exciting, good, hilarious, and novel party and interpersonal experiences. At least it is a superficial impression that prominent members of the entertainment world are part of a hectic social whirl of repeated parties—a kind of manic way of life. Such a way of life, even if satisfactory and productive for show-business people, does not seem to be a particularly productive mode of life for the entire populace. This fact presents the discrepancy of false expectations to large numbers of poeple who may be modeling their life on those in the public limelight. Moreover, significant personal hazards exist for those who adopt behaviors and life styles requiring the euphorigenic or antianxiety action of drugs to continue to function, e.g., in the event such drugs fail to satisfy the escalated search for new "highs," they are progressively more likely to fail to provide a meaning to existence or a sought-after escape.

Some secondary hazards are consequent to society's concern about drug and alcohol taking and to public and school education and persuasion efforts. Such effort "cut both ways" inasmuch as increased attention may serve more to promote use and exploration, and eventually nonadaptive behavior, than to discourage such use. Perhaps more subtle but important is the "labelling" and "instruction" of young people of the areas and modes by which attention, rebellion and deviance can be expressed, e.g., to advertize and dramatize the poolhall or drug-taking as the "trouble in River City," indicates with potentially strong impact those areas that adults are anxious about and how those anxieties can be aggravated, frustrated, or manipulated.

Acknowledgments: Appreciation is expressed to the cooperation of all of the staff of the Research Institute, and especially to secretarial assistance, to DONALD FABER and HEBE GREIZERSTEIN for their review of portions of text, and to EILEEN WILSON for invaluable help with reference materials. Thanks and appreciation beyond any possible recompense are extended to DONNA BALL who tirelessly typed and corrected almost endless text drafts and prepared and edited the extensive bibliography in addition to continuing to fulfill important secretarial duties.

WILLIAM R. MARTIN's stimulation, encouragement, and consultation are gratefully acknowledged.

Abbreviations

EPSP	excitatory post-synaptic potential
e.p.p.	end-plate potential
m.e.p.p.	miniature end-plate potential
m.e.p.c.	miniature end-plate current
IPSP	inhibitory post-synaptic potential

ACh	acetylcholine
GABA	gamma-aminobutyric acid
CNS	central nervous system
Na^+	sodium
K^+	potassium
Ca^{++}	calcium
PTP	post-tetanic potentiation
ATP	adenosine triphosphate
ATPase	adenosine triphosphatase
L-DOPA	*l*, dihydroxyphenylalanine
2,4-DNP	2,4 dinitrophenol
GAD	L-glutamate decarboxylase
GABA-T	4-amino:2-oxoglutamate aminotransferase
H cells	cells exhibiting hyperpolarization
D cells	cells exhibiting depolarization
CA	catecholamine
DA	dopamine
5-HT	5-hydroxytryptamine (serotonin)
E	epinephrine
NE	norepinephrine
AOAA	amino-oxacetic acid
n-DPA	n-dipropyl sodium acetate
AMP	adenosine monophosphate — usually referring to 3', 5' cyclic AMP
GMP	guanosine monophosphate
NADH	reduced nicotinamide adenine dinucleotide
TIQs	tetrahydroisoquinoline alkaloids
PAN	postalcohol nystagmus
Δ^9-THC	Δ^9-tetrahydrocannabinol
ECT	electroconvulsive shock therapy
ADH	alcohol dehydrogenase

ethanol and alcohol are used fairly interchangeably to denote ethyl alcohol

References

Abood, L. G., Ostfeld, A., Biel, J. H.: Structure-activity relationship of 3-piperidyl benzilates with psychotogenic properties (1). Arch. int. Pharmacodyn. **120**, 186—200 (1959)

Abreu, B. E., Emerson, G. A.: Susceptibility to ether anesthesia of mice habituated to alcohol, morphine or cocaine. Anesth. Analg. Curr. Res. **18**, 294 (1939)

Achari, G., Sinha, S. P.: Anti-tremor action of propranolol (Inderal). Jap. J. Pharmacol. **17**, 679—680 (1967)

Achari, G., Sinha, S. P.: Anti-tremor action of a new beta receptor blocking agent (I.C.I. 50, 172). Jap. J. Pharmacol. **18**, 370—371 (1968)

Adams, P. R.: The mechanism by which amylobarbitone and thiopentone block the end-plate response to nicotinic agonists. J. Physiol. (Lond.) **241**, 41 P—42 P (1974)

Adams, P. R., Cash, H. C., Quilliam, J. P.: Extrinsic and intrinsic acetylcholine and barbiturate effects on frog skeletal muscle. Brit. J. Pharmacol. **40**, 552—553 (1970)

Adler, M. W.: Changes in sensitivity to amphetamine in rats with chronic brain lesions. J. Pharmacol. exp. Ther. **134**, 214—221 (1961)

Adler, M. W.: Drug response following brain damage. In: Drugs and Cerebral Function. Springfield, Ill.: Charles C. Thomas 1970

Adler, M. W., Geller, E. B.: Factors to be considered in using brain lesions to study the central sites of action of narcotics. In: Adler, M. W., Samanin, R., Manara, L. (Eds.): Factors Affecting the Action of Narcotics. New York: Raven Press (in press), 1977

Adolph, E. F.: Physiological Regulations. Pennsylvania: Jaques Cattell Press 1943

Adolph, E. F.: The Development of Homeostasis. With special reference to factors of the environment. Proceedings of a Symposium held in Liblice near Prague, Sept. 15—17, 1960. London-New York: Academic Press, Inc. 1960

Adolph, E. F.: Origins of Physiological Regulations. London-New York: Academic Press Inc. 1968

Adriani, J., Morton, R. C.: Drug dependence: important considerations from the anesthesiologists' viewpoint. Anesth. Analg. Curr. Res. **47**, 472 (1968)

Agarwal, S. L., Bose, D.: A study of the role of brain catecholamines in drug induced tremor. Brit. J. Pharmacol. Chemother. **30**, 349—353 (1967)

Ahlquist, R. P., Dille, J. M.: Reactions of alcohol tolerant rabbits to pentobarbital, evipal, ether, amidopyrine and Metrazol. J. Pharmacol. exp. Ther. **70**, 301—308 (1940)

Ahtee, L., Eriksson, K.: Regional distribution of brain 5-hydroxy-tryptamine in rat strains selected for their alcohol intake. Ann. N.Y. Acad. Sci. **215**, 126—134 (1973)

Akera, T., Rech, R. H., Marquis, W. J., Tobin, T., Brody, T. M.: Lack of relationship between brain $(Na^+ + K^+)$-activated adenosine triphosphatase and the development of tolerance to ethanol in rats. J. Pharmacol. exp. Ther. **185**, 594 (1973)

Akil, H., Mayer, D. J., Liebeskind, J. C.: Antagonism of stimulation-produced analgesia by naloxone, a narcotic antagonist. Science **191**, 961—962 (1976)

Alkana, R. L., Parker, E. S., Cohen, H. B., Birch, H., Noble, E. P.: Reversal of ethanol intoxication in humans: an assessment of the efficacy of propranolol. Psychopharmacology **51**, 29—37 (1976)

Allan, F. D., Swinyard, C. A.: Evaluation of tissue tolerance to ethyl alcohol by alterations in electroshock seizure threshold in rats. Anat. Rec. **103**, 419 (1949)

Allison, A. C.: The effects of inhalational anaesthetics on proteins. In: Halsey, M. J., Millar, R. A., Sutton, J. A. (Eds.): Molecular Mechanisms in General Anaesthesia. Edinburgh-London-New York: Churchill Livingstone 1974

Alnaes, E., Rahaminoff, R.: On the role of mitochondria in transmitter release from motor nerve terminals. J. Physiol. (Lond.) **248**, 285—306 (1975)

Alper, M. H., Flacke, W.: The peripheral effects of anesthetics. Ann. Rev. Pharmacol. **9**, 273—296 (1969)

Altenburg, B. M., Michenfelder, J. D., Theye, R. A.: Acute tolerance to thiopental in canine cerebral oxygen consumption studies. Anesthesiology **31**, 443—448 (1969)

Amassian, V. E.: Studies on organization of a somesthetic association area, including a single unit analysis. J. Neurophysiol. **17**, 39—58 (1954)

Andersen, N. B.: Dual effects of general anesthetics on active and passive sodium fluxes and on sympathetic response in toad. In: Cellular Biology and Toxicity of Anesthetics. Baltimore: Williams & Wilkins Co. 1972

Anderson, R. J.: The influence of diphenylhydantoin on auditory and proprioceptive afferent cerebellar pathways. Ph. D. Thesis, Georgetown University 1975

Anderson, R. J., Raines, A.: Suppression by diphenylhydantoin of afferent discharges arising in muscle spindles of the triceps surae of the cat. J. Pharmacol. exp. Ther. **191** (2), 290—299 (1974)

Anton, A. H.: Ethanol and urinary catecholamines in man. Clin. Pharmacol. Ther. **6**, 462—469 (1965)

Armstrong, C. M., Binstock, L.: The effect of several alcohols on the properties of the squid giant axon. J. gen. Physiol. **48**, 265—277 (1964)

Armstrong, D. M., Cogdell, B., Harvey, R. J.: Effects of afferent volleys from the limbs on the discharge patterns of interpositus neurones in cats anaesthetized with α-chloralose. J. Physiol. (Lond.) **248**, 489—517 (1975)

Arrigo, A., Jann, G., Tonali, P.: Some aspects of the action of Valium and Librium on the electrical activity of the rabbit brain. Arch. int. Pharmacodyn. **154**, 364—373 (1965)

Aschan, G., Bergstedt, M., Goldberg, L., Laurell, L.: Positional nystagmus in man during and after alcohol intoxication. Quart. J. Stud. Alcohol **17**, 381—405 (1956)

Askari, A. (ed.): Properties and functions of $(Na^+ + K^+)$-activated adenosinetriphosphatase. Ann. N.Y. Acad. Sci. **242** (1974)

Aston, R.: Quantitative aspects of tolerance and post-tolerance hypersensitivity to pentobarbital in the rat. J. Pharmacol. exp. Ther. **150**, 253—258 (1965)

Aston, R.: Acute tolerance indices for pentobarbital in male and female rats. J. Pharmacol. exp. Ther. **152**, 350—353 (1966 a)

Aston, R.: Latent hypersensitivity to pentobarbital in the rat. Proc. Soc. exp. Biol. (N.Y.) **121**, 623—626 (1966 b)

Aston, R.: Barbiturates, alcohol, and tranquilizers. In: Chemical and Biological Aspects of Drug Dependence. Cleveland, Ohio: CRC Press 1972

Augustine, J. R.: Laboratory studies in acute alcoholics. Canad. med. Ass. J. **96**, 1367—1370 (1967)

Axelrod, J.: Cellular adaptation in the development of tolerance to drugs. Res. Publ. Ass. nerv. ment. Dis. **46**, 247—264 (1968)

Balter, M. B.: An analysis of psychotherapeutic drug consumption in the United States. In: Anglo-American conference on drug abuse: society's reaction—medicine's responsibility. London: Roy. Soc. Med., pp. 58—68, 1973

Balter, M. B., Levine, J.: Character and extent of psychotherapeutic drug usage in the United States. Proceedings of the V World Congress of Psychiatry, pp. 80—88. Amsterdam: Excerpta Medica 1971

Balter, M. B., Levine, J., Manheimer, D. I.: Cross-national study of the extent of anti-anxiety/sedative drug use. New Engl. J. Med. **290**, 769—774 (1974)

Banay-Schwartz, M., Teller, D. N., Gergely, A., Lajtha, A.: The effects of metabolic inhibitors on amino acid uptake and the levels of ATP, Na^+, and K^+ in incubated slices of mouse brain. Brain Res. **71**, 117—131 (1974)

Banna, N. R.: Potentiation of cutaneous inhibition by alcohol. Experientia (Basel) **25**, 619—620 (1969)

Banna, N. R.: Antagonism of barbiturate depression of spinal transmission by catechol. Experientia (Basel) **26**, 1330—1331 (1970)

Banna, N. R., Jabbur, S. J.: Pharmacological studies on inhibition in the cuneate nucleus of the cat. Int. J. Neuropharmacol. **8**, 299—307 (1969)

Barach, A. L.: Homeostasis: A physiologic and psychologic function in man. Perspect. Biol. Med. **17**, 522—528 (1974)

Barany, E. H.: Theoretical note concerning action of drugs on central nervous system. Arch. int. Pharmacodyn. **75**, 222—226 (1947)

Barbizet, J.: Human Memory and Its Pathology. San Francisco: W.H. Freeman and Company 1970

Barker, J. L.: Selective depression of postsynaptic excitation by general anesthetics, pp. 135—156. In: Fink, B.R. (Ed.): Progress in Anesthesiology, Vol. 1, Molecular mechanisms of Anesthesia. New York: Raven Press 1975

Barker, J. L., Gainer, H.: Pentobarbital: selective depression of excitatory postsynaptic potentials. Science **182**, 720—722 (1973)

Barker, J. L., Nicoll, R. A.: The pharmacology and ionic dependency of amino acid responses in the frog spinal cord. J. Physiol. (Lond.) **228**, 259—277 (1973)

Barker, J. L., Nicoll, R. A., Padjen, A.: Studies on convulsants in the isolated frog spinal cord. I. Antagonisms of amino acid responses. J. Physiol. (Lond.) **245**, 521—536 (1975 a)

Barker, J. L., Nicoll, R. A., Padjen, A.: Studies on convulsants in the isolated frog spinal cord. II. Effects on root potentials. J. Physiol. (Lond.) **245**, 537—548 (1975 b)

Barnes, C. D., Moolenaar, G.-M.: Effects of diazepam and picrotoxin on the visual system. Neuropharmacology **10**, 193—201 (1971)

Barnett, A.: Dopamine receptors and their role in brain functions. In: Current Developments in Psychopharmacology. Vol. 1. West Nyack, N.Y.: Spectrum Publications, Inc. 1975

Barnett, A., Fiore, J. W.: Acute tolerance to diazepam in cats and its possible relationship to diazepam metabolism. Europ. J. Pharmacol. **13**, 239—243 (1971 a)

Barnett, A., Fiore, J. W.: Acute tolerance to diazepam in cats: lack of cross-tolerance to methocarbamol and phenobarbital. Europ. J. Pharmacol. **14**, 301—303 (1971 b)

Barnett, A., Fiore, J. W.: Acute tolerance to diazepam in cats. In: Garattini, S., Mussini, E., Randall, L. O. (Eds.): The Benzodiazepines. New York: Raven Press 1973

Barry, H. III.: Prolonged measurements of discrimination between alcohol and nondrug states. J. comp. Physiol. Psychol. **65**, 352—355 (1968)

Barry, H. III.: Classification of drugs according to their discriminable effects in rats. Fed. Proc. **33**, 1814—1824 (1974)
Barry, H. III., Kubena, R. K.: Discriminative stimulus characteristics of alcohol, marihuana and atropine. In: Singh, J. M., Miller, L. H., Lal, H. (Eds.): Vol. 1: Drug Addiction: Experimental Pharmacology. Mount Kisco, N.Y.: Futura 1972
Bartholini, G., Keller, H., Pieri, L., Pletscher, A.: The effect of diazepam on the turnover of cerebral dopamine. In: Garattini, S., Mussini, E., Randall, L. O. (Eds.): The Benzodiazepines. New York: Raven Press 1973
Bauer-Moffett, C., Altman, J.: Ethanol-induced reductions in cerebellar growth of infant rats. Exp. Neurol. **48**, 378—382 (1975)
Beer, M., Chasin, M., Clody, D. H., Vogel, F. R., Horovitz, Z. P.: Cyclic adenosine monophosphate phosphodiesterase in brain: effect on anxiety. Science **176**, 428—431 (1972)
Begleiter, H., Branchey, M. H., Kissin, B.: Effects of ethanol on evoked potentials in the rat. Behav. Biol. **7**, 137—142 (1972)
Begleiter, H., Platz, A.: The effects of alcohol on the central nervous system in humans. In: Kissin, B., Begleiter, H. (Eds.): The Biology of Alcoholism, Vol. 2, pp. 293—343. New York: Plenum Press 1972
Begleiter, H., Porjesz, B., Yerre-Grubstein, C.: Excitability cycle of somatosensory evoked potentials during experimental alcoholization and withdrawal. Psychopharmacologia (Berl.) **37**, 15—21 (1974)
Belenko, S., Woods, S. C.: Physiological correlates of ethanol self-selection by rats. Physiol. Psychol. **1**, 155—157 (1973)
Bell, J. A., Anderson, E.: The influence of semicarbazide induced depletion of gamma-amino butyric acid on presynaptic inhibition. Brain Res. **43**, 161—169 (1972)
Bell, J. A., Anderson, E. G.: Dissociation between amino-oxyacetic acid-induced depression of spinal reflexes and the rise in cord GABA levels. Neuropharmacology **13**, 885—894 (1974)
Belleville, R. E., Fraser, H. F.: Tolerance to some effects of barbiturates. J. Pharmacol. exp. Ther. **120**, 469—474 (1957)
Bennett, P. B., Simon, S., Katz, Y.: High pressures of inert gases and anesthesia mechanisms. In: Fink, B. R. (Ed.): Progress in Anesthesia. New York: Raven Press 1975
Bennion, L. J., Li, T.-K.: Alcohol metabolism in American Indians and Whites. New Engl. J. Med. **294**, 9—13 (1976)
Bergmann, M. C., Klee, M. R., Faber, D. S.: Different sensitivities to ethanol of three early transient voltage clamp currents of Aplysia neurons. Pflügers Arch. **348**, 139—153 (1974)
Berkowitz, B. A., Ngai, S. H., Finck, A. D.: Nitrous oxide "analgesia": resemblance to opiate action. Science **194**, 967—968 (1976)
Bernstein, J., Videla, L., Israel, Y.: Role of the sodium pump in the regulation of liver metabolism in experimental alcoholism. Ann. N.Y. Acad. Sci. **242**, 255—267 (1974)
Berry, C. A.: A study of cortical afterdischarge in the rabbit. Arch. int. Pharmacodyn. **154**, 197—209 (1965)
Bethune, H. C., Burrell, R. H., Culpan, R. H., Ogg, G. J.: Preliminary notes on nitrazepam. N.Z. med. J. **65**, 613—615 (1966)
Bianchi, G. N., Fennessy, M. R., Phillips, J., Everitt, B. S.: Plasma level of diazepam as a therapeutic predictor in anxiety states. Psychopharmacologia (Berl.) **35**, 113—122 (1974)
Biscoe, T. J., Millar, R. A.: The effect of cyclopropane, halothane and ether on sympathetic ganglionic transmission. Brit. J. Anaesth. **38**, 3—12 (1966)
Bjerver, K., Goldberg, L.: Alcohol tolerance in individuals with chronic producer-gas intoxication. Quart. J. Stud. Alcohol **9**, 329—352 (1948)
Blaber, L. C., Gallagher, J. P.: The facilitatory effects of catechol and phenol at the neuromuscular junction of the cat. Neuropharmacology **10**, 153—159 (1971)
Blagoeva, P., Longo, V. G., Masi, I., DeCarolis, A. S.: Alterations in behavior, EEG, and brain cortical GABA content following prolonged administration and subsequent withdrawal of barbital in the rat. Behav. Biol. **7**, 755—760 (1972)
Blank, N. K., Meerschaert, J. R., Rieder, M. J.: Persistent motor neuron discharges of central origin present in the resting state. Neurology (Minneap.) March, 277—281 (1974)
Blaustein, M. P.: Barbiturates block sodium and potassium conductance increases in voltage-clamped lobster axons. J. gen. Physiol. **51**, 293—307 (1968)

Blaustein, M. P., Ector, A. C.: Barbiturate inhibition of calcium uptake by depolarized nerve terminals in vitro. Molec. Pharmacol. **11**, 369—378 (1975)
Bleeker, M., Ford, D. H., Rhines, R. K.: A comparison of ^{131}I-triiodothyronine accumulation and degradation in ethanol treated and control rats. Life Sci. **8**, 267—275 (1969)
Bliding, A.: Effects of different rates of absorption of two benzodiazepins on subjective and objective parameters. Significance for clinical use and risk of abuse. J. clin. Pharmacol. **7**, 201—211 (1974)
Bliss, T. V. P., Richards, C. D.: Some experiments with *in vitro* hippocampal slices. J. Physiol. (Lond.) **214**, 7 P (1971)
Bloedel, J. R., Roberts, W. J.: Functional relationship among neurons of the cerebellar cortex in the absence of anesthesia. J. Neurophysiol. **32**, 75—84 (1969)
Bloom, F. E., Costa, E., Salmoiraghi, G. C.: Anesthesia and the responsiveness of individual neurons of the caudate nucleus of the cat to acetylcholine, norepinephrine, and dopamine administered by microelectrophoresis. J. Pharmacol. exp. Ther. **150**, 244—252 (1965)
Blum, K., Calhoun, W., Wallace, J. E., Merritt, J. H., Geller, I.: Soporific action of ethanol in mice: possible role of biogenic amines. Pharmacol. Biochem. Behav. **1**, 271—276 (1973)
Blum, K., Wallace, J. E., Geller, I.: Synergy of ethanol and putative neurotransmitters: glycine and serine: Science **176**, 292—294 (1972)
Boggan, W. O.: Neuropsychopharmacology of methaqualone. Dept. Psychiatry and Biochem., Med. Uni. of S.C., Charleston, S.C. 29401. In: Program and Abstracts: Society for Neuroscience. Fourth Annual Meeting 1974
Boguslawsky, P. V., Nikander, P.: A difference between high- and low-drinking rats in effects of ethanol on ion movements in cerebral tissue. Acta physiol. scand. **84**, 12 A—13 A (1972)
Bokonjic, N., Trojaborg, W.: The effect of meprobamate on the electroencephalogram during treatment, intoxication, and after abrupt withdrawal. Electroenceph. clin. Neurophysiol. **12**, 177—184 (1960)
Bonfiglio, G., Falli, S., Pacini, A.: Results of the experimentation of new drugs in the therapy of alcoholism. Lav. neuropsichiat. Li. Fascicolo 11 (1972)
Booker, H. E.: Relation of plasma levels to clinical control. In: Woodbury, D. M., Penry, J. K., Schmidt, R. P. (Eds.): Antiepileptic Drugs, pp. 329—334. New York: Raven Press 1972
Bourne, J. P.: Nitrous Oxide in Dentistry. London: Lloyd-Luke 1960
Bowes, H. A.: The role of diazepam (Valium) in emotional illness. Psychosomatics **6**, 336—340 (1965)
Boyd, E. M.: Chlorpromazine tolerance and physical dependence. J. Pharmacol. exp. Ther. **128**, 75—78 (1960)
Boyd, E. S., Boyd, E. H., Brown, L. W.: The effects of some drugs on an evoked response sensitive to tetrahydrocannabinols. J. Pharmacol. exp. Ther. **189**, 748—758 (1974)
Boyd, E. S., Boyd, E. H., Muchmore, J. S., Brown, L. E.: Effects of two tetrahydrocannabinols and of pentobarbital on cortico-cortical evoked responses in the squirrel monkey. J. Pharmac. exp. Ther. **176**, 480—488 (1971)
Bramson, S. M.: Abuse of benzodiazepine tranquilizers. J. Amer. med. Ass. **225**, 749 (1973)
Branchey, M., Rauscher, G., Kissin, B.: Modifications in the response to alcohol following the establishment of physical dependence. Psychopharmacologia (Berl.) **22**, 314—322 (1971)
Brausch, U., Henatsch, H.-D., Student, C., Takano, K.: Effect of diazepam on development of stretch reflex tension. In: Garattini, S., Mussini, E., Randall, L. O. (Eds.): The Benzodiazepines. New York: Raven Press 1973
Brazier, M. A. B.: The electrophysiological effects of barbiturates on the brain. Physiol. Pharmacol. **1**, 219—235 (1963)
Breese, G. R., Cooper, B. R., Prange, A. J., Jr., Cott, J. M., Lipton, M. A.: Interactions of thyrotropin-releasing hormone with centrally acting drugs. In: The Thyroid Axis, Drugs, and Behavior. New York: Raven Press 1974 a
Breese, G. R., Cott, J. M., Cooper, B. R., Prange, A. J., Jr., Lipton, M. A.: Antagonism of ethanol narcosis by thyrotropin releasing hormone. Life Sci. **14**, 1053—1063 (1974 b)
Breese, G. R., Cott, J. M., Cooper, B. R., Prange, A. J., Jr., Lipton, M. A., Plotnikoff, N. P.: Effects of thyrotropin-releasing hormone (TRH) on the actions of pentobarbital and other centrally acting drugs. J. Pharmacol. exp. Ther. **193**, 11—22 (1975)

Breese, G. R., Prange, A. J., Jr., Lipton, M. A.: Pharmacological studies of thyroid-imipramine Interactions in animals. In: The Thyroid Axis, Drugs and Behavior. New York: Raven Press 1974c
Brezenoff, H. E., Mycek, M. J.: Central cholinergic involvement in barbiturate tolerance. Fed. Proc. **34**, 779 (1975)
Brilliant, L.: Nitrous oxide as a psychedelic drug. New Engl. J. Med. **283**, 1522 (1970)
Brimblecombe, R. W., Pinder, R. M.: Tremors and Tremorogenic Agents. Bristol: Scientechnica 1972
Brodie, B. B., Mark, L. C., Lief, P. A., Bernstein, E., Papper, E. M.: Acute tolerance to thiopental. J. Pharmacol. exp. Ther. **102**, 215—218 (1951)
Brody, M. J., Kadowitz, P. J.: Prostaglandins as modulators of the autonomic nervous system. Fed. Proc. **33**, 48—60 (1974)
Brody, T. M.: The uncoupling of oxidative phosphorylation as a mechanism of drug action. Pharmacol. Rev. **7**, 335—363 (1955)
Brody, T. M., Bain, J. A.: Effect of barbiturates on oxidative phosphorylation. Proc. Soc. exp. Biol. (N.Y.) **77**, 50—56 (1951)
Brody, T. M., Bain, J. A.: Barbiturates and oxidative-phosphorylation. J. Pharmacol. exp. Ther. **110**, 148 (1954)
Brooks C. McC., Eccles, J. C.: A study of the effects of anesthesia and asphyxia on the monosynaptic pathway through the spinal cord. J. Neurophysiol. **10**, 349—360 (1947)
Brown, B., Adams, A. J., Haegerstrom-Portnoy, G., Jones, R. T., Flom, M. C.: Effects of alcohol and marijuana on dynamic visual activity: I. Threshold measurements. Percept. Psychophys. **18**, 441—446 (1975)
Brown, C. R., Forrest, W. H., Hayden, J.: The respiratory effects of pentobarbital and secobarbital in clinical doses. J. clin. Pharmacol. **13**, 28—35 (1973)
Brunner, E. A., Chang, S. C., Berman, M. F.: Effects of anesthesia on intermediary metabolism. Ann. Rev. Med. **26**, 391—401 (1975)
Brunner, E. A., Passonneau, J. V.: The effect of inhalational anesthetic agents on brain metabolite levels. In: Cellular Biology and Toxicity of Anesthetics. Baltimore: Williams & Wilkins Co. 1972
Buchtel, H. A., Iosif, G., Marchesi, G. F., Provini, L., Strata, P.: Analysis of the activity evoked in the cerebellar cortex by stimulation of the visual pathways. Exp. Brain Res. **15**, 278—288 (1972)
Buchthal, F., Lenox-Buchthal, M. A.: Phenobarbital: relation of serum concentration to control of seizures. In: Woodbury, D. M., Penry, J. K., Schmidt, R. P. (Eds.): Antiepileptic Drugs. New York: Raven Press 1972, pp. 335—343
Buchthal, F., Svensmark, O., Simonsen, H.: Relation of EEG and seizures to phenobarbital in serum. Arch. Neurol. (Paris) **19**, 567—572 (1968)
Büch, H., Knabe, J., Buzello, W., Rummel, W.: Stereospecificity of anesthetic activity, distribution, inactivation and protein binding of the optical antipodes of two N-methylated barbiturates. J. Pharmacol. exp. Ther. **175**, 709—716 (1970)
Büch, H. P., Schneider-Affeld, F., Rummel, W.: Stereochemical dependence of pharmacological activity in a series of optically active N-methylated barbiturates. Naunyn-Schmiedebergs Arch. Pharmacol. **277**, 191—198 (1973)
Burke, G. W., Anderson, C. W. G.: Response to Librium in individuals with a propensity for addiction. A pilot study. J. Louisiana med. Soc. **114**, 58—60 (1962)
Burns, B. D., Robson, J. G., Welt, P. J.: The effect of nitrous oxide upon sensory thresholds. Canad. Anaesth. Soc. J. **7**, 411—422 (1960)
Busch, G., Henatsch, H.-D., Schulte, F. J.: Elektrophysiologische Analyse der Wirkungen neuroleptischer und tranquilisierender Substanzen (Phenothiazine, Meprobamat) auf die spinalmotorischen Systeme. Arzneimittel-Forsch. **10**, 217—223 (1960)
Butler, T. C., Mahaffee, C., Waddell, W. J.: Phenobarbital: studies of elimination, accumulation, tolerance, and dosage schedules. J. Pharmacol. exp. Ther. **111**, 426—435 (1954)
Cahalan, D., Room, R.: Problem Drinking Among American Men. New Haven, Conn.: College and University Press 1974
Calne, D., Chase, T. N., Barbeau, A. (Eds.): Advances in Neurology, Vol. 9, Dopaminergic Mechanisms. New York: Raven Press 1975

Campbell, M., Fish, B.: Triiodothyronine in schizophrenic children. In: The Thyroid Axis, Drugs, and Behavior. New York: Raven Press 1974
Cannizzaro, G., Nigito, S., Provenzano, P. M., Vitikova, T.: Modification of depressant and disinhibitory action of flurazepam during short-term treatment in the rat. Psychopharmacologia (Berl.) **26**, 173—184 (1972)
Cappell, H., Herman, C. P.: Alcohol and tension reduction. Quart. J. Stud. Alcohol **33**, 33—64 (1972)
Cappell, H., LeBlanc, A. E.: Conditioned aversion by psychoactive drugs: does it have significance for an understanding of drug dependence? Addict. Behav. **1**, 55—64 (1975)
Cardenas, H. L., Ross, D. H.: Morphine induced calcium depletion in discrete regions of rat brain. J. Neurochem. **24**, 487 (1975)
Carew, T. J., Kandel, E. R.: Acquisition and retention of long-term habituation in *Aplysia*: Correlation of behavioral and cellular processes. Science **182**, 1158—1160 (1973)
Carew, T. J., Pinsker, H. M., Kandel, E. R.: Long-term habituation of a defensive withdrawal reflex in Aplysia. Science **175**, 451—454 (1972)
Carlson, H. E., Robbins, J., Murphy, D. L.: The effect of lithium on thyroid iodine release in patients with primary affective disorder. Psychopharmacologia (Berl.) **35**, 249—256 (1974)
Carlsson, A., Engel, J., Svensson, T. H.: Inhibition of ethanol-induced excitation in mice and rats by α-methyl-p-tyrosine. Psychopharmacologia (Berl.) **26**, 307—312 (1972)
Carmichael, E. B., Thompson, W. D.: Effect of repeated administration of Delvinal sodium [5-ethyl-5-(l-methyl-l-butenyl) barbituric acid] to guinea pigs. Proc. Soc. exp. Biol. (N.Y.) **46**, 233—235 (1941)
Carmichael, F. J., Israel, Y.: Effects of ethanol on neurotransmitter release by rat brain cortical slices. J. Pharmacol. exp. Ther. **193**, 824—834 (1975)
Caspers, H., Abele, G.: Hirnelektrische Untersuchungen zur Frage der quantitativen Beziehungen zwischen Blutalkoholgehalt und Alkoholeffekt. Dtsch. Z. ges. gerichtl. Med. **45**, 492—509 (1956)
Cassano, G. B., Ghetti, B., Gliozzi, E., Hansson, E.: Autoradiographic distribution study of "short acting" and "long acting" barbiturates: ^{35}S-thiopentone and ^{14}C-phenobarbitone. Brit. J. Anaesth. **39**, 11—20 (1967)
Catchlove, R. F. H., Krnjević, K., Maretic, H.: Similarity between effects of general anesthetics and dinitrophenol on cortical neurones. Canad. J. Physiol. Pharmacol. **50**, 1111—1114 (1972)
Chalazonitis, N.: Selective actions of volatile anesthetics on synaptic transmission and autorhythmicity in single identifiable neurons. Anesthesiology **28**, 111—122 (1967)
Chamberlain, T. J., Halick, P., Gerard, R. W.: Fixation of experience in the rat spinal cord. J. Neurophysiol. **26**, 662—675 (1963)
Chambers, C. D., Brill, L., Inciardi, J. A.: Barbiturate use, misuse and abuse. J. Drug Issues **2**, 15—20 (1972)
Chambers, C. D., Inciardi, J. A., Siegal, M. A.: Chemical Coping: A Report on Legal Drug Use in the U.S. New York: Halsted Press Publications 1975
Chan, A. W. K.: Putative neurotransmitters in inbred mice: effect of ethanol. Pharmacologist **17**, 198 (1975)
Chanelet, J., Lonchampt, P.: Influence of intra-central microinjections of diazepam on cat spinal cord activity. Fifth Int. Congress on Pharm. July, 1972
Chapman, L. F.: Experimental induction of hangover. Quart. J. Stud. Alcohol Suppl. **5**, 67—86 (1970)
Chappell, J. B., Crofts, A. R.: Calcium ion accumulation and volume changes of isolated liver mitochondria calcium ion-induced swelling. Biochem. J. **95**, 378—386 (1965)
Chase, R.: The suppression of excitatory synaptic responses by ethyl alcohol in the nudibranch mollusc, *tritonia diomedia*. Comp. Biochem. Physiol. **50 C**, 37—40 (1975)
Chein, I.: Psychological functions of drug use. In: Steinberg, H. (Ed.): Scientific Basis of Drug Dependence. London: J & A Churchill, Ltd. 1969a
Chein, I.: Psychological, social and epidemiological factors in drug addiction. Rehabilitating the Narcotic Addict. Vocational Rehabilitation Administration. Washington, D.C.: U.S. Dept. of Health, Education and Welfare, U.S. Government Printing Office 1969b
Chen, C.-S.: A study of the alcohol-tolerance effect and an introduction of a new behavioural technique. Psychopharmacologia (Berl.) **12**, 433—440 (1968)

Cheng,S.-C., Brunner,E.A.: Two neurotransmitters in brain slices. In: Fink,B.R. (Ed.): Progress in anesthesiology, Vol. 1, Molecular Mechanisms of Anesthesia. New York: Raven Press 1975
Cherkin,A.: Mechanisms of general anesthesia by non-hydrogen-bonding molecules. Ann. Rev. Pharmacol. **9**, 259—272 (1969)
Cherkin,A.: Effects of flurothyl on memory processing. In: Drugs and Cerebral Function. Springfield, Ill.: C.C. Thomas, 1970
Chessick,R.D., McFarland,R.L., Clark,R.K., Hammer,M., Bassan,M.I.: The effect of morphine, chlorpromazine, pentobarbital and placebo on anxiety. J. nerv. ment. Dis. **141**, 540—548 (1966)
Chin,J.H., Crankshaw,D.P., Kendig,J.J.: Changes in the dorsal root potential with diazepam and with the analgesics aspirin, nitrous oxide, morphine and meperidine. Neuropharmacology **13**, 305—315 (1974)
Chin,J.H., Smith,C.M.: Effects of some central nervous system depressants on the phasic and tonic stretch reflex. J. Pharmacol. exp. Ther. **136**, 276—283 (1962)
Chin,L., Swinyard,E.A.: Tolerance and withdrawal hyperexcitability induced in mice by chronic administration of phenaglycodal. Proc. Soc. exp. Biol. (N.Y.) **97**, 251—254 (1958)
Chin,L.A., Swinyard,E.A.: Pentylenetetrazol seizure threshold in meprobamate- and phenaglycodol-treated mice. J. Amer. pharma. Ass. **48**, 6—8 (1959)
Christensen,H.D., Lee,I.S.: Anesthetic potency and acute toxicity of optically active disubstituted barbituric acids. Toxicol. appl. Pharmacol. **26**, 495—503 (1973)
Cicero,T.J., Smithloff,B.R.: Alcohol oral self-administration in rats: attempts to elicit excessive intake and dependence. In: Gross,M.M. (Ed.): Alcohol Intoxication and Withdrawal Experimental Studies, Vol. 35, pp. 213—224. New York-London: Plenum Press 1973
Cicero,T.J., Snider,S.R., Perez,V.J., Swanson,L.W.: Physical dependence on and tolerance to alcohol in the rat. Physiol. Behav. **6**, 191—198 (1971)
Cigánek,L.: The EEG response (evoked potential) to light stimulus in man. Electroenceph. clin. Neurophysiol. **13**, 165—172 (1961)
Cigánek,L.: A comparative study of visual, auditory and somatosensory EEG responses in man. Exp. Brain Res. **4**, 118—125 (1967)
Clare,A.W.: Diazepam, alcohol, and barbiturate abuse. Brit. med. J. **4**, 340 (1971)
Clark,D.L., Butler,R.A., Rosner,B.S.: Dissociation of sensation and evoked responses by a general anesthetic in man. J. comp. Physiol. Psychol. **68**, 315—319 (1969)
Clark,D.L., Rosner,B.S.: Neurophysiologic effects of general anesthetics: I. Electroencephalogram and sensory evoked responses in man. Anesthesiology **38**, 564—582 (1973)
Clarke,P.R.F., Eccersley,P.S., Frisby,J.P., Thornton,J.A.: The amnesic effect of diazepam (Valium). Brit. J. Anaesth. **42**, 690—697 (1970)
Cochin,J.: Tolerance to the narcotic analgesics. A long-term phenomenon. In: Psychopharmacology, Sexual Disorders and Drug Abuse. Amsterdam-London: North-Holland 1973; Prague: Avicenum, Czechoslovak Medical Press 1973
Cohen,G.: Alcohol and catecholamine disposition: A role for tetrahydroisoquinoline alkaloids. In: Usdin,E., Snyder,S. (Eds.): Frontiers in Catecholamine Research. London: Pergamon Press 1973a
Cohen,G.: A role for tetrahydroisoquinoline alkaloids as false adrenergic neurotransmitters in alcoholism. Advanc. exp. Biol. **35**, 33 (1973b)
Cohen,G.: Tetrahydroisoquinoline alkaloids: uptake, storage, and secretion by the adrenal medulla and by adrenergic nerves. Ann. N.Y. Acad. of Sci. **215**, 116—119 (1973c)
Cohen,G., Collins,M.: Alkaloids from catecholamines in adrenal tissue: possible role in alcoholism. Science **170**, 1749—1751 (1970)
Cohen,J.B., Changeux,J.P.: The cholinergic receptor protein in its membrane environment. Ann. Rev. Pharmacol. **15**, 83—103 (1975)
Cohen,J.B., Weber,M., Changeux,J.P.: Effects of local anesthetics and calcium on the interaction of cholinergic ligands with the nicotinic receptor protein from torpedo marmorata. Molec. Pharmacol. **10**, 904—932 (1974)
Cohen,P.J., McIntyre,R.: The effects of general anesthesia on respiratory control and oxygen consumption of rat liver mitochondria. In: Fink,B.R. (Ed.): Cellular Biology and Toxicity of Anesthetics. Baltimore: Williams and Wilkins Co. 1972

Coldwell, B. B., Paul, C. J., Thomas, B. H.: Phenobarbital metabolism in ethanol-intoxicated rats. Canad. J. Physiol. Pharmacol. **51**, 458—463 (1973)
Coldwell, B. B., Trenholm, H. L., Thomas, B. H., Charbonneau, S.: The effect of ethanol on phenobarbitone and pentobarbitone absorption into rat blood and brain. J. Pharm. (Lond.) **23**, 947—949 (1971)
Coldwell, B. B., Wiberg, G. S., Trenholm, H. L.: Some effects of ethanol on the toxicity and distribution of barbiturates in rats. Canad. J. Physiol. Pharmacol. **48**, 254—264 (1970)
Collier, H. O. J.: Tolerance, physical dependence and receptors. Advanc. Drug Res. **3**, 171—188 (1966)
Collier, H. O. J.: Humoral transmitters, supersensitivity, receptors and dependence. In: Steinberg, H. (Ed.): Scientific Basis of Drug Dependence. London: J & A Churchill Ltd. 1969
Collier, H. O. J., Hammond, M. D., Schneider, C.: Effects of drugs affecting endogenous amines or cyclic nucleotides on ethanol withdrawal head twitches in mice. Br. J. Pharmac. **58**, 9—16 (1976)
Collins, M. A.: Tetrahydroisoquinoline alkaloids from condensation of alcohol metabolites with norepinephrine: preparative synthesis and potential analysis in nervous tissue by gas chromatography. Ann. N.Y. Acad. Sci. **215**, 92—97 (1973)
Colquhon, D.: Mechanisms of drug action at the voluntary muscle end plate. Ann. Rev. Pharmacol. **15**, 307—325 (1975)
Colton, T., Gosselin, R. E., Smith, R. P.: The tolerance of coffee drinkers to caffeine. Clin. Pharmacol. Ther. **9**, 31—39 (1968)
Conney, A. H., Garren, L.: Contrasting effects of thyroxine on zoxazolamine and hexobarbital metabolism. Biochem. Pharmacol. **6**, 257—262 (1961)
Consolo, S., Garattini, S., Ladinsky, H.: Action of the benzodiazepines on the cholinergic system. Advanc. Biochem. Psychopharm. **14**, 63—80 (1975). In: Mechanism of Action of Benzodiazepines. New York: Raven Press 1975
Cook, L., Sepinwall, J.: Behavioral analysis of the effects and mechanisms of action of benzodiazepines. Advanc. Biochem. Psychopharm. **14**, 1—28 (1975). In: Mechanism of Action of Benzodiazepines. New York: Raven Press 1975
Cooper, S. A., Dretchen, K. L.: Biphasic action of ethanol on contraction of skeletal muscle. Europ. J. Pharmacol. **31**, 232—236 (1975)
Costa, E., Greengard, P. (Eds.): Mechanism of Action of Benzodiazepines. New York: Raven Press 1975
Costa, E., Guidotti, A., Mao, C. C.: Evidence for involvement of GABA in the action of benzodiazepines: studies on rat cerebellum. Advanc. Biochem. Psychopharm. **14**, 113—130 (1975). In: Mechanism of Action of Benzodiazepines. New York: Raven Press 1975
Cox, B., Potkonjak, D.: Effects of drugs on tremor and increase in brain acetylcholine produced by oxotremorine in the rat. Brit. J. Pharmacol. **38**, 171—180 (1970)
Crankshaw, D. P., Raper, C.: The effect of solvents on the potency of chlordiazepoxide, diazepam, medazepam, and nitrazepam. J. Pharm. (Lond.) **23**, 313—321 (1971)
Crawford, J. M.: Anaesthetic agents and the chemical sensitivity of cortical neurones. Neuropharmacology **9**, 31—46 (1970)
Crawford, J. M., Curtis, D. R.: Pharmacological studies on feline Betz cells. J. Physiol. (Lond.) **186**, 121—138 (1966)
Crossland, J., Merrick, A. J.: The effect of anaesthesia on the acetylcholine content of brain. J. Physiol. (Lond.) **125**, 56 (1954)
Crossland, J., Slater, P.: The effect of some drugs on the "free" and "bound" acetylcholine content of rat brain. Brit. J. Pharmacol. **33**, 42 (1968)
Crossland, J., Turnbull, C. J.: Gamma-aminobutyric acid and the barbiturate abstinence syndrome in rats. Neuropharmacology **11**, 733—738 (1972)
Crowley, H. J., Dalton, C., Schallek, W., Sheppard, H.: Benzodiazepine inhibition of cyclic nucleotide phosphodiesterase in the central nervous system (CNS). Fed. Proc. **33**, 493 (1974a)
Crowley, T. J., Chesluk, D., Dilts, S., Hart, R.: Drug and alcohol abuse among psychiatric admissions: a multidrug clinical-toxicologic study. Arch. gen. Psychiat. **30**, 13—20 (1974b)
Cull, J. G., Hardy, R. E.: Language of the drug abuser. In: Types of Drug Abusers and Their Abuses. Springfield, Ill.: Charles C. Thomas 1974

Cullen, D. G., Eger, E. I. II.: The effects of hypoxia and isovolemic anemia on the halothane requirement (MAC) of dogs. Anesthesiology **32**, 28—34 (1970)
Curry, S. H.: Concentration-effect relationships with major and minor tranquilizers. Clin. Pharmacol. Ther. **16**, 192—197 (1974)
Curry, S. H., Norris, H.: Acute tolerance to a sedative in man. Brit. J. Pharmacol. **38**, 450—451 (1970)
Curtis, D. R., Game, C. J. A., Lodge, D.: Benzodiazepines and central glycine receptors. Brit. J. Pharmacol. **56**, 307—311 (1976)
Curtis, D. R., Johnston, G. A. R.: Amino acid transmitters in the mammalian central nervous system. In: Adrian, R. H., Helmreich, E., Holzer, H., Jung, R., Kramer, K., Krayer, O., Lynen, F., Miescher, P. A., Rasmussen, H., Renold, A. E., Trendelenburg, U., Ullrich, K., Vogt, W., Weber, A. (Eds.): Reviews of Physiology. Berlin-Heidelberg-New York: Springer 1974
Cutler, R. W. P., Markowitz, D., Dudzinski, D. S.: The effect of barbiturates on (^{3}H)-GABA transport in rat cerebral cortex slices. Brain Res. **81**, 189—197 (1974)
Daves, D. G. Jr., Belshee, R. B., Anderson, W. R. Jr., Downes, H.: Solution conformations of ethyl-1′-methylbutylbarbituric acids: implications for drug-receptor site interactions. Molec. Pharmacol. **11**, 470—477 (1975)
David, J., Grewal, R. S., Wagle, G. P.: Persistent electroencephalographic changes in rhesus monkeys after single doses of pentobarbital, nitrazepam and imipramine. Psychopharmacologia (Berl.) **35**, 61—75 (1974)
Davidoff, R. A.: The effects of bicuculline on the isolated spinal cord of the frog. Exp. Neurol. **35**, 179—193 (1972)
Davidoff, R. A.: Alcohol and presynaptic inhibition in an isolated spinal cord preparation. Arch. Neurol. **28**, 60—63 (1973)
Davidoff, R. A., Grayson, V., Adair, R.: GABA-transaminase inhibitors and presynaptic inhibition in the amphibian spinal cord. Amer. J. Physiol. **224**, 1230—1234 (1973)
Davidson, N., Rix, K. J. B.: A comparison of the effects of systemic infusion and topical application of ethanol solutions in the rat brain. J. Physiol. (Lond.) **227**, 24—26 P (1972)
Davidson, N., Rix, K. J. B.: Behavioral and clinical studies. In: Davidson, N. (Ed.): The Neurotransmitter Amino Acids. London: Academic Press 1975
Davidson, N., Southwick, C. A. P.: Amino acids and presynaptic inhibition in the rat cuneate nucleus. J. Physiol. (Lond.) **219**, 689—708 (1971)
Davies, J. D., Miller, N. E.: Fear and pain: their effect on self-injection of amobarbital sodium by rats. Science **141**, 1286—1287 (1963)
Davis, H. S., Dillon, W. H., Collins, W. F., Randt, C. T.: The effect of anesthetic agents on evoked central nervous system responses (muscle relaxants, and volatile agents). Anesthesiology **19**, 441—449 (1958)
Davis, V. E.: Neuroamine-derived alkaloids: A possible common denominator in alcoholism and related drug dependencies. Ann. N.Y. Acad. Sci. **215**, 111—115 (1973)
Davis, V. E., Brown, H., Huff, J. A., Cashaw, J. L.: Ethanol-induced alterations of norepinephrine metabolism in man. J. Lab. clin. Med. **69**, 787—799 (1967)
Davis, V. E., Walsh, M. J.: Alcohol, amines, and alkaloids: a possible biochemical basis for alcohol addiction. Science **167**, 1005—1007 (1970a)
Davis, V. E., Walsh, M. J.: Morphine and ethanol physical dependence: a critique of a hypothesis. Science **170**, 1113—1115 (1970b)
Davis, V. E., Walsh, M. J., Yamanaka, Y.: Augmentation of alkaloid formation from dopamine by alcohol and acetaldehyde in vitro. J. Pharmacol. exp. Ther. **174**, 401—412 (1970)
Deitrich, R. A.: Stimulation of rat liver aldehyde dehydrogenase activity by phenobarbital *in vivo*. Pharmacologist **12**, 298 (1970)
Deitrich, R. A., Erwin, V. G.: Involvement of biogenic amine metabolism in ethanol addiction. Fed. Proc. **34**, 1962—1968 (1975)
de Jong, R. H., Nace, R. A.: Nerve impulse conduction and cutaneous receptor responses during general anesthesia. Anesthesiology **28**, 851—855 (1967)
de Jong, R. H., Robles, R., Heavner, J. E.: Suppression of impulse transmission in the cat's dorsal horn by inhalation anaesthetics. Anesthesiology **32**, 440—445 (1970)
Delgado, J. M. R.: Physical Control of the Mind: Toward a Psychocivilized Society. New York: Harper and Row 1969

Deneau, G. A., Klima, M., Wilson, M.: Evaluation of sedative-hypnotic agents for barbiturate-like physical dependence capacity in the dog. Bulletin, Problems of Drug Dependence. National Research Council. Committee on Problems of Drug Dependence. 1968, 1—6 pp.

Deneau, G. A., Yanagita, T., Seevers, M. H.: Psychic dependence studies in self-administration techniques in the rhesus monkey. Committee on Problems of Drug Dependence, NAS-NRC appendix **21**, 4267—4269 (1965)

Deneau, G. A., Yanagita, T., Seevers, M. H.: Self-administration of psychoactive substances by the monkey. A measure of psychological dependence. Psychopharmacologia (Berl.) **16**, 30—48 (1969)

Dent, J. Y.: Apomorphine in the treatment of anxiety states with especial reference to alcoholism. Brit. J. Inebr. **32**, 65—88 (1934)

Devenyi, P., Wilson, M.: Barbiturate abuse and addiction and their relationship to alcohol and alcoholism. Canad. med. Ass. J. **104**, 215—218 (1971a)

Devenyi, P., Wilson, M.: Abuse of barbiturates in an alcoholic population. Canad. med. Ass. J. **104**, 219—221 (1971b)

Dewey, J.: Human Nature and Conduct. New York: The Modern Library (Random House) 1922

Dews, P. B.: The behavioral context of addiction. In: Goldberg, L., Hoffmeister, F. (Eds.): Psychic Dependence, Definition, Assessment in Animals and Man: Theoretical and Clinical Implications. Berlin-Heidelberg-New York: Springer 1973

DiPerri, R., Dravid, A., Schweigerdt, A., Himwich, H. E.: Effects of alcohol on evoked potentials of various parts of the central nervous system of cat. Quart. J. Stud. Alcohol. **29**, 20—37 (1968)

Ditman, K. S.: Evaluation of drugs in the treatment of alcoholics. Quart. J. Stud. Alcohol. (Suppl. **1**) 107—116 (1961)

Ditman, K. S., Bejor, D.: Diazepam (Valium) very high dosage: longitudinal and single case study. West. Med. **7**, 109—110 (1966)

Doggett, N. S., Spencer, P. S. J.: Pharmacological properties of centrally administered ouabain and their modification by other drugs. Brit. J. Pharmacol. **42**, 242—253 (1971)

Doggett, N. S., Spencer, P. S. J., Waite, R.: Hypnotic activity of centrally administered barbiturate and uncouplers of oxidative phosphorylation. Europ. J. Pharmacol. **13**, 23—29 (1970)

Domino, E. F.: Pharmacological actions of a convulsant barbiturate. II. Effects compared with pentobarbital on cerebral cortex and some brain stem systems of the cat. J. Pharmacol. exp. Ther. **119**, 272—283 (1957)

Domino, E. F.: Effects of preanesthetic and anesthetic drugs on visually evoked responses. Anesthesiology **28**, 184—191 (1967)

Domino, E. F.: Neuropsychopharmacologic studies of marihuana, some synthetic and natural THC derivatives in animals and man. Ann. N.Y. Acad. Sci. **191**, 166—191 (1971)

Domino, E. F., Corssen, G., Sweet, R. G.: Effects of various anesthetics on the visually evoked responses in man. Anesth. Analg. Curr. Res. **42**, 735—747 (1963)

Dow, R. S.: Moruzzi, G.: The Physiology and Pathology of the Cerebellum. Minneapolis: University of Minnesota Press 1958

Downes, H., Perry, R. S., Ostlund, R. E., Karler, R.: A study of the excitatory effects of barbiturates. J. Pharmacol. exp. Ther. **175**, 692—699 (1970)

Downes, Y., Williams, J. K.: Effects of a convulsant barbiturate on the spinal monosynaptic pathway. J. Pharmacol. exp. Ther. **168**, 283—289 (1969)

Downs, D. A., Woods, J. H., Llewellyn, M. E.: The behavioural pharmacology of addiction: some conceptual and methodological foci. In: Cappell, H. D., LeBlanc, A. E. (Eds.): Biological and Behavioural Approaches to Drug Dependence. Ontario, Canada: House of Lind 1975

Drachman, D. B.: The role of acetylcholine as a neurotrophic transmitter. In: Trophic Functions of the Neuron. Ann. N.Y. Acad. Sci. **228**, 160—176 (1974)

Dravid, A. R., DiPerri, R., Morillo, A., Himwich, H. E.: Alcohol and evoked potentials in the cat. Nature (Lond.) **200**, 1328—1329 (1963)

Dreyfus, P. M.: Thoughts on the pathophysiology of Wernicke's disease. Ann. N.Y. Acad. Sci. **215**, 367—369 (1973)

Duggan, P. F.: Stimulation of calcium uptake of muscle microsomes by phenothiazines and barbiturates. Europ. J. Pharmacol. **13**, 381—386 (1971)

Dunbar, E.: The light thrown on psychological processes by the action of drugs. Proc. Soc. psych. Res. **19**, 62—77 (1905)

Dundee, J. W.: Alterations in response to somatic pain associated with anesthesia. II. The effect of thiopentone and pentobarbital. Brit. J. Anaesth. **32**, 407—414 (1960)
Dundee, J. W.: Molecular structure-activity relationships of barbiturates. In: Molecular Mechanisms in General Anaesthesia. Edinburgh-London-New York: Churchill Livingstone 1974
Dundee, J. W., Pandit, S. K.: Anterograde amnesic effects of pethidine, hyoscine and diazepam in adults. Brit. J. Pharmacol. **44**, 140—144 (1972)
Dundee, J. W., Price, H. L., Dripps, R. D.: Acute tolerance to thiopentone in man. Brit. J. Anaesth. **28**, 344—352 (1956)
Dysken, M.W., Chan, C. H.: Diazepam withdrawal psychosis: A case report. Amer. J. Psychiatry **134**, 573 (1977)
Ebert, A. G., Yim, G. K., Miya, T. S.: Distribution and metabolism of barbital-^{14}C in tolerant and nontolerant rats. Biochem. Pharmacol. **13**, 1267 (1964)
Eccles, J. C.: Trophic interactions in the mammalian central nervous system. Ann. N.Y. Acad. Sci. **228**, 406—423 (1974)
Eccles, J. C., Faber, D. S., Murphy, J. T., Sabah, N. H., Táboríková, H.: Afferent volleys in limb nerves influencing impulse discharges in cerebellar vortex. I. In mossy fibers and granule cells. Exp. Brain Res. **13**, 15—35 (1971 a)
Eccles, J. C., Faber, D. S., Murphy, J. T., Sabah, N. H., Táboríková, H.: Investigations on integration of mossy fiber inputs to Purkyne cells in the anterior lobe. Exp. Brain Res. **13**, 54—77 (1971 b)
Eccles, J. C., Malcolm, J. L.: Dorsal root potentials of the spinal cord. J. Neurophysiol. **9**, 139—160 (1946)
Eccles, J. C., Schmidt, C. R., Willis, W. D.: Pharmacological studies on presynaptic inhibition. J. Physiol. (Lond.) **168**, 500—530 (1963)
Edgley, R.: Diazepam, nitrazepam, and the N.H.S. Med. J. Aust. **1**, 186—187 (1970)
Edney, S. M., Downes, H.: Contractor effect of barbiturates on smooth muscle. Arch. int. Pharmacodyn. **217**, 180—196 (1975)
Eidelberg, E., Barstow, C. A.: Morphine tolerance and dependence induced by intraventricular injection. Science **174**, 74—76 (1971)
Eidelberg, E., Bond, M. L., Kelter, A.: Effects of alcohol on cerebellar and vestibular neurones. Arch. int. Pharmacodyn. **192**, 213—219 (1971)
Eidelberg, E., Neer, H. M., Miller, M. K.: Anticonvulsant properties of some benzodiazepine derivatives. Neurology (Minneap.) **15**, 223—230 (1965)
Eidelberg, E., Wooley, D. F.: Effects of ethyl alcohol upon spinal cord neurones. Arch. int. Pharmacodyn. **185**, 388—396 (1970)
Ekman, G.: Subjective and objective effects of alcohol as functions of dosage and time. Psychopharmacologia (Berl.) **6**, 399—409 (1964)
Ekman, G., Frankenhauser, M., Goldberg, L., Bjerver, K., Jarpe, G., Myrsten, A. L.: Effects of alcohol intake on subjective and objective variables over a five-hour period. Psychopharmacologia (Berl.) **4**, 28—38 (1963)
Engle, J., Strombom, U., Svensson, T. H., Waldeck, B.: Suppression by α-methyltyrosine of ethanol-induced locomotor stimulation: Partial reversal by L-Dopa. Psychopharmacology **37**, 275—279 (1974)
Erickson, C. K., Burnam, W. L.: Cholinergic alteration of ethanol-induced sleep and death in mice. Ag. Actions **2**, 8—13 (1971)
Erickson, C. K., Graham, D. T.: Alteration of cortical and reticular acetylcholine release by ethanol *in vivo*. J. Pharmacol. exp. Ther. **185**, 583—593 (1973)
Eriksson, C. J.: Ethanol and acetaldehyde metabolism in rat strains genetically selected for their ethanol preference. Biochem. Pharmacol. **22**, 2283—2292 (1973)
Eriksson, K.: Genetic aspects of alcohol drinking behaviour. Int. J. Neurol. **9**, 125—132 (1974)
Eriksson, K.: Alcohol imbibition and behavior: a comparative genetic approach. In: Eleftheriou, B. E. (Ed.): Psychopharmacogenetics, pp. 127—168. New York: Plenum 1975
Erwin, V. G., Deitrich, R. A.: Inhibition of bovine brain aldehyde reductase by anticonvulsant compounds in vitro. Biochem. Pharmacol. **22** 2615—2624 (1973)
Esplin, D. W.: Introduction to drugs acting on the central nervous system. In: Goodman, L. S., Gilman, A. (Eds.): The Pharmacological Basis of Therapeutics, 4. Ed. New York: Macmillan 1970

Essig, C. F.: Focal convulsions during barbiturate abstinence in dogs with cerebrocortical lesions. Psychopharmacologia (Berl.) **3**, 432—437 (1962a)
Essig, C. F.: Convulsive and sham rage behaviors in decorticate dogs during barbiturate withdrawal. Arch. Neurol. **7**, 471—475 (1962b)
Essig, C. F.: Addictive and possible toxic properties of glutethimide. Amer. J. Psychiat. **119**, 993 (1963)
Essig, C. F.: Barbiturate withdrawal convulsions in decerebellate dogs. Int. J. Neuropharmacol. **3**, 453—456 (1964)
Essig, C. F.: Clinical aspects of barbiturate and sedative drug abuse. Amer. J. Hosp. Pharm. **22**, 140—143 (1965)
Essig, C. F.: Barbiturate withdrawal in white rats. Int. J. Neuropharmacol. **5**, 103—110 (1966a)
Essig, C. F.: Newer sedative drugs that can cause states of intoxication and dependence of barbiturate type. J. Amer. med. Ass. **196**, 714—717 (1966b)
Essig, C. F.: Clinical and experimental aspects of barbiturate withdrawal convulsions. Epilepsia **8**, 21—30 (1967)
Essig, C. F.: Addiction to barbiturate and nonbarbiturate sedative drugs. In: The Addictive States, pp. 188—198. Baltimore: Williams & Wilkins Co. 1968a
Essig, C. F.: Possible relation of brain gamma-aminobutyric acid (GABA) to barbiturate abstinence convulsions. Arch. int. Pharmacodyn. **176**, 97—103 (1968b)
Essig, C. F.: The potential risk of addiction to nonbarbiturate sedative and minor tranquilizing drugs. In: Sjoqvist, F., Tottie, M. (Eds.): Symposium on Abuse of Central Stimulants, pp. 89—112. Stockholm: Almqvist and Wiksell 1969
Essig, C. F.: Drug withdrawal convulsions in animals. In: Experimental Models of Epilepsy—A manual for the laboratory worker, Eds.: Purpura, D. P., Penry, J. K., Tower, D., Woodbury, D. M., Walter, R. New York: Raven Press 1972
Essig, C. F., Flanary, H. G.: Convulsions in cats following withdrawal of sodium barbital. Exp. Neurol. **1**, 529—533 (1959)
Essig, C. F., Flanary, H. G.: Convulsive aspects of barbital sodium withdrawal in the cat. Exp. Neurol. **3**, 149—159 (1961)
Essig, C. F., Lam, R. C.: Convulsions and hallucinatory behavior. Arch. Neurol. **18**, 626—632 (1968)
Essig, C. F., Jones, B. E., Lam, R. C.: The effect of pentobarbital on alcohol withdrawal in dogs. Arch. Neurol. **20**, 554—558 (1969)
Evans, W. O., Kline, N. S.: Psychotropic drugs in the year 2000. Springfield, Ill.: Charles C. Thomas 1971
Ewing, J. A., Bakewell, W. E.: Diagnosis and management of depressant drug dependence. Amer. J. Psychiat. **123**, 909—917 (1967)
Ewing, J. A., Rouse, B. A., Mueller, R. A., Mills, K. C.: Alcohol as a euphoriant drug. Searching for a neurochemical basis. Ann. N.Y. Acad. Sci. **273**, 159—166 (1976)
Ewing, J. A., Rouse, B. A., Pellizzari, E. D.: Alcohol sensitivity and ethnic background. Amer. J. Psychiat. **131**, 206—210 (1974)
Faber, D., Greenberg, A.: Effects of cyclic nucleotides on the excitability of Aplysia neurons. Abstract submitted for Symposium on Snail Brain, Sept. 1975, Hungary.
Faber, D. S., Klee, M. R.: Effects of ethanol on collateral inhibition of the goldfish Mauthner cell. Soc. Neurosci. 5th Annual Meeting, November 1975
Faber, D. S., Klee, M. R.: Ethanol suppresses collateral inhibition of the goldfish Mauthner cell. Brain Res. **104**, 347—353 (1976)
Faber, D. S., Klee, M. R.: Actions of ethanol on neuronal membrane properties and synaptic transmission. In: K. Blum (Ed.): Alcohol and Opiates. Neurochemical and Behavioral Mechanisms. New York: Academic Press 1977
Fabre, L. R. Jr., Farmer, R. W., Roach, M. K., Fritchie, G. E., McIsaac, W. M.: Biochemical and metabolic aspects of alcoholism. Ann. N.Y. Acad. Sci. **215**, 346—355 (1973)
Falk, J. L., Samson, H. H.: Schedule-induced physical dependence on ethanol. Pharmacol. Rev. **27**, 449—464 (1976)
Faulconer, A. Jr., Bickford, R. G.: Electroencephalography in Anesthesiology. Springfield, Ill.: Charles C. Thomas 1960
Fehr, K. A., Kalant, H., LeBlanc, A. E.: Residual learning deficit after heavy exposure to cannabis or alcohol in rats. Science **192**, 1249—1251 (1976)

Feltz, P., Rasminsky, M.: A model for the mode of action of GABA on primary afferent terminals: depolarizing effects of GABA applied iontophoretically to neurones of mammalian dorsal root ganglia. Neuropharmacology **13**, 553—563 (1974)

Fencl, V., Koski, G., Papenheimer, J. R.: Factors in cerebrospinal fluid from goats that affect sleep and activity in rats. J. Physiol. (Lond.) **216**, 565—589 (1971)

Fenna, D., Mix, L., Schaefer, O., Gilbert, J. A. L.: Ethanol metabolism in various racial groups. Canad. med. J. **105**, 472—475 (1971)

Findley, J. D., Robinson, W. W., Peregrino, L.: Addiction to secobarbital and chlordiazepoxide in the rhesus monkey by means of self-infusion preference procedure. Psychopharmacologia (Berl.) **26**, 93—114 (1972)

Finer, M. J.: Habituation to chlordiazepoxide in an alcoholic population. J. Amer. med. Ass. **213**, 1342 (1970)

Fink, B. R.: Cellular Biology and Toxicity of Anesthetics. Baltimore: Williams & Wilkins Co. 1972

Fink, B. R.: Progress in Anesthesiology, Vol. 1: Molecular Mechanisms of Anesthesia. New York: Raven Press 1975

Fink, B. R., Haschke, R. H.: Anesthetic effects on cerebral metabolism. Anesthesiology **39**, 199—215 (1973)

Fink, R. D., Knott, D. H., Beard, J. D.: Sedative-hypnotic dependence. Amer. Fam. Phycn **10**, 116—122 (1974)

Fitzgerald, L. A., Thompson, R. F.: Classical conditioning of the hindlimb flexion reflex in the acute spinal cat. Psychon. Sci. **8**, 213—214 (1967)

Flourens, P.: Recherches experimentales sur les propriétés et les functions du système nerveux dans les animaux vertébrés. Paris: Crevot 1824

Floyd, J. B., Jr., Murphy, C. M.: Hallucinations following withdrawal of valium. J. Ky. Med. Ass. **74**, 549—550 (1976)

Folkman, J., Mark, V. H., Ervin, F., Suematsu, Hagiwara, R.: Intracerebral gas anesthesia by diffusion through silicone rubber. Anesthesiology **29**, 419—425 (1968)

Forney, R. B., Hughes, F. W.: Combined Effect of Alcohol and Other Drugs. Springfield, Ill.: Charles C. Thomas 1968

Forrest, W. H. Jr., Bellville, J. W., Brown, B. W. Jr.: The interaction of caffeine with pentobarbital as a nighttime hypnotic. Anesthesiology **36**, 37—41 (1972)

Forte, J. G., Ganser, A. L., Tanisawa, A. S.: The K^+-stimulated ATPase system of microsomal membranes from gastric oxyntic cells. Ann. N.Y. Acad. Sci. **242**, 255—267 (1974)

Frank, G. B.: The effects of anesthetic drugs based on alterations of membrane excitability. In: Cellular Biology and Toxicity of Anesthetics. Baltimore: Williams & Wilkins Co. 1972

Frank, G. B., Jhamandas, K.: Effects of drugs acting alone and in combination on the motor activity of intact mice. Brit. J. Pharmacol. **39**, 696 (1970a)

Frank, G. B., Jhamandas, K.: Effects of general stimulant drugs on the electrical responses of isolated slabs of cat's cerebral cortex. Brit. J. Pharmacol. **39**, 716 (1970b)

Frank, G. B., Jhamandas, K. H.: Interaction between the effects of pentobarbital and procaine in mice: The importance of timing in drug addiction studies. Canad. J. Physiol. Pharmacol. **47**, 493 (1969)

Frankel, D., Khanna, J. M., Kalant, H., LeBlanc, A. E.: Effect of acute and chronic ethanol administration on serotonin turnover in rat brain. Psychopharmacologia (Berl.) **37**, 91—100 (1974)

Franks, C. M.: Alcohol, alcoholism and conditioning: a review of the literature and some theoretical considerations. J. ment. Sci. **104**, 14—33 (1958)

Fraser, H. F.: Tolerance to and physical dependence on opiates, barbiturates and alcohol. Ann. Rev. Med. **8**, 427—440 (1957)

Fraser, H. F.: Criteria for evaluating physical and psychic dependence and overall abuse potential of drugs in man: In: Chemical and Biological Aspects of Drug Dependence. Cleveland, Ohio: CRC Press 1972

Fraser, H. F., Isbell, H.: Abstinence syndrome in dogs after chronic barbiturate medication. J. Pharmacol. exp. Ther. **112**, 261—267 (1954a)

Fraser, H. F., Isbell, H.: Chronic barbiturate intoxication. Arch. intern. Med. **94**, 34—41 (1954b)

Fraser, H. F., Isbell, H., Eisenman, A. J., Wikler, A., Pescor, F. T.: Chronic barbiturate intoxication. Arch. intern. Med. **93**, 34—41 (1954)

Fraser, H. F., Isbell, H., Wikler, A., Belleville, R. E., Essig, C. F., Hill, H. E.: Minimum dose of barbiturates required to produce physical dependence. Fed. Proc. **15**, 423 (1956)

Fraser, H. F., Shaver, M. R., Maxwell, E. S., Isbell, H.: Death due to withdrawal of barbiturates. Ann. intern. Med. **38**, 1319 (1953)

Fraser, H. F., Wikler, A., Essig, C. F., Isbell, H.: Degree of physical dependence induced by secobarbital or pentobarbital. J. Amer. med. Ass. **166**, 126—129 (1958)

Fraser, H. F., Wikler, A., Isbell, H., Johnson, N. K.: Partial equivalence of chronic alcohol and barbiturate intoxications. Quart. J. Stud. Alcohol **18**, 541—551 (1957)

Freedman, D. X.: Drugs and Culture. Triangle **10** (1971)

Fregly, A. R., Bergstedt, M., Graybiel, A.: Relationships between blood alcohol, positional alcohol nystagmus and postural equilibrium. Quart. J. Stud. Alcohol **28**, 11—21 (1967)

French, J. D., Verzeano, M., Magoun, H. W.: A neural basis of the anesthetic state. Arch. Neurol. (Chic.) **69**, 519—529 (1953)

French, S. W.: Effect of chronic ethanol feeding on rat liver phospholipid. J. Nutr. **91**, 292—298 (1967)

French, S. W.: Fragility of liver mitochondria in ethanol-fed rats. Gastroenterology **54**, 1106—1114 (1968)

French, S. W., Morin, R. J.: Mitochondrial injury in experimental chronic ethanol ingestion. In: Sardesai, V. M. (Ed.): Biochemical and Clinical Aspects of Alcohol Metabolism, pp. 123—132. Springfield, Ill.: Charles C. Thomas 1969

French, S. W., Palmer. D. S.: Adrenergic supersensitivity during ethanol withdrawal in the rat. Res. Commun. chem. Path. Pharmacol. **6**, 651—717 (1973)

French, S. W., Sheinbaum, A., Morin, R. J.: Effects of ethanol and a fat-free diet on hepatic mitochondrial fragility and fatty acid composition. Proc. Soc. Exp. Biol. **130**, 781—783 (1969)

French, S. W., Todoroff, T.: Effect of chronic ethanol ingestion and withdrawal on brain mitochondria. Res. Commun. chem. Path. Pharmacol. **2**, 206—215 (1971)

French, S. W.: Track I. A. Metabolism and brain biochemistry and B. Biogenic amines and condensation products. Presentation at National Council on Alcoholism 6th Annual Medical-Scientific Work in Progress on Alcoholism Symposium, April 28—29, 1975, Milwaukee, Wisconsin

Freudenthal, R. I., Martin, J.: Correlation of brain levels of barbiturate enantiomers with reported differenes in duration of sleep. J. Pharmacol. exp. Ther. **193**, 664—668 (1975)

Freund, G.: The effect of ethanol and aging on the transport of α-aminoisobutyric acid into the brain. Brain Res. **46**, 363—368 (1972)

Freund, G.: Alcohol, barbiturate, and bromide withdrawal syndromes in mice. Ann. N.Y. Acad. Sci. **215**, 224—234 (1973a)

Freund, G.: Chronic central nervous system toxicity of alcohol. Ann. Rev. Pharmacol. **13**, 217—227 (1973b)

Frey, H.-H., Kampmann, E.: Tolerance to anticonvulsant drugs. Acta pharmacol. (Kbh.) **22**, 159—171 (1965)

Frisancho, A. R.: Functional adaptation to high altitude hypoxia. Science **187**, 313—319 (1975)

Fuller, J. L.: Strain differences in the effects of chlorpromazine in three strains of mice. Psychopharmacologia (Berl.) **16**, 261—271 (1970)

Furgiuelle, A. R., Aumente, Z., Horovitz, Z. P.: Acute and chronic effects of imipramine and desipramine in normal rats and in rats with lesioned amygdalae. Arch. int. Pharmacodyn. **151**, 170—179 (1964)

Furner, R. L., McCarthy, J. S., Stitzel, R. E., Anders, M. W.: Stereoselective metabolism of the enantiomers of hexobarbital. J. Pharmacol. exp. Ther. **169**, 153—158 (1969)

Fuxe, K., Agnati, L. F., Bolme, P., Hokfelt, T., Lidbrink, P., Ljungdahl, A., Perez, M., Ogren, S.: The possible involvement of GABA mechanisms in the action of benzodiazepines on central catecholamine neurons. Advanc. Biochem. Psychopharmacol. **14**, 45—61 (1975)

Gage, P. W.: The effect of methyl, ethyl and n-propyl alcohol on neuromuscular transmission in the rat. J. Pharmacol. exp. Ther. **150**, 236—243 (1965)

Gage, P. W., McBurney, R. N.: An analysis of the relationship between the current and potential generated by a quantum of acetylcholine in muscle fibers without tranverse tubules. J. Membr. Biol. **12**, 247—272 (1973)

Gage, P. W., McBurney, R. N., Schneider, G. T.: Effects of some aliphatic alcohols on the conductance change caused by a quantum of acetylcholine at the toad end-plate. J. Physiol. (Lond.) **244**, 409—429 (1975)
Gage, P. W., McBurney, R. N., VanHelden, D.: End-plate currents are shortened by octanol: possible role of membrane lipids. Life Sci. **14**, 2277—2283 (1974)
Galindo, A.: Effects of procaine, pentobarbital and halothane on synaptic transmission in the central nervous system. J. Pharmacol. exp. Ther. **169**, 185—195 (1969)
Galindo, A.: Anesthesia and synaptic transmission. In: Cellular Biology and Toxicity of Anesthetics. Baltimore: Williams & Wilkins Co. 1972
Gallagher, J. P., Blaber, L. C.: Catechol, a facilitatory drug that demonstrates only a prejunctional site of action. J. Pharmacol. exp. Ther. **184**, 129—135 (1973)
Gallego, A.: On the effect of ethyl alcohol upon frog nerve. J. cell. comp. Physiol. **31**, 97—106 (1948)
Gangloff, H., Monnier, M.: Tropische Wirkung des Phenobarbitals auf Cortex, Rhinencephalon, Nucleus caudatus, Thalamus und Substantia reticularis des Kaninchens. Naunyn-Schmiedebergs Arch. exp. Path. Pharmak. **231**, 211—218 (1957)
Gangloff, H., Monnier, M.: Effect of phenobarbital on evoked activity following stimulation of cortical and subcortical structures in the unanesthetized rabbit. J. Pharmacol. exp. Ther. **122**, 23A (1958)
Garattini, S., Mussini, E., Randall, L. O.: Eds. The Benzodiazepines. New York: Raven Press 1973
Garg, M., Holland, H. C.: Consolidation and maze learning: The effects of posttrial injections of a depressant drug (pentobarbital sodium). Psychopharmacologia (Berl.) **12**, 127—132 (1968)
Gatz, E. E., Jones, J. R.: Haloperidol antagonism to the hypermetabolic effects of 2,4-dinitrophenol (an in vitro and in vivo correlation). In: Cellular Biology and Toxicity of Anesthetics. Baltimore: Williams & Wilkins Co. 1972
Gebhart, G. F., Mitchell, C. L.: Further studies on the development of tolerance to the analgesic effect of morphine: The role played by the cylinder in the hot plate testing procedure. Arch. int. Pharmacodyn. **191**, 96—103 (1971 a)
Gebhart, G. F., Mitchell, C. L.: The relative contributions to the development of morphine-analgesia tolerance made by the testing cylinder and the plate in the hot plate procedure. Presented at Committee on Problems of Drug Dependence (NAS), Toronto, 16—17 February, 1971 b
Gebhart, G. F., Mitchell, C. L.: The relative contributions of the testing cylinder and the heated plate in the hot plate procedure to the development of tolerance to morphine in rats. Eur. J. Pharmacol. **18**, 56—62 (1972)
Gebhart, G. F., Sherman, A. D., Mitchell, C. L.: The influence of learning on morphine analgesia and tolerance development in rats tested on the hot plate. Psychopharmacologia (Berl.) **27**, 295—304 (1971)
Gebhart, G. F., Sherman, A. D., Mitchell, C. L.: The influence of stress on tolerance development to morphine in rats tested on the hot plate. Arch. int. Pharmacodyn. **197**, 328—337 (1972)
Gehrmann, J. E., Killam, K. F.: Characterization of EEG effects produced by sedative-hypnotic agents using spectral analysis techniques. Fed. Proc. **34**, 779 (1975)
Gergis, S. D., Sokoll, M. D., Cronnelly, R., Dretchen, K. L., Long, J. P.: Changes of acetylcholine release at frog's neuromuscular junction by drugs used during anesthesia. In: Fink, B. R. (Ed.): Progress in anesthesiology, Vol. 1, pp. 181—192. Molecular Mechanisms of Anesthesia. New York: Raven Press 1975
Geschwind, N.: Late changes in the nervous system: an overview. In: Stein, D. G., Rosen, J. J., Butters, N. (Eds.): Plasticity and Recovery of Function in the Central Nervous System. New York: Academic Press 1974
Gessner, P. K.: Induction of a diethyl ether withdrawal syndrome in mice by exposure to ether vapor. Pharmacologist **16**, 304 (1974)
Gey, K. F., Rutishauser, M., Pletscher, A.: Suppression of glycolysis in rat brain in vivo by chlorpromazine, reserpine, and phenobarbital. Biochem. Pharmacol. **14**, 507—514 (1965)
Giacobini, E., Izikowitz, S., Wegmann, A.: Urinary excretion of noradrenalin and adrenaline during acute alcohol intoxication in alcoholic addicts. Experientia (Basel) **16**, 467 (1960 a)
Giacobini, E., Izikowitz, S., Wegmann, A.: Urinary norepinephrine and epinephrine excretion in delirium tremens. Arch. gen. Psychiat. **3**, 289—296 (1960 b)

Gibson, W. R., Doran, W. J., Wood, W. C., Swanson, E. E.: Pharmacology of stereo-isomers of 1-methyl-5-(1-methyl-2-pentynyl)-5-allyl-barbituric acid. J. Pharmacol. exp. Ther. **125**, 23—27 (1959)
Ginsborg, B. L.: Effect of bromide ions on junctional transmission. Nature (Lond.) **218**, 363—365 (1968)
Gissen, A. J., Karis, J. H., Nastuk, W. L.: Effect of halothane on neuromuscular transmission. J. Amer. med. Ass. **197**, 770—774 (1966)
Glagoleva, I. M., Liberman, E. A., Khashaev, Z. K.: The effect of uncouplers of oxidative phosphorylation on the release of acetylcholine from nerve endings. Biofizika **15**, 76—83 (1970)
Glatt, M. M.: Benzodiazepines. Brit. med. J. **1967 II**, 444
Glatt, M. M.: Recent patterns of abuse of and dependence on drugs. Brit. J. Addict. **63**, 111—128 (1968)
Glatt, M. M.: The changing British drug scene. Lancet **1969 II**, 429—430
Glick, S. D.: Changes in drug sensitivity and mechanism of functional recovery following brain damage. In: Stein, D. G., Rosen, J. J., Butters, N. (Eds.): Plasticity and Recovery of Function in the Central Nervous System. New York: Academic Press 1974
Godfraind, J. M., Kawamura, H., Krnjević, K., Putmain, R.: Actions of dinitrophenol and some other metabolic inhibitors on cortical neurones. J. Physiol. (Lond.) **215**, 199—222 (1971)
Godfraind, J. M., Krnjević, K., Pumain, R.: Unexpected features of the action of dinitrophenol on cortical neurones. Nature (Lond.) **228**, 562—564 (1970)
Gogolak, G., Krijzer, F., Stumpf, C.: Action of central depressant drugs on the electrocerebellogram of the rabbit. Naunyn-Schmiedebergs Arch. Pharmacol. **272**, 378—386 (1972)
Gold, B. I., Volicer, L.: Adrenosine triphosphate-derived nucleotide formation in the presence of ethanol. Biochem. Pharmacol. **25**, 1825—1830 (1976)
Goldberg, L.: Quantitative studies on alcohol tolerance in man. The influence of ethyl alcohol on sensory, motor and psychological functions referred to blood alcohol in normal and habituated individuals. Acta physiol. scand. **5**, (Suppl. 16) 1—128 (1943)
Goldberg, L.: Behavioral and physiological effects of alcohol in man. Psychosom. Med. **28**, 570—595 (1966)
Goldberg, M. E., Manian, A. A., Efron, D. H.: A comparative study of certain pharmacologic responses following acute and chronic administrations of chlordiazepoxide. Life Sci. **6** (Part I), 481—491 (1967)
Goldberger, M. E.: Recovery of movement after CNS lesions in monkeys. In: Stein, D. G., Rosen, J. J., Butters, N. (Eds.): Plasticity and Recovery of Function in the Central Nervous System. New York: Academic Press 1974
Goldstein, A., Aronow, L., Kalman, S. M.: Principles of Drug Action. New York: Harper and Row 1968
Goldstein, A., Goldstein, D. B.: Enzyme expansion theory of drug tolerance and physical dependence. Res. Publ. Ass. nerv. ment. Dis. **46**, 265—267 (1968)
Goldstein, A., Judson, B. A.: Alcohol dependence and opiate dependence: lack of relationship in mice. Science **172**, 290—292 (1971)
Goldstein, A., Kaizer, S., Warren, R.: Psychotropic effects of caffeine in man. II. Alertness, psychomotor coordination, and mood. J. Pharmacol. exp. Ther. **150**, 146—151 (1965)
Goldstein, D. B.: Alcohol withdrawal reactions in mice: effects of drugs that modify neurotransmission. J. Pharmacol. exp. Ther. **186**, 1—9 (1973a)
Goldstein, D. B.: Quantitative study of alcohol withdrawal signs in mice. Ann. N.Y. Acad. Sci. **215**, 218—223 (1973b)
Goldstein, D. B.: Rates of onset and decay of alcohol physical dependence in mice. J. Pharmacol. exp. Ther. **190**, 377—383 (1974)
Goldstein, D. B.: Drug dependence as an adaptive response: studies with ethanol in mice. In: Mandell, A. J. (Ed.): Neurobiological Mechanisms of Adaptation and Behavior, Vol. 13, pp. 185—198. New York: Raven Press 1975a
Goldstein, D. B.: Testing the homeostat hypothesis of drug addiction. In: Cappell, H. O., LeBlanc, A. E. (Eds.): Biological and Behavioural Approaches to Drug Dependence. Toronto, Ontario, Canada: Addiction Research Foundation, House of Lind 1975b
Goldstein, D. B., Goldstein, A.: Possible role of enzyme induction and repression in drug tolerance and addiction. Biochem. Pharmacol. **8**, 48 (1961)

Goldstein, D. B., Israel, Y.: Effects of ethanol on mouse brain (Na + K)-activated adenosine triphosphatase. Life Sci. **11**, 957—963 (1972)
Goodman, L. S., Gilman, A.: The Pharmacological Basis of Therapeutics, 4th ed. London-Toronto-New York, 1970 and 5th ed. London-Toronto-New York: Macmillan 1975
Goodwin, D. W.: Is alcoholism hereditary? A review and critique. Arch. gen. Psychiat. **25**, 545—549 (1971)
Goodwin, D. W., Crane, J. B., Guze, S. B.: Alcoholic blackouts: A review and clinical study of 100 alcoholics. Amer. J. Psychiat. **126**, 191—198 (1969a)
Goodwin, D. W., Hill, S. Y.: Short-term memory and the alcoholic blackout. Ann. N.Y. Acad. Sci. **215**, 195—199 (1973)
Goodwin, D. W., Powell, B., Bremer, D., Hoine, H., Stern, J.: Alcohol and recall: state-dependent effects in man. Science **163**, 1358—1360 (1969b)
Gordis, E.: Tolerance to the hypnotic effect of 1-methyl phenobarbitol induced by its nonhypnotic stereoisomer. Biochem. Pharmacol. **20**, 246 (1971)
Gordon, E. B.: Addiction to diazepam (Valium). Brit. med. J. **1967 I**, 112
Gordon, M., Rubia, F. J., Strata, P.: Sensitivity to pentothal of a synapse in the cerebellar cortex. Arch. Fisiol. **68**, 330 (1971)
Gordon, M., Rubia, F. J., Strata, P.: The effect of barbiturate anesthesia on the transmission to the cerebellar cortex. Brain Res. **43**, 677—680 (1972)
Gordon, M., Rubia, F. J., Strata, P.: The effect of pentothal on the activity evoked in the cerebellar cortex. Exp. Brain Res. **17**, 50—62 (1973)
Gorski, R. A.: Barbiturates and sexual differentiation of the brain. In: Zimmermann, E., George, R. (Eds.): Narcotics and the Hypothalamus. New York: Raven Press 1974
Gostomzyk, J. G., Dilger, B., Dilger, K.: Untersuchungen über die arteriovenöse Differenz der Alkoholkonzentration im Blut und ihre Beziehung zum Alkoholgehalt des Gehirns. Z. klin. Chem. klin. Biochem. **7**, 162—166 (1969)
Gotesdam, K. G.: Intragastric self-administration of medazepam. Psychopharmacologia (Berl.) **28**, 87—94 (1973)
Graeff, F. G.: Tryptamine antagonists and punished behavior. J. Pharmacol. exp. Ther. **189**, 344—350 (1974)
Graham, D. T., Erickson, C. K.: Alteration of ethanol-induced CNS depression: ineffectiveness of drugs that modify cholinergic transmission. Psychopharmacologia (Berl.) **34**, 173—180 (1974)
Graham, J. D. P.: Ethanol and the absorption of barbiturate. Toxicol. appl. Pharmacol. **2**, 14—22 (1960)
Green, D. E.: Conformational basis of energy transductions in membrane systems. In: Cellular Biology and Toxicity of Anesthetics. Baltimore: Williams and Wilkins Co. 1972
Green, D. E., MacDonald, M.: Women and Psychoactive Drug Use. Addiction Research Foundation of Ontario, Toronto, Ontario, 1976
Greenberg, R. S., Cohen, G.: Tetrahydroisoquinoline alkaloids: stimulated secretion from the adrenal mudulla. J. Pharmacol. exp. Ther. **184**, 119—128 (1973)
Greenblatt, D. J., Shader, R. I.: Benzodiazepines (first of two parts). New Engl. J. Med. **291**, 1011—1015 (1974a)
Greenblatt, D. J., Shader, R. I.: The price we pay. In: Greenblatt, D. J., Shader, R. L. (Eds.): Benzodiazepines in Clinical Practice, Chap. 13, pp. 263—268. New York: Raven Press 1974b
Greenhouse, D. D., Szumski, A. J.: A technique for studying the effect of ethanol on the isolated rat muscle spindle. Fed. Proc. **31**, 370 (1972)
Greizerstein, H. B.: Effects of rate of absorption and dose on the development of tolerance to ethanol. Presented at the Sixth International Congress of Pharmacology, July 20—25, 1975, Helsinki, Finland.
Greizerstein, H. B., Smith, C. M.: Study of tolerance to ethanol in goldfish. Fifth International Congress on Pharmacology, San Francisco, California 1972
Greizerstein, H. B., Smith, C. M.: Acquired tolerance to ethanol in goldfish. Pharmacologist **15**, 158 (1973a)
Greizerstein, H. B., Smith, C. M.: Development and loss of tolerance to ethanol in goldfish. J. Pharmacol. exp. Ther. **187**, 391—399 (1973b)
Greizerstein, H. B., Smith, C. M.: Ethanol in goldfish: Effect of prior exposure in a test procedure. Psychopharmacologia (Berl.) **38**, 345—349 (1974)

Grenell, R. G.: Alcohols and activity of cerebral neurons. Quart. J. Stud. Alcohol **20**, 421—427 (1959)
Griffin, J. P.: Neurophysiological studies into habituation. In: Horn, G., Hinde, R. A. (Eds.): Short-term changes in neural activity and behaviour, pp. 141—176. London: Cambridge University Press 1970
Griffiths, R. R., Bigelow, G. E., Liebson, I.: Human sedative self-administration: effects of interingestion interval and dose. J. Pharmacol. exp. Ther. **197**, 488—494 (1976)
Gross, M. M., Begleiter, H., Tobin, M., Kissin, B.: Changes in auditory evoked response induced by alcohol. J. nerv. ment. Dis. **143**, 152—156 (1966)
Gross, M. M., Goodenough, D. R., Hastey, J., Lewis, E.: Experimental study of sleep in chronic alcoholics before, during and after four days of heavy drinking with a nondrinking comparison. Ann. N.Y. Acad. Sci. **215**, 254—265 (1973)
Grossie, J., Smith, C. M.: Depression of afferent activity originating in muscle spindles induced by mephenesin, procaine and caramiphen. Arch. int. Pharmacodyn. **159**, 288—298 (1966)
Grossman, S. P.: Behavioral and electroencephalographic effects of microinjections of neurohumors into the midbrain reticular formation. Physiol. Behav. **3**, 777—786 (1968)
Groves, P. M., Thompson, R. F.: A dual-process theory of habituation: Neural mechanisms. In: Habituation, Vol. II. Physiological Substrates. New York-London: Academic Press 1973
Guerrero-Figueroa, R., Gallant, D. M., Guerrero-Figueroa, C., Galant, J.: Electrophysiological analysis of the action of four benzodiazepine derivatives on the central nervous system. In: Garattini, S., Mussini, E., Randall, L. O. (Eds.): The Benzodiazepines. New York: Raven Press 1973
Guerrero-Figueroa, R., Gallant, D. M., Guerrero-Figueroa, C., Rye, M. M.: Electroencephalographic study of diazepam on patients with diagnosis of episodic behavioral disorders. J. clin. Pharmacol. **10**, 57—64 (1970a)
Guerrero-Figueroa, R., Guerrero-Figueroa, E., Sneed, G. A., Kennedy, M. J.: Effects of lorazepam on CNS structures: neurophysiological and behavioral correlations. Curr. ther. Res. clin. Exp. **16**, 137—146 (1974)
Guerrero-Figueroa, R., Rye, M. M., Gallant, D. M., Bishop, M. P.: Electrographic and behavioral effects of diazepam during alcohol withdrawal stage in cats. Neuropharmacology **9**, 143—150 (1970b)
Guerrero-Figueroa, R., Rye, M. M., Heath, R. G.: Effects of two benzodiazepine derivatives on cortical and subcortical epileptogenic tissues in the cat and monkey. 1. Limbic system structures. Curr. ther. Res. **11**, 27—39 (1969a)
Guerrero-Figueroa, R., Rye, M. M., Heath, R. G.: Effects of two benzodiazepine derivatives on cortical and subcortical epileptogenic tissues in the cat and monkey. II. Cortical and centrencephalic structures. Curr. ther. Res. **11**, 40—55 (1969b)
Guha, D., Pradhan, S. N.: Effects of mescaline, (Δ^9)-tetrahydrocannabinol and pentobarbital on the auditory evoked responses in the cat. Neuropharmacology **13**, 755—762 (1974)
Gutman, M., Singer, T. P., Beinhert, H., Casida, J. E.: Reaction sites of rotenone, piericidin A, and amytal in relation to the nonheme iron components of NADH dehydrogenase. Proc. nat. Acad. Sci. (Wash.) **65**, 763 (1970)
Hadji-Dimo, A. A., Ekberg, R., Ingvar, D. H.: Effects of ethanol on EEG and cortical blood flow in the cat. Quart. J. Stud. Alcohol **29**, 828—838 (1968)
Haefely, W., Kulcsar, A., Mohler, H., Pieri, P., Schaffner, R.: Possible involvement of GABA in the central actions of benzodiazepines. Advanc. Biochem. Psychopharmacol. **14**, 131—151 (1975)
Haertzen, C. A.: Addiction Research Center Inventory (ARCI): development of a general drug estimation scale. J. nerv. ment. Dis. **141**, 300—306 (1965)
Haertzen, C. A.: Development of scales based on patterns of drug effects, using the Addiction Research Center Inventory (ARCI). Psychol. Rep. **18**, 163—194 (1966)
Haertzen, C. A., Hill, H. E.: Assessing subjective effects of drugs: an index of carelessness and confusion for use with the Addiction Research Center Inventory (ARCI). J. clin. Psychol. **19**, 407—412 (1963)
Haertzen, C. A., Hill, H. E., Belleville, R. E.: Development of the Addiction Research Center Inventory (ARCI): selection of items that are sensitive to the effects of various drugs. Psychopharmacologia (Berl.) **4**, 155—166 (1963)
Haertzen, C. A., Hooks, N. T.: Dictionary of drug associations to heroin, benzedrine, alcohol, barbiturates and marijuana. J. clin. Psychol. **29**, 115—164 (1973)

Haertzen, C. A., Hooks, N. T., Pross, M.: Drug associations as a measure of habit strength for specific drugs. J. nerv. ment. Dis. **158**, 189—197 (1974)
Haertzen, C. A., Meketon, M. J., Hooks, N. T.: Subjective experiences produced by the withdrawal of opiates. Brit. J. Addict. **65**, 245—255 (1970a)
Haertzen, C. A., Monroe, J. J., Hill, H. E., Hooks, N. T.: Manual for alcoholic scales of the Inventory of Habits and Attitudes (IHA). Psychol. Rep. **25**, 947—973 (1969)
Haertzen, C. A., Monroe, J. J., Hooks, N. T., Hill, H. E.: The language of addiction. Int. J. Addict. **5**, 115—129 (1970b)
Haffner, J. F. W., Mørland, J., Setekleiv, J., Strømsaether, C. E., Danielsen, A., Frivik, P. T., Dybing, F.: Mental and psychomotor effects of diazepam and ethanol. Acta pharmacol. (Kbh.) **32**, 161—178 (1973)
Haggard, H. W.: Devils, Drugs and Doctors. New York: Harper and Brothers 1929
Haley, T. J., Gidley, J. T.: Pharmacological comparison of R(+), S(−) and racemic secobarbital in mice. Europ. J. Pharmacol. **9**, 358—361 (1970)
Hall, G. M., Kirtland, S. J., Baum, H.: The inhibition of mitochondrial respiration by inhalational anaesthetic agents. Brit. J. Anaesth. **45**, 1005—1009 (1973)
Halsey, M. J.: Structure-activity relationships of inhalational anaesthetics. In: Molecular Mechanisms in General Anaesthesia. Edinburgh-London-New York: Churchill Livingstone 1974
Halsey, M. J., Millar, R. A., Sutton, J. A.: Molecular mechanisms in general anaesthesia. Edinburgh-London-New York: Churchill Livingstone 1974
Hammel, H. T., Heller, H. C., Sharp, F. R.: Probing the rostal brainstem of anesthetized unanesthetized, and exercising dogs and hibernating and euthermic ground squirrels. Fed. Proc. **32** 1588—1597 (1973)
Han, Y. H.: Why do chronic alcoholics require more anesthesia? Anesthesiology **30**, 341—342 (1969)
Hanna, S. M.: A case of oxazepam (Serenid D) dependence. Brit. J. Psychiat. **120**, 443—445 (1972)
Hansen, J. E., Claybaugh, J. R.: Ethanol-induced lowering of arterial oxyhemoglobin saturation during hypoxia. Aviation, Space, Environmental Med. **46**, 1123—1127 (1975)
Harrfeldt, 1965 cited by Scheinin, B. (1971)
Harvey, S. C.: Hypnotics and sedatives. 5th ed. The Pharmacological basis of therapeutics. Goodman, L. S., Gilman, A. (Eds.). Macmillan: New York 1975
Hashimoto, T., Shuto, K., Ichikawa, S., Shiozaki, S., Kojima, T., Takahira, H.: Studies on flurazepam (I). Effects of flurazepam on the central nervous system. Pharmacometrics **7**, 381—398 (1973)
Hatch, R. C.: Effect of autonomic blocking agents on development of acute tolerance to thiopental in dogs. Amer. J. vet. Res. **33**, 365—376 (1972)
Hatch, R. C., Currie, R. B., Grieve, G. A.: Feline electroencephalograms and plasma thiopental concentrations associated with clinical stages of anesthesia. Amer. J. vet. Res. **31**, 291—306 (1970)
Haug, J. O.: Pneumonencephalographic evidence of brain damage in chronic alcoholics. Acta psychiat. scand. Suppl. **203**, 135—143 (1968)
Haugen, F. P., Melzack, R.: The effects of nitrous oxide on responses evoked in the brain stem by tooth stimulation. Anesthesiology **18**, 183—195 (1957)
Hawkins, R. D., Kalant, H., Khanna, J. M.: Effects of chronic intake of ethanol on rate of ethanol metabolism. J. Physiol. Pharmacol. **44**, 241—257 (1966)
Hayes, R., Price, D. D., Dubner, R.: Naloxone antagonism as evidence for narcotic mechanisms. Science **196**, 600 (1977)
Heath, R. G.: Pleasure response of human subjects to direct stimulation of the brain: physiologic and psychodynamic considerations. In: Heath, R. G. (Ed.): The Role of Pleasure in Behavior. New York: Harper and Row 1964
Heath, R. G.: Pleasure and brain activity in man. Deep and surface electroencephalograms during orgasm. J. nerv. ment. Dis. **154**, 3—18 (1972)
Heller, A., Harvey, J. A., Hunt, H. F., Roth, L. J.: Effects of lesions in the septal forebrain of the rat on sleeping time under barbiturate. Science **131**, 662—664 (1960)
Henatsch, H. D., Ingvar, D. H.: Chlorphromazin and spastizitat. Arch. Psychiat. Z. Neurol. **195**, 77—93 (1956)
Heston, W. D. W., Erwin, V. G., Anderson, S. M., Robbins, H.: A comparison of the effects of alcohol on mice selectively bred for differences in ethanol sleep-time. Life Sci. **14**, 365—370 (1974)

Hill, H. E.: The social deviant and initial addiction to narcotics and alcohol. Quart. J. Stud. Alcohol **23**, 562—582 (1962)
Hill, H. E., Haertzen, C. A., Wolbach, A. B., Miner, E. J.: The addiction research center inventory: appendix. Psychopharmacologia (Berl.) **4**, 184—205 (1963a)
Hill, H. E., Haertzen, C. A., Wolbach, A. B., Miner, E. J.: The addiction research center inventory: standardization of scales which evaluate subjective effects of morphine, amphetamine, pentobarbital, alcohol, LSD-25, pyrahexyl and chlorpromazine. Psychopharmacologia (Berl.) **4**, 167—183 (1963b)
Hill, R. G., Simmonds, M. A., Straughan, D. W.: Convulsant substances as antagonists of GABA and presynaptic inhibition in cuneate nucleus. Brit. J. Pharmacol. **52**, 117 (1974)
Himwich, H. E., Callison, D. A.: The effects of alcohol on evoked potentials of various parts of the central nervous system of the cat. In: Kissin, B., Begleiter, H. (Eds.): The Biology of Alcoholism, Vol. 2, pp. 67—84. New York: Plenum Press 1972
Himwich, H. E., DiPerri, R., Dravid, A., Schweigerdt, A.: Comparative susceptibility to alcohol of the cortical area and midbrain reticular formation of the cat. Psychosom. Med. **28**, 458—463 (1966)
Hinkley, R. E. Jr., Telser, A. G.: The effects of holothane on microfilamentous systems in cultured neuroblastoma cells, 103—120. In: Fink, B. R. (Ed.): Progress in Anesthesiology, Vol. 1. Molecular Mechanisms of Anesthesia. New York: Raven Press 1975
Ho, A. K. S., Chen, R. C. A., Morrison, J. M.: Interactions of narcotis, narcotic antagonists and ethanol during acute, chronic and withdrawal states. Ann. N.Y. Acad. Sci. **281**, 297—310 (1976)
Ho, A. K. S., Tsai, C. S.: Lithium and ethanol preference communications. J. Pharm. (Lond.) **27**, 58—59 (1975)
Ho, P., McLean, J. R.: Effect of drugs on fighting behavior and motor activity in mice. Fed. Proc. **33**, 465 (1974)
Hobbiger, F.: The mechanism of anticurare action of certain neostigmine analogues. Brit. J. Pharmacol. **7**, 223—236 (1952)
Hocherman, S.: Evidence for a region specifically sensitive to electroanesthesia in the brain stem of the cat. Int. J. Neurosci. **3**, 15—28 (1972)
Hockman, C. H., Livingston, K. E.: Inhibition of reflex vagal bradycardia by diazepam. Neuropharmacology **10**, 307—314 (1971)
Hoffmeister, F.: Negative reinforcing properties of some psychotropic drugs in drug-naive rhesus monkeys. J. Pharmacol. exp. Ther. **192**, 468—477 (1975)
Hogans, A. F., Moreno, O. M., Brodie, D. A.: Effects of ethyl alcohol on EEG and avoidance behavior of chronic electrode monkeys. Amer. J. Physiol. **201**, 434—436 (1961)
Hollister, L. E.: Antianxiety drugs in clinical practice. In: Garattini, S., Mussini, E., Randall, L. O. (Eds.): The Benzodiazepines. New York: Raven Press 1973
Hollister, L. E.: Interactions of Δ^9-tetrahydrocannabinol with other drugs. In: Interactions of drugs of abuse. Ann. N.Y. Acad. Sci. **281**, 212—218 (1976)
Hollister, L. E., Motzenbecker, F. P., Degan, R. O.: Withdrawal reactions from chlordiazepoxide ("Librium"). Psychopharmacologia (Berl.) **2**, 63—68 (1961)
Holtzman, J. L., Thompson, J. A.: Metabolism of R-(+)- and S-(−)-pentobarbital by hepatic microsomes from male rats. Drug Metab. Disposition **3**, 113—117 (1975)
Hong, S. K.: Pattern of cold adaptation in women divers of Korea (ama). Fed. Proc. **32**, 1614—1622 (1973)
Hoogland, D. R., Miya, T. S., Bousquet, W. F.: Metabolism and tolerance studies with chlordiazepoxide-2-^{14}C in the rat. Toxicol. appl. Pharmacol. **9**, 116—123 (1966)
Horn, G.: Changes in neuronal activity and their relationship to behavior. In: Horn, G., Hinde, R. A. (Eds.): Short-term Changes in Neural Activity and Behaviour, pp. 567—602. Great Britain: Cambridge University Press 1970
Horn, G., Hinde, R. A. (Eds.): Short-term Changes in Neural Activity and Behaviour. Great Britain: Cambridge University Press 1970
Horsey, W. J., Akert, K.: The influence of ethyl alcohol on the spontaneous electrical activity of the cerebral cortex and subcortical structures of the cat. Quart. J. Stud. Alcohol **14**, 365—377 (1953)

Houck, D. J.: Effects of alcohols on potentials of lobster axons. Amer. J. Physiol. **216**, 364—367 (1969)
Hug, C. C.: Characteristics and theories related to acute and chronic tolerance development. In: Chemical and Biological Aspects of Drug Dependence, pp. 307—345. Cleveland, Ohio: CRC Press 1972
Hughes, F. W., Forney, R. B., Richards, A. B.: Comparative effect in human subjects of chlordiazepoxide, diazepam, and placebo on mental and physical performance. Clin. Pharmacol. Ther. **6**, 139—145 (1965)
Hupka, A. L., Williams, J. K., Karler, R.: Effects of convulsant barbiturates on vascular smooth muscle. J. Pharm. (Lond.) **21**, 838—844 (1969)
Hurst, P. M., Bagley, S. K.: Acute adaptation to the effects of alcohol. Quart. J. Stud. Alcohol **33**, 358—378 (1972)
Inaba, D. S., Gay, G. R., Newmeyer, J. A., Whitehead, C.: Methaqualone abuse "Luding" out. J. Amer. med. Ass. **224**, 1505—1509 (1973)
Ingle, D.: Enhancement by ethanol of visually evoked responses in the goldfish optic tectum. Exp. Neurol. **33**, 329—342 (1971)
Ingram, L. O.: Adaptation of membrane lipids to alcohols. J. Bacteriol. **125**, 670—678 (1976)
Ingvar, D. H., Lassen, N. A. (Eds.): The Working Brain: The Coupling of Function, Metabolism and Blood Flow in the Brain. Copenhagen: Munksgaard 1975
Inoue, F., Frank, G. B.: Effects of ethyl alcohol on excitability and on neuromuscular transmission in frog skeletal muscle. Brit. J. Pharmacol. **30**, 186—193 (1967)
Irwin, S.: Factors influencing sensitivity of stimulant and depressant drugs affecting (a) locomotor and (b) conditioned avoidance behavior in animals. In: Sarwer-Foner, G. J. (Ed.): The Dynamics of Psychiatric Drug Therapy, pp. 5—28. Springfield: C. C. Thomas 1960
Irwin, S.: A rational approach to drug abuse prevention. Contemp. Drug Probl. **2**, 3—46 (1973)
Irwin, S., Stagg, R. D., Dunbar, E., Govier, W. M.: Methitural, a new intravenous anesthetic: comparison with thiopental in the cat, dog and monkey. J. Pharmacol. exp. Ther. **116**, 317—325 (1956)
Isbell, H., Altschul, S., Kornetsky, C. H., Eisenman, A. J., Flanary, H. G., Fraser, H. F.: Chronic barbiturate intoxication. Arch. Neurol. Psychiat. (Chic.) **64**, 1—28 (1950)
Isbell, H., Fraser, H. F.: Addiction to analgesics and barbiturates. Pharmacol. Rev. **2**, 355—397 (1950)
Isbell, H., Fraser, H. F., Wikler, A., Belleville, R. E., Eisenman, A. J.: An experimental study of the etiology of "rum fits" and delirium tremens. Quart. J. Stud. Alcohol **16**, 1—33 (1955)
Ishikawa, T., Sadanaga, Y., Katsuta, S., Ishiyama, J., Kobayashi, T.: Hippocampal after-discharge and the mode of action of psychotropic drugs. Progr. Brain Res. **21B**, 40—53 (1966)
Israel, M. A., Kuriyama, K.: Effect of *in vivo* ethanol administration on adenosinetriphosphatase. Life Sci. **10**, 591—599 (1971)
Israel, Y.: Cellular effects of alcohol. A Review. Quart. J. Stud. Alcohol **31**, 293—316 (1970)
Israel, Y., Kalant, H., Laufer, I.: Effects of ethanol on Na, K, Mg-stimulated microsomal ATPase activity. Biochem. Pharmacol. **14**, 1803—1814 (1965)
Israel, Y., Kalant, H., LeBlanc, A. E.: Effects of lower alcohols on potassium transport and microsomal adenosine-triphosphatase activity of rat cerebral cortex. Biochem. J. **100**, 27—33 (1966)
Israel, Y., Kalant, H., LeBlanc, E., Bernstein, J. C., Salazar, I.: Changes in cation transport and (Na + K)-activated adenosine triphosphatase produced by chronic administration of ethanol. J. Pharmacol. exp. Ther. **174**, 330—336 (1970)
Israel, Y., Kalant, H., Orrego, H. L., Khanna, J. M., Videla, L., Phillips, J. M.: Experimental alcohol-induced hepatic necrosis: Suppression by propylthiouracil. Proc. nat. Acad. Sci. (Wash.) **72**, 1137—1141 (1975b)
Israel, Y., Mardones, J.: Biological Basis of Alcoholism. New York: Wiley-Interscience 1971
Israel, Y., Salazar, I.: Inhibition of brain microsomal adenosine triphosphatases by general depressants. Arch. Biochem. **122**, 310—317 (1967)
Israel, Y., Videla, L., Bernstein, J.: Liver hypermetabolic state after chronic ethanol consumption: hormonal interrelations and pathogenic implications. Fed. Proc. **34**, 2052—2059 (1975a)
Israel, Y., Videla, L., MacDonald, A., Bernstein, J.: Comparison between the effects produced by ethanol and by thyroid hormones. Biochem. J. **134**, 523—529 (1973)

Israel, M. A., Kuriyama, K., Yoshihawa, K.: Effect of ethanol administration on axoplasmic flow in the brain. Neuropharmacol. **14**, 445—451 (1975c)

Israel-Jacard, Y., Kalant, H.: Effect of ethanol on electrolyte transport and electrogenesis in animal tissues. J. cell. comp. Physiol. **65**, 127—132 (1965)

Itil, T., Gannon, P., Cora, R., Polvan, N., Akpinar, S., Elveris, F., Eskazan, E.: SCH-12,041, A new anti-anxiety agent (quantitative pharmaco-electroencephalography and clinical trials). Phycns. Drug Manual **3**, 26—35 (1971)

Iversen, L. L.: Neuronal uptake processes for amine and amino acid transmitters. In: Callingham, B. A. (Ed.): Drugs and Transport Processes. Baltimore-London-Tokyo: University Park Press 1974

Jacobi, H.: Hemmung des Tremors durch β-adrenolytisch wirksame Substanzen. Naturwissenschaften **54**, 94 (1967)

Jacquet, Y. F., Lajtha, A.: Analgesic tolerance in the perioqueductal gray of the rat following morphine administrations. Fed. Proc. **34**, 786 (1975)

Jaffe, J. H.: Drug addiction and drug abuse. In: Goodman, L. S., Gilman, A. (Eds.): Pharmacological Basis of Therapeutics, 4. Ed. New York: Macmillan 1970

Jaffe, J. H., Sharpless, S. K.: The rapid development of physical dependence on barbiturates. J. Pharmacol. exp. Ther. **150**, 140—145 (1965)

Jaffe, J. H., Sharpless, S. K.: Pharmacological denervation supersensitivity in the central nervous system: A theory of physical dependence. In: The Addictive States, pp. 226—246. Baltimore, Maryland: Williams & Wilkins Co. 1968

James, W.: Varieties of Religious Experience, pp. 298—300. New York-London: Longman's Green 1902

Jarcho, L. W.: Excitability of cortical afferent systems during barbiturate anesthesia. J. Neurophysiol. **12**, 447—457 (1949)

Järnefelt, J.: Inhibition of the brain microsomal adenosinetriphosphatase by depolarizing agents. Biochim. biophys. Acta (Amst.) **48**, 111—116 (1961)

Järnefelt, J.: Lipid requirements of functional membrane structures as indicated by the reversible inactivation of (Na^+-K^+)-ATPase. Biochim. biophys. Acta (Amst.) **266**, 91—96 (1972)

Jasinski, D. R.: Assessment of the dependence liability of opiates and sedative-hypnotics. In: Goldberg, L., Hoffmeister, F. (Eds.): Psychic Dependence. Definition, Assessment in Animals and Man; Theoretical and Clinical Implications. Berlin-Heidelberg-New York: Springer 1973

Jellinek, E. M.: The Disease Concept of Alcoholism. New Haven, Conn.: College and University Press—New Brunswick, N.J.: Hillhouse Press 1960

John, E. R.: State dependent learning. In: Mechanisms of Memory, pp. 67—91. New York: Academic Press 1967

Johnson, F. H., Flagler, E. A.: Hydrostatic pressure reversal of narcosis in tadpoles. Science **112**, 91—92 (1950)

Johnson, F. H., Flagler, E. A.: Activity of narcotized amphibian larvae under hydrostatic pressure. J. cell. Physiol. **37**, 15—25 (1951)

Johnson, G. E., Patel, V. K.: Influence of chronic ethanol consumption of the distribution of thiopental in rats. Int. Congr. Pharmacol. Helsinki 1975

Johnstone, R. E., Kulp, R. A., Smith, T. C.: Effects of acute and chronic ethanol administration on isoflurane requirement in mice. Anesth. Analg. Curr. Res. **54**, 277—281 (1975)

Jones, B. E., Prada, J. A., Martin, W. R.: A method for 1310 assay of physical dependence on sedative drugs in dogs. Psychopharmacology **47**, 7—15 (1976)

Jones, B. M., Vega, A.: Cognitive performance measured on the ascending and descending limb of the blood alcohol curve. Psychopharmacologia (Berl.) **23**, 99—114 (1972)

Jones, R. T., Stone, G. C.: Psychological studies of marijuana and alcohol in man. Psychopharmacologia (Berl.) **18**, 108—117 (1970)

Joyce, C. R. B.: Quantitative estimates of dependence on the symbolic function of drugs. In: Scientific Basis of Drug Dependence. Steinberg, H. (ed.), 271—284. London: Churchill Ltd. 1969

Kaim, S. C.: Benzodiazepines in the treatment of alcohol withdrawal states. In: Garattini, S., Mussini, E., Randall, L. O. (Eds.): The Benzodiazepines, pp. 571—575. New York: Raven Press 1973

Kaim, S. L., Klett, C. J.: Treatment of delirium tremens: a comparative evaluation of four drugs. Quart. J. Stud. Alcohol **33**, 1065—1072 (1972)

Kakihana, R., Brown, D. R., McClearn, G. E., Tabershaw, I. R.: Brain sensitivity to alcohol in inbred mouse strains. Science **154**, 1574—1575 (1966)

Kalant, H.: Pharmacological and behavioral variables in the development of alcohol tolerance. Pharmacology and the Future of Man. Proc. 5th Int. Congr. Pharmacology, San Francisco 1972, Vol. 1, pp. 44—55. Basel: Karger 1973a

Kalant, H.: Biological models of alcohol tolerance and physical dependence. In: Gross, M. (Ed.): Alcohol Intoxication and Withdrawal Experimental Studies, Vol. 35, pp. 3—14. New York-London: Plenum Press 1973b

Kalant, H.: Ethanol and the nervous system experimental neurophysiological aspects. Int. J. Neurol. **9**, 111—120 (1974)

Kalant, H.: Direct effects of ethanol on the nervous system. Fed. Proc. **34**, 1930—1941 (1975)

Kalant, H., Grose, W.: Effects of ethanol and pentobarbital on release of acetylcholine from cerebral cortex slices. J. Pharmacol. exp. Ther. **158**, 386—393 (1967)

Kalant, H., Israel, Y.: Effects of ethanol on active transport of cations. In: Maickel, R. E. (Ed.): Biochemical Factors in Alcoholism, pp. 25—37. Oxford: Pergamon 1967

Kalant, H., Sereny, G., Charlebois, R.: Evaluation of triiodothyronine in the treatment of acute alcohol intoxication. New Engl. J. Med. **267**, 1—6 (1962)

Kalant, H., LeBlanc, A. E., Gibbins, R. J.: Tolerance to, and dependence on, some non-opiate psychotropic drugs. Pharmacol. Rev. **23**, 135—191 (1971)

Kales, A., Bixler, E. O., Tan, T. L., Schart, M. B., Kales, J. D.: Chronic hypnotic drug use. Ineffectiveness, drug-withdrawal insomnia, and dependence. J. Amer. med. Ass. **227**, 513—517 (1974)

Kales, A., Kales, J. D.: Sleep disorders. Recent findings in the diagnosis and treatment of disturbed sleep. New Engl. J. Med. **290**, 487—499 (1974)

Kandel, E. R.: Cellular Basis of Behavior. An Introduction to Behavioral Neurobiology. San Francisco: W. H. Freeman 1976

Kandel, E., Castellucci, V., Pinsker, H., Kupfermann, I.: The role of synaptic plasticity in the short-term modification of behaviour. In: Short-term Changes in Neural Activity and Behaviour, pp. 281—319. London: Cambridge University Press 1970

Karobath, M., Leitich, H.: Antipsychotic drugs and dopamine-stimulated adenylate cyclase prepared from corpus striatum of rat brain. Proc. nat. Acad. Sci. (Wash.) **71**, 2915—2918 (1974)

Kater, R. M. H., Carulli, N., Iber, F. L.: Differences in the rate of ethanol metabolism in recently drinking alcoholic and non drinking subjects. Amer. J. clin. Nutr. **22**, 1608—1617 (1969)

Katz, B., Thesleff, S.: A study of the "desensitization" produced by acetylcholine at the motor end-plate. J. Physiol. (Lond.) **138**, 63—80 (1957)

Kay, D. C., Jasinski, D. R., Eisenstein, R. B., Kelly, O. A.: Quantified human sleep after pentobarbital. Clin. Pharmacol. Ther. **13**, 221—231 (1972)

Kayan, S., Woods, L. A., Mitchell, C. L.: Experience as a factor in the development of tolerance to the analgesic effect of morphine. Europ. J. Pharmacol. **6**, 333—339 (1969)

Keats, A. A., Beecher, H. K.: Pain relief with hypnotic doses of barbiturates and a hypothesis. J. Pharmacol. exp. Ther. **100**, 1—13 (1950)

Keeler, M. H., McCurdy, R. L.: Medical practice without antianxiety drugs. Amer. J. Psychiat. **132**, 654—655 (1975)

Keilty, S. R., Blackwood, S.: Sedation for conservative dentistry. Brit. J. clin. Pract. **23**, 365—367 (1969)

Keller, M., McCormick, M.: A Dictionary of Words about Alcohol. New Brunswick, New Jersey. Rutgers Center of Alcohol Studies 1968

Kellner, R., Collins, A. C., Shulman, R. S., Pathak, D.: The short-term antianxiety effects of Propranolol HCl. J. clin. Pharmacol. **14**, 301—306 (1974)

Kendig, J. J., Cohen, E. N.: Depression of synaptic transmission by stereoisomers of halothane. Abstracts of American Society of Anesthesiologists Annual Meeting, 93 (1971)

Kendig, J. J., Cohen, E. N.: Neural sites of pressure—anesthesia interactions. In: Fink, B. R. (Ed.): Progress in Anesthesiology, Vol. 1. Molecular Mechanisms of Anesthesia. New York: Raven Press 1975

Kendig, J. J., Trudell, J. R., Cohen, E. N.: Effects of pressure and anesthetics on conduction and synaptic transmission. J. Pharmacol. exp. Ther. **195**, 216—224 (1975)

Kennedy, R. D., Galindo, A.: Comparative site of action of various anesthetics at the mammalian myoneural junction. Fed. Proc. **33**, 579 (1974)

Kety, S. S.: Circulation and energy metabolism of the brain. In: Tower, D. B. (Ed.): The Nervous System, Vol. 1, The Basic Neurosciences, pp. 197—205. New York: Raven Press 1975

Khanna, J. M., Kalant, H.: Effect of inhibitors and inducers of drug metabolism on ethanol *in vivo*. Biochem. Pharmacol. **19**, 2033—2041 (1970)

Khanna, J. M., Kalant, H., Bustos, G.: Effects of chronic intake of ethanol on rate of ethanol metabolism. II. influence of sex and of schedule of ethanol administration. Canad. J. Physiol. Pharmacol. **45**, 777—785 (1967)

Kiianmaa, K.: The effect of brain serotonin depletion on alcohol consumption in the rat. Sixth International Congress of Pharmacology, Helsinki, Abstracts, 245 (1975)

Kiianmaa, K., Fuxe, K., Jonsson, G., Ahtee, L.: Evidence for involvement of central neurones in alcohol intake. Increased alcohol consumption after degeneration of the NA pathway to the cortex cerebri. Neurosci. Letters **1**, 41—45 (1975a)

Kiianmaa, K., Fuxe, K., Jonsson, G., Ahtee, L.: Increased alcohol consumption after degeneration of central noradrenaline neurons. First European Neurosciences Meeting, Munich. Abstracts. Exp. Brain Res. (Berl.) **23**, 107 Suppl. (1975b)

Killam, E. K.: Drug action on the brain-stem reticular formation. Pharmacol. Rev. **14**, 175—223 (1962)

Killam, E. K., Matsuzaki, M., Killam, K. F.: Effects of chronic administration of benzodiazepines on epileptic seizures and brain electrical activity in papio papio. In: S. Garattini, E. Mussini, L. O. Randall (Eds.): The Benzodiazepines. New York: Raven Press 1973

Killam, K. F., Brocco, M. J., Robison, C. A.: Evaluation of narcotic and narcotic antagonist interactions in primates. Ann. N.Y. Acad. Sci. **281**, 331—335 (1976)

Killam, K. F., Brody, T. M., Bain, J. A.: Potentiation of barbiturate hypnosis by certain uncoupling agents. Proc. Soc. exp. Biol. (N.Y.) **97**, 744 (1958)

Kissin, B.: The use of psychoactive drugs in the long term treatment of chronic alcoholics. Paper presented at the 20th International Institute on the Prevention and Treatment of Alcoholism, Manchester, England 1974

Kissin, B.: The use of psychoactive drugs in the long-term treatment of chronic alcoholics. Ann. N.Y. Acad. Sci. **252**, 385—395 (1975)

Kissin, B., Begleiter, H.: The Biology of Alcoholism, Vol. 1: Biochemistry. New York: Plenum Press 1971, Vol. 2: Physiology and Behavior. New York: Plenum Press 1972. Vol. 3: Clinical Pathology. New York: Plenum Press 1974

Klee, M. R., Faber, D. S.: Mephenesin blocks early inward currents and strychnine-induced multiple discharges of Aplysia neurons. Pflügers Arch. **346**, 97—106 (1974)

Klee, M. R., Lee, K. C., Park, M. R.: Changes in membrane properties of cat motoneurons due to ethanol. Exp. Brain Res. **23**, 108 (1975a)

Klee, M. R., Faber, D. S., Bergmann, M. C.: Personal communication 1975b

Klein, D. F.: Psychotropic drugs and the regulation of behavioral activation in psychiatric illness. In: Smith, W. L. (Ed.): Drugs and Cerebral Function. Springfield, Ill.: Charles C. Thomas 1970

Kline, N. S., Wren, J. C., Cooper, T. B., Varga, E., Canal, O.: Evaluation of lithium therapy in chronic and periodic alcoholism. Amer. J. med. Sci. **268**, 15—22 (1974)

Klingman, G. I., Goodall, McC: Urinary epinephrine and levarterenol excretion during acute sublethal alcohol intoxication in dogs. J. Pharmacol. exp. Ther. **121**, 313—318 (1957)

Knisely, M. H., Reneau, D. D., Bruley, D. F.: The development and use of equations for predicting the limits on the rates of oxygen supply to the cells of living tissues and organs. J. vasc. Dis. **20**, 1—56 (1969)

Knox, W. H., Perrin, R. G., Sen, A. K.: Effect of chronic administration of ethanol on (Na+K)-activated ATPase activity in six areas of the cat brain. J. Neurochem. **19**, 2881—2884 (1972)

Knutsson, E.: Effects of ethanol on the membrane potential and membrane resistance of frog muscle fibres. Acta physiol. scand. **52**, 242—253 (1961)

Knutsson, E., Katz, S.: The effect of ethanol on the membrane permeability to sodium and potassium ions in frog muscle fibres. Acta pharmacol. (Kbh.) **25**, 54—64 (1967)

Koch-Weser, J., Greenblatt, D. J.: The archaic barbiturate hypnotics. New Engl. J. Med. **291**, 790—791 (1974)

Körlin, D., Larson, B.: Differences in cerebellar potentials evoked by the group I and cutaneous components of the cuneocerebellar tract. In: Andersen, P., Jansen, J.K.S. (Eds.): Excitatory Synaptic Mechanisms. Oslo-Bergen-Tromsö: Universitetsforlaget 1970

Kohli, R.P., Singh, N., Kulshrestha, V.K.: An experimental investigation of dependence liability of methaqualone in rats. Psychopharmacologia (Berl.) **35**, 327—334 (1974)

Kokoski, R.J., Hamner, S., Shiplet, M.: Benzodiazepine tranquilizer abuse in narcotic addict treatment programs: urinalyses as a detection and control measure. 36th meeting of the Committee on Problems of Drug Dependence, Nat. Academy of Sciences, March, 1974. Mexico City, Mexico

Kolmodin, G.M.: The action of ethyl alcohol on the monosynaptic extensor reflex and the multisynaptic reflex. Acta physiol. scand. **29**, Suppl. 106, 530—537 (1953)

Konishi, J., Hickman, C.P. Jr.: Temperature acclimation in the central nervous system of the rainbow trout (Salmo gairdnerii). Comp. Biochem. Physiol. **13**, 433—442 (1964)

Korttila, K., Linnoila, M.: Skills related to driving after intravenous diazepam, flunitrazepam and droperidol. Brit. J. Anaesth. **46**, 961 (1974)

Korttila, K., Linnoila, M.: Recovery and skills related to driving after intravenous sedation: Dose-response relationship with diazepam. Brit. J. Anaesth. **47**, 457—463 (1975a)

Korttila, K., Linnoila, M.: Psychomotor skills related to driving after intramuscular administration of diazepam and meperidine. Anesthesiology **42**, 685—691 (1975b)

Kotchabhakdi, N., Prosser, C.L.: Behavioral electrophysiological, and ultrastructural effects of cooling on the goldfish cerebellum. Abstract, Society for Neuroscience, Houston, Texas, October 8—11 (1972)

Krieglstein, J., Stock, R.: The isolated perfused rat brain as a model for studying drugs acting on the CNS. Psychopharmacologia (Berl.) **35**, 169—177 (1974)

Krnjević, K.: Excitable membranes and anesthetics. In: Cellular Biology and Toxicity of Anesthetics. Baltimore: Williams & Wilkins Co. 1972

Krnjević, K.: Central actions of general anaesthetics. In: Halsey, M.J., Millar, R.A., Sutton, J.A. (Eds.): Molecular Mechanisms in General Anaesthesia. Edinburgh-London-New York: Churchill Livingstone 1974a

Krnjević, K.: Chemical nature of synaptic transmission in vertebrates. Physiol. Rev. **54**, 418—540 (1974b)

Krnjević, K.: Is general anesthesia induced by neuronal asphyxia? In: Fink, B.R. (Ed.): Progress in Anesthesiology, Vol. 1, pp. 93—102, Molecular Mechanisms of Anesthesia. New York: Raven Press 1975

Krnjević, K., Phillis, J.W.: Acetylcholine-sensitive cells in the cerebral cortex. J. Physiol. (Lond.) **166**, 296—327 (1963a)

Krnjević, K., Phillis, J.W.: Pharmacological properties of acetylcholine-sensitive cells in the cerebral cortex. J. Physiol. (Lond.) **166**, 328—350 (1963b)

Krooth, R.S., May, S.R.: A molecular theory of natural and drug induced sleep. Bull. N.Y. Acad. Med. **51**, 1172 (1975)

Krsiak, M.: Behavioral changes and aggressivity evoked by drugs in mice. Res. Commun. chem. Path. Pharmacol. **7**, 237—257 (1974)

Krug, D.C., Sokol, J., Nylander, L.: Inhalation of commerical solvents: a form of defiance among adolescents. In: Harms, E. (Ed.): Drug Addiction in Youth. New York: Pergamon Press 1965 (cited by Hug, 1972)

Kryspin-Exner, K., Demel, I.: The use of tranquilizers in the treatment of mixed drug abuse. Int. J. clin. Pharmacol. **12**, 13—18 (1975)

Kucera, J., Smith, C.M.: Excitation by ethanol of rat muscle spindles. J. Pharmacol. exp. Ther. **179**, 301—311 (1971)

Kucera, J., Smith, C.M.: Muscle afferent outflow during ethanol intoxication. Experientia (Basel) **28**, 908—909 (1972)

Kupfer, D.J., Wyatt, R.J., Synder, F., Davis, J.M.: Chlorpromazine and sleep in psychiatric patients. Arch. gen. Psychiat. **24**, 185—189 (1971)

Kusano, K.: Influence of ionic environment on the relationship between pre- and postsynaptic potentials. J. Neurobiol. **1**, 435—457 (1969)

LaBella, F.S., Pinsky, C.: Opiate receptor: alcohol and acetaldehyde enhance agonist and diminish antagonist binding in vitro. Abstract. Canad. Fed. Biol. Soc., 1976

Lagerspetz, K.: The induction of physiological tolerance to promazine in mice. II. The development of induced tolerance. Ann. Med. exp. Fenn. **41**, 214—219 (1963)
Lahti, R. A., Barsuhn, C.: The effect of minor tranquilizers on stress-induced increases in rat plasma corticosteroids. Psychopharmacologia (Berl.) **35**, 215—220 (1974)
Lake, N., Yarbrough, G. G., Phillis, J. W.: Effects of ethanol on cerebral cortical neurons: Interactions with some putative transmitters. Letters to the Editor, J. Pharm. Pharmac. **25**, 582—584 (1973)
Landauer, A. A., Pocock, D. A., Prott, F. W.: The effect of medazepam and alcohol on cognitive and motor skills used in car driving. Psychopharmacologia (Berl.) **37**, 159—168 (1974)
Larrabee, M. G., Bronk, D. W.: Metabolic requirements of sympathetic neurons. Cold Spr. Harb. quant. Biol. **17**, 245—266 (1952)
Larrabee, M. G., Posternak, J. M.: Selective action of anesthetics on synapses and axons in mammalian sympathetic ganglia. J. Neurophysiol. **15**, 91—114 (1952)
Larrabee, M. G., Ramos, J. B., Bülbring, E.: Effects of anesthetics on oxygen consumption and on synaptic transmission in sympathetic ganglia. Cell. comp. Physiol. **40**, 461—494 (1952)
Larson, M. D., Major, M. A.: The effect of hexobarbital on the duration of the recurrent IPSP on cat motoneurons. Brain Res. **21**, 309—311 (1970)
Larson, P. S., Silvette, H.: Tobacco, Experimental and Clinical Studies. Baltimore: Williams & Wilkins Company 1968
Latham, A., Paul, D. H.: Effects of sodium thiopentone on cerebellar neurone activity. Brain Res. **25**, 212—215 (1971)
Lathers, C. M., Smith, C. M.: Ethanol effects on phasic and static muscle spindle afferent activity in the cat. Fed. Proc. **32**, 730 (1973)
Lathers, C. M., Smith, C. M.: Ethanol effects on muscle spindle afferent activity and spinal reflexes. J. Pharmacol. exp. Ther. **197**, 126—134 (1976)
Laurence, D. R., Webster, R. A.: Tachyphylaxis to the anti-tetanus activity of some phenothiazine compounds. Brit. J. Pharmacol. **16**, 296 (1961)
Lazarewicz, J. W., Haljamäe, H., Hamberger, A.: Calcium metabolism in isolated brain cells and subcellular fractions. J. Neurochem. **22**, 33—45 (1974)
LeBlanc, A. E., Cappell, H. D.: Historical antecedents as determinants of tolerance to and dependence upon psychoactive drugs. In: Cappell, H. D., LeBlanc, A. E. (Eds.): Biological and Behavioural Approaches to Drug Dependence, pp. 43—51. Toronto-Ontario: Addiction Research Foundation, House of Lind 1975
LeBlanc, A. E., Kalant, H.: Ethanol-induced cross tolerance to several homologous alcohols in the rat. Toxicol. appl. Pharmacol. **32**, 123—128 (1975)
LeBlanc, A. E., Gibbins, R. J., Kalant, H.: Behavioral augmentation of tolerance to ethanol in the rat. Psychopharmacologia (Berl.) **30**, 117—122 (1973)
LeBlanc, A. E., Kalant, H., Gibbins, R. J.: Acute tolerance to ethanol in the rat. Psychopharmacologia (Berl.) **41**, 43—46 (1975)
LeBlanc, A. E., Kalant, H., Gibbins, R. J., Berman, N. D.: Acquisition and loss of tolerance to ethanol by the rat. J. Pharmacol. exp. Ther. **168**, 244—250 (1969)
LeBlanc, A. E., Matsunaga, M., Kalant, H.: Effects of frontal polar cortical ablation and cycloheximide on ethanol tolerance in rats. Pharmacol. Biochem. Behav. **4**, 175—179 (1976)
Lee, J. C.: Effect of alcohol injections on the blood-brain barrier. Quart. J. Stud. Alcohol **23**, 4—16 (1962)
Lee, P. K., Cho, M. H., Dobkin, A. B., Curtis, D. A.: Effects of alcoholism, morphinism, and barbiturate resistance on induction and maintenance of general anaesthesia. Can. Anaes. Soc. J. **11**, 354—381 (1964)
Lee-Son, S., Waud, B. E., Waud, D. R.: A comparison of the potencies of a series of barbiturates at the neuromuscular junction and on the central nervous system. J. Pharmacol. exp. Ther. **195**, 251—264 (1975)
Lehninger, A. L.: Mitochondria and calcium ion transport. Biochem. J. **119**, 129—138 (1970)
Leighton, K. M., Jenkins, L. C.: Experimental studies of the central nervous system related to anaesthesia: IV. Effects of pentobarbital placement in caudate nucleus. Canad. Anaesth. Soc. J. **17**, 112—118 (1970)
Lennard, H. L., Epstein, L. J., Bernstein, A., Ransom, D. C.: Mystification and Drug Misuse. San Francisco: Jossey-Bass Inc. 1971

Leslie, C. A., Gottesfeld, Z., Elliott, K. A. C.: Effect of ethanol on entry of some substances into the brains of rats. Canad. J. Physiol. Pharmacol. **49**, 833—840 (1971)
Lester, D.: Self-maintenance of intoxication in the rat. Quart. J. Stud. Alcohol **22**, 223—231 (1961)
Levine, R. L., Hoogenraad, N. J., Kretchmer, N.: A review: biological and clinical aspects of pyrimidine metabolism. Pediat. Res. **8**, 724—734 (1974)
Levy, R. A.: GABA: A direct depolarization action at the mammalian primary afferent terminals. Brain Res. **76**, 155—160 (1974)
Lewis, E. G., Dustman, R. E., Beck, E. C.: The effects of alcohol on visual and somatonsensory evoked responses. Electroenceph. clin. Neurophysiol. **28**, 202—205 (1970)
Libet, B.: Electrical stimulation of cortex in human subjects, and conscious sensory aspects. In: Iggo, A. (Ed.): Handbook of Sensory Physiology, Vol. 2, Somato Sensory System. Berlin-Heidelberg-New York: Springer 1973
Lidbrink, P., Corrodi, H., Fuxe, K.: Benzodiazepines and barbiturates: turnover changes in central 5-hydroxytryptamine pathways. Europ. J. Pharmacol. **26**, 35—40 (1974)
Lidbrink, P., Corrodi, H., Fuxe, K., Olson, L.: The effects of benzodiazepines, meprobamate and barbiturates on central monoamine neurons. In: Garattini, S., Mussini, E., Randall, L. O. (Eds.): The Benzodiazepines. New York: Raven Press 1973
Lieb, J. P., Crandall, P. H.: Differential effects of i.v. diazepam on the spontaneous EEG recorded from the limbic system and cortex of the more epileptogenic hemisphere in temporal lobe epileptics. In: Program and Abstracts: Society for Neuroscience, Fourth Annual Meeting 1974
Lieb, J., Sclabassi, R., Crandall, P., Buchness, R.: Comparison of the action of diazepam and phenobarbital using EEG-derived power spectra obtained from temporal lobe epileptics. Neuropharmacology **13**, 769—783 (1974)
Lin, D. C.: Brain and blood levels of ethanol and acetaldehyde in strains of mice with different preferences for ethanol. Res. Commun. chem. Path. Pharmacol. **11**, 365—371 (1975)
Lin, D. C.: Effect of ethanol on the kinetic parameters of brain ($Na^+ + K^+$)-activated adenosine triphosphatase. Ann. N.Y. Acad. Sci. **273**, 331—337 (1976)
Lingeman, R. R.: Drugs from A to Z: a Dictionary. New York: McGraw-Hill 1969
Linnoila, M., Mattila, M. J.: Drug interaction on psychomotor skills related to driving: diazepam and alcohol. Europ. J. clin. Pharmacol. **5**, 186—194 (1973)
Linnoila, M., Otterstrom, S., Anttila, M.: Serum chlordiazepoxide, diazepam and thioridazine concentrations after the simultaneous ingestion of alcohol or placebo drink. Ann. clin. Res. **6**, 4—6 (1974a)
Linnoila, M., Saario, I., Maki, M.: Effect of treatment with diazepam or lithium and alcohol on psychomotor skills related to driving. Europ. J. clin. Pharmacol. **7**, 337—342 (1974b)
Linnoila, M., Saario, I., Mattila, M. J.: Drug-alcohol interaction on psychomotor skills during subacute treatment with benzodiazepines, flupenthixole, or lithium. Brit. J. clin. Pharmacol. **1**, 176P (1974c)
Linseman, M. A.: Effects of lesions of the ventromedial hypothalamus on naloxone induced morphine withdrawal in rats. Psychopharmacologia. (Berl.) **45**, 271—276 (1976)
Liu, Y., Braud, W. G.: Modification of learning and memory in goldfish through the use of stimulant and depressant drugs. Psychopharmacologia (Berl.) **35**, 99—112 (1974)
Llinás, R.: Mechanisms of supraspinal actions upon spinal cord activities. Pharmacological studies on reticular inhibition of alpha extensor motoneurons. J. Neurophysiol. **27**, 1127—1137 (1964)
Longo, V. G.: Effects of mephenesin on the repetitive discharge of spinal cord interneurones. Arch. int. Pharmacol. Rev. **132**, 222—236 (1961)
Longo, V. G.: Neuropharmacology and Behavior. San Francisco: W.H. Freeman and Company 1972
Longo, V. G., Martin, W. R., Unna, K. R.: A pharmacological study on the Renshaw cell. J. Pharmacol. Exp. Ther. **129**, 61 (1960)
Loomis, T. A., West, T. C.: The influence of alcohol on automobile driving ability. An experimental study for the evaluation of certain medicolegal aspects. Quart. J. Stud. Alcohol **19**, 30—46 (1958)
Louria, D. B.: The Drug Scene. New York: McGraw-Hill 1968

Lous, P.: Phenemal concentration in serum and spinal fluid in epileptics treated with phenemal; preliminary report. Ugesk. loeger **114**, 610—611 (1952)

Low, K.: Changes: the Intoxicant Option in Perspective. Calgary: Lawson Graphics Western Limited 1976

Løyning, Y., Oshima, T., Yokota, T.: Site of action of thiamylal sodium on the monosynaptic spinal reflex pathway in cats. J. Neurophysiol. **27**, 408—427 (1964)

Lynn, E. J., James, M., Dendy, R., Harris, L. A., Walter, R. G.: Non-medical use of nitrous oxide: a preliminary report. Mich. Med. **70**, 203—204 (1971)

MacAndrew, C., Edgerton, R. B.: Drunken Comportment. Chicago: Aldine Publ. Comp. 1969

MacDonnell, M. F., Brown, S. H., Davy, B.: Hyperexcitability in the neural substrate of emotional behavior in cats after alcohol withdrawal. J. Stud. Alcohol **36**, 1480—1492 (1975)

MacDonnell, M. F., Fessock, L.: Some effects of ethanol, amphetamine, disulfiram and p-CPA on seizing of prey in feline predatory attack and on associated motor pathways. Quart. J. Stud. Alcohol **33**, 437—450 (1972)

Madsen, W.: The American Alcoholic—The Nature-Nurture Controversy in Alcoholic Research and Therapy. Springfield, Ill.: Charles C. Thomas 1974

Magoun, H. W.: A neural basis for the anesthetic state. In: Symposium on Sedative and Hypnotic Drugs. Baltimore: Williams & Wilkins Co. 1954

Mallach, H. J., Röseler, P.: Beobachtungen und Untersuchungen über die gemeinsame Wirkung von Alkohol und Kohlenmonoxyd. Arzneimittel-Forsch. **11 (11)**, 1004—1008 (1961)

Malpas, A.: Subjective and objective effects of nitrazepam and amylobarbitone sodium in normal human beings. Psychopharmacologia (Berl.) **27**, 373—378 (1972)

Malpas, A., Rowan, A. J., Joyce, C. R. B., Scott, D. F.: Persistent behavioural and electroencephalographic changes after single doses of nitrazepam and amylobarbitone sodium. Brit. med. J. **1970II**, 762—764

Mandell, A. J. (Ed.): Neurobiological Mechanisms of Adaptation and Behavior. Advances in Biochemical Psychopharmacology, Vol. 13. New York: Raven Press 1975a

Mandell, A. J.: Neurobiological mechanisms of presynaptic metabolic adaptation and their organization: implications for a pathophysiology of the affective disorders. In: Mandell, A. J. (Ed.): Neurobiological Mechanisms of Adaptation and Behavior, Vol. 13, pp. 1—32. New York: Raven Press 1975b

Manthey, A. A.: The effect of calcium on the desensitization of membrane receptors at the neuromuscular junction. J. gen. Physiol. **49**, 963—976 (1966)

Margaria, R.: Exercise at Altitude. New York-London-Milan-Tokyo-Buenos Aires-Amsterdam: Excerpta Medica Foundation 1967

Margules, D. L., Stein, L.: Increase of "antianxiety" activity and tolerance of behavioral depression during chronic administration of oxazepam. Psychopharmacologia (Berl.) **13**, 74—80 (1968)

Marjot, D. H.: Has alcohol psychotomimetic and convulsant properties? Brit. J. Addict. **69**, 295—304 (1974)

Mark, L. C., Papper, E. M., Brodie, B. B., Rovenstine, E. A.: Quantitative pharmacologic studies with pentothal. N.Y. J. Med. **49**, 1546—1549 (1949)

Marley, E., Vane, J. R.: Tryptamine receptors in the central nervous system: Effects of anaesthetics. Nature (Lond.) **198**, 441—444 (1963)

Marsden, C. D., Gimlette, T. M., McCallister, R. G., Owen, D. A. L., Miller, T. N.: Effects of β-adrenergic blockade on finger tremor and achilles reflex time in anxious and thyrotoxic patients. Acta endocr. **57**, 353—362 (1968)

Marshall, W. H.: Observations on subcortical somatic sensory mechanisms of cats under nembutal anesthesia. J. Neurophysiol. **4**, 25—43 (1941)

Marshall, W. H., Woolsey, C. N., Bard, P.: Observations on cortical somatic sensory mechanisms of cat and monkey. J. Neurophysiol. **4**, 1—24 (1941)

Martin, W. R.: Assessment of the dependence producing potentials of narcotic analgesics. International Encyclopedia of Pharmacology and Therapeutics, Sec. 6, Vol. 1, pp. 155—180. Radouco-Thomas, C., Lasagna, L. (Eds.). Glasgow: Pergamon 1966

Martin, W. R.: A homeostatic and redundancy theory of tolerance to and dependence on narcotic analgesics. In: The Addictive States, pp. 206—225. Baltimore: Williams & Wilkins Co. 1968

Martin, W.R.: Pathophysiology of narcotic addiction: possible roles of protracted abstinence in relapse. Drug Abuse—Proceedings of the International Conference, pp. 153—159. Philadelphia: Lea & Febiger 1972
Martin, W.R., Eades, C.G., Thompson, W.O., Thompson, J.A., Flanary, H.G.: Morphine physical dependence in the dog. J. Pharmacol. exp. Ther. **189**, 759—771 (1974a)
Martin, W.R., Jasinski, D.R.: Physiological parameters of morphine dependence in man—tolerance, early abstinence, protracted abstinence. J. psychiat. Res. **7**, 9—17 (1969)
Martin, W.R., Thompson, W.O., Fraser, H.F.: Comparison of graded single intramuscular doses of morphine and pentobarbital in man. Clin. Pharmacol. Ther. **15**, 623—630 (1974b)
Masserman, J.H., Jacobson, L.: Effects of ethyl alcohol on the cerebral cortex and the hypothalamus of the cat. Arch. Neurol. Psychiat. (Chic.) **43**, 334—340 (1940)
Masserman, J.H., Yum, K.S.: An analysis of the influence of alcohol on experimental neuroses in cats. Psychosom. Med. **8**, 36—52 (1946)
Masserman, J.H., Yum, K.S., Nicholson, M.R., Lee, S.: Neurosis and alcohol. Amer. J. Psychol. **101**, 389—395 (1944)
Matsuki, K., Iwamoto, T.: Development of tolerance to tranquilizers in the rat. Jap. J. Pharmacol. **16**, 191—197 (1966)
Matsuki, K., Iwamoto, T.: Development of tolerance to chlorpromazine in the rat. Jap. J. Pharmacol. **18**, 274—277 (1968)
Matsushita, A., Smith, C.M.: Spinal cord function in postischemic rigidity in the rat. Brain Res. **19**, 395—410 (1970a)
Matsushita, A., Smith, C.M.: Muscle relaxants on postischemic spinal rigidity in the rat. Brain Res. **19**, 411—420 (1970b)
Matthews, E.K., Quilliam, J.P.: Effects of central depressant drugs upon acetylcholine release. Brit. J. Pharmacol. **22**, 415—440 (1964)
Matthews, W.D., Connor, J.D.: Effects of diphenylhydantoin and diazepam on hippocampal evoked responses. Neuropharmacology **15**, 181—186 (1976)
Mattson, R.H.: The benzodiazepines. In: Woodbury, D.M., Penry, J.K., Schmidt, R.P. (Eds.): Antiepileptic Drugs, pp. 497—518. New York: Raven Press 1972
Mayer, D., Hayes, R.: Narcotic and stimulation-produced analgesia: tolerance and cross-tolerance. Fed. Proc. **33**, 1644 (1974)
Mayer, D.J., Hayes, R.L.: Stimulation-produced analgesia: development of tolerance and cross-tolerance to morphine. Science **188**, 941—943 (1975)
Mayfield, D.G.: Psychopharmacology of alcohol. I. Affective change with intoxication, drinking behavior and affective state. J. nerv. ment. Dis. **146**, 314—321 (1968a)
Mayfield, D.G.: Psychopharmacology of alcohol II. Effective tolerance in alcohol intoxication. J. nerv. ment. Dis. **146(4)**, 322—327 (1968b)
Maynert, E.W.: Phenobarbital, mephobarbital and metharbital: Absorption, distribution and excretion. In: Woodbury, D.M., Penry, J.K., Schmidt, R.P. (Eds.): Antiepileptic Drugs, pp. 303—310. New York: Raven Press 1972a
Maynert, E.W.: Pharmacology of sedative-hypnotic drugs. In: Zarafonetis, C.J.D. (Ed.): Drug abuse, pp. 199—203. Philadelphia: Lea & Febiger 1972b
Maynert, E.W.: Phenobarbital, mephobarbital, and metharbital: Biotransformation. In: Woodbury, D.M., Penry, J.K., Schmidt, R.P. (Eds.): Antiepileptic Drugs, pp. 311—318. New York: Raven Press 1972c
Maynert, E.W., Klingman, G.I.: Acute tolerance to intravenous anesthetics in dogs. J. Pharmacol. exp. Ther. **128**, 192—200 (1960)
Mazzia, V.D.B., Randt, C.: Amnesia and eye movement in first-stage anesthesia. Arch. Neurol. **14**, 522—525 (1966)
McCabe, E.R., Layne, E.C., Sayler, D.F., Slusher, N., Bessman, S.P.: Synergy of ethanol and a natural soporific—gamma hydroxybutyrate. Science **171**, 404—406 (1971)
McCarthy, J.S., Stitzel, R.E.: Kinetic differences in the microsomal metabolism of the isomers of hexobarbital. J. Pharmacol. exp. Ther. **176**, 772—778 (1971)
McClane, T.K., Martin, W.R.: Subjective and physiologic effects of morphine, pentobarbital, and meprobamate. Clin. Pharmacol. Ther. **20**, 192—198 (1976)
McClearn, G.E.: Genetics and the pharmacology of alcohol. Proc. VI Int. Congress Pharm. **III**, 59—67 (1975)
McDanal, C.E., Owens, D., Bolman, W.M.: Bromide abuse: A continuing problem. Am. J. Psychiat. **131**, 913—915 (1974)

McFarland, R. A., Forbes, W. H.: The metabolism of alcohol in man at high altitudes. Hum. Biol. **8**, 387—398 (1936)
McIlwain, H.: Metabolic adaptation in the brain. Nature (Lond.) **226**, 803—806 (1970)
McLaren, A., Michie, D.: Variability of response in experimental animals. J. Genet. **54**, 440—455 (1956)
McLaughlin, G. T.: A nation tranquilized—a socio-legal analysis of the abuse of sedatives in the United States. Fordham Law Rev. **42**, 725—760 (1974)
Medical Letter **17**, No. 7 (Issue 423), March 28, 1975, New Rochelle, New York
Meech, R. W.: Intracellular calcium injection causes increased potassium conductance in *Aplysia* nerve cells. Comp. Biochem. Physiol. **42 A**, 493—499 (1972)
Meerloo, J. A. M.: Variable individual tolerance for alcohol and drugs: Some dangerous social implications. Postgrad. Med. **22**, 583—590 (1957)
Mellanby, E.: Alcohol: its absorption into and disappearance from the blood under different conditions. (Great Britain Medical Research Committee Special Report Series, No. 31). London: H. M. Stationery Office 1919
Mello, N. K.: Short-term memory function in alcohol addicts during intoxication. In: Gross, M. (Ed.): Alcohol Intoxication and Withdrawal Experimental Studies, Vol. 35, pp. 333—334. New York: Plenum Press 1973a
Mello, N. K.: A review of methods to induce alcohol addiction in animals. Pharmacol. Biochem. and Behav. **1**, 89—101 (1973b)
Mello, N. K.: A semantic aspect of alcoholism. In: Cappel, H. D., LeBlanc, A. E. (Eds.): Biological and Behavioural Approaches to Drug Dependence. Toronto: Addiction Research Foundation of Ontario 1975
Mello, N. K., Mendelson, J. H.: Experimentally induced intoxication in alcoholics: a comparison between programmed and spontaneous drinking. J. Pharmacol. exp. Ther. **173**, 101—116 (1970)
Mendelson, J. H. (Ed.): Experimentally induced chronic intoxication and withdrawal in alcoholics. Quart. J. Stud. Alcohol Suppl. **2**, (1964)
Mendelson, J. H.: Ethanol-1-C^{14} metabolism in alcoholics and non-alcoholics. Science **159**, 319—320 (1968)
Mendelson, J. H.: Biologic concomitants of alcoholism (First of Two Parts). New Engl. J. Med. **283**, 24—32 (1970a)
Mendelson, J. H.: Biologic concomitants of alcoholism (Second of Two Parts) New Engl. J. Med. **283**, 71—81 (1970b)
Mendelson, J. H., Mello, N. K., Corbett, C., Ballard, R.: Puromycin inhibition of ethanol ingestion and liver alcohol dehydrogenase activity in the rat. J. psychiat. Res. **3**, 133—143 (1965)
Mendelson, J. H., Mello, N. K., Solomon, P.: Small group drinking behavior: an experimental study of chronic alcoholics. In: The Addictive States, Baltimore: Williams & Wilkins Co. 1968
Mendelson, J. H., Stein, S., McGuire, M. T.: Comparative psychophysiological studies of alcoholic and nonalcoholic subjects undergoing experimentally induced ethanol intoxication. Psychosom. Med. **28**, 1—12 (1966)
Mendelson, J., Wexler, D., Leiderman, P. H., Solomon, P.: A study of addiction to nonethyl alcohols and other poisonous compounds. Quart. J. Stud. Alcohol **18**, 561—580 (1957)
Merlin, R. G., Kumazawa, T., Honig, C. R.: Reversible interaction between halothane and Ca^{++} on cardiac actomyosin adenosine triphosphatase: mechanism and significance. J. Pharmol. exp. Ther. **190**, 1—14 (1974)
Merlis, S., Koepke, H. H.: The use of oxazepam in elderly patients. Dis. nerv. Syst. **36**, 27—29 (1975)
Merry, J., Marks, V.: The effects of alcohol, barbiturate, and diazepam on hypothalamic/pituitary/adrenal function in chronic alcoholics. Lancet **1972II**, 990—992
Metcalfe, J. C., Hoult, J. R. S., Colley, C. M.: The molecular implications of a unitary hypothesis of anaesthetic action. In: Halsey, M. J., Millar, R. A., Sutton, J. A. (Eds.): Molecular Mechanisms in General Anaesthesia, pp. 145—163. London-New York: Churchill Livingston 1974
Meyer-Lohmann, J., Hagenah, R., Hellweg, C., Benecke, R.: The action of ethyl alcohol on the activity of individual Renshaw cells. Naunyn-Schmiedebergs Arch. Pharmacol. **272**, 131—142 (1972)

Michenfelder, J.D., Theye, R.A.: Cerebral metabolic effects of anesthesia in the dog. In: Fink, B.R. (Ed.): Cellular Biology and Toxicity of Anesthetics. Baltimore: Williams and Wilkins Co. 1972
Miczek, K.A.: Intraspecies aggression in rats: effects of d-amphetamine and chlordiazepoxide. Psychopharmacologia (Berl.) **39**, 275—301 (1974a)
Miczek, K.A.: Intraspecies attack and defense in rats: Effects of amphetamine, chlordiazepoxide and alcohol. Fed. Proc. **33**, 465 (1974b)
Miller, K.W.: Inert gas narcosis and animals under high pressure. Symp. Soc. exp. Biol. **26**, 363—378 (1972)
Miller, K.W.: Inert gas narcosis, the high pressure neurological syndrome, and the critical volume hypothesis. Science **185**, 867—869 (1974)
Miller, K.W.: The pressure reversal of anesthesia and the critical volume hypothesis. In: Fink, B.R. (Ed.): Progress in Anesthesiology, Vol. 1: Molecular Mechanisms of Anesthesia. New York: Raven Press 1975
Miller, L.L., Dolan, M.P.: Effects of alcohol on short term memory as measured by a guessing technique. Psychopharmacologia (Berl.) **35**, 353—364 (1974)
Miller, R.N., Smith, E.E., Hunter, F.E.: Halothane-induced alterations in energy-dependent and energy-independent membrane carrier functions in isolated rat liver mitochondria with some electron microscopic correlations. In: Fink, B.R. (Ed.): Cellular Biology and Toxicity of Anesthetics. Baltimore: Williams & Wilkins Co. 1972
Millichap, J.G.: Testing of anticonvulsants in man. In: Woodbury, D.M., Penry, J.K., Schmidt, R.P. (Eds.): Antiepileptic Drugs, pp. 75—80. New York: Raven Press 1972
Mirsky, I.A., Piker, P., Rosenbaum, M., Lederer, H.: "Adaptation" of the central nervous system to varying concentrations of alcohol in the blood. Quart. J. Stud. Alcohol **2**, 35—45 (1941)
Miyahara, J.T., Esplin, D.W., Zablocka, B.: Differential effects of depressant drugs on presynaptic inhibition. J. Pharmacol. exp. Ther. **154** (1) 119—127 (1966)
Modell, W.: Updating the sleeping pill. Geriatrics **29**, 126—132 (1974)
Mofenson, H.C., Greensher, J.: The unknown poison. Pediatrics **54**, 336—342 (1974)
Mogey, C.A., Young, P.A.: The antagonism of curarizing activity by phenolic substances. Brit. J. Pharmacol. **4**, 359—365 (1949)
Money, K.E., Myles, W.S.: Heavy water nystagmus and effects of alcohol. Nature (Lond.) **247**, 404 (1974)
Morgan, E.P., Phillis, J.W.: The effects of ethanol on acetylcholine release from the brain of unanaesthetized cats. Genet. Pharmacol. **6**, 281—284 (1975)
Morgan, H.W.: Yesterday's Addicts: American Society and Drug Abuse 1865—1920. Norman, Oklahoma: University of Oklahoma Press 1974
Mori, K., Winters, W.D., Spooner, C.E.: Comparison of reticular and cochlear multiple unit activity with auditory evoked responses during various stages induced by anaesthetic agents. Electroenceph. clin. Neurophysiol. **24**, 242—248 (1968)
Morillo, A.: Effects of benzodiazepines upon amygdala and hippocampus of the cat. Int. J. Neuropharmacol. **1**, 353—359 (1962)
Mørland, J., Setekleiv, J., Haffner, J.F.W., Strømsaether, C.E., Danielsen, A., Wethe, G.H.: Combined effects of diazepam and ethanol on mental and psychomotor functions. Acta pharmacol. (Kbh.) **34**, 5—15 (1974)
Mortimer, J.A.: Possible influences of barbiturate anesthesia on cerebellar operation: a computer simulation study. Int. J. Neurosci. **6**, 77—88 (1973)
Moskow, H.A., Pennington, R.C., Knisely, M.H.: Alcohol, sludge, and hypoxic areas of nervous system, liver and heart. Microvasc. Res. **1**, 174—185 (1968)
Muchnik, S., Gage, P.W.: Effect of bromide ions on junctional transmission. Nature (Lond.) **217**, 373—374 (1968)
Mulé, S.J., Brill, H.: Chemical and Biological Aspects of Drug Dependence. Cleveland, Ohio: CRC Press 1972
Müller, W.E., Wollert, V.: High stereospecificity of the benzodiazepine binding site of human serum albumin. Molec. Pharmacol. **11**, 52—60 (1975)
Mullins, L.J.: Anesthesia: an overview. In: Fink, B.R. (Ed.): Progress in Anesthesiology, Vol. 1, pp. 237—242. Molecular Mechanisms of Anesthesia. New York: Raven 1975
Murayama, S.: personal communication (1974)

Murphy, J. T., Sabah, N. H.: Spontaneous firing of cerebellar Purkinje cells in decerebrate and barbiturate anesthesitized cats. Brain Res. **17**, 515—519 (1970)
Murthy, K. S. K., Deshpande, S. S.: Differential sensitivity of primary and secondary spindle afferents to depressant drugs. Brain Res. **79**, 89—99 (1974)
Mycek, M. J., Brezenoff, H. E.: Tolerance to centrally administered phenobarbital. 36th Meeting of the Committee on Problems of Drug Dependence, Nat. Acad. Sci. March, 1974, Mexico City, Mexico
Myers, R. D.: Handbook of drug and chemical stimulation of the brain: behavioural, pharmacological and physiological aspects. New York-Cincinnati-Toronto-London-Melbourne: Van Nostrand Reinhold Company 1974
Myers, R. D., Evans, J. E., Yaksh, T. L.: Ethanol preference in the rat: interactions between brain serotonin and ethanol, acetaldehyde, paraldehyde, 5-HTP and 5-HTOL. Neuropharmacology **11**, 539—549 (1972)
Myers, R. D., Martin, G. E.: The role of cerebral serotonin in the ethanol preference of animals. Ann. N.Y. Acad. Sci. **215**, 135—144 (1973)
Myers, R. D., Veale, W. L.: The determinants of alcohol preference in animals. In: Kissin, B., Begleiter, H. (Eds.): The Biology of Alcoholism, Vol. II, Chap. 6, pp. 131—168. New York: Plenum Press 1972
Myrsten, A. L., Goldberg, L., Neri, A.: Interaction between alcohol and tranquilizing drugs. Report from the Psychological Laboratories, the University of Stockholm, 1—24, 1971
Nagle, D. R.: Anesthetic addiction and drunkenness: a contemporary and historical survey. Int. J. Addict. **3**, 25—39 (1968)
Nahrwold, M. L., Clark, C. R., Cohen, P. J.: Is depression of mitochondrial respiration a predictor of in-vivo anesthetic activity? Anesthesiology **40**, 566—570 (1974)
Nakai, Y., Domino, E. F.: Differential effects of pentobarbital, ethyl alcohol, and chlorpromazine in modifying reticular facilitation of visually evoked responses in the cat. Int. J. Neuropharmacol. **8**, 61—72 (1969)
Nakai, Y., Matsuoka, I., Takaori, S.: Pharmacological analysis of unitary discharges recorded from the inferior colliculus caused by click stimuli in cats. Jap. J. Pharmacol. **15**, 378—385 (1965)
Nakai, Y., Sasa, M., Takaori, S.: Effects of central depressants on the cortical auditory responses in the unrestrained cats. Jap. J. Pharmacol. **16**, 416—422 (1966)
Nakai, Y., Takaori, S.: Effects of central depressants on the cortical auditory responses evoked by repetitive click stimuli in the cat. Jap. J. Pharmacol. **15**, 165—175 (1965)
Nakanishi, T., Norris, F. H.: Effect of diazepam on rat spinal reflexes. J. neurol. Sci. **13**, 189—195 (1971)
Nastuk, W. L.: Activation and inactivation of muscle post-juntional receptors. Fed. Proc. **26**, 1639—1646 (1967)
Nastuk, W. L., Koester, J. D., Gissen, A. J.: The role of divalent cations and certain drugs in the activation and inactivation of muscle postjunctional receptors. In: Cellular Biology and Toxicity of Anesthetics. Baltimore: Williams & Wilkins Co. 1972
Nastuk, W. L., Parsons, R. L.: Factors in the inactivation of postjunctional membrane receptors of frog skeletal muscle. J. gen. Physiol. **56**, 218—249 (1970)
Nelemans, F. A.: The influence of various substances on the contracture of the frog's isolated abdominal muscle. Acta physiol. pharmacol. neerl. **11**, 76—82 (1962)
Newman, H. W.: Alcohol injected intravenously. Some psychological and psychopathological effects in man. Amer. J. Psychiat. **91**, 1343—1352 (1935)
Newman, H. W.: Acute Alcoholic Intoxication. California: Stanford University Press 1941
Newman, H. W., Lehman, A. J.: Nature of acquired tolerance to alcohol. J. Pharmacol. exp. Ther. **62**, 301—306 (1938)
Ngai, S. H., Tseng, D. T. C., Wang, S. C.: Effect of diazepam and other central nervous system depressants on spinal reflexes in cats: A study of the site of action. J. Pharmacol. exp. Ther. **153**, 344—351 (1966)
Nicoll, R. A.: The effects of anaesthetics on synaptic excitation and inhibition in the olfactory bulb. J. Physiol. (Lond.) **223**, 803—814 (1972)
Nicoll, R. A.: The antagonism of amino acid responses by cholinolytic agents in the isolated frog spinal cord. Soc. Neurosci., Fourth Annual Meeting 1974

Nicoll, R. A.: Pentobarbital: action on frog motoneurons. Brain Res. **94**, 1—5 (1975a)
Nicoll, R. A.: Presynaptic action of barbiturates in the frog spinal cord. Proc. nat. Acad. Sci. (Wash.) **72**, 1460—1463 (1975b)
Nicoll, R. A., Eccles, J. C., Oshima, T., Rubia, F.: Prolongation of hippocampal inhibitory postsynaptic potentials by barbiturates. Nature **258**, 625—627 (1975)
Nicoll, R. A., Padjen, A.: Pentylenetetrazol: An antagonist of GABA at primary afferents of the isolated frog spinal cord. Neuropharmacology **15**, 69—71 (1976)
N.I.D.A. (National Institute on Drug Abuse): The CNS Depressant Withdrawal Syndrome and Its Management: An Annotated Bibliography 1950—73. Rockville: National Clearinghouse on Drug Abuse Information 1974
Nishi, S., Minota, S., Karczmar, A. G.: Primary afferent neurons: The ionic mechanism of GABA-mediated depolarization. Neuropharmacology **13**, 215—219 (1974)
Nobes, P.: Intravenous barbiturates for drunkenness. Brit. med. J. **1953I**, 836
Noble, P.: Development of amethystic agents. In: Biomedical Research in Alcohol Abuse Problems, Conference Proceedings. Non-medical Use of Drug Directorate, National Health and Welfare, Canada 1974
Noble, E. P., Alkana, R. L., Parker, E. S.: Ethanol-induced CNS depression and its reversal: a review. Biomed. Res. Session B. In: Proc. 4th Ann. Alcohol Conf. National Institute on Alcohol Abuse and Alcoholism. DHEW 134—170 (1975)
Noble, E. P., Tewari, S.: Protein and ribonucleic acid metabolism in brains of mice following chronic alcohol consumption. Ann. N.Y. Acad. Sci. **215**, 333—345 (1973)
Nunn, J. F.: Mitosis. In: Halsey, M. J., Millar, R. A., Sutton, J. A. (Eds.): Effects of Anaesthetics on Votile Systems. Molecular Mechanisms in General Anaesthesia. Edinburgh-London-New York: Churchill Livingstone 1974
Nurimoto, S., Ogawa, N., Ueki, S.: Effects of psychotropic drugs on hyperemotionality of rats with bilateral ablations of the olfactory bulbs and olfactory tubercles. Jap. J. Pharmacol. **24**, 185—193 (1974)
Nurmand, L. B.: On the cross-tolerance to barbamil. Farmakol. Toksikol. **32**, 41 (1969). Cited by Hug 1972
Oishi, H., Iwahara, S., Yang, K., Yogi, A.: Effects of chlordiazepoxide on passive avoidance responses in rats. Psychopharmacologia (Berl.) **23**, 373—385 (1972)
Okada, K.: Effects of alcohols and actone on the neuromuscular juction of frog. Jap. J. Physiol. **17**, 245—261 (1967)
Okamoto, M., Rosenberg, H. C., Boisse, N. R.: Tolerance characteristics produced during the maximally tolerable chronic pentobarbital dosing in the cat. J. Pharmacol. exp. Ther. **192**, 555—564 (1975)
Oldendorf, W. H.: Drug penetration of the blood-brain barrier. In: Zimmermann, E., George, R. (Eds.): Narcotics and the Hypothalamus. New York: Raven Press 1974
Oliverio, A., Castellano, C., Renzi, P., Sansone, M.: Decreased sensitivity of septal mice to impairment of two-way avoidance by chlorpromazine. Psychopharmacologia (Berl.) **29**, 13—20 (1973)
Oosterfeld, W. J.: Effect of gravity on positional alcohol nystagmus (PAN). Aerosp. Med. **41**, 557 (1970)
Osborn, A. G., Bunker, J. P., Cooper, L. M., Frank, G. S., Hilgard, E. R.: Effects of thiopental sedation on learning and memory. Science **157**, 574—576 (1967)
Ostfeld, A. M.: LSD and JB-318. Arch. gen. Psychiat. **1960**, 390—407
Ostfeld, A. M., Machne, X., Unna, K. R.: The effects of atropine on the electroencephalogram and behavior in man. Pharmacol. exp. Ther. **128**, 265—272 (1960)
Ostfeld, A. M., Visotsky, H., Abood, L., Lebovitz, B. Z.: Studies with a new hallucinogen. Arch. Neurol. Psychiat. **81**, 256—263 (1959)
Oswald, I.: Drugs and sleep. Pharmacol. Rev. **20**, 273—303 (1968)
Oswald, I.: Psychoactive drugs and sleep: withdrawal rebound phenomena. Triangle **10**, 99—104 (1971)
Oswald, I.: Drug research and human sleep. Ann. Rev. Pharmacol. **13**, 243—252 (1973)
Oswald, I., Evans, J. I., Lewis, S. A.: Addictive drugs cause suppression of paradoxical sleep with withdrawal rebound. In: Steinberg, H. (Ed.): Scientific Basis of Drug Dependence, pp. 243—258. London: Churchill Ltd. 1969

Oswald, I., Priest, R. G.: Five weeks to escape the sleeping pill habit. Brit. med. J. **1965 II**, 1093—1099
Overton, D. A.: State-dependent learning produced by depressant and atropine-like drugs. Psychopharmacologia (Berl.) **10**, 6—31 (1966)
Overton, D. A.: Dissociated learning in drug states (state dependent learning). In: Psychopharmacology. A Review of Progress 1957—1967. Efron, D. (Ed.):, pp. 918—930. Washington D.C.: U.S. Government Printing Office 1968
Overton, D. A.: State-dependent learning produced by addicting drugs. In: Fisher, S., Freedman, A. M. (Eds.): Opiate Addiction: Origins and Treatment, pp. 61—75. Washington: Winston & Sons 1973
Overton, D. A.: Experimental methods for the study of state-dependent learning. Fed. Proc. **33**, 1800—1813 (1974)
Padjen, A., Bloom, F.: Problems in the electrophysiological analysis of the site of action of benzodiazepines. Advanc. Biochem. Psychopharm., Vol. 14, pp. 93—102. In: Mechanism of Action of Benzodiazepines. New York: Raven Press 1975
Palmer, K. H., Fowler, M. S., Wall, M. E., Rhodes, L. S., Waddell, W. J., Baggett, B.: The metabolism of R (+)- and RS-pentobarbital. J. Pharmacol. exp. Ther. **170**, 355—363 (1969)
Pappenheimer, J. R.: The sleep factor. Scientific American **235**, 24—29 (1976)
Parbrook, G. D.: Techniques of inhalational analgesia in the postoperative period. Brit. J. Anaesth. **39**, 730—735 (1967 a)
Parbrook, G. D.: The levels of nitrous oxide analgesia. Brit. J. Anaesth. **39**, 974—982 (1967 b)
Pascarelli, E. F.: Methaqualone abuse, the quiet epidemic. J. Amer. med. Ass. **224**, 1512—1514 (1973)
Patch, V. D.: Dangers of diazepam, a street drug. New Engl. J. Med. **290**, 807 (1974)
Patel, G. J., Lal, H.: Reduction in brain γ-aminobutyric acid and in barbital narcosis during ethanol withdrawal. J. Pharmacol. exp. Ther. **186**, 625—629 (1973)
Paton, W. D. M.: A theory of drug action based on the rate of drug-receptor combination. Proc. roy. Soc. B **154**, 21—69 (1961)
Paton, W. D. M.: Unconventional anaesthetic molecules. In: Halsey, M. J., Millar, R. A., Sutton, J. A. (Eds.): Molecular Mechanisms in General Anaesthesia, pp. 48—64. London-New York: Churchill Livingstone 1974
Paton, W. D. M., Speden, R. N.: Uptake of anaesthetics and their action on the central nervous system. Brit. med. Bull. **21**, 44—48 (1965)
Paulson, O. B., Parving, H.-H., Olesen, J., Skinhøj, E.: Influence of carbon monoxide and of hemodilution on cerebral blood flow and blood gases in man. J. appl. Physiol. **35**, 111—116 (1973)
Pearson, R. G., Neal, G. L.: Operator performance as a function of drug, hypoxia, individual, and task factors. Aerosp. Med. **41**, 154—158 (1970)
Pecora, L. J.: Physiologic study of the summating effects ethyl alcohol and carbon monoxide. Amer. industr. Hyg. J. **20**, 235—240 (1959)
Peeke, H. V. S., Herz, M. J.: Habituation. Behavioral Studies, Vol. 1. London-New York: Academic Press 1973 a
Peeke, H. V. S., Herz, M. J.: Habituation. Physiological Substrates, Vol. II. London-New York: Academic Press 1973 b
Penfield, W.: The excitable-cortex in conscious man. Liverpool: Liverpool Univ. Press 1958
Penn, N. W.: Antagonism of barbiturate action by DNA pyrimidines and related cellular constituents. Fed. Proc. **32**, 727 (1973)
Penn, N. W.: Ethanol antagonism by 5-hydroxymethyl cellular compound. Life Sci. **17**, 1055—1061 (1975 a)
Penn, N. W.: Antagonism of barbiturate by DNA pyrimidines and allied compounds. Arch. int. Pharmacodyn. **218**, 156—166 (1975 b)
Pennington, R. C., Knisely, M. H.: Experiments aimed at separating the mechanical circulatory effects of ethanol from specific chemical effects. Ann. N.Y. Acad. Sci. **215**, 356—365 (1973)
Pepeu, G.: The release of acetylcholine from the brain: an approach to the study of central cholinergic mechanisms. Progr. Neurobiol. **2**, 259—288 (1973)
Perman, E. S.: The effect of ethyl alcohol on the secretion from the adrenal medulla in man. Acta physiol. scand. **44**, 241—247 (1958)
Perman, E. S.: The effect of ethyl alcohol on the secretion from the adrenal medulla of the cat. Acta physiol. scand. **48**, 323—328 (1960)

Perman, E.S.: Effect of ethanol and hydration on the urinary excretion of adrenaline and noradrenaline and on the blood sugar of rats. Acta physiol. scand. **51**, 68—74 (1961)
Perrin, R.G., Hockman, C.H., Kalant, H., Livingston, K.E.: Acute effects of ethanol on spontaneous and auditory evoked electrical activity in cat brain. Electroenceph. clin. Neurophysiol. **36**, 19—31 (1974)
Perrin, R.G., Kalant, H., Livingston, K.E.: Electroencephalographic signs of ethanol tolerance and physical dependence in the cat. Electroenceph. clin. Neurophysiol. **39**, 157—162 (1975)
Phillips, B.M., Miya, T.S., Yim, G.K.W.: Studies on the mechanism of meprobamate tolerance on the rat. J. Pharmacol. exp. Ther. **135**, 223—229 (1962)
Phillis, J.W., Jhamandas, K.: The effects of chlorpromazine and ethanol on in vivo release of acetylcholine from the cerebral cortex. Comp. gen. Pharmacol. **2**, 306 (1971)
Pickens, R., Cunningham, M.R., Heston, L.L., Eckert, E., Gustafson, L.K.: Dose preference during pentobarbital self-administration by humans. J. Pharmacol. exp. Ther. (in press) 1977
Pohorecky, L.A.: Effects of ethanol on central and peripheral noradrenergic neurons. J. Pharmacol. exp. Ther. **189**, 380—391 (1974)
Poire, R., Lepoire, P., Rustin, C.: The inhibitory action of diazepam on the electroencephalographic responses evoked by intermittent light stimulation. Its recognition by the electronic integration of Drohocki. Ann. Med. Psychol. **125**, 778—779 (1967)
Polacsek, E., Barnes, T., Turner, N., Hall, R., Weise, C.: Interaction of alcohol and other drugs, 2nd ed. (revised). Toronto: Addiction Research Foundation; 1972
Polc, P., Mohler, H., Haefely, W.: The effect of diazepam on spinal cord activities: Possible sites and mechanisms of action. Naunyn-Schmiedebergs Arch. Pharmacol. **284**, 319—337 (1974)
Pollard, H., Bischoff, S., Schwartz, J.C.: Decreased histamine synthesis in the rat brain by hypnotics and anaesthetics. J. Pharm. (Lond.) **25**, 920—922 (1973)
Pollard, H., Bischoff, S., Schwartz, J.C.: Turnover of histamine in rat brain and its decrease under barbiturate anesthesia. J. Pharmacol. exp. Ther. **190**, 88—90 (1974)
Polzin, R.L., Barnes, C.D.: Effects of diazepam and picrotoxin on dorsal root potentials. Fed. Proc. **33**, 394 (1974)
Polzin, R., Barnes, C.D.: The effect of diazepam and picrotoxin on brainstem evoked dorsal root potentials. Neuropharmacology **15**, 133—137 (1976)
Porsolt, R.D., Joyce, D., Summenfield, A.: Changes in behavior with repeated testing under the influence of drugs: Drug experience interactions. Nature (Lond.) **227**, 286—287 (1970)
Porter, R.: The neurophysiology of movement performance. In: Hunt, C.C. (Ed.): Physiology Series One, Vol. 3, Neurophysiology, pp. 151—183. London: Butterworths 1975
Poser, C.M.: Demyelination in the central nervous system in chronic alcoholism: Central pontine myelinolysis and Marchiafava-Bignami's disease. Ann. N.Y. Acad. Sci. **215**, 373—381 (1973)
Prange, A.J. (Ed.): The Thyroid Axis, Drugs, and Behavior. New York: Raven 1974
Prange, A.J., Breese, G.R., Cott, J.M., Martin, B.R., Cooper, B.R., Wilson, I.C., Plotnikoff, N.P.: Thyrotropin releasing hormone: antagonism of pentobarbital in rodents. Life Sci. **14**, 447—455 (1974a)
Prange, A.J., Wilson, I.C., Lara, P.P., Alltop, L.B., Breese, G.R.: Effects of thyrotropin-releasing hormone in depression. Lancet **1972 II**, 999—1002
Prange, A.J., Wilson, I.C., Lara, P.P., Wilber, J.F., Breese, G.R., Alltop, L.B., Lipton, M.A.: Thyrotropin-releasing hormone: Psychobiological responses of normal women. II. Pituitary-thyroid responses. In: The Thyroid Axis, Drugs, and Behavior. New York: Raven 1974b
Preskorn, S.H., Denner, L.J.: Benzodiazepines and withdrawal psychosis. Report of three cases. J. Amer. med. Ass. **237**, 36—38 (1977)
Press, E., Done, A.K.: Solvent sniffing. Physiologic effects and community control measures for intoxication from the intentional inhalation of organic solvents I. Pediatrics **39**, 451—461 (1967a)
Press, E., Done, A.K.: Solvent sniffing. Physiologic effects and community control measures for intoxication from the intentional inhalation of organic solvents II. Pediatrics **39**, 611—622 (1967b)
Prosser, C.L.: Metabolic and central nervous acclimation of fish to cold. In: Troshin, A.S. (Ed.): The Cell and Environmental Temperature, pp. 375—383. New York: Pergamon 1967
Przybyla, D.C., Wang, S.C.: Locus of central depressant action of diazepam. J. Pharmacol. exp. Ther. **163**, 439—447 (1968)

Puszkin, S., Rubin, E.: Adenosine diphosphate effect on contractility of human muscle actomyosin: Inhibition by ethanol and acetaldehyde. Science **188**, 1319—1320 (1975)
Quastel, D. M. J., Hackett, J. T., Okamoto, K.: Presynaptic action of central depressant drugs. Canad. J. Physiol. Pharmacol. **50**, 279—284 (1972)
Quastel, D. M. J., Linder, T. M.: Pre- and postsynaptic actions of central depressants at the mammalian neuromuscular junction. In: Fink, B. R. (Ed.): Progress in Anesthesiology, Vol. 1, pp. 157—168. Molecular Mechanisms of Anesthesia. New York: Raven Press 1975
Quastel, J. H.: Anesthetics and cerebral fluxes of ions in vitro. In: Fink, B. R. (Ed.): Progress in Anesthesiology, Vol. 1: Molecular Mechanisms of Anesthesia. New York: Raven Press 1975
Quastel, J. H.: Effects of drugs on energy metabolism of the brain and on cerebral transport. In: Iversen, L. L., Iversen, S. D., Snyder, S. H. (Eds.): Handbook of Psychopharmacology, Vol. 5, pp. 1—46. New York: Plenum Press 1975
Quenzer, L. F., Feldman, R. S.: The mechanism of anti-muricidal effects of chlordiazepoxide. Pharmacol. Biochem. Behav. **3**, 567—571 (1975)
Quilliam, J. P.: Paraldehyde and methylpentynol and ganglionic transmission. Brit. J. Pharmacol. Chem. **14**, 277—283 (1959)
Raines, A., Standaert, F. G.: An effect of diphenylhydantoin on posttetanic hyperpolarization of intramedullary nerve terminals. J. Pharmacol. exp. Ther. **156**, 591—597 (1967)
Raines, A., Standaert, F. G.: Effects of anticonvulsant drugs on nerve terminals. Epilepsia **10**, 211—227 (1969)
Ramsey, J. M.: Carbon monoxide, tissue hypoxia, and sensory psychomotor response in hypoxaemic subjects. Clin. Sci. **42**, 619—625 (1972)
Randall, C. L., Lester, D.: Differential effects of ethanol and pentobarbital on sleep time in C57BL and BALB mice. J. Pharmacol. exp. Ther. **188**, 27—33 (1974)
Randall, L. O., Kappell, B.: Pharmacological activity of some benzodiazepines and their metabolites. In: Garattini, S., Mussini, E., Randall, L. O. (Eds.): The Benzodiazepines, New York: Raven 1973
Randall, L. O., Schallek, W.: Pharmacological activity of certain benzodiazepines. In: Efron, D. H. (Ed.): Pharmacology. A Review of Progress (1957—1967), pp. 153—184. USPHS Pub. No. 1836 (1968)
Ransom, B. R., Barker, J. L.: Pentobarbital modulates transmitter effects on mouse spinal neurones grown in tissue culture. Nature (Lond.) **254**, 703—705 (1975)
Ransom, F.: Acquired tolerance for alcohol in the frog's heart. J. Physiol. (Lond.) **53**, 141—146 (1919)
Raskin, N. H., Sokoloff, L.: Ethanol-induced adaptation of alcohol dehydrogenase activity in rat brain. Nature (Lond.) New Biol. **236**, 138—140 (1972)
Ratcliffe, F.: The effect of chronic ethanol administration on the responses to amylobarbitone sodium in the rat. Life Sci. **8**, 1051—1061 (1969) Pt. 1
Rawat, A. K.: Brain levels and turnover rates of presumptive neurotransmitters as influenced by administration and withdrawal of ethanol in mice. J. Neurochem. **22**, 915—922 (1974)
Redmond, G., Cohen, G.: Induction of liver acetaldehyde dehydrogenase: Possible role in ethanol tolerance after exposure to barbiturates. Science **171**, 387—389 (1971)
Redos, J. D., Catravas, G. N., Hunt, W. A.: Ethanol-induced depletion of cerebellar guanosine 3′, 5′-cyclic monophosphate. Science **193**, 58—59 (1976)
Reed, T. E.: Heritability of responses to alcohol in a heterogeneous mouse strain. Paper presented at Ann. Meeting Beh. Genetics Ass., Austin, Texas, March 20—22, 1975
Reggiani, G., Hurlimann, A., Theiss, E.: Some aspects of the experimental and clinical toxicology of chlordiazepoxide. In: Baker, S. B. D., Boissier, J. R., Koll, W. (Eds.): Proceedings of the European Society for the Study of Drug Toxicity. Vol. 9, Toxicity and Side-Effects of Psychotropic Drugs. Amsterdam: Excerpta Medica Foundation 1968
Reich, L. H., Davies, R. K., Himmelhoch, J. M.: Excessive alcohol use in manic-depressive illness. Amer. J. Psychiat. **131**, 83—86 (1974). Also reprinted in Alcohol Hlth and Res. Wld., 26—28 (Spring 1975)
Rementeria, J. L., Bhatt, K.: Withdrawal symptoms in neonates from intrauterine exposure to diazepam. J. Pediatr. **90**, 123—126 (1977)

Report Series 30, No. 1. National Clearinghouse for Drug Abuse Information. The deliberate inhalation of volatile substances. Gamage, J. R., Zerkin, E. L. U.S. Government Printing Office 1974

Reynolds, D. V.: Reduced response to aversive stimuli during focal brain stimulation: electrical analgesia and electrical anesthesia. In: Reynolds, D. V., Sjoberg, A. E. (Eds.): Neuroelectric Research: Electroneuroprosthesis, Electroanesthesia and Nonconvulsive Electrotherapy, pp. 151—167, Springfield, Ill.: Charles C. Thomas 1971

Richards, C. D.: The selective depression of evoked cortical EPSPs by pentobarbitone. J. Physiol. (Lond.) **217**, 41P—43P (1971)

Richards, C. D.: Potentiation and depression of synaptic transmission in the olfactory cortex of the guinea-pig. J. Physiol. (Lond.). **222**, 209—231 (1972a)

Richards, C. D.: On the mechanism of barbiturate anaesthesia. J. Physiol. (Lond.) **227**, 749—767 (1972b)

Richards, C. D.: On the mechanism of halothane anaesthesia. J. Physiol. (Lond.) **233**, 439—456 (1973)

Richards, C. D.: The action of general anesthetics on synaptic transmission within the central nervous system. In: Halsey, M. J., Millar, R. A., Sutton, J. A. (Eds.): Molecular Mechanisms in General Anaesthesia. Edinburgh-London-New York: Churchill Livingstone 1974

Richards, C. D., Russell, W. J., Smaje, J. C.: The action of ether and methoxyflurane on synaptic transmission in isolated preparations of the mammalian cortex. J. Physiol. (Lond.) **248**, 121—142 (1975)

Richardson, E. J.: Alcohol-state dependent learning; acquisition of a spatial discrimination in the goldfish (carassius auratus). Psychol. Rec. **22**, 545—553 (1972)

Richens, A.: Microelectrode studies in the frog isolated spinal cord during depression by general anaesthetic agents. Brit. J. Pharmacol. **36**, 312 (1969)

Rickels, K.: Predictors of response to benzodiazepines in anxious outpatients. In: Garattini, S., Mussini, E., Randall, L. O. (Eds.): The Benzodiazepines, pp. 391—404. New York: Raven Press 1973

Rickels, K., Downing, R. W.: Chlordiazepoxide and hostility in anxious outpatients. Amer. J. Psychiat. **131**, 442—444 (1974)

Riker, W. F. Jr., Okamoto, M.: Pharmacology of motor nerve terminals. Ann. Rev. Pharmacol. **9**, 173—208 (1969)

Ris, M. M., Dietrich, R. A., von Wartburg, J. P.: Inhibition of aldehyde reductase isoenzymes in human and rat brain. Biochem. Pharmacol. **24**, 1865—1869 (1975)

Risberg, J., Ingvar, H.: Increase of blood flow in cortical association areas during memorization and abstract thinking. Eur. Neurol. **6**, 236—241 (1971)

Rix, K. J. B., Davidson, N.: γ-aminobutyric acid in alcohol, barbiturate and morphine dependence: A review. Brit. J. Addict. **72**, 109—115 (1977).

Roach, M. K., Davis, D. L., Pennington, W., Nordyke, E.: Effect of ethanol on the uptake by rat brain synaptosomes of ^{3}H-DL norepinephrine, ^{3}H-5-hydroxytryptamine, ^{3}H-GABA, and ^{3}H-glutamate. Life Sci. **12**, 433—441 (1973)

Roach, M. K., Reese, W. N. Jr.: Effect of ethanol on glucose and amino acid metabolism in brain. Biochem. Pharmacol. **20**, 2805—2812 (1971)

Robın, A., Curry, S., Whelpton, R.: Clinical and biochemical comparison of clorazepate and diazepam. Psychol. Med. **4**, 388—392 (1974)

Robins, L. N.: The Vietnam Drug User Returns. Final Report. Washington, D.C.: U.S. Government Printing Office 1973

Robinson, V.: Victory Over Pain, a History of Anesthesia. New York: Henry Schuman 1946

Robson, J. G., Davenport, H. T., Sugiyama, R.: Differentiation of two types of pain by anesthetics. Anesthesiology **26**, 31—36 (1965)

Rodgers, D. A.: Factors underlying differences in alcohol preference of inbred strains of mice. In: Kissin, B., Begleiter, H. (Eds.): The Biology of Alcoholism, Vol. II, Chap. 5, pp. 107—130. New York: Plenum Press 1972

Röseler, P.: Beobachtungen und Untersuchungen über die gemeinsame Wirkung von Alkohol und Kohlenmonoxyd. (Observations on the interaction of alcohol and carbon monoxide.) Dissertation, Faculty of Medicine of the Free University of Berlin, West Germany, 1961, p. 61

Roots, B. I., Prosser, C. L.: Temperature acclimation and the nervous system of fish. J. exp. Biol. **39**, 617—629 (1962)
Rosadini, G., Rodriguez, G., Siani, C.: Acute alcohol poisoning in man: An experimental electrophysiological study. Psychopharmacologia (Berl.) **35**, 273—285 (1974)
Rose, S. P. R.: Neurochemical correlates of learning and environmental change. In: Horn, G., Hinde, R. A. (Eds.): Short-term Changes in Neural Activity and Behaviour, pp. 517—545. Cambridge, C.R.: Cambridge University Press 1970
Rosenbaum, M.: Adaptation of the central nervous system to varying concentration of alcohol in the blood. Arch. Neurol. Psychiat. (Chic.) **48**, 1010—1012 (1942)
Rosenberg, C. M.: Drug maintenance in the outpatient treatment of chronic alcoholism. Arch. gen. Psychiat. **30**, 373—377 (1974)
Rosenberg, H., Haugaard, N.: The effects of halothane on metabolism and calcium uptake in mitochondria of the rat liver and brain. Anesthesiology **39**, 44—53 (1973)
Rosenberg, H. C., Okamoto, M.: A method for producing maximal pentobarbital dependence in cats: dependency characteristics. In: Singh, J. M., Lal, H. (Eds.): Drug Addiction: Neurobiology and Influences on Behavior, Vol. 3, pp. 89—103, New York: Stratton 1974a
Rosenberg, H. C., Okamoto, M.: Electrophysiology of barbiturate withdrawal in the spinal cord. Fed. Proc. **33**, 528 (1974b)
Rosenfeld, G.: Potentiation of the narcotic action and acute toxicity of alcohol by primary aromatic monoamines. Quart. J. Stud. Alcohol **21**, 584—596 (1960)
Rosenzweig, M. R., Bennett, E. L.: Prospects for application of research on neural mechanisms of learning and memory. In: Neural Mechanisms of Learning and Memory. Cambridge, Mass.: MIT Press 1975
Rosner, B. S.: Recovery of function and localization of function in historical perspective. In: Stein, D. G., Rosen, J. J., Butters, N. (Eds.): Plasticity and Recovery of Function in the Central Nervous System, pp. 1—30. New York: Academic Press 1974
Rosner, B. S., Clark, D. L.: Neurophysiologic effects of general anesthetics: II. Sequential regional actions in the brain. Anesthesiology **39**, 59—81 (1973)
Ross, D. H.: Selective action of alcohol on cerebral calcium levels. Ann. N.Y. Acad. Sci. **273**, 280—294 (1976)
Ross, D. H.: Depletion of regional brain calcium by ethanol and salsolinol: Selective antagonism by naloxone. Soc. Neurosci. Fourth Annual Meeting 1974
Ross, D. H., Medina, M. A., Cardenas, H. L.: Morphine and ethanol: selection depletion of region brain calcium. Science **186**, 63 (1974)
Roth, L. J., Barlow, C. F.: Drugs in the brain. Science **134**, 706—711 (1961)
Roth, S. H.: Anesthesia and pressure: antagonism and enhancement. In: Progress in Anesthesiology. Vol. 1, pp. 405—427. Molecular Mechanisms of Anesthesia. New York: Raven Press 1975
Rubin, E., Lieber, C. S.: Alcoholism, alcohol and drugs. Science **172**, 1097—1102 (1971)
Rubington, E.: The language of "drunks". Quart. J. Stud. Alcohol **32**, 721—740 (1971)
Rummel, W., Schmitz, T.: Die Anticurrarewirkung des Alkohols. Naunyn-Schmiedebergs Arch. exp. Path. Pharmak. **222**, 257—261 (1954)
Rushton, R., Steinberg, H., Tomkiewicz, M.: Effects of chlordiazepoxide alone and in combination with amphetamine on animal and human behavior. In: Garattini, S., Mussini, E., Randall, L. O. (Eds.): The Benzodiazepines. New York: Raven Press 1973
Russell, R. W., Nathan, P. W.: Traumatic amnesia. In: Retrograde Amnesia: Temporal Gradient in Very Long Term Memory Following Electroconvulsive Therapy. Brain **69**, 280—300 (1946)
Russell, R. W., Overstreet, D. H., Cotman, C. W., Carson, V. G., Churchill, L., Dalgish, F. W., Vasquez, B. J.: Experimental tests of hypotheses about neurochemical mechanisms underlying behavioral tolerance to the anticholinesterase, diisopropyl fluorophosphate. J. Pharmacol. exp. Ther. **192**, 73—85 (1975)
Russell, R. W., Steinberg, H.: Effects of nitrous oxide on reactions to "stress". Quart. J. exp. Psychol. **7**, 67—73 (1955)
Ryback, R. S.: Alcohol amnesia: Observations on seven drinking inpatient alcoholics. Quart. J. Stud. Alcohol **31**, 616—632 (1970)
Ryback, R. S.: The continuum and specificity of the effects of alcohol on memory. Quart. J. Stud. Alcohol **32**, 995—1016 (1971)

Ryback, R. S.: Facilitation and inhibition of learning and memory by alcohol. Ann. N.Y. Acad. Sci. **215**, 187—194 (1973)

Ryback, R. S., Ingle, D.: Effect of ethanol and bourbon on Y-maze learning and shock avoidance in the goldfish. Quart. J. Stud. Alcohol (Suppl.) **5**, 136 (1970)

Saarikoski, J.: Effect of cold and stimulation on respiration and on potassium and water contents in brain slices of a hibernator and non-hibernator. Ann. Acad. Sci. Fenn. Ser. A, IV, Biologica **165**, 1—15 (1970)

Sachdev, K. S., Ranjwani, M. W., Joseph, A. D.: Potentiation of the response to acetylcholine on the frog's rectus abdominis by ethyl alcohol. Arch. int. Pharmacodyn. **145**, 36—43 (1963)

Salen, E. B.: Om gengasepoken I sverige 1939—1945: en klinisk och socialmedicinsk oversikt. (On the gengas period in Sweden 1939—1945: a clinical and socio-medical study.) Nord. Med. **1946**, 923—931 and 933—934

Saletu, B.: Classification of psychotropic drugs based on human evoked potentials. Mod. Probl. Pharmacopsychiat. **8**, 258—285 (1974)

Salmoiraghi, G. C., Weight, F.: Micromethods in neuropharmacology: an approach to the study of anesthetics. Anesthesiology **28**, 54—64 (1967)

Salzman, C., Kochansky, G. E., Shader, R. I., Porrino, L. J., Harmatz, J. S., Swett, C. P. Jr.: Chlordiazepoxide-induced hostility in a small group setting. Arch. gen. Psychiat. **31**, 401—405 (1974)

Sansone, M., Messeri, P.: Strain differences on the effects of chlordiazepoxide and chlorpromazine in avoidance behavior of mice. Pharmacol. Res. Commun. **6**, 179—185 (1974)

Sasaki, K., Otani, T.: Accomodation in motoneurons as modified by circumstantial conditions. Jap. J. Physiol. **12**, 383—396 (1962)

Sato, M., Austin, G. M., Yai, H.: Increase in permeability of the postsynaptic membrane to potassium produced by 'nembutal'. Nature (Lond.) **215**, 1506—1508 (1967)

Satoh, M., Takagi, H.: Effect of morphine on the pre and postsynaptic inhibition in the spinal cord. Europ. J. Pharmacol. **14**, 150—154 (1971)

Saubermann, A. J.: Regional and cellular pentobarbital localization in the brain, In: Fink, B. R. (Ed.): Progress in Anesthesiology, Vol. 1, 121—134. Molecular Mechanisms of Anesthesia. New York: Raven Press 1975

Saubermann, A. J., Gallagher, M. L., Hedley-Whyte, J.: Uptake, distribution, and anesthetic effect of pentobarbital-2-^{14}C after intravenous injection into mice. Anesthesiology **40**, 41—51 (1974)

Sauerland, E. K., Harper, R. M.: Effects of ethanol on EEG spectra of the intact brain and isolated forebrain. Exp. Neurol. **27**, 490—496 (1970)

Sauerland, E. K., Mizuno, N., Harper, R. M.: Presynaptic depolarization of trigeminal cutaneous afferent fibers induced by ethanol. Exp. Neurol. **27**, 476—489 (1970)

Schaefer, K. P., Meyer, D. L.: Compensatory mechanisms following labyrinthine lesions in the guinea-pig. A simple model of learning. In: Zippel, H. P. (Ed.): Memory and Transfer of Information, pp. 203—232. New York-London: Plenum Press 1973

Schallek, W., Kuehn, A.: Effects of psychotropic drugs on limbic system of cat. Proc. Soc. exp. Biol. (N.Y.) **105**, 115—117 (1960)

Schallek, W., Kuehn, A.: Effects of trimethadione, diphenylhydantion and chlordiazepoxide on after-discharges in brain of cat. Proc. Soc. exp. Biol. (N.Y.) **112**, 813—817 (1963)

Schallek, W., Kuehn, A., Jew, N.: Effects of chlordiazepoxide (LIBRIUM) and other psychotropic agents on the limbic system of the brain. Ann. N.Y. Acad. Sci. **96**, 303—312 (1962)

Schallek, W., Zabransky, F., Kuehn, A.: Effects of benzodiazepines on central nervous system of the cat. Arch. int. Pharmacodyn. **149**, 467—483 (1964)

Schauf, C. L., Davis, F. A., Marder, J.: Effects of carbamazepine on the ionic conductances of myxicola giant axons. J. Pharmacol. exp. Ther. **189**, 538—543 (1974)

Scheinin, B.: The cross-tolerance between ethanol and general anesthetics: an experimental study on rats. Turku: The Dept. of Anesthesiology and the Institute of Dentistry, University of Turku, Finland 1971

Schenker, V. J., Kissin, B., Maynard, L. S., Schenker, A. C.: Adrenal hormones and amine metabolism in alcoholism. Psychosom. Med. **28**, 564—569 (1966)

Schlan, L., Unna, K. R.: Some effects of myanesin in psychiatric patients. J. Amer. med. Ass. **140**, 672 (1949)

Schlatter, E. K. E., Lal, S.: Treatment of alcoholism with Dent's oral apomorphine method. Quart. J. Stud. Alc. **33**, 430—436 (1972)
Schlosser, W.: Action of diazepam on the spinal cord. Arch. int. Pharmacodyn. **194**, 93—102 (1971)
Schmidt, R. F.: The pharmacology of presynaptic inhibition. Progr. Brain Res. **12**, 119—131 (1964)
Schmidt, R. F.: Die Wirkung von Diazepam ("Valium" Roche) auf synaptische Funktionen des Rückenmarks. Proc. Sixth Int. Congr. Electroenceph. clin. Neurophysiol. (Vienna) **1965**, 627—630
Schmidt, R. F.: Presynaptic inhibition in the vertebrate central nervous system. Ergeb. Physiol. exp. Pharmakol. **63**, 20—101 (1971)
Schmidt, R. F., Vogel, M. E., Zimmerman, M.: Die Wirkung von Diazepam auf die präsynaptische Hemmung und andere Rückenmarksreflexe. Naunyn-Schmiedebergs Arch. exp. Path. Pharmak. **258**, 69—82 (1967)
Schneider, C. W., Evans, S. K., Chenoweth, M. B., Beman, F. L.: Ethanol preference and behavioral tolerance in mice: biochemical and neurophysiological mechanisms. J. comp. Physiol. Psychol. **82**, 466—474 (1973)
Schuberth, J., Sundwall, A.: Biosynthesis and compartmentation of acetylcholine in the brain. In: Psychopharmacology, Sexual Disorders and Drug Abuse. Amsterdam-London: North-Holland 1973; Prague: Avicenum, Czechoslovak Medical Press 1973
Schuckit, M, A., Goodwin, D. A., Winokur, G.: A study of alcoholism in half siblings. Amer. J. Psychiat. **128**, 122—126 (1972)
Schultz, J.: Adenosine 3', 5'-monophosphate in guinea pig cerebral cortical slices: effect of benzodiazepines. J. Neurochem. **22**, 685—690 (1974a)
Schultz, J.: Inhibition of phosphodiesterase activity in brain cortical slices from guinea pig and rat. Pharmacol. Res. Commun. **6**, 335—341 (1974b)
Schultz, J., Hamprecht, B.: Adenosine 3', 5'-monophosphate in cultured neuroblastoma cells: Effect of adenosine, phosphodiesterase inhibitors and benzodiazepines. Naunyn-Schmiedebergs Arch. Pharmacol. **278**, 215—225 (1973)
Schuster, C. R.: Self-administration of drugs. In: Goldberg, L., Hoffmeister, F. (Eds.): Psychic Dependence, Definition, Assessment in Animals and Man—Theoretical and Clinical Implications. New York-Heidelberg-Berlin: Springer 1973
Schuster, C. R., Jonanson, C. E.: The use of animal models for the study of drug dependence. Research Advances in Alcohol and Drug Problems. Gibbons, R. J. (Ed.). New York: Wiley and Sons, Inc. 1974
Schuster, C. R., Thompson, T.: Self administration of and behavioral dependence on drugs. Ann. Rev. Pharmacol. **9**, 483—502 (1969)
Schwartz, A., Lindenmayer, G. E., Allen, J. C.: The sodium-potassium adenosine triphosphatase: pharmacology, physiological and biochemical aspects. Pharmacol. Rev. **27**, 3—134 (1975)
Schwartzburg, M., Lieb, J., Schwartz, A. H.: Methaqualone withdrawal. Arch. gen. Psychiat. **29**, 46—47 (1973)
Seeman, P.: The effects of anesthetics and tranquilizers on cell membranes. In: Cellular Biology and Toxicity of Anesthetics. Baltimore: Williams and Wilkins Co. 1972a
Seeman, P.: The membrane actions of anesthetics and tranquilizers. Pharmacol. Rev. **24**, 583—655 (1972b)
Seeman, P.: The membrane expansion theory of anesthesia. In: Fink, B. R. (Ed.): Progress in Anesthesiology, Vol. 1, pp. 243—252. Molecular Mechanisms of Anesthesia. New York: Raven Press 1975
Seeman, P., Lee, T.: The dopamine-releasing actions of neuroleptics and ethanol. J. Pharmacol. exp. Ther. **190**, 131—140 (1974)
Seevers, M. H.: Morphine and ethanol physical dependence: a critique of a hypothesis. Science **170**, 113—114 (1970)
Seevers, M. H., Deneau, G. A.: Physiological aspects of tolerance and physical dependence. In: Root, W. S., Hofmann, F. G. (Eds.): Physiological Pharmacology, Vol. 1, pp. 565—640. New York: Academic Press 1963
Seguin, J. J., Stavraky, G. W.: The effects of barbiturates on partially isolated regions of the central nervous system. Canad. J. Biochem. Physiol. **35**, 667—680 (1957)

Segundo, J. P., Bell, C. C.: Habituation of single nerve cells in the vertebrate nervous system. In: Horn, G., Hinde, R. A. (Eds.): Short-Term Changes in Neural Activity and Behavior, pp. 77—94. London: Cambridge University Press 1970
Seidel, G.: Verteilung von Pentobarbital, Barbital und Thiopental unter Anthanol. Naunyn-Schmiedebergs Arch. exp. Path. Pharmak. **257**, 221—229 (1967)
Seixas, F. A.: Criteria for the diagnosis of alcoholism. J. Amer. med. Ass. **222**, 207—208 (1972)
Seixas, F. A., Eggleston, S. (Eds.): Alcoholism and the central nervous system. Ann. N.Y. Acad. Sci. **215**, 1—389 (1973)
Selig, J. W.: A possible oxazepam abstinence syndrome. J. Amer. med. Ass. **198**, 951—952 (1966)
Sevitt, I.: The effect of adrenergic beta-receptor blocking drugs on tremor. Practitioner **207**, 677—678 (1971)
Seyama, I., Narahashi, T.: Mechanism of blockade of neuromuscular transmission by pentobarbital. J. Pharmacol. exp. Ther. **192**, 95—104 (1975)
Shagass, C.: Effects of psychotropic drugs on human evoked potentials. Mod. Probl. Pharmacopsychiat. **8**, 238—257 (1974)
Shagass, C., Jones, A. L.: A neurophysiological study of psychiatric patients with alcoholism. Quart. J. Stud. Alcohol **18**, 171—182 (1957)
Shankar, R., Quastel, J. H.: Effects of tetrodotoxin and anaesthetics on brain metabolism and transport during anoxia. Biochem. J. **126**, 851—867 (1972)
Shapovalov, A. I.: Intracellular microelectrode investigation of the effects of anesthetics on transmission of excitation in the spinal cord. Fed. Proc. **23**, T113—T116 (1963)
Sharma, R. P., Stowe, C. M., Good, A. L.: Studies on the distribution and metabolism of thiopental in cattle, sheep, goats and swine. J. Pharmacol. exp. Ther. **172**, 128—137 (1970a)
Sharma, R. P., Stowe, C. M., Good, A. L.: Alteration of thiopental metabolism in phenobarbital-treated calves. Toxicol. appl. Pharmacol. **17**, 400—405 (1970b)
Sharman, D. F.: The formation of some acidic metabolites in the brain, and their subsequent transport. In: Callingham, B. A. (Ed.): Drugs and Transport Processes, Baltimore-London-Tokyo: University Park Press 1974
Sharpless, S.: Hypnotics and sedatives. In: Goodman, L. S., Gilman, A. (Eds.): The Pharmacological Basis of Therapeutics, (Chapters 10, 11), 4th ed. New York: Macmillan 1970
Sharpless, S., Jaffe, J.: Wihtdrawal phenomena as manifestations of disuse supersensitivity, pp. 67—76. In: Steinberg, H. (Ed.): Scientific Basis of Drug Dependence. London: Churchill Ltd. 1969
Shaw, J. A., Donley, P., Morgan, D. W., Robinson, J. A.: Treatment of depression in alcoholics. J. Psychiat. **132**, 6 (1975)
Shideman, F. E., Kelly, A. R., Adams, B. J.: Blood levels of thiopental (Pentothal) following repeated intravenous administration to the dog. (Abstract) Fed. Proc. **7**, 255 (1948)
Shulman, A.: The pharmacology of barbiturates. Med. J. Aust. 1199—1204 (1970)
Shuster, L.: Repression and de-repression of enzyme synthesis as a possible explanation of some aspects of drug action. Nature (Lond.) **189**, 314—315 (1961)
Sidell, B. D., Wilson, F. R., Hazel, J., Prosser, C. L.: Time course of thermal acclimation in goldfish. J. comp. Physiol. **84**, 119—127 (1973)
Sidell, F. R., Pless, J. E.: Ethyl alcohol: blood levels and performance decrements after oral administration to man. Psychopharmacologia (Berl). **19**, 246—261 (1971)
Siemens, A. J., Chan, A. W. K.: Effects of pentobarbital in mice selectively bred for different sensitivities to ethanol. Pharmacologist **17**, 197 (1975)
Siemens, A. J., Chan, A. W. K.: Differential effects of pentobarbital and ethanol in mice. Life Sci. **19**, 581—590 (1976)
Simpson, L. L.: An analysis of the sympathomimetic activity of 6,7-dihydroxy-1,2,3,4-tetrahydroisoquinoline (TIQ). J. Pharmacol. exp. Ther. **192**, 365—371 (1975)
Sinclair, J. D.: Rats learning to work for alcohol. Nature (Lond.) **249**, 590—592 (1974a)
Sinclair, J. D.: Lithium-induced suppression of alcohol drinking by rats. Med. Biol. **52**, 133—136 (1974b)
Singh, J. M.: Clinical signs and development of tolerance to thiopental. Arch. int. Pharmacodyn. **187**, 199—208 (1970)
Singh, J. M.: Comparison between acute and chronic administration of ethyl alcohol on the development of tolerance to pentobarbital. Arch. int. Pharmacodyn. **189**, 123—128 (1971)

Singh, J. M., Fiegenschue, B., Schexnaydre, C.: Development of tolerance to pentobarbital. J. pharm. Sci. **59**, 1020—1022 (1970)
Sitsen, J. M. A., Fresen, J. A.: Contributions to the study of barbiturate derivatives [in Dutch]. III. Farmacologisch gedeelte. Pharm. Weekbl. **109**, 1—10 (1974)
Smart, R. J.: Effects of alcohol on conflict and avoidance behavior. Quart. J. Stud. Alcohol **26**, 187—205 (1965)
Smith, C. M.: Neuromuscular pharmacology: drugs and muscle spindles. Ann. Rev. Pharmacol. **3**, 223—248 (1963)
Smith, C. M.: Relaxants of skeletal muscle. In: Root, W. S., Hofmann, F. G. (Eds.): Physiological Pharmacology, Vol. II, The Nervous System, part B, pp. 1—96. New York: Academic Press 1965
Smith, C. M.: The effects of drugs on the afferent nervous systems. In: Burger, A. (Ed.) Drugs Affecting the Peripheral Nervous System. New York: Marcel Dekker, 1967
Smith, C. M.: Variety of effects resulting from drug action on sensory receptors. Pharmacology and the Future of Man. Proc. 5th Int. Congr. Pharmacology, San Francisco 1972, Vol. 4, 152—166 (1973)
Smith, C. M.: Interactions of drugs of abuse with alcohol. Ann. N.Y. Acad. Sci. **281**, 384—392 (1976)
Smith, C. M., Murayama, S.: Rigidity of spinal origin: quantitative evaluation of agents with muscle relaxant activity in cats. Int. J. Neuropharmacol. **3**, 505—515 (1964)
Smith, E. B.: Physical chemical investigations of the mechanisms of general anaesthesia. In: Halsey, M. J., Millar, R. A., Sutton, J. A. (Eds.): Molecular Mechanisms of General Anesthesia, pp. 112—131. London-New York: Churchill Livingston 1974
Smith, G. M., Beecher, H. K.: Amphetamine sulphate and athletic performance. J. Amer. med. Ass. **170**, 542—557 (1959)
Smith, G. M., Semke, C. W., Beecher, H. K.: Objective evidence of mental effects of heroin, morphine and placebo in normal subjects. J. Pharmacol. exp. Ther. **136**, 53—58 (1962)
Smith, P. B.: Chemical Glimpses of Paradise. Springfield, Ill.: Charles C. Thomas 1972
Smith, S. G., Werner, T. E., Davis, W. M.: Comparison between intravenous and intragastric alcohol self-administration. Physiol. Psychol. **4**, 91—93 (1976)
Smith, W. L.: Drugs and Cerebral Function. Springfield, Ill.: Charles C. Thomas 1970a
Smith, W. L.: Drug response following brain damage. In: Drugs and Cerebral Function. Springfield, Ill.: Charles C. Thomas 1970b
Snyder, S. M., Enna, S. J.: The role of central glycine receptors in the pharmacologic actions of benzodiazepines. Advanc. Biochem. Psychopharm., Vol. 14, pp. 81—91 (1975). In: Mechanism of Action of Benzodiazepines. New York: Raven Press 1975
Soehring, K., Schüppel, R.: Wechselwirkungen zwischen Alkohol und Arzneimitteln. Dtsch. med. Wschr. **91**, 1892—1896 (1966)
Sohn, R. S., Ferrendelli, J. A.: Anticonvulsant drug mechanisms. Arch. Neurol. **33**, 626—629 (1976)
Somjen, G. G.: Effects of ether and thiopental on spinal presynaptic terminals. J. Pharmacol. exp. Ther. **140**, 396—402 (1963)
Somjen, G. G.: Effects of anesthetics on spinal cord of mammals. Anesthesiology **28**, 135—143 (1967)
Somjen, G., Carpenter, D., Henneman, E.: Selective depression of alpha motoneurons of small size by ether. J. Pharmacol. exp. Ther. **148**, 380—385 (1965)
Somjen, G. G., Gill, M.: The mechanism of the blockade of synaptic transmission in the mammalian spinal cord by diethyl ether and by thiopental. J. Pharmacol. exp. Ther. **140**, 19—30 (1963)
Sourbrie, P., Wlodaver, C., Schoonhoed, L., Simon, P., Boissier, J. R.: Preselection of animals in studies of anti-anxiety drugs. Neuropharmacology **13**, 719—728 (1974)
Speden, R. N.: Effect of some volatile anaesthetics on the transmurally stimulated guinea-pig ileum. Brit. J. Pharmacol. **25**, 104—118 (1965)
Spencer, W. A., April, R. S.: Plastic properties of monosynaptic pathways in mammals. In: Horn, G., Hinde, R. A. (Eds.): Short-term Changes in Neural Activity and Behavior, pp. 433—474. Cambridge England: Cambridge University Press 1970
Spencer, W. A., Thompson, R. F., Neilson, D. R. Jr.: Response decrement of the flexion reflex in the acute spinal cat and transient restoration by strong stimuli. J. Neurophysiol. **29**, 221—239 (1966)

Spreng, R. W. E.: Tolserol in acute alcoholism. J. nerv. ment. Dis. **118**, 545—551 (1953)
Squire, L. R., Slater, P. C., Chace, P. M.: Retrograde amnesia: temporal gradient in very long term memory following electroconvulsive therapy. Science **187**, 77—79 (1975)
Stark, L. G., Edmonds, H. L., Keesling, P.: Penicillin-induced epileptogenic foci—I: Time course and the anticonvulsant effects of diphenylhydantoin and diazepam. Neuropharmacology **13**, 261—267 (1974)
Stein, L., Berger, B. D.: Psychopharmacology of 7-chlor-5-(o-chlorophenyl)-1,3-dihydro-3-hydroxy-2H-1,4-benzodiazepin-2-one (lorazepam) in squirrel monkey and rat. Arzneimittel-Forsch. **21**, 1073—1078 (1971)
Stein, L., Wise, C. D., Belluzzi, J. D.: Effects of benzodiazepines on central serotonergic mechanisms. Advanc. Biochem. Psychopharm. Vol. 14, pp. 29—44 (1975). In: Mechanism of Action of Benzodiazepines. New York: Raven Press 1975
Stein, L., Wise, C. D., Berger, B. D.: Antianxiety action of benzodiazepines: decrease in activity of serotonin neurons in the punishment system. In: Garattini, S., Mussini, E., Randall, L. O. (Eds.): The Benzodiazepines. New York: Raven Press 1973a
Stein, R. B., Pearson, K. G., Smith, R. S., Redford, J. B. (Eds.): Control of Posture and Locomotion. New York-London: Plenum Press 1973b
Steinberg, H.: Selective effects of an anaesthetic drug on cognitive behavior. Quart. J. exp. Psychol. **6**, 170—180 (1954)
Steinberg, H.: Changes in time perception induced by an anaesthetic drug. Brit. J. Psychol. **46**, 273—279 (1955)
Steinberg, H.: 'Abnormal behavior' induced by nitrous oxide. Brit. J. Psychol. **47**, 183—194 (1956)
Steinberg, H. (Ed.): Scientific Basis of Drug Dependence, a Symposium. London: Churchill Ltd. 1969
Steinberg, H.: Animal behaviour models in psychopharmacology. In: Porter, R., Birch, J. (Eds.): Chemical Influences on Behaviour, pp. 199—210. London: J. & A. Churchill 1970
Steinberg, H., Tomkiewicz, M.: Long term after-effects of psychoactive drugs on animal behavior. In: Psychopharmacology, Sexual Disorders and Drug Abuse. Amsterdam-London: North-Holland 1973; Prague: Avicenum, Czechoslovak Medical Press 1973
Steiner, C.: Games Alcoholics Play. New York: Grove Press 1971
Steiner, F. A., Hummel, P.: Effects of nitrazepam and phenobarbital on hippocampal and lateral geniculate neurons in the cat. Int. J. Neuropharmacol. **7**, 61—69 (1968)
Stevenson, I. H., Turnbull, M. J.: The sensitivity of the brain to barbiturate during chronic administration and withdrawal of barbitone sodium in the rat. Brit. J. Pharmacol. **39**, 325—333 (1970)
Stevenson, I. H., Turnbull, M. J.: A study of the factors affecting the sleeping time following intracerebroventricular administration of pentobarbitone sodium: Effect of prior administration of centrally active drugs. Brit. J. Pharmacol. **50**, 499—511 (1974)
Stewart, R. D.: The effects of carbon monoxide on humans. Ann. Rev. Pharmacol. **15**, 409—423 (1975)
Stokes, E.: Alcohol-endocrine interrelationships. In: Kissin, B., Begleiter, H. (Eds.): The Biology of Alcoholism. Volume 1: Biochemistry, pp. 397—436. New York: Plenum Press 1971
Story, J. L., Eidelberg, E., French, J. D.: Electrographic changes induced in cats by ethanol intoxication. Arch. Neurol. **5**, 119—124 (1961)
Stratten, W. P., Barnes, C. D.: Diazepam and presynaptic inhibition. Neuropharmacology **10**, 685—696 (1971)
Straw, R. M., Mitchell, C. L.: A comparison of cortical afterdischarge patterns in cat and rabbit. Proc. Soc. exp. Biol. (N.Y.) **121**, 857—861 (1966)
Stumpf, C. M., Chiari, I.: Echte Gewöhnung an Hexobarbital. Naunyn-Schmiedebergs Arch. exp. Path. Pharmak. **251**, 275—287 (1965)
Sugarman, R. C., Cozad, C. P., Zavala, A.: Alcohol-induced degradation of performance on simulated driving tasks. Paper presented at International Automotive Engineering Congress, Detroit, Michigan, January 8—12, 1973. (Society of Automotive Engineers, New York)
Sullivan, G. A.: Dual medication in acute alcoholism. Gen. Practitioner (Kansas City). **9**, 67—69 (1954)
Summerfield, A., Steinberg, H.: Reducing interference in forgetting. Quart. J. exp. Psychol. **9**, 146—154 (1957)

Sun, A. Y.: Alcohol-membrane interaction in the brain: I. Reversibility. Ann. N.Y. Acad. Sci. **273**, 295—302 (1976)
Sun, A. Y., Samorajski, T.: Effect of ethanol on the adenosine triphosphatase and acetylcholinesterase activity in synaptosomes of the guinea-pig brain. J. Neurochem. **17**, 1365—1372 (1970)
Suria, A., Costa, E.: Benzodiazepines and posttetanic potentiation in sympathetic ganglia of the bullfrog. Brain Res. **50**, 235—239 (1973)
Suria, A., Costa, E.: Diazepam inhibition of post-tetanic potentiation in bullfrog sympathetic ganglia: possible role of prostaglandins. J. Pharmacol. exp. Ther. **189**, 690—696 (1974)
Suria, A., Costa, E.: Evidence for GABA involvement in the action of diazepam on presynaptic nerve terminals in bullfrog sympathetic ganglia. Advanc. Biochem. Psychopharm., Vol. 14, pp. 103—112 (1975). In: Mechanism of Action of Benzodiazepines. New York: Raven Press 1975a
Suria, A., Costa, E.: Action of diazepam, dibutyryl cGMP and GABA on presynaptic nerve terminals in bullfrog sympathetic ganglia. Brain Res. **87**, 102—106 (1975b)
Sutton, I., Simmonds, M. A.: Effects of acute and chronic ethanol on the γ-aminobutyric acid system in rat brain. Biochem. Pharmacol. **22**, 1685—1692 (1973)
Sutton, I., Simmonds, M. A.: Effects of acute and chronic pentobarbitone on the γ-aminobutyric acid system in rat brain. Biochem. Pharmacol. **23**, 1801—1808 (1974)
Swanson, D. W., Weddige, R. L., Morse, R. M.: Abuse of prescription drugs. Mayo Clin. Proc. **48**, 359—367 (1973)
Swinyard, E. A., Castellion, A. W.: Anticonvulsant properties of some benzodiazepines J. Pharmacol. exp. Ther. **151**, 369—375 (1966)
Swinyard, E. A., Chin, L., Fingl, E.: Withdrawal hyperexcitability following chronic administration of meprobamate to mice. Science **125**, 739—741 (1957)
Syapin, P. J., Stefanovic, V., Mandel, P., Noble, E. P.: The chronic and acute effects of ethanol on adenosine triphosphatase activity in cultured astroblast and neuroblastoma cells. J. Neuroscience Res. **2**, 147—155 (1976)
Szasz, T.: Ceremonial Chemistry. Garden City, New York: Anchor Press Doubleday 1974
Tabakoff, B., Boggan, W. O.: Effects of ethanol on serotonin metabolism in brain. J. Neurochem. **22**, 759—764 (1974)
Tabakoff, B., Groskopf, W., Anderson, R., Alivisatos, S. G. A.: "Biogenic" aldehyde metabolism relation to pentose shunt activity in brain. Biochem. Pharmac. **23**, 1707—1719 (1974)
Tabakoff, B., Ungar, F., Alivasatos, S. G. A.: Addiction of barbiturates and ethanol: Possible biochemical mechanisms. Advanc. exp. Med. Biol. **35**, 45—55 (1973)
Takaori, S., Nakai, Y., Sasa, M., Shimamoto, K.: Central depressants and evoked click responses with special reference to the reticular formation in the cat. Jap. J. Pharmacol. **16**, 264—275 (1966)
Tamerin, J. S., Weiner, S., Poppen, R., Steinglass, P., Mendelson, J. H.: Alcohol and memory: Amnesia and short term function during experimentally induced intoxication. Amer. J. Psychiat. **127**, 1659—1664 (1971)
Taylor, K. M., Laverty, R.: The interaction of chlordiazepoxide, diazepam, and nitrazepam with cathecholamines and histamine in regions of the rat brain. In: Garattini, S., Mussini, E., Randall, L. O. (Eds.): The Benzodiazepines. New York: Raven Press 1973
Taylor, C. A., Williams, C. H., Wakabayashi, T., Valdivia, E., Harris, R. A., Green, D. E.: The effect of halothane on energized configurational changes in heart mitochondria in situ. In: Fink, B. R. (Ed.): Cellular Biology and Toxicity of Anesthetics. Baltimore: Williams & Wilkins Co. 1972
Teller, D. N., deGuzman, T., Lajtha, A.: The mode of morphine uptake into brain slices. Brain Res. **77**, 121—136 (1974a)
Teller, D. N., deGuzman, T. K., Lajtha, A.: Drug uptake by brain II. Barbiturate uptake alters the transport of amino acids in vitro. In: Buniatian, H. Ch. (Ed.): Problems of Brain Biochemistry, Vol. 9. Armenian Academy of Sciences 1974b
Thadani, P. V., Kulig, B. M., Brown, F. C., Beard, J. D. Acute and chronic ethanol induced alterations in brain norepinephrine metabolites in rats. Pharmacologist **16**, 304 (1974)
Thesleff, S.: The effect of anesthetic agents on skeletal muscle membrane. Acta physiol. scand. **37**, 335—349 (1956)

Thompson,T., Pickens,R.: An experimental analysis of behavioral factors in drug dependence. Fed. Proc. **34**, 1759—1776 (1975)
Thomson,T.D., Turkanis,S.A.: Barbiturate-induced transmitter release at a frog neuromuscular junction. Brit. J. Pharmacol. **48**, 48—58 (1973)
Tirri,R.: Induced tolerance to promazine in mice as a physiological adaptation. Ann. Acad. Sci. Fenn. A. IV **103**, 1—54 (1966)
Traynor,M.E., Woodson,P.B.J., Schlapfer,W.T., Barondes,S.H.: Sustained tolerance to a specific effect of ethanol on posttetanic potentiation in *Aplasia*. Science **193**, 510—511 (1976)
Traynor,M.E., Woodson,P.B.J., Tremblay,J.P., Schlapfer,W.T., Barondes,S.H.: Specific decrease of PTP duration by ethanol at a synapse in *Aplasia* exhibits prolonged tolerance. 5th Annual Meeting of the Society of Neuroscience, 2—6 November 1975
Tseng,T.C., Wang,S.C.: Mechanisms of action of centrally acting muscle relaxants, diazepam and tybamate. J. Pharmacol. exp. Ther. **178**, 350—360 (1971a)
Tseng,T.C., Wang,S.C.: Locus of central depressant action of some benzodiazepine analogues. Proc. Soc. exp. Biol. (N.Y.) **137**, 526—531 (1971b)
Umemoto,M., Olds,M.E.: Effects of chlordiazepoxide, diazepam and chlorpromazine on conditioned emotional behaviour and conditioned neuronal activity in limbic, hypothalamic and geniculate regiona. Neuropharmacology **14**, 413—425 (1975)
Vaillant,G.E.: The natural history of urban narcotic drug addiction—some determinants. In: Steinberg,H. (Ed.): Scientific Basis of Drug Dependence. London: J. & A. Churchill Ltd. 1969
Van Harreveld,A., Foster,R.J., Fasman,G.D.: Effect of diphenyl hydantoin (dilantin) on ether and pentobarbital (nembutal) narcosis. Amer. J. Physiol. **166**, 718—722 (1951)
Van Poznak,A.: The effect of inhalation anesthetics on repetitive activity generated at motor nerve endings. Anesthesiology **28**, 124—127 (1967)
Victor,M.: The pathophysiology of alcoholic epilepsy. In: Wikler,A. (Ed.): The Addictive States. Baltimore: Williams & Wilkins Co. 1968
Victor,M.: The alcohol withdrawal syndrome: theory and pratice. Postgrad. Med. **47**, 68—72 (1970)
Victor,M.: The role of hypomagnesemia and respiratory alkalosis in the genesis of alcohol-withdrawal symptoms. Ann. N.Y. Acad. Sci. **215**, 235—248 (1973)
Victor,M., Adams,R.D.: The effect of alcohol on the nervous system. Res. Publ. Ass. Res. nerv. ment. Dis. **32**, 526—573 (1953)
Victor,M., Adams,R.D., Collins,G.H.: The Wernicke-Korsakoff Syndrome. Philadelphia: Davis 1971
Vieth,J.B., Holm,E., Knopp,P.R.: Electrophysiological studies on the action of MOGADON on central nervous structures of the cat. A comparison with pentobarbital. Arch. int. Pharmacodyn. **171**, 323—339 (1968)
Virtanen,P., Wallgren,H.: Rat brain microsomal lipid labeling in vivo by C-serine after acute and chronic ethanol administration. Abstracts Sixth International Congress of Pharmacol 616 (1975a)
Virtanen,P., Wallgren,H.: Rat brain microsomal total lipid labeling by 14-C-serine after chronic ethanol administration. Exp. Brain Res. **23**, Suppl. (1975b)
Volicer,L., Gold,B.I.: Effect of ethanol on cyclic AMP levels in the rat brain. Life Sci. **13**, 269—280 (1973)
Volicer,L., Gold,B.I.: Interactions of ethanol with cyclic AMP. In: Majchrowicz,E. (Ed.): Biochemical Pharmacology of Ethanol. New York: Plenum Press 1975
Volicer and Hurter: Cited in Redos et al., 1976
Volicer,L., Mirin,R., Gold,B.I.: Effect of acute ethanol administration on the cyclic AMP system in rat brain. J. Stud. Alcohol, **38**, 11—24 (1977a)
Volicer,L., Puri,S.K., Hurter,B.P.: Role of cyclic nucleotides in drug addiction and withdrawal. To be published in: Clinical Aspects of Cyclic Nucleotides, Spectrum Publications, Inc. (1977b)
Von Wartburg,J.P., Ris,M.M., White,T.G.: A possible role of biogenic aldehydes in long-term effects of drugs. In: Psychopharmacology, Sexual Disorders and Drug Abuse. Amsterdam-London: North-Holland 1973; Prague: Avicenum, Czechoslovak Medical Press 1973
Waddell,W.J., Baggett,B.: Anesthetic and lethal activity in mice of the stereoisomers of 5-ethyl-5-(1-methylbutyl) barbituric acid (pentobarbital). Arch. int. Pharmacodyn. **205**, 40—44 (1973)

Wahlström, G.: Differences in anaesthetic properties between the optical antipodes of hexobarbital in the rat. Life Sci. **5**, 1781—1790 (1966). Cited in Waddell and Baggett, 1973
Wahlström, G.: Hexobarbital (enhexymalum MFN) sleeping times and EEG threshold doses as measurements of tolerance to barbiturates in the rat. Acta pharmacol. (Kbh.) **26**, 64—80 (1968a)
Wahlström, G.: Differences in tolerance to hexobarbital (enhexymalum MFN) after barbital (diemalum NFN) pre-treatment during activity or rest. Acta pharmacol. (Kbh.) **26**, 92—104 (1968b)
Wahlström, G., Widerlöv, E.: Interaction and acute cross tolerance between ethanol and hexobarbitone in the rat. J. Pharm. (Lond.) **23**, 58—60 (1971)
Wajda, I. J., Manigault, I., Hudick, J. P.: Dopamine levels in the striatum and the effect of alcohol and reserpine. Biochem. Pharmacol. **26**, 653—655 (1977).
Walker, D. W., Freund, G.: Impairment of shuttle box avoidance learning following prolonged alcohol consumption in rats. Physiol. Behav. **7**, 773—778 (1971)
Walker, D. W., Zornetzer, S. F.: Alcohol withdrawal in mice: Electroencephalographic and behavioral correlates. Electroenceph. clin. Neurophysiol. **36**, 233—243 (1974)
Wall, P. D.: The mechanisms of general anesthesia. Anesthesiology **28**, 46—53 (1967)
Wall, P. D.: Habituation and post-tetanic potentiation in the spinal cord. In: Horn, G., Hinde, R. A. (Eds.): Short-term Changes in Neural Activity and Behaviour, pp. 181—208. London: Cambridge University Press 1970
Wallgren, H.: Effect of ethanol on intracellular metabolism and cerebral function. In: Kissin, B., Begleiter, H. (Eds.): The Biology of Alcoholism, Vol. 1. Biochemistry. New York: Plenum Press 1971
Wallgren, H.: Neurochemical aspects of tolerance to and dependence on ethanol. In: Gross, M. M. (Ed.): Alcohol Intoxication and Withdrawal Experimental Studies I, pp. 15—55. New York-London: Plenum Press 1973
Wallgren, H., Barry, H. III: Actions of Alcohol. Amsterdam-London-New York: Elsevier 1970
Wallgren, H., Lindbohm, R.: Adaptation to ethanol in rats with special reference to brain tissue respiration. Biochem. Pharmacol. **8**, 423—424 (1961)
Wallgren, H., Nikander, P., von Boguslawsky, P., Linkola, J.: Effects of ethanol, tert butanol, and clomethiazole on net movements of sodium and potassium in electrically stimulated cerebral tissue. Acta physiol. scand. **91**, 83—93 (1974)
Wallgren, H., Nikander, P., Virtanen, P.: Ethanol-induced changes in cation-stimulated adenosine triphosphatase activity and lipid-proteolipid labeling of brain microsomes. In: Gross, M. M. (Ed.): Alcohol Intoxication and Withdrawal Experimental Studies—II, pp. 23—36. New York: Plenum Press 1975
Walsh, M. J.: Biogenesis of biologically active alkaloids from amines by alcohol and acetaldehyde. Ann. N.Y. Acad. Sci. **215**, 98—110 (1973)
Warburton, D. M.: Modified sensitivity to pentobarbital in a continuous avoidance situation. Psychopharmacologia (Berl.) **13**, 387 (1968)
Ward, A. A. Jr.: Sodium cyanide: time of appearance of signs as a function of the rate of injection. Proc. Soc. exp. Biol. Med. **64**, 190—193 (1947)
Ward, A. A. Jr., Wheatley, M. D.: Sodium cyanide: sequence of changes of activity induced at various levels of the central nervous system. J. Neuropath. exp. Neurol. **6**, 292—294 (1947)
Waser, P. G., Schaub, E.: The action of some neuro and psychopharmacological agents on the membrane ATPase of cortical synaptosomes. In: Advanc. Cytopharmacol., Vol. 1: First International Symposium on Cell Biology and Cytopharmacology. New York: Raven Press 1971
Watanabe, H., Munakata, H., Chen, S. C., Kasuya, Y.: Effect of L-dopa, adrenergic β-blockers and anticholinergic agents on the tremorine-tremor in mice. Arch. int. Pharmacodyn. **193**, 372—389 (1971)
Waters, D.: Personal communication 1975
Waters, D. H., Okamoto, M.: Increased central excitability in non dependent mice during chronic barbitol dosing. In: Drug Addition: Experimental Pharmacology, Vol. 1. Mount Kisco, New York: Futura 1972
Waters, D. H., Okamoto, M.: Barbital dose and treatment duration: Effects on the incidence and severity of physical dependence. Fed. Proc. **32**, 681 (1973)
Waud, B. E., Cheng, M. C., Waud, D. R.: Comparison of drug-receptor dissociation constants at the mammalian neuromuscular junction in the presence and absence of halothane. J. Pharmacol. exp. Ther. **187**, 40—46 (1973)

Wayner,M.J., Ono,T., Nolley,D.: Effects of ethyl alcohol on central neurons. Pharmacol. Biochem. Behav. **3**, 499—506 (1975)
Weakly,J.N.: Effect of barbiturates on "quantal" synaptic transmission in spinal motoneurones. J. Physiol. (Lond.) **204**, 63—77 (1969)
Weeks,J.R.: Personal communication 1971
Weeks,J.R.: Prostaglandins: Introduction. Fed. Proc. **33**, 37—38 (1974)
Weinryb,I., Chasin,M.: Effects of therapeutic agents on cyclic AMP metabolism in vitro. J. pharmacol. Sci. **61**, 1556—1567 (1972)
Weise,C.E., Busse,S., Hill,R.J.: Solvent Abuse: An Annotated Bibliography with Additional Related Citations. Toronto-Ontario-Canada: Addiction Research Foundation 1973
Weise,C.E., Price,S.F.: The Benzodiazepines—Patterns of Use: An Annotated Bibliography. Toronto, Ontario, Canada: Addiction Research Foundation 1975
Werner,G.: Antidromic activity in motor nerves and its relation to a generator event in nerve terminals. J. Neurophysiol. **24**, 401—413 (1961)
Wesson,D.R., Smith,D.E.: Barbiturates: Their Use, Misuse, and Abuse. New York: Human Sciences Press, 1977
Westmoreland,B.F., Ward,D., Johns,T.R.: The effect of methohexital at the neuromuscular junction. Brain Res. **26**, 465—468 (1971)
White,D.C.: Anaesthetic and enzyme interactions. In: Halsey,M.J., Millar,R.A., Sutton,J.A. (Eds.): Molecular Mechanisms in General Anaesthesia. Edinburgh-London-New York: Churchill Livingstone 1974
Whybrow,P., Ferrell,R.: Thyroid state and human behavior: Contributions from a clinical perspective. In: The Thyroid Axis, Drugs, and Behavior. New York: Raven Press 1974
Wikler,A.: The Addictive States. Baltimore: Williams and Wilkins Co. 1968a
Wikler,A.: Interaction of physical dependence and classical and operant conditioning in the genesis of relapse. In: The Addictive States. Baltimore: Williams and Wilkins Co. 1968b
Wikler,A.: Diagnosis and treatment of drug dependence of the barbiturate type. Amer. J. Psychiat. **125**, 758—765 (1968c)
Wikler,A.: Diagnosis and treatment of drug dependence of the barbiturate type. In: Blachy,P.H. (Ed.): Drug Abuse: Data and Debate, p.283. Springfield, Ill.: Charles C. Thomas 1970
Wikler,A.: Theories related to physical dependence. In: Chemical and Biological Aspects of Drug Dependence. Cleveland, Ohio: CRC Press 1972
Wikler,A.: Dynamics of drug dependence: implications of a conditioning theory for research and treatment. In: Fisher,S., Freedman,A.M. (Eds.): Opiate Addiction: Origins and Treatment, pp.7—21. Washington: V.H. Winston & Sons 1973
Wikler,A.: Theoretical problems in localizing drug actions and origins of withdrawal syndromes in the central nervous system: the glass-eye booby trap. In: Narcotics and the Hypothalamus. New York: Raven Press 1974
Wikler,A., Essig,C.F.: Withdrawal seizures following chronic intoxication with barbiturates and other sedative drugs. In: Neidermeyer,E. (Ed.): Epilepsy, Vol.4: Modern Problems of Pharmacopsychiatry. Basel-New York: Karger 1970
Wikler,A., Frank,K.: Hindlimb reflexes of chronic spinal dogs during cycles of addiction to morphine and methadone. J. Pharmacol. exp. Ther. **94**, 382—400 (1948)
Wikler,A., Pescor,F.T., Fraser,H.F., Isbell,H.: Electroencephalographic changes associated with chronic alcoholic intoxication and the alcohol abstinence syndrome. Amer. J. Psychiat. **113**, 106—114 (1956)
Wilkinson,H.A., Mark,V.H., Wilson,R.: Sleep induced by focal brain suppression using anesthetic gases. Exp. Neurol. **30**, 30—33 (1971)
Williams,J.W., Tada,M., Katz,A.M., Rubin,E.: Effect of ethanol and acetaldehyde on the $(Na^+ + K^+)$-activated adenosine triphosphatase activity of cardiac plasma membranes. Biochem. Pharmacol. **24**, 27—32 (1975)
Willis,J.H.: The natural history of drug dependence: some comparative observations on United Kingdom and United States subjects. In: Steinberg,H. (Ed.): Scientific Basis of Drug Dependence. London: J & A Churchill, Ltd. 1969
Wilson,I.C., Prange,A.J.Jr., Lara,P.P.: L-Triiodothyronine alone and with imipramine in the treatment of depressed women. In: The Thyroid Axis, Drugs, and Behavior. New York: Raven Press 1974a

Wilson, I. C., Prange, A. J. Jr., Lara, P. P., Alltop, L. B., Stikeleather, R. A., Lipton, M. A.: Thyrotropin-releasing hormone: Psychobiological responses of normal women. I. Subjective Experiences. In: The Thyroid Axis, Drugs, and Behavior. New York: Raven Press 1974b
Winger, G. D., Woods, J. H.: The reinforcing property of ethanol in the rhesus monkey: I. Initiation, maintenance, and termination of intravenous ethanol-reinforced responding. Ann. N. Y. Acad. Sci. **215**, 162—175 (1973)
Winick, C. (Ed.): Sociological Aspects of Drug Dependence. Cleveland, Ohio: CRC Press 1974
Winkler, G. F., Young, R. R.: The control of essential tremor by propranol. Trans. Amer. neurol. Ass. **96**, 66—68 (1971)
Winkler, G. F., Young, R. R.: Efficacy of chronic propranolol therapy in action tremors of the familial, senile or essential varieties. New Engl. J. Med. **290**, 984—988 (1974)
Winter, J. C.: Hallucinogens as discriminative stimuli. Fed. Proc. **33**, 1825—1832 (1974)
Winter, J. C.: The stimulus properties of morphine and ethanol. Psychopharmacologia (Berl.) **44**, 209—214 (1975)
Winters, W. D., Mori, K., Spooner, C. E.: The neurophysiology of anesthesia. Anesthesiology **28**, 65—80 (1967a)
Winters, W. D., Mori, K., Spooner, C. E., Kado, R. T.: Correlation of reticular and cochlear multiple unit activity with evoked responses during wakefulness and sleep. Electroenceph. clin. Neurophysiol. **23**, 539—545 (1967b)
Wolff, P. H.: Ethnic differences in alcohol sensitivity. Science **175**, 449—450 (1972)
Wollman, H.: Effects of general anesthetics on cerebral metabolism in man. In: Cellular Biology and Toxicity of Anesthetics. Fink, B. R. (ed.). Baltimore: Williams & Wilkins Co. 1972
Wood, W. G.: Facilitation by dexamethasone of tolerance to ethanol in the rat. Psychopharmacology **52**, 67—72 (1977)
Woodbury, D. M., Esplin, D. W.: Neuropharmacology and neurochemistry of anticonvulsant drugs. Proc. Ass. Res. nerv. ment. Dis. **37**, 24—56 (1959)
Woodbury, D. M., Penry, J. K., Schmidt, R. P.: Antiepileptic drugs. New York: Raven Press 1972
Woodbury, J. W., D'Arrigo, J. S., Eyring, H.: Physiological mechanism of general anesthesia: synaptic blockage, 53—59. In: Fink, B. R. (Ed.): Progress in Anesthesiology, Vol. 1. Molecular Mechanisms of Anesthesia. New York: Raven Press 1975
Woods, J. H., Downs, D. A., Villarreal, J. E.: Changes in operant behavior during deprivation— and antagonist—induced withdrawal states. In: Goldberg, L., Hoffmeister, F. (Eds.): Psychic Dependence, Definition, Assessment in Animals and Man—Theoretical and Clinical Implications. Berlin-Heidelberg-New York: Springer 1973
Woods, J. H., Ikomi, F., Winger, G.: The reinforcing property of ethanol. In: Roach, M. K., McIsaac, W. M., Creaven, P. J. (Eds.): Biological Aspects of Alcoholism, pp. 371—388. Austin: Univ. of Texas Press 1971
Woods, J. H., Schuster, C. F.: Regulation of drug self-administration. In: Harris, R. T., McIsaac, W. M., Schuster, C. F. (Eds.): Drug Dependence, pp. 158—169. Austin: University of Texas 1970
Woods, J. H., Winger, G. D.: A critique of methods for inducing ethanol self-intoxication in animals. In: Mello, N. K., Mendelson, J. H. (Eds.): Recent Advances in Studies of Alcoholism, pp. 413—436. Washington, D.C.: U.S. Government Printing Office 1971
Woodson, P. B. J., Traynor, M. E., Schlapfer, W. T., Barondes, S. H.: Increased membrane fluidity implicated in acceleration of decay of post-tetanic potentiation by alcohols. Nature **260**, 797—799 (1976)
Woody, G. E., O'Brien, C. P., Greenstein, R.: Misuse and abuse of diazepam: an increasingly common medical problem. Int. J. Addict. **10**, 843—848 (1975)
Wren, J. C., Kline, N. S., Cooper, T. B., Varga, E., Canal, O.: Evaluation of lithium therapy in chronic and periodic alcoholism. 3rd Annual Conference of the National Institute on Alcohol Abuse and Alcoholism, Washington, D.C. (1973)
Wulfsohn, N. L.: The benzodiazepines. Clin. Anesth. **10**, 207—232 (1973)
Wyant, G. M., Studney, L. J.: A study of diazepam (Valium) for induction of anaesthesia. Canad. Anaesth. Soc. J. **17**, 166—171 (1970)
Xhenseval, B., Richelle, M.: Behavioural effects of long-term treatment with meprobamate in cats. Int. J. Neuropharmacol. **4**, 1—12 (1965)

Yanagita, T., Takahashi, S.: Development of tolerance to and physical dependence on barbiturates in rhesus monkeys. J. Pharmacol. exp. Ther. **172**, 163—169 (1970)
Yanagita, T., Takahashi, S.: Dependence liability of several sedative-hypnotic agents evaluated in monkeys. J. Pharmacol. exp. Ther. **185**, 307—316 (1973)
Ylikahri, R. H., Huttunen, M. O., Eriksson, C. J. P., Nikkilä, E. A.: Metabolic studies on the pathogenesis of hangover. Europ. J. clin. Invest. **4**, 93—100 (1974)
York, J. L.: Role of brain catecholamines and serotonin in morphine analgesia. Ph.D. thesis, University of Illinois Medical College, Chicago, Illinois, 1972
York, J. L., Winter, J. C.: Assessment of tolerance to barbital by means of drug discrimination procedures. Psychopharmacologia (Berl.) **42**, 283—287 (1975a)
York, J. L., Winter, J. C.: Dissimilar discriminative stimulus properties of barbital and ethanol. Pharmacologist **17**, 198 (1975b)
York, J. L., Winter, J. C.: Long-term effects of barbital on spontaneous activity of rats trained to use the drug as a discriminative stimulus. Psychopharmacologia (Berl.) **42**, 47—50 (1975c)
Young, A. B., Zukin, S. R., Snyder, S. H.: Interaction of benzodiazepines with central nervous glycine receptors: possible mechanism of action. Proc. nat. Acad. Sci. (Wash.) **71**, 2246—2250 (1974)
Zbinden, G., Randall, L. O.: Pharmacology of benzodiazepines: Laboratory and clinical correlations. Advanc. Pharmacol. **5**, 213—291 (1967)
Zilm, D. H., Sellers, E. M.: The effect of propranol on normal physiologic tremor. Electroenceph. clin. Neurophysiol. **41**, 310—313 (1976)
Zilm, D. H., Sellers, E. M., Macleod, A. M., Degani, N.: Propranolol effect on tremor in alcoholic withdrawal. Ann. intern. Med. **83**, 234—235 (1975)
Zinberg, N. E.: "High" states: a beginning study. Washington, D.C.: The Drug Abuse Council Inc. 1974
Zippel, H. P. (Ed.): Memory and Transfer of Information. New York-London: Plenum Press 1973
Zorychta, E.: An intracellular study of the actions of anesthetics on spinal monosynaptic transmission. A Thesis submitted to the Faculty of Graduate Studies and Research in partial fulfillment of the requirements for the degree of Doctor of Philosophy. Montreal, Canada 1974

CHAPTER 2

The Assessment of the Abuse Potentiality of Sedative/Hypnotics (Depressants) (Methods Used in Animals and Man)

H. F. Fraser and D. R. Jasinski

A. Definitions and Scope of Review

In 1964, the WHO Expert Committee on Addiction-Producing Drugs recommended substitution of the term "drug dependence" for the terms "drug addiction" and "drug habituation." The WHO Scientific Group (1964) endorsed this recommendation.

Drug dependence is defined as a state arising from the repeated administration of a drug on a periodic or continuous basis. It includes the attributes "physical" and/or "psychic" dependence. Although the characteristics of dependence vary with the agent involved, it carries no connotation in regard to degree of risk to the public or need for a particular type of control.

Potential for abuse takes into account not only the concept of dependence but also the degree of risk to the public. It is defined as follows: Abuse of a drug exists if its use so harmfully affects the individual and/or society as to require its control (Fraser, 1966). Risks to the public are often reflected in the type and extent of control promulgated. To a considerable extent, however, the degree of control of a drug or substance depends upon whether the "addiction" is socially accepted. In the case of alcohol and tobacco, for example, the controls are limited, since these are socially accepted forms of addiction. Heroin abuse, on the other hand, is not socially accepted, and controls are strict. Thus, it is seen that the term "potential for drug abuse" is quite analogous to the old WHO definition of addiction since it implies harm to the individual and/or society, that is, the drug is capable of inducing psychic or physical pathology.

It would be desirable to have screening procedures and tests in animals and man for sedatives and hypnotics which would predict not only "drug dependence" but also, if possible, the extent to which they might be abused and harmful to society. In carrying out such evaluations, the pharmacologist studying both animals and man determines (1) the degree of physical dependence an unknown agent induces as compared with a control drug—that is, the extent to which a chronically intoxicated subject will manifest a withdrawal illness when the drug under test is discontinued abruptly and (2) degree of psychic dependence or the extent to which the unknown agent as compared to a control drug reinforces drug-seeking behavior.

For convenience in presentation the drugs covered in this review fall in three main categories:

1. Barbituratelike agents, which would include such drugs as barbital, secobarbital, pentobarbital, amobarbital, and phenobarbital, and nonbarbiturates such as glutethimide or methaqualone which are hypnotic CNS depressants. Drugs in this

category induce physical dependence, euphoria, and are toxic in that they cause impaired coordination and judgment, respiratory depression, and in very large doses, drunkenness, coma, and death.

2. Minor tranquilizers such as meprobamate, chlordiazepoxide (Librium), and diazepam (Valium). Drugs in this category induce physical dependence and euphoria. However, their toxic effects overall are much less than those of barbituratelike agents.

3. Major tranquilizers such as chlorpromazine. They induce neither physical nor psychic dependence and overall are less toxic than the barbiturates.

It is realized that to some extent this classification is arbitrary. For example, barbiturates are employed clinically not only as hypnotics but also as sedatives. Phenobarbital which accounts for at least half of the clinical use of all barbiturates is used primarily as a sedative and antiepileptic agent, not as an hypnotic, and many clinicians employ it in a manner similar to that of the minor tranquilizers, that is, as an antianxiety agent.

The approach taken in this review will be to evaluate for each classification the procedures available in animals and man with a discussion of their sophistication and reliability for predicting to what extent an unknown agent will be abused by man.

B. Assessment of Barbituratelike Agents

I. Tests for Physical Dependence in Animals

1. Dog

SEEVERS and TATUM (1931) demonstrated that if dogs were chronically and continuously intoxicated with a high dosage of barbital, abrupt withdrawal of such medication was followed by an abstinence syndrome which was characterized by acute loss of weight and grand mal convulsions. FRASER and ISBELL (1954) confirmed these observations, and in addition, demonstrated that the dog also developed in many instances behavior which resembled the delirium observed in man and described this as "canine delirium." On the basis of these observations the oldest and best established method for evaluating the abuse potential of a new sedative/hypnotic has been to use dogs that are clinically intoxicated with sodium barbital as a control, attempt to intoxicate other dogs to an equivalent amount of the experimental agent, and then abruptly withdraw the animals after 60–90 days of intoxication and observe the two groups for abstinence signs. Using this procedure, it is feasible to evaluate in a quantitative manner the incidence of grand mal convulsions, delirium, the extent to which food and water intake is decreased, and the amount of weight loss. In addition, milder signs of withdrawal such as nervousness, apprehension, and tremulousness are observable (DENEAU et al., 1968). This method is generally referred to as a direct addiction procedure.

A second method widely used is the substitution technique (DENEAU and WEISS, 1968). This method has been described in detail by DENEAU and WEISS as follows:

> Mongrel dogs, averaging 15 kg in body weight, were used. Prior to the experiment, the dogs were conditioned by treatment with distemper and rabies vaccines and appropriate anthelmintics. Sodium barbital was chosen as the primary drug of dependence because it has a duration of action of approximately 24 h—a desirable feature, in that dosing once daily only is required in order to maintain the animals under continuous drug effect. An initial dose of 80 mg/kg was

Table 1. The incidence of convulsions and delirium in relation to the duration of treatment with 100 mg/kg barbital sodium

Months of treatment	Percent of dogs with convulsions	Percent of dogs with delirium
1–3	20	0
4–6	40	20
7–12	90	40
13–24	100	100

administered daily for 2 weeks. The dose was then raised to 100 mg/kg and maintained at that level. Although higher doses of sodium barbital can be tolerated, we purposely avoided the development of maximal physiological dependence because of the potential danger of a lethal outcome should a test drug not be an effective substitute.

It was of basic importance to develop an objective point-score system to assess the severity of the barbiturate abstinence signs. After the various abstinence signs were observed, definitions of 4 grades of severity were established. These were: slight, intermediate, severe, and very severe; and the grades of 1–4 were assigned.

Prior to substitution testing, range-finding experiments were conducted in normal dogs, to ascertain each test drug's potency and duration of action. The dose which produced light anesthesia in normal dogs was employed in the substitution test. A frequency of administration was selected which would not result in cumulative drug effects, but did maintain the animals under continuous drug effect. If the supply of test drug was sufficient, it was administered to 6 dogs for a period of 6 days. All treatment was then abruptly discontinued. The animals were observed continuously and graded throughout the substitution and withdrawal periods.

It was observed that the intensity of abstinence was dependent upon the duration of physical dependence. This is well illustrated in Table 1.

The convulsions were grand mal type and stopped spontaneously within a few seconds to 4 min. Convulsions seldom occurred prior to the 40th h of withdrawal. After 2 years of treatment, the number of convulsions in 11 dogs which occurred during a 5-day withdrawal ranged from 5–20. Each dog showed his own characteristic type of delirium which would recur every time the treatment was withdrawn. After the extent of the various signs had been determined, they were given numerical grades of severity as follows:

1. (Mild) Dog fails to respond or responds abnormally to the observer.
2. (Intermediate) Dog displays poor judgment or aggressiveness, or maintains a strange posture for minutes, or barks at no apparent object.
3. (Severe) Purposeless jumping, unprovoked attacks on other dogs or observers, moving the head and eyes as if seeing an imaginary object.
4. (Very Severe) Continuous bizarre behavior, searching for imaginary object."

In addition to the major signs of abstinence described above, many other signs appeared, many of which occur earlier during withdrawal and are more consistent indicators of abstinence than the major signs. These other signs include: tremor, sleeplessness, nervousness, mydriasis, tachycardia, fever, and weight loss.*

The results with unknown drugs that effectively substitute in dogs which were dependent on sodium barbital are summarized in Table 2. In each case, withdrawal of all treatment following the substitution period resulted in the appearance of a typical abstinence syndrome except after substitution of carisoprodol. The start of

* Certain of these minor symptoms have been quantitatively scored to measure degree of physical dependence [Jones, B.E., Prada, J.A. and Martin, W.R.: Psychopharmacologia (Berl.) **47**, 7—15 (1976)].

Table 2. Drugs which substituted for 100 mg/kg sodium barbital in physiologically dependent dogs

Drug	Dose	Dosage interval
Phenobarbital sodium	30 mg/kg	12 h
Pentobarbital sodium	15 mg/kg	6 h
Chloral hydrate	500 mg/kg	12 h
Paraldehyde	1 cc/kg	12 h
Ethyl alcohol	8 mg/kg	12 h
Bromural	450 mg/kg	6 h
Carisoprodol	200 mg/kg	6 h
Meprobamate	150 mg/kg	6 h
Chlordiazepoxide	100 mg/kg	12 h
Glutethimide	125 mg/kg	6 h
Methyprylon	80 mg/kg	6 h
Methaqualone	300 mg/kg	12 h

the abstinence syndrome corresponded to the duration of action of a particular drug; convulsions and/or delirium occurred earlier following withdrawal of the shorter-acting drugs.

2. Monkey

The monkey has also been utilized for evaluating physical dependence on barbiturates by the substitution technique which has been described by YANAGITA and TAKAHASHI (1970, 1973). It may be described as follows:

Physical dependence on barbital was produced in monkeys within three months by repeated oral administration of 75 mg/kg of barbital once daily at the beginning, then twice daily. These animals were withdrawn for 15–18 h prior to testing. When an intermediate to severe grade of withdrawal signs appeared, single doses of a test drug were administered and suppression of barbital withdrawal signs was observed for the following 5 to 24 h.

The experiment was controlled by administering only the vehicle during substitution to parallel chronically intoxicated monkeys. The severity of withdrawal signs was evaluated by the double blind procedure in accordance with the "grades for withdrawal signs in barbiturate-dependent monkeys" which are presented in Table 3. If the substitution of the unknown drug for sodium barbital is associated with continued intoxication and if there are typical signs of barbiturate withdrawal when a substitute drug is abruptly withdrawn, it is assumed to have equivalent dependence characteristics to that of sodium barbital.

Table 3. Grades of withdrawal signs in barbiturate-dependent monkeys

Mild	Intermediate	Severe
Apprehension	Aggravated tremor	Convulsions
Hyperirritability	Muscle rigidity	Delirium: hallucinatory behavior, nystagmus, dissociation from environment
Mild tremor	Impaired motor activities	Hyperthermia (> 1.5° C)
Anorexia	Retching or vomiting	
Piloerection	Weight loss (10%)	

3. Rat

CROSSLAND and LEONARD (1963) first reported that the white rat will develop physical dependence on sodium barbital if the fluid intake is restricted to solutions containing increasing concentrations of sodium barbital. Convulsions occurred in such rats following the withdrawal of sodium barbital.

ESSIG (1966) reported on another study using the method of CROSSLAND and LEONARD (1963) which may be summarized as follows: Male Wistar rats weighing between 360 and 486 g were used in the study. They were maintained on dry Purina lab chow pellets, and fluid intake was limited to tap water solution of barbital. The sodium barbital concentration was 2 mg/ml at the beginning and was increased in 1 mg/ml increments. Five rats in Group I were intoxicated for 111 days, and the final barbital concentration received was 6 mg/ml. Thirteen rats in Group II were intoxicated 159 days, and the final barbital concentration received was 5 mg/ml.

Both groups and their controls were studied 6 days before barbital was abruptly withdrawn and up to 23 days thereafter. Observations for activity, rectal temperature, body weight and fluid as well as food consumption were made at noon each day. Activity was recorded during 5 min in a cage balanced to activate a mercury switch and an electronic counter when the animal moved. Rectal temperatures were determined by means of a telethermometer and the probe was inserted 3.5 cm during

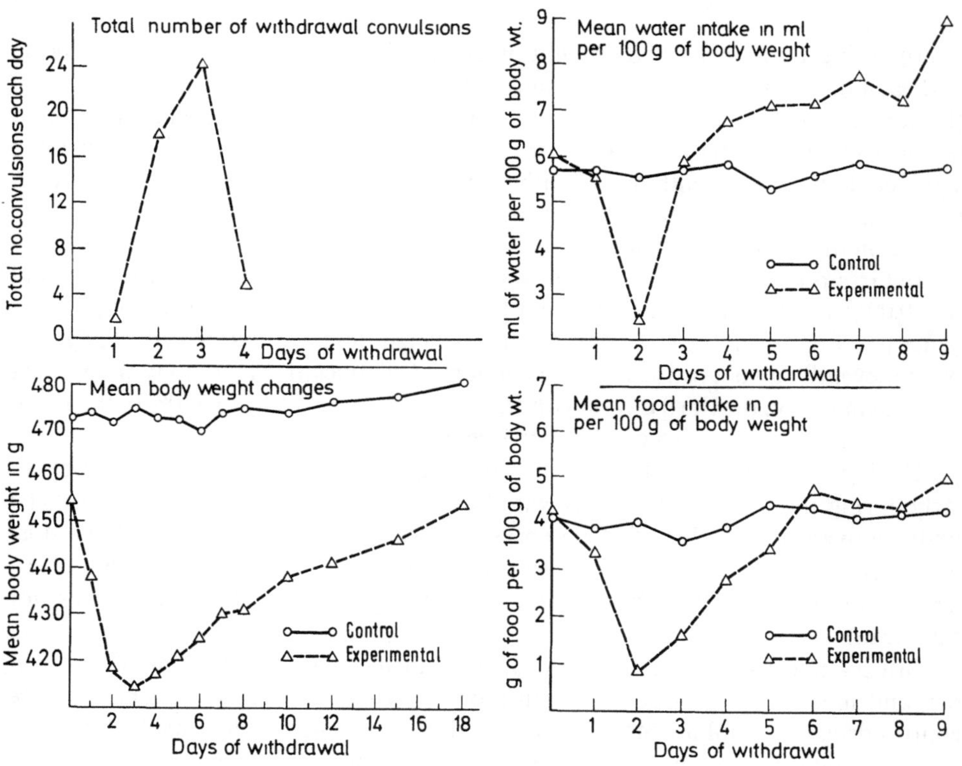

Fig. 1. Effect of sodium barbital withdrawal in rats Group II. [Reproduced from ESSIG (1958) with the permission of the publisher]

a 3-min interval. In addition to these measurements, the rats were observed around the clock for the occurrence of convulsions during the last day of addiction and for the first $4^1/_2$ days of withdrawal.

The results of these experiments with rats were as follows: During intoxication control rats gained weight but both Group I and II rats had a moderate loss of weight (a mean of 54 g for Group I and 25 g for Group II). During the later stages of intoxication rats were somnolent and waddled as they walked.

Most of the abstinence convulsions were associated with loss of the upright posture, and there were generalized tonic-clonic seizures of less than a minute's duration. Occasionally such convulsions lacked a clonic phase, and infrequently a rat would remain upright while exhibiting generalized shaking movements. Partial convulsions were observed in which the forelegs were raised off the floor, and while the rat remained balanced on its hindlegs, the head, face, and forelegs underwent movements (see Fig. 1).

The first abstinence convulsions occurred from 5.4–48.8 (27.2 mean) h after the barbital was replaced with tap water. However, the exact time at which each rat drank the last barbital is unknown. The last abstinence convulsions occurred from 48.8–88 (66.6 mean) h after barbital was discontinued. All 5 experimental rats in Group I convulsed, and the number of seizures varied from 3–64 (23.4 mean). The 13 experimental rats in Group II were intoxicated at a slower rate and to a lower final dose level than Group I. Nine of these 13 rats developed from 1–11 (5.4 mean) convulsions.

In both groups, mean body weight differences between the last day of addiction and each day of withdrawal were compared with the variations in body weights of the control rats on the same days. In Group I, mean weight loss reached significant levels ($P<0.02$) on the first day of withdrawal. The decrease in weight remained significant until the seventh day of withdrawal ($P<0.05$). Based on means, the maximum weight loss developed by the fourth day of withdrawal was 44 g or 12.5% of the weight on the last day of addiction.

In Group II the mean weight loss was significant on the first day of withdrawal ($P<0.001$) and remained at that level of significance through the fifth day of abstinence (see Fig. 1). On the sixth day, the mean weight loss was marked ($P<0.02$), and by the seventh day of abstinence it was no longer significant. Mean body weight loss was maximum on the third day of withdrawal when it was 39.8 g or 8.7% of the mean body weight on the last day of addiction (see Fig. 1).

During withdrawal food and water consumption were greatly reduced, particularly during the first 2–3 days after withdrawal (see Fig. 1). During withdrawal the body temperature which had been depressed during intoxication returned to a normal range. Activity measurements were elevated for some of the rats during abstinence; however, there was much variation in the groups and this change was not significant.

MILLER and FRASER (unpublished data) confirmed the observations of ESSIG in rats and in addition conducted substitution experiments using secobarbital as the control drug when unknowns were evaluated. It was observed that satisfactory dose response could be obtained for secobarbital using only body weight, a nonspecific parameter, as an index of intensity of abstinence. The authors also chronically intoxicated dogs with progressively increasing dosages of sodium barbital in the drinking water for 3–6 months and observed them for signs of abstinence according to the procedure of DENEAU and WEISS (1968). It was the opinion of these investigators that the white rat provides a simpler and more economical method for evaluating sedative-hypnotic drugs for their degree of physical dependence as compared with the dog.

4. Mouse

WATERS and OKAMOTO (1972) found that the laboratory mouse can be made physically dependent on barbital and that degrees of dependence can be uncovered by a PTZ seizure threshold procedure. Furthermore, the occurrence of behavioral and excitability changes in withdrawing animals correlates well with the time course of the blood barbital elimination curves.

GOLDSTEIN (1972) made mice physically dependent on alcohol by 3 days of continuous intoxication produced by inhalation of alcohol vapor. Elimination of alcohol was reduced and stabilized by the concurrent administration of pyrazole. The severity of the withdrawal reaction which followed removal of the mice from the inhalation chamber was quantitatively assessed by scoring the intensity of the convulsions which were elicited by hand lifting the mice.

5. Cat

ROSENBERG and OKAMOTO (1974) made cats physically dependent on sodium pentobarbital by administering sodium pentobarbital through an implanted intragastric catheter twice daily in increasing doses as tolerated for 22–40 days. On withdrawal of pentobarbital a characteristic abstinence syndrome developed, the intensity of which paralleled the rate of elimination of pentobarbital from the animal. Frequency of grand mal convulsions on withdrawal was found to be a good indicator of degree of dependence.

II. Tests for Psychic Dependence on Barbiturates in Animals

The above tests for dependence on barbiturates as well as other drugs in animals speak only to physical dependence-producing properties of the drug. Another important factor is the primary reinforcing property of the agent which may be related to psychic dependence. In physical dependence tests in animals the drugs are administered on a schedule over which the animal has no control. Sedative/hypnotic abusers, on the other hand, control the dosage as well as frequency of administration.

1. Rat

WEEKS (1962) described a technique for self-administration of drugs to the rat which is more analogous to drug-seeking behavior in man. Female rats weighing 200–250 g were used. They could move freely about their cages, carrying a light weight saddle strapped behind the forelegs. The saddle was connected by a sprocket chain to a small swivel and a stuffing box to permit injection of the drug. Intravenous injections were made through a polyethylene cannula passed down the jugular vein into the right heart. A lever was put into the cage, which, when pressed, caused injection of 10 mg/kg of morphine as the sulfate. After experiencing drug effects a few times by chance lever pressing, rats regularly responded. When they had been made dependent on morphine sulfate their frequency of response to lever pressing depended upon the concentration of the morphine sulfate solution.

2. Monkey

DENEAU et al. (1969) extended this technique to the monkey. They demonstrated that a naive monkey would on a trial and error basis inject a morphine solution in preference to saline using an intravenously implanted polyethylene catheter that discharged directly into the right auricle of the heart. These authors utilized a special harness that was attached to the monkey's back. The procedure permits a drug to be injected at a rate which can be controlled by the monkey by pressing a lever, or arbitrarily controlled by the experimenter. If the monkey does not voluntarily inject the test drug by a trial and error pressure on the lever, then an injection by the animal is facilitated by taping a raisin or a small piece of candy to the lever with transparent cellulose tape. In his efforts to obtain the raisin, the monkey will activate the injector. If test drug is positively rewarding, a small number of trials will teach the monkey that lever-pressing is associated with favorable subjective effects. On the other hand, if the subjective effects of the drugs are aversive, the monkey will not attempt to extricate the raisin even though he will eat a raisin when it is hand-fed to him. In the application of this technique it was very soon learned that there was a distinct difference among monkeys in respect to the ease with which they would voluntarily self-inject drugs.

When this self-injection technique was applied to barbiturates, they were found to be more reinforcing than narcotics in the monkey (YANAGITA and TAKAHASHI, 1973). For example, in the case of pentobarbital, 5 monkeys initiated and maintained self-administration of 3 mg/kg doses of pentobarbital. After 1 week all monkeys were taking a dosage that would make them comatose. As soon as they recovered from a coma they promptly attempted to again press the lever switch. As time elapsed the monkeys became tolerant and were able to increase their daily dosage and a maximum weekly dose of 420 mg/kg of pentobarbital was attained. All monkeys discontinued self-injection for 30–45 min each morning and afternoon while they consumed a larger than average meal. The monkeys maintained a good physical condition and gained weight throughout the experimental procedure. In this respect the monkeys resembled man since apparently barbiturates do not interfere with food intake; in fact, man gains weight in analogous tests (ISBELL et al., 1950).

Prolonged voluntary abstinence never occurred and if pentobarbital was abruptly discontinued either deliberately by the investigator or because of failure of the injection mechanism, the animals showed signs of abstinence or withdrawal within 46 h. The syndrome was characterized by extreme restlessness, tremor, grand mal convulsions, and hallucinogenic behavior.

DENEAU (personal communication) in parallel intravenous experiments evaluated phenobarbital. It was available to 6 monkeys in a dose of 7.5 mg/kg. All 6 monkeys took it to some extent but never to points of becoming ataxic. None of the monkeys developed any predictable pattern of self-administration. There were no signs of abstinence when phenobarbital was eventually discontinued. This is in sharp contrast to the results with pentobarbital described above.

The technique of self-administration of drugs has been criticized on the basis that it involves the injection of such drugs into the right auricle, a procedure not available to man. In addition it permits the injection of drugs which are quite insoluble in water. Furthermore, in the case of barbiturates, abuse is usually by an oral route since protracted or continuous attempts to abuse barbiturates by injection are usually self-terminated by abscesses and inflammatory reactions at the site of injection. To overcome these difficulties YANAGITA and TAKAHASHI (1973) described a technique which permitted the chronic programmed or self-administration of drugs to the monkey via a catheter that was implanted in the stomach. Implantation of the catheter in the stomach minimizes the effect of gustatory sensations on drug-seeking. One complication of self-ingesting drugs via a catheter to the stomach is that the monkey may have difficulty associating pressing the lever and injecting the drug into the stomach with rewarding subjective effects since there would be an expected delay in the onset of such effects. YANAGITA and TAKAHASHI (1973) overcame a reluctance on the part of monkeys to self-ingest drugs by the oral route by "conditioning" them to the subjective effects of "Spa" (1-2-diphenyl-1-dimethyl-amino ethane hydrochloride, a potent stimulant drug, with morphine characteristics that has a high abuse potential).

One must classify the "Spa" monkeys as being hypersensitive to the subjective effects of drugs, and as a consequence, they may exaggerate to some extent the actual drug-seeking behavior that is induced by the agents which are administered subsequent to it. The applicability of such experiments to man is uncertain, but if the mildly reinforcing agent is taken by the monkey after the highly reinforcing agent has been administered, it demonstrates that the milder reinforcing agent is not aversive. There is also considerable human epidemiologic evidence that exposure to any given addictive agent makes such individuals more prone to abuse another drug as compared to individuals not exposed to the initial agent. For example, persons who smoke tobacco are much more prone to smoke marijuana as compared to persons who do not smoke tobacco (R.G. SMART and D. FEJER, unpublished observations).

III. Evaluation of Physical and Psychic Dependence Tests Conducted in Animals for their Capacity to Predict Abuse Potential of Barbiturates in Man

It appears that most barbiturates and pharmacologically related drugs, such as glutethimide, will provoke withdrawal symptoms when they are abruptly withdrawn following long continuous intoxication. As far as is known, all barbiturates are capable of substituting for sodium barbital. It is of interest that carisoprodol and tybamate (FELDMAN and MULINOS, 1966; COLMORE and MOORE, 1967) also effectively substituted for barbital; however, these drugs did not demonstrate dependence in direct addiction tests in dogs since signs of abstinence were not observed when they were withdrawn abruptly after long-term administration in large doses. Carisoprodol has been administered in high dosage chronically to man and then abruptly withdrawn, and no signs of physical dependence were observed (FRASER et al., 1961 b). This inconsistency shows the limitations of the suppression test for depressants. We should not necessarily consider all drugs that satisfactorily substitute for barbital as dependence-inducing agents or rely on only one procedure to predict dependency.

The studies of DENEAU (personal communication) in which pentobarbital and phenobarbital were compared intravenously in monkeys for their acceptability by self-administration in animals indicate possibilities for differentiating the relative abuse of barbituratelike drugs.

IV. Tests for Physical Dependence in Man

The experimental basis for physical dependence tests for barbituratelike activity are the classical studies of ISBELL et al. (1950, 1955). These investigators demonstrated in man that the chronic administration of continuously intoxicating doses of barbiturates and of alcohol induced a moderate degree of tolerance and a high degree of physical dependence which was associated with an abstinence syndrome when barbiturates and alcohol were abruptly withdrawn. The abstinence syndrome of abrupt withdrawal of alcohol and barbiturates differs considerably from that observed following abrupt withdrawal of opiates and may be described as follows:

There is a progressive decrease in signs of intoxication over a period of 6–15 h, depending upon the length of action of the barbiturate or alcohol and the degree of intoxication. When signs of intoxication disappear, the patient begins to show anxiety, nervousness, coarse tremor of the hands and face, progressive weakness, loss of appetite, nausea, and even vomiting, and acute loss of weight and insomnia. These symptoms may be followed after 24 h or more by grand mal convulsions, auditory or visual hallucinations and in some patients by a frank delirium. Hyperpyrexia and death may ensue for certain patients withdrawn abruptly from barbiturates (FRASER et al., 1953). In the case of alcoholic delirium, as many as 15% of the patients, if inadequately treated, may die (VICTOR, 1958).

Dependence has been experimentally demonstrated in man with secobarbital, pentobarbital, and sodium amytal (ISBELL et al., 1950; FRASER et al., 1954). Withdrawal syndromes analogous to those described for animals are observed. Further evidence for the consistency of barbiturates in the induction of physical dependence is the observation at Lexington that pentobarbital can be satisfactorily used routinely to withdraw patients regardless of the level of dependence. Over the past 15 years hundreds of such cases have been so treated. In addition, FRASER et al. (1957) chronically intoxicated 3 patients with high doses of secobarbital (1.06, 1.6, and 1.62 average grams daily). Then an equivalent dose of pentobarbital was substituted for 7 consecutive days, following which the patients were again stabilized on secobarbital. All the patients remained intoxicated and had no symptoms or signs of abstinence from barbiturates when pentobarbital was substituted for secobarbital. The transition from pentobarbital to secobarbital was equally smooth. FRASER et al. (1958) observed the incidence of withdrawal convulsions and/or delirium in patients chronically administered various dosages of secobarbital and pentobarbital for 32–365 days. The results of this study are summarized in Table 4. Other abstinence changes included a sharp drop in the eosinophil count and an increase in serum uric acid and nonprotein nitrogen. The effects of convulsions per se on serum uric and nonprotein nitrogen levels have been studied in psychiatric patients receiving electro-convulsive

Table 4. Summary of data on relationship of dosage of secobarbital or pentobarbital to intensity of physical dependence

Patients			Daily dose of barbiturate g	Days of intoxication in hospital	Number of patients having symptoms		
Total Number	Number receiving						
	Seco-barbital	Pento-barbital			Con-vulsions	Delirium	Minor symptoms of significant degree
18	16	2	0.9–2.2	32–144	14	12	18
5	5		0.8	42–57	1	0	5
18	18		0.6	35–57	2	0	9
18	10	8	0.4	90	0	0	1
2	1	1	0.2	365	0	0	0

therapy and comparable elevations of uric acid and nonprotein nitrogen in such patients were observed. It was concluded, therefore, that their elevation during barbiturate withdrawal was nonspecific and was a consequence of the convulsions.

FRASER et al. (1954) studied the effects of a cycle of barbiturate dependence on the EEG in 14 patients who were chronically and maximally intoxicated and withdrawn. The daily dose to induce chronic intoxication ranged from 0.6–2.6 g of secobarbital or pentobarbital. Electroencephalograms consisted of bipolar and monopolar tracings from frontal, temporal, parietal, and occipital electrodes. An analysis of 157 records made in this study is summarized in Table 5.

These observations confirmed those reported by ISBELL et al. (1950) and are in accordance with those of WIKLER et al. (1955) and WULFF (1959).

ESSIG and FRASER (1958) studied the effects on the EEG of 0.4 g of sodium secobarbital (10 subjects) and sodium pentobarbital (8 subjects). Concurrently, clinical observations were made. With this dosage schedule the following results were reported:

1. In contrast to doses of over 0.6 g per day, the withdrawal of 0.4 g daily of secobarbital or pentobarbital after 3 months was not followed by psychotic or convulsive manifestations in any of 18 nonepileptic subjects.

Table 5. Distribution of EEG patterns during barbiturate addiction, withdrawal, and recovery

Phase of addiction cycle	Normal EEGs %	Abnormal EEGs, %				
		Random spikes	Fast waves	Mixed fast and slow waves	Slow waves	Par-oxysmal activity
Chronic intoxication	1.9	9.8	13.8	47.0	27.5	—
Acute abstinence period						
1st–3rd day	7.2	10.4	—	8.8	33.3	40.3
4th–8th day	23.3	20.0	—	—	36.7	20.0
Recovery period (9th–83rd day)	72.7	9.1	—	—	13.7	4.5

2. Seven of 18 individuals demonstrated some evidence of EEG tolerance to 0.4 g daily of secobarbital or pentobarbital during 90 days.

3. During withdrawal from this same barbiturate regimen, 5 of 18 subjects developed paroxysmal EEG discharges which later disappeared and were not associated with observable clinical abnormalities.

V. Tests for Psychic Dependence in Man

The measurement of the ability of barbituratelike agents to produce psychic dependence in man has had only limited systematic study. In contrast to animal studies which utilize the behavioral measure of self-injection, the primary reinforcing capacity of barbituratelike agents in man has been estimated by measurement of subjective effects of single graded doses. The preliminary approach used to measure psychic dependence on barbiturates employed a method analogous to that used for measuring the primary reinforcing properties of morphinelike agents. This consisted of specific measures of subjective and behavioral effects which were correlated with an objective measure of facilitation of postrotatory nystagmus. Utilizing the single and chronic dose opiate questionnaires, facilitation of postrotatory nystagmus, and change in pupil diameter measured photographically, MARTIN et al. (1962, 1974) demonstrated that graded single doses of morphine sulfate (8, 16, and 32 mg) and pentobarbital sodium (150, 200, and 250 mg) administered intramuscularly under double blind conditions in nontolerant narcotic addicts produced two distinct syndromes as measured by these parameters. Pentobarbital produced a dose-related increase in the duration of postrotational nystagmus and symptoms of sleepiness and drunkenness. In contrast, morphine produced a dose-related increase in the signs scratching, relaxation, coasting, talkativeness, conjunctival injection, and pupillary constriction. The symptoms which were dose related for morphine included talkativeness and turning of the stomach. With the larger doses subjects readily differentiated pentobarbital from morphine. As measured by "liking" scores both drugs produced euphoria; however, in this study there was no evidence of dose relationship for the "liking" scores produced by pentobarbital.

Studies by HILL et al. (1963) and HAERTZEN (1966) indicated that the subjective effects of pentobarbital assessed in the same population were not distinctively characterized by euphoria, at least of a morphine type. These authors developed and utilized the Addiction Research Center Inventory (ARCI) a 550 item structured questionnaire consisting of a series of statements to be answered "yes" or "no." The content of these statements was originally developed with the use of sentence completion and other association techniques employing male addicts under drug and no-drug conditions. Of importance from the viewpoint of assessing barbituratelike agents, HAERTZEN (1966) found that pentobarbital as well as alcohol and chlorpromazine produced most characteristically feelings of lethargy, weakness, and loss of energy. The items in this scale, Pentobarbital-Chlorpromazine-Alcohol Group scale (PCAG), differentiated these three drugs from other drugs in the series which included morphine, amphetamine, LSD, pyrahexyl, and placebo. On the other hand, morphine and DL-amphetamine induced a characteristic type of effects for these two classes of agents as exemplified by a factor-analytic derived scale, the Morphine-Benzedrine Group (MBG), characterized by feelings of elation, well-being, enhanced self-image, and increased energy. MBG scores have been used as measures of "euphoria" induced by these drugs. Both morphine and barbiturates produce an internal sensation of pleasantness, but there is a qualitative difference in the characteristics of the euphoria induced by morphine as compared to barbiturates in that morphine induces more subjective effects of energy or stimulation.

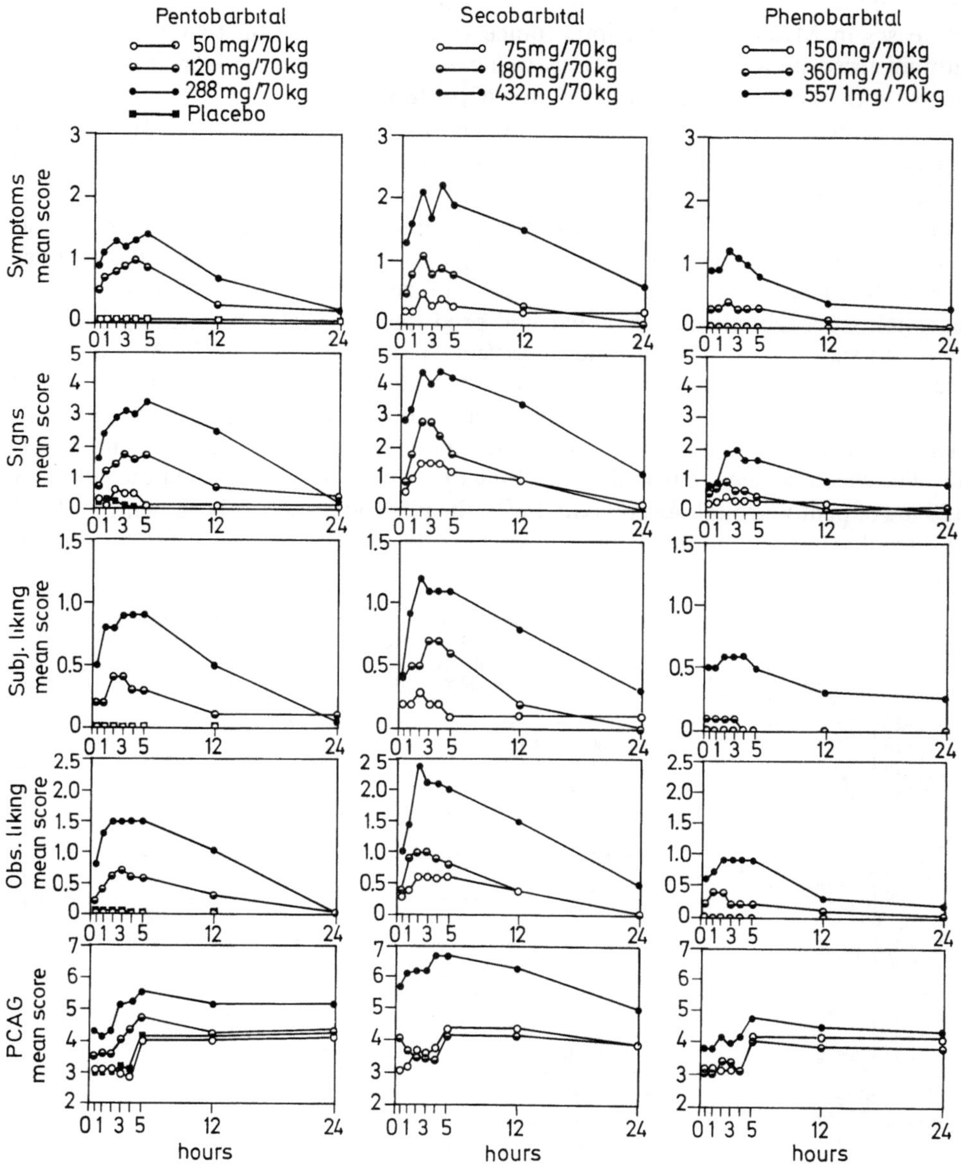

Fig. 2. Time-action curves for responses to intramuscularly administered pentobarbital, secobarbital, phenobarbital, and placebo as measured by PCAG scale scores, symptom and "liking" scale scores from subjects' single dose opiate questionnaire, and sign scores and observers' "liking" scores from observers' single dose opiate questionnaire

McCLANE and MARTIN (1976) compared the effects of 50, 150, 250, and 300 mg of pentobarbital sodium and 12 and 24 mg of morphine sulfate administered intramuscularly under double blind conditions in a crossover study. In addition to the single dose opiate questionnaire, the investigators utilized a subset of items from the PCAG and MBG scales of HAERTZEN. Again, subjects and observers distinguished pentobarbital from morphine sulfate. Pentobarbital but not morphine in these dosages produced dose-related increases in PCAG scores. Morphine produced dose-related increases in MBG scores. Of importance was the observation that pentobarbital administered under these conditions was capable of producing increases in MBG scores with maximum effects with 150 mg of pentobarbital.

Subsequently, D. R. JASINSKI (unpublished observations) conducted studies using the methods devised by MARTIN and his coworkers described above, to measure the effects of other barbituratelike agents. The first study compared the effects of pentobarbital, 50, 120, and 288 mg/70 kg; secobarbital 75, 180, and 432 mg/70 kg; and phenobarbital 140, 360, and 557.1 mg/70 kg; all administered intramuscularly under double blind conditions. Subjects identified all three drugs as barbituratelike. Pentobarbital, secobarbital, and phenobarbital produced dose-related increases in symptom and sign scores, subjects' and observers' "liking" scores, and PCAG scores as well as dose-related increases in postrotatory nystagmus (Figs. 2, 3). Relative potencies on each of these measures indicated that secobarbital was equipotent to pentobarbital, while phenobarbital was one-fifth to one-seventh as potent as pentobarbital. The effects of phenobarbital in 557.1 mg doses were no greater than the effects of 120 mg of pentobarbital. As compared with the placebo response, both pentobarbital and secobarbital produced significant MBG scale scores; however, phenobarbital in

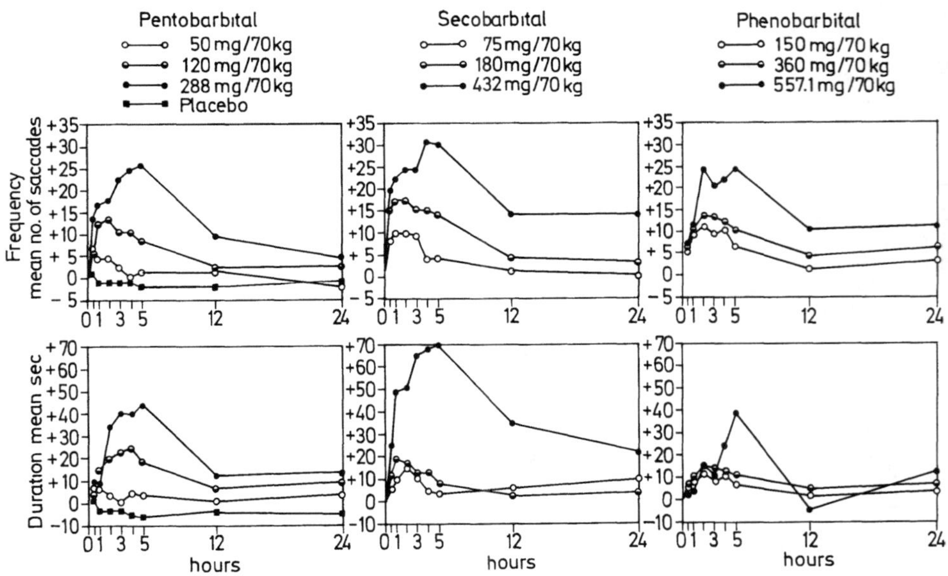

Fig. 3. Time-action curves for facilitation of duration and frequency of postrotatory nystagmus by pentobarbital, secobarbital, phenobarbital, and placebo. Each point represents mean change in frequency or duration from mean of two predrug controls. Nystagmus was recorded with a polygraph. Data were obtained in a crossover study employing 10 subjects

these doses did not produce significant MBG scores as compared with placebo. JASINSKI (unpublished observations) also compared the effects of pentobarbital 60, 120, and 240 mg and secobarbital 60, 120, and 240 mg administered orally under double blind conditions. Again, the responses were quite similar as measured by similar increases in dose-related scale scores for all the above-described parameters. Again, secobarbital was found to be equipotent to pentobarbital. The 240 mg dose of pentobarbital and secobarbital produced significant MBG responses.

Studies to date indicate that the potential for inducing psychic dependence (primary reinforcing capacity) of new barbituratelike agents may possibly be assessed by measurement of specific subjective effects, behavioral effects, and an objective measure of intoxication assessed concurrently–i.e., enhancement of postrotational nystagmus. These effects then could be related to similarly measured effects of a standard agent such as pentobarbital administered as a control in a crossover experiment.

VI. Evaluation of Physical and Psychic Tests in Man for Predicting the Abuse Potential of Barbituratelike Drugs

In order to induce experimental dependence with barbiturates, it is necessary to intoxicate such patients around the clock for approximately 45 or more days. Such a procedure is hazardous to the individual and should not be undertaken, certainly, as a routine test procedure. Acute withdrawal is also hazardous, since such serious complications as grand mal convulsions and delirium may ensue.

However, a panel of experts (Workshop on Sedatives-Hypnotics, jointly sponsored by the Bureau of Narcotics and Dangerous Drugs and the Food and Drug Administration, Washington, D.C. 1969) agreed that short-term, direct addiction preference tests might be made comparing an unknown drug with a known agent that is abused, such as secobarbital.

The direct addiction and the substitution procedures described for barbiturates will, in a qualitative manner, demonstrate those depressants that have a potential for producing a similar physical dependence. However, these procedures in themselves will not determine the extent to which patients will deliberately seek out one barbiturate in preference to another. For example, although phenobarbital accounts for at least half of all the sales of barbiturates in the United States, there is an extremely low incidence of abuse of phenobarbital characterized by drug-seeking behavior, even though it is fully capable of inducing physical dependence. On the other hand, there is a higher abuse incidence of secobarbital, pentobarbital, and amobarbital or combinations of secobarbital and amobarbital and of glutethimide. These observations suggest that tests other than substitution and/or direct addiction in the case of barbiturates are necessary to determine the relative abuse of depressants.

Although tests for psychic dependence have been evaluated in a systematic manner by two routes of administration and only in single doses, these tests indicate that the procedures utilized have a potential for assessing psychic dependence. This possibility is demonstrated by the finding of significant "liking" scores for pentobarbital and secobarbital and by positive "MBG" scores (euphoria scores) for these drugs also.

To avoid the hazards to human life involved in substitution and/or direct addiction procedures and yet develop information as to the relative abuse of a new drug, the following methods for evaluating barbituratelike drugs in man are suggested:

A new barbituratelike agent in single oral doses would be compared with a standard drug such as pentobarbital or secobarbital. In preliminary experiments, an attempt would be made to establish equivalent intoxicating doses for the standard and the unknown. Then, with a placebo as the negative control, the routine-type study of dose response for single doses would be carried out using 8–10 subjects sophisticated in drug abuse. Emphasis would be placed on the quality of the subjective effects, including identification of the class of drug and the degree of "liking" for the unknown as compared to the standard and the frequency of responses of "sleepy" and "drunken", characteristics associated with the barbiturates. The single dose questionnaire of Fraser et al. (1961a) for subjects and observers and the PCAG and the MBG scales of Haertzen (1966) and Jasinski et al. (1968) could be utilized. For objective evaluations of degree of intoxication, the frequency and duration of nystagmus (Martin et al., 1974) would be employed.

Upon conclusion of the single dose study, a pilot experiment would be set up in an attempt to establish a chronic oral dosage of the unknown and the standard that would maintain a moderate degree of intoxication for 10 days. Should this procedure be safe and feasible, a routine crossover designed experiment would be carried out with 8–10 subjects receiving the unknown and the standard drug for 10 days with an interval of 10 days of placebo between drugs. A daily written questionnaire would be administered to both the patients and the observers to ascertain the degree of attractiveness or aversiveness of the standard and the unknown utilizing methods parallel to those of the chronic dosage opiate questionnaire of Fraser et al. (1961a). An experiment of this type would not ascertain degree of physical dependence because the interval of intoxication is too short; it would primarily indicate acceptability or the primary reinforcing characteristics of the agents under test.

The initial experiment to test, in part, the validity of such a method would be a comparision of secobarbital and phenobarbital with both drugs orally administered in single doses and then multiple doses daily for 10 days using a crossover double blind design.

C. Assessment of Minor Tranquilizers

I. Tests for Physical Dependence in Animals

1. Dog

Essig (1958) studied the physical dependence capacity of orally administered meprobamate in the dog. The details are as follows: Five dogs, weighing between 8.6 and 13.4 kg, were given 1.6 g of meprobamate orally at 9:00–10:00 a.m. and 3.00–4:00 p.m., daily. The total daily dose was increased to 4.8 and then to 5.2 g during the subsequent 75–130 days. At that time a dose was added at midnight. The total duration of intoxication varied from 124–188 days. The total daily dose of meprobamate ranged between 8.0 and 8.8 g. Withdrawal was abrupt, and the dogs were observed continuously until death occurred or abstinence signs subsided. The results were as follows: One dog had a convulsion on the 56th day of intoxication while receiving 2.4 g of meprobamate daily. This animal died on the 65th day of intoxication. Convulsions were not observed after the third dose was added. Otherwise, the dogs ate well, remained ambulatory, and maintained their usual physical appearance throughout the period of intoxication. When meprobamate was abruptly withdrawn in the remaining 4 dogs after 124–188 days of intoxication, repeated convulsions and death were observed in 3 dogs and an induced convulsion was observed in the fourth dog.

2. Monkey

YANAGITA and TAKAHASHI (1970) evaluated the physical dependence capacity of meprobamate, diazepam, and chlordiazepoxide in the monkey. The tests conducted were similar to those for evaluation of dependence characteristics of barbiturates and included administration of single doses, substitution of the experimental drug for barbital in chronically intoxicated monkeys, and direct addiction procedures. The results were as follows: Single doses of diazepam and meprobamate induced a significant CNS depression. In the case of chlordiazepoxide a dose of 40 mg/kg provoked a grade 4 CNS depression (the monkey cannot climb but can sit up). It was not feasible, by increasing the dose, to increase the degree of CNS depression. In substitution tests for barbital an adequate oral dose of meprobamate, diazepam, and chlordiazepoxide suppressed signs of the dependence in animals dependent on barbital. Likewise, when direct addiction procedures were carried out with these 3 agents all monkeys showed signs of abstinence and in many instances both hallucinatory behavior and convulsions were observed when drugs were abruptly discontinued. Thus there are tests available in animals for demonstrating the physical dependence characteristics of the minor tranquilizers.

II. Tests for Psychic Dependence in Animals

Monkey

YANAGITA and TAKAHASHI (1973) have used the monkey also for evaluating the psychic dependence (drug-seeking behavior) of the minor tranquilizers. In particular, they studied diazepam and chlordiazepoxide and compared their results with these agents with those observed with alcohol, pentobarbital, chloroform, and chlorpromazine. These drugs are self-administered by monkeys intravenously using an intracardiac catheter and orally using a gastric-implanted catheter. YANAGITA and TAKAHASHI evaluated chlordiazepoxide for drug-seeking behavior by intracardiac and intragastric routes and diazepam by the intravenous route and compared them with pentobarbital, alcohol, and chloroform. These agents induced definite drug-seeking behavior and monkeys administered them until they were comatose. On the other hand, the monkey would not bar press for intravenously administered chlorpromazine. It is of interest that neither diazepam nor chlordiazepoxide were self-administered to the extent that anesthesia was induced although the animals became ataxic. To some extent chlordiazepoxide could be differentiated from diazepam since the degree of intoxication with chlordiazepoxide was less than that observed with diazepam. In addition, after 4 weeks of self-administration the monkeys gradually decreased their daily intake of chlordiazepoxide. Two monkeys that were conditioned to the positive effects of drugs initiated and maintained intragastric self-administration of chlordiazepoxide at the unit dose of 100 mg/kg for over 8 weeks. They developed no marked depressive effects.

FINDLEY et al. (1972) compared the self-infusion of chlordiazepoxide and secobarbital in the rhesus monkey by means of a forced choice preference procedure. The observations were as follows: Monkeys were exposed to a 24-h continuous experimental procedure which provided a periodic forced choice between the self-infusion intracardiacally of secobarbital *versus* saline and of chlordiazepoxide *versus* saline. Monkeys showed a preference for both drugs over saline. Following a period of intake of both drugs, a gradual shift in preference from chlordiazepoxide to secobarbital was observed over a period of some 60 days. This forced preference procedure affords another method for evaluating in the monkey relative preference among drugs. The results observed are somewhat different from those of YANAGITA and TAKAHASHI (1973) in that in their experiments there was a tendency of the monkeys to reduce the dose of self-administered chlordiazepoxide over time. The difference between the two approaches may be related to the penalty (shock) involved in the studies by FINDLEY et al. (1972).

III. Evaluation of Physical and Psychic Dependence Tests in Animals of Minor Tranquilizers for their Capacity to Predict Abuse Potential in Man

In respect to tests for physical dependence there is a good agreement between observations in animals and those thus far described for man. Table 6 summarizes observations in the rhesus monkey (YANAGITA and TAKAHASHI, 1973). Regarding tests for psychic dependence in animals there have been important advances in the evaluation of drug-seeking behavior for tranquilizers and other CNS depressants. There is good evidence of a differentiation of the relative abuse liability of several agents in this category; for example, chlorpromazine is not reinforcing while alcohol, pentobarbital, and chloroform are. Chlordiazepoxide and diazepam are less reinforcing than pentobarbital but are more reinforcing than chlorpromazine (see Table 6). Although these tests in animals are by and large consistent with clinical observations (chloroform which has been easily obtainable for many years has not been abused), there has been as yet only limited experimental observations in man to confirm the animal studies. (The reader is also referred to points noted previously under a parallel heading for evaluating tests for barbiturates.)

IV. Tests for Physical Dependence in Man

There have been several clinical reports of physical dependence on meprobamate, chlordiazepoxide, and diazepam. For example, LEMERE (1956) reported the case of a patient who had ingested 6.4 g of meprobamate daily for 1 month and had a convulsion 10 h after discontinuation of the drug. BARSA and KLINE (1956) noted a single convulsion in 6 of 25 schizophrenic patients following withdrawal of meprobamate

Table 6. Summary of dependence liability tests with sedative-hypnotic agents in rhesus monkeys

Drugs	Complete suppression doses in cross-physical dependence test mg/kg	Physical dependence-producing test[a]	Self-administration test[a]	Overt signs of drug effect during self-administration
Pentobarbital-Na	25 (i.v.)	Positive	Markedly reinforcing (i.v., i.g.)[e]	Coma
Alcohol	4000 (p.o.)	Positive	Markedly reinforcing (i.g.)	Coma
Chloroform	Subanesthetic dose (inhal.)	Untested	Markedly reinforcing[b] (inhal.)	Coma
Meprobamate	200 (p.o.)	Positive	Untested	Untested
Diazepam	5 (p.o.)	Positive	Moderately reinforcing (i.v.)	Ataxia
Chlordiazepoxide	20 (p.o.)	Positive	Mildly reinforcing (i.v., i.g.)	Ataxia
Oxazolam	20 (p.o.)	Positive	Mildly reinforcing (i.g.)	Sedation
Benzoctamine	Not suppressed	Negative	Mildly reinforcing[c] (i.g.)	Sedation
Perlapine	Not suppressed	Negative	Not reinforcing (i.g.)	None
Chlorpromazine	Not suppressed	Negative	Not reinforcing[d] (i.v.)	None

[a] For details, see text.
[b] Quoted from a previous study (YANAGITA and TAKAHASHI, 1970).
[c] Reinforcement observed in only 1 of 6 monkeys.
[d] Quoted from a previous study (DENEAU et al., 1969).
[e] Route given in parentheses.
Abbreviations used: i.v., intravenous; p.o., oral; i.g., intragastric; inhal., inhalation.

(2.4 g daily for 9 months). TUCKER and WILENSKY (1957) reported one grand mal seizure following discontinuation of the drug in 2 of 32 psychotic patients who had received 1.6 g of meprobamate, increasing to 4.8 g daily for 3 months.

The question of whether physical dependence is induced by the chronic administration of benzodiazepines in high dosage experimentally in man has been studied by several investigators. HOLLISTER et al. (1961) administered high daily doses of chlordiazepoxide (300–600 mg, 8–20 times the usual therapeutic dosage) for several months to patients. Chlordiazepoxide was abruptly discontinued and a placebo substituted. Ten of the 11 patients developed new symptoms which included depression, aggravation of psychosis, agitation, insomnia, loss of appetite, and nausea between 2 and 8 days following abrupt withdrawal. One patient, not in the withdrawal study, had a seizure 12 days after discontinuation of a 300 mg daily dose. An additional case of grand mal seizure was observed in a mentally retarded child after discontinuation of chlordiazepoxide in the dose range of 40–60 mg (PILKINGTON, 1961). In the clinical study conducted on 25 alcoholic patients, chlordiazepoxide was administered for 14 days at 50 mg t.i.d. and then withdrawn suddenly. No withdrawal symptoms were observed (BURKE and ANDERSON, 1962). This observation suggests that chlordiazepoxide needs to be administered in very large doses and for a considerable period of time for physical dependence to develop.

HOLLISTER et al. (1963) reported severe withdrawal reactions in patients after abrupt discontinuation of diazepam (up to 120 mg per day).

V. Tests for Psychic Dependence in Man

The authors are not aware of any systematic studies in man evaluating relative reinforcing properties induced by minor tranquilizers such as meprobamate, chlordiazepoxide, and diazepam nor of a comparison of these agents with barbiturates.

To evaluate the relative abuse potential of a new drug proposed as a tranquilizer, single dose tests and chronic administration for 10 days could be carried out in a parallel manner to that described for evaluating a new barbituratelike drug. In the case of a new tranquilizer, however, one might wish to use two positive controls, namely a barbiturate and a tranquilizer considered to have similar effects.

VI. Evaluation of Experimental Tests in Man of Physical and Psychic Dependence for their Capacity to Predict the Abuse Potential of the Minor Tranquilizers

As yet the studies conducted are insufficient to establish methods for predicting the relative abuse of an unknown minor tranquilizer type of drug.

D. Assessment of Major Tranquilizers

I. Assessment in Animals

Agents belonging to this class include the phenothiazines, the reserpine alkaloids, and butyrophenones. All these drugs have in common antipsychotic effects which separate them from other classes of depressants. They apparently do not elicit self-administration in animals.

BOYD (1960) administered to 32 rats progressively increasing doses of chlorpromazine intramuscularly for 40 weeks. A control group of 32 rats received injections of saline. The age of the rats when the experiment began was 41–52 days and their body weight was 134 ± 19 g. The initial daily dose of chlorpromazine was 1 mg/kg, and this was progressively increased to 200 mg/kg by the end of 40 weeks. This dosage schedule was quite toxic since the chlorpromazine group attained an average weight of only 300 g; whereas, the control group after 40 weeks averaged nearly 500 g. Both groups of rats were observed for 40 days after saline and chlorpromazine were discontinued. The chlorpromazine rats had a mild diarrhea for 2 weeks but gradually gained weight to an average of about 375 g. In addition, the chlorpromazine group on withdrawal showed a marked increase in locomotor activity (up to 300%) which persisted to 200% by the 40th day. On withdrawal of saline the control group showed no significant change in any parameter observed.

It is difficult to interpret from these experiments with rats as to whether they demonstrate signs of physical dependence since the dosage schedule of chlorpromazine was so toxic and the results observed during withdrawal could be interpreted in part as the sequelae of a toxic state.

HOFFMEISTER and GOLDBERG (1973) compared chlorpromazine, imipramine, morphine, and D-amphetamine by self-administration in cocaine-dependent rhesus monkeys and found that chlorpromazine exhibited aversive properties.

The authors have found no studies indicative of physical dependence on reserpine in animals.

II. Assessment in Man

We are not aware of any clinical studies that indicate compulsive drug-seeking behavior on the part of patients who chronically ingest or even inject chlorpromazine. In fact, one of the chief complaints of clinicians using chlorpromazine therapeutically is that when patients are discharged from the hospital for outpatient treatment they often discontinue taking chlorpromazine. HAERTZEN (1966) has selected items from the MMPI which measure some of the subjective effects produced by chlorpromazine. He observed that pentobarbital, chlorpromazine, and alcohol had a great many common effects and he grouped the items that measure these effects together to form a Pentobarbital-Chlorpromazine-Alcohol Group scale (PCAG). This scale includes such items as "I feel more tired than usual" and "I feel drowsy" and the items of the Morphine-Benzedringe Group (MBG) or the "euphoria" scale.

FRASER and ISBELL (1956) studied the effects of chlorpromazine and reserpine alone and the combination of each with morphine in single oral and intramuscular doses in nontolerant former opiate addicts. Chlorpromazine's miotic effect added to that of morphine and prolonged some of the subjective effects. Chlorpromazine alone, either orally or intramuscularly, produced a feeling of lethargy associated with drowsiness. None of the 8 subjects liked the effects. These authors tested chlorpromazine and reserpine for their capacity to substitute for morphine in patients dependent on 240 mg of morphine sulfate daily and observed that it was no more effective than a placebo in relieving morphine abstinence.

There is general agreement, therefore, that the drugs thus far used clinically as major tranquilizers do not possess any significant abuse potential. Whether new drugs of the major tranquilizer class will induce abuse is unknown.

E. Summary

1. Sedative-hypnotics and the minor tranquilizers cause (a) incoordination, impaired thinking and judgment, and uninhibited behavior; (b) death from excessive dosage; and (c) psychic and/or physical dependence leading to compulsive drug-seeking behavior. The harm to the individual or to society is not equal for all of these drugs or even within a given class. There are decided differences among drugs in a class. For example, the abuse of the barbiturate secobarbital is very high; whereas, phenobarbital, even though it is capable of inducing physical dependence, is low despite its extensive clinical use.

2. Screening procedures for abuse liability of the sedative-hypnotics conducted in animals and man have been critically reviewed. The classes of drugs covered include (a) barbituratelike agents, (b) the minor tranquilizers such as meprobamate, chlordiazepoxide, and diazepam, and (c) the major tranquilizers such as chlorpromazine and reserpine. Tests for "physical" and "psychic" dependence were discussed separately.

3. The reliability of the direct addiction tests for physical dependence in animals and man for predicting subsequent abuse in man is high. However, substitution tests may not be specific, since carisoprodol and tybamate will substitute for sodium barbital but do not induce physical dependence in direct addiction procedures. It

would be prudent, therefore, to include direct addiction studies on animals and/or man to comprehensively evaluate the ability of a sedative-hypnotic to produce physical dependence. Substitution and direct addiction tests in man are hazardous and for this reason studies in animals are preferable.

4. Tests for psychic dependence on barbituratelike agents and minor tranquilizers in animals have been well validated and the degree of reinforcing properties is well correlated with their clinical abuse. A limited series of drug-seeking studies on barbiturates have been carried out in man which suggest a potential for assessing psychic dependence in man.

5. The major tranquilizers, such as chlorpromazine, do not induce dependence and whether new drugs of this type will possess dependence characteristics is unknown.

References

Barsa, J. F., Kline, N. W.: Use of meprobamate in the treatment of psychotic patients. Amer. J. Psychiat. **112**, 1023—1025 (1956)

Boyd, E. M.: Chlorpromazine tolerance and physical dependence. J. Pharmacol. exp. Ther. **128**, 75—78 (1960)

Burke, G. W., Anderson, C. W. G.: Response to Librium in individuals with a propensity for addiction: A pilot study. J. La med. Soc. **114**, 58—60 (1962)

Colmore, J. P., Moore, J. P.: Lack of dependence and withdrawal symptoms in healthy volunteers given high doses of tybamate. J. clin. Pharmacol. **7**, 319—323 (1967)

Crossland, J., Leonard, B. E.: Barbiturate withdrawal convulsions in the rat. Biochem. Pharmacol. **12**, Suppl. 103 (1963)

Deneau, G. A., Klima, M., Wilson, M.: Evaluation of sedative hypnotic agents for barbiturate-like physiological dependence capacity in the dog. Reported to the Committee on Problems of Drug Dependence, National Research Council, Indianapolis, Ind., 1968

Deneau, G. A., Weiss, S.: A substitution technique for determining barbiturate-like physiological dependence capacity in the dog. Pharmakopsychiatr. Neuropsychopharmacol. **1**, 270—275 (1968)

Deneau, G. A., Yanagita, T., Seevers, M. H.: Self-administration of psycho-active substances by the monkey. Psychopharmacologia (Berl.) **16**, 30—48 (1969)

Essig, C. F.: Withdrawal convulsions in dogs following chronic meprobamate intoxication. Arch. Neurol. Psychiat. (Chic.) **80**, 414—417 (1958)

Essig, C. F.: Barbiturate withdrawal in white rats. Int. J. Neuropharmacol. **5**, 103—107 (1966)

Essig, C. F., Fraser, H. F.: Electroencephalographic changes in man during use and withdrawal of barbiturates in moderate dosage. Electroenceph. clin. Neurophysiol. **10**, 649—659 (1958)

Feldman, H. S., Mulinos, M. G.: Lack of addiction from high doses of meprobamate. J. clin. Pharmacol. **6**, 354—360 (1966)

Findley, J. D., Robinson, W. W., Peligrinio, L.: Addiction to secobarbital and chlordiazepoxide in the rhesus monkey by means of a self-infusion preference procedure. Psychopharmacologia (Berl.) **26**, 93—114 (1972)

Fraser, H. F.: Methods for testing for narcotic addiction in animals and man and their efficacy for predicting human abuse. In: Mantegazza, P., Piccinini, F. (Eds.): Methods in drug evaluation, Amsterdam: North-Holland Publishing Co., 1966

Fraser, H. F., Essig, C. F., Wolbach, A. B.: Evaluation of carisoprodol and phenyramidol for addictiveness. Bull. Narcot. **13**, 3—7 (1961 b)

Fraser, H. F., Isbell, H.: Abstinence syndrome in dogs after chronic barbiturate administration. J. Pharmacol. exp. Ther. **112**, 261—267 (1954)

Fraser, H. F., Isbell, H.: Chlorpromazine and reserpine. (A) Effects of each with morphine. (B) Failure of each in treatment of abstinence from morphine. Arch. Neurol. Psychiat. (Chic.) **76**, 257—262 (1956)

Fraser, H. F., Isbell, H., Eisenman, A. J., Wikler, A., Pescor, F. T.: Chronic barbiturate intoxication. Further studies. Arch. intern. Med. **94**, 34—41 (1954)

Fraser, H. F., Shaver, M. R., Maxwell, E. S., Isbell, H.: Death due to withdrawal of barbiturates. Report of a case. Ann. intern. Med. **38**, 1319—1325 (1953)

Fraser, H. F., Van Horn, G. D., Martin, W. R., Wolbach, A. B., Isbell, H.: Methods for evaluating addiction liability. (A) "Attitude" of opiate addicts toward opiate-like drugs. (B) A short-term "direct" addiction test. J. Pharmacol. exp. Ther. **133**, 371—387 (1961 a)

Fraser, H. F., Wikler, A., Essig, C. F., Isbell, H.: Degree of physical dependence induced by secobarbital and pentobarbital. J. Amer. med. Ass. **166**, 126—129 (1958)

Fraser, H. F., Wikler, A., Isbell, H., Johnson, N. K.: Partial equivalence of chronic alcohol and barbiturate intoxications. Quart. J. Stud. Alcohol **18**, 541—551 (1957)

Goldstein, D. B.: An animal model for testing effects of drugs on alcohol withdrawal. J. Pharmacol. exp. Ther. **183**, 14—22 (1972)

Haertzen, C. A.: Development of scales based on patterns of drug effects, using the Addiction Research Center Inventory (ARCI). Psychol. Rep. **18**, 163—194 (1966)

Hill, H. E., Haertzen, C. A., Wolbach, A. B., Jr., Miner, E. J.: The Addiction Reserach Center Inventory: Appendix. I. Items comprising empirical scales for seven drugs. II. Items which do not differentiate placebo from any drug condition. Psychopharmacologia (Berl.) **4**, 184—205 (1963)

Hoffmeister, F., Goldberg, S. R.: A comparison of chlorpromazine, imipramine, morphine, and D-amphetamine self-administration in cocaine-dependent rhesus monkeys. J. Pharmacol. exp. Ther. **187**, 8—14 (1973)

Hollister, L. E., Bennett, J. L., Kimbell, I., Jr., Savage, C., Overall, J. E.: Diazepam in newly admitted schizophrenics. Dis. nerv. Syst. **24**, 746—750 (1963)

Hollister, L. E., Motzenbecker, F. P., Degan, R. O.: Withdrawal reaction with chlordiazepoxide ("Librium"). Psychopharmacologia (Berl.) **2**, 63—68 (1961)

Isbell, H., Altschul, S., Kornetsky, C. H., Eisenman, A. J., Flanary, H. G., Fraser, H. F.: Chronic barbiturate intoxication. An experimental study. Arch. Neurol. Psychiat. (Chic.) **64**, 1—28 (1950)

Isbell, H., Fraser, H. F., Wikler, A., Belleville, R. E., Eisenman, A. J.: An experimental study of the etiology of "rum fits" and delirium tremens. Quart. J. Stud. Alcohol **16**, 1—33 (1955)

Jasinski, D. R., Martin, W. R., Sapira, J. D.: Antagonism of the subjective, pupillary and respiratory depressant effects of cyclazocine by naloxone. Clin. Pharmacol. Ther. **9**, 215—222 (1968)

Lemere, F.: Habit-forming properties of meprobamate. Arch. Neurol. Psychiat. (Chic.) **76**, 205—206 (1956)

Martin, W. R., Fraser, H. F., Isbell, H.: A comparison of the effects of intramuscularly administered pentobarbital sodium and morphine sulfate in man. Fed. Proc. **21**, 326 (1962)

Martin, W. R., Thompson, W. O., Fraser, H. F.: Comparison of graded single intramuscular doses of morphine and pentobarbital in man. Clin. Pharmacol. Ther. **15**, 623—630 (1974)

McClane, T. K., Martin, W. R.: Subjective and physiologic effects of morphine, pentobarbital, and meprobamate. Clin. Pharmacol. Ther. **20**, 192—198 (1976)

Pilkington, T. L.: Comparative effects of Librium and taractan on behavior disorders of mentally retarded children. Dis. nerv. Syst. **22**, 573—575 (1961)

Rosenberg, H. C., Okamoto, M.: A method for producing maximal pentobarbital dependency in cats: dependency characteristics. In: Singh, J. M., Lal, H. (Eds.): Neurobiology and influences on behavior. Miami: Miami Symposium 1974.

Seevers, M. H., Tatum, A. L.: Chronic experimental barbital poisoning. J. Pharmacol. exp. Ther. **42**, 217—231 (1931)

Tucker, K., Wilensky, H.: A clinical evaluation of meprobamate therapy in a chronic schizophrenic population. Amer. J. Psychiat. **113**, 698—703 (1957)

Victor, M.: Alcohol and nutritional diseases of the nervous system. J. Amer. med. Ass. **167**, 65—71 (1958)

Waters, D. H., Okamoto, M.: Increased central excitability in nondependent mice during chronic barbital dosing. In: Singh, J. M., Miller, L., Lal, H. (Eds.): Drug addiction: Experimental pharmacology, Vol. 1. Mount Kisco, New York: Futura Publication Co. 1972

Weeks, J. R.: Experimental morphine addiction: Method for automatic intravenous injections in unrestrained rats. Science **138**, 143—144 (1962)

Wikler, A., Fraser, H.F., Isbell, H., Pescor, F.T.: Electroencephalograms during cycles of addiction to barbiturates in man. Electroenceph. clin. Neurophysiol. **7**, 1—14 (1955)
Wulff, M.H.: The barbiturate withdrawal syndrome. A clinical and electroencephalographic study. Electroenceph. clin. Neurophysiol. Suppl. 14 (1959)
World Health Organization: Expert Committee on Addiction-Producing Drugs, 13th report, Wld Hlth Org. techn. Rep. Ser. No. 273 (1964)
Yanagita, T., Takahashi, S.: Development of tolerance to and physical dependence on barbiturates in rhesus monkey. J. Pharmacol. exp. Ther. **172**, 163—169 (1970)
Yanagita, T., Takahashi, S.: Dependence liability of several sedative-hypnotic agents evaluated in monkeys. J. Pharmacol. exp. Ther. **185**, 307—316 (1973)

CHAPTER 3

Clinical Aspects of Alcohol Dependence*

N. K. MELLO and J. H. MENDELSON

A. Introduction

I. Toward a Definition of Alcoholism

Traditionally, there has been relatively poor agreement concerning an adequate definition of alcoholism. Although a definition involving excessive drinking that results in injury to an individual's health, adequate social function, or both would be generally accepted, considerable variation continues to exist in the formulation of more precise definitions with concomitant criteria for differential diagnosis and implications for treatment (WHO Report, 1955; JELLINEK, 1960; KELLER, 1962; MENDELSON and STEIN, 1966a; PLAUT, 1967; CAHALAN, 1970; National Council on Alcoholism, 1972).

One limitation of those definitions of alcoholism which involve primarily *social* criteria lies in the enormous variation of acceptable drinking habits both within and between countries. The inconsistency of culturally determined standards of acceptable drinking presents a particular problem for the biomedical researcher concerned with selecting equivalent subjects so that comparisons can be made between laboratories in different areas. This problem can be resolved by defining alcoholism in terms of objective pharmacologic criteria of addiction, tolerance, and physical dependence, rather than in terms of its social consequences. Physical dependence, i.e., the alcohol withdrawal syndrome, is by far the most objective and unambiguous criterion of alcohol addiction. However, an emphasis upon physical dependence as the defining criterion of alcoholism is necessarily restricted to the extreme end stage of this disease process which differs in a number of respects from antecedent heavy drinking. Tolerance for alcohol may be rapidly induced and precede the development of physical dependence. Several techniques are now available for the measurement of tolerance (KALANT et al., 1971).

II. The Concept of Addiction

General acceptance of the notion that alcoholism is a form of addiction has occurred only within the last decade. Although the addiction concept of alcoholism derives from the observations of clinicians in the early 19th century, these ideas only recently culminated in unambiguous description of alcohol withdrawal following cessation of drinking (VICTOR and ADAMS, 1953; ISBELL et al., 1955; MENDELSON, 1964). At one

*This review was initially prepared in 1974 with support from the Intramural Laboratory of Alcohol Research, NIAAA, ADAMHA. Completion of this review was supported by Grant No. DA 01676-01 from the National Institute on Drug Abuse, ADAMHA.

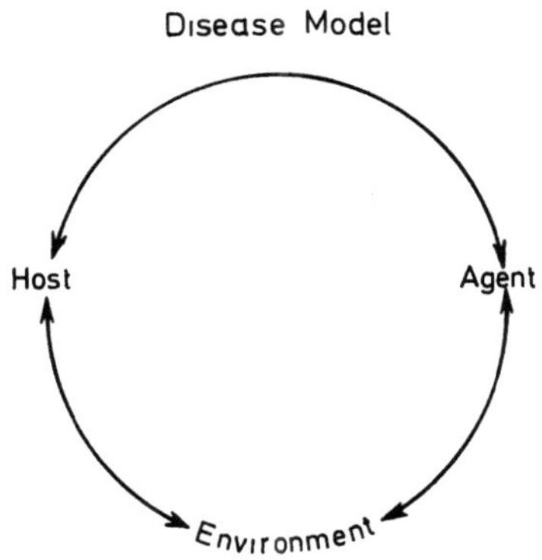

Fig. 1. Schematic diagram of disease model of alcoholism which shows interaction between host, agent (alcohol) and environment. (Reprinted from MENDELSON and MELLO, 1969.)

time, it was thought that the alcohol withdrawal syndrome reflected intercurrent illness or vitamin and nutritional deficiency rather than the effects of cessation of drinking. This interpretation of alcohol withdrawal phenomena was challenged by the clinical observations of VICTOR and ADAMS (1953) and the experimental studies of ISBELL et al. (1955) on the effects of alcohol intoxication in morphine addicts. Subsequently, it was shown experimentally that withdrawal signs and symptoms occur in healthy, well-nourished alcoholics solely as a function of cessation of drinking (MENDELSON, 1964). These data justified consideration of alcoholism as a disease within the context of the addictive disorders. This transition from the time-honored tendency to view alcoholism as a type of moral transgression and social deviancy has done much to promote the systematic study of alcoholism within the framework of biomedical research.

Since alcoholism is a form of addiction, it is useful to think of this disorder in terms of a model generally applied in medical science. A schematic diagram of this model is shown in Figure 1. The disease model of alcoholism assumes that the expression of the disorder is dependent upon an *interaction* between a host, the agent of the disease, and the environment in which the disease occurs. Disease processes can rarely be explained on the basis of any specific factor within each of these three categories. An analogy can be drawn between alcoholism and infectious disease in which an agent can be identified but the expression of the disease is more closely related to host resistance factors and environmental variables than to the presence or even to the virulence of a given agent (MENDELSON and MELLO, 1969).

The crucial determinants of the development of alcohol addiction are unknown and the nature of the addictive process remains a matter of conjecture. The variety of theories which have been advanced to account for the development of alcoholism are as numerous as the medical and scientific disciplines which have become concerned with the problem. Thus far, no single theory has proved adequate to explain the complex multilevel symptoms which we collectively term alcoholism. Most probably, alcoholism derives from as many diverse factors in the individual and in the environment as can be postulated to contribute to the development of any behavior disorder. However, alcoholism does differ from most other behavior disorders, such as depression or schizophrenia, in that we can isolate, describe, and define the nature of the agent, i.e., alcohol, that is essential for the expression of the disease process. Consequently, it is possible to observe the effects of the agent, alcohol, upon any measurable behavioral, physiologic, or biochemical variable as a function of dosage through time. Moreover, it is possible to compare measures of the effects of alcohol with observations on the same subject obtained prior to and following exposure to alcohol. The experimental induction of an addictive disorder in encapsulated form by administration of a drug provides a very useful model for studying behavioral and biological correlations in man and other species.

In alcoholism and in most other addictive disorders, the presence of the drug is a *necessary* but not a sufficient condition for the expression of the behavioral disorder. The factors which determine a particular individual's susceptibility to alcohol addiction, given a prolonged pattern of heavy drinking, remain to be determined. Since individuals afflicted with alcoholism are essentially heterogenous in terms of their sociocultural background, psychodynamics, and drinking patterns (cf. MELLO, 1972), we have argued that reconstruction of those events presumed to precipitate the development of alcoholism is probably not an effective strategy for modifying the disease process (MENDELSON and MELLO, 1969). Eventual modification of drinking behavior and the medical complications of alcoholism may come from an understanding of the variables which contribute to the maintenance of drinking behavior rather than an elucidation of etiologic factors. Even the relationship between physical dependence and drug-seeking behavior remains to be clarified. Research on opiate narcotics has consistently shown that the condition of physical dependence is not invariably associated with compulsive drug use and that behavioral addiction need not be accompanied by physical dependence (cf. JAFFE, 1970 for review). A comparable situation appears to exist in the realm of alcohol abuse.

In summary, alcoholism has been shown to be a form of addiction (ISBELL et al., 1955; MENDELSON, 1964). The alcoholic individual fills the classic pharmacologic criteria of addiction, i.e., tolerance and dependence. The addiction model of alcoholism is schematically illustrated in Figure 2. The alcoholic exhibits *tolerance* in that he must ingest progressively larger quantities of alcohol through time in order to produce a change in feelings and behavior which had previously been attained with smaller doses of alcohol. *Physical dependence* upon alcohol is demonstrated by the

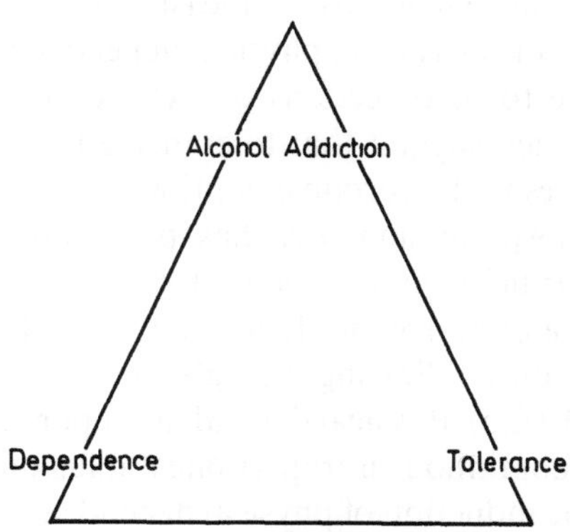

Fig. 2. Schematic description of pharmacologic basis of alcoholism, i.e., alcohol addiction is defined by presence of tolerance for and physical dependence upon alcohol

fact that the abrupt cessation of drinking may result in dramatic withdrawal signs and symptoms to be described in Subsection C of this review (p. 621).

B. Tolerance

After repeated exposure to a drug, a given dose of that drug tends to produce a decreasing effect or alternatively more of the drug is required to produce the same effect in the addicted individual. Alcoholism is characterized by those aspects of tolerance which are common to the other addictive disorders, i.e., *behavioral tolerance, pharmacologic tolerance*, and *cross tolerance* to other potentially addictive agents such as barbiturates, hypnotics, and sedatives (SEEVERS and DENEAU, 1963). These three aspects of tolerance will be discussed in the remainder of this section. Some theoretical models proposed to account for tolerance and physical dependence will be described in Subsection D (p. 635). A more detailed examination of the phenomena comprising tolerance may be found in several recent reviews (cf., COCHIN, 1971; KALANT et al., 1971; SHUSTER, 1971; SMITH, 1971).

It was once thought that tolerance and physical dependence were sequential and inseparable aspects of the same underlying addictive process. Support for this notion was primarily based on the findings that opioid antagonists and other agents can prevent the development of *both* tolerance and physical dependence (SMITH, 1971). SEEVERS and DENEAU have suggested that since an antagonist prevents the depressive effects of the drug, a lack of neuronal depression per se could account for the absence of physical dependence (1963). However, narcotic antagonists prevent both stimulant and depressant effects of narcotics, and the relative contribution of each of these variables to the development of physical dependence is unknown.

Our current understanding of the relationship between tolerance and physical dependence suggests that although physical dependence is invariably accompanied by the induction of tolerance, drug tolerance may occur independently of physical dependence (COCHIN, 1966; SMITH, 1971; JAFFE, 1970; GOLDSTEIN et al., 1968). This argument is based in part on differences in the rate of development believed to exist between tolerance and physical dependence. Several investigators have suggested that tolerance may develop far more rapidly than physical dependence, for both alcohol (KALANT et al., 1971; MENDELSON, et al., 1965) and for narcotics (KALANT et al., 1971; SHUSTER, 1971). Since methods for measuring tolerance may be somewhat more sensitive than those for measuring physical dependence, the alleged differences in time course may prove to be reflections of methodologic differences. There are now considerable data which suggest that tolerance and physical dependence may develop at comparable rates under certain conditions.

Acute tolerance to morphine analgesia has been reported to develop in rats within 3–4 h of continuous infusion (COX et al., 1968) and to persist for as long as 6 months after a single dose of morphine (KORNETSKY and BAIN, 1968). Induction of physical dependence in mouse following a single i.p. injection of alcohol has been reported by GOLDSTEIN (1972a). Physical dependence upon morphine in rat has been produced by two days of continuous intraperitoneal infusion (TEIGER, 1974). A comparable time course for the induction of physical dependence upon alcohol in mouse, using an inhalation technique, has been reported by GOLDSTEIN (1972a, 1974). MARTIN has induced physical dependence to morphine in dog in 8 h of continuous administration (MARTIN and EADES, 1961, 1964; MARTIN et al., 1974). Finally, WAY and coworkers have presented evidence that tolerance and physical dependence to morphine may develop at the same rate in mouse (1969). Physical dependence was measured by precipitated abstinence, a technique which appears to facilitate objective measurement of the early development of physical dependence. The need for a continuing re-evaluation of the concepts of tolerance and physical dependence is indicated by these data.

The mechanisms underlying the induction of tolerance for alcohol are unknown. However, considerable data converge to suggest that the adaptive processes subserving tolerance may occur in the central nervous system rather than at a metabolic level (cf., MENDELSON and MELLO, 1969). Although a variety of techniques have been developed for the measurement of tolerance in animals (KALANT et al., 1971), tolerance has been somewhat difficult to measure clinically. Ambiguities in clinical measurements and the possibility of rapid induction of acute tolerance make it less valuable for the differential diagnosis of alcohol addiction than the appearance of withdrawal signs and symptoms.

I. Behavioral Tolerance

Prolonged drinking may produce qualitatively as well as quantitatively different effects in the alcohol addict and the normal individual. We have found that many alcohol addicts can drink between four-fifths and 1 quart of bourbon per day without signs of gross inebriation (e.g., significant impairment of cognitive processes and motor function) (MELLO and MENDELSON, 1970a, 1972; MELLO, 1972, 1973a). An early study of behavioral tolerance in the alcoholic involved a comparison between global intoxication ratings of alcoholic and nonalcoholic subjects given equivalent doses of alcohol (JETTER, 1938). It was found that the percentage of alcoholics rated as intoxicated according to these criteria was dramatically lower than the percentage of normal drinkers that were rated as intoxicated. One of the first comprehensive studies comparing the effects of alcohol on performance in alcoholics and nonalcoholics was performed by GOLDBERG in 1943. The evidence for behavioral tolerance in alcoholics obtained in these studies has been confirmed in many laboratories (cf., reviews by CARPENTER, 1962; WALLGREN and BARRY, 1970b; MELLO, 1972).

A second approach to the study of behavioral tolerance has been to compare the performance of the alcoholic during sobriety and inebriation. TALLAND and his associates have examined the effects of a period of sustained intoxication on tasks which involve attention and motor skills in chronic alcoholics (TALLAND et al., 1964a, b; TALLAND, 1966). It was found that blood alcohol levels as high as 200 mg/100 ml did not significantly affect simple manual dexterity tasks. Similarly, no significant decrement in performance occurred on vigilance and reaction tasks in which an alcoholic subject was required to respond by pressing a button whenever he detected the occurrence of a visual (TALLAND et al., 1964b) or an auditory signal (DOCTER et al., 1966) despite blood alcohol levels approaching 200 mg/100 ml. Even in a complex attention task in which subjects were required to identify sequences of 3 consecutive odd or 3 consecutive even numbers, acute doses of alcohol ranging between 0.5 and 1.0 g per kilogram produced no serious disruption of performance (TALLAND, 1966). More recently, we found that when alcoholic subjects were required to make a specified number of consecutive correct responses on a simple visual vigilance task in order to earn alcohol, response efficiency was not correlated with blood alcohol levels. Moreover, response efficiency tended to improve in 9 of the 14 subjects studied during the course of the experiment (MELLO et al., 1968).

Alcoholics, trained to perform a visual matching-to-sample task in order to earn alcohol reinforcement, were able to perform as accurately as during baseline at blood alcohol levels in excess of 200 mg/100 ml (MELLO, 1973a). There were no significant changes in daily average performance accuracy in comparison to baseline following an acute dose of alcohol (3.0–3.5 mg/kg) or 10–12 days of chronic drinking.

II. Pharmacologic Tolerance

One indication of pharmacologic tolerance in the alcoholic is that consistent consumption of as much as a quart of bourbon per day may result in unexpectedly low levels of alcohol in the blood as measured by an enzymatic method (cf., MELLO and MENDELSON, 1970a, 1972). Although alcohol tolerance is striking in alcohol addicts, the magnitude of tolerance for alcohol is less, on a comparative dosage basis, than is seen in opiate or barbiturate addiction. For example, barbiturate addicts may continue to function at dose levels which are 20–30 times greater than the average dose for induction of sleep. Heroin addicts show similar increases in self-medication as a function of tolerance. It is well known that heroin addicts may go through withdrawal in order to reduce their daily maintenance dosage requirements by reducing drug tolerance (GOLDSTEIN et al., 1968). The lethal level of alcohol dosage for the alcohol addict remains close to the level generally observed. Blood alcohol concentrations between 500 and 600 mg/100 ml may result in severe respiratory depression and death.

The dramatic behavioral tolerance for alcohol shown by heavily intoxicated alcoholics cannot be accounted for by a more rapid or effective capacity to metabolize alcohol, since the critical behavioral measures are taken at high blood alcohol concentrations (200 mg/100 ml). Moreover, a number of studies have shown that the rate of ethanol metabolism in alcoholics and nonalcoholics is not significantly different (CARPENTER, 1962; NEWMAN, 1941; JACOBSEN, 1952; HARGER and HULPIEU, 1956). More recently, a comparison of the rates of alcohol metabolism in alcoholic and nonalcoholic individuals, abstinent for 3 weeks, by measuring C^{14} carbon dioxide output following administration of an acute dose of carbon 14-labeled ethanol showed that there are no significant differences in rates of alcohol metabolism (MENDELSON, 1968a).

It has been difficult to compare the effects of prolonged heavy alcohol ingestion in alcoholics and nonalcoholics, because nonalcoholic individuals are unable to sustain a high intake of alcohol without developing acute gastrointestinal disorders. However, it has been shown that alcohol ingestion does induce an increased rate of alcohol metabolism in alcoholics *and* controls as measured by $^{14}CO_2$ output following C^{14}-labeled ethanol ingestion (MENDELSON et al., 1965). Moreover, the rates of ethanol metabolism may increase as a function both of the amount and the duration of alcohol ingestion (MENDELSON et al., 1965). The alcohol-induced increase in rates of ethanol metabolism observed in man (MENDELSON et al., 1965) and animals (HAWKINS et al., 1966) is consistent with the finding that chronic drug administration may induce an increased activity of those enzymes which catalyze specific drug metabolism.

A second possibility is the occurrence of de novo synthesis of enzymes following administration of certain substrates or hormones. The increased rate of ethanol metabolism in experimental animals following chronic alcohol administration has been correlated with an increase in hepatic alcohol dehydrogenase activity (HAWKINS et al., 1966). The functional significance of the liver microsomal enzyme systems remains to be determined (cf., LIEBER et al., 1971; ISSELBACHER and CARTER, 1971).

At present, there is only one pharmacologic manipulation which has been shown to be somewhat effective in *accelerating* the rate of *ethanol metabolism* with concomitant reduction of the acute intoxicating effects of alcohol. In 1937, it was first shown that the administration of fructose can increase the rate of ethanol metabolism in man (CARPENTER and LEE, 1937). This finding has been confirmed by a number of investigators; however, the mechanism by which fructose enhances ethanol metabolism is unclear (PLETSCHER et al., 1952; LUNDQUIST and WOLTHERS, 1958; TYGSTRUP et al., 1965), and this technique has had little clinical application. There are no other specific agents currently available which appreciably enhance rates of ethanol metabolism with a concomitant reduction of the acute intoxicating effects of alcohol.

In contrast, *reduction* of the rate of *ethanol metabolism* can be accomplished both pharmacologically by the administration of pyrazole, and through the reduction of food intake. Pyrazole slows the rate of ethanol metabolism by inhibition of liver ADH (THEORELLE and YONETANI, 1963; LESTER et al., 1968). The mechanism of relative nutritional deficiency in contributing to reduction in the rate of ethanol metabolism is not understood. However, studies with experimental animals have unequivocally demonstrated a 50% reduction in the alcohol metabolism rate of fasted as contrasted to adequately nourished rats (LEBRETON, 1936; OWENS and MARSHALL, 1955). We have consistently observed that blood alcohol levels increase as a function of reduced food intake in alcohol addicts (MELLO and MENDELSON, 1970a, 1972). A constant level of alcohol was administered (5.0 ml/kg q4h) to alcohol addicts and caloric intake reduced from 1800 to 450 calories over 4 days. There was a progressive increase in blood alcohol levels (MENDELSON, 1970).

Both experimentally programmed (MENDELSON, 1970) and spontaneous reduction of food intake during intoxication (MELLO and MENDELSON, 1970a) result in an increase in observed and subjective intoxication which parallels the increase in blood alcohol levels. Several of our subjects have volunteered that they deliberately stop eating in order to increase their intoxication level. This may be an effective way for the alcohol addict to counteract the effects of tolerance and intensify and prolong a drinking experience for less money.

Since there is no good evidence that significant alterations in metabolic rates can adequately account for the degree of behavioral tolerance observed in alcoholics, the adaptive processes subserving behavioral tolerance probably occur in the central nervous system. In this connection, it is of interest that although some, as yet, unspecified change in central nervous system processes may underlie the phenomena of addiction, alcohol is not metabolized by brain (WALLGREN and LINDBOHN, 1961). RASKIN and his associates have shown that small or trace amounts of alcohol are catabolized by neural tissue (RASKIN and SOKOLOFF, 1968). It is therefore possible that ethanol degradation in crucial regional areas of the CNS could be related to tolerance. Confirmation of this hypothesis will require far more extensive exploration once appropriate methods become available.

The liver is the site of virtually all ethanol metabolism. Recently, the status of alcohol dehydrogenase as the major enzyme responsible for catalyzing 90–98% of the initial oxidation of ethanol to acetaldehyde in liver (WESTERFELD, 1961) has been challenged by research on microsomal oxidase systems (cf., ORME-JOHNSON and ZIEGLER, 1965; LIEBER and DECARLI, 1968; LIEBER et al., 1971; ISSELBACHER and CARTER, 1971). The relative importance of this microsomal oxidase system in ethanol metabolism remains to be determined since other data suggest that this system does not have an important role in alcohol oxidation in vivo (TEPHLY et al., 1969). Moreover, it appears that mutual induction of the microsomal enzyme system is not important in accounting for cross-tolerance (KALANT et al., 1971). However, recent studies carried out on rates of acetaldehyde catabolism in alcohol addicts and controls indicate that the role of the microsomal oxidase system may be of major importance in ethanol metabolism (LIEBER, 1974).

There have been recent data which suggest that the rate-limiting step in ethanol metabolism is not solely dependent upon hepatic ADH activity. GOLDSTEIN and associates have suggested that the notion that metabolism proceeds at a constant rate independent of the concentration of ethanol is open to question (GOLDSTEIN et al., 1968). According to their calculations, ADH is only half saturated at moderate blood alcohol levels and is not fully saturated even at very high blood alcohol levels. Consequently, the availability of hepatic NAD may be the major factor regulating the zero order kinetics of ethanol oxidation and the ratios of NAD to reduced nicotinamide-adenine dinucleotide (NADH) should be a crucial factor affecting ethanol metabolism rates. There is considerable experimental support for this hypothesis and in vivo metabolism of alcohol has been clearly shown to be associated with a significant decrease in hepatic NAD-NADH ratios (FORSANDER et al., 1958; SMITH and NEWMAN, 1959; REBOUCAS and ISSELBACHER, 1961; RAIHA and OURA, 1962). A more complete discussion of the biochemical pharmacology of alcohol metabolism appears elsewhere (LUNDQUIST, 1971; MENDELSON, 1968b, 1970, 1971; VON WARTBURG, 1971; WALLGREN and BARRY, 1970a).

III. Cross Tolerance

Cross tolerance refers to the general phenomena of reduced responsivity to drugs other than the primary addicting agent. It has been shown that alcoholics may metabolize a number of drugs more rapidly than nonalcoholics (KATER et al., 1968, 1969). There are a number of reports that alcohol addicts undergoing surgery require far larger doses of anesthesia to induce a surgical level of anesthesia than nonaddict patients (BLOOMQUIST, 1959; LEE et al., 1964; HAN, 1969). Similarly, the alcohol addict may show cross-tolerance for certain other central nervous system depressants such as barbiturates, sedatives, and hypnotics (cf., review by SEEVERS and DENEAU, 1963; KALANT et al., 1971); however, there is no cross-dependence between alcohol and opiate narcotics (GOLDSTEIN et al., 1968). Finally, the alcoholic shows tolerance for many toxic alcohols and is able to ingest these in quantities that would be fatal for nonalcoholics (MENDELSON et al., 1957). It should be emphasized that the phenomenon of cross-tolerance occurs in the sober alcohol addict. Under conditions of intoxication, a number of drugs might show synergism or facilitation with alcohol and the alcoholic individual might metabolize these drugs more slowly than normal subjects (RUBIN et al., 1970).

Since alcoholics do show cross-tolerance to many other central nervous system depressants which involve diverse metabolic pathways, this is further evidence in support of the notion that the adaptive processes subserving addiction occur in the central nervous system.

C. Physical Dependence

I. The Alcohol Withdrawal Syndromes in Historical Perspective

The association between alcohol abuse and the abstinence syndromes have been recognized from the beginning of man's recorded history. ZILBORG and HENRY wrote in their scholarly *History of Medical Psychology:* "It is highly probable that there were frequent instances of delirium tremens attending the Bacchanalian Orgies. Herodotus held the Scyphians responsible for teaching Cleomenes to drink and for his excessive indulgences which caused him to be mentally disordered for a time." These same authors quoted an observation of Hippocrates, "If the patient be in the prime of life, and ... if from drinking he has trembling hands, it may be well to announce beforehand, either delirium or convulsion" (ZILBORG and HENRY, 1941).

THOMAS SUTTON has been credited by ZILBORG and HENRY for devising a specific diagnostic classification of delirium tremens as well as inventing the nosologic term (SUTTON, 1813). However, there is some doubt about both priorities. In 1786, LETTSOM provided a good description of this variant of the alcohol abstinence syndrome. Moreover, in 1813, the same year that SUTTON published his Tracts, PEARSON employed the nomenclature "delirium tremens" in his paper in the *Edinborough Medical and Surgical Journal* (LETTSOM, 1787; SUTTON, 1813; PEARSON, 1813).

The clinical descriptions provided by THOMAS SUTTON remain the earliest astute and incisive reports of the phenomenology of delirium tremens (SUTTON, 1813). SUTTON's recognition of the latency between cessation of drinking and the onset of delirium tremens is highly consistent with the systematic observations of VICTOR and ADAMS which were made 140 years later (VICTOR and ADAMS, 1953). SUTTON advocated treatment of this condition with opium in large doses (SUTTON, 1813). This concept would be rejected by contemporary clinicians on pharmacologic grounds alone because there is no cross-tolerance between opiate analgesics and ethanol (GOLDSTEIN et al., 1968). Claims of therapeutic efficacy, even by experts, were not gratuitously accepted in the early 19th century. In 1828, BURROWS wrote that he "treated such patients by opiates and without any narcotic at all; and they have, by both modes, recovered in the time this disease usually occupies" (BURROWS, 1828). Thus, BURROWS can be credited with having first observed that the natural history of the abstinence syndrome is a most important factor to consider in assessing the success of drug therapy; a point which has been strongly re-emphasized by VICTOR over 100 years later (VICTOR, 1966).

In view of the long history of medical documentation of the association between cessation of drinking and appearance of the abstinence syndrome, it may seem surprising that such great controversy existed among medical specialists concerning the causal basis of delirium tremens as late as the mid-twentieth century. The reason for this was not necessarily ignorance or denial of the role of alcohol as a predeterminant condition, but an increased awareness of the role of multiple factors in the causation of any illness. Clinicians who observed the natural history of alcohol-related disorders found that their patients had poor nutritional intake during drinking and were susceptible to many toxic, metabolic, and infectious diseases coincident with alcohol abuse. It is, therefore, not surprising to find that many scientists and clinicians rejected what they consider to be an oversimplistic explanation of the causation of delirium tremens and entertained more complex hypotheses of multifactorial geneses of the abstinence syndrome (i.e., nutritional factors.)

Although there is now good evidence that the alcohol abstinence syndrome can be initiated in man as a function of cessation of drinking or significant reduction of alcohol intake following a period of chronic consumption without other intercurrent disease, the multifactorial model should not be completely abandoned. There is now sufficient evidence to indicate that the alcohol withdrawal syndrome is not "an all or none" phenomenon. Recent data acquired under carefully controlled research ward conditions indicate that alcoholics may show evidence of psychomotor agitation and autonomic dysfunction (phenomena which herald more severe withdrawal states), even when they are drinking relatively large amounts of alcohol and have significantly high blood alcohol levels (MELLO and MENDELSON, 1970a, 1972). Withdrawal phenomena may appear when there is a relative reduction of alcohol intake, and a relative fall of blood alcohol levels (MELLO and MENDELSON, 1970a, 1972). Thus, those clinicians who maintain that the withdrawal syndrome may occur when individuals are still drinking, are probably correct in their observations but incorrect in their assumption that withdrawal has to be considered as a total or absolute process.

In summary, the history of documentation and explanation of alcohol dependence has progressed from anecdotal observation of a correlation between cessation of alcohol use and appearance of the withdrawal states, to more complex multifactorial theories, and currently a return to more unitary conceptualizations based upon detailed observations of pathophysiologic processes antecedent to and during the alcohol withdrawal syndrome.

II. Basic Phenomenology of the Alcohol Withdrawal Syndrome

1. Clinical Description

Even after 20 years, it would be difficult to improve upon the careful clinical descriptions of the sequelae to cessation of drinking reported by VICTOR and ADAMS in 1953. Their observations form the basis for the material to be presented in this section. Their classic paper provided the first distinction between the several clinical syndromes which may follow cessation of drinking, i.e., tremulousness, alcoholic epilepsy or "rum fits", and delirium tremens. They differentiated these three abstinence syndromes in terms of the time of onset following cessation of drinking; the exclusive defining symptoms and signs; and the relative severity of the disorder. Their findings are summarized in Figure 3 and Table 1.

It is important to recall that in 1953, it was not generally acknowledged that the tremulous, hallucinatory, epileptic, and delirious states so frequently seen as "alco-

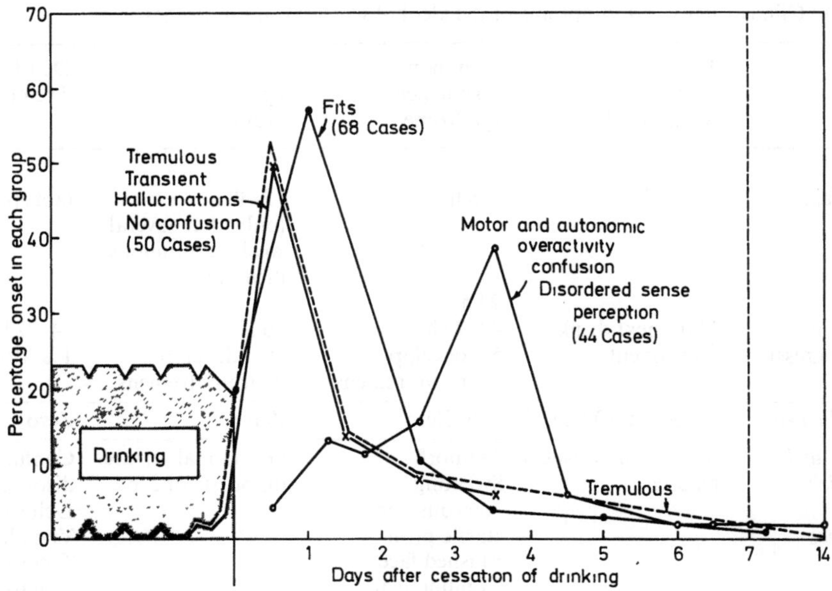

Fig. 3. Summary of temporal relationship of onset of acute neurologic disturbances to cessation of drinking in alcohol addicts. (Reprinted from VICTOR and ADAMS, 1953.)

holic complications" did represent an abstinence syndrome (cf., VICTOR and ADAMS, 1953, pp. 551—552). In discussing their observations of 266 consecutive patients admitted for alcohol-related illness to the Boston City Hospital, VICTOR and ADAMS remarked that "It is difficult to escape the conclusion that the clinical states under discussion depend for their production, not only upon the effects of prolonged exposure to alcohol, but temporally on abstinence from the drug" (p. 550). They supported this conclusion by analogy with withdrawal from barbiturate intoxication and further explain that "This analogy can be drawn not only in regard to the individual variations in susceptibility to the drugs and the acquisition of tolerance, but also in regard to the intoxicating effects and the orderly sequence of the tremulous and delirious symptoms that follow withdrawal" (p. 551).

Clinical evidence that the syndrome which we now term the alcohol withdrawal syndrome represented a true abstinence syndrome received its first experimental validation in the studies of ISBELL et al., in 1955. Former morphine addicts were given 9–16 ounces of alcohol per day in divided doses for 48–87 days. Upon withdrawal of alcohol, these subjects showed withdrawal signs and symptoms (ISBELL et al., 1955). It was not until 1964 that a comparable study was carried out with well-nourished alcoholics given programmed dosage of alcohol (q4h) for 24 consecutive days with a maximum dosage of 40 ounces per day (MENDELSON, 1964). These clinical and experimental data combined to support the addiction concept of alcoholism (cf., Subsect. A, pp. 613–616).

There is a frequent tendency to confuse the term *delirium tremens* with the milder form of the alcohol abstinence syndrome and to use delirium tremens to refer to all complications following cessation of drinking. In accordance with the distinction

Table 1. Clinical signs and symptoms of the alcohol withdrawal syndromes[a]

	Early or partial withdrawal	Common abstinence syndrome	Alcoholic epilepsy or "rum fits"	Delirium Tremens
Time course				
Onset (after last drink)	8–16 h[d]	6–8 h	12–48 h (only after several weeks continuous drinking)	(abrupt) 73–96 h
Peak		24 h		
Usual duration	Until next drink	48–72 h	36 h	24–72 h
Symptom progression	if abstinent →	5% develop delirium tremens	30% develop delirium tremens	1% Mortality[f]
Blood alcohol levels	0–100 mg/100 ml[b,d]	↓ to Zero	Zero	Zero
Defining signs and symptoms which differentiate the 4 syndromes	Mild tremulousness Nausea	Tremor[g] Sweating Nervous and startle prone Flushed face Conjunctival injection Mild tachycardia[g]	Grand mal seizures (in bursts of 2–6)	Confusion and disorientation Delusions Vivid hallucinations Tremor Agitation Insomnia Autonomic hyperactivity Fever Sweating Tachycardia Mydriasis
Additional common signs and symptoms	Early morning drink "Anxiety"[d] Agitation[d] Tachycardia[d]	Nausea[g] Vomiting[g] Anorexia[g] Nystagmus[c] Hyperreflexia[c] Mild disorientation[g] Nightmares[g], Illusions[g], Hallucinations[g] } 25% visual (83.4%) auditory, mixed } 16.6% Insomnia[g] Sleep fragmentation[e,g]		Distractible Suggestible
Estimated incidence in alcoholics		80%		5%

[a] All data from VICTOR and ADAMS, 1953 and VICTOR, 1966 unless otherwise indicated.
[b] Reported by ISBELL et al., 1955.
[c] Reported by MENDELSON, 1964.
[d] Reported by MELLO and MENDELSON, 1970a, 1972.
[e] Reported by MELLO and MENDELSON, 1970b.
[f] Reported by TAVEL et al., 1961.
[g] Also observed during intoxication by the authors.

proposed by VICTOR and ADAMS (1953), we wish to emphasize that delirium tremens is a distinct and specific syndrome which occurs relatively infrequently and much later than the usual alcohol abstinence syndrome (cf., Fig. 3; Table 1, Column 4). However, delirium tremens is by far the most severe complication of alcohol withdrawal characterized by profound confusion and disorientation, delusions, and vivid hallucinations in contrast to the common abstinence syndrome (cf., Table 1). Delirium tremens may occur independently of or following apparent recovery of the common abstinence syndrome (VICTOR and ADAMS, 1953). It was once estimated that 15% of persons with delirium tremens, so defined, die as a direct consequence of this illness (VICTOR and ADAMS, 1953). As a result of improvements in medical management generally, mortality following delirium tremens is now estimated at about 1% (TAVEL et al., 1961). Methods for treatment of the alcohol withdrawal syndromes will be discussed in Subsection F of this review (p. 651).

The condition known as alcoholic epilepsy or *rum fits* is distinguished both from the common abstinence syndrome and from delirium tremens by the occurrence of grand mal seizures (usually in bursts of 2–6; Table 1, Column 3). Seizures occur during alcohol withdrawal in persons with otherwise normal clinical EEGs (VICTOR and BRAUSCH, 1967; VICTOR, 1968). Single or multiple seizures may occur over a period of a few hours. About one-third of the alcoholics who develop *rum fits* proceed to delirium tremens after termination of the seizures.

The occurrence of seizures during drug withdrawal is, of course, not unique to alcohol addiction. Severe seizures during withdrawal are commonly seen following prolonged intoxication with barbiturates (FRASER et al., 1954), meprobamate (HAIZLIP and EWING, 1958), chloral hydrate, and paraldehyde (KALINOWSKY, 1942). On the other hand, the heroin abstinence syndrome is rarely associated with convulsions (cf., JAFFE, 1970). It was once thought that alcohol had anticonvulsant properties which were useful in the clinical management of epilepsy (KALINOWSKY, 1942). It has now been shown that alcoholics with idiopathic epilepsy or with epilepsy secondary to head trauma have more frequent and severe seizures during drinking (VICTOR and BRAUSCH, 1967). There is no evidence for the idea that rum fits represent a latent epilepsy which is unmasked or precipitated by alcohol (VICTOR and BRAUSCH, 1967; VICTOR, 1968). The mechanism by which withdrawal from alcohol may lead to rum fits is unknown (cf., review by VICTOR, 1968; discussion Subsection C.III, pp. 631–635).

In the sequential development of the alcohol withdrawal syndromes, a period of heavy intoxication may be punctuated by *early* or *partial withdrawal* signs and symptoms attendant upon the abrupt fall in blood alcohol levels which usually occur during sleep. These early withdrawal signs and symptoms were first noted by VICTOR and ADAMS (1953) and are schematically indicated by the notches in the alcohol administration portion of Figure 3. These early withdrawal symptoms are usually characterized by a mild tremulousness and some general feelings of discomfort, "nervousness", and possible nausea (Table 1, Column 1). The alcohol addict usually combats partial withdrawal symptoms with an "early morning drink."

Our understanding of the determinants of alcohol abstinence is complicated by this observation that alcoholics may experience withdrawal symptoms during a drinking episode if their blood alcohol level falls rapidly enough, even though the fall may be relatively small. These data suggest the futility of trying to establish a critical blood alcohol level for the initiation of withdrawal signs and symptoms since a small decrease from an initial value of 300 mg/100 ml may evoke partial withdrawal phenomena which are just as severe as a small decrease from an initial level of 100 mg/100 ml. A relative fall of about 100 mg/100 ml in blood alcohol levels within 24 h appears to be sufficient to produce tremulousness, sweating, tachycardia and agitation, even when the blood alcohol level is above 100 mg/100 ml (MELLO and MENDELSON, 1970a, 1972).

Upon complete cessation of drinking, the symptoms associated with early withdrawal become more pronounced and severe and comprise part of the total clinical picture of the *common abstinence syndrome* (Table 1, Column 2). The common alcohol abstinence syndrome is initiated within 6–8 h following cessation of drinking and usually reaches its peak after about 24 h. This syndrome is usually self-limiting and lasts for a maximum of 72 h. This is, by far, the most common complication of alcohol abstinence and usually responds to appropriate medical management (cf., Subsect. F, p. 651).

The alcoholic in withdrawal has coarse tremors of the extremities which are most obvious during intentional movements. There are also frequently tremors of the tongue and of the trunk. Sweating is usually profuse and patients often present with a flushed face, conjunctival injection, and a mild tachycardia, and describe themselves as being nervous, shaky, and restless. VICTOR has emphasized that the instability and fluctuations of tremor make it somewhat unreliable as a criterion by which to judge the efficacy of therapeutic intervention (VICTOR, 1966). Patients frequently report nausea and some vomiting during alcohol withdrawal. Many subjects exhibit nystagmus and hyperreflexia upon neurologic examination. Subjects are usually alert and well oriented, although some may show some mild degree of disorientation with regard to time.

A summary of the various signs and symptoms of the alcohol withdrawal syndrome that have been emphasized by various investigators has recently been presented by GROSS and coworkers (GROSS et al., 1971b). On the basis of this survey, a daily clinical evaluation scale was developed (GROSS et al., 1971a) which was then applied to 100 consecutive male alcoholic admissions to a general psychiatric hospital and the results were factor-analyzed (GROSS et al., 1971b). It was found that three symptom clusters accounted for 66% of the variance. These factors were:

1. Nausea, tinnitus, visual disturbances, pruritus, parasthesias, muscle pain, hallucinations (tactile, visual, and auditory), and agitation
2. Tremor, sweating, depression, and anxiety
3. Level of consciousness, quality of contact, disturbance of gait, and nystagmus.

These factors accounted for 27, 19, and 20% of the total variance, respectively. The authors point out that each factor contains one element of what they define as the basic triad of hallucinations, tremor, and clouded sensorium. This factor-analytic approach does provide some quantitative impression of the typical groupings of the various types of symptoms. However, these data do not take into account the *temporal* course of symptom development that has been emphasized by VICTOR and ADAMS (1953).

Most observers report insomnia in their clinical sample (cf., GROSS and GOODENOUGH, 1968; GROSS et al., 1966, 1971c, 1973; JOHNSON et al., 1970), although we have seen exceptions to this general finding (MELLO and MENDELSON, 1970b; WOLIN and MELLO, 1973). Many alcoholics in withdrawal experience a transitory hallucinosis which may be accompanied by illusions or nightmares. There does not appear to be a progression from illusions to nightmares to hallucinations, and hallucinations also occur during intoxication (WOLIN and MELLO, 1973).

The factors which contribute to the severity and duration of a particular withdrawal episode have not been determined. It was once thought that the mode of cessation of drinking, i.e., gradual reduction of intake as opposed to abrupt cessation, might influence the expression of withdrawal signs. Indeed, a "tapering off procedure" has long been advanced as an effective treatment for the chronic inebriate (WILLIAMS, 1931). However, we have consistently observed withdrawal signs and symptoms following cessation of *programmed* drinking in which alcohol intake was tapered over 3 days (MENDELSON, 1964; MELLO and MENDELSON, 1970a). Even after *spontaneous* alcohol access, those subjects that gradually tapered their drinking at the end of the drinking period did not always avoid withdrawal signs and symptoms. This is because the behavioral act of increasing or decreasing alcohol intake is not the sole determinant of blood alcohol levels, but rather a complex series of factors including tolerance, nutritional status, and intercurrent illness combine to determine the effective dosage of alcohol at any one time (cf., this review, pp. 619). Consequently, we have argued that insofar as withdrawal symptoms are related to the absolute concentration of alcohol in the blood and to the rate of fall in blood alcohol levels, an alcoholic cannot effectively manipulate either variable by his drinking behavior alone (MELLO and MENDELSON, 1970a, 1972).

When considered as independent factors, neither volume of alcohol ingested nor blood alcohol levels can reliably predict the severity and duration of withdrawal signs and symptoms. In a comparison between spontaneous and programmed drinking (in which alcohol was administered in divided doses once every 4 h), the same subjects drank more, ate less, achieved higher blood alcohol levels, and tolerated alcohol better during the spontaneous unrestricted drinking paradigm. Moreover, the spontaneous drinking paradigm consistently produced the most severe, varied, and prolonged withdrawal signs and symptoms. Since subjects drank for 20 days in each condition, these data suggested that the pattern of drinking is more important than the duration of drinking in determining the severity of withdrawal signs and symptoms. Since all subjects drank more alcohol and maintained higher blood alcohol levels during the free choice than during the programmed drinking period, it is somewhat difficult to isolate the contribution of the pattern of drinking from the overall volume ingested in precipitating the observed withdrawal syndromes. However, in two subjects who drank similar quantities of alcohol during the two periods, both showed the most varied, severe, and prolonged withdrawal signs and symptoms after spontaneous drinking (MELLO and MENDELSON, 1970a).

2. Temporal Development of Alcohol Dependence

Very little is known about the critical contributing factors or the time course of development of the alcohol withdrawal syndrome. It has generally been assumed that physical dependence upon alcohol develops after some number of years of heavy drinking, and it is of course impossible to confirm or disconfirm this impression by direct experimental observation. The only method available to examine these issues is the analysis of self-report data. We have frequently commented on the unreliability

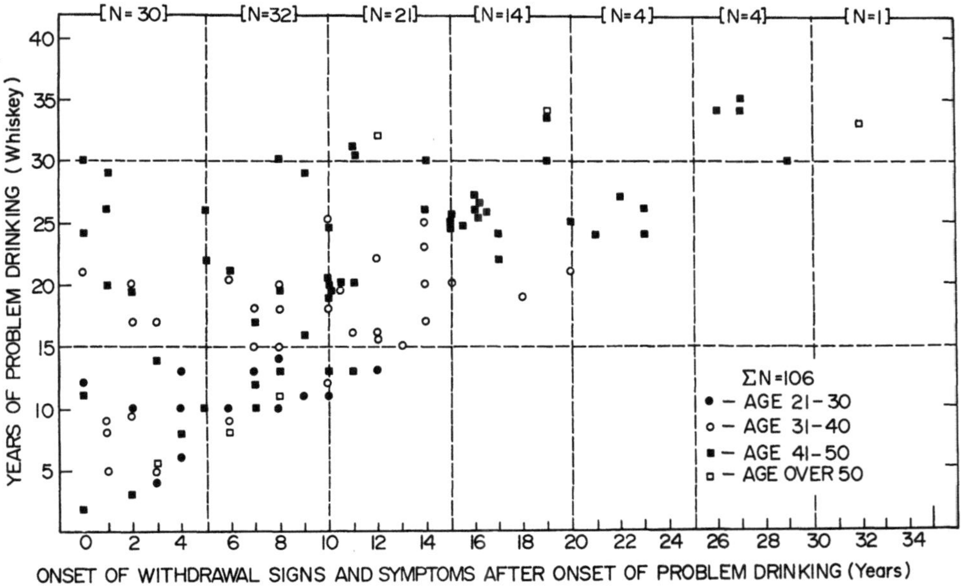

Fig. 4. Self-report data from 106 alcohol addicts describing latency between initiation of problem drinking (whiskey) and development of withdrawal signs and symptoms. Circles and squares indicate age of subject at time of interview (lower right). (Reprinted from MELLO and MENDELSON, 1976.)

of self-report data obtained from sober alcohol addicts and all of the self-evident reservations apply equally in the present case. However, we felt it might be useful to examine the case records of 129 alcohol addicts studied over a 4 year period on the clinical research ward of the Laboratory of Alcohol Research, National Institute on Alcohol Abuse and Alcoholism, ADAMHA at the Saint Elizabeths Hospital in Washington, D.C. Our goal was to determine if there were discernible relationships between the time of onset of withdrawal signs and symptoms as a function of years of problem drinking or reported amount or type of alcohol consumed each day (MELLO and MENDELSON, 1976).

Upon admission to the clinical research ward, a number of medical and psychologic screening tests were performed on each subject, and his drinking and withdrawal history were evaluated in a standard interview. A copy of the drinking and withdrawal history interview form is attached as Appendix I. Answers to this questionnaire were compared for consistency with responses given during a second medical interview and data are presented for 129 subjects[1] from a total sample of 133 subjects (MELLO and MENDELSON, 1976).

The reported time of onset of withdrawal signs and symptoms after the onset of problem drinking is shown in Figure 4. The latency of development of withdrawal signs and symptoms, after the subject identified drinking as a problem, is shown on the abcissa. Data are presented for 106 whiskey drinkers and their age at the time of

[1] The 8 subjects described previously have not been included in this report (MELLO and MENDELSON, 1970a).

interview is indicated at the lower right of Figure 4. The withdrawal onset latency is plotted against the number of years of problem drinking reported at the time of the clinical interview. Consequently, all subjects reporting 10 years of problem drinking must have a withdrawal syndrome onset latency of less than 10 years. The ordinate of Figure 4 shows the duration of each subject's drinking history (1–40 years) at the time of interview.

Fifty percent of this sample reported that withdrawal signs and symptoms began after at least 5 years and within 15 years of problem drinking (cf., cols. 2, 3, Fig. 4). A total of 23 subjects or 22% of this sample reported at least 15 years of problem drinking before the onset of withdrawal signs and symptoms (Fig. 4). Early development of withdrawal signs and symptoms during the first 5 years of problem drinking was reported by 30 subjects or 28% of this sample. Twelve of these subjects (11% of total sample) reported that withdrawal signs and symptoms began within the first year of heavy drinking. These data indicate that a minimum of 2 years of heavy drinking usually preceded the development of withdrawal signs and symptoms and, for the vast majority (72%) at least 5 or more years of heavy whiskey drinking was required.

It might be expected that the consistent consumption of a lower concentration of beverage alcohol, such as wine or beer, could result in a postponement of the development of withdrawal signs and symptoms. In Figure 5, withdrawal latency data are presented for 17 subjects who drank wine or beer rather than whiskey. The onset of withdrawal signs and symptoms is displayed as a function of years after the onset of problem drinking. The age and the beverage choice of each subject is shown at the lower right of Figure 5.

The meaningfulness of percentage comparisons between whiskey and wine and beer drinkers is limited by the small sample size. However, the number of wine and beer drinkers who reported at least 15 years of heavy drinking before the onset of withdrawal signs and symptoms was 35%, a 13% increase over the whiskey drinkers. Fifty-three percent (53%) reported at least 5 years of heavy drinking prior to the onset of withdrawal signs and symptoms and this figure is equivalent to that for the whiskey drinkers. However, within the first 5 years of problem drinking, only 2 subjects reported withdrawal signs and symptoms. Therefore, the frequency of early onset of withdrawal signs and symptoms is 17% less in the wine and beer drinkers. These data do offer some tentative support for the notion that the abuse of wine and beer may result in some delay in the development of physical dependence in comparison to the abuse of whiskey. However, it is apparent that alcohol addiction does result if sufficient alcohol is consumed in any beverage form, regardless of the alcohol concentration.

It is obvious that these data relating the latency of withdrawal syndrome onset to years of heavy drinking can be meaningfully interpreted only if the volume of alcohol consumed is also considered. Figure 6 presents data for 94 of the 106 whiskey drinking subjects described in Figure 4. Only subjects who gave the most consistent estimates of daily consumption were used. These subjects were spree drinkers and their daily intake estimates were based upon spree consumption during a usual drinking episode.

The development of physical dependence is probably a function of both volume consumed and duration of drinking. It might be expected that those whiskey drink-

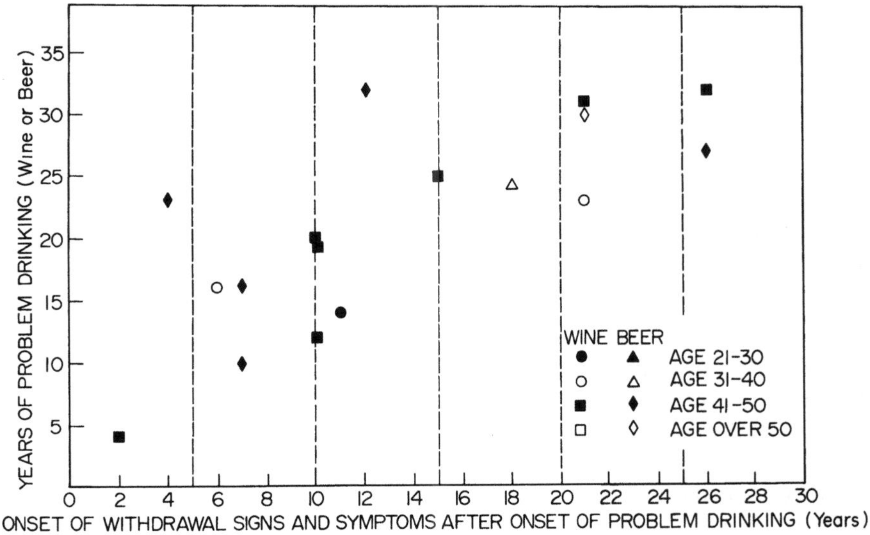

Fig. 5. Self-report data from 17 alcohol addicts describing latency between initiation of problem drinking of wine or beer and development of withdrawal signs and symptoms. Circles and squares indicate age of subject at time of interview (lower right). (Reprinted from MELLO and MENDELSON, 1976.)

ers who consumed only 1 pint of alcohol per day would develop withdrawal signs and symptoms somewhat later than those subjects who consumed 4/5 quart or more than 1 quart per day. However, the data presented in Figure 6 indicate that the majority of subjects developed withdrawal signs and symptoms during the first 10 years of heavy drinking, independently of the absolute amount of alcohol that they drink during sprees. Fifty-three (53) of the 94 subjects drank between 4/5 and 1 quart of whiskey per day. Thirty-seven percent (37%) of these individuals did not develop withdrawal signs until after at least 10 years of heavy drinking. Sixty-two percent (62%) required between 5 and 10 years of heavy drinking before the development of withdrawal signs and symptoms. Twenty-six percent (26%) of these subjects did develop withdrawal signs and symptoms within the first 5 years of drinking. Approximately the same temporal distribution was observed for the 16 persons who drank 1 pint of whiskey per day. No clear distribution pattern emerged for those subjects who reported drinking $1\frac{1}{2}$ quarts or more of whiskey per day. It was somewhat surprising to see that those subjects who claimed to drink 1 quart of whiskey or more per day also reported that withdrawal signs and symptoms began over 25 years after the onset of problem drinking.

These findings appear to testify to an enormous interindividual variability in resistance to the induction of physical dependence upon alcohol. Since the absolute volume of alcohol consumed over time is also a function of the frequency of drinking sprees, and since drinking spree frequency could not be assessed with any degree of reliability from a retrospective report, these data must be interpreted with some caution.

Data presented in Figure 7 indicate that the absolute volume of wine and beer consumed per day during a drinking spree also did not predict the time of onset of

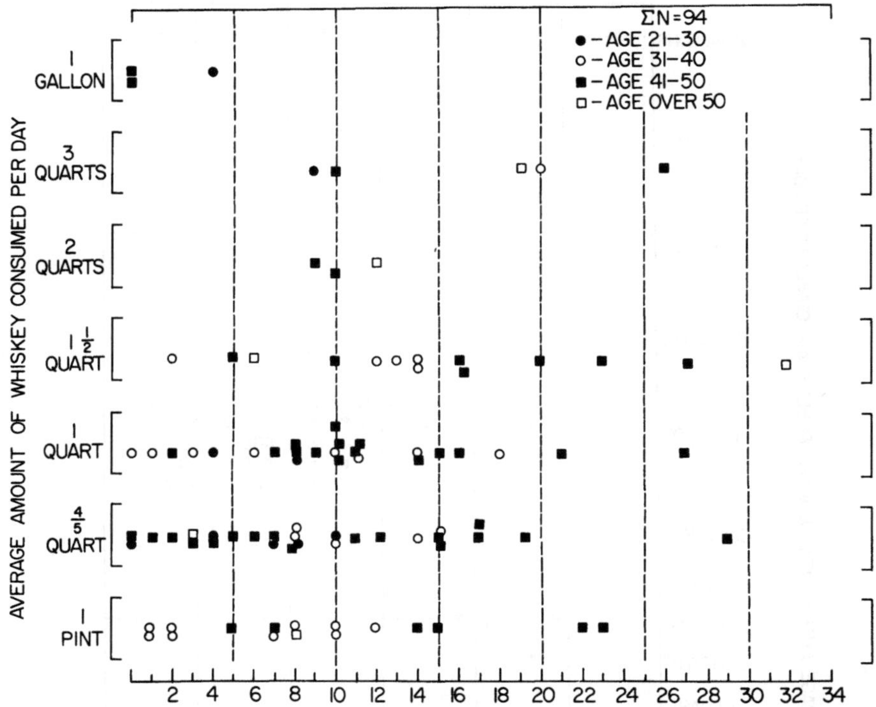

Fig. 6. Self-report data from 94 alcohol addicts describing relationship between average amount of whiskey consumed each day during a drinking spree, and time between onset of problem drinking and development of withdrawal signs and symptoms. Circles and squares indicate age of subject at time of interview (upper right). (Reprinted from MELLO and MENDELSON, 1976.)

withdrawal signs and symptoms with any degree of reliability. Some subjects who reported consuming about 1 pint per day and some subjects who reported consuming about 2 gallons per day of wine or beer were both able to drink for 15–20 years before the onset of signs of physical dependence. Once again, the limitations of retrospective reports of drinking behavior should be emphasized. However, insofar as these data are valid reflections of actual drinking patterns in alcohol addicts, it would appear that the onset of physical dependence upon alcohol is a complex interaction of both the years of drinking and the average volume of ingestion which varies greatly from individual to individual. Those factors which are most important in accounting for differential susceptibility to or resistance to addiction to alcohol remain to be determined (MELLO and MENDELSON, 1976).

III. Some Attempts to Account for the Alcohol Withdrawal Syndrome

The determinants of the alcohol withdrawal syndrome remain largely unspecified and the biological mechanisms underlying the expression of the alcohol withdrawal phenomena are unknown. Before proceeding to a discussion of the theoretical models of alcohol withdrawal and the development of physical dependence, we will first consider several hypotheses concerning the development of alcohol withdrawal signs

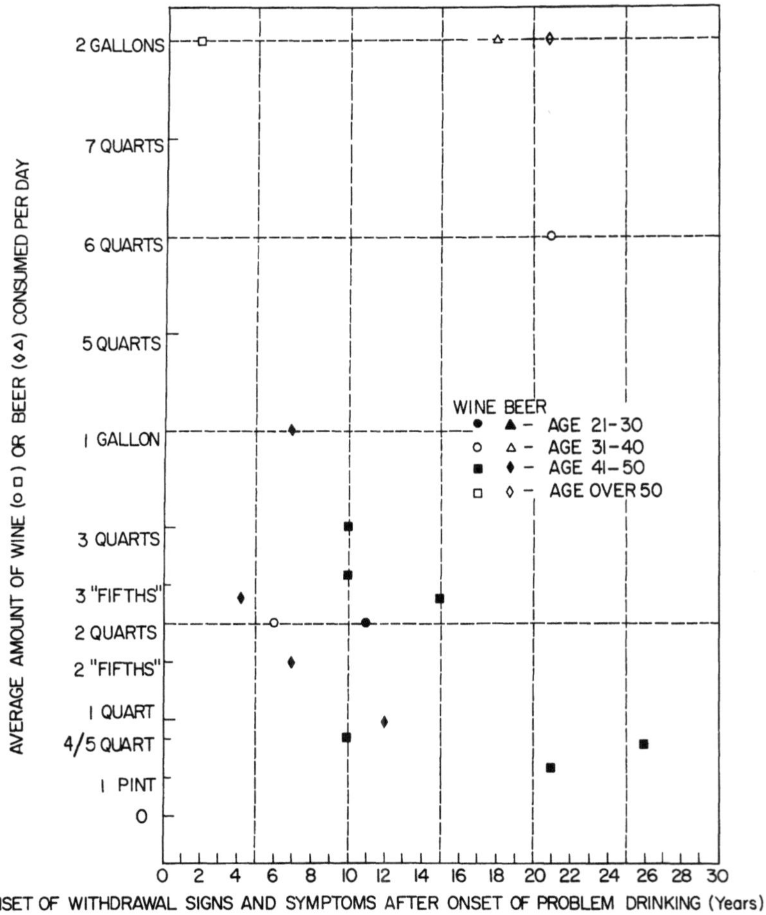

Fig. 7. Self-report data from 16 alcohol addicts describing relationship between average amount of beer or wine consumed each day during drinking spree, and time between onset of problem drinking and development of withdrawal signs and symptoms. Circles and squares indicate age of subject at time of interview (middle right). (Reprinted from MELLO and MENDELSON, 1976.)

and symptoms that have been empirically derived from clinical observations. Each of these hypotheses shares the notion that the psychomotor and autonomic hyperactivity which characterize the alcohol withdrawal syndromes, reflect a heightened central nervous system excitability, and each attempts to account for this CNS hyperexcitability.

WOLFE and his associates have recently suggested that the physiologic basis of the alcohol withdrawal syndrome may be related to the development of a significant respiratory alkalosis which begins as early as 8 h following cessation of drinking and lasts for 24–60 h (WOLFE et al., 1969; WOLFE and VICTOR, 1971). This hypothesis was developed following the observation that alcoholics in withdrawal who showed photo myoclonus (involuntary clonic movements in response to stroboscopic stimulation) had significantly lower serum magnesium levels than alcoholics who did not show a photo myoclonic response and the acute hypomagnesemia appeared to be correlated with a severe alkalosis (WOLFE et al., 1969; WOLFE and VICTOR, 1971). Observations of alcoholics in alcohol withdrawal following a period of experimentally induced intoxication revealed a comparable pattern of alkalosis associated with hyperventilation. All subjects had tachypnea with respiratory rates as high as 25/min. Calculation of bicarbonate values indicated that the abnormal rise in arterial pH (from 7.40–7.49) was a pure respiratory alkalosis and not a metabolic alkalosis. Those subjects showing the greatest increase in arterial pH, with a concomitant decrease in pCO_2 (from 36—30 mm/HG) had the most severe withdrawal signs and symptoms (WOLFE et al., 1969).

These findings were confirmed and extended in an additional series of 31 patients, 9 of whom had classical delirium tremens (WOLFE and VICTOR, 1971). In this series, the 9 patients with seizures associated with alcohol withdrawal showed the most severe respiratory alkalosis. Both the occurrence of withdrawal seizures and the occurrence of hallucinations temporally coincided with the peak respiratory alkalosis in an additional 13 patients. The bicarbonate values calculated from pH and pCO_2 data were either decreased or in the normal range, indicating a purely respiratory alkalosis. Arterial pH values in the delirium tremens group were only slightly higher than in the group with seizure disorders. In patients with delirium tremens, the pCO_2 values had normalized prior to the onset of symptoms and then decreased again. These patients consistently showed decreased serum magnesium levels which followed a time course similar to the respiratory alkalosis.

These data are of particular interest in view of several other clinical syndromes in which increased neural excitability is associated with a respiratory alkalosis. Symptoms very similar to those of alcohol withdrawal may develop in patients with chronic obstructive lung disease who are mechanically ventilated to combat their state of respiratory acidosis. During the course of hyperventilation treatment, the pCO_2 decreases and arterial pH increases, with the subsequent development of the symptoms, e.g., disorientation, hallucinations, tremor, hyperreflexia, seizures, and hyperpyrexia (ADDINGTON et al., 1966; HAMILTON and GROSS, 1963; KILBURN, 1966; ROTHERAN et al., 1964). These symptoms are rapidly relieved by decreasing the mechanical ventilation which produces a concomitant increase in pCO_2 and decrease in pH (ROTHERAM et al., 1964). A respiratory alkalosis has also been associated with tremor, confusion, delirium, and seizures in salicylism and the hyperventilation syndrome (cf., WOLFE and VICTOR, 1971).

Most other efforts to account for the CNS hyperexcitability associated with alcohol withdrawal can be subsumed under the general rubric of the "toxic agent hypothesis." The general idea is that alcohol withdrawal may be accounted for by an unspecified toxic agent induced by alcohol and eliminated more slowly than alcohol and with a maximal effect following cessation of drinking (cf., KAIM, 1971).

One candidate for the role of the toxic agent has been *acetaldehyde*, a by-product of the first step in alcohol metabolism (TRUITT and WALSH, 1971). Ethanol is oxidized to acetaldehyde which, in turn, is oxidized to acetate or acetyl coenzyme A. However, the rate of acetaldehyde metabolism is faster than the rate of oxidation of ethanol to acetaldehyde (LUBIN and WESTERFELD, 1945). In a study of serial determinations of acetaldehyde concentration during chronic experimentally induced intoxication, no dose or dose-time relationships were found between blood ethanol concentrations and blood acetaldehyde concentrations during intoxication or withdrawal. No descending dose or dose-time relationships were found between blood ethanol and blood acetaldehyde concentrations immediately following cessation of drinking (MAJCHROWICZ and MENDELSON, 1970, 1971 a).

A second candidate for the role of the toxic agent is *methanol* which is known to accumulate in blood as a result of the competitive inhibition by alcohol of alcohol dehydrogenase, the enzyme which metabolizes both methanol and alcohol. Gas chromatographic determination of blood levels of methanol and ethanol in chronic alcoholic subjects, during a 10–15 day period of spontaneous alcohol intake, showed a progressive increase in blood methanol levels with ascending levels of blood alcohol. It was not possible to establish any definitive relationship between the degree of intoxication observed and the accumulation of methanol in blood. Following cessation of drinking, the blood methanol disappearance lagged behind the linear disappearance of blood alcohol by approximately 6–8 h (MAJCHROWICZ and MENDELSON, 1971 a, b). The time course of blood methanol decay precedes the usual maximum development of withdrawal signs and symptoms. On the basis of temporal association alone, it is unlikely that methanol catabolism contributes to the alcohol withdrawal syndrome following cessation of alcohol ingestion. Moreover, the symptoms associated with methanol poisoning and with alcohol withdrawal are quite dissimilar. The alcoholic in withdrawal shows a striking and significant respiratory alkalosis (WOLFE et al., 1969; WOLFE and VICTOR, 1971). However, methanol poisoning is associated with a metabolic acidosis which is usually found in severe or fatal cases associated with relatively high blood methanol levels (BENNETT et al., 1953; ROE, 1955; COOPER and KINI, 1962).

A third variable which has been associated with alcohol withdrawal signs and symptoms is the relative level of *serum magnesium*. Although low serum magnesium levels are correlated with the onset and severity of alcohol withdrawal states, severe alcohol withdrawal is not invariably associated with abnormal serum magnesium levels (MENDELSON et al., 1969). Serum magnesium levels may not reflect a significant deficit in total body magnesium. Poor food intake during prolonged drinking, combined with an enhanced urinary secretion of magnesium during rising blood alcohol levels, may produce relative states of magnesium deficiency in chronic alcoholics. The abrupt significant fall in serum magnesium levels which may accompany cessation of drinking, is associated with a transient decrease in other serum electrolytes, especially potassium, and coincides with the onset of neuromuscular hyperexcitability. The transient fall and subsequent rise in serum magnesium and in potassium probably reflect a shift in the extracellular-intracellular electrolyte pools associated with changes in acid base equilibrium. The abrupt fall in serum magnesium levels during early withdrawal may occur when there is no evidence of poor dietary intake or enhanced urinary excretion of magnesium during drinking. These findings appear to challenge a simplistic explanation of the genesis of alcohol withdrawal syndrome in terms of magnesium shifts alone (MENDELSON et al., 1969).

A fourth group of substances which have been related to alcohol dependence are the alkaloidlike compounds which may be formed during ethanol metabolism. Studies by COHEN and COLLINS (1970) and DAVIS and WALSH (1970) suggested that a biologically active alkaloid might be synthesized in vivo as a function of condensation of acetaldehyde, produced during ethanol catabolism, and endogenous catechol and biogenic amines. At the core of both of these theories is the observation that acetaldehyde may interact chemically with endogenously present catechol and biogenic amines to produce intermediary compounds which are further transposed to molecules which closely resemble biologically active opiates such as morphine. COHEN and COLLINS (1970) showed that tetrahydroisoquinoline alkaloids can be formed in adrenal tissue if optimal concentrations of acetaldehyde are present. DAVIS and WALSH (1970) suggested that endogenous formation of a dopamine-acetaldehyde-derived alcohol was theoretically possible in organisms which were metabolizing large quantities of ethanol. The work of these investigators was based upon studies carried out as early as 1967 by SMITH and GITLOW (1967) who demonstrated ethanol catabolism could induce a shift in the catabolism of catecholamines from oxidative to reductive pathways.

The notion that physical dependence upon a number of centrally acting drugs, such as ethanol and opiates, may occur via a common biochemical pathway is an attractive one. Unfortunately, this hypothesis is not consistent with both clinical and pharmacologic data and a detailed critique has been presented by SEEVERS (1970). First, there are several differences between the withdrawal syndromes observed following alcohol and opiate abuse (SEEVERS, 1970). In contrast to alcoholics, opiate addicts rarely show severe tremor, convulsions, fever, disorientation, or hallucinations. Opiate withdrawal is characterized by severe anxiety, abdominal pain, anorexia, diarrhea, lacrimation and nasal discharge, alternate chills and sweating, muscle spasm, and tachycardia (JAFFE, 1970). Moreover, there is no evidence of specific cross-dependence or cross-tolerance between morphine and alcohol (SEEVERS, 1970). Finally, SEEVERS (1970) argues that the DAVIS and WALSH (1970) hypothesis is untenable since morphine antagonists do not also antagonize alcohol intoxication or intoxication with drugs which are cross dependent with alcohol such as barbiturates and sedative hypnotics.

There are additional data which suggest that formation of an opiatelike compound cannot be explained as a function of acetaldehyde formation during ethanol catabolism. The removal of acetaldehyde which is produced by ethanol oxidation is extremely rapid (LUBIN and WESTERFELD, 1945) and no dose-response relationships have been found between alcohol and acetaldehyde (MAJCHROWICZ and MENDELSON, 1970, 1971a). These findings, considered in the context of the DAVIS and WALSH (1970) hypothesis, would suggest that social drinkers consuming small amounts of alcohol should be as susceptible to alcohol withdrawal signs as persons who consume very large quantities of alcohol. Thus, the kinetics of acetaldehyde metabolism, the clinical phenomenology of the alcohol and opiate withdrawal states, and the pharmacologic data on cross-tolerance and cross-dependence argue against the hypothesis that a biologically active alkaloid is the basis of physical dependence on ethanol. More recently, COHEN et al. (1972) have emphasized that biologically active alkaloids which may be produced during ethanol catabolism may act as false transmitters in the central nervous system. This hypothesis remains to be further explored in animal and human studies.

D. Theoretical Models of Withdrawal Syndromes

A number of models have been proposed to explain the appearance of withdrawal signs and symptoms following cessation of use of addictive agents, including ethanol. The most comprehensive theories attempt to account for both tolerance and physical dependence within a unitary construct, since there is good evidence that most centrally acting agents which have addictive potency also produce tolerance. Although tolerance may develop more rapidly than physical dependence, these are usually regarded as parallel processes (cf., pp. 616–622).

The degree of tolerance observed in addicts appears to be drug specific. For example, GRAHAM (1972) has noted that "... the dose of drug consumable with impunity by the fully dependent person increases approximately for alcohol × 2, for barbiturates × 4, for heroin × 20, for amphetamine × 100 over that consumable by the nontolerant person" (p. 83). It appears that the lower the tolerance factor, the more severe the withdrawal syndrome. For example, the psychomotor agitation observed in the most severe alcohol withdrawal syndrome, delirium tremens, is significantly greater than that seen in amphetamine withdrawal. The barbiturate withdrawal syndrome resembles the ethanol withdrawal syndrome more closely than the heroin abstinence syndrome (cf., FRASER et al., 1957; JAFFE, 1970).

Tolerance may be related to physical dependence insofar as tolerance limits the drug dosage an individual can self-administer and thereby affects the time over which physical dependence can develop. Alcohol has a low tolerance factor and alcohol dependence in man usually develops slowly over the course of several years (cf., pp. 627–631). Tolerance to heroin may develop very rapidly and man may become physically dependent on heroin within several months. The differences between rate of tolerance development for alcohol and barbiturates are probably due to metabolic factors. Barbiturate use results in a very rapid induction of drug catabolizing enzymes (JAFFE, 1970), whereas the metabolic changes which result in enhanced ethanol catabolism are relatively small (MENDELSON et al., 1965).

To date, no unified theory of tolerance and physical dependence has satisfactorily explained the relationship between these two phenomena, and moreover, no existing theory adequately accounts for physical dependence alone. There follows a description of three of the major theories of drug dependence—the denervation supersensitivity hypothesis, the receptor induction theory, and the enzyme de-repression theory.

I. The Disuse Supersensitivity Hypothesis

One of the earliest models suggested to explain the process of tolerance and physical dependence was derived from the work carried out by CANNON and ROSENBLUETH (1949) on autonomic effector processes following destruction of preganglionic cells and postganglionic axones. These investigators observed a decreased threshold in autonomic effectors following denervation and termed the phenomenon "denervation supersensitivity." Autonomic denervation supersensitivity has never been fully explained, although a number of mechanisms have been suggested, such as loss of inactivation of adrenergic transmitters via tissue uptake processes as a consequence of denervation (SHARPLESS, 1964).

JAFFE and SHARPLESS (1968) extended the denervation supersensitivity model to withdrawal phenomena observed following cessation of chronic use of centrally acting drugs. They pointed out that withdrawal signs often indicate an exaggeration of a physiologic response which is suppressed or inhibited by the drug which produces physical dependence. For example, administration of morphine to man produces miosis, but mydriasis is usually seen in the morphine abstinence syndrome.

A crucial notion embedded in the disuse supersensitivity model is that drugs which induce dependence have significant excitatory effects, but these excitatory effects are masked by other more prominent depressive actions of the drug. Both excitatory and inhibitory pathways regulate neural activity in the central nervous system. Inhibition of excitation could occur via a direct depressant effect of the drug on those same mechanisms which are responsible for the expression of excitation *or* through activation of inhibitory pathways in the central nervous system. As drug use continues, it is postulated that the degree of latent hyperexcitability increases, even though it is masked by the depressive effect of the agent. However, when the drug is withdrawn, the masked or latent hyperexcitability or hypersensitivity becomes prominent and withdrawal signs and symptoms occur. Moreover, as the degree of latent hypersensitivity increases during chronic drug use, it becomes necessary for more drug to be utilized in order to suppress such hypersensitivity and thereby achieve the depressive action of the drug. Thus, tolerance and physical dependence develop simultaneously and presumably the degree of tolerance should be correlated with the degree of physical dependence.

SEEVERS and DENEAU (1962) have stressed that there is no universal concordance between the development of tolerance and physical dependence, and those processes are not based upon a common neural mechanism. They have pointed out that amphetamine injection produces no significant physical dependence in man, yet significant tolerance for the drug does develop. However, their argument is not supported by clinical observations which suggest that although amphetamine users do not develop psychomotor signs of withdrawal following cessation of amphetamine use, they do develop significant behavioral problems which are consistent with a withdrawal state. In addition, WAY and his associates have reported data which suggest that there is a common neural basis for the development of tolerance and physical dependence for morphine. In studies with rodents, a close temporal continuity between development of and disappearance of physical dependence and tolerance for morphine has been observed (WAY et al., 1968). Similar studies have not been carried out in experimental animals with ethyl alcohol. In clinical studies with alcohol addicts, it is extremely difficult to establish any dose-time relationships between volume and frequency of ethanol intake, and propensity to develop the alcohol withdrawal syndrome (MELLO and MENDELSON, 1976; see also pages 627–631).

In general the disuse hypersensitivity model goes somewhat beyond a restatement of behavioral phenomenology in proposing a common mechanism for tolerance and physical dependence. The most novel aspect of the model is that it suggests that the *excitable* actions of drugs are as important as their depressant actions. However, differentiation of excitation versus depressant effects is extremely difficult in behavioral studies and is only in the very earliest stages of development in neurophysiologic and neurochemical investigations. Excitatory effects of alcohol on cerebral neurons were first reported by GRENELL (1959).

II. Receptor Induction

A theory of tolerance and physical dependence which is consistent with the denervation hypersensitivity model has been proposed by COLLIER (1965). He theorizes that drugs which can induce physical dependence may also have the property of inducing an increase in receptor sites for the drug in the brain. A necessary condition for this theory is the existence of specific receptor sites for neural cells which could bind with an addictive drug. Support for this notion has been advanced by the work of GOLDSTEIN and colleagues (1971) and PERT and SNYDER (1973). An opiate specific protein binding site in CNS tissue has recently been reported, and interpreted as direct evidence of an opiate receptor (PERT and SNYDER, 1973).

COLLIER's theory proposes that such receptors are not finite in number but may increase as a function of drug exposure. As the organism is exposed to more drug through time, more receptor sites evolve as a function of drug induction. Therefore, it is necessary for the organism to employ more drug in order to bind with the newly formed receptor sites, and this process would explain, in part, the phenomenon of tolerance. Upon cessation of drug use, a large number of receptor sites would be present which would not be occupied by a drug molecule, and COLLIER suggests that the presence of a large number of such unoccupied receptor sites would be associated with an excitation process which would be expressed behaviorally as a withdrawal syndrome. Thus, the increased number of receptor sites following drug removal would provide the basis for physical dependence. COLLIER speculated that receptor site processes associated with 5-hydroxytryptamine, acetylcholine, and central catecholamines might be related to a drug-induced receptor system (COLLIER, 1965).

III. Enzyme De-Repression

Another theoretical construct for explaining tolerance and physical dependence has been advanced by SHUSTER (1961) and by GOLDSTEIN and GOLDSTEIN (1968). These theories are based upon developments in biochemistry which have occurred during the past decade on repression and de-repression of enzyme biosynthesis. A schematic depiction of this theory is provided in Figure 8. In Figure 8, Substance C, which might be a central catechol or biogenic amine, is synthesized from a series of precursors (B) which are available in the neural tissue or blood. The synthesis of C from B is catalyzed by the enzyme E. A centrally acting drug, such as ethanol or morphine, may inhibit the enzyme E, thus decreasing the net rate of synthesis of the excitatory substance C from the precursors B. However, it is also proposed that the rate of

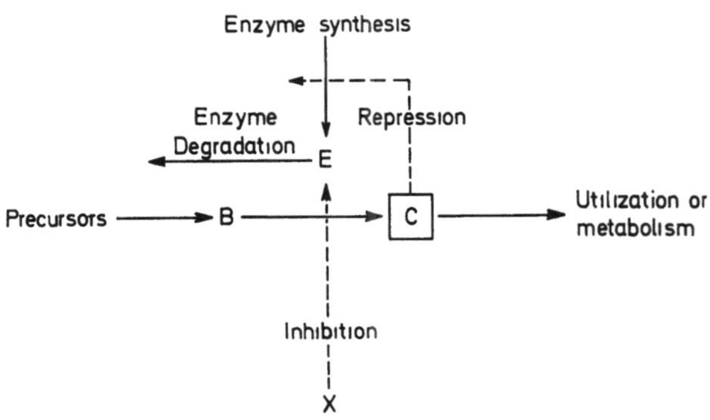

Fig. 8. Schematic representation of enzyme repression theory of tolerance and physical dependence. (From GOLDSTEIN et al., 1968.)

enzyme synthesis is controlled by the concentration of the excitatory substance, C. Drug inhibition of E would result in an increased synthesis of the enzyme because the concentration of C would be diminished. As more enzyme was synthesized, more drug would be required to inhibit the enzyme. This process would be associated with an increased degree of tolerance by the organism.

As the organism became increasingly tolerant, more and more drug would be necessary to produce an inhibition of enzyme E through time. However, when the drug was withdrawn, a large quantity of enzyme E would be present, and given an adequate concentration of precursors, a large amount of excitatory substance C would be synthesized. This process would be associated with the withdrawal state as a consequence of an overabundance of the excitatory substance C. However, the duration of the withdrawal syndrome would be limited as a function of the time necessary for the increased concentration of substance C to repress the synthesis of enzyme E. Following such repression after cessation of drug use, both tolerance and physical dependence would disappear. As with most other theories, the specific entities involved, particularly the enzyme and the excitatory substance, remain a matter of conjecture.

IV. Conclusions

A number of other theories have been proposed which incorporate variants of the notion that drugs which have addictive potency also have excitatory as well as depressant activities. Some theories stress the role of the specific properties of the drug molecule while others emphasize the nature of receptor mechanisms which interact with such drugs. An example of the latter is the concept of pharmacologic redundancy as proposed by MARTIN (1970). Other theories have highlighted the importance of long-term immunologic responses in the role of tolerance and physical dependency such as that described by COCHIN and KORNETSKY (1964). Still others have stressed the importance of learned behaviors in the expression of the withdrawal state (SCHUSTER, 1970). All of these conceptualizations are useful for developing hypotheses which can be tested in laboratory research, but none to date has provided a model, sufficiently comprehensive to explain the neural mechanisms involved in physical dependence upon alcohol (MENDELSON, 1971).

KALANT (1973) has recently presented a concise and comprehensive review of biological models of alcohol tolerance and physical dependence. He has categorized a series of known properties of alcohol tolerance and physical dependence which should be explained by an adequate model and concludes that none of the models which have been devised to account for alcohol dependence or dependence upon other drugs is satisfactory (KALANT et al., 1971) . The lack of an adequate model is unquestionably due to our limited state of knowledge of the effects of alcohol on neurochemical and neurophysiologic processes and the basic neural mechanisms subserving behavior. In particular, much more information is necessary concerning complex factors such as the physiochemical properties of neural membranes and the manner in which they control and program the logic of central nervous system information processes before an understanding of drug effects on the brain can be synthesized into a valid model.

E. Experimental Animal Models of Alcohol Withdrawal [2]

The limitations on experiments that can be performed on human subjects have necessarily restricted progress in examining the central nervous system mechanisms involved in the phenomena of alcohol addiction. The importance of central nervous system alterations in alcoholism have been illustrated by behavioral tolerance for alcohol, which cannot be explained by metabolic factors, and by physical dependence upon alcohol, as described on pages 616–618 and 622–627. Until recently, research on the biological basis of alcohol addiction has been hampered by the lack of adequate experimental animal models of alcoholism. For many years, investigators have concentrated upon developing techniques for inducing *preference* for alcohol in animals. There has been an unfortunate tendency to equate a transitory alcohol preference with alcohol addiction, even though no withdrawal signs and symptoms occurred upon cessation of drinking. Removal of the factors which accelerated alcohol preference, e.g., noxious stimuli, is usually accompanied by a decrease in alcohol intake (cf., MARDONES, 1960; LESTER, 1966, LESTER and FREED, 1973; MELLO, 1968, 1973b; MYERS and VEALE, 1972; WOODS and WINGER, 1971 for reviews).

Recently, several groups of investigators have succeeded in devising techniques to induce physical dependence upon alcohol in animals using oral, intragastric, intravenous, and inhalation routes of administration. Representative studies are summarized in Table 2. Both forced administration and self-selection procedures have been effective in producing alcohol abstinence signs in several species.

Development of criteria for evaluating the adequacy and potential applicability of the various techniques for inducing physical dependence upon ethanol has proceeded rather slowly. It is obvious that demonstration of *physical dependence* per se must be unequivocally attributed to the administration and removal of alcohol and not to the confounding effects of nutritional deficiency or toxic drug effects (cf., OGATA et al., 1972). Evidence concerning the induction of behavioral and pharmacologic *tolerance* should be obtained whenever possible. It has also been argued that levels of ethanol intake should be sufficient to produce *intoxication* as indicated both by behavioral measures and blood alcohol levels (WOODS and WINGER, 1971). Moreover, an adequate model of addiction should show that alcohol functions as a *reinforcer*, i.e., the rate of operant responding for alcohol increases and is sustained (WOODS et al., 1971; WOODS and WINGER, 1971). Addictive drug-seeking behavior is presumably maintained by its consequences and it is the identification and analysis of the complex of drug-related reinforcers which present a major challenge to the behavioral pharmacologist.

The rationale for applying behavioral techniques to induce addictive drinking derives from the argument that a situation in which an animal drinks alcohol as a form of *motivated* behavior in order to avoid pain or to obtain a reward is most analogous to the human condition. Operant conditioning techniques permit evaluation of the efficacy of positive and aversive stimulus control in maintaining alcohol consumption. Moreover, operant techniques can require "drinking" as the contingent response.

[2] Portions of this section have appeared previously in MELLO, 1973b.

Implicit in these comments is the notion that voluntary selfselection techniques for inducing alcohol addiction may eventually prove more valuable than forced administration techniques. Although the rapid induction of physical dependence by intravenous or nasogastric alcohol administration provides an important tool for assessing the end product of addiction, the analysis of developmental correlates of addiction may be obscured by an accelerated induction time. Only a behavioral technique designed to produce alcohol self-administration would permit examination of those factors which contribute to the development of the addictive process as a function of time. Ideally, a behavioral method should permit the identification and subsequent manipulation of the environmental determinants that affect the acquisition and maintenance of addictive drinking. The effects of a variety of pharmacologic interventions could then be studied.

The following description of animal models of alcohol will be organized in terms of routes of administration of alcohol, i.e., oral and intragastric administration, intravenous administration, and administration via an inhalation technique—since these determine the rapidity with which an effective alcohol dose is achieved. Moreover, the route of administration influences the duration and stability of the drug effect.

I. Oral and Intragastric Administration

FALK and coworkers have recently reported the successful application of a behavioral technique, *schedule-induced polydipsia*, in producing physical dependence in the rat (FALK et al., 1972). Schedule-induced polydipsia refers to a situation in which rats will consume as much as $^1/_2$ of their body weight in water within a few hours, when dry food is presented intermittently (FALK, 1961). In this procedure, rats reduced to 80% of their free feeding weight were exposed to a schedule in which a food pellet was delivered automatically, every 2 min, during 1-h sessions. There were six 1-h sessions in each 24-h day. Following establishment of polydipsic drinking with water, an ethanol concentration of 1% was introduced and increased in 1% increments every 6–8 days until 5 or 6% solutions were reached. This procedure resulted in an alcohol intake of between 11 and 15 g/ETOH/kg/day. Blood alcohol levels were maintained above 100 mg/100 ml and usually ranged between 150 and 300 mg/100 ml. Following 3-months exposure to this polydipsia schedule, alcohol was abruptly withdrawn and animals consistently showed evidence of physical dependence including hyperactivity, tremor, spasticity, and audiogenically induced tonic-clonic seizures (FALK et al., 1972).

The polydipsia paradigm for inducing physical dependence upon alcohol in the rat as reported by FALK et al. (1972) meets the evaluative criteria previously described insofar as animals ingested more alcohol through time, showed evidence of intoxication, and evidence of physical dependence upon alcohol withdrawal. The polydipsia technique has the particular advantage of maintaining high levels of alcohol ingestion in the presence of adequate food intake.

LESTER (1961) was the first to show that schedule-induced polydipsia was an effective procedure for inducing alcohol intoxication in rats, however, he did not report evidence of physical dependence. High levels of alcohol intake in rats have been reported by several investigators using variations on a polydipsia technique,

Table 2 Methods used to induce physical dependence on alcohol in animals

Investigator and year	Procedures				Results				
	Technique	Species	Dose	Days of exposure	Other variables	Physical depend.	Enhanced metabolic rate	Intoxication (bal range)	ETOH as a reinf
FALK et al., 1972	Polydipsia	Rat	13 g/kg/day	90		Yes[a]	Yes	100–300 mg/100 ml	Yes
LESTER, 1961	Polydipsia	Rat		2		No[a]	No data	100–200 mg/100 ml	Yes
OGATA et al , 1972	Polydipsia	Mouse	14–24 mg/g/day	7–14		No[a]	No data	73–279 mg/100 ml	Yes
MELLO and MENDELSON, 1971a	Polydipsia	Rhesus monkey	3 g/kg/day	90–195		No[a]	No data	50 mg/100 ml	No
WOODS and WINGER, 1971	Polyidpsia	Rhesus monkey	4–7 g/kg/day	No data		No[a]	No data	+200 mg/100 ml	Yes
FREUND, 1969	ETOH-liquid diet	Mouse	0.51 ml abs. ETOH	4	35% Weight reduction	Yes[a]	No data	100–650 mg/100 ml	Yes
OGATA et al , 1972	ETOH-liquid diet	Mouse	15–18 mg/g	4	35% Weight reduction	Yes[a]	No data	178–499 mg/100 ml	Yes
BRANCHEY et al , 1971	ETOH-liquid diet	Rat	4 26–4.33 ml abs. ETOH	4–21	33% Weight reduction	Yes[a]	No data	No data	No
LIEBER and DECARLI, 1973	ETOH-liquid diet	Rat	14–16 g/day	10–35		Yes[a]	Yes	No data	NA
PIEPER et al., 1972	ETOH-liquid diet	Chimpanzee	2–8 g/kg/day	42–70		Yes[a]	Yes	50–500 mg/100 ml	Yes
PIEPER and SKEEN, 1972	ETOH-liquid diet	Rhesus monkey	5–7 g/kg/day	40		Yes[a]	Yes	+150 mg/100 ml	No
RUBIN and LIEBER, 1973, 1974	ETOH-liquid diet	Baboon	4 5–8.3 g/kg/day	9 months–4 years		Yes[a]	No data	No data	NA
MENDELSON and MELLO, 1973	ETOH-sole fluid	Infant rhesus monkey	4–10 g/kg/day	730	Peer and maternal deprivation	No[a]	No	50–200 mg/100 ml	No
CICERO et al., 1971	ETOH-sole fluid	Infant rats	5 6–7 3 g/kg	133		Yes[a]	Yes	No data	NA

MAJCHROWICZ, 1973	Intubation	Rat	9–14 g/kg/day	5–7		Yes[a]	Yes	+600 mg/100 ml	NA
ELLIS and PICK, 1969, 1970b	N.G intubation	Rhesus monkey	4–8 g/kg/day	10–18		Yes[a]	Yes	100–500 mg/100 ml	NA
ELLIS and PICK, 1970a	N G intubation	Beagle dog	3–7 g/kg/day	14–56		Yes[a]	No data	100–500 mg/100 ml	NA
ESSIG and LAM, 1968	Intragas infus	Beagle dog	4–4 5 ml/kg/day (40%)	40–54		Yes[a]	No data	No data	NA
YANAGITA et al , 1969	Intragas. self-infus.	Rhesus monkey	3.2–7.5 g/kg/day	35		Yes[a]	No data	No data	Yes
GOLDSTEIN and PAL, 1971	Inhalation ETOH vapor	Mouse	11 mg/l air	1–3	Pyrazole	Yes[a]	No data	147–186 mg/100 ml	NA
GOLDSTEIN, 1972a				1–3	(1 0 mmol/kg/day)				
FRENCH and MORRIS, 1972	Inhalation ETOH vapor	Rat	1.4 mg/l air	14		Yes[a]	No data	Not detectable	NA
MELLO and MENDELSON, 1971b	Drinking to avoid shock	Rhesus monkey	2.5 g/kg/day	70–700		No[a]	No data	30–70 mg/100 ml	No
DENEAU et al., 1969	Intravenous self-administration	Rhesus monkey	8 6 g/kg/day	120+		Yes[a]	No data	No data	Yes
WOODS et al., 1971	Intravenous self-administration	Rhesus monkey	6–8 g/kg/day	90–360		Yes[a]	No	300–400 mg/100 ml	Yes
MCQUARRIE and FINGL, 1958	Gavage	Mouse	5.4 g/kg/day	14	Electroconvulsive seizure threshold	Yes[b]	No data	No data	No
GIBBINS et al., 1971	Intubation and ETOH-sole fluid	Rat	3–7 g/kg/day	37	startle threshold to electric shock	Yes[b]	No data	104–200 mg/100 ml	No
RATCLIFFE, 1972	ETOH-sole fluid	Rat	No data	35–49	audiogenic and drug induced withdrawal seizures	Yes[b]	No data	No data	No

[a] Assessed by observation. [b] Assessed by a threshold measure NA = not applicable. (Reprinted from MELLO, 1973b).

but none has produced physical dependence (HOLMAN and MYERS, 1968; MEISCH and PICKENS, 1968; MEISCH, 1969).

The use of schedule-induced polydipsia procedures has not been effective in producing alcohol addiction in mouse (OGATA et al., 1972). Although mice consumed alcohol in quantities comparable to that of addicted rats (15 mg/g/day), the period of exposure to these conditions (7–14 days) was considerably shorter than that used in rat (cf., FALK et al., 1972; OGATA et al., 1972).

Efforts to develop behavioral methods to induce addictive drinking in rhesus monkey have also been unsuccessful. The major impediment appears to be the aversive taste of alcohol which is difficult to mask effectively. A second factor is the inevitable delay between alcohol ingestion and intoxication. It requires about 2 h for a 25% alcohol solution, administered via a nasogastric tube, to reach peak absorption as indicated by enzymatic blood alcohol determinations (MELLO, 1971). The rate of ethanol metabolism in rhesus monkey is similar to the metabolic rate in man, i.e., about 25–30 mg/100 ml per hour (cf., ELLIS and PICK, 1969; MENDELSON, 1968b; WESTERFELD and SCHULMAN, 1959; WOODS et al., 1971). Consequently, if alcohol consumption is only associated with an aversive taste, and never with intoxication, monkeys might never discover the potential reinforcing properties of oral alcohol ingestion. In contrast, the immediacy of the effects obtained from intravenous alcohol administration is undoubtedly a critical factor in maintaining self-administration.

Monkeys exposed to a polydipsia procedure during single daily 3-h sessions only consumed between 2.5 and 5 g/kg/day and showed no signs of gross intoxication. There was no evidence of physical dependence following approximately 10 months of continuous access to gradually increasing concentrations of alcohol (5–15%) (MELLO and MENDELSON, 1971a). The potential utility of a single session polydipsia paradigm is greatly limited, since maximal levels of alcohol intake were induced for only 3 h each day and alcohol ingestion during the remaining 21 h was optional. It appears that maintenance of a consistent, high blood alcohol level is important for the induction of physical dependence. It has been reported that induction of blood alcohol levels of 400 mg/100 ml for only 3 h each day, using an intravenous self-selection paradigm, is not sufficient to produce physical dependence (WINGER et al., 1970). Production of a frequent high alcohol intake with a behavioral technique involving food reinforcement is also restricted by rate of food satiation.

The use of relatively low alcohol concentrations (2.5 w/v) in a polydipsia paradigm has been more effective in inducing severe intoxication in the monkey (WOODS and WINGER, 1971). Total volumes of alcohol consumption exceeding 1000 ml over a 24-h period were not unusual in monkeys with a prior history of intravenous alcohol administration and alcohol doses as high as 7.1 g/kg were self-administered (WOODS et al., 1971; WOODS and WINGER, 1971). It was found that monkeys with an intravenous alcohol administration history drank more alcohol in a 2.5–4-h polydipsia paradigm than naive controls during the initial 3–4 weeks of exposure to this schedule. Subsequently, the naive controls reached about the same level of intake as the alcohol-experienced animals. About 5 out of 6 animals showed signs of intoxication and blood alcohol levels ranged between 150 and 200 mg/100 ml (WOODS, personal communication). Signs of physical dependence were not observed in these monkeys under these conditions.

The optimal parameters of schedule-induced ethanol consumption to produce self-intoxication remain to be determined. These data suggest that a procedure involving multiple polydipsia sessions in which large volumes of low alcohol concentrations are consumed might prove most effective. However, it could also be argued that monkeys would have to consume very large volumes of a low concentration alcohol solution before intoxication levels would be reached.

In a paradigm which involved using a consummatory (lick) response as the operant response, rhesus monkeys were trained to drink in order to avoid a noxious shock (MELLO and MENDELSON, 1971 b). Both bourbon and ethanol solutions were presented in concentrations ranging between 5 and 25%. A 6-h Sidman avoidance period (SIDMAN, 1953) was alternated with a 6-h rest period during which no shocks occurred. Experiments were run 24 h a day, 7 days a week. Each monkey learned to drink to avoid shock at a rate sufficient to avoid virtually all possible shocks. However, the amount of fluid consumed did not remain stable across concentrations but decreased linearly as a function of increasing alcohol concentrations even though the rate of response was the same. It was found that monkeys had learned a dual avoidance response in which it was possible to postpone the occurrence of a noxious shock by making a lick response, and to avoid consuming an aversive fluid by modulating the duration of the lick response. As the ethanol concentration was increased, the mode of the lick duration distributions was shifted toward shorter lick durations which presumably resulted in smaller amounts of fluid dispensed per lick.

The apparatus was then modified so that only discrete licks of a specified duration were effective in postponing shock. Monkeys were run for 60 days on a 10% alcohol solution and lick duration requirements were increased in 50 ms increments. Each monkey's lick duration shifted toward longer durations in accordance with the programmed lick duration requirement. However, despite the increase in lick durations, the volume of alcohol consumed did not increase. Monkeys drank about 2.5 g/kg per day and blood alcohol levels ranged between 30 and 70 mg/100 ml. No monkey showed evidence of intoxication or of physical dependence upon removal of alcohol. These data further testify to the monkey's dislike of alcohol and suggest that monkeys learned to control the amount of fluid dispensed by manipulating the displacement of the ball valve (MELLO and MENDELSON, 1971 b). Studies in which *both* lick duration and lick displacement were specified as part of the avoidance schedule have yielded comparable data.

A more consistently effective approach to inducing alcohol dependence has been an oral forced administration procedure which employs a nutritionally adequate *liquid diet* combined with alcohol. The liquid diet, in which some percentage of calories are replaced by ethanol, is the only nutrient available to the animal. This procedure has been used to induce alcohol dependence in mouse (FREUND, 1969; OGATA et al., 1972), rat (BRANCHEY et al., 1971; LIEBER and DECARLI, 1973), rhesus monkey (PIEPER and SKEEN, 1972), chimpanzee (PIEPER et al., 1972), and baboon (RUBIN and LIEBER, 1973, 1974).

The first application of this procedure in rodent was described by FREUND in 1969. Food-depleted mice, given a *liquid diet* in which 35% of the calories came from alcohol, showed gross intoxication and evidence of physical dependence within 4 days (FREUND, 1969). The drastic (65% ad libitum) weight reduction of these mice induces a higher effective dosage of alcohol because of the concomitant reduction in the rate of alcohol metabolism. It has been shown in man and in animals that the rate of alcohol metabolism is reduced by as much as 50% in fasted, as contrasted to well-fed organisms (LEBRETON, 1936; SMITH and NEWMAN, 1959; FORSANDER et al., 1965; OWENS and MARSHALL, 1955; LELOIR and MUNOZ, 1938; MENDELSON, 1970). FREUND's basic observations have been replicated (OGATA et al., 1972). However, the potentially confounding effects of severe weight reduction in the FREUND mouse model of physical dependence have introduced some serious questions about the adequacy of that model (cf., OGATA et al., 1972). During the 4 days of alcohol administration, the mice, already reduced to 65% of their free-feeding weight, lost an additional 10% of their weight.

In order to determine if *equivalent* alcohol consumption with an adequate diet would be effective in producing physical dependence, a naive group of mice were exposed to a variation of the polydipsia procedure first described by FALK (1961) (OGATA et al., 1972). This procedure was effective in inducing the consumption of large quantities of fluid each day, which approached 40% of total body weight. Alcohol intake in the polydipsia paradigm was between 0.58 and 0.71 ml of absolute alcohol or 14–24 mg/g. This volume exceeded that observed with the replication of the FREUND procedure, i.e., 0.45–0.46 ml of absolute alcohol or 15–18 mg/g. Average blood alcohol levels in the polydipsia procedure ranged between 73 and 279 mg/100 ml, a level significantly lower than that observed in the nutritionally deprived FREUND replication animals (178–499 mg/100 ml). Despite sustained high alcohol intake for periods of 7–14 days, no polydipsia mouse showed evidence of physical dependence during a 10-h observation period after the substitution of water for alcohol. Consequently, consumption of alcohol in amounts exceeding 15 mg/g or 0.50 ml of absolute alcohol per day for prolonged periods of time is not a necessary and sufficient condition to produce withdrawal signs in mice upon removal of alcohol (OGATA et al., 1972).

Forced administration of a liquid diet combined with alcohol has proven very useful in chronic studies with primates. RUBIN and LIEBER (1973, 1974) have developed the first convincing animal model of alcohol-induced cirrhosis using this procedure. Thirteen baboons were maintained on a nutritious liquid diet in which alcohol accounted for 50% of the total calories in the experimental group, yielding an alcohol dose ranging between 4.8 and 8.3 g/kg per day. Over the course of 9 months–4 years, the entire range of alcohol-associated liver injuries were produced. All baboons developed fatty liver; 4 baboons developed alcoholic hepatitis, and 2 baboons developed alcoholic cirrhosis. Cirrhosis of the liver is one of the most serious medical complications of alcoholism, and development of an animal model should help to clarify the nature of this potentially lethal disorder. One significant implication of these data is the demonstration that adequate nutrition is not a protection against alcohol-induced liver damage. Moreover alcohol, rather than any congeners or impurities, appears to be responsible for the development of cirrhosis.

Rhesus monkeys and young chimpanzees have also been addicted to alcohol using a liquid diet (PIEPER and SKEEN, 1972; PIEPER et al., 1972). One to 7-month-old chimpanzees (Pan-Troglodytes) were given a liquid diet with 45% of the calories from ethanol, 4–5 times daily at standard feeding times. Chimpanzees maintained normal weight gains and consumed alcohol in doses from 2 g/kg to 8 g/kg. Blood alcohol levels ranged between 50 and 300 mg/100 ml with peaks as high as 500 mg/100 ml depending on the concentration of alcohol in the liquid diet. During the alcohol administration period, mild tremulousness, hyperreflexia, and irritability were observed prior to the morning feeding (a 9-h abstinent interval) when blood alcohol levels fell below 100–150 mg/100 ml. After 6–10 weeks on this regimen, alcohol was abruptly withdrawn and withdrawal symptoms observed included hyperreflexia, irritability, photophobia, rapid respiration, sweaty palms and feet, spastic rigidity, and, in some instances, tonic and clonic convulsions resulting in death in 1 animal. An induced increase in the rate of ethanol metabolism, comparable to that observed in man (MENDELSON et al., 1965), was also found in chimpanzees. Liver biopsies indicated reversible fatty infiltration comparable to changes observed in man following ethanol ingestion (PIEPER et al., 1972).

Oral self-administration of alcohol solutions can also be achieved by restricting the animals' total fluid intake to alcohol. This procedure has not been effective in mature rats (MENDELSON and MELLO, 1964). Young rats maintained on a 7% alcohol solution from weaning at 21 days to 154 days of age developed alcohol tolerance and signs of physical dependence (CICERO et al., 1971). Moreover, rats subsequently showed preference for alcohol, even when given a choice between alcohol and a saccharin solution. CICERO et al. interpret these data to suggest that exposure to alcohol during an early and critical stage in CNS development may be a critical factor in inducing tolerance and physical dependence. This interpretation is logically appealing but empirically, the effect may not generalize to all species. Infant rhesus monkeys did not develop physical dependence after prolonged exposure to ethanol as the only fluid in an otherwise normal diet (MENDELSON and MELLO, 1973). In order to examine the addictive potential of early experience with alcohol, 4 infant monkeys were provided with alcohol continuously from within 2 h of birth. Newborn monkeys were immediately separated from their mothers and subsequently maintained under conditions of maternal and peer deprivation. Four monkeys were given increasing concentrations of alcohol as their only fluid source and 4 control monkeys were maintained under identical conditions and given a free choice between comparable concentrations of alcohol and water. During the first 3 months of life, the experimental monkeys consumed between 2 and 5 g/kg of ethanol with blood alcohol levels ranging between 50 and 80 mg/100 ml. During the next 6 months, alcohol intake increased and varied between 4 and 10 g/kg and blood alcohol levels of 200 mg/100 ml were occasionally observed. This pattern was sustained for the remainder of the first year and stabilized during the second year with alcohol consumption of between 5 and 8 g/kg. No monkey showed evidence of physical dependence when withdrawn from alcohol. Subsequent assessment of blood alcohol disappearance curves after acute nasogastric intubation of 2 g/kg of 25% ethanol showed no significant differences from control monkeys, raised under identical conditions with access to water. Examination of alcohol selection in comparison to water showed that no monkey had developed a preference for alcohol (MENDELSON and MELLO, 1973).

Forced alcohol administration procedures have proved to be consistently effective in producing alcohol addiction in monkeys (ELLIS and PICK, 1969, 1971) and in dogs (ESSIG and LAM, 1968, 1971; ELLIS and PICK, 1970a). ELLIS and PICK (1969) were first to report that *nasogastric intubation* of alcohol (25%) in 2 or 3 divided doses of 4–8 g/kg is effective in producing physical dependence upon alcohol in rhesus monkeys within 10–18 days. These monkeys showed tremor, spasticity, hyperreflexia, mydriasis, and clonic-tonic convulsions which could be suppressed with ethanol. Monkeys also exhibited an alcohol-induced increase in metabolic rate (ELLIS and PICK, 1969, 1971).

Gastric intubation of alcohol, in a range of 9–14 g/kg over a 5–7 day period, has also been shown to produce physical dependence in rats. Alcohol was administered every 6 h to ensure consistent high blood alcohol levels. The reliability of the procedure, in terms of the total number of animals showing positive signs of physical dependence, is greatest when alcohol administration is maintained for at least 7 days (MAJCHROWICZ, 1973).

ESSIG and LAM (1968) were the first to report alcohol dependence in the dog following prolonged administration via a surgically implanted gastric cannula. The 8 out of 12 dogs that survived the alcohol administration regimen did show definite signs of physical dependence including tremulousness, tonic extension of the extremities, and seizures. Some animals who did not exhibit convulsions appeared to be attending to nonexistent visual stimuli, described by the authors as hallucinations, and they reported a disruption of sleep during the withdrawal period (ESSIG and LAM, 1968, 1971). These animals lost an average of 1.6 kg during the total period of intoxication.

The effect of intragastric administration of alcohol in monkeys has recently been examined by YANAGITA and colleagues (1969). An intragastric catheter was implanted via a nasogastric route. All monkeys were maintained for 5 or 6 weeks using either voluntary or forced ingestion of alcohol. The reinforcing properties of alcohol were apparently dependent upon the prior history of reinforcement as has also been reported with intravenous alcohol administration (WOODS et al., 1971). Naive monkeys did not initiate self-administration and programmed alcohol was presented at 1.0 or 2.0 g/kg/infusion every 3 h or every 6 h for 5 or 6 weeks. After only 5 days of intragastric administration of 2.8 g/kg of alcohol every 3 h, mild abstinence signs were observed. However, a brief withdrawal period did not result in increases in lever-pressing behavior for alcohol.

Three monkeys that were familiar with intravenous self-administration procedures administered intragastric doses of alcohol ranging from 2.8–7.5 g/kg, and showed severe abstinence syndromes. It appears that the critical distinguishing variable is experience with drug self-administration rather than experience with alcohol since 2 naive monkeys failed to initiate self-administration even *after* programmed administration of alcohol at a dose level of 8.0 g/kg/day for periods of 2 months (YANAGITA et al., 1969). Comparable data have been reported for initiation of intravenous self-selection (DENEAU et al., 1969).

II. Intravenous Alcohol Administration

YANAGITA et al. (1965) were the first to report the successful addiction of monkeys to alcohol, using a paradigm in which a monkey could lever press to activate an automatic intravenous injection apparatus. Their original observations have been subsequently confirmed and extended (DENEAU et al., 1969; WINGER et al., 1970; WOODS et al., 1971; WOODS and WINGER, 1971; WINGER and WOODS, 1973). Consistent delivery of 6–8 g/kg/day for 10 weeks or more appears to be sufficient to produce signs of physical dependence. Upon discontinuation of alcohol, an abstinence syndrome appeared within 6 h characterized by tremor, vomiting, "hallucinatory" behavior, and convulsions (DENEAU et al., 1969).

The initiation of intravenous administration of alcohol does not occur as consistently with alcohol as with cocaine and opiate narcotics. In one study, only 11 of 27 monkeys initiated responding for alcohol when exposed for 10 or more days.

Consequently, it was necessary to elicit high response rates with the use of another consistently reinforcing pharmacologic agent, i.e., cocaine (WOODS et al., 1971). This procedure proved to be effective in establishing a high response rate which then persisted once ethanol was substituted for the initial drug. Both cocaine and pentobarbital have been used to establish ethanol reinforced self-administration. The factors which make ethanol a less potent reinforcer than these other psychotropic agents are unclear.

Once self-administration is established, intravenous alcohol is a reinforcer and monkeys will self-administer to the point of intoxication. The degree of intoxication produced with this technique results in severe motor incoordination, stupor, and occasionally light anesthesia (DENEAU et al., 1969). During the course of ethanol self-administration, monkeys show loss of weight, minimal food intake and general ill health associated with malnutrition (DENEAU et al., 1969).

Within a dose range of 0.05–0.20 g/kg/injection, the overall intake within a 3-h period appears to be independent of the dose delivered with each injection. Moreover, if a presession infusion of alcohol is introduced and increased from 1–3 g/kg, response frequency for ethanol reinforcement is reduced correspondingly (WOODS et al., 1971). Over a period of 6 months, pharmacologic tolerance as reflected in changed elimination rates of alcohol has not been observed (WOODS et al., 1971).

There is a striking similarity between patterns of intravenous self-selection of alcohol by the rhesus monkey and spontaneous drinking patterns observed in human alcohol addicts (cf., WOODS et al., 1971; DENEAU et al., 1969; MELLO and MENDELSON, 1972; NATHAN et al., 1970, 1971). Human alcohol addicts, given an opportunity to work for alcohol at a simple operant task, consistently alternate drinking episodes of 3–6 days with relatively abstinent work periods of 2–3 days. These abstinent periods are usually associated with partial withdrawal signs and symptoms (MELLO and MENDELSON, 1972; NATHAN et al., 1970, 1971). Alcohol self-administration in monkeys is also punctuated by periods of spontaneous abstinence which are associated with withdrawal signs (DENEAU et al., 1969; WOODS et al., 1971).

The factors which determine these episodes of self-imposed abstinence are difficult to specify. Self-imposed abstinence occurs only when monkeys are given 24-h access to alcohol and never when ethanol access is limited to 3 h per day (WOODS et al., 1971). However, only 3 h a day of intoxication (400 mg/100 ml) is not sufficient to produce physical dependence (WINGER et al., 1970). Termination of self-administration has not been consistently associated with any particular number of days or amount of alcohol taken by monkeys on a 24-h access paradigm.

The cyclicity of alcohol self-administration, both in animal and human addicts, raises the question of the relationship of physical dependence to drug selection. Since responding for alcohol under conditions in which physical dependence cannot be produced (i.e., the 3-h access paradigm) is more consistent than in the 24-h access paradigm (WINGER et al., 1970; WINGER and WOODS, 1973), it appears that the momentary reinforcing effects of alcohol are most important in the maintenance of responding. The question of the extent to which forced administration techniques for producing physical dependence on alcohol result in subsequent motivated drinking behavior has not been examined. It is not known whether an animal thus addicted will elect to consume large quantities of alcohol once the initial induction of physical dependence and recovery from an abstinence episode is past. There have been no studies of oral alcohol consumption in rhesus monkeys during alcohol withdrawal. Information is not available concerning the persistence of alcohol dependence. The ambiguous relationship of physical dependence to subsequent drug selection has been acknowledged repeatedly by behavioral pharmacologists, but, as yet, there has been no empirical resolution of these issues (MELLO and MENDELSON, 1971a, b, 1976; MELLO, 1972, 1973b; WOODS et al., 1971; WOODS and WINGER, 1971, WINGER and WOODS, 1973).

The extent to which the reinforcing properties of a drug may be enhanced following withdrawal from the drug have not been established. It has repeatedly been shown that monkeys will self administer i.v. doses of alcohol and opiates at levels below that required to produce physical dependence (WOODS et al., 1971; WOODS and SCHUSTER, 1970). However, in the narcotic addicted monkey, it appears that physical dependence does maintain responding, presumably to avoid withdrawal signs (WOODS and SCHUSTER, 1970; DENEAU et al., 1969).

III. Alcohol Administration via Inhalation

In 1971, GOLDSTEIN and PAL reported that alcohol dependence could be produced by exposing mice to a situation in which they inhaled alcohol vapor at concentrations between 10 and 16 mg/liter. Maintenance of blood alcohol levels of 180 mg/100 ml for 3 days was found sufficient to produce withdrawal signs of tremor, convulsions on handling, tail lift, startle reaction to noise, and spontaneous seizures (GOLDSTEIN and PAL, 1971).

In order to ensure stable high blood alcohol levels, the alcohol inhalation technique was combined with daily doses of pyrazole (1.0 mmol/kg I.P.). Pyrazole reduces the rate of alcohol metabolism (LESTER et al., 1968; GOLDBERG and RYDBERG, 1969) even more effectively than starvation, i.e., by an estimated 70% (GOLDSTEIN and PAL, 1971). Since pyrazole has toxic side effects on hepatic function (LIEBER et al., 1970), evaluation of the adequacy of this technique for inducing a true model of physical dependence is complicated. Mice exposed to alcohol vapor alone do show withdrawal signs and symptoms but the fluctuations in blood alcohol concentrations achieved without pyrazole seemed to contraindicate its omission to the investigators (GOLDSTEIN and PAL, 1971; GOLDSTEIN, 1972a). All animals exposed to the inhalation technique lost weight rapidly and those mice treated with pyrazole and maintained at moderate blood alcohol levels (140 mg/100 ml) lost as much as 17% in 6 days (GOLDSTEIN, 1972a). Moreover, in mice exposed only to alcohol, coma and death occurred at blood alcohol levels between 200–300 mg/100 ml; a level which is rarely lethal in mice (cf., OGATA et al., 1972; SMITH and HAYASHIDA, 1970; FREUND, 1969). Solubilization of ethanol exerts osmolar pressure which leads to a relative dehydration. Control studies to examine the possible role of the dehydrating effect of high levels of alcohol vapor on nasal mucosa with attendant susceptibility to infection have not been reported.

FRENCH and MORRIS (1972) have reported that low concentrations of ethanol vapor (1.4 mg of alcohol per liter of air), insufficient to produce detectable blood ethanol levels, also produce signs of physical dependence in mice after 1–2 weeks of exposure via inhalation. The time course of alcohol withdrawal was comparable to that observed in other animal models (FRENCH and MORRIS, 1972). These data challenge the accepted notion that sustained exposure to high ethanol doses with attendant CNS depression is essential for the development of alcohol dependence. The necessary and sufficient dose-time parameters for inducing alcohol dependence through various routes of administration are not currently available in the literature.

GOLDSTEIN (1972a) has subsequently shown that the intensity of withdrawal convulsions (as rated by a special scoring system) increased as a function of the duration of exposure to alcohol vapor and concomitant blood alcohol levels. Moreover, mild withdrawal signs appeared following a single injection of alcohol, with or without pyrazole, while mice were still intoxicated. Ethanol as well as barbiturates and benzodiazepines were effective in suppressing the alcohol withdrawal reaction whereas phenothiazines and chlormethiazole were not (GOLDSTEIN, 1972b).

Observations that injection of minute quantities of ethanol into the lateral ventricle in rat and monkey resulted in marked increases in oral alcohol preferences (MYERS, 1963; MYERS et al., 1972a; MYERS et al., 1972b) have not been replicated by other investigators (CICERO and SMITHLOFF, 1973; FRIEDMAN and LESTER, 1975). This controversial area has been reviewed by FRIEDMAN and LESTER (1975) who conclude that methodologic differences and species variations cannot explain the recurrent failures to confirm the original findings. It now appears that intraventricular infusion of ethanol does not reliably increase alcohol preference and consequently does not produce an alcohol-addicted animal.

It is apparent that although several techniques now exist to induce physical dependence upon alcohol, our understanding of the behavioral aspects of alcohol self-administration is severely limited. The comparability of withdrawal signs and symptoms in different species is summarized in Table 3. Although a number of theories have been developed concerning the biological bases of addiction (see pp. 635–640), none has yet been sufficient to account for these phenomena.

F. Treatment of the Alcohol Withdrawal Syndrome

Successful treatment of patients with alcohol withdrawal signs and symptoms is basically dependent upon the application of the best principles of contemporary medical care. Although a number of pharmacotherapies appear to have some efficacy, as will be discussed below, treatment of the alcohol withdrawal syndrome has not been significantly advanced by development of new psychotropic agents. The most important factor which determines favorable therapeutic outcome in the treatment of withdrawal states has been the application of judicious diagnostic and therapeutic procedures based upon continuous assessment of the patient's condition rather than utilization of a standardized formulary procedure. Any approach which advocates a specific routine method for treatment of the withdrawal syndrome should be viewed with suspicion. The following principles should be considered in treatment of all patients and in the critical evaluation of new therapeutic procedures.

I. The Natural History of the Alcohol Withdrawal Syndrome

In detailed clinical studies VICTOR and ADAMS (1953) found that 15% of all patients exhibiting withdrawal states showed significant remission of signs and symptoms within 24 h, and over 80% of all patients had remission or recovery within 3 days. Thus the withdrawal state, similar to virtually every other clinical condition, is time-limited and claims for efficacy of any specific therapy must take into account that spontaneous remission occurs in most patients in a relatively short duration of time. Knowledge of this fact should not impede adequate treatment as will be described below, but should preclude overzealous use of physical and pharmacologic agents in the early treatment of the disorder.

Table 3. Comparison of alcohol withdrawal signs observed in different species

	Man	Monkey	Chimpanzee	Dog	Rat	Mouse
Motor hyperactivity						
Tremor	X	X	X	X	X	X
Muscle fasciculation	X	X	--	X	--	--
Hyperreflexia	X	X	X	X	--	--
Spasticity	--	X	X	X	X	--
Rigidity	--	X	X	X	--	--
CNS hyperactivity						
Nystagmus	X	--	--	--	--	--
Tonic-clonic-seizures	X	X	X	X	X	X
Convulsions on handling	NA	NA	NA	NA	--	X
Fatal convulsions	X	X	X	X	X	X
Tail beating and arching	NA	NA	NA	NA	--	X
Tail lift	NA	NA	NA	NA	--	X
Lethargy	--	--	--	--	--	X
Hyperactivity	X	--	--	--	X	X
Irritability	X	X	X	--	--	
Autonomic hyperactivity						
Mydriasis/photophobia	X	X	X	--	--	--
Tachycardia	X	--	X	--	--	--
Respiratory rate	X	--	X	--	--	--
Temperature	X	--	X	--	--	--
Sweating	X	--	X	--	--	--
Salivation	X	X	--	--	--	--
Gastrointestinal signs						
Retching	X	X	--	--	--	--
Vomiting	X	X	--	--	--	--
Gross behavioral changes						
Spontaneous fear reactions	X	X	--	X	--	--
Startle to noise	X	--	--	--	X	X
Decreased response to noise	--	--	X	--	--	--
Exaggerated postures	--	X	--	X	--	--
Intense scratching	--	X	--	--	--	--
Stereotyped movements	--	--	--	--	--	X
Sleep disturbances	X	--	--	X	--	--

observed = X.
not reported = - - .
not applicable = NA .
Man: VICTOR and ADAMS, 1953; WOLFE and VICTOR, 1971.
Monkey: DENEAU et al., 1969; ELLIS and PICK, 1970b, 1971.
Chimpanzee: PIEPER et al., 1972.
Dog: ELLIS and PICK, 1970a; ESSIG and LAM, 1968, 1971.
Rat: FALK et al., 1972.
Mouse: FREUND, 1969; GOLDSTEIN, 1972a; OGATA et al., 1972.
(Reprinted from MELLO, 1973b).

II. Intercurrent Illness

Initiation of treatment of patients should always begin with the careful search for occult trauma. Any examination which produces equivocal findings of bone injury should be corroborated by appropriate x-ray examination. Evidence of head trauma, particularly in disoriented or confused patients, mandates a skull x-ray series. Soft tissue and internal organ trauma are also not unusual concomitants of alcohol abuse and alcohol withdrawal. Careful palpation of the abdomen is especially important to detect spleen or hepatic damage and similar palpation of the flanks should be carried out to determine if renal traumatic injury has occurred.

Intercurrent illness involving pulmonary, gastrointestinal, hematologic, and peripheral nervous system function is not uncommon in patients during alcohol withdrawal. Individuals who consume large amounts of alcohol usually suffer from some nutritional inadequacy as a function of poor dietary intake and their resistance to infectious disease is often reduced. The importance of adequate diagnosis and treatment of intercurrent illness has been highlighted in comprehensive discussions of the hospital treatment of patients with alcohol-related illness (WALLERSTEIN, 1957). A complete description of the various disorders frequently seen in patients with alcohol related problems including gastritis, pancreatitis, tuberculosis, and cirrhosis is not attempted here since this material is available in existing textbooks of medicine (HARRISON, 1970). The role of nutritional inadequacy in peripheral nervous system disorders in alcohol withdrawal states has been discussed in detail by VICTOR (1966). However, it is useful to reemphasize for both the therapist concerned with providing the best care for patients and the investigator exploring the efficacy of new interventions for the treatment of alcohol withdrawal states that intercurrent illness affecting a number of organ systems may often be present, and the severity and duration of such disorders interact with the overall clinical picture of the withdrawal syndrome. In essence, patients exhibiting the alcohol withdrawal state require the most diligent and careful diagnostic evaluations both for initiation of treatment and during the course of medical management.

III. Disorders of Electrolyte Acid Base and Water Balance

A number of studies have demonstrated that patients in alçohol withdrawal may have significant derangements in electrolyte and water balance (BEARD and KNOTT, 1968; OGATA et al., 1968; KNOTT and BEARD, 1970). Some patients may have severe dehydration associated with alcohol-induced diuresis and profuse sweating. However, other patients may have a state of overhydration and still other patients may appear peripherally dehydrated, but actually have an undetected increase in intracellular water. It is therefore extremely important to carry out a careful clinical assessment of the patient's state of hydration and to administer intravenous fluids in a most careful manner. Patients who are severely dehydrated should be treated with intravenous fluids. But patients who can take fluids orally without recurrent vomiting and who have normal renal function should be given frequent oral fluid intake in best tolerated volume in preference to use of i.v. fluids.

Derangements in electrolyte status, in particular magnesium, have also been well documented in patients during the withdrawal state (FLINK et al., 1954; KLINGMAN et al., 1955; MENDELSON et al., 1959). Although patients may have low serum magnesium levels during alcohol withdrawal, it should be emphasized that not all patients, even in severe withdrawal, have a magnesium deficit. Parenteral magnesium replacement therapy should not be a routine measure and should be undertaken only after determination of serum magnesium levels reveal abnormally low values. Initiation of magnesium replacement therapy should be followed by careful monitoring of serum magnesium levels and by careful observation of the patient's clinical status. Magnesium deficiency may occur as a function of poor dietary intake by alcoholics when they are drinking or by enhanced excretion of magnesium as a consequence of the action of ethanol on mechanisms subserving the conservation of magnesium by the kidney. Magnesium levels may also be affected by changes in acid base balance as will be described below. A number of complex factors may regulate magnesium homeostasis and a simplistic formulation of the relationship between alcohol withdrawal and magnesium balance is not possible.

The dynamics of acid base balance in patients exhibiting withdrawal syndromes may be extraordinarily variable. Patients who are dehydrated, sweating, and have a history of poor dietary intake may have a relative state of metabolic acidosis. Other patients may develop respiratory alkalosis. WOLFE et al. (1969) have shown that patients in alcohol withdrawal may have increased respiratory rates. Increased respiration is often associated with hypocapnia and clinical signs of respiratory alkalosis. Respiratory alkalosis, in turn, is often associated with elevated arterial pH levels, and WOLFE et al. (1969) have been able to establish a correlation between increased arterial pH and a decrease in serum magnesium levels. Serum magnesium levels may fall during alkalosis as a function of the effect of magnesium on cell-buffering processes, and magnesium shifts from the intravascular to the intracellular compartment, a phenomenon which occurs without any deficit in total body magnesium.

In summary, patients in alcohol withdrawal may be dehydrated or overhydrated and may have relative increases or decreases of electrolytes in intravascular and intracellular body fluids. They may exhibit states of acidosis or alkalosis depending upon status of dietary history, intercurrent illness, or physiologic response mechanisms which occur during the withdrawal syndrome. Determination of the presence of any specific abnormality necessitates precise clinical and laboratory diagnostic procedures.

IV. The Use of Psychotropic Agents in the Treatment of the Alcohol Withdrawal Syndrome

Although psychotropic agents have been used in the treatment of alcohol withdrawal syndromes for almost 20 years, there are still many questions concerning the relative efficacy of specific drugs. Following a critical evaluation of the existing literature, VICTOR (1966) concluded that there was no evidence that any of the new psychotropic agents were effective in preventing delirium tremens. VICTOR also pointed out that no psychotropic drug had been shown to be better than paraldehyde, an agent used in the treatment of alcohol withdrawal for many decades. In fact, there are data which indicate that paraldehyde may be safer and more effective than certain psychotropic agents such as promazine. THOMAS and FREEDMAN (1964) reported a series of 39 patients diagnosed as having delirium tremens. In this series 6 of the 17 patients who received promazine died, whereas only 1 of the 22 patients who received paraldehyde succumbed. THOMAS and FREEDMAN also found that those patients who did not show an initial response to promazine (during the first 2 days of treatment) appeared to develop more severe signs and symptoms during the course of the withdrawal state than those patients treated with paraldehyde. In other words, patients who received promazine for periods of time longer than 2 days after onset of withdrawal developed an accentuation of the withdrawal syndrome.

SERENY and KALANT (1965) carried out a rigorous double blind study of the efficacy of placebo, chlordiazepoxide, and promazine in a series of 58 hospitalized patients with signs and symptoms of acute alcohol withdrawal. These investigators concluded that both promazine and chlordiazepoxide improved sleep and reduced severity of sweating. However, they also observed that grand mal seizures occurred in four patients treated with promazine. None of the patients treated with chlordiazepoxide developed seizure disorders or delirium.

One of the most comprehensive studies for determining the efficacy of a number of psychotropic compounds in the treatment of alcohol withdrawal was carried out in a collaborative study by the Veterans Administration Hospital (KAIM et al., 1969). Five hundred and thirty-seven patients were studied in 23 Veterans Administration Hospitals. The major outcome criterion measured in this study was the degree of progression of mild withdrawal states to either seizure disorders or overt delirium tremens. The findings revealed that chlordiazepoxide was superior to chlorpromazine, hydroxyzine, thiamine and placebo for preventing both delirium tremens and seizure disorders. Thus chlordiazepoxide has emerged as the safest and most efficacious drug for use in the treatment of the alcohol withdrawal syndromes.

V. Nonspecific Pharmacotherapies in the Treatment of Alcohol Withdrawal

During the course of many decades, a host of pharmacotherapies have been advocated for the treatment of alcohol withdrawal without an adequate rationale for their utilization. Many bizzare drug cocktails have been concocted for administration to patients during acute alcohol withdrawal, and it is likely that some of the agents employed enhanced rather than reduced the severity of the disorder. For example it has been argued that alcoholics may have some evidence of hypothyroidism and thus thyroid preparations may assist in treatment. This notion has been critically reviewed and discredited by the studies of KALANT and his associates (1962). It has also been postulated that alcoholics may have adrenal exhaustion during the withdrawal syndrome and hence adrenocortical preparations might be of value. This hypothesis was critically reviewed by MENDELSON and STEIN (1966b), and in controlled studies of both alcoholics and nonalcoholics during experimentally induced ethanol intoxication and withdrawal, no evidence for adrenocortical exhaustion or insufficiency was found to be associated with the withdrawal state. WALDER and his associates (1969) have advocated rapid detoxification of individuals during acute alcohol intoxication with hemodialysis. This procedure is not only expensive and difficult, but may also be hazardous in terms of precipitating the withdrawal syndrome. There is no evidence that the withdrawal syndrome is caused by any unique metabolite of ethanol or any toxic derivative associated with alcohol catabolism. A rapid decrease in blood alcohol levels during dialysis would abruptly induce withdrawal rather than diminish the possibility of the occurrence of the syndrome.

In conclusion, the application of specific pharmacotherapies should be based upon knowledge of the pathophysiologic basis of the withdrawal syndrome. Administration of fluids, vitamins, antibiotics, and other agents to treat specific intercurrent illness may be necessary and should be based upon appropriate clinical examinations and laboratory studies. Diagnosis and classification of the behavior disorders associated with alcohol withdrawal should follow the principles of a comprehensive mental status examination (GROSS, 1967). Careful clinical investigation utilizing appropriate research design and methods is a mandatory prrequisite prior to any general application of new treatment modalities, in particular drug therapy, for the alcohol withdrawal syndrome.

Appendix

Social and drinking history questionnaire

DEPARTMENT OF HEALTH, EDUCATION, AND WELFARE
PUBLIC HEALTH SERVICE
HEALTH SERVICES AND MENTAL HEALTH ADMINISTRATION
NATIONAL INSTITUTE OF MENTAL HEALTH

NIAAA

SOCIAL AND DRINKING HISTORY

NAME

DATE | UNIT NO

INTERVIEWER

Age | Weight | Race | Birthplace | Parent's Birthplace: Father / Mother

Residence at time of admission

Education

Current Job

Previous Job

Marital Status

Number of Marriages ______

No of Children

Religion

FAMILY DRINKING PATTERNS
(Check appropriate column for each):

	Social	Heavy	Alcoholic	Abstinen
Wife				
Mother				
Father				
Sibling				

Age of First Drink | Beverage First Drunk | Onset of Heavy Drinking Age ______ | Duration of Drinking ______ Years | Preferred Beverage

Frequency of Drinking: ☐ Daily ☐ Weekend ☐ Spree
Starts Drinking ______
Finishes Drinking

Amount of Alcohol per Day

Pattern of Drinking Within a Day

Usually Drinks: ☐ Alone ☐ With Friends ☐ With Strangers ☐ At Home ☐ At Bar of Club ☐ In Street or Park ☐ Other *(Specify)*

Any Change in Drinking Pattern? ☐ No ☐ Yes

If Yes, specify what change and when

Memory Problems

Intercurrent Illness During Drinking? ☐ No ☐ Yes If Yes, specify illness

Use of
Toxic Beverage *(Specify)*: ______
Drugs *(Specify)*: ______

Withdrawal Symptoms

Withdrawal Symptoms? ☐ No ☐ Yes *(Specify)*:

☐ Tremulous
☐ Sweating
☐ Insomnia
☐ Nausea
☐ Vomiting
☐ Hallucinosis
☐ Visual
☐ Auditory
☐ Seizures
☐ Delirium

Withdrawal Symptoms Usually Last ______ Days

Withdrawal Symptoms First Began ______ Years Ago

Usual Severity ☐ Mild ☐ Moderate ☐ Seve

During Withdrawal Usually ☐ At Home ☐ Hospital Emergency Ward ☐ Jail Equivalent *(Specify)*:

Withdrawal Usually After What Kind of Drinking?

Attitude About Withdrawal Symptoms

MH-180-3
Rev 4-71

CONTINUE ON REVERSE SIDE

FORM APPROVED
OMB NO 68-R1208

ttitude About Own Drinking

On Behavior

On Feelings

On Sleep Pattern

On Food Intake

SUAL FFECT F LCOHOL:

While Drinking	Elevated +	Diminished −	No Change 0
Mood			
Sexuality			
Aggression			
Socialization			
Guilt			
Memory			
Self Esteem			
Anxiety			
Emotionality			
Sleep			

Are effects constant throughout drinking spree? *(Describe):*

Is more alcohol required to produce these effects than previously?

☐ No ☐ Yes

Expectancy About Effects of Alcohol During This Study

Drinking Problem Attributed to a Specific Initiating Event?

sually Begins a Drinking Episode Because:

Usually Terminates a Drinking Episode Because:

hen was the Longest Period of Voluntary Abstinence? Why begun and Why ended?

DDITIONAL COMMENTS

-180-3 (Back)
v 4-71

HEW-Lex

References

Addington, M. W., Kettel, L. J., Lugell, D. W.: Alkalosis due to mechanical hyperventilation in patients with chronic hypercapnea. Ann. Rev. Resp. Dis. **93**, 736—741 (1966)
Beard, J. D., Knott, D. H.: Fluid and electrolyte balance during acute withdrawal in chronic alcoholic patients. J. Amer. med. Ass. **204**, 135—139 (1968)
Bennett, I. L., Cary, F. H., Mitchel, G. L., Cooper, M. H.: Acute methyl alcohol poisoning: A review based on experiences in an outbreak of 323 cases. Medicine **32**, 431—463 (1953)
Bloomquist, E. R.: Addiction, addicting drugs, and the anesthesiologist. J. Amer. med. Ass. **171**, 518—523 (1959)
Branchey, M., Rauscher, G., Kissin, B.: Modifications in the response to alcohol following the establishment of physical dependence. Psychopharmacologia (Berl.) **22**, 314—322 (1971)
Burrows, G. M.: Commentaries on the causes, forms and symptoms and treatment, moral and medical of insanity, p. 332. London, 1828
Cahalan, D.: Problem Drinkers. San Francisco: Jossey-Bass Inc., 1970
Cannon, W. B., Rosenblueth, A.: The Supersensitivity of Denervated Structures: A Law of Denervation. New York: Macmillan 1949
Carpenter, J. A.: Effects of alcohol on some psychological processes. Quart. J. Stud. Alcohol **23**, 274—314 (1962)
Carpenter, T. M., Lee, R. D.: The effects of glucose on the metabolism of ethyl alcohol in man. J. Pharmacol. exp. Ther. **60**, 264—285 (1937)
Cicero, T. J., Smithloff, B. R.: Alcohol oral self-administration in rats: Attempts to elicit excessive intake and dependence. In: Gross, M. (Ed.): Alcohol Intoxication and Withdrawal: Experimental Studies, pp. 213—224. New York: Plenum Press 1973
Cicero, T. J., Snider, S. R., Perez, V. J., Swanson, L. W.: Physical dependence on and tolerance to alcohol in the rat. Physiol. Behav. **6**, 191—198 (1971)
Cochin, J.: The pharmacology of addiction to narcotics. In: Martin, G. J., Kisch, B. (Eds.): Enzymes in Mental Health, pp. 27—42. Philadelphia: J. B. Lippincott Co. 1966
Cochin, J.: Role of possible immune mechanisms in the development of tolerance. In: Clouet, D. H. (Ed.): Narcotic Drugs, Biochemical Pharmacology, pp. 432—448. New York-London: Plenum Press 1971
Cochin, J., Kornetsky, C.: Development and loss of tolerance to morphine in the rat after single and multiple injections. J. Pharmacol. exp. Ther. **145**, 1—10 (1964)
Cohen, G., Collins, M.: Alkaloids from catecholamines in adrenal tissue: Possible role in alcoholism. Science **167**, 1749—1751 (1970)
Cohen, G., Mytilineou, C., Barrett, R. E.: 6,7-Dihydroxytetrahydroisoquinoline: Uptake and storage by peripheral sympathetic nerve of the rat. Science **175**, 1269—1272 (1972)
Collier, H. O. J.: A general theory of the genesis of drug dependence by induction of receptors. Nature (Lond.) **205**, 181—182 (1965)
Cooper, J. R., Kini, M. M.: Biochemical aspects of methanol poisoning. Biochem. Pharmacol. **11**, 405—416 (1962)
Cox, B. M., Ginsburg, M., Osman, O. H.: Acute tolerance to narcotic analgesic drugs in rats. Brit. J. Pharmacol. **33**, 245—256 (1968)
Davis, V. E., Walsh, M. J.: Alcohol, amines and alkaloids: A possible biochemical basis for alcohol addiction. Science **167**, 1005—1006 (1970)
Deneau, G., Yanagita, T., Seevers, M. H.: Self administration of psychoactive substances by the monkey. Psychopharmacologia (Berl.) **16**, 30—48 (1969)
Docter, R. G., Naitoh, P., Smith, J. C.: Electroencephalographic changes and vigilance behavior during experimentally-induced intoxication with alcoholic subjects. Psychosom. Med. **28**, 605—615 (1966)
Ellis, F. W., Pick, J. R.: Ethanol-induced withdrawal reactions in rhesus monkey. Pharmacologist **11**, 256 (1969)
Ellis, F. W., Pick, J. R.: Evidence of ethanol dependence in dogs. Fed. Proc. **29**, 649 Abs. (1970a)
Ellis, F. W., Pick, J. R.: Experimentally-induced ethanol dependence in rhesus monkeys. J. Pharmacol. exp. Ther. **175**, 88—93 (1970b)
Ellis, F. W., Pick, J. R.: Ethanol intoxication and dependence in rhesus monkeys. In: Mello, N. K., Mendelson, J. H. (Eds.): Recent Advances in Studies of Alcoholism, pp. 401—412. Washington, D.C.: U.S. Govt. Printing Office Publ. No. (HSM) 71-9045, 1971

Essig, C. F., Lam, R. C.: Convulsions and hallucinatory behavior following alcohol withdrawal in the dog. Arch. Neurol. (Paris) **18**, 626—632 (1968)
Essig, C. F., Lam, R. C.: The alcohol abstinence syndrome in dogs and its treatment with pentobarbital. In: Mello, N. K., Mendelson, J. H. (Eds.): Recent Advances in Studies of Alcoholism, pp. 437—452. Washington, D.C.: U.S. Govt. Printing Office Publ. No. (HSM) 71-9045, 1971
Falk, J. L.: Production of polydipsia in normal rats by an intermittent food schedule. Science **133**, 195—196 (1961)
Falk, J. L., Samson, H. H., Winger, G.: Behavioral maintenance of high concentrations of blood ethanol and physical dependence in the rat. Science **177**, 811—813 (1972).
Flink, E. B., Stutzman, F. L., Anderson, A. R., Lontig, T., Frasier, R.: Magnesium deficiency after prolonged parenteral fluid administration and after chronic alcoholism complicated by delirium tremens. J. Lab. clin. Med. **43**, 169—183 (1954)
Forsander, O. A., Raiha, N., Salaspuro, M., Maenpaa, P.: Influence of ethanol on the liver metabolism of fed and starved rats. Biochem. J. **94**, 259—265 (1965)
Forsander, O. A., Raiha, N., Sumalainen, H.: Alkoholoxydation und Bildung von Acetoacetat in normaler und glykogenarmer intakter Rattenleber. Hoppe-Seylers Z. physiol. Chem. **312**, 243—248 (1958)
Fraser, H. F., Isbell, H., Eisenman, A. J., Wikler, A., Pescor, F. T.: Chronic barbiturate intoxication. Arch. intern. Med. **94**, 34—41 (1954)
Fraser, H. F., Wikler, A., Isbell, H., Johnson, N. K.: Partial equivalence of chronic alcohol and barbiturate intoxications. Quart. J. Stud. Alcohol **18**, 541—551 (1957)
French, S. W., Morris, J. R.: Ethanol dependence in the rat induced by non-intoxicating levels of ethanol. Res. Commun. chem. Path. Pharmacol. **4**, 221—233 (1972)
Freund, G.: Alcohol withdrawal syndrome in mice. Arch. Neurol. (Paris) **21**, 315—320 (1969)
Friedman, H. J., Lester, D.: Intraventricular ethanol and ethanol intake: A behavioral and radiographic study. Pharmacol. Biochem. Behav. **3**, 393—401 (1975)
Gibbins, R. J., Kalant, M., LeBlanc, A. E., Clark, W.: The effects of chronic administration of ethanol on startle thresholds in rats. Psychopharmacologia (Berl.) **19**, 95—104 (1971)
Goldberg, L.: Quantitative studies on alcohol tolerance in man. The influence of ethyl alcohol on sensory, motor and psychological functions referred to blood alcohol in normal and habituated individuals. Acta physiol. scand. **5** (Suppl. 16), 1—128 (1943)
Goldberg, L., Rydberg, U.: Inhibition of ethanol metabolism in vivo by administration of pyrazole. Biochem. Pharmacol. **18**, 1749—1762 (1969)
Goldstein, A., Aronow, L., Kalman, S. M.: Principles of drug action. New York: Harper & Row 1968
Goldstein, A., Goldstein, D. B.: Enzyme expansion theory of drug tolerance and physical dependence. In: A. Wikler (Ed.): The Addictive States. Res. Publ. Ass. nerv. ment. Dis. Vol. 46, pp. 265—267. Baltimore: Williams & Wilkins 1968
Goldstein, A., Lowney, L. L., Pal, B. K.: Stereospecific and nonspecific interactions of the morphine congener levorphanol in sub-cellular fractions of mouse brain. Proc. nat. Acad. Sci. (Wash.) **68**, 1742—1747 (1971)
Goldstein, D. B.: Relationship of alcohol dose to intensity of withdrawal signs in mice. J. Pharmacol. exp. Ther. **180** (2), 203—215 (1972a)
Goldstein, D. B.: An animal model for testing effects of drugs on alcohol withdrawal reactions. J. Pharmacol. exp. Ther. **183** (1), 14—22 (1972b)
Goldstein, D. B.: Rates of onset and decay of alcohol physical dependence in mice. J. Pharmacol. exp. Ther. **190**, 377—383 (1974)
Goldstein, D. B., Pal, N.: Alcohol dependence produced in mice by inhalation of ethanol; Grading the withdrawal reaction. Science **172**, 288—290 (1971)
Graham, J. D. P.: Recent theories on the pharmacological basis of tolerance and dependence. Brit. J. Addict. **67**, 83—87 (1972)
Grenell, R. G.: Alcohols and activity of cerebral neurons. Quart. J. Stud. Alcohol **20**, 421—427 (1959)
Gross, M. M.: Management of acute alcohol withdrawal states. Quart. J. Stud. Alcohol **28**, 655—666 (1967)
Gross, M. M., Goodenough, D. R.: Sleep disturbances in the acute alcoholic psychoses. In: Cole, J. O. (Ed.): Clinical Research in Alcoholism, pp. 132—147. A.P.A. Psychiatric Research Report 24, 1968

Gross, M. M., Goodenough, D. R., Hastey, J., Lewis, E.: Experimental study of sleep in chronic alcoholics before, during and after four days of heavy drinking, with a non-drinking comparison. In: Seixas, F. A., Eggleston, S. (Eds.): Alcoholism and the Central Nervous System. Ann. N.Y. Acad. Sci. **215**, 254—265 (1973)
Gross, M. M., Goodenough, D. R., Hastey, J. M., Rosenblatt, S. M., Lewis, E.: Sleep disturbances in alcohol intoxication and withdrawal. In: Mello, N. K., Mendelson, J. H. (Eds.): Recent Advances in Studies of Alcoholism, pp. 317—397. Washington, D.C.: U.S. Govt. Printing Office Publ. No. (HSM) 71-9045, 1971c
Gross, M. M., Goodenough, D. R., Tobin, M., Halpert, E., Lepore, D., Perstein, A., Sirota, M., Dibianco, J., Fuller, R., Kishner, I.: Sleep disturbances and hallucinations in the acute alcoholic psychoses. J. nerv. ment. Dis. **142**, 493—514 (1966)
Gross, M. M., Rosenblatt, S. M., Chartoff, S., Herman, A., Schacter, E., Sheinkin, D., Broman, M.: Evaluation of the acute alcoholic psychoses and related states. The daily clinical course rating scale. Quart. J. Stud. Alcohol **32** (3), 611—619 (1971a)
Gross, M. M., Rosenblatt, S. M., Malenowski, B., Broman, M., Lewis, E.: A factor analytic study of the clinical phenomena in the acute alcohol withdrawal syndromes. Alkohologia **2** (1), 1—7 (1971b)
Haizlip, T. M., Ewing, J. A.: Meprobamate habituation: A controlled clinical study. New Engl. J. Med. **258**, 1181—1186 (1958)
Hamilton, J. D., Gross, N. J.: Unusual neurological and cardiovascular complications of respiratory failure. Brit. med. J. **II**, 1092—1096 (1963)
Han, J. Y.: Why do chronic alcoholics require more anesthesia? Anesthesiology **30**, 341—342 (1969)
Harger, R. N., Hulpieu, H. R.: The pharmacology of alcohol. In: Thompson, G. N. (Ed.): Alcoholism, pp. 103—232. Springfield, Illinois: Charles C. Thomas 1956
Harrison's Principles of Internal Medicine (6th Ed.). New York: McGraw Hill 1970
Hawkins, R. D., Kalant, H., Khanna, J. M.: Effects of chronic intake of ethanol on rate of ethanol metabolism. Canad. J. Physiol. Pharmacol. **44**, 241—257 (1966)
Holman, R. B., Myers, R. D.: Ethanol consumption under conditions of psychogenic polydipsia. Physiol. Behav. **3**, 369—371 (1968)
Isbell, H., Fraser, H., Wikler, A., Belleville, R., Eisenman, A.: An experimental study of the etiology of rum fits and delirium tremens. Quart. J. Stud. Alcohol **16**, 1—33 (1955)
Isselbacher, K. J., Carter, E. A.: Effect of alcohol on liver and intestinal function. In: Mello, N. K., Mendelson, J. H. (Eds.): Recent Advances in Studies of Alcoholism, pp. 42—58. Washington, D.C.: U.S. Govt. Printing Office Publ. No. (HSM) 71-9045, 1971
Jacobsen, E.: The metabolism of ethyl alcohol. Pharmacol. Rev. **4**, 107—135 (1952)
Jaffe, J. H.: Drug addiction and drug abuse. In: Goodman, L. S., Gilman, A. (Eds.): The Pharmacological Basis of Therapeutics, pp. 276—313. New York: Macmillan Co. 1970
Jaffe, J. H., Sharpless, S. K.: Pharmacological denervation supersensitivity in the central nervous system: A theory of physical dependence. In: Wikler, A. (Ed.): The Addictive States, Proceedings of the Association for Research in Nervous and Mental Disease, Vol. 46, pp. 226—243. Baltimore: Williams & Wilkins Co. 1968
Jellinek, E. M.: The Disease Concept of Alcoholism. Highland Park, New Jersey: Hillhouse Press 1960
Jetter, W. W.: Studies in alcohol. II. Experimental feeding of alcohol to non-alcoholic individuals. Amer. J. med. Sci. **196**, 487—493 (1938)
Johnson, L. C., Burdick, J. A., Smith, J.: Sleep during alcohol intake and withdrawal in the chronic alcoholic. Arch. gen. Psychiat. **22**, 406—418 (1970)
Kaim, S.: Drug treatment of the alcohol withdrawal syndrome. In: Mello, N. K., Mendelson, J. H. (Eds.): Recent Advances in Studies of Alcoholism, pp. 767—780. Washington, D.C.: U.S. Govt. Printing Office Publ. No. (HSM) 71-9045, 1971
Kaim, S. C., Klett, C. J., Rothfeld, B.: Treatment of the acute alcohol withdrawal state: A comparison of four drugs. Amer. J. Psychiat. **125**, 1640—1646 (1969)
Kalant, H.: Absorption diffusion, distribution and elimination of ethanol: Effects on biological membranes. In: Kissin, B., Begleiter, H.: The Biology of Alcoholism, Vol. I, Biochemistry, pp. 1—62. New York: Plenum Press 1971

Kalant,H.: Biological models of alcohol tolerance and physical dependence. In: Gross,M.M. (Ed.): Alcohol Intoxication and Withdrawal: Experimental Studies, Advances in Experimental Medicine, Biology, Vol. 35, pp. 3—14. New York: Plenum Press 1973

Kalant,H., LeBlanc,A.E., Gibbins,R.J.: Tolerance to, and dependence on, some non-opiate psychotropic drugs. Pharmacol. Rev. **23**, 135—191 (1971)

Kalant,H., Sereny,G., Charlebois,R.: Evaluation of tri-iodethryonine in the treatment of acute alcoholic intoxication. New Engl. J. Med. **267**, 1—8 (1962)

Kalinowsky,L.B.: Convulsions in non-epileptic patients on withdrawal from barbiturates, alcohol and other drugs. Arch. Neurol. Psychiat. (Chic.) **48**, 946—956 (1942)

Kater,R.M.H., Tobon,F., Iber,F.L.: Increased rate of tolbutamide metabolism in alcoholic patients. J. Amer. med. Ass. **207** (2), 363—365 (1969)

Kater,R.M.H., Zeive,D., Tobon,F., Roggin,G.M., Iber,F.L.: Heavy drinking accelerates drugs' breakdown in liver. J. Amer. med. Ass. **206**, 1709 (1968)

Keller,M.: The definition of alcoholism and the estimation of its prevalence. In: Pittman,D.J., Snyder,C.R. (Eds.): Society, Culture and Drinking Patterns, pp. 310—329. New York: Wiley 1962

Kilburn,K.H.: Shock, seizures and coma with alkalosis during mechanical ventilation. Ann. intern. Med. **65**, 977—984 (1966)

Klingman,W.O., Suter,C., Green,R., Robinson,I.: Role of alcoholism and magnesium deficiency in convulsions. Trans. Amer. neurol. Ass. **80**, 162—165 (1955)

Knott,D.H., Beard,J.D.: Diagnosis and therapy of acute withdrawal from alcohol. In: Masserman,J.H. (Ed.): Current Psychiatric Therapies, Vol. 10, pp. 145—153. New York: Grune & Stratton 1970

Kornetsky,C., Bain,G.: Morphine: Single-dose tolerance. Science **162**, 1011—1012 (1968)

Le Breton,E.: Influence du jeune sur la vitesse d'oxydation de l'alcool ethylique chez le rat blanc. C.R. Soc. Biol. (Paris) **122**, 330—332 (1936)

Lee,T.K., Cho,M.H., Dobkin,A.B.: Effects of alcoholism, morphinism and barbiturate resistance on induction and maintenance of general anesthesia. Canad. Anaesth. Soc. J. **11**, 354—381 (1964)

Leloir,L.F., Munoz,J.M.: Ethyl alcohol metabolism in animal tissues. Biochem. J. **32**, 299—307 (1938)

Lester,D.: Self-maintenance of intoxication in the rat. Quart. J. Stud. Alcohol **22**, 223—231 (1961)

Lester,D.: Self-selection of alcohol by animals, human variation, and the etiology of alcoholism. Quart. J. Stud. Alcohol **27**, 395—438 (1966)

Lester,D., Freed,E.X.: Criteria for an animal model of alcoholism. Pharmacol. Biochem. Behav. **1**, 103—107 (1973)

Lester,D., Keobosky,W.Z., Felzenberg,F.: Effect of pyrazoles and other compounds on alcohol metabolism. Quart. J. Stud. Alcohol **29**, 449—454 (1968)

Lettsom,J.C.: Some remarks on the effects of Lignum Quassil Amare. Memoirs Med. Soc. London **1**, 151—165 (1787)

Lieber,C.S.: Hepatic and intestinal adaptation and injury in alcoholism. Paper presented at meeting of Biomedical Research in Alcohol Abuse Problems, sponsored by Non-Medical Use of Drugs Directorate, Health and Welfare of Canada. Halifax, Nova Scotia. August 29 and 30, 1974

Lieber,C.S., DeCarli,L.M.: Ethanol dependence and tolerance: A nutritionally controlled experimental model in the rat. Res. Commun. chem. Path. Pharmacol. **6** (3), 983—991 (1973)

Lieber,C.S., DeCarli,L.M.: Ethanol oxidation by hepatic microsomes: Adaptive increase after ethanol feeding. Science **162**, 917—918 (1968)

Lieber,C.S., Rubin,E., DeCarli,L.M.: Chronic and acute effects of ethanol on hepatic metabolism of ethanol, lipids and drugs: Correlation with ultrastructural changes. In: Mello,N.K., Mendelson,J.H. (Eds.): Recent Advances in Studies of Alcoholism, pp. 3—41. Washington, D.C.: U.S. Govt. Printing Office Publ. No. (HSM) 71-9045, 1971

Lieber,C.S., Rubin,E., DeCarli,L.M., Misra,P., Gang,H.: Effects of pyrazole on hepatic function and structure. Lab. Invest. **22**, 615—621 (1970)

Lubin,M., Westerfeld,W.W.: The metabolism of acetaldehyde. J. Biol. Chem. **161**, 503—512 (1945)

Lundquist, F.: The metabolism of ethanol. In: Israel, Y., Mardones, J. (Eds.): Biological Basis of Alcoholism. New York: Wiley-Interscience 1971
Lundquist, F., Wolthers, H.: The influence of fructose on the kinetics of alcohol elimination in man. Acta pharmacol. (Kbh.) **14**, 290—294 (1958)
Majchrowicz, E.: Induction of physical dependence on alcohol and the associated metabolic and behavioral changes in the rat. Pharmacologist **15**, 159 (1973)
Majchrowicz, E., Mendelson, J. H.: Blood concentrations of acetaldehyde and ethanol in chronic alcoholics. Science **168**, 1100—1102 (1970)
Majchrowicz, E., Mendelson, J. H.: Blood levels of acetaldehyde and methanol during chronic ethanol ingestion and withdrawal. In: Mello, N. K., Mendelson, J. H. (Eds.): Recent Advances in Studies of Alcoholism, pp. 199—216. Washington, D.C.: U.S. Govt. Printing Office Publ. No. (HSM) 71-9045, 1971a
Majchrowicz, E., Mendelson, J. H.: Blood methanol concentrations during experimentally-induced alcohol intoxication in alcoholics. J. Pharmacol. exp. Ther. **179**, 293—300 (1971b)
Mardones, J.: Experimentally-induced changes in the free selection of ethanol. In: Pfeiffer, C. C., Smythies, J. R. (Eds.): International Review of Neurobiology, Vol. 2, pp. 41—76. New York: Academic Press 1960
Martin, W. R.: Pharmacological redundancy as an adaptive mechanism in the central nervous system. Fed. Proc. **29**, 13—18 (1970)
Martin, W. R., Eades, C. G.: Demonstration of tolerance and physical dependence in the dog following a short-term infusion of morphine. J. Pharmacol. exp. Ther. **133**, 262—270 (1961)
Martin, W. R., Eades, C. G.: A comparison between acute and chronic physical dependence in the chronic spinal dog. J. Pharmacol. exp. Ther. **146**, 385—394 (1964)
Martin, W. R., Eades, C. G., Thompson, W. O., Thompson, J. A., Flanary, H. G.: Morphine physical dependence in the dog. J. Pharmacol. exp. Ther. **189**, 759—771 (1974)
McQuarrie, D. G., Fingl, E.: Effect of single doses and chronic administration of ethanol on experimental seizures in mice. J. Pharmacol. exp. Ther. **124**, 264—271 (1958)
Meisch, R.: Increased rate of ethanol self-administration as a function of experience. Reports from the Research Laboratories, Dept. of Psychiatry, University of Minnesota, No. PR-69-3, 1969
Meisch, R., Pickens, R.: Oral self-administration of ethanol by the rat. Psychonomic Society Meetings in St. Louis, Missouri 1968
Mello, N. K.: Some aspects of the behavioral pharmacology of alcohol. In: Efron, D. H., Cole, J. O., Levine, J., Wittenborn, J. R. (Eds.): Psychopharmacology: A Review of Progress 1957—1967, pp. 787—809. Washington, D.C.: U.S. Govt. Printing Office PHS Publ. No. 1863, 1968
Mello, N. K.: Alcohol effects on delayed matching-to-sample performance by rhesus monkey. Physiol. Behav. **7**, 77—101 (1971)
Mello, N. K.: Behavioral studies of alcoholism. In: Kissin, B., Begleiter, H. (Eds.): The Biology of Alcoholism: Vol. II, Physiology and Behavior, pp. 219—291. New York: Plenum Publishing Co. 1972
Mello, N. K.: Short-term memory function in alcohol addicts during intoxication. In: Gross, M. M. (Ed.): Alcohol Intoxication and Withdrawal: Experimental Studies, Advances in Experimental Medicine and Biology, Vol. 35, pp. 333—344. New York: Plenum Press 1973a
Mello, N. K.: A review of methods to induce alcohol addiction in animals. Pharmacol. Biochem. Behav. **1**, 89—101 (1973b)
Mello, N. K., Mendelson, J. H.: Experimentally induced intoxication in alcoholics: A comparison between programmed and spontaneous drinking. J. Pharmacol. exp. Ther. **173**, 101—116 (1970a)
Mello, N. K., Mendelson, J. H.: Behavioral studies of sleep patterns in alcoholics during intoxication and withdrawal. J. Pharmacol. exp. Ther. **175**, 94—112 (1970b)
Mello, N. K., Mendelson, J. H.: Evaluation of a polydipsia technique to induce alcohol consumption in monkeys. Physiol. Behav. **7**, 827—836 (1971a)
Mello, N. K., Mendelson, J. H.: The effects of drinking to avoid shock on alcohol intake in primates. In: Roach, M. K., McIsaac, W. M., Creaven, P. J. (Eds.): Biological Aspects of Alcohol, pp. 313—340. Austin: University of Texas Press 1971b

Mello, N. K., Mendelson, J. H.: Drinking patterns during work-contingent and non-contingent alcohol acquisition. Psychosom. Med. **34** (2), 139—164 (1972)

Mello, N. K., Mendelson, J. H.: Alcoholism: A biobehavioral disorder. In American Handbook of Psychiatry, pp. 371—403. New York: Basic Books, 1975

Mello, N. K., Mendelson, J. H.: The development of alcohol dependence: A clinical study. McLean Hosp. J. **1** (2), 64—88 (1976)

Mello, N. K., McNamee, H. B., Mendelson, J. H.: Drinking patterns of chronic alcoholics: Gambling and motivation for alcohol. In: Cole, J. O. (Ed.): Clinical Research in Alcoholism, Psychiatric Research Report No. 24, pp. 83—118. Amer. Psychiat. Ass., Washington, D.C. 1968

Mendelson, J. H. (Ed.): Experimentally induced chronic intoxication and withdrawal in alcoholics. Quart. J. Stud. Alcohol, Suppl. No. 2, 1964

Mendelson, J. H.: Ethanol-1-C^{14} metabolism in alcoholics and non-alcoholics. Science **159**, 319—320 (1968a)

Mendelson, J. H.: Biochemical pharmacology of alcohol. In: Efron, D. H. Cole, J. O., Levine, J., Wittenborn, J. R. (Eds.): Psychopharmacology: A Review of Progress 1957—1967, pp. 769—785. Washington, D.C.: U.S. Govt. Printing Office PHS Publ. No. 1836, 1968b

Mendelson, J. H.: Biologic concomitants of alcoholism. New Engl. J. Med. **283**, 24—32; 71—81 (1970)

Mendelson, J. H.: Biochemical mechanisms of alcohol addiction. In: Kissin, B., Begleiter, H. (Eds.): The Biology of Alcoholism. Vol. I, Biochemistry, pp. 513—544. New York: Plenum Press 1971

Mendelson, J. H., Mello, N. K.: Ethanol and whiskey drinking patterns in rats under free-choice and forced-choice conditions. Quart. J. Stud. Alcohol **25**, 1—25 (1964)

Mendelson, J. H., Mello, N. K.: A disease as an organizer for biochemical research: Alcoholism. In: Mandell, A. J., Mandell, M. P. (Eds.): Psychochemical Research in Man, pp. 379—403. New York: Academic Press 1969

Mendelson, J. H., Mello, N. K.: Studies of the development of alcohol addiction in infant monkeys. In: Seixas, F., Eggleston, S. (Eds.): Alcoholism and the Central Nervous System. Ann. N.Y. Acad. Sci. **215**, 145—161 (1973)

Mendelson, J. H., Ogata, M., Mello, N. K.: Effects of alcohol ingestion and withdrawal on magnesium states of alcoholics: Clinical and experimental findings. Ann. N.Y. Acad. Sci. **162** (2), 918—933 (1969)

Mendelson, J. H., Stein, S.: The definition of alcoholism. In: Mendelson, J. H. (Ed.): Alcoholism, International Psychiatry Clinics, pp. 3—16. Boston: Little, Brown & Co. 1966a

Mendelson, J. H., Stein, S.: Serum cortisol levels in alcoholic and nonalcoholic subjects during experimentally induced ethanol intoxication. Psychosom. Med. **28**, 616—626 (1966b)

Mendelson, J. H., Stein, S., Mello, N. K.: Effects of experimentally induced intoxication on metabolism of ethanol-1-C^{14} in alcoholic subjects. Metabolism **14**, 1255—1266 (1965)

Mendelson, J. H., Wexler, D., Kubzansky, P., Leiderman, H., Solomon, P.: Serum magnesium in delirium tremens and alcoholic hallucinosis. J. nerv. ment. Dis. **128**, 352—357 (1959)

Mendelson, J. H., Wexler, D., Leiderman, P., Solomon, P.: A study of addiction to nonethyl alcohols and other poisonous compounds. Quart. J. Stud. Alcohol **18**, 561—580 (1957)

Myers, R. D.: Alcohol consumption in rats: Effects of intracranial injections of ethanol. Science **142**, 240—241 (1963)

Myers, R. D., Evans, J. E., Yaksh, T. L.: Ethanol preference in the rat: Interactions between brain serotonin and ethanol, acetaldehyde, paraldehyde, 5-HTP and 5-HTOL. Neuropharmacology **11**, 539—549 (1972a)

Myers, R. D., Veale, W. L.: The determinants of alcohol preference in animals. In: Kissin, B., Begleiter, H., (Eds.): Biology of Alcoholism. Vol. II, Physiology and Behavior, pp. 131—168. New York: Plenum Press 1972

Myers, R. D., Veale, W. L., Yaksh, T. L.: Preference for ethanol in the rhesus monkey following chronic infusion of ethanol into the cerebral ventricles. Physiol. Behav. **8**, 431—435 (1972b)

Nathan, P. E., O'Brien, J. S., Lowenstein, L. M.: Operant studies of chronic alcoholism: Interaction of alcohol and alcoholics. In: Roach, M. K., McIsaac, W. M., Creaven, P. J. (Eds.): Biological Aspects of Alcohol, pp. 341—370. Austin: University of Texas 1971

Nathan, P. E., Titler, N. A., Lowenstein, L. M., Solomon, P., Rossi, A. M.: Behavioral analysis of chronic alcoholism. Arch. gen. Psychiat. **22**, 419—430 (1970)

National Council on Alcoholism Criteria for the Diagnosis of Alcoholism: Amer. J. Psychiat. **129** (2), 127—135 (1972)
Newman, H. W.: Acquired tolerance to ethyl alcohol. Quart. J. Stud. Alcohol **2**, 453—463 (1941)
Ogata, H., Ogata, F., Mendelson, J. H., Mello, N. K.: A comparison of techniques to induce alcohol dependence in mouse. J. Pharmacol. exp. Ther. **180** (2), 216—230 (1972)
Ogata, M., Mendelson, J. H., Mello, N. K.: Electrolytes and osmolality in alcoholics during experimentally induced intoxication. Psychosom. Med. **30**, 463—488 (1968)
Orme-Johnson, W. H., Ziegler, D. M.: Alcohol mixed function oxidase activity of mammalian liver microsomes. Biochem. biophys. Res. Commun. **21**, 78—82 (1965)
Owens, A. H., Marshall, E. K., Jr.: The metabolism of ethyl alcohol in the rat. J. Pharmacol. exp. Ther. **115**, 360—370 (1955)
Pearson, S. B.: Observations on brain fever; delirium tremens. Edinb. med. Surg. J. **9**, 326—332 (1813)
Pert, C. B., Snyder, S. H.: Opiate receptor: Demonstration in nervous tissue. Science **179**, 1011—1014 (1973)
Pieper, W. A., Skeen, M. J.: Induction of physical dependence on ethanol in rhesus monkeys using oral acceptance techniques. Life Sci. **11**, 989—997 (1972)
Pieper, W. A., Skeen, M. J., McClure, H. M., Bourne, P. G.: The chimpanzee as an animal model for investigating alcoholism. Science **176**, 71—73 (1972)
Plaut, T. F. A.: Alcohol Problems: A report to the nation. Cooperative Commission on the Study of Alcoholism. New York: Oxford University Press 1967
Pletscher, A., Bernstein, A., Stabu, H.: Beschleunigung des Alkoholabbaus durch Fructose beim Menschen. Experientia (Basel) **8**, 307—308 (1952)
Raiha, N., Oura, E.: Effect of ethanol oxidation on levels of pyridine nucleotides in liver and yeast. Proc. Soc. exp. Biol. (N.Y.) **109**, 908—910 (1962)
Raskin, N. H., Sokoloff, L.: Brain alcohol dehydrogenase. Science **162**, 131—132 (1968)
Ratcliffe, F.: Ethanol dependence in the rat: Its production and characteristics. Arch. int. Pharmacodyn. **196**, 146—156 (1972)
Reboucas, G., Isselbacher, K. J.: Studies on the pathogenesis of the ethanol-induced fatty liver. I. Synthesis and oxidation of fatty acids by the liver. J. clin. Invest. **40**, 1355—1362 (1961)
Roe, E.: The metabolism and toxicity of methanol. Pharmacol. Rev. **17**, 399—412 (1955)
Rotheran, E. B., Safar, P., Robin, E. D.: CNS disorder during mechanical ventilation. J. Amer. med. Ass. **189**, 933—996 (1964)
Rubin, E., Gang, H., Misra, P., Lieber, C. S.: Inhibition of drug metabolism by acute ethanol intoxication: A hepatic microsomal mechanism. Amer. J. Med. **49**, 800—806 (1970)
Rubin, E., Lieber, C. S.: Experimental alcoholic hepatitis: A new primate model. Science **182**, 712—713 (1973)
Rubin, E., Lieber, C. S.: Fatty liver, alcoholic hepatitis and cirrhosis produced by alcohol in primates. New Engl. J. Med. **209** (3), 128—135 (1974)
Schuster, C. R.: Psychological approaches to opiate dependence and self-administration by laboratory animals. Fed. Proc. **29**, 2—5 (1970)
Seevers, M. H.: Morphine and ethanol physical dependence: A critique of a hypothesis. Science **170**, 1113—1114 (1970)
Seevers, M. H., Deneau, G. A.: A critique of the dual action hypothesis of morphine physical dependence. Arch. int. Pharmacodyn. **140**, 514—520 (1962)
Seevers, M. H., Deneau, G. A.: Physiological aspects of tolerance and physical dependence. In: Root, W. S., Hofmann, F. G. (Eds.): Physiological Pharmacology, pp. 565—640. New York: Academic Press 1963
Sereny, G., Kalant, H.: Comparative clinical evaluation of chlordiazepoxide and promazine in treatment of alcohol withdrawal syndrome. Brit. med. J. **I**, 92—97 (1965)
Sharpless, S. K.: Reorganization of function in the nervous system—use and disuse. Ann. Rev. Physiol. **26**, 357—388 (1964)
Shuster, L.: Repression and de-repression of enzyme synthesis as a possible explanation of some aspects of drug addiction. Nature (Lond.) **189**, 314—315 (1961)
Shuster, L.: Tolerance and physical dependence. In: Clouet, D. H. (Ed.): Narcotic Drugs, Biochemical Pharmacology, pp. 408—423. New York-London: Plenum Press 1971

Sidman, M.: Avoidance conditioning with brief shock and no exteroceptive warning signal. Science **118**, 157—158 (1953)
Smith, A. A.: Inhibitors of tolerance development. In: Clouet, D. H. (Ed.): Narcotic Drugs, Biochemical Pharmacology, pp. 424—431. New York-London: Plenum Press 1971
Smith, A. A., Gitlow, S.: Effect of disulfiram and ethanol on the catabolism of norepinephrine in man. In: Maickel, R. P. (Ed.): Biochemical Factors in Alcoholism, pp. 33—59. Oxford: Pergamon 1967
Smith, A. A., Hayashida, K.: Blockade or reversal by propranolol of the narcotic and respiratory depression induced in mice by ethanol. Fed. Proc. **29** (Abstr. No. 2267), 649 (1970)
Smith, M. E., Newman, H. W.: The rate of ethanol metabolism in fed and fasting animals. J. biol. Chem. **234**, 1544 (1959)
Sutton, T.: Tracts on Delirium Tremens, on Peridonitis and Other Inflammatory Afflictions. London: Thomas Underwood 1813
Talland, G. A.: Effects of alcohol on performance in continuous attention tasks. Psychosom. Med. **28**, 596—604 (1966)
Talland, G. A., Mendelson, J. H., Ryack, P.: Experimentally induced chronic intoxication and withdrawal in alcoholics. Pt. 4, Tests of Motor Skills. Quart. J. Stud. Alcohol Suppl. No. **2**, 53—73 (1964 a)
Talland, G. A., Mendelson, J. H., Ryack, P.: Experimentally induced chronic intoxication and withdrawal in alcoholics. Pt. 5, Tests of Attention. Quart. J. Stud. Alcohol Suppl. No. **2**, 74—86 (1964 b)
Tavel, M. E., Davidson, W., Batterton, T. D.: A critical analysis of mortality associated with delirium tremens: Review of 39 fatalities in a 9-year period. Amer. J. med. Sci. **242**, 18—29 (1961)
Teiger, D. G.: Induction of physical dependence on morphine, codeine and meperidine in the rat by continuous infusion. J. Pharmacol. exp. Ther. **190**, 408—415 (1974)
Tephly, T. R., Tinelli, F., Watkins, W. D.: Alcohol metabolism: Role of microsomal oxidation in vivo. Science **166**, 627—628 (1969)
Theorelle, H., Yonetani, T.: Liver alcohol dehydrogenase-DPN-pyrazole complex: A model of ternary intermediate and the enzyme reaction. Biochem. Z. **338**, 537—553 (1963)
Thomas, D. W., Freedman, D. X.: Treatment of the alcohol withdrawal syndrome: Comparison of promazine and paraldehyde. J. Amer. med. Ass. **188**, 316—318 (1964)
Truitt, E. B., Walsh, J. J.: The role of acetaldehyde in the actions of ethanol. In: Kissin, B., Begleiter, H. (Eds.): The Biology of Alcoholism: Vol. 1, Biochemistry, pp. 161—195. New York: Plenum Press 1971
Tygstrup, N., Winkler, K., Lundquist, F.: The mechanism of the fructose effect on the ethanol metabolism of the human liver. J. clin. Invest. **44**, 817—830 (1965)
Victor, M.: Treatment of alcoholic intoxication and the withdrawal syndrome. A syndrome analysis of the use of drugs and other forms of therapy. Psychosom. Med. **28**, No. 3 (II), 636—650 (1966)
Victor, M.: The pathophysiology of alcoholic epilepsy. In: The Addictive States. Res. Publ. Ass. nerv. ment. Dis., Vol. 46, pp. 431—454. Baltimore: Williams & Wilkins Co. 1968
Victor, M., Adams, R. D.: The effect of alcohol on the nervous system. Res. Publ. Ass. nerv. ment. Dis. **32**, 526—573 (1953)
Victor, M., Brausch, C.: The role of abstinence in the genesis of alcoholic epilepsy. Epilepsia **8**, 1—20 (1967)
von Wartburg, J. P.: The metabolism of alcohol in normals and alcoholics: Enzymes. In: Kissin, B., Begleiter, H. (Eds.): The Biology of Alcoholism: Vol. I, Biochemistry, pp. 63—102. New York: Plenum Press 1971
Walder, A. I., Redding, J. S., Faillace, L., Sternburg, R. W.: Rapid detoxification of the acute alcoholic with hemodialysis. Surgery **66**, 201—207 (1969)
Wallerstein, R. S.: Hospital Treatment of Alcoholism. New York: Basic Books, Inc. 1957
Wallgren, H., Barry, H.: Actions of Alcohol. Vol. I. Biochemical and Physiological Aspects. Amsterdam: Elsevier Publishing Co. 1970 a
Wallgren, H., Barry, H.: Actions of Alcohol. Vol. II. Chronic and Clinical Aspects. Amsterdam: Elsevier Publishing Co. 1970 b
Wallgren, H., Lindbohn, R.: Adaptation to ethanol in rats with special reference to brain tissue respiration. Biochem. Pharmacol. **8**, 423—424 (1961)

Way, E. L., Loh, H. H., Shen, F.: Morphine tolerance, physical dependence and synthesis of brain 5-OH tryptamine. Science **162**, 1290—1292 (1968)
Way, E. L., Loh, H. H., Shen, F.: Simultaneous quantitative assessment of morphine tolerance and physical dependence. J. Pharmacol. exp. Ther. **167**, 1—8 (1969)
Westerfeld, W. W.: The intermediary metabolism of alcohol. Amer. J. clin. Nutr. **9**, 426—431 (1961)
Westerfeld, W. W., Schulman, M. P.: Metabolism and caloric value of alcohol. J. Amer. med. Ass. **170**, 197 (1959)
WHO Expert Committee on Mental Health and on Alcohol Symposium: The craving for alcohol. Quart. J. Stud. Alcohol **16**, 34—66 (1955)
Williams, I. E.: The emergency treatment of the alcoholic. J. nerv. ment. Dis. **74**, 161—172 (1931)
Winger, G. D., Ikomi, F., Woods, J. H.: Intravenous ethanol self-administration in rhesus monkeys. Proceedings of the 32nd Annual Conference of the Committee on Problems of Drug Dependence, pp. 6598—6605. Washington, D.C.: NAS-NRC, 1970
Winger, G. D., Woods, J. H.: The reinforcing property of ethanol in the rhesus monkey: I. Initiation, maintenance and termination of intravenous ethanol-reinforced responding. In: Seixas, F. A., Eggleston, S. (Eds.): Alcoholism and the Central Nervous System. Ann. N.Y. Acad. Sci. **215**, 162—175 (1973)
Wolfe, S. M., Mendelson, J. H., Ogata, M., Victor, M., Marshall, W., Mello, N. K.: Respiratory alkalosis and alcohol withdrawal. Trans. Ass. Amer. Phycns **82**, 344—352 (1969)
Wolfe, S. M., Victor, M.: The physiological basis of the alcohol withdrawal syndrome. In: Mello, N. K., Mendelson, J. H. (Eds.): Recent Advances in Studies of Alcoholism, pp. 188—199. Washington, D.C.: U.S. Govt. Printing Office, Publ. No. (HSM) 71-9045, 1971
Wolin, S. J., Mello, N. K.: The effects of alcohol on dreams and hallucinations in alcohol addicts. In: Seixas, F. A., Eggleston, S. (Eds.): Ann. N.Y. Acad. Sci. **215**, 266—302 (1973)
Woods, J. H., Ikomi, F. I., Winger, G.: The reinforcing properties of ethanol. In: Roach, M. K., McIsaac, W. M., Creaven, P. J. (Eds.): Biological Aspects of Alcoholism, pp. 371—388. Austin: University of Texas Press 1971
Woods, J. H., Schuster, C. R.: Regulation of drug self administration. In: Harris, R. T., McIsaac, W. M., Schuster, C. R. (Eds.): Drug Dependence, pp. 158—169. Austin: University of Texas Press 1970
Woods, J. H., Winger, G.: A critique of methods for inducing ethanol self intoxication in animals. In: Mello, N. K., Mendelson, J. H. (Eds.): Recent Advances in Studies of Alcoholism, pp. 413—436. Washington, D.C.: U.S. Govt. Printing Office Publ. No. (HSM) 71-9045, 1971
Yanagita, T., Ando, K., Takahashi, D. V. M., Ishida, K.: Self-administration of barbiturates, alcohol (intragastric) and CNS stimulants (intravenous) in monkeys. Proceedings, Committee on Problems of Drug Dependence, pp. 6039—6051. Washington, D.C.: NAS-NRC, 1969
Yanagita, T., Deneau, G. A., Seevers, M. H.: Evaluation of pharmacologic agents in the monkey by long-term intravenous self or programmed administration. 23rd International Congress of Physiological Sciences, Abs. 66. Tokyo 1965
Zilborg, G., Henry, G. W.: A history of medical psychology. New York: W. W. Norton & Co. 1941

CHAPTER 4

Abuse of Non-Narcotic Analgesics

P. KIELHOLZ and D. LADEWIG

A. Incidence of the Abuse of Non-Narcotic Analgesics

After reports had first been published from Switzerland referring to the abuse of antipyretic analgesics containing phenacetin (KIELHOLZ, 1954; HORRISBERGER et al., 1958; BATTEGAY, 1958) and to the possibility of a causal connection between such abuse and chronic renal diseases (SPÜHLER and ZOLLINGER, 1963), numerous authors subsequently drew attention to the problem of the misuse of non-narcotic analgesics (cf. GSELL et al., 1968). The misuse of non-narcotic analgesics may lead to genuine dependence (EDDY et al., 1965; JAFFE, 1970), which on theoretical grounds might be assigned to the barbiturate type of drug dependence. An enquiry which was carried out from 1955 to 1969 over the whole of Switzerland revealed that the number of first admittances to psychiatric hospitals of patients suffering from dependence on non-narcotic analgesics had trebled (KIELHOLZ, 1968; KIELHOLZ and BATTEGAY, 1967). Evidence showed that the abuse of analgesics had increased not only in urban districts but also in rural areas. Although an initial enquiry undertaken over the period 1945–1954 (MÜLLER and KIELHOLZ, 1967; LADEWIG, 1973) had disclosed that the abuse of analgesics was significantly higher among workers engaged in certain special occupations, such as the watch and textile industries, this difference later ceased to be significant. Patients dependent upon analgesics are predominantly women, the ratio of women to men being 4:1. The age at which the incidence is greatest (i.e., the so-called age of manifestation) is the same for both sexes, namely, 35–45 years. Compared with the normal population, such patients tend to be more frequently either divorced or unmarried.

With regard to their occupation, women who abuse analgesics are more often encountered among "housewives" and "gainfully employed housewives," and men among "laborers" and "craftsmen." These patients presented for clinical treatment after an average of 5–7 years, during which time their consumption of an antipyretic analgesic had averaged between 8–20 tablets daily (KIELHOLZ and LADEWIG, 1975).

B. Symptomatology and Etiology of the Abuse of Non-Narcotic Analgesics

Most of the antipyretic analgesics abusively employed consist of mixed preparations which, as reported by WOODBURY (1970), contain analgesic and antipyretic substances such as:

1. Salicylates
2. Para-aminophenol derivatives, e.g., phenacetin, paracetamol, acetanilide

3. Pyrazolone derivatives, e.g., antipyrine, aminophenazone, or phenylbutazone
4. Caffeine
5. Hypnotic

Some of the combinations commonly used in this connection are: (1) propyphenazone 150 mg, phenacetin 250 mg, pyrithyldione 50 mg, caffeine 50 mg; (2) isobutyllallylbarbituric acid 50 mg, aminophenazone 125 mg, caffeine 25 mg, and (3) aminophenazone 100 mg, antipyrine 100 mg, phenacetin 250 mg, caffeine 50 mg.

Distinctive features exhibited by these patients upon examination are their dirty greyish-brown skin complexion and their cyanotic lips (Fig. 1). During the stage of intoxication, the patients are befuddled, dysarthric, and ataxic, their psychomotor status being characterized either by disinhibition and enhanced drive or by retardation and apathy. They bump into objects and may even fall, thus sustaining typical injuries. Hematomas can usually be seen on the face, on the shoulders, and on the pelvis. Diagnosis is facilitated by the presence of signs such as nystagmus, a characteristic ash-grey, café-au-lait colored skin, and hematomas, as well as by a history of accidents which the patient has had either at work or while driving. The basic mood in these cases is anxious, sub-depressive, dysphoric, or—less often—irritable and hypomanic. Following years of such drug abuse, the patient's memory may also be impaired. The personality may display varying degrees of "deprivation."

The reason given by patients for consuming excessive quantities of non-narcotic analgesics is that they suffer from painful syndromes of predominantly psychosomatic origin, such as headaches, pains in the nuchal region, in the limbs, or in the back, gastrointestinal spasms, cardiac oppression, dysmenorrhea, or feelings of pressure which assume various forms.

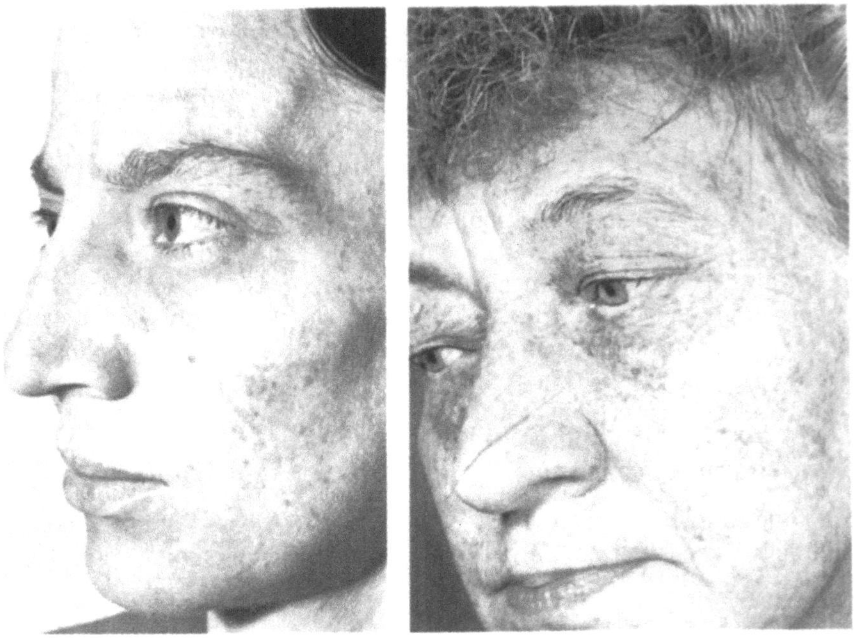

Fig. 1. Typical discoloration of the skin resulting from the abuse of analgesics

When the chronic consumer of analgesics describes the type of headache to which he is subject, he confines himself to generalities, the description usually being vague, long-winded, and, depending on the degree and duration of the intoxication, either verbose or monosyllabic. Such descriptions frequently contain references to changes in the intensity and location of the pain or to alterations in the symptomatology. The picture presented by the patient must also be interpreted to some extent in terms of his or her reaction to the type of interrogation used, to the manner in which the interview is conducted, and to the physician's own attitude.

The headache as described may be either unilateral or bilateral. Though rather difficult to classify accurately, the headache has been identified as migraine in 15% of cases, as vasomotor cephalalgia in 55%, due to muscular tension in 18% (i.e., "myalgic" headache, or in some cases "tension headache"), and in 12% as chronic post-traumatic neuralgic headache or as headache due to diseases of the nose, sinuses, teeth, ears, temperomandibular joint, or pharynx. Patients most commonly complain of a feeling of pressure over the eyes or forehead or of a sensation such as would be produced by a helmet pressing on their head. Less often they report pain in the nuchal or occipital region. In patients of this type seen by us, headache is usually not the only painful symptom encountered. A study of the case histories of patients complaining of headache reveals a high incidence of emotional tension (see below), as well as previous physical illnesses in 28% of the women and 32% of the men.

In addition to headaches, other painful symptoms may be present; these are liable to occur in any of the muscles, in the limbs, in the neck and shoulder region, or in the back. Occasionally an abdominal syndrome consisting of a loss of appetite amounting sometimes to frank anorexia (an increase in appetite is less common), as well as nausea, the sensation of a lump in the throat or of choking or dryness of the throat, vomiting, heartburn, epigastric tenderness, gastrointestinal spasms, gaseous distension, constipation (possibly alternating with diarrhea), and tenesmus is seen. In certain instances the condition becomes such that even surgery is performed; following the operation, the patient obtains temporary relief from the symptoms, which later recur and are associated with increased headaches. More frequent than these abdominal symptoms is the occurrence of headaches associated with a variety of complaints which include tiredness, impairment of performance, sleep disorders, hypoesthesia or hyperesthesia, pruritus, tremor, jitteriness, mental unrest, sweating, shivering, impotence, frigidity, and frequency of urination. In the general symptomatology of these syndromes, sleep disorders such as difficulty in getting to sleep, light sleep, and early awakening constitute cardinal signs. Less often encountered are cardiopulomary syndromes marked by a sensation of pressure and/or stabbing pains in the region of the heart, precordial anxiety, pseudoangina, tachycardia, extrasystoles, a tendency to syncope, dizziness, feelings of pressure and constriction in the chest, and dyspnea, as well as endocrine symptoms such as disturbances in the water balance, hypoglycemia, loss of weight, and menstrual disorders. In female patients, headache—in conjunction with the above-mentioned categories of symptoms—is an early manifestation of systemic psychosomatic diseases. The female patients tend to have personalities that are predominantly of the asthenic, sensitive type, i.e., they take life very seriously and feel uncertain of themselves. Owing to their emotional weakness and sensitivity, they are afraid of injury to their feelings, which they therefore strive to bottle up. This tendency to introversion is often further intensified by the attitude which others adopt towards them. When they do voice their intuitive opinions, anxieties, and apprehensions, they are not adequately understood. Consequently, they become increasingly withdrawn and isolated. The personalities of the male patients are likewise largely of the asthenic, sensitive, egocentric type which, though passive, is characterized by the desire to assert itself. Ambition and the urge toward self-assertion induce such men to aim at high goals, but the discrepancy between their goals and their actual performance results in feelings of failure, in resignation and dysphoria, in overcompensating pompousness, and in increasing emotional encapsulation coupled with the feeling of not being understood by others.

Though in such cases we may suspect various conflict situations, we learn little about these at first. It is often not until after weeks of intensive interviewing that the patients begin to thaw out. As they are progressively weaned from the drug, they show less tendency to suppress their feelings, to deny their difficulties, and to explain away their unplesant experiences. We are then informed by the patients that the headaches had the habit of worsening after they had suffered losses or disappointments, and that a change of environment, a pleasant experience, or success at work caused the pains to disappear temporarily.

Underlying the psychosomatic disturbances as described above are long-standing, conscious or unconscious, emotional tensions. In women these take the form of persistent tensions due to thwarted yearnings for love, sexual satisfaction, or recognition, i.e., tensions which are usually attributable to lack of affection, tenderness, appreciation, and sexual fulfilment, to chronic emotional upsets, or to the husband's infidelity or alcoholism. Alternatively, they may arise from emotional overstrain, from inability to cope with the household chores and with the upbringing of the children, from dissatisfaction with the job in which the patient is employed, or from financial worries. The strain of having to perform work which taxes the patient's inner resistance rapidly leads to autonomic nervous disturbances and to organic manifestations of functional disorders. In men, on the other hand, the main causative factors are conflicts occurring in connection with their occupations, ambitions, rivalries, the fear of failure at work, or simply anticipatory, competitive, or existential anxiety as such.

It is quite understandable that, to combat their psychosomatic symptoms, these patients resort to analgesics and hypnotics, which are on sale everywhere and are claimed to be completely safe and harmless. The persons who take them certainly have no desire to become dependent on drugs, nor do they consciously wish to escape from reality; they are merely trying to treat themselves so that they can keep going and live free of the symptoms that afflict them. In much the same way as hypnotics are taken in order to combat insomnia, so drugs are consumed by these individuals in response to anticipatory anxiety, i.e., in the hope that they will ward off pain and discomfort.

Very gradually the patient becomes habituated to the drug and slowly has to increase the dose so as to obtain the desired analgesic affect. The intervals between doses become shorter and the dosage becomes progressively higher. Often only after prolonged habituation does the patient suddenly find that, in addition to its analgesic properties, the drug also relieves tiredness, exerts a reviving and stimulant effect, and apparently helps to improve his performance. This effect—which serves to combat weariness, apathy, and general listlessness and which as a rule makes itself felt only secondarily—is the one that the patient can no longer do without (BUCHER, 1962). The insidious feature of the phenacetin-containing combinations is that when consumed in a fairly high dosage, they themselves give rise to headache and so prompt the patient to take increasing doses. Such patients thus become caught up in a vicious circle, which leads to intoxication and may ultimately result in genuine dependence.

When antipyretic analgesics are discontinued, withdrawal symptoms set in within 12–48 h, persisting for 10 days in the acute phase of abstinence and for 2–4 months in the protracted withdrawal period. Among the chief symptoms encountered are headache, excessive wakefulness, mental unrest, sleep disturbances, fine tremor, muscular twitching, rheumatic pains in the limbs, mood disorders marked by anxiety and depression, pre-collapse, diarrhea alternating with constipation, vomiting, and, finally, withdrawal deliria, hallucinations, and grand mal seizures. As in the case of barbiturates, the severity of the withdrawal symptoms—including especially the occurrence of epileptic seizures and deliria—seems to be dose-dependent.

During the withdrawal of analgesics, we carried out six electroencephalographic examinations at standardized intervals over a period of 6–10 weeks. Immediately after the withdrawal, unspecific generalized abnormalities, which are usually of mild to moderate severity and are rarely of an episodic character, can be detected in the EEG. The EEG pattern generally soon shows a tendency towards normalization, as evidenced by stabilization of the alpha frequency and a decrease in theta, delta, and beta waves. In one-third of the patients, however, the electrical activity of the brain showed only mild abnormalities in the first EEG recordings made after withdrawal of the drug, but underwent a deterioration during the following 7–10 days—a deterioration marked by an increase in slow waves, asymmetry, localized focal abnormalities, or even paroxysmal electrical manifestations. Except in one case, in which two recurrences of abnormalities were observed, the third and fourth examinations generally revealed a normal EEG. These phenomena are probably attributable to the fact that the withdrawal of a preparation which the patient has been taking for years results in disruption of the equilibrium which the electrophysiologic functions of the brain have gradually established, the disruption persisting until such time as a new balance is attained.

The possibility has also been examined that a connection may exist between the abuse of phenacetin-containing combinations and the occurrence of renal damage (DUBACH et al., 1967; DUBACH, 1968; GILMAN, 1964). In the light of recent epidemiological studies and clinical observations it appears more and more probable that there is indeed a connection between renal lesions and the abuse of such combined preparations. The lesions found have chiefly taken the form of interstitial nephritis and necrosis of the renal papillae. An enquiry undertaken throughout the whole of Switzerland disclosed that 70% of all patients suffering from dependence on analgesics who had been treated in Swiss medical clinics and polyclinics displayed pathologic renal findings. In addition, evidence of hemotoxic effects—i.e., accelerated breakdown of erythrocytes, anemia, and formation of methemoglobin and sulphemoglobin—was obtained in 30% of such cases. It will not be possible, however, to reach any final conclusions in this connection until detailed epidemiologic investigations have been carried out. In some of our findings, significant correlations between the amount of intake and personality abnormalities demonstrated by a personality inventory could be proven (LADEWIG et al., 1973).

References

Battegay, R.: Aktuelle Aspekte der Analgeticasucht. Nervenarzt **29**, 467, (1958)

Bucher, K.: Ursachen und Konsequenzen des Arzneimittelverbrauchs. Schweiz. Apoth.-Ztg. **100**, 190, 1962

Dubach, U. C.: Mortalitätsentwicklung für Nierenleiden in der Schweiz. Schweiz. med. Wschr. **40**, 1542, (1968)

Dubach, U. C., Minder, F., Gsell, O.: An Epidemiological Study of Analgesic Abuse, Proc. 3rd Int. Congr. Nephrol. Washington, p. 300 (1966). Basel-New York: S. Karger 1967

Eddy, N. B., Halbach, H., Isbell, H., Seevers, M. H.: Drug dependence: Its significance and characteristics. Bull. WHO 37, (1965)

Gilman, A.: Analgesic nephrotoxicity a pharmacological analysis. Amer. J. Med. **36**, 167—173 (1964)

Gsell, O., Dubach, U. C., Raillard-Peucker, U.: Phenazetinabusus und Nierenleiden. Dtsch. med. Wschr. **94**, 101 (1968)

Horrisberger, B., Grandjean, E., Lanz, F.: Untersuchungen über den Medikamentenmißbrauch in einem Großbetrieb der schweiz. Uhrenindustrie, Schweiz. med. Wschr. **88**, 920 (1958)

Jaffe, J. H.: Drug addiction and drug abuse. In: Goodman, L. S., Gilman, A. (Eds.): The Pharmacological Basis of Therapeutics, pp. 276—313. London: Macmillan 1970

Kielholz, P.: Aetiologie und Therapie der Analgetica- und Hypnoticasucht. Schweiz. med. Wschr. **84**, 1214 (1954)

Kielholz, P.: Gesamtschweizerische Enquête über die Häufigkeit des Medikamentenmißbrauchs, Bericht der von der Schweiz. Sanitätsdirektoren-Konferenz eingesetzten Kommission zur Behandlung des Problems des Medikamentenmißbrauchs. Schweiz. Ärzteztg. **49**, 1077 (1968)

Kielholz, P., Battegay, R.: Vergleichende Untersuchungen über die Genese und den Verlauf der Drogenabhängigkeit und des Alkoholismus. Schweiz. med. Wschr. **97**, 893 (1967)

Kielholz, P., Ladewig, D.: Arzneimittelmißbrauch. In: IKS (Eds.): Jubiläumsschrift zum 75jährigen Bestehen der interkantonalen Vereinbarung über die Kontrolle der Heilmittel, Bern 1975, pp. 137—144

Ladewig, D.: Gesamtschweizerische Enquête über die Häufigkeit des Medikamenten- und Drogenmißbrauchs. Schweiz. Ärzteztg. **28**, 971—974 (1973)

Ladewig, D., Dubach, U. C., Ehrensperger, Th., Hobi, V., Miest, P.: Epidemiologische, Psychiatrische und Psychologische Aspekte des Analgetika-Mißbrauchs. Beitr. Int. Symp. Probleme des Phenacetin-Abusus, Wien 1973. Facta Publication, Wien 1973, 275—286

Müller, T., Kielholz, P.: Erhebung über Ausmaß, Verbreitung und Prophylaxe des Medikamenten- insbesondere des Analgeticmißbrauchs in der Schweiz. Bull. Eidg. Gesundheitsamt B 5 (1967)

Spühler, O., Zollinger, H. U.: Die chronische interstitielle Nephritis. Z. klin. Med. **151**, 1 (1963)

Woodbury, D. M.: Analgesic-antipyretics, anti-inflammatory agents and inhibitors of uric acid synthesis. In: Goodman, L. S., Gilman, A. (Eds.): The Pharmacological Basis of Therapeutics, pp. 314—347. London: Macmillan 1970

Author Index

Page numbers in *italics* refer to bibliography

Subject Index

Reviews of Physiology, Biochemistry and Pharmacology

formerly
Ergebnisse der Physiologie, biologischen Chemie und experimentellen Pharmakologie

This series presents rapid and comprehensive information on topical problems and research in progress over the entire range of physiology, biochemistry, and pharmacology. An international group of editors is responsible for inviting experts in these fields to submit contributions. Every year three to four volumes are published. The language of publication is English.

Volume 73
W. Hasselbach:
Hans Hermann Weber, 1896-1974
E. de Robertis:
Synaptic Receptor Proteins. Isolation and Reconstruction in Artificial Membranes
A. Melander, L. E. Ericson, F. Sunder and U. Westgren:
Intrathyroidal Amines in the Regulation of Thyroid Activity
J. Haase, S. Cleveland and H.-G. Ross:
Problems of Postsynaptic Autogenous and Recurrent Inhibition in the Mammalian Spinal Cord
I. S. Kulaev:
Biochemistry of Inorganic Polyphosphates

Volume 74
F. E. Bloom:
The Role of Cyclic Nucleotides in Central Synaptic Function
S. Silbernagl, E. C. Foulkes and P. Deetjen:
Renal Transport of Amino Acids

Volume 75
E. Hofmann:
The Significance of Phosphofructokinase to the Regulation of Carbohydrate Metabolism
H. Grunicke, B. Puschendorf and H. Werchau:
Mechanism of Action of Distamycin A and Other Antibiotics with Antiviral Activity
A. E. Lambert:
The Regulation of Insulin Secretion

Volume 76
H. Hilz and P. Stone:
Poly (ADP-Ribose) and ADP-Ribosylation of Proteins
W. Wuttke:
Neuroendocrine Mechanisms in Reproductive Physiology
F. Ellendorff:
Evaluation of Extrahypothalamic Control of Reproductive Physiology
A. M. Rappaport and J. H. Schneiderman:
The Function of the Hepatic Artery

Volume 77
K. Starke:
Regulation of Noradrenaline Release by Presynaptic Receptor Systems
P. Ward and E. Becker:
Biology of Leukotaxis
W. A. Grunewald and W. Sowa:
Capillary Structures and O_2 Supply to Tissue

Volume 78
B. Deuticke:
Properties and Structural Basis of Simple Diffusion Pathways in the Erythrocyte Membrane
R. K. Crane:
The Gradient Hypothesis and Other Models of Carrier – Mediated Active Transport

Volume 79
J. M. Ritchie, R. B. Rogart:
The Binding of Saxitoxin and Tetrodotoxin to Excitable Tissue
K. Sato:
The Physiology, Pharmacology, and Biochemistry of the Eccrine Sweat Gland
G. Sachs:
H Transport by Non-Electrogenic Gastric ATPase as a Model for Acid Reaction

Springer-Verlag
Berlin
Heidelberg
New York

Handbuch der experimentellen Pharmakologie/ Handbook of Experimental Pharmacology

Heffter-Heubner, New Series

Volume 4:
General Pharmacology

Volume 10:
Die Pharmakologie anorganischer Anionen

Volume 11:
Lobelin und Lobeliaalkaloide

Volume 12:
Morphin und morphinähnlich wirkende Verbindungen

Volume 14: Part 1
The Adrenocortical Hormones I
Part 2
The Adrenocortical Hormones II
Part 3
The Adrenocortical Hormones III

Volume 16:
Erzeugung von Krankheitszuständen durch das Experiment/ Experimental Production of Diseases
Part 1
Blut (In Vorbereitung)
Part 2
Atemwege
Part 3
Heart and Circulation
Part 4
Niere, Nierenbecken, Blase
Part 5
Liver
Part 6
Schilddrüse
(In Vorbereitung)
Part 7
Zentralnervensystem
Part 8
Stütz- und Hartgewebe
Part 9
Infektionen I
Part 10
Infektionen II
Part 11A
Infektionen III
Part 11B
Infektionen IV
Part 12
Tumoren I
Part 13
Tumoren II
Part 14
Tumoren III
(In preparation)
Part 15
Kohlenhydratstoffwechsel, Fieber/ Carbohydrate Metabolism, Fever

Volume 17: Part 1
Ions alcalino-terreux I
Part 2
Ions alcalino-terreux II

Volume 18: Part 1
Histamine
Part 2
Anti-Histaminics

Volume 19:
5-Hydroxytryptamine and Related Indolealkylamines

Volume 20: Part 1
Pharmacology of Fluorides I
Part 2
Pharmacology of Fluorides II

Volume 21:
Beryllium

Volume 22: Part 1
Die Gestagene I
Part 2
Die Gestagene II

Volume 23:
Neurohypophysial Hormones and Similar Polipeptides

Springer-Verlag
Berlin
Heidelberg
New York

MIX
Papier aus verantwortungsvollen Quellen
Paper from responsible sources
FSC® C105338

If you have any concerns about our products,
you can contact us on
ProductSafety@springernature.com

In case Publisher is established outside the EU,
the EU authorized representative is:
Springer Nature Customer Service Center GmbH
Europaplatz 3, 69115 Heidelberg, Germany

Printed by Libri Plureos GmbH
in Hamburg, Germany